ECOLOGY
From Individuals to Ecosystems

ECOLOGY
From Individuals to Ecosystems

MICHAEL BEGON

School of Biological Sciences,
The University of Liverpool, Liverpool, UK

COLIN R. TOWNSEND

Department of Zoology, University of Otago, Dunedin, New Zealand

JOHN L. HARPER

Chapel Road, Brampford Speke, Exeter, UK

FOURTH EDITION

Blackwell
Publishing

BLACKWELL PUBLISHING
350 Main Street, Malden, MA 02148-5020, USA
9600 Garsington Road, Oxford OX4 2DQ, UK
550 Swanston Street, Carlton, Victoria 3053, Australia

First edition published 1986
Second edition published 1990
Third edition published 1996
Fourth edition published 2006 by Blackwell Publishing Ltd

7   ©2011

Library of Congress Cataloging-in-Publication Data

Begon, Michael
  Ecology : from individuals to ecosystems / Michael Begon, Colin R.
Townsend, John L. Harper.—4th ed.
      p. cm.
  Includes bibliographical references and index.
  ISBN: 978-1-4051-1117-1 (paperback : alk. paper)
  1. Ecology. I. Townsend, Colin R. II. Harper, John L. III. Title.
  QH54.B416   2005
  577—dc22

                                        2005004136

A catalogue record for this title is available from the British Library.

Set in 9.5 / 12 Dante MT
by Graphicraft Ltd, Hong Kong
Printed and bound in Singapore
by C.O.S. Printers Pte Ltd

The publisher's policy is to use permanent paper from mills that operate a sustainable forestry policy, and which has been manufactured from pulp processed using acid-free and elementary chlorine-free practices. Furthermore, the publisher ensures that the text paper and cover board used have met acceptable environmental accreditation standards.

For further information on
Blackwell Publishing, visit our website:
www.blackwellpublishing.com

Contents

Preface

A science for everybody – but not an easy science

This book is about the distribution and abundance of different types of organism, and about the physical, chemical but especially the biological features and interactions that determine these distributions and abundances.

Unlike some other sciences, the subject matter of ecology is apparent to everybody: most people have observed and pondered nature, and in this sense most people are ecologists of sorts. But ecology is not an easy science. It must deal explicitly with three levels of the biological hierarchy – the organisms, the populations of organisms, and the communities of populations – and, as we shall see, it ignores at its peril the details of the biology of individuals, or the pervading influences of historical, evolutionary and geological events. It feeds on advances in our knowledge of biochemistry, behavior, climatology, plate tectonics and so on, but it feeds back to our understanding of vast areas of biology too. If, as T. H. Dobzhansky said, 'Nothing in biology makes sense, except in the light of evolution', then, equally, very little in evolution, and hence in biology as a whole, makes sense except in the light of ecology.

Ecology has the distinction of being peculiarly confronted with uniqueness: millions of different species, countless billions of genetically distinct individuals, all living and interacting in a varied and ever-changing world. The challenge of ecology is to develop an understanding of very basic and apparent problems, in a way that recognizes this uniqueness and complexity, but seeks patterns and predictions within this complexity rather than being swamped by it. As L. C. Birch has pointed out, Whitehead's recipe for science is never more apposite than when applied to ecology: seek simplicity, but distrust it.

Nineteen years on: applied ecology has come of age

This fourth edition comes fully 9 years after its immediate predecessor and 19 years after the first edition. Much has changed – in ecology, in the world around us, and even (strange to report!) in we authors. The Preface to the first edition began: 'As the cave painting on the front cover of this book implies, ecology, if not the oldest profession, is probably the oldest science', followed by a justification that argued that the most primitive humans had to understand, as a matter of necessity, the dynamics of the environment in which they lived. Nineteen years on, we have tried to capture in our cover design both how much and how little has changed. The cave painting has given way to its modern equivalent: urban graffiti. As a species, we are still driven to broadcast our feelings graphically and publicly for others to see. But simple, factual depictions have given way to urgent statements of frustration and aggression. The human subjects are no longer mere participants but either perpetrators or victims.

Of course, it has taken more than 19 years to move from man-the-cave-painter to man-the-graffiti-artist. But 19 years ago it seemed acceptable for ecologists to hold a comfortable, objective, not to say aloof position, in which the animals and plants around us were simply material for which we sought a scientific understanding. Now, we must accept the immediacy of the environmental problems that threaten us and the responsibility of ecologists to come in from the sidelines and play their full part in addressing these problems. Applying ecological principles is not only a practical necessity, but also as scientifically challenging as deriving those principles in the first place, and we have included three new 'applied' chapters in this edition, organized around the three sections of the book: applications at the level of individual organisms and of single-species populations, of species inter-actions, and of whole communities and ecosystems. But we remain wedded to the belief that environmental action can only ever be as sound as the ecological principles on which it is based. Hence, while the remaining chapters are still largely about the principles themselves rather than their application, we believe that the *whole* of this book is aimed at improving preparedness for addressing the environmental problems of the new millennium.

Ecology's ecological niche

We would be poor ecologists indeed if we did not believe that the principles of ecology apply to all facets of the world around us and all aspects of human endeavor. So, when we wrote the first edition of *Ecology*, it was a generalist book, designed to overcome the opposition of all competing textbooks. Much more recently, we have been persuaded to use our 'big book' as a springboard to produce a smaller, less demanding text, *Essentials of Ecology* (also published by Blackwell Publishing!), aimed especially at the first year of a degree program and at those who may, at that stage, be taking the only ecology course they will ever take.

This, in turn, has allowed us to engineer a certain amount of 'niche differentiation'. With the first years covered by *Essentials*, we have been freer to attempt to make this fourth edition an up-to-date guide to ecology *now* (or, at least, when it was written). To this end, the results from around 800 studies have been newly incorporated into the text, most of them published since the third edition. None the less, we have shortened the text by around 15%, mindful that for many, previous editions have become increasingly overwhelming, and that, clichéd as it may be, less is often more. We have also consciously attempted, while including so much modern work, to avoid bandwagons that seem likely to have run into the buffers by the time many will be using the book. Of course, we may also, sadly, have excluded bandwagons that go on to fulfil their promise.

Having said this, we hope, still, that this edition will be of value to all those whose degree program includes ecology and all who are, in some way, practicing ecologists. Certain aspects of the subject, particularly the mathematical ones, will prove difficult for some, but our coverage is designed to ensure that wherever our readers' strengths lie – in the field or laboratory, in theory or in practice – a balanced and up-to-date view should emerge.

Different chapters of this book contain different proportions of descriptive natural history, physiology, behavior, rigorous laboratory and field experimentation, careful field monitoring and censusing, and mathematical modeling (a form of simplicity that it is essential to seek but equally essential to distrust). These varying proportions to some extent reflect the progress made in different areas. They also reflect intrinsic differences in various aspects of ecology. Whatever progress is made, ecology will remain a meeting-ground for the naturalist, the experimentalist, the field biologist and the mathematical modeler. We believe that all ecologists should to some extent try to combine all these facets.

Technical and pedagogical features

One technical feature we have retained in the book is the incor-poration of marginal es as signposts throughout the text. These, we hope, will serve a number of purposes. In the first place, they constitute a series of subheadings highlighting the detailed struc-ture of the text. However, because they are numerous and often informative in their own right, they can also be read in sequence along with the conventional subheadings, as an outline of each chapter. They should act too as a revision aid for students – indeed, they are similar to the annotations that students themselves often add to their textbooks. Finally, because the marginal notes generally summarize the take-home message of the paragraph or paragraphs that they accompany, they can act as a continuous assessment of comprehension: if you can see that the signpost is the take-home message of what you have just read, then you have understood. For this edition, though, we have also added a brief summary to each chapter, that, we hope, may allow readers to either orient and prepare themselves before they embark on the chapter or to remind themselves where they have just been.

So: to summarize and, to a degree, reiterate some key features of this fourth edition, they are:

- marginal notes throughout the text
- summaries of all chapters
- around 800 newly-incorporated studies
- three new chapters on applied ecology
- a reduction in overall length of around 15%
- a dedicated website (www.blackwellpublishing.com/begon), twinned with that for *Essentials of Ecology*, including inter-active mathematical models, an extensive glossary, copies of artwork in the text, and links to other ecological sites
- an up-dating and redrawing of all artwork, which is also avail-able to teachers on a CD-ROM for ease of incorporation into lecture material.

Acknowledgements

Finally, perhaps the most profound alteration to the construction of this book in its fourth edition is that the revision has been the work of two rather than three of us. John Harper has very rea-sonably decided that the attractions of retirement and grand-fatherhood outweigh those of textbook co-authorship. For the two of us who remain, there is just one benefit: it allows us to record publicly not only what a great pleasure it has been to have collaborated with John over so many years, but also just how much we learnt from him. We cannot promise to have absorbed or, to be frank, to have accepted, every one of his views; and we hope in particular, in this fourth edition, that we have not strayed too far from the paths through which he has guided us. But if readers recognize any attempts to stimulate and inspire rather than simply to inform, to question rather than to accept, to respect our readers rather than to patronize them, and to avoid unquestioning obedience to current reputation while acknowledging our debt to the masters of the past, then they will have identified John's intellectual legacy still firmly imprinted on the text.

In previous editions we thanked the great many friends and colleagues who helped us by commenting on various drafts of the text. The effects of their contributions are still strongly evident in the present edition. This fourth edition was also read by a series of reviewers, to whom we are deeply grateful. Several remained anonymous and so we cannot thank them by name, but we are delighted to be able to acknowledge the help of Jonathan Anderson, Mike Bonsall, Angela Douglas, Chris Elphick, Valerie Eviner, Andy Foggo, Jerry Franklin, Kevin Gaston, Charles Godfray, Sue Hartley, Marcel Holyoak, Jim Hone, Peter Hudson, Johannes Knops, Xavier Lambin, Svata Louda, Peter Morin, Steve Ormerod, Richard Sibly, Andrew Watkinson, Jacob Weiner, and David Wharton. At Blackwell, and in the production stage, we were particularly helped and encouraged by Jane Andrew, Elizabeth Frank, Rosie Hayden, Delia Sandford and Nancy Whilton.

This book is dedicated to our families – by Mike to Linda, Jessica and Robert, and by Colin to Laurel, Dominic, Jenny and Brennan, and especially to the memory of his mother, Jean Evelyn Townsend.

Mike Begon
Colin Townsend

Introduction: Ecology and its Domain

Definition and scope of ecology

The word 'ecology' was first used by Ernest Haeckel in 1869. Paraphrasing Haeckel we can describe ecology as the scientific study of the interactions between organisms and their environment. The word is derived from the Greek *oikos*, meaning 'home'. Ecology might therefore be thought of as the study of the 'home life' of living organisms. A less vague definition was suggested by Krebs (1972): 'Ecology is the scientific study of the interactions that determine the distribution and abundance of organisms'. Notice that Krebs' definition does not use the word 'environment'; to see why, it is necessary to define the word. The environment of an organism consists of all those factors and phenomena outside the organism that influence it, whether these are physical and chemical (abiotic) or other organisms (biotic). The 'interactions' in Krebs' definition are, of course, interactions with these very factors. The environment therefore retains the central position that Haeckel gave it. Krebs' definition has the merit of pinpointing the ultimate subject matter of ecology: the distribution and abundance of organisms – *where* organisms occur, *how many* occur there, and *why*. This being so, it might be better still to define ecology as:

the scientific study of the distribution and abundance of organisms and the interactions that determine distribution and abundance.

As far as the subject matter of ecology is concerned, 'the distribution and abundance of organisms' is pleasantly succinct. But we need to expand it. The living world can be viewed as a biological hierarchy that starts with subcellular particles, and continues up through cells, tissues and organs. Ecology deals with the next three levels: the individual *organism*, the *population* (consisting of individuals of the same species) and the *community* (consisting of a greater or lesser number of species populations). At the level of the organism, ecology deals with how individuals are affected by (and how they affect) their environment. At the level of the population, ecology is concerned with the presence or absence of particular species, their abundance or rarity, and with the trends and fluctuations in their numbers. Community ecology then deals with the composition and organization of ecological communities. Ecologists also focus on the pathways followed by energy and matter as these move among living and nonliving elements of a further category of organization: the *ecosystem*, comprising the community together with its physical environment. With this in mind, Likens (1992) would extend our preferred definition of ecology to include 'the interactions between organisms and the transformation and flux of energy and matter'. However, we take energy/matter transformations as being subsumed in the 'interactions' of our definition.

There are two broad approaches that ecologists can take at each level of ecological organization. First, much can be gained by building from properties at the level below: physiology when studying organismal ecology; individual clutch size and survival probabilities when investigating the dynamics of individual species populations; food consumption rates when dealing with interactions between predator and prey populations; limits to the similarity of coexisting species when researching communities, and so on. An alternative approach deals directly with properties of the level of interest – for example, niche breadth at the organismal level; relative importance of density-dependent processes at the population level; species diversity at the level of community; rate of biomass production at the ecosystem level – and tries to relate these to abiotic or biotic aspects of the environment. Both approaches have their uses, and both will be used in each of the three parts of this book: Organisms; Species Interactions; and Communities and Ecosystems.

Explanation, description, prediction and control

At all levels of ecological organization we can try to do a number of different things. In the first place we can try to *explain* or *understand*. This is a search for knowledge in the pure scientific tradition. In order to do this, however, it is necessary first to *describe*. This, too, adds to our knowledge of the living world. Obviously, in order to understand something, we must first have a description of whatever it is that we wish to understand. Equally, but less obviously, the most valuable descriptions are those carried out with a particular problem or 'need for understanding' in mind. All descriptions are selective: but undirected description, carried out for its own sake, is often found afterwards to have selected the wrong things.

Ecologists also often try to *predict* what will happen to an organism, a population, a community or an ecosystem under a particular set of circumstances: and on the basis of these predictions we try to *control* the situation. We try to minimize the effects of locust plagues by predicting when they are likely to occur and taking appropriate action. We try to protect crops by predicting when conditions will be favorable to the crop and unfavorable to its enemies. We try to maintain endangered species by predicting the conservation policy that will enable them to persist. We try to conserve biodiversity to maintain ecosystem 'services' such as the protection of chemical quality of natural waters. Some prediction and control can be carried out without explanation or understanding. But confident predictions, precise predictions and predictions of what will happen in unusual circumstances can be made only when we can explain what is going on. Mathematical modeling has played, and will continue to play, a crucial role in the development of ecology, particularly in our ability to predict outcomes. But it is the real world we are interested in, and the worth of models must always be judged in terms of the light they shed on the working of natural systems.

It is important to realize that there are two different classes of explanation in biology: proximal and ultimate explanations. For example, the present distribution and abundance of a particular species of bird may be 'explained' in terms of the physical environment that the bird tolerates, the food that it eats and the parasites and predators that attack it. This is a *proximal* explanation. However, we may also ask how this species of bird comes to have these properties that now appear to govern its life. This question has to be answered by an explanation in evolutionary terms. The *ultimate* explanation of the present distribution and abundance of this bird lies in the ecological experiences of its ancestors. There are many problems in ecology that demand evolutionary, ultimate explanations: 'How have organisms come to possess particular combinations of size, developmental rate, reproductive output and so on?' (Chapter 4), 'What causes predators to adopt particular patterns of foraging behavior?' (Chapter 9) and 'How does it come about that coexisting species are often similar but rarely the same?' (Chapter 19). These problems are as much part of modern ecology as are the prevention of plagues, the protection of crops and the preservation of rare species. Our ability to control and exploit ecosystems cannot fail to be improved by an ability to explain and understand. And in the search for understanding, we must combine both proximal and ultimate explanations.

Pure and applied ecology

Ecologists are concerned not only with communities, populations and organisms *in nature*, but also with manmade or human-influenced environments (plantation forests, wheat fields, grain stores, nature reserves and so on), and with the consequences of human influence *on* nature (pollution, overharvesting, global climate change). In fact, our influence is so pervasive that we would be hard pressed to find an environment that was totally unaffected by human activity. Environmental problems are now high on the political agenda and ecologists clearly have a central role to play: a sustainable future depends fundamentally on ecological understanding and our ability to predict or produce outcomes under different scenarios.

When the first edition of this text was published in 1986, the majority of ecologists would have classed themselves as pure scientists, defending their right to pursue ecology for its own sake and not wishing to be deflected into narrowly applied projects. The situation has changed dramatically in 20 years, partly because governments have shifted the focus of grant-awarding bodies towards ecological applications, but also, and more fundamentally, because ecologists have themselves responded to the need to direct much of their research to the many environmental problems that have become ever more pressing. This is recognized in this new edition by a systematic treatment of ecological applications – each of the three sections of the book concludes with an applied chapter. We believe strongly that the application of ecological theory must be based on a sophisticated understanding of the pure science. Thus, our ecological application chapters are organized around the ecological understanding presented in the earlier chapters of each section.

Part 1
Organisms

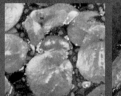

Introduction

We have chosen to start this book with chapters about organisms, then to consider the ways in which they interact with each other, and lastly to consider the properties of the communities that they form. One could call this a 'constructive' approach. We could though, quite sensibly, have treated the subject the other way round – starting with a discussion of the complex communities of both natural and manmade habitats, proceeding to *de*construct them at ever finer scales, and ending with chapters on the characteristics of the individual organisms – a more analytical approach. Neither is 'correct'. Our approach avoids having to describe community patterns before discussing the populations that comprise them. But when we start with individual organisms, we have to accept that many of the environmental forces acting on them, especially the species with which they coexist, will only be dealt with fully later in the book.

This first section covers individual organisms and populations composed of just a single species. We consider initially the sorts of correspondences that we can detect between organisms and the environments in which they live. It would be facile to start with the view that every organism is in some way ideally fitted to live where it does. Rather, we emphasize in Chapter 1 that organisms frequently are as they are, and live where they do, because of the constraints imposed by their evolutionary history. All species are absent from almost everywhere, and we consider next, in Chapter 2, the ways in which environmental conditions vary from place to place and from time to time, and how these put limits on the distribution of particular species. Then, in Chapter 3, we look at the resources that different types of organisms consume, and the nature of their interactions with these resources.

The particular species present in a community, and their abundance, give that community much of its ecological interest. Abundance and distribution (variation in abundance from place to place) are determined by the balance between birth, death, immigration and emigration. In Chapter 4 we consider some of the variety in the schedules of birth and death, how these may be quantified, and the resultant patterns in 'life histories': lifetime profiles of growth, differentiation, storage and reproduction. In Chapter 5 we examine perhaps the most pervasive interaction acting within single-species populations: intraspecific competition for shared resources in short supply. In Chapter 6 we turn to movement: immigration and emigration. Every species of plant and animal has a characteristic ability to disperse. This determines the rate at which individuals escape from environments that are or become unfavorable, and the rate at which they discover sites that are ripe for colonization and exploitation. The abundance or rarity of a species may be determined by its ability to disperse (or migrate) to unoccupied patches, islands or continents. Finally in this section, in Chapter 7, we consider the application of the principles that have been discussed in the preceding chapters, including niche theory, life history theory, patterns of movement, and the dynamics of small populations, paying particular attention to restoration after environmental damage, biosecurity (resisting the invasion of alien species) and species conservation.

Chapter 1

Organisms in their Environments: the Evolutionary Backdrop

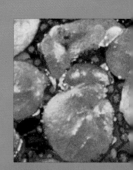

1.1 Introduction: natural selection and adaptation

From our definition of ecology in the Preface, and even from a layman's understanding of the term, it is clear that at the heart of ecology lies the relationship between organisms and their environments. In this opening chapter we explain how, fundamentally, this is an evolutionary relationship. The great Russian–American biologist Theodosius Dobzhansky famously said: 'Nothing in biology makes sense, except in the light of evolution'. This is as true of ecology as of any other aspect of biology. Thus, we try here to explain the processes by which the properties of different sorts of species make their life possible in particular environments, and also to explain their failure to live in other environments. In mapping out this evolutionary backdrop to the subject, we will also be introducing many of the questions that are taken up in detail in later chapters.

The phrase that, in everyday speech, is most commonly used to describe the match between organisms and environment is: 'organism X is adapted to' followed by a description of where the organism is found. Thus, we often hear that 'fish are adapted to live in water', or 'cacti are adapted to live in conditions of drought'. In everyday speech, this may mean very little: simply that fish have characteristics that allow them to live in water (and perhaps exclude them from other environments) or that cacti have characteristics that allow them to live where water is scarce. The word 'adapted' here says nothing about how the characteristics were acquired.

For an ecologist or evolutionary biologist, however, 'X is adapted to live in Y' means that environment Y has provided forces of natural selection that have affected the life of X's ancestors and so have molded and specialized the evolution of X. 'Adaptation' means that genetic change has occurred.

the meaning of adaptation

Regrettably, though, the word 'adaptation' implies that organisms are matched to their present environments, suggesting 'design' or even 'prediction'. But organisms have not been designed for, or fitted to the present: they have been molded (by *natural selection*) by past environments. Their characteristics reflect the successes and failures of ancestors. They appear to be apt for the environments that they live in at present only because present environments tend to be similar to those of the past.

The theory of evolution by natural selection is an ecological theory. It was first elaborated by Charles Darwin (1859), though its essence was also appreciated by a contemporary and correspondent of Darwin's, Alfred Russell Wallace (Figure 1.1). It rests on a series of propositions.

evolution by natural selection

1 The individuals that make up a population of a species are *not identical*: they vary, although sometimes only slightly, in size, rate of development, response to temperature, and so on.

2 Some, at least, of this variation is *heritable*. In other words, the characteristics of an individual are determined to some extent by its genetic make-up. Individuals receive their genes from their ancestors and therefore tend to share their characteristics.

3 All populations have the *potential* to populate the whole earth, and they would do so if each individual survived and each individual produced its maximum number of descendants. But they do not: many individuals die prior to reproduction, and most (if not all) reproduce at a less than maximal rate.

4 Different ancestors leave *different numbers of descendants*. This means much more than saying that different individuals produce different numbers of offspring. It includes also the chances of survival of offspring to reproductive age, the survival and reproduction of the progeny of these offspring, the survival and reproduction of their offspring in turn, and so on.

5 Finally, the number of descendants that an individual leaves depends, not entirely but crucially, on *the interaction between the characteristics of the individual and its environment.*

Figure 1.1   (a) Charles Darwin, 1849 (lithograph by Thomas H. Maguire; courtesy of The Royal Institution, London, UK/Bridgeman Art Library). (b) Alfred Russell Wallace, 1862 (courtesy of the Natural History Museum, London).

In any environment, some individuals will tend to survive and reproduce better, and leave more descendants, than others. If, because of this, the heritable characteristics of a population change from generation to generation, then evolution by natural selection is said to have occurred. This is the sense in which nature may loosely be thought of as *selecting*. But nature does not select in the way that plant and animal breeders select. Breeders have a defined end in view – bigger seeds or a faster racehorse. But nature does not *actively* select in this way: it simply sets the scene within which the evolutionary play of differential survival and reproduction is played out.

fitness: it's all relative

The fittest individuals in a population are those that leave the greatest number of descendants. In practice, the term is often applied not to a single individual, but to a typical individual or a type. For example, we may say that in sand dunes, yellow-shelled snails are fitter than brown-shelled snails. *Fitness*, then, is a relative not an absolute term. The fittest individuals in a population are those that leave the greatest number of descendants *relative* to the number of descendants left by other individuals in the population.

When we marvel at the diversity of complex specializations, there is a temptation to regard each case as an

evolved perfection?
no

example of evolved perfection. But this would be wrong. The evolutionary process works on the genetic variation that is available. It follows that natural selection is unlikely to lead to the evolution of perfect, 'maximally fit' individuals. Rather, organisms come to match their environments by being 'the fittest available' or 'the fittest yet': they are not 'the best imaginable'. Part of the lack of fit arises because the present properties of an organism have not all originated in an environment similar in every respect to the one in which it now lives. Over the course of its evolutionary history (its phylogeny), an organism's remote ancestors may have evolved a set of characteristics – evolutionary 'baggage' – that subsequently constrain future evolution. For many millions of years, the evolution of vertebrates has been limited to what can be achieved by organisms with a vertebral column. Moreover, much of what we now see as precise matches between an organism and its environment may equally be seen as constraints: koala bears live successfully on *Eucalyptus* foliage, but, from another perspective, koala bears cannot live without *Eucalyptus* foliage.

1.2 Specialization within species

The natural world is not composed of a continuum of types of organism each grading into the next: we recognize boundaries between one type of organism and another. Nevertheless, within what we recognize as *species* (defined below), there is often considerable variation, and some of this is heritable. It is on such intraspecific variation, after all, that plant and animal breeders (and natural selection) work.

Since the environments experienced by a species in different parts of its range are themselves different (to at least some extent), we might expect natural selection to have favored different variants of the species at different sites. The word '*ecotype*' was first coined for plant populations (Turesson, 1922a, 1922b) to describe genetically determined differences between populations within a species that reflect local matches between the organisms and their environments. But evolution forces the characteristics of populations to diverge from each other only if: (i) there is sufficient heritable variation on which selection can act; and (ii) the forces favoring divergence are strong enough to counteract the mixing and hybridization of individuals from different sites. Two populations will not diverge completely if their members (or, in the case of plants, their pollen) are continually migrating between them and mixing their genes.

Local, specialized populations become differentiated most conspicuously amongst organisms that are immobile for most of their lives. Motile organisms have a large measure of control over the environment in which they live; they can recoil or retreat from a lethal or unfavorable environment and actively seek another. Sessile, immobile organisms have no such freedom. They must live, or die, in the conditions where they settle. Populations of sessile organisms are therefore exposed to forces of natural selection in a peculiarly intense form.

This contrast is highlighted on the seashore, where the intertidal environment continually oscillates between the terrestrial and the aquatic. The fixed algae, sponges, mussels and barnacles all meet and tolerate life at the two extremes. But the mobile shrimps, crabs and fish track their aquatic habitat as it moves; whilst the shore-feeding birds track their terrestrial habitat. The mobility of such organisms enables them to match their environments to themselves. The immobile organism must match itself to its environment.

1.2.1 Geographic variation within species: ecotypes

The sapphire rockcress, *Arabis fecunda*, is a rare perennial herb restricted to calcareous soil outcrops in western Montana (USA) – so rare, in fact, that there are just 19 existing populations separated into two groups ('high elevation' and 'low elevation') by a distance of around 100 km. Whether there is local adaptation is of practical importance for conservation: four of the low elevation populations are under threat from spreading urban areas and may require reintroduction from elsewhere if they are to be sustained. Reintroduction may fail if local adaptation is too marked. Observing plants in their own habitats and checking for differences between them would not tell us if there was local adaptation in the evolutionary sense. Differences may simply be the result of immediate responses to contrasting environments made by plants that are essentially the same. Hence, high and low elevation plants were grown together in a 'common garden', eliminating any influence of contrasting immediate environments (McKay *et al.*, 2001). The low elevation sites were more prone to drought; both the air and the soil were warmer and drier. The low elevation plants in the common garden were indeed significantly more drought tolerant (Figure 1.2).

On the other hand, local selection by no means always overrides hybridization. For example, in a study of *Chamaecrista fasciculata*, an annual legume from disturbed habitats in eastern North America, plants were grown in a common garden that were derived from the 'home' site or were transplanted from distances of 0.1, 1, 10, 100, 1000 and 2000 km (Galloway & Fenster, 2000). The study was replicated three times: in Kansas, Maryland and northern Illinois. Five characteristics were measured: germination, survival, vegetative biomass, fruit production and the number of fruit produced per seed planted. But for all characters in all replicates there was little or no evidence for local adaptation except at the very furthest spatial scales (e.g. Figure 1.3). There is 'local adaptation' – but it's clearly not *that* local.

the balance between local adaptation and hybridization

We can also test whether organisms have evolved to become specialized to life in their local environment in *reciprocal transplant* experiments: comparing their performance when they are grown 'at home' (i.e. in their original habitat) with their performance 'away' (i.e. in the habitat of others). One such experiment (concerning white clover) is described in the next section.

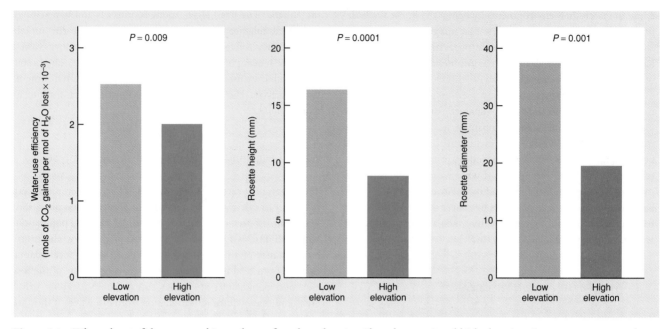

Figure 1.2   When plants of the rare sapphire rockcress from low elevation (drought-prone) and high elevation sites were grown together in a common garden, there was local adaptation: those from the low elevation site had significantly better water-use efficiency as well as having both taller and broader rosettes. (From McKay *et al.*, 2001.)

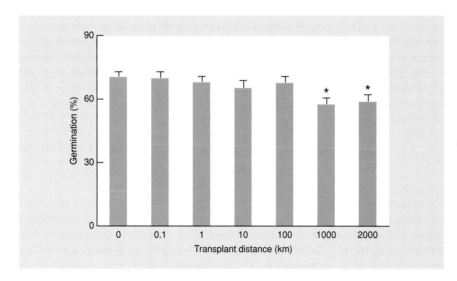

Figure 1.3   Percentage germination of local and transplanted *Chamaecrista fasciculata* populations to test for local adaptation along a transect in Kansas. Data for 1995 and 1996 have been combined because they do not differ significantly. Populations that differ from the home population at $P < 0.05$ are indicated by an asterisk. Local adaptation occurs at only the largest spatial scales. (From Galloway & Fenster, 2000.)

1.2.2   Genetic polymorphism

transient
polymorphisms

On a finer scale than ecotypes, it may also be possible to detect levels of variation *within* populations. Such variation is known as polymorphism. Specifically, genetic polymorphism is 'the occurrence together in the same habitat of two or more discontinuous forms of a species

in such proportions that the rarest of them cannot merely be maintained by recurrent mutation or immigration' (Ford, 1940). Not all such variation represents a match between organism and environment. Indeed, some of it may represent a mismatch, if, for example, conditions in a habitat change so that one form is being replaced by another. Such polymorphisms are called transient. As all communities are always changing, much polymorphism that we observe in nature may be transient, representing the extent to which the genetic response of populations to environmental change will always be out of step with the environment and unable to anticipate changing circumstances – this is illustrated in the peppered moth example below.

the maintenance of polymorphisms

Many polymorphisms, however, are actively maintained in a population by natural selection, and there are a number of ways in which this may occur.

1 *Heterozygotes may be of superior fitness*, but because of the mechanics of Mendelian genetics they continually generate less fit homozygotes within the population. Such 'heterosis' is seen in human sickle-cell anaemia where malaria is prevalent. The malaria parasite attacks red blood cells. The sickle-cell mutation gives rise to red cells that are physiologically imperfect and misshapen. However, sickle-cell heterozygotes are fittest because they suffer only slightly from anemia and are little affected by malaria; but they continually generate homozygotes that are either dangerously anemic (two sickle-cell genes) or susceptible to malaria (no sickle-cell genes). None the less, the superior fitness of the heterozygote maintains both types of gene in the population (that is, a polymorphism).

2 *There may be gradients of selective forces* favoring one form (morph) at one end of the gradient, and another form at the other. This can produce polymorphic populations at intermediate positions in the gradient – this, too, is illustrated below in the peppered moth study.

3 *There may be frequency-dependent selection* in which each of the morphs of a species is fittest when it is rarest (Clarke & Partridge, 1988). This is believed to be the case when rare color forms of prey are fit because they go unrecognized and are therefore ignored by their predators.

4 *Selective forces may operate in different directions within different patches* in the population. A striking example of this is provided by a reciprocal transplant study of white clover (*Trifolium repens*) in a field in North Wales (UK). To determine whether the characteristics of individuals matched local features of their environment, Turkington and Harper (1979) removed plants from marked positions in the field and multiplied them into clones in the common environment of a greenhouse. They then transplanted samples from each clone into the place in the sward of vegetation from which it had originally been taken (as a control), and also to the places from where all the others had been taken (a transplant). The plants were allowed to grow for a year before they were removed, dried and weighed. The mean weight of clover plants transplanted back into their home sites was 0.89 g but at away sites it was only 0.52 g, a statistically highly significant difference. This provides strong, direct evidence that clover clones in the pasture had evolved to become specialized such that they performed best in their local environment. But all this was going on within a single population, which was therefore polymorphic.

In fact, the distinction between local ecotypes and polymorphic populations is not always a clear one. This is illustrated by another study in North Wales, where there was a gradation in habitats at the margin between maritime cliffs and grazed pasture, and a common species, creeping bent grass (*Agrostis stolonifera*), was present in many of the habitats. Figure 1.4 shows a map of the site and one of the transects from which plants were sampled. It also shows the results when plants from the sampling points along this transect were grown in a common garden. The

no clear distinction between local ecotypes and a polymorphism

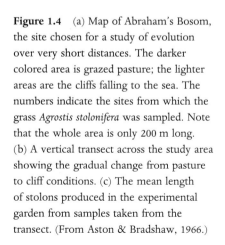
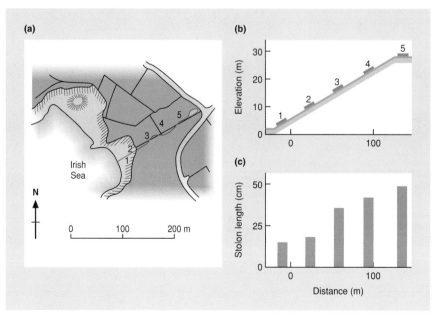

Figure 1.4 (a) Map of Abraham's Bosom, the site chosen for a study of evolution over very short distances. The darker colored area is grazed pasture; the lighter areas are the cliffs falling to the sea. The numbers indicate the sites from which the grass *Agrostis stolonifera* was sampled. Note that the whole area is only 200 m long. (b) A vertical transect across the study area showing the gradual change from pasture to cliff conditions. (c) The mean length of stolons produced in the experimental garden from samples taken from the transect. (From Aston & Bradshaw, 1966.)

plants spread by sending out shoots along the ground surface (stolons), and the growth of plants was compared by measuring the lengths of these. In the field, cliff plants formed only short stolons, whereas those of the pasture plants were long. In the experimental garden, these differences were maintained, even though the sampling points were typically only around 30 m apart – certainly within the range of pollen dispersal between plants. Indeed, the gradually changing environment along the transect was matched by a gradually changing stolon length, presumably with a genetic basis, since it was apparent in the common garden. Thus, even though the spatial scale was so small, the forces of selection seem to outweigh the mixing forces of hybridization – but it is a moot point whether we should describe this as a small-scale series of local ecotypes or a polymorphic population maintained by a gradient of selection.

1.2.3 Variation within a species with manmade selection pressures

It is, perhaps, not surprising that some of the most dramatic examples of local specialization within species (indeed of natural selection in action) have been driven by manmade ecological forces, especially those of environmental pollution. These can provide rapid change under the influence of powerful selection pressures. *Industrial melanism*, for example, is the phenomenon in which black or blackish forms of species have come to dominate populations in industrial areas. In the dark individuals, a dominant gene is typically responsible for producing an excess of the black pigment melanin. Industrial melanism is known in most industrialized countries and more than 100 species of moth have evolved forms of industrial melanism.

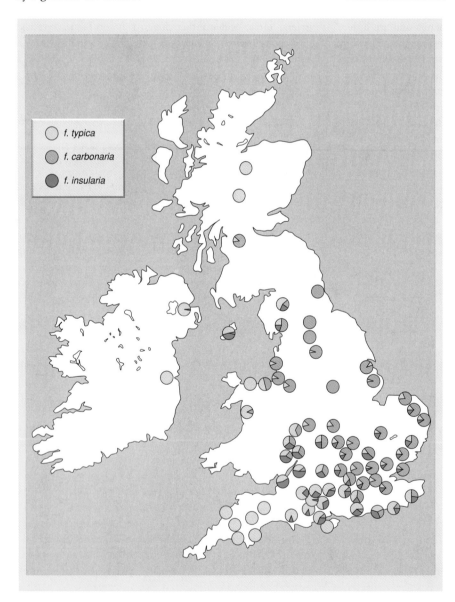

Figure 1.5 Sites in Britain where the frequencies of the pale (*forma typica*) and melanic forms of *Biston betularia* were recorded by Kettlewell and his colleagues. In all more than 20,000 specimens were examined. The principal melanic form (*forma carbonaria*) was abundant near industrial areas and where the prevailing westerly winds carry atmospheric pollution to the east. A further melanic form (*forma insularia*, which looks like an intermediate form but is due to several different genes controlling darkening) was also present but was hidden where the genes for *forma carbonaria* were present. (From Ford, 1975.)

industrial melanism
in the peppered
moth

The earliest recorded species to evolve in this way was the peppered moth (*Biston betularia*); the first black specimen in an otherwise pale population was caught in Manchester (UK) in 1848. By 1895, about 98% of the Manchester peppered moth population was melanic. Following many more years of pollution, a large-scale survey of pale and melanic forms of the peppered moth in Britain recorded more than 20,000 specimens between 1952 and 1970 (Figure 1.5). The winds in Britain are predominantly westerlies, spreading industrial pollutants (especially smoke and sulfur dioxide) toward the east. Melanic forms were concentrated toward the east and were completely absent from the unpolluted western parts of England and Wales, northern Scotland and Ireland. Notice from the figure, though, that many populations were polymorphic: melanic and nonmelanic forms coexisted. Thus, the polymorphism seems to be a result both of environments changing (becoming more polluted) – to this extent the polymorphism is transient – and of there being a gradient of selective pressures from the less polluted west to the more polluted east.

The main selective pressure appears to be applied by birds that prey on the moths. In field experiments, large numbers of melanic and pale ('typical') moths were reared and released in equal numbers. In a rural and largely unpolluted area of southern England, most of those captured by birds were melanic. In an industrial area near the city of Birmingham, most were typicals (Kettlewell, 1955). Any idea, however, that melanic forms were favored simply because they were camouflaged against smoke-stained backgrounds in the polluted areas (and typicals were favored in unpolluted areas because they were camouflaged against pale backgrounds) may be only part of the story. The moths rest on tree trunks during the day, and nonmelanic moths are well hidden against a background of mosses and lichens. Industrial pollution has not just blackened the moths' background; sulfur dioxide, especially, has also destroyed most of the moss and lichen on the tree trunks. Thus, sulfur dioxide pollution may have been as important as smoke in selecting melanic moths.

In the 1960s, industrialized environments in Western Europe and the United States started to change again, as oil and electricity began to replace coal, and legislation was passed to impose smoke-free zones and to reduce industrial emissions of sulfur dioxide. The frequency of melanic forms then fell back to near pre-Industrial levels with remarkable speed (Figure 1.6). Again, there was transient polymorphism – but this time while populations were *en route* in the other direction.

1.3 Speciation

It is clear, then, that natural selection can force populations of plants and animals to change their character – to evolve. But none of the examples we have considered has involved the evolution of

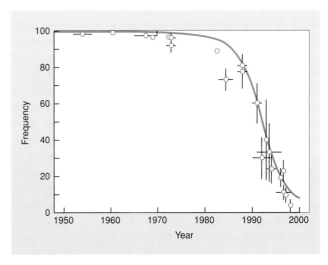

Figure 1.6 Change in the frequency of the *carbonaria* form of the peppered moth *Biston betularia* in the Manchester area since 1950. Vertical lines show the standard error and the horizontal lines show the range of years included. (After Cook *et al.*, 1999.)

a new species. What, then, justifies naming two populations as different species? And what is the process – 'speciation' – by which two or more new species are formed from one original species?

1.3.1 What do we mean by a 'species'?

Cynics have said, with some truth, that a species is what a competent taxonomist regards as a species. On the other hand, back in the 1930s two American biologists, Mayr and Dobzhansky, proposed an empirical test that could be used to decide whether two populations were part of the same species or of two different species. They recognized organisms as being members of a single species if they could, at least potentially, breed together in nature to produce fertile offspring. They called a species tested and defined in this way a *biological species* or *biospecies*. In the examples that we have used earlier in this chapter we know that melanic and normal peppered moths can mate and that the offspring are fully fertile; this is also true of plants from the different types of *Agrostis*. They are all variations within species – not separate species.

biospecies: the Mayr–
Dobzhansky test

In practice, however, biologists do not apply the Mayr–Dobzhansky test before they recognize every species: there is simply not enough time or resources, and in any case, there are vast portions of the living world – most microorganisms, for example – where an absence of sexual reproduction makes a strict interbreeding criterion inappropriate. What is more important is that the test recognizes a crucial element in the evolutionary process that we have met already in considering specialization

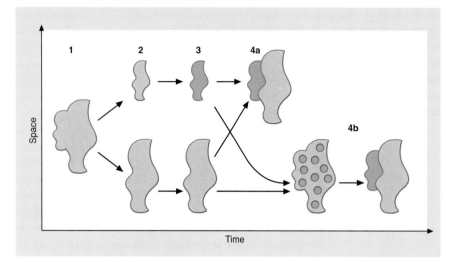

Figure 1.7 The orthodox picture of ecological speciation. A uniform species with a large range (1) differentiates (2) into subpopulations (for example, separated by geographic barriers or dispersed onto different islands), which become genetically isolated from each other (3). After evolution in isolation they may meet again, when they are either already unable to hybridize (4a) and have become true biospecies, or they produce hybrids of lower fitness (4b), in which case evolution may favor features that prevent interbreeding between the 'emerging species' until they are true biospecies.

within species. If the members of two populations are able to hybridize, and their genes are combined and reassorted in their progeny, then natural selection can never make them truly distinct. Although natural selection may tend to force a population to evolve into two or more distinct forms, sexual reproduction and hybridization mix them up again.

orthodox ecological speciation

'Ecological' speciation is speciation driven by divergent natural selection in distinct subpopulations (Schluter, 2001). The most orthodox scenario for this comprises a number of stages (Figure 1.7). First, two subpopulations become geographically isolated and natural selection drives genetic adaptation to their local environments. Next, as a *by-product* of this genetic differentiation, a degree of reproductive isolation builds up between the two. This may be 'pre-zygotic', tending to prevent mating in the first place (e.g. differences in courtship ritual), or 'post-zygotic': reduced viability, perhaps inviability, of the offspring themselves. Then, in a phase of '*secondary contact*', the two subpopulations re-meet. The hybrids between individuals from the different subpopulations are now of low fitness, because they are literally neither one thing nor the other. Natural selection will then favor any feature in either subpopulation that *reinforces* reproductive isolation, especially pre-zygotic characteristics, preventing the production of low-fitness hybrid offspring. These breeding barriers then cement the distinction between what have now become separate species.

allopatric and sympatric speciation

It would be wrong, however, to imagine that all examples of speciation conform fully to this orthodox picture (Schluter, 2001). First, there may never be secondary contact. This would be pure 'allopatric' speciation (that is, with all divergence occurring in subpopulations in *different* places). Second, there is clearly room for considerable variation in the relative importances of pre-zygotic and post-zygotic

mechanisms in both the allopatric and the secondary-contact phases.

Most fundamentally, perhaps, there has been increasing support for the view that an allopatric phase is not necessary: that is, 'sympatric' speciation is possible, with subpopulations diverging despite not being geographically separated from one another. Probably the most studied circumstance in which this seems likely to occur (see Drès & Mallet, 2002) is where insects feed on more than one species of host plant, and where each requires specialization by the insects to overcome the plant's defenses. (Consumer resource defense and specialization are examined more fully in Chapters 3 and 9.) Particularly persuasive in this is the existence of a continuum identified by Drès and Mallet: from populations of insects feeding on more than one host plant, through populations differentiated into 'host races' (defined by Drès and Mallet as sympatric subpopulations exchanging genes at a rate of more than around 1% per generation), to coexisting, closely related species. This reminds us, too, that the origin of a species, whether allopatric or sympatric, is a process, not an event. For the formation of a new species, like the boiling of an egg, there is some freedom to argue about when it is completed.

The evolution of species and the balance between natural selection and hybridization are illustrated by the extraordinary case of two species of sea gull. The lesser black-backed gull (*Larus fuscus*) originated in Siberia and colonized progressively to the west, forming a chain or *cline* of different forms, spreading from Siberia to Britain and Iceland (Figure 1.8). The neighboring forms along the cline are distinctive, but they hybridize readily in nature. Neighboring populations are therefore regarded as part of the same species and taxonomists give them only 'subspecific' status (e.g. *L. fuscus graellsii*, *L. fuscus fuscus*). Populations of the gull have, however, also spread east from Siberia, again forming a cline of freely hybridizing forms. Together, the populations spreading east and west encircle the northern hemisphere. They meet and overlap

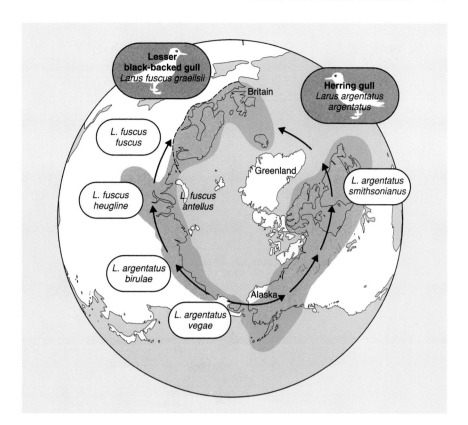

Figure 1.8 Two species of gull, the herring gull and the lesser black-backed gull, have diverged from a common ancestry as they have colonized and encircled the northern hemisphere. Where they occur together in northern Europe they fail to interbreed and are clearly recognized as two distinct species. However, they are linked along their ranges by a series of freely interbreeding races or subspecies. (After Brookes, 1998.)

in northern Europe. There, the eastward and westward clines have diverged so far that it is easy to tell them apart, and they are recognized as two different *species*, the lesser black-backed gull (*L. fuscus*) and the herring gull (*L. argentatus*). Moreover, the two species do not hybridize: they have become true biospecies. In this remarkable example, then, we can see how two distinct species have evolved from one primal stock, and that the stages of their divergence remain frozen in the cline that connects them.

1.3.2 Islands and speciation

Darwin's finches

We will see repeatedly later in the book (and especially in Chapter 21) that the isolation of islands – and not just land islands in a sea of water – can have a profound effect on the ecology of the populations and communities living there. Such isolation also provides arguably the most favorable environment for populations to diverge into distinct species. The most celebrated example of evolution and speciation on islands is the case of Darwin's finches in the Galápagos archipelago. The Galápagos are volcanic islands isolated in the Pacific Ocean about 1000 km west of Ecuador and 750 km from the island of Cocos, which is itself 500 km from Central America. At more than 500 m above sea level the vegetation is open grassland. Below this

is a humid zone of forest that grades into a coastal strip of desert vegetation with some endemic species of prickly pear cactus (*Opuntia*). Fourteen species of finch are found on the islands. The evolutionary relationships amongst them have been traced by molecular techniques (analyzing variation in 'microsatellite' DNA) (Figure 1.9) (Petren *et al.*, 1999). These accurate modern tests confirm the long-held view that the family tree of the Galápagos finches radiated from a single trunk: a single ancestral species that invaded the islands from the mainland of Central America. The molecular data also provide strong evidence that the warbler finch (*Certhidea olivacea*) was the first to split off from the founding group and is likely to be the most similar to the original colonist ancestors. The entire process of evolutionary divergence of these species appears to have happened in less than 3 million years.

Now, in their remote island isolation, the Galápagos finches, despite being closely related, have radiated into a variety of species with contrasting ecologies (Figure 1.9), occupying ecological niches that elsewhere are filled by quite unrelated species. Members of one group, including *Geospiza fuliginosa* and *G. fortis*, have strong bills and hop and scratch for seeds on the ground. *G. scandens* has a narrower and slightly longer bill and feeds on the flowers and pulp of the prickly pears as well as on seeds. Finches of a third group have parrot-like bills and feed on leaves, buds, flowers and fruits, and a fourth group with a parrot-like bill (*Camarhynchus*

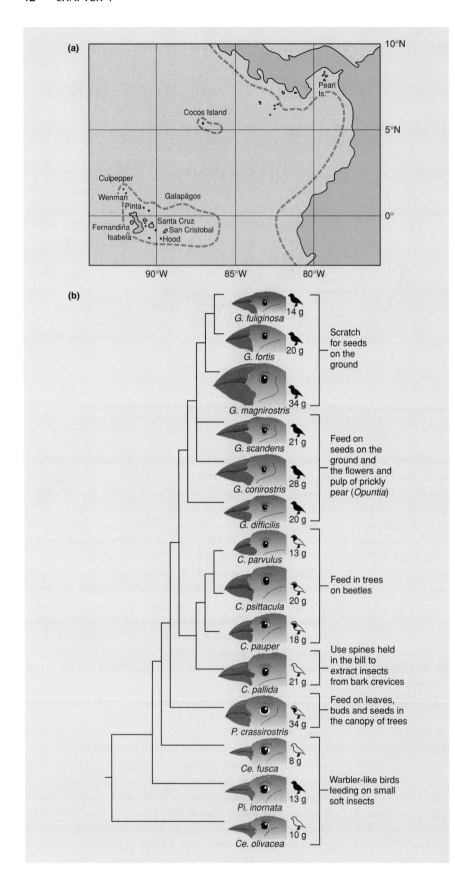

Figure 1.9 (a) Map of the Galápagos Islands showing their position relative to Central America; on the equator 5° equals approximately 560 km. (b) A reconstruction of the evolutionary history of the Galápagos finches based on variation in the length of microsatellite deoxyribonucleic acid (DNA). The feeding habits of the various species are also shown. Drawings of the birds are proportional to actual body size. The maximum amount of black coloring in male plumage and the average body mass are shown for each species. The *genetic distance* (a measure of the genetic difference) between species is shown by the length of the horizontal lines. Notice the great and early separation of the warbler finch (*Certhidea olivacea*) from the others, suggesting that it may closely resemble the founders that colonized the islands. C, *Camarhynchus*; Ce, *Certhidea*; G, *Geospiza*; P, *Platyspiza*; Pi, *Pinaroloxias*. (After Petren et al., 1999.)

psittacula) has become insectivorous, feeding on beetles and other insects in the canopy of trees. A so-called woodpecker finch, *Camarhynchus* (*Cactospiza*) *pallida*, extracts insects from crevices by holding a spine or a twig in its bill, while yet a further group includes the warbler finch, which flits around actively and collects small insects in the forest canopy and in the air. Isolation – both of the archipelago itself and of individual islands within it – has led to an original evolutionary line radiating into a series of species, each matching its own environment.

1.4   Historical factors

Our world has not been constructed by someone taking each species in turn, testing it against each environment, and molding it so that every species finds its perfect place. It is a world in which species live where they do for reasons that are often, at least in part, accidents of history. We illustrate this first by continuing our examination of islands.

1.4.1   Island patterns

Many of the species on islands are either subtly or profoundly different from those on the nearest comparable area of mainland. Put simply, there are two main reasons for this.

1   The animals and plants on an island are limited to those types having ancestors that managed to disperse there, although the extent of this limitation depends on the isolation of the island and the intrinsic dispersal ability of the animal or plant in question.
2   Because of this isolation, as we saw in the previous section, the rate of evolutionary change on an island may often be fast enough to outweigh the effects of the exchange of genetic material between the island population and related populations elsewhere.

Thus, islands contain many species unique to themselves ('*endemics*' – species found in only one area), as well as many differentiated 'races' or 'subspecies' that are distinguishable from mainland forms. A few individuals that disperse by chance to a habitable island can form the nucleus of an expanding new species. Its character will have been colored by the particular genes that were represented among the colonists – which are unlikely to be a perfect sample of the parent population. What natural selection can do with this *founder population* is limited by what is in its limited sample of genes (plus occasional rare mutations). Indeed much of the deviation among populations isolated on islands appears to be due to a *founder effect* – the chance composition of the pool of founder genes puts limits and constraints on what variation there is for natural selection to act upon.

The *Drosophila* fruit-flies of Hawaii provide a further spectacular example of species formation on islands. The Hawaiian chain of islands (Figure 1.10) is volcanic in origin, having been formed gradually over the last 40 million years, as the center of the Pacific tectonic plate moved steadily over a 'hot spot' in a southeasterly direction (Niihau is the most ancient of the islands, Hawaii itself the most recent). The richness of the Hawaiian *Drosophila* is spectacular: there are probably about 1500 *Drosophila* spp. worldwide, but at least 500 of these are found only in the Hawaiian islands.

Of particular interest are the 100   Hawaiian *Drosophila*
or so species of 'picture-winged' *Drosophila*. The lineages through which these species have evolved can be traced by analyzing the banding patterns on the giant chromosomes in the salivary glands of their larvae. The evolutionary tree that emerges is shown in Figure 1.10, with each species lined up above the island on which it is found (there are only two species found on more than one island). The historical element in 'what lives where' is plainly apparent: the more ancient species live on the more ancient islands, and, as new islands have been formed, rare dispersers have reached them and eventually evolved in to new species. At least some of these species appear to match the same environment as others on different islands. Of the closely related species, for example, *D. adiastola* (species 8) is only found on Maui and *D. setosimentum* (species 11) only on Hawaii, but the environments that they live in are apparently indistinguishable (Heed, 1968). What is most noteworthy, of course, is the power and importance of isolation (coupled with natural selection) in generating new species. Thus, island biotas illustrate two important, related points: (i) that there is a historical element in the match between organisms and environments; and (ii) that there is not just one perfect organism for each type of environment.

1.4.2   Movements of land masses

Long ago, the curious distributions of species between continents, seemingly inexplicable in terms of dispersal over vast distances, led biologists, especially Wegener (1915), to suggest that the continents themselves must have moved. This was vigorously denied by geologists, until geomagnetic measurements required the same, apparently wildly improbable explanation. The discovery that the tectonic plates of the earth's crust move and carry with them the migrating continents, reconciles geologist and biologist (Figure 1.11b–e). Thus, whilst major evolutionary developments were occurring in the plant and animal kingdoms, populations were being split and separated, and land areas were moving across climatic zones.

Figure 1.12 shows just one example   large flightless birds
of a major group of organisms (the
large flightless birds), whose distributions begin to make sense only in the light of the movement of land masses. It would be

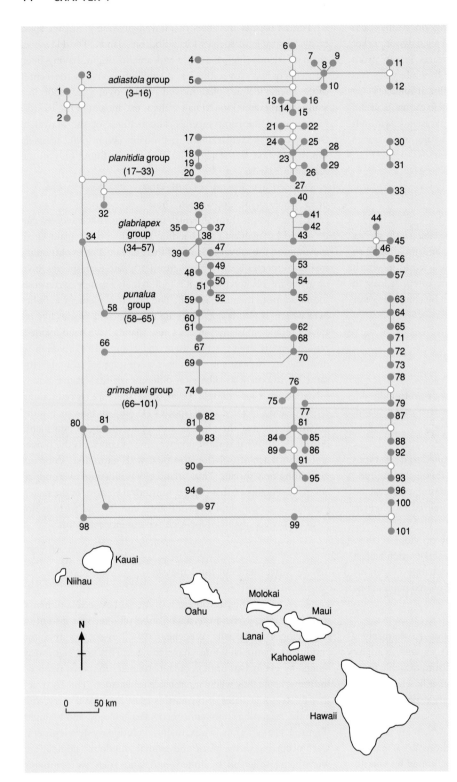

Figure 1.10 An evolutionary tree linking the picture-winged *Drosophila* of Hawaii, traced by the analysis of chromosomal banding patterns. The most ancient species are *D. primaeva* (species 1) and *D. attigua* (species 2), found only on the island of Kauai. Other species are represented by solid circles; hypothetical species, needed to link the present day ones, are represented by open circles. Each species has been placed above the island or islands on which it is found (although Molokai, Lanai and Maui are grouped together). Niihau and Kahoolawe support no *Drosophila*. (After Carson & Kaneshiro, 1976; Williamson, 1981.)

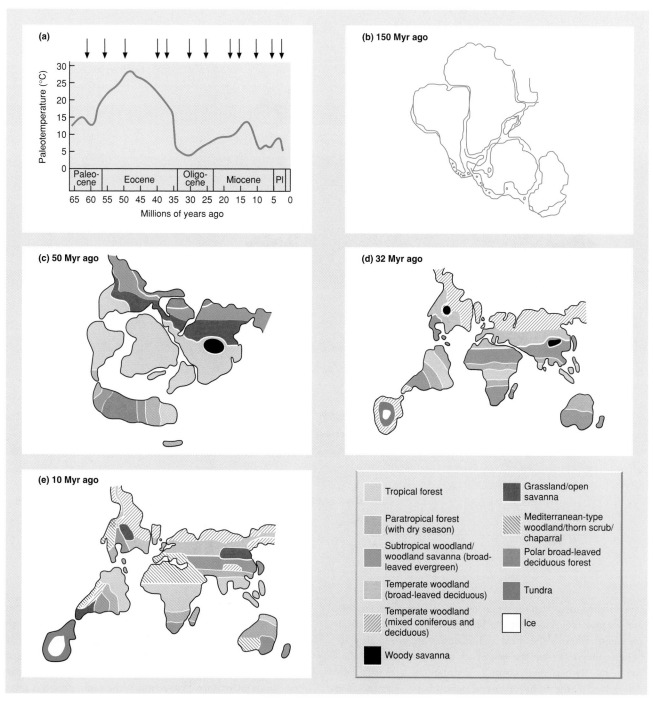

Figure 1.11   (a) Changes in temperature in the North Sea over the past 60 million years. During this period there were large changes in sea level (arrows) that allowed dispersal of both plants and animals between land masses. (b–e) Continental drift. (b) The ancient supercontinent of Gondwanaland began to break up about 150 million years ago. (c) About 50 million years ago (early Middle Eocene) recognizable bands of distinctive vegetation had developed, and (d) by 32 million years ago (early Oligocene) these had become more sharply defined. (e) By 10 million years ago (early Miocene) much of the present geography of the continents had become established but with dramatically different climates and vegetation from today; the position of the Antarctic ice cap is highly schematic. (Adapted from Norton & Sclater, 1979; Janis, 1993; and other sources).

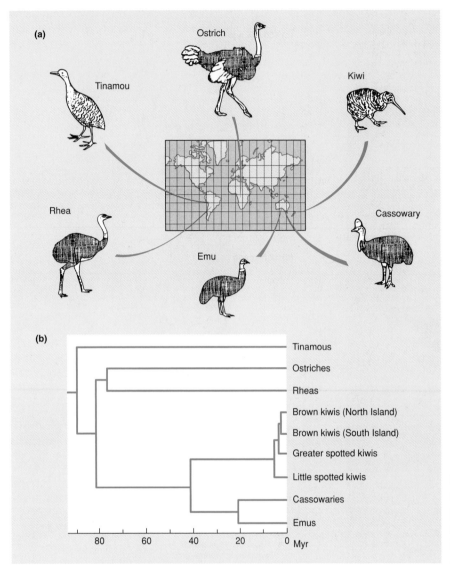

Figure 1.12   (a) The distribution of terrestrial flightless birds. (b) The phylogenetic tree of the flightless birds and the estimated times (million years, Myr) of their divergence. (After Diamond, 1983; from data of Sibley & Ahlquist.)

unwarranted to say that the emus and cassowaries are where they are because they represent the best match to Australian environments, whereas the rheas and tinamous are where *they* are because they represent the best match to South American environments. Rather, their disparate distributions are essentially determined by the prehistoric movements of the continents, and the subsequent impossibility of geographically isolated evolutionary lines reaching into each others' environment. Indeed, molecular techniques make it possible to analyze the time at which the various flightless birds started their evolutionary divergence (Figure 1.12). The tinamous seem to have been the first to diverge and became evolutionarily separate from the rest, the *ratites*. Australasia next split away from the other southern continents, and from the latter, the ancestral stocks of ostriches and rheas were subsequently separated when the Atlantic opened up between Africa

and South America. Back in Australasia, the Tasman Sea opened up about 80 million years ago and ancestors of the kiwi are thought to have made their way, by island hopping, about 40 million years ago across to New Zealand, where divergence into the present species happened relatively recently. An account of the evolutionary trends amongst mammals over much the same period is given by Janis (1993).

1.4.3   Climatic changes

Changes in climate have occurred on shorter timescales than the movements of land masses (Boden *et al.*, 1990; IGBP, 1990). Much of what we see in the present distribution of species represents phases in a recovery from past climatic shifts. Changes in climate during the Pleistocene ice ages, in particular, bear a lot of the responsibility for the present patterns of distribution of plants and animals. The extent of these climatic and biotic changes is only beginning to be unraveled as the technology for discovering, analyzing and dating biological remains becomes more sophisticated (particularly by the analysis of buried pollen samples). These methods increasingly allow us to determine just how much of the present distribution of organisms represents a precise local match to present environments, and how much is a fingerprint left by the hand of history.

the Pleistocene glacial cycles . . .

Techniques for the measurement of oxygen isotopes in ocean cores indicate that there may have been as many as 16 glacial cycles in the Pleistocene, each lasting for about 125,000 years (Figure 1.13a). It seems that each glacial phase may have lasted for as long as 50,000–100,000 years, with brief intervals of 10,000–20,000 years when the temperatures rose close to those we experience today. This suggests that it is present floras and faunas that are unusual, because they have developed towards the end of one of a series of unusual catastrophic warm events!

During the 20,000 years since the peak of the last glaciation, global temperatures have risen by about 8°C, and the rate at which vegetation has changed over much of this period has been detected by examining pollen records. The woody species that dominate pollen profiles at Rogers Lake in Connecticut (Figure 1.13b) have arrived in turn: spruce first and chestnut most recently. Each new arrival has added to the number of the species present, which has increased continually over the past 14,000-year period. The same picture is repeated in European profiles.

As the number of pollen records has increased, it has become possible not only to plot the changes in vegetation

. . . from which trees are still recovering

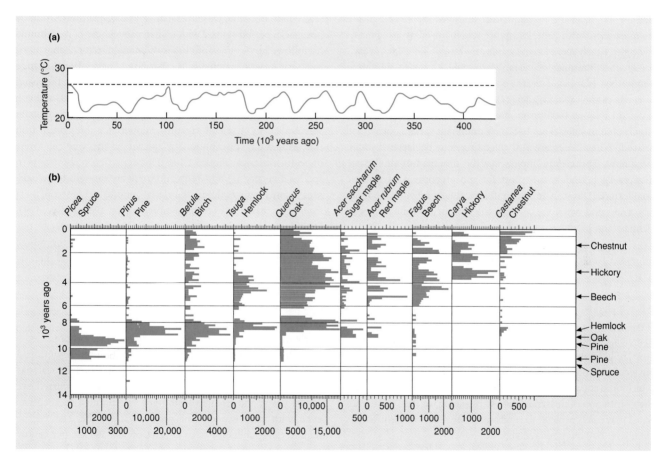

Figure 1.13   (a) An estimate of the temperature variations with time during glacial cycles over the past 400,000 years. The estimates were obtained by comparing oxygen isotope ratios in fossils taken from ocean cores in the Caribbean. The dashed line corresponds to the ratio 10,000 years ago, at the start of the present warming period. Periods as warm as the present have been rare events, and the climate during most of the past 400,000 years has been glacial. (After Emiliani, 1966; Davis, 1976.) (b) The profiles of pollen accumulated from late glacial times to the present in the sediments of Rogers Lake, Connecticut. The estimated date of arrival of each species in Connecticut is shown by arrows at the right of the figure. The horizontal scales represent pollen influx: 10^3 grains cm^{-2} year^{-1}. (After Davis *et al.*, 1973.)

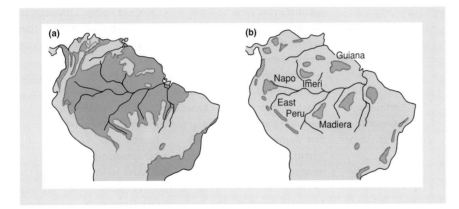

Figure 1.14 (a) The present-day distribution of tropical forest in South America. (b) The possible distribution of tropical forest refuges at the time when the last glaciation was at its peak, as judged by present-day hot spots of species diversity within the forest. (After Ridley, 1993.)

at a point in space, but to begin to map the movements of the various species as they have spread across the continents (see Bennet, 1986). In the invasions that followed the retreat of the ice in eastern North America, spruce was followed by jack pine or red pine, which spread northwards at a rate of 350–500 m year^{-1} for several thousands of years. White pine started its migration about 1000 years later, at the same time as oak. Hemlock was also one of the rapid invaders (200–300 m year^{-1}), and arrived at most sites about 1000 years after white pine. Chestnut moved slowly (100 m year^{-1}), but became a dominant species once it had arrived. Forest trees are still migrating into deglaciated areas, even now. This clearly implies that the timespan of an average interglacial period is too short for the attainment of floristic equilibrium (Davis, 1976). Such historical factors will have to be borne in mind when we consider the various patterns in species richness and biodiversity in Chapter 21.

'history' on a smaller scale

'History' may also have an impact on much smaller space and time scales. Disturbances to the benthic (bottom dwelling) community of a stream occurs when high discharge events (associated with storms or snow melt) result in a very small-scale mosaic of patches of scour (substrate loss), fill (addition of substrate) and no change (Matthaei *et al.*, 1999). The invertebrate communities associated with the different patch histories are distinctive for a period of months, within which time another high discharge event is likely to occur. As with the distribution of trees in relation to repeating ice ages, the stream fauna may rarely achieve an equilibrium between flow disturbances (Matthaei & Townsend, 2000).

changes in the tropics

The records of climatic change in the tropics are far less complete than those for temperate regions. There is therefore the temptation to imagine that whilst dramatic climatic shifts and ice invasions were dominating temperate regions, the tropics persisted in the state we know today. This is almost certainly wrong. Data from a variety of sources indicate that there were abrupt fluctuations in postglacial climates in Asia and Africa. In continental monsoon areas (e.g. Tibet, Ethiopia, western Sahara and subequatorial Africa) the postglacial period started with an extensive phase of high humidity followed by a series of phases of intense aridity (Zahn, 1994). In South America, a picture is emerging of vegetational changes that parallel those occurring in temperate regions, as the extent of tropical forest increased in warmer, wetter periods, and contracted, during cooler, drier glacial periods, to smaller patches surrounded by a sea of savanna. Support for this comes from the present-day distribution of species in the tropical forests of South America (Figure 1.14). There, particular 'hot spots' of species diversity are apparent, and these are thought to be likely sites of forest refuges during the glacial periods, and sites too, therefore, of increased rates of speciation (Prance, 1987; Ridley, 1993). On this interpretation, the present distributions of species may again be seen as largely accidents of history (where the refuges were) rather than precise matches between species and their differing environments.

how will global warming compare?

Evidence of changes in vegetation that followed the last retreat of the ice hint at the consequence of the global warming (maybe 3°C in the next 100 years) that is predicted to result from continuing increases in atmospheric carbon dioxide (discussed in detail in Sections 2.9.1 and 18.4.6). But the scales are quite different. Postglacial warming of about 8°C occurred over 20,000 years, and changes in the vegetation failed to keep pace even with this. But current projections for the 21st century require range shifts for trees at rates of 300–500 km per century compared to typical rates in the past of 20–40 km per century (and exceptional rates of 100–150 km). It is striking that the only precisely dated extinction of a tree species in the Quaternary, that of *Picea critchfeldii*, occurred around 15,000 years ago at a time of especially rapid postglacial warming (Jackson & Weng, 1999). Clearly, even more rapid change in the future could result in extinctions of many additional species (Davis & Shaw, 2001).

1.4.4   Convergents and parallels

analogous and
homologous structures

A match between the nature of organisms and their environment can often be seen as a similarity in form and behavior between organisms living in a similar environment, but belonging to different phyletic lines (i.e. different branches of the evolutionary tree). Such similarities also undermine further the idea that for every environment there is one, and only one, perfect organism. The evidence is particularly persuasive when the phyletic lines are far removed from each other, and when similar roles are played by structures that have quite different evolutionary origins, i.e. when the structures are *analogous* (similar in superficial form or function) but not *homologous* (derived from an equivalent structure in a common ancestry). When this is seen to occur, we speak of *convergent evolution*. Many flowering plants and some ferns, for example, use the support of others to climb high in the canopies of vegetation, and so gain access to more light than if they depended on their own supporting tissues. The ability to climb has evolved in many different families, and quite different organs have become modified into climbing structures (Figure 1.15a): they are analogous structures but not homologous. In other plant species the same organ has been modified into quite different structures with quite different roles: they are therefore homologous, although they may not be analogous (Figure 1.15b).

Other examples can be used to show the *parallels* in evolutionary pathways within separate groups that have radiated after they were isolated from each other. The classic example of such parallel evolution is the radiation amongst the placental and marsupial mammals. Marsupials arrived on the Australian continent in the Cretaceous period (around 90 million years ago), when the only other mammals present were the curious egg-laying monotremes (now represented only by the spiny anteaters (*Tachyglossus aculeatus*) and the duckbill platypus (*Ornithorynchus anatinus*)). An evolutionary process of radiation then occurred that in many

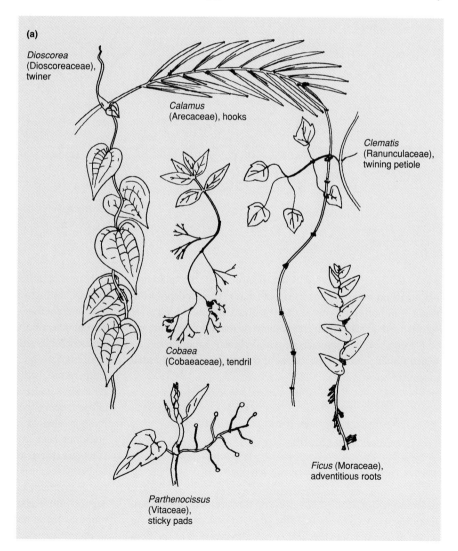

Figure 1.15 A variety of morphological features that allow flowering plants to climb. (a) Structural features that are analogous, i.e. derived from modifications of quite different organs, e.g. leaves, petioles, stems, roots and tendrils.

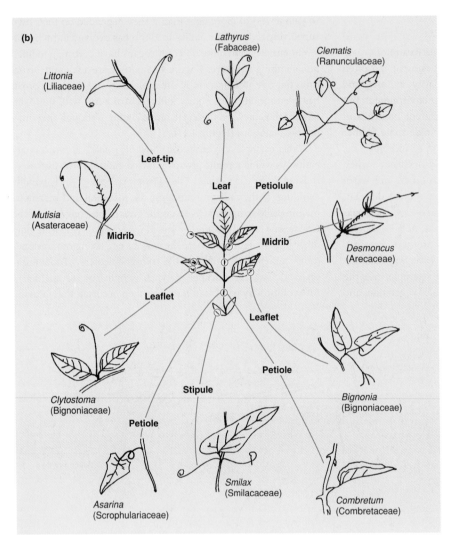

Littonia
(Liliaceae)

Lathyrus
(Fabaceae)

Clematis
(Ranunculaceae)

Leaf-tip

Leaf   **Petiolule**

Mutisia
(Asateraceae)

Midrib

Midrib

Desmoncus
(Arecaceae)

Leaflet

Leaflet

Petiole

Clytostoma
(Bignoniaceae)

Stipule

Bignonia
(Bignoniaceae)

Petiole

Asarina
(Scrophulariaceae)

Smilax
(Smilacaceae)

Combretum
(Combretaceae)

Figure 1.15 (*continued*) (b) Structural features that are homologous, i.e. derived from modifications of a single organ, the leaf, shown by reference to an idealized leaf in the center of the figure. (Courtesy of Alan Bryant.)

ways accurately paralleled what occurred in the placental mammals on other continents (Figure 1.16). The subtlety of the parallels in both the form of the organisms and their lifestyle is so striking that it is hard to escape the view that the environments of placentals and marsupials provided similar opportunities to which the evolutionary processes of the two groups responded in similar ways.

1.5   The match between communities and their environments

1.5.1   Terrestrial biomes of the earth

Before we examine the differences and similarities between communities, we need to consider the larger groupings, 'biomes', in which biogeographers recognize marked differences in the flora and fauna of different parts of the world. The number of biomes that are distinguished is a matter of taste. They certainly grade into one another, and sharp boundaries are a convenience for cartographers rather than a reality of nature. We describe eight terrestrial biomes and illustrate their global distribution in Figure 1.17, and show how they may be related to annual temperature and precipitation (Figure 1.18) (see Woodward, 1987 for a more detailed account). Apart from anything else, understanding the terminology that describes and distinguishes these biomes is necessary when we come to consider key questions later in the book (especially in Chapters 20 and 21). Why are there more species in some communities than in others? Are some communities more stable in their composition than others, and if so why? Do more productive environments support more diverse communities? Or do more diverse communities make more productive use of the resources available to them?

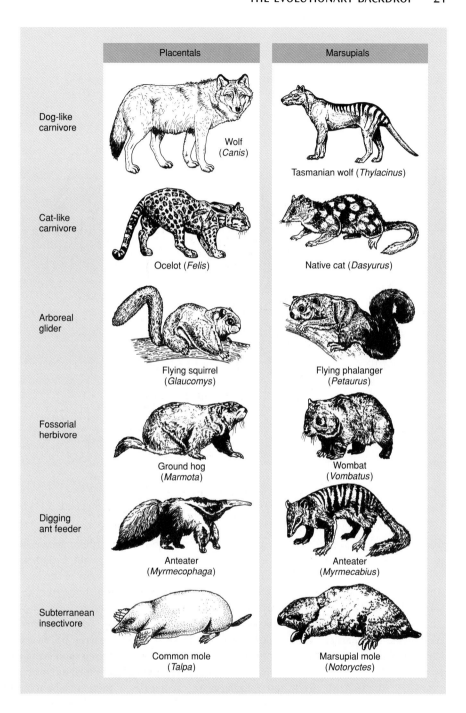

	Placentals	Marsupials
Dog-like carnivore	Wolf (*Canis*)	Tasmanian wolf (*Thylacinus*)
Cat-like carnivore	Ocelot (*Felis*)	Native cat (*Dasyurus*)
Arboreal glider	Flying squirrel (*Glaucomys*)	Flying phalanger (*Petaurus*)
Fossorial herbivore	Ground hog (*Marmota*)	Wombat (*Vombatus*)
Digging ant feeder	Anteater (*Myrmecophaga*)	Anteater (*Myrmecabius*)
Subterranean insectivore	Common mole (*Talpa*)	Marsupial mole (*Notoryctes*)

Figure 1.16 Parallel evolution of marsupial and placental mammals. The pairs of species are similar in both appearance and habit, and usually (but not always) in lifestyle.

tundra

Tundra (see Plate 1.1, facing p. 84) occurs around the Arctic Circle, beyond the tree line. Small areas also occur on sub-Antarctic islands in the southern hemisphere. 'Alpine' tundra is found under similar conditions but at high altitude. The environment is characterized by the presence of permafrost – water permanently frozen in the soil – while liquid water is present for only short periods of the year. The typical flora includes lichens, mosses, grasses, sedges and dwarf trees. Insects are extremely seasonal in their activity, and the native bird and mammal fauna is enriched by species that migrate from warmer latitudes in the summer. In the colder areas, grasses and sedges disappear, leaving nothing rooted in the permafrost. Ultimately, vegetation that consists only of lichens and mosses gives way, in its turn, to the polar desert. The number of species of higher plants (i.e. excluding mosses and lichens) decreases

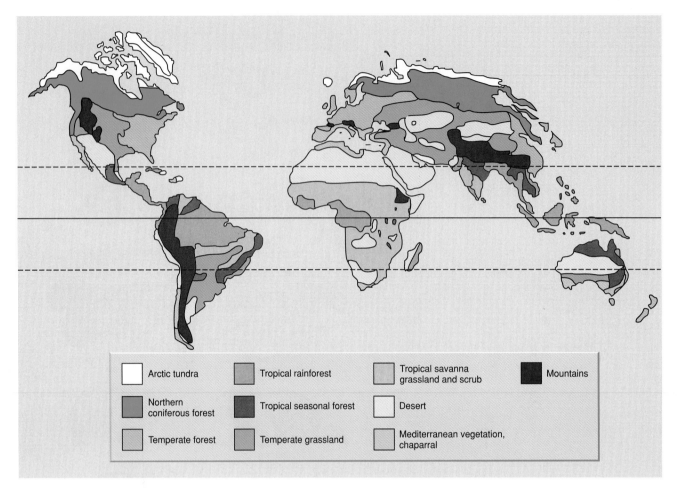

Figure 1.17   World distribution of the major biomes of vegetation. (After Audesirk & Audesirk, 1996.)

from the Low Arctic (around 600 species in North America) to the High Arctic (north of 83°, e.g. around 100 species in Greenland and Ellesmere Island). In contrast, the flora of Antarctica contains only two native species of vascular plant and some lichens and mosses that support a few small invertebrates. The biological productivity and diversity of Antarctica are concentrated at the coast and depend almost entirely on resources harvested from the sea.

taiga

Taiga or northern coniferous forest (see Plate 1.2, facing p. 84) occupies a broad belt across North America and Eurasia. Liquid water is unavailable for much of the winter, and plants and many of the animals have a conspicuous winter dormancy in which metabolism is very slow. Generally, the tree flora is very limited. In areas with less severe winters, the forests may be dominated by pines (*Pinus* species, which are all evergreens) and deciduous trees such as larch (*Larix*), birch (*Betula*) or aspens (*Populus*), often as mixtures of species. Farther north, these species give way to single-species forests of spruce (*Picea*) covering immense areas. The overriding environmental constraint in

northern spruce forests is the presence of permafrost, creating drought except when the sun warms the surface. The root system of spruce can develop in the superficial soil layer, from which the trees derive all their water during the short growing season.

Temperate forests (see Plate 1.3, between pp. 84 and 85) range from the mixed conifer and broad-leaved forests of much of North America and northern central Europe (where there may be 6 months of freezing temperatures), to the moist dripping forests of broad-leaved evergreen trees found at the biome's low latitude limits in, for example, Florida and New Zealand. In most temperate forests, however, there are periods of the year when liquid water is in short supply, because potential evaporation exceeds the sum of precipitation and water available from the soil. Deciduous trees, which dominate in most temperate forests, lose their leaves in the fall and become dormant. On the forest floor, diverse floras of perennial herbs often occur, particularly those that grow quickly in the spring before the new tree foliage has developed. Temperate forests also

temperate forests

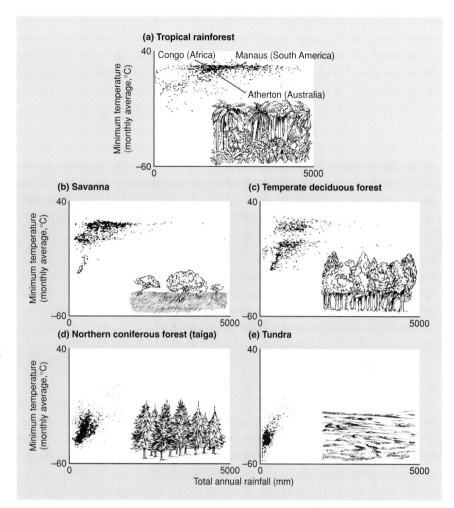

Figure 1.18 The variety of environmental conditions experienced in terrestrial environments can be described in terms of their annual rainfall and mean monthly minimum temperatures. The range of conditions experienced in: (a) tropical rainforest, (b) savanna, (c) temperate deciduous forest, (d) northern coniferous forest (taiga), and (e) tundra. (After Heal *et al.*, 1993; © UNESCO.)

provide food resources for animals that are usually very seasonal in their occurrence. Many of the birds of temperate forests are migrants that return in spring but spend the remainder of the year in warmer biomes.

grassland

Grassland occupies the drier parts of temperate and tropical regions. Temperate grassland has many local names: the steppes of Asia, the prairies of North America, the pampas of South America and the veldt of South Africa. Tropical grassland or savanna (see Plate 1.4, between pp. 84 and 85) is the name applied to tropical vegetation ranging from pure grassland to some trees with much grass. Almost all of these temperate and tropical grasslands experience seasonal drought, but the role of climate in determining their vegetation is almost completely overridden by the effects of grazing animals that limit the species present to those that can recover from frequent defoliation. In the savanna, fire is also a common hazard in the dry season and, like grazing animals, it tips the balance in the vegetation against trees and towards grassland. None the less, there is typically a seasonal glut of food, alternating with shortage, and as a consequence

the larger grazing animals suffer extreme famine (and mortality) in drier years. A seasonal abundance of seeds and insects supports large populations of migrating birds, but only a few species can find sufficiently reliable resources to be resident year-round.

Many of these natural grasslands have been cultivated and replaced by arable annual 'grasslands' of wheat, oats, barley, rye and corn. Such annual grasses of temperate regions, together with rice in the tropics, provide the staple food of human populations worldwide. At the drier margins of the biome, many of the grasslands are 'managed' for meat or milk production, sometimes requiring a nomadic human lifestyle. The natural populations of grazing animals have been driven back in favor of cattle, sheep and goats. Of all the biomes, this is the one most coveted, used and transformed by humans.

Chaparral or *maquis* occurs in Mediterranean-type climates (mild, wet winters and summer drought) in Europe, California and northwest Mexico, and in a few small areas in Australia, Chile and South Africa. Chaparral develops in regions with less rainfall than temperate grasslands and is dominated mainly by a

chaparral

drought-resistant, hard-leaved scrub of low-growing woody plants. Annual plants are also common in chaparral regions during the winter and early spring, when rainfall is more abundant. Chaparral is subject to periodic fires; many plants produce seeds that will only germinate after fire while others can quickly resprout because of food reserves in their fire-resistant roots.

desert

Deserts (see Plate 1.5, between pp. 84 and 85) are found in areas that experience extreme water shortage: rainfall is usually less than about 25 cm year^{-1}, is usually very unpredictable and is considerably less than potential evaporation. The desert biome spans a very wide range of temperatures, from hot deserts, such as the Sahara, to very cold deserts, such as the Gobi in Mongolia. In their most extreme form, the hot deserts are too arid to bear any vegetation; they are as bare as the cold deserts of Antarctica. Where there is sufficient rainfall to allow plants to grow in arid deserts, its timing is always unpredictable. Desert vegetation falls into two sharply contrasted patterns of behavior. Many species have an opportunistic lifestyle, stimulated into germination by the unpredictable rains. They grow fast and complete their life history by starting to set new seed after a few weeks. These are the species that can occasionally make a desert bloom. A different pattern of behavior is to be long-lived with sluggish physiological processes. Cacti and other succulents, and small shrubby species with small, thick and often hairy leaves, can close their stomata (pores through which gas exchange takes place) and tolerate long periods of physiological inactivity. The relative poverty of animal life in arid deserts reflects the low productivity of the vegetation and the indigestibility of much of it.

tropical rainforest

Tropical rainforest (see Plate 1.6, between pp. 84 and 85) is the most productive of the earth's biomes – a result of the coincidence of high solar radiation received throughout the year and regular and reliable rainfall. The productivity is achieved, overwhelmingly, high in the dense forest canopy of evergreen foliage. It is dark at ground level except where fallen trees create gaps. Often, many tree seedlings and saplings remain in a suppressed state from year to year and only leap into action if a gap forms in the canopy above them. Apart from the trees, the vegetation is largely composed of plant forms that reach up into the canopy vicariously; they either climb and then scramble in the tree canopy (vines and lianas, including many species of fig) or grow as epiphytes, rooted on the damp upper branches. Most species of both animals and plants in tropical rain forest are active throughout the year, though the plants may flower and ripen fruit in sequence. Dramatically high species richness is the norm for tropical rainforest, and communities rarely if ever become dominated by one or a few species. The diversity of rainforest trees provides for a corresponding diversity of resources for herbivores, and so on up the food chain. Erwin (1982) estimated that there are 18,000 species of beetle in 1 ha of Panamanian rainforest (compared with only 24,000 in the whole of the United States and Canada!).

aquatic biomes?

All of these biomes are terrestrial. Aquatic ecologists could also come up with a set of biomes, although the tradition has largely been a terrestrial one. We might distinguish springs, rivers, ponds, lakes, estuaries, coastal zones, coral reefs and deep oceans, among other distinctive kinds of aquatic community. For present purposes, we recognize just two aquatic biomes, *marine* and *freshwater*. The oceans cover about 71% of the earth's surface and reach depths of more than 10,000 m. They extend from regions where precipitation exceeds evaporation to regions where the opposite is true. There are massive movements within this body of water that prevent major differences in salt concentrations developing (the average concentration is about 3%). Two main factors influence the biological activity of the oceans. Photosynthetically active radiation is absorbed in its passage through water, so photosynthesis is confined to the surface region. Mineral nutrients, especially nitrogen and phosphorus, are commonly so dilute that they limit the biomass that can develop. Shallow waters (e.g. coastal regions and estuaries) tend to have high biological activity because they receive mineral input from the land and less incident radiation is lost than in passage through deep waters. Intense biological activity also occurs where nutrient-rich waters from the ocean depths come to the surface; this accounts for the concentration of many of the world's fisheries in Arctic and Antarctic waters.

Freshwater biomes occur mainly on the route from land drainage to the sea. The chemical composition of the water varies enormously, depending on its source, its rate of flow and the inputs of organic matter from vegetation that is rooted in or around the aquatic environment. In water catchments where the rate of evaporation is high, salts leached from the land may accumulate and the concentrations may far exceed those present in the oceans; brine lakes or even salt pans may be formed in which little life is possible. Even in aquatic situations liquid water may be unavailable, as is the case in the polar regions.

Differentiating between biomes allows only a very crude recognition of the sorts of differences and similarities that occur between communities of organisms. Within biomes there are both small- and large-scale patterns of variation in the structure of communities and in the organisms that inhabit them. Moreover, as we see next, what characterizes a biome is not necessarily the particular species that live there.

1.5.2   The 'life form spectra' of communities

We pointed out earlier the crucial importance of geographic isolation in allowing populations to diverge under selection. The geographic distributions of species, genera, families and even higher taxonomic categories of plants and animals often reflect this geographic divergence. All species of lemurs, for example, are found on the island of Madagascar and nowhere else. Similarly,

230 species in the genus *Eucalyptus* (gum tree) occur naturally in Australia (and two or three in Indonesia and Malaysia). The lemurs and the gum trees occur where they do because they evolved there – not because these are the only places where they could survive and prosper. Indeed, many *Eucalyptus* species grow with great success and spread rapidly when they have been introduced to California or Kenya. A map of the natural world distribution of lemurs tells us quite a lot about the evolutionary history of this group. But as far as its relationship with a biome is concerned, the most we can say is that lemurs happen to be one of the constituents of the tropical rainforest biome in Madagascar.

Similarly, particular biomes in Australia include certain marsupial mammals, while the *same* biomes in other parts of the world are home to their placental counterparts. A map of biomes, then, is not usually a map of the distribution of species. Instead, we recognize different biomes and different types of aquatic community from the *types* of organisms that live in them. How can we describe their similarities so that we can classify, compare and map them? In addressing this question, the Danish biogeographer Raunkiaer developed, in 1934, his idea of 'life forms', a deep insight into the ecological significance of plant forms (Figure 1.19). He then used the spectrum of life forms present in different types of vegetation as a means of describing their ecological character.

Plants grow by developing new shoots from the buds that lie at the apices (tips) of existing shoots and in the leaf axils. Within the buds, the meristematic cells are the most sensitive part of the whole shoot – the 'Achilles' heel' of plants. Raunkiaer argued that the ways in which these buds are protected in different plants are powerful indicators of the hazards in their environments and may be used to define the different plant forms (Figure 1.19). Thus, trees expose their buds high in the air, fully exposed to the wind, cold and drought; Raunkiaer called them *phanerophytes* (Greek *phanero*, 'visible'; *phyte*, 'plant'). By contrast, many perennial herbs form cushions or tussocks in which buds are borne above ground but are protected from drought and cold in the dense mass of old leaves and shoots (*chamaephytes*: 'on the ground plants'). Buds are even better protected when they are formed at or in the soil surface (*hemicryptophytes*: 'half hidden plants') or on buried dormant storage organs (bulbs, corms and rhizomes – *cryptophytes*: 'hidden plants'; or *geophytes*: 'earth plants'). These allow the plants to make rapid growth and to flower before they die back to a dormant state. A final major category consists of annual plants that depend wholly on dormant seeds to carry their populations through seasons of drought and cold (*therophytes*: 'summer plants'). Therophytes are the plants of deserts (they make up nearly 50% of the flora of Death Valley, USA), sand dunes and repeatedly disturbed habitats. They also include the annual weeds of arable lands, gardens and urban wastelands.

But there is, of course, no vegetation that consists entirely of one growth form. All vegetation contains a mixture, a spectrum, of Raunkiaer's life forms. The composition of the spectrum in any particular habitat is as good a shorthand description of its vegetation as ecologists have yet managed to devise. Raunkiaer compared these with a 'global spectrum' obtained by sampling from a compendium of all species known and described in his time (the *Index Kewensis*), biased by the fact that the tropics were, and still are, relatively unexplored. Thus, for example, we recognize a chaparral type of vegetation when we see it in Chile, Australia, California or Crete because the life form spectrums are similar. Their detailed taxonomies would only emphasize how different they are.

Faunas are bound to be closely tied to floras – if only because most herbivores are choosy about their diet. Terrestrial carnivores range more widely than their herbivore prey, but the distribution of herbivores still gives the carnivores a broad vegetational allegiance. Plant scientists have tended to be keener on classifying floras than animal scientists on classifying faunas, but one interesting attempt to classify faunas compared the mammals of forests in Malaya, Panama, Australia and Zaire (Andrews *et al.*, 1979). They were classified into carnivores, herbivores, insectivores and mixed feeders, and these categories were subdivided into those that were aerial (mainly bats and flying foxes), arboreal (tree dwellers), scansorial (climbers) or small ground mammals (Figure 1.20). The comparison reveals some strong contrasts and similarities. For example, the ecological diversity spectra for the Australian and Malayan forests were very similar despite the fact that their faunas are taxonomically very distinct – the Australian mammals are marsupials and the Malaysian mammals are placentals.

1.6 The diversity of matches within communities

Although a particular type of organism is often characteristic of a particular ecological situation, it will almost inevitably be only part of a diverse community of species. A satisfactory account, therefore, must do more than identify the similarities between organisms that allow them to live in the same environment – it must also try to explain why species that live in the same environment are often profoundly different. To some extent, this 'explanation' of diversity is a trivial exercise. It comes as no surprise that a plant utilizing sunlight, a fungus living on the plant, a herbivore eating the plant and a parasitic worm living in the herbivore should all coexist in the same community. On the other hand, most communities also contain a variety of different species that are all constructed in a fairly similar way and all living (at least superficially) a fairly similar life. There are several elements in an explanation of this diversity.

1.6.1 Environments are heterogeneous

There are no homogeneous environments in nature. Even a continuously stirred culture of microorganisms is heterogeneous

Raunkiaer's classification

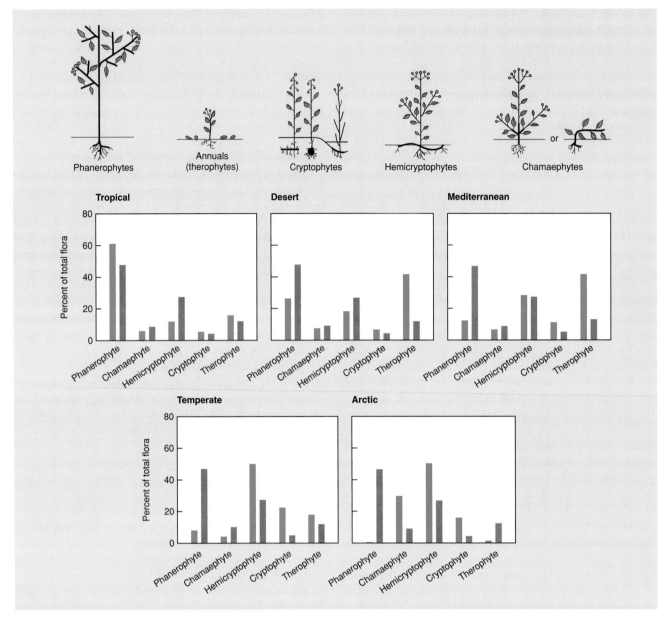

Figure 1.19   The drawings above depict the variety of plant forms distinguished by Raunkiaer on the basis of where they bear their buds (shown in color). Below are life form spectrums for five different biomes. The colored bars show the percentage of the total flora that is composed of species with each of the five different life forms. The gray bars are the proportions of the various life forms in the world flora for comparison. (From Crawley, 1986.)

because it has a boundary – the walls of the culture vessel – and cultured microorganisms often subdivide into two forms: one that sticks to the walls and the other that remains free in the medium.

The extent to which an environment is heterogeneous depends on the scale of the organism that senses it. To a mustard seed, a grain of soil is a mountain; and to a caterpillar, a single leaf may represent a lifetime's diet. A seed lying in the shadow of a leaf may be inhibited in its germination while a seed lying outside that shadow germinates freely. What appears to the human observer as a homogeneous environment may, to an organism within it, be a mosaic of the intolerable and the adequate.

There may also be gradients in space (e.g. altitude) or gradients in time, and the latter, in their turn, may be rhythmic (like

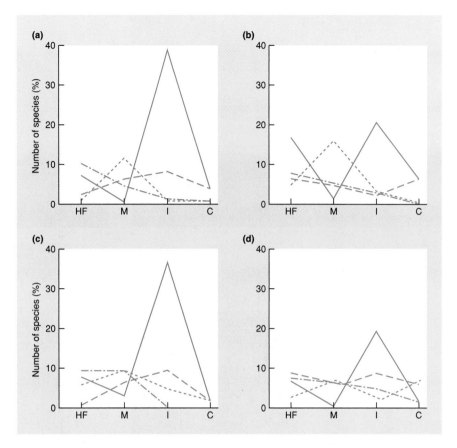

Figure 1.20 The percentages of forest mammals in various locomotory and feeding habitat categories in communities in: (a) Malaya, all forested areas (161 species), (b) Panama dry forest (70 species), (c) Australia, Cape York forest (50 species), and (d) Zaire, Irangi forest (96 species). C, carnivores; HF, herbivores and fructivores; I, insectivores; M, mixed feeders; (——) aerial; (-----) arboreal; (– – –) scansorial; (—·—) small ground mammals. (After Andrews *et al.*, 1979.)

daily and seasonal cycles), directional (like the accumulation of a pollutant in a lake) or erratic (like fires, hailstorms and typhoons).

Heterogeneity crops up again and again in later chapters – in part because of the challenges it poses to organisms in moving from patch to patch (Chapter 6), in part because of the variety of opportunities it provides for different species (Chapters 8 and 19), and in part because heterogeneity can alter communities by interrupting what would otherwise be a steady march to an equilibrium state (Chapters 10 and 19).

1.6.2 Pairs of species

As we have already noted, the existence of one type of organism in an area immediately diversifies it for others. Over its lifetime, an organism may increase the diversity of its environment by contributing dung, urine, dead leaves and ultimately its dead body. During its life, its body may serve as a place in which other species find homes. Indeed, some of the most strongly developed matches between organisms and their environment are those in which one species has developed a dependence upon another. This is the case in many relationships between consumers and their foods. Whole syndromes of form, behavior and metabolism constrain the

animal within its narrow food niche, and deny it access to what might otherwise appear suitable alternative foods. Similar tight matches are characteristic of the relationships between parasites and their hosts. The various interactions in which one species is consumed by another are the subject matter of Chapters 9–12.

Where two species have evolved a mutual dependence, the fit may be even tighter. We examine such 'mutualisms' in detail in Chapter 13. The association of nitrogen-fixing bacteria with the roots of leguminous plants, and the often extremely precise relationships between insect pollinators and their flowers, are two good examples.

When a population has been exposed to variations in the physical factors of the environment, for example a short growing season or a high risk of frost or drought, a once-and-for-all tolerance may ultimately evolve. The physical factor cannot itself change or evolve as a result of the evolution of the organisms. By contrast, when members of two species interact, the change in each produces alterations in the life of the other, and each may generate selective forces that direct the evolution of the other. In such a coevolutionary process the interaction between two species may continually escalate. What we then see in nature may be pairs of species that have driven each other into ever narrowing ruts of specialization – an ever closer match.

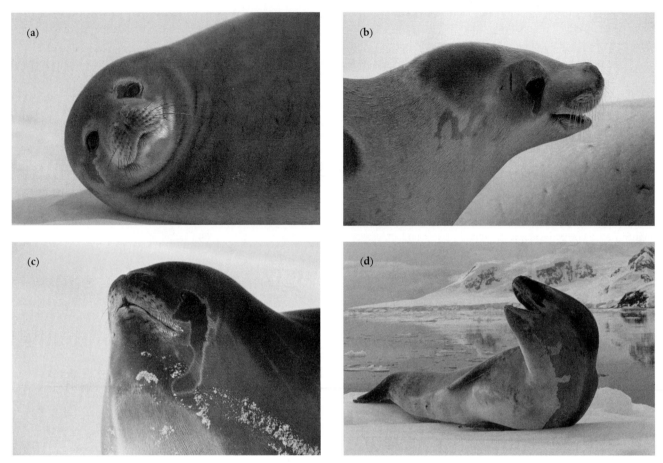

Figure 1.21   Antarctic seals, similar species that coexist: (a) the Weddell seal, *Leptonychotes weddellii* (© Imageshop – zefa visual media uk ltd / Alamy), (b) the crab-eater seal *Lobodon carcinophagus* (© Bryan & Cherry Alexander Photography / Alamy), (c) the Ross seal, *Omatophoca rossii* (© Chris Sattlberger / Science Photo Library), and (d) the leopard seal, *Hydrurga leptonyx* (© Kevin Schafer / Alamy).

1.6.3   Coexistence of similar species

While it is no surprise that species with rather different roles coexist within the same community, it is also generally the case that communities support a variety of species performing apparently rather similar roles. The Antarctic seals are an example. It is thought that the ancestral seals evolved in the northern hemisphere, where they are present as Miocene fossils, but one group of seals moved south into warmer waters and probably colonized the Antarctic in the Late Miocene or Early Pliocene (about 5 million years ago). When they entered the Antarctic, the Southern Ocean was probably rich in food and free from major predators, as it is today. It was within this environment that the group appears to have undergone radiative evolution (Figure 1.21). For example, the Weddell seal feeds primarily on fish and has unspecialized dentition; the crab-eater seal feeds almost exclusively on krill and its teeth are suited to filtering these from the sea water; the Ross seal has small, sharp teeth and feeds mainly on pelagic squid; and the leopard seal has large, cusped, grasping teeth and feeds on a wide variety of foods, including other seals and, in some seasons, penguins.

Do these species compete with one another? Do competing species need to be different if they are to coexist? If so, how different do they need to be: is there some limit to their similarity? Do species like the seals interact with one another at the present time, or has evolution in the past led to the absence of such interactions in contemporary communities? We return to these questions about coexisting, similar species in Chapter 8.

Even at this stage, though, we may note that coexisting species, even when apparently very similar, commonly differ in subtle ways – not simply in their morphology or physiology but also in their responses to their environment and the role they play within the community of which they are part. The 'ecological niches' of such species are said to be differentiated from one another. The concept of the ecological niche is itself explained in the next two chapters.

Summary

'Nothing in biology makes sense, except in the light of evolution'. We try in this chapter to illustrate the processes by which the properties of different sorts of species make their life possible in particular environments.

We explain what is meant by evolutionary adaptation and by the theory of evolution by natural selection, an ecological theory first elaborated by Charles Darwin in 1859. Through natural selection, organisms come to match their environments by being 'the fittest available' or 'the fittest yet': they are not 'the best imaginable'.

Adaptive variation within species can occur at a range of levels: all represent a balance between local adaptation and hybridization. Ecotypes are genetically determined variants between populations within a species that reflect local matches between the organisms and their environments. Genetic polymorphism is the occurrence together in the same habitat of two or more distinct forms. Dramatic examples of local specialization have been driven by manmade ecological forces, especially those of environmental pollution.

We describe the process of speciation by which two or more new species are formed from one original species and explain what we mean by a 'species', especially a biospecies. Islands provide arguably the most favorable environment for populations to diverge into distinct species.

Species live where they do for reasons that are often accidents of history. We illustrate this by examining island patterns, the movements of land masses over geological time, climatic changes especially during the Pleistocene ice ages (and we compare this with predicted changes consequent on current global warming) and the concepts of convergent and parallel evolution.

The various terrestrial biomes of the earth are reviewed and their aquatic equivalents touched on briefly. Raunkiaer's concept of life form spectra, in particular, emphasizes that ecological communities may be fundamentally very similar even when taxonomically quite distinct.

All communities comprise a diversity of species: a diversity of matches to the local environment. Environmental heterogeneity, interactions between predators and prey, parasites and hosts and mutualists, and the coexistence of similar species all contribute to this.

Chapter 2
Conditions

2.1 Introduction

In order to understand the distribution and abundance of a species we need to know its history (Chapter 1), the resources it requires (Chapter 3), the individuals' rates of birth, death and migration (Chapters 4 and 6), their interactions with their own and other species (Chapters 5 and 8–13) and the effects of environmental conditions. This chapter deals with the limits placed on organisms by environmental conditions.

conditions may be altered – but not consumed

A condition is as an abiotic environmental factor that influences the functioning of living organisms. Examples include temperature, relative humidity, pH, salinity and the concentration of pollutants. A condition may be modified by the presence of other organisms. For example, temperature, humidity and soil pH may be altered under a forest canopy. But unlike resources, conditions are not consumed or used up by organisms.

For some conditions we can recognize an optimum concentration or level at which an organism performs best, with its activity tailing off at both lower and higher levels (Figure 2.1a). But we need to define what we mean by 'performs best'. From an evolutionary point of view, 'optimal' conditions are those under which individuals leave most descendants (are fittest), but these are often impossible to determine in practice because measures of fitness should be made over several generations. Instead, we more often measure the effect of conditions on some key property like the activity of an enzyme, the respiration rate of a tissue, the growth rate of individuals or their rate of reproduction. However, the effect of variation in conditions on these various properties will often not be the same; organisms can usually survive over a wider range of conditions than permit them to grow or reproduce (Figure 2.1a).

The precise shape of a species' response will vary from condition to condition. The generalized form of response, shown in Figure 2.1a, is appropriate for conditions like temperature and pH

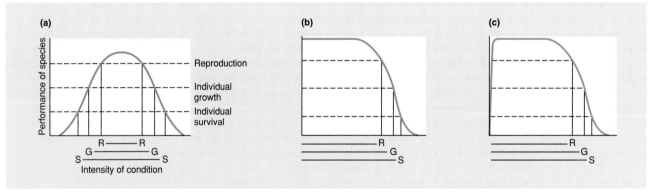

Figure 2.1 Response curves illustrating the effects of a range of environmental conditions on individual survival (S), growth (G) and reproduction (R). (a) Extreme conditions are lethal; less extreme conditions prevent growth; only optimal conditions allow reproduction. (b) The condition is lethal only at high intensities; the reproduction–growth–survival sequence still applies. (c) Similar to (b), but the condition is required by organisms, as a resource, at low concentrations.

in which there is a continuum from an adverse or lethal level (e.g. freezing or very acid conditions), through favorable levels of the condition to a further adverse or lethal level (heat damage or very alkaline conditions). There are, though, many environmental conditions for which Figure 2.1b is a more appropriate response curve: for instance, most toxins, radioactive emissions and chemical pollutants, where a low-level intensity or concentration of the condition has no detectable effect, but an increase begins to cause damage and a further increase may be lethal. There is also a different form of response to conditions that are toxic at high levels but essential for growth at low levels (Figure 2.1c). This is the case for sodium chloride – an essential resource for animals but lethal at high concentrations – and for the many elements that are essential micronutrients in the growth of plants and animals (e.g. copper, zinc and manganese), but that can become lethal at the higher concentrations sometimes caused by industrial pollution.

In this chapter, we consider responses to temperature in much more detail than other conditions, because it is the single most important condition that affects the lives of organisms, and many of the generalizations that we make have widespread relevance. We move on to consider a range of other conditions, before returning, full circle, to temperature because of the effects of other conditions, notably pollutants, on global warming. We begin, though, by explaining the framework within which each of these conditions should be understood here: the ecological niche.

2.2 Ecological niches

The term *ecological niche* is frequently misunderstood and misused. It is often used loosely to describe the sort of place in which an organism lives, as in the sentence: 'Woodlands are the niche of woodpeckers'. Strictly, however, where an organism lives is its *habitat*. A niche is not a place but an idea: a summary of the organism's tolerances and requirements. The habitat of a gut microorganism would be an animal's alimentary canal; the habitat of an aphid might be a garden; and the habitat of a fish could be a whole lake. Each habitat, however, provides many different niches: many other organisms also live in the gut, the garden or the lake – and with quite different lifestyles. The word *niche* began to gain its present scientific meaning when Elton wrote in 1933 that the niche of an organism is its mode of life 'in the sense that we speak of trades or jobs or professions in a human community'. The niche of an organism started to be used to describe how, rather than just where, an organism lives.

niche dimensions   The modern concept of the niche was proposed by Hutchinson in 1957 to address the ways in which tolerances and requirements interact to define the conditions (this chapter) and resources (Chapter 3) needed by an individual or a species in order

to practice its way of life. Temperature, for instance, limits the growth and reproduction of all organisms, but different organisms tolerate different ranges of temperature. This range is one *dimension* of an organism's ecological niche. Figure 2.2a shows how species of plants vary in this dimension of their niche: how they vary in the range of temperatures at which they can survive. But there are many such dimensions of a species' niche – its tolerance of various other conditions (relative humidity, pH, wind speed, water flow and so on) and its need for various resources. Clearly the real niche of a species must be *multi*dimensional.

It is easy to visualize the early stages of building such a multidimensional niche. Figure 2.2b illustrates the way in which two niche dimensions (temperature and salinity) together define a two-dimensional area that is part of the niche of a sand shrimp. Three dimensions, such as temperature, pH and the availability of a particular food, may define a three-dimensional niche volume (Figure 2.2c). In fact, we consider a niche to be an *n-dimensional hypervolume*, where *n* is the number of dimensions that make up the niche. It is hard to imagine (and impossible to draw) this more realistic picture. None the less, the simplified three-dimensional version captures the idea of the ecological niche of a species. It is defined by the boundaries that limit where it can live, grow and reproduce, and it is very clearly a concept rather than a place. The concept has become a cornerstone of ecological thought.

the *n*-dimensional hypervolume

Provided that a location is characterized by conditions within acceptable limits for a given species, and provided also that it contains all the necessary resources, then the species can, potentially, occur and persist there. Whether or not it does so depends on two further factors. First, it must be able to reach the location, and this depends in turn on its powers of colonization and the remoteness of the site. Second, its occurrence may be precluded by the action of individuals of other species that compete with it or prey on it.

Usually, a species has a larger ecological niche in the absence of competitors and predators than it has in their presence. In other words, there are certain combinations of conditions and resources that can allow a species to maintain a viable population, but only if it is not being adversely affected by enemies. This led Hutchinson to distinguish between the *fundamental* and the *realized* niche. The former describes the overall potentialities of a species; the latter describes the more limited spectrum of conditions and resources that allow it to persist, even in the presence of competitors and predators. Fundamental and realized niches will receive more attention in Chapter 8, when we look at interspecific competition.

fundamental and realized niches

The remainder of this chapter looks at some of the most important condition dimensions of species' niches, starting with temperature; the following chapter examines resources, which add further dimensions of their own.

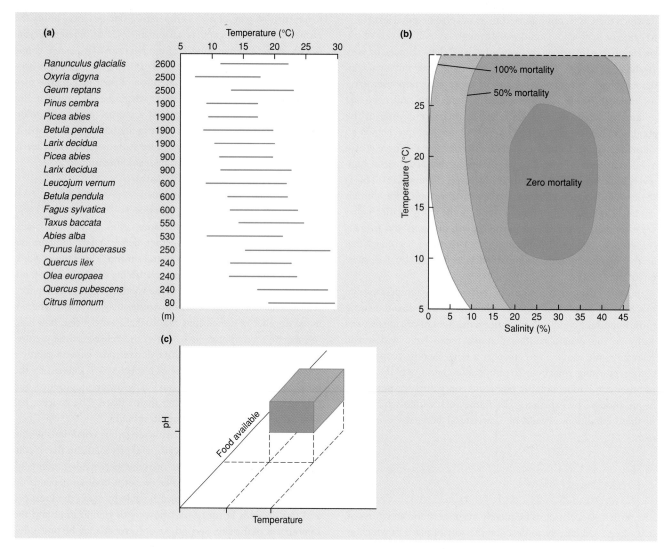

Figure 2.2 (a) A niche in one dimension. The range of temperatures at which a variety of plant species from the European Alps can achieve net photosynthesis of low intensities of radiation (70 W m^{-2}). (After Pisek *et al.*, 1973.) (b) A niche in two dimensions for the sand shrimp (*Crangon septemspinosa*) showing the fate of egg-bearing females in aerated water at a range of temperatures and salinities. (After Haefner, 1970.) (c) A diagrammatic niche in three dimensions for an aquatic organism showing a volume defined by the temperature, pH and availability of food.

2.3 Responses of individuals to temperature

2.3.1 What do we mean by 'extreme'?

It seems natural to describe certain environmental conditions as 'extreme', 'harsh', 'benign' or 'stressful'. It may seem obvious when conditions are 'extreme': the midday heat of a desert, the cold of an Antarctic winter, the salinity of the Great Salt Lake. But this only means that these conditions are extreme *for us*, given our particular physiological characteristics and tolerances.

To a cactus there is nothing extreme about the desert conditions in which cacti have evolved; nor are the icy fastnesses of Antarctica an extreme environment for penguins (Wharton, 2002). It is too easy and dangerous for the ecologist to assume that all other organisms sense the environment in the way we do. Rather, the ecologist should try to gain a worm's-eye or plant's-eye view of the environment: to see the world as others see it. Emotive words like harsh and benign, even relativities such as hot and cold, should be used by ecologists only with care.

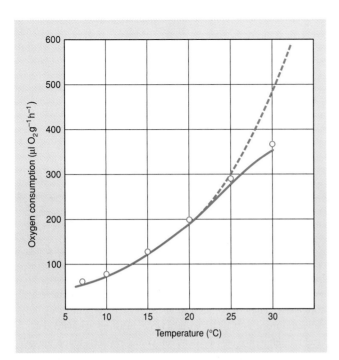

Figure 2.3 The rate of oxygen consumption of the Colorado beetle (*Leptinotarsa decemineata*), which doubles for every 10°C rise in temperature up to 20°C, but increases less fast at higher temperatures. (After Marzusch, 1952.)

2.3.2 Metabolism, growth, development and size

exponential effects of temperature on metabolic reactions

Individuals respond to temperature essentially in the manner shown in Figure 2.1a: impaired function and ultimately death at the upper and lower extremes (discussed in Sections 2.3.4 and 2.3.6), with a functional range between the extremes, within which there is an optimum. This is accounted for, in part, simply by changes in metabolic effectiveness. For each 10°C rise in temperature, for example, the rate of biological enzymatic processes often roughly doubles, and thus appears as an exponential curve on a plot of rate against temperature (Figure 2.3). The increase is brought about because high temperature increases the speed of molecular movement and speeds up chemical reactions. The factor by which a reaction changes over a 10°C range is referred to as a Q_{10}: a rough doubling means that $Q_{10} \approx 2$.

effectively linear effects on rates of growth and development

For an ecologist, however, effects on individual chemical reactions are likely to be less important than effects on rates of growth (increases in mass), on rates of development (progression through lifecycle stages) and on final body size, since, as we shall discuss much more fully in Chapter 4, these tend

to drive the core ecological activities of survival, reproduction and movement. And when we plot rates of growth and development of whole organisms against temperature, there is quite commonly an extended range over which there are, at most, only slight deviations from linearity (Figure 2.4).

day-degree concept

When the relationship between growth or development *is* effectively linear, the temperatures experienced by an organism can be summarized in a single very useful value, the number of 'day-degrees'. For instance, Figure 2.4c shows that at 15°C (5.1°C above a development threshold of 9.9°C) the predatory mite, *Amblyseius californicus*, took 24.22 days to develop (i.e. the proportion of its total development achieved each day was 0.041 (= 1/24.22)), but it took only 8.18 days to develop at 25°C (15.1°C above the same threshold). At both temperatures, therefore, development required 123.5 day-degrees (or, more properly, 'day-degrees above threshold'), i.e. 24.22 × 5.1 = 123.5, and 8.18 × 15.1 = 123.5. This is also the requirement for development in the mite at other temperatures within the nonlethal range. Such organisms cannot be said to require a certain length of time for development. What they require is a combination of time and temperature, often referred to as 'physiological time'.

temperature–size rule

Together, the rates of growth and development determine the final size of an organism. For instance, for a given rate of growth, a faster rate of development will lead to smaller final size. Hence, if the responses of growth and development to variations in temperature are not the same, temperature will also affect final size. In fact, development usually increases more rapidly with temperature than does growth, such that, for a very wide range of organisms, final size tends to decrease with rearing temperature: the 'temperature–size rule' (see Atkinson *et al.*, 2003). An example for single-celled protists (72 data sets from marine, brackish and freshwater habitats) is shown in Figure 2.5: for each 1°C increase in temperature, final cell volume decreased by roughly 2.5%.

These effects of temperature on growth, development and size may be of practical rather than simply scientific importance. Increasingly, ecologists are called upon to predict. We may wish to know what the consequences would be, say, of a 2°C rise in temperature resulting from global warming (see Section 2.9.2). Or we may wish to understand the role of temperature in seasonal, interannual and geographic variations in the productivity of, for example, marine ecosystems (Blackford *et al.*, 2004). We cannot afford to assume exponential relationships with temperature if they are really linear, nor to ignore the effects of changes in organism size on their role in ecological communities.

'universal temperature dependence'?

Motivated, perhaps, by this need to be able to extrapolate from the known to the unknown, and also simply by a wish to discover fundamental organizing principles governing the world

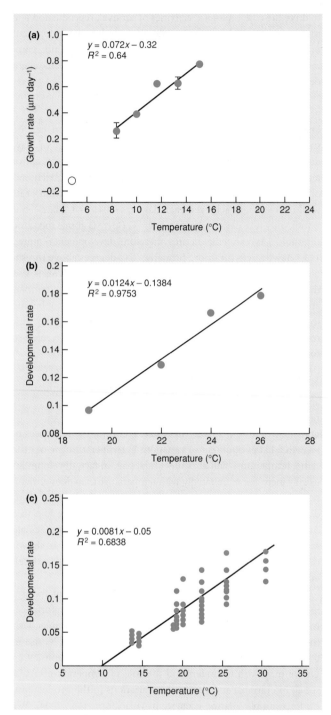

Figure 2.4 Effectively linear relationships between rates of growth and development and temperature. (a) Growth of the protist *Strombidinopsis multiauris*. (After Montagnes *et al.*, 2003.) (b) Egg development in the beetle *Oulema duftschmidi*. (After Severini *et al.*, 2003.) (c) Egg to adult development in the mite *Amblyseius californicus*. (After Hart *et al.*, 2002.) The vertical scales in (b) and (c) represent the proportion of total development achieved in 1 day at the temperature concerned.

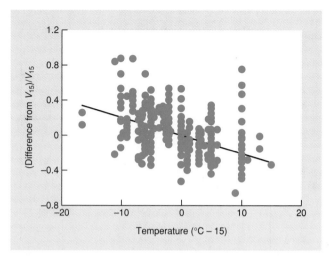

Figure 2.5 The temperature–size rule (final size decreases with increasing temperature) illustrated in protists (65 data sets combined). The horizontal scale measures temperature as a deviation from 15°C. The vertical scale measures standardized size: the difference between the cell volume observed and the cell volume at 15°C, divided by cell volume at 15°C. The slope of the mean regression line, which must pass through the point (0,0), was −0.025 (SE, 0.004); the cell volume decreased by 2.5% for every 1°C rise in rearing temperature. (After Atkinson *et al.*, 2003.)

around us, there have been attempts to uncover universal rules of temperature dependence, for metabolism itself and for development rates, linking all organisms by scaling such dependences with aspects of body size (Gillooly *et al.*, 2001, 2002). Others have suggested that such generalizations may be *over*simplified, stressing for example that characteristics of whole organisms, like growth and development rates, are determined not only by the temperature dependence of individual chemical reactions, but also by those of the availability of resources, their rate of diffusion from the environment to metabolizing tissues, and so on (Rombough, 2003; Clarke, 2004). It may be that there is room for coexistence between broad-sweep generalizations at the grand scale and the more complex relationships at the level of individual species that these generalizations subsume.

2.3.3   Ectotherms and endotherms

Many organisms have a body temperature that differs little, if at all, from their environment. A parasitic worm in the gut of a mammal, a fungal mycelium in the soil and a sponge in the sea acquire the temperature of the medium in which they live. Terrestrial organisms, exposed to the sun and the air, are different because they may acquire heat directly by absorbing solar radiation or be cooled by the latent heat of evaporation of water (typical

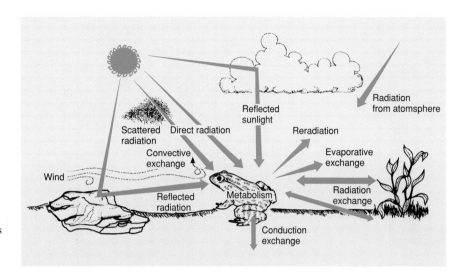

Figure 2.6 Schematic diagram of the avenues of heat exchange between an ectotherm and a variety of physical aspects of its environment. (After Tracy, 1976; from Hainsworth, 1981.)

pathways of heat exchange are shown in Figure 2.6). Various fixed properties may ensure that body temperatures are higher (or lower) than the ambient temperatures. For example, the reflective, shiny or silvery leaves of many desert plants reflect radiation that might otherwise heat the leaves. Organisms that can move have further control over their body temperature because they can seek out warmer or cooler environments, as when a lizard chooses to warm itself by basking on a hot sunlit rock or escapes from the heat by finding shade.

Amongst insects there are examples of body temperatures raised by controlled muscular work, as when bumblebees raise their body temperature by shivering their flight muscles. Social insects such as bees and termites may combine to control the temperature of their colonies and regulate them with remarkable thermostatic precision. Even some plants (e.g. *Philodendron*) use metabolic heat to maintain a relatively constant temperature in their flowers; and, of course, birds and mammals use metabolic heat almost all of the time to maintain an almost perfectly constant body temperature.

An important distinction, therefore, is between *endotherms* that regulate their temperature by the production of heat within their own bodies, and *ectotherms* that rely on external sources of heat. But this distinction is not entirely clear cut. As we have noted, apart from birds and mammals, there are also other taxa that use heat generated in their own bodies to regulate body temperature, but only for limited periods; and there are some birds and mammals that relax or suspend their endothermic abilities at the most extreme temperatures. In particular, many endothermic animals escape from some of the costs of endothermy by hibernating during the coldest seasons: at these times they behave almost like ectotherms.

endotherms: temperature regulation – but at a cost

Birds and mammals usually maintain a constant body temperature between

35 and 40°C, and they therefore tend to lose heat in most environments; but this loss is moderated by insulation in the form of fur, feathers and fat, and by controlling blood flow near the skin surface. When it is necessary to increase the rate of heat loss, this too can be achieved by the control of surface blood flow and by a number of other mechanisms shared with ectotherms like panting and the simple choice of an appropriate habitat. Together, all these mechanisms and properties give endotherms a powerful (but not perfect) capability for regulating their body temperature, and the benefit they obtain from this is a constancy of near-optimal performance. But the price they pay is a large expenditure of energy (Figure 2.7), and thus a correspondingly large requirement for food to provide that energy. Over a certain temperature range (the thermoneutral zone) an endotherm consumes energy at a basal rate. But at environmental temperatures further and further above or below that zone, the endotherm consumes more and more energy in maintaining a constant body temperature. Even in the thermoneutral zone, though, an endotherm typically consumes energy many times more rapidly than an ectotherm of comparable size.

The responses of endotherms and ectotherms to changing temperatures, then, are not so different as they may at first appear to be. Both are at risk of being killed by even short exposures to very low temperatures and by more prolonged exposure to moderately low temperatures. Both have an optimal environmental temperature and upper and lower lethal limits. There are also costs to both when they live at temperatures that are not optimal. For the ectotherm these may be slower growth and reproduction, slow movement, failure to escape predators and a sluggish rate of search for food. But for the endotherm, the maintenance of body temperature costs energy that might have been used to catch more prey, produce and nurture more offspring or escape more predators. There are also costs of insulation (e.g. blubber in whales, fur in mammals) and even costs of changing the insulation between

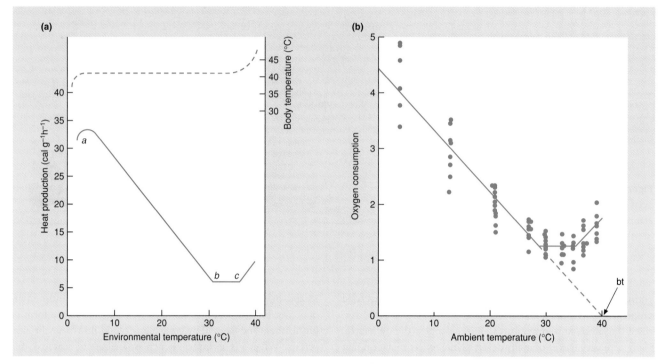

Figure 2.7   (a) Thermostatic heat production by an endotherm is constant in the thermoneutral zone, i.e. between *b*, the lower critical temperature, and *c*, the upper critical temperature. Heat production rises, but body temperature remains constant, as environmental temperature declines below *b*, until heat production reaches a maximum possible rate at a low environmental temperature. Below *a*, heat production and body temperature both fall. Above *c*, metabolic rate, heat production and body temperature all rise. Hence, body temperature is constant at environmental temperatures between *a* and *c*. (After Hainsworth, 1981.) (b) The effect of environmental temperature on the metabolic rate (rate of oxygen consumption) of the eastern chipmunk (*Tamias striatus*). bt, body temperature. Note that at temperatures between 0 and 30°C oxygen consumption decreases approximately linearly as the temperature increases. Above 30°C a further increase in temperature has little effect until near the animal's body temperature when oxygen consumption increases again. (After Neumann, 1967; Nedgergaard & Cannon, 1990.)

seasons. Temperatures only a few degrees higher than the metabolic optimum are liable to be lethal to endotherms as well as ectotherms (see Section 2.3.6).

ectotherms and
endotherms coexist:
both strategies 'work'

It is tempting to think of ectotherms as 'primitive' and endotherms as having gained 'advanced' control over their environment, but it is difficult to justify this view. Most environments on earth are inhabited by mixed communities of endothermic and ectothermic animals. This includes some of the hottest – e.g. desert rodents and lizards – and some of the coldest – penguins and whales together with fish and krill at the edge of the Antarctic ice sheet. Rather, the contrast, crudely, is between the high cost–high benefit strategy of endotherms and the low cost–low benefit strategy of ectotherms. But their coexistence tells us that both strategies, in their own ways, can 'work'.

2.3.4 Life at low temperatures

The greater part of our planet is below 5°C: 'cold is the fiercest and most widespread enemy of life on earth' (Franks *et al.*, 1990). More than 70% of the planet is covered with seawater: mostly deep ocean with a remarkably constant temperature of about 2°C. If we include the polar ice caps, more than 80% of earth's biosphere is permanently cold.

By definition, all temperatures below the optimum are harmful, but there is usually a wide range of such temperatures that cause no physical damage and over which any effects are fully reversible. There are, however, two quite distinct types of damage at low temperatures that can be lethal, either to tissues or to whole organisms: chilling and freezing. Many organisms are damaged by exposure to temperatures that are low but above freezing point – so-called

chilling injury

'chilling injury'. The fruits of the banana blacken and rot after exposure to chilling temperatures and many tropical rainforest species are sensitive to chilling. The nature of the injury is obscure, although it seems to be associated with the breakdown of membrane permeability and the leakage of specific ions such as calcium (Minorsky, 1985).

Temperatures below 0°C can have lethal physical and chemical consequences even though ice may not be formed. Water may 'supercool' to temperatures at least as low as −40°C, remaining in an unstable liquid form in which its physical properties change in ways that are bound to be biologically significant: its viscosity increases, its diffusion rate decreases and its degree of ionization of water decreases. In fact, ice seldom forms in an organism until the temperature has fallen several degrees below 0°C. Body fluids remain in a supercooled state until ice forms suddenly around particles that act as nuclei. The concentration of solutes in the remaining liquid phase rises as a consequence. It is very rare for ice to form within cells and it is then inevitably lethal, but the freezing of extracellular water is one of the factors that prevents ice forming within the cells themselves (Wharton, 2002), since water is withdrawn from the cell, and solutes in the cytoplasm (and vacuoles) become more concentrated. The effects of freezing are therefore mainly osmoregulatory: the water balance of the cells is upset and cell membranes are destabilized. The effects are essentially similar to those of drought and salinity.

freeze-avoidance and freeze-tolerance

Organisms have at least two different metabolic strategies that allow survival through the low temperatures of winter. A 'freeze-avoiding' strategy uses low-molecular-weight polyhydric alcohols (polyols, such as glycerol) that depress both the freezing and the supercooling point and also 'thermal hysteresis' proteins that prevent ice nuclei from forming (Figure 2.8a, b). A contrasting 'freeze-tolerant' strategy, which also involves the formation of polyols, encourages the formation of extracellular ice, but protects the cell membranes from damage when water is withdrawn from the cells (Storey, 1990). The tolerances of organisms to low temperatures are not fixed but are preconditioned by the experience of temperatures in their recent past. This process is called *acclimation* when it occurs in the laboratory and *acclimatization* when it occurs naturally. Acclimatization may start as the weather becomes colder in the fall, stimulating the conversion of almost the entire glycogen reserve of animals into polyols (Figure 2.8c), but this can be an energetically costly affair: about 16% of the carbohydrate reserve may be consumed in the conversion of the glycogen reserves to polyols.

acclimation and acclimatization

The exposure of an individual for several days to a relatively low temperature can shift its whole temperature response downwards along the temperature scale. Similarly, exposure to a high temperature can shift the temperature response upwards. Antarctic springtails (tiny

arthropods), for instance, when taken from 'summer' temperatures in the field (around 5°C in the Antarctic) and subjected to a range of acclimation temperatures, responded to temperatures in the range +2°C to −2°C (indicative of winter) by showing a marked drop in the temperature at which they froze (Figure 2.9); but at lower acclimation temperatures still (−5°C, −7°C), they showed no such drop because the temperatures were themselves too low for the physiological processes required to make the acclimation response.

Acclimatization aside, individuals commonly vary in their temperature response depending on the stage of development they have reached. Probably the most extreme form of this is when an organism has a dormant stage in its life cycle. Dormant stages are typically dehydrated, metabolically slow and tolerant of extremes of temperature.

2.3.5 Genetic variation and the evolution of cold tolerance

Even within species there are often differences in temperature response between populations from different locations, and these differences have frequently been found to be the result of genetic differences rather than being attributable solely to acclimatization. Powerful evidence that cold tolerance varies between geographic races of a species comes from a study of the cactus, *Opuntia fragilis*. Cacti are generally species of hot dry habitats, but *O. fragilis* extends as far north as 56°N and at one site the lowest extreme minimum temperature recorded was −49.4°C. Twenty populations were sampled from diverse localities in northern USA and Canada, and were tested for freezing tolerance and ability to acclimate to cold. Individuals from the most freeze-tolerant population (from Manitoba) tolerated −49°C in laboratory tests and acclimated by 19.9°C, whereas plants from a population in the more equable climate of Hornby Island, British Columbia, tolerated only −19°C and acclimated by only 12.1°C (Loik & Nobel, 1993).

There are also striking cases where the geographic range of a crop species has been extended into colder regions by plant breeders. Programs of deliberate selection applied to corn (*Zea mays*) have expanded the area of the USA over which the crop can be profitably grown. From the 1920s to the 1940s, the production of corn in Iowa and Illinois increased by around 24%, whereas in the colder state of Wisconsin it increased by 54%.

If deliberate selection can change the tolerance and distribution of a domesticated plant we should expect natural selection to have done the same thing in nature. To test this, the plant *Umbilicus rupestris*, which lives in mild maritime areas of Great Britain, was deliberately grown outside its normal range (Woodward, 1990). A population of plants and seeds was taken from a donor population in the mild-wintered habitat of Cardiff in the west and introduced in a cooler environment at an altitude of

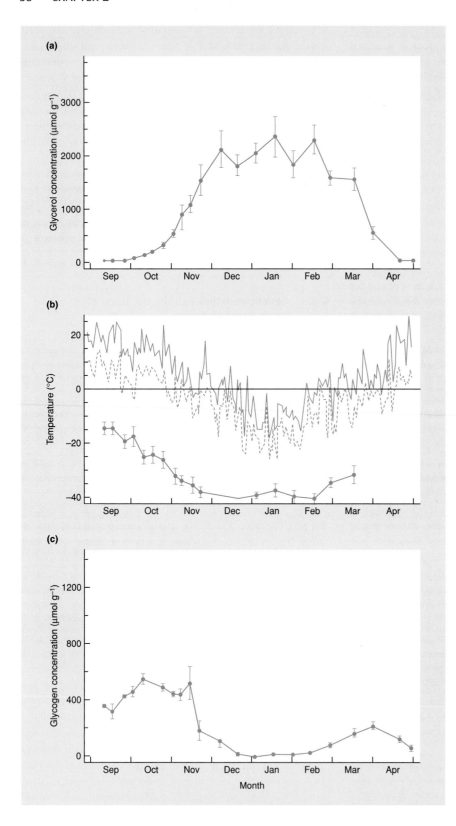

Figure 2.8 (a) Changes in the glycerol concentration per gram wet mass of the freeze-avoiding larvae of the goldenrod gall moth, *Epiblema scudderiana*. (b) The daily temperature maxima and minima (above) and whole larvae supercooling points (below) over the same period. (c) Changes in glycogen concentration over the same period. (After Rickards *et al.*, 1987.)

Figure 2.9 Acclimation to low temperatures. Samples of the Antarctic springtail *Cryptopygus antarcticus* were taken from field sites in the summer (*c*. 5°C) on a number of days and their supercooling point (at which they froze) was determined either immediately (●) or after a period of acclimation (●) at the temperatures shown. The supercooling points of the controls themselves varied because of temperature variations from day to day, but acclimation at temperatures in the range +2 to −2°C (indicative of winter) led to a drop in the supercooling point, whereas no such drop was observed at higher temperatures (indicative of summer) or lower temperatures (too low for a physiological acclimation response). Bars are standard errors. (After Worland & Convey, 2001.)

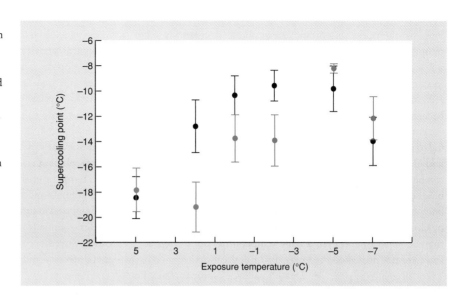

157 m in Sussex in the south. After 8 years, the temperature response of seeds from the donor and the introduced populations had diverged quite strikingly (Figure 2.10a), and subfreezing temperatures that kill in Cardiff (−12°C) were then tolerated by 50% of the Sussex population (Figure 2.10b). This suggests that past climatic changes, for example ice ages, will have changed the temperature tolerance of species as well as forcing their migration.

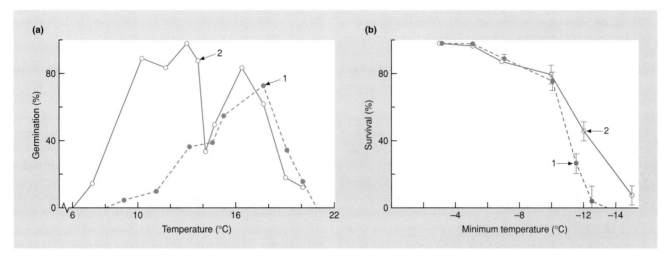

Figure 2.10 Changes in the behavior of populations of the plant *Umbilicus rupestris*, established for a period of 8 years in a cool environment in Sussex from a donor population in a mild-wintered area in South Wales (Cardiff, UK). (a) Temperature responses of seed germination: (1) responses of samples from the donor population (Cardiff) in 1978, and (2) responses from the Sussex population in 1987. (b) The low-temperature survival of the donor population at Cardiff, 1978 (1) and of the established population in Sussex, 1987 (2). (After Woodward, 1990.)

2.3.6   Life at high temperatures

Perhaps the most important thing about dangerously high temperatures is that, for a given organism, they usually lie only a few degrees above the metabolic optimum. This is largely an unavoidable consequence of the physicochemical properties of most enzymes (Wharton, 2002). High temperatures may be dangerous because they lead to the inactivation or even the denaturation of enzymes, but they may also have damaging indirect effects by leading to dehydration. All terrestrial organisms need to conserve water, and at high temperatures the rate of water loss by evaporation can be lethal, but they are caught between the devil and the deep blue sea because evaporation is an important means of reducing body temperature. If surfaces are protected from evaporation (e.g. by closing stomata in plants or spiracles in insects) the organisms may be killed by too high a body temperature, but if their surfaces are not protected they may die of desiccation.

high temperature and water loss

Death Valley, California, in the summer, is probably the hottest place on earth in which higher plants make active growth. Air temperatures during the daytime may approach 50°C and soil surface temperatures may be very much higher. The perennial plant, desert honeysweet (*Tidestromia oblongifolia*), grows vigorously in such an environment despite the fact that its leaves are killed if they reach the same temperature as the air. Very rapid transpiration keeps the temperature of the leaves at 40–45°C, and in this range they are capable of extremely rapid photosynthesis (Berry & Björkman, 1980).

Most of the plant species that live in very hot environments suffer severe shortage of water and are therefore unable to use the latent heat of evaporation of water to keep leaf temperatures down. This is especially the case in desert succulents in which water loss is minimized by a low surface to volume ratio and a low frequency of stomata. In such plants the risk of overheating may be reduced by spines (which shade the surface of a cactus) or hairs or waxes (which reflect a high proportion of the incident radiation). Nevertheless, such species experience and tolerate temperatures in their tissues of more than 60°C when the air temperature is above 40°C (Smith *et al.*, 1984).

fire

Fires are responsible for the highest temperatures that organisms face on earth and, before the fire-raising activities of humans, were caused mainly by lightning strikes. The recurrent risk of fire has shaped the species composition of arid and semiarid woodlands in many parts of the world. All plants are damaged by burning but it is the remarkable powers of regrowth from protected meristems on shoots and seeds that allow a specialized subset of species to recover from damage and form characteristic fire floras (see, for example, Hodgkinson, 1992).

Decomposing organic matter in heaps of farmyard manure, compost heaps and damp hay may reach very high temperatures. Stacks of damp hay are heated to temperatures of 50–60°C by

the metabolism of fungi such as *Aspergillus fumigatus*, carried further to approximately 65°C by other thermophilic fungi such as *Mucor pusillus* and then a little further by bacteria and actinomycetes. Biological activity stops well short of 100°C but autocombustible products are formed that cause further heating, drive off water and may even result in fire. Another hot environment is that of natural hot springs and in these the microbe *Thermus aquaticus* grows at temperatures of 67°C and tolerates temperatures up to 79°C. This organism has also been isolated from domestic hot water systems. Many (perhaps all) of the extremely thermophilic species are prokaryotes. In environments with very high temperatures the communities contain few species. In general, animals and plants are the most sensitive to heat followed by fungi, and in turn by bacteria, actinomycetes and archaebacteria. This is essentially the same order as is found in response to many other extreme conditions, such as low temperature, salinity, metal toxicity and desiccation.

thermal vents and other hot environments

An ecologically very remarkable hot environment was first described only towards the end of the last century. In 1979, a deep oceanic site was discovered in the eastern Pacific at which fluids at high temperatures ('smokers') were vented from the sea floor forming thin-walled 'chimneys' of mineral materials. Since that time many more vent sites have been discovered at mid-ocean crests in both the Atlantic and Pacific Oceans. They lie 2000–4000 m below sea level at pressures of 200–400 bars (20–40 MPa). The boiling point of water is raised to 370°C at 200 bars and to 404°C at 400 bars. The superheated fluid emerges from the chimneys at temperatures as high as 350°C, and as it cools to the temperature of seawater at about 2°C it provides a continuum of environments at intermediate temperatures.

Environments at such extreme pressures and temperatures are obviously extraordinarily difficult to study *in situ* and in most respects impossible to maintain in the laboratory. Some thermophilic bacteria collected from vents have been cultured successfully at 100°C at only slightly above normal barometric pressures (Jannasch & Mottl, 1985), but there is much circumstantial evidence that some microbial activity occurs at much higher temperatures and may form the energy resource for the warm water communities outside the vents. For example, particulate DNA has been found in samples taken from within the 'smokers' at concentrations that point to intact bacteria being present at temperatures very much higher than those conventionally thought to place limits on life (Baross & Deming, 1995).

There is a rich eukaryotic fauna in the local neighborhood of vents that is quite atypical of the deep oceans in general. At one vent in Middle Valley, Northeast Pacific, surveyed photographically and by video, at least 55 taxa were documented of which 15 were new or probably new species (Juniper *et al.*, 1992). There can be few environments in which so complex and specialized a community depends on so localized a special condition. The closest known vents with similar conditions are 2500 km distant. Such communities add a further list to the planet's record of species richness. They present tantalizing problems in evolution and daunting problems for the technology needed to observe, record and study them.

2.3.7 Temperature as a stimulus

We have seen that temperature as a condition affects the rate at which organisms develop. It may also act as a stimulus, determining whether or not the organism starts its development at all. For instance, for many species of temperate, arctic and alpine herbs, a period of chilling or freezing (or even of alternating high and low temperatures) is necessary before germination will occur. A cold experience (physiological evidence that winter has passed) is required before the plant can start on its cycle of growth and development. Temperature may also interact with other stimuli (e.g. photoperiod) to break dormancy and so time the onset of growth. The seeds of the birch (*Betula pubescens*) require a photoperiodic stimulus (i.e. experience of a particular regime of day length) before they will germinate, but if the seed has been chilled it starts growth without a light stimulus.

2.4 Correlations between temperature and the distribution of plants and animals

2.4.1 Spatial and temporal variations in temperature

Variations in temperature on and within the surface of the earth have a variety of causes: latitudinal, altitudinal, continental, seasonal, diurnal and microclimatic effects and, in soil and water, the effects of depth.

Latitudinal and seasonal variations cannot really be separated. The angle at which the earth is tilted relative to the sun changes with the seasons, and this drives some of the main temperature differentials on the earth's surface. Superimposed on these broad geographic trends are the influences of altitude and 'continentality'. There is a drop of 1°C for every 100 m increase in altitude in dry air, and a drop of 0.6°C in moist air. This is the result of the 'adiabatic' expansion of air as atmospheric pressure falls with increasing altitude. The effects of continentality are largely attributable to different rates of heating and cooling of the land and the sea. The land surface reflects less heat than the water, so the surface warms more quickly, but it also loses heat more quickly. The sea therefore has a moderating, 'maritime' effect on the temperatures of coastal regions and especially islands; both daily and seasonal variations in temperature are far less marked than at more inland, continental locations at the same latitude. Moreover, there are comparable effects within land masses: dry, bare areas like deserts suffer greater daily and seasonal extremes of temperature

than do wetter areas like forests. Thus, global maps of temperature zones hide a great deal of local variation.

It is much less widely appreciated that on a smaller scale still there can be a great deal of microclimatic variation. For example, the sinking of dense, cold air into the bottom of a valley at night can make it as much as 30°C colder than the side of the valley only 100 m higher; the winter sun, shining on a cold day, can heat the south-facing side of a tree (and the habitable cracks and crevices within it) to as high as 30°C; and the air temperature in a patch of vegetation can vary by 10°C over a vertical distance of 2.6 m from the soil surface to the top of the canopy (Geiger, 1955). Hence, we need not confine our attention to global or geographic patterns when seeking evidence for the influence of temperature on the distribution and abundance of organisms.

microclimatic variation

Long-term temporal variations in temperature, such as those associated with the ice ages, were discussed in the previous chapter. Between these, however, and the very obvious daily and seasonal changes that we are all aware of, a number of medium-term patterns have become increasingly apparent. Notable amongst these are the El Niño-Southern Oscillation (ENSO) and the North Atlantic Oscillation (NAO) (Figure 2.11) (see Stenseth *et al.*, 2003). The ENSO originates in the tropical Pacific Ocean off the coast of South America and is an alternation (Figure 2.11a) between a warm (El Niño) and a cold (La Niña) state of the water there, though it affects temperature, and the climate generally, in terrestrial and marine environments throughout the whole Pacific basin (Figure 2.11b; for color, see Plate 2.1, between pp. 84 and 85) and beyond. The NAO refers to a north–south alternation in atmospheric mass between the subtropical Atlantic and the Arctic (Figure 2.11c) and again affects climate in general rather than just temperature (Figure 2.11d; for color, see Plate 2.2, between pp. 084 and 85). Positive index values (Figure 2.11c) are associated, for example, with relatively warm conditions in North America and Europe and relatively cool conditions in North Africa and the Middle East. An example of the effect of NAO variation on species abundance, that of cod, *Gadus morhua*, in the Barents Sea, is shown in Figure 2.12.

ENSO and NAO

2.4.2 Typical temperatures and distributions

There are very many examples of plant and animal distributions that are strikingly correlated with some aspect of environmental temperature even at gross taxonomic and systematic levels (Figure 2.13). At a finer scale, the distributions of many species closely match maps of some aspect of temperature. For example, the northern limit of the distribution of wild madder plants (*Rubia peregrina*) is closely correlated with the position of the January 4.5°C

isotherms

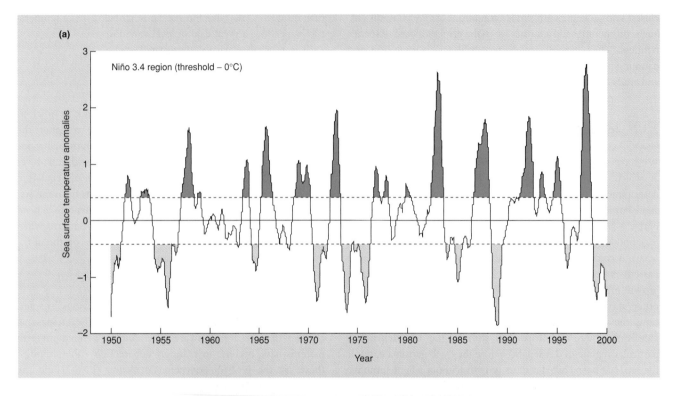

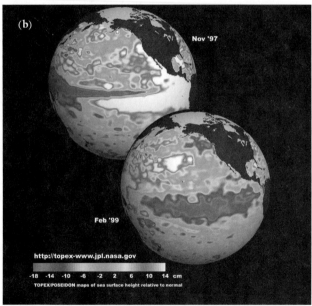

Figure 2.11   (a) The El Niño–Southern Oscillation (ENSO) from 1950 to 2000 as measured by sea surface temperature anomalies (differences from the mean) in the equatorial mid-Pacific. The El Niño events (> 0.4°C above the mean) are shown in dark color, and the La Niña events (> 0.4°C below the mean) are shown in pale color. (Image from http://www.cgd.ucar.edu/cas/catalog/ climind/Nino_3_3.4_indices.html.) (b) Maps of examples of El Niño (November 1997) and La Niña (February 1999) events in terms of sea height above average levels. Warmer seas are higher; for example, a sea height 15–20 cm below average equates to a temperature anomaly of approximately 2–3°C. (Image from http://topex-www.jpl.nasa.gov/science/images/el-nino-la-nina.jpg.) (For color, see Plate 2.1, between pp. 84 and 85.)

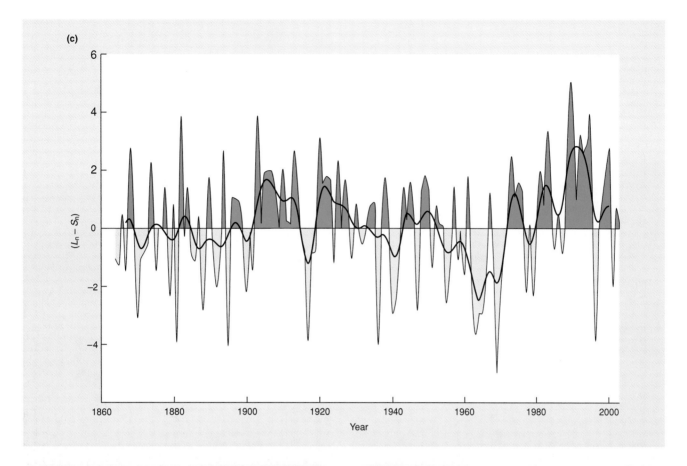

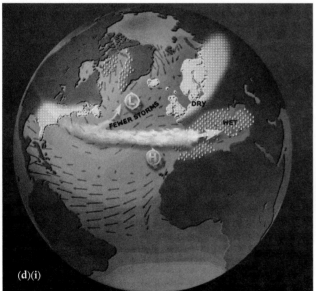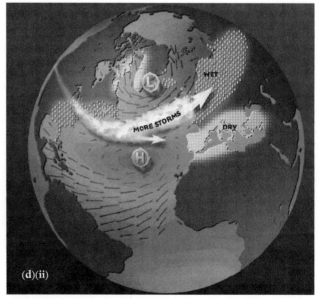

Figure 2.11 (*continued*) (c) The North Atlantic Oscillation (NAO) from 1864 to 2003 as measured by the normalized sea-level pressure difference ($L_n - S_n$) between Lisbon, Portugal and Reykjavik, Iceland. (Image from http://www.cgd.ucar.edu/~jhurrell/ nao.stat.winter.html#winter.) (d) Typical winter conditions when the NAO index is positive or negative. Conditions that are more than usually warm, cold, dry or wet are indicated. (Image from http://www.ldeo.columbia.edu/NAO/.) (For color, see Plate 2.2, between pp. 84 and 85.)

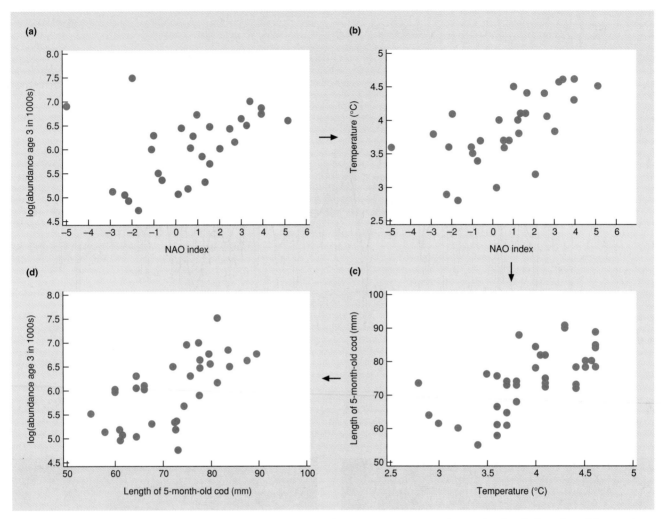

Figure 2.12 (a) The abundance of 3-year-old cod, *Gadus morhua*, in the Barents Sea is positively correlated with the value of the North Atlantic Oscillation (NAO) index for that year. The mechanism underlying this correlation is suggested in (b–d). (b) Annual mean temperature increases with the NAO index. (c) The length of 5-month-old cod increases with annual mean temperature. (d) The abundance of cod at age 3 increases with their length at 5 months. (After Ottersen *et al.*, 2001.)

isotherm (Figure 2.14a; an isotherm is a line on a map joining places that experience the same temperature – in this case a January mean of 4.5°C). However, we need to be very careful how we interpret such relationships: they can be extremely valuable in predicting where we might and might not find a particular species; they may suggest that some feature related to temperature is important in the life of the organisms; but they do not prove that temperature *causes* the limits to a species' distribution. The literature relevant to this and many other correlations between temperature and distribution patterns is reviewed by Hengeveld (1990), who also describes a more subtle graphical procedure. The minimum temperature of the coldest month and the maximum temperature of the hottest month are estimated for many places within and

outside the range of a species. Each location is then plotted on a graph of maximum against minimum temperature, and a line is drawn that optimally discriminates between the presence and absence records (Figure 2.14b). This line is then used to define the geographic margin of the species distributions (Figure 2.14c). This may have powerful predictive value, but it still tells us nothing about the underlying forces that cause the distribution patterns.

One reason why we need to be cautious about reading too much into correlations of species distributions with maps of temperature is that the temperatures measured for constructing isotherms for a map are only rarely those that the organisms experience. In nature an organism may choose to lie in the sun or hide

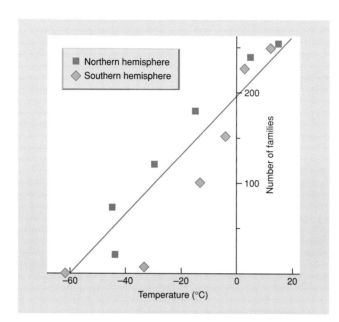

Figure 2.13   The relationship between absolute minimum temperature and the number of families of flowering plants in the northern and southern hemispheres. (After Woodward, 1987, who also discusses the limitations to this sort of analysis and how the history of continental isolation may account for the odd difference between northern and southern hemispheres.)

in the shade and, even in a single day, may experience a baking midday sun and a freezing night. Moreover, temperature varies from place to place on a far finer scale than will usually concern a geographer, but it is the conditions in these 'microclimates' that will be crucial in determining what is habitable for a particular species. For example, the prostrate shrub *Dryas octopetala* is restricted to altitudes exceeding 650 m in North Wales, UK, where it is close to its southern limit. But to the north, in Sutherland in Scotland, where it is generally colder, it is found right down to sea level.

2.4.3   Distributions and extreme conditions

For many species, distributions are accounted for not so much by average temperatures as by occasional extremes, especially occasional lethal temperatures that preclude its existence. For instance, injury by frost is probably the single most important factor limiting plant distribution. To take one example: the saguaro cactus (*Carnegiea gigantea*) is liable to be killed when temperatures remain below freezing for 36 h, but if there is a daily thaw it is under no threat. In Arizona, the northern and eastern edges of the cactus' distribution correspond to a line joining places where on occasional days it fails to thaw. Thus, the saguaro is absent where there are occasionally lethal conditions – an individual need only be killed once.

Similarly, there is scarcely any crop   you only die once
that is grown on a large commercial
scale in the climatic conditions of its wild ancestors, and it is well known that crop failures are often caused by extreme events, especially frosts and drought. For instance, the climatic limit to the geographic range for the production of coffee (*Coffea arabica* and *C. robusta*) is defined by the 13°C isotherm for the coldest month of the year. Much of the world's crop is produced in the highland microclimates of the São Paulo and Paraná districts of

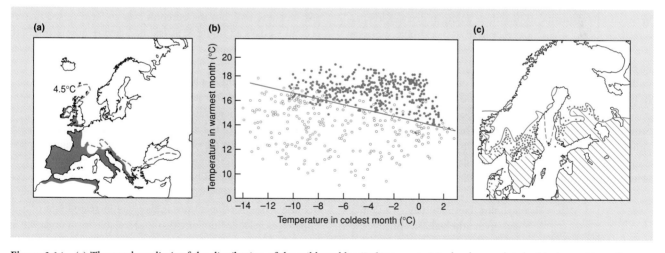

Figure 2.14   (a) The northern limit of the distribution of the wild madder (*Rubia peregrina*) is closely correlated with the position of the January 4.5°C isotherm. (After Cox *et al.*, 1976.) (b) A plot of places within the range of *Tilia cordat* (●), and outside its range (○) in the graphic space defined by the minimum temperature of the coldest month and the maximum temperature of the warmest month. (c) Margin of the geographic range of *T. cordata* in northern Europe defined by the straight line in (b). ((b, c) after Hintikka, 1963; from Hengeveld, 1990.)

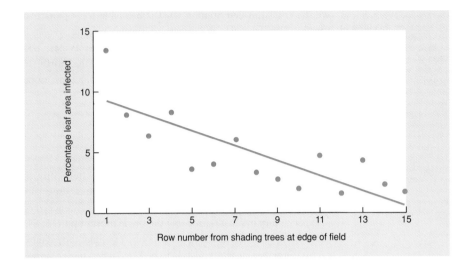

Figure 2.15   The incidence of southern corn leaf blight (*Helminthosporium maydis*) on corn growing in rows at various distances from trees that shaded them. Wind-borne fungal diseases were responsible for most of this mortality (Harper, 1955). (From Lukens & Mullany, 1972.)

Brazil. Here, the average minimum temperature is 20°C, but occasionally cold winds and just a few hours of temperature close to freezing are sufficient to kill or severely damage the trees (and play havoc with world coffee prices).

2.4.4   Distributions and the interaction of temperature with other factors

Although organisms respond to each condition in their environment, the effects of conditions may be determined largely by the responses of other community members. Temperature does not act on just one species: it also acts on its competitors, prey, parasites and so on. This, as we saw in Section 2.2, was the difference between a fundamental niche (where an organism *could* live) and a realized niche (where it *actually* lived). For example, an organism will suffer if its food is another species that cannot tolerate an environmental condition. This is illustrated by the distribution of the rush moth (*Coleophora alticolella*) in England. The moth lays its eggs on the flowers of the rush *Juncus squarrosus* and the caterpillars feed on the developing seeds. Above 600 m, the moths and caterpillars are little affected by the low temperatures, but the rush, although it grows, fails to ripen its seeds. This, in turn, limits the distribution of the moth, because caterpillars that hatch in the colder elevations will starve as a result of insufficient food (Randall, 1982).

disease

The effects of conditions on disease may also be important. Conditions may favor the spread of infection (winds carrying fungal spores), or favor the growth of the parasite, or weaken the defenses of the host. For example, during an epidemic of southern corn leaf blight (*Helminthosporium maydis*) in a corn field in Connecticut, the plants closest to the trees that were shaded for the longest periods were the most heavily diseased (Figure 2.15).

competition

Competition between species can also be profoundly influenced by environmental conditions, especially temperature. Two stream salmonid fishes, *Salvelinus malma* and *S. leucomaenis*, coexist at intermediate altitudes (and therefore intermediate temperatures) on Hokkaido Island, Japan, whereas only the former lives at higher altitudes (lower temperatures) and only the latter at lower altitudes (see also Section 8.2.1). A reversal, by a change in temperature, of the outcome of competition between the species appears to play a key role in this. For example, in experimental streams supporting the two species maintained at 6°C over a 191-day period (a typical high altitude temperature), the survival of *S. malma* was far superior to that of *S. leucomaenis*; whereas at 12°C (typical low altitude), both species survived less well, but the outcome was so far reversed that by around 90 days all of the *S. malma* had died (Figure 2.16). Both species are quite capable, alone, of living at either temperature.

temperature and humidity

Many of the interactions between temperature and other physical conditions are so strong that it is not sensible to consider them separately. The relative humidity of the atmosphere, for example, is an important condition in the life of terrestrial organisms because it plays a major part in determining the rate at which they lose water. In practice, it is rarely possible to make a clean distinction between the effects of relative humidity and of temperature. This is simply because a rise in temperature leads to an increased rate of evaporation. A relative humidity that is acceptable to an organism at a low temperature may therefore be unacceptable at a higher temperature. Microclimatic variations in relative humidity can be even more marked than those involving temperature. For instance, it is not unusual for the relative humidity to be almost 100% at ground level amongst dense vegetation and within the soil, whilst the air immediately above, perhaps 40 cm away, has a relative humidity

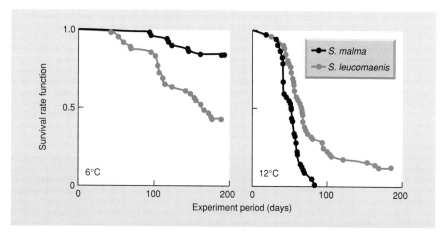

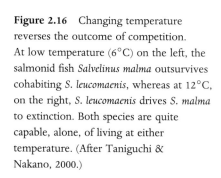

Figure 2.16 Changing temperature reverses the outcome of competition. At low temperature (6°C) on the left, the salmonid fish *Salvelinus malma* outsurvives cohabiting *S. leucomaenis*, whereas at 12°C, on the right, *S. leucomaenis* drives *S. malma* to extinction. Both species are quite capable, alone, of living at either temperature. (After Taniguchi & Nakano, 2000.)

of only 50%. The organisms most obviously affected by humidity in their distribution are those 'terrestrial' animals that are actually, in terms of the way they control their water balance, 'aquatic'. Amphibians, terrestrial isopods, nematodes, earthworms and molluscs are all, at least in their active stages, confined to microenvironments where the relative humidity is at or very close to 100%. The major group of animals to escape such confinement are the terrestrial arthropods, especially insects. Even here though, the evaporative loss of water often confines their activities to habitats (e.g. woodlands) or times of day (e.g. dusk) when relative humidity is relatively high.

2.5 pH of soil and water

The pH of soil in terrestrial environments or of water in aquatic ones is a condition that can exert a powerful influence on the distribution and abundance of organisms. The protoplasm of the root cells of most vascular plants is damaged as a direct result of toxic concentrations of H^+ or OH^- ions in soils below pH 3 or above pH 9, respectively. Further, indirect effects occur because soil pH influences the availability of nutrients and/or the concentration of toxins (Figure 2.17).

Increased acidity (low pH) may act in three ways: (i) directly, by upsetting osmoregulation, enzyme activity or gaseous exchange across respiratory surfaces; (ii) indirectly, by increasing the concentration of toxic heavy metals, particularly aluminum (Al^{3+}) but also manganese (Mn^{2+}) and iron (Fe^{3+}), which are essential plant nutrients at higher pHs; and (iii) indirectly, by reducing the quality and range of food sources available to animals (e.g. fungal growth is reduced at low pH in streams (Hildrew *et al.*, 1984) and the aquatic flora is often absent or less diverse). Tolerance limits for pH vary amongst plant species, but only a minority are able to grow and reproduce at a pH below about 4.5.

In alkaline soils, iron (Fe^{3+}) and phosphate (PO_4^{3+}), and certain trace elements such as manganese (Mn^{2+}), are fixed in relatively

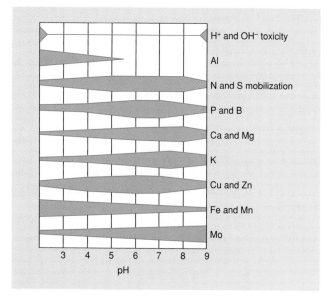

Fgiure 2.17 The toxicity of H^+ and OH^- to plants, and the availability to them of minerals (indicated by the widths of the bands) is influenced by soil pH. (After Larcher, 1980.)

insoluble compounds, and plants may then suffer because there is too little rather than too much of them. For example, calcifuge plants (those characteristic of acid soils) commonly show symptoms of iron deficiency when they are transplanted to more alkaline soils. In general, however, soils and waters with a pH above 7 tend to be hospitable to many more species than those that are more acid. Chalk and limestone grasslands carry a much richer flora (and associated fauna) than acid grasslands and the situation is similar for animals inhabiting streams, ponds and lakes.

Some prokaryotes, especially the Archaebacteria, can tolerate and even grow best in environments with a pH far outside the range tolerated by eukaryotes. Such environments are rare, but occur in volcanic lakes and geothermal springs where they are dominated by sulfur-oxidizing bacteria whose pH optima lie between 2 and 4 and which cannot grow at neutrality (Stolp, 1988). *Thiobacillus ferroxidans* occurs in the waste from industrial metal-leaching processes and tolerates pH 1; *T. thiooxidans* cannot only tolerate but can grow at pH 0. Towards the other end of the pH range are the alkaline environments of soda lakes with pH values of 9–11, which are inhabited by cyanobacteria such as *Anabaenopsis arnoldii* and *Spirulina platensis*; *Plectonema nostocorum* can grow at pH 13.

2.6   Salinity

For terrestrial plants, the concentration of salts in the soil water offers osmotic resistance to water uptake. The most extreme saline conditions occur in arid zones where the predominant movement of soil water is towards the surface and cystalline salt accumulates. This occurs especially when crops have been grown in arid regions under irrigation; salt pans then develop and the land is lost to agriculture. The main effect of salinity is to create the same kind of osmoregulatory problems as drought and freezing and the problems are countered in much the same ways. For example, many of the higher plants that live in saline environments (halophytes) accumulate electrolytes in their vacuoles, but maintain a low concentration in the cytoplasm and organelles (Robinson *et al.*, 1983). Such plants maintain high osmotic pressures and so remain turgid, and are protected from the damaging action of the accumulated electrolytes by polyols and membrane protectants.

Freshwater environments present a set of specialized environmental conditions because water tends to move into organisms from the environment and this needs to be resisted. In marine habitats, the majority of organisms are isotonic to their environment so that there is no net flow of water, but there are many that are hypotonic so that water flows out from the organism to the environment, putting them in a similar position to terrestrial organisms. Thus, for many aquatic organisms the regulation of body fluid concentration is a vital and sometimes an energetically expensive process. The salinity of an aquatic environment can have an important influence on distribution and abundance, especially in places like estuaries where there is a particularly sharp gradient between truly marine and freshwater habitats.

The freshwater shrimps *Palaemonetes pugio* and *P. vulgaris*, for example, co-occur in estuaries on the eastern coat of the USA at a wide range of salinities, but the former seems to be more tolerant of lower salinities than the latter, occupying some habitats from which the latter is absent. Figure 2.18 shows the mechanism likely to be underlying this (Rowe, 2002). Over the low salinity range (though not at the effectively lethal lowest salinity) metabolic expenditure was significantly lower in *P. pugio*. *P. vulgaris* requires far more energy simply to maintain itself, putting it at a severe disadvantage in competition with *P. pugio* even when it is able to sustain such expenditure.

2.6.1   Conditions at the boundary between the sea and land

Salinity has important effects on the distribution of organisms in intertidal areas but it does so through interactions with other conditions – notably exposure to the air and the nature of the substrate.

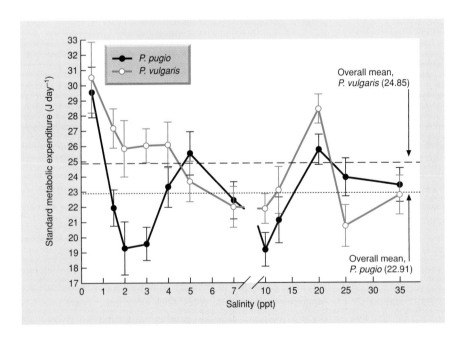

Figure 2.18   Standard metabolic expenditure (estimated through minimum oxygen consumption) in two species of shrimp, *Palaemonetes pugio* and *P. vulgaris*, at a range of salinities. There was significant mortality of both species over the experimental period at 0.5 ppt (parts per thousand), especially in *P. vulgaris* (75% compared with 25%). (After Rowe, 2002.)

Algae of all types have found suitable habitats permanently immersed in the sea, but permanently submerged higher plants are almost completely absent. This is a striking contrast with submerged freshwater habitats where a variety of flowering plants have a conspicuous role. The main reason seems to be that higher plants require a substrate in which their roots can find anchorage. Large marine algae, which are continuously submerged except at extremely low tides, largely take their place in marine communities. These do not have roots but attach themselves to rocks by specialized 'holdfasts'. They are excluded from regions where the substrates are soft and holdfasts cannot 'hold fast'. It is in such regions that the few truly marine flowering plants, for example sea grasses such as *Zostera* and *Posidonia*, form submerged communities that support complex animal communities.

| algae and higher plants | Most species of higher plants that root in seawater have leaves and shoots that are exposed to the atmosphere for a large part of the tidal cycle, such |

as mangroves, species of the grass genus *Spartina* and extreme halophytes such as species of *Salicornia* that have aerial shoots but whose roots are exposed to the full salinity of seawater. Where there is a stable substrate in which plants can root, communities of flowering plants may extend right through the intertidal zone in a continuum extending from those continuously immersed in full-strength seawater (like the sea grasses) through to totally nonsaline conditions. Salt marshes, in particular, encompass a range of salt concentrations running from full-strength seawater down to totally nonsaline conditions.

Higher plants are absent from intertidal rocky sea shores except where pockets of soft substrate may have formed in crevices. Instead, such habitats are dominated by the algae, which give way to lichens at and above the high tide level where the exposure to desiccation is highest. The plants and animals that live on rocky sea shores are influenced by environmental conditions in a very profound and often particularly obvious way by the extent to which they tolerate exposure to the aerial environment and the forces of waves and storms. This expresses itself in the *zonation* of the organisms, with different species at different heights up the shore (Figure 2.19).

| | zonation |

The extent of the intertidal zone depends on the height of tides and the slope of the shore. Away from the shore, the tidal rise and fall are rarely greater than 1 m, but closer to shore, the shape of the land mass can funnel the ebb and flow of the water to produce extraordinary spring tidal ranges of, for example, nearly 20 m in the Bay of Fundy (between Nova Scotia and New Brunswick, Canada). In contrast, the shores of the Mediterranean Sea

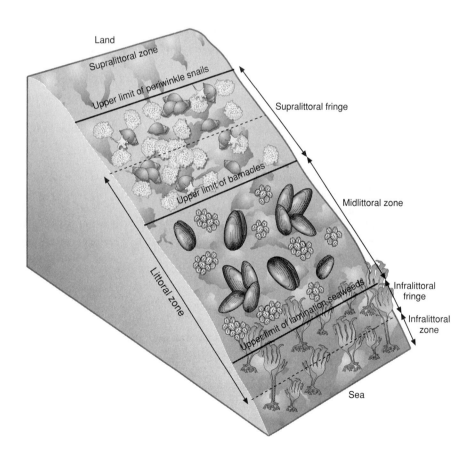

Figure 2.19 A general zonation scheme for the seashore determined by relative lengths of exposure to the air and to the action of waves. (After Raffaelli & Hawkins, 1996.)

experience scarcely any tidal range. On steep shores and rocky cliffs the intertidal zone is very short and zonation is compressed.

To talk of 'zonation as a result of exposure', however, is to oversimplify the matter greatly (Raffaelli & Hawkins, 1996). In the first place, 'exposure' can mean a variety, or a combination of, many different things: desiccation, extremes of temperature, changes in salinity, excessive illumination and the sheer physical forces of pounding waves and storms (to which we turn in Section 2.7). Furthermore, 'exposure' only really explains the *upper* limits of these essentially marine species, and yet zonation depends on them having lower limits too. For some species there can be *too little exposure* in the lower zones. For instance, green algae would be starved of blue and especially red light if they were submerged for long periods too low down the shore. For many other species though, a lower limit to distribution is set by competition and predation (see, for example, the discussion in Paine, 1994). The seaweed *Fucus spiralis* will readily extend lower down the shore than usual in Great Britain whenever other competing midshore fucoid seaweeds are scarce.

2.7   Physical forces of winds, waves and currents

In nature there are many forces of the environment that have their effect by virtue of the force of physical movement – wind and water are prime examples.

In streams and rivers, both plants and animals face the continual hazard of being washed away. The average velocity of flow generally increases in a downstream direction, but the greatest danger of members of the benthic (bottom-dwelling) community being washed away is in upstream regions, because the water here is turbulent and shallow. The only plants to be found in the most extreme flows are literally 'low profile' species like encrusting and filamentous algae, mosses and liverworts. Where the flow is slightly less extreme there are plants like the water crowfoot (*Ranunculus fluitans*), which is streamlined, offering little resistance to flow and which anchors itself around an immovable object by means of a dense development of adventitious roots. Plants such as the free-floating duckweed (*Lemna* spp.) are usually only found where there is negligible flow.

The conditions of exposure on sea shores place severe limits on the life forms and habits of species that can tolerate repeated pounding and the suction of wave action. Seaweeds anchored on rocks survive the repeated pull and push of wave action by a combination of powerful attachment by holdfasts and extreme flexibility of their thallus structure. Animals in the same environment either move with the mass of water or, like the algae, rely on subtle mechanisms of firm adhesion such as the powerful organic glues of barnacles and the muscular feet of limpets. A comparable diversity of morphological specializations is to be found amongst the invertebrates that tolerate the hazards of turbulent, freshwater streams.

2.7.1   Hazards, disasters and catastrophes: the ecology of extreme events

The wind and the tides are normal daily 'hazards' in the life of many organisms. The structure and behavior of these organisms bear some witness to the frequency and intensity of such hazards in the evolutionary history of their species. Thus, most trees withstand the force of most storms without falling over or losing their living branches. Most limpets, barnacles and kelps hold fast to the rocks through the normal day to day forces of the waves and tides. We can also recognize a scale of more severely damaging forces (we might call them 'disasters') that occur occasionally, but with sufficient frequency to have contributed repeatedly to the forces of natural selection. When such a force recurs it will meet a population that still has a genetic memory of the selection that acted on its ancestors – and may therefore suffer less than they did. In the woodlands and shrub communities of arid zones, fire has this quality, and tolerance of fire damage is a clearly evolved response (see Section 2.3.6).

When disasters strike natural communities it is only rarely that they have been carefully studied before the event. One exception is cyclone 'Hugo' which struck the Caribbean island of Guadeloupe in 1994. Detailed accounts of the dense humid forests of the island had been published only recently before (Ducrey & Labbé, 1985, 1986). The cyclone devastated the forests with mean maximum wind velocities of 270 km h^{-1} and gusts of 320 km h^{-1}. Up to 300 mm of rain fell in 40 h. The early stages of regeneration after the cyclone (Labbé, 1994) typify the responses of long-established communities on both land or sea to massive forces of destruction. Even in 'undisturbed' communities there is a continual creation of gaps as individuals (e.g. trees in a forest, kelps on a sea shore) die and the space they occupied is recolonized (see Section 16.7). After massive devastation by cyclones or other widespread disasters, recolonization follows much the same course. Species that normally colonize only natural gaps in the vegetation come to dominate a continuous community.

In contrast to conditions that we have called 'hazards' and 'disasters' there are natural occurrences that are enormously damaging, yet occur so rarely that they may have no lasting selective effect on the evolution of the species. We might call such events 'catastrophes', for example the volcanic eruption of Mt St Helens or of the island of Krakatau. The next time that Krakatau erupts there are unlikely to be any genes persisting that were selected for volcano tolerance!

2.8   Environmental pollution

A number of environmental conditions that are, regrettably, becoming increasingly important are due to the accumulation of toxic by-products of human activities. Sulfur dioxide emitted from power stations, and metals like copper, zinc and lead, dumped

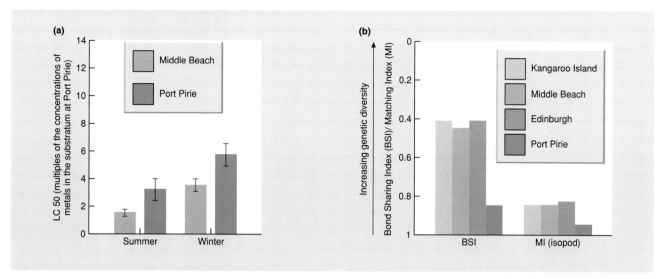

Figure 2.20 The response of the marine isopod, *Platynympha longicaudata*, to pollution around the largest lead smelting operation in the world, Port Pirie, South Australia. (a) Tolerance, both summer and winter, was significantly higher ($P < 0.05$) than for animals from a control (unpolluted) site, as measured by the concentration in food of a combination of metals (lead, copper, cadmium, zinc and manganese) required to kill 50% of the population (LC50). (b) Genetic diversity at Port Pirie was significantly lower than at three unpolluted sites, as measured by two indices of diversity based on RAPDs (random amplified polymorphic DNA). (After Ross *et al.*, 2002.)

around mines or deposited around refineries, are just some of the pollutants that limit distributions, especially of plants. Many such pollutants are present naturally but at low concentrations, and some are indeed essential nutrients for plants. But in polluted areas their concentrations can rise to lethal levels. The loss of species is often the first indication that pollution has occurred, and changes in the species richness of a river, lake or area of land provide bioassays of the extent of their pollution (see, for example, Lovett Doust *et al.*, 1994).

rare tolerators

Yet it is rare to find even the most inhospitable polluted areas entirely devoid of species; there are usually at least a few individuals of a few species that can tolerate the conditions. Even natural populations from unpolluted areas often contain a low frequency of individuals that tolerate the pollutant; this is part of the genetic variability present in natural populations. Such individuals may be the only ones to survive or colonize as pollutant levels rise. They may then become the founders of a tolerant population to which they have passed on their 'tolerance' genes, and, because they are the descendants of just a few founders, such populations may exhibit notably low genetic diversity overall (Figure 2.20). Moreover, species themselves may differ greatly in their ability to tolerate pollutants. Some plants, for example, are 'hyperaccumulators' of heavy metals – lead, cadmium and so on – with an ability not only to tolerate but also to accumulate much higher concentrations than the norm (Brooks, 1998). As a result, such plants may have an important role to play in 'bioremediation' (Salt *et al.*, 1998), removing pollutants from the soil so that

eventually other, less tolerant plants can grow there too (discussed further in Section 7.2.1).

Thus, in very simple terms, a pollutant has a twofold effect. When it is newly arisen or is at extremely high concentrations, there will be few individuals of any species present (the exceptions being naturally tolerant variants or their immediate descendants). Subsequently, however, the polluted area is likely to support a much higher density of individuals, but these will be representatives of a much smaller range of species than would be present in the absence of the pollutant. Such newly evolved, species-poor communities are now an established part of human environments (Bradshaw, 1987).

Pollution can of course have its effects far from the original source (Figure 2.21). Toxic effluents from a mine or a factory may enter a watercourse and affect its flora and fauna for its whole length downstream. Effluents from large industrial complexes can pollute and change the flora and fauna of many rivers and lakes in a region and cause international disputes.

A striking example is the creation of 'acid rain' – for example that falling in Ireland and Scandinavia from industrial activities in other countries. Since the Industrial Revolution, the burning of fossil fuels and the consequent emission to the atmosphere of various pollutants, notably sulfur dioxide, has produced a deposition of dry acidic particles and rain that is essentially dilute sulfuric acid. Our knowledge of the pH tolerances of diatom species enables an approximate pH history of a lake to be constructed. The history of the acidification of lakes is often

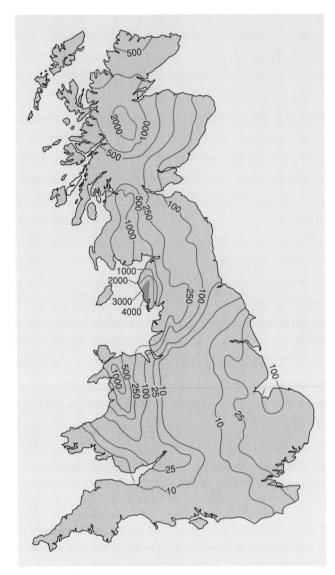

Figure 2.21 An example of long-distance environmental pollution. The distribution in Great Britain of fallout of radioactive caesium (Bq m^{-2}) from the Chernobyl nuclear accident in the Soviet Union in 1986. The map shows the persistence of the pollutant on acid upland soils where it is recycled through soils, plants and animals. Sheep in the upland areas contained more caesium-137 (^{137}Cs) in 1987 and 1988 (after recycling) than in 1986. ^{137}Cs has a half-life of 30 years! On typical lowland soils it is more quickly immobilized and does not persist in the food chains. (After NERC, 1990.)

recorded in the succession of diatom species accumulated in lake sediments (Flower *et al.*, 1994). Figure 2.22, for example, shows how diatom species composition has changed in Lough Maam, Ireland – far from major industrial sites. The percentage of various diatom species at different depths reflects the flora present at various times in the past (four species are illustrated). The age of layers of sediment can be determined by the radioactive decay of lead-210 (and other elements). We know the pH tolerance of the diatom species from their present distribution and this can be used to reconstruct what the pH of the lake has been in the past. Note how the waters acidified since about 1900. The diatoms *Fragilaria virescens* and *Brachysira vitrea* have declined markedly during this period while the acid-tolerant *Cymbella perpusilla* and *Frustulia rhomboides* increased after 1900.

2.9   Global change

In Chapter 1 we discussed some of the ways in which global environments have changed over the long timescales involved in continental drift and the shorter timescales of the repeated ice ages. Over these timescales some organisms have failed to accommodate to the changes and have become extinct, others have migrated so that they continue to experience the same conditions but in a different place, and it is probable that others have changed their nature (evolved) and tolerated some of the changes. We now turn to consider global changes that are occurring in our own lifetimes – consequences of our own activities – and that are predicted, in most scenarios, to bring about profound changes in the ecology of the planet.

2.9.1   Industrial gases and the greenhouse effect

A major element of the Industrial Revolution was the switch from the use of sustainable fuels to the use of coal (and later, oil) as a source of power. Between the middle of the 19th and the middle of the 20th century the burning of fossil fuels, together with extensive deforestation, added about 9×10^{10} tonnes of carbon dioxide (CO_2) to the atmosphere and even more has been added since. The concentration of CO_2 in the atmosphere before the Industrial Revolution (measured in gas trapped in ice cores) was about 280 ppm, a fairly typical interglacial 'peak' (Figure 2.23), but this had risen to around 370 ppm by around the turn of the millennium and is still rising (see Figure 18.22).

Solar radiation incident on the earth's atmosphere is in part reflected, in part absorbed, and part is transmitted through to the earth's surface, which absorbs and is warmed by it. Some of this absorbed energy is radiated back to the atmosphere where atmospheric gases, mainly water vapor and CO_2 absorb about 70% of it. It is this trapped reradiated energy that heats the atmosphere in what is called the 'greenhouse effect'. The greenhouse effect was of course part of the normal environment before the Industrial Revolution and carried responsibility for some of the environmental warmth before industrial activity started to enhance it. At that time, atmospheric water vapor was responsible for the greater portion of the greenhouse effect.

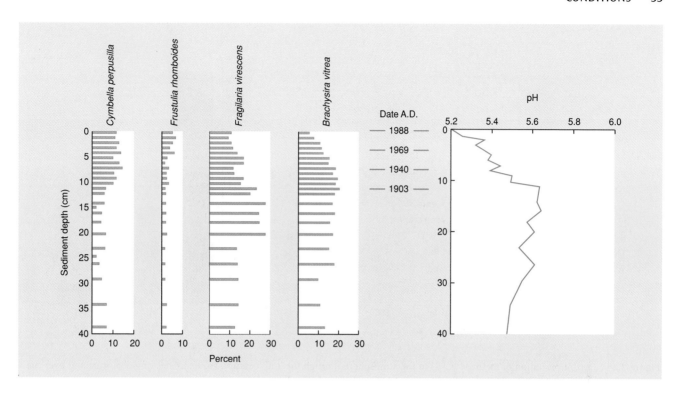

Figure 2.22 The history of the diatom flora of an Irish lake (Lough Maam, County Donegal) can be traced by taking cores from the sediment at the bottom of the lake. The percentage of various diatom species at different depths reflects the flora present at various times in the past (four species are illustrated). The age of the layers of sediment can be determined by the radioactive decay of lead-210 (and other elements). We know the pH tolerance of the diatom species from their present distribution and this can be used to reconstruct what the pH of the lake has been in the past. Note how the waters have been acidified since about 1900. The diatoms *Fragilaria virescens* and *Brachysira vitrea* have declined markedly during this period, while the acid-tolerant *Cymbella perpusilla* and *Frustulia rhomboides* have increased. (After Flower *et al.*, 1994.)

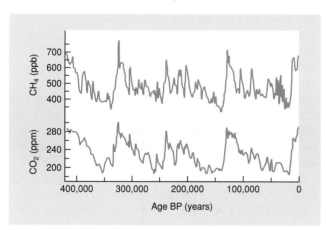

Figure 2.23 Concentrations of CO_2 and methane (CH_4) in gas trapped in ice cores from Vostok, Antarctica deposited over the past 420,000 years. Estimated temperatures are very strongly correlated with these. Thus, transitions between glacial and warm epochs occurred around 335,000, 245,000, 135,000 and 18,000 years ago. BP, before present; ppb, parts per billion; ppm, parts per million. (After Petit *et al.*, 1999; Stauffer, 2000.)

In addition to the enhancement of greenhouse effects by increased CO_2, other trace gases have increased markedly in the atmosphere, particularly methane (CH_4) (Figure 2.24a; and compare this with the historical record in Figure 2.23), nitrous oxide (N_2O) and the chlorofluorocarbons (CFCs, e.g. trichlorofluoromethane (CCl_3F) and dichlorodifluoromethane (CCl_2F_2)). Together, these and other gases contribute almost as much to enhancing the greenhouse effect as does the rise in CO_2 (Figure 2.24b). The increase in CH_4 is not all explained but probably has a microbial origin in intensive agriculture on anaerobic soils (especially increased rice production) and in the digestive process of ruminants (a cow produces approximately 40 litres of CH_4 each day); around 70% of its production is anthropogenic (Khalil, 1999). The effect of the CFCs from refrigerants, aerosol propellants and so on is potentially great, but international agreements at least appear to have halted further rises in their concentrations (Khalil, 1999).

It should be possible to draw up a balance sheet that shows how the CO_2 produced by human activities translates into the changes in concentration in the atmosphere. Human activities

CO₂ – but not only CO₂

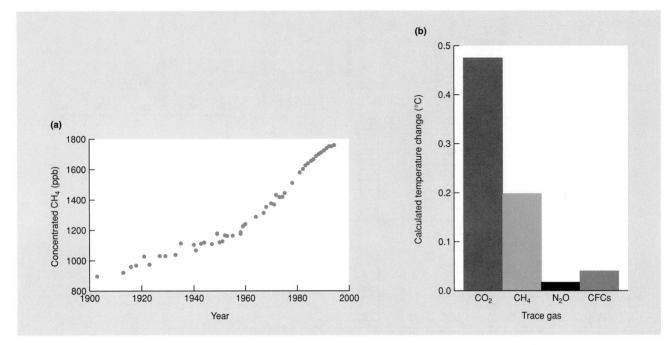

Figure 2.24   (a) Concentration of methane (CH_4) in the atmosphere through the 20th century. (b) Estimates of global warming over the period 1850–1990 caused by CO_2 and other major greenhouse gases. (After Khalil, 1999.)

release 5.1–7.5×10^9 metric tons of carbon to the atmosphere each year. But the increase in atmospheric CO_2 (2.9×10^9 metric tons) accounts for only 60% of this, a percentage that has remained remarkably constant for 40 years (Hansen *et al.*, 1999). The oceans absorb CO_2 from the atmosphere, and it is estimated that they may absorb 1.8–2.5×10^9 metric tons of the carbon released by human activities. Recent analyses also indicate that terrestrial vegetation has been 'fertilized' by the increased atmospheric CO_2, so that a considerable amount of extra carbon has been locked up in vegetation biomass (Kicklighter *et al.*, 1999). This softening of the blow by the oceans and terrestrial vegetation notwithstanding, however, atmospheric CO_2 and the greenhouse effect are increasing. We return to the question of global carbon budgets in Section 18.4.6.

2.9.2   Global warming

We started this chapter discussing temperature, moved through a number of other environmental conditions to pollutants, and now return to temperature because of the effects of those pollutants on global temperatures. It appears that the present air temperature at the land surface is $0.6 \pm 0.2°C$ warmer than in preindustrial times (Figure 2.25), and temperatures are predicted to continue to rise by a further 1.4–$5.8°C$ by 2100 (IPCC, 2001). Such changes will probably lead to a melting of the ice caps, a consequent rising of sea level and large changes in the pattern of

global climates and the distribution of species. Predictions of the extent of global warming resulting from the enhanced greenhouse effect come from two sources: (i) predictions based on sophisticated computer models ('general circulation models') that simulate the world's climate; and (ii) trends detected in measured data sets, including the width of tree rings, sea-level records and measures of the rate of retreat of glaciers.

Not surprisingly, different global circulation models differ in their predictions of the rise in global temperature that will result from predicted

a 3–4°C rise in the next 100 years

increases in CO_2. However, most model predictions vary only from 2.3 to 5.2°C (most of the variation is accounted for by the way in which the effects of cloud cover are modeled), and a projected rise of 3–$4°C$ in the next 100 years seems a reasonable value from which to make projections of ecological effects (Figure 2.26).

But temperature regimes are, of course, only part of the set of conditions that determine which organisms live where. Unfortunately, we can place much less faith in computer projections of rainfall and evaporation because it is very hard to build good models of cloud behavior into a general model of climate. If we consider only temperature as a relevant variable, we would project a 3°C rise in temperature giving London (UK) the climate of Lisbon (Portugal) (with an appropriate vegetation of olives, vines, *Bougainvillea* and semiarid scrub). But with more reliable rain it would be nearly subtropical, and with a little less it might qualify for the status of an arid zone!

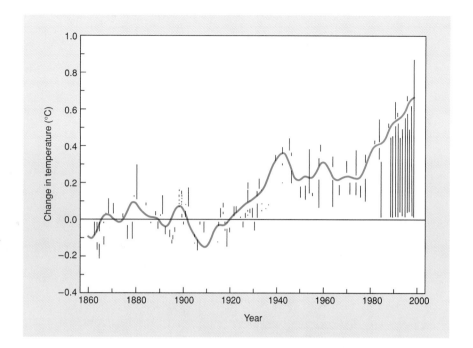

Figure 2.25 Global annual surface temperature variations from 1860 to 1998. The bars show departures from the mean at the end of the 19th century. The curve is a moving average obtained using a 21-year filter. Mean global temperatures are now higher than at any time since 1400. (After Saunders, 1999.)

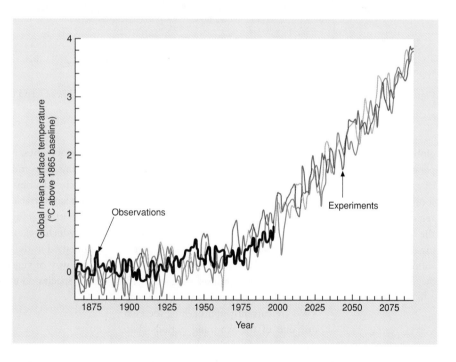

Figure 2.26 The rise in global mean surface temperature projected by the global coupled model (i.e. both the oceans and the atmosphere are modeled) for climate variability and change in use at the Geophysical Fluid Dynamics Laboratory, Princeton, USA. Observed increases in greenhouse gases are used for the period 1865–1990 (and clearly the projections match closely the observed trend in temperature); thereafter, greenhouse gases are assumed to increase at 1% per year. Since the model simulates the global behavior of the oceans and atmosphere, the precise behavior depends on the initial state of the system. The three 'experiments' were started from different states. (After Delworth *et al.*, 2002.)

the global distribution of climate change

Also, global warming is not evenly distributed over the surface of the earth. Figure 2.27 shows the measured global change in the trends of surface temperature over the 46 years from 1951 to 1997. Areas of North America (Alaska) and Asia experienced rises of 1.5–2°C in that period, and these places are predicted to continue experiencing the fastest warming in the first half of the present century. In some regions the temperature has apparently not changed (New York, for example) and should not change greatly in the next 50 years. There are also some areas, notably Greenland and the northern Pacific Ocean, where surface temperatures have fallen.

We have emphasized, too, that the distribution of many organisms is determined by occasional extremes rather than by average conditions. Computer modeled projections imply that

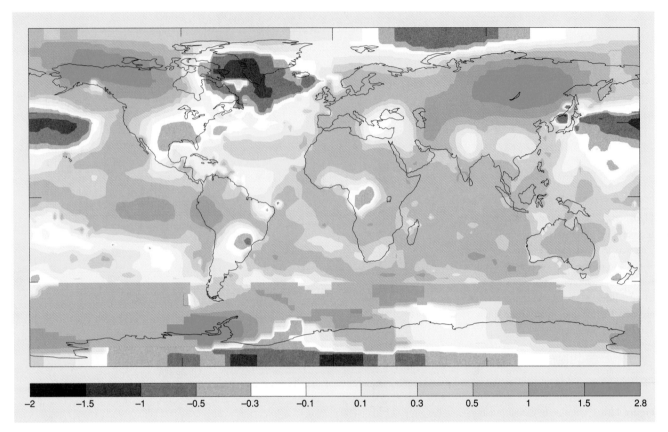

Figure 2.27 Change in the surface temperature of the globe expressed as the linear trend over 46 years from 1951 to 1997. The bar below gives the temperatures in °C. (From Hansen *et al.*, 1999.)

global climatic change will also bring greater variance in temperature. Timmerman *et al.* (1999), for example, modeled the effect of greenhouse warming on the ENSO (see Section 2.4.1). They found that not only was the mean climate in the tropical Pacific region predicted to move towards that presently represented by the (warmer) El Niño state, but that interannual variability was also predicted to increase and that variability was predicted to be more skewed towards unusually cold events.

can the biota keep up with the pace? Global temperatures have changed naturally in the past, as we have seen. We are currently approaching the end of one of the warming periods that started around 20,000 years ago, during which global temperatures have risen by about 8°C. The greenhouse effect adds to global warming at a time when temperatures are already higher than they have been for 400,000 years. Buried pollen gives us evidence that North American forest boundaries have migrated north at rates of 100–500 m year^{-1} since the last ice age. However, this rate of advance has not been fast enough to keep pace with postglacial warming. The rate of warming forecast to result from the greenhouse effect is 50–100 times faster than postglacial warming. Thus, of all the types of environmental pollution caused by human activities, none may have such profound effects as global warming. We must expect latitudinal and altitudinal changes to species' distributions and widespread extinctions as floras and faunas fail to track and keep up with the rate of change in global temperatures (Hughes, 2000). What is more, large tracts of land over which vegetation might advance and retreat have been fragmented in the process of civilization, putting major barriers in the way of vegetational advance. It will be very surprising if many species do not get lost on the journey.

Summary

A condition is an abiotic environmental factor that influences the functioning of living organisms. For most, we can recognize an optimum level at which an organism performs best. Ultimately, we should define 'performs best' from an evolutionary point of view, but in practice we mostly measure the effect of conditions on some key property like the activity of an enzyme or the rate of reproduction.

The ecological niche is not a place but a summary of an organism's tolerances of conditions and requirements for resources. The modern concept – Hutchinson's *n*-dimensional hypervolume – also distinguishes fundamental and realized niches.

Temperature is discussed in detail as a typical, and perhaps the most important, condition. Individuals respond to temperature with impaired function and ultimately death at upper and lower extremes, with a functional range between the extremes, within which there is an optimum, although these responses may be subject to evolutionary adaptation and to more immediate acclimatization.

The rates of biological enzymatic processes often increase exponentially with temperature (often $Q_{10} \approx 2$), but for rates of growth and development there are often only slight deviations from linearity: the basis for the day-degree concept. Because development usually increases more rapidly with temperature than does growth, final size tends to decrease with rearing temperature. Attempts to uncover universal rules of temperature dependence remain a matter of controversy.

We explain the differences between endotherms and ectotherms but also the similarities between them, ultimately, in their responses to a range of temperatures.

We examine variations in temperature on and within the surface of the earth with a variety of causes: latitudinal, altitudinal, continental, seasonal, diurnal and microclimatic effects, and, in soil and water, the effects of depth. Increasingly, the importance of medium-term temporal patterns have become apparent. Notable amongst these are the El Niño–Southern Oscillation (ENSO) and the North Atlantic Oscillation (NAO).

There are very many examples of plant and animal distributions that are strikingly correlated with some aspect of environmental temperature but these do not prove that temperature directly causes the limits to a species' distribution. The temperatures measured are only rarely those that the organisms experience. For many species, distributions are accounted for not so much by average temperatures as by occasional extremes; and the effects of temperature may be determined largely by the responses of other community members or by interactions with other conditions.

A range of other environmental conditions are also discussed: the pH of soil and water, salinity, conditions at the boundary between sea and land, and the physical forces of winds, waves and currents. Hazards, disasters and catastrophes are distinguished.

A number of environmental conditions are becoming increasingly important due to the accumulation of toxic by-products of human activities. A striking example is the creation of 'acid rain'. Another is the effect of industrial gases on the greenhouse effect and consequent effects on global warming. A projected rise of $3–4°C$ in the next 100 years seems a reasonable value from which to make projections of ecological effects, though global warming is not evenly distributed over the surface of the earth. This rate is 50–100 times faster than postglacial warming. We must expect latitudinal and altitudinal changes to species' distributions and widespread extinctions of floras and faunas.

Chapter 3
Resources

3.1 Introduction

what are resources?

According to Tilman (1982), all things consumed by an organism are resources for it. But consumed does not simply mean 'eaten'. Bees and squirrels do not eat holes, but a hole that is occupied is no longer available to another bee or squirrel, just as an atom of nitrogen, a sip of nectar or a mouthful of acorn are no longer available to other consumers. Similarly, females that have already mated may be unavailable to other mates. All these things have been consumed in the sense that the stock or supply has been reduced. Thus, resources are entities required by an organism, the quantities of which can be reduced by the activity of the organism.

organisms may
compete for
resources

Green plants photosynthesize and obtain both energy and matter for growth and reproduction from inorganic materials. Their resources are solar radiation, carbon dioxide (CO_2), water and mineral nutrients. 'Chemosynthetic' organisms, such as many of the Archaebacteria, obtain energy by oxidizing methane, ammonium ions, hydrogen sulfide or ferrous iron; they live in environments such as hot springs and deep sea vents and use resources that were much more abundant during early phases of life on earth. All other organisms use as their food resource the bodies of other organisms. In each case, what has been consumed is no longer available to another consumer. The rabbit eaten by an eagle is no longer available to another eagle. The quantum of solar radiation absorbed and photosynthesized by a leaf is no longer available to another leaf. This has an important consequence: organisms may *compete* with each other to capture a share of a limited resource – a topic that will occupy us in Chapter 5.

A large part of ecology is about the assembly of inorganic resources by green plants and the reassembly of these packages at each successive stage in a web of consumer–resource inter-actions. In this chapter we start with the resources of plants and focus especially on those most important in photosynthesis: radiation and CO_2. Together, plant resources fuel the growth of individual plants, which, collectively, determine the *primary productivity* of whole areas of land (or volumes of water): the rate, per unit area, at which plants produce biomass. Patterns of primary productivity are examined in Chapter 17. Relatively little space in this chapter is given to food as a resource for animals, simply because a series of later chapters (9–12) is devoted to the ecology of predators, grazers, parasites and saprotrophs (the consumers and decomposers of dead organisms). This chapter then closes where the previous chapter began: with the ecological niche, adding resource dimensions to the condition dimensions we have met already.

3.2 Radiation

Solar radiation is the only source of energy that can be used in metabolic activities by green plants. It comes to the plant as a flux of radiation from the sun, either directly having been diffused to a greater or lesser extent by the atmosphere, or after being reflected or transmitted by other objects. The direct fraction is highest at low latitudes (Figure 3.1). Moreover, for much of the year in temperate climates, and for the whole of the year in arid climates, the leaf canopy in terrestrial communities does not cover the land surface, so that most of the incident radiation falls on bare branches or on bare ground.

the fate of radiation

When a plant intercepts radiant energy it may be reflected (with its wavelength unchanged), transmitted (after some wavebands have been filtered out) or absorbed. Part of the fraction that is absorbed may raise the plant's temperature and be reradiated at much longer wavelengths; in terrestrial plants, part may contribute latent heat of evaporation of water and so power the transpiration

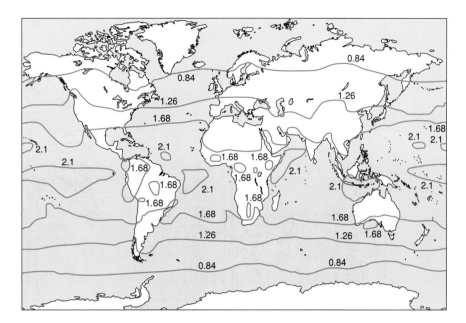

Figure 3.1 Global map of the solar radiation absorbed annually in the earth–atmosphere system: from data obtained with a radiometer on the Nimbus 3 meteorological satellite. The units are J cm^{-2} min^{-1}. (After Raushke *et al.*, 1973.)

stream. A small part may reach the chloroplasts and drive the process of photosynthesis (Figure 3.2).

radiant energy must be captured or is lost forever

Radiant energy is converted during photosynthesis into energy-rich chemical compounds of carbon, which will subsequently be broken down in respiration (either by the plant itself or by organisms that consume it). But unless the radiation is captured and chemically fixed at the instant it falls on the leaf, it is irretrievably lost for photosynthesis. Radiant energy that has been fixed in photosynthesis passes just once through the world. This is in complete contrast to an atom of nitrogen or carbon or a molecule of water that may cycle repeatedly through endless generations of organisms.

photosynthetically active radiation

Solar radiation is a resource continuum: a spectrum of different wavelengths. But the photosynthetic apparatus is able to gain access to energy in only a restricted band of this spectrum. All green plants depend on chlorophyll and other pigments for the photosynthetic fixation of carbon, and these pigments fix radiation in a waveband between roughly 400 and 700 nm. This is the band of 'photosynthetically active radiation' (PAR). It corresponds broadly with the range of the spectrum visible to the human eye that we call 'light'. About 56% of the radiation incident on the earth's surface lies outside the PAR range and is thus unavailable as a resource for green plants. In other organisms there are pigments, for example bacteriochlorophyll in bacteria, that operate in photosynthesis outside the PAR range of green plants.

3.2.1 Variations in the intensity and quality of radiation

photoinhibition at high intensities

A major reason why plants seldom achieve their intrinsic photosynthetic capacity is that the intensity of radiation varies continually (Figure 3.3). Plant morphology and physiology that are optimal for photosynthesis at one intensity of radiation will usually be inappropriate at another. In terrestrial habitats, leaves live in a radiation regime that varies throughout the day and the year, and they live in an environment of other leaves that modifies the quantity and quality of radiation received. As with all resources, the supply of radiation can vary both systematically (diurnal, annual) and unsystematically. Moreover, it is not the case simply that the intensity of radiation is a greater or lesser proportion of a maximum value at which photosynthesis would be most productive. At high intensities, *photoinhibition* of photosynthesis may occur (Long *et al.*, 1994), such that the rate of fixation of carbon decreases with increasing radiation intensity. High intensities of radiation may also lead to dangerous overheating of plants. Radiation is an essential resource for plants, but they can have too much as well as too little.

systematic variations in supply

Annual and diurnal rhythms are systematic variations in solar radiation (Figure 3.3a, b). The green plant experiences periods of famine and glut in its radiation resource every 24 h (except near the poles) and seasons of famine and glut every year (except in the tropics). In aquatic habitats, an additional

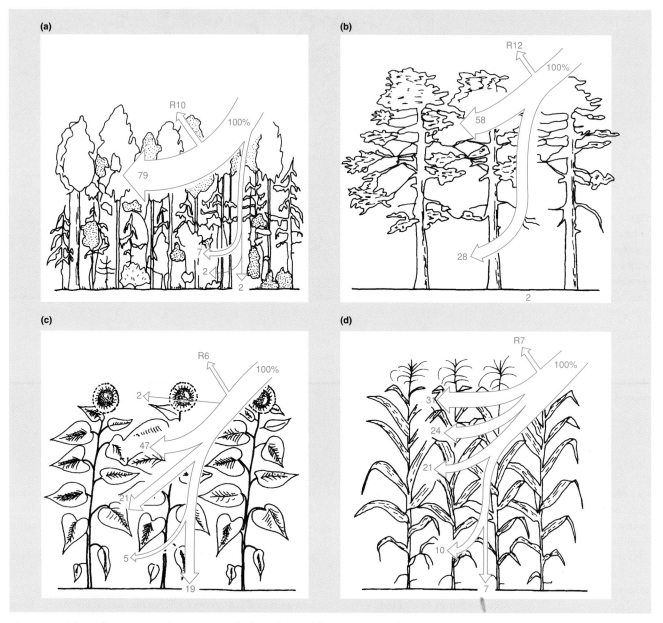

Figure 3.2 The reflection (R) and attenuation of solar radiation falling on various plant communities. The arrows show the percentage of incident radiation reaching various levels in the vegetation. (a) A boreal forest of mixed birch and spruce; (b) a pine forest; (c) a field of sunflowers; and (d) a field of corn (maize). These figures represent data obtained in particular communities and great variation will occur depending on the stage of growth of the forest or crop canopy, and on the time of day and season at which the measurements are taken. (After Larcher, 1980, and other sources.)

systematic and predictable source of variation in radiation intensity is the reduction in intensity with depth in the water column (Figure 3.3c), though the extent of this may vary greatly. For example, differences in water clarity mean that seagrasses may grow on solid substrates as much as 90 m below the surface in the relatively unproductive open ocean, whereas macrophytes in fresh waters rarely grow at depths below 10 m (Sorrell *et al.*, 2001), and

often only at considerably shallower locations, in large part because of differences in concentrations of suspended particles and also phytoplankton (see below).

The way in which an organism reacts to systematic, predictable variation in the supply of a resource reflects both its present physiology and its past evolution. The seasonal shedding of leaves by deciduous trees in temperate regions in part reflects the annual

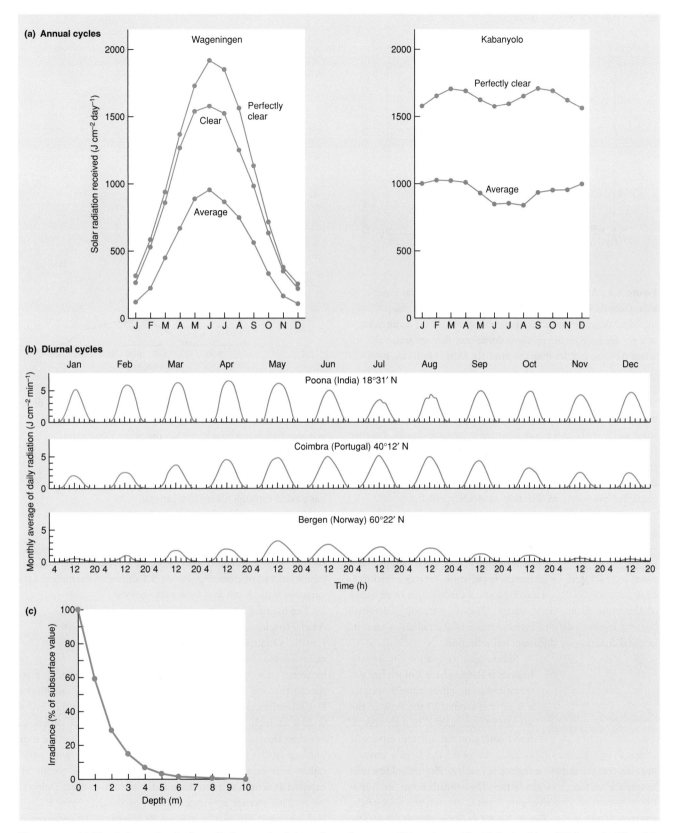

Figure 3.3 (a) The daily totals of solar radiation received throughout the year at Wageningen (the Netherlands) and Kabanyolo (Uganda). (b) The monthly average of daily radiation recorded at Poona (India), Coimbra (Portugal) and Bergen (Norway). ((a, b) after de Wit, 1965, and other sources.) (c) Exponential diminution of radiation intensity in a freshwater habitat (Burrinjuck Dam, Australia). (After Kirk, 1994.)

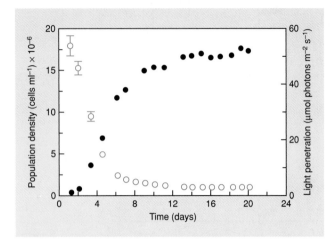

Figure 3.4 As population density (●) of the unicellular green alga, *Chlorella vulgaris*, increased in laboratory culture, this increased density reduced the penetration of light (○; its intensity at a set depth). Bars are standard deviations; they are omitted when they are smaller than the symbols. (After Huisman, 1999.)

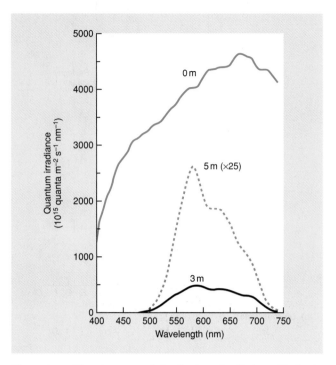

Figure 3.5 Changing spectral distribution of radiation with depth in Lake Burley Griffin, Australia. Note that photosynthetically active radiation lies broadly within the range 400–700 nm. (After Kirk, 1994.)

rhythm in the intensity of radiation – they are shed when they are least useful. In consequence, an evergreen leaf of an understory species may experience a further systematic change, because the seasonal cycle of leaf production of overstory species determines what radiation remains to penetrate to the understory. The daily movement of leaves in many species also reflects the changing intensity and direction of incident radiation.

shade: a resource-depletion zone

Less systematic variations in the radiation environment of a leaf are caused by the nature and position of neighboring leaves. Each canopy, each plant and each leaf, by intercepting radiation, creates a resource-depletion zone (RDZ) – a moving band of shadow over other leaves of the same plant, or of others. Deep in a canopy, shadows become less well defined because much of the radiation loses its original direction by diffusion and reflection.

attenuation with depth, and plankton density, in aquatic habitats

Submerged vegetation in aquatic habitats is likely to have a much less systematic shading effect, simply because it is moved around by the flow of the water in which it lives, though vegetation floating on the surface, especially of ponds or lake, inevitably has a profound and largely unvarying effect on the radiation regime beneath it. Phytoplankton cells nearer the surface, too, shade the cells beneath them, such that the reduction of intensity with depth is greater, the greater the phytoplankton density. Figure 3.4, for example, shows the decline in light penetration, measured at a set depth in a laboratory system, as a population of the unicellular green alga, *Chlorella vulgaris*, built up over a 12-day period (Huisman, 1999).

The composition of radiation that has passed through leaves in a canopy, or through a body of water, is also altered. It may be less useful photosynthetically because the PAR component has been reduced – though such reductions may also, of course, prevent photoinhibition and overheating. Figure 3.5 shows an example for the variation with depth in a freshwater habitat.

variations in quality as well as quantity

The major differences amongst terrestrial species in their reaction to systematic variations in the intensity of radiation are those that have evolved between 'sun species' and 'shade species'. In general, plant species that are characteristic of shaded habitats use radiation at low intensities more efficiently than sun species, but the reverse is true at high intensities (Figure 3.6). Part of the difference between them lies in the physiology of the leaves, but the morphology of the plants also influences the efficiency with which radiation is captured. The leaves of sun plants are commonly exposed at acute angles to the midday sun (Poulson & DeLucia, 1993). This spreads an incident beam of radiation over a larger leaf area, and effectively reduces its intensity. An intensity of radiation that is superoptimal for photosynthesis when it strikes a leaf at 90° may therefore be optimal for a leaf inclined at an acute angle. The leaves of sun plants are often superimposed into

sun and shade species

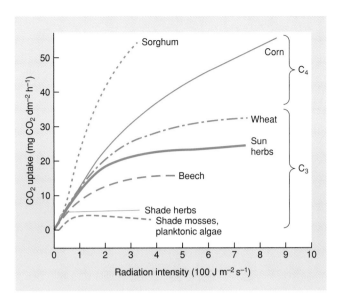

Figure 3.6 The response of photosynthesis to light intensity in various plants at optimal temperatures and with a natural supply of CO_2. Note that corn and sorghum are C_4 plants and the remainder are C_3 (the terms are explained in Sections 3.3.1 and 3.3.2). (After Larcher, 1980, and other sources.)

a multilayered canopy. In bright sunshine even the shaded leaves in lower layers may have positive rates of net photosynthesis. Shade plants commonly have leaves held near to the horizontal and in a single-layered canopy.

sun and shade leaves

In contrast to these 'strategic' differences, it may also happen that as a plant grows, its leaves develop differently as a 'tactical' response to the radiation environment in which it developed. This often leads to the formation of 'sun leaves' and 'shade leaves' within the canopy of a single plant. Sun leaves are typically smaller, thicker, have more cells per unit area, denser veins, more densely packed chloroplasts and a greater dry weight per unit area of leaf. These tactical maneuvers, then, tend to occur not at the level of the whole plant, but at the level of the individual leaf or even its parts. Nevertheless, they take time. To form sun or shade leaves as a tactical response, the plant, its bud or the developing leaf must sense the leaf's environment and respond by growing a leaf with an appropriate structure. For example, it is impossible for the plant to change its form fast enough to track the changes in intensity of radiation between a cloudy and a clear day. It can, however, change its rate of photosynthesis extremely rapidly, reacting even to the passing of a fleck of sunlight. The rate at which a leaf photosynthesizes also depends on the demands that are made on it by other vigorously growing parts. Photosynthesis may be reduced, even though conditions are otherwise ideal, if there is no demanding call on its products.

In aquatic habitats, much of the variation between species is accounted for by differences in photosynthetic pigments, which contribute significantly to the precise wavelengths of radiation that can be utilized (Kirk, 1994). Of the three types of pigment – chlorophylls, carotenoids and biliproteins – all photosynthetic plants contain the first two, but many algae also contain biliproteins; and within the chlorophylls, all higher plants have chlorophyll *a* and *b*, but many algae have only chlorophyll *a* and some have chlorophyll *a* and *c*. Examples of the absorption spectra of a number of pigments, the related contrasting absorption spectra of a number of groups of aquatic plants, and the related distributional differences (with depth) between a number of groups of aquatic plants are illustrated in Figure 3.7. A detailed assessment of the evidence for direct links between pigments, performance and distribution is given by Kirk (1994).

pigment variation in aquatic species

3.2.2 Net photosynthesis

The rate of photosynthesis is a gross measure of the rate at which a plant captures radiant energy and fixes it in organic carbon compounds. However, it is often more important to consider, and very much easier to measure, the net gain. Net photosynthesis is the increase (or decrease) in dry matter that results from the difference between gross photosynthesis and the losses due to respiration and the death of plant parts (Figure 3.8).

Net photosynthesis is negative in darkness, when respiration exceeds photosynthesis, and increases with the intensity of PAR. The *compensation point* is the intensity of PAR at which the gain from gross photosynthesis exactly balances the respiratory and other losses. The leaves of shade species tend to respire at lower rates than those of sun species. Thus, when both are growing in the shade the net photosynthesis of shade species is greater than that of sun species.

the compensation point

There is nearly a 100-fold variation in the *photosynthetic capacity* of leaves (Mooney & Gulmon, 1979). This is the rate of photosynthesis when incident radiation is saturating, temperature is optimal, relative humidity is high, and CO_2 and oxygen concentrations are normal. When the leaves of different species are compared under these ideal conditions, the ones with the highest photosynthetic capacity are generally those from environments where nutrients, water and radiation are seldom limiting (at least during the growing season). These include many agricultural crops and their weeds. Species from resource-poor environments (e.g. shade plants, desert perennials, heathland species) usually have low photosynthetic capacity – even when abundant resources are provided. Such patterns can be understood by noting that photosynthetic capacity, like all capacity, must be 'built'; and the investment in building

photosynthetic capacity

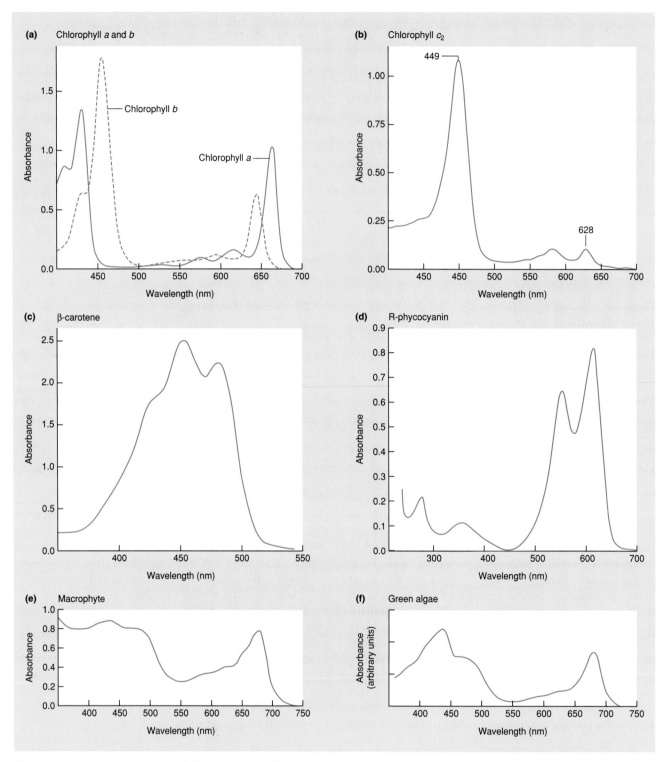

Figure 3.7 (a) Absorption spectra of chlorophylls *a* and *b*. (b) Absorption spectrum of chlorophyll *c₂*. (c) Absorption spectrum of β-carotene. (d) Absorption spectrum of the biliprotein, R-phycocyanin. (e) Absorption spectrum of a piece of leaf of the freshwater macrophyte, *Vallisneria spiralis*, from Lake Ginnindera, Australia. (f) Absorption spectrum of the planktonic alga *Chlorella pyrenoidos* (green).

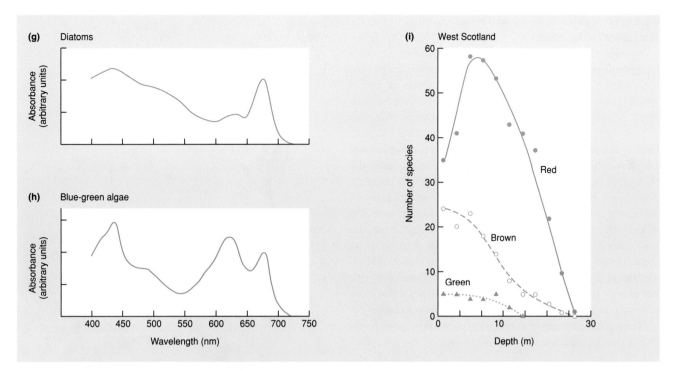

Figure 3.7 *(continued)* (g–h) Absorption spectra of the planktonic algae *Navicula minima* (diatom) and *Synechocystis* sp. (blue-green). (i) The numbers of species of benthic red, green and brown algae at various depths (and in various light regimes) off the west coast of Scotland (56–57°N). (After Kirk, 1994; data from various sources.)

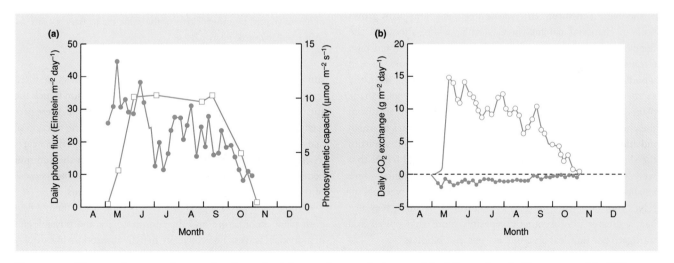

Figure 3.8   The annual course of events that determined the net photosynthetic rate of the foliage of maple (*Acer campestre*) in 1980. (a) Variations in the intensity of PAR (●), and changes in the photosynthetic capacity of the foliage (□) appearing in spring, rising to a plateau and then declining through late September and October. (b) The daily fixation of carbon dioxide (CO_2) (○) and its loss through respiration during the night (●). The annual total gross photosynthesis was 1342 g CO_2 m^{-2} and night respiration was 150 g CO_2 m^{-2}, giving a balance of 1192 g CO_2 m^{-2} net photosynthesis. (After Pearcy *et al.*, 1987.)

capacity is only likely to be repaid if ample opportunity exists for that capacity to be utilized.

Needless to say, ideal conditions in which plants may achieve their photosynthetic capacity are rarely present outside a physiologist's controlled environment chamber. In practice, the rate at which photosynthesis actually proceeds is limited by conditions (e.g. temperature) and by the availability of resources other than radiant energy. Leaves seem also to achieve their maximal photosynthetic rate only when the products are being actively withdrawn (to developing buds, tubers, etc.). In addition, the photosynthetic capacity of leaves is highly correlated with leaf nitrogen content, both between leaves on a single plant and between the leaves of different species (Woodward, 1994). Around 75% of leaf nitrogen is invested in chloroplasts. This suggests that the availability of nitrogen as a resource may place strict limits on the ability of plants to garner CO_2 and energy in photosynthesis. The rate of photosynthesis also increases with the intensity of PAR, but in most species ('C$_3$ plants' – see below) reaches a plateau at intensities of radiation well below that of full solar radiation.

The highest efficiency of utilization of radiation by green plants is 3–4.5%, obtained from cultured microalgae at low intensities of PAR. In tropical forests values fall within the range 1–3%, and in temperate forests 0.6–1.2%. The approximate efficiency of temperate crops is only about 0.6%. It is on such levels of efficiency that the energetics of all communities depend.

3.2.3 Sun and shade plants of an evergreen shrub

A number of the general points above are illustrated by a study of the evergreen shrub, *Heteromeles arbutifolia*. This plant grows both in chaparral habitats in California, where shoots in the upper crown are consistently exposed to full sunlight and high temperatures, especially during the dry season, and also in woodland habitats, where the plant grows both in open sites and in the shaded understory (Valladares & Pearcy, 1998). Shade plants from the understory were compared with sun plants from the chaparral, where they received around seven times as much radiation ('photon flux density', PFD). Compared to those from the shade (Figure 3.9 and Table 3.1a), sun plants had leaves that were inclined at a much steeper angle to the horizontal, were smaller but thicker, and were borne on shoots that were themselves shorter (smaller internode distances). The sun leaves also had a greater photosynthetic capacity (more chlorophyll and nitrogen) per unit leaf area but not per unit biomass.

The 'architectural' consequences of these differences (Table 3.1b) were first that shade plants had a much greater 'projection efficiency' in the summer, but a much lower efficiency in the winter. Projection efficiency expresses the degree to which the effective leaf area is reduced by being borne at an angle other than right angles to the incident radiation. Thus, the more angled leaves of sun plants absorbed the direct rays of the overhead summer

Figure 3.9 Computer reconstructions of stems of typical sun (a, c) and shade (b, d) plants of the evergreen shrub *Heteromeles arbutifolia*, viewed along the path of the sun's rays in the early morning (a, b) and at midday (c, d). Darker tones represent parts of leaves shaded by other leaves of the same plant. Bars = 4 cm. (After Valladares & Pearcy, 1998.)

sun over a wider leaf area than the more horizontal shade plant leaves, but the more sideways rays of the winter sun struck the sun plant leaves at closer to a right angle. Furthermore, these projection efficiencies can themselves be modified by the fraction of leaf area subject to self-shading, giving rise to 'display efficiencies'. These were higher in shade than in sun plants, in the summer because of the higher projection efficiency, but in the winter because of the relative absence of self-shading in shade plants.

Whole plant physiological properties (Table 3.1b), then, reflect both plant architecture and the morphologies and physiologies of individual leaves. The efficiency of light absorption, like display efficiency, reflects both leaf angles and self-shading. Hence, absorption efficiency was consistently higher for shade than for sun plants, though the efficiency for sun plants was significantly higher in winter compared to summer. The effective leaf ratio (the light absorption efficiency per unit of biomass) was then massively greater for shade than for sun plants (as a result of their thinner leaves), though again, somewhat higher for the latter in winter.

Table 3.1 (a) Observed differences in the shoots and leaves of sun and shade plants of the shrub *Heteromeles arbutifolia*. Standard deviations are given in parentheses; the significance of differences are given following analysis of variance. (b) Consequent whole plant properties of sun and shade plants. (After Valladares & Pearcy, 1998.)

(a)

	Sun		Shade		P
Internode distance (cm)	1.08	(0.06)	1.65	(0.02)	< 0.05
Leaf angle (degrees)	71.3	(16.3)	5.3	(4.3)	< 0.01
Leaf surface area (cm²)	10.1	(0.3)	21.4	(0.8)	< 0.01
Leaf blade thickness (μm)	462.5	(10.9)	292.4	(9.5)	< 0.01
Photosynthetic capacity, area basis (μmol CO_2 m^{-2} s^{-1})	14.1	(2.0)	9.0	(1.7)	< 0.01
Photosynthetic capacity, mass basis (μmol CO_2 kg^{-1} s^{-1})	60.8	(10.1)	58.1	(11.2)	NS
Chlorophyll content, area basis (mg m^{-2})	280.5	(15.3)	226.7	(14.0)	< 0.01
Chlorophyll content, mass basis (mg g^{-1})	1.23	(0.04)	1.49	(0.03)	< 0.05
Leaf nitrogen content, area basis (g m^{-2})	1.97	(0.25)	1.71	(0.21)	< 0.05
Leaf nitrogen content, mass basis (% dry weight)	0.91	(0.31)	0.96	(0.30)	NS

(b)

	Sun plants		Shade plants	
	Summer	Winter	Summer	Winter
E_P	0.55[a]	0.80[b]	0.88[b]	0.54[a]
E_D	0.33[a]	0.38[a, b]	0.41[b]	0.43[b]
Fraction self-shaded	0.22[a]	0.42[b]	0.47[b]	0.11[a]
$E_{A, direct\ PFD}$	0.28[a]	0.44[b]	0.55[c]	0.53[c]
LAR$_c$ (cm² g^{-1})	7.1[a]	11.7[b]	20.5[c]	19.7[c]

E_P, projection efficiency; E_D, display efficiency; E_A, absorption efficiency; LAR$_e$, effective leaf area ratio; NS, not significant.
Letter codes indicate groups that differed significantly in analyses of variance ($P < 0.05$).

Overall, therefore, despite receiving only one-seventh of the PFD of sun plants, shade plants reduced the differential in the amount absorbed to one-quarter, and reduced the differential in their daily rate of carbon gain to only a half. Shade plants successfully counterbalanced their reduced photosynthetic capacity at the leaf level with enhanced light-harvesting ability at the whole plant level. The sun plants can be seen as striking a compromise between maximizing whole plant photosynthesis on the one hand while avoiding photoinhibition and overheating of individual leaves on the other.

3.2.4 Photosynthesis or water conservation? Strategic and tactical solutions

stomatal opening

In fact, in terrestrial habitats especially, it is not sensible to consider radiation as a resource independently of water. Intercepted radiation does not result in photosynthesis unless there is CO_2 available, and the prime route of entry of CO_2 is through open stomata. But if the stomata are open to the air, water will evaporate through them. If water is lost faster than it can be gained, the leaf (and the plant) will sooner or later wilt and eventually die. But in most terrestrial communities, water is, at least sometimes, in short supply. Should a plant conserve water at the expense of present photosynthesis, or maximize photosynthesis at the risk of running out of water? Once again, we meet the problem of whether the optimal solution involves a strict strategy or the ability to make tactical responses. There are good examples of both solutions and also compromises.

Perhaps the most obvious strategy that plants may adopt is to have a short life and high photosynthetic activity during periods when water is abundant, but remain dormant as seeds during the rest of the year, neither photosynthesizing nor transpiring (e.g. many desert annuals, annual weeds and most annual crop plants).

short active interludes in a dormant life

leaf appearance and
structure

Second, plants with long lives may produce leaves during periods when water is abundant and shed them during droughts (e.g. many species of *Acacia*). Some shrubs of the Israeli desert (e.g. *Teucrium polium*) bear finely divided, thin-cuticled leaves during the season when soil water is freely available. These are then replaced by undivided, small, thick-cuticled leaves in more drought-prone seasons, which in turn fall and may leave only green spines or thorns (Orshan, 1963): a sequential polymorphism through the season, with each leaf morph being replaced in turn by a less photosynthetically active but more water-tight structure.

Next, leaves may be produced that are long lived, transpire only slowly and tolerate a water deficit, but which are unable to photosynthesize rapidly even when water is abundant (e.g. evergreen desert shrubs). Structural features such as hairs, sunken stomata and the restriction of stomata to specialized areas on the lower surface of a leaf slow down water loss. But these same morphological features reduce the rate of entry of CO_2. Waxy and hairy leaf surfaces may, however, reflect a greater proportion of radiation that is not in the PAR range and so keep the leaf temperature down and reduce water loss.

*physiological
strategies*

Finally, some groups of plants have evolved particular physiologies: C_4 and crassulacean acid metabolism (CAM). We consider these in more detail in Sections 3.3.1–3.3.3. Here, we simply note that plants with 'normal' (i.e. C_3) photosynthesis are wasteful of water compared with plants that possess the modified C_4 and CAM physiologies. The water-use efficiency of C_4 plants (the amount of carbon fixed per unit of water transpired) may be double that of C_3 plants.

The viability of alternative strategies to solve a common problem is nicely illustrated by the trees of seasonally dry tropical forests and woodlands (Eamus, 1999). These communities are found naturally in Africa, the Americas, Australia and India, and as a result of human interference elsewhere in Asia. But whereas, for example, the savannas of Africa and India are dominated by deciduous species, and the Llanos of South America are dominated by evergreens, the savannas of Australia are occupied by roughly equal numbers of species from four groups (Figure 3.10a): evergreens (a full canopy all year), deciduous species (losing all leaves for at least 1 and usually 2–4 months each year), semideciduous species (losing around 50% or more of their leaves each year) and brevideciduous species (losing only about 20% of their leaves). At the ends of this continuum, the deciduous species avoid drought in the dry season (April–November in Australia) as a result of their vastly reduced rates of transpiration (Figure 3.10b), but the evergreens maintain a positive carbon balance throughout the year (Figure 3.10c), whereas the deciduous species make no net photosynthate at all for around 3 months.

The major tactical control of the rates of both photosynthesis and water loss is through changes in stomatal 'conductance' that may occur rapidly during the course of a day and allow a very rapid response to immediate water shortages. Rhythms of stomatal opening and closure may ensure that the above-ground parts of the plant remain more or less watertight except during controlled periods of active photosynthesis. These rhythms may

*coexisting alternative
strategies in
Australian savannas*

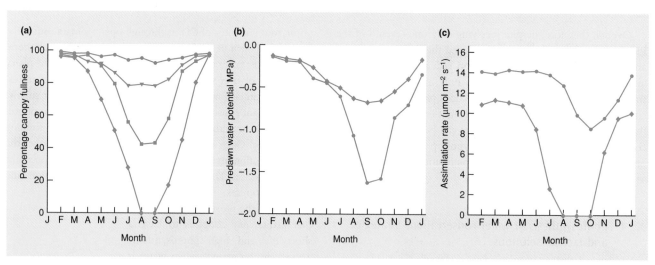

Figure 3.10   (a) Percentage canopy fullness for deciduous (♦), semideciduous (■), brevideciduous (▼) and evergreen (●) trees Australian savannas throughout the year. (Note that the southern hemisphere dry season runs from around April to November.) (b) Susceptibility to drought as measured by increasingly negative values of 'predawn water potential' for deciduous (♦) and evergreen (●) trees. (c) Net photosynthesis as measured by the carbon assimilation rate for deciduous (♦) and evergreen (●) trees. (After Eamus, 1999.)

be diurnal or may be quickly responsive to the plant's internal water status. Stomatal movement may even be triggered directly by conditions at the leaf surface itself – the plant then responds to desiccating conditions at the very site, and at the same time, as the conditions are first sensed.

3.3 Carbon dioxide

the rise in global levels — The CO_2 used in photosynthesis is obtained almost entirely from the atmosphere, where its concentration has risen from approximately 280 $\mu l\, l^{-1}$ in 1750 to about 370 $\mu l\, l^{-1}$ today and is still increasing by 0.4–0.5% $year^{-1}$ (see Figure 18.22). In a terrestrial community, the flux of CO_2 at night is upwards, from the soil and vegetation to the atmosphere; on sunny days above a photosynthesizing canopy, there is a downward flux.

variations beneath a canopy — Above a vegetation canopy, the air becomes rapidly mixed. However, the situation is quite different within and beneath canopies. Changes in CO_2 concentration in the air within a mixed deciduous forest in New England were measured at various heights above ground level during the year (Figure 3.11a) (Bazzaz & Williams, 1991). Highest concentrations, up to around 1800 $\mu l\, l^{-1}$, were measured near the surface of the ground, tapering off to around 400 $\mu l\, l^{-1}$ at 1 m above the ground. These high values near ground level were achieved in the summer when high temperatures allowed the rapid decomposition of litter and soil organic matter. At greater heights within the forest, the CO_2 concentrations scarcely ever (even in winter) reached the value of 370 $\mu l\, l^{-1}$ which is the atmospheric concentration of bulk air measured at the Mauna Loa laboratory in Hawaii (see Figure 18.22). During the winter months, concentrations remained virtually constant through the day and night at all heights. But in the summer, major diurnal cycles of concentration developed that reflected the interaction between the production of CO_2 by decomposition and its consumption in photosynthesis (Figure 3.11b).

That CO_2 concentrations vary so widely within vegetation means that plants growing in different parts of a forest will experience quite different CO_2 environments. Indeed the lower leaves on a forest shrub will usually experience higher CO_2 concentrations than its upper leaves, and seedlings will live in environments richer in CO_2 than mature trees.

In aquatic environments, variations in CO_2 concentration can be just as striking, especially when water mixing is limited, for example during the summer 'stratification' of lakes, with layers of warm water towards the surface and colder layers beneath (Figure 3.12).

variations in aquatic habitats . . .

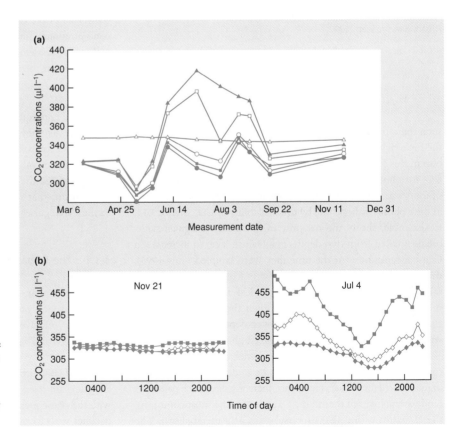

Figure 3.11 (a) CO_2 concentrations in a mixed deciduous forest (Harvard Forest, Massachusetts, USA) at various times of year at five heights above ground: ▲, 0.05 m; □, 0.20 m; ■, 3.00 m; ○, 6.00 m; ●, 12.00 m. Data from the Mauna Loa CO_2 observatory (△) are given on the same axis for comparison. (b) CO_2 concentrations for each hour of the day (averaged over 3–7-day periods) on November 21 and July 4. (After Bazzaz & Williams, 1991.)

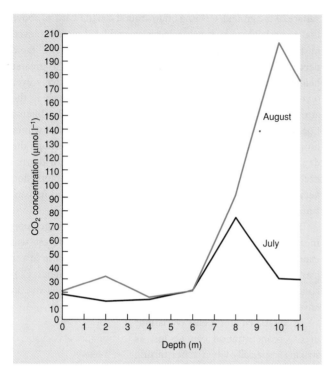

Figure 3.12 Variation in CO_2 concentration with depth in Lake Grane Langsø, Denmark in early July and again in late August after the lake becomes stratified with little mixing between the warm water at the surface and the colder water beneath. (After Riis & Sand-Jensen, 1997.)

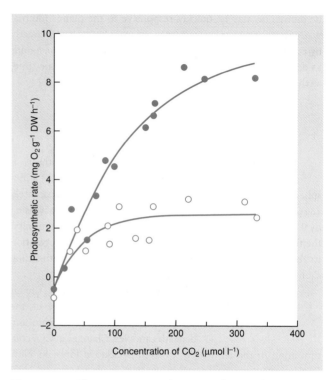

Figure 3.13   The increase (to a plateau) in photosynthetic rate with artificially manipulated CO_2 concentrations in moss, *Sphagnum subsecundum*, taken from depths of 9.5 m (●) and 0.7 m (○) in Lake Grane Langsø, Denmark, in early July. These concentrations – and hence the rates of photosynthesis – are much higher than those occurring naturally (see Figure 3.12). (After Riis & Sand-Jensen, 1997.)

| ... setting a limit on photosynthetic rates |

Also, in aquatic habitats, dissolved CO_2 tends to react with water to form carbonic acid, which in turn ionizes, and these tendencies increase with pH, such that 50% or more of inorganic carbon in water may be in the form of bicarbonate ions. Many aquatic plants can utilize carbon in this form, but since it must ultimately be reconverted to CO_2 for photosynthesis, this is likely to be less useful as a source of inorganic carbon, and in practice, many plants will be limited in their photosynthetic rate by the availability of CO_2. Figure 3.13, for example, shows the response of the moss, *Sphagnum subsecundum*, taken from two depths in a Danish lake, to increases in CO_2 concentration. At the time they were sampled (July 1995), the natural concentrations in the waters from which they were taken (Figure 3.12) were 5–10 times less than those eliciting maximum rates of photosynthesis. Even the much higher concentrations that occurred at the lower depths during summer stratification would not have maximized photosynthetic rate.

One might expect a process as fundamental to life on earth as carbon fixation in photosynthesis to be underpinned by a single unique biochemical pathway. In fact, there are three such pathways (and variants within them): the C_3 pathway (the most common), the C_4 pathway and CAM (crassulacean acid metabolism). The ecological consequences of the different pathways are profound, especially as they affect the reconciliation of photosynthetic activity and controlled water loss (see Section 3.2.4). Even in aquatic plants, where water conservation is not normally an issue, and most plants use the C_3 pathway, there are many CO_2-concentrating mechanisms that serve to enhance the effectiveness of CO_2 utilization (Badger *et al.*, 1997).

3.3.1   The C_3 pathway

In this, the Calvin–Benson cycle, CO_2 is fixed into a three-carbon acid (phosphoglyceric acid) by the enzyme Rubisco, which is present in massive amounts in the leaves (25–30% of the total leaf nitrogen). This same enzyme can also act as an oxygenase, and this activity (photorespiration) can result in a wasteful release of CO_2 – reducing by about one-third the net amounts of CO_2 that are fixed. Photorespiration increases with temperature with the consequence that the overall efficiency of carbon fixation declines with increasing temperature.

The rate of photosynthesis of C_3 plants increases with the intensity of radiation, but reaches a plateau. In many species, particularly shade species, this plateau occurs at radiation intensities far below that of full solar radiation (see Figure 3.6). Plants with C_3 metabolism have low water-use efficiency compared with C_4 and CAM plants (see below), mainly because in a C_3 plant, CO_2 diffuses rather slowly into the leaf and so allows time for a lot of water vapor to diffuse out of it.

3.3.2 The C_4 pathway

In this, the Hatch–Slack cycle, the C_3 pathway is present but it is confined to cells deep in the body of the leaf. CO_2 that diffuses into the leaves via the stomata meets mesophyll cells containing the enzyme phosphoenolpyruvate (PEP) carboxylase. This enzyme combines atmospheric CO_2 with PEP to produce a four-carbon acid. This diffuses, and releases CO_2 to the inner cells where it enters the traditional C_3 pathway. PEP carboxylase has a much greater affinity than Rubisco for CO_2. There are profound consequences.

First, C_4 plants can absorb atmospheric CO_2 much more effectively than C_3 plants. As a result, C_4 plants may lose much less water per unit of carbon fixed. Furthermore, the wasteful release of CO_2 by photorespiration is almost wholly prevented and, as a consequence, the efficiency of the overall process of carbon fixation does not change with temperature. Finally, the concentration of Rubisco in the leaves is a third to a sixth of that in C_3 plants, and the leaf nitrogen content is correspondingly lower. As a consequence of this, C_4 plants are much less attractive to many herbivores and also achieve more photosynthesis per unit of nitrogen absorbed.

One might wonder how C_4 plants, with such high water-use efficiency, have failed to dominate the vegetation of the world, but there are clear costs to set against the gains. The C_4 system has a high light compensation point and is inefficient at low light intensities; C_4 species are therefore ineffective as shade plants. Moreover, C_4 plants have higher temperature optima for growth than C_3 species: most C_4 plants are found in arid regions or the tropics. In North America, C_4 dicotyledonous species appear to be favored in sites of limited water supply (Figure 3.14) (Stowe & Teeri, 1978), whereas the abundance of C_4 monocotyledonous species is strongly correlated with maximum daily temperatures during the growing season (Teeri & Stowe, 1976). But these correlations are not universal. More generally, where there are mixed populations of C_3 and C_4 plants, the proportion of C_4 species tends to fall with elevation on mountain ranges, and in seasonal climates it is C_4 species that tend to dominate the vegetation in the hot dry seasons and C_3 species in the cooler wetter seasons. The few C_4 species that extend into temperate regions (e.g. *Spartina* spp.) are found in marine or other saline environments where osmotic conditions may especially favor species with efficient water use.

Perhaps the most remarkable feature of C_4 plants is that they do not seem to use their high water-use efficiency in faster shoot growth, but instead devote a greater fraction of the plant body to a well-developed root system. This is one of the hints that the rate of carbon assimilation is not the major limit to their growth, but that the shortage of water and/or nutrients matters more.

3.3.3 The CAM pathway

Plants with a crassulacean acid metabolism (CAM) pathway also use PEP carboxylase with its strong power of concentrating CO_2. In contrast to C_3 and C_4 plants, though, they open their stomata and fix CO_2 at night (as malic acid). During the daytime the stomata are closed and the CO_2 is released within the leaf and fixed by Rubisco. However, because the CO_2 is then at a high concentration within the leaf, photorespiration is prevented, just as it is in plants using the C_4 pathway. Plants using the CAM photosynthetic pathway have obvious advantages when water is in short supply, because their stomata are closed during the daytime when evaporative forces are strongest. The system is now known in a wide variety of families, not just the Crassulaceae. This appears to be a highly effective means of water conservation, but CAM species have not come to inherit the earth. One cost to CAM plants is the problem of storing the malic acid that is formed at night: most CAM plants are succulents with extensive water-storage tissues that cope with this problem.

In general, CAM plants are found in arid environments where strict stomatal control of daytime water is vital for survival (desert succulents) and where CO_2 is in short supply during the daytime, for example in submerged aquatic plants, and in photosynthetic organs that lack stomata (e.g. the aerial photosynthetic roots of orchids). In some CAM plants, such as *Opuntia basilaris*, the stomata remain closed both day and night during drought. The CAM process then simply allows the plant to 'idle' – photosynthesizing only the CO_2 produced internally by respiration (Szarek *et al.*, 1973).

A taxonomic and systematic survey of C_3, C_4 and CAM photosynthetic systems is given by Ehleringer and Monson (1993). They describe the very strong evidence that the C_3 pathway is evolutionarily primitive and, very surprisingly, that the C_4 and CAM systems must have arisen repeatedly and independently during the evolution of the plant kingdom.

3.3.4 The response of plants to changing atmospheric concentrations of CO_2

Of all the various resources required by plants, CO_2 is the only one that is increasing on a global scale. This rise is strongly correlated with the increased rate of consumption of fossil fuels

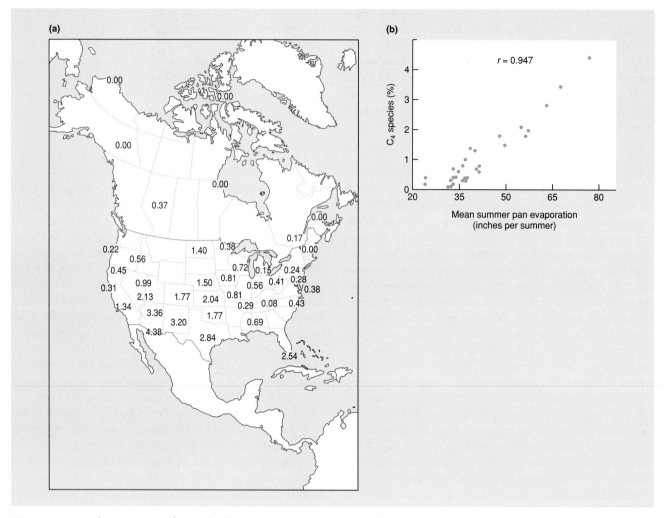

Figure 3.14   (a) The percentage of native C_4 dicot species in various regions of North America. (b) The relationship between the percentage of native C_4 species in 31 geographic regions of North America, and the mean summer (May–October) pan evaporation – a climatic indicator of plant/water balance. Regions for which appropriate climatic data were unavailable were excluded, together with south Florida, where the peculiar geography and climate may explain the aberrant composition of the flora. (After Stowe & Teeri, 1978.)

and the clearing of forests. As Loladze (2002) points out, while consequential changes to global climate may be controversial in some quarters, marked increases in CO_2 concentration itself are not. Plants now are experiencing around a 30% higher concentration compared to the pre-industrial period – effectively instantaneous on geological timescales; trees living now may experience a doubling in concentration over their lifetimes – effectively an instantaneous change on an *evolutionary* timescale; and high mixing rates in the atmosphere mean that these are changes that will affect *all* plants.

changes in geological time

There is also evidence of large-scale changes in atmospheric CO_2 over much longer timescales. Carbon balance models suggest that during the Triassic, Jurassic and Cretaceous periods, atmospheric concentrations of CO_2 were four to eight times greater than at present, falling after the Cretaceous from between 1400 and 2800 $\mu l\ l^{-1}$ to below 1000 $\mu l\ l^{-1}$ in the Eocene, Miocene and Pliocene, and fluctuating between 180 and 280 $\mu l\ l^{-1}$ during subsequent glacial and interglacial periods (Ehleringer & Monson, 1993).

The declines in CO_2 concentration in the atmosphere after the Cretaceous may have been the primary force that favored the evolution of plants with C_4 physiology (Ehleringer *et al.*, 1991), because at low concentrations of CO_2, photorespiration places C_3 plants at a particular disadvantage. The steady rise in CO_2 since the Industrial Revolution is therefore a partial return to pre-Pleistocene conditions and C_4 plants may begin to lose some of their advantage.

what will be the
consequences of
current rises?

When other resources are present at adequate levels, additional CO_2 scarcely influences the rate of photosynthesis of C_4 plants but increases the rate of C_3 plants. Indeed, artificially increasing the CO_2 concentration in greenhouses is a commercial technique to increase crop (C_3) yields. We might reasonably predict dramatic increases in the productivity of individual plants and of whole crops, forests and natural communities as atmospheric concentrations of CO_2 continue to increase. In the 1990s alone, results from more than 2700 studies on free-air CO_2 enrichment (FACE) experiments were published, and it is clear that, for example, doubling CO_2 concentration generally stimulates photosynthesis and increases agricultural yield by an average of 41% (Loladze, 2002). However, there is also much evidence that the responses may be complicated (Bazzaz, 1990). For example, when six species of temperate forest tree were grown for 3 years in a CO_2-enriched atmosphere in a glasshouse, they were generally larger than controls, but the CO_2 enhancement of growth declined even within the relatively short timescale of the experiment (Bazzaz et al., 1993).

Moreover, there is a general tendency for CO_2 enrichment to change the composition of plants, and in particular to reduce nitrogen concentration in above-ground plant tissues – around 14% on average under CO_2 enhancement (Cotrufo et al., 1998). This in turn may have indirect effects on plant–animal interactions, because insect herbivores may then eat 20–80% more foliage to maintain their nitrogen intake and fail to gain weight as fast (Figure 3.15).

CO_2 and nitrogen and
micronutrient
composition

CO_2 enhancement may also reduce concentrations in plants of other essential nutrients and micronutrients (Figure 3.16) (see Section 3.5), contributing in turn to 'micronutrient malnutrition', which diminishes the health and economy of more than one-half of the world's human population (Loladze, 2002).

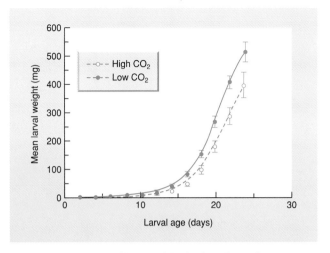

Figure 3.15   Growth of larvae of the buckeye butterfly (*Junonia coenia*) feeding on *Plantago lanceolata* that had been grown at ambient and elevated CO_2 concentrations. (After Fajer, 1989.)

3.4   Water

The volume of water that becomes incorporated in higher plants during growth is infinitesimal in comparison to the volume that flows through the plant in the transpiration stream. Nevertheless, water is a critical resource. Hydration is a necessary condition for metabolic reactions to proceed, and because no organism is completely watertight its water content needs continual replenishment. Most terrestrial animals drink free water and also generate some from the metabolism of food and body materials; there are extreme cases in which animals of arid zones may obtain all their water from their food.

3.4.1   Roots as water foragers

For most terrestrial plants, the main source of water is the soil and they gain access to it through a root system. We proceed here

Figure 3.16   Changes in the concentrations of nutrients in plant material grown at twice-ambient atmospheric CO_2 concentrations, based on 25 studies on leaves of a variety of plants (colored bars) and five studies of wheat grains (gray bars). Black lines indicate the standard errors. (After Loladze, 2002.)

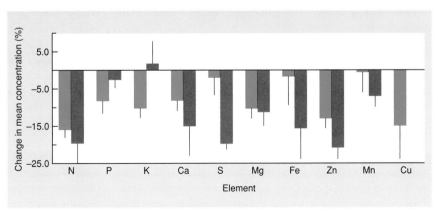

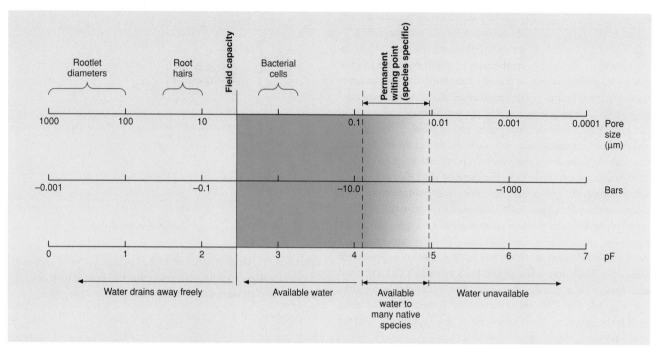

Figure 3.17   The status of water in the soil, showing the relationship between three measures of water status: (i) pF, the logarithm of the height (cm) of the column of water that the soil would support; (ii) water status expressed as atmospheres or bars; (iii) the diameter of soil pores that remain water-filled. The size of water-filled pores may be compared in the figure with the sizes of rootlets, root hairs and bacterial cells. Note that for most species of crop plant the permanent wilting point is at approximately −15 bars (−1.5 × 10⁶ Pa), but in many other species it reaches −80 bars (−8 ×10⁶ Pa), depending on the osmotic potentials that the species can develop.

(and in the next section on plant nutrient resources) on the basis of plants simply having 'roots'. In fact, most plants do not have roots – they have mycorrhizae: associations of fungal and root tissue in which both partners are crucial to the resource-gathering properties of the whole. Mycorrhizae, and the respective roles of the plants and the fungi, are discussed in Chapter 13.

It is not easy to see how roots evolved by the modification of any more primitive organ (Harper *et al.*, 1991), yet the evolution of the root was almost certainly the most influential event that made an extensive land flora and fauna possible. Once roots had evolved they provided secure anchorage for structures the size of trees and a means for making intimate contact with mineral nutrients and water within the soil.

field capacity and the permanent wilting point
Water enters the soil as rain or melting snow and forms a reservoir in the pores between soil particles. What happens to it then depends on the size of the pores, which may hold it by capillary forces against gravity. If the pores are wide, as in a sandy soil, much of the water will drain away until it reaches some impediment and accumulates as a rising watertable or finds its way into streams or rivers. The water held by soil pores against the force of gravity is called the 'field capacity' of the soil. This is the upper limit of the water that a freely drained soil will retain.

There is a less clearly defined lower limit to the water that can be used in plant growth (Figure 3.17). This is determined by the ability of plants to extract water from the narrower soil pores, and is known as the 'permanent wilting point' – the soil water content at which plants wilt and are unable to recover. The permanent wilting point does not differ much between the plant species of mesic environments (i.e. with a moderate amount of water) or between species of crop plants, but many species native to arid regions can extract significantly more water from the soil.

As a root withdraws water from the soil pores at its surface, it creates water-depletion zones around it. These determine gradients of water potential between the interconnected soil pores. Water flows along the gradient into the depleted zones, supplying further water to the root. This simple process is made much more complex because the more the soil around the roots is depleted of water, the more resistance there is to water flow. As the root starts to withdraw water from the soil, the first water that it obtains is from the wider pores because they hold the water with weaker capillary forces. This leaves only the narrower, more tortuous water-filled paths through which flow can occur, and so the resistance to water flow increases. Thus, when the root draws water from the soil very rapidly, the resource depletion zone (RDZ; see Section 3.2.1) becomes sharply defined and water can move across it only slowly. For this reason, rapidly transpiring plants may wilt in a soil that contains abundant water. The fineness and degree of ramification of the root system through the soil then become important in determining the access of the plant to the water in the soil reservoir.

roots and the dynamics of water depletion zones

Water that arrives on a soil surface as rain or as melting snow does not distribute itself evenly. Instead, it tends to bring the surface layer to field capacity, and further rain extends this layer further and further down into the soil profile. This means that different parts of the same plant root system may encounter water held with quite different forces, and indeed the roots can move water between soil layers (Caldwell & Richards, 1986). In arid areas, where rainfall is in rare, short showers, the surface layers may be brought to field capacity whilst the rest of the soil stays at or below wilting point. This is a potential hazard in the life of a seedling that may, after rain, germinate in the wet surface layers lying above a soil mass that cannot provide the water resource to support its further growth. A variety of specialized dormancy-breaking mechanisms are found in species living in such habitats, protecting them against too quick a response to insufficient rain.

The root system that a plant establishes early in its life can determine its responsiveness to future events. Where most water is received as occasional showers on a dry substrate, a seedling with a developmental program that puts its early energy into a deep taproot will gain little from subsequent showers. By contrast, a program that determines that the taproot is formed early in life may guarantee continual access to water in an environment in which heavy rains fill a soil reservoir to depth in the spring, but there is then a long period of drought.

3.4.2 Scale, and two views of the loss of plant water to the atmosphere

There are two very different ways in which we can analyze and explain the loss of water from plants to the atmosphere. Plant physiologists going back at least to Brown and Escombe in 1900 have emphasized the way in which the behavior of the stomata determines the rate at which a leaf loses water. It now seems obvious that it is the frequency and aperture of pores in an otherwise mainly waterproof surface that will control the rate at which water diffuses from a leaf to the outside atmosphere. But micrometeorologists take a quite different viewpoint, focusing on vegetation as a whole rather than on the single stoma, leaf or plant. Their approach emphasizes that water will be lost by evaporation only if there is latent heat available for this evaporation. This may be from solar radiation received directly by the transpiring leaves or as 'advective' energy, i.e. heat received as solar radiation elsewhere but transported in moving air. The micrometeorologists have developed formulae for the rate of water loss that are based entirely on the weather: wind speed, solar radiation, temperature and so

on. They wholly ignore both the species of plants and their physiology, but their models nevertheless prove to be powerful predictors of the evaporation of water from vegetation that is not suffering from drought. Neither approach is right or wrong: which to use depends on the question being asked. Large-scale, climatically based models, for example, are likely to be the most relevant in predicting the evapotranspiration and photosynthesis that might occur in areas of vegetation as a result of global warming and changes in precipitation (Aber & Federer, 1992).

3.5 Mineral nutrients

It takes more than light, CO_2 and water to make a plant. Mineral resources are also needed. The mineral resources that the plant must obtain from the soil (or, in the case of aquatic plants, from the surrounding water) include macronutrients (i.e. those needed in relatively large amounts) – nitrogen (N), phosphorus (P), sulfur (S), potassium (K), calcium (Ca), magnesium (Mg) and iron (Fe) – and a series of trace elements – for example, manganese (Mn), zinc (Zn), copper (Cu), boron (B) and molybdenum (Mo) (Figure 3.18). (Many of these elements are also essential to animals, although it is more common for animals to obtain them in organic form in their food than as inorganic chemicals.) Some plant groups have special requirements. For example, aluminum is a necessary nutrient for some ferns, silicon for diatoms and selenium for certain planktonic algae.

macronutrients and trace elements

Green plants do not obtain their mineral resources as a single package. Each element enters the plant independently as an ion or a molecule, and each has its own characteristic properties of absorption in the soil and of diffusion, which affect its accessibility to the plant even before any selective processes of uptake occur at the root membranes. All green plants require all of the 'essential' elements listed in Figure 3.18, although not in the same proportion, and there are some quite striking differences between the mineral compositions of plant tissues of different species and between the different parts of a single plant (Figure 3.19).

Many of the points made about water as a resource, and about roots as extractors of this resource, apply equally to mineral nutrients. Strategic differences in developmental programs can be recognized between the roots of different species (Figure 3.20a), but it is the ability of root systems to override strict programs and be opportunistic that makes them effective exploiters of the soil. Most roots elongate before they produce laterals, and this ensures that exploration precedes exploitation. Branch roots usually emerge on radii of the parent root, secondary roots radiate from these primaries and tertiaries from the secondaries. These rules reduce the chance that two branches of the same root will forage in the same soil particle and enter each other's RDZs.

roots as foragers

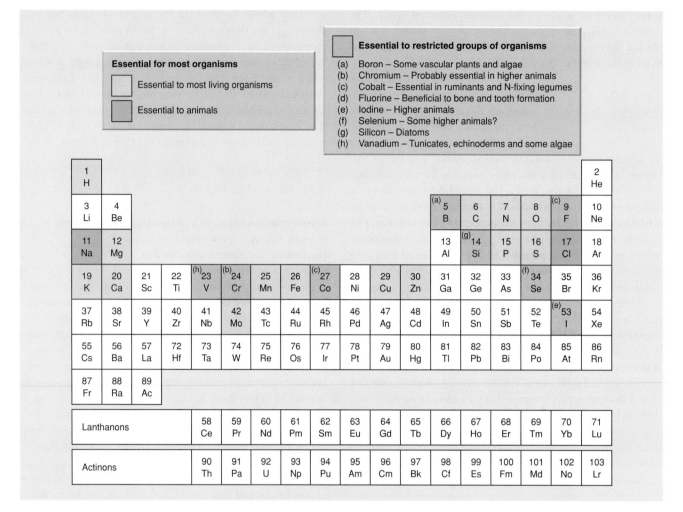

Figure 3.18   Periodic table of the elements showing those that are essential resources in the life of various organisms.

Roots pass through a medium in which they meet obstacles and encounter heterogeneity – patches of nutrient that vary on the same scale as the diameter of a root itself. In 1 cm of growth, a root may encounter a boulder, pebbles and sand grains, a dead or living root, or the decomposing body of a worm. As a root passes through a heterogeneous soil (and all soils are heterogeneous seen from a 'root's-eye view'), it responds by branching freely in zones that supply resources, and scarcely branching in less rewarding patches (Figure 3.20b). That it can do so depends on the individual rootlet's ability to react on an extremely local scale to the conditions that it meets.

interactions between foraging for water and nutrients

There are strong interactions between water and nutrients as resources for plant growth. Roots will not grow freely into soil zones that lack available water, and so nutrients in these zones will not be exploited. Plants deprived of essential minerals make less growth and may then fail to reach volumes of soil that contain available water. There are similar interactions between mineral resources. A plant starved of nitrogen makes poor root growth and so may fail to 'forage' in areas that contain available phosphate or indeed contain more nitrogen.

Of all the major plant nutrients, nitrates move most freely in the soil solution and are carried from as far away from the root surface as water is carried. Hence nitrates will be most mobile in soils at or near field capacity, and in soils with wide pores. The RDZs for nitrates will then be wide, and those produced around neighboring roots will be more likely to overlap. Competition can then occur – even between the roots of a single plant.

The concept of RDZs is important not only in visualizing how one organism influences the resources available to another, but also in understanding how the architecture of the root system affects the capture of these resources. For a plant growing in an environment in which water moves freely to the root surface, those nutrients that are freely in solution will move with the water. They will then be most effectively captured by wide ranging, but not

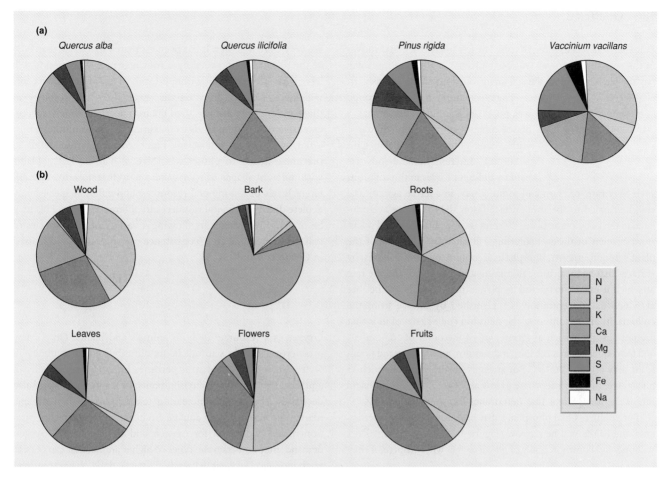

Figure 3.19 (a) The relative concentration of various minerals in whole plants of four species in the Brookhaven Forest, New York. (b) The relative concentration of various minerals in different tissues of the white oak (*Quercus alba*) in the Brookhaven Forest. Note that the differences between species are much less than between the parts of a single species. (After Woodwell *et al.*, 1975).

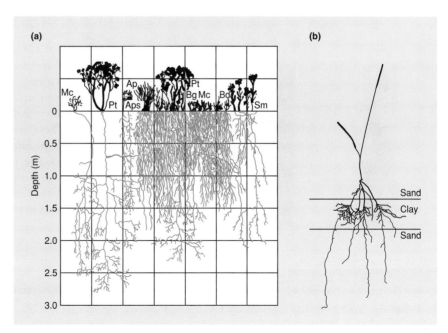

Figure 3.20 (a) The root systems of plants in a typical short-grass prairie after a run of years with average rainfall (Hays, Kansas). Ap, *Aristida purpurea*; Aps, *Ambrosia psilostachya*; Bd, *Buchloe dactyloides*; Bg, *Bouteloua gracilis*; Mc, *Malvastrum coccineum*; Pt, *Psoralia tenuiflora*; Sm, *Solidago mollis*. (After Albertson, 1937; Weaver & Albertson, 1943.) (b) The root system developed by a plant of wheat grown through a sandy soil containing a layer of clay. Note the responsiveness of root development to the localized environment that it encounters. (Courtesy of J.V. Lake.)

intimately branched, root systems. The less freely that water moves in the soil, the narrower will be the RDZs, and the more it will pay the plant to explore the soil intensively rather than extensively.

variations between nutrients in their freedom of movement

The soil solution that flows through soil pores to the root surface has a biased mineral composition compared with what is potentially available. This is because different mineral ions are held by different forces in the soil. Ions such as nitrate, calcium and sodium may, in a fertile agricultural soil, be carried to the root surface faster than they are accumulated in the body of the plant. By contrast, the phosphate and potassium content of the soil solution will often fall far short of the plant's requirements. Phosphate is bound on soil colloids by surfaces that bear calcium, aluminum and ferric ions, and the rate at which it can be extracted by plants then depends on the rate at which its concentration is replenished by release from the colloids. In dilute solutions, the diffusion coefficients of ions that are not absorbed, such as nitrate, are of the order of 10^{-5} cm^2 s^{-1}, and for cations such as calcium, magnesium, ammonium and potassium they are 10^{-7} cm^2 s^{-1}. For strongly absorbed anions such as phosphate, the coefficients are as low as 10^{-9} cm^2 s^{-1}. The diffusion rate is the main factor that determines the width of an RDZ.

For resources like phosphate that have low diffusion coefficients, the RDZs will be narrow (Figure 3.21); roots or root hairs will only tap common pools of resource (i.e. will compete) if they are very close together. It has been estimated that more than 90% of the phosphate absorbed by a root hair in a 4-day period will have come from the soil within 0.1 mm of its surface. Two roots will therefore only draw on the same phosphate resource in this period if they are less than 0.2 mm apart. A widely spaced, extensive root system tends to maximize access to nitrate, whilst a narrowly spaced, intensively branched root system tends to maximize access to phosphates (Nye & Tinker, 1977). Plants with different shapes of root system may therefore tolerate different levels of soil mineral resources, and different species may deplete different mineral resources to different extents. This may be of great importance in allowing a variety of plant species to cohabit in the same area (coexistence of competitors is discussed in Chapters 8 and 19).

3.6   Oxygen

Oxygen is a resource for both animals and plants. Only a few prokaryotes can do without it. Its diffusibility and solubility in water are very low and so it becomes limiting most quickly in aquatic and waterlogged environments. Its solubility in water also decreases rapidly with increasing temperature. When organic matter decomposes in an aquatic environment, microbial respiration makes a demand for oxygen and this 'biological oxygen demand' may constrain the types of higher animal that can persist. High biological oxygen demands are particularly characteristic of still waters into which leaf litter or organic pollutants are deposited and they become most acute during periods of high temperature.

Because oxygen diffuses so slowly in water, aquatic animals must either maintain a continual flow of water over their respiratory surfaces (e.g. the gills of fish), or have very large surface areas relative to body volume (e.g. many aquatic crustacea have large feathery appendages), or have specialized respiratory pigments or a slow respiration rate (e.g. the midge larvae that live in still and nutrient-rich waters), or continually return to the surface to breathe (e.g. whales, dolphins, turtles and newts).

The roots of many higher plants fail to grow into waterlogged soil, or die if the water table rises after they have penetrated deeply. These reactions may be direct responses to oxygen deficiency or responses to the accumulation of gases such as hydrogen sulfide, methane and ethylene, which are produced by microorganisms engaged in anaerobic decomposition. Even if roots do not die when starved of oxygen, they may cease to absorb mineral nutrients so that the plants suffer from mineral deficiencies.

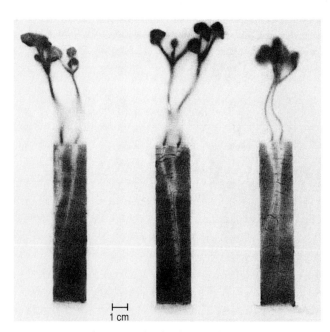

Figure 3.21   Radioautograph of soil in which seedlings of mustard have been grown. The soil was supplied with radioactively labeled phosphate ($^{32}PO_4^-$) and the zones that have been depleted by the activity of the roots show up clearly as white. (After Nye & Tinker, 1977.)

3.7   Organisms as food resources

Autotrophic organisms (green plants and certain bacteria) assimilate inorganic resources into packages of organic

autotrophs and heterotrophs

molecules (proteins, carbohydrates, etc.). These become the resources for *heterotrophic* organisms (decomposers, parasites, predators and grazers), which take part in a chain of events in which each consumer of a resource becomes, in turn, a resource for another consumer. At each link in this food chain the most obvious distinction is between saprotrophs and predators (defined broadly).

Saprotrophs – bacteria, fungi and detritivorous animals (see Chapter 11) – use other organisms, or parts of other organisms, as food but only after they have died, or they consume another organism's waste or secretory products.

saprotrophs, predators, grazers and parasites

Predators use other living organisms, or parts of other living organisms, as food. True predators predictably kill their prey. Examples include a mountain lion consuming a rabbit but also consumers that we may not refer to as predators in everyday speech: a water flea consuming phytoplankton cells, a squirrel eating an acorn, and even a pitcherplant drowning a mosquito. *Grazing* can also be regarded as a type of predation, but the food (prey) organism is not killed; only part of the prey is taken, leaving the remainder with the potential to regenerate. Grazers feed on (or from) many prey during their lifetime. True predation and grazing are discussed in detail in Chapter 9. *Parasitism*, too, is a form of predation in which the consumer usually does not kill its food organism; but unlike a grazer, a parasite feeds from only one or a very few host organisms in its lifetime (see Chapter 12).

specialists and generalists

An important distinction amongst animal consumers is whether they are specialized or generalized in their diet. Generalists (*polyphagous* species) take a wide variety of prey species, though they very often have clear preferences and a rank order of what they will choose when there are alternatives available. Specialists may consume only particular parts of their prey though they range over a number of species. This is most common among herbivores because, as we shall see, different parts of plants are quite different in their composition. Thus, many birds specialize on eating seeds though they are seldom restricted to a particular species. Other specialists, however, may feed on only a narrow range of closely related species or even just a single species (when they are said to be *monophagous*). Examples are caterpillars of the cinnabar moth (which eat the leaves, flower buds and very young stems of species of ragwort, *Senecio*) and many species of host-specific parasites.

Many of the resource-use patterns found among animals reflect the different lifespans of the consumer and what it consumes. Individuals of long-lived species are likely to be generalists: they cannot depend on one food resource being available throughout their life. Specialization is increasingly likely if a consumer has a short lifespan. Evolutionary forces can then shape the timing of the consumer's food demands to match the

timetable of its prey. Specialization also allows the evolution of structures that make it possible to deal very efficiently with particular resources – this is especially the case with mouthparts. A structure like the stylet of an aphid (Figure 3.22) can be interpreted as an exquisite product of the evolutionary process that has given the aphid access to a valuable food resource – or as an example of the ever-deepening rut of specialization that has constrained what aphids can feed on. The more specialized the food resource required by an organism, the more it is constrained to live in patches of that resource *or* to spend time and energy in searching for it among a mixture of resources. This is one of the costs of specialization.

3.7.1 The nutritional content of plants and animals as food

As a 'package' of resources, the body of a green plant is quite different from the body of an animal. This has a tremendous effect on the value of these resources as potential food (Figure 3.23). The most important contrast is that plant cells are bounded by walls of cellulose, lignin and/or other structural materials. It is these cell walls that give plant material its high fiber content. The presence of cell walls is also largely responsible for the high fixed carbon content of plant tissues and the high ratio of carbon to other important elements. For example, the carbon : nitrogen (C : N) ratio of plant tissues commonly exceeds 40 : 1, in contrast to the ratios of approximately 10 : 1 in bacteria, fungi and animals. Unlike plants, animal tissues contain no structural carbohydrate or fiber component but are rich in fat and, in particular, protein.

C : N ratios in animals and plants

The various parts of a plant have very different compositions (Figure 3.23) and so offer quite different resources. Bark, for example, is largely composed of dead cells with corky and lignified walls and is quite useless as a food for most herbivores (even species of 'bark beetle' specialize on the nutritious cambium layer just beneath the bark, rather than on the bark itself). The richest concentrations of plant proteins (and hence of nitrogen) are in the meristems in the buds at shoot apices and in leaf axils. Not surprisingly, these are usually heavily protected with bud scales and defended from herbivores by thorns and spines. Seeds are usually dried, packaged reserves rich in starch or oils as well as specialized storage proteins. And the very sugary and fleshy fruits are resources provided by the plant as 'payment' to the animals that disperse the seeds. Very little of the plants' nitrogen is 'spent' on these rewards.

different plant parts represent very different resources . . .

The dietary value of different tissues and organs is so different that it is no surprise to find that most small herbivores are specialists – not only on particular species or plant groups, but on particular plant parts: meristems, leaves, roots, stems, etc. The smaller the herbivore, the finer is the scale of heterogeneity of

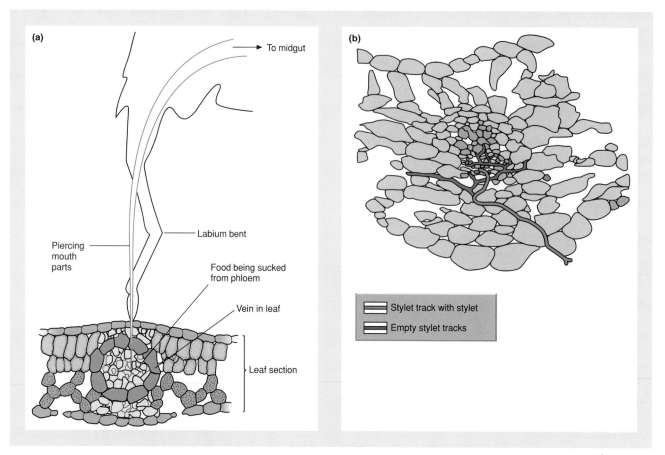

Figure 3.22   The stylet of an aphid penetrating the host tissues and reaching the sugar-rich phloem cells in the leaf veins. (a) Aphid mouthparts and cross-section of a leaf. (b) A stylet, showing its circuitous path through a leaf. (After Tjallingii & Hogen Esch, 1993.)

the plant on which it may specialize. Extreme examples can be found in the larvae of various species of oak gall wasps, some of which may specialize on young leaves, some on old leaves, some on vegetative buds, some on male flowers and others on root tissues.

... but the composition of all herbivores is remarkably similar

Although plants and their parts may differ widely in the resources they offer to potential consumers, the composition of the bodies of different herbivores is remarkably similar. In terms of the content of protein, carbohydrate, fat, water and minerals per gram there is very little to choose between a diet of caterpillars, cod or venison. The packages may be differently parceled (and the taste may be different), but the contents are essentially the same. Carnivores, then, are not faced with problems of digestion (and they vary rather little in their digestive apparatus), but rather with difficulties in finding, catching and handling their prey (see Chapter 9).

Differences in detail aside, herbivores that consume living plant material – and saprotrophs that consume dead plant material –

all utilize a food resource that is rich in carbon and poor in protein. Hence, the transition from plant to consumer involves a massive burning off of carbon as the C : N ratio is lowered. This is the realm of ecological stoichiometry (Elser & Urabe 1999): the analysis of constraints and consequences in ecological interactions of the mass balance of multiple chemical elements (particularly the ratios of carbon to nitrogen and of carbon to phosphorus – see Sections 11.2.4 and 18.2.5). The main waste products of organisms that consume plants are carbon-rich compounds: CO_2, fiber, and in the case of aphids, for example, carbon-rich honeydew dripping from infested trees. By contrast, the greater part of the energy requirements of carnivores is obtained from the protein and fats of their prey, and their main excretory products are in consequence nitrogenous.

The differential in C : N ratios between plants and microbial decomposers also means that the long-term effects of CO_2 enhancement (see Section 3.3.4) are not as straightforward as might be imagined (Figure 3.24): that is, it is not necessarily the case that plant

C : N ratios and the effects of CO_2 enhancement

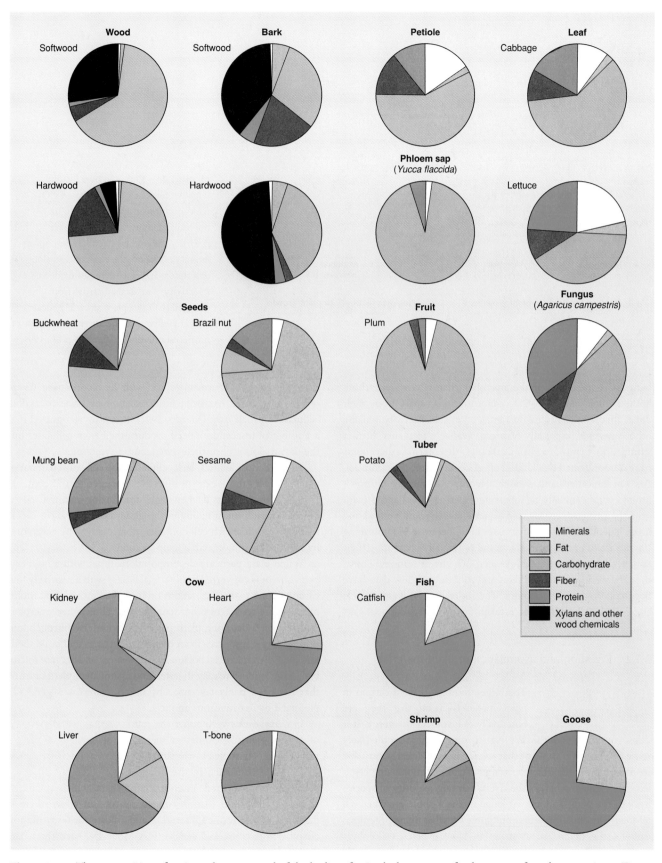

Figure 3.23 The composition of various plant parts and of the bodies of animals that serve as food resources for other organisms. (Data from various sources.)

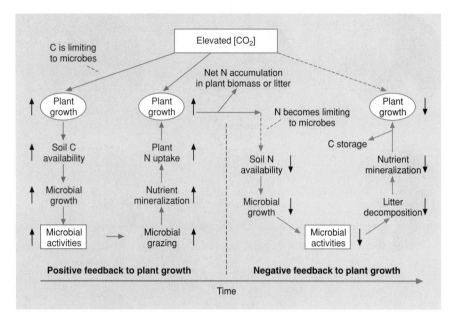

Figure 3.24 Potential positive and negative feedback from elevated CO_2 concentrations to plant growth, to microbial activity and back to plant growth. The arrows between descriptors indicate causation; the black arrows alongside descriptors indicate increases or decreases in activity. The dashed arrow from elevated $[CO_2]$ to plant growth indicates that any effect may be absent as a result of nutrient-limitation. (After Hu *et al.*, 1999.)

biomass is increased. If the microbes themselves are carbon-limited, then increased CO_2 concentrations, apart from their direct effects on plants, might stimulate microbial activity, making other nutrients, especially nitrogen, available to plants, further stimulating plant growth. Certainly, short-term experiments have demonstrated this kind of effect on decomposer communities. On the other hand, though, decomposers may be nitrogen-limited, either initially or following a period of enhanced plant growth during which nitrogen accumulates in plant biomass and litter. Then, microbial activity would be depressed, diminishing the release of nutrients to plants and potentially preventing their enhanced growth in spite of elevated CO_2 concentrations. These, though, are longer term effects and to date very few data have been collected to detect them. The more general issue of local and global 'carbon budgets' is taken up again in Section 18.4.6.

3.7.2 Digestion and assimilation of plant material

cellulases, which most animals lack

The large amounts of fixed carbon in plant materials mean that they are potentially rich sources of energy. It is other components of the diet (e.g. nitrogen) that are more likely to be limiting. Yet most of that energy is only directly available to consumers if they have enzymes capable of mobilizing cellulose and lignins, whereas the overwhelming majority of species in both the plant and animal kingdoms lack these enzymes. Of all the many constraints that put limits on what living organisms can do, the failure of so many to have evolved cellulolytic enzymes is a particular evolutionary puzzle. It may be that gut-inhabiting, cellulolytic prokaryotes have so readily formed

intimate, 'symbiotic' relationships with herbivores (see Chapter 13) that there has been little selection pressure to evolve cellulases of their own (Martin, 1991). It is now recognized that a number of insects do indeed produce their own cellulases but the vast majority nevertheless depend on symbionts.

Because most animals lack cellulases, the cell wall material of plants hinders the access of digestive enzymes to the contents of plant cells. The acts of chewing by the grazing mammal, cooking by humans and grinding in the gizzard of birds allow digestive enzymes to reach cell contents more easily. The carnivore, by contrast, can more safely gulp its food.

When plant parts are decomposed, material with a high carbon content is converted to microbial bodies with a relatively low carbon content – the limitations on microbial growth and multiplication are resources other than carbon. Thus, when microbes multiply on a decaying plant part, they withdraw nitrogen and other mineral resources from their surroundings and build them into their own microbial bodies. For this reason, and because microbial tissue is more readily digested and assimilated, plant detritus that has been richly colonized by microorganisms is generally preferred by detritivorous animals.

In herbivorous vertebrates the rate of energy gain from different dietary resources is determined by the structure of the gut – in particular, the balance between a well-stirred anterior chamber in which microbial fermentation occurs (AF), a connecting tube in which there is digestion but no fermentation (D), and a posterior fermentation chamber, the colon and cecum (PF). Models of such three-part digestive systems (Alexander, 1991) suggest that large AF, small D and small PF (e.g. the ruminant) would give near-optimal gains

the gut structures of herbivorous vertebrates

from poor-quality food, and that large PF, as in horses, is more appropriate for food with less cell wall material and more cell contents. For very high-quality food (a very high proportion of cell contents and little cell wall material) the optimum gut has long D and no AF or PF.

Elephants, lagomorphs and some rodents eat their own feces and so double the distance traveled by the food resource through the digestive system. This allows further fermentation and digestion but may also allow time for dietary deficiencies (e.g. of vitamins) to be made good by microbial synthesis. These issues are picked up again in Section 13.5.

3.7.3 Physical defenses

coevolution

All organisms are potentially food resources for others and so it is not surprising that many organisms have evolved physical, chemical, morphological and/or behavioral defenses that reduce the chance of an encounter with a consumer and/or increase the chance of surviving such an encounter. But the interaction does not necessarily stop there. A better defended food resource itself exerts a selection pressure on consumers to overcome that defense; though in overcoming that defense, rather than the defenses of other species, the consumer is likely to become relatively specialized on that resource – which is then under particular pressure to defend itself against that particular consumer, and so on. A continuing interaction can therefore be envisaged in which the evolution of both the consumer and the organism consumed depend crucially on the evolution of the other: a coevolutionary 'arms race' (Ehrlich & Raven, 1964), which, in its most extreme form, has a coadapted pair of species locked together in perpetual struggle.

Of course, the resources of green plants (and of autotrophs in general) are not alive and cannot therefore evolve defenses. Coevolution is also not possible between decomposer organisms and their dead food resources, although bacteria, fungi and detritivorous animals will often have to contend with the residual effects of physical and, in particular, chemical defenses in their food.

spines

Simple spines can be an effective deterrent. The spiny leaves of holly are not eaten by oak eggar moth larvae (*Lasiocampa quercus*), but if the spines are removed the leaves are eaten readily. No doubt a similar result would be achieved with foxes as predators and de-spined hedgehogs as prey. In many small planktonic invertebrates that live in lakes, the development of spines, crests and other appendages that reduce their vulnerability to predation can be induced by a predator's presence. Thus, for example, spine development in the offspring of brachionid rotifers, including *Keratella cochlearis*, is promoted if their mother was cultured in a medium conditioned by the predatory rotifer, *Asplachna priodonta* (Stemberger & Gilbert, 1984; Snell, 1998). At a smaller scale still, many plant surfaces are clothed in epidermal hairs (trichomes) and in some species these develop thick secondary walls to form strong hooks or points that may trap or impale insects.

shells

Any feature that increases the energy a consumer spends in discovering or handling a food item – the thick shell of a nut or the fibrous cone on a pine – is a defense if, as a consequence, the consumer eats less of it. The green plant uses none of its energetic resources in running away and so may have relatively more available to invest in energy-rich defense structures. Moreover, most green plants are probably relatively overprovided with energy resources, making it cheap to build shells around seeds and woody spines on stems – mainly out of cellulose and lignin – and so protecting the real riches: the scarce resources of nitrogen, phosphorus, potassium, etc. in the embryos and meristems.

seeds: dissipation or protection

Seeds are most at risk to predators when they have just ripened and are still attached, in a cone or ovary, to the parent plant, but their value is literally dissipated as soon as the capsule opens and the seeds are shed. The poppies illustrate this point. The seeds of wild poppies are shed through a series of pores at the apex of the capsule as it waves in the wind. Two of the species, *Papaver rhoeas* and *P. dubium*, open these pores as soon as the seed is ripe and the capsules are often empty by the following day. Two other species, *P. argemone* and *P. hybridum*, have seeds that are large relative to the size of the capsule pores and dispersal is a slow process over the fall and winter months. The capsules of these species are defended by spines. The cultivated poppy (*P. somniferum*) by contrast, has been selected by humans not to disperse its seeds – the capsule pores do not open. Birds can therefore be a serious pest of the cultivated poppy; they tear open the capsules to reach an oil- and protein-rich reward. Humans, in fact, have selected most of their crops to retain rather than disperse their seeds and these represent sitting targets for seed-eating birds.

3.7.4 Chemical defenses

secondary chemicals: protectants?

The plant kingdom is very rich in chemicals that apparently play no role in the normal pathways of plant biochemistry. These 'secondary' chemicals range from simple molecules like oxalic acid and cyanide to the more complex glucosinolates, alkaloids, terpenoids, saponins, flavonoids and tannins (Futuyma, 1983). Many of these have been shown to be toxic to a wide range of potential consumers. For example, populations of white clover, *Trifolium repens*, are commonly polymorphic for the ability to release hydrogen cyanide when the tissues are attacked. Plants that lack the ability to generate hydrogen cyanide are eaten by slugs and snails: the cyanogenic forms are nibbled but then rejected. Many researchers have assumed that protection against consumers has provided the selective pressure favoring the production of such chemicals. Many others, however, have questioned whether the selective force of herbivory is powerful enough for this (their production may be costly to the plants in terms of essential nutrients) and have pointed to other properties that they possess: for example as protectants against ultraviolet radiation (Shirley, 1996). None the less, in the few cases where selection experiments have been carried out, plants reared in the presence of consumers have evolved enhanced defenses against these enemies, relative to control plants reared in the absence of consumers (Rausher, 2001). Later, in Chapter 9 when we look in more detail at the *interaction* between predators and their prey, we will look at the costs and benefits of prey (especially plant) defense to both the prey itself and its consumers. Here, we focus more on the nature of those defenses.

apparency theory

If the attentions of herbivores select for plant defensive chemicals, then equally, those chemicals will select for adaptations in herbivores that can overcome them: a classic coevolutionary 'arms race'. This, though, suggests that plants should become ever more noxious and herbivores ever more specialized, leaving unanswered the question of why there are so many generalist herbivores, capable of feeding from many plants (Cornell & Hawkins, 2003). An answer has been suggested by 'apparency theory' (Feeny, 1976; Rhoades & Cates, 1976). This is based on the observation that noxious plant chemicals can be classified broadly into two types: (i) toxic (or qualitative) chemicals, which are poisonous even in small quantities; and (ii) digestion-reducing (or quantitative) chemicals, which act in proportion to their concentration. Tannins are an example of the second type. They bind proteins, rendering tissues such as mature oak leaves relatively indigestible. The theory further supposes that toxic chemicals, by virtue of their specificity, are likely to be the foundation of an arms race, requiring an equally simple and specific response from a herbivore; whereas chemicals that make plants generally indigestible are much more difficult to overcome.

Apparency theory then proposes that relatively short-lived, ephemeral plants (said to be 'unapparent') gain a measure of protection from consumers because of the unpredictability of their appearance in space and time. They therefore need to invest less in defense than predictable, long-lived ('apparent') species like forest trees. Moreover, the apparent species, precisely because they are apparent for long, predictable periods to a large number of herbivores, should invest in digestion-reducing chemicals that, while costly, will afford them broad protection; whereas unapparent plants should produce toxins since it is only likely to pay a few specialist species to coevolve against them.

Apparency theory, incorporating ideas on coevolution, therefore makes a number of predictions (Cornell & Hawkins, 2003). The most obvious is that more unapparent plants are more likely to be protected by simple, toxic compounds than by more complex, digestion-inhibiting compounds. This can even be seen in the changing balance of chemical defense in some plants as the season progresses. For example, in the bracken fern (*Pteridium aquilinum*), the young leaves that push up through the soil in spring are less apparent to potential herbivores than the luxuriant foliage in late summer. The young leaves are rich in cyanogenic glucosinolates, whilst the tannin content steadily increases in concentration to its maximum in mature leaves (Rhoades & Cates, 1976).

A more subtle prediction of the theory is that specialist herbivores, having invested evolutionarily in overcoming particular chemicals, should perform best when faced with those chemicals (compared to chemicals they would not normally encounter); whereas generalists, having invested in performing well when faced with a wide range of chemicals, should perform least well when faced with chemicals that have provoked coevolutionary responses from specialists. This is supported by an analysis of a wide range of data sets for insect herbivores fed on artificial diets with added chemicals (892 insect/chemical combinations) shown in Figure 3.25.

Furthermore, plants are predicted to differ in their chemical defenses not only from species to species but also within an individual plant. 'Optimal defense theory' predicts that the more important an organ or tissue is for an organism's fitness, the better protected it will be; and in the present context, it predicts that more important plant parts should be protected by *constitutive chemicals* (produced all the time), whereas less important parts should rely on *inducible chemicals*, only produced in response to damage itself, and hence with far lower fixed costs to the plants (McKey, 1979; Strauss *et al.*, 2004). This is confirmed, for example, by a study of wild radish, *Raphanus sativus*, in which plants were either subjected to herbivory by caterpillars of the butterfly, *Pieris rapae*, or left as unmanipulated controls (Strauss *et al.*, 2004). Petals (and all parts of the flower) are known in this insect-pollinated plant to be highly important to fitness. Concentrations of protective glucosinolates were twice as high in petals as in undamaged leaves, and these levels were maintained constitutively, irrespective of whether the petals were damaged by the caterpillars (Figure 3.26). Leaves, on the other hand, have a much less direct influence on fitness: high levels of leaf damage can be sustained without any measurable effect on reproductive output. Constitutive levels of glucosinolates, as already noted, were low; but if the leaves were damaged the (induced) concentrations were even higher than in the petals.

optimal defense theory: constitutive and inducible defenses

Similar results were found for the brown seaweed, *Sargassum filipendula*, where the holdfast at its base was the most valuable tissue: without it the plant would be cast adrift in the water (Taylor *et al.*, 2002). This was protected by costly constitutive, quantitative chemicals, whereas the much less valuable youngest stipes (effectively stems) near the tip of the plant were protected only by toxic chemicals induced by grazing.

Plate 1.1 Arctic tundra, Greenland. (Courtesy of J. A. Vickery.)

(a) 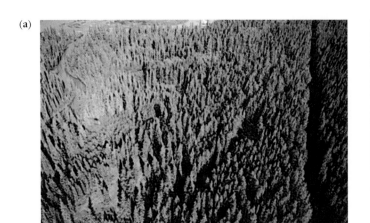 (b)

Plate 1.2 Coniferous forest: (a) aerial view of conifer forest, Alberta, Canada (© Planet Earth Pictures/Martin King) and (b) a pine forest in the fall, Sweden (© Planet Earth Pictures/Jan Tove Johansson).

(a)  (b)

Plate 1.3 Temperate forest: (a) mixed woodland in the fall, North Carolina, USA (© The Image Bank/Arthur Mayerson) and (b) late summer in Beechwood, Harburn, Scotland (© Ecoscene/Wilkinson).

Plate 1.4 Savanna. (a) Huge herds of wildebeest and common zebra seen from Naabi Hill, Serengeti, Tanzania. (b) Grassland savanna with scattered trees. Common zebra and wildebeest in the western corridor of the Serengeti, Tanzania. (© Images of Africa/David Keith Jones.)

(a)  **(b)**

Plate 1.5 Desert: (a) summertime and (b) spring flowers in Namaqualand, western South Africa. (© Planet Earth Pictures/ J. MacKinnon.)

(a)

(b)

Plate 1.6 Rainforest. (a, b) Impenetrable forest in southwest Uganda. (© Images of Africa/David Keith Jones.)

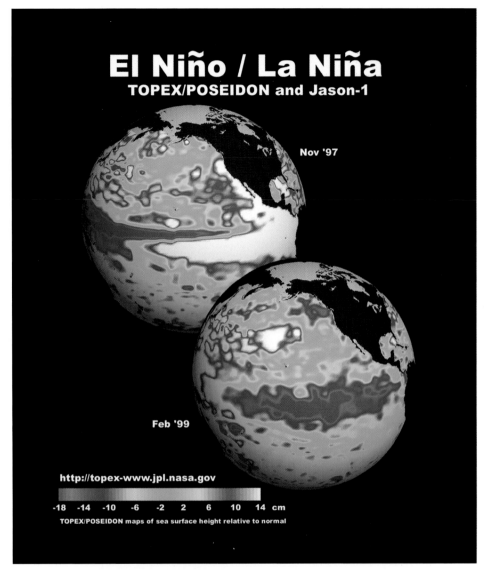

Plate 2.1 Maps of examples of El Niño (November 1997) and La Niña (February 1999) events in terms of sea height above average levels. Warmer seas are higher; for example, a sea height 15–20 cm below average equates to a temperature anomaly of approximately 2–3°C. (Image from http://topex-www.jpl.nasa.gov/science/images/el-nino-la-nina.jpg.) (See Figure 2.11. Courtesy of NASA JPL-Caltech)

Plate 2.2 Typical winter conditions when the NAO index is positive or negative. Conditions that are more than usually warm (red), cold (blue), dry (orange) or wet (turquoise) are indicated. (Image from http://www.ldeo.columbia.edu/NAO/.) (See Figure 2.11.)

(d)

(e)

Plate 4.1 Modular plants (on the left) and animals (on the right), showing the underlying parallels in the various ways they may be constructed. (*previous page*) (a) Modular organisms that fall to pieces as they grow: duckweed (*Lemna* sp.) and *Hydra* sp. (b) Freely branching organisms in which the modules are displayed as individuals on 'stalks': a vegetative shoot of a higher plant (*Lonicera japonica*) with leaves (feeding modules) and a flowering shoot, and a hydroid colony (*Obelia*) bearing both feeding and reproductive modules. (c) Stoloniferous organisms in which colonies spread laterally and remain joined by 'stolons' or rhizomes: a single plant of strawberry (*Fragaria*) spreading by means of stolons, and a colony of the hydroid *Tubularia crocea*. (*above*) (d) Tightly packed colonies of modules: a tussock of the spotted saxifrage (*Saxifraga bronchialis*), and a segment of the hard coral *Turbinaria reniformis*. (e) Modules accumulated on a long persistent, largely dead support: an oak tree (*Quercus robur*) in which the support is mainly the dead woody tissues derived from previous modules, and a gorgonian coral in which the support is mainly heavily calcified tissues from earlier modules.

((a) left, © Visuals Unlimited/John D. Cunningham; right, © Visuals Unlimited/Larry Stepanowicz; (b) left, © Visuals Unlimited; right, © Visuals Unlimited/Larry Stepanowicz; (c) left, © Visuals Unlimited/Science VU; right, © Visuals Unlimited/John D. Cunningham; (d) left, © Visuals Unlimited/Gerald and Buff Corsi; right, © Visuals Unlimited/Dave B. Fleetham; (e) left, © Visuals Unlimited/Silwood Park; right, © Visuals Unlimited/Daniel W. Gotshall.)

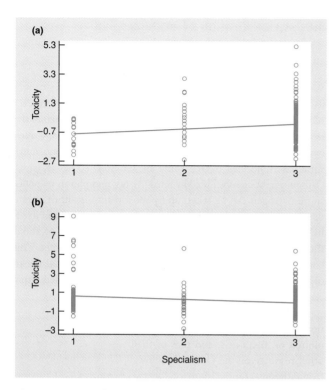

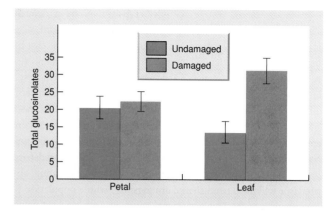

Figure 3.25 Combining data from a wide range of published studies, herbivores were split into three groups: 1, specialists (feeding from one or two plant families), 2, oligophages (3–9 families) and 3, generalists (more than nine families). Chemicals were split into two groups: (a) those that are, and (b) those that are not, found in the normal hosts of specialists and oligophages. With increasing specialization, (a) herbivores suffered decreased mortality on chemicals that have not provoked a coevolutionary response from specialist herbivores, but (b) suffered higher mortality on chemicals that have not provoked such a response. Regressions: (a) $y = 0.33x - 1.12$; $r^2 = 0.032$; $t = 3.25$; $P = 0.0013$; (b) $y = 0.93 - 0.36x$; $r^2 = 0.049$; $t = -4.35$; $P < 0.00001$. (After Cornell & Hawkins, 2003.)

Figure 3.26 Concentrations of glucosinolates (μg mg^{-1} dry mass) in the petals and leaves of wild radish, *Raphanus sativus*, either undamaged or damaged by caterpillars of *Pieris rapae*. Bars are standard errors. (After Strauss *et al.*, 2004.)

butterfly before) will vomit violently after eating one, and once it recovers will reject all others on sight. In contrast, monarchs reared on cabbage are edible (Brower & Corvinó, 1967).

Chemical defenses are not equally effective against all consumers. Indeed, what is unacceptable to most animals may be the chosen, even unique, diet

one man's poison is another man's meat

of others. It is, after all, an inevitable consequence of having evolved resistance to a plant's defenses that a consumer will have gained access to a resource unavailable to most (or all) other species. For example, the tropical legume *Dioclea metacarpa* is toxic to almost all insect species because it contains a nonprotein amino acid, L-canavanine, which insects incorporate into their proteins in place of arginine. But a species of bruchid beetle, *Caryedes brasiliensis*, has evolved a modified tRNA synthetase that distinguishes between L-canavanine and arginine, and the larvae of these beetles feed solely on *D. metacarpa* (Rosenthal *et al.*, 1976).

3.7.5 Crypsis, aposematism and mimicry

An animal may be less obvious to a predator if it matches its background,

crypsis

or possesses a pattern that disrupts its outline, or resembles an inedible feature of its environment. Straightforward examples of such *crypsis* are the green coloration of many grasshoppers and caterpillars, and the transparency of many planktonic animals that inhabit the surface layers of oceans and lakes. More dramatic cases are the sargassum fish (*Histrio pictus*), whose body outline mimics the sargassum weed in which it is found, or the caterpillar of the viceroy butterfly (*Limenitis archippus*) that resembles a bird dropping. Cryptic animals may be highly palatable, but their morphology and color (and their choice of the appropriate background) reduce the likelihood that they will be used as a resource.

animal defenses

Animals have more options than plants when it comes to defending themselves, but some still make use of chemicals. For example, defensive secretions of sulfuric acid of pH 1 or 2 occur in some marine gastropod groups, including the cowries. Other animals that can tolerate the chemical defenses of their plant food, store and use them in their own defense. A classic example is the monarch butterfly (*Danaus plexippus*), whose caterpillars feed on milkweeds (*Asclepias* spp.). Milkweeds contain secondary chemicals, cardiac glycosides, which affect the vertebrate heartbeat and are poisonous to mammals and birds. Monarch caterpillars can store the poison, and it is still present in the adults, which in consequence are completely unacceptable to bird predators. A naive blue jay (*Cyanocitta cristata*) (i.e. one that has not tried a monarch

aposematism

Whilst crypsis may be a defense strategy for a palatable organism, noxious or dangerous animals often seem to advertize the fact by bright, conspicuous colors and patterns. This phenomenon is referred to as *aposematism*. The monarch butterfly, discussed above, is aposematically colored, as is its caterpillar, which actually sequesters the defensive cardiac glucosinolates from its food. The usual evolutionary argument for this runs as follows: conspicuous coloration will be favored because noxious prey will be recognized (memorized) as such by experienced predators, and thus will be protected, whereas the costs of 'educating' the predator will have been shared amongst the whole population of conspicuous prey. This argument, however, leaves unanswered the question of how conspicuous, noxious prey arose in the first place, since when initially rare, they seem likely to be repeatedly eliminated by naive (i.e. 'uneducated') predators (Speed & Ruxton, 2002). One possible answer is that predators and prey have coevolved: in each generation – from an original mixture of conspicuous and inconspicuous, noxious and edible prey – conspicuous edible prey are eliminated, and, with conspicuous prey therefore becoming disproportionately noxious, predators evolve an increased wariness for conspicuous prey (Sherratt, 2002).

Batesian and Müllerian mimicry

The adoption of memorable body patterns by distasteful prey also immediately opens the door for deceit by other species, because there will be a clear evolutionary advantage to a palatable prey, 'the mimic', if it looks like an unpalatable species, 'the model' (Batesian mimicry). Developing the story of the monarch butterfly a little further, the adult of the palatable viceroy butterfly mimics the distasteful monarch, and a blue jay that has learned to avoid monarchs will also avoid viceroys. There will also be an advantage to aposematically colored, distasteful prey in looking like one another (Müllerian mimicry), though many unanswered questions remain as to where exactly Batesian mimicry ends and Müllerian mimicry begins, in part because there are more theoretical viewpoints than impeccable data sets that might distinguish between them (Speed, 1999).

By living in holes (e.g. millipedes and moles) animals may avoid stimulating the sensory receptors of predators, and by 'playing dead' (e.g. the opossum *Didelphis virginiana* and African ground squirrels) animals may fail to stimulate a killing response. Animals that withdraw to a prepared retreat (e.g. rabbits and prairie dogs to their burrows, snails to their shells), or which roll up and protect their vulnerable parts by a tough exterior (e.g. armadillos and pill millipedes), reduce their chance of capture but stake their lives on the chance that the attacker will not be able to breach their defenses. Other animals seem to try to bluff themselves out of trouble by threat displays. The startle response of moths and butterflies that suddenly expose eye-spots on their wings is one example. No doubt the most common behavioral response of an animal in danger of being preyed upon is to flee.

3.8 A classification of resources, and the ecological niche

We have seen that every plant requires many distinct resources to complete its life cycle, and most plants require the same set of resources, although in subtly different proportions. Each of the resources has to be obtained independently of the others, and often by quite different uptake mechanisms – some as ions (potassium), some as molecules (CO_2), some in solution, some as gases. Carbon cannot be substituted by nitrogen, nor phosphorus by potassium. Nitrogen can be taken up by most plants as either nitrate or ammonium ions, but there is no substitute for nitrogen itself. In complete contrast, for many carnivores, most prey of about the same size are wholly interchangeable as articles of diet. This contrast between resources that are individually *essential* for an organism, and those that are *substitutable*, can be extended into a classification of resources taken in pairs (Figure 3.27).

zero net growth isoclines

In this classification, the concentration or quantity of one resource is plotted on the *x*-axis, and that of the other resource on the *y*-axis. We know that different combinations of the two resources will support different growth rates for the organism in question (this can be individual growth or population growth). Thus, we can join together points (i.e. combinations of resources) with the same growth rates, and these are therefore contours or 'isoclines' of equal growth. In Figure 3.27, line B in each case is an isocline of *zero* net growth: each of the resource combinations on these lines allows the organism just to maintain itself, neither increasing nor decreasing. The A isoclines, then, with less resources than B, join combinations giving the same *negative* growth rate; whilst the C isoclines, with more resources than B, join combinations giving the same *positive* growth rate. As we shall see, the shapes of the isoclines vary with the nature of the resources.

3.8.1 Essential resources

Two resources are said to be *essential* when neither can substitute for the other. Thus, the growth that can be supported on resource 1 is absolutely dependent on the amount available of resource 2 and vice versa. This is denoted in Figure 3.27a by the isoclines running parallel to both axes. They do so because the amount available of one resource defines a maximum possible growth rate, irrespective of the amount of the other resource. This growth rate is achieved unless the amount available of the other resource defines an even lower growth rate. It will be true for nitrogen and potassium as resources in the growth of green plants, and for two obligate hosts in the life of a parasite or pathogen that are required to alternate in its life cycle (see Chapter 12).

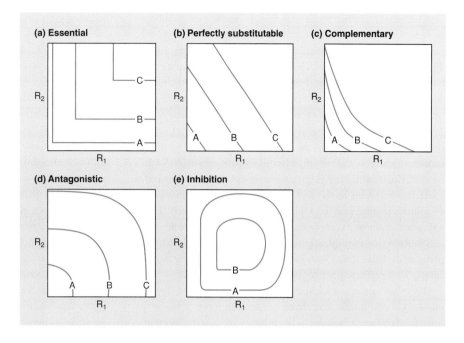

Figure 3.27 Resource-dependent growth isoclines. Each of the growth isoclines represents the amounts of two resources (R_1 and R_2) that would have to exist in a habitat for a population to have a given growth rate. Because this rate increases with resource availability, isoclines further from the origin represent higher population growth rates – isocline A has a negative growth rate, isocline B a zero growth rate and isocline C a positive growth rate. (a) Essential resources; (b) perfectly substitutable; (c) complementary; (d) antagonistic; and (e) inhibition. (After Tilman, 1982.)

3.8.2 Other categories of resource

Two resources are said to be *perfectly substitutable* when either can wholly replace the other. This will be true for seeds of wheat or barley in the diet of a farmyard chicken, or for zebra and gazelle in the diet of a lion. Note that we do not imply that the two resources are as good as each other. This feature (perfectly substitutable but not necessarily as good as each other) is included in Figure 3.27b by the isoclines having slopes that do not cut both axes at the same distance from the origin. Thus, in Figure 3.27b, in the absence of resource 2, the organism needs relatively little of resource 1, but in the absence of resource 1 it needs a relatively large amount of resource 2.

|complementary resources| Substitutable resources are defined as *complementary* if the isoclines bow inwards towards the origin (Figure 3.27c). This shape means that a species requires less of two resources when taken together than when consumed separately. A good example is human vegetarians combining beans and rice in their diet. The beans are rich in lysine, an essential amino acid poorly represented in rice, whilst rice is rich in sulfur-containing amino acids that are present only in low abundance in beans.

|antagonistic resources| A pair of substitutable resources with isoclines that bow away from the origin are defined as *antagonistic* (Figure 3.27d). The shape indicates that a species requires proportionately more resource to maintain a given rate of increase when two resources are consumed together than when consumed separately. This could arise, for example, if the resources contain different toxic compounds that act synergistically (more than just additively) on their consumer. For example, D, L-pipecolic acid and djenkolic acid (two defensive chemicals found in certain seeds) had no significant effect on the growth of the seed-eating larva of a bruchid beetle if consumed separately, but they had a pronounced effect if taken together (Janzen *et al.*, 1977).

Finally, Figure 3.27e illustrates the phenomenon of *inhibition* at high |inhibition| resource levels for a pair of essential resources: resources that are essential but become damaging when in excess. CO_2, water and mineral nutrients such as iron are all required for photosynthesis, but each is lethal in excess. Similarly, light leads to increased growth rates in plants through a broad range of intensities, but can inhibit growth at very high intensities. In such cases, the isoclines form closed curves because growth decreases with an increase in resources at very high levels.

3.8.3 Resource dimensions of the ecological niche

In Chapter 2 we developed the concept of the ecological niche as an *n*-dimensional hypervolume. This defines the limits within which a given species can survive and reproduce, for a number (*n*) of environmental factors, including both conditions and resources. Note, therefore, that the zero growth isoclines in Figure 3.27 define niche boundaries in two dimensions. Resource combinations to one side of line B allow the organisms to thrive – but to the other side of the line the organisms decline.

The resource dimensions of a species' niche can sometimes be represented in a manner similar to that adopted for conditions, with lower and upper limits within which a species can thrive. Thus, a predator may only be able to detect and handle prey between lower and upper limits of size. For other resources, such as mineral nutrients for plants, there may be a lower limit below which individuals cannot grow and reproduce but an upper limit may not exist (Figure 3.27a–d). However, many resources must be viewed as discrete entities rather than continuous variables. Larvae of butterflies in the genus *Heliconius* require *Passiflora* leaves to eat; those of the monarch butterfly specialize on plants in the milkweed family; and various species of animals require nest sites with particular specifications. These resource requirements cannot be arranged along a continuous graph axis labeled, for example, 'food plant species'. Instead, the food plant or nest-site dimension of their niches needs to be defined simply by a restricted list of the appropriate resources.

Together, then, conditions and resources define a species' niche. We turn in the next chapter to look in more detail at the most fundamental responses of organisms to those conditions and resources: their patterns of growth, survival and reproduction.

Summary

Resources are entities required by an organism, the quantities of which can be reduced by the activity of the organism. Hence, organisms may compete with each other to capture a share of a limited resource.

Autotrophic organisms (green plants and certain bacteria) assimilate inorganic resources into packages of organic molecules (proteins, carbohydrates, etc.). These become the resources for heterotrophic organisms, which take part in a chain of events in which each consumer of a resource becomes, in turn, a resource for another consumer.

Solar radiation is the only source of energy that can be used in metabolic activities by green plants. Radiant energy is converted during photosynthesis into energy-rich chemical compounds of carbon, which will subsequently be broken down in respiration. But the photosynthetic apparatus is able to gain access to energy only in the waveband of 'photosynthetically active radiation'. We examine variations in the intensity and quality of radiation, and the responses of plants to such variations. We examine, too, the strategic and tactical solutions adopted by plants to resolve the conflicts between photosynthesis and water conservation.

Carbon dioxide is also essential for photosynthesis. We examine variations in its concentration, and their consequences, including global rises over time and those at the smallest spatial scales. There are three pathways to carbon fixation in photosynthesis: C_3, C_4 and CAM. The differences between the different pathways and the ecological consequences of them are explained.

Water is a critical resource for all organisms. For plants, we examine how roots 'forage' for water, and the dynamics of resource depletion zones around roots, for water and for mineral nutrients. Mineral nutrients, broadly divisible into macronutrients and trace elements, each enter a plant independently as an ion or a molecule, and have their own characteristic properties of absorption in the soil and of diffusion, which affect their accessibility to a plant.

Oxygen is a resource for both animals and plants. It becomes limiting most quickly in aquatic and waterlogged environments, and when organic matter decomposes in an aquatic environment, microbial respiration may so deplete oxygen as to constrain the types of higher animal that can persist.

Amongst heterotrophs, we explain the distinctions between saprotrophs, predators, grazers and parasites, and between specialists and generalists.

The carbon : nitrogen ratio of plant tissues commonly exceeds greatly that in bacteria, fungi and animals. The main waste products of organisms that consume plants are therefore carbon-rich compounds. By contrast, the main excretory products of carnivores are nitrogenous. The various parts of a plant have very different compositions. Hence, most small herbivores are specialists. The composition of the bodies of different herbivores is remarkably similar.

Most of the energy sources potentially available to herbivores comprise cellulose and lignins, but most animals lack cellulases – an evolutionary puzzle. We explain how, in herbivorous vertebrates, the rate of energy gain from different dietary resources is determined by the structure of the gut.

Living resources are typically defended: physically, by chemicals, or by crypsis, aposematism or mimicry. This may lead to a coevolutionary arms races between the consumer and the consumed.

Apparency theory and optimal defense theory seek to make sense of the distribution of different protective chemicals, especially those that are constitutive and those that are induced, in different plant species and plant parts.

Taking resources in pairs, plots for the consumers of zero net growth isoclines allow resource pairs to be classified as essential, perfectly substitutable, complementary, antagonistic or displaying inhibition. The zero net growth isoclines themselves define a boundary of a species' ecological niche.

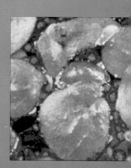

Chapter 4
Life, Death and Life Histories

4.1 Introduction: an ecological fact of life

In this chapter we change the emphasis of our approach. We will not be concerned so much with the interaction between individuals and their environment, as with the numbers of individuals and the processes leading to changes in the number of individuals.

In this regard, there is a fundamental ecological fact of life:

$$N_{\text{now}} = N_{\text{then}} + B - D + I - E. \tag{4.1}$$

This simply says that the numbers of a particular species presently occupying a site of interest (N_{now}) is equal to the numbers previously there (N_{then}), plus the number of births between then and now (B), minus the number of deaths (D), plus the number of immigrants (I), minus the number of emigrants (E).

This defines the main aim of ecology: to describe, explain and understand the distribution and abundance of organisms. Ecologists are interested in the number of individuals, the distributions of individuals, the demographic processes (birth, death and migration) that influence these, and the ways in which these demographic processes are themselves influenced by environmental factors.

4.2 What is an individual?

4.2.1 Unitary and modular organisms

Our 'ecological fact of life', though, implies by default that all individuals are alike, which is patently false on a number of counts. First, almost all species pass through a number of *stages* in their life cycle: insects metamorphose from eggs to larvae, sometimes to pupae, and then to adults; plants pass from seeds to seedlings to photosynthesizing adults; and so on. The different stages are likely to be influenced by different factors and to have different rates of migration, death and of course reproduction.

Second, even within a stage, individuals can differ in 'quality' or 'condition'. The most obvious aspect of this is size, but it is also common, for example, for individuals to differ in the amount of stored reserves they possess.

individuals differ in their life cycle stage and their condition

Uniformity amongst individuals is especially unlikely, moreover, when organisms are *modular* rather than *unitary*. In unitary organisms, form is highly determinate: that is, barring aberrations, all dogs have four legs, all squid have two eyes, etc. Humans are perfect examples of unitary organisms. A life begins when a sperm fertilizes an egg to form a zygote. This implants in the wall of the uterus, and the complex processes of embryonic development commence. By 6 weeks the fetus has a recognizable nose, eyes, ears and limbs with digits, and accidents apart, will remain in this form until it dies. The fetus continues to grow until birth, and then the infant grows until perhaps the 18th year of life; but the only changes in form (as opposed to size) are the relatively minor ones associated with sexual maturity. The reproductive phase lasts for perhaps 30 years in females and rather longer in males. This is followed by a phase of senescence. Death can intervene at any time, but for surviving individuals the succession of phases is, like form, entirely predictable.

unitary organisms

In modular organisms (Figure 4.1), on the other hand, neither timing nor form is predictable. The zygote develops into a unit of construction (a module, e.g. a leaf with its attendant length of stem), which then produces further, similar modules. Individuals are composed of a highly variable number of such modules, and their program of development is strongly dependent on their interaction with their environment. The product is almost always branched, and except for a juvenile phase, effectively immobile. Most plants are modular and are certainly the most obvious group of modular organisms. There are, however, many important groups of modular animals

modular organisms

(a)

(b)

(c)

(d)

(e)

Figure 4.1   Modular plants (on the left) and animals (on the right), showing the underlying parallels in the various ways they may be constructed. *(opposite page)* (a) Modular organisms that fall to pieces as they grow: duckweed (*Lemna* sp.) and *Hydra* sp. (b) Freely branching organisms in which the modules are displayed as individuals on 'stalks': a vegetative shoot of a higher plant (*Lonicera japonica*) with leaves (feeding modules) and a flowering shoot, and a hydroid colony (*Obelia*) bearing both feeding and reproductive modules. (c) Stoloniferous organisms in which colonies spread laterally and remain joined by 'stolons' or rhizomes: a single plant of strawberry (*Fragaria*) spreading by means of stolons, and a colony of the hydroid *Tubularia crocea*. *(above)*   (d) Tightly packed colonies of modules: a tussock of the spotted saxifrage (*Saxifraga bronchialis*), and a segment of the hard coral *Turbinaria reniformis*. (e) Modules accumulated on a long persistent, largely dead support: an oak tree (*Quercus robur*) in which the support is mainly the dead woody tissues derived from previous modules, and a gorgonian coral in which the support is mainly heavily calcified tissues from earlier modules. (For color, see Plate 4.1, between pp. 84 and 85.)

((a) left, © Visuals Unlimited/John D. Cunningham; right, © Visuals Unlimited/Larry Stepanowicz; (b) left, © Visuals Unlimited; right, © Visuals Unlimited/Larry Stepanowicz; (c) left, © Visuals Unlimited/Science VU; right, © Visuals Unlimited/John D. Cunningham; (d) left, © Visuals Unlimited/Gerald and Buff Corsi; right, © Visuals Unlimited/Dave B. Fleetham; (e) left, © Visuals Unlimited/Silwood Park; right, © Visuals Unlimited/Daniel W. Gotshall.

(indeed, some 19 phyla, including sponges, hydroids, corals, bryozoans and colonial ascidians), and many modular protists and fungi. Reviews of the growth, form, ecology and evolution of a wide range of modular organisms may be found in Harper *et al.* (1986a), Hughes (1989), Room *et al.* (1994) and Collado-Vides (2001).

Thus, the potentialities for individual difference are far greater in modular than in unitary organisms. For example, an individual of the annual plant *Chenopodium album* may, if grown in poor or crowded conditions, flower and set seed when only 50 mm high. Yet, given more ideal conditions, it may reach 1 m in height, and produce 50,000 times as many seeds as its depauperate counterpart. It is modularity and the differing birth and death rates of plant parts that give rise to this plasticity.

In the growth of a higher plant, the fundamental module of construction above ground is the leaf with its axillary bud and the attendant internode of the stem. As the bud develops and grows, it produces further leaves, each bearing buds in their axils. The plant grows by accumulating these modules. At some stage in the development, a new sort of module appears, associated with reproduction (e.g. the flowers in a higher plant), ultimately giving rise to new zygotes. Modules that are specialized for reproduction usually cease to give rise to new modules. The roots of a plant are also modular, although the modules are quite different (Harper *et al.*, 1991). The program of development in modular organisms is typically determined by the proportion of modules that are allocated to different roles (e.g. to reproduction or to continued growth).

4.2.2   Growth forms of modular organisms

A variety of growth forms and architectures produced by modular growth in animals and plants is illustrated in Figure 4.1 (for color, see Plate 4.1, between pp. 84 and 85). Modular organisms may broadly be divided into those that concentrate on vertical growth, and those that spread their modules laterally, over or in a substrate. Many plants produce new root systems associated with a laterally extending stem: these are the rhizomatous and stoloniferous plants. The connections between the parts of such plants may die and rot away, so that the product of the original zygote becomes represented by physiologically separated parts. (Modules with the potential for separate existence are known as 'ramets'.) The most extreme examples of plants 'falling to pieces' as they grow are the many species of floating aquatics like duckweeds (*Lemna*) and the water hyacinth (*Eichhornia*). Whole ponds, lakes or rivers may be filled with the separate and independent parts produced by a single zygote.

Trees are the supreme example of plants whose growth is concentrated vertically. The peculiar feature distinguishing trees and shrubs from most herbs is the connecting system linking modules together and connecting them to the root system. This does not rot away, but thickens with wood, conferring perenniality. Most of the structure of such a woody tree is dead, with a thin layer of living material lying immediately below the bark. The living layer, however, continually regenerates new tissue, and adds further layers of dead material to the trunk of the tree, which solves, by the strength it provides, the difficult problem of obtaining water and nutrients below the ground, but also light perhaps 50 m away at the top of the canopy.

We can often recognize two or more levels of modular construction. The strawberry is a good example of this: leaves are repeatedly developed from a bud, but these leaves are arranged into rosettes. The strawberry plant grows: (i) by adding new leaves to a rosette; and (ii) by producing new rosettes on stolons grown from the axils of its rosette leaves. Trees also exhibit modularity at several levels: the leaf with its axillary bud, the whole shoot on which the leaves are arranged, and the whole branch systems that repeat a characteristic pattern of shoots.

modules within modules

Many animals, despite variations in their precise method of growth and reproduction, are as 'modular' as any plant. Moreover, in corals, for example, just like many plants, the individual may exist as a physiologically integrated whole, or may be split into a number of colonies – all part of one individual, but physiologically independent (Hughes *et al.*, 1992).

4.2.3   What is the size of a modular population?

In modular organisms, the number of surviving zygotes can give only a partial and misleading impression of the 'size' of the population. Kays and Harper (1974) coined the word 'genet' to describe the 'genetic individual': the product of a zygote. In modular organisms, then, the distribution and abundance of genets (individuals) is important, but it is often more useful to study the distribution and abundance of modules (ramets, shoots, tillers, zooids, polyps or whatever): the amount of grass in a field available to cattle is not determined by the number of genets but by the number of leaves (modules).

4.2.4   Senescence – or the lack of it – in modular organisms

There is also often no programed senescence of whole modular organisms – they appear to have perpetual somatic youth. Even in trees that accumulate their dead stem tissues, or gorgonian corals that accumulate old calcified branches, death often results from becoming too big or succumbing to disease rather than from programed senescence. This is illustrated for three types of coral in

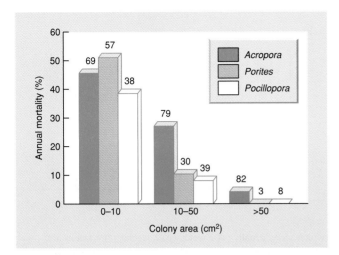

Figure 4.2   The mortality rate declines steadily with colony size (and hence, broadly, age) in three coral taxa from the reef crest at Heron Island, Great Barrier Reef (sample sizes are given above each bar). (After Hughes & Connell, 1987; Hughes *et al.*, 1992.)

the Great Barrier Reef in Figure 4.2. Annual mortality declined sharply with increasing colony size (and hence, broadly, age) until, amongst the largest, oldest colonies, mortality was virtually zero, with no evidence of any increase in mortality at extreme old age (Hughes & Connell, 1987).

At the modular level, things are quite different. The annual death of the leaves on a deciduous tree is the most dramatic

example of senescence – but roots, buds, flowers and the modules of modular animals all pass through phases of youth, middle age, senescence and death. The growth of the individual genet is the combined result of these processes. Figure 4.3 shows that the age structure of shoots of the sedge *Carex arenaria* is changed dramatically by the application of NPK fertilizer, even when the total number of shoots present is scarcely affected by the treatment. The fertilized plots became dominated by young shoots, as the older shoots that were common on control plots were forced into early death.

4.2.5   Integration

For many rhizomatous and stoloniferous species, this changing age structure is in turn associated with a changing level to which the connections between individual ramets remain intact. A young ramet may benefit from the nutrients flowing from an older ramet to which it is attached and from which it grew, but the pros and cons of attachment will have changed markedly by the time the daughter is fully established in its own right and the parent has entered a postreproductive phase of senescence (a comment equally applicable to unitary organisms with parental care) (Caraco & Kelly, 1991).

The changing benefits and costs of integration have been studied experimentally in the pasture grass *Holcus lanatus*, by comparing the growth of: (i) ramets that were left with a physiological connection to their parent plant, and in the same pot, so that parent and daughter might compete (unsevered,

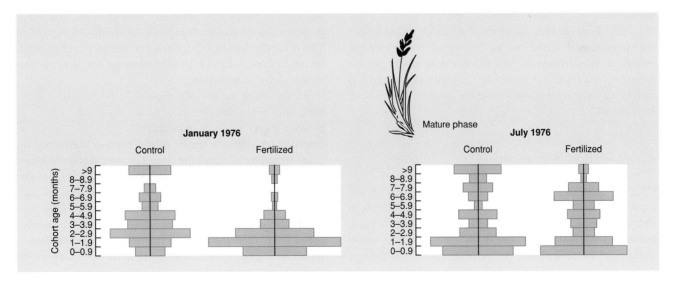

Figure 4.3   The age structure of shoots in clones of the sand sedge *Carex arenaria* growing on sand dunes in North Wales, UK. Clones are composed of shoots of different ages. The effect of applying fertilizer is to change this age structure. The clones become dominated by young shoots and the older shoots die. (After Noble *et al.*, 1979.)

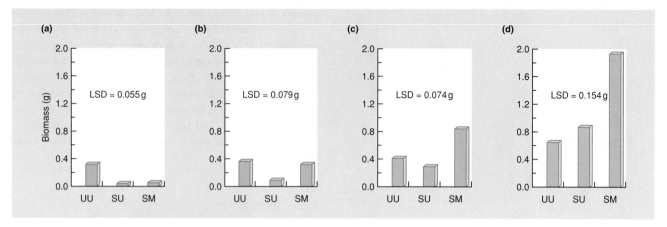

Figure 4.4   The growth of daughter ramets of the grass *Holcus lanatus*, which were initially (a) 1 week, (b) 2 weeks, (c) 4 weeks and (d) 8 weeks old, and were then grown on for a further 8 weeks. LSD, least significant difference, which needs to be exceeded for two means to be significantly different from each other. For further discussion, see text. (After Bullock *et al.*, 1994a.)

unmoved: UU); (ii) ramets that had their connection severed but were left in the same pot so competition was possible (severed, unmoved: SU); and (iii) ramets that had their connection severed and were repotted in their parent's soil, but after the parent had been removed, so no competition was possible (SM) (Figure 4.4). These treatments were applied to daughter ramets of various ages, which were then examined after a further 8 weeks' growth. For the youngest daughters (Figure 4.4a) attachment to the parent significantly enhanced growth (UU > SU), but competition with the parent had no apparent effect (SU ≈ SM). For slightly older daughters (Figure 4.4b), growth could be depressed by the parent (SU < SM), but physiological connection effectively negated this (UU > SU; UU ≈ SM). For even older daughters, the balance shifted further still: physiological connection to the parent was either not enough to fully overcome the adverse effects of the parent's presence (Figure 4.4c; SM > UU > SU) or eventually appeared to represent a drain on the resources of the daughter (Figure 4.4d; SM > SU > UU).

4.3 Counting individuals

If we are going to study birth, death and modular growth seriously, we must quantify them. This means counting individuals and (where appropriate) modules. Indeed, many studies concern themselves not with birth and death but with their consequences, i.e. the total number of individuals present and the way these numbers vary with time. Such studies can often be useful none the less. Even with unitary organisms, ecologists face enormous technical problems when they try to count what is happening to populations in nature. A great many ecological questions remain unanswered because of these problems.

It is usual to use the term *population* to describe a group of individuals of one species under investigation. What actually constitutes a population, though, will vary from species to species and from study to study. In some cases, the boundaries of a population are readily apparent: the sticklebacks occupying a small lake are the 'stickleback population of the lake'. In other cases, boundaries are determined more by an investigator's purpose or convenience: it is possible to study the population of lime aphids inhabiting one leaf, one tree, one stand of trees or a whole woodland. In yet other cases – and there are many of these – individuals are distributed continuously over a wide area, and an investigator must define the limits of a population arbitrarily. In such cases, especially, it is often more convenient to consider the *density* of a population. This is usually defined as 'numbers per unit area', but in certain circumstances 'numbers per leaf', 'numbers per host' or some other measure may be appropriate.

To determine the size of a population, one might imagine that it is possible simply to count individuals, especially for relatively small, isolated habitats like islands and relatively large individuals like deer. For most species, however, such 'complete enumerations' are impractical or impossible: observability – our ability to observe every individual present – is almost always less than 100%. Ecologists, therefore, must almost always *estimate* the number of individuals in a population rather than count them. They may estimate the numbers of aphids on a crop, for example, by counting the number on a representative sample of leaves, then estimating the number of leaves per square meter of ground, and from this estimating the number of aphids per square meter. For plants and animals living on the ground surface, the sample unit is generally a small area known as a quadrat (which is also the name given to the

what is a population?

determining population size

square or rectangular device used to demarcate the boundaries of the area on the ground). For soil-dwelling organisms the unit is usually a volume of soil; for lake dwellers a volume of water; for many herbivorous insects the unit is one typical plant or leaf, and so on. Further details of sampling methods, and of methods for counting individuals generally, can be found in one of many texts devoted to ecological methodology (e.g. Brower *et al.*, 1998; Krebs, 1999; Southwood & Henderson, 2000).

For animals, especially, there are two further methods of estimating population size. The first is known as capture–recapture. At its simplest, this involves catching a random sample of a population, marking individuals so that they can be recognized subsequently, releasing them so that they remix with the rest of the population and then catching a further random sample. Population size can be estimated from the proportion of this second sample that bear a mark. Roughly speaking, the proportion of marked animals in the second sample will be high when the population is relatively small, and low when the population is relatively large. Data sets become much more complex – and methods of analysis become both more complex and much more powerful – when there are a whole sequence of capture-recapture samples (see Schwarz & Seber, 1999, for a review).

The final method is to use an index of abundance. This can provide information on the relative size of a population, but by itself usually gives little indication of absolute size. As an example, Figure 4.5 shows the effect on the abundance of leopard frogs (*Rana*

pipiens) in ponds near Ottawa, Canada, of the number of occupied ponds and the amount of summer (terrestrial) habitat in the vicinity of the pond. Here, frog abundance was estimated from the 'calling rank': essentially compounded from whether there were no frogs, 'few', 'many' or 'very many' frogs calling on each of four occasions. Despite their shortcomings, even indices of abundance can provide valuable information.

Counting births can be more difficult even than counting individuals. The formation of the zygote is often   counting births
regarded as the starting point in the life of an individual. But it is a stage that is often hidden and extremely hard to study. We simply do not know, for most animals and plants, how many embryos die before 'birth', though in the rabbit at least 50% of embryos are thought to die in the womb, and in many higher plants it seems that about 50% of embryos abort before the seed is fully grown and mature. Hence, it is almost always impossible in practice to treat the formation of a zygote as the time of birth. In birds we may use the moment that an egg hatches; in mammals when an individual ceases to be supported within the mother on her placenta and starts to be supported outside her as a suckling; and in plants we may use the germination of a seed as the birth of a seedling, although it is really only the moment at which a developed embryo restarts into growth after a period of dormancy. We need to remember that half or more of a population will often have died before they can be recorded as born!

Counting deaths poses as many problems. Dead bodies do not linger long in nature. Only the skeletons of   counting deaths
large animals persist long after death. Seedlings may be counted and mapped one day and gone without trace the next. Mice, voles and soft-bodied animals such as caterpillars and worms are digested by predators or rapidly removed by scavengers or decomposers. They leave no carcasses to be counted and no evidence of the cause of death. Capture–recapture methods can go a long way towards estimating deaths from the loss of marked individuals from a population (they are probably used as often to measure survival as abundance), but even here it is often impossible to distinguish loss through death and loss through emigration.

4.4 Life cycles

To understand the forces determining the abundance of a population, we need to know the phases of the constituent organisms' lives when these forces act most significantly. For this, we need to understand the sequences of events that occur in those organisms' life cycles. A highly simplified, generalized life history (Figure 4.6a) comprises birth, followed by a prereproductive period, a period of reproduction, perhaps a postreproductive period, and then death as a result of senescence (though of course other forms of mortality may intervene at any time). The variety of life cycles is also

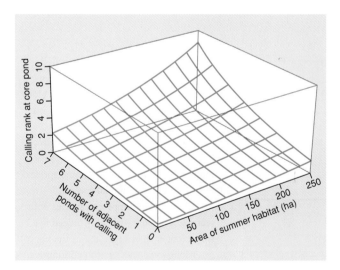

Figure 4.5 The abundance (calling rank) of leopard frogs in ponds increases significantly with both the number of adjacent ponds that are occupied and the area of summer habitat within 1 km of the pond. Calling rank is the sum of an index measured on four occasions, namely: 0, no individuals calling; 1, individuals can be counted, calls not overlapping; 2, calls of < 15 individuals can be distinguished with some overlapping; 3, calls of ≥ 15 individuals. (After Pope *et al.*, 2000.)

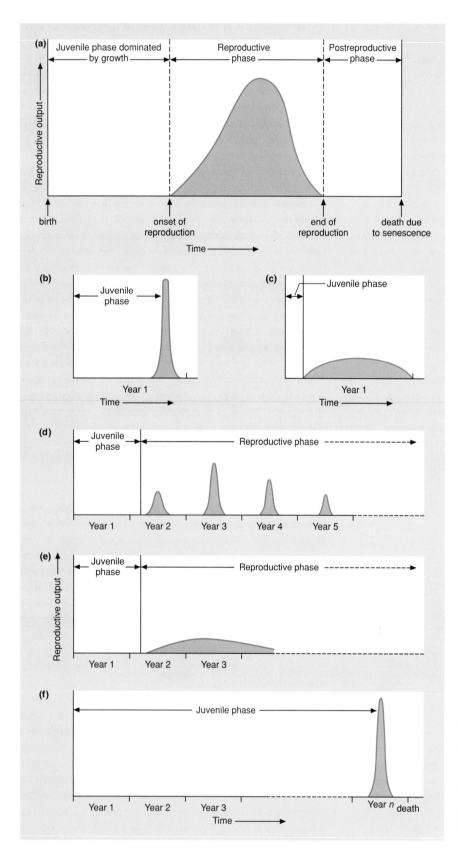

Figure 4.6 (a) An outline life history for a unitary organism. Time passes along the horizontal axis, which is divided into different phases. Reproductive output is plotted on the vertical axis. The figures below (b–f) are variations on this basic theme. (b) A semelparous annual species. (c) An iteroparous annual species. (d) A long-lived iteroparous species with seasonal breeding (that may indeed live much longer than suggested in the figure). (e) A long-lived species with continuous breeding (that may again live much longer than suggested in the figure). (f) A semelparous species living longer than a year. The pre-reproductive phase may be a little over 1 year (a biennial species, breeding in its second year) or longer, often much longer, than this (as shown).

summarized diagrammatically in Figure 4.6, although there are many life cycles that defy this simple classification. Some organisms fit several or many generations within a single year, some have just one generation each year (annuals), and others have a life cycle extended over several or many years. For all organisms, though, a period of growth occurs before there is any reproduction, and growth usually slows down (and in some cases stops altogether) when reproduction starts.

Whatever the length of their life cycle, species may, broadly, be either *semelparous* or *iteroparous* (often referred to by plant scientists as monocarpic and polycarpic). In semelparous species, individuals have only a single, distinct period of reproductive output

semelparous and iteroparous life cycles

in their lives, prior to which they have largely ceased to grow, during which they invest little or nothing in survival to future reproductive events, and after which they die. In iteroparous species, an individual normally experiences several or many such reproductive events, which may in fact merge into a single extended period of reproductive activity. During each period of reproductive activity the individual continues to invest in future survival and possibly growth, and beyond each it therefore has a reasonable chance of surviving to reproduce again.

For example, many annual plants are semelparous (Figure 4.6b): they have a sudden burst of flowering and seed set, and then they die. This is commonly the case among the weeds of arable crops. Others, such as groundsel (*Senecio vulgaris*), are iteroparous (Figure 4.6c): they continue to grow and produce new flowers and seeds through the season until they are killed by the first lethal frost of winter. They die with their buds on.

There is also a marked seasonal rhythm in the lives of many long-lived iteroparous plants and animals, especially in their reproductive activity: a period of reproduction once per year (Figure 4.6d). Mating (or the flowering of plants) is commonly triggered by the length of the photoperiod (see Section 2.3.7) and usually makes sure that young are born, eggs hatch or seeds are ripened when seasonal resources are likely to be abundant. Here, though, unlike annual species, the generations overlap and individuals of a range of ages breed side by side. The population is maintained in part by survival of adults and in part by new births.

In wet equatorial regions, on the other hand, where there is very little seasonal variation in temperature and rainfall and scarcely any variation in photoperiod, we find species of plants that are in flower and fruit throughout the year – and continuously breeding species of animal that subsist on this resource (Figure 4.6e). There are several species of fig (*Ficus*), for instance, that bear fruit continuously and form a reliable year-round food supply for birds and primates. In more seasonal climates, humans are unusual in also breeding continuously throughout the year, though numbers of other species, cockroaches, for example, do so in the stable environments that humans have created.

Amongst long-lived (i.e. longer than annual) semelparous plants (Figure 4.6f), some are strictly biennial

the variety of life cycles

– each individual takes two summers and the intervening winter to develop, but has only a single reproductive phase, in its second summer. An example is the white sweet clover, *Melilotus alba*. In New York State, this has relatively high mortality during the first growing season (whilst seedlings were developing into established plants), followed by much lower mortality until the end of the second summer, when the plants flowered and survivorship decreased rapidly. No plants survive to a third summer. Thus, there is an overlap of two generations at most (Klemow & Raynal, 1981). A more typical example of a semelparous species with overlapping generations is the composite *Grindelia lanceolata*, which may flower in its third, fourth or fifth years. But whenever an individual does flower, it dies soon after.

A well-known example of a semelparous animal with overlapping generations (Figure 4.6f) is the Pacific salmon *Oncorhynchus nerka*. Salmon are spawned in rivers. They spend the first phase of their juvenile life in fresh water and then migrate to the sea, often traveling thousands of miles. At maturity they return to the stream in which they were hatched. Some mature and return to reproduce after only 2 years at sea; others mature more slowly and return after 3, 4 or 5 years. At the time of reproduction the population of salmon is composed of overlapping generations of individuals. But all are semelparous: they lay their eggs and then die; their bout of reproduction is terminal.

There are even more dramatic examples of species that have a long life but reproduce just once. Many species of bamboo form dense clones of shoots that remain vegetative for many years: up to 100 years in some species. The whole population of shoots, from the same and sometimes different clones, then flowers simultaneously in a mass suicidal orgy. Even when shoots have become physically separated from each other, the parts still flower synchronously.

In the following sections we look at the patterns of birth and death in some of these life cycles in more detail, and at how these patterns are quantified. Often, in order to monitor and examine changing patterns of mortality with age or stage, a *life table* is used. This allows a *survivorship curve* to be constructed, which traces the decline in numbers, over time, of a group of newly born or newly emerged individuals or modules – or it can be thought of as a plot of the probability, for a representative newly born individual, of surviving to various ages. Patterns of birth amongst individuals of different ages are often monitored at the same time as life tables are constructed. These patterns are displayed in *fecundity schedules*.

4.5   Annual species

Annual life cycles take approximately 12 months or rather less to complete (Figure 4.6b, c). Usually, every individual in a population breeds during one particular season of the year, but then dies before the same season in the next year. Generations are therefore said to be discrete, in that each generation is distinguishable from every other; the only overlap of generations is between breeding adults and their offspring during and immediately after the breeding season. Species with discrete generations need not be annual, since generation lengths other than 1 year are conceivable. In practice, however, most are: the regular annual cycle of seasonal climates provides the major pressure in favor of synchrony.

4.5.1   Simple annuals: cohort life tables

A life table and fecundity schedule are set out in Table 4.1 for the annual plant *Phlox drummondii* in Nixon, Texas (Leverich & Levin, 1979). The life table is known as a cohort life table, because a single cohort of individuals (i.e. a group of individuals born within the same short interval of time) was followed from birth to the death of the last survivor. With an annual species like *Phlox*, there is no other way of constructing a life table. The life cycle of *Phlox* was divided into a number of age classes. In other cases, it is more appropriate to divide it into stages (e.g. insects with eggs, larvae, pupae, etc.) or into size classes. The number

in the *Phlox* population was recorded on various occasions before germination (i.e. when the plants were seeds), and then again at regular intervals until all individuals had flowered and died. The advantage of using age classes is that it allows an observer to look in detail at the patterns of birth and mortality *within* stages (e.g. the seedling stage). The disadvantage is an individual's age is not necessarily the best, nor even a satisfactory, measure of its biological 'status'. In many long-lived plants, for instance, individuals of the same age may be reproducing actively, or growing vegetatively but not reproducing, or doing neither. In such cases, a classification based on developmental stages (as opposed to ages) is clearly appropriate. The decision to use age classes in *Phlox* was based on the small number of stages, the demographic variation within each and the synchronous development of the whole population.

The first column of Table 4.1 sets out the various classes (in this case, age classes). The second column, a_x, then lists the major part of the raw data: it gives the total number of individuals surviving to the start of each class (a_0 individuals in the initial class, a_{63} in the following one (which started on day 63), and so on). The problem with any a_x column is that its information is specific to one population in 1 year, making comparisons with other populations and other years very difficult. The data have therefore been standardized, next, in a column of l_x values. This is headed by an l_0 value of 1.000, and all succeeding figures have been brought into line accordingly (e.g. $l_{124} = 1.000 \times 295/$

the columns of a life table

Table 4.1   A cohort life table for *Phlox drummondii*. The columns are explained in the text. (After Leverich & Levin, 1979.)

Age interval (days) $x - x'$	Number surviving to day x a_x	Proportion of original cohort surviving to day x l_x	Proportion of original cohort dying during interval d_x	Mortality rate per day q_x	$Log_{10} l_x$	Daily killing power k_x	F_x	m_x	$l_x m_x$
0–63	996	1.000	0.329	0.006	0.00	0.003	–	–	–
63–124	668	0.671	0.375	0.013	−0.17	0.006	–	–	–
124–184	295	0.296	0.105	0.007	−0.53	0.003	–	–	–
184–215	190	0.191	0.014	0.003	−0.72	0.001	–	–	–
215–264	176	0.177	0.004	0.002	−0.75	0.001	–	–	–
264–278	172	0.173	0.005	0.002	−0.76	0.001	–	–	–
278–292	167	0.168	0.008	0.004	−0.78	0.002	–	–	–
292–306	159	0.160	0.005	0.002	−0.80	0.001	53.0	0.33	0.05
306–320	154	0.155	0.007	0.003	−0.81	0.001	485.0	3.13	0.49
320–334	147	0.148	0.043	0.025	−0.83	0.011	802.7	5.42	0.80
334–348	105	0.105	0.083	0.106	−0.98	0.049	972.7	9.26	0.97
348–362	22	0.022	0.022	1.000	−1.66	–	94.8	4.31	0.10
362–	0	0.000	–	–	–	–	–	–	–
							2408.2		2.41

$$R_0 = \sum l_x m_x = \frac{\sum F_x}{a_0} = 2.41.$$

996 = 0.296). Thus, whilst the a_0 value of 996 is peculiar to this set of data, all studies have an l_0 value of 1.000, making all studies comparable. The l_x values are best thought of as the proportion of the original cohort surviving to the start of a stage or age class.

To consider mortality more explicitly, the proportion of the original cohort dying during each stage (d_x) is computed in the next column, being simply the difference between successive values of l_x; for example $d_{124} = 0.296 - 0.191 = 0.105$. The stage-specific mortality *rate*, q_x, is then computed. This considers d_x as a fraction of l_x. Furthermore, the variable length of the age classes makes it sensible to convert the q_x values to 'daily' rates. Thus, for instance, the fraction dying between days 124 and 184 is $0.105/0.296 = 0.355$, which translates, on the basis of *compound* 'interest', into a daily rate or fraction, q_{124}, of 0.007. q_x may also be thought of as the average 'chance' or probability of an individual dying during an interval. It is therefore equivalent to $(1 - p_x)$ where p refers to the probability of survival.

The advantage of the d_x values is that they can be summed: thus, the proportion of the cohort dying in the first 292 days (essentially the prereproductive stage) was $d_0 + d_{63} + d_{124} \ldots + d_{278}$ (= 0.840). The disadvantage is that the individual values give no real idea of the intensity or importance of mortality during a particular stage. This is because the d_x values are larger the more individuals there are, and hence the more there are available to die. The q_x values, on the other hand, are an excellent measure of the intensity of mortality. For instance, in the present example it is clear from the q_x column that the mortality rate increased markedly in the second period; this is not clear from the d_x column. The q_x values, however, have the disadvantage that, for example, summing the values over the first 292 days gives no idea of the mortality rate over that period.

The advantages are combined, however, in the next column of the life table, which contains k_x values (Haldane, 1949; Varley & Gradwell, 1970). k_x is defined simply as the difference between successive values of $\log_{10}a_x$ *or* successive values of $\log_{10}l_x$ (they amount to the same thing), and is sometimes referred to as a 'killing power'. Like q_x values, k_x values reflect the intensity or rate of mortality (as Table 4.1 shows); but unlike summing the q_x values, summing k_x values is a legitimate procedure. Thus, the killing power or k value for the final 28 days is $(0.011 \times 14) + (0.049 \times 14) = 0.84$, which is also the difference between -0.83 and -1.66 (allowing for rounding errors). Note too that like l_x values, k_x values are standardized, and are therefore appropriate for comparing quite separate studies. In this and later chapters, k_x values will be used repeatedly.

k values

4.5.2 Fecundity schedules and basic reproductive rates

The fecundity schedule in Table 4.1 (the final three columns) begins with a column of raw data, F_x: the total number of seeds produced during each period. This is followed in the next column by m_x: the individual fecundity or birth rate, i.e. the mean number of seeds produced per surviving individual. Although the reproductive season for the *Phlox* population lasts for 56 days, each individual plant is semelparous. It has a single reproductive phase during which all of its seeds develop synchronously (or nearly so). The extended reproductive season occurs because different individuals enter this phase at different times.

Perhaps the most important summary term that can be extracted from a life table and fecundity schedule is the basic reproductive rate, denoted by R_0. This is the mean number of offspring (of the first stage in the life cycle – in this case seeds) produced per original individual by the end of the cohort. It therefore indicates, in annual species, the overall extent by which the population has increased or decreased over that time. (As we shall see below, the situation becomes more complicated when generations overlap or species breed continuously.)

There are two ways in which R_0 can be computed. The first is from the formula:

the basic reproductive rate, R_0

$$R_0 = \sum F_x/a_0, \qquad (4.2)$$

i.e. the total number of seeds produced during one generation divided by the original number of seeds ($\sum F_x$ means the sum of the values in the F_x column). The more usual way of calculating R_0, however, is from the formula:

$$R_0 = \sum l_x m_x, \qquad (4.3)$$

i.e. the sum of the number of seeds produced per original individual during each of the stages (the final column of the fecundity schedule). As Table 4.1 shows, the basic reproductive rate is the same, whichever formula is used.

The age-specific fecundity, m_x (the fecundity per surviving individual), demonstrates the existence of a preproductive period, a gradual rise to a peak and then a rapid decline. The reproductive output of the whole population, F_x, parallels this pattern to a large extent, but also takes into account the fact that whilst the age-specific fecundity was changing, the size of the population was gradually declining. This combination of fecundity and survivorship is an important property of F_x values, shared by the basic reproductive rate (R_0). It makes the point that actual reproduction depends both on reproductive potential (m_x) and on survivorship (l_x).

In the case of the *Phlox* population, R_0 was 2.41. This means that there was a 2.41-fold increase in the size of the population over one generation. If such a value were maintained from generation to generation, the *Phlox* population would grow ever larger and soon cover the globe. Thus, a balanced and realistic picture of the life and death of *Phlox*, or any other species, can only emerge from several or many years' data.

4.5.3 Survivorship curves

The pattern of mortality in the *Phlox* population is illustrated in Figure 4.7a using both q_x and k_x values. The mortality rate was fairly high at the beginning of the seed stage but became very low towards the end. Then, amongst the adults, there was a period where the mortality rate fluctuated about a moderate level, followed finally by a sharp increase to very high levels during the last weeks of the generation. The same pattern is shown in a different form in Figure 4.7b. This is a survivorship curve, and follows the decline of $\log_{10}l_x$ with age. When the mortality rate is roughly constant, the survivorship curve is more or less straight; when the rate increases, the curve is convex; and when the rate decreases, the curve is concave. Thus, the curve is concave towards the end of the seed stage, and convex towards the end of the generation. Survivorship curves are the most widely used way of depicting patterns of mortality.

the logarithmic scale in survivorship curves

The *y*-axis in Figure 4.7b is logarithmic. The importance of using logarithms in survivorship curves can be seen by imagining two investigations of the same population. In the first, the whole population is censused: there is a decline in one time interval from 1000 to 500 individuals. In the second, samples are taken, and over the same time interval this index of density declines from 100 to 50. The two cases are biologically identical, i.e. the rate or probability of death per individual over the time interval (the *per capita* rate) is the same. The slopes of the two logarithmic survivorship curves reflect this: both would be −0.301. But on simple linear scales the slopes would differ. Logarithmic survivorship curves therefore have the advantage of being standardized from study to study, just like the 'rates' q_x, k_x and m_x. Plotting numbers on a logarithmic scale will also indicate when per capita rates of increase are identical. 'Log numbers' will therefore often be used in preference to 'numbers' when numerical change is being plotted.

4.5.4 A classification of survivorship curves

Life tables provide a great deal of data on specific organisms. But ecologists search for generalities: patterns of life and death that we can see repeated in the lives of many species. A useful set of survivorship curves was developed long ago by Pearl (1928) whose three types generalize what we know about the way in which the risks of death are distributed through the lives of different organisms (Figure 4.8). Type I describes the situation in which mortality is concentrated toward the end of the maximum lifespan. It is perhaps most typical of humans in developed countries and their carefully tended zoo animals and pets. Type II is a straight line that describes a constant mortality rate from birth to maximum age. It describes, for instance, the survival of seeds buried in the soil. Type III indicates extensive early mortality, but a high rate of subsequent survival. This is typical of species that produce many offspring. Few survive initially, but once individuals reach a critical size, their risk of death remains low and more or less constant. This appears to be the most common survivorship curve among animals and plants in nature.

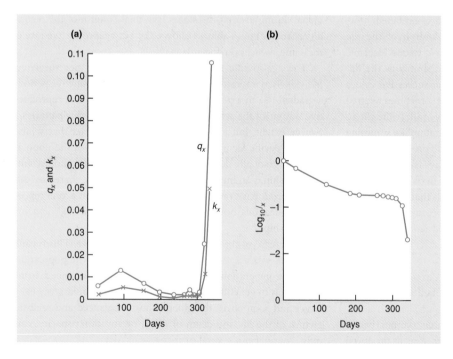

Figure 4.7 Mortality and survivorship in the life cycle of *Phlox drummondii*. (a) The age-specific daily mortality rate (q_x) and daily killing power (k_x). (b) The survivorship curve: $\log_{10} l_x$ plotted against age. (After Leverich & Levin, 1979.)

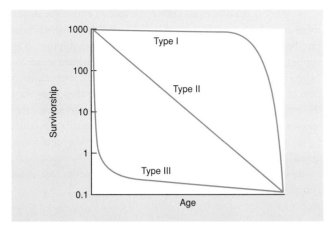

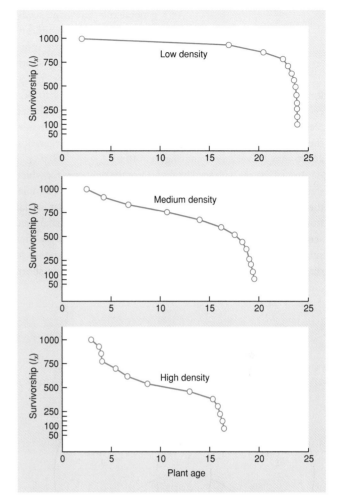

Figure 4.8 A classification of survivorship curves. Type I (convex) – epitomized perhaps by humans in rich countries, cosseted animals in a zoo or leaves on a plant – describes the situation in which mortality is concentrated at the end of the maximum lifespan. Type II (straight) indicates that the probability of death remains constant with age, and may well apply to the buried seed banks of many plant populations. Type III (concave) indicates extensive early mortality, with those that remain having a high rate of survival subsequently. This is true, for example, of many marine fish, which produce millions of eggs of which very few survive to become adults. (After Pearl, 1928; Deevey, 1947.)

Figure 4.9 Survivorship curves (l_x, where $l_0 = 1000$) for the sand-dune annual plant *Erophila verna* monitored at three densities: high (initially 55 or more seedlings per 0.01 m^2 plot); medium (15–30 seedlings per plot); and low (1–2 seedlings per plot). The horizontal scale (plant age) is standardized to take account of the fact that each curve is the average of several cohorts, which lasted different lengths of time (around 70 days on average). (After Symonides, 1983.)

These types of survivorship curve are useful generalizations, but in practice, patterns of survival are usually more complex. Thus, in a population of *Erophila verna*, a very short-lived annual plant inhabiting sand dunes, survival can follow a type I curve when the plants grow at low densities; a type II curve, at least until the end of the lifespan, at medium densities; and a type III curve in the early stages of life at the highest densities (Figure 4.9).

4.5.5 Seed banks, ephemerals and other not-quite-annuals

Using *Phlox* as an example of an annual plant has, to a certain extent, been misleading, because the group of seedlings developing in 1 year is a true cohort: it derives entirely from seed set by adults in the previous year. Seeds that do not germinate in 1 year will not survive till the next. In most 'annual' plants this is not the case. Instead, seeds accumulate in the soil in a buried *seed bank*. At any one time, therefore, seeds of a variety of ages are likely to occur together in the seed bank, and when they germinate the seedlings will also be of varying ages (age being the length of time since the seed was first produced). The formation of something comparable to a seed bank is rarer amongst animals, but there are

examples to be seen amongst the eggs of nematodes, mosquitoes and fairy shrimps, the gemmules of sponges and the statocysts of bryozoans.

Note that species commonly referred to as 'annual', but with a seed bank (or animal equivalent), are not strictly annual species at all, even if they progress from germination to reproduction within 1 year, since some of the seeds destined to germinate each year will already be more than 12 months old. All we can do, though, is bear this fact in mind, and note that it is just one example of real organisms spoiling our attempts to fit them neatly into clear-cut categories.

the species
composition of seed
banks

As a general rule, dormant seeds, which enter and make a significant contribution to seed banks, are more common in annuals and other short-lived plant species than they are in longer lived species, such that short-lived species tend to predominate in buried seed banks, even when most of the established plants above them belong to much longer lived species. Certainly, the species composition of seed banks and the mature vegetation above may be very different (Figure 4.10).

Annual species with seed banks are not the only ones for which the term annual is, strictly speaking, inappropriate. For example, there are many annual plant species living in deserts that are far from seasonal in their appearance. They have a substantial buried seed bank, with germination occurring on rare occasions after substantial rainfall. Subsequent development is usually rapid, so that the period from germination to seed production is short. Such plants are best described as semelparous *ephemerals*.

A simple annual label also fails to fit species where the majority of individuals in each generation are annual, but where a small number postpone reproduction until their second summer. This applies, for example, to the terrestrial isopod *Philoscia muscorum* living in northeast England (Sunderland *et al.*, 1976). Approximately 90% of females bred only in the first summer after they were born; the other 10% bred only in their second summer. In some other species, the difference in numbers between those that reproduce in their first or second years is so slight that the description *annual–biennial* is most appropriate.

In short, it is clear that annual life cycles merge into more complex ones without any sharp discontinuity.

4.6 Individuals with repeated breeding seasons

Many species breed repeatedly (assuming they survive long enough), but nevertheless have a specific breeding season. Thus, they have overlapping generations (see Figure 4.6d). Amongst the more obvious examples are temperate-region birds living for more than 1 year, some corals, most trees and other iteroparous perennial plants. In these, individuals of a range of ages breed side by side. None the less, some species in this category, some grasses for example, and many birds, live for relatively short periods.

4.6.1 Cohort life tables

Constructing a cohort life table for species that breed repeatedly is more difficult than constructing one for an annual species. A cohort must be recognized and followed (often for many years), even though the organisms within it are coexisting and intermingling with organisms from many other cohorts, older and younger. This was possible, though, as part of an extensive study of red deer (*Cervus elaphus*) on the small island of Rhum, Scotland (Lowe, 1969). The deer live for up to 16 years, and the females (hinds) are capable of breeding each year from their fourth summer onwards. In 1957, Lowe and his coworkers made a very careful count of the total number of deer on the island, including the total number of calves (less than 1 year old). Lowe's cohort consisted of the deer that were calves in 1957. Thus, each year from 1957 to 1966, every one of the deer that was discovered that had died from natural causes, or had been shot under the rigorously controlled conditions of this Nature Conservancy Council reserve, was examined and aged reliably by examining tooth replacement, eruption and wear. It was therefore possible to identify those dead deer that had been calves in 1957; and by 1966, 92% of this cohort had been observed dead and their age at death therefore determined. The life table for this cohort of hinds (or the 92% sample of it) is presented in Table 4.2; the survivorship curve is shown in Figure 4.11. There appears to be a fairly consistent increase in the risk of mortality with age (the curve is convex).

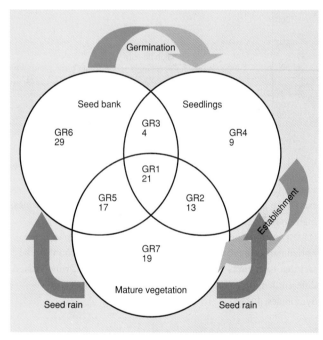

Figure 4.10 Species recovered from the seed bank, from seedlings and from mature vegetation in a coastal grassland site on the western coast of Finland. Seven species groups (GR1–GR7) are defined on the basis of whether they were found in only one, two, or all three stages. GR3 (seed bank and seedlings only) is an unreliable group of species that are mostly incompletely identified; in GR5 there are many species difficult to identify as seedlings that may more properly belong to GR1. None the less, the marked difference in composition, especially between the seed bank and the mature vegetation, is readily apparent. (After Jutila, 2003.)

Table 4.2 Cohort life table for red deer hinds on the island of Rhum that were calves in 1957. (After Lowe, 1969.)

Age (years) x	Proportion of original cohort surviving to the beginning of age-class x l_x	Proportion of original cohort dying during age-class x d_x	Mortality rate q_x
1	1.000	0	0
2	1.000	0.061	0.061
3	0.939	0.185	0.197
4	0.754	0.249	0.330
5	0.505	0.200	0.396
6	0.305	0.119	0.390
7	0.186	0.054	0.290
8	0.132	0.107	0.810
9	0.025	0.025	1.000

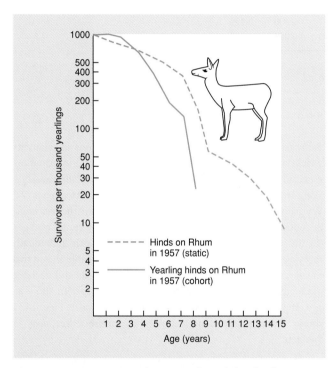

Figure 4.11 Two survivorship curves for red deer hinds on the island of Rhum. As explained in the text, one is based on the cohort life table for the 1957 calves and therefore applies to the post-1957 period; the other is based on the static life table of the 1957 population and therefore applies to the pre-1957 period. (After Lowe, 1969.)

4.6.2 Static life tables

The difficulties of constructing a cohort life table for an organism with overlapping generations are eased somewhat when the organism is sessile. In such a case, newly arrived or newly emerged individuals can be mapped, photographed or even marked in some way, so that they (or their exact location) can be recognized whenever the site is revisited subsequently. Taken overall, however, practical problems have tended to deter ecologists from constructing cohort life tables for long-lived iteroparous organisms with overlapping generations, even when the individuals are sessile. But there is an alternative: the construction of a static life table. As will become clear, this alternative is seriously flawed – but it is often better than nothing at all.

An interesting example emerges from Lowe's study of red deer on Rhum. As has already been explained, a large proportion of the deer that died from 1957 to 1966 could be aged reliably. Thus, if, for example, a fresh corpse was examined in 1961 and was found to be 6 years old, it was known that in 1957 the deer was alive and 2 years old. Lowe was therefore eventually able to reconstruct the age structure of the 1957 population: age structures are the basis for static life tables. Of course, the age structure of the 1957 population could have been ascertained by shooting and examining large numbers of deer in 1957; but since the ultimate aim of the project was the enlightened conservation of the deer, this method would have been somewhat inappropriate. (Note that Lowe's results did not represent the total numbers alive in 1957, because a few carcasses must have decomposed or been eaten before they could be discovered and examined.) Lowe's raw data for red deer hinds are presented in column 2 of Table 4.3.

Remember that the data in Table 4.3 refer to ages in 1957. They can be used as a basis for a life table, but only if it is assumed that there had been no year-to-year variation prior to 1957 in either the total number of births or the age-specific survival rates. In other words, it must be assumed that the 59 6-year-old deer alive in 1957 were the survivors of 78 5-year-old deer alive in 1956, who were themselves the survivors of 81 4-year olds in 1955, and so on. Or, in short, that the data in Table 4.3 are the same as would have been obtained if a single cohort *had* been followed.

Age (years) x	Number of individuals observed of age x a_x	l_x	d_x	q_x	Smoothed l_x	d_x	q_x
1	129	1.000	0.116	0.116	1.000	0.137	0.137
2	114	0.884	0.008	0.009	0.863	0.085	0.097
3	113	0.876	0.251	0.287	0.778	0.084	0.108
4	81	0.625	0.020	0.032	0.694	0.084	0.121
5	78	0.605	0.148	0.245	0.610	0.084	0.137
6	59	0.457	0.047	–	0.526	0.084	0.159
7	65	0.504	0.078	0.155	0.442	0.085	0.190
8	55	0.426	0.232	0.545	0.357	0.176	0.502
9	25	0.194	0.124	0.639	0.181	0.122	0.672
10	9	0.070	0.008	0.114	0.059	0.008	0.141
11	8	0.062	0.008	0.129	0.051	0.009	0.165
12	7	0.054	0.038	0.704	0.042	0.008	0.198
13	2	0.016	0.008	0.500	0.034	0.009	0.247
14	1	0.080	−0.023	–	0.025	0.008	0.329
15	4	0.031	0.015	0.484	0.017	0.008	0.492
16	2	0.016	–	–	0.009	0.009	1.000

Table 4.3 A static life table for red deer hinds on the island of Rhum, based on the reconstructed age structure of the population in 1957. (After Lowe, 1969.)

static life tables: flawed but sometimes useful, none the less

Having made these assumptions, the l_x, d_x and q_x columns were constructed. It is clear, however, that the assumptions are false. There were actually more animals in their seventh year than in their sixth year, and more in their 15th year than in their 14th year. There were therefore 'negative' deaths and meaningless mortality rates. The pitfalls of constructing such static life tables (and equating age structures with survivorship curves) are amply illustrated.

Nevertheless, the data can be useful. Lowe's aim was to provide a *general* idea of the population's age-specific survival rate prior to 1957 (when culling of the population began). He could then compare this with the situation after 1957, as illustrated by the cohort life table previously discussed. He was more concerned with general trends than with the particular changes occurring from 1 year to the next. He therefore 'smoothed out' the variations in numbers between ages 2–8 and 10–16 years to give a steady decline during both of these periods. The results of this process are shown in the final three columns of Table 4.3, and the survivorship curve is plotted in Figure 4.11. A general picture does indeed emerge: the introduction of culling on the island appears to have decreased overall survivorship significantly, overcoming any possible compensatory decreases in natural mortality.

Notwithstanding this successful use of a static life table, the interpretation of static life tables generally, and the age structures from which they stem, is fraught with difficulty: usually, age structures offer no easy short cuts to understanding the dynamics of populations.

4.6.3 Fecundity schedules

Static fecundity schedules, i.e. age-specific variations in fecundity within a particular season, can also provide useful information, especially if they are available from successive breeding seasons. We can see this for a population of great tits (*Parus major*) in Wytham Wood, near Oxford, UK (Table 4.4), where the data could be obtained only because the individual birds could be aged (in this case, because they had been marked with individually recognizable leg-rings soon after hatching). The table shows that mean fecundity rose to a peak in 2-year-old birds and declined gradually thereafter. Indeed, most iteroparous species show an age- or stage-related pattern of fecundity. For instance, Figure 4.12 shows the age-dependent fecundity of moose (*Alces alces*) in Sweden.

4.6.4 The importance of modularity

The sedge *Carex bigelowii*, growing in a lichen heath in Norway, illustrates the difficulties of constructing any sort of life table for organisms that are not only iteroparous with overlapping generations but are also modular (Figure 4.13). *Carex bigelowii* has an extensive underground rhizome system that produces tillers (aerial shoots) at intervals along its length as it grows. It grows by producing a lateral meristem in the axil of a leaf belonging to a 'parent' tiller. This lateral is completely dependent on the parent tiller at first, but is potentially capable of developing into a vegetative parent tiller itself, and also of flowering, which it does

Table 4.4 Mean clutch size and age of great tits in Wytham Wood, near Oxford, UK. (After Perrins, 1965.)

Age (years)	1961		1962		1963	
	Number of birds	Mean clutch size	Number of birds	Mean clutch size	Number of birds	Mean clutch size
Yearlings	128	7.7	54	8.5	54	9.4
2	18	8.5	43	9.0	33	10.0
3	14	8.3	12	8.8	29	9.7
4			5	8.2	9	9.7
5			1	8.0	2	9.5
6					1	9.0

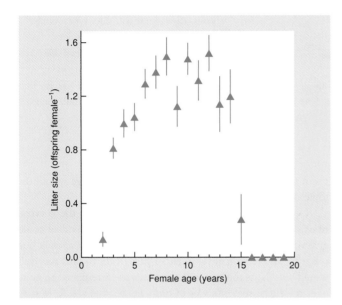

Figure 4.12 Age-dependent reproduction (average litter size) in a population of moose (*Alces alces*) in Sweden (means with standard errors). (After Ericsson *et al.*, 2001.)

when it has produced a total of 16 or more leaves. Flowering, however, is always followed by tiller death, i.e. the tillers are semelparous although the genets are iteroparous.

Callaghan (1976) took a number of well-separated young tillers, and excavated their rhizome systems through progressively older generations of parent tillers. This was made possible by the persistence of dead tillers. He excavated 23 such systems containing a total of 360 tillers, and was able to construct a type of static life table (and fecundity schedule) based on the growth stages (Figure 4.13). There were, for example, 1.04 dead vegetative tillers (per m²) with 31–35 leaves. Thus, since there were also 0.26 tillers in the next (36–40 leaves) stage, it can be assumed that a total of 1.30 (i.e. 1.04 + 0.26) living vegetative tillers entered the 31–35 leaf stage. As there were 1.30 vegetative tillers and 1.56 flowering tillers in the 31–35 leaf stage, 2.86 tillers must have survived from the 26–30 stage. It is in this way that the

life table – applicable not to individual genets but to tillers (i.e. modules) – was constructed.

There appeared to be no new establishment from seed in this particular population (no new genets); tiller numbers were being maintained by modular growth alone. However, a 'modular growth schedule' (*laterals*), analogous to a fecundity schedule, has been constructed.

Note finally that stages rather than age classes have been used here – something that is almost always necessary when dealing with modular iteroparous organisms, because variability stemming from modular growth accumulates year upon year, making age a particularly poor measure of an individual's chances of death, reproduction or further modular growth.

4.7 Reproductive rates, generation lengths and rates of increase

4.7.1 Relationships between the variables

In the previous section we saw that the life tables and fecundity schedules drawn up for species with overlapping generations are at least superficially similar to those constructed for species with discrete generations. With discrete generations, we were able to compute the basic reproductive rate (R_0) as a summary term describing the overall outcome of the patterns of survivorship and fecundity. Can a comparable summary term be computed when generations overlap?

Note immediately that previously, for species with discrete generations, R_0 described two separate population parameters. It was the number of offspring produced on average by an individual over the course of its life; but it was also the multiplication factor that converted an original population size into a new population size, one generation hence. With overlapping generations, when a cohort life table is available, the basic reproductive rate can be calculated using the same formula:

$$R_0 = \sum l_x m_x, \tag{4.4}$$

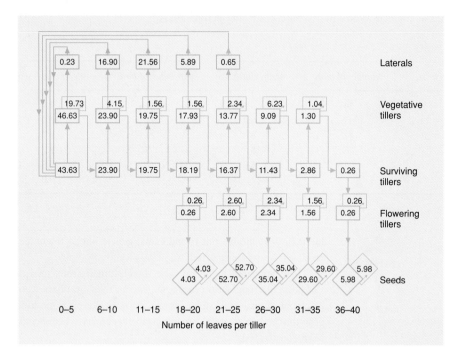

Figure 4.13 A reconstructed static life table for the modules (tillers) of a *Carex bigelowii* population. The densities per m² of tillers are shown in rectangular boxes, and those of seeds in diamond-shaped boxes. Rows represent tiller types, whilst columns depict size classes of tillers. Thin-walled boxes represent dead tiller (or seed) compartments, and arrows denote pathways between size classes, death or reproduction. (After Callaghan, 1976.)

and it still refers to the average number of offspring produced by an individual. But further manipulations of the data are necessary before we can talk about the rate at which a population increases or decreases in size – or, for that matter, about the length of a generation. The difficulties are much greater still when only a static life table (i.e. an age structure) is available (see below).

We begin by deriving a general relationship that links population size, the rate of population increase, and time – but which is not limited to measuring time in terms of generations. Imagine a population that starts with 10 individuals, and which, after successive intervals of time, rises to 20, 40, 80, 160 individuals and so on. We refer to the initial population size as N_0 (meaning the population size when no time has elapsed). The population size after one time interval is N_1, after two time intervals it is N_2, and in general after t time intervals it is N_t. In the present case, $N_0 = 10$, $N_1 = 20$, and we can say that:

$$N_1 = N_0R, \tag{4.5}$$

the fundamental net reproductive rate, R

where R, which is 2 in the present case, is known as the *fundamental net reproductive rate* or the *fundamental net per capita rate of increase*. Clearly, populations will increase when $R > 1$, and decrease when $R < 1$. (Unfortunately, the ecological literature is somewhat divided between those who use 'R' and those who use the symbol λ for the same parameter. Here we stick with R, but we sometimes

use λ in later chapters to conform to standard usage within the topic concerned.)

R combines the birth of new individuals with the survival of existing individuals. Thus, when $R = 2$, each individual could give rise to two offspring but die itself, or give rise to only one offspring and remain alive: in either case, R (birth plus survival) would be 2. Note too that in the present case R remains the same over the successive intervals of time, i.e. $N_2 = 40 = N_1R$, $N_3 = 80 = N_2R$, and so on. Thus:

$$N_3 = N_1R \times R = N_0R \times R \times R = N_0R^3, \tag{4.6}$$

and in general terms:

$$N_{t+1} = N_tR, \tag{4.7}$$

and:

$$N_t = N_0R^t. \tag{4.8}$$

Equations 4.7 and 4.8 link together population size, rate of increase and time; and we can now link these in turn with R_0, the basic reproductive rate, and with the generation length (defined as lasting T intervals of time). In Section 4.5.2, we saw that R_0 is the multiplication factor that converts one population size to another population size, one generation later, i.e. T time intervals later. Thus:

R, R_0 and T

$$N_T = N_0 R_0. \tag{4.9}$$

But we can see from Equation 4.8 that:

$$N_T = N_0 R^T. \tag{4.10}$$

Therefore:

$$R_0 = R^T, \tag{4.11}$$

or, if we take natural logarithms of both sides:

$$\ln R_0 = T \ln R. \tag{4.12}$$

r, the intrinsic rate of natural increase

The term $\ln R$ is usually denoted by r, the *intrinsic rate of natural increase*. It is the rate at which the population increases in size, i.e. the change in population size per individual per unit time. Clearly, populations will increase in size for $r > 0$, and decrease for $r < 0$; and we can note from the preceding equation that:

$$r = \ln R_0 / T. \tag{4.13}$$

Summarizing so far, we have a relationship between the average number of offspring produced by an individual in its lifetime, R_0, the increase in population size per unit time, r ($= \ln R$), and the generation time, T. Previously, with discrete generations (see Section 4.5.2), the unit of time *was* a generation. It was for this reason that R_0 was the same as R.

4.7.2 Estimating the variables from life tables and fecundity schedules

In populations with overlapping generations (or continuous breeding), r is the intrinsic rate of natural increase that the population has the *potential* to achieve; but it will only actually achieve this rate of increase if the survivorship and fecundity schedules remain steady over a long period of time. If they do, r will be approached gradually (and thereafter maintained), and over the same period the population will gradually approach a stable age structure (i.e. one in which the proportion of the population in each age class remains constant over time; see below). If, on the other hand, the fecundity and survivorship schedules alter over time – as they almost always do – then the rate of increase will continually change, and it will be impossible to characterize in a single figure. Nevertheless, it can often be useful to characterize a population in terms of its potential, especially when the aim is to make a comparison, for instance comparing various populations of the same species in different environments, to see which environment appears to be the most favorable for the species.

The most precise way to calculate r is from the equation:

$$\sum e^{-rx} l_x m_x = 1, \tag{4.14}$$

where the l_x and m_x values are taken from a cohort life table, and e is the base of natural logarithms. However, this is a so-called 'implicit' equation, which cannot be solved directly (only by iteration, usually on a computer), and it is an equation without any clear biological meaning. It is therefore customary to use instead an approximation to Equation 4.13, namely:

$$r \approx \ln R_0 / T_c, \tag{4.15}$$

where T_c is the *cohort generation time* (see below). This equation shares with Equation 4.13 the advantage of making explicit the dependence of r on the reproductive output of individuals (R_0) and the length of a generation (T). Equation 4.15 is a good approximation when $R_0 \approx 1$ (i.e. population size stays approximately constant), or when there is little variation in generation length, or for some combination of these two things (May, 1976).

We can estimate r from Equation 4.15 if we know the value of the cohort generation time T_c, which is the average length of time between the birth of an individual and the birth of one of its own offspring. This, being an average, is the sum of all these birth-to-birth times, divided by the total number of offspring, i.e.:

$$T_c = \sum x l_x m_x / \sum l_x m_x$$

or

$$T_c = \sum x l_x m_x / R_0. \tag{4.16}$$

This is only approximately equal to the true generation time T, because it takes no account of the fact that some offspring may themselves develop and give birth during the reproductive life of the parent.

Thus Equations 4.15 and 4.16 allow us to calculate T_c, and thus an approximate value for r, from a cohort life table of a population with either overlapping generations or continuous breeding. In short, they give us the summary terms we require. A worked example is set out in Table 4.5, using data for the barnacle *Balanus glandula*. Note that the precise value of r, from Equation 4.14, is 0.085, compared to the approximation 0.080; whilst T, calculated from Equation 4.13, is 2.9 years compared to $T_c = 3.1$ years. The simpler and biologically transparent approximations are clearly satisfactory in this case. They show that since r was somewhat greater than zero, the population would have increased in size, albeit rather slowly, if the schedules had remained steady. Alternatively, we may say that, as judged by this cohort life table, the barnacle population had a good chance of continued existence.

Table 4.5 A cohort life table and a fecundity schedule for the barnacle *Balanus glandula* at Pile Point, San Juan Island, Washington (Connell, 1970). The computations for R_0, T_c and the approximate value of r are explained in the text. Numbers marked with an asterisk were interpolated from the survivorship curve.

Age (years) x	a_x	l_x	m_x	$l_x m_x$	$x l_x m_x$
0	1,000,000	1.000	0	0	
1	62	0.0000620	4,600	0.285	0.285
2	34	0.0000340	8,700	0.296	0.592
3	20	0.0000200	11,600	0.232	0.696
4	15.5*	0.0000155	12,700	0.197	0.788
5	11	0.000110	12,700	0.140	0.700
6	6.5*	0.0000065	12,700	0.082	0.492
7	2	0.0000020	12,700	0.025	0.175
8	2	0.0000020	12,700	0.025	0.200
				1.282	3.928

$$R_0 = 1.282; \quad T_c = \frac{3.928}{1.282} = 3.1; \quad r \approx \frac{\ln R_0}{T_c} = 0.08014.$$

4.7.3 The population projection matrix

A more general, more powerful, and therefore more useful method of analyzing and interpreting the fecundity and survival schedules of a population with overlapping generations makes use of the population projection matrix (see Caswell, 2001, for a full exposition). The word 'projection' in its title is important. Just like the simpler methods above, the idea is not to take the cur-

rent state of a population and forecast what *will* happen to the population in the future, but to project forward to what *would* happen if the schedules remained the same. Caswell uses the analogy of the speedometer in a car: it provides us with an invaluable piece of information about the car's current state, but a reading of, say, 80 km h^{-1} is simply a projection, not a serious forecast that we will actually have traveled 80 km in 1 hour's time.

The population projection matrix acknowledges that most life cycles comprise a sequence of distinct classes with different rates of fecundity and survival: life cycle stages, perhaps, or size classes, rather than simply different ages. The resultant patterns can be summarized in a 'life cycle graph', though this is not a graph in the everyday sense but a flow diagram depicting the transitions from class to class over each step in time. Two examples are shown in Figure 4.14 (see also Caswell, 2001). The first (Figure 4.14a) indicates a straightforward sequence of classes where, over each time step, individuals in class *i* may: (i) survive and remain in that class (with probability p_i); (ii) survive and grow and/or develop into the next class (with probability g_i); and (iii) give birth to m_i newborn individuals into the youngest/smallest class. Moreover, as Figure 4.14b shows, a life cycle graph can also depict a more complex life cycle, for example with both sexual reproduction (here, from reproductive class 4 into 'seed' class 1) and vegetative growth of new modules (here, from 'mature module' class 3 to 'new module' class 2). Note that the notation here is slightly different from that in life tables like Table 4.1 above. There the focus was on age classes, and the passage of time inevitably meant the passing of individuals from one age class to the next: *p* values therefore referred to survival from one age class to the next. Here, by contrast, an individual

life cycle graphs

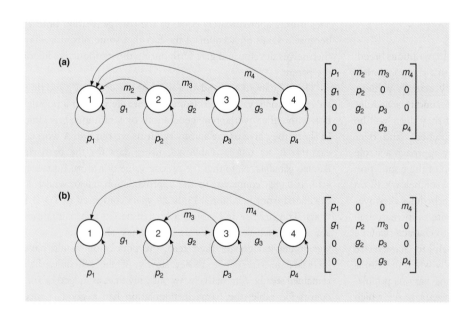

Figure 4.14 Life cycle graphs and population projection matrices for two different life cycles. The connection between the graphs and the matrices is explained in the text. (a) A life cycle with four successive classes. Over one time step, individuals may survive within the same class (with probability p_i), survive and pass to the next class (with probability g_i) or die, and individuals in classes 2, 3 and 4 may give birth to individuals in class 1 (with per capita fecundity m_i). (b) Another life cycle with four classes, but in this case only reproductive class 4 individuals can give birth to class 1 individuals, but class 3 individuals can 'give birth' (perhaps by vegetative growth) to further class 2 individuals.

need not pass from one class to the next over a time step, and it is therefore necessary to distinguish survival within a class (*p* values here) from passage and survival into the next class (*g* values).

elements of
the matrix

The information in a life cycle graph can be summarized in a population projection matrix. Such matrices are shown alongside the graphs in Figure 4.14. The convention is to contain the elements of a matrix within square brackets. In fact, a projection matrix is itself always 'square': it has the same number of columns as rows. The rows refer to the class number at the endpoint of a transition: the columns refer to the class number at the start. Thus, for instance, the matrix element in the third row of the second column describes the flow of individuals from the second class into the third class. More specifically, then, and using the life cycle in Figure 4.14a as an example, the elements in the main diagonal from top left to bottom right represent the probabilities of surviving and remaining in the same class (the *p*s), the elements in the remainder of the first row represent the fecundities of each subsequent class into the youngest class (the *m*s), while the *g*s, the probabilities of surviving and moving to the next class, appear in the subdiagonal below the main diagonal (from 1 to 2, from 2 to 3, etc).

Summarizing the information in this way is useful because, using standard rules of matrix manipulation, we can take the numbers in the different classes (n_1, n_2, etc.) at one point in time (t_1), expressed as a 'column vector' (simply a matrix comprising just one column), *pre*-multiply this vector by the projection matrix, and generate the numbers in the different classes one time step later (t_2). The mechanics of this – that is, where each element of the new column vector comes from – are as follows:

$$\begin{bmatrix} p_1 & m_2 & m_3 & m_4 \\ g_1 & p_2 & 0 & 0 \\ 0 & g_2 & p_3 & 0 \\ 0 & 0 & g_3 & p_4 \end{bmatrix} \times \begin{bmatrix} n_{1,t1} \\ n_{2,t1} \\ n_{3,t1} \\ n_{4,t1} \end{bmatrix} = \begin{bmatrix} n_{1,t2} \\ n_{2,t2} \\ n_{3,t2} \\ n_{4,t2} \end{bmatrix}$$

$$= \begin{bmatrix} (n_{1,t1} \times p_1) + (n_{2,t1} \times m_2) + (n_{3,t1} \times m_3) + (n_{4,t1} \times m_4) \\ (n_{1,t1} \times g_1) + (n_{2,t1} \times p_2) + (n_{3,t1} \times 0) + (n_{4,t1} \times 0) \\ n_{1,t1} \times 0) + (n_{2,t1} \times g_2) + (n_{3,t1} \times p_3) + (n_{4,t1} \times 0) \\ (n_{1,t1} \times 0) + (n_{2,t1} \times 0) + (n_{3,t1} \times g_3) + (n_{4,t1} \times p_4) \end{bmatrix}$$

Thus, the numbers in the first class, n_1, are the survivors from that class one time step previously plus those

determining *R* from
a matrix

born into it from the other classes, and so on. Figure 4.15 shows this process repeated 20 times (i.e. for 20 time steps) with some hypothetical values in the projection matrix shown as an inset in the figure. It is apparent that there is an initial (transient) period in which the proportions in the different classes alter, some increasing and others decreasing, but that after about nine time steps, all classes grow at the same exponential rate (a straight line on a logarithmic scale), and so therefore does the whole population. The *R* value is 1.25. Also, the proportions in the different classes are constant: the population has achieved a stable class structure with numbers in the ratios 51.5 : 14.7 : 3.8 : 1.

Hence, a population projection matrix allows us to summarize a potentially complex array of survival, growth and reproductive processes, and characterize that population succinctly by determining the per capita rate of increase, *R*, implied by the matrix. But crucially, this 'asymptotic' *R* can be determined directly, without the need for a simulation, by application of the methods of matrix algebra, though these are quite beyond our scope here

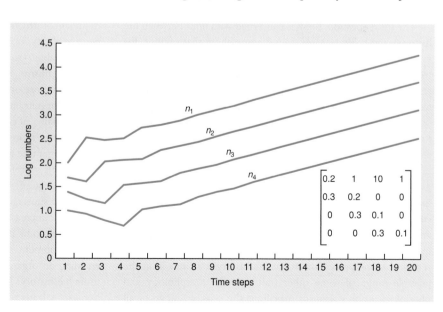

Figure 4.15 A population growing according to the life cycle graph shown in Figure 4.14a, with parameter values as shown in the insert here. The starting conditions were 100 individuals in class 1 ($n_1 = 100$), 50 in class 2, 25 in class 4 and 10 in class 4. On a logarithmic (vertical) scale, exponential growth appears as a straight line. Thus, after about 10 time steps, the parallel lines show that all classes were growing at the same rate ($R = 1.25$) and that a stable class structure had been achieved.

(but see Caswell, 2001). Moreover, such algebraic analysis can also indicate whether a simple, stable class structure will indeed be achieved, and what that structure will be. It can also determine the importance of each of the different components of the matrix in generating the overall outcome, R – a topic to which we return in Section 14.3.2.

4.8   Life history evolution

An organism's life history is its lifetime pattern of growth, differentiation, storage and reproduction; and we have seen something in the preceding sections about the variety of patterns life histories may take and what the consequences may be in terms of population rates of increase. But can we understand how different species' life histories have evolved? In fact, there are at least three different types of question that are commonly asked about the evolution of life histories.

three types of question

The first is concerned with individual life history traits. Why is it that swifts, for example, usually produce clutches of three eggs, when other birds produce larger clutches, and the swifts themselves are physiologically capable of doing so? Can we establish that *this* clutch size is ultimately the most productive, i.e. the fittest in evolutionary terms, and what is it about this particular clutch size that makes it so?

The second question is concerned with links between life history traits. Why is it, for example, that the ratio between age at maturity and average lifespan is often roughly constant within a group of organisms but markedly different between groups (e.g. mammals 1.3, fish 0.45)? What is the basis for the link between these two traits within a group of related organisms? What is the basis for differences amongst groups?

The third question, then, is concerned with links between life histories and habitats. How does it come about that orchids, for example, produce vast numbers of tiny seeds when tropical *Mora* trees produce just a few enormous ones? Can the difference be related directly to differences in the habitats that they occupy, or to any other differences between them?

In short, the study of the evolution of life histories is a search for patterns – and for explanations for those patterns. We must remember, however, that every life history, and every habitat, is unique. In order to find ways in which life histories might be grouped, classified and compared, we must find ways of describing them that apply to all life histories and all habitats. Only then can we search for associations between one life history trait and another or between life history traits and features of the habitats in which the life histories are found. It is also important to realize that the possession of one life history trait may limit the possible range of some other trait, and the morphology and physiology of an organism may limit the possible range of all its life history traits. The most that natural selection can do is to favor, in a particular environment with its many, often conflicting demands, the life history that has been most (not 'perfectly') successful, overall, at leaving descendants in the past.

None the less, most of the successes in the search for an understanding of life history evolution have been based on the idea of optimization: establishing that observed combinations of life history traits are those with the highest fitness (Stearns, 2000). It is also important to note, however, that there are alternative approaches – one long-established, two others more recent – that certainly have much to recommend them in theory, even if their explanatory powers to date have been limited compared to the optimization approach (Stearns, 2000). The first is 'bet-hedging': the idea that when fitness fluctuates, it may be most important to minimize the setbacks from periods of low fitness rather than evolving to a single optimum (Gillespie, 1977). The second acknowledges that the fitness of any life history cannot be seen in isolation: it depends on the life histories of other individuals in the population, such that fitness of a life history is 'frequency dependent' – dependent on the proportions of that and other life histories in the population (e.g. Sinervo *et al.*, 2000). The third, then, includes an explicit consideration of the dynamics of the population concerned, rather than making the usual simplifying assumption of population stability (e.g. Ranta *et al.*, 2000). Here, though, we focus on the optimization approach.

optimization and other approaches to understanding life history evolution

4.8.1   Components of life histories

What are the most important components of any organism's life history? Individual size is perhaps the most apparent aspect. As we have seen, it is particularly variable in organisms with a modular construction. Large size may increase an organism's competitive ability, or increase its success as a predator or decrease its vulnerability to predation, and hence increase the survival of larger organisms. Stored energy and/or resources will also be of benefit to those organisms that pass through periods of reduced or irregular nutrient supply (probably true of most species at some time). Finally, of course, larger individuals within a species usually produce more offspring. Size, however, can increase some risks: a larger tree is more likely to be felled in a gale, many predators exhibit a preference for larger prey, and larger individuals typically require more resources and may therefore be more prone to a shortage of them. Hence it is easy to see why detailed studies are increasingly confirming an intermediate, not a maximum, size to be optimal (Figure 4.16).

Development is the progressive differentiation of parts, enabling an organism to do different things at different stages in its life history. Hence rapid development can increase fitness because it leads to the rapid initiation of reproduction. As we have seen, reproduction itself may occur in one terminal burst

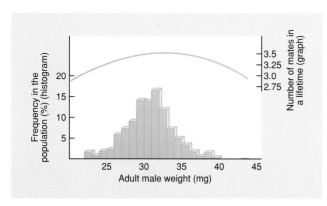

Figure 4.16 For adult male damselflies, *Coenagrion puella*, the predicted optimum size (weight) is intermediate (upper graph), and corresponds closely to the modal size class in the population (histogram below). The upper graph takes this form because mating rate decreases with weight, whereas lifespan increases with weight (mating rate = 1.15 − 0.018 weight, $P < 0.05$; lifespan = 0.21 − 0.44 weight, $P < 0.05$; $n = 186$). (After Thompson, 1989.)

(semelparity) or as a series of repeated events (iteroparity). Amongst iteroparous organisms, variation is possible in the number of separate clutches of offspring, and all organisms can vary in the number of offspring in a clutch.

The individual offspring can themselves vary in size. Large newly emerged or newly germinated offspring are often better competitors, better at obtaining nutrients and better at surviving in extreme environments. Hence, they often have a better chance of surviving to reproduce themselves.

Combining all of this detail, life histories are often described in terms of a composite measure of reproductive activity known as 'reproductive allocation' (also often called 'reproductive effort'). This is best defined as the proportion of the available resource

input that is allocated to reproduction over a defined period of time; but it is far easier to define than it is to measure. Figure 4.17 shows an example involving the allocation of nitrogen, a crucial resource in this case. In practice, even the better studies usually monitor only the allocation of energy or just dry weight to various structures at a number of stages in the organism's life cycle.

4.8.2 Reproductive value

Natural selection favors those individuals that make the greatest proportionate contribution to the future of the population to which they belong. All life history components affect this contribution, ultimately through their effects on fecundity and survival. It is necessary, though, to combine these effects into a single currency so that different life histories may be judged and compared. A number of measures of fitness have been used. All the better ones have made use of both fecundity and survival schedules, but they have done so in different ways, and there has often been marked disagreement as to which of them is the most appropriate. The intrinsic rate of natural increase, r, and the basic reproductive rate, R_0 (see above) have had their advocates, as has 'reproductive value' (Fisher, 1930; Williams, 1966), especially reproductive value at birth (Kozlowski, 1993; de Jong, 1994). For an exploration of the basic patterns in life histories, however, the similarities between these various measures are far more important than the minor differences between them. We concentrate here on reproductive value.

Reproductive value is described in some detail in Box 4.1. For most purposes though, these details can be ignored as long as it is remembered that: (i) reproductive value at a given age or stage is the sum of the current reproductive output and the residual (i.e. future) reproductive value (RRV); (ii) RRV combines expected future

reproductive value described in words

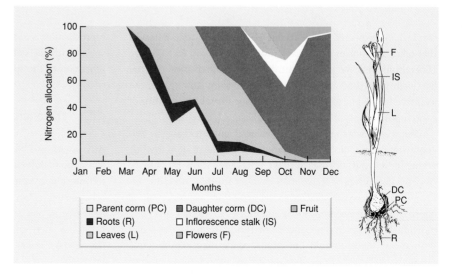

Figure 4.17 Percentage allocation of the crucial resource nitrogen to different structures throughout the annual cycle of the perennial plant *Sparaxis grandiflora* in South Africa, where it sets fruit in the southern hemisphere spring (September–December). The plant grows each year from a corm, which it replaces over the growing season, but note the development of reproductive parts at the expense of roots and leaves toward the end of the growing season. The plant parts themselves are illustrated to the right for a plant in early spring. (After Ruiters & McKenzie, 1994.)

Box 4.1   Reproductive value

The reproductive value of an individual of age x (RV_x) is the currency by which the worth of a life history in the hands of natural selection may be judged. It is defined in terms of the life-table statistics discussed earlier. Specifically:

$$RV_x = \sum_{y=x}^{y=y_{max}} \left(\frac{l_y}{l_x} \cdot m_y \cdot R^{x-y} \right)$$

where m_x is the birth rate of the individual in age-class x; l_x is the probability that the individual will survive to age x; R is the net reproductive rate of the whole population per unit time (the time unit here being the age interval); and Σ means 'the sum of'.

To understand this equation, it is easiest to split RV_x into its two components:

$$RV_x = m_x + \sum_{y=x+1}^{y=y_{max}} \left(\frac{l_y}{l_x} \cdot m_y \cdot R^{x-y} \right).$$

Here, m_x, the individual's birth rate at its current age, can be thought of as its *contemporary reproductive output*. What remains is then the *residual reproductive value* (Williams, 1966): the sum of the 'expectations of reproduction' at all subsequent ages, modified in each case by R^{x-y} for reasons described below. The 'expectation of reproduction' for age class y is $(l_y/l_x \cdot (m_y))$, i.e. it is the birth rate of the individual should it reach that age (m_y), discounted by the probability of it doing so given that it has already reached stage x (l_y/l_x).

Reproductive value takes on its simplest form where the overall population size remains approximately constant. In such cases, $R = 1$ and can be ignored. The reproductive value of an individual is then simply its total lifetime expectation of reproductive output (from its current age class and from all subsequent age classes).

However, when the population consistently increases or decreases, this must be taken into account. If the population increases, then $R > 1$ and $R^{x-y} < 1$ (because $x < y$). Hence, the terms in the equation are reduced by R^{x-y} the larger the value of y (the further into the future we go), signifying that future (i.e. 'residual') reproduction adds relatively little to RV_x, because the proportionate contribution to a growing population made by a given reproductive output in the future is relatively small – whereas the offspring from present or early reproduction themselves have an early opportunity to contribute to the growing population. Conversely, if the population decreases, then $R < 1$ and $R^{x-y} > 1$, and the terms in the equation are successively increased, reflecting the greater proportionate contribution of future reproduction.

In any life history, the reproductive values at different ages are intimately connected, in the sense that when natural selection acts to maximize reproductive value at one age, it constrains the values of the life table parameters – and thus reproductive value itself – for subsequent ages. Hence, strictly speaking, natural selection acts ultimately to maximize reproductive value *at birth*, RV_0 (Kozlowski, 1993). (Note that there is no contradiction between this and the fact that reproductive value is typically low at birth (Figure 4.18). Natural selection can discriminate only between those options available at that stage.)

survival and expected future fecundity; (iii) this is done in a way that takes account of the contribution of an individual to future generations, relative to the contributions of others; and (iv) the life history favored by natural selection from amongst those available in the population will be the one for which the sum of contemporary output and RRV is highest.

The way in which reproductive value changes with age in two contrasting populations is illustrated in Figure 4.18. It is low for young individuals when each of them has only a low probability of surviving to reproductive maturity; but for those that do survive, it then increases steadily as the age of first reproduction is approached, as it becomes more and more certain that surviving individuals will reach reproductive maturity. Reproductive value is then low again for old individuals, since their reproductive output is likely to have declined, and their expectation of future reproduction is even lower. The detailed rise and fall, of course, varies with the detailed age- or stage-specific birth or mortality schedules of the species concerned.

4.8.3   Trade-offs

Any organism's life history must, of necessity, be a compromise allocation of the resources that are available to it. Resources devoted to one trait are unavailable to others. A 'trade-off' is a negative relationship between two life history characteristics in which increases in one are associated with decreases in the other as a result of such compromises. For instance, Douglas fir trees (*Pseudotsuga menziesii*) benefit both from reproducing and from growing (since, amongst other things, this enhances future

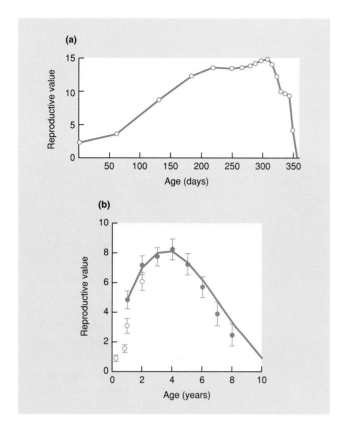

Figure 4.18 Reproductive value generally rises and then falls with age, as explained in the text. (a) The annual plant *Phlox drummondii*, described earlier in the chapter. (After Leverich & Levin, 1979.) (b) The sparrowhawk, *Accipiter nisus*, in southern Scotland. Solid symbols (±1 SE) refer to breeders only; open symbols include nonbreeders. (After Newton & Rothery, 1997.) Note that in both cases the vertical scale is arbitrary, in the sense that the rate of increase (R) for the whole population was not known, and a value therefore had to be assumed.

reproduction), but the more cones they produce the less they grow (Figure 4.19a). Male fruit-flies benefit both from a long period of reproductive activity and from a high frequency of matings, but the higher their level of reproductive activity earlier in life the sooner they die (Figure 4.19b).

trade-offs are not easy to observe

Yet it would be quite wrong to think that such negative correlations abound in nature, only waiting to be observed. On the contrary, we cannot generally expect to see trade-offs by simply observing correlations in natural populations (Lessells, 1991). In the first place, if there is just one clearly optimal way of combining, say, growth and reproductive output, then all individuals may approximate closely to this optimum and a population would then lack the variation in these traits necessary for a trade-off to be seen. Moreover, if there

is variation between individuals in the amount of resource they have at their disposal, then there is likely to be a positive, not a negative, correlation between two apparently alternative processes – some individuals will be good at everything, others consistently awful. For instance, in Figure 4.20, the aspic vipers (*Vipera aspis*) in the best condition produced larger litters but also recovered from breeding more rapidly, ready to breed again.

Two approaches have sought to overcome these problems and hence allow the investigation of the nature of trade-off curves. The first is based on comparisons of individuals differing *genetically*, where different genotypes are thought likely to give rise to different allocations of resources to alternative traits. Genotypes can be compared in two ways: (i) by a breeding experiment, in which genetically contrasting groups are bred and then compared; or (ii) by a selection experiment, in which a population is subjected to a selection pressure to alter one trait, and associated changes in other traits are then monitored. For example, in one selection experiment, populations of the Indian meal moth, *Plodia interpunctella*, that evolved increased resistance to a virus having been infected with it for a number of generations, exhibited an associated decrease (negative correlation) in their rate of development (Boots & Begon, 1993). Overall, however, the search for genetic correlations has generated more zero and positive than negative correlations (Lessells, 1991), and it has therefore had only limited success to date in measuring trade-offs, despite receiving strong support from its adherents by virtue of its direct approach to the underlying basis for selective differentials between life histories (Reznick, 1985; Rose *et al.*, 1987).

genetic comparisons

The alternative approach is to use experimental manipulation to reveal a trade-off directly from a negative *phenotypic* correlation. The *Drosophila* study in Figure 4.19b is an example of this. The great advantage of experimental manipulation over simple observation is that individuals are assigned experimental treatments at random rather than differing from one another, for instance, in the quantity of resource that they have at their disposal. This contrast is illustrated in Figure 4.21, which shows two sets of data for the bruchid beetle *Callosobruchus maculatus* in which fecundity and longevity were correlated. Simple observation of an unmanipulated population gave rise to a positive correlation: the 'better' individuals both lived longer and laid more eggs. When fecundity varied, however, not as a result of differing resource availability, but because access to mates and/or egg-laying sites was manipulated, a trade-off (negative correlation) was revealed.

experimental manipulations

However, this contrast between experimental manipulation ('good') and simple observation ('bad') is not always straightforward (Bell & Koufopanou, 1986; Lessells, 1991). Some manipulations suffer from much the same problems as simple observations. For instance, if clutch size is manipulated by giving supplementary food, then improvements in other traits are to be expected as well.

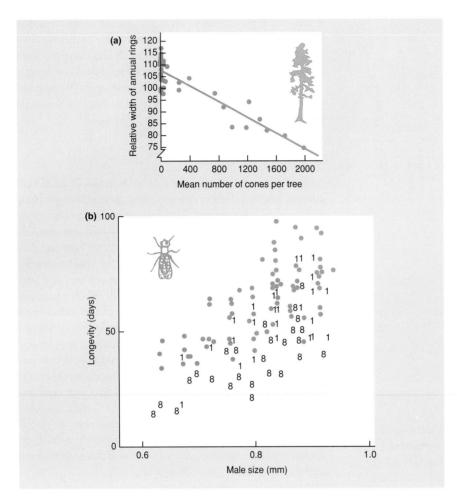

Figure 4.19 Life history trade-offs. (a) The negative correlation between cone crop size and annual growth increment for a population of Douglas fir trees, *Pseudotsuga menziesii*. (After Eis *et al.*, 1965.) (b) The longevity of male fruit-flies (*Drosophila melanogaster*) generally increases with size (thorax length). However, longevity was reduced in males provided with one virgin and seven mated females per day (1) compared with those provided with eight mated famales (●), because of the increase in courtship activity, and reduced still further in males provided with eight virgins per day (8). (After Partridge & Farquhar, 1981.)

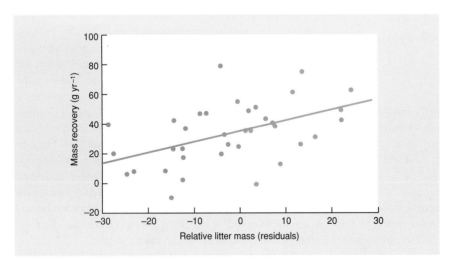

Figure 4.20 Female aspic vipers (*Vipera aspis*) that produced larger litters ('relative' litter mass because total female mass was taken into account) also recovered more rapidly from reproduction (not 'relative' because mass recovery was not affected by size) ($r = 0.43$; $P = 0.01$). (After Bonnet *et al.*, 2002.)

It is important that the manipulation should alter the target trait and nothing else. On the other hand, simple observation may be acceptable if based on the results of a 'natural' experiment. For example, it is likely that as a result of 'mast seeding' (see Section 9.4) the population of fir trees in Figure 4.19a produced large and small crops of cones in response to factors other than resource availability, and that the negative correlation therefore genuinely represented an underlying trade-off.

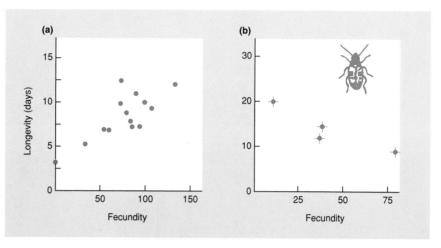

Figure 4.21 (a) The (positive) phenotypic correlation in an unmanipulated population between adult longevity and fecundity in female bruchid beetles, *Callosobruchus maculatus*. (b) The (negative) trade-off between the same traits when access to mates and/or egg-laying sites was manipulated. Points are means of four treatments with standard errors. (After Lessells, 1991; from K. Wilson, unpublished data.)

In the final analysis, though, it is generally agreed that trade-offs are widespread and important. The problems arise in revealing and hence quantifying them.

4.8.4 The cost of reproduction

Most attention has been directed at trade-offs that reveal an apparent 'cost of reproduction' (CR). Here, 'cost' is used in a particular way to indicate that an individual, by increasing its current allocation to reproduction, is likely to decrease its survival and/or its rate of growth, and therefore decrease its potential for reproduction in future. This is shown by the fir trees and fruit-flies in Figure 4.19 and by the beetles in Figure 4.21. The costs of reproduction can be shown even more easily with plants. All good gardeners know, for example, that to prolong the life of perennial flowering herbs, the ripening seed heads should

be removed, since these compete for resources that may be available for improved survivorship and even better flowering next year. When ragwort plants (*Senecio jacobaea*) of a given size are compared at the end of a season, it is only those that have made the smallest reproductive allocation that survive (Figure 4.22).

Thus, individuals that delay reproduction, or restrain their reproduction to a level less than the maximum, may grow faster, grow larger or have an increased quantity of resources available for maintenance, storage and, ultimately, future reproduction. Any 'cost' incurred by contemporary reproduction, therefore, is likely to contribute to a decrease in residual reproductive value (RRV). Yet, as we have noted, natural selection favors the life history with the highest available *total* reproductive value: the sum of two quantities, one of which (contemporary reproductive output) tends to go up as the other (RRV) goes down. Trade-offs involving the cost of reproduction are at the heart of the evolution of any life history.

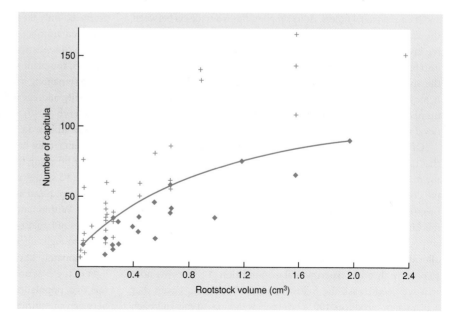

Figure 4.22 The cost of reproduction in ragwort (*Senecio jacobaea*). The line divides plants that survive (♦), from those that have died by the end of the season (+). There are no surviving plants above and to the left of the line. For a given size (rootstock volume), only those that have made the smallest reproductive allocation (number of capitula) survive, although larger plants are able to make a larger allocation and still survive. (After Gilman & Crawley, 1990.)

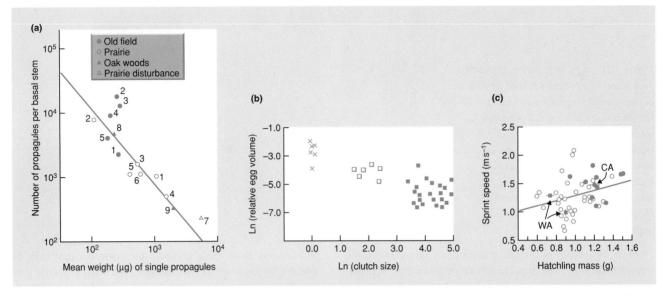

Figure 4.23 Evidence for a trade-off between the number of offspring produced in a clutch by a parent and the individual fitness of those offspring. (a) A negative correlation between the number of propagules per stem produced by goldenrod plants (*Solidago*) and the weight of single propagules. The species are: 1, *S. nemoralis*; 2, *S. graminifolia*; 3, *S. canadensis*; 4, *S. speciosa*; 5, *S. missouriensis*; 6, *S. gigantea*; 7, *S. rigida*; 8, *S. caesia*; and 9, *S. rugosa*, taken from a variety of habitats as shown. (After Werner & Platt, 1976.) (b) A negative correlation between clutch size and egg volume amongst Hawaiian species of *Drosophila*, developing either on a relatively poor and restricted resource, flower pollen (×), or on bacteria within decaying leaves (□) or on especially unpredictable but productive resources – yeasts within rotten fruits, barks and stems (■). (After Montague *et al.*, 1981; Stearns, 1992.) (c) The mass and sprint speed of California hatchlings of the lizard, *Sceloporus occidentalis*, are lower from eggs with some yolk removed (○) than from unmanipulated control eggs (●). Also shown are the means for the control California hatchlings (CA) (fewer, larger eggs) and those for two samples from Washington (WA) (more, smaller eggs). (After Sinervo, 1990.)

4.8.5   The number and fitness of offspring

A second key trade-off is that between the number of offspring and their individual fitness. At its simplest, this is a trade-off between the size and number of offspring, within a given total reproductive investment. That is, a reproductive allocation can be divided into fewer, larger offspring or more, smaller offspring. However, the size of an egg or seed is only an index of its likely fitness. It may be more appropriate to look for a trade-off between the number of offspring and, say, their individual survivorship or developmental rate.

Of the few genetic correlations between egg size and number that have so far been examined (dominated by domestic poultry), the majority have been negative, as expected (Lessells, 1991). Negative correlations have also been observed in simple cross-species or cross-population comparisons (e.g. Figure 4.23a, b), although it is unlikely in such cases that the individuals from different species or populations are making precisely the same total reproductive allocation. Moreover, this type of trade-off is especially difficult to observe through experimental manipulation. To see why, note that we need to ask the following kind of question. Given that a plant, say, produces 100 seeds each weighing 10 mg and each

with a 5% chance of developing to reproductive maturity, what would be the seed size, and what would be the chance of developing to maturity, if an identical plant receiving identical resources produced only 80 seeds? Clearly, it would be invalid to manipulate seed number by altering the provision of resources; and even if 20 seeds were removed at or close to their point of production, the plant would be limited in its ability to alter the size of the remaining seeds, and their subsequent survivorship would not really address the question originally posed.

Sinervo (1990), though, did manipulate the size of the eggs of an iguanid lizard (*Sceloporus occidentalis*) by removing yolk from them after they have been produced, giving rise to healthy but smaller offspring than unmanipulated eggs. These smaller hatchlings had a slower sprint speed (Figure 4.23c) – probably an indication of a reduced ability to avoid predators, and hence of a lower fitness. Within natural populations, this species produces smaller clutches of larger eggs in California than in Washington (typically seven to eight eggs with an average weight of 0.65 g, as against approximately 12 eggs weighing 0.4 g; Figure 4.23c). Thus, in the light of the experimental manipulations, the comparison between the two populations does indeed appear to reflect a trade-off between the number of offspring and their individual fitness.

4.9   Options sets, fitness contours and a classification of habitats

We turn next to another of our original life history questions – are there patterns linking particular types of life history to particular types of habitat? To address this question, we introduce two further concepts. We do so in the context of the cost of reproduction, since the trade-offs associated with it are the most fundamental – but the same principles apply to any trade-off.

4.9.1   Options sets and fitness contours

An *options set* describes the whole range of combinations of two life history traits that an organism is capable of exhibiting. Hence,

it reflects the organism's underlying physiology. Here, for purposes of illustration, we use present reproduction, m_x, and growth (as a potentially important indicator of RRV) (Figure 4.24). The options set therefore describes, for any given level of present reproduction, the range of growth increments that the organism can achieve, and for any given growth increment, the range of levels of present reproduction that the organism can achieve. The outer boundary of the options set represents the trade-off curve. For any point on that boundary, the organism can only increase m_x by making a compensatory reduction in growth and vice versa.

An options set may be convex-outwards (Figure 4.24a), implying in the present case that a level of present reproduction only slightly less than the maximum none the less allows a considerable amount of growth. Alternatively, the set may be concave-outwards (Figure 4.24b), implying that substantial growth

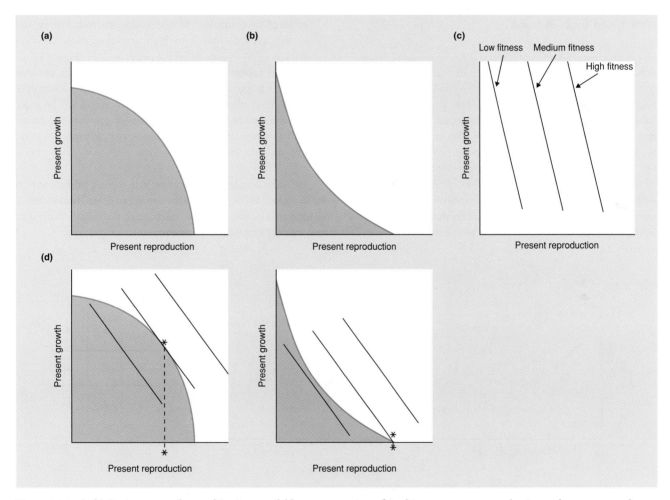

Figure 4.24   (a, b) Options sets – the combinations available to an organism of, in this case, present reproduction and present growth. As explained in the text, the outer boundary of the options set is a trade-off curve: (a) is convex-outwards, (b) concave-outwards. (c) Fitness contours linking combinations of present reproduction and present growth that have equal fitness in a given habitat. Hence, contours further from the origin have greater fitness. (d) The point in an options set with the greatest fitness is the one that reaches the highest fitness contour. This point, and the (optimal) value for present reproduction giving rise to it, are marked with an asterisk. (After Sibly & Calow, 1983.)

can only be achieved with a level of present reproduction considerably less than the maximum.

A *fitness contour*, then, is a line joining, in this case, combinations of m_x and growth for which fitness (reproductive value) is constant (Figure 4.24c). Contours further away from the origin therefore represent combinations with greater fitness. As described below, the shapes of fitness contours reflect not the organism's intrinsic properties but the habitat in which it lives.

The combination of traits, amongst those available, that has the highest fitness determines the direction of natural selection. Natural selection therefore favors the point in the options set (on the trade-off curve) that reaches the highest fitness contour (indicated by the asterisks in Figure 4.24d, e). Since different options sets imply different types of organism, and different shapes of fitness contour imply different types of habitat, they can be used together as a guide to where and when different types of life history might be found.

4.9.2   Habitats: a classification

Each organism's habitat is unique, but if a pattern linking habitats and life histories is to be established, habitats must be classified in terms that apply to them all. Moreover, they must be described and classified from the point of view of the organism concerned, rather than whether *we* feel the habitat is patchy or homogeneous, harsh or benign. Thus, when we say that the shapes of fitness contours reflect an organism's habitat, we mean that they reflect the effect of the habitat on that particular organism or the response of that organism to the habitat.

A number of classifications of habitat types have been proposed (e.g. Schaffer, 1974; Grime *et al.*, 1988; Silvertown *et al.*, 1993), but a review of these is beyond our scope here. Instead, we classify habitats by focusing on fitness contours, and hence on the ways in which present reproduction and growth combine to determine fitness in different types of habitat (following Levins, 1968; Sibly & Calow, 1983).

For established individuals (i.e. not newly emerged or newly born offspring) two contrasting habitat types can be recognized.

1 High CR (cost of reproduction) habitats, in which any reduced growth that results from present reproduction has a significant negative effect on RRV, and hence on fitness. Thus, similar fitness can be achieved by combining high reproduction with low growth or low reproduction with high growth. Fitness contours therefore run diagonally with a negative slope (Figure 4.25a). *high and low CR habitats: a comparative classification*

2 Low CR habitats, in which RRV is little affected by the level of present growth. Fitness is thus essentially determined by the level of present reproduction alone and will be much the same whatever the level of present growth. The fitness contours therefore run approximately vertically (parallel to the 'growth' axis; Figure 4.25a).

This classification is comparative. In practice, a habitat can only be described as 'high CR' relative to some other habitat that is, comparatively, low CR. The purpose of the classification is to contrast habitats with one another.

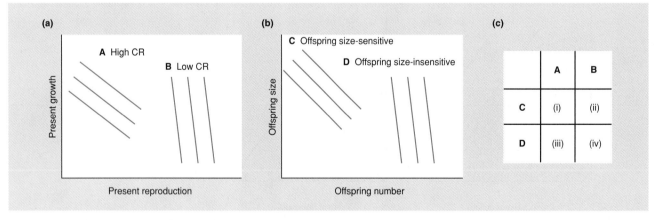

Figure 4.25   A demographic classification of habitats. (a) Habitats of established individuals can be either: (A) relatively high CR (fitness contours indicate that residual reproductive value rises sharply with increased growth resulting from decreased present reproduction) or (B) relatively low CR (fitness contours largely reflect the level of present reproduction). (b) Habitats of recently produced offspring can be either: (C) relatively offspring size-sensitive or (D) relatively offspring size-insensitive. Larger offspring size is assumed to imply a smaller number of them (for a given reproductive allocation). Hence, for example, in (D), fitness largely reflects the number of offspring – not their individual size. (c) By combining these two contrasting pairs, the habitat of one organism over its whole life, compared with another organism, can be of four basic types, arbitrarily referred to as (i)–(iv) in the figure.

habitat types can
arise for a variety of
reasons

Moreover a habitat can be of a particular type for a variety of reasons. Habitats can be relatively high CR for at least two reasons.

1 When there is intense competition amongst established individuals (see Chapter 5), with only the best competitors surviving and reproducing, present reproduction may be costly because it reduces growth and hence substantially reduces competitive ability in the future, and thus reduces RRV. Red deer stags, where only the best competitors can hold a harem of females, are a good example of this.

2 Whenever diminutive adults are particularly susceptible to an important source of mortality from a predator or some abiotic factor, present reproduction may be costly because it maintains adults in these vulnerable size classes. For instance, mussels on the seashore may, through reproductive restraint, outgrow predation by both crabs and eider ducks.

On the other hand, habitats can be relatively low CR for at least three different reasons.

1 Much mortality may be indiscriminate and unavoidable, so that any increase in size caused by reproductive restraint is likely to be worthless in future. For instance, when temporary ponds dry out, most individuals die irrespective of their size or condition.

2 The habitat may be so benign and competition-free for established individuals that all of them have a high probability of surviving, and a large future reproductive output, irrespective of any present lack of reproductive restraint. This is true, at least temporarily, for the first colonists to arrive in a newly arisen habitat.

3 A habitat may be low CR simply because there are important sources of mortality to which the largest individuals are especially prone. Thus, restrained present reproduction, by leading to greater size, may give rise to *lowered* survival in future. For instance, in the Amazon, avian predators preferentially prey upon the largest individuals of certain fish species.

related classification
of habitats for newly
born offspring

A related classification of habitats for newly born offspring can also be constructed. Again, there are two contrasting types (Figure 4.25b), assuming that, for a given reproductive allocation, larger offspring can be produced only if there are fewer of them.

1 'Offspring size-sensitive' habitats, in which the reproductive value of individual offspring rises significantly with size (as above, either because of competition amongst offspring, or because of important sources of mortality to which small offspring are especially vulnerable). An increase in size implies a significant rise up the fitness contours.

2 'Offspring size-insensitive' habitats, in which the reproductive value of individual offspring is little affected by their size (as above, because of indiscriminate mortality, or because of superabundant resources, or because there are sources of mortality to which larger individuals are more prone). An increase in size implies a negligible move up the fitness contours.

Together, clearly, the two contrasting pairs can be combined into four types of habitat (Figure 4.25c).

4.10 Reproductive allocation and its timing

4.10.1 Reproductive allocation

If we assume initially that all options sets are convex-outwards, then we can see that relatively low CR habitats should favor a higher reproductive allocation, whilst relatively high CR habitats favor a lower reproductive allocation (Figure 4.26a). This pattern can be seen in three populations of the dandelion *Taraxacum officinale*. The populations were composed of a number of distinct clones that belonged to one or other of four biotypes (A–D). The habitats of the populations varied from a footpath (the habitat in which adult mortality was most indiscriminate – 'lowest CR') to an old, stable pasture (the habitat with most adult competition – 'highest CR'); the third site was intermediate between the other two. In line with predictions, the biotype that predominated in the footpath site (A) made the greatest reproductive allocation (whichever site it was obtained from), whilst the biotype that predominated in the old pasture (D) made the lowest reproductive allocation (Figure 4.26b, c). Biotypes B and C were appropriately intermediate with respect both to their site occupancies and their reproductive allocations.

4.10.2 Age at maturity

Since relatively high CR habitats should favor low reproductive allocations, maturity (the onset of sexual reproduction) should be relatively delayed in such habitats but should occur at a relatively large size (in deferring maturity at any given time, an organism is making a reproductive allocation of zero). These ideas are supported by a study of guppies, *Poecilia reticulata*, a small fish species in Trinidad (Table 4.6). The same work also provides support for the patterns of reproductive allocation discussed above, and for patterns of variation in offspring size, discussed in Section 4.11. The guppies live in small streams that can be divided into two contrasting types. In one, their main predator is a cichlid fish, *Crenicichla alta*, which eats mostly large, sexually mature guppies. In the other, the main predator is a killifish, *Rivulus hartii*, which prefers small, juvenile guppies. The *Crenicichla* sites

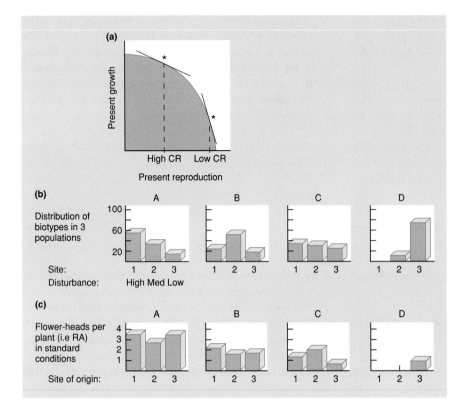

Figure 4.26 (a) Options sets and fitness contours (see Figure 4.25) suggest that relatively high CR habitats should favor relatively small reproductive allocations. (b) The distribution of four biotypes (A–D) of the dandelion *Taraxacum officinale*, amongst three populations subject to low, medium and high levels of disturbance (i.e. habitats ranging from relatively high CR to relatively low CR). (c) The reproductive allocations (RAs) of the different biotypes from the different sites of origin, showing that biotype A, which predominates in the relatively low CR habitat, has a relatively large RA, and so on. ((b, c) after Solbrig & Simpson, 1974.)

Table 4.6 A comparison of guppies (*Poecilia reticulata*) from relatively low CR, offspring size-insensitive sites (*Crenicichla* – predation concentrated on larger, adult fish) and relatively high CR, offspring size-sensitive sites (*Rivulus* – predation concentrated on small, juvenile fish). In the former, the guppies (male and female) mature earlier and smaller, make a larger reproductive allocation (shorter interlitter interval, higher percentage effort) and produce smaller offspring (and more of them). This is true both for natural populations from contrasting sites (left) and comparing a population introduced to a *Rivulus* site with its unmanipulated control (right). (After Reznick et al., 1982, 1990.)

	Reznick (1982)			Reznick et al. (1990)		
	Crenicichla		Rivulus	Control (Crenicichla)		Introduction (Rivulus)
Male age at maturity (days)	51.8	$P < 0.01$	58.8	48.5	$P < 0.01$	58.2
Male size at maturity (mg wet)	87.7	$P < 0.01$	99.7	67.5	$P < 0.01$	76.1
Female age at first birth (days)	71.5	$P < 0.01$	81.9	85.7	$P < 0.05$	92.3
Female size at first birth (mg wet)	218.0	$P < 0.01$	270.0	161.5	$P < 0.01$	185.6
Size of litter 1	5.2	$P < 0.01$	3.2	4.5	$P < 0.05$	3.3
Size of litter 2	10.9	NS	10.2	8.1	NS	7.5
Size of litter 3	16.1	NS	16.0	11.4	NS	11.5
Offspring weight (mg dry) litter 1	0.84	$P < 0.01$	0.99	0.87	$P < 0.10$	0.95
Offspring weight litter 2	0.95	$P < 0.05$	1.05	0.90	$P < 0.05$	1.02
Offspring weight litter 3	1.03	$P < 0.01$	1.17	1.10	NS	1.17
Interlitter interval (days)	22.8	NS	25.0	24.5	NS	25.2
Reproductive effort (%)	25.1	$P < 0.05$	19.2	22.0	NS	18.5

NS, not significant.

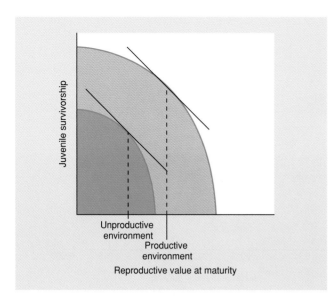

Figure 4.27 Age and size at maturity in productive and unproductive environments. When reproductive value at maturity is traded off against juvenile survivorship, the trade-off curve at the edge of the options set in the productive environment lies beyond that in the unproductive environment, and earlier maturity at a larger size is predicted.

are therefore relatively low CR, and, as predicted, the guppies there mature earlier and at a smaller size. They also make a larger reproductive allocation (left-hand columns in Table 4.6) (Reznick, 1982). Moreover, when 200 guppies were introduced from a *Crenicichla* to a *Rivulus* site, and lived there for 11 years (30–60 generations), not only did the phenotypes in the field come to resemble those of other high CR (*Rivulus*) sites, but it was also clear that these differences had evolved and were heritable, since they were also discernable under laboratory conditions (right-hand columns in Table 4.5) (Reznick *et al.*, 1990).

moving beyond the classification of habitats

Understanding age at maturity, however, requires us to move beyond our simple classification of habitats. For example, the favored age and size at maturity can be thought of as being governed by a trade-off between juvenile (pre-maturity) survival and reproductive value at maturity (for a review, see Stearns, 1992). Deferring maturity to a larger size increases the reproductive value at maturity, but it does so at the expense of decreased juvenile survival, since the juvenile phase is, by definition, extended when maturity is deferred. With this trade-off in mind, we can ask, for example, how age and size at maturity might differ between a 'productive' environment, with abundant food, and an 'unproductive' one in which individuals are poorly nourished. If increased food availability increases both the rate of growth (i.e. the size at a given age) and juvenile survival (i.e. the probability of reaching a given age), then the options set in the productive environment will extend beyond that in the unproductive environment, whatever the shape of the trade-off curve (Figure 4.27). Organisms in more productive environments should then mature both earlier and at a larger size. This has been observed commonly in *Drosophila melanogaster*: flies growing at 27°C with abundant food at moderate densities start to reproduce at 11 days, weighing 1.0 mg, whereas crowded, poorly nourished flies start after 15 days or more, weighing 0.5 mg (Stearns, 1992). Note here that we are comparing the immediate responses of individuals to their environments, rather than comparing two quite independent populations or species. We return to this point in Section 4.13.

4.10.3 Semelparity

Returning to a comparison of high and low CR habitats, it is clear that semelparity is most likely to evolve in the latter (Figure 4.28a). This is borne out in a detailed way by work on two species of *Lobelia* living on Mount Kenya (Figure 4.28b). These are long-lived herbaceous species: both take from 40 to 60 years even to mature, following which the semelparous *L. telekii* dies, whereas the iteroparous *L. keniensis* breeds only every 7–14 years. Young (1990) and Young and Augspurger (1991) showed that in drier *Lobelia* sites, probabilities of adult survival are smaller and periods between reproductive events are longer, i.e. the drier sites are lower CR. Semelparity would only actually be favored there, however, if semelparous plants also gained a sufficient reproductive advantage by diverting more resources to reproduction and less to future survival. In fact, there appears to be a close correspondence between the geographic boundary between the semelparous and the iteroparous species and the boundary where the balance of advantage swings from one to the other reproductive strategy (Figure 4.28b).

If we now relax our assumption that all options sets are convex-outwards, it is apparent that semelparity is especially likely to evolve in organisms with options sets that are concave-outwards, i.e. where even a low level of present reproduction leads to a considerable drop in, say, future survival, but increases from low to moderate levels have little influence on survival (see Figure 4.24). This is the likely explanation for why many species of salmon display suicidal semelparity. Reproduction for them demands a dangerous and effortful upstream migration from the sea to their spawning grounds, but the risks and extra costs are associated with the 'act' of reproduction and are largely independent of the magnitude of the reproductive allocation.

4.11   The size and number of offspring

According to the classification of Section 4.9, the division of a given reproductive allocation into a smaller number of larger offspring is expected in relatively offspring size-sensitive habitats. Support

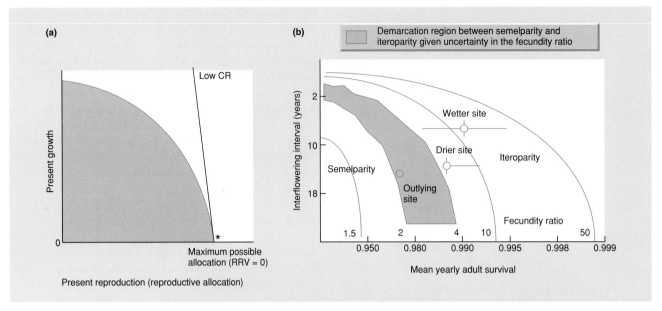

Figure 4.28 (a) Relatively low CR habitats (near-vertical fitness contours) are more likely to give rise to semelparity (maximum reproductive allocation: nothing kept in reserve). RRV, residual reproductive value. (b) For *Lobelia* spp. on Mount Kenya, habitats become lower CR as interflowering interval increases (down the axis) and mean yearly adult survival decreases. Given that the semelparous *L. telekii* sets approximately four times the weight of seed set by the iteroparous *L. keniensis*, habitats can be predicted to favor either semelparity (bottom left) or iteroparity (upper right), with a region of uncertainty between the two. Three study populations of *L. keniensis* have, as predicted, either habitat characteristics favoring iteroparity, or, in the case of an outlying site, characteristics in the region of uncertainty. (After Young, 1990; Stearns, 1992.)

for this is provided by the observations and experiments on guppies described previously (see Table 4.6): offspring size was larger where predation was most concentrated on the smaller juveniles; and also by examples in Figure 4.23 where offspring size was larger in habitats where competition was likely to be most intense – goldenrods in prairies (as opposed to more temporary old field habitats) and *Drosophila* on pollen (as opposed to rich but unpredictable sources of yeast).

4.11.1 The number of offspring: clutch size

The offspring number and fitness trade-off, however, is perhaps best not viewed in isolation. Rather, if we combine it with the CR trade-off, we can turn to another of our types of life history question and ask: 'How is it that particular clutch sizes, or particular sizes of seed crop, have been favored?'

the Lack clutch size

Lack (1947b) concentrated on the trade-off between offspring number and fitness and proposed that natural selection will favor not the largest clutch size but a compromise clutch size, which, by balancing the number produced against their subsequent survival, leads to the maximum number surviving to maturity. This has come to be known as the 'Lack clutch size' (Figure 4.29a). A number of attempts, especially with birds and to a lesser extent with insects, have been made to test the validity of this proposal by adding eggs to or removing them from natural clutches or broods, determining which clutch size is ultimately the most productive, and comparing this with the normal clutch size. Many of these have suggested that Lack's proposal is wrong: the clutch size most commonly observed 'naturally' is not the most productive. Experimental increases in clutch size, in particular, often lead to apparent increases in productivity (Godfray, 1987; Lessells, 1991; Stearns, 1992). Nevertheless, as is so often the case, Lack's proposal, whilst wrong in detail, has been immensely important in directing ecologists towards an understanding of clutch size. A number of reasons for the lack of fit are now apparent and two are particularly important.

First, many of the studies are likely to have made an inadequate assessment of the fitness of individual off-spring. It is not enough to add two eggs to a bird's normal clutch of four and note that six apparently healthy birds hatch, develop and fledge from the nest. How well do they survive the following winter? How many chicks do they have themselves? For example, in a long-term study of great tits (*Parus major*) near Oxford, UK,

beyond the Lack clutch size

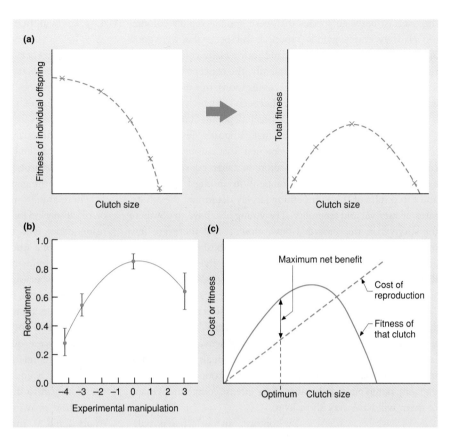

Figure 4.29 (a) The 'Lack clutch size'. If the fitness of each individual offspring decreases as total clutch size increases, then the total fitness of a clutch (the product of number and individual fitness) must be maximized at some intermediate ('Lack') clutch size. (b) The mean observed number of young recruited per nest ± SE relative to experimental manipulations (additions to, or removals from, clutches) in great tits. The curve is the polynomial RECRUITMENT ~ EXPTL MANIP + (EXPTL MANIP)^2. (After Pettifor *et al.*, 2001.) (c) However, if there is also a cost of reproduction, then the 'optimum' clutch size is that where the net fitness is greatest, i.e. here, where the distance between the cost line and the 'benefit' (total clutch) curve is greatest. (After Charnov & Krebs, 1974.)

whereas 'addition' nests were more immediately productive (10.96) than control nests (8.68), which were more productive than removal nests (5.68), recruitment (i.e. survival of offspring to become breeding adults themselves) was highest from the unmanipulated clutches (Figure 4.29b).

Second, perhaps the most important omission from Lack's proposal is any consideration of the cost of reproduction. Natural selection will favor a lifetime pattern of reproduction that gives rise to the greatest fitness overall. A large and apparently productive clutch may extract too high a price in terms of RRV. The favored clutch size will then be less than what appears to be the most productive in the short term (Figure 4.29c). Few studies have been sufficiently detailed to allow the cost of reproduction to be taken into account in assessing an optimal clutch size. In one, bank vole females (*Clethrionomys glareolus*) were treated with gonadotropin hormones, inducing them to increase their reproductive allocation to a larger litter (Oksanen *et al.*, 2002). Treated females were considerably more productive at the time the litters were born, and a very small but none the less significant increase was maintained in the number of offspring surviving to the following winter. However, the treated females also paid a significant cost for their increased reproductive efforts: higher mortality during nursing, decreased body mass gain and a decreased probability

of producing a subsequent litter. A second study, on kestrels, is discussed below (see Section 4.13).

4.12 *r* and *K* selection

Some of the predictions of the previous sections can be brought together in a scheme that has been particularly influential in the search for life history patterns. This is the concept of *r* and *K* selection, originally propounded by MacArthur and Wilson (1967; MacArthur, 1962) and elaborated by Pianka (1970) (but, see Boyce, 1984). The letter *r* refers to the intrinsic rate of natural increase (above) and indicates that *r*-selected individuals have been favored for their ability to reproduce rapidly (i.e. have a high *r* value). The letter *K* will not be introduced properly until intraspecific competition is discussed fully in the next chapter, but for now we need note only that it refers to the size ('carrying capacity') of a crowded population, limited by competition. Thus, *K*-selected individuals have been favored for their ability to make a large proportional contribution to a population that remains near that carrying capacity. The concept is therefore based on there being two contrasting types of habitat: *r*-selecting and *K*-selecting. It originally emerged (MacArthur & Wilson, 1967) from the contrast between species that were good at rapidly colonizing relatively 'empty' islands (r species), and species that were good at maintaining themselves on islands once many colonizers had reached there (K species). Subsequently, the concept was applied much more generally. Like all generalizations, this dichotomy is an oversimplification – but one that has been immensely productive.

K selection

A K-selected population lives in a habitat that imposes few random environmental fluctuations on it. As a consequence, a crowded population of fairly constant size is established. There is intense competition amongst the adults, and the results of this competition largely determine the adults' rates of survival and fecundity. The young also have to compete for survival in this crowded environment, and there are few opportunities for the young to become established as breeding adults themselves. In short, the population lives in a habitat that, because of intense competition, is both high CR and offspring size-sensitive.

The predicted characteristics of these K-selected individuals are therefore larger size, deferred reproduction, iteroparity (i.e. more extended reproduction), a lower reproductive allocation and larger (and thus fewer) offspring. The individuals will generally invest in attributes that increase survival (as opposed to reproduction); but in practice (because of the intense competition) many of them will have very short lives.

r selection

By contrast, an r-selected population lives in a habitat that is either unpredictable in time or short lived. Intermittently, the population experiences benign periods of rapid population growth, free from competition (either when the environment fluctuates into a favorable period, or when a site has been newly colonized). But these benign periods are interspersed with malevolent periods of unavoidable mortality (either in an unpredictable, unfavorable phase, or when an ephemeral site has been fully exploited or disappears). The mortality rates of both adults and juveniles are therefore highly variable and unpredictable, and they are frequently independent of population density and of the size and condition of the individuals concerned. In short, the habitat is both low CR and offspring size-insensitive.

The predicted characteristics of r-selected individuals are therefore smaller size, earlier maturity, possibly semelparity, a larger reproductive allocation and more (and thus smaller) offspring. The individuals will invest little in survivorship, but their actual survival will vary considerably depending on the (unpredictable) environment in which they find themselves.

The scheme is thus a special case of the general classification of habitats in Figure 4.25c. Note, therefore, first, that adult and offspring habitats need not be linked in the way the r/K scheme envisages, and second, that the life history characteristics associated with the r/K scheme can arise for all sorts of reasons beyond its scope (e.g. predation of diminutive adults as opposed to intense competition amongst adults).

4.12.1 Evidence for the r/K concept

The r/K concept can certainly be useful in describing some of the general differences between taxa. For instance, amongst plants it is possible to draw up a number of very broad and general relationships (Figure 4.30). Trees, in relatively K-selecting woodland habitats (relatively constant and predictable), exhibit long life, delayed maturity, large seed size, low reproductive allocation, large individual size and a very high frequency of iteroparity. Whilst in more disturbed, open, r-selecting habitats, plants tend to conform to the general syndrome of r characteristics.

There are also many cases in which populations of a species, or of closely related species, have been compared, and the correspondence with the r/K scheme has been good. For instance, this is true of a study of *Typha* (cattail or reed mace) populations (Table 4.7). Individuals of a southerly species, *T. domingensis*, and a northerly species, *T. angustifolia*, were taken from sites in Texas and North Dakota, respectively, and were grown side by side under the same conditions. In addition, certain aspects of the habitats, with long and short growing seasons, in which these species are found were quantified. It is clear from Table 4.7 that the former were relatively K-selecting and the latter relatively r-selecting. It is equally clear that the species inhabiting these sites conform to the r/K scheme. *T. angustifolia* (which naturally has a short growing season) matures earlier (trait 1), is smaller (traits 2 and 3), makes a larger reproductive allocation (traits 3 and 6) and produces more and smaller offspring (traits 4 and 5) than does *T. domingensis* (long growing season).

There are, then, examples that fit the r/K scheme. Stearns (1977), however, in an extensive review of the data available at that time, found that of 35 thorough studies, 18 conformed to

the scheme explains much – but leaves as much unexplained

the scheme whilst 17 did not. We might regard this as a damning criticism of the r/K concept, since it undoubtedly shows that the explanatory powers of the scheme are limited. On the other hand, a 50% success rate is hardly surprising given the number of additional factors already described (or to be described) that further our understanding of life history patterns. It is therefore equally possible to regard it as very satisfactory that a relatively simple concept can help make sense of a large proportion of the multiplicity of life histories. Nobody, though, can regard the r/K scheme as the whole story.

4.13 Phenotypic plasticity

A life history is not a fixed property that an organism exhibits irrespective of the prevailing environmental conditions. An observed life history is the result of long-term evolutionary forces, but also of the more immediate responses of an organism to the environment in which it is and has been living. This

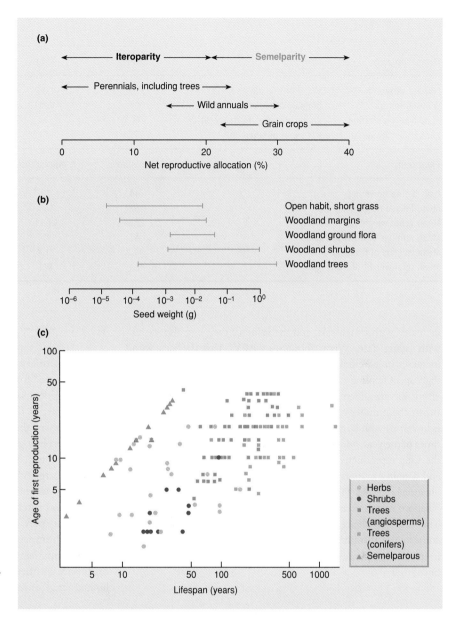

Figure 4.30 Broadly speaking, plants show some conformity with the *r/K* scheme. For example, trees in relatively *K*-selecting woodland habitats: (a) have a relatively high probability of being iteroparous and a relatively small reproductive allocation; (b) have relatively large seeds; and (c) are relatively long lived with relatively delayed reproduction. (After Harper, 1977; following Salisbury, 1942; Ogden, 1968; Harper & White, 1974.)

ability of a single genotype to express itself in different ways in different environments is known as phenotypic plasticity.

One of the most important questions we need to ask about phenotypic plasticity is the extent to which it represents a response by which an organism allocates resources differently in different environments such that it maximizes its fitness in each. The alternative would be that the response represented a degree of inevitable or uncontrolled damage or stunting by the environment (Lessells, 1991). Note, especially, that if phenotypic plasticity is governed by natural selection, then it is just as valid to seek patterns linking different environments and the different responses to them by a single individual, as it is to seek patterns linking the habitats and the life histories of genetically different individuals.

In some cases at least, the appropriateness of a plastic response seems clear. For example, kestrels (predatory birds) in the Netherlands vary in the quality of their territory, the size of their clutch and the date on which they lay it (Daan *et al.*, 1990). The differences appear not to be genetically determined but to be an example of phenotypic plasticity. Is each combination of clutch size and laying date optimal in its own territory?

The optimal combination is, as usual, the one with the highest total reproductive value – the value of the present clutch plus the parent's RRV. The value of the present clutch clearly

Habitat property	Measured by	Growing season	
		Short	Long
Climate variability	$s^2/\bar{x}$ frost-free days per year	3.05	1.56
Competition	Biomass above ground (g m^{-2})	404	1336
Annual recolonization	Winter rhizome mortality (%)	74	5
Annual density variation	$s^2/\bar{x}$ shoot numbers m^{-2}	2.75	1.51
Plant traits		T. angustifolia	T. domingensis
Days before flowering		44	70
Mean foliage height (cm)		162	186
Mean genet weight (g)		12.64	14.34
Mean number of fruits per genet		41	8
Mean weights of fruits (g)		11.8	21.4
Mean total weight of fruits (g)		483	171

Table 4.7 Life history traits of two *Typha* (cattail) species, along with properties of the habitats in which they grow. '$s^2/\bar{x}$' refers to the variance : mean ratio, a measure of variability. The cattails conform to the *r/K* scheme. (After McNaughton, 1975.)

increases with clutch size, and the value of each egg also varies with laying date. What, though, of RRV? This declines with increases in 'parental effort' (i.e. the number of hours per day in flight spent hunting in order to raise a clutch of chicks), and parental effort, in turn, decreases with increases in the 'quality' of a territory: the number of prey caught per hunting hour. Thus, RRV is lower: (i) with larger clutches; (ii) at particular, less productive times of the year; and (iii) in lower quality territories. On this basis, the total reproductive value of each combination of clutch size and date in each territory could be computed, the optimal combination predicted (Figure 4.31a) and the predicted and actual combinations compared in territories of different quality (Figure 4.31b). The correspondence is impressive. Each individual apparently comes close to optimizing its clutch size and laying date as an immediate response to the environment (territory) in which it finds itself.

4.14 Phylogenetic and allometric constraints

The life histories that natural selection favors (and we observe) are not selected from an unlimited supply but are constrained by the phylogenetic or taxonomic position that organisms occupy. For example, in the entire order Procellariiformes (albatrosses, petrels, fulmars) the clutch size is one, and the birds are 'prepared' for this morphologically by having only a single brood patch with which they can incubate this one egg (Ashmole, 1971). A bird might produce a larger clutch, but this is bound to be a waste unless it exhibits concurrent changes in all the processes in the development of the brood patch. Albatrosses are therefore prisoners of their evolutionary past, as are all organisms. Their life histories can evolve to only a limited number of options, and the organisms are therefore confined to a limited range of habitats.

organisms are prisoners of their evolutionary past

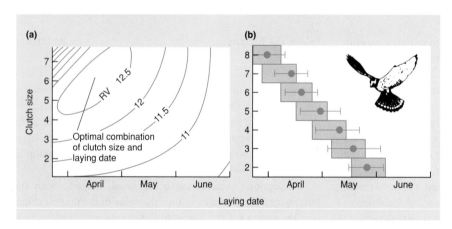

Figure 4.31 Phenotypic plasticity in the combination of clutch size and laying date in the kestrel, *Falco tinnunculus*, in the Netherlands. (a) Within particular territories (in this case, one of high quality) the expected, optimal combination is that with the highest total reproductive value (for calculation, see text). (b) Predicted (rectangles) and observed (points with standard deviations) combinations for territories varying in quality from high (left) to low (right). (After Daan *et al.*, 1990; Lessells, 1991.)

It follows from these 'phylogenetic' constraints that caution must be exercised when life histories are compared. The albatrosses, as a group, may be compared with other types of birds in an attempt to discern a link between the typical albatross life history and the typical albatross habitat. The life histories and habitats of two albatross species might reasonably be compared. But if an albatross species is compared with a distantly related bird species, then care must be taken to distinguish between differences attributable to habitat (if any) and those attributable to phylogenetic constraints.

4.14.1 Effects of size and allometry

One element of phylogenetic constraint is that of size. Figure 4.32a shows the relationship between time to maturity and size (weight)

in a wide range of organisms from viruses to whales. Note first that particular groups of organisms are confined to particular size ranges. For instance, unicellular organisms cannot exceed a certain size because of their reliance on simple diffusion for the transfer of oxygen from their cell surface to their internal organelles. Insects cannot exceed a certain size because of their reliance on unventilated tracheae for the transfer of gases to and from their interiors. Mammals, being endothermic, must exceed a certain size, because at smaller sizes the relatively large body surface would dissipate heat faster than the animal could produce it, and so on.

The second point to note is that time to maturity and size are strongly correlated. In fact, as Figure 4.32a–c illustrates, size is strongly correlated with many life history components. Since the sizes of organisms are constrained by phylogenetic position, these other life history components will be constrained too.

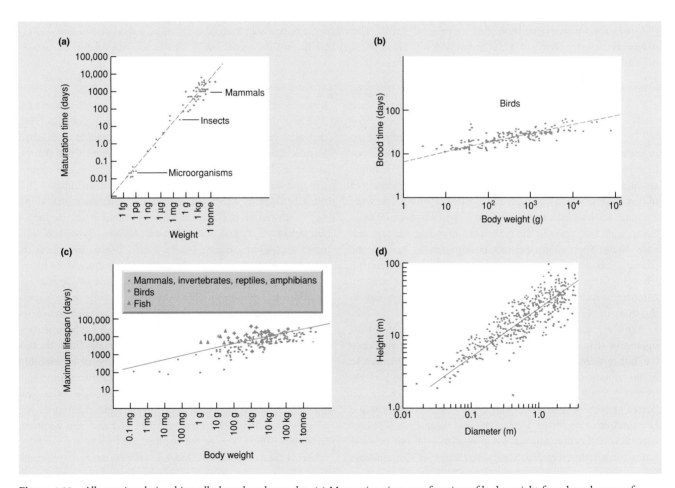

Figure 4.32 Allometric relationships, all plotted on log scales. (a) Maturation time as a function of body weight for a broad range of animals. (b) Brood time as a function of maternal body weight in birds. (c) Maximum lifespan as a function of adult body weight for a broad range of animals. (After Blueweiss *et al.*, 1978.) (d) The allometric relationship between tree height and trunk diameter 1.525 m from the ground, for 576 individual 'record' trees representing nearly every American species. (After McMahon, 1973.)

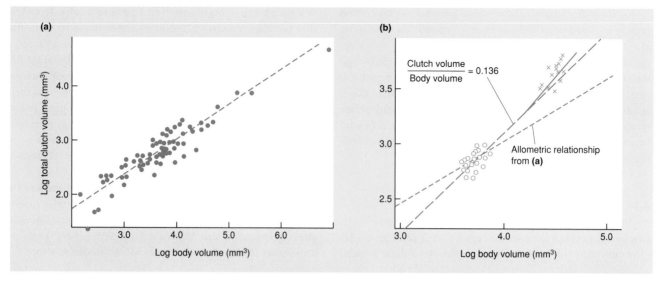

Figure 4.33   Allometric relationships between total clutch volume and body volume in female salamanders. (a) The overall relationship for 74 salamander species, using one mean value per species ($P < 0.01$). (b) The relationships within a population of *Ambystoma tigrinum* ($\times$) ($P < 0.01$), and within a population of *A. opacum* ($\circ$) ($P < 0.05$). The allometric relationship from (a) is shown as a short dashed line: *A. opacum* conforms closely to it; *A. tigrinum* does not. However, they both lie on an isometric line along which clutch volume is 13.6% of body volume (— — —). (After Kaplan & Salthe, 1979.)

allometry defined

An allometric relationship (see Gould, 1966) is one in which a physical or physiological property of an organism alters relative to the size of the organism. For example, in Figure 4.33a an increase in size (actually volume) amongst salamander species leads to a decrease in the *proportion* of that volume which is allocated to a clutch of young. Likewise, in Figure 4.32b an increase in weight amongst bird species is associated with a decrease in the time spent brooding eggs *per unit body weight*. Such allometric relationships can be ontogenetic (changes occurring as an organism develops) or phylogenetic (changes that are apparent when related taxa of different sizes are compared), and it is the latter that are particularly important in the study of life histories (Figures 4.32 and 4.33).

Why are there allometric relationships? Briefly, if similar organisms that differed in size retained a geometric similarity (i.e. if they were *isometric*), then all surface areas would increase as the square of linear size, whilst all volumes and weights increased as the cube. An increase in size would then lead to decreases in the length : area ratios, decreases in length : volume ratios and, most important, decreases in area : volume ratios. Almost every bodily function depends for its efficiency on one of these ratios (or a ratio related to them). A change in size amongst isometric organisms would therefore lead to a change in efficiency.

why are there allometric relationships?

For example, the transfer of heat, or water, or nutrients, either within an organism or between an organism and its environment, takes place across a surface, which has an area. The amount of heat produced or water required, however, depends on the volume of the organ or organism concerned. Hence, changes in area : volume ratios resulting from changes in size are bound to lead to changes in the efficiency of transfer per unit volume. Thus, if efficiency is to be maintained, this must be done by allometric alterations. Exact allometric slopes vary from system to system and from taxon to taxon (for further discussion see Gould, 1966; Schmidt-Nielsen, 1984; and, in a more ecological context, Peters, 1983). What, though, is the significance of allometry in the study of life histories?

The usual approach to the ecological study of life histories has been to compare the life histories of two or more populations (or species or groups), and to seek to understand the differences between them by reference to their environments. It must be clear by now, however, that taxa can also differ because they lie at different points on the same allometric relationship, or because they are subject to different phylogenetic constraints generally. It is therefore important to disentangle 'ecological' differences from allometric and phylogenetic differences (see Harvey & Pagel, 1991; Harvey, 1996; and also the summary in Stearns, 1992). This is not because the former are 'adaptive' whereas the latter are not. Indeed, we have seen, for example, that the basis for allometric relationships is a matching of organisms of different sizes to their respective environments. Rather, it is a question of the evolutionary responses of a species to its habitat being limited by constraints that have themselves evolved.

comparing
salamanders:
dangerous if
allometries are
ignored

These ideas are illustrated in Figure 4.33a, which shows the allometric relationship between clutch volume and body volume for salamanders generally. Figure 4.33b then shows the same relationships in outline; but superimposed upon it are the allometric relationships within populations for two salamander species, *Ambystoma tigrinum* and *A. opacum* (Kaplan & Salthe, 1979). If the species' means are simply compared, without reference to the general salamander allometry, then the species are seen to have the same ratio of clutch volume : body volume (0.136). This seems to suggest that the species' life histories 'do not differ', and that there is therefore 'nothing to explain' – but any such suggestion would be wrong. *A. opacum* conforms closely to the general salamander relationship. *A. tigrinum*, on the other hand, has a clutch volume which is almost twice as large as would be expected from that relationship. Within the allometric constraints of being a salamander, *A. tigrinum* is making a much greater reproductive allocation than *A. opacum*; and it would be reasonable for an ecologist to look at their respective habitats and seek to understand why this might be so.

In other words, it is reasonable to compare taxa from an 'ecological' point of view as long as the allometric relationship linking them at a higher taxonomic level is known (Clutton-Brock & Harvey, 1979). It will then be their respective deviations from the relationship that form the basis for the comparison. Problems arise, though, when allometric relationships are unknown (or ignored). Without the general salamander allometry in Figure 4.33a, the two species would have seemed similar when in fact they are different. Conversely, two other species might have seemed different when in fact they were simply conforming to the same allometric relationship. Comparisons oblivious to allometries are clearly perilous, but regrettably, ecologists *are* frequently oblivious to allometries. Typically, life histories have been compared, and attempts have been made to explain the differences between them, in terms of habitat differences. As previous sections have

shown, these attempts have often been successful. But they have also often been unsuccessful, and unrecognized allometries undoubtedly go some way towards explaining this.

4.14.2 Effects of phylogeny

The approach used with the salamanders, of comparing species or other groups in terms of their deviations from an allometric relationship that links them, has been applied successfully to a number of larger assemblages. By removing the effects of size, the approach searches for phylogenetic relationships beyond those associated with size. For example, Figure 4.34 shows, for a number of mammal species, that 'relative' age at first reproduction increases as 'relative' life expectancy increases (i.e. relative to a value expected on the basis of an underlying allometry). This shows a powerful relationship between these two life history characters once the confounding effects of size have been removed. It also reveals underlying similarities between species of very different sizes: elephants and otters, and mice and warthogs.

removing the
confounding effects
of size

A further impression of the strength of the influence of phylogeny can be gained from analyses like those in Table 4.8 (Read & Harvey, 1989). A nested analysis of variance has been applied to the variation in seven life history traits amongst a large number of mammal species. This has led to the determination of the percentage of the total variance attributable to: (i) differences between species within genera; (ii) differences between genera within families; and so on. Species vary very little within genera; genera vary little within families. Far and away the largest part of the variance, for all the traits, is accounted for by differences between orders within the mammalian class as a whole. This emphasizes that in simply comparing two species from different orders, we are in essence comparing those orders (which probably diverged many millions of years ago) rather than the species themselves. It does not mean, however, that comparing species

Table 4.8 When nested analyses of variance are performed on data sets for a number of life history traits from a large number of mammal species, the percentage of the variance is greatest at the highest taxonomic level (orders within the class) and least at the lowest level (species within genera). (After Read & Harvey, 1989.)

Trait	Species within genera	Genera within families	Families within orders	Orders within the class
Gestation length	2.4	5.8	21.1	70.7
Age at weaning	8.4	11.5	18.9	61.6
Age at maturity	10.7	7.2	26.7	55.4
Interlitter interval	6.6	13.5	16.1	63.8
Maximum lifespan	9.7	10.1	12.4	67.8
Neonatal weight	2.9	5.5	26.6	64.9
Adult weight	2.9	7.5	21.0	68.5

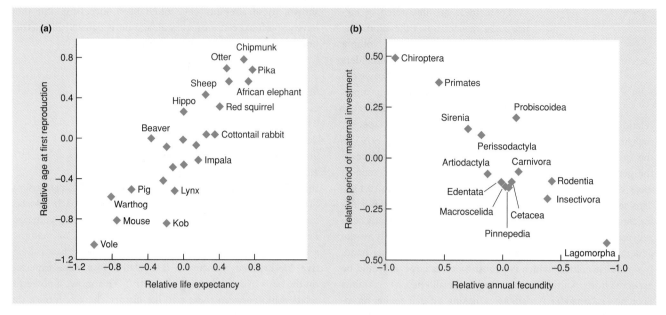

Figure 4.34 After the effects of size have been removed, age at first reproduction increases with life expectancy at birth for 24 species of mammals. 'Relative' refers to the deviation from the underlying allometric relationship linking the character in question to organism size. (After Harvey & Zammuto, 1985.)

in the same genus, say, is only 'scratching the surface'. Even when two species are very similar in their life histories and habitats, if one makes a greater reproductive allocation and also lives in a habitat that is lower CR, then this allows us to build a pattern linking the two.

a role, still, for habitat? . . . Moreover, the strength of these relationships at the higher taxonomic levels does not mean that attempts to relate life histories to lifestyles and habitats should be abandoned even there, since lifestyles and habitats are also constrained by an organism's size and phylogenetic position. There may still, therefore, at these higher levels, be patterns linking habitats and life histories rooted in natural selection. For example, insects (small size, many offspring, high reproductive allocation, frequent semelparity) have been described as relatively *r*-selected, compared to mammals (large size, few offspring, etc. – relatively *K*-selected) (Pianka, 1970). Such differences could be dismissed as being 'no more' than the product of an ancient evolutionary divergence (Stearns, 1992). But, as we have stressed, an organism's habitat reflects its own responses to its environment. Hence a mammal and an insect living side by side are also almost certain to experience very different habitats. The larger, homeothermic, behaviorally sophisticated, longer lived mammal is likely to maintain a relatively constant population size, subject to frequent competition, and be relatively immune from environmental catastrophes and uncertainties. The smaller, poikilothermic, behaviorally unsophisticated, shorter lived insect, by contrast, is likely to live a relatively opportunistic life, with a

high probability of unavoidable death. Insects and mammals are prisoners of their evolutionary past in their range of habitats just as they are in their range of life histories – and the *r/K* scheme provides a reasonable (although certainly not perfect) summary of the patterns linking the two.

The same point is illustrated in a more quantitative way by an application of the 'phylogenetic-subtraction' method (Harvey & Pagel, 1991) to patterns of covariation in 10 life history traits of mammals (Stearns, 1983). In the unmanipulated data, the pattern to be expected under the influence of *r/K* selection was pronounced: it accounted for 68% of the covariation. This *r/K* influence was reduced to about 42% when the effects of weight were removed, and was reduced still further when the trait values were replaced in the analysis by their deviations from the mean value for the family to which the species belonged (33%), or their order (32%). In the first place, this reaffirms the importance of both size and phylogeny, since each clearly accounted for much of the interspecific variation. It is also significant that the *r/K* pattern remained clearly visible even after these other effects had been removed. But the strength of the pattern in the unmanipulated data set cannot simply be dismissed as an artefact arising out of phylogenetic naivety. It may well be that important differences in habitat, too, are associated with an organism's size, or its order or family.

It is undoubtedly true that life history ecology cannot proceed oblivi- *. . . yes!* ous to phylogenetic and allometric constraints. Yet it would be unhelpful to see phylogeny as an alternative explanation to habitat in seeking to understand life histories. Phylogeny sets limits to an organism's life history and to its habitat. But the essentially ecological task of relating life histories to habitats remains the most fundamental challenge.

Summary

Ecologists are interested in the numbers of individuals, the distributions of individuals, the demographic processes (birth, death and migration) that influence these, and the ways in which these demographic processes are themselves influenced by environmental factors.

Not all individuals are alike, especially amongst modular, as opposed to unitary, organisms. The growth forms of modular organisms are described, as well the nature and ecological importance of senescence and physiological integration in modular organisms. Ecology necessarily involves counting individuals or modules. A population is a group of individuals of one species, though what constitutes a population will vary from study to study. It is often most convenient to consider the density as opposed to the size of a population. Methods of estimating population size or density are described briefly.

We explain the variety of patterns of life cycle, including the distinction between semelparous and iteroparous species. Basic methods of quantification of these include life tables, survivorship curves and fecundity schedules. For annual species, cohort life tables can be constructed, the elements of which are described. A summary term of this and a fecundity schedule is the basic reproductive rate, R_0. The survivorship curves that emerge from a life table can be classified into three broad types. However, a variety of features, including seed banks, mean that there are many not-quite-annual species.

For individuals with repeated breeding seasons, it may also be possible to construct cohort life tables; a static life table is an imperfect alternative that must be interpreted with caution.

We explain how basic reproductive rates, R_0, generation lengths and population rates of increase are interrelated when generations overlap, leading to definitions of the fundamental net reproductive rate, R, and the intrinsic rate of natural increase, r ($= \ln R$). We explain, too, how these may be estimated from life tables and fecundity schedules, and move on to describe the population projection matrix, a more powerful method of analyzing and interpreting fecundity and survival schedules when generations overlap.

Three different types of question that are commonly asked about the evolution of life histories are described. Most answers to these questions have been based on the idea of optimization. The components of life histories, and their ecological importance, are also described: size, development rate, semel- or iteroparity, clutch size, offspring size and some composite measures – reproductive allocation and especially reproductive value.

Trade-offs are central to an understanding of life history evolution, though they may be difficult to observe in practice. Key trade-offs are those that reveal an apparent 'cost of reproduction' in terms of a decrease in residual reproductive value. Another is that between the number and fitness of offspring.

To address the question of whether there are patterns linking particular types of life history to particular types of habitat, the concepts of options sets and fitness contours are introduced, leading to a general, comparative classification of habitats. Armed with this, light is thrown on patterns in reproductive allocation and its timing, the optimal size and number of offspring. We explain the concept of r and K selection, its limitations and the evidence for it. We explain, too, that patterns in the phenotypic plasticity of life histories may equally be governed by natural selection.

Finally, the effects of phylogenetic and allometric constraints on the evolution of life histories are discussed – especially the effects of size – but end with the conclusion that the essentially ecological task of relating life histories to habitats remains the most fundamental challenge.

Chapter 5
Intraspecific Competition

5.1 Introduction

Organisms grow, reproduce and die (Chapter 4). They are affected by the conditions in which they live (Chapter 2), and by the resources that they obtain (Chapter 3). But no organism lives in isolation. Each, for at least part of its life, is a member of a population composed of individuals of its own species.

a definition of competition

Individuals of the same species have very similar requirements for survival, growth and reproduction; but their combined demand for a resource may exceed the immediate supply. The individuals then compete for the resource and, not surprisingly, at least some of them become deprived. This chapter is concerned with the nature of such intraspecific competition, its effects on the competing individuals and on populations of competing individuals. We begin with a working definition: 'competition is an interaction between individuals, brought about by a shared requirement for a resource, and leading to a reduction in the survivorship, growth and/or reproduction of at least some of the competing individuals concerned'. We can now look more closely at competition.

Consider, initially, a simple hypothetical community: a thriving population of grasshoppers (all of one species) feeding on a field of grass (also of one species). To provide themselves with energy and material for growth and reproduction, grasshoppers eat grass; but in order to find and consume that grass they must use energy. Any grasshopper might find itself at a spot where there is no grass because some other grasshopper has eaten it. The grasshopper must then move on and expend more energy before it takes in food. The more grasshoppers there are, the more often this will happen. An increased energy expenditure and a decreased rate of food intake may all decrease a grasshopper's chances of survival, and also leave less energy available for development and reproduction. Survival and reproduction determine a grasshopper's contribution to the next generation. Hence, the more intraspecific competitors for food a grasshopper has, the less its likely contribution will be.

As far as the grass itself is concerned, an isolated seedling in fertile soil may have a very high chance of surviving to reproductive maturity. It will probably exhibit an extensive amount of modular growth, and will probably therefore eventually produce a large number of seeds. However, a seedling that is closely surrounded by neighbors (shading it with their leaves and depleting the water and nutrients of its soil with their roots) will be very unlikely to survive, and if it does, will almost certainly form few modules and set few seeds.

We can see immediately that the ultimate effect of competition on an individual is a decreased contribution to the next generation compared with what would have happened had there been no competitors. Intraspecific competition typically leads to decreased rates of resource intake per individual, and thus to decreased rates of individual growth or development, or perhaps to decreases in the amounts of stored reserves or to increased risks of predation. These may lead, in turn, to decreases in survivorship and/or decreases in fecundity, which together determine an individual's reproductive output.

5.1.1 Exploitation and interference

In many cases, competing individuals do not interact with one another directly.

exploitation

Instead, individuals respond to the level of a resource, which has been depressed by the presence and activity of other individuals. The grasshoppers were one example. Similarly, a competing grass plant is adversely affected by the presence of close neighbors, because the zone from which it extracts resources (light, water, nutrients) has been overlapped by the 'resource depletion zones' of these neighbors, making it more difficult to extract those resources. In such cases, competition may be described as

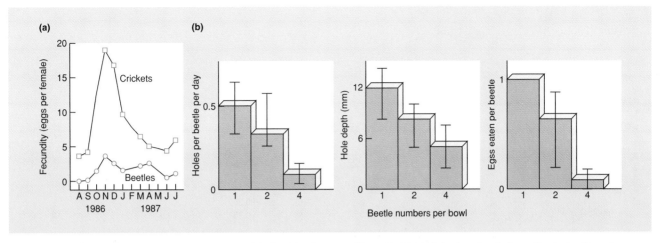

Figure 5.1 Intraspecific competition amongst cave beetles (*Neapheanops tellkampfi*). (a) Exploitation. Beetle fecundity is significantly correlated (*r* = 0.86) with cricket fecundity (itself a good measure of the availability of cricket eggs – the beetles' food). The beetles themselves reduce the density of cricket eggs. (b) Interference. As beetle density in experimental arenas with 10 cricket eggs increased from 1 to 2 to 4, individual beetles dug fewer and shallower holes in search of their food, and ultimately ate much less (*P* < 0.001 in each case), in spite of the fact that 10 cricket eggs was sufficient to satiate them all. Means and standard deviations are given in each case. (After Griffith & Poulson, 1993.)

exploitation, in that each individual is affected by the amount of resource that remains after that resource has been exploited by others. Exploitation can only occur, therefore, if the resource in question is in limited supply.

interference
In many other cases, competition takes the form of *interference*. Here individuals interact directly with each other, and one individual will actually prevent another from exploiting the resources within a portion of the habitat. For instance, this is seen amongst animals that defend territories (see Section 5.11) and amongst the sessile animals and plants that live on rocky shores. The presence of a barnacle on a rock prevents any other barnacle from occupying that same position, even though the supply of food at that position may exceed the requirements of several barnacles. In such cases, space can be seen as a resource in limited supply. Another type of interference competition occurs when, for instance, two red deer stags fight for access to a harem of hinds. Either stag, alone, could readily mate with all the hinds, but they cannot both do so since matings are limited to the 'owner' of the harem.

Thus, interference competition may occur for a resource of real value (e.g. space on a rocky shore for a barnacle), in which case the interference is accompanied by a degree of exploitation, or for a surrogate resource (a territory, or ownership of a harem), which is only valuable because of the access it provides to a real resource (food, or females). With exploitation, the intensity of competition is closely linked to the level of resource present and the level required, but with interference, intensity may be high even when the level of the real resource is not limiting.

In practice, many examples of competition probably include elements of both exploitation and interference. For instance, adult cave beetles, *Neapheanops tellkampfi*, in Great Onyx Cave, Kentucky, compete amongst themselves but with no other species and have only one type of food – cricket eggs, which they obtain by digging holes in the sandy floor of the cave. On the one hand, they suffer indirectly from exploitation: beetles reduce the density of their resource (cricket eggs) and then have markedly lower fecundity when food availability is low (Figure 5.1a). But they also suffer directly from interference: at higher beetle densities they fight more, forage less, dig fewer and shallower holes and eat far fewer eggs than could be accounted for by food depletion alone (Figure 5.1b).

5.1.2 One-sided competition

Whether they compete through exploitation or interference, individuals within a species have many fundamental features in common, using similar resources and reacting in much the same way to conditions. None the less, intraspecific competition may be very one sided: a strong, early seedling will shade a stunted, late one; an older and larger bryozoan on the shore will grow over a smaller and younger one. One example is shown in Figure 5.2. The overwinter survival of red deer calves in the resource-limited population on the island of Rhum, Scotland (see Chapter 4) declined sharply as the population became more crowded, but those that were smallest at birth were by far the most likely to die. Hence, the ultimate effect of competition is

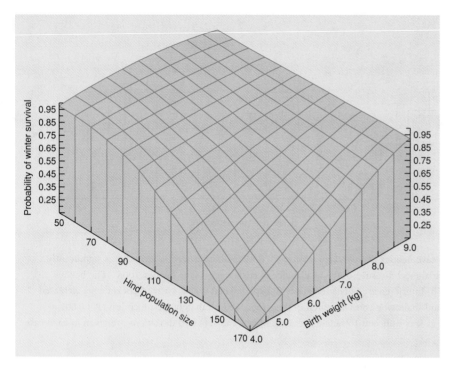

Figure 5.2   Those red deer that are smallest when born are the least likely to survive over winter when, at higher densities, survival declines. (After Clutton-Brock *et al.*, 1987.)

far from being the same for every individual. Weak competitors may make only a small contribution to the next generation, or no contribution at all. Strong competitors may have their contribution only negligibly affected.

Finally, note that the likely effect of intraspecific competition on any individual is greater the more competitors there are. The effects of intraspecific competition are thus said to be density dependent. We turn next to a more detailed look at the density-dependent effects of intraspecific competition on death, birth and growth.

5.2   Intraspecific competition, and density-dependent mortality and fecundity

Figure 5.3 shows the pattern of mortality in the flour beetle *Tribolium confusum* when cohorts were reared at a range of densities. Known numbers of eggs were placed in glass tubes with 0.5 g of a flour–yeast mixture, and the number of individuals that survived to become adults in each tube was noted. The same data have been expressed in three ways, and in each case the resultant curve has been divided into three regions. Figure 5.3a describes the relationship between density and the *per capita* mortality rate – literally, the mortality rate 'per head', i.e. the probability of an individual dying or the proportion that died between the egg and adult stages. Figure 5.3b describes how the number that died prior to the adult stage changed with density; and Figure 5.3c describes the relationship between density and the numbers that survived.

Throughout region 1 (low density) the mortality rate remained constant as density was increased (Figure 5.3a). The numbers dying and the numbers surviving both rose (Figure 5.3b, c) (not surprising, given that the numbers 'available' to die and survive increased), but the proportion dying remained the same, which accounts for the straight lines in region 1 of these figures. Mortality in this region is said to be density *in*dependent. Individuals died, but the chance of an individual surviving to become an adult was not changed by the initial density. Judged by this, there was no intraspecific competition between the beetles at these densities. Such density-independent deaths affect the population at all densities. They represent a baseline, which any density-dependent mortality will exceed.

In region 2, the mortality rate increased with density (Figure 5.3a): there was density-dependent mortality. The numbers dying continued to rise with density, but unlike region 1 they did so more than proportionately (Figure 5.3b). The numbers surviving also continued to rise, but this time less than proportionately (Figure 5.3c). Thus, over this range, increases in egg density continued to lead to increases in the total number of surviving adults. The mortality rate had increased, but it 'undercompensated' for increases in density.

In region 3, intraspecific competition was even more intense. The increasing mortality rate 'overcompensated' for any increase in density, i.e. over this range, the more eggs there were present, the fewer adults survived: an increase in the initial number of eggs led to an even

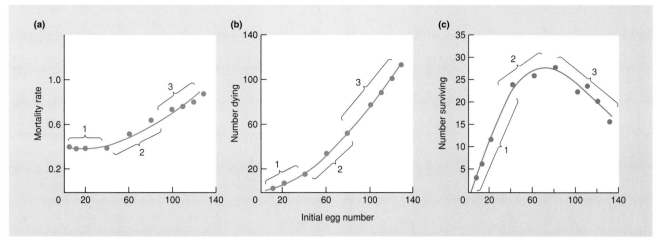

Figure 5.3 Density-dependent mortality in the flour beetle *Tribolium confusum*: (a) as it affects mortality rate, (b) as it affects the numbers dying, and (c) as it affects the numbers surviving. In region 1 mortality is density independent; in region 2 there is undercompensating density-dependent mortality; in region 3 there is overcompensating density-dependent mortality. (After Bellows, 1981.)

greater proportional increase in the mortality rate. Indeed, if the range of densities had been extended, there would have been tubes with no survivors: the developing beetles would have eaten all the available food before any of them reached the adult stage.

exactly compensating density dependence

A slightly different situation is shown in Figure 5.4. This illustrates the relationship between density and mortality in young trout. At the lower densities there was undercompensating density dependence, but at higher densities mortality never overcompensated. Rather, it compensated exactly for any increase in density: any rise in the number of fry was matched by an exactly equivalent rise in the

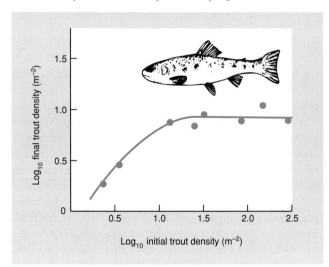

Figure 5.4 An exactly compensating density-dependent effect on mortality: the number of surviving trout fry is independent of initial density at higher densities. (After Le Cren, 1973.)

mortality rate. The number of survivors therefore approached and maintained a constant level, irrespective of initial density.

The patterns of density-dependent fecundity that result from intraspecific competition are, in a sense, a mirror-image of those for mortality (Figure 5.5). Here, though, the per capita birth rate falls as intraspecific competition intensifies. At low enough densities, the birth rate may be density independent (Figure 5.5a, lower densities). But as density increases, and the effects of intraspecific competition become apparent, birth rate initially shows undercompensating density dependence (Figure 5.5a, higher densities), and may then show exactly compensating density dependence (Figure 5.5b, throughout; Figure 5.5c, lower densities) or overcompensating density dependence (Figure 5.5c, higher densities).

intraspecific competition and fecundity

Thus, to summarize, irrespective of variations in over- and undercompensation, the essential point is a simple one: at appropriate densities, intraspecific competition can lead to density-dependent mortality and/or fecundity, which means that the death rate increases and/or the birth rate decreases as density increases. Thus, whenever there is intraspecific competition, its effect, whether on survival, fecundity or a combination of the two, is density dependent. However, as subsequent chapters will show, there are processes other than intraspecific competition that also have density-dependent effects.

5.3 Density or crowding?

Of course, the intensity of intraspecific competition experienced by an individual is not really determined by the density of the population as a whole. The effect on an individual is determined,

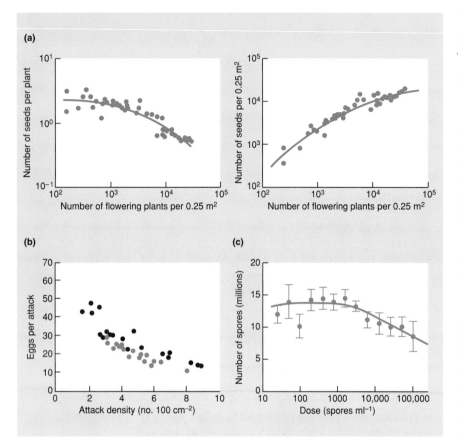

Figure 5.5 (a) The fecundity (seeds per plant) of the annual dune plant *Vulpia fasciculata* is constant at the lowest densities (density independence, left). However, at higher densities, fecundity declines but in an undercompensating fashion, such that the total number of seeds continues to rise (right). (After Watkinson & Harper, 1978.) (b) Fecundity (eggs per attack) in the southern pine beetle, *Dendroctonus frontalis*, in East Texas declines with increasing attack density in a way that compensates more or less exactly for the density increases: the total number of eggs produced was roughly 100 per 100 cm^2, irrespective of attack density over the range observed (●, 1992; ●, 1993). (After Reeve *et al.*, 1998.) (c) When the planktonic crustacean *Daphnia magna* was infected with varying numbers of spores of the bacterium *Pasteuria ramosa*, the total number of spores produced per host in the next generation was independent of density (exactly compensating) at the lower densities, but declined with increasing density (overcompensating) at the higher densities. Standard errors are shown. (After Ebert *et al.*, 2000.)

rather, by the extent to which it is crowded or inhibited by its immediate neighbors.

One way of emphasizing this is by noting that there are actually at least three different meanings of 'density' (see Lewontin & Levins, 1989, where details of calculations and terms can be found). Consider a population of insects, distributed over a population of plants on which they feed. This is a typical example of a very general phenomenon – a population (the insects in this case) being distributed amongst different patches of a resource (the plants). The density would usually be calculated as the number of insects (let us say 1000) divided by the number of plants (say 100), i.e. 10 insects per plant. This, which we would normally call simply the 'density', is actually the 'resource-weighted density'. However, it gives an accurate measure of the intensity of competition suffered by the insects (the extent to which they are crowded) only if there are exactly 10 insects on every plant and every plant is the same size.

three meanings of density

Suppose, instead, that 10 of the plants support 91 insects each, and the remaining 90 support just one insect. The resource-weighted density would still be 10 insects per plant. But the average density experienced by the insects would be 82.9 insects per plant. That is, one adds

up the densities experienced by each of the insects (91 + 91 + 91 . . . + 1 + 1) and divides by the total number of insects. This is the 'organism-weighted density', and it clearly gives a much more satisfactory measure of the intensity of competition the insects are likely to suffer.

However, there remains the further question of the average density of insects experienced by the plants. This, which may be referred to as the 'exploitation pressure', comes out at 1.1 insects per plant, reflecting the fact that most of the plants support only one insect.

What, then, is the density of the insect? Clearly, it depends on whether you answer from the perspective of the insect or the plant – but whichever way you look at it, the normal practice of calculating the resource-weighted density and calling it the 'density' looks highly suspect. The difference between resource- and organism-weighted densities is illustrated for the human population of a number of US states in Table 5.1 (where the 'resource' is simply land area). The organism-weighted densities are so much larger than the usual, but rather unhelpful, resource-weighted densities essentially because most people live, crowded, in cities (Lewontin & Levins, 1989).

The difficulties of relying on density to characterize the potential intensity of intraspecific competition are particularly

Table 5.1 A comparison of the resource- and organism-weighted densities of five states, based on the 1960 USA census, where the 'resource patches' are the counties within each state. (After Lewontin & Levins, 1989.)

State	*Resource-weighted density* (km^{-2})	*Organism-weighted density* (km^{-2})
Colorado	44	6,252
Missouri	159	6,525
New York	896	48,714
Utah	28	684
Virginia	207	13,824

acute with sessile, modular organisms, because, being sessile, they compete almost entirely only with their immediate neighbors, and being modular, competition is directed most at the modules that are closest to those neighbors. Thus, for instance, when silver birch trees (*Betula pendula*) were grown in small groups, the sides of individual trees that interfaced with neighbors typically had a lower 'birth' and higher death rate of buds (see Section 4.2); whereas on sides of the same trees with no interference, bud birth rate was higher, death rate lower, branches were longer and the form approached that of an open-grown individual (Figure 5.6). Different modules experience different intensities of competition, and quoting the density at which an individual was growing would be all but pointless.

density: a convenient expression of crowding

Thus, whether mobile or sessile, different individuals meet or suffer from different numbers of competitors. Density, especially resource-weighted density, is an abstraction that applies to the population as a whole but need not apply to any of the individuals within it. None the less, density may often be the most convenient way of expressing the degree to which individuals are crowded – and it is certainly the way it has usually been expressed.

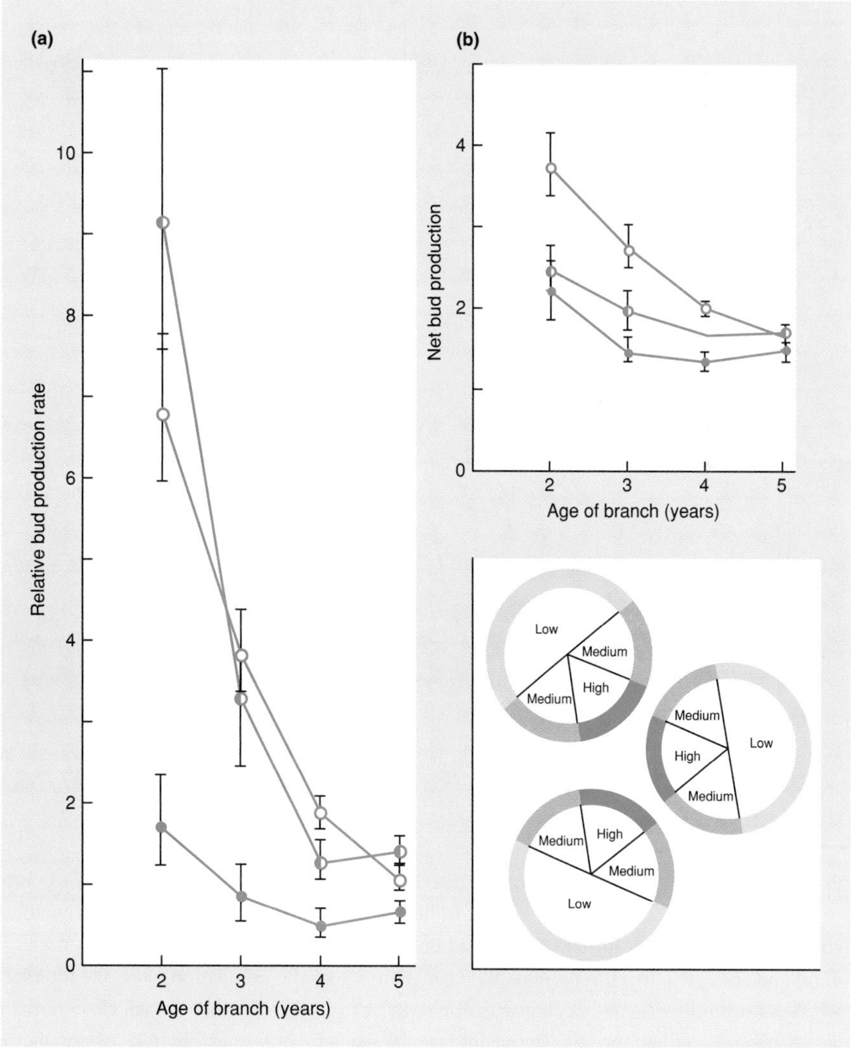

Figure 5.6 Mean relative bud production (new buds per existing bud) for silver birch trees (*Betula pendula*), expressed (a) as gross bud production and (b) as net bud production (birth minus death), in different interference zones. These zones are themselves explained in the inset. ●, high interference; ◐, medium; ○, low. Bars represent standard errors. (After Jones & Harper, 1987.)

5.4   Intraspecific competition and the regulation of population size

There are, then, typical patterns in the effects of intraspecific competition on birth and death (see Figures 5.3–5.5). These generalized patterns are summarized in Figures 5.7 and 5.8.

5.4.1   Carrying capacities

Figure 5.7a–c reiterates the fact that as density increases, the per capita birth rate eventually falls and the per capita death rate eventually rises. There must, therefore, be a density at which these curves cross. At densities below this point, the birth rate exceeds

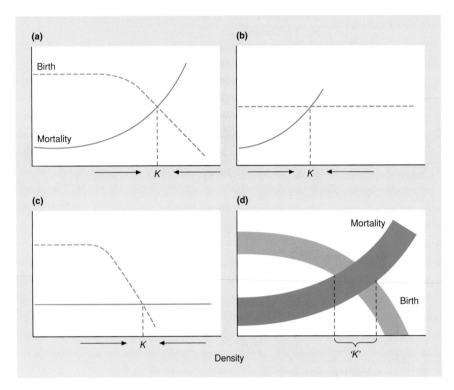

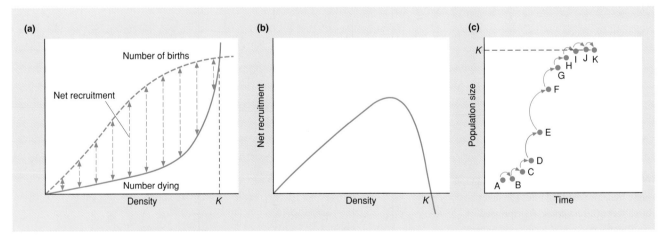

Figure 5.7   Density-dependent birth and mortality rates lead to the regulation of population size. When both are density dependent (a), or when either of them is (b, c), their two curves cross. The density at which they do so is called the carrying capacity (*K*). Below this the population increases, above it the population decreases: *K* is a stable equilibrium. However, these figures are the grossest of caricatures. The situation is closer to that shown in (d), where mortality rate broadly increases, and birth rate broadly decreases, with density. It is possible, therefore, for the two rates to balance not at just one density, but over a broad range of densities, and it is towards this broad range that other densities tend to move.

Figure 5.8   Some general aspects of intraspecific competition. (a) Density-dependent effects on the numbers dying and the number of births in a population: net recruitment is 'births minus deaths'. Hence, as shown in (b), the density-dependent effect of intraspecific competition on net recruitment is a domed or 'n'-shaped curve. (c) A population increasing in size under the influence of the relationships in (a) and (b). Each arrow represents the change in size of the population over one interval of time. Change (i.e. net recruitment) is small when density is low (i.e. at small population sizes: A to B, B to C) and is small close to the carrying capacity (I to J, J to K), but is large at intermediate densities (E to F). The result is an 'S'-shaped or sigmoidal pattern of population increase, approaching the carrying capacity.

the death rate and the population increases in size. At densities above the crossover point, the death rate exceeds the birth rate and the population declines. At the crossover density itself, the two rates are equal and there is no net change in population size. This density therefore represents a stable equilibrium, in that all other densities will tend to approach it. In other words, intraspecific competition, by acting on birth rates and death rates, can regulate populations at a stable density at which the birth rate equals the death rate. This density is known as the *carrying capacity* of the population and is usually denoted by *K* (Figure 5.7). It is called a carrying capacity because it represents the population size that the resources of the environment can just maintain ('carry') without a tendency to either increase or decrease.

real populations lack simple carrying capacities

However, whilst hypothetical populations caricatured by line drawings like Figures 5.7a–c can be characterized by a simple carrying capacity, this is not true of any natural population. There are unpredictable environmental fluctuations; individuals are affected by a whole wealth of factors of which intraspecific

competition is only one; and resources not only affect density but respond to density as well. Hence, the situation is likely to be closer to that depicted in Figure 5.7d. Intraspecific competition does not hold natural populations to a predictable and unchanging level (the carrying capacity), but it may act upon a very wide range of starting densities and bring them to a much narrower range of final densities, and it therefore tends to keep density within certain limits. It is in this sense that intraspecific competition may be said typically to be capable of regulating population size. For instance, Figure 5.9 shows the fluctuations within and between years in populations of the brown trout (*Salmo trutta*) and the grasshopper, *Chorthippus brunneus*. There are no simple carrying capacities in these examples, but there are clear tendencies for the 'final' density each year ('late summer numbers' in the first case, 'adults' in the second) to be relatively constant, despite the large fluctuations in density within each year and the obvious potential for increase that both populations possess.

In fact, the concept of a population settling at a stable carrying capacity, even in caricatured populations, is relevant only to situations in which density dependence is not strongly overcompensating. Where there is overcompensation, cycles or even

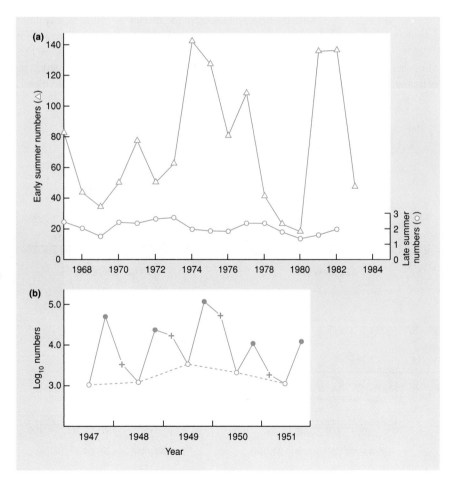

Figure 5.9 Population regulation in practice. (a) Brown trout (*Salmo trutta*) in an English Lake District stream. △, numbers in early summer, including those newly hatched from eggs; ○, numbers in late summer. Note the difference in vertical scales. (After Elliott, 1984.) (b) The grasshopper, *Chorthippus brunneus*, in southern England. ●, eggs; +, nymphs; ○, adults. Note the logarithmic scale. (After Richards & Waloff, 1954.) There are no definitive carrying capacities, but the 'final' densities each year ('late summer' and 'adults') are relatively constant despite large fluctuations within years.

chaotic changes in population size may be the result. We return to this point later (see Section 5.8).

5.4.2 Net recruitment curves

An alternative general view of intraspecific competition is shown in Figure 5.8a, which deals with numbers rather than rates. The difference there between the two curves ('births minus deaths' or 'net recruitment') is the net number of additions expected in the population during the appropriate stage or over one interval of time. Because of the shapes of the birth and death curves, the net number of additions is small at the lowest densities, increases as density rises, declines again as the carrying capacity is approached and is then negative (deaths exceed births) when the initial density exceeds K (Figure 5.8b). Thus, total recruitment into a population is small when there are few individuals available to give birth, and small when intraspecific competition is intense. It reaches a peak, i.e. the population increases in size most rapidly, at some intermediate density.

peak recruitment occurs at intermediate densities

The precise nature of the relationship between a population's net rate of recruitment and its density varies with the detailed biology of the species concerned (e.g. the trout, clover plants, herring and whales in Figure 5.10a–d). Moreover, because recruitment is affected by a whole multiplicity of factors, the data points rarely fall exactly on any single curve. Yet, in each case in Figure 5.10, a domed curve is apparent. This reflects the general nature of density-dependent birth and death whenever there is intraspecific competition. Note also that one of these (Figure 5.10b) is modular: it describes the relationship between the leaf area index (LAI) of a plant population (the total leaf area being borne per unit area of ground) and the population's growth rate (modular birth minus modular death). The growth rate is low when there are few leaves, peaks at an intermediate LAI, and is then low again at a high LAI, where there is much mutual shading and competition and many leaves may be consuming more in respiration than they contribute through photosynthesis.

5.4.3 Sigmoidal growth curves

In addition, curves of the type shown in Figure 5.8a and b may be used to suggest the pattern by which a population might increase from an initially very small size (e.g. when a species colonizes a previously unoccupied area). This is illustrated in Figure 5.8c. Imagine a small population, well below the carrying capacity of its environment (point A). Because the population is small, it increases in size only slightly during one time interval, and only reaches point B. Now, however, being larger, it increases in size more rapidly during the next time interval (to point C), and even more during the next (to point D). This process continues until the population passes beyond the peak of its net recruitment curve (Figure 5.8b). Thereafter, the population increases in size less and less with each time interval until the population reaches its

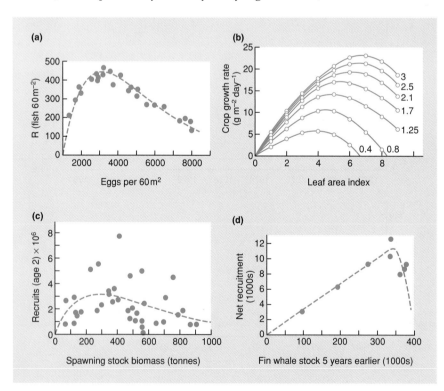

Figure 5.10 Some dome-shaped net-recruitment curves. (a) Six-month old brown trout, *Salmo trutta*, in Black Brows Beck, UK, between 1967 and 1989. (After Myers, 2001; following Elliott, 1994.) (b) The relationship between crop growth rate of subterranean clover, *Trifolium subterraneum*, and leaf area index at various intensities of radiation (kJ cm^{-2} day^{-1}). (After Black, 1963.) (c) 'Blackwater' herring, *Clupea harengus*, from the Thames estuary between 1962 and 1997. (After Fox, 2001.) (d) Estimates for the stock of Antarctic fin whales. (After Allen, 1972.)

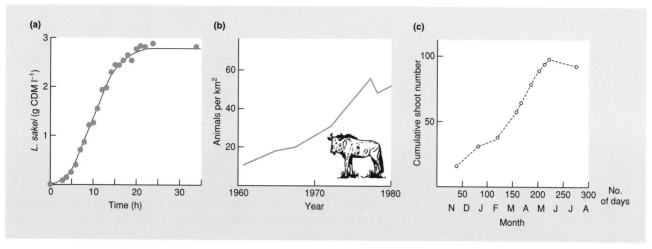

Figure 5.11   Real examples of S-shaped population increase. (a) The bacterium *Lactobacillus sakei* (measured as grams of 'cell dry mass' (CDM) per liter) grown in nutrient broth. (After Leroy & de Vuyst, 2001.) (b) The population of wildebeest *Connochaetes taurinus*, of the Serengeti region of Tanzania and Kenya seems to be leveling off after rising from a low density caused by the disease rinderpest. (After Sinclair & Norton-Griffiths, 1982; Deshmukh, 1986.) (c) The population of shoots of the annual *Juncus gerardi* in a salt marsh habitat on the west coast of France. (After Bouzille *et al.*, 1997.)

carrying capacity (*K*) and ceases completely to increase in size. The population might therefore be expected to follow an S-shaped or 'sigmoidal' curve as it rises from a low density to its carrying capacity. This is a consequence of the hump in its recruitment rate curve, which is itself a consequence of intraspecific competition.

Of course, Figure 5.8c, like the rest of Figure 5.8, is a gross simplification. It assumes, apart from anything else, that changes in population size are affected only by intraspecific competition. Nevertheless, something akin to sigmoidal population growth can be perceived in many natural and experimental situations (Figure 5.11).

Intraspecific competition will be obvious in certain cases (such as overgrowth competition between sessile organisms on a rocky shore), but this will not be true of every population examined. Individuals are also affected by predators, parasites and prey, competitors from other species, and the many facets of their physical and chemical environment. Any of these may outweigh or obscure the effects of intraspecific competition; or the effect of these other factors at one stage may reduce the density to well below the carrying capacity for all subsequent stages. Nevertheless, intraspecific competition probably affects most populations at least sometimes during at least one stage of their life cycle.

5.5   Intraspecific competition and density-dependent growth

Intraspecific competition, then, can have a profound effect on the number of individuals in a population; but it can have an equally

profound effect on the individuals themselves. In populations of unitary organisms, rates of growth and rates of development are commonly influenced by intraspecific competition. This necessarily leads to density-dependent effects on the composition of a population. For instance, Figure 5.12a and b shows two examples in which individuals were typically smaller at higher densities. This, in turn, often means that although the numerical size of a population is regulated only approximately by intraspecific competition, the total biomass is regulated much more precisely. This, too, is illustrated by the limpets in Figure 5.12b.

5.5.1   The law of constant final yield

Such effects are particularly marked in modular organisms. For example, when carrot seeds (*Daucus carrota*) were sown at a range of densities, the yield per pot at the first harvest (29 days) increased with the density of seeds sown (Figure 5.13). After 62 days, however, and even more after 76 and 90 days, yield no longer reflected the numbers sown. Rather it was the same over a wide range of initial densities, especially at higher densities where competition was most intense. This pattern has frequently been noted by plant ecologists and has been called the 'law of constant final yield' (Kira *et al.*, 1953). Individuals suffer density-dependent reductions in growth rate, and thus in individual plant size, which tend to compensate exactly for increases in density (hence the constant final yield). This suggests, of course, that there are limited resources available for plant growth, especially at high densities, which is borne out in Figure 5.13 by the higher (constant) yields at higher nutrient levels.

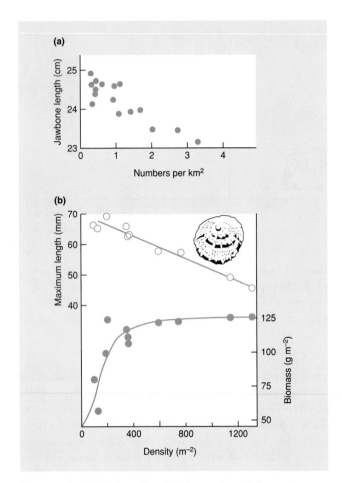

Figure 5.12 (a) Jawbone length indicates that reindeer grow to a larger size at lower densities. (After Skogland, 1983.) (b) In populations of the limpet *Patella cochlear*, individual size declines with density leading to an exact regulation of the population's biomass. (After Branch, 1975.)

Yield is density (d) multiplied by mean weight per plant ($\bar{w}$). Thus, if yield is constant (c):

$$d\bar{w} = c, \tag{5.1}$$

and so:

$$\log d + \log \bar{w} = \log c \tag{5.2}$$

and:

$$\log \bar{w} = \log c - 1 \cdot \log d \tag{5.3}$$

and thus, a plot of log mean weight against log density should have a slope of −1.

Data on the effects of density on the growth of the grass *Vulpia fasciculata* are shown in Figure 5.14, and the slope of the curve

towards the end of the experiment does indeed approach a value of −1. Here too, as with the carrot plants, individual plant weight at the first harvest was reduced only at very high densities – but as the plants became larger, they interfered with each other at successively lower densities.

The constancy of the final yield is a result, to a large extent, of the modularity of plants. This was clear when perennial rye grass (*Lolium perenne*) was sown at a 30-fold range of densities (Figure 5.15). After 180 days some genets had died; but the range of final tiller (module) densities was far narrower than that of genets (individuals). The regulatory powers of intraspecific competition were operating largely by affecting the number of modules per genet rather than the number of genets themselves.

constant yield and modularity

5.6 Quantifying intraspecific competition

Every population is unique. Nevertheless, we have already seen that there are general patterns in the action of intraspecific competition. In this section we take such generalizations a stage further. A method will be described, utilizing k values (see Chapter 4) to summarize the effects of intraspecific competition on mortality, fecundity and growth. Mortality will be dealt with first. The method will then be extended for use with fecundity and growth.

A k value was defined by the formula:

use of k values

$$k = \log (\text{initial density}) - \log (\text{final density}), \tag{5.4}$$

or, equivalently:

$$k = \log (\text{initial density/final density}). \tag{5.5}$$

For present purposes, 'initial density' may be denoted by B, standing for 'numbers *before* the action of intraspecific competition', whilst 'final density' may be denoted by A, standing for 'numbers *after* the action of intraspecific competition'. Thus:

$$k = \log (B/A). \tag{5.6}$$

Note that k increases as mortality rate increases.

Some examples of the effects of intraspecific competition on mortality are shown in Figure 5.16, in which k is plotted against log B. In several cases, k is constant at the lowest densities. This is an indication of density independence: the proportion surviving is not correlated with initial density. At higher densities, k increases with initial density; this indicates density dependence. Most importantly,

plots of k against log density

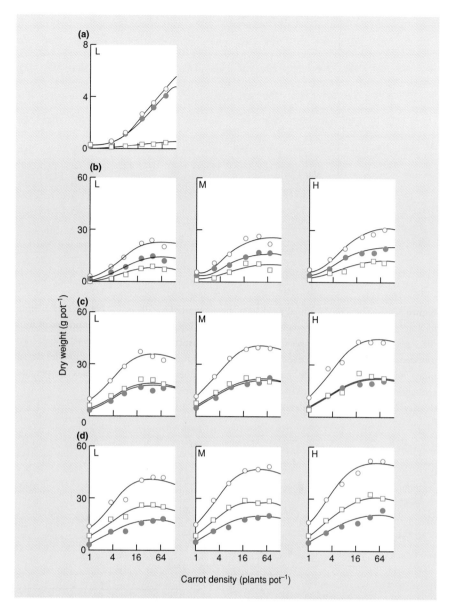

Figure 5.13 The relationship between yield per pot and sowing density in carrots (*Dacaus carrota*) at four harvests ((a) 29 days after sowing, (b) 62 days, (c) 76 days, and (d) 90 days) and at three nutrient levels (low, medium and high: L, M and H), given to pots weekly after the first harvest. Points are means of three replicates, with the exception of the lowest density (9) and the first harvest (9). □, root weight, ●, shoot weight, ○, total weight. The curves were fitted in line with theoretical yield–density relationships, the details of which are unimportant in this context. (After Li *et al.*, 1996.)

however, the way in which k varies with the logarithm of density indicates the precise nature of the density dependence. For example, Figure 5.16a and b describes, respectively, situations in which there is under- and exact compensation at higher densities. The exact compensation in Figure 5.16b is indicated by the slope of the curve (denoted by b) taking a constant value of 1 (the mathematically inclined will see that this follows from the fact that with exact compensation A is constant). The undercompensation that preceded this at lower densities, and which is seen in Figure 5.16a even at higher densities, is indicated by the fact that b is less than 1.

scramble and contest

Exact compensation ($b = 1$) is often referred to as pure contest competition, because there are a constant number of winners (survivors) in the competitive process. The term was initially proposed by Nicholson (1954), who contrasted it with what he called pure scramble competition. Pure scramble is the most extreme form of overcompensating density dependence, in which all competing individuals are so adversely affected that none of them survive, i.e. $A = 0$. This would be indicated in Figure 5.16 by a b value of infinity (a vertical line), and Figure 5.16c is an example in which this is the case. More common, however, are examples in which competition is scramble-*like*, i.e. there is considerable but not total overcompensation ($b \gg 1$). This is shown, for instance, in Figure 5.16d.

Plotting k against $\log B$ is thus an informative way of depicting the effects of intraspecific competition on mortality. Variations in the slope of the curve (b) give a clear indication

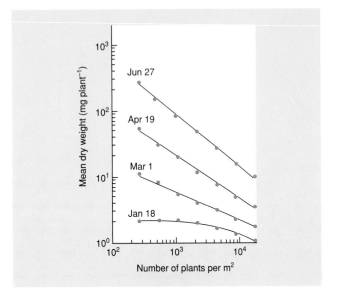

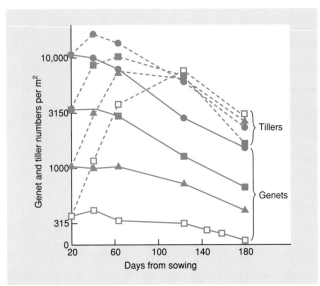

Figure 5.14 The 'constant final yield' of plants illustrated by a line of slope −1 when log mean weight is plotted against log density in the dune annual, *Vulpia fasciculata*. On January 18, particularly at low densities, growth and hence mean dry weight were roughly independent of density. But by June 27, density-dependent reductions in growth compensated exactly for variations in density, leading to a constant yield. (After Watkinson, 1984.)

Figure 5.15 Intraspecific competition in plants often regulates the number of modules. When populations of rye grass (*Lolium perenne*) were sown at a range of densities, the range of final tiller (i.e. module) densities was far narrower than that of genets. (After Kays & Harper, 1974.)

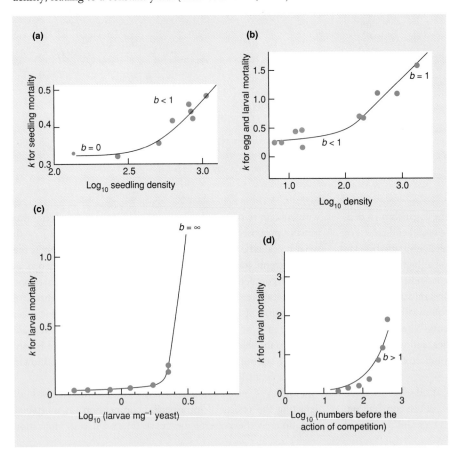

Figure 5.16 The use of k values for describing patterns of density-dependent mortality. (a) Seedling mortality in the dune annual, *Androsace septentrionalis*, in Poland. (After Symonides, 1979.) (b) Egg mortality and larval competition in the almond moth, *Ephestia cautella*. (After Benson, 1973a.) (c) Larval competition in the fruit-fly, *Drosophila melanogaster*. (After Bakker, 1961.) (d) Larval mortality in the moth, *Plodia interpunctella*. (After Snyman, 1949.)

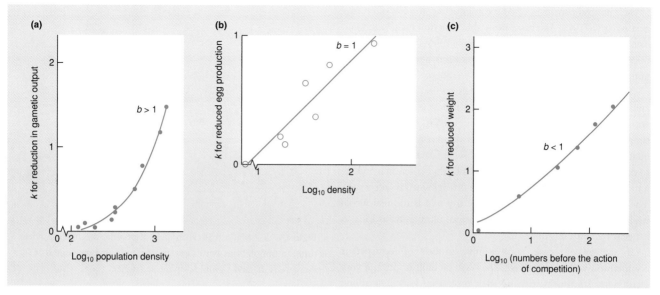

Figure 5.17   The use of *k* values for describing density-dependent reductions in fecundity and growth. (a) Fecundity in the limpet *Patella cochlear* in South Africa. (After Branch, 1975.) (b) Fecundity in the cabbage root fly, *Eriosichia brassicae*. (After Benson, 1973b.) (c) Growth in the shepherd's purse plant, *Capsella bursa-pastoris*. (After Palmblad, 1968.)

of the manner in which density dependence changes with density. The method can also be extended to fecundity and growth.

For fecundity, it is necessary to think of *B* as the 'total number of offspring that *would* have been produced had there been no intraspecific competition', i.e. if each reproducing individual had produced as many offspring as it would have done in a competition-free environment. *A* is then the total number of offspring *actually* produced. (In practice, *B* is usually estimated from the population experiencing the least competition – not necessarily competition-free.) For growth, *B* must be thought of as the total biomass, or total number of modules, that would have been produced had all individuals grown as if they were in a competition-free situation. *A* is then the total biomass or total number of modules actually produced.

Figure 5.17 provides examples in which *k* values are used to describe the effects of intraspecific competition on fecundity and growth. The patterns are essentially similar to those in Figure 5.16. Each falls somewhere on the continuum ranging between density independence and pure scramble, and their position along that continuum is immediately apparent. Using *k* values, all examples of intraspecific competition can be quantified in the same terms. With fecundity and growth, however, the terms 'scramble' and especially 'contest' are less appropriate. It is better simply to talk in terms of exact, over- and undercompensation.

5.7   Mathematical models: introduction

The desire to formulate general rules in ecology often finds its expression in the construction of mathematical or graphical models. It may seem surprising that those interested in the natural living world should spend time reconstructing it in an artificial mathematical form; but there are several good reasons why this should be done. The first is that models can crystallize, or at least bring together in terms of a few parameters, the important, shared properties of a wealth of unique examples. This simply makes it easier for ecologists to think about the problem or process under consideration, by forcing us to try to extract the essentials from complex systems. Thus, a model can provide a 'common language' in which each unique example can be expressed; and if each can be expressed in a common language, then their properties relative to one another, and relative perhaps to some ideal standard, will be more apparent.

These ideas are more familiar, perhaps, in other contexts. Newton never laid hands on a perfectly frictionless body, and Boyle never saw an ideal gas – other than in their imaginations – but Newton's Laws of Motion and Boyle's Law have been of immeasurable value to us for centuries.

Perhaps more importantly, however, models can actually shed light on the real world that they mimic. Specific examples below will make this apparent. Models can, as we shall see, exhibit properties that the system being modeled had not previously been known to possess. More commonly, models make it clear how the behavior of a population, for example, depends on the properties of the individuals that comprise it. That is, models allow us to see the likely consequences of any assumptions that we choose to make – 'If it were the case that only juveniles migrate, what would this do to the dynamics of their populations?' – and so on. Models can do this because mathematical methods are designed precisely to allow a set of assumptions to be followed through to their natural conclusions. As a consequence, models often suggest what would be the most profitable experiments to carry out or observations to make – 'Since juvenile migration rates appear to be so important, these should be measured in each of our study populations'.

These reasons for constructing models are also criteria by which any model should be judged. Indeed, a model is only useful (i.e. worth constructing) if it does perform one or more of these functions. Of course, in order to perform them a model must adequately describe real situations and real sets of data, and this 'ability to describe' or 'ability to mimic' is itself a further criterion by which a model can be judged. However, the crucial word is 'adequate'. The only perfect description of the real world is the real world itself. A model is an adequate description, ultimately, as long as it performs a useful function.

In the present case, some simple models of intraspecific competition will be described. They will be built up from a very elementary starting point, and their properties (i.e. their ability to satisfy the criteria described above) will then be examined. Initially, a model will be constructed for a population with discrete breeding seasons.

5.8   A model with discrete breeding seasons

5.8.1   Basic equations

In Section 4.7 we developed a simple model for species with discrete breeding seasons, in which the population size at time t, N_t, altered in size under the influence of a fundamental net reproductive rate, R. This model can be summarized in two equations:

$$N_{t+1} = N_t R \tag{5.7}$$

and:

$$N_t = N_0 R^t. \tag{5.8}$$

| no competition: exponential growth |

The model, however, describes a population in which there is no competition. R is constant, and if $R > 1$, the population will continue to increase in size indefinitely ('exponential growth', shown in Figure 5.18). The first step is therefore to modify the equations by making the net reproductive rate subject to intraspecific competition. This is done in Figure 5.19, which has three components.

At point A, the population size is very small (N_t is virtually zero). Competition is therefore negligible, and the actual net reproductive rate is adequately defined by an unmodified R. Thus, Equation 5.7 is still appropriate, or, rearranging the equation:

$$N_t/N_{t+1} = 1/R. \tag{5.9}$$

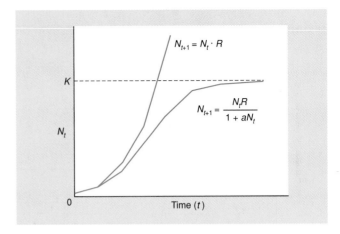

Figure 5.18   Mathematical models of population increase with time, in populations with discrete generations: exponential increase (left) and sigmoidal increase (right).

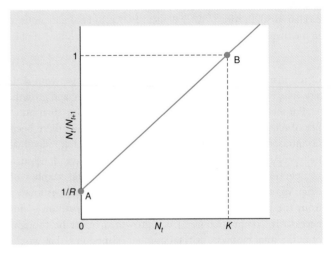

Figure 5.19   The simplest, straight-line way in which the inverse of generation increase (N_t/N_{t+1}) might rise with density (N_t). For further explanation, see text.

At point B, by contrast, the population size (N_t) is very much larger and there is a significant amount of intraspecific competition, such that the net reproductive rate has been so modified by competition that the population can collectively do no better than replace itself each generation, because 'births' equal 'deaths'. In other words, N_{t+1} is simply the same as N_t, and N_t/N_{t+1} equals 1. The population size at which this occurs is, by definition, the carrying capacity, K (see Figure 5.7).

The third component of Figure 5.19 is the straight line joining point A to point B and extending beyond it. This

| incorporating competition |

describes the progressive modification of the actual net reproductive rate as population size increases; but its straightness is simply an assumption made for the sake of expediency, since all straight lines are of the simple form: $y = (\text{slope}) x + (\text{intercept})$. In Figure 5.19, N_t/N_{t+1} is measured on the y-axis, N_t on the x-axis, the intercept is $1/R$ and the slope, based on the segment between points A and B, is $(1 - 1/R)/K$. Thus:

$$\frac{N_t}{N_{t+1}} = \frac{1 - \dfrac{1}{R}}{K} \cdot N_t + \frac{1}{R} \qquad (5.10)$$

or, rearranging:

$$N_{t+1} = \frac{N_t R}{1 + \dfrac{(R-1)N_t}{K}}. \qquad (5.11)$$

a simple model of intraspecific competition

For further simplicity, $(R - 1)/K$ may be denoted by a giving:

$$N_{t+1} = \frac{N_t R}{(1 + aN_t)}. \qquad (5.12)$$

This is a model of population increase limited by intraspecific competition. Its essence lies in the fact that the unrealistically constant R in Equation 5.7 has been replaced by an actual net reproductive rate, $R/(1 + aN_t)$, which decreases as population size (N_t) increases.

which comes first – a or K?

We, like many others, derived Equation 5.12 as if the behavior of a population is jointly determined by R and K, the per capita rate of increase and the population's carrying capacity – a is then simply a particular combination of these. An alternative point of view is that a is meaningful in its own right, measuring the per capita susceptibility to crowding: the larger the value of a, the greater the effect of density on the actual rate of increase in the population (Kuno, 1991). Now the behavior of a population is seen as being jointly determined by two properties of the individuals within it – their intrinsic per capita rate of increase and their susceptibility to crowding, R and a. The carrying capacity of the population ($K = (R - 1)/a$) is then simply an outcome of these properties. The great advantage of this viewpoint is that it places individuals and populations in a more realistic biological perspective. Individuals come first: individual birth rates, death rates and susceptibilities to crowding are subject to natural selection and evolve. Populations simply follow: a population's carrying capacity is just one of many features that reflect the values these individual properties take.

properties of the simplest model

The properties of the model in Equation 5.12 may be seen in Figure 5.19 (from which the model was derived) and Figure 5.18 (which shows a hypothetical population increasing in size over time in conformity with the model). The population in Figure 5.18 describes an S-shaped curve over time. As we saw earlier, this is a desirable quality of a model of intraspecific competition. Note, however, that there are many other models that would also generate such a curve. The advantage of Equation 5.12 is its simplicity.

The behavior of the model in the vicinity of the carrying capacity can best be seen by reference to Figure 5.19. At population sizes that are less than K the population will increase in size; at population sizes that are greater than K the population size will decline; and at K itself the population neither increases nor decreases. The carrying capacity is therefore a stable equilibrium for the population, and the model exhibits the regulatory properties classically characteristic of intraspecific competition.

5.8.2 What *type* of competition?

It is not yet clear, however, just exactly what type or range of competition this model is able to describe. This can be explored by tracing the relationship between k values and $\log N$ (as in Section 5.6). Each generation, the potential number of individuals produced (i.e. the number that would be produced if there were no competition) is $N_t R$. The actual number produced (i.e. the number that survive the effects of competition) is $N_t R/(1 + aN_t)$.

Section 5.6 established that:

$$k = \log (\text{number produced}) - \log (\text{number surviving}). \qquad (5.13)$$

Thus, in the present case:

$$k = \log N_t R - \log N_t R/(1 + aN_t), \qquad (5.14)$$

or, simplifying:

$$k = \log(1 + aN_t). \qquad (5.15)$$

Figure 5.20 shows a number of plots of k against $\log_{10} N_t$ with a variety of values of a inserted into the model. In every case, the slope of the graph approaches and then attains a value of 1. In other words, the density dependence always begins by undercompensating and then compensates perfectly at higher values of N_t. The model is therefore limited in the type of competition that it can produce, and all we have been able to say so far is that *this type* of competition leads to very tightly controlled regulation of populations.

5.8.3 Time lags

One simple modification that we can make is to relax the assumption that populations respond instantaneously to changes

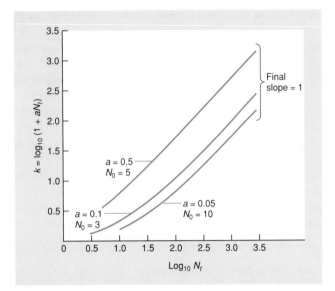

Figure 5.20   The intraspecific competition inherent in Equation 5.13. The final slope of k against $\log_{10}N_t$ is unity (exact compensation), irrespective of the starting density N_0 or the constant $a\ (= (R-1)/K)$.

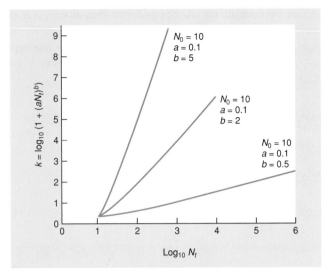

Figure 5.21   The intraspecific competition inherent in Equation 5.19. The final slope is equal to the value of b in the equation.

in their own density, i.e. that present density determines the amount of resource available to a population and this in turn determines the net reproductive rate within the population. Suppose instead that the amount of resource available is determined by the density one time interval previously. To take a specific example, the amount of grass in a field in spring (the resource available to cattle) might be determined by the level of grazing (and hence, the density of cattle) in the previous year. In such a case, the reproductive rate itself will be dependent on the density one time interval ago. Thus, since in Equations 5.7 and 5.12:

$$N_{t+1} = N_t \times \text{reproductive rate}, \tag{5.16}$$

Equation 5.12 may be modified to:

$$N_{t+1} = \frac{N_t R}{1 + aN_{t-1}}. \tag{5.17}$$

time lags provoke population fluctuations

There is a time lag in the population's response to its own density, caused by a time lag in the response of its resources. The behavior of the modified model is as follows:

$R < 1.33$: direct approach to a stable equilibrium
$R > 1.33$: damped oscillations towards that equilibrium.

In comparison, the original Equation 5.12, without a time lag, gave rise to a direct approach to its equilibrium for all values of R. The

time lag has provoked the fluctuations in the model, and it can be assumed to have similar, destabilizing effects on real populations.

5.8.4   Incorporating a range of competition

A simple modification of Equation 5.12 of far more general importance was originally suggested by Maynard Smith and Slatkin (1973) and was discussed in detail by Bellows (1981). It alters the equation to:

$$N_{t+1} = \frac{N_t R}{1 + (aN_t)^b}. \tag{5.18}$$

The justification for this modification may be seen by examining some of the properties of the revised model. For example, Figure 5.21 shows plots of k against $\log N_t$, analogous to those in Figure 5.20: k is now $\log_{10}[1 + (aN_t)^b]$. The slope of the curve, instead of approaching 1 as it did previously, now approaches the value taken by b in Equation 5.18. Thus, by the choice of appropriate values, the model can portray undercompensation ($b < 1$), perfect compensation ($b = 1$), scramble-like overcompensation ($b > 1$) or even density independence ($b = 0$). This model has the generality that Equation 5.12 lacks, with the value of b determining the type of density dependence that is being incorporated.

Another desirable quality that Equation 5.18 shares with other good models is an ability to throw fresh light on the real world. By sensible analysis of the population dynamics generated by the equation,

the dynamic pattern? it's R and b

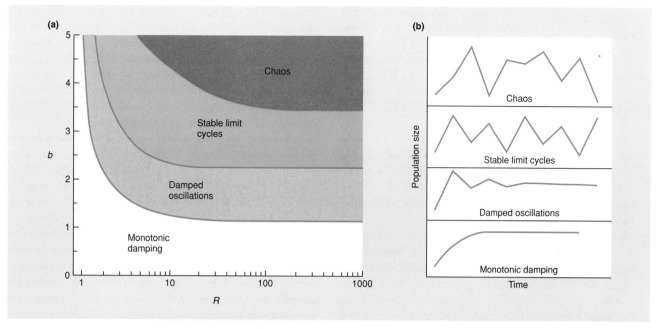

Figure 5.22   (a) The range of population fluctuations (themselves shown in (b)) generated by Equation 5.19 with various combinations of *b* and *R* inserted. (After May, 1975a; Bellows, 1981.)

it is possible to draw guarded conclusions about the dynamics of natural populations. The mathematical method by which this and similar equations may be examined is set out and discussed by May (1975a), but the results of the analysis (Figure 5.22) can be appreciated without dwelling on the analysis itself. Figure 5.22b shows the various patterns of population growth and dynamics that Equation 5.18 can generate. Figure 5.22a sets out the conditions under which each of these patterns occurs. Note first that the pattern of dynamics depends on two things: (i) *b*, the precise type of competition or density dependence; and (ii) *R*, the effective net reproductive rate (taking density-independent mortality into account). By contrast, *a* determines not the type of pattern, but only the level about which any fluctuations occur.

As Figure 5.22a shows, low values of *b* and/or *R* lead to populations that approach their equilibrium size without fluctuating at all ('monotonic damping'). This has already been hinted at in Figure 5.18. There, a population behaving in conformity with Equation 5.12 approached equilibrium directly, irrespective of the value of *R*. Equation 5.12 is a special case of Equation 5.18 in which *b* = 1 (perfect compensation); Figure 5.22a confirms that for *b* = 1, monotonic damping is the rule whatever the effective net reproductive rate.

As the values of *b* and/or *R* increase, the behavior of the population changes first to damped oscillations gradually approaching equilibrium, and then to 'stable limit cycles' in which the population fluctuates around an equilibrium level, revisiting the same two, four or even more points time and time again. Finally, with large values of *b* and *R*, the population fluctuates in an apparently irregular and chaotic fashion.

5.8.5   Chaos

Thus, a model built around a density-dependent, supposedly regulatory process (intraspecific competition) can lead to a very wide range of population dynamics. If a model population has even a moderate fundamental net reproductive rate (and the ability to leave 100 (= *R*) offspring in the next generation in a competition-free environment is not unreasonable), and if it has a density-dependent reaction which even moderately overcompensates, then far from being stable, it may fluctuate widely in numbers without the action of any extrinsic factor. The biological significance of this is the strong suggestion that even in an environment that is wholly constant and predictable, the intrinsic qualities of a population and the individuals within it may, by themselves, give rise to population dynamics with large and perhaps even chaotic fluctuations. The consequences of intraspecific competition are clearly not limited to 'tightly controlled regulation'.

This leads us to two important conclusions. First, time lags, high reproductive rates and overcompensating density dependence are capable (either alone or in combination) of producing all types of fluctuations in population density, without invoking any extrinsic cause. Second, and equally important, this has been made apparent by the analysis of mathematical models.

key characteristics of
chaotic dynamics

In fact, the recognition that even simple ecological systems may contain the seeds of chaos has led to chaos itself becoming a topic of interest amongst ecologists (Schaffer & Kot, 1986; Hastings *et al.*, 1993; Perry *et al.*, 2000). A detailed exposition of the nature of chaos is not appropriate here, but a few key points should be understood.

1 The term 'chaos' may itself be misleading if it is taken to imply a fluctuation with absolutely no discernable pattern. Chaotic dynamics do not consist of a sequence of random numbers. On the contrary, there are tests (although they are not always easy to put into practice) designed to distinguish chaotic from random and other types of fluctuations.
2 Fluctuations in chaotic ecological systems occur between definable upper and lower densities. Thus, in the model of intraspecific competition that we have discussed, the idea of 'regulation' has not been lost altogether, even in the chaotic region.
3 Unlike the behavior of truly regulated systems, however, two similar population trajectories in a chaotic system will not tend to converge on ('be attracted to') the same equilibrium density or the same limit cycle (both of them 'simple' attractors). Rather, the behavior of a chaotic system is governed by a 'strange attractor'. Initially, very similar trajectories will diverge from one another, exponentially, over time: chaotic systems exhibit 'extreme sensitivity to initial conditions'.
4 Hence, the long-term future behavior of a chaotic system is effectively impossible to predict, and prediction becomes increasingly inaccurate as one moves further into the future. Even if we appear to have seen the system in a particular state before – and know precisely what happened subsequently last time – tiny (perhaps immeasurable) initial differences will be magnified progressively, and past experience will become of increasingly little value.

Ecology must aim to become a predictive science. Chaotic systems set us some of the sternest challenges in prediction. There has been an understandable interest, therefore, in the question 'How often, if ever, are ecological systems chaotic?' Attempts to answer this question, however, whilst illuminating, have certainly not been definitive.

Takens' theorem:
reconstructing the
attractor

Most recent attempts to detect chaos in ecological systems have been based on a mathematical advance known as *Takens' theorem*. This says, in the context of ecology, that even when a system comprises a number of interacting elements, its characteristics (whether it is chaotic, etc.) may be deduced from a time series of abundances of just one of those elements (e.g. one species). This is called 'reconstructing the attractor'. To be more specific: suppose, for example, that a system's behavior is determined by interactions between four elements (for simplicity, four species). First, one expresses the abundance of just one of those species at time t, N_t, as a function of the sequence of abundances at *four* successive previous time points: N_{t-1}, N_{t-2}, N_{t-3}, N_{t-4} (the same number of 'lags' as there are elements in the original system). Then, the attractor of this lagged system of abundances is an accurate reconstruction of the attractor of the original system, which determines its characteristics.

In practice, this means taking a series of abundances of, say, one species and finding the 'best' model, in statistical terms, for predicting N_t as a function of lagged abundances, and then investigating this reconstructed attractor as a means of investigating the nature of the dynamics of the underlying system. Unfortunately, ecological time series (compared, say, to those of physics) are particularly short and particularly noisy. Thus, methods for identifying a 'best' model and applying Takens' theorem, and for identifying chaos in ecology generally, have been 'the focus of continuous methodological debate and refinement' (Bjørnstad & Grenfell, 2001), one consequence of which is that any suggestion of a suitable method in a textbook such as this is almost certainly doomed to be outmoded by the time it is first read.

Notwithstanding these technical difficulties, however, and in spite of occasional demonstrations of apparent chaos in artificial laboratory environments (Costantino *et al.*, 1997), a consensus view has grown that chaos is not a dominant pattern of dynamics in natural ecological systems. One trend, therefore, has been to seek to understand why chaos might *not* occur in nature, despite its being generated readily by ecological models. For example, Fussmann and Heber (2002) examined model populations embedded in food webs and found that as the webs took on more of the characteristics observed in nature (see Chapter 20) chaos became less likely.

Thus, the potential importance of chaos in ecological systems is clear. From a fundamental point of view, we need to appreciate that if we have a relatively simple system, it may nevertheless generate complex, chaotic dynamics; and that if we observe complex dynamics, the underlying explanation may nevertheless be simple. From an applied point of view, if ecology is to become a predictive and manipulative science, then we need to know the extent to which long-term prediction is threatened by one of the hallmarks of chaos – extreme sensitivity to initial conditions. The key practical question, however – 'how common is chaos?' – remains largely unanswered.

how common – or
important – is chaos?

5.9 Continuous breeding: the logistic equation

The model derived and discussed in Section 5.8 was appropriate for populations that have discrete breeding seasons and can therefore be described by equations growing in discrete steps, i.e. by 'difference' equations. Such models are not appropriate, however, for those populations in which birth and death are continuous. These are best described by models of continuous growth, or 'differential' equations, which will be considered next.

r, the intrinsic rate of natural increase

The net rate of increase of such a population will be denoted by dN/dt (referred to in speech as 'dN by dt'). This represents the 'speed' at which a population increases in size, N, as time, t, progresses. The increase in size of the whole population is the sum of the contributions of the various individuals within it. Thus, the average rate of increase per individual, or the 'per capita rate of increase' is given by $dN/dt(1/N)$. But we have already seen in Section 4.7 that in the absence of competition, this is the definition of the 'intrinsic rate of natural increase', r. Thus:

$$\frac{dN}{dt}\left(\frac{1}{N}\right) = r \qquad (5.19)$$

and:

$$\frac{dN}{dt} = rN. \qquad (5.20)$$

A population increasing in size under the influence of Equation 5.20, with $r > 0$, is shown in Figure 5.23. Not surprisingly, there is unlimited, 'exponential' increase. In fact, Equation 5.20 is the continuous form of the exponential difference Equation 5.8, and as discussed in Section 4.7, r is simply $\log_e R$. (Mathematically adept readers will see that Equation 5.20 can be obtained by differentiating Equation 5.8.) R and r are clearly measures of the same commodity: 'birth plus survival' or 'birth minus death'; the difference between R and r is merely a change of currency.

For the sake of realism, intraspecific competition must obviously be added to Equation 5.20. This can be achieved

the logistic equation

most simply by a method exactly equivalent to the one used in Figure 5.19, giving rise to:

$$\frac{dN}{dt} = rN\left(\frac{K - N}{K}\right). \qquad (5.21)$$

This is known as the logistic equation (coined by Verhulst, 1838), and a population increasing in size under its influence is shown in Figure 5.23.

The logistic equation is the continuous equivalent of Equation 5.12, and it therefore has all the essential characteristics of Equation 5.12, and all of its shortcomings. It describes a sigmoidal growth curve approaching a stable carrying capacity, but it is only one of many reasonable equations that do this. Its major advantage is its simplicity. Moreover, whilst it was possible to incorporate a range of competitive intensities into

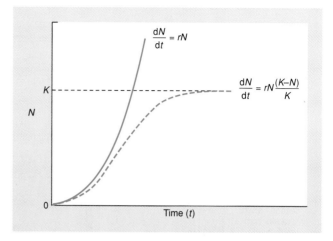

Figure 5.23 Exponential (——) and sigmoidal (- - -) increase in density (N) with time for models of continuous breeding. The equation giving sigmoidal increase is the logistic equation.

Equation 5.12, this is by no means easy with the logistic equation. The logistic is therefore doomed to be a model of perfectly compensating density dependence. Nevertheless, in spite of these limitations, the equation will be an integral component of models in Chapters 8 and 10, and it has played a central role in the development of ecology.

5.10 Individual differences: asymmetric competition

5.10.1 Size inequalities

Until now, we have focused on what happens to the whole population or the average individual within it. Different individuals, however, may respond to intraspecific competition in very different ways. Figure 5.24 shows the results of an experiment in which flax (*Linum usitatissimum*) was sown at three densities, and harvested at three stages of development, recording the weight of each plant individually. This made it possible to monitor the effects of increasing amounts of competition not only as a result of variations in sowing density, but also as a result of plant growth (between the first and the last harvests). When intraspecific competition was at its least intense (at the lowest sowing density after only 2 weeks' growth) the individual plant weights were distributed symmetrically about the mean. When competition was at its most intense, however, the distribution was strongly skewed to the left: there were many very small individuals and a few large ones. As the intensity of competition gradually increased, the degree of skewness increased as well. Decreased size – but increased skewness in size – is also seen to

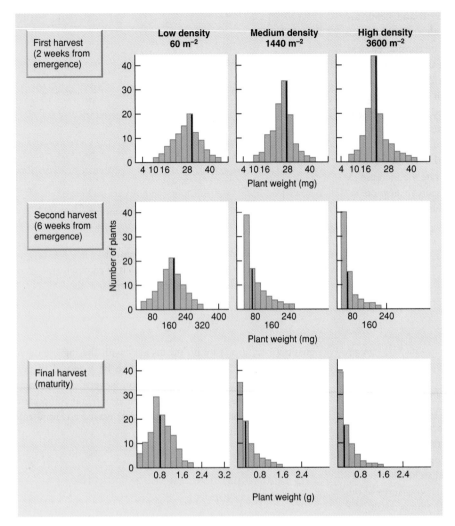

Figure 5.24 Competition and a skewed distribution of plant weights. Frequency distributions of individual plant weights in populations of flax (*Linum usitatissimum*), sown at three densities and harvested at three ages. The black bar is the mean weight. (After Obeid *et al.*, 1967.)

be associated with increased density (and presumably competition) in cod (*Gadus morhua*) living off the coast of Norway (Figure 5.25).

the inadequacy of the average

More generally, we may also say that increased competition increased the degree of size inequality within the population, i.e. the extent to which total biomass was unevenly distributed amongst the different individuals (Weiner, 1990). Rather similar results have been obtained from a number of other populations of animals (Uchmanski, 1985) and plants (Uchmanski, 1985; Weiner & Thomas, 1986). Typically, populations experiencing the most intense competition have the greatest size inequality and often have a size distribution in which there are many small and a few large individuals. Characterizing a population by an arbitrary 'average' individual can obviously be very misleading under such circumstances, and can divert attention from the fact that intraspecific competition is a force affecting individuals, even though its effects may often be detected in whole populations.

5.10.2 Preempting resources

An indication of the way in which competition can exaggerate underlying inequalities in a population comes from observations on a natural, crowded population of the woodland annual *Impatiens pallida* in southeastern Pennsylvania. Over an 8-week period, growth was very much faster in large than in small plants – in fact, small plants did not grow at all (Figure 5.26a). This increased significantly the size inequality within the population (Figure 5.26b). Thus, the smaller a plant was initially, the more it was affected by neighbors. Plants that established early preempted or 'captured' space, and subsequently were little affected by intraspecific competition. Plants that emerged later entered a universe in which most of the available space had already been preempted; they were therefore greatly affected by intraspecific competition. Competition was asymmetric: there was a hierarchy. Some individuals were affected far more than others, and small initial differences were transformed by competition into much larger differences 8 weeks later.

If competition is asymmetric because superior competitors preempt resources, then competition is most likely to be asymmetric when it occurs for resources that are most liable to be preempted. Specifically, competition amongst plants for light, in which a superior competitor can overtop and shade an inferior, might be expected to lend itself far more readily to preemptive resource capture than competition for soil nutrients or water, where the roots of even a very inferior competitor will have more immediate access to at least some of the available resources than

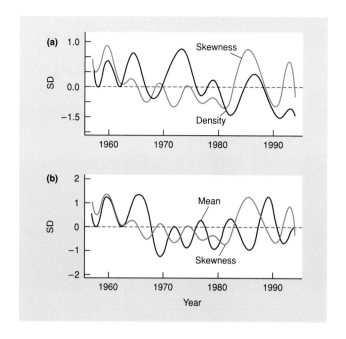

Figure 5.25 *(right)* Values of skewness (in the frequency distribution of lengths) and density (a) and of skewness and mean length (b) are expressed as standard deviations from mean values for the years 1957–94 for cod (*Gadus morhua*) from the Skagerrak, off the coast of Norway. Despite marked fluctuations from year to year, much of it the result of variations in weather, skewness was clearly greatest at high densities ($r = 0.58$, $P < 0.01$) when lengths were smallest ($r = -0.45$, $P < 0.05$), that is, when competition was most intense. (After Lekve *et al.*, 2002.)

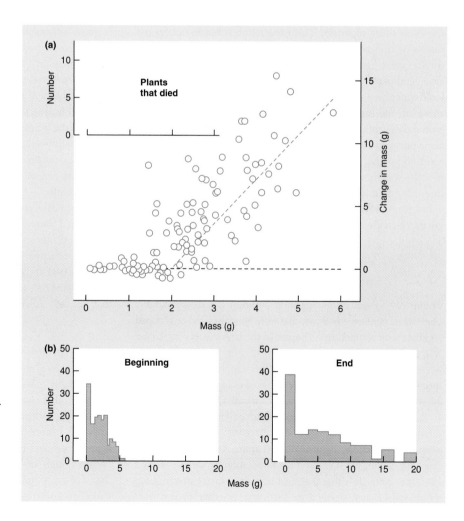

Figure 5.26 Asymmetric competition in a natural population of *Impatiens pallida*. (a) The increase in mass of survivors of different sizes over an 8-week period, and the distribution of initial sizes of those individuals that died over the same period. The horizontal axis is the same in each case. (b) The distribution of individual weights at the beginning (Gini coefficient, a measure of inequality, 0.39) and the end of this period (Gini coefficient, 0.48). (After Thomas & Weiner, 1989.)

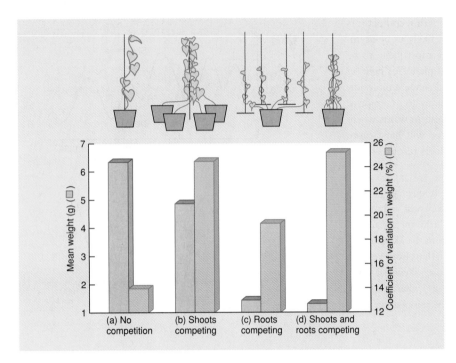

Figure 5.27   When morning glory vines competed, root competition was most effective in reducing mean plant weight (treatments significantly different, $P < 0.01$, for all comparisons except (c) with (d)), but shoot competition was most effective in increasing the degree of size inequality, as measured by the coefficient of variation in weight (significant differences between treatments (a) and (b), $P < 0.05$, and (a) and (d), $P < 0.01$). (After Weiner, 1986.)

the roots of its superiors. This expectation is borne out by the results of an experiment in which morning glory vines (*Ipomoea tricolor*) were grown as single plants in pots ('no competition'), as several plants rooted in their own pots but with their stems intertwined on a single stake ('shoots competing'), as several plants rooted in the same pot, but with their stems growing up their own stakes ('roots competing') and as several plants rooted in the same pot with their stems intertwined on one stake ('shoots and roots competing') (Figure 5.27). Despite the fact that root competition was more intense than shoot competition, in the sense that it led to a far greater decrease in the mean weight of individual plants, it was shoot competition for light that led to a much greater increase in size inequality.

skews and other hierarchies

Skewed distributions are one possible manifestation of hierarchical, asymmetric competition, but there are many others. For instance, Ziemba and Collins (1999) studied competition amongst larval salamanders (*Ambystoma tigrinum nebulosum*) that were either isolated or grouped together with competitors. The size of the largest surviving larvae was unaffected by competition ($P = 0.42$) but the smallest larvae were much smaller ($P < 0.0001$). This emphasizes that intraspecific competition is not only capable of exaggerating individual differences, it is also greatly affected by individual differences.

Asymmetric competition was observed on a much longer timescale in a population of the herbaceous perennial *Anemone hepatica* in Sweden (Figure 5.28) (Tamm, 1956). Despite the crops of seedlings that entered the population between 1943 and 1952, it is quite clear that the most important factor determining which individuals survived to 1956 was whether or not they were established in 1943. Of the 30 individuals that had reached

large or intermediate size by 1943, 28 survived until 1956, and some of these had branched. By contrast, of the 112 plants that were either small in 1943 or appeared as seedlings subsequently, only 26 survived to 1956, and not one of these was sufficiently well established to have flowered. Similar patterns can be observed in tree populations. The survival rates, the birth rates and thus the fitnesses of the few established adults are high; those of the many seedlings and saplings are comparatively low.

These considerations illustrate a final, important general point: asymmetries tend to reinforce the regulatory powers of intraspecific competition. Tamm's established plants were successful competitors year after year, but his small plants and seedlings were repeatedly unsuccessful. This guaranteed a near constancy in the number of established plants between 1943 and 1956. Each year there was a near-constant number of 'winners', accompanied by a variable number of 'losers' that not only failed to grow, but usually, in due course, died.

asymmetry enhances regulation

5.11   Territoriality

Territoriality is one particularly important and widespread phenomenon that results in asymmetric intraspecific competition. It occurs when there is active interference between individuals, such that a more or less exclusive area, the territory, is defended against intruders by a recognizable pattern of behavior.

Individuals of a territorial species that fail to obtain a territory often make no contribution whatsoever to future generations. Territoriality, then,

territoriality is a contest

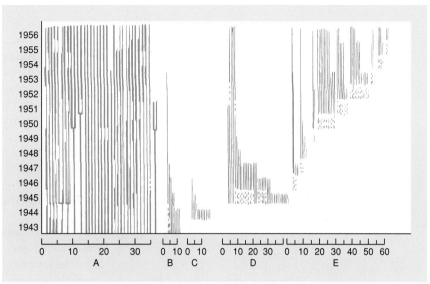

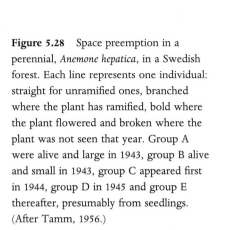

Figure 5.28 Space preemption in a perennial, *Anemone hepatica*, in a Swedish forest. Each line represents one individual: straight for unramified ones, branched where the plant has ramified, bold where the plant flowered and broken where the plant was not seen that year. Group A were alive and large in 1943, group B alive and small in 1943, group C appeared first in 1944, group D in 1945 and group E thereafter, presumably from seedlings. (After Tamm, 1956.)

is a 'contest'. There are winners (those that come to hold a territory) and losers (those that do not), and at any one time there can be only a limited number of winners. The exact number of territories (winners) is usually somewhat indeterminate in any one year, and certainly varies from year to year, depending on environmental conditions. Nevertheless, the contest nature of territoriality ensures, like asymmetric competition generally, a comparative constancy in the number of surviving, reproducing individuals. One important consequence of territoriality, therefore, is population regulation, or more particularly, the regulation of the number of territory holders. Thus, when territory owners die, or are experimentally removed, their places are often rapidly taken by newcomers. For instance, in great tit (*Parus major*) populations, vacated woodland territories are reoccupied by birds coming from hedgerows where reproductive success is noticeably lower (Krebs, 1971).

Some have felt that the regulatory consequences of territoriality must themselves be the root cause underlying the evolution of territorial behavior – territoriality being favored because the population as a whole benefitted from the rationing effects, which guaranteed that the population did not overexploit its resources (e.g. Wynne-Edwards, 1962). However, there are powerful and fundamental reasons for rejecting this 'group selectionist' explanation (essentially, it stretches evolutionary theory beyond reasonable limits): the ultimate cause of territoriality must be sought within the realms of natural selection, in some advantage accruing to the individual.

benefits and costs of territoriality

Any benefit that an individual does gain from territoriality, of course, must be set against the costs of defending the territory. In some animals this defense involves fierce combat between competitors, whilst in

others there is a more subtle mutual recognition by competitors of one another's keep-out signals (e.g. song or scent). Yet, even when the chances of physical injury are minimal, territorial animals typically expend energy in patrolling and advertizing their territories, and these energetic costs must be exceeded by any benefits if territoriality is to be favored by natural selection (Davies & Houston, 1984; Adams, 2001).

Praw and Grant (1999), for example, investigated the costs and benefits to convict cichlid fish (*Archocentrus nigrofasciatus*) of defending food patches of different sizes. As patch size increased, the amount of food eaten by a patch defender increased (the benefit; Figure 5.29a), but the frequency of chasing intruders (the cost; Figure 5.29b) also increased. Evolution should favor an intermediate patch (territory) size at which the trade-off between costs and benefits is optimized, and indeed, the growth rate of defenders was greatest in intermediate-sized patches (Figure 5.29c).

On the other hand, explaining territoriality only in terms of a net benefit to the territory owner is rather like history always being written by the victors. There is another, possibly trickier question, which seems not to have been answered – could those individuals without a territory not do better by challenging the territory owners more often and with greater determination?

Of course, describing territoriality in terms of just 'winners' and 'losers' is an oversimplification. Generally, there are first, second and a range of consolation prizes – not all territories are equally valuable. This has been demonstrated in an unusually striking way in a study of oystercatchers (*Haematopus ostralegus*) on the Dutch coast, where pairs of birds defend both nesting territories on the salt marsh and feeding territories on the mudflats (Ens *et al.*, 1992). For some birds (the 'residents'), the feeding territory is simply an extension of the

not simply winners and losers

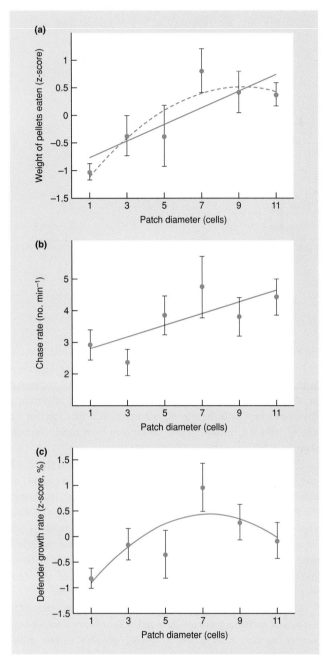

Figure 5.29 Optimal territory size in the convict cichlid fish, *Archocentrus nigrofasciatus*. (a) As patch (territory) size increased, the amount of food eaten by a territory defender (standardized z score) increased but leveled off at the largest sizes (solid line, linear regression: $r^2 = 0.27$, $P = 0.002$; dashed line, quadratic regression: $r^2 = 0.33$, $P = 0.003$). (b) As patch (territory) size increased, the chase rate of territory defenders increased (linear regression: $r^2 = 0.68$, $P < 0.0001$). (c) As patch (territory) size increased, the growth rate of territory defenders (standardized z score) was highest at intermediate-sized territories (quadratic regression: $r^2 = 0.22$, $P = 0.028$). (After Praw & Grant, 1999.)

nesting territory: they form one spatial unit. For other pairs, however (the 'leapfrogs'), the nesting territory is further inland and hence separated spatially from the feeding territory (Figure 5.30a). Residents fledge many more offspring than do leapfrogs (Figure 5.30b), because they deliver far more food to them (Figure 5.30c). From an early age, resident chicks follow their parents onto the mudflats, taking each prey item as soon as it is captured. Leapfrog chicks, however, are imprisoned on their nesting territory prior to fledging; all their food has to be flown in. It is far better to have a resident than a leapfrog territory.

5.12   Self-thinning

We have seen throughout this chapter that intraspecific competition can influence the number of deaths, the number of births and the amount of growth within a population. We have illustrated this largely by looking at the end results of competition. But in practice, the effects are often progressive. As a cohort ages, the individuals grow in size, their requirements increase and they therefore compete at a greater and greater intensity. This in turn tends gradually to increase their risk of dying. But if some individuals die, then the density and the intensity of competition are decreased – which affects growth, which affects competition, which affects survival, which affects density, and so on.

5.12.1   Dynamic thinning lines

The patterns that emerge in growing, crowded cohorts of individuals were originally the focus of particular attention in plant populations. For example, perennial rye grass (*Lolium perenne*) was sown at a range of densities, and samples from each density were harvested after 14, 35, 76, 104 and 146 days (Figure 5.31a). Figure 5.31a has the same logarithmic axes – density and mean plant weight – as Figure 5.14. It is most important to appreciate the difference between the two. In Figure 5.14, each line represented a separate yield–density relationship at different ages of a cohort. Successive points along a line represent different initial sowing densities. In Figure 5.31, each line itself represents a different sowing density, and successive points along a line represent populations of this initial sowing density at different ages. The lines are therefore trajectories that follow a cohort through time. This is indicated by arrows, pointing from many small, young individuals (bottom right) to fewer, larger, older individuals (top left).

Mean plant weight (at a given age) was always greatest in the lowest density populations (Figure 5.31a). It is also clear that the highest density populations were the first to suffer substantial mortality. What is most noticeable, however, is that eventually, in all cohorts, density declined and mean plant weight increased in unison: populations progressed along roughly the same straight

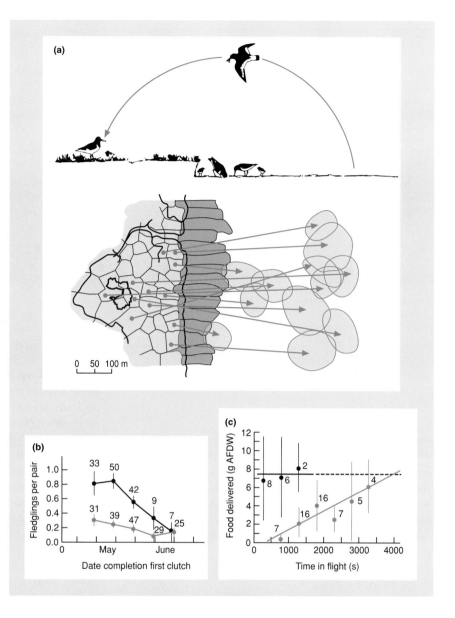

Figure 5.30 (a) A coastal area in the Netherlands providing both nesting and feeding territories for oystercatchers. In 'resident' territories (dark shading), nesting and feeding areas are adjacent and chicks can be taken from one to the other at an early age. 'Leapfrogs', however, have separate nesting and feeding territories (light shading) and food has to be flown in until the chicks fledge. (b) Residents (●) fledge more chicks than leapfrogs (●). (c) Residents (●) deliver more food per tide (grams of ash-free dry weight (g AFDW), with standard deviations) than leapfrogs (●). The latter deliver more, the more effort (in flying) they expend, but still cannot match the residents. (After Ens *et al.*, 1992.)

line. The populations are said to have experienced *self-thinning* (i.e. a progressive decline in density in a population of growing individuals), and the line that they approached and then followed is known as a *dynamic thinning line* (Weller, 1990).

The lower the sowing density, the later was the onset of self-thinning. In all cases, though, the populations initially followed a trajectory that was almost vertical, i.e. there was little mortality. Then, as they neared the thinning line, the populations suffered increasing amounts of mortality, so that the slopes of all the self-thinning trajectories gradually approached the dynamic thinning line and then progressed along it. Note also that Figure 5.31 has been drawn, following convention, with log density on the *x*-axis and log mean weight on the *y*-axis. This is not meant to imply that density is the independent variable on

which mean weight depends. Indeed, it can be argued that mean weight increases naturally during plant growth, and this determines the decrease in density. The most satisfactory view is that density and mean weight are wholly interdependent.

Plant populations (if sown at sufficiently high densities) have repeatedly been found to approach and then follow

the −3/2 power law

a dynamic thinning line. For many years, all such lines were widely perceived as having a slope of roughly −3/2, and the relationship was often referred to as the '−3/2 power law' (Yoda *et al.*, 1963; Hutchings, 1983), since density (*N*) was seen as related to mean weight ($\bar{w}$) by the equation:

$$\log \bar{w} = \log c - 3/2 \log N \qquad (5.22)$$

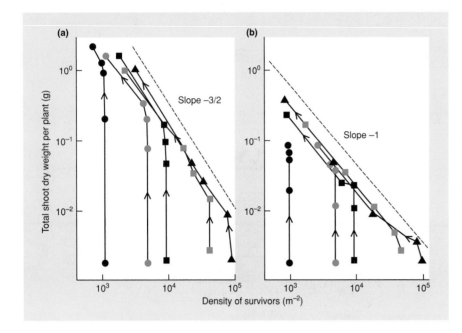

Figure 5.31 Self-thinning in *Lolium perenne* sown at five densities: 1000 (●), 5000 (●), 10,000 (■), 50,000 (■) and 100,000 (▲) 'seeds' m⁻², in: (a) 0% shade and (b) 83% shade. The lines join populations of the five sowing densities harvested on five successive occasions. They therefore indicate the trajectories, over time, that these populations would have followed. The arrows indicate the directions of the trajectories, i.e. the direction of self-thinning. For further discussion, see text. (After Lonsdale & Watkinson, 1983.)

or:

$$\bar{w} = c\,N^{-3/2} \qquad (5.23)$$

where c is constant.

Note, however, that there are statistical problems in using Equations 5.22 and 5.23 to estimate the slope of the relationship (Weller, 1987). In particular, since $\bar{w}$ is usually estimated as B/N, where B is the total biomass per unit area, $\bar{w}$ and N are inevitably correlated, and any relationship between them is, to a degree, spurious. It is therefore preferable to use the equivalent relationships, lacking autocorrelation:

$$\log B = \log c - 1/2 \log N \qquad (5.24)$$

or:

$$B = c\,N^{-1/2}. \qquad (5.25)$$

5.12.2 Species and population boundary lines

In fact, in many cases where biomass–density relationships have been documented, it is not a single cohort that has been followed over time, but a series of crowded populations at different densities (and possibly different ages) that have been compared. In such cases, it is more correct to speak of a *species boundary line* – a line beyond which combinations of density and mean weight appear not to be possible for that species (Weller, 1990). Indeed, since what is possible for a species will vary with the environment in which it is living, the species boundary line will itself subsume a whole series of *population* boundary lines, each of which defines the limits of a particular population of that species in a particular environment (Sackville Hamilton *et al.*, 1995).

Thus, a self-thinning population should approach and then track its population boundary line, which, as a trajectory, we would call its dynamic thinning line – but this need not also be its species boundary line. For example, the light regime, soil fertility, spatial arrangement of seedlings, and no doubt other factors may all alter the boundary line (and hence the dynamic thinning line) for a particular population (Weller, 1990; Sackville Hamilton *et al.*, 1995). Soil fertility, for example, has been found in different studies to alter the slope of the thinning line, the intercept, neither, or both (Morris, 2002).

dynamic thinning and boundary lines need not be the same

The influence of light is also worth considering in more detail, since it highlights a key feature of thinning and boundary lines. A slope of roughly −3/2 means that mean plant weight is increasing faster than density is decreasing, and hence that total biomass is increasing (a slope of −1/2 on a total biomass–density graph). But eventually this must stop: total biomass cannot increase indefinitely. Instead, the thinning line might be expected to change to a slope of −1: that is, loss through mortality is exactly balanced by the growth of survivors, such that the total biomass remains constant (a horizontal line on a total biomass–density graph). This can be seen when populations of *Lolium perenne* (Figure 5.31b) were grown at low light intensities. A boundary (and thinning line) with a slope of −1 was apparent

thinning slopes of −1

at much lower densities than it would otherwise be. Clearly, the light regime can alter the population boundary line. This also emphasizes, however, that boundary lines with negative slopes *steeper* than −1 (whether or not they are exactly −3/2) imply limits to the allowable combinations of plant densities and mean weights that set in *before* the maximum biomass from an area of land has been reached. Possible reasons are discussed below.

5.12.3   A single boundary line for all species?

Intriguingly, when the thinning and boundary lines of all sorts of plants are plotted on the same figure, they all appear to have approximately the same slope and also to have intercepts (i.e. values of c in Equation 5.24) falling within a narrow range (Figure 5.32). To the lower right of the figure are high-density populations of small plants (annual herbs and perennials with short-lived

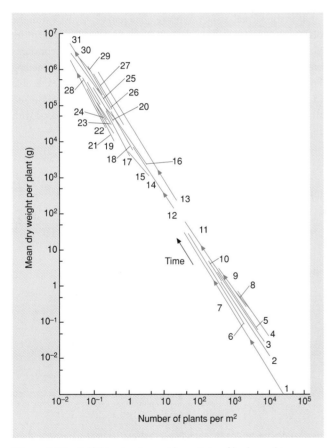

Figure 5.32 Self-thinning in a wide variety of herbs and trees. Each line is a different species, and the line itself indicates the range over which observations were made. The arrows, drawn on representative lines only, indicate the direction of self-thinning over time. The figure is based on Figure 2.9 of White (1980), which also gives the original sources and the species names for the 31 data sets.

shoots), whilst to the upper left are sparse populations of very large plants, including coastal redwoods (*Sequoia sempervirens*), the tallest known trees. Fashions change in science as in everything else. At one time, ecologists looked at Figure 5.32 and saw uniformity – all plants marching in −3/2 time (e.g. White, 1980), with variations from the norm seen as either 'noise' or as only of minor interest. Subsequently, serious doubt was cast on the conformity of individual slopes to −3/2, and on the whole idea of a single, ideal thinning line (Weller, 1987, 1990; Zeide, 1987; Lonsdale, 1990). There really is no contradiction, though. On the one hand, the lines in Figure 5.32 occupy a very much smaller portion of the graph than one would expect by chance alone. There *is* apparently some fundamental phenomenon linking this whole spectrum of plant types: not an invariable 'rule' but an underlying trend. On the other hand, the variations between the lines are real and important and in as much need of explanation as the general trend.

5.12.4   A geometric basis for self-thinning

We proceed, therefore, by examining possible bases for the general trend, and then enquiring why different species or populations might display their own variations on this common theme. Two broad types of explanation for the trend have been proposed. The first (and for many years the only one) is geometric; the second is based on resource allocation in plants of different sizes.

The geometric argument runs as follows. In a growing cohort of plants, as the mass of the population increases, the leaf area index (L, the leaf area per unit area of land) does not keep on increasing. Instead, beyond a certain point, it remains constant irrespective of plant density (N). It is, in fact, precisely beyond this point that the population follows the dynamic thinning line. We can express this by writing:

$$L = \lambda N = \text{constant} \qquad (5.26)$$

where λ is the mean leaf area per surviving plant. However, the leaf area of individual plants increases as they grow, and so too therefore does their mean, λ. It is reasonable to expect λ, because it is an area, to be related to linear measurements of a plant, such as stem diameter, D, by a formula of the following type:

$$\lambda = aD^2 \qquad (5.27)$$

where a is a constant. Similarly, it is reasonable to expect mean plant weight, $\bar{w}$, to be related to D by:

$$\bar{w} = bD^3 \qquad (5.28)$$

where b is also a constant. Putting Equations 5.26–5.28 together, we obtain:

$$\bar{w} = b(L/a)^{3/2} \cdot N^{-3/2} \qquad (5.29)$$

This is structurally equivalent to the −3/2 power law in Equation 5.23, with the intercept constant, c, given by $b(L/a)^{3/2}$.

It is apparent, therefore, why thinning lines might generally be expected to have slopes of approximately −3/2. Moreover, if the relationships in Equations 5.27 and 5.28 were roughly the same for all plant species, and if all plants supported roughly the same leaf area per unit area of ground (L), then the constant c would be approximately the same for all species. On the other hand, suppose that L is not quite constant for some species (see Equation 5.26), or that the powers in Equations 5.27 and 5.28 are not exactly 2 or 3, or that the constants in these equations (a and b) either vary between species or are not actually constants at all. Thinning lines will then have slopes that depart from −3/2, and slopes and intercepts that vary from species to species. It is easy to see why, according to the geometric argument, there is a broad similarity in the behavior of different species, but also why, on closer examination, there are variations between species and no such thing as a single, 'ideal' thinning line.

complications of the geometric argument

Furthermore, contrary to the simple geometric argument, the yield–density relationship in a growing cohort need not depend only on the numbers that die and the way the survivors grow. We have seen (see Section 5.10) that competition is frequently highly asymmetric. If those that die in a cohort are predominantly the very smallest individuals, then density (individuals per unit area) will decline more rapidly as the cohort grows than it would otherwise do, and the slope will be shallower, especially in the early stages of self-thinning. This idea is supported by a comparison of self-thinning in normal *Arabidopsis thaliana* plants with self-thinning in mutants that overexpress phytochrome A, greatly reducing their shade tolerance, and making competition amongst them more asymmetric (Figure 5.33a).

It seems possible, too, to use departures from the assumptions built into Equations 5.26–5.29 to explain at least some of the variations from a 'general' −3/2 rule. Osawa and Allen (1993) estimated a number of the parameters in these equations from data on the growth of individual plants of mountain beech (*Nothofagus solandri*) and red pine (*Pinus densiflora*). They estimated, for instance, that the exponents in Equations 5.27 and 5.28 were not 2 and 3, but 2.08 and 2.19 for mountain beech, and 1.63 and 2.41 for red pine. These suggest thinning slopes of −1.05 in the first case and −1.48 in the second, which compare quite remarkably well with the observed slopes of −1.06 and −1.48 (Figure 5.33b). The similarities between the estimates and observations for the intercept constants were equally impressive. These results show, therefore, that thinning lines with slopes other than −3/2 can occur, but can be explicable in terms of the detailed biology of the species concerned – and that even when slopes of −3/2 do occur, they may do so, as with red pine, for the 'wrong' reason (−2.41/1.63 rather than −3/2).

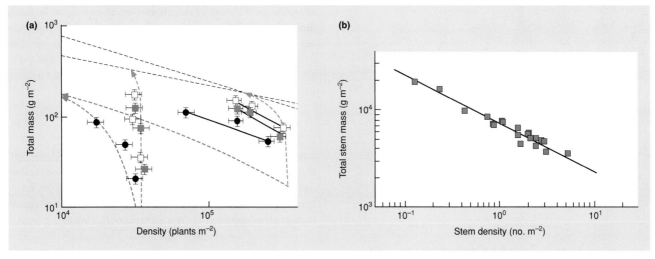

Figure 5.33 (a) The relationship between total biomass and density in two wild-type strains of *Arabidopsis thaliana* plants (□ and ▪) and a phytochrome A overexpressing mutant strain (●) 15, 22 and 33 days after sowing (bottom to top). Data points give means (±1 SE, n = 3). In each case, the strains were sown at two initial densities; solid black regression lines are shown in each case for the higher density. The steeper dotted black line has a slope of −1/2 (indicative of −3/2 self-thinning) and the shallow line a slope of −1/3 (indicative of −4/3 self-thinning). Model trajectories for asymmetric (———) and symmetric (− − −) competition are also shown. The mutant strain exhibited shallower thinning lines, indicative of more asymmetric competition. (After Stoll *et al.*, 2002.) (b) The species boundary line for populations of red pine, *Pinus densiflora* (slope = −1.48) from northern Japan. (After Osawa & Allen, 1993.)

5.12.5 A resource-allocation basis for thinning boundaries

The increasing recognition that a variety of slope values might be expected, even on the geometric argument, along with the statistical difficulties of estimating slopes, has left the way open for alternative explanations for the underlying trend itself. Enquist *et al.* (1998) made use of the much more general model of West *et al.* (1997), which considered the most effective architectural designs of organisms (not just plants) for distributing acquired resources throughout those organisms. This suggested that the rate of resource use per individual, u, should be related to mean plant weight, $\bar{w}$, according to the equation:

$$u = a\bar{w}^{3/4} \qquad (5.30)$$

where a is a constant. Indeed, Enquist *et al.* (1998) were also able to find empirical support for this relationship.

−4/3 or −3/2?

They then argued that plants have evolved to make full use of the resources available, and so if S is the rate of resource supply per unit area and N_{max} the maximum allowable density of plants, then:

$$S = N_{max}u \qquad (5.31)$$

or, from Equation 5.30:

$$S = aN_{max}\bar{w}^{3/4}. \qquad (5.32)$$

But when the plants have arrived at an equilibrium with the rate of resource supply, S should itself be constant. Hence:

$$\bar{w} = bN_{max}^{-4/3}, \qquad (5.33)$$

where b is another constant. In short, the expected slope of a population boundary on this argument is −4/3 rather than −3/2.

Enquist and colleagues themselves considered the available data to be more supportive of their prediction of a slope of −4/3 than the more conventional −3/2. This has not, however, been the conclusion drawn either from previous data surveys or from the analysis of subsequent experiments (e.g. Figure 5.33a; Stoll *et al.*, 2002). In part, the discrepancy may have arisen because the geometric argument is focused on light acquisition, and the data collected to test it have likewise been focused on above-ground plant parts (photosynthetic or support tissue); whereas Enquist *et al.*'s is a much more general resource-acquisition argument, and at least some of their data were based on overall plant weights (leaves, shoots and roots). Related to this, Enquist *et al.*'s data sets were focused on maximum densities of large numbers of species, whereas other analyses have focused on the self-thinning *process*,

which occurs largely before the overall resource-determined limit has been reached. Again, therefore, there may be no contradiction between the two approaches.

5.12.6 Self-thinning in animal populations

Animals, whether they are sessile or mobile, must also 'self-thin', insofar as growing individuals within a cohort increasingly compete with one another and reduce their own density. There is nothing linking all animals quite like the shared need for light interception that links all plants, so there is even less likelihood of a general self-thinning 'law' for animals. On the other hand, crowded sessile animals can, like plants, be seen as needing to pack 'volumes' beneath an approximately constant area, and mussels, for example, have been found to follow a thinning line with a slope of −1.4, and barnacles a line with a slope of −1.6 (Hughes & Griffiths, 1988). Moreover, self-thinning in the gregarious tunicate, *Pyura praeputialis*, on the coast of Chile was found to follow a slope of only −1.2; but when the analysis was modified to acknowledge that rocky shore invertebrates are more 'three-dimensional' than plants, and may fit more than one layer into a fully occupied area (as opposed to the constant leaf area index of plants), then the estimated slope was −1.5 (Figure 5.34a).

For mobile animals, it has been suggested that the relationship between metabolic rate and body size could generate thinning lines with slopes of −4/3 (Begon *et al.*, 1986). However, the generality of this is probably even more questionable than the 'rules' in plants, given variations in resource supply, variations in the coefficients in the underlying relationships, and the possibilities of self-thinning depending on, say, territorial behavior rather than simply food availability (Steingrimsson & Grant, 1999). Nonetheless, evidence of self-thinning in animals is increasingly reported, especially in fish, even if the basis for it remains uncertain (e.g. Figure 5.34b).

Plants are not so consistent in their pattern of self-thinning as was once thought. It may be that animals are not much less bound than plants by 'general' self-thinning rules.

Summary

Intraspecific competition is defined and explained. Exploitation and interference are distinguished, and the commonly one-sided nature of competition is emphasized.

We describe the effects of intraspecific competition on rates of mortality and fecundity, distinguishing under-, over- and exactly compensating density dependence. We explain, however, that density itself is usually just a convenient expression of crowding or shortage of resources.

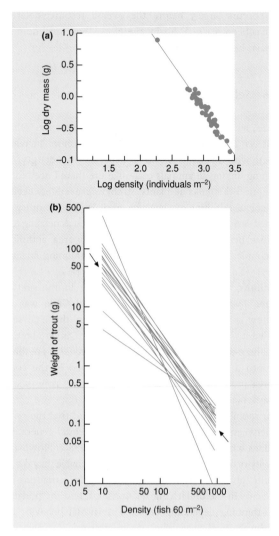

Figure 5.34   (a) Self-thinning in the gregarious tunicate, *Pyura praeputialis*, where density has been modified to include an 'effective area' which incorporates the number of layers in the tunicate colonies. The estimated slope is −1.49 (95% CI −1.59 to −1.39, P < 0.001). (After Guiñez & Castilla, 2001.) (b) Dynamic thinning lines for 23 year-classes of sea trout, *Salmo trutta*, from an English Lake District stream, with the position of the mean regression line (slope = −1.35) indicated by the arrows (After Elliott, 1993.)

These effects at the individual level lead in turn to patterns, and regulatory tendencies, at the population level. The carrying capacity is defined and its limitations are explained, along with the domed nature of net recruitment curves and the sigmoidal nature of population growth curves.

We describe the effects of intraspecific competition on rates of growth, explaining the 'law of constant final yield', especially in modular organisms.

The use of k values in quantifying intraspecific competition is described, and scramble and contest competition are distinguished.

We introduce the use of mathematical models in ecology generally, then go on to develop a model of a population with discrete breeding seasons subject to intraspecific competition. The model illustrates the tendency of time lags to provoke population fluctuations and that different types of competition may lead to different types of population dynamics, including patterns of deterministic chaos – the nature and importance of which are themselves explained. A model with continuous breeding is also developed, leading to the logistic equation.

The importance of individual differences in generating asymmetries in competition is explained, as is the importance of competition in generating individual differences. Asymmetries tend to enhance regulation; territoriality is a particularly important example of this.

The progressive effects of competition on growth and mortality may often be interlinked in the process of self-thinning, which has been a particular focus in plant populations. We explain the nature of dynamic thinning lines and the −3/2 power law when single cohorts are followed, and also species and population boundary lines when a series of crowded populations is observed at different densities. We address the question of whether there is a single boundary line for all species.

We explain how two broad types of explanation for the consistent trend amongst species have been proposed: those based on geometry and those based on resource allocation in plants of different sizes.

Finally we examine self-thinning in animal populations and conclude that plants are not so consistent in their pattern of self-thinning as was once thought, while animals are not much less bound than plants by 'general' self-thinning rules.

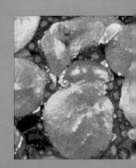

Chapter 6
Dispersal, Dormancy and Metapopulations

6.1 Introduction

All organisms in nature are where we find them because they have moved there. This is true for even the most apparently sedentary of organisms, such as oysters and redwood trees. Their movements range from the passive transport that affects many plant seeds to the apparently purposeful actions of many mobile animals. *Dispersal* and *migration* are used to describe aspects of the movement of organisms. The terms are defined for groups of organisms, although it is of course the individual that moves.

the meanings of 'dispersal' and 'migration'

Dispersal is most often taken to mean a spreading of individuals away from others, and is therefore an appropriate description for several kinds of movements: (i) of plant seeds or starfish larvae away from each other and their parents; (ii) of voles from one area of grassland to another, usually leaving residents behind and being counterbalanced by the dispersal of other voles in the other direction; and (iii) of land birds amongst an archipelago of islands (or aphids amongst a mixed stand of plants) in the search for a suitable habitat.

Migration is most often taken to mean the mass directional movements of large numbers of a species from one location to another. The term therefore applies to classic migrations (the movements of locust swarms, the intercontinental journeys of birds) but also to less obvious examples like the to and fro movements of shore animals following the tidal cycle. Whatever the precise details of dispersal in particular cases, it will be useful in this chapter to divide the process into three phases: *starting, moving* and *stopping* (South *et al.*, 2002) or, put another way, *emigration, transfer* and *immigration* (Ims & Yoccoz, 1997). The three phases differ (and the questions we ask about them differ) both from a behavioral point of view (what triggers the initiation and cessation of movement?, etc.) and from a demographic point of view (the distinction between loss and gain of individuals, etc.).

The division into these phases also emphasizes that dispersal can refer to the process by which individuals, in leaving, escape from the immediate environment of their parents and neighbors; but it can also often involve a large element of discovery or even exploration. It is useful, too, to distinguish between *natal dispersal* and *breeding dispersal* (Clobert *et al.*, 2001). The former refers to the movement between the natal area (i.e. where the individual was born) and where breeding first takes place. This is the only type of dispersal possible in a plant. Breeding dispersal is movement between two successive breeding areas.

6.2 Active and passive dispersal

Like most biological categories, the distinction between active and passive dispersers is blurred at the edges. Passive dispersal in air currents, for example, is not restricted to plants. Young spiders that climb to high places and then release a gossamer thread that carries them on the wind are then passively at the mercy of air currents; i.e. 'starting' is active but moving itself is effectively passive. Even the wings of insects are often simply aids to what is effectively passive movement (Figure 6.1).

6.2.1 Passive dispersal: the seed rain

Most seeds fall close to the parent and their density declines with distance from that parent. This is the case for wind-dispersed seeds and also for those that are ejected actively by maternal tissue (e.g. many legumes). The eventual destination of the dispersed offspring is determined by the original location of the parent and by the relationship relating disperser density to distance from parent, but the detailed microhabitat of that destination is left to chance. Dispersal is nonexploratory; discovery is a matter of chance. Some animals have essentially this same type of dispersal. For

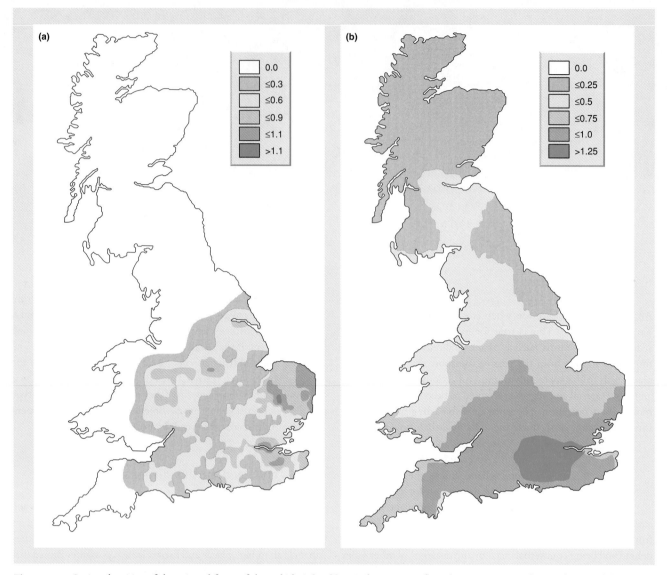

Figure 6.1 Spring densities of the winged form of the aphid, *Aphis fabae*, in large part reflect their carriage on the wind. (a) *A. fabae* eggs are found on spindle plants and their distribution in the UK over winter reflects that of the plants ($\log_{10}$ geometric mean number of eggs per 100 spindle buds). (b) But by spring, although the highest densities are in spindle regions, the aphids have dispersed on the wind over the whole country ($\log_{10}$ geometric mean aerial density). (After Compton, 2001; from Cammell *et al.*, 1989.)

example, the dispersal of most pond-dwelling organisms without a free-flying stage depends on resistant wind-blown structures (e.g. gemmules of sponges, cysts of brine shrimps).

The density of seeds is often low immediately under the parent, rises to a peak close by and then falls off steeply with distance (Figure 6.2a). However, there are immense practical problems in studying seed dispersal (i.e. in following the seeds), and these become increasingly irresolvable further from the source. Greene and Calogeropoulos (2001) liken any assertion that 'most seeds travel short distances' to a claim that most lost keys and contact lenses fall close to streetlights. Certainly, the very few

studies of long-distance dispersal that have been carried out suggest that seed density declines only very slowly at larger distances from the original source (Figure 6.2b), and even a few long-distance dispersers may be crucial in either invasion or recolonization dispersal (see Section 6.3.1).

6.2.2 Passive dispersal by a mutualistic agent

Uncertainty of destination may be reduced if an active agent of dispersal is involved. The seeds of many herbs of the woodland

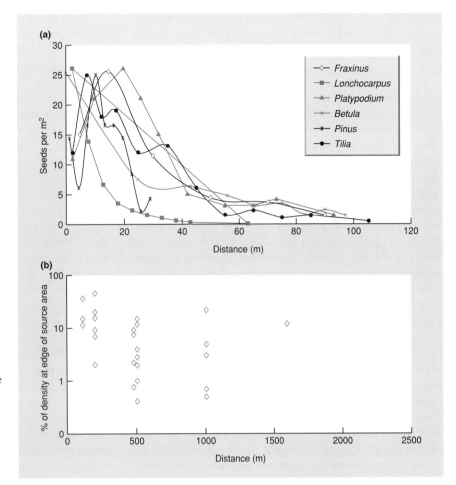

Figure 6.2 (a) The density of wind-dispersed seeds from solitary trees within forests. The studies had a reasonable density of sampling points, there were no nearby conspecific trees and the source tree was neither in a clearing nor at the forest edge. (b) Observed long-distance dispersal up to 1.6 km of wind dispersed seeds from a forested source area. (After Greene & Calogeropoulos, 2001, where the original data sources may also be found.)

floor have spines or prickles that increase their chance of being carried passively on the coats of animals. The seeds may then be concentrated in nests or burrows when the animal grooms itself. The fruits of many shrubs and lower canopy trees are fleshy and attractive to birds, and the seed coats resist digestion in the gut. Where the seed is dispersed to is then somewhat less certain, depending on the defecating behavior of the bird. It is usually presumed that such associations are 'mutualistic' (beneficial to both parties – see Chaper 13): the seed is dispersed in a more or less predictable fashion; the disperser consumes either the fleshy 'reward' or a proportion of the seeds (those that it finds again).

There are also important examples in which animals are dispersed by an active agent. For instance, there are many species of mite that are taken very effectively and directly from dung pat to dung pat, or from one piece of carrion to another, by attaching themselves to dung or carrion beetles. They usually attach to a newly emerging adult, and leave again when that adult reaches a new patch of dung or carrion. This, too, is typically mutualistic: the mites gain a dispersive agent, and many of them attack and eat the eggs of flies that would otherwise compete with the beetles.

6.2.3 Active discovery and exploration

Many other animals cannot be said to explore, but they certainly control their settlement ('stopping', see Section 6.1.1) and cease movement only when an acceptable site has been found. For example, most aphids, even in their winged form, have powers of flight that are too weak to counteract the forces of prevailing winds. But they control their take-off from their site of origin, they control when they drop out of the windstream, and they make additional, often small-scale flights if their original site of settlement is unsatisfactory. In a precisely analogous manner, the larvae of many river invertebrates make use of the flowing column of water for dispersing from hatching sites to appropriate microhabitats ('invertebrate drift') (Brittain & Eikeland, 1988). The dispersal of aphids in the wind and of drifting invertebrates in streams, therefore, involves 'discovery', over which they have some, albeit limited, control.

Other animals explore, visiting many sites before returning to a favored suitable one. For example, in contrast to their drifting larvae, most adults of freshwater insects depend on flight for upstream dispersal and movement from stream to stream. They explore and, if successful, discover, suitable sites within which to lay their eggs: starting, moving and stopping are all under active control.

6.2.4   Clonal dispersal

In almost all modular organisms (see Section 4.2.1), an individual genet branches and spreads its parts around it as it grows. There is a sense, therefore, in which a developing tree or coral actively disperses its modules into, and explores, the surrounding environment. The interconnections of such a clone often decay, so that it becomes represented by a number of dispersed parts. This may result ultimately in the product of one zygote being represented by a clone of great age that is spread over great distances. Some clones of the rhizomatous bracken fern (*Pteridium aquilinum*) were estimated to be more than 1400 years old and one extended over an area of nearly 14 ha (Oinonen, 1967).

guerrillas and phalanx-formers   We can recognize two extremes in a continuum of strategies in clonal dispersal (Lovett Doust & Lovett Doust, 1982; Sackville Hamilton *et al.*, 1987). At one extreme, the connections between modules are long and the modules themselves are widely spaced. These have been called 'guerrilla' forms, because they give the plant, hydroid or coral a character like that of a guerrilla army. Fugitive and opportunist, they are constantly on the move, disappearing from some territories and penetrating into others. At the other extreme are 'phalanx' forms, named by analogy with the phalanxes of a Roman army, tightly packed with their shields held around them. Here, the connections are short and the modules are tightly packed, and the organisms expand their clones slowly, retain their original site occupancy for long periods, and neither penetrate readily amongst neighboring plants nor are easily penetrated by them.

Even amongst trees, it is easy to see that the way in which the buds are placed gives them a guerrilla or a phalanx type of growth form. The dense packing of shoot modules in species like cypresses (*Cupressus*) produces a relatively undispersed and impenetrable phalanx canopy, whilst many loose-structured, broad-leaved trees (*Acacia*, *Betula*) can be seen as guerrilla canopies, bearing buds that are widely dispersed and shoots that interweave with the buds and branches of neighbors. The twining or clambering lianas in a forest are guerrilla growth forms *par excellence*, dispersing their foliage and buds over immense distances, both vertically and laterally.

The way in which modular organisms disperse and display their modules affects the ways in which they interact with their neighbors. Those with a guerrilla form will continually meet and compete with other species and other genets of their own kind. With a phalanx structure, however, most meetings will be between modules of a single genet. For a tussock grass or a cypress tree, competition must occur very largely between parts of itself.

Clonal growth is most effective as a means of dispersal in aquatic environments. Many aquatic plants fragment easily, and the parts of a single clone become independently dispersed because they are not dependent on the presence of roots to maintain their water relations. The major aquatic weed problems of the world are caused by plants that multiply as clones and fragment and fall to pieces as they grow: duckweeds (*Lemna* spp.), the water hyacinth (*Eichhornia crassipes*), Canadian pond weed (*Elodea Canadensis*) and the water fern *Salvinia*.

6.3   Patterns of distribution: dispersion

The movements of organisms affect the spatial pattern of their distribution (their *dispersion*) and we can recognize three main patterns of dispersion, although they too form part of a continuum (Figure 6.3).

Random dispersion occurs when there is an equal probability of an organism occupying any point in space (irrespective of the position of any others). The result is that individuals are unevenly distributed because of chance events. random, regular and aggregated distributions

Regular dispersion (also called a *uniform* or *even* distribution or *overdispersion*) occurs either when an individual has a tendency to avoid other individuals, or when individuals that are especially close to others die. The result is that individuals are more evenly spaced than expected by chance.

Aggregated dispersion (also called a *contagious* or *clumped* distribution or *underdispersion*) occurs either when individuals tend to be attracted to (or are more likely to survive in) particular parts of the environment, or when the presence of one individual

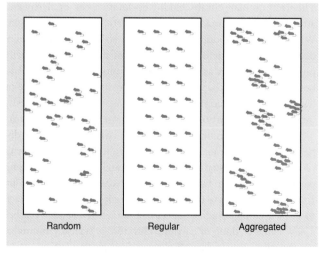

Figure 6.3   Three generalized spatial patterns that may be exhibited by organisms across their habitats.

attracts, or gives rise to, another close to it. The result is that individuals are closer together than expected by chance.

How these patterns appear to an observer, however, and their relevance to the life of other organisms, depends on the spatial scale at which they are viewed. Consider the distribution of an aphid living on a particular species of tree in a woodland. At a large scale, the aphids will appear to be aggregated in particular parts of the world, i.e. in woodlands as opposed to other types of habitat. If samples are smaller and taken only in woodlands, the aphids will still appear to be aggregated, but now on their host tree species rather than on trees in general. However, if samples are smaller still (25 cm^2, about the size of a leaf) and are taken within the canopy of a single tree, the aphids might appear to be randomly distributed over the tree as a whole. At an even smaller scale (c. 1 cm^2) we might detect a regular distribution because individual aphids on a leaf avoid one another.

6.3.1 Patchiness

fine- and coarse-grained environments

In practice, the populations of all species are patchily distributed at some scale or another, but it is crucial to describe dispersion at scales that are relevant to the lifestyle of the organisms concerned. MacArthur and Levins (1964) introduced the concept of environmental *grain* to make this point. For example, the canopy of an oak–hickory forest, from the point of view of a bird like the scarlet tanager (*Piranga olivacea*) that forages indiscriminately in both oaks and hickories, is *fine grained*: i.e. it is patchy, but the birds experience the habitat as an oak–hickory mixture. The habitat is *coarse grained*, however, for defoliating insects that attack either oaks or hickories preferentially: they experience the habitat one patch at a time, moving from one preferred patch to another (Figure 6.4).

Patchiness may be a feature of the physical environment: islands surrounded by water, rocky outcrops in a moorland, and so on. Equally important, patchiness may be created by the activities of organisms themselves; by their grazing, the deposition of dung, trampling or by the local depletion of water and mineral resources. Patches in the environment that are created by the activity of organisms have life histories. A gap created in a forest by a falling tree is colonized and grows up to contain mature trees, whilst other trees fall and create new gaps. The dead leaf in a grassland area is a patch for colonization by a succession of fungi and bacteria that eventually exhaust it as a resource, but new dead leaves arise and are colonized elsewhere.

Patchiness, dispersal and scale are tied intimately together. A framework that is useful in thinking about this distinguishes between local and landscape scales (though what is 'local' to a worm is very different from what is local to the bird that eats it) and between *turnover* and *invasion* dispersal (Bullock et al., 2002). Turnover dispersal at the local scale describes the movement into

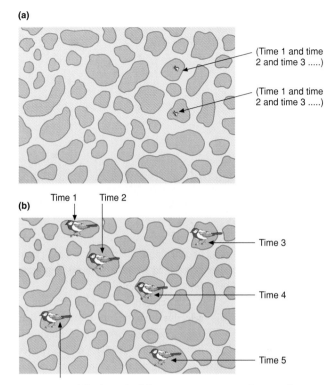

Figure 6.4 The 'grain' of the environment must be seen from the perspective of the organism concerned. (a) An organism that is small or moves little is likely to see the environment as coarse-grained: it experiences only one habitat type within the environment for long periods or perhaps all of its life. (b) An organism that is larger or moves more may see the same environment as fine-grained: it moves frequently between habitat types and hence samples them in the proportion in which they occur in the environment as a whole.

a gap from occupied habitat immediately surrounding the gap; whereas that gap may also be invaded or colonized by individuals moving in from elsewhere in the surrounding community. At the landscape scale, similarly, dispersal may be part of an on-going turnover of extinction and recolonization of occupiable patches within an otherwise unsuitable habitat matrix (e.g. islands in a stream: 'metapopulation dynamics' – see Section 6.9, below), or dispersal may result in the invasion of habitat by a 'new' species expanding its range.

6.3.2 Forces favoring aggregations (in space and time)

The simplest evolutionary explanation for the patchiness of populations is that organisms aggregate when and where they find resources and conditions that favor reproduction and survival. These resources and conditions are usually patchily distributed in both space and time. It pays (and has paid in evolutionary time)

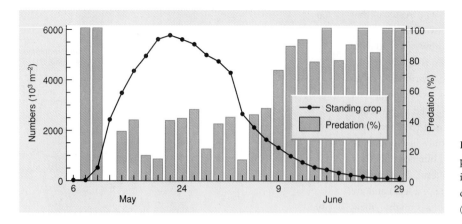

Figure 6.5 Changes in the density of a population of 13-year periodical cicadas in northwestern Arkansas in 1985, and changes in the percentage eaten by birds. (After Williams *et al.*, 1993.)

to disperse to these patches when and where they occur. There are, however, other specific ways in which organisms may gain from being close to neighbors in space and time.

aggregation and the selfish herd

An elegant theory identifying a selective advantage to individuals that aggregate with others was suggested by Hamilton (1971) in his paper 'Geometry for the selfish herd'. He argued that the risk to an individual from a predator may be lessened if it places another potential prey individual between itself and the predator. The consequence of many individuals doing this is bound to be an aggregation. The 'domain of danger' for individuals in a herd is at the edge, so that an individual would gain an advantage if its social status allowed it to assimilate into the center of a herd. Subordinate individuals might then be forced into the regions of greater danger on the edge of the flock. This seems to be the case in reindeer (*Rangifer tarandus*) and woodpigeons (*Columba palumbus*), where a newcomer may have to join the herd or flock at its risky perimeter and can only establish itself in a more protected position within the flock after social interaction (Murton *et al.*, 1966). Individuals may also gain from living in groups if this helps to locate food, give warning of predators or if it pays for individuals to join forces in fighting off a predator (Pulliam & Caraco, 1984).

The principle of the selfish herd as described for the aggregation of organisms in space is just as appropriate for the synchronous appearance of organisms in time. The individual that is precocious or delayed in its appearance, outside the norm for its population, may be at greater risk from predators than those conformist individuals that take part in 'flooding the market' thereby diluting their own risk. Amongst the most remarkable examples of synchrony are the 'periodic cicadas' (insects), the adults of which emerge simultaneously after 13 or 17 years of life underground as nymphs. Williams *et al.* (1993) studied the mortality of populations of 13-year periodic cicadas that emerged in northwestern Arkansas in 1985. Birds consumed almost all of the standing crop of cicadas when the density was low, but only 15–40% when the cicadas reached peak density. Predation then rose to near 100%

as the cicada density fell again (Figure 6.5). Equivalent arguments apply to the many species of tree, especially in temperate regions, that have synchronous 'mast' years (see Section 9.4).

6.3.3 Forces diluting aggregations: density-dependent dispersal

There are also strong selective pressures that can act *against* aggregation in space or time. In some species a group of individuals may actually concentrate a predator's attention (the opposite effect to the 'selfish herd'). However, the foremost diluting forces are certain to be the more intense competition suffered by crowded individuals (see Chapter 5) and the direct interference between such individuals even in the absence of a shortage of resources. One likely consequence is that the highest rates of dispersal will be away from the most crowded patches: density-dependent emigration dispersal (Figure 6.6) (Sutherland *et al.*, 2002), though as we shall see below, density-dependent dispersal is by no means a general rule.

Overall, though, the types of distribution over available patches found in nature are bound to be compromises between forces attracting individuals to disperse towards one another and forces provoking individuals to disperse away from one another. As we shall see in a later chapter, such compromises are conventionally crystallized in the 'ideal free' and other theoretical distributions (see Section 9.6.3).

6.4   Patterns of migration

6.4.1   Tidal, diurnal and seasonal movements

Individuals of many species move *en masse* from one habitat to another and back again repeatedly during their life. The timescale involved may be hours, days, months or years. In some cases, these movements have the effect of maintaining the

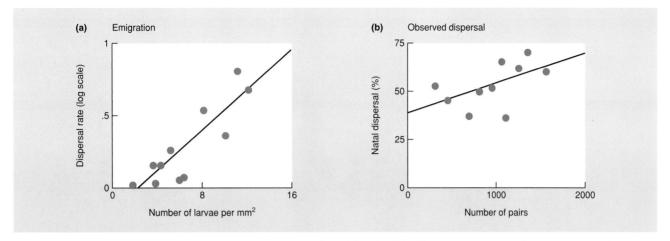

Figure 6.6   Density-dependent dispersal. (a) The dispersal rates of newly hatched blackfly (*Simulium vittatum*) larvae increase with increasing density. (Data from Fonseca & Hart, 1996.) (b) The percentage of juvenile male barnacle geese, *Branta leucopsis*, dispersing from breeding colonies on islands in the Baltic Sea to non-natal breeding locations increased as density increased. (Data from van der Juegd 1999.) (After Sutherland *et al.*, 2002.)

organism in the same type of environment. This is the case in the movement of crabs on a shoreline: they move with the advance and retreat of the tide. In other cases, diurnal migration may involve moving between two environments: the fundamental niches of these species can only be satisfied by alternating life in two distinct habitats within each day of their lives. For example, some planktonic algae both in the sea and in lakes descend to the depths at night but move to the surface during the day. They accumulate phosphorus and perhaps other nutrients in the deeper water at night before returning to photosynthesize near the surface during daylight hours (Salonen *et al.*, 1984). Other species aggregate into tight populations during a resting period and separate from each other when feeding. For example, most land snails rest in confined humid microhabitats by day, but range widely when they search for food by night.

Many organisms make seasonal migrations – again, either tracking a favorable habitat or benefitting from different, complementary habitats. The altitudinal migration of grazing animals in mountainous regions is one example. The American elk (*Cervus elaphus*) and mule deer (*Odocoileus hemionus*), for instance, move up into high mountain areas in the summer and down to the valleys in the winter. By migrating seasonally the animals escape the major changes in food supply and climate that they would meet if they stayed in the same place. This can be contrasted with the 'migration' of amphibians (frogs, toads, newts) between an aquatic breeding habitat in spring and a terrestrial environment for the remainder of the year. The young develop (as tadpoles) in water with a different food resource from that which they will later eat on land. They will return to the same aquatic habitat for mating, aggregate into dense populations for a time and then separate to lead more isolated lives on land.

6.4.2   Long-distance migration

The most remarkable habitat shifts are those that involve traveling very long distances. Many terrestrial birds in the northern hemisphere move north in the spring when food supplies become abundant during the warm summer period, and move south to savannas in the fall when food becomes abundant only after the rainy season. Both are areas in which seasons of comparative glut and famine alternate. Migrants then make a large contribution to the diversity of a local fauna. Of the 589 species of birds (excluding seabirds) that breed in the Palaearctic (temperate Europe and Asia), 40% spend the winter elsewhere (Moreau, 1952). Of those species that leave for the winter, 98% travel south to Africa. On an even larger scale, the Arctic tern (*Sterna paradisaea*) travels from its Arctic breeding ground to the Antarctic pack ice and back each year – about 10,000 miles (16,100 km) each way (although unlike many other migrants it can feed on its journey).

The same species may behave in different ways in different places. All robins (*Erithacus rubecula*) leave Finland and Sweden in winter, but on the Canary Islands the species is resident the whole year-round. In most of the intervening countries, a part of the population migrates and a part remains resident. Such variations are in some cases associated with clear evolutionary divergence. This is true of the knot (*Calidris canutus*), a species of small wading bird mostly breeding in remote areas of the Arctic tundras and 'wintering' in the summers of the southern hemisphere. At least five subspecies appear to have diverged in the Late Pleistocene (based on genetic evidence from the sequencing of mitochondrial DNA), and these now have strikingly different patterns of distribution and migration (Figure 6.7).

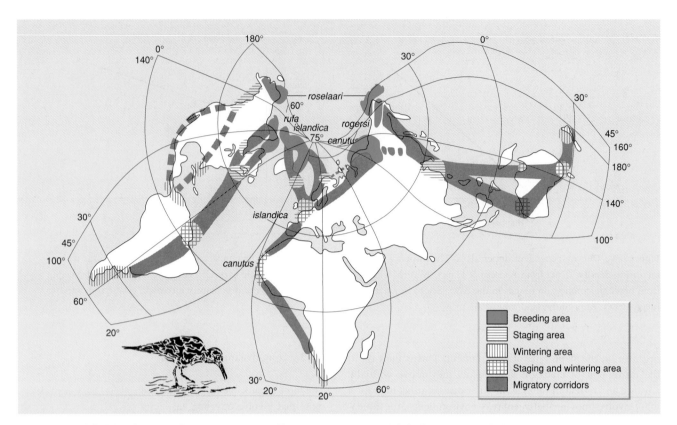

Figure 6.7 Global distribution and migration pattern of knots (*Calidris* spp.). Solid shading indicates the breeding areas; horizontally striped spots indicate the stop-over areas, used only during south- and northward migration; the cross-hatched spots indicate the areas used both as stop-over and wintering sites; and the vertically striped spots designate areas used only for wintering. The gray shaded corridors indicate proven migration routes; the broken-shaded corridors indicate tentative migration routes suggested in the literature. (After Piersma & Davidson, 1992.)

Long-distance migration is a feature of many other groups too. Baleen whales in the southern hemisphere move south in summer to feed in the food-rich waters of the Antarctic. In winter they move north to breed (but scarcely to feed) in tropical and subtropical waters. Caribou (*Rangifer tarandus*) travel several hundred kilometers per year from northern forests to the tundra and back. In all of these examples, an individual of the migrating species may make the return journey several times.

eels and salmon

Many long-distance migrants, however, make only one return journey during their lifetime. They are born in one habitat, make their major growth in another habitat, but then return to breed and die in the home of their infancy. Eels and migratory salmon provide classic examples. The European eel (*Anguilla anguilla*) travels from European rivers, ponds and lakes across the Atlantic to the Sargasso Sea, where it is thought to reproduce and die (although spawning adults and eggs have never actually been caught there). The American eel (*Anguilla rostrata*) makes a comparable journey from areas ranging between the Guianas in the south, to southwest Greenland in the north. Salmon make

a comparable transition, but from a freshwater egg and juvenile phase to mature as a marine adult. The fish then returns to freshwater sites to lay eggs. After spawning, all Pacific salmon (*Oncorhynchus nerka*) die without ever returning to the sea. Many Atlantic salmon (*Salmo salar*) also die after spawning, but some survive to return to the sea and then migrate back upstream to spawn again.

6.4.3 'One-way only' migration

In some migratory species, the journey for an individual is on a strictly one-way ticket. In Europe, the clouded yellow (*Colias croceus*), red admiral (*Vanessa atalanta*) and painted lady (*Vanessa cardui*) butterflies breed at both ends of their migrations. The individuals that reach Great Britain in the summer breed there, and their offspring fly south in autumn and breed in the Mediterranean region – the offspring of these in turn come north in the following summer.

Most migrations occur seasonally in the life of individuals or of populations. They usually seem to be triggered by some external seasonal phenomenon (e.g. changing day length), and sometimes also by an internal physiological clock. They are often preceded by quite profound physiological changes such as the accumulation of body fat. They represent strategies evolved in environments where seasonal events like rainfall and temperature cycles are reliably repeated from year to year. There is, however, a type of migration that is tactical, forced by events such as overcrowding, and appears to have no cycle or regularity. These are most common in environments where rainfall is not seasonally reliable. The economically disastrous migration plagues of locusts in arid and semiarid regions are the most striking examples.

6.5 Dormancy: migration in time

An organism gains in fitness by dispersing its progeny as long as the progeny are more likely to leave descendants than if they remained undispersed. Similarly, an organism gains in fitness by delaying its arrival on the scene, so long as the delay increases its chances of leaving descendants. This will often be the case when conditions in the future are likely to be better than those in the present. Thus, a delay in the recruitment of an individual to a population may be regarded as 'migration in time'.

Organisms generally spend their period of delay in a state of *dormancy*. This relatively inactive state has the benefit of conserving energy, which can then be used during the period following the delay. In addition, the dormant phase of an organism is often more tolerant of the adverse environmental conditions prevailing during the delay (i.e. tolerant of drought, extremes of temperature, lack of light and so on). Dormancy can be either *predictive* or *consequential* (Müller, 1970). Predictive dormancy is initiated in advance of the adverse conditions, and is most often found in predictable, seasonal environments. It is generally referred to as 'diapause' in animals, and in plants as 'innate' or 'primary' dormancy (Harper, 1977). Consequential (or 'secondary') dormancy, on the other hand, is initiated in response to the adverse conditions themselves.

6.5.1 Dormancy in animals: diapause

Diapause has been most intensively studied in insects, where examples occur in all developmental stages. The common field grasshopper *Chorthippus brunneus* is a fairly typical example. This annual species passes through an *obligatory* diapause in its egg stage, where, in a state of arrested development, it is resistant to the cold winter conditions that would quickly kill the nymphs and adults. In fact, the eggs require a long cold period before development can start again (around 5 weeks at 0°C, or rather longer at a slightly higher temperature) (Richards & Waloff, 1954). This ensures that the eggs are not affected by a short, freak period of warm winter weather that might then be followed by normal, dangerous, cold conditions. It also means that there is an enhanced synchronization of subsequent development in the population as a whole. The grasshoppers 'migrate in time' from late summer to the following spring.

Diapause is also common in species with more than one generation per year. For instance, the fruit-fly *Drosophila obscura* passes through four generations per year in England, but enters diapause during only one of them (Begon, 1976). This *facultative* diapause shares important features with obligatory diapause: it enhances survivorship during a predictably adverse winter period, and it is experienced by *resistant* diapause adults with arrested gonadal development and large reserves of stored abdominal fat. In this case, synchronization is achieved not only during diapause but also prior to it. Emerging adults react to the short daylengths of the fall by laying down fat and entering the diapause state; they recommence development in response to the longer days of spring. Thus, by relying, like many species, on the utterly predictable *photoperiod* as a cue for seasonal development, *D. obscura* enters a state of predictive diapause that is confined to those generations that inevitably pass through the adverse conditions.

Consequential dormancy may be expected to evolve in environments that are relatively unpredictable. In such circumstances, there will be a disadvantage in responding to adverse conditions only after they have appeared, but this may be outweighed by the advantages of: (i) responding to favorable conditions *immediately* after they reappear; and (ii) entering a dormant state only if adverse conditions *do* appear. Thus, when many mammals enter hibernation, they do so (after an obligatory preparatory phase) in direct response to the adverse conditions. Having achieved 'resistance' by virtue of the energy they conserve at a lowered body temperature, and having periodically emerged and monitored their environment, they eventually cease hibernation whenever the adversity disappears.

6.5.2 Dormancy in plants

Seed dormancy is an extremely widespread phenomenon in flowering plants. The young embryo ceases development whilst still attached to the mother plant and enters a phase of suspended activity, usually losing much of its water and becoming dormant in a desiccated condition. In a few species of higher plants, such as some mangroves, a dormant period is absent, but this is very much the exception – almost all seeds are dormant when they are shed from the parent and require special stimuli to return them to an active state (germination).

Dormancy in plants, though, is not confined to seeds. For example, as the sand sedge *Carex arenaria* grows, it tends to accumulate dormant buds along the length of its predominantly linear rhizome. These may remain alive but dormant long after the shoots with which they were produced have died, and they have been found in numbers of up to 400–500 m^{-2} (Noble *et al.*, 1979). They play a role analogous to the bank of dormant seeds produced by other species.

Indeed, the very widespread habit of deciduousness is a form of dormancy displayed by many perennial trees and shrubs. Established individuals pass through periods, usually of low temperatures and low light levels, in a leafless state of low metabolic activity.

innate, enforced and induced dormancy

Three types of dormancy have been distinguished.

1 *Innate dormancy* is a state in which there is an absolute requirement for some special external stimulus to reactivate the process of growth and development. The stimulus may be the presence of water, low temperature, light, photoperiod or an appropriate balance of near- and far-red radiation. Seedlings of such species tend to appear in sudden flushes of almost simultaneous germination. Deciduousness is also an example of innate dormancy.

2 *Enforced dormancy* is a state imposed by external conditions (i.e. it is consequential dormancy). For example, the Missouri goldenrod *Solidago missouriensis* enters a dormant state when attacked by the beetle *Trirhabda canadensis*. Eight clones, identified by genetic markers, were followed prior to, during and after a period of severe defoliation. The clones, which varied in extent from 60 to 350 m^2 and from 700 to 20,000 rhizomes, failed to produce any above-ground growth (i.e. they were dormant) in the season following defoliation and had apparently died, but they reappeared 1–10 years after they had disappeared, and six of the eight bounced back strongly within a single season (Figure 6.8). Generally, the progeny of a single plant with enforced dormancy may be dispersed in time over years, decades or even centuries. Seeds of *Chenopodium album* collected from archeological excavations have been shown to be viable when 1700 years old (Ødum, 1965).

3 *Induced dormancy* is a state produced in a seed during a period of enforced dormancy in which it acquires some new requirement before it can germinate. The seeds of many agricultural and horticultural weeds will germinate without a light stimulus when they are released from the parent; but after a period of enforced dormancy they require exposure to light before they will germinate. For a long time it was a puzzle that soil samples taken from the field to the laboratory would quickly generate huge crops of seedlings, although these same seeds had failed to germinate in the field. It was a simple idea of genius that prompted Wesson and Wareing (1969) to collect soil samples from the field at night and bring them to the laboratory in darkness. They obtained large crops of seedlings from the soil only when the samples were exposed to light. This type of induced dormancy is responsible for the accumulation of large populations of seeds in the soil. In nature

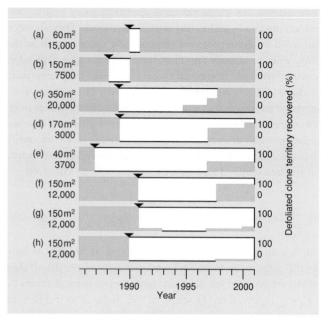

Figure 6.8 The histories of eight Missouri goldenrod (*Solidago missouriensis*) clones (rows a–h). Each clone's predefoliation area (m^2) and estimated number of ramets is given on the left. The panels show a 15-year record of the presence (shading) and absence of ramets in each clone's territory. The arrowheads show the beginning of dormancy, initiated by a *Trirhabda canadensis* eruption and defoliation. Reoccupation of entire or major segments of the original clone's territory by postdormancy ramets is expressed as the percentage of the original clone's territory. (After Morrow & Olfelt, 2003.)

they germinate only when they are brought to the soil surface by earthworms or other burrowing animals, or by the exposure of soil after a tree falls.

Seed dormancy may be induced by radiation that contains a relatively high ratio of far-red (730 nm) to near-red (approximately 660 nm) wavelengths, a spectral composition characteristic of light that has filtered through a leafy canopy. In nature, this must have the effect of holding sensitive seeds in the dormant state when they land on the ground under a canopy, whilst releasing them into germination only when the overtopping plants have died away.

Most of the species of plants with seeds that persist for long in the soil are annuals and biennials, and they are mainly weedy species – opportunists waiting (literally) for an opening. They largely lack features that will disperse them extensively in space. The seeds of trees, by contrast, usually have a very short expectation of life in the soil, and many are extremely difficult to store artificially for more than 1 year. The seeds of many tropical trees are particularly short lived: a matter of weeks or even days. Amongst trees, the most striking longevity is seen in those that retain the seeds in cones or pods on the tree until they are released after fire (many species of *Eucalyptus* and *Pinus*). This phenomenon of *serotiny* protects the seeds against risks on the ground until fire creates an environment suitable for their rapid establishment.

6.6   Dispersal and density

Density-dependent emigration was identified in Section 6.3.3 as a frequent response to overcrowding. We turn now to the more general issue of the density dependence of dispersal and also to the evolutionary forces that may have led to any density dependences that are apparent. In doing so, it is important to bear in mind the point made earlier (see Section 6.1.1): that 'effective' dispersal (from one place to another) requires emigration, transfer *and* immigration. The density dependences of the three need not be the same.

6.6.1   Inbreeding and outbreeding

Much of this chapter is devoted to the demographic or ecological consequences of dispersal, but there are also important genetic and evolutionary consequences. Any evolutionary 'consequence' is, of course, then a potentially important selective force favoring particular patterns of dispersal or indeed the tendency to disperse at all. In particular, when closely related individuals breed, their offspring are likely to suffer an 'inbreeding depression' in fitness (Charlesworth & Charlesworth, 1987), especially as a result of the expression in the phenotype of recessive deleterious alleles. With limited dispersal, inbreeding becomes more likely, and inbreeding avoidance is thus a force favoring dispersal. On the other hand, many species show local adaptation to their immediate environment (see Section 1.2). Longer distance dispersal may therefore bring together genotypes adapted to different local environments, which on mating give rise to low-fitness offspring adapted to neither habitat. This is called 'outbreeding depression', resulting from the break-up of coadapted combinations of genes – a force acting *against* dispersal. The situation is complicated by the fact that inbreeding depression is most likely amongst populations that normally outbreed, since inbreeding itself will purge populations of their deleterious recessives. None the less, natural selection can be expected to favor a pattern of dispersal that is in some sense intermediate – maximizing fitness by avoiding both inbreeding and outbreeding depression, though these will clearly be by no means the only selective forces acting on dispersal.

Certainly, there are several examples in plants of inbreeding and outbreeding depression when pollen is transferred from either close or distant donors, and in some cases both effects can be demonstrated in a single experiment. For example, when larkspur (*Delphinium nelsonii*) offspring were generated by hand pollinating with pollen brought from 1, 3, 10 and 30 m to the receptor flowers (Figure 6.9), both inbreeding and outbreeding depression in fitness were apparent.

6.6.2   Avoiding kin competition

In fact, inbreeding avoidance is not the only force likely to favor natal dispersal of offspring away from their close relatives. Such

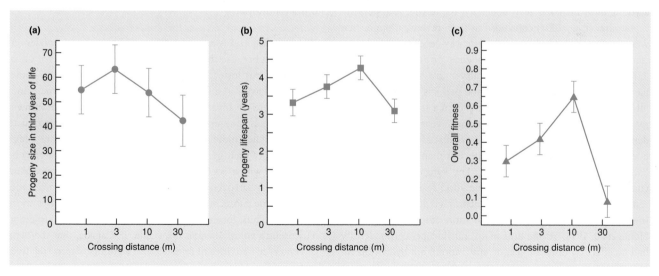

Figure 6.9 Inbreeding and outbreeding depression in *Delphinium nelsonii*: (a) progeny size in the third year of life, (b) progeny lifespan and (c) the overall fitness of progeny cohorts were all lower when progeny were the result of crosses with pollen taken close to (1 m) or far from (30 m) the receptor plant. Bars show standard errors. (After Waser & Price, 1994.)

dispersal will also be favored because it decreases the likelihood of competitive effects being directed at close kin. This was explained in a classic modeling paper by Hamilton and May (1977; see also Gandon & Michalakis, 2001), who demonstrated that even in very stable habitats, all organisms will be under selective pressure to disperse some of their progeny. Imagine a population in which the majority of organisms have a stay-at-home, nondispersive genotype O, but in which a rare mutant genotype, X, keeps some offspring at home but commits others to dispersal. The disperser X will suffer no competition in its own patch from O-type individuals but will compete against O-type individuals in their home patches. Disperser X will direct much of its competitive effects at non-kin (with genotype O), while O directs all of its competition at kin (also with genotype O). X will therefore increase in frequency in the population. On the other hand, if the majority of the population are type X, whilst O is the rare mutant, O will still do worse than X, since O can never displace any of the Xs from their patches but has itself to contend with several or many dispersers in its own patch. Dispersal is therefore said to be an evolutionarily stable strategy (ESS) (Maynard Smith, 1972; Parker, 1984). A population of nondispersers will evolve towards the ubiquitous possession of a dispersive tendency; but a population of dispersers will be under no selective pressure to lose that tendency. Hence, the avoidance of both inbreeding and kin competition seem likely to give rise to higher emigration rates at higher densities, when these forces are most intense.

There is indeed evidence for kin competition playing a role in driving offspring away from their natal habitat (Lambin *et al.*, 2001), but much of it is indirect. For example, in the California mouse, *Peromyscus californicus*, mean dispersal distance increased with increasing litter size in males and, in females, with increasing numbers of sisters in the litter (Ribble, 1992). The more kin a young individual was surrounded by, the further it dispersed.

Lambin *et al.* (2001) concluded in their review, though, that whereas there is plentiful evidence for density-dependent emigration (see Section 6.3.3), there is little evidence for density-dependent 'effective' dispersal (emigration, transfer *and* immigration), in part at least because immigration (and perhaps transfer) may be inhibited at high densities. For example, in a study of kangaroo rats, *Dipodomys spectabilis*, over several years during which density varied, dispersal was monitored first after juveniles had become independent of their parents, but then again after they had survived to breed themselves. The kangaroo rats occupy complex burrow systems containing food reserves, and these remain more or less constant in number: high densities therefore mean a saturated environment and more intense competition (Jones *et al.*, 1988). At the time of juvenile independence, density had no effect on dispersal (i.e. on emigration); but by first breeding, dispersal rates (i.e. *effective* dispersal rates) were *lower* at higher densities (inverse density dependence) (Figure 6.10). In males, this was mainly because they moved less between juvenile independence and breeding. In females, it occurred mainly because

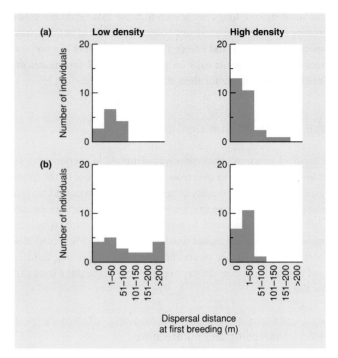

Figure 6.10 Inverse density-dependent effective dispersal in the kangaroo rat, *Dipodomys spectabilis*: (a) males, (b) females. Natal dispersal distances were greater at low than at high densities. (After Jones, 1988.)

their survival rate in new patches was lower at high densities (Jones, 1988).

6.6.3 Philopatry

Effective dispersal is not straightforwardly density dependent at least in part because there are also selective forces in favor of *not* dispersing, but instead showing so-called philopatry or 'home-loving' behavior (Lambin *et al.*, 2001). This can come about because there are advantages of inhabiting a familiar environment; or individuals may cooperate with (or at least be prepared to tolerate) related individuals in the natal habitat that share a high proportion of their genes; or individuals that *do* disperse may be confronted with a 'social fence' of aggression or intolerance from groups of *un*related individuals (Hestbeck, 1982). These forces, too, may become more intense as the environment becomes more saturated. Thus, for example, Lambin and Krebs (1993) found in Townsend's voles, *Microtus townsendii*, in Canada, that the nests or centers of activities of females that were first degree relatives (mother–daughters, littermate sisters) were closer than those that were second degree relatives (nonlittermate sisters, aunt–nieces), which were closer than those that were more distantly related, which in turn were closer than those not related at all.

And in a study of Belding's ground squirrels, *Spermophilus beldingi*, even when females dispersed, they tended to settle near their sisters (Nunes *et al.*, 1997). Moreover, there are examples of fitness being higher when close kin are nearby. For instance, Lambin and Yoccoz (1998) manipulated the relatedness of groups of breeding females of Townsend's vole, mimicking either a situation where the population had experienced philopatric recruitment followed by high survival ('high kinship'), or where the population had experienced either low philopatric recruitment or high mortality of recruits ('low kinship'). Survival of pups, especially early in their life, was significantly higher in the high kinship than in the low kinship treatment.

Overall, then, the relationship between dispersal and density will depend, just like all other adaptations, on evolved compromises to conflicting forces, and also on which aspect of dispersal (emigration, effective dispersal, etc.) is the focus of attention. It is no surprise either that, as we shall see below, the balance of advantage works out differently for different groups: males and females, old and young, and so on. Such variation also argues against broad generalizations suggesting that dispersal is 'typically' at presaturation densities (i.e. before resource limitation is intense) or for that matter at saturation densities (Lidicker, 1975).

6.7 Variation in dispersal within populations

6.7.1 Dispersal polymorphism

One source of variability in dispersal within populations is a somatic polymorphism amongst the progeny of a single parent. This is typically associated with habitats that are variable or unpredictable. A classic example is the desert annual plant *Gymnarrhena micrantha*. This bears very few (one to three) large seeds (achenes) in flowers that remain unopened below the soil surface, and these seeds germinate in the original site of the parent. The root system of the seedling may even grow down through the dead parent's root channel. But the same plants also produce above-ground, smaller seeds with a feathery pappus, and these are wind dispersed. In very dry years only the undispersed underground seeds are produced, but in wetter years the plants grow vigorously and produce a large number of seeds above ground, which are released to the hazards of dispersal (Koller & Roth, 1964).

There are very many examples of such seed dimorphism amongst the flowering plants. Both the dispersed and the 'stay at home' seeds will, in their turn, produce both dispersed and 'stay at home' progeny. Moreover, the 'stay at home' seed is often produced from self-pollinated flowers below ground or from unopened flowers, whereas the seeds that are dispersed are more often the product of cross-fertilization. Hence, the tendency to disperse is coupled with the possession of new, recombinant ('experimental') genotypes, whereas the 'stay at home' progeny are more likely to be the product of self-fertilization.

A dimorphism of dispersers and nondispersers is also a common phenomenon amongst aphids (winged and wingless progeny). As this differentiation occurs during the phase of population growth when reproduction is parthenogenetic, the winged and wingless forms are genetically identical. The winged morphs are clearly more capable of dispersing to new habitats, but they also often have longer development times, lower fecundity, shorter lifespans and hence a reduced intrinsic rate of increase (Dixon, 1998). It is perhaps no surprise, therefore, that aphids may modify the proportions of winged and wingless morphs in immediate response to the environments in which they find themselves. The pea aphid, *Acyrthosiphon pisum*, for example, produces more winged morphs in the presence of predators (Figure 6.11), presumably as an escape response from an adverse environment.

dispersal dimorphisms

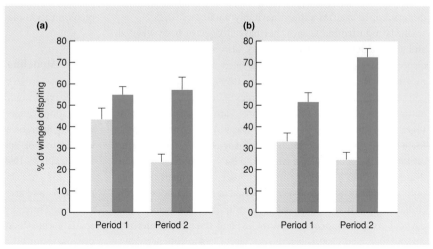

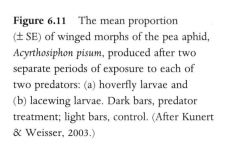

Figure 6.11 The mean proportion (± SE) of winged morphs of the pea aphid, *Acyrthosiphon pisum*, produced after two separate periods of exposure to each of two predators: (a) hoverfly larvae and (b) lacewing larvae. Dark bars, predator treatment; light bars, control. (After Kunert & Weisser, 2003.)

6.7.2 Sex-related differences

Males and females often differ in their liability to disperse. Differences are especially strong in some insects, where it is the male that is usually the more active disperser. For example, in the winter moth (*Operophtera brumata*), the female is wingless whilst the male is free-flying. In a seminal paper, Greenwood (1980) contrasted the sex-biased dispersal of birds and mammals. Amongst birds it is usually the females that are the main dispersers, but amongst mammals it is usually the males. Evolutionary explanations for a sex bias have emphasized on the one hand the advantages of a sex bias in its own right as a means of minimizing inbreeding, but also that details of the mating system may generate asymmetries in the costs and benefits of dispersal and philopatry in the two sexes (Lambin *et al.*, 2001). Thus, in birds, competition for territories is typically most intense amongst males. They, therefore, have most to gain from philopatry in terms of being familiar with their natal habitat, whereas the dispersing (and often monogamous) females may gain from exercising a choice of mate amongst the males. In mammals, the (often polygamous) males may compete more often for mates than for territories, and they therefore have most to gain by dispersing to areas with the largest number of defensible females.

6.7.3 Age-related differences

Much dispersal is natal dispersal, i.e. dispersal by juveniles before they reproduce for the first time. In many taxa this is constitutional: we have already noted that seed dispersal in plants is, by its nature, natal dispersal. Likewise, many marine invertebrates have a sessile adult (reproductive) stage and rely on their larvae (obviously pre-reproductive) for dispersal. On the other hand, most insects have a sessile larval stage and rely on the reproductive adults for dispersal. Here, for iteroparous species, dispersal is most often something that occurs throughout the adult life, before and after the first breeding episode; but for semelparous species, dispersal once again is almost inevitably natal.

Birds and mammals, once they have fledged or been weaned and are independent of their mothers, also have the potential to disperse throughout the rest of their lives. None the less, most dispersal here, too, is natal (Wolff, 1997). Indeed, age-biases and sex-biases in dispersal, and the forces of inbreeding-avoidance, competition-avoidance and philopatry, are all tied intimately together in the patterns of dispersal observed in mammals. Thus, for example, in an experiment with gray-tailed voles, *Microtus canicaudus*, 87% of juvenile males and 34% of juvenile females dispersed within 4 weeks of initial capture at low densities, but only 16% and 12%, respectively, dispersed at high densities (Wolff *et al.*, 1997). There was massive juvenile dispersal, which was particularly pronounced in the males; and the inverse density dependence, and especially the very high rates at low densities,

argue in favor of inbreeding-avoidance as a major force shaping the pattern.

6.8 The demographic significance of dispersal

The ecological fact of life identified in Section 4.1 emphasized that dispersal can have a potentially profound effect on the dynamics of populations. In practice, however, many studies have paid little attention to dispersal. The reason often given is that emigration and immigration are approximately equal, and they therefore cancel one another out. One suspects, though, that the real reason is that dispersal is usually extremely difficult to quantify.

The nature of the role of dispersal in population dynamics depends on how we think of populations. The simplest view sees a population as a collection of individuals distributed more or less continuously over a stretch of more or less suitable habitat, such that the population is a single, undivided entity. Dispersal is then a process contributing to either the increase (immigration) or decrease (emigration) in the population. Many populations, however, are in fact *meta*populations; that is, collections of *sub*populations.

metapopulations and subpopulations

We noted in Section 6.3.1 the ubiquity of patchiness in ecology and the importance of dispersal in linking patches to one another. A subpopulation, then, occupies a habitable patch in the landscape, and it corresponds, in isolation, to the simple view of a population described above. But the dynamics of the metapopulation as a whole is determined in large part by the rate of extinction of individual subpopulations, and the rate of colonization – by dispersal – of habitable but uninhabited patches. Note, however, that just because a species occupies more than one habitable site, each of which supports a population, this does not mean that those populations comprise a metapopulation. As we shall discuss more fully below, 'classic' metapopulation status is conferred only when extinction and recolonization play a major role in the overall dynamics.

6.8.1 Modeling dispersal: the distribution of patches

The ways in which dispersal intervenes in the dynamics of populations can be envisaged, or indeed modeled mathematically, in three different ways (see Kareiva, 1990; Keeling, 1999). The first is an 'island' or 'spatially implicit' approach (Hanski & Simberloff, 1997; Hanski, 1999). Here, the key feature is that a proportion of the individuals leave their home patches and enter a pool of dispersers and are then redistributed amongst patches, usually at random. Thus, these models do not place patches at any specific spatial location. All patches may lose or gain individuals through dispersal, but all are, in a sense, equally distant from all other patches. Many metapopulation models, including the earliest (Levins' model, see below), come into this category, and despite their simplicity (real patches *do* have a location in space) they have provided important insights, in part because their simplicity makes them easier to analyze.

In contrast, spatially *explicit* models acknowledge that the distances between patches vary, as do therefore the chances of them exchanging individuals through dispersal. The earliest such models, developed in population genetics, were linear 'stepping stones', where dispersal occurred only between adjacent patches in the line (Kimura & Weiss, 1964). More recently, spatially explicit approaches have often involved 'lattice' models in which patches are arranged on a (usually) square grid, and patches exchange dispersing individuals with 'neighboring' patches – perhaps the four with which they share a side, or the eight with which they make any contact at all, including the diagonals (Keeling, 1999). Of course, despite being spatially explicit, such models are still caricatures of patch arrangements in the real world. They are none the less useful in highlighting new dynamic patterns that appear as soon as space is incorporated explicitly: not only spatial patterns (see, for example, Section 10.5.6), but also altered temporal dynamics, including, for example, the increased probability of extinction of whole spatially explicit metapopulations as habitat is destroyed (Figure 6.12). Further spatially explicit models are also spatially 'realistic' (see Hanski, 1999) in that they include information about the actual geometry of fragmented landscapes. One of these, the 'incidence function model' (Hanski, 1994b), is utilized below (Section 6.9.4).

Finally, the third approach treats space not as patchy at all but as continuous and homogeneous, and usually models dispersal as part of a reaction–diffusion system, where the dynamics at any given location in space are captured by the 'reaction', and dispersal is added as separate 'diffusion' terms. The approach has been more useful in other areas of biology (e.g. developmental biology) than it has in ecology. None the less, the mathematical understanding of such systems is strong, and they are particularly good at demonstrating how spatial variation (i.e. patchiness) can be generated, internally, within an intrinsically homogeneous system (Kareiva, 1990; Keeling, 1999).

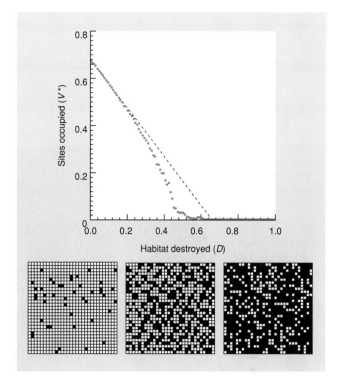

Figure 6.12 In a series of models, as an increasing fraction of habitat is destroyed (left to right on the *x*-axis), the fraction of available sites occupied (*y*-axis) declines until the whole population is effectively extinct (no sites occupied). The diagonal dotted line shows the relationship for a spatially implicit model in which all sites are equally connected. The dots show output from a spatially explicit lattice model: values are the means of five replicates (the model is probabilistic: each run is slightly different). Three examples of the lattice are shown below, with 0.05, 0.40 and 0.70 of the patches destroyed (black). With little habitat destruction (towards the left), an explicit spatial structure makes negligible difference as the remaining patches are well connected to other patches. But with more habitat destruction, patches in the lattice become increasingly isolated and unlikely to be recolonized, and many more of them remain unoccupied than in the spatially implicit model. (After Bascompte & Sole, 1996.)

6.8.2 Dispersal and the demography of single populations

The studies that *have* looked carefully at dispersal have tended to bear out its importance. In a long-term and intensive investigation of a population of great tits, *Parus major*, near Oxford, UK, it was observed that 57% of breeding birds were immigrants rather than born in the population (Greenwood *et al.*, 1978). In a population of the Colorado potato beetle, *Leptinotarsa decemlineata*, in Canada, the average emigration rate of newly emerged adults was 97% (Harcourt, 1971). This makes the rapid spread of the beetle in Europe in the middle of the last century easy to understand (Figure 6.13).

A profound effect of dispersal on the dynamics of a population was seen in a study of *Cakile edentula*, a summer annual plant growing on the sand dunes of Martinique Bay, Nova Scotia. The population was concentrated in the middle of the dunes, and declined towards both the sea and the land. Only in the area towards the sea, however, was seed production high enough and mortality sufficiently low for the population to maintain itself year after year. At the middle and landward sites, mortality exceeded seed production. Hence, one might have expected the population

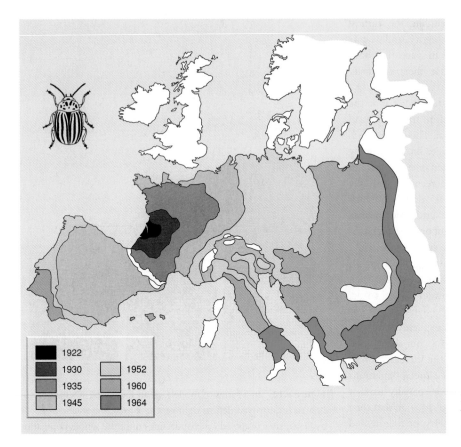

Figure 6.13 Spread of the Colorado beetle (*Leptinotarsa decemlineata*) in Europe. (After Johnson, 1967.)

Legend:
- 1922
- 1930
- 1935
- 1945
- 1952
- 1960
- 1964

to become extinct (Figure 6.14). But the distribution of *Cakile* did not change over time. Instead, large numbers of seeds from the seaward zone dispersed to the middle and landward zones. Indeed, more seeds were dispersed into and germinated in these two zones than were produced by the residents. The distribution and abundance of *Cakile* were directly due to the dispersal of seeds in the wind and the waves.

Probably the most fundamental consequence of dispersal for the dynamics of single populations, though, is the regulatory effect of density-dependent emigration (see Section 6.3.3). Locally, all that was said in Chapter 5 regarding density-dependent mortality applies equally to density-dependent emigration. Globally, of course, the consequences of the two may be quite different. Those that die are lost forever and from everywhere. With emigration, one population's loss may be another's gain.

6.8.3   Invasion dynamics

the importance of eccentric dispersers

In almost every aspect of life, there is a danger in imagining that what is usual and 'normal' is in fact universal, and that what is unusual or eccentric can safely be dismissed or ignored. Every statistical distribution has a tail, however, and those that occupy the tail are as real as the conformists that outnumber them. So it is with dispersal. For many purposes, it is reasonable to characterize dispersal rates and distances in terms of what is typical. But especially when the focus is on the spread of a species into a habitat that it has previously not been occupied, those propagules dispersing furthest may be of the greatest importance. Neubert and Caswell (2000), for example, analyzed the rate of spread of two species of plants, *Calathea ovandensis* and *Dipsacus sylvestris*. In both cases they found that the rate of spread was strongly dependent on the maximum dispersal distance, whereas variations in the pattern of dispersal at lesser distances had little effect.

This dependence of invasion on rare long-distance dispersers means, in turn, that the probability of a species invading a new habitat may have far more to do with the proximity of a source population (and hence the opportunity to invade) than it does on the performance of the species once an initial bridgehead has been established. For instance, the invasion of 116 patches of lowland heath vegetation in southern England by scrub and tree species was studied for the period from 1978 to 1987 (Figure 6.15) and also from 1987 to 1996 (Nolan *et al.*, 1998; Bullock *et al.*, 2002). There were four types of heath – dry, humid, wet and mire – and with

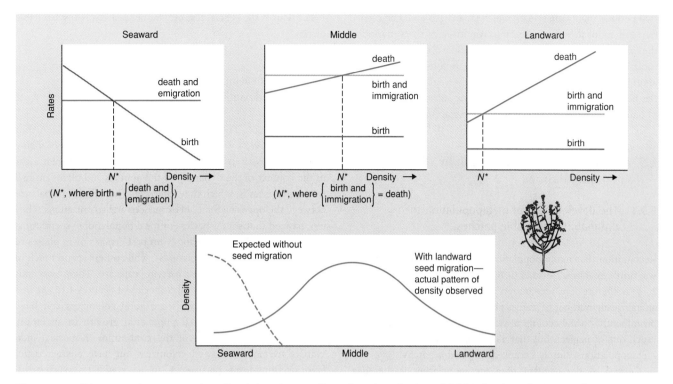

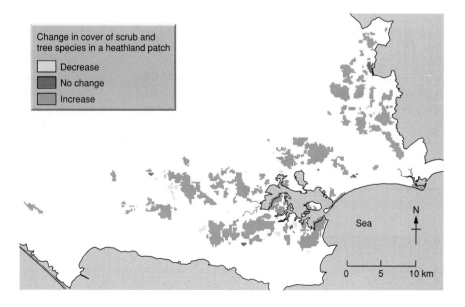

Figure 6.14 Diagrammatic representation of variations in mortality and seed production of *Cakile edentula* in three areas along an environmental gradient from open sand beach (seaward) to densely vegetated dunes (landward). In contrast to other areas, seed production was prolific at the seaward site. Births, however, declined with plant density, and where births and deaths were equal, an equilibrium population density can be envisaged, N^*. In the middle and landward sites, deaths always exceeded births resulting from local seeds, but populations persisted there because of the landward drift of the majority of seed produced by plants on the beach (seaward site). Thus, the sum of local births plus immigrating seeds can balance mortality in the middle and landward sites, resulting in equilibria at appropriate densities. (After Keddy, 1982; Watkinson, 1984.)

Figure 6.15 The invasion (i.e. increase in abundance) of most of the 116 patches of lowland heath in Dorset, UK, by scrub and tree species between 1978 and 1987. Coastland is to the south and the county boundary to the east. (After Bullock *et al.*, 2002.)

two periods, eight data sets on which an analysis could be carried out. For six of these, a significant proportion of the variation in the loss of heath to invading species could be accounted for. The most important explanatory variables were those describing the abundance of scrub and tree species in the vegetation bordering the heath patches. Invasions, and thus the subsequent dynamics of patches, were being driven by initiating acts of dispersal.

6.9   Dispersal and the demography of metapopulations

6.9.1   The development of metapopulation theory: uninhabited habitable patches

Recognition that many populations are in fact metapopulations was firmly established around 1970, but there was a delay of around 20 years before that recognition was translated into action and an increasing number of studies placed metapopulation dynamics prominently on the ecological stage. Now, the danger is not so much one of neglect, but that all populations are thought of as metapopulations, simply because the world is patchy.

Central to the concept of a metapopulation is the idea, emphasized by Andrewartha and Birch back in 1954, that habitable patches might be uninhabited simply because individuals have failed to disperse into them. To establish that this is so, we need to be able to identify habitable sites that are not inhabited. Only very rarely has this been attempted. One method involves identifying characteristics of habitat patches to which a species is restricted and then determining the distribution and abundance of similar patches in which the species might be expected to occur. The water vole (*Arvicola terrestris*) lives in river banks, and in a survey of 39 sections of river bank in North Yorkshire, UK, 10 contained breeding colonies of voles (core sites), 15 were visited by voles but they did not breed there (peripheral sites) and 14 were apparently never used or visited. A 'principle component' analysis was used to characterize the core sites, and on the basis of these characteristics a further 12 unoccupied or peripheral sites were identified that should have been suitable for breeding voles (i.e. habitable sites). Apparently, about 30% of habitable sites were uninhabited by voles because they were too isolated to be colonized or in some cases suffered high levels of predation by mink (Lawton & Woodroffe, 1991).

Habitable patches can also be identified for a number of rare butterfly species because the larvae feed on only one or a few patchily distributed plant species. Thomas *et al.* (1992) found that the patches that remained uninhabited were small and isolated from the sources of dispersal: the butterfly *Phlebejus argus* was able to colonize virtually all habitable sites less than 1 km from existing populations. Indeed, the habitability of some of the isolated (previously uninhabited) sites was established when the butterfly was successfully introduced (Thomas & Harrison,

1992). This is the crucial test of whether a site is really habitable or not.

6.9.2   The development of metapopulation theory: islands and metapopulations

The classic book, *The Theory of Island Biogeography* by MacArthur and Wilson (1967), was an important catalyst in radically changing ecological theory in general. The authors developed their ideas in the context of the dynamics of the animals and plants on real (maritime) islands, which they interpreted as reflecting a balance between the opposing forces of extinctions and colonizations. They emphasized that some species (or local populations) spend most of their time either recovering from past crashes or in phases of invasion of new territories (islands), while others spend much of their time at or around their carrying capacity. These two ends of a continuum are the *r*- and *K*-species of Section 4.12. At one extreme (*r*-species), individuals are good colonizers and have characteristics favoring rapid population growth in an empty habitat. At the other end of the continuum (*K*-species) individuals are not such good colonizers but have characteristics favoring long-term persistence in a crowded environment. *K*-species therefore have relatively low rates of both colonization and extinction, whereas *r*-species have relatively high rates. These ideas are developed further in the discussion of island biogeography in Chapter 21.

At about the same time as MacArthur and Wilson's book was published, a simple model of 'metapopulation' dynamics was proposed by Levins (1969, 1970). Like MacArthur and Wilson, he sought to incorporate into ecological thinking the essential patchiness of the world around us. MacArthur and Wilson were more concerned with whole communities of species, and envisaged a 'mainland' that could provide a regular source of colonists for the islands. Levins focused on populations of a single species and awarded none of his patches special mainland status. Levins introduced the variable $p(t)$, the fraction of habitat patches occupied at time t, reflecting an acceptance that not all habitable patches are always inhabited.

The rate of change in the fraction of occupied habitat (patches, p) is given in Levins' model as:

Levins' model

$$dp/dt = mp(1 - p) - \mu p, \qquad (6.1)$$

in which μ is the rate of local extinction of patches and m is the rate of recolonization of empty patches. That is, the rate of recolonizations increases both with the fraction of empty patches prone to recolonization $(1 - p)$ and with the fraction of occupied patches able to provide colonizers, p, whereas the rate of extinctions increases simply with the fraction of patches prone to extinction, p. Rewriting this equation, Hanski (1994a) showed that it is structurally identical to the logistic equation (see Section 5.9):

$$\mathrm{d}p/\mathrm{d}t = (m - \mu)\, p \,\{1 - p/[1 - (m/\mu)]\}. \qquad (6.2)$$

Hence, as long as the intrinsic rate of recolonization exceeds the intrinsic rate of extinction $((m - \mu) > 0)$, the total metapopulation will reach a stable equilibrium, with a fraction, $1 - (\mu/m)$, of the patches occupied.

extinctions and colonizations in subpopulations: a stable metapopulation

The most fundamental message from taking a metapopulation perspective, then, which emerges from even the simplest models, is that a metapopulation can persist, stably, as a result of the balance between random extinctions and recolonizations even though none of the local populations are stable in their own right. An example of this is shown in Figure 6.16, where within a persistent, highly fragmented metapopulation of the Glanville fritillary butterfly (*Melitaea cinxia*) in Finland, even the largest local populations had a high probability of declining to extinction within 2 years. To re-state the message another way: if we wish to understand the long-term persistence of a population, or indeed that population's dynamics, then we may need to look beyond the local rates of birth and death (and what determines them), or even the local rates of immigration and emigration. If the population as a whole functions as a metapopulation, then the rates of subpopulation extinction and colonization may be of at least comparable importance.

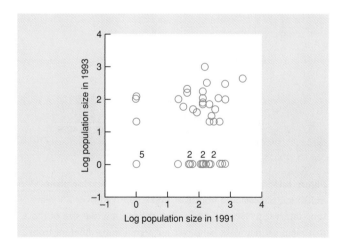

Figure 6.16 Comparison of the local population sizes in June 1991 (adults) and August 1993 (larvae) of the Glanville fritillary butterfly (*Melitaea cinxia*) on Åland island in Finland. Multiple data points are indicated by numbers. Many 1991 populations, including many of the largest, had become extinct by 1993. (After Hanski *et al.*, 1995.)

6.9.3 When is a population a metapopulation?

Two necessary features of a metapopulation have already been established here: that individual subpopulations have a realistic chance of experiencing both extinction *and* recolonization. To this we can add a third, which has been implicit in the discussion thus far. The dynamics of the various subpopulations should be largely independent, i.e. not synchronous. There would, after all, be little hope of stability if when one subpopulation went extinct they all did. Rather, asynchrony guarantees that as one goes extinct (or even declines), there are likely to be others that are thriving and generating dispersers, promoting the 'rescue effect' (Brown & Kodric-Brown, 1977) of the former by the latter.

Some metapopulations may conform to the 'classic' concept, in which all the subpopulations have a realistic (and roughly equal) chance of extinction, but in other cases there may be significant variation in either the size or quality of individual patches. Thus, patches may be divided into 'sources' (donor patches) and 'sinks' (receiver patches) (Pulliam, 1988). In source patches at equilibrium, the number of births exceeds the number of deaths, whereas in sink patches the reverse is true. Hence, source populations support one or more sink populations within a metapopulation. The persistence of the metapopulation depends not only on the overall balance between extinction and recolonization, as in the simple model, but also on the balance between sources and sinks.

sources and sinks

In practice, of course, there is likely to be a continuum of types of metapopulation: from collections of nearly identical local populations, all equally prone to extinction, to metapopulations in which there is great inequality between local populations, some of which are effectively stable in their own right. This contrast is illustrated in Figure 6.17 for the silver-studded blue butterfly (*Plejebus argus*) in North Wales.

Just because a population is patchily distributed, however, this does not necessarily make it a metapopulation (Harrison & Taylor, 1997; Bullock *et al.*, 2002). First, a population may be patchily distributed, but dispersal between the patches may be so great that the dynamics of the individual patches are no longer independent: a single population, albeit occupying a heterogeneous habitat. Alternatively, patches may be so isolated from one another that dispersal between them is negligible: a series of effectively separate populations.

Finally, and perhaps most commonly, all patches may simply have a negligible chance of extinction, at least on observable timescales. This means that their dynamics may be influenced by birth, death, immigration and emigration – but not to any significant degree by extinction or recolonization. This last category comes closest to being a true metapopulation, and there can be little doubt that the title has been given to many patchy populations fitting this description. Of course, there can be a danger in being overprotective of the purity of definitions.

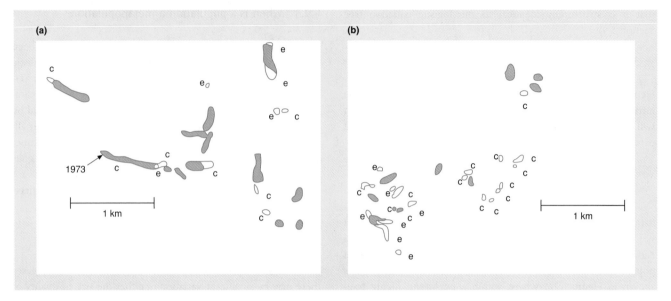

Figure 6.17   Two metapopulations of the silver-studded blue butterfly (*Plejebus argus*) in North Wales: (a) in a limestone habitat in the Dulas Valley, where there was a large number of persistent (often larger) local populations amongst smaller, much more ephemeral local populations; (b) in a heathland habitat at South Stack Cliffs, where the proportion of smaller and ephemeral populations was much greater. Filled outlines, present in both 1983 and 1990; open outlines, not present at both times; e, present only in 1983 (presumed extinction); c, present only in 1990 (presumed colonization). (After Thomas & Harrison, 1992.)

What harm can there be if, as interest in the metapopulation concept grows, the term itself is extended to a wider variety of ecological scenarios? Perhaps none – and the spread of the term's usage to populations originally beyond its reach may, in any case, be unstoppable. But a word, like any other signal, is only effective if the receiver understands what the sender intends. At the very least, care should be taken by users of the term to confirm whether the extinction and recolonization of patches has been established.

metapopulations of plants? remember the seed bank

The problem of identifying metapopulations is especially apparent for plants (Husband & Barrett, 1996; Bullock *et al.*, 2002). There is no doubt that many plants inhabit patchy environments, and apparent extinctions of local populations may be common. This is shown in Figure 6.18 for the annual aquatic plant *Eichhornia paniculata*, living in temporary ponds and ditches in arid regions in northeast Brazil. However, the applicability of the idea of recolonization following a genuine extinction is questionable in any plant species that has a buried seed bank. In *E. paniculata*, for instance, the heavy seeds almost always drop in the immediate vicinity of the parent rather than being dispersed to other patches. 'Extinctions' are typically the result of the catastrophic loss of habitat (note in Figure 6.18 that the chance of extinction has effectively nothing to do with the previous population size) and 'recolonizations' are almost always simply the result of the germination of seeds following habitat restoration. Recolon-

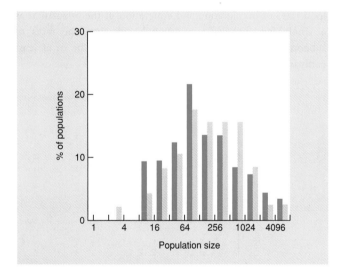

Figure 6.18   Of 123 populations of the annual aquatic plant *Eichhornia paniculata* in northeast Brazil observed over a 1-year time interval, 39% went extinct, but the mean initial size of those that went extinct (dark bars) was not significantly different from those that did not (pale bars). (Mann-Whitney U = 1925, P > 0.3.) (After Husband & Barrett, 1996.)

ization by dispersal, a prerequisite for a true metapopulation, is extremely rare.

Moreover, as Bullock *et al.* (2002) point out, of the plant studies that have documented patch extinctions and colonizations, the vast majority have been in recently emerged patches (the early stages of succession, see Chapter 16). Extinctions mostly occur when the vegetation in a patch develops to a state where it is no longer suitable for the plant species in question, and that patch is therefore also not suitable for recolonization by the same species. This is 'habitat tracking' (Harrison & Taylor, 1997) rather than the repeated extinction and recolonization of the same habitat that is central to the concept of a metapopulation.

6.9.4 Metapopulation dynamics

Levins' simple model does not take into account the variation in size of patches, their spatial locations, nor the dynamics of populations within individual patches. Not surprisingly, models that do take all these highly relevant variables into account become mathematically complex (Hanski, 1999). Nevertheless, the nature and consequences of some of these modifications can be understood without going into the details of the mathematics.

For example, imagine that the habitat patches occupied by a metapopulation vary in size and that large patches support larger local populations. This allows persistence of the metapopulation, with lower rates of colonization than would otherwise be the case, as a result of the lowered rates of extinction on the larger patches (Hanski & Gyllenberg, 1993). Indeed, the greater the variation in patch size, the more likely it is that the metapopulation will persist, other things being equal. Variations in the size of local populations may, alternatively, be the result of variations in patch quality rather than patch size: the consequences would be broadly the same.

The probability of extinction of local populations typically declines as local population size increases (Hanski, 1991). Moreover, as the fraction of patches occupied by the metapopulation, p, increases, there should on average be more migrants, more immigration into patches, and hence larger local populations (confirmed, for example, for the Glanville fritillary – Hanski *et al.*, 1995). Thus, the extinction rate, μ, should arguably not be constant as it is in the simple model, but should decline as p increases. Models incorporating this effect (Hanski, 1991; Hanski & Gyllenberg, 1993) often give rise to an intermediate unstable threshold value of p. Above the threshold, the sizes of local populations are sufficiently large, and their rate of extinction sufficiently low, for the metapopulation to persist at a relatively high fraction of patches, as in the simple model. Below the threshold, however, the average size of local populations is too low and their rate of extinction hence too high. The metapopulation declines either to an alternative stable equilibrium at $p = 0$ (extinction of the whole metapopulation) or to one in which p is low, where essentially only the most favorable patches are occupied.

alternative stable equilibria?

Different metapopulations of the same species might therefore be

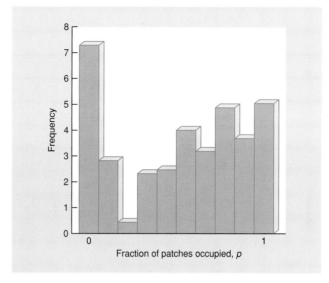

Figure 6.19 The bimodal frequency distribution of patch occupancy (proportion of habitable patches occupied, p) amongst different metapopulations of the Glanville fritillary (*Melitaea cinxia*) on Åland island in Finland. (After Hanski *et al.*, 1995.)

expected to occupy either a high or a low fraction of their habitable patches (the alternative stable equilibria) but not an intermediate fraction (close to the threshold). Such a bimodal distribution is indeed apparent for the Glanville fritillary in Finland (Figure 6.19). In addition, these alternative equilibria have potentially profound implications for conservation (see Chapter 15), especially when the lower equilibrium occurs at $p = 0$, suggesting that the threat of extinction for any metapopulation may increase or decline quite suddenly as the fraction of habitable patches occupied moves below or above some threshold value.

One study drawing many of the preceding threads together examined the dynamics of a presumed metapopulation of a small mammal, the American pika *Ochotona princeps*, in California (Moilanen *et al.*, 1998). (The qualifier 'presumed' is necessary because dispersal between habitat patches was itself presumed rather than actually observed (see Clinchy *et al.*, 2002).) The overall metapopulation could itself be divided into northern, middle and southern networks, and the patch occupancy in each was determined on four occasions between 1972 and 1991 (Figure 6.20a). These purely spatial data were used alongside more general information on pika biology, to provide parameter values for Hanski's (1994b) incidence function model (see Section 6.8.1). This was then used to simulate the overall dynamics of each of the networks, with a realistic degree of stochastic variation incorporated, starting from the observed situation in 1972 and either treating the entire metapopulation as a single entity (Figure 6.20b) or simulating each of the networks in isolation (Figure 6.20c).

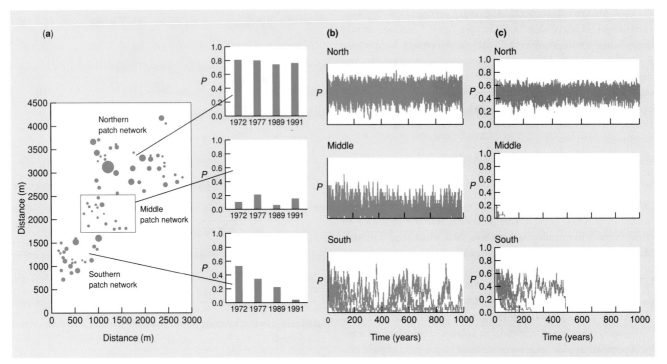

Figure 6.20   The metapopulation dynamics of the American pika, *Ochotona princeps*, in Bodie, California. (a) The relative positions and approximate sizes of the habitable patches, and the occupancies in the northern, middle and southern networks of patches in 1972, 1977, 1989 and 1991. (b) The temporal dynamics of the three networks, with the entire metapopulation treated as a single entity, using Hanski's (1994b) incidence function model. Ten replicate simulations are shown, each starting with the actual data in 1972. (c) Equivalent simulations to (b) but with each of the networks simulated in isolation. (After Moilanen *et al.*, 1998.)

The data themselves (Figure 6.20a) show that the northern network maintained a high occupancy throughout the study period, the middle network maintained a more variable and much lower occupancy, while the southern network suffered a steady and substantial decline. The output from the incidence function model (Figure 6.20b) was very encouraging in mirroring accurately these patterns in temporal dynamics despite being based only on spatial data. In particular, the southern network was predicted to collapse periodically to overall extinction but to be rescued by the middle network acting, despite its low occupancy, as a stepping stone from the much more buoyant northern network. This interpretation is supported by the results when the three networks are simulated in isolation (Figure 6.20c). The northern network remains at a stable high occupancy; but the middle network, starved of migrants from the north, rapidly and predictably crashes; and the southern network, while not so unstable, eventually suffers the same fate. On this view, then, within the metapopulation as a whole, the northern network is a source and the middle and southern networks are sinks. Thus, there is no need to invoke any environmental change to explain the decline in the southern network; such declines are predicted even in an unchanging environment.

Even more fundamentally, these results illustrate how whole metapopulations can be stable when their individual subpopulations are not. Moreover, the comparison of the northern and middle networks, both stable but at very different occupancies, shows how occupancy may depend on the size of the pool of dispersers, which itself may depend on the size and number of the subpopulations.

Finally, these simulations direct us to a theme that recurs throughout this book. Simple models (and one's own first thoughts) often focus on equilibria attained in the long term. But in practice such equilibria may rarely be reached. In the present case, stable equilibria can readily be generated in simple metapopulation models, but the observable dynamics of a species may often have more to do with the 'transient' behavior of its metapopulations, far from equilibrium. To take another example, the silver-spotted skipper butterfly (*Hesperia comma*) declined steadily in Great Britain from a widespread distribution over most calcareous hills in 1900, to 46 or fewer refuge localities (local populations) in 10 regions by the early 1960s (Thomas & Jones, 1993). The probable reasons were changes in land use – increased ploughing of unimproved grasslands and reduced stocking with

equilibria may rarely be reached

domestic grazing animals – and the virtual elimination of rabbits by myxomatosis with its consequent profound vegetational changes. Throughout this nonequilibrium period, rates of local extinction generally exceeded those of recolonization. In the 1970s and 1980s, however, the reintroduction of livestock and the recovery of the rabbits led to increased grazing and the number of suitable habitats increased again. Recolonization exceeded local extinction – but the spread of the skipper remained slow, especially into localities isolated from the 1960s refugia. Even in southeast England, where the density of refugia was greatest, it is predicted that the abundance of the butterfly will increase only slowly – and remain far from equilibrium – for at least 100 years.

Summary

We distinguish between dispersal and migration, and within dispersal between emigration, transfer and immigration.

Various categories of active and passive dispersal are described, including especially passive dispersal in the seed rain and the guerrilla and phalanx strategies of clonal dispersers.

Random, regular and aggregated distributions are explained, and the importance of scale and patchiness in the perception of such distributions is emphasized, especially in the context of environmental 'grain'. Forces favoring and diluting aggregations are elaborated, including the theory of the selfish herd and density-dependent dispersal.

We describe some of the main patterns of migration at a range of scales – tidal, diurnal, seasonal and intercontinental – including those that recur repeatedly and those that occur just once.

We examine dormancy as migration in time in both animals (especially diapause) and plants. The importance of photoperiod in the timing of dormancy is emphasized.

The relationship between dispersal and density is examined in detail. The roles of in- and outbreeding in driving density dependences are explained, including especially the importance of avoiding kin competition on the one hand and the attractions of philopatry on the other.

We describe a variety of types of variation in dispersal within populations: polymorphisms and sex- and age-related differences.

We turn to the demographic significance of dispersal and introduce the concept of the metapopulation composed of a number of subpopulations. Dispersal can be incorporated into the dynamics of populations, and modeled, in three different ways: (i) an 'island' or 'spatially implicit' approach; (ii) a spatially explicit approach that acknowledges that the distances between patches vary; and (iii) an approach treating space as continuous and homogeneous.

Probably the most fundamental consequence of dispersal for the dynamics of single populations is the regulatory effect of density-dependent emigration. It is important also, though, to recognize the importance of rare long-distance dispersers in invasion dynamics.

Metapopulation theory developed from the earlier concept of the uninhabited habitable patch. Its origin as a concept in its own right was the Levins' model, which established the most fundamental message: that a metapopulation can persist, stably, as a result of the balance between random extinctions and recolonizations, even though no subpopulations are stable in their own right.

Not all patchily distributed populations are metapopulations, so we address the question 'When is a population a metapopulation?', which may be particularly problematic with plant populations.

Finally, we explore the dynamics of metapopulations, emphasizing especially the likely importance of alternative stable equilibria.

Chapter 7

Ecological Applications at the Level of Organisms and Single-Species Populations: Restoration, Biosecurity and Conservation

7.1 Introduction

environmental problems resulting from human population growth . . .

The expanding human population (Figure 7.1) has created a wide variety of environmental problems. Our species is not unique in depleting and contaminating the environment but we are certainly unique in using fire, fossil fuels and nuclear fission to provide the energy to do work. This power generation has had far-reaching consequences for the state of the land, aquatic ecosystems and the atmosphere, with dramatic repercussions for global climate (see Chapter 2). Moreover, the energy generated has provided people with the power to transform landscapes (and waterscapes) through urbanization, industrial agriculture, forestry, fishing and mining. We have polluted land and water, destroyed large areas of almost all kinds of natural habitat, overexploited living resources, transported organisms around the world with negative consequences for native ecosystems, and driven a multitude of species close to extinction.

An understanding of the scope of the problems facing us, and the means to counter and solve these problems, depends absolutely on a proper grasp of ecological fundamentals. In the first section of this book we have dealt with the ecology of individual organisms, and of populations of organisms of single species (population interactions will be the subject of the second section). Here we switch attention to how this knowledge can be turned to advantage by resource managers. At the end of the second and third sections of the book we will address, in a similar manner, the application of ecological knowledge at the level of population interactions (Chapter 15) and then of communities and ecosystems (Chapter 22).

. . . require the application of ecological knowledge, . . .

Individual organisms have a physiology that fits them to tolerate particular ranges of physicochemical conditions and dictates their need for specific resources (see Chapters 2 and 3). The occurrence and distribution of species therefore depends fundamentally on their physiological ecology and, for animals, their behavioral repertoire too. These facts of ecological life are encapsulated in the concept of the niche (see Chapter 2). We have observed that species do not occur everywhere that conditions and resources are

. . . niche theory, . . .

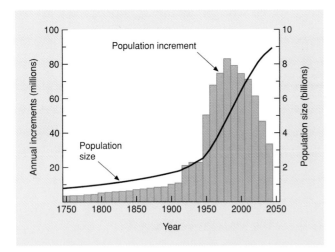

Figure 7.1 Growth in size of the world's human population since 1750 and predicted growth until 2050 (solid line). The histograms represent decadal population increments. (After United Nations, 1999.)

appropriate for them. However, management strategies often rely on an ability to predict where species might do well, whether we wish to restore degraded habitats, predict the future distribution of invasive species (and through biosecurity measures prevent their arrival), or conserve endangered species in new reserves. Niche theory therefore provides a vital foundation for many management actions. We deal with this in Section 7.2.

. . . life history theory . . .   The life history of a species (see Chapter 4) is another basic feature that can guide management. For example, whether organisms are annuals or perennials, with or without dormant stages, large or small, or generalists or specialists may influence their likelihood of being a successful part of a habitat restoration project, a problematic invader or a candidate for extinction and therefore worthy of conservation priority. We turn to these ideas in Section 7.3.

A particularly influential feature of the behavior of organisms, whether animals or plants, is their pattern of movement and dispersion (see Chapter 6). Knowledge of animal migratory behavior can be especially important in attempts to restore damaged habitats, predict and prioritize invaders, and design conservation reserves. This is covered in Section 7.4.

. . . and the dynamics of small populations   Conservation of endangered species requires a thorough understanding of the dynamics of small populations. In Section 7.5 we deal with an approach called population viability analysis (PVA), an assessment of extinction probabilities that depends on knowledge of life tables (see Chapter 4, in particular Section 4.6), population rates of increase (see Section 4.7), intraspecific competition (see Chapter 5), density dependence (see Section 5.2), carrying capacities (see Section 5.3) and, in some cases, metapopulation structure (if the endangered species occurs in a series of linked subpopulations – see Section 6.9). As we shall see in Part 2 of this book (and particularly in the synthesis provided in Chapter 14), the determination of abundance, and thus the likelihood of extinction of a population, depends not only on intrinsic properties of individual species (birth and death rates, etc.) but also on interactions with other species in their community (competitors, predators, parasites, mutualists, etc.). However, PVA usually takes a more simplistic approach and does not deal explicitly with these complications. For this reason, the topic is dealt with in the present chapter.

the challenge of global climate change   One of the biggest future challenges to organisms, ecologists and resource managers is global climate change (see Section 2.9). Attempts to mitigate predicted changes to climate have an ecological dimension (e.g. plant more trees to soak up some of the extra carbon dioxide produced by the burning of fossil fuels), although mitigation must also focus on the economic and sociopolitical dimensions of the problem. This is discussed in Chapter 22, because the relevant issues relate to ecosystem functioning. However, in the current

chapter we deal with the way we can use knowledge about the ecology of individual organisms to predict and manage the consequences of global climate change such as the spread of disease and weeds (see Section 7.6.1) and the positioning of conservation reserves (see Section 7.6.2).

Given the pressing environmental problems we face, it is not surprising that a large number of ecologists now perform research that is applied (i.e. aimed directly at such problems) and then publish it in specialist scientific journals. But to what extent is this work assimilated and used by resource managers? Questionnaire assessments by two applied journals, *Conservation Biology* (Flashpohler *et al.*, 2000) and the *Journal of Applied Ecology* (Ormerod, 2003), revealed that 82 and 99% of responding authors, respectively, made management recommendations in their papers. Of these, it is heartening to note that more than 50% of respondents reported that their work had been taken up by managers. For papers published between 1999 and 2001 in the *Journal of Applied Ecology*, for example, the use of findings by managers most commonly involved planning aimed at species and habitats of conservation importance, pest species, agroecosystems, river regulation and reserve design (Ormerod, 2003).

7.2 Niche theory and management

7.2.1 Restoration of habitats impacted by human activities

using knowledge of species niches . . .   The term 'restoration ecology' can be used, rather unhelpfully, to encompass almost every aspect of applied ecology (recovery of overexploited fisheries, removal of invaders, revegetation of habitat corridors to assist endangered species, etc.) (Ormerod, 2003). We restrict our consideration here to restoration of landscapes and waterscapes whose physical nature has been affected by human activities, dealing specifically with mining, intensive agriculture and water abstraction from rivers.

. . . to reclaim contaminated land, . . .   Land that has been damaged by mining is usually unstable, liable to erosion and devoid of vegetation. Tony Bradshaw, the father of restoration ecology, noted that the simple solution to land reclamation is the reestablishment of vegetation cover, because this will stabilize the surface, be visually attractive and self-sustaining, and provide the basis for natural or assisted succession to a more complex community (Bradshaw, 2002). Candidate plants for reclamation are those that are tolerant of the toxic heavy metals present; such species are characteristic of naturally metalliferous soils (e.g. the Italian serpentine endemic *Alyssum bertolonii*) and have fundamental niches that incorporate the extreme conditions. Moreover, of particular value are ecotypes (genotypes within a species having different fundamental niches

– see Section 1.2.1) that have evolved resistance in mined areas. Antonovics and Bradshaw (1970) were the first to note that the intensity of selection against intolerant genotypes changes abruptly at the edge of contaminated areas, and populations on contaminated areas may differ sharply in their tolerance of heavy metals over distances as small as 1.5 m (e.g. sweet vernal grass, *Anthoxanthum odoratum*). Subsequently, metaltolerant grass cultivars were selected for commercial production in the UK for use on neutral and alkaline soils contaminated by lead or zinc (*Festuca rubra* cv 'Merlin'), acidic lead and zinc wastes (*Agrostis capillaris* cv 'Goginan') and acidic copper wastes (*A. capillaris* cv 'Parys') (Baker, 2002).

... to improve contaminated soil, ...

Since plants lack the ability to move, many species that are characteristic of metalliferous soils have evolved biochemical systems for nutrient acquisition, detoxification and the control of local geochemical conditions (in effect, they help create the conditions appropriate to their fundamental niche). *Phytoremediati* involves placing such plants in contaminated soil with the aim of reducing the concentrations of heavy metals and other toxic chemicals. It can take a variety of forms (Susarla *et al.*, 2002). *Phytoaccumulation* occurs when the contaminant is taken up by the plants but is not degraded rapidly or completely; these plants, such as the herb *Thlaspi caerulescens* that hyperaccumulates zinc, are harvested to remove the contaminant and then replaced. *Phytostabilization*, on the other hand, takes advantage of the ability of root exudates to precipitate heavy metals and thus reduce bioavailability. Finally, *phytotransformation* involves elimination of a contaminant by the action of plant enzymes; for example, hybrid poplar trees *Populus deltoides x nigra* have the remarkable ability to degrade TNT (2,4,6-trinitrotoluene) and show promise in the restoration of munition dump areas. Note that microorganisms are also used for remediation in polluted situations.

Sometimes the aim of land managers is to restore the landscape for the benefit of a particular species. The European hare *Lepus europaeus* provides a case in point. The hare's fundamental niche includes landscapes created over the centuries by human activity. Hares are most common in farmed areas but populations have declined where agriculture has become too intensive and the species is now protected. Vaughan *et al.* (2003) used a farm postal survey (1050 farmers responded) to investigate the relationships between hare abundance and current land management. Their aim was to establish key features of the two most significant niche dimensions for hares, namely resource availability (crops eaten by hares) and habitat availability, and then to propose management action to maintain and restore landscapes beneficial to the species. Hares were more common on arable farms, especially on those growing wheat or beet, and where fallow land was present (areas not currently used for crops). They were less common on pasture farms, but the abundance of hares increased if 'improved' grass (ploughed, sown with a grass mixture and fertilized), some arable crops or woodland were present (Table 7.1). To increase the distribution and abundance of hares, Vaughan *et al.*'s (2003) recommendations include the provision on all farms of forage and year-round cover (from foxes *Vulpes vulpes*), the provision of woodland, improved grass and arable crops on pasture farms, and of wheat, beet and fallow land on arable farms.

... to restore landscape for a declining mammal ...

... and to restore river flow for native fish

One of the most pervasive of human influences on river ecosystems has been

Variable	Variable description	Arable farms	Pasture farms
Wheat	Wheat *Triticum aestivum* (no, yes)	***	–
Barley	Barley (no, yes)	**	–
Cereal	Other cereals (no, yes)	NS	–
Spring	Any cereal grown in spring? (no, yes)	*	–
Maize	Maize (no, yes)	NS	–
Rape	Oilseed rape *Brassica napus* (no, yes)	**	–
Legume	Peas/beans/clover *Trifolium* sp. (no, yes)	**	–
Linseed	Flax *Linum usitatissimum* (no, yes)	NS	–
Horticulture	Horticultural crops (no, yes)	NS	–
Beet	Beet *Beta vulgaris* (no, yes)	***	–
Arable	Arable crops present (see above; no, yes)	–	**
Grass	Grass (including ley, nonpermanent) (no, yes)	NS	–
Type grass	Ley, improved, semi-improved, unimproved	NS	***
Fallow	Set aside/fallow (no, yes)	***	–
Woods	Woodland/orchard (no, yes)	NS	*

Table 7.1 Habitat variables potentially determining the abundance of hares (estimated from the frequency of hare sightings), analyzed separately for arable and pasture farms. Analysis was not performed for variables where fewer than 10% of farmers responded (–). For those variables that were significantly related to whether or not hares were seen by farmers (*, $P < 0.05$; **, $P < 0.01$; ***, $P < 0.001$), the variable descriptor associated with most frequent sightings are shown in bold. (After Vaughan *et al.*, 2003.)

NS, not significant.

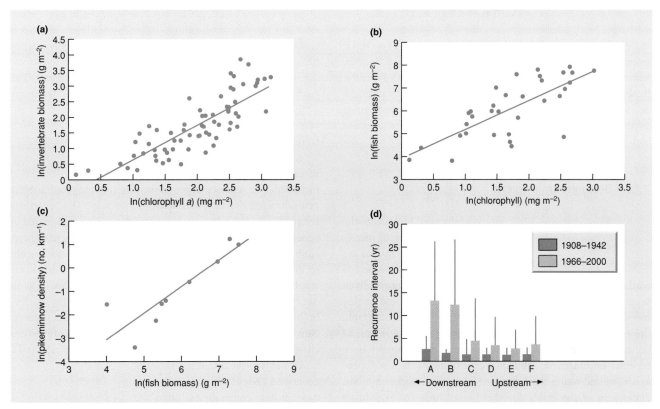

Figure 7.2   Interrelationships among biological parameters measured in a number of reaches of the Colorado River in order to determine the ultimate causes of the declining distribution of Colorado pikeminnows. (a) Invertebrate biomass versus algal biomass (chlorophyll *a*). (b) Prey fish biomass versus algal biomass. (c) Pikeminnow density versus prey fish biomass (from catch rate per minute of electrofishing). (d) Mean recurrence intervals in six reaches of the Colorado River (for which historic data were available) of discharges necessary to produce widespread stream bed mobilization and to remove silt and sand that would otherwise accumulate, during recent (1966–2000) and preregulation periods (1908–42). Lines above the histograms show maximum recurrence intervals. (After Osmundson *et al.*, 2002.)

the regulation of discharge, and river restoration often involves reestablishing aspects of the natural flow regime. Water abstraction for agricultural, industrial and domestic use has changed the hydrographs (discharge patterns) of rivers both by reducing discharge (volume per unit time) and altering daily and seasonal patterns of flow. The rare Colorado pikeminnow, *Ptychocheilus lucius*, is a piscivore (fish-eater) that is now restricted to the upper reaches of the Colorado River. Its present distribution is positively correlated with prey fish biomass, which in turn depends on the biomass of invertebrates upon which the prey fish depend, and this, in its turn, is positively correlated with algal biomass, the basis of the food web (Figure 7.2a–c). Osmundson *et al.* (2002) argue that the rarity of pikeminnows can be traced to the accumulation of fine sediment (reducing algal productivity) in downstream regions of the river. Fine sediment is not part of the fundamental niche of pikeminnows. Historically, spring snowmelt often produced flushing discharges with the power to mobilize the bed of the stream and remove much of the silt and sand that would otherwise accumulate. As a result of river regulation, however,

the mean recurrence interval of such discharges has increased from once every 1.3–2.7 years to only once every 2.7–13.5 years (Figure 7.2d), extending the period of silt accumulation.

High discharges can influence fish in other ways too by, for example, maintaining side channels and other elements of habitat heterogeneity, and by improving substrate conditions for spawning (all elements of the fundamental niche of particular species). Managers must aim to incorporate ecologically influential aspects of the natural hydrograph of a river into river restoration efforts, but this is easier said than done. Jowett (1997) describes three approaches commonly used to define minimum discharges: historic flow, hydraulic geometry and habitat assessment. The first of these assumes that some percentage of the mean discharge is needed to maintain a 'healthy' river ecosystem: 30% is often used as a rule of thumb. Hydraulic methods relate discharge to the hydraulic geometry of stream channels (based on multiple measurements of river cross-sections); river depth and width begin to decline sharply at discharges less than a certain percentage of mean discharge (10% in some rivers) and this inflection point is sometimes used as a basis for setting a minimum discharge. Finally, habitat assessment methods are based on discharges that meet specified ecological criteria, such as a critical amount of food-producing habitat for particular fish species. Managers need to beware the simplified assumptions inherent in these various approaches because, as we saw with the pikeminnows, the integrity of a river ecosystem may require something other than setting a minimum discharge, such as infrequent but high flushing discharges.

7.2.2   Dealing with invasions

a technique for displaying species niches . . .

It is not straightforward to visualize the multidimensional niche of a species when more than three dimensions are involved (see Chapter 2). However, a mathematical technique called *ordination* (discussed more fully in Section 16.3.2) allows us to simultaneously analyze and display species and multiple environmental variables on the same graph, the two dimensions of which combine the most important of the niche dimensions. Species with similar niches appear close together on the graph. Influential environmental factors appear as arrows indicating their direction of increase within the two dimensions of the graph. Marchetti and Moyle (2001) used an ordination method called canonical correspondence analysis to describe how a suite of fish species – 11 native and 14 invaders – are related to environmental factors at multiple sites in a regulated stream in California (Figure 7.3). It is clear that the native and invasive species occupy different parts of the niche space: most

of the native species occurred in places associated with higher mean discharge (m³ s⁻¹), good canopy cover (higher levels of percent shade), lower concentrations of plant nutrients (lower conductivity, µS), cooler temperatures (°C) and less pool habitat in the stream (i.e. greater percent of fast flowing, shallow riffle habitat). This combination of variables reflects the natural condition of the stream.

The pattern for introduced species was generally the opposite: invaders were favored by the present combination of conditions where water regulation had reduced discharge and increased the representation of slower flowing pool habitat, riparian vegetation had been removed leading to higher stream temperatures, and nutrient concentrations had been increased by agricultural and domestic runoff. Marchetti and Moyle (2001) concluded that restoration of more natural flow regimes is needed to limit the advance of invaders and halt the continued downward decline of native fish in this part of the western USA. It should not be imagined, however, that invaders inevitably do less well in 'natural' flow regimes. Invasive brown trout (*Salmo trutta*) in New Zealand streams seem to do better in the face of high discharge events than some native galaxiid fish (Townsend, 2003).

. . . shows why native fish are replaced by invaders

Of the invader taxa responsible for economic losses, fish are a relatively insignificant component. Table 7.2 breaks down the tens of thousands of exotic invaders in the USA into a variety of taxonomic groups. Among these, the yellow star thistle (*Centaurea solstitalis*) is a crop weed that now dominates more than 4 million ha in California, resulting in the total loss of once productive grassland. Rats are estimated to destroy US$19 billion of stored grains nationwide per year, as well as causing fires (by gnawing electric wires), polluting foodstuffs, spreading diseases and preying on native species. The red fire ant (*Solenopsis invicta*) kills poultry, lizards, snakes and ground-nesting birds; in Texas alone, its estimated damage to livestock, wildlife and public health is put at about $300 million per year, and a further $200 million is spent on control. Large populations of the zebra mussel (*Dreissena polymorpha*) threaten native mussels and other fauna, not only by reducing food and oxygen availability but by physically smothering them. The mussels also invade and clog water intake pipes, and millions of dollars need to be spent clearing them from water filtration and hydroelectric generating plants. Overall, pests of crop plants, including weeds, insects and pathogens, engender the biggest economic costs. However, imported human disease organisms, particularly HIV and influenza viruses, cost $7.5 billion to treat and result in 40,000 deaths per year. (See Pimentel *et al.*, 2000, for further details and references.)

a diversity of invaders and their economic costs

The alien plants of the British Isles illustrate a number of points about invaders and the niches they fill

species niches and the prediction of invasion success

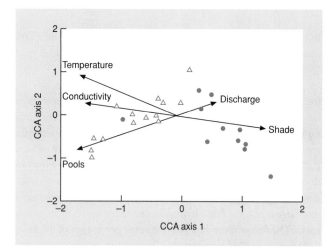

Figure 7.3   Plot of results of canonical correspondence analysis (first two CCA axes) showing native species of fish (●), introduced invader species (△) and five influential environmental variables (arrows represent the correlation of the physical variables with the canonical axes). (After Marchetti & Moyle, 2001.)

Table 7.2 Estimated annual costs (billions of US$) associated with invaders in the United States. Taxonomic groups are ordered in terms of the total costs associated with them. (After Pimentel *et al.*, 2000.)

Type of organism	Number of invaders	Major culprits	Loss and damage	Control costs	Total costs
Microbes (pathogens)	> 20,000	Crop pathogens	32.1	9.1	41.2
Mammals	20	Rats and cats	37.2	NA	37.2
Plants	5,000	Crop weeds	24.4	9.7	34.1
Arthropods	4,500	Crop pests	17.6	2.4	20.0
Birds	97	Pigeons	1.9	NA	1.9
Molluscs	88	Asian clams, Zebra mussels	1.2	0.1	1.3
Fishes	138	Grass carp, etc.	1.0	NA	1.0
Reptiles, amphibians	53	Brown tree snake	0.001	0.005	0.006

NA, not available.

(Godfray & Crawley, 1998). Species whose niches encompass areas where people live and work are more likely to be transported to new regions, where they will tend to be deposited in habitats like those where they originated. Thus more invaders are found in disturbed habitats close to transport centers and fewer are found in remote mountain areas (Figure 7.4a). Moreover, more invaders arrive from nearby locations (e.g. Europe) or from remote locations whose climate (and therefore the invader's niche) matches that found in Britain (Figure 7.4b). Note the small number of alien plants from tropical environments; these species usually lack the frost-hardiness required to survive the British winter. Shea and Chesson (2002) use the phrase *niche opportunity* to describe the potential provided in a given region for invaders to succeed – in terms of a high availability of resources and appropriate physico-chemical conditions (coupled with a lack or scarcity of natural enemies). They note that human activities often disrupt conditions

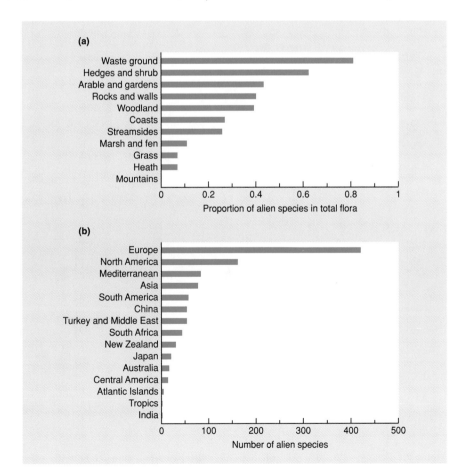

Figure 7.4 The alien flora of the British Isles: (a) according to community type (note the large number of aliens in open, disturbed habitats close to human settlements) and (b) by geographic origin (reflecting proximity, trade and climatic similarity). (After Godfray & Crawley, 1998.)

in ways that provide niche opportunities for invaders – river regulation is a case in point. Not all invaders cause obvious ecological harm or economic loss; indeed some ecologists distinguish exotic species that establish without significant consequences from those they consider 'truly invasive' – whose populations expand 'explosively' in their new environment, with significant impacts for indigenous species. Managers need to differentiate among potential new invaders both according to their likelihood of establishing should they arrive in a new region (largely dependent on their niche requirements) and in relation to the probability of having dramatic consequences in the receiving community (dealt with in Chapter 22). Management strategies to get rid of invading pests usually require an understanding of the dynamics of interacting populations and are covered in Chapter 17.

7.2.3 Conservation of endangered species

The conservation of species at risk often involves establishing protected areas and sometimes the translocation of individuals to new locations. Both approaches should be based on considerations of the niche requirements of the species concerned.

The overwintering habitat in Mexico is absolutely critical for the monarch butterfly (*Danaus plexippus*), which breeds in southern Canada and the eastern United States. The butterflies form dense colonies in oyamel (*Abies religiosa*) forests on 11 separate mountains in central Mexico. A group of experts was assembled to define objectives, assess and analyze the available data, and to produce alternative feasible solutions to the problem of maximizing the protection of overwintering habitat while minimizing the inclusion of valuable land for logging (Bojorquez-Tapia *et al.*, 2003). As in many areas of applied ecology, ecological and economic criteria had to be judged together. The critical dimensions of the butterfly's overwintering niche include relatively warm and humid conditions (permitting survival and conservation of energy for the return north) and the availability of streams (resource) from which the butterflies drink on clear, hot days. The majority of known colony sites are in forests on moderately steep slopes, at high elevation (>2890 m), facing towards the south or southwest, and within 400 m of streams (Figure 7.5). According to the degree to which locations in central Mexico matched the optimal habitat features, and taking into account the desire to mimimize

niche ecology and the selection of conservation reserves

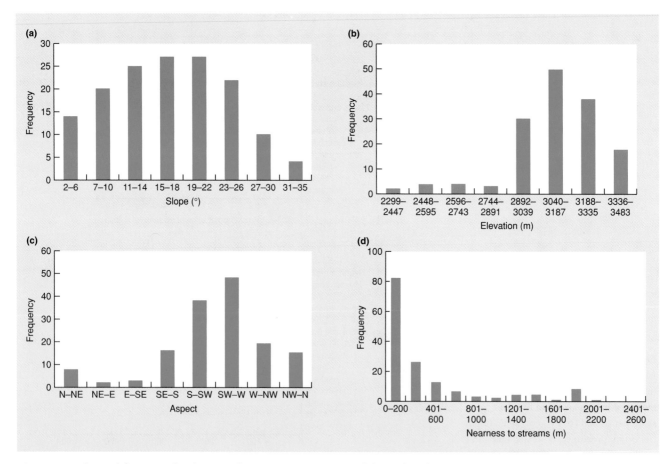

Figure 7.5 Observed frequency distributions of 149 overwintering monarch butterfly colonies in central Mexico in relation to: (a) slope, (b) elevation, (c) aspect and (d) proximity to a stream. (After Bojorquez-Tapia *et al.*, 2003.)

Figure 7.6 Optimal distribution in the mountains of central Mexico of overwintering monarch butterfly reserves (colored areas) according to three scenarios: (a) area constraint of 4500 ha, (b) area constraint of 16,000 ha, and (c) no area constraint (area included is 21,727 ha). The orange lines are the boundaries between river catchment areas. Scenario (c) was accepted by the authorities for the design of Mexico's 'Monarch Butterfly Biosphere Reserve'. (After Bojorquez-Tapia *et al.*, 2003.)

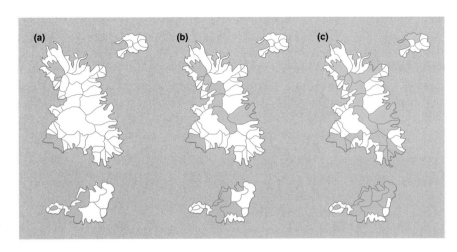

the inclusion of prime logging habitat, a geographic information system (GIS) was then used to delineate three scenarios. These differed according to the area the government might be prepared to set aside for monarch butterfly conservation (4500 ha, 16,000 ha or no constraint) (Figure 7.6). The experts preferred the no-constraint scenario, which called for 21,727 ha of reserves (Figure 7.6c), and despite the fact that their recommendation was the most expensive it was accepted by the authorities.

Unraveling the fundamental niche of species that have been driven to extreme rarity may not be straight-forward. The takahe (*Porphyrio hochstetteri*), a giant rail, is one of only two remaining species of the guild of large, flightless herbivorous birds that dominated the prehuman New Zealand landscape (Figure 7.7). Indeed, it was also believed to be

present distributions do not always coincide with optimal niche conditions

Figure 7.7 The location of fossil bones of the takahe in the South Island of New Zealand. (After Trewick & Worthy, 2001.)

extinct until the discovery in 1948 of a small population in the remote and climatically extreme Murchison Mountains in the southeast of South Island (Figure 7.7). Since then intense conservation efforts have involved habitat management, captive breeding, wild releases into the Murchison Mountains and nearby ranges, and translocation to offshore islands that lack the mammals introduced by people that are now widespread on the mainland (Lee & Jamieson, 2001). Some ecologists argued that because takahe are grassland specialists (tall tussocks in the genus *Chionochloa* are their most important food) and adapted to the alpine zone they would not fare well outside this niche (Mills *et al.*, 1984). Others pointed to fossil evidence that the species was once widespread and occurred mainly at altitudes below 300 m (often in coastal areas – Figure 7.7) where they were associated with a mosaic of forest, shrublands and grasslands. These ecologists argued that takahe might be well suited for life on offshore islands that are free of mammalian invaders. It turned out that the sceptics were wrong in thinking that translocated island populations would not become self-sustaining (takahe have been successfully introduced to four islands), but they seem to have been right that islands would not provide an optimal habitat: island birds have poorer hatching and fledging success than mountain birds (Jamieson & Ryan, 2001). The fundamental niche of takahe probably encompasses a large part of the landscape of South Island, but the species became confined to a much narrower realized niche by people who hunted them, and by mammalian invaders such as red deer (*Cervus elaphus scoticus*) that compete with them for food and stoats (*Mustella erminea*) that prey upon them. The current distributions of species like takahe, which have been driven very close to extinction, may provide misleading information about niche requirements. It is likely that neither the Murchison Mountains nor offshore islands (with pasture rather than tussock grasses) coincide with the optimal set of conditions and resources of the takahe's fundamental niche. Historical reconstructions of the ranges of endangered species may help managers identify the best sites for reserves.

7.3   Life history theory and management

We saw in Chapter 4 that particular combinations of ecological traits help determine lifetime patterns of fecundity and survival, which in turn determine the distribution and abundance of species in space and time. In this section we consider whether particular traits can be of use to managers concerned with restoration, biosecurity and the risk of extinction of rare species.

7.3.1   Species traits as predictors for effective restoration

Pywell *et al.* (2003) assembled the results of 25 published experiments dealing with the restoration of species-rich grasslands from land that had previously been 'improved' for pasture or used for arable farming. They wished to relate plants' performances to their life histories. On the basis of the results of the

using knowledge of species traits . . .

. . . to restore grassland, . . .

first 4 years of restoration, they calculated a performance index for commonly sown grasses (13 species) and forbs (45 species; forbs are defined as herbaceous plants that are not grass-like). The index, calculated for each of the 4 years, was based on the proportion of quadrats (0.4×0.4 m or larger) that contained the species in treatments where that species was sown. Their life history analysis included 38 plant traits, including longevity of seeds in the seed bank, seed viability, seedling growth rate, life form and life history strategy (e.g. competitiveness, stress tolerance, colonization ability (ruderality)) (Grime *et al.*, 1988) and the timing of life cycle events (germination, flowering, seed dispersal). The best performing grasses included *Festuca rubra* and *Trisetum flavescens* (performance indexes averaged for the 4 years of 0.77); and among the forbs *Leucanthemum vulgare* (0.50) and *Achillea mellefolium* (0.40) were particularly successful. Grasses, which showed few relationships between species traits and performance (only ruderality was positively correlated), consistently outperformed the forbs. Within the forbs, good establishment was linked to colonization ability, percent germination of seeds, fall germination, vegetative growth, seed bank longevity and habitat generalism, while competitive ability and seedling growth rate became increasingly important determinants of success with time (Table 7.3). Stress tolerators, habitat specialists and species of infertile habitats performed badly (partly reflecting the high residual nutrient availability in many restored grasslands). Pywell *et al.* (2003) argue that restoration efficiency could be increased by only sowing species with the identified ecological traits. However, because this would lead to uniformity amongst restored grasslands, they also suggest that desirable but poorly performing species could be assisted by phased introduction several years after restoration begins, when environmental conditions are more favorable for their establishment.

7.3.2   Species traits as predictors for setting biosecurity priorities

A number of species have invaded widely separated places on the planet (e.g. the shrub *Lantana camara* (Figure 7.8), the starling *Sturnus vulgaris* and the rat *Rattus rattus*) prompting the question of whether successful invaders share traits that raise the odds of successful invasion (Mack *et al.*, 2000). Were it possible to produce a list of traits associated with invasion success, managers would be in a good position to assess the risks of establishment, and thus to prioritize potential invaders and devise appropriate biosecurity

. . . to set priorities for dealing with invasive species . . .

Table 7.3 Ecological traits of forbs that showed a significant relationship with plant performance in years 1–4 after sowing in grassland restoration experiments. The sign shows whether the relationship was positive or negative. (After Pywell *et al.*, 2003.)

Trait	n	Year 1	Year 2	Year 3	Year 4
Ruderality (colonization ability)	39	+ *	NS	NS	NS
Fall germination	42	+ *	NS	NS	NS
Germination (%)	43	+ **	+ *	+ *	NS
Seedling growth rate	21	NS	+ *	+ **	+ *
Competitive ability	39	+ *	+ **	+ ***	+ ***
Vegetative growth	36	+ **	+ *	+ *	+ *
Seed bank longevity	44	+ *	+ *	+ *	+ *
Stress tolerance	39	– **	– **	– ***	– ***
Generalist habitat	45	+ **	+ **	+ **	+ **

*, P < 0.05; **, P < 0.01; ***, P < 0.001; *n*, number of species in analysis; NS, not significant.

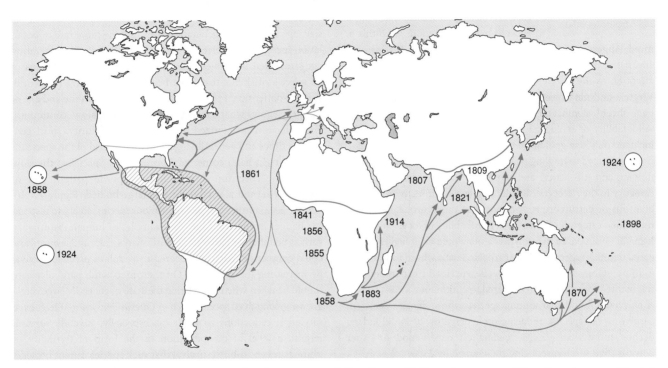

Figure 7.8 The shrub *Lantana camara*, an example of a very successful invader, was deliberately transported from its native range (shaded area) to widely dispersed subtropical and tropical locations where it spread and increased to pest proportions. (After Cronk & Fuller, 1995.)

procedures (Wittenberg & Cock, 2001). The success of some invasive taxa has an element of predictability. Of 100 or so introduced pine species in the USA, for example, the handful that have successfully encroached into native habitats are characterized by small seeds, a short interval between successive large seed crops and a short juvenile period (Rejmanek & Richardson, 1996). In New Zealand there is a similarly precise record of successes and failures of attempted bird introductions. Sol and Lefebvre (2000) found that invasion success increased with introduction effort (number of attempts and number of individuals since European colonization), which is not surprising. Invasion success was also higher for nidifugous species whose young are not fed by their

parents (such as game birds), species that do not migrate and, in particular, birds with relatively large brains. The relationship with brain size was partly a consequence of nidifugous species having large brains but probably also reflects greater behavioral flexibility; the successful invaders have more reports in the international literature of adopting novel food or feeding techniques (mean for 28 species 1.96, SD 3.21) than the unsuccessful species (mean for 48 species 0.58, SD 1.01).

Despite indications of predictability of invasion success for some taxa, related to high fecundity (e.g. pine seed production) and broad niches (e.g. bird behavioral flexibility), exceptions to the 'rules' are common and there are many more cases where no relationships have been found, prompting Williamson (1999) to wonder whether invasions are any more predictable than earthquakes. The best predictor of invasion success is previous success as an invader elsewhere. However, even this provides invasion managers with useful pointers for prioritizing potential invaders to their regions.

7.3.3 Species traits as predictors for conservation and harvest management priorities

. . . and to set priorities for conservation of endangered species

Managers would be better able to prioritize species for conservation intervention if it were possible to predict, on the basis of species traits, those most at risk of extinction. With this in mind, Angermeier (1995) analyzed the traits of the 197 historically native freshwater fish in Virginia, USA, paying particular attention to the characteristics of the 17 species now extinct in Virginia and nine more considered at risk because their ranges have shrunk significantly. Of particular interest was the greater vulnerability of ecological specialists. Thus species whose niche included only one geological type (of several present in Virginia), those restricted to flowing water (as opposed to occurring in both flowing and still water) and those that included only one food category in their diet (i.e. wholly piscivorous, insectivorous, herbivorous or detritivorous as opposed to omnivorous on two or more food categories) had a higher probability of local extinction. It might be supposed that top predators would be at higher risk of extinction than species at lower trophic levels whose food supply is more stable. In a study of beetle species in experimentally fragmented forest habitat (compared to continuous forest) Davies *et al.* (2000) found that among species whose density declined, carnivores (10 species, reducing on average by 70%) did indeed decline more than species feeding on dead wood or other detritus (five species, reducing on average by 25%).

large body size and extinction risk are often correlated

A pattern that has repeatedly emerged is that extinction risk tends to be highest for species with a large body size. Figure 7.9 illustrates this for Australian marsupials that have gone extinct within the last 200 years or are currently considered endangered. Some geographic regions (e.g. arid compared to mesic zone) and some taxa (e.g. potoroos, bettongs, bandicoots and bilbies) have experienced higher extinction/endangerment rates than others, but the strongest relationship is between body size and risk of extinction (Cardillo & Bromham, 2001). Recall that body size is part of a common life history syndrome (essentially r/K) that associates large size, late maturity and small reproductive allocation (see Section 4.12).

Cortes (2002) has explored the relationship between body size, age at maturity, generation time and the finite rate of population increase λ (referred to in Section 4.7 as R), by generating age-structured life tables (see Chapter 4) for 41 populations of 38 species of sharks that have been studied around the world. A three-dimensional plot of λ against generation time and age at maturity shows what Cortes (2002) calls a 'fast–slow' continuum, with species characterized by early age at maturity, short generation times and generally high λ at the fast end of the spectrum (bottom right hand corner of Figure 7.10a). Species at the slow end of the spectrum displayed the opposite pattern (left of Figure 7.10a) and also tended to be large bodied (Figure 7.10b). Cortes (2002) further assessed the various species' ability to respond to changes in survival (due, for example, to human disturbance such as pollution or harvesting). 'Fast' sharks, such as *Sphyrna tiburo*, could compensate for a 10% decrease in adult or juvenile survival by increasing the birth rate. On the other hand, particular care should be taken when considering the state of generally large, slow-growing, long-lived species, such as *Carcharhinus leucas*. Here, even moderate reductions to adult or, especially, juvenile survival require a level of compensation in the form of fecundity or immediately post-birth survival that such species cannot provide.

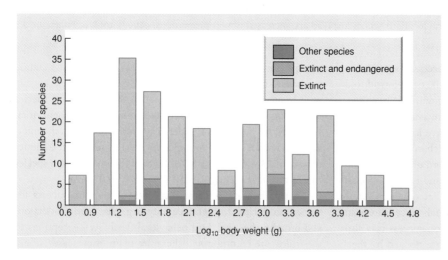

Figure 7.9 Body size frequency distribution of the Australian terrestrial marsupial fauna including 25 species that have gone extinct in the last 200 years (dark orange). Sixteen species currently considered endangered are shown in gray. (After Cardillo & Bonham, 2001.)

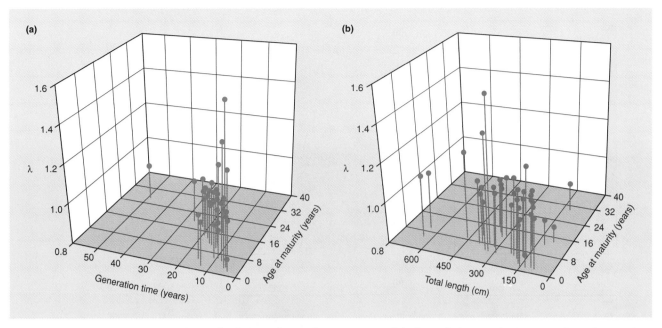

Figure 7.10 Mean population growth rates λ of 41 populations from 38 species of shark in relation to: (a) age at maturity and generation time and (b) age at maturity and total body length. (After Cortes, 2002.)

Skates and rays (Rajidae) provide a graphic illustration of Cortes' warning. Of the world's 230 species, only four are known to have undergone local extinctions and significant range reduction (Figure 7.11a–d). These are among the largest of their group (Figure 7.11e) and Dulvy and Reynolds (2002) propose that seven further species, each as large or larger than the locally extinct taxa, should be prioritized for careful monitoring.

7.4 Migration, dispersion and management

7.4.1 Restoration and migratory species

using knowledge of animal movements . . .

. . . to restore harvested fish species, . . .

Species that spend part of their time in one habitat (or region) and part in another (see Section 6.4) can be badly affected by human activities that influence the ability to move between them. The declining populations of river herrings (*Alosa pseudoharengus* and *A. aestivalis*) in the northeastern USA provide a case in point. These species are anadromous: adults ascend coastal rivers to spawn in lakes between March and July and the young fish remain in fresh water for 3–7 months before migrating to the ocean. Yako *et al.* (2002) sampled river herrings three times per week from June to December in the Santuit River downstream of Santuit Pond, which contains the only spawning habitat in the catchment. They identified periods of migration as either 'peak' (>1000 fish week⁻¹)

or 'all' (>30 fish week⁻¹, obviously including the 'peak'). By simultaneously measuring a range of physicochemical and biotic variables, they aimed to identify factors that could predict the timing of juvenile migration (Figure 7.12). They determined that peaks of migration were most likely to occur during the new moon and when the density of important zooplankton prey was low (*Bosmina* spp.). All migration periods, taken together (30 to 1000+), tended to occur when water visibility was low and during decreased periods of rainfall. It is not unusual for changes in the moon phase to influence animal behavior by acting as cues for life cycle events; in the herrings' case, migration near to the new moon phase, when nights are dark, may reduce the risk of becoming prey to piscivorous fish and birds. The trough in availability of the herrings' preferred food may also play a role in promoting migration, and this could be exacerbated by murky water interfering with the foraging of the visually hunting fish. Predictive models such as the one for river herrings can help managers identify periods when river discharge needs to be maintained to coincide with migration.

Populations of flying squirrels (*Pteromys volans*) have declined dramatically since the 1950s in Finland, mainly because of habitat loss, habitat fragmentation and reduced habitat connectivity associated with intensive forestry practices. Areas of natural forest are now separated by clear-cut and regenerating areas. The core breeding habitat of the flying squirrels only occupies a few hectares, but individuals, particularly males, move to and from

. . . to restore habitat for a declining squirrel population . . .

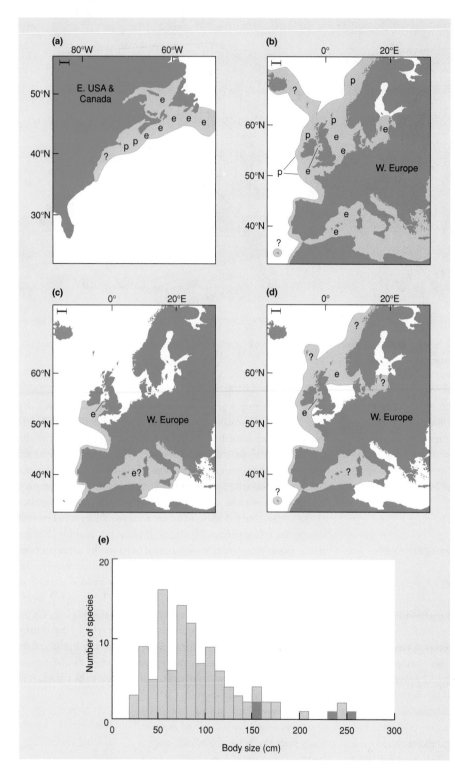

Figure 7.11 Historic distribution of four locally extinct skates in the northwest and northeast Atlantic: (a) barndoor skate *Dipturus laevis*, (b) common skate *D. batis*, (c) white skate *Rostroraja alba* and (d) long-nose skate *D. oxyrhinchus*. e, area of local extinction; e?, possible local extinction; p, present in recent fisheries surveys; ?, no knowledge of status; scale bar represents 150 km. (e) Frequency distribution of skate body size – the four locally extinct species are dark orange. (After Dulvy & Reynolds, 2002.)

this core for temporary stays in a much larger 'dispersal' area (1–3 km²), and juveniles permanently disperse within this range (Section 6.7 dealt with within-population variation in dispersal). Reunanen *et al.* (2000) compared the landscape structure around known flying squirrel home ranges (63 sites) with randomly chosen areas (96 sites) to determine the forest patterns that favor the squirrels. They first established that landscape patch types could be divided into optimal breeding habitat (mixed spruce–

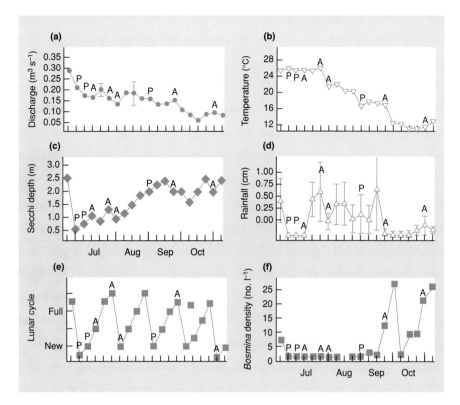

Figure 7.12 Variation in physical and biotic variables in the Santuit River, USA during the migratory period of river herrings: (a) discharge, (b) temperature, (c) Secchi disc depth (low values indicate poor light transmission because of high turbidity), (d) rainfall, (e) lunar cycle and (f) *Bosmina* density. P denotes 'peak' periods of migration (>1000 fish week⁻¹), P and A (>30 fish week⁻¹) together denote 'all' periods of migration. (After Yako *et al.*, 2002.)

deciduous forests), dispersal habitat (pine and young forests) and unsuitable habitat (young sapling stands, open habitats, water). Figure 7.13 shows the amount and spatial arrangement of the breeding habitat and dispersal habitat for examples of a typical flying squirrel site and a random forest site. Overall, flying squirrel landscapes contained three times more suitable breeding habitat within a 1 km radius than random landscapes. Squirrel landscapes also contained about 23% more dispersal habitat than random landscapes but, more significantly, squirrel dispersal habitat was much better connected (fewer fragments per unit area) than random landscapes. Reunanen *et al.* (2000) recommend that forest managers should restore and maintain a deciduous mixture, particularly in spruce-dominated forests, for optimal breeding habitat. But of particular significance in the context of dispersal behavior, they need to ensure good physical connectivity between the optimal squirrel breeding and dispersal habitats.

7.4.2 Predicting the spread of invaders

... and to predict the spread of invaders

A broad scale approach to preventing the arrival of potential invaders is to identify major 'migration' pathways, such as hitchhiking in the mail or cargos and on aircraft or in ships, and to manage the risks associated with these (Wittenberg & Cock, 2001). The Great Lakes of North America have been invaded by more

than 145 alien species, many arriving in the ballast water of ships. For example, a whole series of recent invaders (including fish, mussels, amphipods, cladocerans and snails) originated from the other end of an important trade route in the Black and Caspian Seas (Ricciardi & MacIsaac, 2000). A ballasted ocean freighter before taking on cargo in the Great Lakes may discharge 3 million liters of ballast water that contain various life stages of many plant and animal taxa (and even the cholera bacterium *Vibrio cholerae*) that originate where the ballast water was taken aboard. One solution is to make the dumping of ballast water in the open ocean compulsory rather than voluntary (this is now the case for the Great Lakes). Other possible methods involve filter systems when loading ballast water, and on-board treatment by ultra-violet irradiation or waste heat from the ship's engines.

The most damaging invaders are not simply those that arrive in a new part of the world; the subsequent pattern and speed of their spread is also significant to managers. Zebra mussels (*Dreissena polymorpha*) have had a devastating effect (see Section 7.2.2) since arriving in North America via the Caspian Sea/Great Lakes trade route. Range expansion quickly occurred throughout commercially navigable waters, but overland dispersal into inland lakes, mainly attached to recreational boats, has been much slower (Kraft & Johnson, 2000). Geographers have developed so-called 'gravity' models to predict human dispersal patterns based on distance to and attractiveness of destination points, and Bossenbroek *et al.* (2001) adopted the technique to predict the spread of zebra mussels through the inland lakes of Illinois, Indiana,

Breeding habitats   **Breeding plus dispersal habitats**

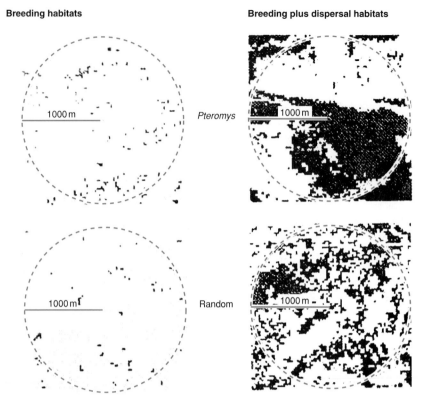

Figure 7.13 The spatial arrangement of patches (dark) of breeding habitat (left hand panels) and breeding plus dispersal habitat (right hand panels) in a typical landscape containing flying squirrels (*Pteromys*) (top panels) and a random forest location (bottom panels). This flying squirrel landscape contains 4% breeding habitat and 52.4% breeding plus dispersal habitat, compared with 1.5 and 41.5% for the random landscape. Dispersal habitat in the squirrel landscape is much more highly connected (fewer fragments per unit area) than in the random landscape. (After Reunanen *et al.*, 2000.)

Michigan and Wisconsin (364 counties in all). The model has three steps involving (i) the probability of a boat traveling to a zebra mussel source; (ii) the probability of the same boat making a subsequent outing to an uncolonized lake; and (iii) the probability of zebra mussels becoming established in the uncolonized lake.

1 Uninfested boats travel to an already colonized lake or boat ramp and inadvertently pick up mussels. The number of boats, T, that travel from county i to a lake or boat ramp, j, is estimated as:

$$T_{ij} = A_i O_i W_j \, c_{ij}^{-\alpha}$$

where A_i is a correction factor that ensures all boats from county i reach some lake, O_i is the number of boats in county i, W_j is the attractiveness of location j, c_{ij} is the distance from county i to location j and α is a distance coefficient.

2 Infested boats travel to an uncolonized lake and release mussels. The number of infested boats P_i consists of those boaters that travel from county i to a source of zebra mussels,

summed for each county over the total number of zebra mussel sources. T_{iu}, then, is the number of infested boats that travel from county i to an uncolonized lake u:

$$T_{iu} = A_i P_i W_u c_{iu}^{-\alpha}.$$

The total number of infested boats that arrive at a given uncolonized lake is summed over all the counties (Q_u).

3 The probability that transported individuals will establish a new colony depends on lake physicochemistry (i.e. key elements of the mussel's fundamental niche) and stochastic elements. In the model, a new colony is recruited if Q_u is greater than a colonization threshold of f.

To generate a probabilistic distribution of zebra mussel-colonized lakes, 2000 trials were run for 7 years and the number of colonized lakes for each county was estimated by summing the individual colonization probabilities for each lake in the county. The results, shown in Figure 7.14, are highly correlated with the pattern of colonization that actually occurred up to 1997, giving confidence

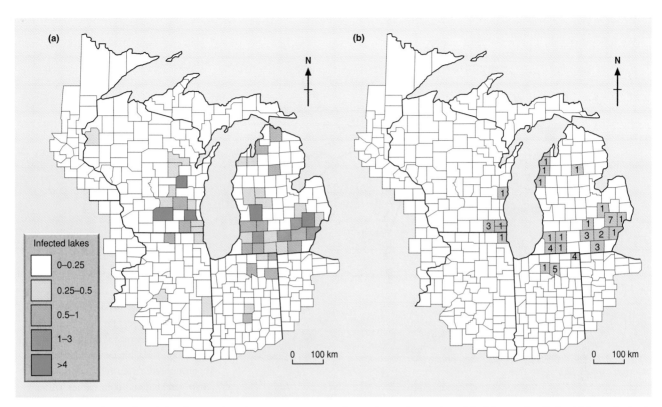

Figure 7.14   (a) The predicted distribution (based on 2000 iterations of a stochastic 'gravity' model of dispersal) of inland lakes colonized by zebra mussels in 364 counties in the USA; the large lake in the middle is Lake Michigan, one of the Great Lakes of North America. (b) The actual distribution of colonized lakes as of 1997. (After Bossenbroek *et al.*, 2001.)

in the predictions of the model. However, areas of central Wisconsin and western Michigan were predicted to be colonized, but no colonies have so far been documented. Bossenbroek *et al.* (2001) suggest that invasion may be imminent in these locations, which should therefore be the focus of biosecurity efforts and education campaigns.

Of course invaders do not all rely on human agency; many disperse by their own devices. The red fire ant (*Solenopsis invicta*) has spread rapidly through much of southern USA with dramatic economic consequences (see Section 7.2.2). The species, which originated in Argentina, occurs in two distinct social forms. The single-queened (monogyne) form and the multiple-queened (polygyne) form differ in their patterns of reproduction and modes of dispersal. The queens from monogyne colonies take part in mating flights and found colonies independently, whereas the queens from polygyne colonies are adopted into established nests after mating. As a result, the monogyne populations spread three orders of magnitude more quickly than their polygyne counterparts (Holway & Suarez, 1999). The ability of managers to prioritize potentially problematic invaders and to devise strategies to counter their spread can be expected to be improved by a thorough understanding of the invaders' behavior.

7.4.3   Conservation of migratory species

An understanding of the behavior of species at risk can also assist managers to devise conservation strategies. Sutherland (1998) describes an intriguing case where the knowledge of migratory and dispersal behavior has proved critical. A scheme was devised to alter the migration route of the lesser white-fronted geese (*Anser erythropus*) from southeastern Europe, where they tend to get shot, to spend their winters in the Netherlands. A population of captive barnacle geese (*Branta leucopsis*) breeds in Stockholm Zoo but overwinters in the Netherlands. Some were taken to Lapland where they nested and were given lesser white-fronted goose eggs to rear. The young geese then flew with their adopted parents to the Netherlands for the winter, but next spring the lesser white-fronted geese returned to Lapland and bred with conspecifics there, subsequently returning again to the Netherlands. Another example involves the reintroduction of captive-reared *Phascogale tapoatafa*, a carnivorous marsupial. Soderquist (1994) found that if males and females were released together, the males dispersed and females could not

using behavioral ecology . . .

. . . to conserve endangered species . . .

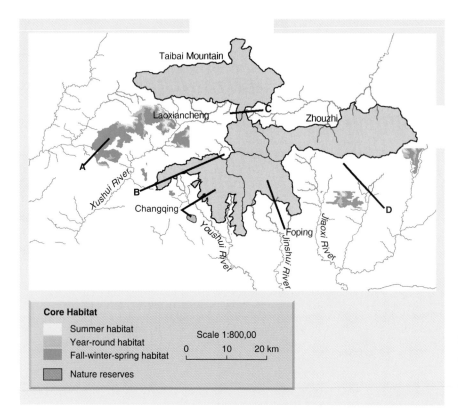

Figure 7.15 Core panda habitats (A–D), each of which caters for the year-round needs of the elevational migration of giant pandas in China's Qinling Province. Superimposed are current nature reserves (cross-hatched) and their names. (After Loucks *et al.*, 2003.)

find a mate. Much more successful was a 'ladies first' release scheme; this allowed the females to establish a home range before males came and joined them.

. . . and to design nature reserves

Where migrating species are concerned, the design of nature reserves must take account of their seasonal movements. The Qinling Province in China is home to approximately 220 giant pandas (*Ailuropoda melanoleuca*), representing about 20% of the wild population of one of the world's most imperiled mammals. Of particular significance is the fact that pandas in this region are elevational migrants, needing both low and high elevation habitat to survive, but current nature reserves do not cater for this. Pandas are extreme dietary specialists, primarily consuming a few species of bamboo. In Qinling Province, from June to September pandas eat *Fargesia spathacea*, which grows from 1900 to 3000 m. But as colder weather sets in, they travel to lower elevations and from October to May they feed primarily on *Bashania fargesii*, which grows from 1000 to 2100 m. Loucks *et al.* (2003) used a combination of satellite imagery, fieldwork and GIS analysis to identify a landscape to meet the long-term needs of the species. The process for selecting potential habitat first excluded areas lacking giant pandas, forest block areas that were smaller than 30 km² (the minimum area needed to support a pair of giant pandas over the short term) and forest with roads, settlements or plantation forests. Figure 7.15 maps summer habitat (1900–3000 m; *F. spathacea* present), fall/winter/ spring habitat (1400–2100 m; *B. fargesii* present) and a small amount of year-round habitat (1900–2100 m, both bamboo species present) and identifies four areas of core panda habitat (A–D) that provide for the migrational needs of the pandas. Superimposed on Figure 7.15 are the current nature reserves; disturbingly, they cover only 45% of the core habitat. Loucks *et al.* (2003) recommend that the four core habitat areas they have identified should be incorporated into a reserve network. Moreover, they note the importance of promoting linkage between the zones, because extinction in any one area (and in all combined) is more likely if the populations are isolated from each other (see Section 6.9, which deals with metapopulation behavior). Thus, they also identify two important linkage zones for protection, between areas A and B where steep topography means few roads exist, and between B and D across high elevation forests.

7.5 Dynamics of small populations and the conservation of endangered species

Extinction has always been a fact of life, but the arrival on the scene of humans has injected some novelty into the list of its causes. Overexploitation by hunting was probably the first of these, but more recently a large array of other impacts have been brought to bear, including habitat destruction, introduction of exotic pest species and pollution. Not surprisingly, conservation

of the world's remaining species has come to assume great importance. Here we deal with the conservation of species populations,
leaving the management of communities and ecosystems to
Chapter 22.

7.5.1 The scale of the problem

To judge the scale of the problem facing conservation managers we need to know the total number of species that occur
in the world, the rate at which these are going extinct and how
this rate compares with that of prehuman times. Unfortunately,
there are considerable uncertainties in our estimates of all these
things.

how many species
on earth?

About 1.8 million species have so
far been named (Alonso *et al.*, 2001), but
the real number is very much larger.
Estimates have been derived in a vari-
ety of ways (see May, 1990). One approach is based on a general
observation that for every temperate or boreal mammal or bird
(taxa where most species have now apparently been described)
there are approximately two tropical counterparts. If this is
assumed also to hold for insects (where there are many undescribed
species), the grand total would be around 3–5 million. Another
approach uses information on the rate of discovery of new
species to project forward, group by group, to a total estimate of
up to 6–7 million species in the world. A third approach is based
on a species size–species richness relationship, taking as its starting point that as one goes down from terrestrial animals whose
characteristic linear dimensions are a few meters to those of
around 1 cm, there is an approximate empirical rule that for each
10-fold reduction in length there are 100 times the number of
species. If the observed pattern is arbitrarily extrapolated down
to animals of a characteristic length of 0.2 mm, we arrive at a global
total of around 10 million species of terrestrial animals. A fourth
approach is based on estimates of beetle species richness (more
that 1000 species recorded in one tree) in the canopies of tropical
trees (about 50,000 species), and assumptions about the proportion
of nonbeetle arthropods that will also be present in the canopy
plus other species that do not occupy the canopy; this yields an
estimate of about 30 million tropical arthropods. The uncertainties in estimating global species richness are profound and our best
guesses range from 3 to 30 million or more.

modern and historic
extinction rates
compared

An analysis of recorded extinctions
during the modern period of human
history shows that the majority have
occurred on islands and that birds and
mammals have been particularly badly
affected (Figure 7.16). The percentage of extant species involved
appears at first glance to be quite small, and moreover, the
extinction rate appears to have dropped in the latter half of the
20th century, but how good are these data?

Once again, these estimates are bedevilled by uncertainty. First,
the data are much better for some taxa and in some places than
others, so the patterns in Figure 7.16 must be viewed with a good
deal of scepticism. For example, there may be serious underestimates even for the comparatively well-studied birds and mammals
because many tropical species have not received the careful
attention needed for fully certified extinction. Second, a very large
number of species have gone unrecorded and we will never
know how many of these have become extinct. Finally, the drop
in recorded extinctions in the latter half of the 20th century
may signal some success for the conservation movement. But it
might equally well be due to the convention that a species is only
denoted extinct if it has not been recorded for 50 years. Or it may
indicate that many of the most vulnerable species are already
extinct. Balmford *et al.* (2003) suggest that all our attention
should not be focused on extinction rates, but that a more meaningful view of the scale of the problem of species at risk will come
from the long-term assessment of changes (often significant
reductions) in the relative abundance of species (which have not
yet gone extinct) or of their habitats.

a human-induced
mass extinction?

An important lesson from the fossil
record is that the vast majority, probably all, of extant species will become
extinct eventually – more than 99% of species that ever existed
are now extinct (Simpson, 1952). However, given that individual
species are believed, on average, to have lasted for 1–10 million
years (Raup, 1978), and if we take the total number of species on
earth to be 10 million, we would predict that only an average
of between 100 and 1000 species (0.001–0.01%) would go extinct
each century. The current observed rate of extinction of birds
and mammals of about 1% per century is 100–1000 times this
'natural' background rate. Furthermore, the scale of the most
powerful human influence, that of habitat destruction, continues
to increase and the list of endangered species in many taxa is alarm-
ingly long (Table 7.4). We cannot be complacent. The evidence,
whilst inconclusive because of the unavoidable difficulty of making accurate estimates, suggests that our children and grandchil-
dren may live through a period of species extinction comparable
to the five 'natural' mass extinctions evident in the geological record
(see Chapter 21).

7.5.2 Where should we focus conservation effort?

Several categories of risk of species
extinction have been defined (Mace
& Lande, 1991). A species can be
described as *vulnerable* if there is considered to be a 10% probability of extinction within 100 years, *endangered* if the probability
is 20% within 20 years or 10 generations, whichever is longer, and
critically endangered if within 5 years or two generations the risk of
extinction is at least 50% (Figure 7.17). Based on these criteria,

classification of the
threat to species

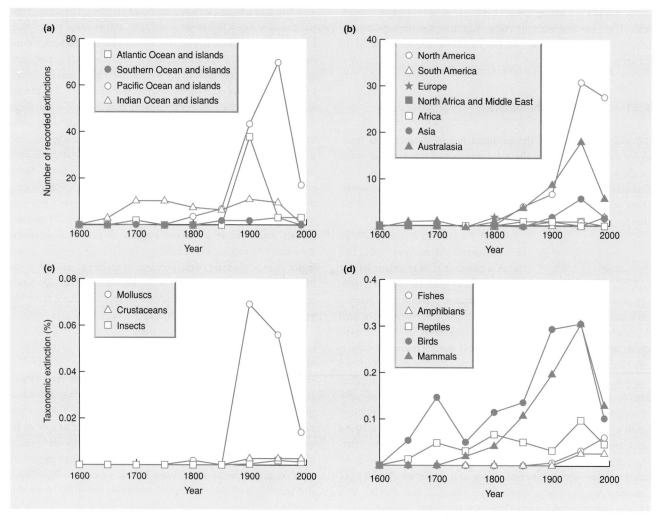

Figure 7.16 Trends in recorded animal species extinctions since 1600, for which a date is known, in: (a) the major oceans and their islands, (b) major continental areas, for (c) invertebrates and (d) vertebrates. (After Smith *et al.*, 1993.)

43% of vertebrate species have been classified as threatened (i.e. they fell into one of the above categories) (Mace, 1994).

On the basis of these definitions, both governments and non-governmental organizations have produced threatened species lists (the basis of analyses like that shown in Table 7.4). Clearly, these lists provide a starting point for setting priorities for developing plans to manage individual species. However, resources for conservation are limited and spending the most money on species with the highest extinction probabilities will be a false economy if a particular highly ranked species would require a huge recovery effort but with little chance of success (Possingham *et al.*, 2002). As in all areas of applied ecology, conservation priorities have both ecological and economic dimensions. In desperate times, painful decisions have to be made about priorities. Wounded soldiers arriving at field hospitals in the First World War were subjected to a

triage evaluation: priority 1 – those who were likely to survive but only with rapid intervention; priority 2 – those who were likely to survive without rapid intervention; priority 3 – those who were likely to die with or without intervention. Conservation managers are often faced with the same kind of choices and need to demonstrate some courage in giving up on hopeless cases, and prioritizing those species where something can be done.

Species that are at high risk of extinction are almost always rare. Nevertheless, rare species, just by virtue of their rarity, are not necessarily at risk of extinction. It is clear that many, probably most, species are naturally rare. The population dynamics of such species may follow a characteristic pattern. For example, out of a group of four species of *Calochortus* lilies in California, one is abundant and three

many species are naturally rare

Table 7.4 The current numbers and percentages of named animal and plant species in major taxa judged to be threatened with extinction. The higher values associated with plants, birds and mammals may reflect our greater knowledge of these taxa. (After Smith *et al.*, 1993.)

Taxons	Number of threatened species	Approximate total species	Percentage threatened
Animals			
Molluscs	354	10^5	0.4
Crustaceans	126	4.0×10^3	3
Insects	873	1.2×10^6	0.07
Fishes	452	2.4×10^4	2
Amphibians	59	3.0×10^3	2
Reptiles	167	6.0×10^3	3
Birds	1,029	9.5×10^3	11
Mammals	505	4.5×10^3	11
Total	3,565	1.35×10^6	0.3
Plants			
Gymnosperms	242	758	32
Monocotyledons	4,421	5.2×10^4	9
Monocotyledons: palms	925	2,820	33
Dicotyledons	17,474	1.9×10^5	9
Total	22,137	2.4×10^5	9

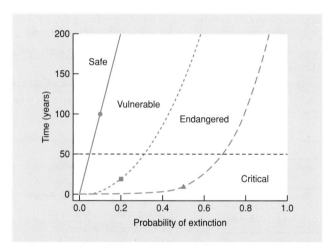

Figure 7.17 Levels of threat as a function of time and probability of extinction. (After Akçakaya, 1992.)

are rare (Fiedler, 1987). The rare species have larger bulbs but produce fewer fruits per plant and have a smaller probability of survival to reproductive age than the common species. The rare species can all be categorized as climax species that are restricted to unusual soil types, whilst the common one is a colonizer of disturbed habitats. Rare taxa may generally have a tendency towards asexual reproduction, lower overall reproductive effort and poorer dispersal abilities (Kunin & Gaston, 1993). In the absence of human interference there is no reason to expect that the rarer types would be substantially more at risk of extinction.

However, while some species are born rare, others have rarity thrust upon them. The actions of humans have undoubtedly reduced the abundance and range of many species (including naturally rare species). A review of the factors responsible for recorded vertebrate extinctions shows that habitat loss, overexploitation and invasions of exotic species are all of considerable significance, although habitat loss was less prominent in the case of reptiles and overexploitation less important in the case of fishes (Table 7.5). As far as threatened extinctions are concerned, habitat loss is most commonly the major hazard, whilst risk of overexploitation remains very high, especially for mammals and reptiles. The probability of extinction may be enhanced in small populations for two different reasons related to genetics (Section 7.5.3) and population dynamics (Section 7.5.4). We deal with them in turn.

other species have rarity thrust upon them

7.5.3 Genetics of small populations: significance for species conservation

Theory tells conservation biologists to beware genetic problems in small populations that may arise through loss of genetic variation. Genetic variation is determined primarily by the joint action of natural selection and genetic drift (where the frequency of genes in a population is determined by chance rather than evolutionary advantage). The relative importance of genetic drift is higher in small, isolated populations, which as a consequence are expected

possible genetic problems in small populations

to lose genetic variation. The rate at which this happens depends on the effective population size (N_e). This is the size of the 'genetically idealized' population to which the actual population (N) is equivalent in genetic terms. As a first approximation, N_e is equal to or less than the number of breeding individuals. N_e is usually less, often much less, than N, for a number of reasons (detailed formulae can be found in Lande & Barrowclough, 1987):

1 If the sex ratio is not 1 : 1; for example with 100 breeding males and 400 breeding females $N = 500$, but $N_e = 320$.
2 If the distribution of progeny from individual to individual is not random; for instance if 500 individuals each produce one individual for the next generation on average $N = 500$, but if the variance in progeny production is 5 (with random variation this would be 1), then $N_e = 100$.
3 If population size varies from generation to generation, then N_e is disproportionately influenced by the smaller sizes; for example for the sequence 500, 100, 200, 900 and 800, mean $N = 500$, but $N_e = 258$.

loss of evolutionary potential

The preservation of genetic diversity is important because of the long-term evolutionary potential it provides. Rare forms of a gene (alleles), or combinations of alleles, may confer no immediate advantage but could turn out to be well suited to changed environmental conditions in the future. Small populations that have lost rare alleles, through genetic drift, have less potential to adapt.

risk of inbreeding depression

A more immediate potential problem is inbreeding depression. When populations are small there is a tendency for individuals breeding with one another to be related. Inbreeding reduces the heterozygosity of the offspring far below that of the population as a whole. More important, though, is that all populations carry recessive alleles which are deleterious, or even lethal, when homozygous. Individuals that are forced to breed with close relatives are more likely to produce offspring where the harmful alleles are derived from both parents so the deleterious effect is expressed. There are many examples of inbreeding depression – breeders have long been aware of reductions in fertility, survivorship, growth rates and resistance to disease – although high levels of inbreeding may be normal and nondeleterious in some animal species (Wallis, 1994) and many plants.

magic genetic numbers?

How many individuals are needed to maintain genetic variability? Franklin (1980) suggested that an effective population size of about 50 would be unlikely to suffer from inbreeding depression, whilst 500–1000 might be needed to maintain longer term evolutionary potential (Franklin & Frankham, 1998). Such rules of thumb should be applied cautiously and, bearing in mind the relationship between N_e and N, the minimum population size N should probably be set an order of magnitude higher than N_e (5000–12,500 individuals) (Franklin & Frankham, 1998).

It is interesting that no example of extinction due to genetic problems is found in Table 7.5. Perhaps inbreeding depression has

Group	Percentage due to each cause*					
	Habitat loss	Overexploitation†	Species introduction	Predators	Other	Unknown
Extinctions						
Mammals	19	23	20	1	1	36
Birds	20	11	22	0	2	37
Reptiles	5	32	42	0	0	21
Fishes	35	4	30	0	4	48
Threatened extinctions						
Mammals	68	54	6	8	12	–
Birds	58	30	28	1	1	–
Reptiles	53	63	17	3	6	–
Amphibians	77	29	14	–	3	–
Fishes	78	12	28	–	2	–

Table 7.5 Review of the factors responsible for recorded extinctions of vertebrates and an assessment of risks currently facing species categorized globally as endangered, vulnerable or rare by the International Union for the Conservation of Nature (IUCN). (After Reid & Miller, 1989.)

* The values indicated represent the percentage of species that are influenced by the given factor. Some species may be influenced by more than one factor, thus, some rows may exceed 100%.
† Overexploitation includes commercial, sport, and subsistence hunting and live animal capture for any purpose.

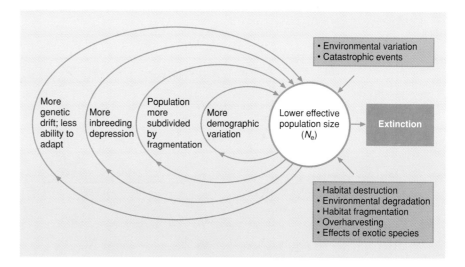

Figure 7.18 Extinction vortices may progressively lower population sizes leading inexorably to extinction. (After Primack, 1993.)

occurred, although undetected, as part of the 'death rattle' of some dying populations (Caughley, 1994). Thus, a population may have been reduced to a very small size by one or more of the processes described above and this may have led to an increased frequency of matings amongst relatives and the expression of deleterious recessive alleles in offspring, leading to reduced survivorship and fecundity, causing the population to become smaller still – the so-called extinction vortex (Figure 7.18).

genetic effects and the persistence of a rare plant

Evidence of a role of genetic effects in population persistence was reported in a study of 23 local populations of the rare plant *Gentianella germanica* in grasslands in the Jura mountains (Swiss–German border). Fischer and Matthies (1998) found a negative correlation between reproductive performance and population size (Figure 7.19a–c). Furthermore, population size decreased between 1993 and 1995 in most of the studied populations, but population size decreased more rapidly in the smaller populations (Figure 7.19d). These results are consistent with the hypothesis that genetic effects resulted in a reduction in fitness in the small populations. However, they may equally have been caused by differences in local habitat conditions (small populations may be small because they have low fecundity resulting from low-quality habitat) or because of the disruption of plant–pollinator interactions (small populations may have low fecundity because of low frequencies of visitation by pollinators). To determine whether genetic differences were, indeed, responsible, seeds from each population were grown under standard conditions in a common garden experiment. After 17 months, there were significantly more flowering plants and more flowers (per planted seed) from seeds from large populations than those from small populations. We can conclude that genetic effects are of importance for population persistence in this rare species and need to be considered when developing a conservation management strategy.

7.5.4   Uncertainty and the risk of extinction: the population dynamics of small populations

Much of conservation biology is a crisis discipline. Managers are inevitably confronted with too many problems and too few resources. Should they focus attention on the various forces that bring species to extinction and attempt to persuade governments to act to reduce their prevalence; or should they restrict activities to identifying areas of high species richness where reserves can be set up and protected (see Section 22.4); or should they identify species at most risk of extinction and work out ways of keeping them in existence? Ideally, we should do all these things. However, the greatest pressure is often in the area of species preservation. For example, the remaining populations of pandas in China or of yellow-eyed penguins (*Megadyptes antipodes*) in New Zealand has become so small that if nothing is done the species may become extinct within a few years or decades. Responding to the crisis requires that we devote scarce resources to identifying some special solutions; more general approaches may have to be put on the back-burner.

The dynamics of small populations are governed by a high level of uncertainty, whereas large populations can be described as being governed by the law of averages (Caughley, 1994). Three kinds of uncertainty or variation can be identified that are of particular importance to the fate of small populations.

three kinds of uncertainty for small populations . . .

1   Demographic uncertainty: random variations in the number of individuals that are born male or female, or in the number that happen to die or reproduce in a given year or in the quality (genotypic/phenotypic) of the individuals in terms of survival/reproductive capacities can matter very much to the fate of small populations. Suppose a breeding pair produces a

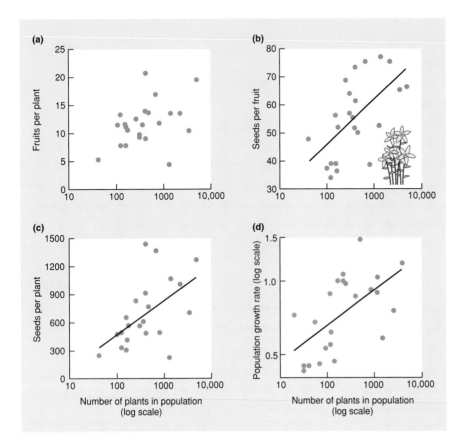

Figure 7.19 Relationships for 23 populations of *Gentianella germanica* between population size and (a) mean number of fruits per plant, (b) mean number of seeds per fruit and (c) mean number of seeds per plant. (d) The relationship between the population growth rate from 1993 to 1995 (ratio of population sizes) and population size (in 1994). (From Fischer & Matthies, 1998.)

clutch consisting entirely of females – such an event would go unnoticed in a large population but it would be the last straw for a species down to its last pair.

2 Environmental uncertainty: unpredictable changes in environmental factors, whether 'disasters' (such as floods, storms or droughts of a magnitude that occur very rarely – see Chapter 2) or more minor (year to year variation in average temperature or rainfall), can also seal the fate of a small population. Even where the average rainfall of an area is known accurately, because of records going back centuries, we cannot predict whether next year will be average or extreme, nor whether we are in for a number of years of particularly dry conditions. A small population is more likely than a large one to be reduced by adverse conditions to zero (extinction), or to numbers so low that recovery is impossible (quasi-extinction).

3 Spatial uncertainty: many species consist of an assemblage of subpopulations that occur in more or less discrete patches of habitat (habitat fragments). Since the subpopulations are likely to differ in terms of demographic uncertainty, and the patches they occupy in terms of environmental uncertainty, the patch dynamics of extinction and local recolonization can be expected to have a large influence on the chance of extinction of the metapopulation (see Section 6.9).

To illustrate some of these ideas, take the demise in North America of the heath hen (*Tympanychus cupido cupido*) (Simberloff, 1998). This bird was once extremely common from Maine to Virginia. Being tasty and easy to shoot (and also susceptible to introduced cats and affected by conversion of its grassland habitat to farmland), by 1830 it had disappeared from the mainland and was only found on the island of Martha's Vineyard. In 1908 a reserve was established for the remaining 50 birds and by 1915 the population had increased to several thousand. However, 1916 was a bad year. Fire (a disaster) eliminated much of the breeding ground, there was a particularly hard winter coupled with an influx of goshawks (*Accipiter gentilis*) (environmental uncertainty), and finally poultry disease arrived on the scene (another disaster). At this point, the remnant population was likely to have become subject to demographic uncertainty; for example, of the 13 birds remaining in 1928 only two were females. A single bird was left in 1930 and the species became extinct in 1932.

The heath hen provides one example of a relatively recent *global* extinction. At a different scale, *local* extinctions of small populations in insular habitat patches are common events for diverse taxa, often being in the range of 10–20% per year (Figure 7.20). Such extinctions are also observed on true islands. The detailed

... illustrated by the heath hen

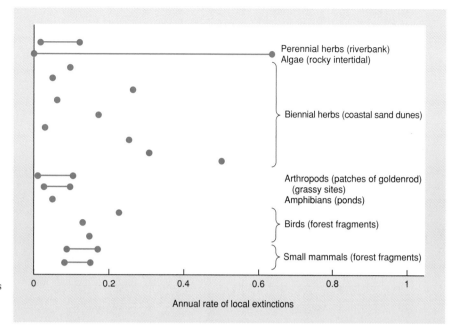

Figure 7.20 Fractions of local populations in habitat patches becoming extinct each year. (After Fahrig & Merriam, 1994.)

records from 1954 to 1969 of birds breeding on Bardsey Island, a small island (1.8 km²) off the west coast of Great Britain, revealed that 16 species bred every year, two of the original species disappeared, 15 flickered in and out, whilst four were initially absent but became regular breeders (Diamond, 1984). We can build a picture of frequent local extinctions, which in some cases are countered by recolonization from the mainland or other islands. Examples such as these provide a rich source of information about the factors affecting the fate of small populations in general. The understanding gained is entirely applicable to species in danger of global extinction, since a global extinction is nothing more nor less than the final local extinction. Thus, of the high-risk factors associated with local extinctions, habitat or island area is probably the most pervasive (Figure 7.21). No doubt the main reason for the vulnerability of populations in small areas is the fact that the populations themselves are small. A local extinction of an endemic species on a remote island is precisely equivalent to a global extinction, since recolonization is impossible. This is a principal reason for the high rates of global extinction on islands (see Figure 7.16).

7.5.5 Population viability analysis: the application of theory to management

trying to determine the minimum viable population . . .

The focus of population viability analysis (PVA) differs from many of the population models developed by ecologists (such as those discussed in Chapters 5, 10 and 14) because an aim of PVA is to predict extreme events (such as extinction) rather that central tendencies

such as mean population sizes. Given the environmental circumstances and life history characteristics of a particular rare species, what is the chance it will go extinct in a specified period? Alternatively, how big must its population be to reduce the chance of extinction to an acceptable level? These are frequently the crunch questions in conservation management. The ideal classical approach of experimentation, which might involve setting up and monitoring for several years a number of populations of various sizes, is unavailable to those concerned with species at risk, because the situation is usually too urgent and there are inevitably too few individuals to work with. How then are we to decide what constitutes the minimum viable population (MVP)? Three approaches will be discussed in turn: (i) a search for patterns in evidence already gathered in long-term studies (Section 7.5.5.1); (ii) subjective assessment based on expert knowledge (Section 7.5.5.2); and (iii) the development of population models, both general (Section 7.5.5.3) and specific to particular species of interest (Section 7.5.5.4). All the approaches have their limitations, which we will explore by looking at particular examples. But first it should be noted that the field of PVA has largely moved away from the simple estimation of extinction probabilities and times to extinction, to focus on the comparison of likely outcomes (in terms of extinction probabilities) of alternative management strategies.

7.5.5.1 Clues from long-term studies of biogeographic patterns

Data sets such as the one displayed in Figure 7.22 are unusual because they depend on a long-term commitment

. . . from biogeographic data . . .

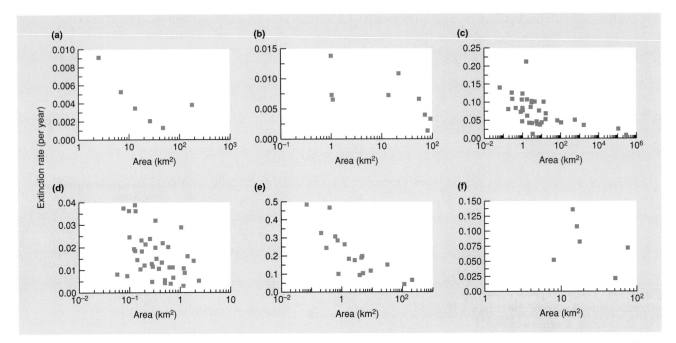

Figure 7.21 Percentage extinction rates as a function of habitat area for: (a) zooplankton in lakes in the northeastern USA lakes, (b) birds of the Californian Channel Islands, (c) birds on northern European islands, (d) vascular plants in southern Sweden, (e) birds on Finnish islands and (f) birds on the islands in Gatun Lake, Panama. (Data assembled by Pimm, 1991.)

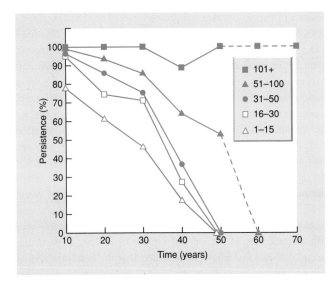

Figure 7.22 The percentage of populations of bighorn sheep in North America that persists over a 70-year period reduces with initial population size. (After Berger, 1990.)

to monitoring a number of populations – in this case, bighorn sheep in desert areas of North America. If we set an arbitrary definition of the necessary MVP as one that will give at least a 95% probability of persistence for 100 years, we can explore the data on the fate of bighorn populations to provide an approximate

answer. Populations of fewer than 50 individuals all went extinct within 50 years whilst only 50% of populations of 51–100 sheep lasted for 50 years. Evidently, for our MVP we require more than 100 individuals; in the study, such populations demonstrated close to 100% success over the maximum period studied of 70 years.

A similar analysis of long-term records of birds on the Californian Channel Islands indicates an MVP of between 100 and 1000 pairs of birds (needed to provide a probability of persistence of between 90 and 99% for the 80 years of the study) (Table 7.6).

Studies such as these are rare and valuable. The long-term data are available because of the extraordinary interest people have in hunting (bighorn sheep) and ornithology (Californian birds). Their value for conservation, however, is limited because they deal with species that

. . . is a risky approach

Table 7.6 The relationship for a variety of bird species on the Californian Channel Islands between initial population size and probability of populations persisting. (After Thomas, 1990.)

Population size (pairs)	Time period (years)	Percentage persisting
1–10	80	61
10–100	80	90
100–1000	80	99
1000+	80	100

are generally not at risk. It is at our peril that we use them to produce recommendations for management of endangered species. There will be a temptation to report to a manager 'if you have a population of more than 100 pairs of your bird species you are above the minimum viable threshold'. Indeed, such a statement would not be without value. But, it will only be a safe recommendation if the species of concern and the ones in the study are sufficiently similar in their vital statistics, and if the environmental regimes are similar, something that it would rarely be safe to assume.

7.5.5.2   Subjective expert assessment

in the minds of experts: decision analysis

Information that may be relevant to a conservation crisis exists not only in the scientific literature but also in the minds of experts. By bringing experts together in conservation workshops, well-informed decisions can be reached (we have already considered an example of this approach in the selection of overwintering reserves for monarch butterflies – Section 7.2.3). To illustrate the strengths and weaknesses of the approach when estimating extinction probability, we take as our example the results of a workshop concerning the Sumatran rhinoceros (*Dicerorhinus sumatrensis*).

the case of the Sumatran rhinoceros

The species persists only in small, isolated subpopulations in an increasingly fragmented habitat in Sabah (East Malaysia), Indonesia and West Malaysia, and perhaps also in Thailand and Burma. Unprotected habitat is threatened by timber harvest, human resettlement and hydroelectric development. There are only a few designated reserves, which are themselves subject to exploitation, and only two individuals were held in captivity at the time of the workshop.

The vulnerability of the Sumatran rhinoceros, the way this vulnerability varies with different management options and the most appropriate management option given various criteria-were assessed by a technique known as *decision analysis*. A decision tree is shown in Figure 7.23, based on the estimated probabilities of the species becoming extinct within a 30-year period (equivalent to approximately two rhinoceros generations). The tree was constructed in the following way. The two squares are decision points: the first distinguishes between intervention on the rhinoceros' behalf and nonintervention (status quo); the second distinguishes the various management options. For each option, the line branches at a small circle. The branches represent alternative scenarios that might occur, and the numbers on each branch indicate the probabilities estimated for the alternative scenarios. Thus, for the status quo option, there was estimated to be a

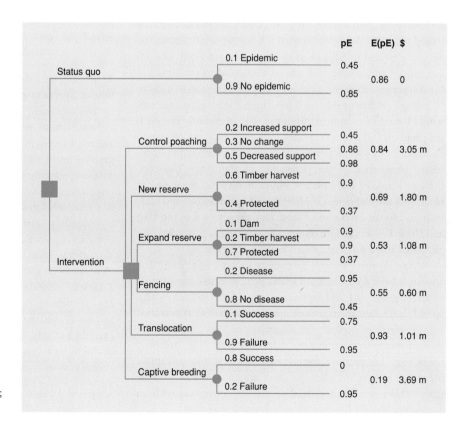

Figure 7.23 Decision tree for management of the Sumatran rhinoceros. ■, decision points; ●, random events. Probabilities of random events are estimated for a 30-year period; pE, probability of species extinction within 30 years; E(pE), expected value of pE for each alternative. Costs are present values of 30-year costs discounted at 4% per year; m, million. (After Maguire *et al.*, 1987.)

probability of 0.1 that a disease epidemic will occur in the next 30 years, and hence, a probability of 0.9 that no epidemic will occur.

If there is an epidemic, the probability of extinction (pE) is estimated to be 0.95 (i.e. 95% probability of extinction in 30 years), whereas with no epidemic the pE is 0.85. The overall estimate of species extinction for an option, E(pE), is then given by:

E(pE) = probability of first option × pE for first option +
        probability of second option × pE for second option,

which, in the case of the status quo option, is:

E(pE) = (0.1 × 0.95) + (0.9 × 0.85) = 0.86.

The values of pE and E(pE) for the intervention options are calculated in a similar way. The final column in Figure 7.23 then lists the estimated costs of the various options.

evaluating management options
We will consider here two of the interventionist management options in a little more detail. The first is to fence an area in an existing or new reserve, and to manage the resulting high density of rhinoceroses with supplemental feeding and veterinary care. Disease here is a major risk: the probability of an epidemic was estimated to be higher than in the status quo option (0.2 as opposed to 0.1) because the density would be higher. Moreover, the pE if there was an epidemic was considered to be higher (0.95), because animals would be transferred from isolated subpopulations to the fenced area. On the other hand, if fencing were successful, the pE was expected to fall to 0.45, giving an overall E(pE) of 0.55. The fenced area would cost around US$60,000 to establish and $18,000 per year to maintain, giving a 30-year total of $0.60 million.

For the establishment of a captive breeding program, animals would have to be captured from the wild, increasing the pE if the program failed to an expected 0.95. However, the pE would clearly drop to 0 if the program succeeded (in terms of the continued persistence of the population in captivity). The cost, though, would be high, since it would involve the development of facilities and techniques in Malaysia and Indonesia (around $2.06 million) and the extension of those that already exist in the USA and Great Britain ($1.63 million). The probability of success was estimated to be 0.8. The overall E(pE) is therefore 0.19.

Where do these various probability values come from? The answer is from a combination of data, the educated use of data, educated guesswork and experience with related species. Which would be the best management option? The answer depends on what criteria are used in the judgement of 'best'. Suppose we wanted simply to minimize the chances of extinction, irrespective of cost. The proposal for best option would then be captive breeding. In practice, though, costs are most unlikely to be ignored. We would then need to identify an option with an acceptably low E(pE) but with an acceptable cost.

strengths of subjective expert assessment...
The subjective expert assessment approach has much to commend it. It makes use of available data, knowledge and experience in a situation when a decision is needed and time for further research is unavailable. Moreover, it explores the various options in a systematic manner and does not duck the regrettable but inevitable truth that unlimited resources will not be available.

... and weaknesses
However, it also runs a risk. In the absence of all necessary data, the recommended best option may simply be wrong. With the benefit of hindsight (and in all probability some rhinoceros experts who were not part of the workshop would have suggested this alternative outcome), we can now report that about $2.5 million have been spent catching Sumatran rhinoceroses; three died during capture, six died postcapture, and of the 21 rhinoceroses now in captivity only one has given birth and she was pregnant when captured (data of N. Leader-Williams reported in Caughley, 1994). Leader-Williams suggests that $2.5 million could have been used effectively to protect 700 km² of prime rhinoceros habitat for nearly two decades. This could in theory hold a population of 70 Sumatran rhinoceroses which, with a rate of increase of 0.06 per individual per year (shown by other rhinoceros species given adequate protection), might give birth to 90 calves during that period.

7.5.5.3 A general mathematical model of population persistence time

a general modeling approach...
At its simplest, the likely persistence time of a population, T, can be expected to be influenced by its size, N, its intrinsic rate of increase, r, and the variance in r resulting from variation in environmental conditions through time, V. Demographic uncertainty is only expected to be influential in very small populations; persistence time increases from a low level with population size when numbers are tiny, but approaches infinity at a relatively small population (dashed curve in Figure 7.24).

Various researchers have manipulated the mathematics of population growth, allowing for uncertainty in the expression of the intrinsic rate of increase, to provide an explicit estimate, T, of mean time to extinction as a function of carrying capacity, K (briefly reviewed by Caughley, 1994). Making a number of approximations (e.g. that demographic uncertainty is inconsequential, and that r is constant except if the population is at the carrying capacity when r is zero), Lande (1993) has produced one of the more accessible equations:

$$T = \frac{2}{Vc}\left(\frac{K^c - 1}{c} - \ln K\right)$$

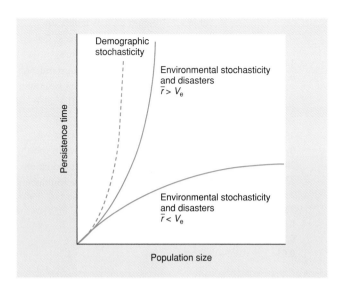

Figure 7.24 Relationships between population persistence time and population size, both on arbitrary scales, when the population is subject to demographic uncertainty or to environmental uncertainty/disasters. (After Lande, 1993.)

where:

$$c = 2r/V - 1,$$

r is the intrinsic rate of increase and V is the variance in r resulting from variation in environmental conditions through time.

This equation is the basis for the solid curves in Figure 7.24, which indicate that mean time to extinction is higher for larger maximum population sizes (K), for greater intrinsic rates of population growth and when environmental influences on the expression of r are smaller. In contrast to earlier claims that random disasters pose a greater threat than smaller environmental variation, it turns out that what really matters is the relationship between the mean and variance of r (Lande, 1993; Caughley, 1994). The relationship between persistence time and population size curves sharply upwards (i.e. is influential only for small or intermediate population sizes) if the mean rate of increase is greater than the variance, whereas if the variance is greater than the mean the relationship is convex – so that, even at large population sizes, environmental uncertainty still has an influence on likely persistence time. This all makes intuitive good sense but can it be put to practical use?

... put to the test   In their study of the Tana River crested mangabey (*Cercocebus galeritus galeritus*) in Kenya, Kinnaird and O'Brien (1991) used a similar equation to estimate the population size (K) needed to provide a 95% probability of persistence for 100 years. This endangered primate is confined to the floodplain forest of a single river where it declined in numbers from 1200

to 700 between 1973 and 1988 despite the creation of a reserve. Its naturally patchy habitat has become progressively more fragmented through agricultural expansion. The model parameters, estimated on the basis of some real population data, were taken to be $r = 0.11$ and $V = 0.20$. The latter was particularly uncertain because only a few years' data were available. Substituting in the model yielded a MVP of 8000. Using the standard rule of thumb described earlier, that to avoid genetic problems an effective population size of 500 individuals is needed, an actual population of about 5000 individuals was indicated. Given the available habitat it was concluded that the mangabeys could not attain a population size of 5000–8000. Moreover, Kinnaird and O'Brien think it unlikely that this naturally rare and restricted species ever has. Either the data were deficient (e.g. environmental variation in r may be smaller than estimated if they are able to undergo dietary shifts in response to habitat change) or the model is too general to be much use in specific cases. The latter is likely to be true. However, this is not to deny the value of ecologists continuing to search for generalizations about processes underlying the problems facing conservation managers.

7.5.5.4 Simulation models: population viability analysis (PVA)

Simulation models provide an alternative, more specific way of gauging viability. Usually, these encapsulate survivorships and reproductive rates in age-structured populations. Random variations in these elements or in K can be employed to represent the impact of environmental variation, including that of disasters of specified frequency and intensity. Density dependence can be introduced where required, as can population harvesting or supplementation. In the more sophisticated models, every individual is treated separately in terms of the probability, with its imposed uncertainty, that it will survive or produce a certain number of offspring in the current time period. The program is run many times, each giving a different population trajectory because of the random elements involved. The outputs, for each set of model parameters used, include estimates of population size each year and the probability of extinction during the modeled period (the proportion of simulated populations that go extinct).

Koalas (*Phascolarctos cinereus*) are regarded as near-threatened nationally, with populations in different parts of Australia varying from secure to vulnerable or extinct. The primary aim of the national management strategy is to retain viable populations throughout their natural range (ANZECC, 1998). Penn *et al.* (2000) used a widely available demographic forecasting tool, known as VORTEX (Lacey, 1993), to model two populations in Queensland, one thought to be declining (at Oakey), the other secure (at Springsure). Koala breeding commences at 2 years in

Variable	Oakey	Springsure
Maximum age	12	12
Sex ratio (proportion male)	0.575	0.533
Litter size of 0 (%)	57.00 (± 17.85)	31.00 (± 15.61)
Litter size of 1 (%)	43.00 (± 17.85)	69.00 (± 15.61)
Female mortality at age 0	32.50 (± 3.25)	30.00 (± 3.00)
Female mortality at age 1	17.27 (± 1.73)	15.94 (± 1.59)
Adult female mortality	9.17 (± 0.92)	8.47 (± 0.85)
Male mortality at age 0	20.00 (± 2.00)	20.00 (± 2.00)
Male mortality at age 1	22.96 (± 2.30)	22.96 (± 2.30)
Male mortality at age 2	22.96 (± 2.30)	22.96 (± 2.30)
Adult male mortality	26.36 (± 2.64)	26.36 (± 2.64)
Probability of catastrophe	0.05	0.05
Multiplier, for reproduction	0.55	0.55
Multiplier for survival	0.63	0.63
% males in breeding pool	50	50
Initial population size	46	20
Carrying capacity, K	70 (± 7)	60 (± 6)

Table 7.7 Values used as inputs for simulations of koala populations at Oakey (declining) and Springsure (secure), Australia. Values in brackets are standard deviations due to environmental variation; the model procedure involves the selection of values at random from the range. Catastrophes are assumed to occur with a certain probability; in years when the model selects a catastrophe, reproduction and survival are reduced by the multipliers shown. (After Penn *et al.*, 2000.)

females and 3 years in males. The other demographic values used in the two PVAs were derived from extensive knowledge of the two populations and are shown in Table 7.7. Note how the Oakey population had somewhat higher female mortality and fewer females producing young each year. The Oakey population was modeled from 1971 and the Springsure population from 1976 (when the first estimates of density were available) and the model trajectories were indeed declining and stable, respectively. Over the modeled period (Figure 7.25), the probability of extinction of the Oakey population was 0.380 (i.e. 380 out of 1000 iterations went extinct) while that for Springsure was 0.063. Managers concerned with critically endangered species do not usually have the luxury of monitoring populations to check the accuracy of their predictions. In contrast, Penn *et al.* (2000) were able to compare the predictions of their PVAs with real population trajectories, because the koala populations have been continuously monitored since the 1970s (Figure 7.25). The predicted trajectories were close to the actual population trends, particularly for the Oakey population, and this gives added confidence to the modeling approach.

The predictive accuracy of VORTEX and other simulation modeling tools was also found to be high for 21 long-term animal data sets by Brook *et al.* (2000). How can such modeling be put to management use? Local governments in New South Wales are obliged both to prepare comprehensive koala management plans and to ensure that developers survey for potential koala habitat when a building application affects an area greater than 1 ha. Penn *et al.* (2000) argue that PVA modeling can be used to determine whether any effort made to protect habitat is likely to be rewarded by a viable population.

Overall numbers of African elephants (*Loxodonta africana*) are in decline and few populations are expected to survive over the next few decades outside high-security areas, mainly because

the case of the African elephant – necessary size of reserves?

of habitat loss and poaching for ivory. For their simulation models Armbruster and Lande (1992) chose to represent the elephant population in twelve 5-year age classes through discrete 5-year time steps. Values for age-specific survivorship and density-dependent reproductive rates were derived from a thorough data set from Tsavo National Park in Kenya, because its semiarid nature has the general characteristics of land planned for game reserves now, and in the future. Environmental stochasticity, perhaps most appropriately viewed as disasters, was modeled as drought events affecting sex- and age-specific survivorship – again realistic data from Tsavo were used, based on a mild drought cycle of approximately 10 years superimposed on a more severe 50-year drought and an even more severe 250-year drought cycle. Table 7.8 gives the survivorship of females under 'normal' conditions and the three drought conditions. The relationship between habitat area and the probability of extinction was examined in 1000-year simulations with and without a culling regime. At least 1000 replicates were performed for each model with many more (up to 30,000) to attain acceptable statistical confidence in the smaller extinction probabilities associated with larger habitat areas. Extinctions were taken to have occurred when no individuals remained or when only a single sex was represented.

The results imply that an area of 1300 km² (500 sq. miles) is required to yield a 99% probability of persistence for 1000 years (Figure 7.26). This conservative outcome was chosen because of

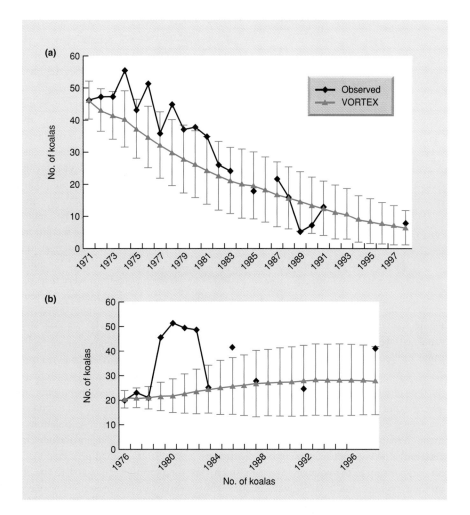

Figure 7.25   Observed koala population trends (♦) compared with trajectories (▲ ± 1 SD) predicted by 1000 iterations of VORTEX at (a) Oakey and (b) Springsure, USA. (After Penn *et al.*, 2000.)

Table 7.8   Survivorship for 12 elephant age classes in normal years (occur in 47% of 5-year periods), and in years with 10-year droughts (41% of 5-year periods), 50-year and 250-year droughts (10 and 2% of 5-year periods, respectively). (After Armbruster & Lande, 1992.)

| | *Female survivorship* | | | |
Age class (years)	Normal years	10-year droughts	50-year droughts	250-year droughts
0–5	0.500	0.477	0.250	0.01
5–10	0.887	0.877	0.639	0.15
10–15	0.884	0.884	0.789	0.20
15–20	0.898	0.898	0.819	0.20
20–25	0.905	0.905	0.728	0.20
25–30	0.883	0.883	0.464	0.10
30–35	0.881	0.881	0.475	0.10
35–40	0.875	0.875	0.138	0.05
40–45	0.857	0.857	0.405	0.10
45–50	0.625	0.625	0.086	0.01
50–55	0.400	0.400	0.016	0.01
55–60	0.000	0.000	0.000	0.00

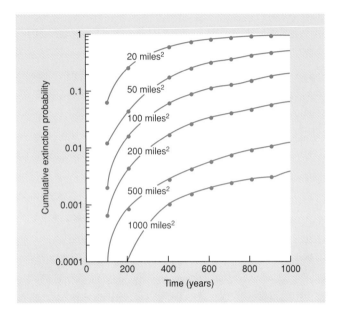

Figure 7.26 Cumulative probability of elephant population extinction over 1000 years for six habitat areas (without culling). (After Armbruster & Lande, 1992.)

the difficulty of reestablishing viable populations in isolated areas where extinctions have occurred and because of the elephant's long generation time (about 31 years). In fact, the authors recommend to managers an even more conservative minimum area of 2600 km² (1000 sq. miles) for reserves. The data are least reliable for survivorship in the youngest age class and for the long-term drought regime, and a 'sensitivity analysis' shows extinction probability to be particularly sensitive to slight variations in these parameters. Of the parks and game reserves in Central and Southern Africa, only 35% are larger than 2600 km².

the case of the royal catchfly: management of an endangered plant

Many aspects of the life history of plants present particular challenges for simulation modeling, including seed dormancy, highly periodic recruitment and clonal growth (Menges, 2000). However, as with endangered animals, different management scenarios can be usefully simulated in PVAs. The royal catchfly, *Silene regia*, is a long-lived iteroparous prairie perennial whose range has shrunk dramatically. Menges and Dolan (1998) collected demographic data for up to 7 years from 16 midwestern USA populations (adult population sizes of 45–1302) subject to different management regimes. The species has high survivorship, slow growth, frequent flowering and nondormant seeds, but very episodic recruitment (most populations in most years fail to produce seedlings). Matrices, such as that illustrated in Table 7.9, were produced for individual populations and years. Multiple simulations were then run for every matrix to determine the finite rate of increase (λ; see Section 4.7) and the probability of extinction in 1000 years. Figure 7.27 shows the median finite rate of increase for the 16 populations, grouped into cases where particular management regimes were in place, for years when recruitment of seedlings occurred and for years when it did not. All sites where λ was greater than 1.35 when recruitment took place were managed by burning and some by mowing as well; none of these were predicted to go extinct during the modeled period. On the other hand, populations with no management, or whose management did not include fire, had lower values for λ and all except two had predicted extinction probabilities (over 1000 years) of from 10 to 100%. The obvious management recommendation is to use prescribed burning to provide opportunities for seedling recruitment. Low establishment rates of seedlings in the field may be due to frugivory by rodents or ants and/or competition for light with established vegetation (Menges & Dolan, 1998) – burnt

Table 7.9 An example of a projection matrix for a particular *Silene regia* population from 1990 to 1991, assuming recruitment. Numbers represent the proportion changing from the stage in the column to the stage in the row (bold values represent plants remaining in the same stage). 'Alive undefined' represents individuals with no size or flowering data, usually as a result of mowing or herbivory. Numbers in the top row are seedlings produced by flowering plants. The finite rate of increase λ for this population is 1.67. The site is managed by prescribed burning. (After Menges & Dolan, 1998.)

	Seedling	*Vegetative*	*Small flowering*	*Medium flowering*	*Large flowering*	*Alive undefined*
Seedling	–	–	5.32	12.74	30.88	–
Vegetative	0.308	**0.111**	0	0	0	0
Small flowering	0	0.566	**0.506**	0.137	0.167	0.367
Medium flowering	0	0.111	0.210	**0.608**	0.167	0.300
Large flowering	0	0	0.012	0.039	**0.667**	0.167
Alive undefined	0	0.222	0.198	0.196	0	**0.133**

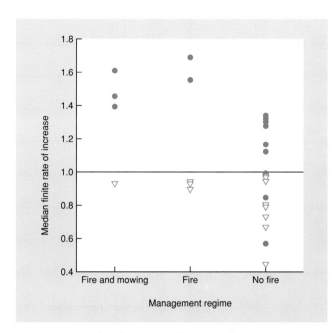

Figure 7.27 Median finite rates of increase of *Silene regia* populations as a function of management regime, for years with seedling recruitment (●) and without (▽). Unburned management regimes include just mowing, herbicide use or no management.

areas probably reduce one or both of these negative effects. While management regime was by far the best predictor of persistence, it is of interest that populations with higher genetic diversity also had higher median values for λ.

value and limits of population viability analysis

In an ideal world, a PVA would enable us to produce a specific and reliable recommendation for an endangered species of the population size, or reserve area, that would permit persistence for a given period with a given level of probability. But this is rarely achievable because the biological data are hardly ever good enough. The modelers know this and it is important that conservation managers also appreciate it. Within the inevitable constraints of lack of knowledge and lack of time and opportunity to gather data, the model building exercise is no more nor less than a rationalization of the problem and quantification of ideas. Moreover, even though such models produce quantitative outputs, common sense tells us to trust the results only in a qualitative fashion. Nevertheless, the examples above show how, on the basis of ecological theory discussed in Chapters 4–6, we can construct models that allow us to make the very best use of available data and may well give us the confidence to make a choice between various possible management options and to identify the relative importance of factors that put a population at risk (Reed *et al.*, 2003). The sorts of management interventions that may then be recommended include translocation of individuals

to augment target populations, creating larger reserves, raising the carrying capacity by artificial feeding, restricting dispersal by fencing, fostering of young (or cross-fostering of young by related species), reducing mortality by controlling predators or poachers, or through vaccination and, of course, habitat preservation.

7.5.6 Conservation of metapopulations

We noted in Section 7.5.4 that local extinctions are common events. It follows that conservation biologists need to be aware of the critical importance of recolonization of habitat fragments

adding metapopulation structure

if fragmented populations are to persist. Thus, we need to pay particular attention to the relationships amongst landscape elements, including dispersal corridors, in relation to the dispersal characteristics of focal species (Fahrig & Merriam, 1994).

Westphal *et al.* (2003) built a stochastic patch occupancy model for the critically endangered southern emu-wren (*Stipiturus malachurus intermedius*) (based on realistic extinction and recolonization matrices) and then used a

the case of the southern emu-wren: comparing the cost of different strategies

technique known as stochastic dynamic modeling to find optimal solutions for its future management. The metapopulation in the Mount Lofty ranges of South Australia occurs in six remaining patches of dense swamp habitat (Figure 7.28). Emu-wrens are poor flyers and interpatch corridors of appropriate vegetation are likely to be important for metapopulation persistence. The management strategies that Westphal *et al.* (2003) evaluated were the enlargement of existing patches, linking patches via corridors and creating a new patch (Figure 7.28). The 'cost' of each strategy was standardized to be equivalent to 0.9 ha of revegetated area. The optimization modeling checked among individual management actions, and also compared a variety of management scenario trajectories (e.g. first build a corridor from the largest patch to its neighbor, then, in the next time period, enlarge the largest patch; then create a new patch, etc.), to find those that reduced the 30-year extinction risk to the greatest extent.

Optimal metapopulation management decisions depended on the current state of the population. For example, if only the two smallest patches were occupied, the optimal single action would be to enlarge one of them (patch 2; strategy E2). However, when only a more extinction-resistant large patch is occupied, connecting it to neighboring patches is optimal (strategy C5). The best of these fixed strategies reduced the 30-year extinction probabilities by up to 30%. On the other hand, the optimal state-dependent strategies, where chains of different actions were taken over successive time periods, reduced extinction probabilities by 50–80% compared to no-management models. The optimal scenario trajectories varied according to the starting state of the metapopulation and are shown in Figure 7.29.

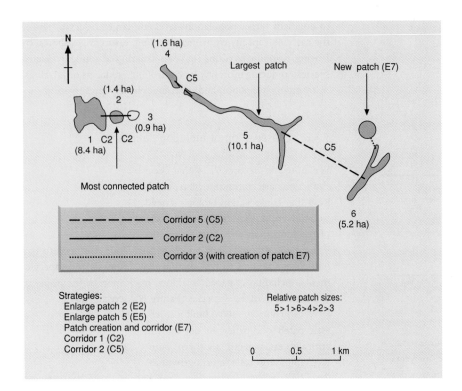

Figure 7.28 The southern emu-wren metapopulation, showing the size and location of patches and corridors. For further details, see text. (After Westphal et al., 2003.)

These results hold a number of lessons for conservation managers. First, optimal decisions are highly state dependent, relying on knowledge of patch occupancy and a good understanding of extinction and recolonization rates. Second, the sequence of actions is critical and recognition of an optimal sequence can only be underpinned by an approach such as stochastic dynamic modeling (Clark & Mangel, 2000); it will be hard to come up with simple rules of thumb about metapopulation management. Most important of all is the point that funds available for conservation will always be limited and tools such as these should help achieve the optimal use of scarce resources.

set up for key species may be in the wrong places and species currently appropriate for restoration projects may no longer succeed. Moreover, each region of the world is likely to be subject to a new set of invaders, pests and diseases.

Political approaches to the mitigation of climate change focus on international efforts to reduce emissions and to augment ecological sinks (e.g. by increasing the amount of the world's surface that is forest). We deal with these aspects in Chapter 22. Here we focus on predicting the effects of climate change on the spread of diseases and other invasive species (Section 7.6.1) and deciding where to locate nature reserves in a changing world (Section 7.6.2).

7.6   Global climate change and management

climate change models predict a shifting geographic template of abiotic factors . . .

Given predicted increases in carbon dioxide and other greenhouse gases, temperature is expected to increase by between 1.4 and 5.8°C from 1990 levels by the year 2100 (IPCC, 2001). The effects of such increases can be expected to be profound, through a melting of glaciers and ice caps and the consequent raising of sea level, and more generally through large-scale changes to the global climate. Alterations to temperature, and other aspects of climate, provide a shifting physicochemical template upon which species' niches will be superimposed in future. In other words, nature reserves already

7.6.1   Predicting the spread of diseases and other invaders in a changing world

We are only at an early stage of projected trends in global climate change but already there is evidence of responses by the flora and fauna. Thus, shoot production and flowering of a variety of plants is happening earlier, many birds, butterflies and amphibians are breeding earlier, and shifts in species' ranges have been detected both polewards and towards higher altitudes (Walther et al., 2002; Parmesan & Yohe, 2003). We can expect much more dramatic changes in the potential ranges of both native and invasive species in the coming century.

. . . which will be reflected in new patterns of invader risk

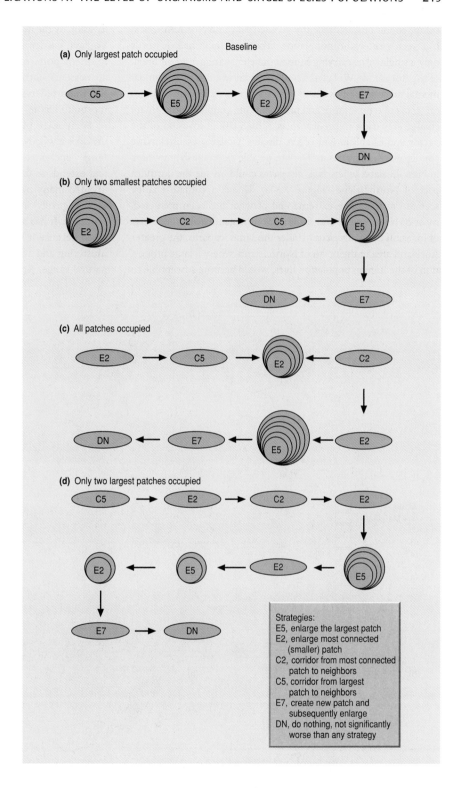

Figure 7.29 The optimal management scenario trajectories for different starting configurations of the southern emu-wren metapopulation. Each circle represents one action. Concentric circles show the repeated execution of a strategy before the next strategy is implemented. Note that each trajectory ends with a do-nothing action when the metapopulation state is such that lack of management action does not produce a probability of extinction that is significantly worse than any active strategy. (After Westphal *et al.*, 2003.)

the case of mosquitoes and dengue fever

Dengue fever is a potentially fatal viral disease currently limited to tropical and subtropical countries where its mosquito vectors occur. No mosquito species currently in New Zealand appears to be capable of carrying the disease. Worldwide, the two most important vectors are *Aedes aegypti* and *A. albopictus*. Both have been intercepted at New Zealand's borders and the latter, which is tolerant of somewhat colder conditions, has recently invaded Italy and North America. If a vector mosquito population becomes established, it needs only a single virus-carrying human traveler to trigger an outbreak of the disease. de Wet *et al.* (2001) used knowledge of the fundamental niches of the two mosquito species in their natural ranges (in terms of temperature and precipitation), coupled with climate change scenarios, to predict areas of high risk of invasion of the vector and establishment of the disease. Under present climatic conditions, *A. aegypti* is unlikely to be able to establish anywhere in New Zealand whereas *A. albopictus* could invade the northern part of North Island (Figure 7.30a). Under a climate change scenario at the more extreme end of what has been predicted, most of North Island and some of South Island would be at risk of invasion by *A. albopictus*. Under the same scenario, the greater Auckland area in the north of North Island, where a large proportion of the human population lives, would become susceptible to

invasion by the more efficient virus vector *A. aegypti* (Figure 7.30b). Vigilant border surveillance is vital, with most emphasis on northern ports of entry, in particular Auckland (with 75% of air passenger arrivals, 74% of bulk shipping cargo and 50% of the imported tyres that provide a prime transport route for mosquito larvae) (Hearnden *et al.*, 1999).

Spiny acacia (*Acacia nilotic* subspecies *indica*) is a woody legume whose native range encompasses parts of Africa and extends as far east as India. It has invaded many parts of the world, including Australia where it was originally introduced for shade, fodder and ornamental purposes. It has spread widely and is now considered a noxious weed because it reduces pasture production and impedes stock mustering and access to water. On the basis of conditions in its natural range, Kriticos *et al.* (2003) first determined the species'

the case of invasive acacias

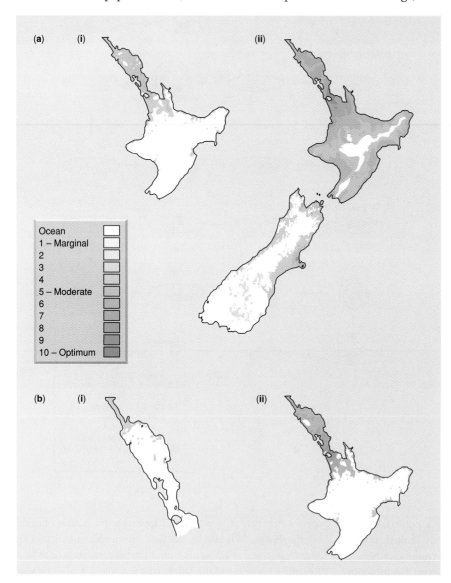

Figure 7.30 Dengue fever risk maps for: (a) *Aedes albopictus* for (i) present climatic conditions and (ii) for a high-range climate change scenario for 2100, and (b) *A. aegypti* for (i) a high-range climate change scenario for (i) 2050 and (ii) 2100. (After de Wet *et al.*, 2001.)

fundamental niche in terms of lower and upper tolerance limits and optima for temperature and moisture, and the thresholds for cold stress, heat stress, dry stress and wet stress (water-logging). They then modeled the invasive potential of spiny acacia under two climate change scenarios. Both assumed a 2°C temperature rise, coupled with either a 10% increase or 10% decrease in rainfall, because there is considerable uncertainty about the effects of global change on precipitation in Australia (Figure 7.31). The actual current distribution of spiny acacia is widespread within the range indicated by the model, but it has not yet spread to all predicted areas. When climate change is taken into account, its eventual invaded range should be much greater, particularly because the plant is expected to become more efficient in its use of water as a result of a fertilization effect of increased atmospheric carbon dioxide. Thus elevated atmospheric concentration can have both indirect effects, via climate change, and direct effects on the performance and distribution of plants (Volk *et al.*, 2000). Further spread of this species should be containable because trees can be physically removed and the spread of seeds (in stock feces) can be prevented as long as animals are not moved indiscriminately. A crucial component in containing the invasion will be raising public awareness of the weed and how to control it (Kriticos *et al.*, 2003).

7.6.2 Managing endangered species

global climate change: will nature reserves be in the right place?

Temperature and moisture also strongly influence the life cycle of butterflies. Beaumont and Hughes (2002) used the approach applied to spiny acacia above to predict the effect of climate change on the distribution of 24 Australian butterfly species. Under even a moderate set of future conditions (temperature increase of 0.8–1.4°C by 2050), the distributions of 13 of the species decreased by more than 20%. Most at risk are those, such as *Hypochrysops halyetus*, that not only have specialized food-plant requirements but also depend on the presence of ants for a mutualistic relationship. The models suggest that *H. halyetus*, which is restricted to coastal heathland in Western Australia, will lose 58–99% of its current climatic range. Moreover, less than 27% of its predicted future distribution occurs in locations that it

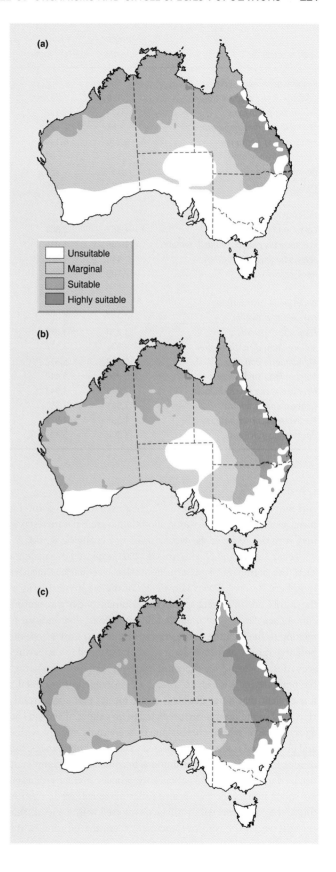

Figure 7.31 (*right*) The predicted distribution of spiny acacia in Australia on the basis of (a) current climate, (b) a scenario with an average 2°C increase and a 10% increase in precipitation, and (c) a 2°C increase and a 10% decrease in precipitation. The predicted distributions in (b) and (c) also assume an increased efficiency of water use by spiny acacia because of a fertilizing effect of increased atmospheric carbon dioxide. (After Kriticos *et al.*, 2003.)

Species category	Current climate	+1.0°C −10% rain	+2.0°C −10% rain	+2.0°C −15% rain
Restricted to the reserve				
Cephalocereus columna-trajani	138	27	0	0
Ferocactus flavovirens	317	532	100	55
Mammillaria huitzilopochtli	68	21	0	0
Mammillaria pectinifera	5,130	1,124	486	69
Pachycereus hollianus	175	87	0	0
Polaskia chende	157	83	76	41
Polaskia chichipe	387	106	10	0
Intermediate distribution				
Coryphantha pycnantha	1,367	2,881	1,088	807
Echinocactus platyacanthus f. grandis	1,285	1,046	230	1,148
Ferocactus haematacanthus	340	1,979	1,220	170
Pachycereus weberi	2,709	3,492	1,468	1,012
Widespread distribution				
Coryphantha pallida	10,237	5,887	3,459	2,920
Ferocactus recurvus	3,220	3,638	1,651	151
Mammillaria dixanthocentron	9,934	7,126	5,177	3,162
Mammillaria polyedra	10,118	5,512	3,473	2,611
Mammillaria sphacelata	3,956	5,440	2,803	2,580
Neobuxbaumia macrocephala	2,846	4,943	3,378	1,964
Neobuxbaumia tetetzo	2,964	1,357	519	395
Pachycereus chrysacanthus	1,395	1,929	872	382
Pachycereus fulviceps	3,306	5,405	2,818	1,071

Table 7.10 The potential core distributions (km²) of cacti under current climatic conditions and for three climate change scenarios for Mexico. Species in the first category of cacti are currently completely restricted to the 10,000 km² Tehuacán–Cuicatlán Biosphere Reserve. Those in the second category have a current range more or less equally distributed within and outside the reserve. The current ranges of species in the final category extend widely beyond the reserve boundaries. (After Téllez-Valdés & Dávila-Aranda, 2003.)

currently occupies. This result highlights a general point for managers: regional conservation efforts and current nature reserves may turn out to be in the wrong place in a changing world.

Téllez-Valdés and Dávila-Aranda (2003) explored this issue for cacti, the dominant plant form in Mexico's Tehuacán-Cuicatlán Biosphere Reserve. From knowledge of the biophysical basis of the distribution of current species and assuming one of three future climate scenarios, they predicted future species' distributions in relation to the location of the reserve. Table 7.10 shows how the potential ranges of species contracted or expanded in the various scenarios. Focusing on the most extreme scenario (an average temperature increase of 2.0°C and a 15% reduction in rainfall), it is evident that more than half of the species that are currently restricted to the reserve are predicted to go extinct. A second category of cacti, whose current ranges are almost equally within and outside the reserve, are expected to contract their ranges, but in such a way that their distributions become almost completely confined to the reserve. A final category, whose current distributions are much more widespread, also suffer range contraction but in future they are expected to still be distributed within and outside the reserve. In the case of these cacti, then, the location of the reserve seems to cater adequately for potential range changes.

We noted above that the performance of the butterfly *Hypochrysops halyetus* depends not just on its own physiology and behavior but also on a mutualistic interaction with ants. Moreover, while cactus distributions are fundamentally dependent on appropriate physicochemical conditions, they are also certain to be influenced by competition for resources with other plants and by their interactions with the species that feed upon them. We now turn our attention, in the second section of the book, to the ecology of interacting populations.

Summary

Ecologists and managers need to identify effective ways to apply ecological knowledge to deal with the wide range of environmental problems that confront us all. In this chapter we discuss ecological applications of theory and knowledge at the level of individual organisms and of single populations. This is the first of a trio of chapters; the others will address, in a similar manner, the application of the fundamentals of ecology at the level of population interactions (Chapter 15) and of communities and ecosystems (Chapter 22).

Management strategies often rely on an ability to predict where species might do well, whether we wish to revegetate contaminated land, restore degraded animal habitats, predict the future distribution of invasive species (and through biosecurity measures prevent their arrival) or conserve endangered species in new reserves. We describe how our understanding of niche theory provides a vital foundation for many management actions.

The life history of a species is another basic feature that can guide management. Particular combinations of ecological traits help determine lifetime patterns of fecundity and survival, which in turn determine the distribution and abundance of species in space and time. We consider whether particular traits (such as seed size, growth rate, longevity and behavioral flexibility) can be of use to managers concerned with the likelihood of a species being a successful part of a habitat restoration project, a problematic invader or a candidate for extinction and therefore worthy of conservation priority. Body size turns out to be a particularly important indicator of extinction risk.

A particularly influential feature of the behavior of organisms, whether animals or plants, is their pattern of movement and dispersion. Knowledge of migratory behavior and dispersion behavior in a patchy environment can underpin attempts to restore damaged and suboptimal habitats and in the design of conservation reserves. Moreover, a detailed understanding of patterns of species transmission by human agency permits us to predict and counter the spread of invaders.

Conservation of endangered species requires a thorough understanding of the dynamics of small populations. Theory tells conservation biologists to beware genetic problems in small populations, which needs to be taken into account when devising conservation management plans. Small populations are also subject to particular demographic risks that make extinction more likely. We focus on an approach called population viability analysis (PVA) – an assessment of extinction probabilities that depends on knowledge of life tables, population rates of increase, intraspecific competition, density dependence, carrying capacities and, when appropriate, metapopulation structure. Careful analysis of populations of particular species at risk can be used to suggest management approaches with the greatest chance of ensuring their persistence.

One of the biggest future challenges to organisms, ecologists and resource managers is global climate change. We deal with the way we can use knowledge about the ecology of individual organisms, coupled with predicted global changes in patterns in physicochemical conditions across the face of the globe, to predict and manage the spread of disease-carrying organisms and other invaders, and to determine the appropriate positioning of conservation reserves.

Part 2
Species Interactions

Introduction

The activity of any organism changes the environment in which it lives. It may alter conditions, as when the transpiration of a tree cools the atmosphere, or it may add or subtract resources from the environment that might have been available to other organisms, as when that tree shades the plants beneath it. In addition, though, organisms interact when individuals enter into the lives of others. In the following chapters (8–15) we consider the variety of these interactions between individuals of different species. We distinguish five main categories: competition, predation, parasitism, mutualism and detritivory, although like most biological categories, these five are not perfect pigeon-holes.

In very broad terms, 'competition' is an interaction in which one organism consumes a resource that would have been available to, and might have been consumed by, another. One organism deprives another, and, as a consequence, the other organism grows more slowly, leaves fewer progeny or is at greater risk of death. The act of deprivation can occur between two members of the same species or between individuals of different species. We have already examined *intra*specific competition in Chapter 5. We turn to *inter*specific competition in Chapter 8.

Chapters 9 and 10 deal with various aspects of 'predation', though we have defined predation broadly. We have combined those situations in which one organism eats another and kills it (such as an owl preying on mice), and those in which the consumer takes only part of its prey, which may then regrow to provide another bite another day (grazing). We have also combined herbivory (animals eating plants) and carnivory (animals eating animals). In Chapter 9 we examine the nature of predation, i.e. what happens to the predator and what happens to the prey, paying particular attention to herbivory because of the subtleties that characterize the response of a plant to attack. We also discuss the behavior of predators. Then, in Chapter 10, we examine the 'consequences of consumption' in terms of the dynamics of predator and prey populations. This is the part of ecology that has the most obvious relevance to those concerned with the management of natural resources: the efficiency of harvesting (whether of fish, whales, grasslands or prairies) and the biological and chemical control of pests and weeds – themes that we take up in Chapter 15.

Most of the processes in this section involve genuine *inter*actions between organisms of different species. However, when dead organisms (or dead parts of organisms) are consumed – decomposition and detritivory – the affair is far more one-sided. None the less, as we describe in Chapter 11, these processes themselves incorporate competition, parasitism, predation and mutualism: microcosms of all the major ecological processes (except photosynthesis).

Chapter 12, 'Parasitism and Disease', deals with a subject that in the past was often neglected by ecologists – and by ecology texts. Yet more than half of all species are parasites, and recent years have seen much of that past neglect rectified. Parasitism itself has blurred edges, particularly where it merges into predation. But whereas a predator usually takes all or part of many individual prey, a parasite normally takes its resources from one or a very few hosts, and (like many grazing predators) it rarely kills its hosts immediately, if at all.

Whereas the earlier chapters of this section deal largely with conflict between species, Chapter 13 is concerned with mutualistic interactions, in which both organisms experience a net benefit. None the less, as we shall see, conflict often lies at the heart of mutualistic interactions too: each participant exploiting the other, such that the *net* benefit arises only because, overall, gains exceed losses. Like parasitism, the ecology of mutualism has often been neglected. Again, though, this neglect has been unwarranted: the greater part of the world's biomass is composed of mutualists.

Ecologists have often summarized interactions between organisms by a simple code that represents each one of the pair of interacting organisms by a '+', a '−' or a '0', depending on how it is affected by the interaction. Thus, a predator–prey (including a herbivore–plant) interaction, in which the predator benefits and the prey is harmed, is denoted by + −, and a parasite–host interaction is also clearly + −. Another straightforward case is mutualism, which, overall, is obviously + +; whereas if organisms do not interact at all, we can denote this by 0 0 (sometimes called 'neutralism'). Detritivory must be denoted by + 0, since the detritivore itself benefits, while its food (dead already) is unaffected. The general term applied to + 0 interactions is 'commensalism', but paradoxically this term is not usually used for detritivores. Instead, it is reserved for cases, allied to parasitism, in which one organism (the 'host') provides resources or a home for another organism, but in which the host itself suffers no tangible ill effects. Competition is usually described as a − − interaction, but it is often impossible to establish that both organisms are harmed. Such asymmetric interactions may then approximate to a − 0 classification, generally referred to as 'amensalism'. True cases of amensalism may occur when one organism produces its ill effect (for instance a toxin) whether or not the potentially affected organism is present.

Although the earlier chapters in this section deal with these various interactions largely in isolation, members of a population are subject simultaneously to many such interactions, often of all conceivable types. Thus, the abundance of a population is determined by this range of interactions (and indeed environmental conditions and the availability of resources) all acting in concert. Attempts to understand variations in abundance therefore demand an equally wide ranging perspective. We adopt this approach in Chapter 14.

Finally in this section, we discuss in Chapter 15 applications of the principles elaborated in the preceding chapters. Our focus is on pest control and the management of natural resources. With the former, the pest species is either a competitor or a predator of desirable species (for example food crops), and we are either predators of the pest ourselves or we manipulate its natural predators to our advantage (biological control). With the latter, again, we are predators of a living, natural resource (harvestable trees in a forest, fish in the sea), but the challenge for us is to establish a stable and sustainable relationship with the prey, guaranteeing further valuable harvests for generations to come.

Chapter 8
Interspecific Competition

8.1 Introduction

The essence of interspecific competition is that individuals of one species suffer a reduction in fecundity, growth or survivorship as a result of resource exploitation or interference by individuals of another species. This competition is likely to affect the population dynamics of the competing species, and the dynamics, in their turn, can influence the species' distributions and their evolution. Of course, evolution, in *its* turn, can influence the species' distributions and dynamics. Here, we concentrate on the effects of competition on populations of species, whilst Chapter 19 examines the role of interspecific competition (along with predation and parasitism) in shaping the structure of ecological communities. There are several themes introduced in this chapter that are taken up and discussed more fully in Chapter 20. The two chapters should be read together for a full coverage of interspecific competition.

8.2 Some examples of interspecific competition

a diversity of examples of competition . . . There have been many studies of interspecific competition between species of all kinds. We have chosen six initially, to illustrate a number of important ideas.

8.2.1 Competition between salmonid fishes

. . . between salmonid fishes, . . . *Salvelinus malma* (Dolly Varden charr) and *S. leucomaenis* (white-spotted charr) are morphologically similar and closely related fishes in the family Salmonidae. The two species are found together in many streams on Hokkaido Island in Japan, but Dolly Varden are distributed at higher altitudes (further upstream) than white-spotted charr, with a zone of overlap at intermediate altitudes. In streams where one species happens to be absent, the other expands its range, indicating that the distributions may be maintained by competition (i.e. each species suffers, and is thus excluded from certain sites, in the presence of the other species). Water temperature, an abiotic factor with profound consequences for fish ecology (discussed already in Section 2.4.4), increases downstream.

By means of experiments in artificial streams, Taniguchi and Nakano (2000) showed that when either species was tested alone, higher temperatures led to increased aggression. But this effect was reversed for Dolly Varden when in the presence of white-spotted charr (Figure 8.1a). Reflecting this, at the higher temperature, Dolly Varden were suppressed from obtaining favorable foraging positions when white-spotted charr were present, and they suffered lower growth rates (Figure 8.1b, c) and a lower probability of survival.

Thus, the experiments lend support to the idea that Dolly Varden and white-spotted charr compete: one species, at least, suffers directly from the presence of the other. They coexist in the same river, but on a finer scale their distributions overlap very little. Specifically, the white-spotted charr appear to outcompete and exclude Dolly Varden from downstream locations in the latter's range. The reason for the upper boundary of white-spotted charr remains unknown as they did not suffer from the presence of Dolly Varden at the lower temperature.

8.2.2 Competition between barnacles

The second study concerns two species of barnacle in Scotland: *Chthamalus stellatus* and *Balanus balanoides* (Figure 8.2) (Connell, 1961). These are frequently found together on the same Atlantic rocky shores of northwest . . . between barnacles, . . .

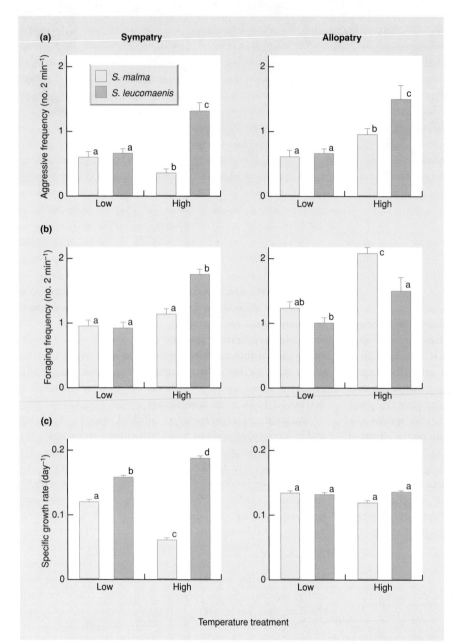

Figure 8.1 (a) Frequency of aggressive encounters initiated by individuals of each fish species during a 72-day experiment in artificial stream channels with two replicates each of 50 Dolly Varden (*Salvelinus malma*) or 50 white-spotted charr (*S. leucomaenis*) alone (allopatry) or 25 of each species together (sympatry). (b) Foraging frequency. (c) Specific growth rate in length. Different letters indicate that the means are significantly different from each other. (From Taniguchi & Nakano, 2000.)

Europe. However, adult *Chthamalus* generally occur in an intertidal zone that is higher up the shore than that of adult *Balanus*, even though young *Chthamalus* settle in considerable numbers in the *Balanus* zone. In an attempt to understand this zonation, Connell monitored the survival of young *Chthamalus* in the *Balanus* zone. He took successive censuses of mapped individuals over the period of 1 year and, most importantly, he ensured at some sites that young *Chthamalus* that settled in the *Balanus* zone were kept free from contact with *Balanus*. In contrast with the normal pattern, such individuals survived well, irrespective of the intertidal level. Thus, it seemed that the usual cause of mortality in young *Chthamalus* was not the increased submergence times of the lower zones, but competition from *Balanus* in those zones. Direct observation confirmed that *Balanus* smothered, undercut or crushed *Chthamalus*, and the greatest *Chthamalus* mortality occurred during the seasons of most rapid *Balanus* growth. Moreover, the few *Chthamalus* individuals that survived 1 year of *Balanus* crowding were much smaller than uncrowded ones, showing, since smaller barnacles produce fewer offspring, that interspecific competition was also reducing fecundity.

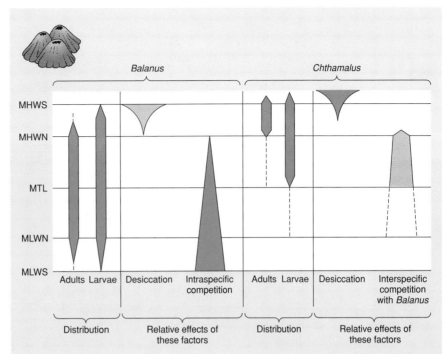

Figure 8.2 The intertidal distribution of adults and newly settled larvae of *Balanus balanoides* and *Chthamalus stellatus*, with a diagrammatic representation of the relative effects of desiccation and competition. Zones are indicated to the left: from MHWS (mean high water, spring) down to MLWS (mean low water, spring); MTL, mean tide level; N, neap. (After Connell, 1961.)

Thus, *Balanus* and *Chthamalus* compete. They coexist on the same shore but, like the fish in the previous section, on a finer scale their distributions overlap very little. *Balanus* outcompetes and excludes *Chthamalus* from the lower zones; but *Chthamalus* can survive in the upper zones where *Balanus*, because of its comparative sensitivity to desiccation, cannot.

8.2.3   Competition between bedstraws (*Galium* spp.)

... between
bedstraws, ...

A. G. Tansley, one of the greatest of the 'founding fathers' of plant ecology, studied competition between two species of bedstraw (Tansley, 1917). *Galium hercynicum* is a species which grows naturally in Great Britain at acidic sites, whilst *G. pumilum* is confined to more calcareous soils. Tansley found in experiments that as long as he grew them alone, both species would thrive on both the acidic soil from a *G. hercynicum* site and the calcareous soil from a *G. pumilum* site. Yet, if the species were grown together, only *G. hercynicum* grew successfully in the acidic soil and only *G. pumilum* grew successfully in the calcareous soil. It seems, therefore, that when they grow together the species compete, and that one species wins, whilst the other loses so badly that it is competitively excluded from the site. The outcome depends on the habitat in which the competition occurs.

8.2.4   Competition between *Paramecium* species

The fourth example comes from the classic work of the great Russian ecologist G. F. Gause, who studied competition in laboratory experiments using three species of the protozoan *Paramecium* (Gause, 1934, 1935). All three species grew well alone, reaching stable carrying capacities in tubes of liquid medium. There, *Paramecium* consumed bacteria or yeast cells, which themselves lived on regularly replenished oatmeal (Figure 8.3a).

... between
Paramecium
species, ...

When Gause grew *P. aurelia* and *P. caudatum* together, *P. caudatum* always declined to the point of extinction, leaving *P. aurelia* as the victor (Figure 8.3b). *P. caudatum* would not normally have starved to death as quickly as it did, but Gause's experimental procedure involved the daily removal of 10% of the culture and animals. Thus, *P. aurelia* was successful in competition because near the point where its population size leveled off, it was still increasing by 10% per day (and able to counteract the enforced mortality), whilst *P. caudatum* was only increasing by 1.5% per day (Williamson, 1972).

By contrast, when *P. caudatum* and *P. bursaria* were grown together, neither species suffered a decline to the point of extinction – they coexisted. But, their stable densities were much lower than when grown alone (Figure 8.3c), indicating that they were in competition with one another (i.e. they 'suffered'). A closer

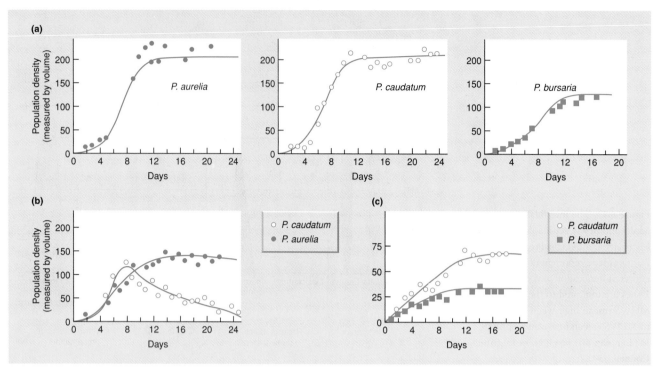

Figure 8.3 Competition in *Paramecium*. (a) *P. aurelia*, *P. caudatum* and *P. bursaria* all establish populations when grown alone in culture medium. (b) When grown together, *P. aurelia* drives *P. caudatum* towards extinction. (c) When grown together, *P. caudatum* and *P. bursaria* coexist, although at lower densities than when alone. (After Clapham, 1973; from Gause, 1934.)

look, however, revealed that although they lived together in the same tubes, they were, like Taniguchi and Nakano's fish and Connell's barnacles, spatially separated. *P. caudatum* tended to live and feed on the bacteria suspended in the medium, whilst *P. bursaria* was concentrated on the yeast cells at the bottom of the tubes.

8.2.5 Coexistence amongst birds

... among birds ... Ornithologists are well aware that closely related species of birds often coexist in the same habitat. For example, five *Parus* species occur together in English broad-leaved woodlands: the blue tit (*P. caeruleus*), the great tit (*P. major*), the marsh tit (*P. palustris*), the willow tit (*P. montanus*) and the coal tit (*P. ater*). All have short beaks and hunt for food chiefly on leaves and twigs, but at times on the ground; all eat insects throughout the year, and also seeds in winter; and all nest in holes, normally in trees. However, the closer we look at the details of the ecology of such coexisting species, the more likely we will find ecological differences – for example, in precisely where within the trees they feed, in the size of their insect prey and the hardness of the seeds they take. Despite their similarities, we may be tempted to conclude that the tit species compete but coexist by eating slightly different resources

in slightly different ways. However, a scientifically rigorous approach to determine the current role of competition requires the removal of one or more of the competing species and monitoring the responses of those that remain. Martin and Martin (2001) did just this in a study of two very similar species: the orange-crowned warbler (*Vermivora celata*) and virginia's warbler (*V. virginiae*) whose breeding territories overlap in central Arizona. On plots where one of the two species had been removed, the remaining orange-crowned or virginia's warblers fledged between 78 and 129% more young per nest, respectively. The improved performance was due to improved access to preferred nest sites and consequent decreased losses of nestlings to predators. In the case of virginia's warblers, but not orange-crowned warblers, feeding rate also increased in plots from which the other species was removed (Figure 8.4).

8.2.6 Competition between diatoms

The final example is from a laboratory investigation of two species of fresh-water diatom: *Asterionella formosa* and ... and between diatoms
Synedra ulna (Tilman *et al.*, 1981). Both these algal species require silicate in the construction of their cell walls. The investigation was

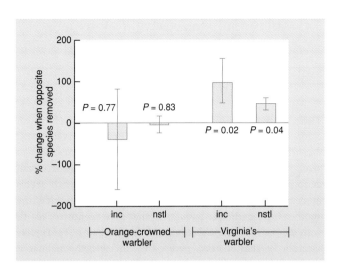

Figure 8.4 (*right*) Percentage difference in feeding rates (mean ± SE) at orange-crowned warbler and virginia's warbler nests on plots where the other species had been experimentally removed. Feeding rates (visits per hour to the nest with food) were measured during incubation (inc) (rates of male feeding of incubating females on the nest) and during the nestling period (nstl) (nestling feeding rates by both parents combined). *P* values are from *t*-tests of the hypothesis that each species fed at higher rates on plots from which the other had been removed. This hypothesis was supported for virginia's warblers but not orange-crowned warblers. (After Martin & Martin, 2001.)

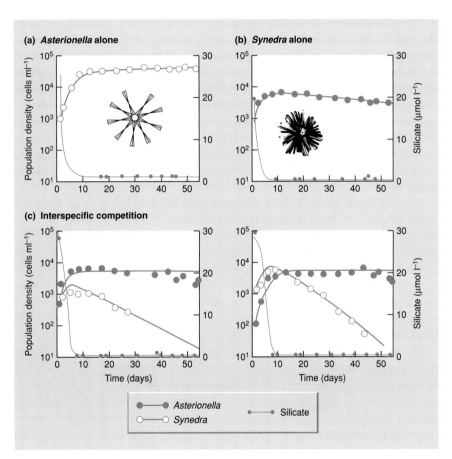

Figure 8.5 Competition between diatoms. (a) *Asterionella formosa*, when grown alone in a culture flask, establishes a stable population and maintains a resource, silicate, at a constant low level. (b) When *Synedra ulna* is grown alone it does the same, but maintains silicate at an even lower level. (c) When grown together, in two replicates, *Synedra* drives *Asterionella* to extinction. (After Tilman *et al.*, 1981.)

unusual because at the same time as population densities were being monitored, the impact of the species on their limiting resource (silicate) was being recorded. When either species was cultured alone in a liquid medium to which resources were continuously being added, it reached a stable carrying capacity whilst maintaining the silicate at a constant low concentration (Figure 8.5a, b). However, in exploiting this resource, *Synedra* reduced the silicate concentration to a lower level than did *Asterionella*. Hence, when the two species were grown together, *Synedra* maintained the concentration at a level that was too low for the survival and reproduction of *Asterionella*. *Synedra* therefore competitively excluded *Asterionella* from mixed cultures (Figure 8.5c).

8.3 Assessment: some general features of interspecific competition

8.3.1 Unraveling ecological and evolutionary aspects of competition

These examples show that individuals of different species can compete. This is hardly surprising. The field experiments with barnacles and warblers also show that different species do compete in nature (i.e. there was a measurable interspecific reduction in abundance and/or fecundity and/or survivorship). It seems, moreover, that competing species may either exclude one another from particular habitats so that they do not coexist (as with the bedstraws, the diatoms and the first pair of *Paramecium* species), or may coexist, perhaps by utilizing the habitat in slightly different ways (e.g. the barnacles and the second pair of *Paramecium* species).

But what about the story of the coexisting tits? Certainly the five bird species coexist and utilize the habitat in slightly different ways. But does this have anything to do with competition? It may do. It may be that the five species of tit coexist as a result of evolutionary responses to interspecific competition. This requires some further explanation. When two species compete, individuals of one or both species may suffer reductions in fecundity and/or survivorship, as we have seen. The fittest individuals of each species may then be those that (relatively speaking) escape competition because they utilize the habitat in ways that differ most from those adopted by individuals of the other species. Natural selection will then favor such individuals, and eventually the population may consist entirely of them. The two species will evolve to become more different from one another than they were previously; they will compete less, and thus will be more likely to coexist.

coexisting competitors or the 'ghost of competition past'? . . .
The trouble with this as an explanation for the tit story is that there is no proof. We need to beware, in Connell's (1980) phrase, of uncritically invoking the 'ghost of competition past'. We cannot go back in time to check whether the species ever competed more than they do now. A plausible alternative interpretation is that the species have, in the course of their evolution, responded to natural selection in different but entirely independent ways. They are distinct species, and they have distinctive features. But they do not compete now, nor have they ever competed; they simply happen to be different. If all this were true, then the coexistence of the tits would have nothing to do with competition. Alternatively again, it may be that competition in the past eliminated a number of other species, leaving behind only those that are different in their utilization of the habitat: we can still see the hand of the ghost of competition past, but acting as an ecological force (eliminating species) rather than an evolutionary one (changing them).

The tit story, therefore, and the difficulties with it, illustrate two important general points. The first is that we must pay careful, and separate, attention to both the ecological and the evolutionary effects of interspecific competition. The ecological effects are, broadly, that species may be eliminated from a habitat by competition from individuals of other species; or, if competing species coexist, that individuals of at least one of them suffer reductions in survival and/or fecundity. The evolutionary effects appear to be that species differ more from one another than they would otherwise do, and hence compete less (but see Section 8.9).

The second point, though, is that there are profound difficulties in invoking competition as an explanation for observed patterns, and especially in invoking it as an evolutionary explanation. An experimental manipulation (for instance, the removal of one or more species) can, as we have seen with the warblers, indicate the presence of current competition if it leads to an increase in the fecundity or survival or abundance of the remaining species. But negative results would be equally compatible with the past elimination of species by competition, the evolutionary avoidance of competition in the past, and the independent evolution of noncompeting species. In fact, for many sets of data, there are no easy or agreed methods of distinguishing between these explanations (see Chapter 19). Thus, in the remainder of this chapter (and in Chapter 19) when examining the ecological and, especially, the evolutionary effects of competition, we will need to be more than usually cautious.

8.3.2 Exploitation and interference competition and allelopathy

For now, though, what other general features emerge from our examples? As with intraspecific competition, a basic distinction can be made between interference and exploitation competition (although elements of both may be found in a single interaction) (see Section 5.1.1). With exploitation, individuals interact with each other indirectly, responding to a resource level that has been depressed by the activity of competitors. The diatom work provides a clear example of this. By contrast, Connell's barnacles provide an equally clear example of interference competition. *Balanus*, in particular, directly and physically interfered with the occupation by *Chthamalus* of limited space on the rocky substratum.

Interference, on the other hand, is not always as direct as this. Amongst plants, it has often been claimed that interference occurs through the production and release into the environment of chemicals that are toxic to other species but not to the producer (known as allelopathy). There is no doubt that chemicals with such properties can be extracted from plants, but establishing a role for them in nature or that they have evolved *because of* their allelopathic effects, has proved difficult. For example, extracts from more than 100 common agricultural weeds have been reported to have allelopathic potential against crop species (Foy & Inderjit, 2001), but the studies generally involved unnatural laboratory bioassays rather than realistic field experiments. In a similar manner, Vandermeest *et al.* (2002) showed in the laboratory that an extract from American chestnut leaves (*Castanea dentata*) suppressed germination of the shrub rosebay rhododendron (*Rhododendron maximum*). The American chestnut was the most common overstory tree in the USA's eastern deciduous forest until ravaged by chestnut blight (*Cryphonectria parasitica*). Vandermeest *et al.* concluded that the expansion of rhododendron thickets throughout the 20th century may have been due as much to the cessation of the chestnut's allelopathic influence as to the more commonly cited invasion of canopy openings following blight, heavy logging and fire. However, their hypothesis cannot be tested. Amongst competing tadpole species, too, water-borne inhibitory products have been implicated as a means of interference (most notably, perhaps, an alga produced in the feces of the common frog, *Rana temporaria*, inhibiting the natterjack toad, *Bufo calamita* (Beebee, 1991; Griffiths *et al.*, 1993)), but here again their importance in nature is unclear (Petranka, 1989). Of course, the production by fungi and bacteria of allelopathic chemicals that inhibit the growth of potentially competing microorganisms is widely recognized – and exploited in the selection and production of antibiotics.

8.3.3 Symmetric and asymmetric competition

interspecific competition is frequently highly asymmetric

Interspecific competition (like intra-specific competition) is frequently highly asymmetric – the consequences are often not the same for both species. For instance, with Connell's barnacles, *Balanus* excluded *Chthamalus* from their zone of potential overlap, but any effect of *Chthamalus* on *Balanus* was negligible: *Balanus* was limited by its own sensitivity to desiccation. An analogous situation is provided by two species of cattail (reedmace) in ponds in Michigan; *Typha latifolia* occurs mostly in shallower water whilst *T. angustifolia* occurs in deeper water. When grown together (in sympatry) in artificial ponds, the two species mirror their natural distributions, with *T. latifolia* mainly occupying depth zones from 0 to 60 cm below the water surface and *T. angustifolia* mainly from 60 to 90 cm (Grace & Wetzel, 1998). When grown on its own (allopatry), the depth distribution of *T. angustifolia* shifts markedly towards shallower depths. In contrast, *T. latifolia* shows only a minor shift towards greater depth in the absence of interspecific competition.

On a broader front, it seems that highly asymmetric cases of interspecific competition (where one species is little affected)

generally outnumber symmetric cases (e.g. Keddy & Shipley, 1989). The more fundamental point, however, is that there is a continuum linking the perfectly symmetric competitive cases to strongly asymmetric ones. Asymmetric competition results from the differential ability of species to occupy higher positions in a competitive hierarchy. In plants, for example, this may result from height differences, with one species able to completely over-top another and preempt access to light (Freckleton & Watkinson, 2001). In a similar vein, Dezfuli *et al.* (2002) have argued that asymmetric competition might be expected between parasite species that occupy sequential positions in the gut of their host, with a stomach parasite reducing resources and adversely influencing an intestinal parasite further downstream, but not vice versa. Asymmetric competition is especially likely where there is a very large difference in the size of competing species. Reciprocal exclusion experiments have shown that grazing ungulates (domestic sheep and Spanish ibex *Capra pyrenaica*) reduce the abundance of the herbivorous beetle *Timarcha lugens* in Spanish scrubland by exploitation competition (and partly by incidental predation). However, there was no effect of beetle exclusion on ungulate performance (Gomez & Gonzalez-Megias, 2002).

8.3.4 Competition for one resource may influence competition for another

Finally, it is worth noting that competition for one resource often affects the ability of an organism to exploit another resource. For example, Buss (1979) showed that in interactions between species of bryozoa (colonial, modular animals), there appears to be an interdependence between competition for space and for food. When a colony of one species contacts a colony of another species, it interferes with the self-generated feeding currents upon which bryozoans rely (competition for space affects feeding). But a colony short of food will, in turn, have a greatly reduced ability to compete for space (by overgrowth).

Comparable examples are found amongst rooted plants. If one species invades the canopy of another and deprives it of light, the suppressed species will suffer directly from the reduction in light energy that it obtains, but this will also reduce its rate of root growth, and it will therefore be less able to exploit the supply of water and nutrients in the soil. This in turn will reduce its rate of shoot and leaf growth. Thus, when plant species compete, repercussions flow backwards and forwards between roots and shoots (Wilson, 1988a). A number of workers have attempted to separate the effects of canopy and root competition by an experimental design in which two species are grown: (i) alone; (ii) together; (iii) in the same soil, but with their canopies separated; and (iv) in separate soil with their canopies intermingling. One example is a study of

root and shoot competition

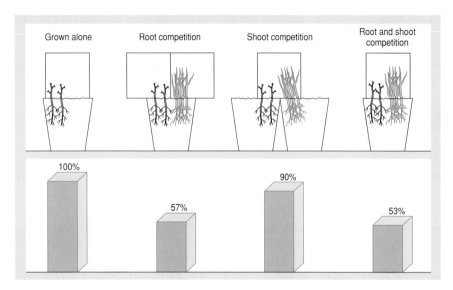

Figure 8.6   Root and shoot competition between maize and pea plants. Above are the experimental plants used, below are the dry weights of pea plants after 46 days as a percentage of those achieved when grown alone. (Data from Semere & Froud-Williams, 2001.)

maize (*Zea mays*) and pea plants (*Pisum sativum*) (Semere & Froud-Williams, 2001). In full competition, with roots and shoots intermingling, the biomass production of maize and peas respectively (dry matter per plant, 46 days after sowing) was reduced to 59 and 53% of the 'control' biomass when the species were grown alone. When only the roots intermingled, pea plant biomass production was still reduced to 57% of the control value, but when just the shoots intermingled, biomass production was only reduced to 90% of the control (Figure 8.6). These results indicate, therefore, that soil resources (mineral nutrients and water) were more limiting than light, a common finding in the literature (Snaydon, 1996). They also support the idea of root and shoot competition combining to generate an overall effect, in that the overall reduction in plant biomass (to 53%) was close to the product of the root-only and shoot-only reductions (90% of 57% is 51.3%).

8.4   Competitive exclusion or coexistence?

The results of experiments such as those described here highlight a critical question in the study of the ecological effects of interspecific competition: what are the general conditions that permit the coexistence of competitors, and what circumstances lead to competitive exclusion? Mathematical models have provided important insights into this question.

8.4.1   A logistic model of interspecific competition

The 'Lotka–Volterra' model of interspecific competition (Volterra, 1926; Lotka, 1932) is an extension of the logistic equation described in Section 5.9. As such, it incorporates all of the logistic's shortcomings, but a useful model can none the less be constructed, shedding light on the factors that determine the outcome of a competitive interaction.

The logistic equation:

$$\frac{dN}{dt} = rN\frac{(K-N)}{K} \tag{8.1}$$

contains, within the brackets, a term responsible for the incorporation of intraspecific competition. The basis of the Lotka–Volterra model is the replacement of this term by one which incorporates both intra- and interspecific competition.

The population size of one species can be denoted by N_1, and that of a second species by N_2. Their carrying capacities and intrinsic rates of increase are K_1, K_2, r_1 and r_2, respectively.

Suppose that 10 individuals of species 2 have, between them, the same competitive, inhibitory effect on species 1 as does a single individual of species 1. The total competitive effect on species 1 (intra- and interspecific) will then be equivalent to the effect of $(N_1 + N_2/10)$ species 1 individuals. The constant (1/10 in the present case) is called a competition coefficient and is denoted by α_{12} ('alpha-one-two'). It measures the per capita competitive effect on species 1 of species 2. Thus, multiplying N_2 by α_{12} converts it to a number of 'N_1-equivalents'. (Note that $\alpha_{12} < 1$ means that individuals of species 2 have less inhibitory effect on individuals of species 1 than individuals of species 1 have on others of their own species, whilst $\alpha_{12} > 1$ means that individuals of species 2 have a greater inhibitory effect on individuals of species 1 than do the species 1 individuals themselves.)

α: the competition coefficient

Lotka–Volterra model: a logistic model for two species

The crucial element in the model is the replacement of N_1 in the bracket of the logistic equation with a term signifying 'N_1 plus N_1-equivalents', i.e.:

$$\frac{dN_1}{dt} = r_1 N_1 \frac{(K_1 - (N_1 + \alpha_{12} N_2))}{K_1} \qquad (8.2)$$

or:

$$\frac{dN_1}{dt} = r_1 N_1 \frac{(K_1 - N_1 - \alpha_{12} N_2)}{K_1} \qquad (8.3)$$

and in the case of the second species:

$$\frac{dN_2}{dt} = r_2 N_2 \frac{(K_2 - N_2 - \alpha_{21} N_1)}{K_2}. \qquad (8.4)$$

These two equations constitute the Lotka–Volterra model.

behavior of the Lotka–Volterra model is investigated using 'zero isoclines'

To appreciate the properties of this model, we must ask the question: when (under what circumstances) does each species increase or decrease in abundance? In order to answer this, it is necessary to construct diagrams in which all possible combinations of species 1 and species 2 abundance can be displayed (i.e. all possible combinations of N_1 and N_2). These will be diagrams (Figures 8.7 and 8.9), with N_1 plotted on the horizontal axis and N_2 plotted on the vertical axis, such that there are low numbers of both species towards the bottom left, high numbers of both species towards the top right, and so on. Certain combinations of N_1 and N_2 will give rise to increases in species 1 and/or species 2, whilst other combinations will give rise to decreases in species 1 and/or species 2. Crucially, there

must also therefore be 'zero isoclines' for each species (lines along which there is neither an increase nor a decrease), dividing the combinations leading to increase from those leading to decrease. Moreover, if a zero isocline is drawn first, there will be combinations leading to an increase on one side of it, and combinations leading to a decrease on the other.

In order to draw a zero isocline for species 1, we can use the fact that on the zero isocline $dN_1/dt = 0$ (by definition), that is (from Equation 8.3):

$$r_1 N_1 (K_1 - N_1 - \alpha_{21} N_2) = 0. \qquad (8.5)$$

This is true when the intrinsic rate of increase (r_1) is zero, and when the population size (N_1) is zero, but – much more importantly in the present context – it is also true when:

$$K_1 - N_1 - \alpha_{21} N_2 = 0, \qquad (8.6)$$

which can be rearranged as:

$$N_1 = K_1 - \alpha_{21} N_2. \qquad (8.7)$$

In other words, everywhere along the straight line which this equation represents, $dN_1/dt = 0$. The line is therefore the zero isocline for species 1; and since it is a straight line it can be drawn by finding two points on it and joining them. Thus, in Equation 8.7, when:

$$N_1 = 0, N_2 = \frac{K_1}{\alpha_{12}} \text{ (point A, Figure 8.7a)} \qquad (8.8)$$

and when:

$$N_2 = 0, N_1 = K \text{ (point B, Figure 8.7a)}, \qquad (8.9)$$

and joining them gives the zero isocline for species 1. Below and to the left of this, the numbers of both species are relatively low, and species 1, subjected to only weak competition, increases in abundance (the arrows in the figure, representing this increase, point from left to right, since N_1 is on the horizontal axis). Above and to the right of the line, the numbers are high, competition is strong and species 1 decreases in abundance (arrows from right to left). Based on an equivalent derivation, Figure 8.7b has combinations leading to an increase and decrease in species 2, separated by a species 2 zero isocline, with arrows, like the N_2 axis, running vertically.

Finally, in order to determine the outcome of competition in this model, it is necessary to fuse Figures 8.7a and b, allowing the behavior of a joint population to be predicted. In doing this, it should be noted that the arrows in Figure 8.7 are actually vectors – with a strength as well as a direction – and that to determine

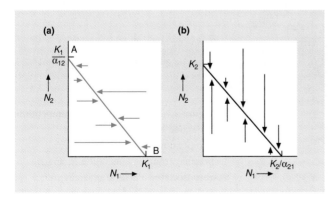

Figure 8.7 The zero isoclines generated by the Lotka–Volterra competition equations. (a) The N_1 zero isocline: species 1 increases below and to the left of it, and decreases above and to the right of it. (b) The equivalent N_2 zero isocline.

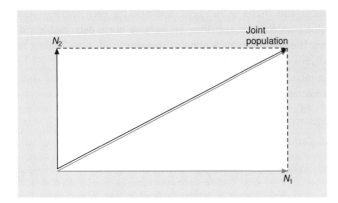

Figure 8.8 Vector addition. When species 1 and 2 increase in the manner indicated by the N_1 and N_2 arrows (vectors), the joint population increase is given by the vector along the diagonal of the rectangle, generated as shown by the N_1 and N_2 vectors.

two zero isoclines can be arranged relative to one another, and the outcome of competition will be different in each case. The different cases can be defined and distinguished by the intercepts of the zero isoclines. For instance, in Figure 8.9a:

$$\frac{K_1}{\alpha_{12}} > K_2 \quad \text{and} \quad K_1 > \frac{K_2}{\alpha_{21}} \tag{8.10}$$

i.e.:

$$K_1 > K_2\alpha_{12} \quad \text{and} \quad K_1\alpha_{21} > K_2. \tag{8.11}$$

The first inequality ($K_1 > K_2\alpha_{12}$) indicates that the inhibitory intraspecific effects that species 1 can exert on itself are greater than the interspecific effects that species 2 can exert on species 1. The second inequality, however, indicates

strong interspecific competitors outcompete weak interspecific competitors

that species 1 can exert more of an effect on species 2 than species 2 can on itself. Species 1 is thus a strong interspecific competitor, whilst species 2 is a weak interspecific competitor; and as the vectors in Figure 8.9a show, species 1 drives species 2 to extinction and attains its own carrying capacity. The situation is

four ways in which the two zero isoclines can be arranged

the behavior of a joint N_1, N_2 population, the normal rules of vector addition should be applied (Figure 8.8).

Figure 8.9 shows that there are, in fact, four different ways in which the

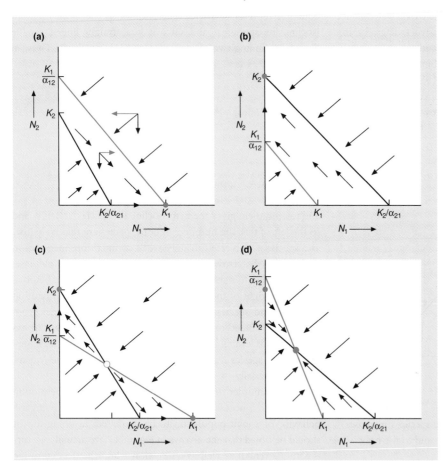

Figure 8.9 The outcomes of competition generated by the Lotka–Volterra competition equations for the four possible arrangements of the N_1 and N_2 zero isoclines. Vectors, generally, refer to joint populations, and are derived as indicated in (a). The solid circles show stable equilibrium points. The open circle in (c) is an unstable equilibrium point. For further discussion, see the text.

reversed in Figure 8.8b. Hence, Figures 8.8a and b describe cases in which the environment is such that one species invariably outcompetes the other.

In Figure 8.9c:

$$K_2 > \frac{K_1}{\alpha_{12}} \quad \text{and} \quad K_1 > \frac{K_2}{\alpha_{21}} \tag{8.12}$$

i.e.:

$$K_2\alpha_{12} > K_1 \quad \text{and} \quad K_1\alpha_{21} > K_2. \tag{8.13}$$

when interspecific competition is more important than intraspecific, the outcome depends on the species' densities

Thus, individuals of both species compete more strongly with individuals of the other species than they do amongst themselves. This will occur, for example, when each species produces a substance that is toxic to the other species but is harmless to itself, or when each species is aggressive towards or even preys upon individuals of the other species, more than individuals of its own species. The consequence, as the figure shows, is an unstable equilibrium combination of N_1 and N_2 (where the isoclines cross), and two stable points. At the first of these stable points, species 1 reaches its carrying capacity with species 2 extinct; whilst at the second, species 2 reaches its carrying capacity with species 1 extinct. Which of these two outcomes is actually attained is determined by the initial densities: the species which has the initial advantage will drive the other species to extinction.

Finally, in Figure 8.9d:

$$\frac{K_1}{\alpha_{12}} > K_2 \quad \text{and} \quad \frac{K_2}{\alpha_{21}} > K_1 \tag{8.14}$$

i.e.:

$$K_1 > K_2\alpha_{12} \quad \text{and} \quad K_2 > K_1\alpha_{21}. \tag{8.15}$$

when interspecific competition is less important than intraspecific, the species coexist

In this case, both species have less competitive effect on the other species than they have on themselves. The outcome, as Figure 8.9d shows, is a stable equilibrium combination of the two species, which all joint populations tend to approach.

Overall, therefore, the Lotka–Volterra model of interspecific competition is able to generate a range of possible outcomes: the predictable exclusion of one species by another, exclusion dependent on initial densities, and stable coexistence. Each of these possibilities will be discussed in turn, alongside the results of laboratory and field investigations. We will see that the three outcomes from the model correspond to biologically reasonable

circumstances. The model, therefore, in spite of its simplicity and its failure to address many of the complexities of the dynamics of competiton in the real world, serves a useful purpose.

Before we move on, however, one particular shortcoming of the Lotka–Volterra model is worth noting. The outcome of competition in the model depends on the Ks and the αs, but not on the rs, the intrinsic rates of increase. These determine the speed with which the outcome is achieved but not the outcome itself. This, though, seems to be a result peculiar to competition between only two species, since in models of competition between three or more species, the Ks, αs and rs combine to determine the outcome (Strobeck, 1973).

Ks, αs and rs

8.4.2 The Competitive Exclusion Principle

Figure 8.9a and b describes cases in which a strong interspecific competitor invariably outcompetes a weak interspecific competitor. It is useful to consider this situation from the point of view of niche theory (see Sections 2.2 and 3.8). Recall that the niche of a species in the absence of competition from other species is its *fundamental* niche (defined by the combination of conditions and resources that allow the species to maintain a viable population). In the presence of competitors, however, the species may be restricted to a *realized* niche, the precise nature of which is determined by which competing species are present. This distinction stresses that interspecific competition reduces fecundity and survival, and that there may be parts of a species' fundamental niche in which, as a result of interspecific competition, the species can no longer survive and reproduce successfully. These parts of its fundamental niche are absent from its realized niche. Thus, returning to Figures 8.9a and b, we can say that the weak interspecific competitor lacks a realized niche when in competition with the stronger competitor. The real examples of interspecific competition previously discussed can now be re-examined in terms of niches.

fundamental and realized niches

In the case of the diatom species, the fundamental niches of both species were provided by the laboratory regime (they both thrived when alone). Yet when *Synedra* and *Asterionella* competed, *Synedra* had a realized niche whilst *Asterionella* did not: there was competitive exclusion of *Asterionella*. The same outcome was recorded when Gause's *P. aurelia* and *P. caudatum* competed; *P. caudatum* lacked a realized niche and was competitively excluded by *P. aurelia*. When *P. caudatum* and *P. bursaria* competed, on the other hand, both species had realized niches, but these niches were noticeably different: *P. caudatum* living and feeding on the bacteria in the medium, *P. bursaria* concentrating on the yeast cells on the bottom of the

coexisting competitors often exhibit a differentiation of their realized niches

tube. Coexistence was therefore associated with a differentiation of realized niches, or a 'partitioning' of resources.

In the *Galium* experiments, the fundamental niches of both species included both acidic and calcareous soils. In competition with one another, however, the realized niche of *G. hercynicum* was restricted to acidic soils, whilst that of *G. pumilum* was restricted to calcareous ones – there was reciprocal competitive exclusion. Neither habitat allowed niche differentiation, and neither habitat fostered coexistence.

Amongst Taniguchi and Nakano's salmonid fishes, the fundamental niches of each species extended over a broad range in altitude (and temperature) but both were restricted to a smaller realized niche (Dolly Varden at higher altitudes and white-spotted charr at lower altitudes).

Similarly, amongst Connell's barnacles, the fundamental niche of *Chthamalus* extended down into the *Balanus* zone, but competition from *Balanus* restricted *Chthamalus* to a realized niche higher up the shore. In other words, *Balanus* competitively excluded *Chthamalus* from the lower zones, but for *Balanus* itself, even its fundamental niche did not extend up into the *Chthamalus* zone: its sensitivity to desiccation prevented it surviving even in the absence of *Chthamalus*. Hence, overall, the coexistence of these species was also associated with a differentiation of realized niches.

the Competitive Exclusion Principle

The pattern that has emerged from these examples has also been uncovered in many others, and has been elevated to the status of a principle: the *Competitive Exclusion Principle* or 'Gause's Principle'. It can be stated as follows: if two competing species coexist in a stable environment, then they do so as a result of niche differentiation, i.e. differentiation of their realized niches. If, however, there is no such differentiation, or if it is precluded by the habitat, then one competing species will eliminate or exclude the other. Thus exclusion occurs when the realized niche of the superior competitor completely fills those parts of the inferior competitor's fundamental niche that are provided by the habitat.

difficulty proving and, especially, disproving the Principle

When there *is* coexistence of competitors, a differentiation of realized niches is sometimes seen to arise from current competition (an 'ecological' effect), as with the barnacles. Often, however, the niche differentiation is believed to have arisen either as a result of the past elimination of those species without realized niches (leaving behind only those exhibiting niche differentiation – another ecological effect) or as an *evolutionary* effect of competition. In either case, present competition may be negligible or at least impossible to detect. Consider again the coexisting tits. The species coexist and exhibit differentiation of their realized niches. But we do not know whether they compete now, or have ever competed in the past, or whether other species have been competitively excluded in the

past. It is impossible to say with certainty whether the Competitive Exclusion Principle was relevant. If the species do actually compete currently, or if other species are being or have been competitively excluded, then the Principle is relevant in the strictest sense. If they competed only in the past, and that competition has led to their niche differentiation, then the Principle is relevant, but only if it is extended from applying to the coexistence of 'competitors' to the coexistence of 'species that are *or have ever been* competitors'. Of course, if the species have never competed, then the Principle is of no relevance here. Clearly, interspecific competition cannot be studied by the mere documentation of present interspecific differences.

With Martin and Martin's warblers, on the other hand, the two species competed and coexisted, and the Competitive Exclusion Principle would suggest that this was a result of niche differentiation. But, whilst reasonable,

niche differentiation and interspecific competition: a pattern and a process not always linked

this is by no means proven, since such differentiation was neither observed nor shown to be effective. Thus, when two competitors coexist, it is often difficult to establish positively that there is niche differentiation. Worse still, it is impossible to prove the absence of it. When ecologists fail to find differentiation, this might simply mean that they have looked in the wrong place or in the wrong way. Clearly, there can be very real methodological problems in establishing the pertinence of the Competitive Exclusion Principle in any particular case.

The Competitive Exclusion Principle has become widely accepted because: (i) there is much good evidence in its favor; (ii) it makes intuitive good sense; and (iii) there are theoretical grounds for believing in it (the Lotka–Volterra model). But there will always be cases in which it has not been positively established; and as Section 8.5 will make plain, there are many other cases in which it simply does not apply. In short, interspecific competition is a process that is often associated, ecologically and evolutionarily, with a particular pattern (niche differentiation), but interspecific competition and niche differentiation (the process and the pattern) are not inextricably linked. Niche differentiation can arise through other processes, and interspecific competition need not lead to a differentiation of niches.

8.4.3 Mutual antagonism

Figure 8.9c, derived from the Lotka–Volterra model, describes a situation in which interspecific competition is, for both species, a more powerful force than intraspecific competition. This is known as mutual antagonism.

An extreme example of such a situation is provided by work on two species of flour beetle: *Tribolium confusum* and *T. castaneum* (Park, 1962). Park's experiments in the 1940s,

reciprocal predation in flour beetles

Table 8.1 Reciprocal predation (a form of mutual antagonism) between two species of flour beetle, *Tribolium confusum* and *T. castaneum*. Both adults and larvae eat both eggs and pupae. In each case, and overall, the preference of each species for its own or the other species is indicated. Interspecific predation is more marked than intraspecific predation. (After Park *et al.*, 1965.)

	'Predator'	'Shows a preference for . . .'
Adults eating eggs	T. confusum	T. confusum
	T. castaneum	T. confusum
Adults eating pupae	T. confusum	T. castaneum
	T. castaneum	T. confusum
Larvae eating eggs	T. confusum	T. castaneum
	T. castaneum	T. castaneum
Larvae eating pupae	T. confusum	T. castaneum
	T. castaneum	T. confusum
Overall	T. confusum	T. castaneum
	T. castaneum	T. confusum

Table 8.2 Competition between *Tribolium confusum* and *T. castaneum* in a range of climates. One species is always eliminated and climate alters the outcome, but at intermediate climates the outcome is nevertheless probable rather than definite. (After Park, 1954.)

	Percentage wins	
Climate	T. confusum	T. castaneum
Hot–moist	0	100
Temperate–moist	14	86
Cold–moist	71	29
Hot–dry	90	10
Temperate–dry	87	13
Cold–dry	100	0

Table 8.2 shows that this was indeed the case with Park's flour beetles. There was always only one winner, and the balance between the species changed with climatic conditions. Yet at all intermediate climates *the outcome was probable rather than definite*. Even the inherently inferior competitor occasionally achieved a density at which it could outcompete the other species.

8.5 Heterogeneity, colonization and preemptive competition

At this point it is necessary to sound a loud note of caution. It has been assumed in this chapter until now that the environment is sufficiently constant for the outcome of competition to be determined by the competitive abilities of the competing species. In reality, though, such situations are far from universal. Environments are usually a patchwork of favorable and unfavorable habitats; patches are often only available temporarily; and patches often appear at unpredictable times and in unpredictable places. Even when interspecific competition occurs, it does not necessarily continue to completion. Systems do not necessarily reach equilibrium, and superior competitors do not necessarily have time to exclude their inferiors. Thus, an understanding of interspecific competition itself is not always enough. It is often also necessary to consider how interspecific competition is influenced by, and interacts with, an inconstant or unpredictable environment. To put it another way: *K*s and αs alone may determine an equilibrium, but in nature, equilibria are very often not achieved. Thus, the speed with which an equilibrium is approached becomes important. That is, as we have already noted in Section 8.4.1 in another context, not only *K*s and αs, but *r*s too play their part.

1950s and 1960s were amongst the most influential in shaping ideas about interspecific competition. He reared the beetles in simple containers of flour, which provided fundamental and often realized niches for the eggs, larvae, pupae and adults of both species. There was certainly exploitation of common resources by the two species; but in addition, the beetles preyed upon each other. The larvae and adults ate eggs and pupae, cannibalizing their own species as well as attacking the other species, and their propensity for doing so is summarized in Table 8.1. The important point is that taken overall, beetles of both species ate more individuals of the other species than they did of their own. Thus, a crucial mechanism in the interaction of these competing species was reciprocal predation (i.e. mutual antagonism), and it is easy to see that both species were more affected by inter- than intraspecific predation.

Figure 8.9c, the Lotka–Volterra model, suggests that the consequences of mutual antagonism are essentially the same whatever the exact mechanism. Because species are affected more by inter- than intraspecific competition, the outcome is strongly dependent on the relative abundances of the competing species. The small amount of interspecific aggression displayed by a rare species will have relatively little effect on an abundant competitor; but the large amount of aggression displayed by an abundant species might easily drive a rare species to local extinction. Moreover, if abundances are finely balanced, a small change in relative abundance will be sufficient to shift the advantage from one species to the other. The outcome of competition will then be unpredictable – either species could exclude the other, depending on the exact densities that they start with or attain.

8.5.1   Unpredictable gaps: the poorer competitor is a better colonizer

'Gaps' of unoccupied space occur unpredictably in many environments. Fires, landslips and lightning can create gaps in woodlands; storm-force seas can create gaps on the shore; and voracious predators can create gaps almost anywhere. Invariably, these gaps are recolonized. But the first species to do so is not necessarily the one that is best able to exclude other species in the long term. Thus, so long as gaps are created at the appropriate frequency, it is possible for a 'fugitive' species and a highly competitive species to coexist. The fugitive species tends to be the first to colonize gaps; it establishes itself, and it reproduces. The other species tends to be slower to invade the gaps, but having begun to do so, it outcompetes and eventually excludes the fugitive from that particular gap.

fugitive annuals
and competitive
perennials

This outline sketch has been given some quantitative substance in a simulation model in which the 'fugitive' species is thought of as an annual plant and the superior competitor as a perennial (Crawley & May, 1987). The model is one of a growing number that combine temporal and spatial dynamics by having interactions occur within individual cells of a two-dimensional lattice, but also having movement between cells (see also Inghe, 1989; Dytham, 1994; Bolker *et al.*, 2003). In this model, each cell can either be empty or occupied by either a single individual of the annual or a single ramet of the perennial. Each 'generation', the perennial can invade cells adjacent to those it already occupies, and it does so irrespective of whether those cells support an annual (a reflection of the perennial's competitive superiority), but individual ramets of the perennial may also die. The annual, however, can colonize any empty cell, which it does through the deposition of randomly dispersed 'seed', the quantity of which reflects the annual's abundance. Putting details aside, the annual can coexist with its superior competitor, providing the product (cE^*) of the annual's fecundity (c) and the equilibrium proportion of empty cells (E^*) is sufficiently great (Figure 8.10), i.e. as long as the annual is a sufficiently good colonizer and there are sufficient opportunities for it to do so. Indeed, the greater cE^*, the more the balance in the equilibrium mixture shifts towards the annual (Figure 8.10).

coexistence of a
competitive mussel
and a fugitive sea
palm

An example is provided by the coexistence of the sea palm *Postelsia palmaeformis* (a brown alga) and the mussel *Mytilus californianus* on the coast of Washington (Paine, 1979). *Postelsia* is an annual that must re-establish itself each year in order to persist at a site. It does so by attaching to the bare rock, usually in gaps in the mussel bed created by wave action. However, the mussels themselves slowly encroach on these gaps, gradually filling them and precluding colonization by

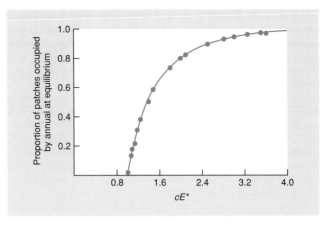

Figure 8.10   In a spatial lattice, a model fugitive annual plant can coexist with a competitively superior perennial provided $cE^* > 1$ (where c is the annual's fecundity and E^* the equilibrium proportion of empty cells in the lattice). For larger values, the fraction of cells occupied by the annual increases with cE^*. (After Crawley & May, 1987.)

Postelsia. Paine found that these species coexisted only at sites in which there was a relatively high average rate of gap formation (about 7% of surface area per year), and in which this rate was approximately the same each year. Where the average rate was lower, or where it varied considerably from year to year, there was (either regularly or occasionally) a lack of bare rock for colonization. This led to the overall exclusion of *Postelsia*. At the sites of coexistence, on the other hand, although *Postelsia* was eventually excluded from each gap, these were created with sufficient frequency and regularity for there to be coexistence in the site as a whole.

8.5.2   Unpredictable gaps: the preemption of space

When two species compete on equal terms, the result is usually predictable. But in the colonization of unoccupied space, competition is rarely even handed. Individuals of one species are likely to arrive, or germinate from the seed bank, in advance of individuals of another species. This, in itself, may be enough to tip the competitive balance in favor of the first species. If space is preempted by different species in different gaps, then this may allow coexistence, even though one species would always exclude the other if they competed 'on equal terms'.

first come, first
served

For instance, Figure 8.11 shows the results of a competition experiment between the annual grasses *Bromus madritensis* and *B. rigidus*, which occur together in Californian rangelands (Harper, 1961). When they were sown simultaneously in an equiproportional mixture, *B. rigidus* contributed overwhelmingly to the biomass of the mixed population. But, by delaying the introduction of *B. rigidus*

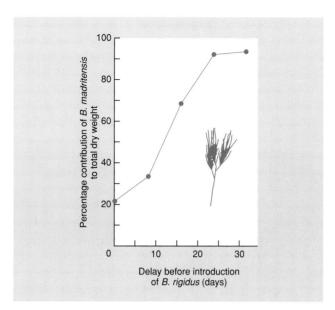

Figure 8.11 The effect of timing on competition. *Bromus rigidus* makes an overwhelming contribution to the total dry weight per pot after 126 days growth when sown at the same time as *B. madritensis*. But, as the introduction of *B. rigidus* is delayed, its contribution declines. Total yield per pot was unaffected by delaying the introduction of *B. rigidus*. (After Harper, 1961.)

into the mixtures, the balance was tipped decisively in favour of *B. madritensis*. It is therefore quite wrong to think of the outcome of competition as being always determined by the inherent competitive abilities of the competing species. Even an 'inferior' competitor can exclude its superior if it has enough of a head start. This can foster coexistence when repeated colonization occurs in a changing or unpredictable environment.

8.5.3 Fluctuating environments

paradox of the plankton

The balance between competing species can be shifted repeatedly, in fact, and coexistence therefore fostered, simply as a result of environmental change. This was the argument used by Hutchinson (1961) to explain the 'paradox of the plankton' – the paradox being that numerous species of planktonic algae frequently coexist in simple environments with little apparent scope for niche differentiation. Hutchinson suggested that the environment, although simple, was continually changing, particularly on a seasonal basis. Thus, although the environment at any one time would tend to promote the exclusion of certain species, it would alter and perhaps even favor these same species before exclusion occurred. In other words, the equilibrium outcome of a competitive interaction may not be of paramount importance

if the environment typically changes long before the equilibrium can be reached.

8.5.4 Ephemeral patches with unpredictable lifespans

Many environments, by their very nature, are not simply variable but ephemeral. Amongst the more obvious examples are decaying corpses (carrion), dung, rotting fruit and fungi,

coexistence of the strong with the fast . . .

and temporary ponds. But note too that a leaf or an annual plant can be seen as an ephemeral patch, especially if it is palatable to its consumer for only a limited period. Often, these ephemeral patches have an unpredictable lifespan – a piece of fruit and its attendant insects, for instance, may be eaten at any time by a bird. In these cases, it is easy to imagine the coexistence of two species: a superior competitor and an inferior competitor that reproduces early.

One example concerns two species of pulmonate snail living in ponds in northeastern Indiana. Artificially altering the density of one or other species in the field showed that the fecundity of *Physa gyrina* was significantly reduced by interspecific competition from *Lymnaea elodes*, but the effect was not reciprocated. *L. elodes* was clearly the superior competitor when competition continued throughout the summer. Yet *P. gyrina* reproduced earlier and at a smaller size than *L. elodes*, and in the many ponds that dried up by early July it was often the only species to have produced resistant eggs in time. The species therefore coexisted in the area as a whole, in spite of *P. gyrina*'s apparent inferiority (Brown, 1982). Among frogs and toads, on the other hand, the competitively superior tadpoles of *Scaphiopus holbrooki* are even more successful when ponds dry up because they have shorter larval periods than weaker competitors such as *Hyla chrysoscelis* (Wilbur, 1987).

. . . but not always

8.5.5 Aggregated distributions

A more subtle, but more generally applicable path to the coexistence of a superior and an inferior competitor on a patchy and ephemeral resource is based on the idea that the two species may have independent, aggregated (i.e.

a clumped superior competitor adversely affects itself and leaves gaps for its inferior

clumped) distributions over the available patches. This would mean that the powers of the superior competitor were mostly directed against members of its own species (in the high-density clumps), but that this aggregated superior competitor would be absent from many patches – within which the inferior competitor could escape competition. An inferior competitor may then be able to coexist with a superior competitor that would rapidly exclude it

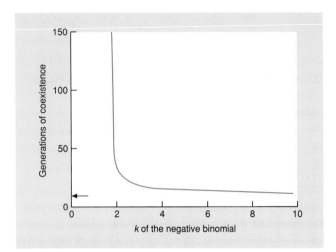

Figure 8.12 When two species compete on a continuously distributed resource, one species would exclude the other in approximately 10 generations (as indicated by the arrow). However, with these same species on a patchy and ephemeral resource, the number of generations of coexistence increases with the degree of aggregation of the competitors, as measured by the parameter *k* of the 'negative binomial' distribution. Values above 5 are effectively random distributions; values below 5 represent increasingly aggregated distributions. (After Atkinson & Shorrocks, 1981.)

from a continuous, homogeneous environment. Certainly it can do so in models (see, for example, Atkinson & Shorrocks, 1981; Kreitman *et al.*, 1992; Dieckmann *et al.*, 2000). For instance, a simulation model (Figure 8.12) shows that the persistence of such coexistence between competitors increases with the degree of aggregation (as measured by the parameter *k* of the 'negative binomial' distribution) until, at high levels of aggregation, coexistence is apparently permanent, although this has nothing to do with any niche differentiation. Since many species have aggregated distributions in nature, these results may be applicable widely.

Note, however, that whilst such coexistence of competitors has nothing to do with niche differentiation, it is linked to it by a common theme – that of species competing more frequently and intensively intraspecifically than they do interspecifically. Niche differentiation is one means by which this can occur, but temporary aggregations can give rise to the same phenomenon – even for the inferior competitor.

In seeking to justify the applicability of these models to the real world, however, one question in particular needs to be answered: are two similar species really likely to have independent distributions over available patches of resource? The question has been addressed through an examination of a large number of data sets from Diptera, especially drosophilid flies – where eggs are laid, and larvae develop, in ephemeral patches (fruits, fungi, flowers, etc.). In fact, there was little evidence for independence

in the aggregations of coexisting species (Shorrocks *et al.*, 1990; see also Worthen & McGuire, 1988). However, computer simulations suggest that whilst a positive association between species (i.e. a tendency to aggregate in the same patches) does make coexistence more difficult, the *level* of association and aggregation actually found would still generally lead to coexistence, whereas there would be exclusion in a homogeneous environment (Shorrocks & Rosewell, 1987).

The importance of aggregation for coexistence has been further supported by another spatially explicit model based on a two-dimensional lattice of cells (see Section 8.5.1), each of which could be occupied by one of five species of grass: *Agrostis stolonifera*, *Cynosurus cristatus*, *Holcus lanatus*, *Lolium perenne* and *Poa trivialis* (Silvertown *et al.*, 1992). The model was a 'cellular automaton', in which each cell can exist in a limited number of discrete states (in this case, which species was in occupancy), with the state of each cell determined at each time step by a set of rules. In this case, the rules were based on the cell's current state, the state of the neighboring cells and the probability that a species in a neighboring cell would replace its current occupant. These replacement rates of each species by each other species were themselves based on field observations (Thórhallsdóttir, 1990).

grasses in a cellular automaton

If the initial arrangement of the species over the grid was random (no aggregation), the three competitively inferior species were quickly driven to extinction, and of the survivors, *Agrostis* (greater than 80% cell occupancy) rapidly dominated *Holcus*. If, however, the initial arrangement was five equally broad single-species bands across the landscape, the outcome changed dramatically: (i) competitive exclusion was markedly delayed even for the worst competitors (*Cynosurus* and *Lolium*); (ii) *Holcus* sometimes occupied more than 60% of the cells, at a time (600 time steps) where, with an initially random arrangement, it would have been close to extinction; and (iii) the outcome itself depended largely on which species started next to each other, and hence, initially competed with each other.

There is no suggestion, of course, that natural communities of grasses exist as broad single-species bands – but, neither are we likely to find communities with species mixed at random, such that there is no spatial organization to be taken into account. The model emphasizes the dangers of ignoring aggregations (because they shift the balance towards intra- rather than interspecific competition, and hence promote coexistence), but also the dangers of ignoring the juxtaposition of aggregations, since these too may serve to keep competitive subordinates away from their superiors.

Despite a rich body of theory and models, there are few experimental studies that directly address the impact of spatial patterns on population dynamics. Stoll and Prati (2001) performed experiments with real plants in a study that had much in common with Silvertown's

plants in a field experiment

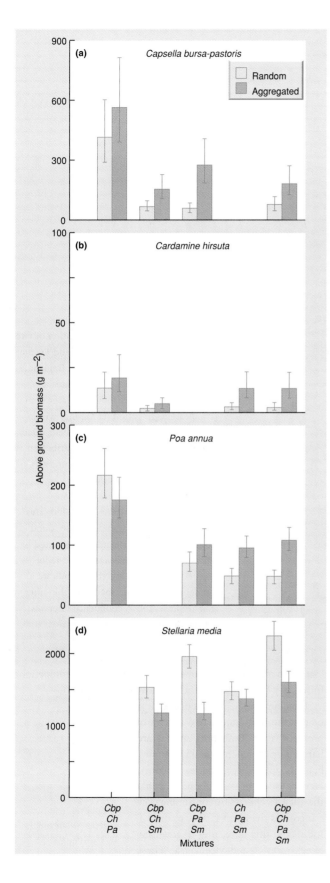

theoretical treatment. They tested the hypothesis that intraspecific aggregation can promote coexistence and thus maintain high species richness in experimental communities of four annual terrestrial plants: *Capsella bursa-pastoris*, *Cardamine hirsuta*, *Poa annua* and *Stellaria media*. *Stellaria* is known to be the superior competitor among these species. Replicate three- and four-species mixtures were sown at high density, and the seeds were either placed completely at random or seeds of each species were aggregated in subplots within the experimental areas. Intraspecific aggregation decreased the performance of the superior *Stellaria* in the mixtures, whereas in all but one case aggregation improved the performance of the three inferior competitors (Figure 8.13).

More generally, the success of 'neighborhood' approaches (Pacala, 1997) in the study of plant competition, where the focus is on the competition experienced by individuals in local patches, rather than densities averaged out over whole populations, argues again in favor of the importance of acknowledging spatial heterogeneity. Coomes *et al.* (2002), for example, investigated competition between two species of sand-dune plant, *Aira praecox* and *Erodium cicutarium*, in northwest England. The smaller plant, *Aira*, tended to be aggregated even at the smallest spatial scales, whereas *Erodium* was moderately aggregated in patches of 30 and 50 mm radius but, if anything, was evenly spaced within 10 mm radius patches (Figure 8.14a). The two species, though, were negatively associated with one another at the smallest spatial scale (Figure 8.14b), indicating that *Aira* tended to occur in small, single-species clumps. *Aira* was therefore much less liable to competition from *Erodium* than would be the case if they were distributed at random, justifying the application by Coomes *et al.* of simulation models of competition where local responses were explicitly incorporated.

Repeatedly in this section, then, the heterogeneous nature of the environment can be seen to have fostered coexistence without there being a marked differentiation of niches. A realistic view of interspecific competition, therefore, must acknowledge that it often proceeds not in isolation, but under the influence of, and within the constraints of, a patchy, impermanent or unpredictable world. Furthermore, the heterogeneity need not be in the temporal or spatial dimensions that we have discussed so far. Individual variation in competitive ability

heterogeneity often stabilizes

Figure 8.13 (*left*) The effect of intraspecific aggregation on above-ground biomass (mean ± SE) of four plant species grown for 6 weeks in three- and four-species mixtures (four replicates of each). The normally competitively superior *Stellaria media* (Sm) did consistently less well when seeds were aggregated than when they were placed at random. In contrast, the three competitively inferior species – *Capsella bursa-pastoris* (Cbp), *Cardamine hirsuta* (Ch) and *Poa annua* (Pa) – almost always performed better when the seeds had been aggregated. Note the different scales on the vertical axes. (From Stoll & Prati, 2001.)

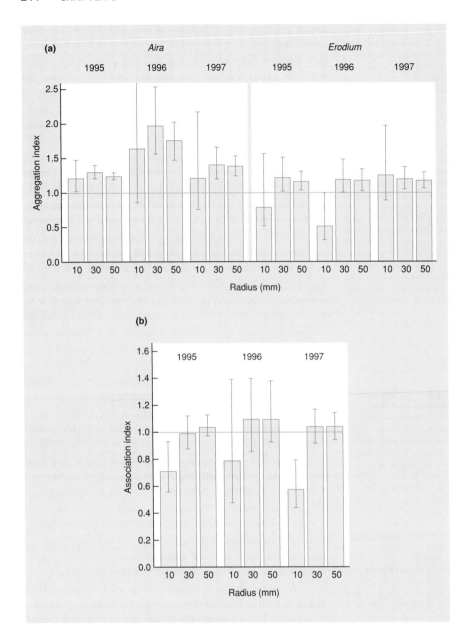

Figure 8.14 (a) Spatial distribution of two sand-dune species, *Aira praecox* and *Erodium cicutarium* at a site in northwest England. An aggregation index of 1 indicates a random distribution. Indices greater than 1 indicate aggregation (clumping) within patches with the radius as specified; values less than 1 indicate a regular distribution. Bars represent 95% confidence intervals. (b) The association between *Aira* and *Erodium* in each of the 3 years. An association index greater than 1 indicates that the two species tended to be found together more than would be expected by chance alone in patches with the radius as specified; values less than 1 indicate a tendency to find one species or the other. Bars represent 95% confidence intervals. (After Coomes *et al.*, 2002.)

within species can also foster stable coexistence in cases where a superior nonvariable competitor would otherwise exclude an inferior nonvariable species (Begon & Wall, 1987). This reinforces a point that recurs throughout this text: heterogeneity (spatial, temporal or individual) can have a stabilizing influence on ecological interactions.

8.6   Apparent competition: enemy-free space

Another reason for being cautious in our discussion of competition is the existence of what Holt (1977, 1984) has called

'apparent competition', and what others have called 'competition for enemy-free space' (Jeffries & Lawton, 1984, 1985).

Imagine a single species of predator or parasite that attacks two species of prey (or host). Both prey species are harmed by the enemy, and the enemy benefits from both species of prey. Hence, the increase in abundance that the enemy achieves by consuming prey 1 increases the harm it does to prey 2. Indirectly, therefore, prey 1 adversely affects prey 2 and vice versa. These

two prey species attacked by a predator are, in essence, indistinguishable from two consumer species competing for a resource

Figure 8.15 In terms of the signs of their interactions, all of the following are indistinguishable from one another: (a) two species interfering directly (interference competition); (b) two species consuming a common resource (exploitation competition); (c) two species being attacked by a common predator ('apparent competition' for 'enemy-free space'); and (d) two species linked by a third which is a competitor of one and a mutualist of the other. (——), direct interactions; (- - - - -), indirect interactions; arrows indicate positive influences, circles indicate negative influences. (After Holt, 1984; Connell, 1990.)

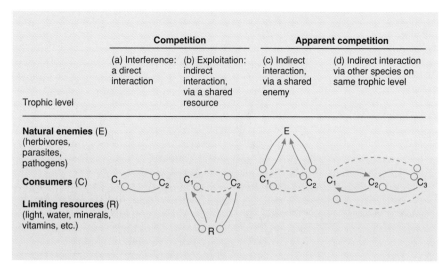

interactions are summarized in Figure 8.15, which shows that from the point of view of the two prey species, the signs of the interactions are indistinguishable from those that would apply in the indirect interaction of two species competing for a single resource (exploitation competition). In the present case there appears to be no limiting resource. Hence, the term 'apparent competition'.

evidence for apparent competition . . .

. . . in two caterpillars sharing a parasitoid, . . .

In an experiment involving a parasitoid (the ichneumonid wasp *Venturia canescens*) and two caterpillar hosts (*Plodia interpunctella* and *Ephestia kuehniella*), Bonsall and Hassell (1997) allowed free passage of the parasitoid between the host species but kept the hosts apart to avoid the possibility of resource competition between them. When the experimental chambers contained just a single host species together with the parasitoid, both the parasite and host persisted and exhibited damped oscillations in population size, tending towards a stable equilibrium (Figure 8.16). But when the system was run with both host species, the parasitoid had a greater impact on the species with the lower intrinsic rate of increase (*E. kuehniella*). This host showed increasing population oscillations and invariably went extinct. By means of their elegant experimental design, Bonsall and Hassell were able to demonstrate the effect of apparent competition in a situation where resource competition between the caterpillar species was ruled out.

While the term 'apparent competition' is entirely appropriate, it is sometimes useful to think of 'enemy-free space' as the limiting resource for which prey (or host) species compete. This is because the persistence of prey species 1 will be favored by avoiding attacks from the predator, which we know also attacks prey 2. Clearly, prey 1 can achieve this by occupying a habitat, or adopting a form or a behavioral pattern, that is sufficiently different from that of prey 2. In short, 'being different' (i.e. niche differentiation) will once again favor coexistence – but it will do so because it diminishes apparent competition or competition for enemy-free space.

A rare experimental demonstration of apparent competition for enemy-free space involves two groups of prey living on subtidal rocky reefs at Santa Catalina Island, California. The first comprises three species of mobile gastropods, *Tegula aureotincta*, *T. eiseni* and *Astraea undosa*; the second comprises sessile bivalves, dominated by the clam *Chama arcana*. Both groups were preyed upon by a lobster (*Panulirus interruptus*), an octopus (*Octopus bimaculatus*) and a whelk (*Kelletia kelletii*), although these predators showed a marked preference for the bivalves. In areas characterized by large boulders and much crevice space ('high relief') there were high densities of bivalves and predators, but only moderate densities of gastropods; whereas in low relief areas largely lacking crevice space ('cobble fields') there were apparently no bivalves, only a few predators but high densities of gastropods.

. . . in gastropods, bivalves and their predators . . .

The densities of the two prey groups were inversely correlated, but there was little in their feeding biology to suggest that they were competing for a shared food resource. On the other hand, when bivalves were experimentally introduced into cobble-field areas, the number of predators congregating there increased, the mortality rates of the gastropods increased (often observably associated with lobster or octopus predation) and the densities of the gastropods declined (Figures 8.17a, b). Experimental manipulation of the (mobile) gastropods proved impossible, but cobble sites with high densities of gastropods supported higher densities of predators, and had higher mortality rates of experimentally added bivalves than did sites with relatively low densities of gastropods (Figure 8.17c). On the rare high relief sites without *Chama* bivalves, predator densities were lower, and gastropod densities higher, than was normally the case (Figure 8.17d). It seems clear that each prey group adversely affected the other through an

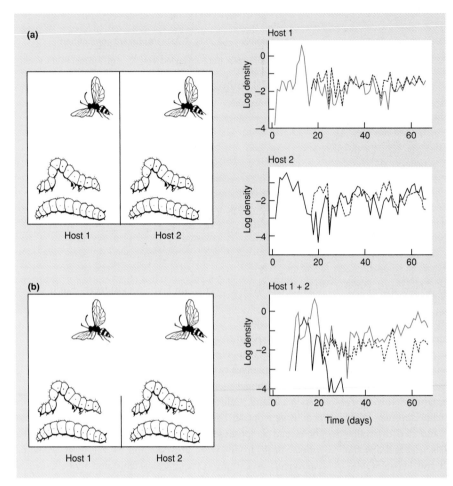

Figure 8.16 Parasite-mediated apparent competition via a parasitoid wasp *Venturia canescens* that lays eggs in two caterpillar host species. The experimental setups are illustrated on the left and the population dynamics of the parasitoid (dashed black lines) and host species (host 1 *Plodia interpunctella* (orange lines); host 2 *Ephestia kuehniella* (black lines)) on the right. (a) When only a single host was present, the parasitoid and host coexisted with stable dynamics. (b) When the parasitoid had access to both hosts, host 2 showed diverging oscillations and went extinct. (From Hudson & Greenman, 1998, after Bonsall & Hassell, 1997.)

increased number of predators, and hence increased predator-induced mortality.

... and in leaf-mining flies sharing parasitoids in a tropical forest

An experiment with a similar aim involved removing a common leaf-mining fly (*Calycomyza* sp.) and its host plant *Lepidaploa tortuosa* (Asteracea) in replicate sites in a tropical forest community in Belize, Central America. Other leaf-mining fly species that shared natural enemies (parasitoid wasps) with *Calycomyza*, but whose host plants were different, demonstrated reduced parasitism and increased abundance (a year later) in the removal sites than in the control sites (Morris *et al.*, 2004). These results support predictions of apparent competition, involving a shared natural enemy, in a situation where interspecific competition among the fly species for host plants could not occur.

To complete the picture, there is another indirect interaction between two species that qualifies for the term 'apparent competition' (Figure 8.15d), where species 1 and 2 have negative impacts on one another, and species 2 and 3 have positive (mutualistic) impacts (see Chapter 13). Species 1 and 3 then have indirect negative impacts on one another without sharing a common resource or, for that matter, a common predator. They exhibit apparent competition, although not for enemy-free space (Connell, 1990).

reappraisal of plant competition

The examples mentioned so far concern apparent competition in animals. Connell (1990) carried out a particularly revealing reappraisal of 54 published plant examples of field experiments on 'competition', where the original authors had claimed to have demonstrated conventional interspecific competition in 50. A closer look revealed that, in many of these, insufficient information had been collected to distinguish between conventional competition and apparent competition; and in a number of others the information was available – but was ambiguous. For example, one study showed that removal of *Artemisia* bushes from a large site in Arizona led to much better growth of 22 species of herb than was observed in either undisturbed sites or sites where *Artemisia* was removed from narrow 3 m strips. This was originally interpreted in terms of greatly

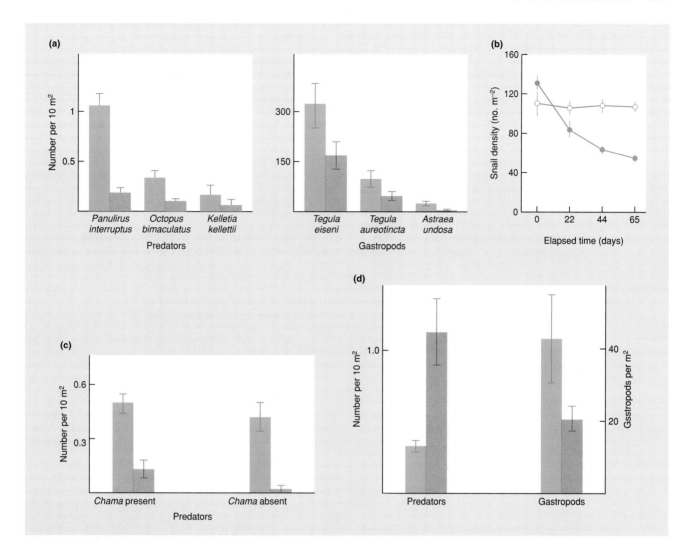

Figure 8.17 Evidence for apparent competition for predator-free space at Santa Catalina Island, USA. (a) Predator density (number per 10 m², with standard errors) and gastropod mortality increased (number of 'newly dead' shells per site, with standard errors) when bivalves were added to gastropod-dominated cobble sites (colored bars) relative to controls (gray bars). (b) This led to a decline in gastropod density (standard error bars shown). (c) Predator density was higher (number per 10 m², with standard errors) at high (colored bars) than at low (gray bars) gastropod-density cobble sites, both in the presence and absence of *Chama*. (d) Densities of predators were lower (number per 10 m², with standard errors) and densities of gastropods higher (number per m², with standard errors) at high-relief sites without *Chama* (colored bars) than at those with (gray bars). (After Schmitt, 1987.)

reduced exploitative competition for water in the former case (Robertson, 1947). However, the herbs in the larger site also experienced greatly reduced grazing pressure from deer, rodents and insects, for which the *Artemisia* bushes were not only a source of food but a place of shelter, too. The outcome is therefore equally likely to have resulted from reduced apparent competition.

distinguishing pattern and process

This emphasizes that the relative neglect of apparent competition in the past has been unwarranted, but also re-emphasizes that the distinction is important within interspecific competition between pattern on the one hand, and process or mechanism on the other. In the past, patterns of niche differentiation, and also of increased abundance of one species in the absence of another, have been interpreted as evidence of competition too readily. Now we can see that such patterns can arise through a wide variety of processes, and that a proper understanding requires that we distinguish between them – not only discriminating between conventional and apparent competition, but also specifying mechanisms within, say, conventional competition (a point to which we return in Section 8.10).

8.7 Ecological effects of interspecific competition: experimental approaches

field and laboratory experiments

Notwithstanding the important interactions between competition and environmental heterogeneity, and the complications of apparent competition, a great deal of attention has been focused on conventional competition itself. We have already noted the difficulties in interpreting merely observational evidence (but see Freckleton & Watkinson, 2001), and it is for this reason that many studies of the ecological effects of interspecific competition have taken an experimental approach. For example, we have seen manipulative field experiments involving barnacles (see Section 8.2.2), birds (see Section 8.2.5), cattails (see Section 8.3.3) and snails (see Section 8.5.4), where the density of one or both species was altered (usually reduced). The fecundity, the survivorship, the abundance or the resource utilization of the remaining species was subsequently monitored. It was then compared either with the situation prior to the manipulation, or, far better, with a comparable control plot in which no manipulation had occurred. Such experiments have consistently provided valuable information, but they are typically easier to perform on some types of organism (e.g. sessile organisms) than they are on others.

The second type of experimental evidence has come from work carried out under artificial, controlled (often laboratory) conditions. Again, the crucial element has usually been a comparison between the responses of species living alone and their responses when in combination. Such experiments have the advantage of being comparatively easy to perform and control, but they have two major disadvantages. The first is that species are examined in environments that are different from those they experience naturally. The second is the simplicity of the environment: it may preclude niche differentiation because niche dimensions are missing that would otherwise be important. Nevertheless, these experiments can provide useful clues to the likely effects of competition in nature.

8.7.1   Longer term experiments

The most direct way of discovering the outcome of competition between two species in the laboratory, or under other controlled conditions, is to put them together and leave them to it. However, since even the most one-sided competition is likely to take a few generations (or a reasonable period of modular growth) before

it is completed, this direct approach is easier, and has been more frequently used, in some species than in others. It has most frequently been applied to insects (such as the flour beetle example in Section 8.4.3) and microorganisms (such as the *Paramecium* example in Section 8.2.4). Note that neither higher plants, nor vertebrates, nor large invertebrates, lend themselves readily to this approach (although a plant example is discussed in Section 8.10.1). We must be aware that this may bias our view of the nature of interspecific competition.

8.7.2   Single-generation experiments

Given these problems, the alternative 'laboratory' approach, especially with plants (although the methods have occasionally been used with animals), has generally been to follow populations over just a single generation, comparing 'inputs' and 'outputs'. A number of experimental designs have been used.

In 'substitutive' experiments, the effect of varying the proportion of each of two species is explored whilst keeping overall density constant (de Wit, 1960). Thus, at an overall density of say 200 plants, a series of mixtures would be set up: 100 of species A with 100 of species B, 150 A and 50 B, 0 A and 200 B, and so on. At the end of the experimental period, the amount of seed or the biomass of each species in each mixture would be monitored. Such replacement series may then be established at a range of total densities. In practice, however, most workers have used only a single total density, and this has led to considerable criticism of the design since it means that the effect of competition over several generations – when total density would inevitably alter – cannot be predicted (see Firbank & Watkinson, 1990).

None the less, replacement series have provided valuable insights into the nature of interspecific competition and the factors influencing its intensity (Firbank & Watkinson, 1990). An early, influential study was that of de Wit *et al.* (1966) on competition between the grass *Panicum maximum* and the legume *Glycine javanica*, which often form mixtures in Australian pastures. *Panicum* acquires its nitrogen only from the soil, but *Glycine* acquires part of its nitrogen from the air, by nitrogen fixation, through its root association with the bacterium *Rhizobium* (see Section 13.10.1). The competitors were grown in replacement series with and without an inoculation of *Rhizobium*, and the results are given both as replacement diagrams and as 'relative yield totals' (Figure 8.18). The relative yield of a species in a mixture is the ratio of its yield in the mixture to its yield alone in the replacement series, removing any absolute yield differences between species and referring both to the same scale. The relative yield total of a particular mixture is then the sum of the two relative yields. It is fairly clear from the replacement series (Figure 8.18a)

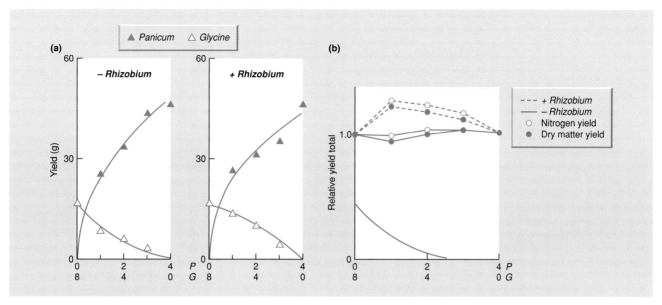

Figure 8.18 A substitutive experiment on interspecific competition between *Panicum maximum* (P), and *Glycine javanica* (G), in the presence and absence of *Rhizobium*: (a) replacement diagrams; (b) relative yield totals. (After de Wit *et al.*, 1966.)

that both species, but especially *Glycine*, fared better (were less affected by interspecific competition) in the presence than in the absence of *Rhizobium*. This is clearer still, however, from the relative yield totals (Figure 8.18b), which never departed significantly

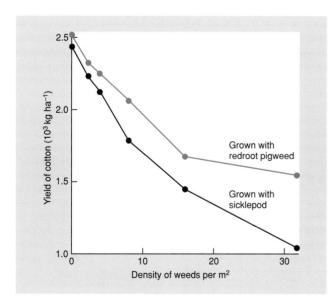

Figure 8.19 An 'additive design' competition experiment: the yield of cotton produced from stands planted at constant density, infested with weeds (either sicklepod or redroot pigweed) at a range of densities. (After Buchanan *et al.*, 1980.)

from 1 in the absence of *Rhizobium*, but consistently exceeded 1 in its presence. This suggested that niche differentiation was not possible without *Rhizobium* (a second species could only be accommodated by a compensatory reduction in the output of the first) and that niche differentiation occurred in its presence (the species yielded more between them than either could alone).

A second popular approach in the past has been the use of an 'additive' design, in which one species (typically a crop) is sown at a constant density, along with a range of densities of a second species (typically a weed). The justification for this is that it mimics the natural situation of a crop infested by a weed, and it therefore provides information on the likely effect on the crop of various levels of infestation (Firbank & Watkinson, 1990). A problem with additive experiments, however, is that overall density and species proportion are changed simultaneously. It has therefore proved difficult to separate the effect of the weed itself on crop yield from the simple effect of increasing total density (crop plus weed). An example is shown in Figure 8.19, describing the effects of two weeds, sicklepod (*Cassia obtusifolia*) and redroot pigweed (*Amaranthus retroflexus*), on the yield of cotton grown in Alabama (Buchanan *et al.*, 1980). As weed density increased, so cotton yield decreased, and this effect of interspecific competition was always more pronounced with sicklepod than with redroot pigweed.

In substitutive designs the proportions of competitors are varied but total density is held constant, whilst in

additive experiments

response surface analysis

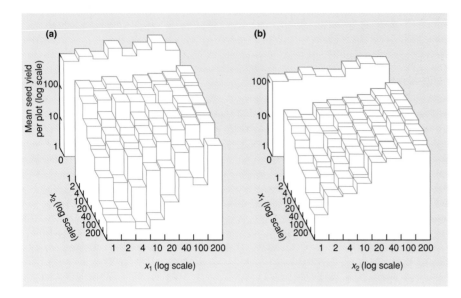

Figure 8.20 The response surface of competition, as indicated by seed production per pot, between (a) *Phleum arenarium* and (b) *Vulpia fasciculata* sown alone and in mixtures over a range of densities and frequencies. x_1 and x_2 are the sowing densities of Phleum and Vulpia, respectively. (After Law & Watkinson, 1987.)

additive designs the proportions are varied but the density of one competitor is held constant. It is perhaps not surprising, therefore, and certainly welcome, that a 'response surface analysis' has been proposed and applied, in which two species are grown alone and in mixtures at a wide range of densities and proportions (Figure 8.20) (Firbank & Watkinson, 1985; Law & Watkinson, 1987; Bullock *et al.*, 1994b; although the last of these deals with clones of the same species). Overall, these studies suggest that good equations for describing the competitive effect of one species (A) on another (B) are, for mortality:

$$N_A = N_{i,A}[1 + m(N_{i,A} + \beta N_{i,B})]^{-1} \qquad (8.16)$$

and for fecundity:

$$Y_A = N_A R_A [1 + a(N_A + \alpha N_B)]^{-b}, \qquad (8.17)$$

which can both be seen to be related to Equation 5.17 (see Section 5.8.1 – basic model of intraspecific competition) and Equation 5.12 (see Section 8.4.1 – incorporation of interspecific competition). Thus, $N_{i,A}$ and $N_{i,B}$ are the initial numbers of species A and B; N_A and N_B are the numbers of species A and B after mortality; Y_A is the yield (seeds or biomass) of species A; m and a are susceptibilities to crowding; β and α are competition coefficients; R_A is the basic reproductive rate of species A (and hence, $N_A R_A$ is the yield in the absence of competition); and b determines the type of density dependence (assumed equal to 1 for mortality – perfect compensation). Data like those shown in Figure 8.20, obtained over a single generation, can thus be used to fit values (by computer program) to the parameters in Equations 8.16 and 8.17, and the equations in their turn can be used to predict the

outcome of competition between the species over many generations – which is not possible with either substitutive or additive designs.

On the other hand, Law and Watkinson (1987) found that they could obtain an improved fit to their response surfaces, especially for one of their species, if they used an equation in which competition coefficients were not fixed, but varied with both frequency and density – although the meaning of this in terms of 'plant behavior' is not clear. Hence, response surface analyses, in revealing the potential complexities in interactions between competing species, also reveal that knowing or predicting dynamical outcomes may be only part of the story. It may also be necessary to understand the underlying mechanisms (see Section 8.10).

8.8 Evolutionary effects of interspecific competition

8.8.1 Natural experiments

We have seen that interspecific competition is commonly studied by an experimenter comparing species alone and in combination. Nature, too, often provides information of this sort: the distribution of certain potentially competing species is such that they sometimes occur together (sympatry) and sometimes occur alone (allopatry). These 'natural experiments' can provide additional information about interspecific competition, and especially about evolutionary effects, since the differences between sympatric and allopatric populations are often of long standing. The attractions of natural experiments are first that they

pros and cons of natural experiments

are natural – they are concerned with organisms living in their natural habitats – and second, that they can be 'carried out' simply by observation – no difficult or impracticable experimental manipulations are required. They have the disadvantage, however, of lacking truly 'experimental' and 'control' populations. Ideally, there should be only one difference between the populations: the presence or absence of a competitor species. In practice, though, populations typically differ in other ways too, simply because they exist in different locations. Natural experiments should therefore always be interpreted cautiously.

competitive release and character displacement

Evidence for competition from natural experiments usually comes either from niche expansion in the absence of a competitor (known as *competitive release*) or simply from a difference in the realized niche of a species between sympatric and allopatric populations. If this difference is accompanied by morphological changes, then the effect is referred to as *character displacement*. On the other hand, physiological, behavioral and morphological traits are all equally likely to be involved in competitive interactions and to be reflections of a species' realized niche. One difference may be that morphological distinctions are most obviously the result of evolutionary change, but as we shall see, physiological and behavioral 'characters' are also liable to 'competitive displacement'.

gerbils in Israel: competitive release

One example of natural competitive release is provided by work on two gerbilline rodents living in the coastal sand dunes of Israel (Abramsky & Sellah, 1982). In northern Israel, the protrusion of the Mt Carmel ridge towards the sea separates the narrow coastal strip into two isolated areas, north and south. *Meriones tristrami* is a gerbil that has colonized Israel from the north. It now occurs, associated with the dunes, throughout the length of the coast, including the areas both north and south of Mt Carmel. *Gerbillus allenbyi* is another gerbil, also associated with the dunes and feeding on similar seeds to *M. tristrami*; but this species has colonized Israel from the south and has not crossed the Mt Carmel ridge. To the north of Mt Carmel, where *M. tristrami* lives alone, it is found on sand as well as other soil types. However, south of Mt Carmel it occupies several soil types but not the coastal sand dunes. Here, only *G. allenbyi* occurs on dunes.

This appears to be a case of competitive exclusion and competitive release: exclusion of *M. tristrami* by *G. allenbyi* from the sand to the south of Mt Carmel; release of *M. tristrami* to the north. Is this present day competitive exclusion, however, or an evolutionary effect? Abramsky and Sellah set up a number of plots south of Mt Carmel from which *G. allenbyi* was removed, and they compared the densities of *M. tristrami* in these plots with those in a number of similar control plots. They monitored the plots for 1 year, but the abundance of *M. tristrami* remained essentially unchanged. It seems that south of Mt Carmel, *M. tristrami* has

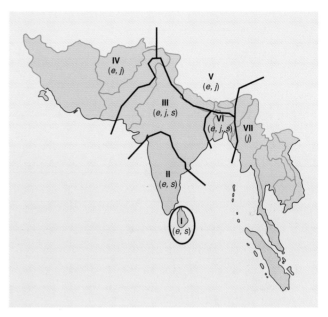

Figure 8.21   Native geographic ranges (I–VII) of *Herpestes javanicus* (*j*), *H. edwardsii* (*e*) and *H. smithii* (*s*). (From Simberloff *et al.*, 2000.)

evolved to select those habitats in which it avoids competition with *G. allenbyi*, and that even in the absence of *G. allenbyi* it retains this genetically fixed preference. Note, though, as ever, that this interpretation, because it invokes the ghost of competition past, may be sound and sensible – but it is not established fact.

A case of apparent morphological character displacement comes from work on Indian mongooses. In the western parts of its natural range, the small Indian mongoose (*Herpestes javanicus*) coexists with one or two slightly larger species in the same

morphological character displacement . . . . . . in Indian mongooses . . .

genus (*H. edwardsii* and *H. smithii*), but these species are absent in the eastern part of its range (Figure 8.21). Simberloff *et al.* (2000) examined size variation in the upper canine tooth, the animal's principal prey-killing organ (note that female mongooses are smaller than males). In the east where it occurs alone (area VII in Figure 8.21), both males and females have larger canines than in the western areas (III, V, VI) where it coexists with the larger species (Figure 8.22). This is consistent with the view that where similar but larger predators are present, the prey-catching apparatus of *H. javanicus* has been selected for reduced size. This is likely to reduce the strength of competition with other species in the genus because smaller predators tend to take smaller prey than larger predators. Where *H. javanicus* occurs in isolation, its canine teeth are much larger.

It is of particular interest that the small Indian mongoose was introduced about a century ago to many islands outside its

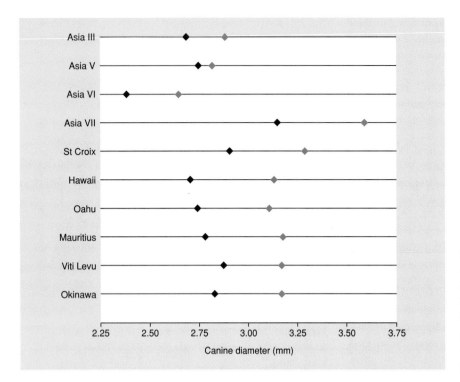

Figure 8.22 Maximum diameter (mm) of the upper canine for *Herpestes javanicus* in its native range (data only for areas III, V, VI and VII from Figure 8.21) and introduced range. Black symbols represent mean female size and colored symbols represent mean male size. (From Simberloff *et al.*, 2000.)

native range (often as part of a naive attempt to control introduced rodents). In these places, the larger competitor mongoose species are absent. Within 100–200 generations the small Indian mongoose has increased in size (Figure 8.22), so that the sizes of island individuals are now intermediate between those in the region of origin (where they coexisted with other species and were small) and those in the east where they occur alone. On the islands they show variation consistent with 'ecological release' from competition with larger species.

... and in three-spined sticklebacks in Canada

A further example concerns populations of the originally marine three-spined stickleback, *Gasterosteus aculeatus*, living in freshwater lakes in British Columbia, Canada, having apparently been left behind either following uplifting of the land after deglaciation, around 12,500 years ago, or after the subsequent rise and fall of sea levels around 11,000 years ago (Schluter & McPhail, 1992, 1993). As a result of this 'double invasion', some lakes now support two species of *G. aculeatus* (although they have not, as yet, been given their own specific names), whilst others support only one. Wherever there are two species, one is always 'limnetic', the other 'benthic'. The first concentrates its feeding on plankton in the open water and has correspondingly long (and closely spaced) gill rakers that seive the plankton from the stream of ingested water. The second, with much shorter gill rakers, concentrates on larger prey that it consumes largely from vegetation or sediments (Figure 8.23b). Wherever there is only one species in a lake, however, this species exploits both food

resources and is morphologically intermediate (Figure 8.23a). Presumably, either ecological character displacement has evolved since the second invasion, and this has promoted the coexistence of the species pairs, or it was a necessary prerequisite for the second invasion to be successful. Genetic evidence, based on

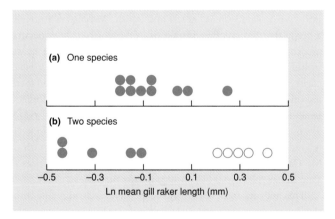

Figure 8.23 Character displacement in three-spined sticklebacks (*Gasterosteus aculeatus*). In small lakes in coastal British Columbia supporting two stickleback species (b), the gill rakers of the benthic species (●) are significantly shorter than those of the limnetic species (○), whilst those species of sticklebacks that occupy comparable lakes alone (a) are intermediate in length. Lengths of gill rakers have been adjusted to take account of species differences in overall size. (After Schluter & McPhail, 1993.)

analyses of mitochondrial DNA of several species pairs, supports the idea of repeated patterns of adaptive radiation within individual lakes (Rundle *et al.*, 2000).

If character displacement has ultimately been caused by competition, then the effects of competition should decline with the degree of displacement. Brook sticklebacks (*Culaea inconstans*) that are sympatric in Canadian lakes with ninespine sticklebacks (*Pungitius pungitius*) possess significantly shorter gill rakers, longer jaws and deeper bodies than allopatric brook sticklebacks. Gray and Robinson (2002) view allopatric brook sticklebacks as predisplacement phenotypes and sympatric brook sticklebacks as postdisplacement phenotypes. When each phenotype was separately placed in enclosures in the presence of ninespine sticklebacks, the allopatric (predisplacement) brook sticklebacks grew significantly less well than their sympatric (postdisplacement) counterparts (Figure 8.24). This is consistent with the hypothesis that competition is reduced when divergence between competing species occurs.

mud snails: a classic example of character displacement? Two final, plausible examples of character displacement are provided by work on mud snails in Finland (*Hydrobia ulvae* and *H. ventrosa*) and giant rhinoceros beetles in Southeast Asia (*Chalcosoma caucasus* and *C. atlas*). When the two mud snail species live apart, their sizes are more or less identical; but when they coexist they are always different in size (Figure 8.25a)

(Saloniemi, 1993) and they tend to consume different food particle sizes (Fenchel 1975). The beetles display a similar morphological pattern (Figure 8.25b) (Kawano, 2002). These data, therefore, strongly suggest character displacement, allowing coexistence. However, even an apparently exemplary example such as that of the mud snails is open to serious question (Saloniemi, 1993). In Finland, the sympatric and allopatric habitats were not identical: *H. ulvae* and *H. ventrosa* coexisted in sheltered water bodies rarely affected by tidal action, *H. ulvae* was found alone in relatively exposed tidal mudflats and salt marshes, and *H. ventrosa* was found alone in nontidal lagoons and pools. Moreover, *H. ulvae* naturally grows larger in less tidal habitats, and *H. ventrosa* may grow less well in this habitat. This alone could account for the size differences between sympatry and allopatry in these species. This emphasizes the major problem with natural experiments such as those that seem to demonstrate character displacement: sympatric and allopatric populations can occur in different environmental conditions over which the observer has no control. Sometimes it will be these environmental differences, rather than competition, that have led to the character displacement.

8.8.2 Experimenting with natural experiments

niche divergence in clover–grass competition Sometimes, as we have already seen with the gerbils, natural experiments may themselves provide an opportunity for a further – and more informative – experimental manipulation. In one such case, niche divergence was sought in clover, *Trifolium repens*, as a result of its having to compete with the grass *Lolium perenne* (Turkington & Mehrhoff, 1990). Clover was examined from two sites: (i) a 'two-species' site, in which clover achieved a ground coverage of 48% and the grass achieved a coverage of 96% (the two added together exceed 100% because their leaves can overlap); and (ii) a site in which clover achieved 40% coverage, but *L. perenne* covered only 4% (effectively a 'clover-alone' site). A total of three transplant (into the other site) and three re-plant (back into the home site) experiments were carried out (described and numbered in Figure 8.26a). *T. repens*, from both sites, was planted in: (i) plots at the two-species site cleared of *T. repens* only; (ii) plots at the two-species site cleared of both *T. repens* and *L. perenne*; and (iii) plots at the clover-alone site cleared of *T. repens*. The extent of competitive suppression or release was assessed from the amount of growth achieved by the different plantings of *T. repens*. From this the extent of the evolution of niche divergence between 'clover-alone' and 'two-species' *T. repens* was deduced, as was that between *T. repens* and *L. perenne*.

The *T. repens* population from the two-species site had indeed apparently diverged from the *L. perenne* population with which it was coexisting (and with which it may otherwise have competed strongly), and had diverged too from the clover-alone

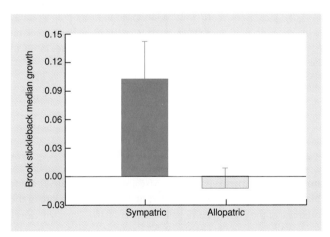

Figure 8.24 Means (with standard error) of group-median growth (natural log of the final mass of fish in each enclosure divided by the initial mass of the group) for sympatric brook sticklebacks representing postdisplacement phenotypes (dark orange bar) and allopatric brook sticklebacks representing predisplacement phenotypes (light bar), both reared in the presence of ninespine sticklebacks. In competition with ninespine sticklebacks, growth was significantly greater for postdisplacement versus predisplacement phenotypes (*P* = 0.012). (After Gray & Robinson, 2002.)

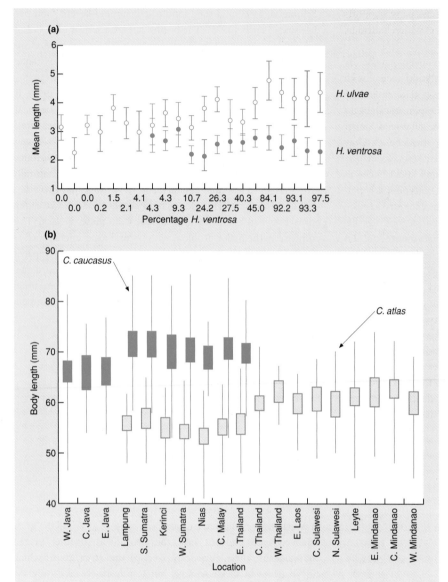

Figure 8.25 Character displacement in body size. (a) Mud snails in Finland (average lengths of *Hydrobia ulvae* and *H. ventrosa*, arranged in order of increasing percentage of *H. ventrosa*). (After Saloniemi, 1993.) (b) Giant rhinoceros beetles in southeast Asia (average lengths of *Chalcosoma caucasus* and *C. atlas*). (After Kawano, 2002.) In each case in allopatry, body sizes overlap broadly, but in sympatry body sizes are significantly different.

population (Figure 8.26b). When the two-species site was cleared of *T. repens* only, the *re*-planted *T. repens* grew better than the *trans*planted clover-alone plants (treatments 1 and 4, respectively; $P = 0.086$, close to significance), suggesting that the transplanted plants were competing more with the resident *L. perenne*. Moreover, when *L. perenne* was also removed, this made no difference to the two-species *T. repens* (treatments 4 and 5; $P > 0.9$), but led to a large increase in the growth of the clover-alone plants (treatments 1 and 2; $P < 0.005$). Also, when *L. perenne* was removed, the clover-alone plants grew better than the two-species ones (treatments 2 and 5; $P < 0.05$) – all of which suggests that only the clover-alone plants were released from competition by the absence of *L. perenne*. Finally, at the clover-alone site, the two-species clover plants grew no better than they had at their home site (treatments 4 and 6; $P > 0.7$), whereas the clover-alone plants grew far better than they had at the two-species site in the presence of the grass (treatments 1 and 3; $P < 0.05$). Thus, the clover from the two-species population hardly competes with the *L. perenne* with which it coexists, whereas the clover-alone population would do – and does so if transplanted to the two-species site.

8.8.3 Selection experiments

The most direct way of demonstrating the evolutionary effects of competition within a pair of competing species is for the experimenter to induce these

direct demonstrations of evolutionary effects of competition have been rare

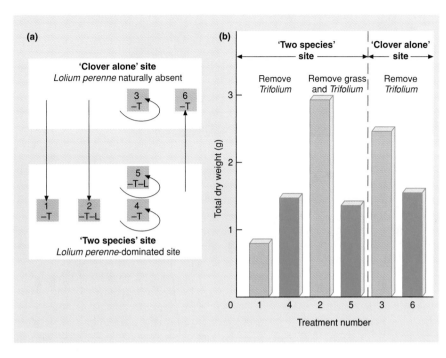

Figure 8.26 (a) Experimental design to test for the evolution of *Trifolium repens* (T) in competition with *Lolium perenne* (L). Indigenous populations of *T. repens*, and sometimes also *L. perenne*, were removed. *Trifolium repens* was removed from the base of the arrow and transplanted, or replanted, at the head of the arrow. Treatment numbers are consistent with the usage of Connell (1980). (b) The results of this experiment are in terms of the total plot dry weight achieved by *T. repens* in the various treatments. Significance levels for comparisons between pairs of treatments are given in the text. (After Turkington & Mehrhoff, 1990.)

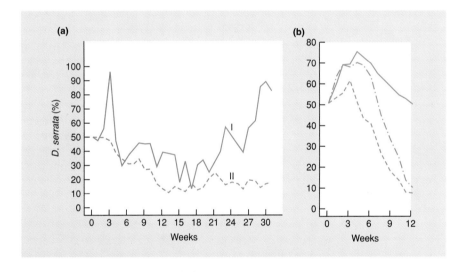

Figure 8.27 Apparent evolution of competitive ability in *Drosophila serrata*. (a) Of two experimental populations coexisting (and competing) with *D. nebulosa*, one (I) increased markedly in frequency after around week 20. (b) Individuals from this population did better in further competition with *D. nebulosa* ((———), mean of five populations) than did individuals from population II ((– – –), mean of five), or individuals from a stock not previously subjected to interspecific competition ((—·—), mean of five). (After Ayala, 1969.)

effects – to impose the selection pressure (competition) and observe the outcome. Surprisingly perhaps, there have been very few successful experiments of this type. In some cases, a species has responded to the selection pressure applied by a second, competitor species by apparently increasing its 'competitive ability', in the sense of increasing its frequency within a joint population. An example of this with two species of *Drosophila* is shown in Figure 8.27. Such results, however, tell us nothing about the means by which such apparent increases were achieved (e.g. whether it was as a result of niche differentiation).

To find an example of a demonstrable increase in niche differentiation giving rise to coexistence of competitors in a selection experiment, we must turn away from interspecific competition in the strictest sense to competition between three types of the same bacterial species, *Pseudomonas fluorescens*, which behave as separate species because they reproduce asexually (Rainey & Trevisano, 1998). The three types are named 'smooth' (SM), 'wrinkly spreader' (WS) and 'fuzzy spreader' (FS) on the basis of the morphology of their colonies plated out on solid medium. In liquid medium, they also occupy quite different parts of the culture vessel (Figure 8.28a). In vessels that were continually shaken, so that no separate niches for the different types could be established, an initially pure culture of SM individuals retained its purity (Figure 8.28b). But in the absence

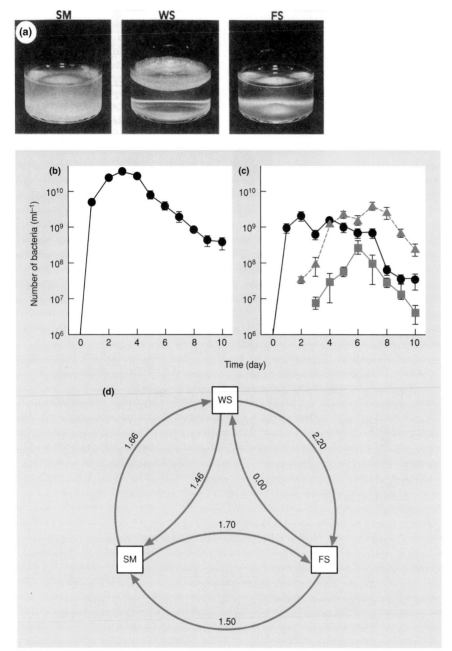

Figure 8.28   (a) Pure cultures of three types of the bacterium, *Pseudomonas fluorescens* (smooth, SM, wrinkly spreader, WS, and fuzzy spreader, FS) concentrate their growth in different parts of a liquid culture vessel. (b) In shaken culture vessels, pure SM cultures are maintained. Bars represent standard errors. (c) But in unshaken, initially pure SM cultures (●), WS (▲) and FS (■) mutants arise, invade and establish. Bars represent standard errors. (d) The competitive abilities (relative rates of increase) when an initially rare type (foot of the arrow) invades a pure colony of another type (head of the arrow). Hence, values >1 indicate an ability to invade (superior competitor when rare) and values <1 an inability. (After Rainey & Trevisano, 1998. Reproduced by permission of Nature.)

of shaking, mutant WS and FS types invaded and established (Figure 8.28c). Furthermore, it was possible to determine the competitive abilities of the different types, when rare, to invade pure cultures of the other types (Figure 8.28d). Five of six possible invasions are favored. The exception – WS repels the invasion of FS – is unlikely to lead to the elimination of FS, because FS can invade cultures of SM, and SM can invade cultures of WS. In general, however, the experimental selection of increased niche differentiation amongst competing species appears to be either frustratingly elusive or sadly neglected.

8.9   Niche differentiation and similarity amongst coexisting competitors

It might be imagined that scientific progress is made by providing answers to questions. In fact, progress often consists of replacing one question with another, more pertinent, more challenging question. In this section, we deal with an area where this is the case: the questions of *how* different coexisting competitors are, and how different coexisting species need to be if competition is not to eliminate one of them.

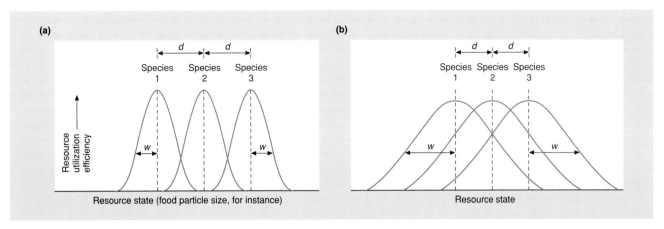

(a)

Species 1   Species 2   Species 3

Resource utilization efficiency

Resource state (food particle size, for instance)

(b)

Species 1 Species 2 Species 3

Resource state

Figure 8.29 Resource-utilization curves for three species coexisting along a one-dimensional resource spectrum. d is the distance between adjacent curve peaks and w is the standard deviation of the curves. (a) Narrow niches with little overlap ($d > w$), i.e. relatively little interspecific competition. (b) Broader niches with greater overlap ($d < w$), i.e. relatively intense interspecific competition.

The Lotka–Volterra model predicts the stable coexistence of competitors in situations where interspecific competition is, for both species, less significant than intraspecific competition. Niche differentiation will obviously tend to concentrate competitive effects more within species than between them. The Lotka–Volterra model, and the Competitive Exclusion Principle, therefore imply that *any* amount of niche differentiation will allow the stable coexistence of competitors. Hence, in an attempt to discover whether this was 'true', the question 'do competing species need to be different in order to coexist stably?' greatly exercised the minds of ecologists during the 1940s (Kingsland, 1985).

how much niche differentiation is needed for coexistence?

It is easy to see now, however, that the question is badly put, since it leaves the precise meaning of 'different' undefined. We have seen examples in which the coexistence of competitors is apparently associated with some degree of niche differentiation, but it seems that if we look closely enough, all coexisting species will be found to be different – without this having anything to do with competition. A more pertinent question, therefore, would be 'is there a minimum amount of niche differentiation that has to be exceeded for stable coexistence?' That is, is there a limit to the similarity of coexisting species?

One influential attempt to answer this question for exploitative competition, based on variants of the Lotka–Volterra model, was initiated by MacArthur and Levins (1967) and developed by May (1973). With hindsight, their approach is certainly open to question (Abrams, 1983). Nevertheless, we can learn most about the 'limiting similarity problem' by first examining their approach and then looking at the objections to it. Here, as so often, the models can be instructive without being 'right'.

Imagine three species competing for a resource that is unidimensional and is distributed continuously; food size is a clear example. Each species has its own realized niche in this single dimension,

a simple model provides a simple answer . . .

which can be visualized as a resource-utilization curve (Figure 8.29). The consumption rate of each species is highest at the center of its niche and tails off to zero at either end, and the more the utilization curves of adjacent species overlap, the more the species compete. Indeed, by assuming that the curves are 'normal' distributions (in the statistical sense), and that the different species have similarly shaped curves, the competition coefficient (applicable to both adjacent species) can be expressed by the following formula:

$$\alpha = e^{-d^2/4w^2} \tag{8.18}$$

where w is the standard deviation (or, roughly, 'relative width') of the curves, and d is the distance between the adjacent peaks. Thus, α is very small when there is considerable separation of adjacent curves ($d/w \gg 1$; Figure 8.29a), and approaches unity as the curves themselves approach one another ($d/w < 1$; Figure 8.29b).

How much overlap of adjacent utilization curves is compatible with stable coexistence? Assume that the two peripheral species have the same carrying capacity (K_1, representing the suitability of the available resources for species 1 and 3) and consider the coexistence, in between them on the resource axis, of another species (carrying capacity K_2). When d/w is low (α is high and the species are similar) the conditions for coexistence are extremely restrictive in terms of the $K_1 : K_2$ ratio; but these restrictions lift rapidly as d/w approaches and exceeds unity (Figure 8.30). In other words, coexistence is possible when d/w

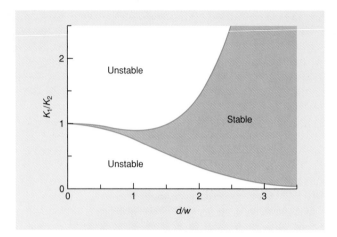

Figure 8.30 The range of habitat favorabilities (indicated by the carrying capacities K_1 and K_2, where $K_1 = K_3$) that permit a three-species equilibrium community with various degrees of niche overlap (d/w). (After May, 1973.)

is low, but only if the suitabilities of the environment for the different species are extremely finely balanced. Furthermore, if the environment is assumed to vary, then the fluctuations will lead to variations in the $K_1 : K_2$ ratio, and coexistence will now only be possible if there is a broad range of $K_1 : K_2$ ratios leading to stability, i.e. if, roughly, $d/w > 1$.

... that is almost certainly wrong

This model, then, suggests that there *is* a limit to the similarity of coexisting competitors, and that the limit is given by the condition $d/w > 1$. Are these the correct answers? In fact, it seems most unlikely that there is a universal limit to similarity, or even a widely applicable one that we could express in such a simple way as $d/w > 1$. Abrams (1976, 1983), amongst others, has emphasized that models with competition in several dimensions, with alternative utilization curves and so on, all lead to alternative limits to similarity, and often to much lower values of d/w being compatible with robust, stable coexistence. In other words, '$d/w > 1$' is a property of one type of model analysis, but not of others, and thus, almost certainly, not of nature as a whole. Furthermore, we have already seen that because of environmental heterogeneity, apparent competition and so on, exploitative competition and any niche differentiation associated with it are not necessarily the whole story when it comes to the coexistence of competitors. This too argues against the idea of a universal limit.

On the other hand, the most general messages from the early models still seem valid, namely: (i) in the real world, with all its intrinsic variability, there *are* likely to be limits to the similarity of coexisting exploitative competitors; and (ii) these limits will reflect not only the differences between species, but also the variability within them, the nature of the resource, the nature of the utilization curve and so on.

But is even the limiting similarity question the best question to ask? We want to understand the extent of niche differentiation amongst coexisting

the answer? it depends

species. If species were always packed as tightly together as they could be, then presumably they would differ by the minimum (limiting) amount. But why should they be? We return, once again, to the distinction between the ecological and the evolutionary consequences of competition (Abrams, 1990). The ecological effects are that species with 'inappropriate' niches will be eliminated (or repelled if they try to invade), and the limiting similarity question implicitly concerned itself with this: how many species can be 'packed in'? But coexisting competitors may also evolve. Do we generally observe the ecological effects, or the combined ecological and evolutionary effects? Do they differ? We cannot attempt to answer the first question without answering the second, and the answer to that seems to be, perhaps inevitably, 'it depends'. Different models, based on different underlying mechanisms in the competitive process, predict that evolution will lead to more widely spaced niches, or to more closely packed niches or to much the same disposition of niches as those predicted by ecological processes alone (Abrams, 1990).

Two points, therefore, emerge from this discussion. The first is that it has been entirely theoretical. This is a reflection of the second point, which is that we have seen progress – but in terms of successive questions superseding their predecessors rather than actually answering them. Data provide answers – what we have seen is a refinement of questions. The latest stage in this appears to be that attempts to answer questions regarding 'niche similarity' may need to be postponed until we know more about resource distributions, utilization curves and, more generally, the mechanisms underlying exploitative competition. It is to these that we now turn.

8.10   Niche differentiation and mechanisms of exploitation

In spite of all the difficulties of making a direct connection between interspecific competition and niche differentiation, there is no doubt that niche differentiation is often the basis for the coexistence of species.

niche differentiation ...

There are a number of ways in which niches can be differentiated. One is resource partitioning or, more generally, differential resource utilization. This can be observed when species living in precisely the same habitat nevertheless utilize different resources.

... easy to imagine in animals, less easy in plants ...

Since the majority of resources for animals are individuals of other species (of which there are literally millions of types), or parts of individuals, there is no difficulty, in principle, in imagining how competing animals might partition resources amongst themselves. Plants, on the other hand, all have very similar requirements for the same potentially limited resources (see Chapter 3), and there is much less apparent scope for resource partitioning (but see below).

In many cases, the resources used by ecologically similar species are separated spatially. Differential resource utilization will then express itself as either microhabitat differentiation between the species (e.g. different species of fish feeding at different depths), or even a difference in geographic distribution. Alternatively, the availability of the different resources may be separated in time; for example, different resources may become available at different times of the day or in different seasons. Differential resource utilization may then express itself as a temporal separation between the species.

... based on resources and conditions

The other major way in which niches can be differentiated is on the basis of conditions (Wilson, 1999). Two species may use precisely the same resources, but if their ability to do so is influenced by environmental conditions (as it is bound to be), and if they respond differently to those conditions, then each may be competitively superior in different environments. This too can express itself as a microhabitat differentiation, or a difference in geographic distribution or a temporal separation, depending on whether the appropriate conditions vary on a small spatial scale, a large spatial scale or over time. Of course, in a number of cases (especially with plants) it is not easy to distinguish between conditions and resources (see Chapter 3). Niches may then be differentiated on the basis of a factor (such as water) which is both a resource and a condition.

spatial and temporal separation

There are many examples of the separation of competing species in space or time involving both animals and plants. For example, tadpoles of two anuran species in New Jersey, USA (*Hyla crucifer* and *Bufo woodhousii*), have their feeding periods offset by around 4–6 weeks each year, apparently, though not certainly, associated with differential responses to environmental conditions rather than seasonal changes in resources (Lawler & Morin, 1993). Two coexisting species of spiny mice in rocky deserts in Israel partition activity on a diel basis: *Acomys cahirinus* is nocturnal and *A. russatus* is diurnal, although the latter becomes nocturnal if its congener is removed (Jones *et al.*, 2001). Two phloem-feeding bark beetles, *Ips duplicatus* and *I. typographus*, on Norway spruce trees, in Norway, are separated in their feeding sites on a small spatial scale by trunk diameter, although the reason for this is not at all clear (Schlyter & Anderbrandt, 1993). But, it is amongst plants and other sessile organisms, because of their limited scope for differential resource utilization at the same location and instant, that spatial and temporal separation are likely to be of particular significance (see Harper, 1977). Although, as ever, it is one thing to show that

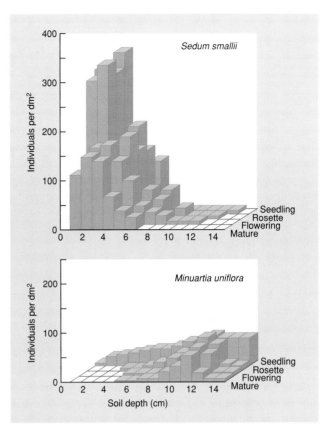

Figure 8.31 The zonation of individuals, according to soil depth, of two annual plants, *Sedum smallii* and *Minuartia uniflora* at four stages of the life cycle. (After Sharitz & McCormick, 1973.)

species differ in their spatial or temporal distribution – it is quite another to prove that this has anything to do with competition. The cattails in Section 8.3.3 provide one example of competing plants separated spatially. Another is shown in Figure 8.31, concerning the annuals *Sedum smallii* and *Minuartia uniflora* that dominate the vegetation growing on granite outcrops in southeastern USA. The adult plants exhibit an especially clear spatial zonation associated with soil depth (itself strongly correlated with soil moisture), and further experimental results reinforce the idea that it is competition rather than mere differences in tolerance that gives rise to this zonation.

Describing the outcome of competition, however – 'one species coexists with or excludes another' – and even associating this with niche differentiation, whether based on resources themselves, or conditions or merely differences in space or time, actually provides us with rather little understanding of the competitive process. For this, as we have seen repeatedly in this chapter, we may need to focus more on the mechanisms of exploitation. *How*, precisely, does one species outexploit and outcompete another? *How* can two consumers coexist on two limiting resources, when both resources are absolutely essential to both consumers?

the need to consider resource dynamics

Furthermore, as Tilman (1990) has pointed out, whilst monitoring the population dynamics of two competing species may give us some powers of prediction for the next time they compete, it will give us very little help in predicting how each would fare against a third species. Whereas, if we understood the dynamics of the interaction of all the species with their shared limiting resources, then we might be able to predict the outcome of exploitative competition between any given pair of the species. We therefore turn now to some attempts to explain the coexistence of species competing for limiting resources that explicitly consider not only the dynamics of the competing species but also the dynamics of the resources themselves. Rather than going into details, we examine the outlines of models and some major conclusions.

8.10.1 Exploitation of a single resource

a model for a single resource . . .

Tilman (1990) shows, for a number of models, what we have already seen demonstrated empirically in Section 8.2.6, that when two species compete exploitatively for a single limiting resource the outcome is determined by which species is able, in its exploitation, to reduce the resource to the lower equilibrium concentration, R^*. (Satisfyingly, for apparent competition the reverse is true: the prey or host able to support the *greatest* abundance of predators or parasites is the winner (see, for example, Begon & Bowers, 1995) – a prediction we have seen borne out in Figure 8.17.)

Different models, based on varying details in the mechanism of exploitation, give rise to different formulae for R^*, but even the simplest model is revealing, giving:

$$R_i^* = m_i C_i / (g_i - m_i). \tag{8.19}$$

Here m_i is the mortality or loss rate of consumer species i; C_i is the resource concentration at which species i attains a rate of growth and reproduction per unit biomass (relative rate of increase, RRI) equal to half its maximal RRI (C_i is thus highest in consumers that require the most resource in order to grow rapidly); and g_i is the maximum RRI achievable by species i. This suggests that successful exploitative competitors (low R_i^*) are those that combine resource-utilization efficiency (low C_i), low rates of loss (low m_i) and high rates of increase (high g_i). On the other hand, it may not be possible for an organism to combine, say, low C_i and high g_i. A plant's growth will be most enhanced by putting its matter and energy into leaves and photosynthesis – but to enhance its nutrient-utilization efficiency it would have to put these into roots. A lioness will be best able to subsist at low densities of prey by being fleet-footed and maneuverable – but this may be difficult if she is often heavily pregnant. Understanding successful exploitative competitiveness, therefore, may require us ultimately to understand how organisms trade off features giving rise to low values of R^* against features that enhance other aspects of fitness.

. . . tested on grasses

A rare test of these ideas is provided by Tilman's own work on terrestrial plants competing for nitrogen (Tilman & Wedin, 1991a, 1991b). Five grass species were grown alone in a range of experimental conditions that gave rise in turn to a range of nitrogen concentrations. Two species, *Schizachyrium scoparium* and *Andropogon gerardi*, consistently reduced the nitrate and ammonium concentrations in soil solutions to lower values than those achieved by the other three species (in all soils but those with the very highest nitrogen levels). Of these three other species, one, *Agrostis scabra*, left behind higher concentrations than the other two, *Agropyron repens* and *Poa pratensis*. Then, when *A. scabra* was grown with *A. repens*, *S. scoparium* and *A. gerardi*, the results, especially at low nitrogen concentrations where nitrogen was most likely to be limiting, were very much in line with the exploitative competition theory (Figure 8.32). The species that could reduce nitrogen to the lowest concentration always won – *A. scabra* was always competitively displaced. A similar result has been obtained for the nocturnal, insectivorous gecko *Hemidactylus frenatus*, an invader of urban habitats across the Pacific basin, where it is responsible for population declines of the native gecko *Lepidodactylus lugubris*. Petren and Case (1996) established that insects are a limiting resource for both. The invader is capable of depleting insect resources in experimental enclosures to lower levels than the native gecko, and the latter suffers reductions in body condition, fecundity and survivorship as a result.

Returning to Tilman's grasses, the five species were chosen from various points in a typical old-field successional sequence in Minnesota (Figure 8.33a), and it is clear that the better competitors for nitrogen are found later in the sequence. These species, and *S. scoparium* and *A. gerardi* in particular, had higher root allocations, but lower above-ground vegetative growth rates and reproductive allocations (e.g. Figure 8.33b). In other words, they achieved their low values of R^* by the high resource-utilization efficiency given to them by their roots (low C_i, Equation 8.19), even though they appeared to have paid for this through a reduction in growth and reproductive rates (lower g_i). In fact, over all the species, a full 73% of the variance in the eventual soil nitrate concentration was explained by variations in root mass (Tilman & Wedin, 1991a). This successional sequence (see Section 16.4 for a much fuller discussion of succession) therefore appears to be one in which fast growers and reproducers are replaced by efficient and powerful exploiters and competitors.

8.10.2 Exploitation of two resources

a model for two resources – the zero net isocline: a niche boundary

Tilman (1982, 1986; see also Section 3.8) has also considered what happens when two competitors compete for two resources. Beginning with *intra*specific

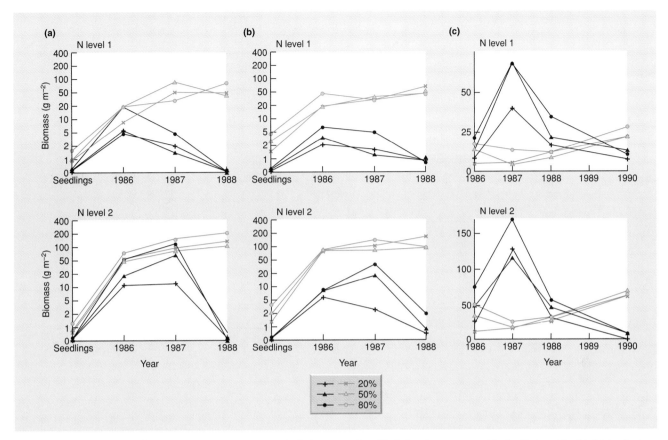

Figure 8.32 The results of competition experiments in which *Agrostis scabra* (black lines) was competitively displaced by (a) *Schizachyrium scoparium*, (b) *Andropogon gerardi* and (c) *Agropyron repens* (orange lines, at each of two nitrogen levels (both low) and whether it represented 20, 50 or 80% of the initial seed sown. In each case, *A. scabra* had lower values of *R** for nitrate and ammonium (see text). Displacement was least rapid in (c) where the differential was least marked. (After Tilman & Wedin, 1991b.)

competition, we can define a zero net growth isocline for a single species utilizing two essential resources (see Section 3.8). This isocline is the boundary between resource combinations that allow the species to survive and reproduce, and resource combinations that do not (Figure 8.34), and therefore represents the boundary of the species' niche in these two dimensions. For present purposes, we can ignore the complications of overcompensation, chaos, etc. and assume that intraspecific competition brings the population to a stable equilibrium. Here, however, the equilibrium has two components: both population size and the resource levels should remain constant. Population size is constant (by definition) at all points on the isocline, and Tilman established that there is only one point on the isocline where resource levels are also constant (point *S** in Figure 8.34). This point, which is the two-resource equivalent of *R** for one resource, represents a balance between the consumption of the resources by the consumer (taking the resource concentrations towards the bottom left of the figure) and the natural renewal of the resources (taking the concentrations towards the top right). Indeed, in the absence of the consumer, resource renewal would take the resource concentrations to the 'supply point', shown in the figure.

To move from intra- to *inter*specific competition, it is necessary to superimpose the isoclines of two species on the same diagram (Figure 8.35). The two species will presumably have different consumption rates, but there will still be a single supply point. The outcome depends on the position of this supply point.

In Figure 8.35a, the isocline of species A is closer to both axes than that of species B. There are three regions in which the supply point might be found. If it was in region 1, below the isoclines of both species, then there would never be sufficient resources for either species and neither would survive. If it was in region 2, below the isocline of species B but above that of species A, then species B would be unable to survive and the system would equilibrate on the isocline of species A. If the supply point was in region 3, then this system too would equilibrate on the isocline of species A. Analogous to the one-resource case, species A would competitively exclude species B because of its ability to exploit both resources down to levels at which species B could not survive. Of course, the outcome would be reversed if the positions of the isoclines were reversed.

a superior and an inferior competitor

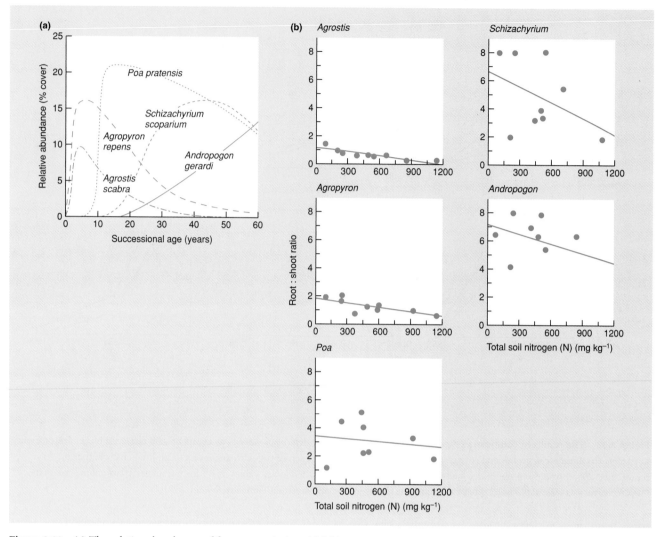

Figure 8.33 (a) The relative abundances of five grasses during old-field successions at Cedar Creek Natural History Area, Minnesota, USA. (b) The root : shoot ratios were generally higher in the later successional species and declined as soil nitrogen increased. (After Tilman & Wedin, 1991a.)

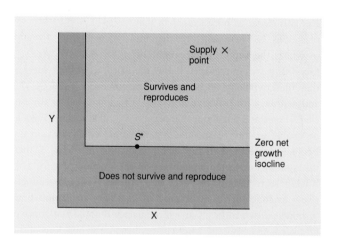

Figure 8.34 (*left*) The zero net growth isocline of a species potentially limited by two resources (X and Y), divides resource combinations on which the species can survive and reproduce, from those on which it cannot. The isocline is rectangular in this case because X and Y are essential resources (see Section 3.8.1). Point *S** is the only point on the isocline at which there is also no net change in resource concentrations (consumption and resource renewal are equal and opposite). In the absence of the consumer, resource renewal would take the resource concentrations to the 'supply point' shown.

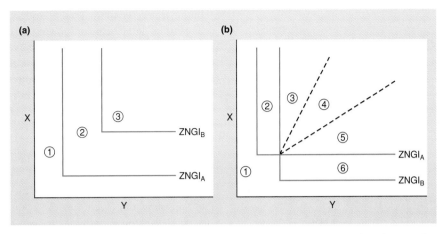

Figure 8.35   (a) Competitive exclusion: the isocline (zero net growth isocline, ZNGI) of species A lies closer to the resource axes than the isocline of species B. If the resource supply point is in region 1, then neither species can exist. But if the resource supply point is in regions 2 or 3, then species A reduces the resource concentrations to a point on its own isocline (where species B cannot survive and reproduce): species A excludes species B. (b) Potential coexistence of two competitors limited by two essential resources. The isoclines of species A and B overlap, leading to six regions of interest. With supply points in region 1 neither species can exist; with points in regions 2 and 3, species A excludes species B; and with points in regions 5 and 6, species B excludes species A. Region 4 contains supply points lying between the limits defined by the two dashed lines. With supply points in region 4 the two species coexist. For further discussion, see text.

*coexistence –
dependent on the
ratio of resource
levels at the supply
point*

In Figure 8.35b the isoclines of the two species overlap, and there are six regions in which the supply point might be found. Points in region 1 are below both isoclines and would allow neither species to exist; those in region 2 are below the isocline of species B and would only allow species A to exist; and those in region 6 are below the isocline of species A and would only allow species B to exist. Regions 3, 4 and 5 lie within the fundamental niches of both species. However, the outcome of competition depends on which of these regions the supply point is located in.

The most crucial region in Figure 8.35b is region 4. For supply points here, the resource levels are such that species A is more limited by resource X than by resource Y, whilst species B is more limited by Y than X. However, species A consumes more X than Y, whilst species B consumes more Y than X. Because each species consumes more of the resource that more limits its own growth, the system equilibrates at the intersection of the two isoclines, and this equilibrium is stable: the species coexist.

*subtle niche
differentiation – each
species consumes
more of the resource
that more limits its
own growth*

This is niche differentiation, but of a subtle kind. Rather than the two species exploiting different resources, species A disproportionately limits itself by its exploitation of resource X, whilst species B disproportionately limits itself by its exploitation of resource Y. The result is the coexistence of competitors. By contrast, for supply points in region 3, both species are more limited by Y than X. But species A can reduce the level of

Y to a point on its own isocline below species B's isocline, where species B cannot exist. Conversely, for supply points in region 5, both species are more limited by X than Y, but species B depresses X to a point below species A's isocline. Thus, in regions 3 and 5, the supply of resources favors one species or the other, and there is competitive exclusion.

It seems then that two species can compete for two resources and coexist as long as two conditions are met.

1   The habitat (i.e. the supply point) must be such that one species is more limited by one resource, and the other species more limited by the other resource.
2   Each species must consume more of the resource that more limits its own growth. Thus, it is possible, in principle, to understand coexistence in competing plants on the basis of differential resource utilization. The key seems to be an explicit consideration of the dynamics of the resources as well as the dynamics of the competing species. As with other cases of coexistence by niche differentiation, the essence is that intraspecific competition is, for both species, a more powerful force than interspecific competition.

The best evidence for the validity of the model comes from Tilman's own experimental laboratory work on competition between the diatoms *Asterionella formosa* and *Cyclotella meneghiniana* (Tilman, 1977). For both species, Tilman observed directly the consumption rates and the isoclines for both phosphate and silicate. He then used these to predict the outcome of competition with a range of resource supply points (Figure 8.35). Finally, he ran a number of competition experiments with a variety of supply

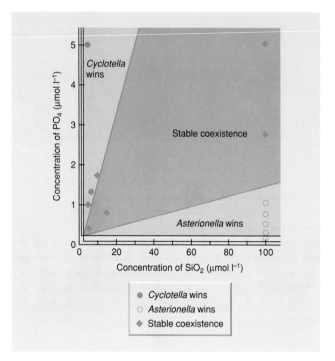

Figure 8.36 The observed isoclines and consumption vectors of two diatom species, *Asterionella formosa* and *Cyclotella meneghiniana*, were used to predict the outcome of competition between them for silicate and phosphate. The predictions were then tested in a series of experiments, the outcomes of which are depicted by the symbols explained in the key. Most experiments confirmed the predictions, with the exception of two lying close to the regional boundary. (After Tilman, 1977, 1982.)

points, and the results of these are illustrated in Figure 8.36. In most cases the results confirmed the predictions. In the two that did not, the supply points were very close to the regional boundary. The results are therefore encouraging. However, it has proved extremely difficult to transfer this approach from the laboratory, where supply points can be manipulated, to natural populations, where they cannot, and where even the estimation of supply points has proved practically impossible (Sommer, 1990). There is considerable need for consolidation and extension from work on other types of plants and animals.

8.10.3   Exploitation of more than two resources

the more limiting resources there are, the more species may coexist

We have seen how two diatom species may coexist in the laboratory on two shared limiting resources. In fact, Tilman's resource competition theory predicts that the diversity of coexisting species should be proportional to the total number of resources in a system that are at physiological

limiting levels: the more limiting resources, the more coexisting competitors. Interlandi and Kilham (2001) tested this hypothesis directly in three lakes in the Yellowstone region of Wyoming, USA using an index (Simpson's index) of the species diversity of phytoplankton there (diatoms and other species). If one species exists on its own, the index equals 1; in a group of species where biomass is strongly dominated by a single species, the index will be close to 1; when two species exist at equal biomass, the index is 2; and so on. According to resource competition theory, this index should therefore increase in direct proportion to the number of resources limiting growth. The spatial and temporal patterns in phytoplankton diversity in the three lakes for 1996 and 1997 are shown in Figure 8.37. The principal limiting resources for phytoplankton growth are nitrogen, phosphorus, silicon and light. These parameters were measured at the same depths and times that the phytoplankton were sampled, and it was noted where and when any of the *potential* limiting factors *actually* occurred at levels below threshold limits for growth. Consistent with resource competition theory, species diversity increased as the number of resources at physiologically limiting levels increased (Figure 8.38).

These results suggest that even in the highly dynamic environments of lakes where equilibrium conditions are rare, resource competition plays a role in continuously structuring the phytoplankton community. It is heartening that the results of experiments performed in the artificial world of the laboratory are echoed here in the much more complex natural environment.

Our survey of interspecific competition has concluded with a realization that we need to understand much more about the mechanisms underlying the interactions between consumers and their resources. If these resources are alive, then we normally refer to such interactions as predation; and if they were alive once, but are now dead, we refer to them as detritivory. It would seem, therefore, that the distinction normally made between competition and predation is, in a very real sense, an artificial one (Tilman, 1990). None the less, having dealt with competition here, we turn next, in a separate series of chapters, to predation and detritivory.

Summary

In interspecific competition, individuals of one species suffer a reduction in fecundity, growth or survivorship as a result of resource exploitation or interference by individuals of another species. Competing species may exclude one another from particular habitats so that they do not coexist, or may coexist, perhaps by utilizing the habitat in slightly different ways. Interspecific competition is frequently highly asymmetric.

Although species may not be competing now, their ancestors may have done so in the past. Species may have evolved characteristics that ensure they compete less, or not at all, with other species. Moreover, species whose niches appear differentiated may have evolved independently and, in fact, never have

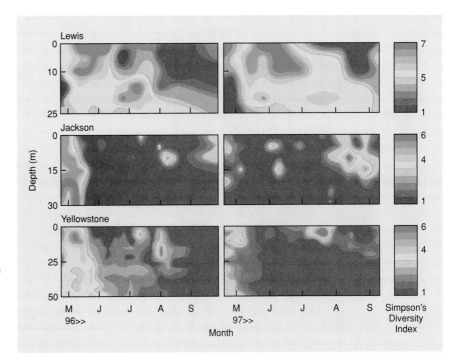

Figure 8.37 Variation in phytoplankton species diversity (Simpson's index) with depth in 2 years in three large lakes in the Yellowstone region, USA. Shading indicates depth–time variation in a total of 712 discrete samples: dark orange areas denote high species diversity, and gray areas denote low species diversity. (After Interlandi & Kilham, 2001.)

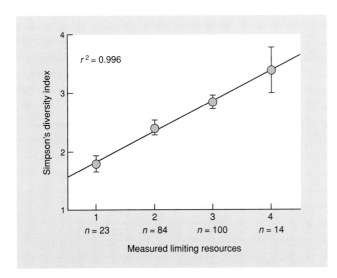

Figure 8.38 Phytoplankton diversity (Simpson's index; mean ± SE) associated with samples with different numbers of measured limiting resources. It was possible to perform this analysis on 221 samples from those displayed in Figure 6.14. The number of samples (n) in each limiting resource class is shown. (From Interlandi & Kilham, 2001.)

But negative results would be equally compatible with the past elimination of species by competition, the evolutionary avoidance of competition in the past, and the independent evolution of noncompeting species.

Mathematical models, most notably the Lotka–Volterra model, have provided important insights into the circumstances that permit the coexistence of competitors, and those that lead to competitive exclusion. However, the simplified assumptions of the Lotka–Volterra model limit its applicability to real situations in nature. We know from other models and experiments that the outcome of interspecific competition can be strongly influenced by heterogeneous, inconstant or unpredictable environments. Coexistence of a superior and an inferior competitor on a patchy and ephemeral resource can occur if the two species have independent, aggregated distributions over the available patches.

We describe the range of approaches used to study both the ecological and evolutionary effects of interspecific competition, paying particular attention to experiments in the laboratory or field (e.g. substitutive, additive, response surface analysis) and natural experiments (e.g. comparing niche dimensions of species in sympatry and allopatry). The important question of whether a minimum amount of niche differentiation is required for stable coexistence is much easier to pose than answer.

The chapter concludes by acknowledging the need to consider not just the population dynamics of the competing populations but also the dynamics of the resources for which they are competing, if we wish to achieve a full understanding of interspecific competition and species coexistence.

competed, now or historically. An experimental manipulation (for instance, the removal of one or more species) can indicate the presence of current competition if it leads to an increase in the fecundity or survival or abundance of the remaining species.

Chapter 9
The Nature of Predation

9.1 Introduction: the types of predators

Consumers affect the distribution and abundance of the things they consume and vice versa, and these effects are of central importance in ecology. Yet, it is never an easy task to determine what the effects are, how they vary and why they vary. These topics will be dealt with in this and the next few chapters. We begin here by asking 'What is the nature of predation?', 'What are the effects of predation on the predators themselves and on their prey?' and 'What determines where predators feed and what they feed on?' In Chapter 10, we turn to the consequences of predation for the dynamics of predator and prey populations.

definition of predation

Predation, put simply, is consumption of one organism (the prey) by another organism (the predator), in which the prey is alive when the predator first attacks it. This excludes detritivory, the consumption of dead organic matter, which is discussed in its own right in Chapter 11. Nevertheless, it is a definition that encompasses a wide variety of interactions and a wide variety of 'predators'.

taxonomic and functional classifications of predators

There are two main ways in which predators can be classified. Neither is perfect, but both can be useful. The most obvious classification is 'taxonomic': carnivores consume animals, herbivores consume plants and omnivores consume both (or, more correctly, prey from more than one trophic level – plants and herbivores, or herbivores and carnivores). An alternative, however, is a 'functional' classification of the type already outlined in Chapter 3. Here, there are four main types of predator: true predators, grazers, parasitoids and parasites (the last is divisible further into microparasites and macroparasites as explained in Chapter 12).

true predators

True predators kill their prey more or less immediately after attacking them; during their lifetime they kill several or many different prey individuals, often consuming prey in their entirety. Most of the more obvious carnivores like tigers, eagles, coccinellid beetles and carnivorous plants are true predators, but so too are seed-eating rodents and ants, plankton-consuming whales, and so on.

grazers

Grazers also attack large numbers of prey during their lifetime, but they remove only part of each prey individual rather than the whole. Their effect on a prey individual, although typically harmful, is rarely lethal in the short term, and certainly never predictably lethal (in which case they would be true predators). Amongst the more obvious examples are the large vertebrate herbivores like sheep and cattle, but the flies that bite a succession of vertebrate prey, and leeches that suck their blood, are also undoubtedly grazers by this definition.

parasites

Parasites, like grazers, consume parts of their prey (their 'host'), rather than the whole, and are typically harmful but rarely lethal in the short term. Unlike grazers, however, their attacks are concentrated on one or a very few individuals during their life. There is, therefore, an intimacy of association between parasites and their hosts that is not seen in true predators and grazers. Tapeworms, liver flukes, the measles virus, the tuberculosis bacterium and the flies and wasps that form mines and galls on plants are all obvious examples of parasites. There are also many plants, fungi and microorganisms that are parasitic on plants (often called 'plant pathogens'), including the tobacco mosaic virus, the rusts and smuts and the mistletoes. Moreover, many herbivores may readily be thought of as parasites. For example, aphids extract sap from one or a very few individual plants with which they enter into intimate contact. Even caterpillars often rely on a single plant for their development. Plant pathogens, and animals parasitic on animals, will be dealt with together in Chapter 12. 'Parasitic' herbivores, like aphids and caterpillars, are dealt with here and in the next chapter, where we group them together with true predators, grazers and parasitoids under the umbrella term 'predator'.

parasitoids

The parasitoids are a group of insects that belong mainly to the order Hymenoptera, but also include many Diptera. They are free-living as adults, but lay their eggs in, on or near other insects (or, more rarely, in spiders or woodlice). The larval parasitoid then develops inside or on its host. Initially, it does little apparent harm, but eventually it almost totally consumes the host and therefore kills it. An adult parasitoid emerges from what is apparently a developing host. Often, just one parasitoid develops from each host, but in some cases several or many individuals share a host. Thus, parasitoids are intimately associated with a single host individual (like parasites), they do not cause immediate death of the host (like parasites and grazers), but their eventual lethality is inevitable (like predators). For parasitoids, and also for the many herbivorous insects that feed as larvae on plants, the rate of 'predation' is determined very largely by the rate at which the adult females lay eggs. Each egg is an 'attack' on the prey or host, even though it is the larva that hatches from the egg that does the eating.

Parasitoids might seem to be an unusual group of limited general importance. However, it has been estimated that they account for 10% or more of the world's species (Godfray, 1994). This is not surprising given that there are so many species of insects, that most of these are attacked by at least one parasitoid, and that parasitoids may in turn be attacked by parasitoids. A number of parasitoid species have been intensively studied by ecologists, and they have provided a wealth of information relevant to predation generally.

In the remainder of this chapter, we examine the nature of predation. We will look at the effects of predation on the prey individual (Section 9.2), the effects on the prey population as a whole (Section 9.3) and the effects on the predator itself (Section 9.4). In the cases of attacks by true predators and parasitoids, the effects on prey individuals are very straightforward: the prey is killed. Attention will therefore be placed in Section 9.2 on prey subject to grazing and parasitic attack, and herbivory will be the principal focus. Apart from being important in its own right, herbivory serves as a useful vehicle for discussing the subtleties and variations in the effects that predators can have on their prey.

Later in the chapter we turn our attention to the behavior of predators and discuss the factors that determine diet (Section 9.5) and where and when predators forage (Section 9.6). These topics are of particular interest in two broad contexts. First, foraging is an aspect of animal behavior that is subject to the scrutiny of evolutionary biologists, within the general field of 'behavioral ecology'. The aim, put simply, is to try to understand how natural selection has favored particular patterns of behavior in particular circumstances (how, behaviorally, organisms match their environment). Second, the various aspects of predatory behavior can be seen as components that combine to influence the population

dynamics of both the predator itself and its prey. The population ecology of predation is dealt with much more fully in the next chapter.

9.2 Herbivory and individual plants: tolerance or defense

The effects of herbivory on a plant depend on which herbivores are involved, which plant parts are affected, and the timing of attack relative to the plant's development. In some insect–plant interactions as much as 140 g, and in others as little as 3 g, of plant tissue are required to produce 1 g of insect tissue (Gavloski & Lamb, 2000a) – clearly some herbivores will have a greater impact than others. Moreover, leaf biting, sap sucking, mining, flower and fruit damage and root pruning are all likely to differ in the effect they have on the plant. Furthermore, the consequences of defoliating a germinating seedling are unlikely to be the same as those of defoliating a plant that is setting its own seed. Because the plant usually remains alive in the short term, the effects of herbivory are also crucially dependent on the response of the plant. Plants may show tolerance of herbivore damage or resistance to attack.

9.2.1 Tolerance and plant compensation

Plant compensation is a term that refers to the degree of tolerance exhibited by plants. If damaged plants have greater fitness than their undamaged counterparts, they have *overcompensated*, and if they have lower fitness, they have *undercompensated* for herbivory (Strauss & Agrawal, 1999). Individual plants can compensate for the effects of herbivory in a variety of ways. In the first place, the removal of shaded leaves (with their normal rates of respiration but low rates of photosynthesis; see Chapter 3) may improve the balance between photosynthesis and respiration in the plant as a whole. Second, in the immediate aftermath of an attack from a herbivore, many plants compensate by utilizing reserves stored in a variety of tissues and organs or by altering the distribution of photosynthate within the plant. Herbivore damage may also lead to an increase in the rate of photosynthesis per unit area of surviving leaf. Often, there is compensatory regrowth of defoliated plants when buds that would otherwise remain dormant are stimulated to develop. There is also, commonly, a reduced death rate of surviving plant parts. Clearly, then, there are a number of ways in which individual plants compensate for the effects of herbivory (discussed further in Sections 9.2.3–9.2.5). But perfect compensation is rare. Plants are usually harmed by herbivores even though the compensatory reactions tend to counteract the harmful effects.

individual plants can compensate for herbivore effects

9.2.2 Defensive responses of plants

plants make
defensive
responses . . .

The evolutionary selection pressure exerted by herbivores has led to a variety of plant physical and chemical defenses that resist attack (see Sections 3.7.3 and 3.7.4). These may be present and effective continuously (constitutive defense) or increased production may be induced by attack (inducible defence) (Karban et al., 1999). Thus, production of the defensive hydroxamic acid is induced when aphids (*Rhopalosiphum padi*) attack the wild wheat *Triticum uniaristatum* (Gianoli & Niemeyer, 1997), and the prickles of dewberries on cattle-grazed plants are longer and sharper than those on ungrazed plants nearby (Abrahamson, 1975). Particular attention has been paid to rapidly inducible defenses, often the production of chemicals within the plant that inhibit the protease enzymes of the herbivores. These changes can occur within individual leaves, within branches or throughout whole tree canopies, and they may be detectable within a few hours, days or weeks, and last a few days, weeks or years; such responses have now been reported in more than 100 plant–herbivore systems (Karban & Baldwin, 1997).

. . . or do they?

There are, however, a number of problems in interpreting these responses (Schultz, 1988). First, are they 'responses' at all, or merely an incidental consequence of regrowth tissue having different properties from that removed by the herbivores? In fact, this issue is mainly one of semantics – if the metabolic responses of a plant to tissue removal happen to be defensive, then natural selection will favor them and reinforce their use. A further problem is much more substantial: are induced chemicals actually defensive in the sense of having an ecologically significant effect on the herbivores that seem to have induced them? Finally, and of most significance, are they truly defensive in the sense of having a measurable, positive impact on the plant making them, especially after the costs of mounting the response have been taken into account?

are herbivores
really adversely
affected? . . .

Fowler and Lawton (1985) addressed the second problem – 'are the responses harmful to the herbivores?' – by reviewing the effects of rapidly inducible plant defenses and found little clear-cut evidence that they are effective against insect herbivores, despite a widespread belief that they were. For example, they found that most laboratory studies revealed only small adverse effects (less than 11%) on such characters as larval development time and pupal weight, with many studies that claimed a larger effect being flawed statistically, and they argued that such effects may have negligible consequences for field populations. However, there are also a number of cases, many of which have been published since Fowler and Lawton's review, in which the plant's responses do seem to be genuinely harmful to the herbivores. When larch trees were defoliated by the larch budmoth, *Zeiraphera diniana*, the survival and adult fecundity of the moths were reduced throughout the succeeding 4–5 years as a combined result of delayed leaf production, tougher leaves, higher fiber and resin concentration and lower nitrogen levels (Baltensweiler et al., 1977). Another common response to leaf damage is early abscission ('dropping off') of mined leaves; in the case of the leaf-mining insect *Phyllonorycter* spp. on willow trees (*Salix lasiolepis*), early abscission of mined leaves was an important mortality factor for the moths – that is, the herbivores were harmed by the response (Preszler & Price, 1993). As a final example, a few weeks of grazing on the brown seaweed *Ascophyllum nodosum* by snails (*Littorina obtusata*) induces substantially increased concentrations of phlorotannins (Figure 9.1a), which reduce further snail grazing (Figure 9.1b). In this case, simple clipping of the plants did not have the same effect as the herbivore. Indeed, grazing by another herbivore, the isopod *Idotea granulosa*, also failed to induce the chemical defense. The snails can stay and feed on the same plant for long time periods (the isopods are much more mobile), so that induced responses that take time to develop can still be effective in reducing damage by snails.

. . . and do plants
really benefit?

The final question – 'do plants benefit from their induced defensive responses?' – has proved the most difficult to answer and only a few well designed field studies have been performed (Karban et al., 1999). Agrawal (1998) estimated lifetime fitness of wild radish plants (*Raphanus sativus*) (as number of seeds produced multiplied by seed mass) assigned to one of three treatments: grazed plants (subject to grazing by the caterpillar of *Pieris rapae*), leaf damage controls (equivalent amount of biomass removed using scissors) and overall controls (undamaged). Damage-induced responses, both chemical and physical, included increased concentrations of defensive glucosinolates and increased densities of trichomes (hair-like structures). Earwigs (*Forficula* spp.) and other chewing herbivores caused 100% more leaf damage on the control and artificially leaf-clipped plants than on grazed plants and there were 30% more sucking green peach aphids (*Myzus persicae*) on the control and leaf-clipped plants (Figure 9.2a, b). Induction of resistance, caused by grazing by the *P. rapae* caterpillars, significantly increased the lifetime index of fitness by more than 60% compared to the control. However, leaf damage control plants (scissors) had 38% lower fitness than the overall controls, indicating the negative effect of tissue loss without the benefits of induction (Figure 9.2c).

This fitness benefit to wild radish occurred only in environments containing herbivores; in their absence, an induced defensive response was inappropriate and the plants suffered reduced fitness (Karban et al., 1999). A similar fitness benefit has been shown in a field experiment involving wild tobacco (*Nicotiana attenuata*) (Baldwin, 1998). A specialist consumer of wild tobacco, the caterpillar of *Manduca sexta*, is remarkable in that it not only induces an accumulation of secondary metabolites and proteinase inhibitors when it feeds on wild tobacco, but it also induces the plants to

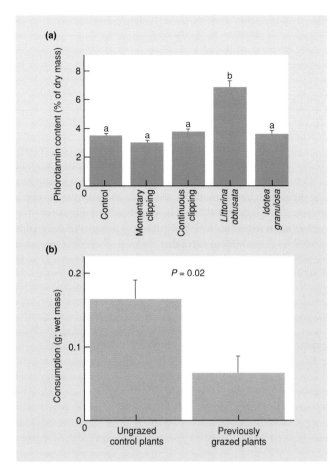

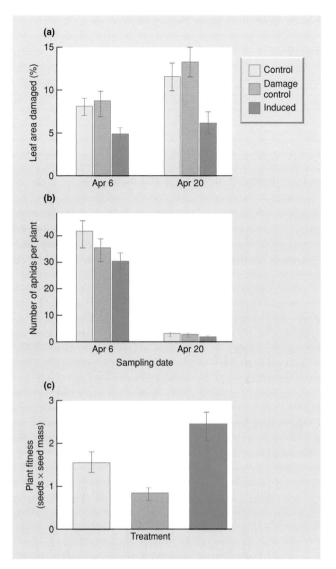

Figure 9.1 (a) Phlorotannin content of *Ascophyllum nodosum* plants after exposure to simulated herbivory (removing tissue with a hole punch) or grazing by real herbivores of two species. Means and standard errors are shown. Only the snail *Littorina obtusata* had the effect of inducing increased concentrations of the defensive chemical in the seaweed. Different letters indicate that means are significantly different ($P < 0.05$). (b) In a subsequent experiment, the snails were presented with algal shoots from the control and snail-grazed treatments in (a); the snails ate significantly less of plants with a high phlorotannin content. (After Pavia & Toth 2000.)

release volatile organic compounds that attract the generalist predatory bug *Geocoris pallens*, which feeds on the slow moving caterpillars (Kessler & Baldwin, 2004). Using molecular techniques, Zavala *et al.* (2004) were able to show that in the absence of herbivory, plant genotypes that produced little or no proteinase inhibitor grew faster and taller and produced more seed capsules than inhibitor-producing genotypes. Moreover, naturally occurring genotypes from Arizona that lacked the ability to produce proteinase inhibitors were damaged more, and sustained greater *Manduca* growth, in a laboratory experiment, compared with Utah inhibitor-producing genotypes (Glawe *et al.*, 2003).

Figure 9.2 (a) Percentage of leaf area consumed by chewing herbivores and (b) number of aphids per plant, measured on two dates (April 6 and April 20) in three field treatments: overall control, damage control (tissue removed by scissors) and induced (caused by grazing of caterpillars of *Pieris rapae*). (c) The fitness of plants in the three treatments calculated by multiplying the number of seeds produced by the mean seed mass (in mg). (After Agrawal, 1998.)

It is clear from the wild radish and wild tobacco examples that the evolution of inducible (plastic) responses involves significant costs to the plant. We may expect inducible responses to be favored by selection only when past herbivory is a reliable predictor of future risk of herbivory *and* if the likelihood of herbivory is not constant (constant herbivory should select for a fixed defensive phenotype that is best for that set of conditions) (Karban *et al.*, 1999). Of course, it is not only the costs of inducible defenses that can be set against fitness benefits. Constitutive defenses, such as spines, trichomes or defensive chemicals (particularly in the families Solanaceae and Brassicaceae), also have costs that have been measured (in phenotypes or genotypes lacking the defense) in terms of reductions in growth or the production of flowers, fruits or seeds (see review by Strauss *et al.*, 2002).

9.2.3 Herbivory, defoliation and plant growth

timing of herbivory is crucial

Despite a plethora of defensive structures and chemicals, herbivores still eat plants. Herbivory can stop plant growth, it can have a negligible effect on growth rate, and it can do just about anything in between. Plant compensation may be a general response to herbivory or may be specific to particular herbivores. Gavloski and Lamb (2000b) tested these alternative hypotheses by measuring the biomass of two cruciferous plants *Brassica napus* and *Sinapis alba* in response to 0, 25 and 75% defoliation of seedling plants by three herbivore species with biting and chewing mouthparts – adult flea beetles *Phyllotreta cruciferae* and larvae of the moths *Plutella xylostella* and *Mamestra configurata*. Not surprisingly, both plant species compensated better for 25% than 75% defoliation. However, although defoliated to the same extent, both plants tended to compensate best for defoliation by the moth *M. configurata* and least for the beetle *P. cruciferae* (Figure 9.3). Herbivore-specific compensation may reflect plant responses to slightly different patterns of defoliation or different chemicals in saliva that suppress growth in contrasting ways (Gavloski & Lamb, 2000b).

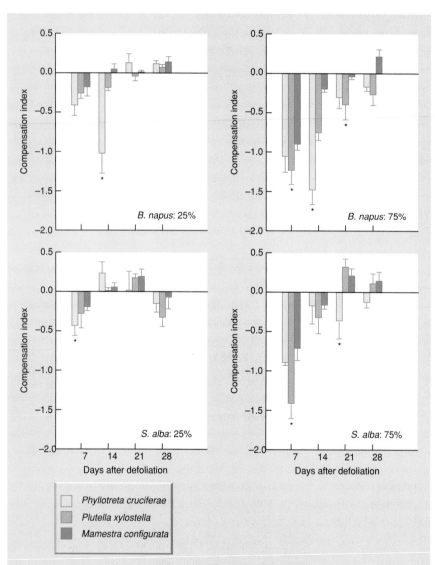

Figure 9.3 Compensation of leaf biomass (mean ± SE: ($\log_e$ biomass defoliated plant) – ($\log_e$ of mean for control plants)) of *Brassica napus* and *Sinapis alba* seedlings with 25 or 75% defoliation by three species of insect (see key) in a controlled environment. On the vertical axis, zero equates to perfect compensation, negative values to undercompensation and positive values to overcompensation. Mean biomasses of defoliated plants that differ significantly from corresponding controls are indicated by an asterisk. (After Gavloski & Lamb, 2000b.)

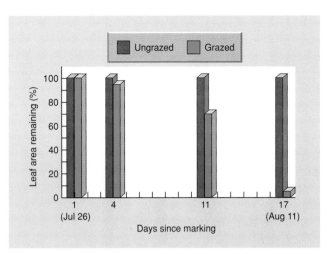

Figure 9.4 The survivorship of leaves on waterlily plants grazed by the waterlily leaf beetle was much lower than that on ungrazed plants. Effectively, all leaves had disappeared at the end of 17 days, despite the fact that 'snapshot' estimates of loss rates to grazing on grazed plants during this period suggested only around a 13% loss. (After Wallace & O'Hop, 1985.)

In the example above, compensation, which was generally complete by 21 days after defoliation, was associated with changes in root biomass consistent with the maintenance of a constant shoot : root ratio. Many plants compensate for herbivory in this way by altering the distribution of photosynthate in different parts of the plant. Thus, for example, Kosola *et al.* (2002) found that the concentration of soluble sugars in the young (white) fine roots of poplars (*Populus canadensis*) defoliated by gypsy moth caterpillars (*Lymantria dispar*) was much lower than in undefoliated trees. Older roots (>1 month in age), on the other hand, showed no significant effect of defoliation.

Often, there is considerable difficulty in assessing the real extent of defoliation, refoliation and hence net growth. Close monitoring of waterlily leaf beetles (*Pyrrhalta nymphaeae*) grazing on waterlilies (*Nuphar luteum*) revealed that leaves were rapidly removed, but that new leaves were also rapidly produced. More than 90% of marked leaves on grazed plants had disappeared within 17 days, while marked leaves on ungrazed plants were still completely intact (Figure 9.4). However, simple counts of leaves on grazed and ungrazed plants only indicated a 13% loss of leaves to the beetles, because of new leaf production on grazed plants.

grasses are particularly tolerant of grazing

The plants that seem most tolerant of grazing, especially vertebrate grazing, are the grasses. In most species, the meristem is almost at ground level amongst the basal leaf sheaths, and this major point of growth (and regrowth) is therefore usually protected from grazers' bites. Following defoliation, new leaves are produced using either stored carbohydrates or the photosynthate of surviving leaves, and new tillers are also often produced.

Grasses do not benefit directly from their grazers' attentions. But it is likely that they are helped by grazers in their competitive interactions with other plants (which are more strongly affected by the grazers), accounting for the predominance of grasses in so many natural habitats that suffer intense vertebrate grazing. This is an example of the most widespread reason for herbivory having a more drastic effect on grazing-intolerant species than is initially apparent – the interaction between herbivory and plant competition (the range of possible consequences of which are discussed by Pacala & Crawley, 1992; see also Hendon & Briske, 2002). Note also that herbivores can have severe nonconsumptive effects on plants when they act as vectors for plant pathogens (bacteria, fungi and especially viruses) – what the herbivores take from the plant is far less important than what they give it! For instance, scolytid beetles feeding on the growing twigs of elm trees act as vectors for the fungus that causes Dutch elm disease. This killed vast numbers of elms in northeastern USA in the 1960s, and virtually eradicated them in southern England in the 1970s and early 1980s.

9.2.4 Herbivory and plant survival

Generally, it is more usual for herbivores to increase a plant's susceptibility to mortality than to kill it outright. For example, although the flea beetle *Altica sublicata* reduced the growth rate of the sand-dune willow *Salix cordata* in both 1990 and 1991 (Figure 9.5), significant mortality as a result of drought stress only occurred in 1991. Then, however, susceptibility was strongly influenced by the herbivore: 80% of plants died in a high herbivory treatment (eight beetles per plant), 40% died at four beetles per plant, but none of the beetle-free control plants died (Bach, 1994).

mortality: the result of an interaction with another factor?

Repeated defoliation can have an especially drastic effect. Thus, a single defoliation of oak trees by the gypsy moth (*Lymantria dispar*) led to only a 5% mortality rate whereas three successive heavy defoliations led to mortality rates of up to 80% (Stephens, 1971). The mortality of established plants, however, is not necessarily associated with massive amounts of defoliation. One of the most extreme cases where the removal of a small amount of plant has a disproportionately profound effect is ring-barking of trees, for example by squirrels or porcupines. The cambial tissues and the phloem are torn away from the woody xylem, and the carbohydrate supply link between the leaves and the roots is broken. Thus, these pests of forestry plantations often kill young trees whilst removing very little tissue. Surface-feeding slugs can also do more damage to newly established grass populations than might be expected from the quantity of material they consume (Harper, 1977). The slugs chew through

repeated defoliation or ring-barking can kill

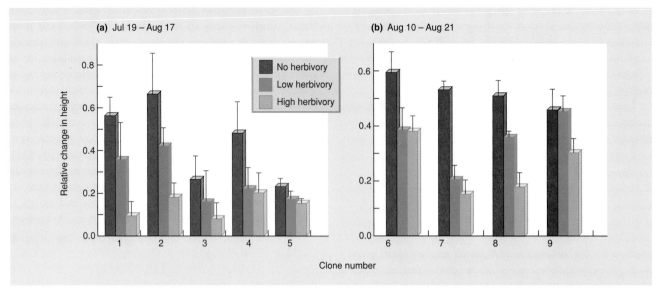

Figure 9.5   Relative growth rates (changes in height, with standard errors) of a number of different clones of the sand-dune willow, *Salix cordata*, (a) in 1990 and (b) in 1991, subjected either to no herbivory, low herbivory (four flea beetles per plant) or high herbivory (eight beetles per plant). (After Bach, 1994.)

the young shoots at ground level, leaving the felled leaves uneaten on the soil surface but consuming the meristematic region at the base of shoots from which regrowth would occur. They therefore effectively destroy the plant.

Predation of seeds, not surprisingly, has a predictably harmful effect on individual plants (i.e. the seeds themselves). Davidson *et al.* (1985) demonstrated dramatic impacts of seed-eating ants and rodents on the composition of seed banks of 'annual' plants in the deserts of southwestern USA and thus on the make up of the plant community.

9.2.5   Herbivory and plant fecundity

herbivores affect plant growth . . .
. . . indirectly by reducing seed production . . .

The effects of herbivory on plant fecundity are, to a considerable extent, reflections of the effects on plant growth: smaller plants bear fewer seeds. However, even when growth appears to be fully compensated, seed production may nevertheless be reduced because of a shift of resources from reproductive output to shoots and roots. This was the case in the study shown in Figure 9.3 where compensation in growth was complete after 21 days but seed production was still significantly lower in the herbivore-damaged plants. Moreover, indirectly through its effect on leaf area, or by directly feeding on reproductive structures, herbivory can affect floral traits (corolla diameter, floral tube length, flower number) and have an adverse impact on pollination and seed set (Mothershead &

Marquis, 2000). Thus experimentally 'grazed' plants of *Oenothera macrocarpa* produced 30% fewer flowers and 33% fewer seeds.

Plants may also be affected more directly, by the removal or destruction of flowers, flower buds or seeds. Thus, caterpillars of the large blue butterfly *Maculinea rebeli* feed only in the flowers and on the fruits of the rare plant *Gentiana cruciata*, and the number of seeds per fruit (70 compared to 120) is reduced where this specialist herbivore occurs (Kery *et al.*, 2001). Many studies, involving the artificial exclusion or removal of seed predators, have shown a strong influence of predispersal seed predation on recruitment and the density of attacked species. For example, seed predation was a significant factor in the pattern of increasing abundance of the shrub *Haplopappus squarrosus* along an elevational gradient from the Californian coast, where predispersal seed predation was higher, to the mountains (Louda, 1982); and restriction of the crucifer *Cardamine cordifolia* to shaded situations in the Rocky Mountains was largely due to much higher levels of predispersal seed predation in unshaded locations (Louda & Rodman, 1996).

It is important to realize, however, that many cases of 'herbivory' of reproductive tissues are actually mutualistic, benefitting both the herbivore and the plant (see Chapter 13). Animals that 'consume' pollen and nectar usually transfer pollen inadvertently from plant to plant in the process; and there are many fruit-eating animals that also confer a net benefit on both the parent

. . . and directly by removing reproductive structures

much pollen and fruit herbivory benefits the plant

plant and the individual seed within the fruit. Most vertebrate fruit-eaters, in particular, either eat the fruit but discard the seed, or eat the fruit but expel the seed in the feces. This disperses the seed, rarely harms it and frequently enhances its ability to germinate.

Insects that attack fruit or developing fruit, on the other hand, are very unlikely to have a beneficial effect on the plant. They do nothing to enhance dispersal, and they may even make the fruit less palatable to vertebrates. However, some large animals that normally kill seeds can also play a part in dispersing them, and they may therefore have at least a partially beneficial effect. There are some 'scatter-hoarding' species, like certain squirrels, that take nuts and bury them at scattered locations; and there are other 'seed-caching' species, like some mice and voles, that collect scattered seeds into a number of hidden caches. In both cases, although many seeds are eaten, the seeds are dispersed, they are hidden from other seed predators and a number are never relocated by the hoarder or cacher (Crawley, 1983).

Herbivores also influence fecundity in a number of other ways. One of the most common responses to herbivore attack is a delay in flowering. For instance, in longer lived semelparous species, herbivory frequently delays flowering for 1 year or more, and this typically increases the longevity of such plants since

death almost invariably follows their single burst of reproduction (see Chapter 4). *Poa annua* on a lawn can be made almost immortal by mowing it at weekly intervals, whereas in natural habitats, where it is allowed to flower, it is commonly an annual – as its name implies.

Generally, the timing of defoliation is critical in determining the effect on plant fecundity. If leaves are removed before inflorescences are formed, then the extent to which fecundity is depressed clearly depends on the extent to which the plant is able to compensate. Early defoliation of a plant with sequential leaf production may have a negligible effect on fecundity; but where defoliation takes place later, or where leaf production is synchronous, flowering may be reduced or even inhibited completely. If leaves are removed after the inflorescence has been formed, the effect is usually to increase seed abortion or to reduce the size of individual seeds.

the timing of herbivory is critical

An example where timing is important is provided by field gentians (*Gentianella campestris*). When herbivory on this biennial plant is simulated by clipping to remove half its biomass (Figure 9.6a), the outcome depends on the timing of the clipping (Figure 9.6b). Fruit production was much increased over controls if clipping

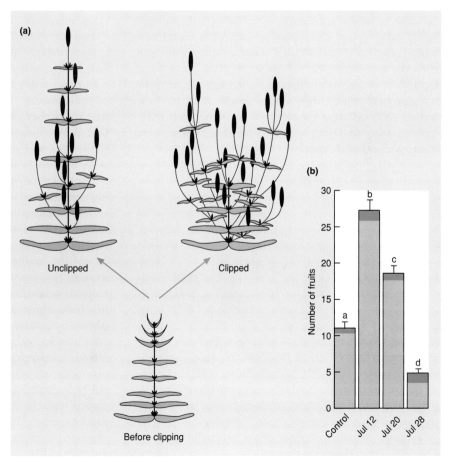

Figure 9.6 (a) Clipping of field gentians to simulate herbivory causes changes in the architecture and numbers of flowers produced. (b) Production of mature (open histograms) and immature fruits (black histograms) of unclipped control plants and plants clipped on different occasions from July 12 to 28, 1992. Means and standard errors are shown and all means are significantly different from each other (*P* < 0.05). Plants clipped on July 12 and 20 developed significantly more fruits than unclipped controls. Plants clipped on July 28 developed significantly fewer fruits than controls. (After Lennartsson *et al.*, 1998).

occurred between 1 and 20 July, but if clipping occurred later than this, fruit production was less in the clipped plants than in the unclipped controls. The period when the plants show compensation coincides with the time when damage by herbivores normally occurs.

9.2.6　A postscript: antipredator chemical defenses in animals

animals also defend themselves

It should not be imagined that antipredator chemical defenses are restricted to plants. A variety of constitutive animal chemical defenses were described in Chapter 3 (see Section 3.7.4), including plant defensive chemicals sequestered by herbivores from their food plants (see Section 3.7.4). Chemical defenses may be particularly important in modular animals, such as sponges, which lack the ability to escape from their predators. Despite their high nutritional value and lack of physical defenses, most marine sponges appear to be little affected by predators (Kubanek *et al.*, 2002). In recent years, several triterpene glycosides have been extracted from sponges, including from *Ectyoplasia ferox* in the Caribbean. In a field study, crude extracts of refined triterpene glycosides from this sponge were presented in artificial food substrates to natural assemblages of reef fishes in the Bahamas. Strong antipredatory affects were detected when compared to control substrates (Figure 9.7). It is of interest that the triterpene glycosides also adversely affected competitors of the sponge, including 'fouling' organisms that overgrow them (bacteria, invertebrates and algae) and other sponges (an example of allelopathy – see Section 8.3.2). All these enemies were apparently deterred by surface contact with the chemicals rather than by water-borne effects (Kubanek *et al.*, 2002).

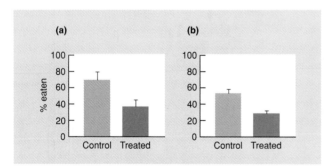

Figure 9.7　Results of field assays assessing antipredatory effects of compounds from the sponge *Ectyoplasia ferox* against natural assemblages of reef fish in the Bahamas. Means (+ SE) are shown for percentages of artificial food substrates eaten in controls (containing no sponge extracts) in comparison with: (a) substrates containing a crude sponge extract (*t*-test, $P = 0.036$) and (b) substrates containing triterpene glycosides from the sponge ($P = 0.011$). (After Kubanek *et al.*, 2002.)

9.3　The effect of predation on prey populations

Returning now to predators in general, it may seem that since the effects of predators are harmful to individual prey, the immediate effect of predation on a population of prey must also be predictably harmful. However, these effects are not always so predictable, for one or both of two important reasons. In the first place, the individuals that are killed (or harmed) are not always a random sample of the population as a whole, and may be those with the lowest potential to contribute to the population's future. Second, there may be compensatory changes in the growth, survival or reproduction of the surviving prey: they may experience reduced competition for a limiting resource, or produce more offspring, or other predators may take fewer of the prey. In other words, whilst predation is bad for the prey that get eaten, it may be good for those that do not. Moreover, predation is least likely to affect prey dynamics if it occurs at a stage of the prey's life cycle that does not have a significant effect, ultimately, on prey abundance.

To deal with the second point first, if, for example, plant recruitment is not limited by the number of seeds produced, then insects that reduce seed production are unlikely to have an important effect on plant abundance (Crawley, 1989). For instance, the weevil *Rhinocyllus conicus* does not reduce recruitment of the nodding thistle, *Carduus nutans*, in southern France despite inflicting seed losses of over 90%. Indeed, sowing 1000 thistle seeds per square meter also led to no observable increase in the number of thistle rosettes. Hence, recruitment appears not to be limited by the number of seeds produced; although whether it is limited by subsequent predation of seeds or early seedlings, or the availability of germination sites, is not clear (Crawley, 1989). (However, we have seen in other situations (see Section 9.2.5) that predispersal seed predation can profoundly affect seedling recruitment, local population dynamics and variation in relative abundance along environmental gradients and across microhabitats.)

predation may occur at a demographically unimportant stage

compensatory reactions amongst survivors

The impact of predation is often limited by compensatory reactions amongst the survivors as a result of reduced intraspecific competition. Thus, in a classic experiment in which large numbers of woodpigeons (*Columba palumbus*) were shot, the overall level of winter mortality was not increased, and stopping the shooting led to no increase in pigeon abundance (Murton *et al.*, 1974). This was because the number of surviving pigeons was determined ultimately not by shooting but by food availability, and so when shooting reduced density, there were compensatory reductions in intraspecific competition and in natural mortality, as well as density-dependent immigration of birds moving in to take advantage of unexploited food.

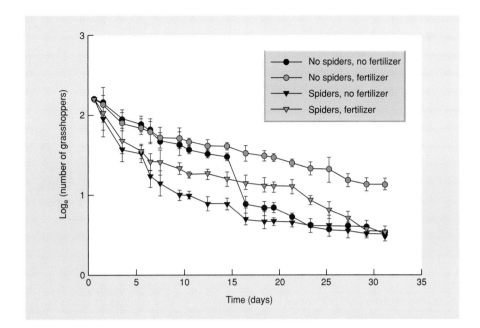

Figure 9.8 Trajectories of numbers of grasshoppers surviving (mean ± SE) for fertilizer and predation treatment combinations in a field experiment involving caged plots in the Arapaho Prairie, Nebraska, USA. (After Oedekoven & Joern, 2000.)

effects ameliorated
by reduced
competition

Indeed, whenever density is high enough for intraspecific competition to occur, the effects of predation on a population should be ameliorated by the consequent reductions in intraspecific competition. Outcomes of predation may, therefore, vary with relative food availability. Where food quantity or quality is higher, a given level of predation may not lead to a compensatory response because prey are not food-limited. This hypothesis was tested by Oedekoven and Joern (2000) who monitored grasshopper (*Ageneotettix deorum*) survivorship in caged prairie plots subject to fertilization (or not) to increase food quality in the presence or absence of lycosid spiders (*Schizocoza* spp.). With ambient food quality (no fertilizer, black symbols), spider predation and food limitation were compensatory: the same numbers of grasshoppers were recovered at the end of the 31-day experiment (Figure 9.8). However, with higher food quality (nitrogen fertilizer added, colored symbols), spider predation reduced the numbers surviving compared to the no-spider control: a noncompensatory response. Under ambient conditions after spider predation, the surviving grasshoppers encountered more food per capita and lived longer as a result of reduced competition. However, grasshoppers were less food-limited when food quality was higher so that after predation the release of additional per capita food did not promote survivorship (Oedekoven & Joern, 2000).

predatory attacks are
often directed at the
weakest prey

Turning to the nonrandom distribution of predators' attention within a population of prey, it is likely, for example, that predation by many large carnivores is focused on the old (and infirm), the young (and naive) or the sick. For instance, a study in the Serengeti found that cheetahs and wild dogs killed a disproportionate number from the younger age classes of Thomson's gazelles (Figure 9.9a), because: (i) these young animals were easier to catch (Figure 9.9b); (ii) they had lower stamina and running speeds; (iii) they were less good at outmaneuvering the predators (Figure 9.9c); and (iv) they may even have failed to recognize the predators (FitzGibbon & Fanshawe, 1989; FitzGibbon, 1990). Yet these young gazelles will also have been making no reproductive contribution to the population, and the effects of this level of predation on the prey population will therefore have been less than would otherwise have been the case.

Similar patterns may also be found in plant populations. The mortality of mature *Eucalyptus* trees in Australia, resulting from defoliation by the sawfly *Paropsis atomaria*, was restricted almost entirely to weakened trees on poor sites, or to trees that had suffered from root damage or from altered drainage following cultivation (Carne, 1969).

Taken overall, then, it is clear that the step from noting that individual prey are harmed by individual predators to demonstrating that prey abundance

difficulties of
demonstrating effects
on prey populations

is adversely affected is not an easy one to take. Of 28 studies in which herbivorous insects were experimentally excluded from plant communities using insecticides, 50% provided evidence of an effect on plants at the population level (Crawley, 1989). As Crawley noted, however, such proportions need to be treated cautiously. There is an almost inevitable tendency for 'negative' results (no population effect) to go unreported, on the grounds of there being 'nothing' to report. Moreover, the exclusion studies often took 7 years or more to show any impact on the plants: it may be that many of the 'negative' studies were simply given up too early.

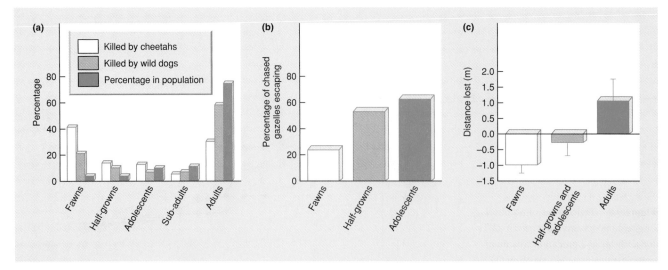

Figure 9.9   (a) The proportions of different age classes (determined by tooth wear) of Thomson's gazelles in cheetah and wild dog kills is quite different from their proportions in the population as a whole. (b) Age influences the probability for Thomson's gazelles of escaping when chased by cheetahs. (c) When prey (Thomson's gazelles) 'zigzag' to escape chasing cheetahs, prey age influences the mean distance lost by the cheetahs. (After FitzGibbon & Fanshawe, 1989; FitzGibbon, 1990.)

Many more recent investigations have shown clear effects of seed predation on plant abundance (e.g. Kelly & Dyer, 2002; Maron et al., 2002).

9.4   Effects of consumption on consumers

consumers often need to exceed a threshold of consumption

The beneficial effects that food has on individual predators are not difficult to imagine. Generally speaking, an increase in the amount of food consumed leads to increased rates of growth, development and birth, and decreased rates of mortality. This, after all, is implicit in any discussion of intraspecific competition amongst consumers (see Chapter 5): high densities, implying small amounts of food per individual, lead to low growth rates, high death rates, and so on. Similarly, many of the effects of migration previously considered (see Chapter 6) reflect the responses of individual consumers to the distribution of food availability. However, there are a number of ways in which the relationships between consumption rate and consumer benefit can be more complicated than they initially appear. In the first place, all animals require a certain amount of food simply for maintenance and unless this threshold is exceeded the animal will be unable to grow or reproduce, and will therefore be unable to contribute to future generations. In other words, low consumption rates, rather than leading to a small benefit to the consumer, simply alter the rate at which the consumer starves to death.

At the other extreme, the birth, growth and survival rates of individual consumers cannot be expected to rise indefinitely as food availability is increased. Rather, the consumers become satiated. Consumption rate eventually reaches a plateau, where it becomes independent of the amount of food available, and benefit to consumers therefore also reaches a plateau. Thus, there is a limit to the amount that a particular consumer population can eat, a limit to the amount of harm that it can do to its prey population at that time, and a limit to the extent by which the consumer population can increase in size. This is discussed more fully in Section 10.4.

consumers may become satiated

The most striking example of whole populations of consumers being satiated simultaneously is provided by the many plant species that have mast years. These are occasional years in which there is synchronous production of a large volume of seed, often across a large geographic area, with a dearth of seeds produced in the years in between (Herrera et al., 1998; Koenig & Knops, 1998; Kelly et al., 2000). This is seen particularly often in tree species that suffer generally high intensities of seed predation (Silvertown, 1980) and it is therefore especially significant that the chances of escaping seed predation are likely to be much higher in mast years than in other years. Masting seems to be especially common in the New Zealand flora (Kelly, 1994) where it has also been reported for tussock grass species (Figure 9.10). The individual predators of seeds are satiated in mast years, and the populations of predators cannot increase in abundance rapidly enough to exploit the glut. This

mast years and the satiation of seed predators

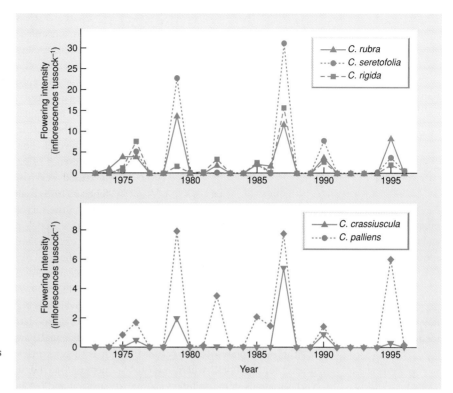

Figure 9.10 The flowering rate for five species of tussock grass (*Chionochloa*) between 1973 and 1996 in Fiordland National Park, New Zealand. Mast years are highly synchronized in the five species, seemingly in response to high temperatures in the previous season, when flowering is induced. (After McKone *et al.*, 1998.)

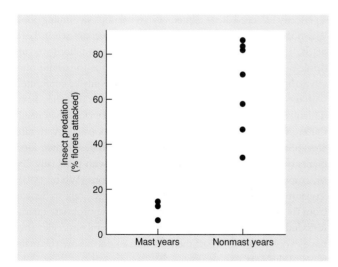

Figure 9.11 Insect predation on florets of *Chionochloa pallens* in mast (*n* = 3) and nonmast years (*n* = 7) from 1988 to 1997 at Mount Hutt, New Zealand. A mast year is defined here as one with greater than 10 times as many florets produced per tussock than in the previous year. The significant difference in insect damage supports the hypothesis that the function of masting is to satiate seed predators. (After McKone *et al.*, 1998.)

is illustrated in Figure 9.11 where the percentage of florets of the grass *Chionochloa pallens* attacked by insects remains below 20% in mast years but ranges up to 80% or more in nonmast years. The fact that *C. pallens* and four other species of *Chionochloa* show strong synchrony in masting is likely to result in an increased benefit to each species in terms of escaping seed predation in mast years.

On the other hand, the production of a mast crop makes great demands on the internal resources of a plant. A spruce tree in a mast year averages 38% less annual growth than in other years, and the annual ring increment in forest trees may be reduced by as much during a mast year as by a heavy attack of defoliating caterpillars. The years of seed famine are therefore essentially years of plant recovery.

As well as illustrating the potential importance of predator satiation, the example of masting highlights a further point relating to timescales. The seed predators are unable to extract the maximum benefit from (or do the maximum harm to) the mast crop because their generation times are too long. A hypothetical seed predator population that could pass through several generations during a season would be able to increase exponentially and explosively on the mast crop and destroy it. Generally speaking, consumers with relatively short generation times tend to closely track fluctuations in the quantity or abundance of their food or

a consumer's numerical response is limited by its generation time . . .

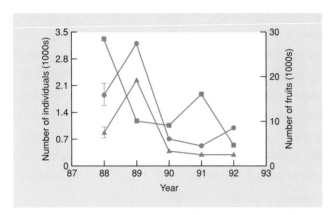

Figure 9.12 Fluctuations in the fruit production of *Asphodelus* (■) and the number of *Capsodes* nymphs (●) and adults (▲) at a study site in the Negev desert, Israel. (After Ayal, 1994.)

prey, whereas consumers with relatively long generation times take longer to respond to increases in prey abundance, and longer to recover when reduced to low densities.

... as illustrated by desert interactions

The same phenomenon occurs in desert communities, where year-to-year variations in precipitation can be both considerable and unpredictable, leading to similar year-to-year variation in the productivity of many desert plants. In the rare years of high productivity, herbivores are typically at low abundance following one or more years of low plant productivity. Thus, the herbivores are likely to be satiated in such years, allowing plant populations to add considerably to their reserves, perhaps by augmenting their buried seed banks or their underground storage organs (Ayal, 1994). The example of fruit production by *Asphodelus ramosus* in the Negev desert in Israel in shown in Figure 9.12. The mirid bug, *Capsodes infuscatus*, feeds on *Asphodelus*, exhibiting a particular preference for the developing flowers and young fruits. Potentially, therefore, it can have a profoundly harmful effect on the plant's fruit production. But it only passes through one generation per year. Hence, its abundance tends never to match that of its host plant (Figure 9.12). In 1988 and 1991, fruit production was high but mirid abundance was relatively low: the reproductive output of the mirids was therefore high (3.7 and 3.5 nymphs per adult, respectively), but the proportion of fruits damaged was relatively low (0.78 and 0.66). In 1989 and 1992, on the other hand, when fruit production had dropped to much lower levels, the proportion of fruits damaged was much higher (0.98 and 0.87) and the reproductive output was lower (0.30 nymphs per adult in 1989; unknown in 1992). This suggests that herbivorous insects, at least, may have a limited ability to affect plant population dynamics in desert communities, but that the potential is much greater for the dynamics of herbivorous insects to be affected by their food plants (Ayal, 1994).

food quality rather than quantity can be of paramount importance

Chapter 3 stressed that the quantity of food consumed may be less important than its quality. In fact, food quality, which has both positive aspects (like the concentrations of nutrients) and negative aspects (like the concentrations of toxins), can only sensibly be defined in terms of the effects of the food on the animal that eats it; and this is particularly pertinent in the case of herbivores. For instance, we saw in Figure 9.8 how even in the presence of predatory spiders, enhanced food quality led to increased survivorship of grasshoppers. Along similar lines, Sinclair (1975) examined the effects of grass quality (protein content) on the survival of wildebeest in the Serengeti of Tanzania. Despite selecting protein-rich plant material (Figure 9.13a), the wildebeest consumed food in the dry season that contained well below the level of protein necessary even for maintenance (5–6% of crude protein); and to judge by the depleted fat reserves of dead males (Figure 9.13b), this was an important cause of mortality. Moreover, it is highly relevant that the protein requirements of females during late pregnancy and lactation (December–May in the wildebeest) are three to four times higher than the normal. It is therefore clear that the shortage of high-quality food (and not just food shortage *per se*) can have a drastic effect on the growth, survival and fecundity of a consumer. In the case of herbivores especially, it is possible for an animal to be apparently surrounded by its food whilst still experiencing a food shortage. We can see the problem if we imagine that we ourselves are provided with a perfectly balanced diet – diluted in an enormous swimming pool. The pool contains everything we need, and we can see it there before us, but we may very well starve to death before we can drink enough water to extract enough nutrients to sustain ourselves. In a similar fashion, herbivores may frequently be confronted with a pool of available nitrogen that is so dilute that they have difficulty processing enough material to extract what they need. Outbreaks of herbivorous insects may then be associated with rare elevations in the concentration of available nitrogen in their food plants (see Section 3.7.1), perhaps associated with unusually dry or, conversely, unusually waterlogged conditions (White, 1993). Consumers obviously need to acquire resources – but, to benefit from them fully they need to acquire them in appropriate quantities and in an appropriate form. The behavioral strategies that have evolved in the face of the pressures to do this are the main topic of the next two sections.

9.5   Widths and compositions of diets

Consumers can be classified as either monophagous (feeding on a single prey type), oligophagous (few prey types) or polyphagous (many prey types). An equally useful distinction is

range and classification of diet widths

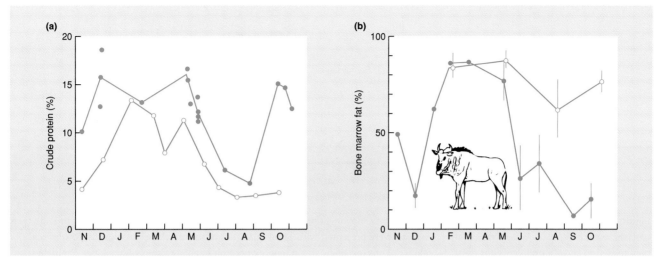

Figure 9.13   (a) The quality of food measured as percentage crude protein available to (○) and eaten by (●) wildebeest in the Serengeti during 1971. Despite selection ('eaten' > 'available'), the quality of food eaten fell during the dry season below the level necessary for the maintenance of nitrogen balance (5–6% of crude protein). (b) The fat content of the bone marrow of the live male population (○) and those found dead from natural causes (●). Vertical lines, where present, show 95% confidence limits. (After Sinclair, 1975.)

between specialists (broadly, monophages and oligophages) and generalists (polyphages). Herbivores, parasitoids and true predators can all provide examples of monophagous, oligophagous and polyphagous species. But the distribution of diet widths differs amongst the various types of consumer. True predators with specialized diets do exist (for instance the snail kite *Rostrahamus sociabilis* feeds almost entirely on snails of the genus *Pomacea*), but most true predators have relatively broad diets. Parasitoids, on the other hand, are typically specialized and may even be monophagous. Herbivores are well represented in all categories, but whilst grazing and 'predatory' herbivores typically have broad diets, 'parasitic' herbivores are very often highly specialized. For instance, Janzen (1980) examined 110 species of beetle that feed as larvae inside the seeds of dicotyledonous plants in Costa Rica ('parasitizing' them) and found that 83 attacked only one plant species, 14 attacked only two, nine attacked three, two attacked four, one attacked six and one attacked eight of the 975 plants in the area.

9.5.1   Food preferences

preference is defined by comparing diet with 'availability'

It must not be imagined that polyphagous and oligophagous species are indiscriminate in what they choose from their acceptable range. On the contrary, some degree of preference is almost always apparent. An animal is said to exhibit a preference for a particular type of food when the proportion of that type in the animal's diet is higher than its proportion in the animal's environment. To measure food preference in nature, therefore, it is necessary not only to examine the animal's diet (usually by the analysis of gut contents) but also to assess the 'availability' of different food types. Ideally, this should be done not through the eyes of the observer (i.e. not by simply sampling the environment), but through the eyes of the animal itself.

A food preference can be expressed in two rather different contexts. There can be a preference for items that are the most valuable amongst those available *or* for items that provide an integral part of a mixed and balanced diet. These will be referred to as ranked and balanced preferences, respectively. In the terms of Chapter 3 (Section 3.8), where resources were classified, individuals exhibit ranked preferences in discriminating between resource types that are 'perfectly substitutable' and exhibit balanced preferences between resource types that are 'complementary'.

Ranked preferences are usually seen most clearly amongst carnivores. For instance, Figure 9.14 shows two examples in which carnivores actively selected prey items that were the most profitable in terms of energy intake per unit time spent dealing with (or

ranked preferences predominate when food items can be classified on a single scale . . .

'handling') prey. Results such as these reflect the fact that a carnivore's food often varies little in composition (see Section 3.7.1), but may vary in size or accessibility. This allows a single measure (like 'energy gained per unit handling time') to be used to characterize food items, and it therefore allows food items to be ranked. In other words, Figure 9.14 shows consumers exhibiting an active preference for food of a high rank.

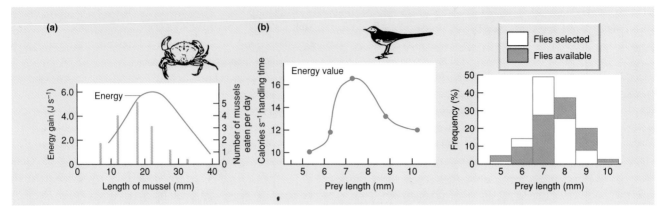

Figure 9.14 Predators eating 'profitable' prey, i.e. predators showing a preponderance in their diet for those prey items that provide them with the most energy. (a) When crabs (*Carcinus maenas*) were presented with equal quantities of six size classes of mussels (*Mytilus edulis*), they tended to show a preference for those providing the greatest energy gain (energy per unit handling time). (After Elner & Hughes, 1978.) (b) Pied wagtails (*Motacilla alba yarrellii*) tended to select, from scatophagid flies available, those providing the greatest energy gain per unit handling time. (After Davies, 1977; Krebs, 1978.)

... but many consumers show a combination of ranked and balanced preferences

For many consumers, however, especially herbivores and omnivores, no simple ranking is appropriate, since none of the available food items matches the nutritional requirements of the consumer. These requirements can therefore only be satisfied either by eating large quantities of food, and eliminating much of it in order to get a sufficient quantity of the nutrient in most limited supply (for example aphids and scale insects excrete vast amounts of carbon in honeydew to get sufficient nitrogen from plant sap), or by eating a combination of food items that between them match the consumer's requirements. In fact, many animals exhibit both sorts of response. They select food that is of generally high quality (so the proportion eliminated is minimized), but they also select items to meet specific requirements. For instance, sheep and cattle show a preference for high-quality food, selecting leaves in preference to stems, green matter in preference to dry or old material, and generally selecting material that is higher in nitrogen, phosphorus, sugars and gross energy, and lower in fiber, than what is generally available. In fact, all generalist herbivores appear to show rankings in the rate at which they eat different food plants when given a free choice in experimental tests (Crawley, 1983).

mixed diets can be favored for a variety of reasons

On the other hand, a balanced preference is also quite common. For instance, the plate limpet, *Acmaea scutum*, selects a diet of two species of encrusting microalgae that contains 60% of one species and 40% of the other, almost irrespective of the proportions in which they are available (Kitting, 1980). Whilst caribou, which survive on lichen through the winter, develop a sodium deficiency by the spring that they overcome by drinking seawater, eating urine-contaminated snow and gnawing shed antlers (Staaland *et al.*, 1980). We have only to look at ourselves to see an example in which 'performance' is far better on a mixed diet than on a pure diet of even the 'best' food.

There are two other important reasons why a mixed diet may be favored. First, consumers may accept low-quality items simply because, having encountered them, they have more to gain by eating them (poor as they are) than by ignoring them and continuing to search. This is discussed in detail in Section 9.5.3. Second, consumers may benefit from a mixed diet because each food type may contain a different undesirable toxic chemical. A mixed diet would then keep the concentrations of all of these chemicals within acceptable limits. It is certainly the case that toxins can play an important role in food preference. For instance, dry matter intake by Australian ringtail possums (*Pseudocheirus peregrinus*) feeding on *Eucalyptus* tree leaves was strongly negatively correlated with the concentration of sideroxylonal, a toxin found in *Eucalyptus* leaves, but was not related to nutritional characteristics such as nitrogen or cellulose (Lawler *et al.*, 2000).

Overall, however, it would be quite wrong to give the impression that all preferences have been clearly linked with one explanation or another. For example, Thompson (1988) reviewed the relationship between the oviposition preferences of phytophagous insects and the performance of their offspring on the selected food plants in terms of growth, survival and reproduction. A number of studies have shown a good association (i.e. females preferentially oviposit on plants where their offspring perform best), but in many others the association is poor. In such cases there is generally no shortage of explanations for the apparently unsuitable behavior, but these explanations are, as yet, often just untested hypotheses.

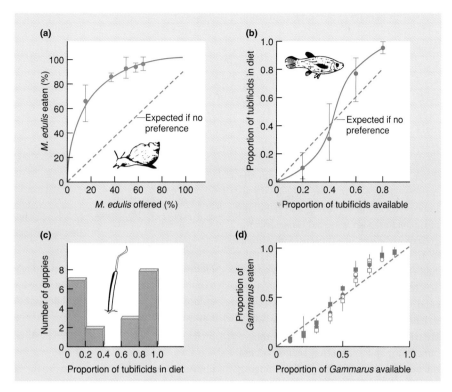

Figure 9.15 Switching. (a) A lack of switching: snails exhibit a consistent preference amongst the mussels *Mytilus edulis* and *M. californianus*, irrespective of their relative abundance (means plus standard errors). (After Murdoch & Stewart-Oaten, 1975.) (b) Switching by guppies fed on tubificids and fruit-flies: they take a disproportionate amount of whichever prey type is the more available (means and total ranges). (After Murdoch *et al.*, 1975.) (c) Preferences shown by the individual guppies in (b) when offered equal amounts of the two prey types: individuals were mostly specialists on one or other type. (d) Switching by sticklebacks fed mixtures of *Gammarus* and *Artemia*: overall they take a disproportionate amount of whichever is more available. However, in the first series of trials, with *Gammarus* availability decreasing (closed symbols), first-day trialists (■) tended to take more *Gammarus* than third-day trialists (●), whereas with *Gammarus* availability increasing, firsts (□) tended to take less *Gammarus* than thirds (○). The effects of learning are apparent. (After Hughes & Croy, 1993.)

9.5.2 Switching

switching involves a preference for food types that are common

The preferences of many consumers are fixed; in other words, they are maintained irrespective of the relative availabilities of alternative food types. But many others switch their preference, such that food items are eaten disproportionately often when they are common and are disproportionately ignored when they are rare. The two types of preference are contrasted in Figure 9.15. Figure 9.15a shows the fixed preference exhibited by predatory shore snails when they were presented with two species of mussel prey at a range of proportions. The line in Figure 9.15a has been drawn on the assumption that they exhibited the same preference at all proportions. This assumption is clearly justified: irrespective of availability, the predatory snails showed the same marked preference for the thin-shelled, less protected *Mytilus edulis*, which they could exploit more effectively. By contrast,

Figure 9.15b shows what happened when guppies (fish) were offered a choice between fruit-flies and tubificid worms as prey. The guppies clearly switched their preference, and consumed a disproportionate number of the more abundant prey type.

There are a number of situations in which switching can arise. Probably the most common is where different types of prey are found in different microhabitats, and the consumers concentrate on the most profitable microhabitat. This was the case for the guppies in Figure 9.15b: the fruit-flies floated at the water surface whilst the tubificids were found at the bottom. Switching can also occur (Bergelson, 1985) in the following situations:

when might switching arise?

1   When there is an increased probability of orientating toward a common prey type, i.e. consumers develop a 'search image' for abundant food (Tinbergen, 1960) and concentrate on their 'image' prey to the relative exclusion of nonimage prey.

2 When there is an increased probability of pursuing a common
   prey type.
3 When there is an increased probability of capturing a common
   prey type.
4 When there is an increased efficiency in handling a common
   prey type.

In each case, increasingly common prey generate increased
interest and/or success on the part of the predator, and hence an
increased rate of consumption. For instance, switching occurred
in the 15-spined stickleback, *Spinachia spinachia*, feeding on the
crustaceans *Gammarus* and *Artemia* as alternative prey (Figure 9.15d)
as a result of learned improvements in capturing and handling
efficiencies, especially of *Gammarus*. Fish were fed *Gammarus* for
7 days, which was then replaced in the diet, in 10% steps, with *Artemia*
until the diet was 100% *Artemia*. This diet was then maintained
for a further 7 days, when the process was reversed back down
to 100% *Gammarus*. Each 'step' itself lasted 3 days, on each of
which the fish were tested. The learning process is apparent in
Figure 9.15d in the tendency for first-day trialists to be more
influenced than third-day trialists by the previous dietary mix.

Interestingly, switching in a population often seems to be a
consequence not of individual consumers gradually changing
their preference, but of the proportion of specialists changing. Figure
9.15c shows this for the guppies. When the prey types were equally
abundant, individual guppies were not generalists – rather, there
were approximately equal numbers of fruit-fly and tubificid
specialists.

a plant that 'switches' It may come as a surprise that a
plant may show behavior akin to
switching. The northern pitcher plant
Sarracenia purpurea lives in nutrient-poor bogs and fens, circumstances that are thought to favor carnivory in plants. Carnivorous
plants such as pitcher plants invest an excess of carbon (captured
in photosynthesis) in specialist organs for capturing invertebrate
prey (effectively nitrogen-capturing structures). Figure 9.16 shows
how relative size of the pitcher keel responded to nitrogen addition to plots in Molly Bog in Vermont, USA. The more nitrogen
that was applied, the larger the relative keel size – this corresponds
to an increase in size of the noncarnivorous keel of the pitcher
and a decrease in size of the prey-catching tube. Thus, with
increasing nitrogen levels, the capacity for carnivory decreased
while maximum photosynthesis rates increased. In effect, the plants
switched effort from nitrogen to carbon capture when more
nitrogen was available in their environment.

9.5.3 The optimal foraging approach to diet width

diet width and
evolution Predators and prey have undoubtedly
influenced one another's evolution.
This can be seen in the distasteful or

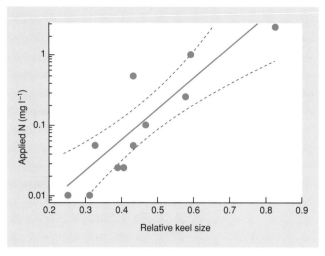

Figure 9.16 The relationship between relative keel size of pitchers
of *Sarracenia purpurea* and nitrogen added as aerial spray in plots
at Molly Bog, Vermont. Dotted lines indicate 95% confidence
intervals. A larger relative keel size corresponds to a reduced
investment in organs of prey capture. (After Ellison & Gotelli, 2002.)

poisonous leaves of many plants, in the spines of hedgehogs and
in the camouflage coloration of many insect prey; and it can be
seen in the stout ovipositors of wood wasps, the multichambered
stomachs of cattle and the silent approach and sensory excellence
of owls. Such specialization makes it clear, though, that no predator
can possibly be capable of consuming all types of prey. Simple
design constraints prevent shrews from eating owls (even though
shrews are carnivores) and prevent humming-birds from eating
seeds.

Even within their constraints, however, most animals consume a narrower range of food types than they are morphologically
capable of consuming. In trying to understand what determines
a consumer's actual diet within its wide potential range, ecologists
have increasingly turned to *optimal foraging theory*. The aim of
optimal foraging theory is to predict the foraging strategy to be
expected under specified conditions. It generally makes such predictions on the basis of a number of assumptions:

1 The foraging behavior that is
   exhibited by present-day animals is
   the one that has been favored by
   natural selection in the past but
   also most enhances an animal's fitness at present. assumptions inherent
in optimal foraging
theory
2 High fitness is achieved by a high net rate of energy intake
   (i.e. gross energy intake minus the energetic costs of obtaining that energy).
3 Experimental animals are observed in an environment to which
   their foraging behavior is suited, i.e. it is a natural environment
   very similar to that in which they evolved, or an experimental
   arena similar in essential respects to the natural environment.

These assumptions will not always be justified. First, other aspects of an organism's behavior may influence fitness more than optimal foraging does. For example, there may be such a premium on the avoidance of predators that animals forage at a place and time where the risk from predators is lower, and in consequence gather their food less efficiently than is theoretically possible (see Section 9.5.4). Second, and just as important, for many consumers (particularly herbivores and omnivores) the efficient gathering of energy may be less critical than of some other dietary constituent (e.g. nitrogen), or it may be of prime importance for the forager to consume a mixed and balanced diet. In such cases, the value of existing optimal foraging theory is limited. However, in circumstances where the energy maximization premise can be expected to apply, optimal foraging theory offers a powerful insight into the significance of the foraging 'decisions' that predators make (for reviews see Stephens & Krebs, 1986; Krebs & Kacelnik, 1991; Sih & Christensen, 2001).

theoreticians are omniscient mathematicians – the foragers need not be

Typically, optimal foraging theory makes predictions about foraging behavior based on mathematical models constructed by ecological theoreticians who are omniscient ('all knowing') as far as their model systems are concerned. The question therefore arises: is it necessary for a real forager to be equally omniscient and mathematical, if it is to adopt the appropriate, optimal strategy? The answer is 'no'. The theory simply says that if there is a forager that in some way (in any way) manages to do the right thing in the right circumstances, then this forager will be favored by natural selection; and if its abilities are inherited, these should spread, in evolutionary time, throughout the population.

Optimal foraging theory does not specify precisely how the forager should make the right decisions, and it does not require the forager to carry out the same calculations as the modeler. Later we consider another group of 'mechanistic' models (see Section 9.6.2) that attempt to show how a forager, given that it is not omniscient, might nevertheless manage to respond by 'rules of thumb' to limited environmental information and thereby exhibit a strategy that is favored by natural selection. But it is optimal foraging theory that predicts the nature of the strategy that should be so favored.

The first paper on optimal foraging theory (MacArthur & Pianka, 1966) sought to understand the determination of diet 'width' (the range of food types eaten by an animal) within a habitat. Subsequently, the model was developed into a more rigorous algebraic form, notably by Charnov (1976a). MacArthur and Pianka argued that to obtain food, any predator must expend time and energy, first in searching for its prey and then in handling it (i.e. pursuing, subduing and consuming it). Whilst searching, a predator is likely to encounter a wide variety of food items. MacArthur and Pianka therefore saw diet width as depending on the responses of predators once they had encountered prey.

Generalists pursue (and may then subdue and consume) a large proportion of the prey types they encounter; specialists continue searching except when they encounter prey of their specifically preferred type.

The 'problem' for any forager is this: if it is a specialist, then it will only pursue profitable prey items, but it

to pursue or not pursue?

may expend a great deal of time and energy searching for them. Whereas if it is a generalist, it will spend relatively little time searching, but it will pursue both more and less profitable types of prey. An optimal forager should balance the pros and cons so as to maximize its overall rate of energy intake. MacArthur and Pianka expressed the problem as follows: given that a predator already includes a certain number of profitable items in its diet, should it expand its diet (and thereby decrease its search time) by including the next most profitable item as well?

We can refer to this 'next most profitable' item as the ith item. E_i/h_i is then the profitability of the item, where E_i is its energy content, and h_i its handling time. In addition, $\bar{E}/\bar{h}$ is the average profitability of the 'present' diet (i.e. one that includes all prey types that are more profitable than i, but does not include prey type i itself), and $\bar{s}$ is the average search time for the present diet. If a predator does pursue a prey item of type i, then its expected rate of energy intake is E_i/h_i. But if it ignores this prey item, whilst pursuing all those that are more profitable, then it can expect to search for a further $\bar{s}$, following which its expected rate of energy intake is $\bar{E}/\bar{h}$. The total time spent in this latter case is $\bar{s} + \bar{h}$, and so the overall expected rate of energy intake is $\bar{E}/(\bar{s} + \bar{h})$. The most profitable, optimal strategy for a predator will be to pursue the ith item if, and only if:

$$E_i/h_i \geq \bar{E}/(\bar{s} + \bar{h}). \tag{9.1}$$

In other words, a predator should continue to add increasingly less profitable items to its diet as long as Equation 9.1 is satisfied (i.e. as long as this increases its overall rate of energy intake). This will serve to maximize its overall rate of energy intake, $\bar{E}/(\bar{s} + \bar{h})$.

This optimal diet model leads to a number of predictions.

1 Predators with handling times that are typically short compared to their search times should be gener-

searchers should be generalists

alists, because in the short time it takes them to handle a prey item that has already been found, they can barely begin to search for another prey item. (In terms of Equation 9.1: E_i/h_i is large (h_i is small) for a wide range of prey types, whereas $\bar{E}/(\bar{s} + \bar{h})$ is small ($\bar{s}$ is large) even for broad diets.) This prediction seems to be supported by the broad diets of many insectivorous birds feeding in trees and shrubs. Searching is always moderately time consuming, but handling the minute insects takes negligible time and is almost always successful. A bird, therefore, has something to gain and virtually nothing to lose by consuming an item once found, and overall profitability is maximized by a broad diet.

handlers should be specialists

2 By contrast, predators with handling times that are long relative to their search times should be specialists. That is, if $\bar{s}$ is always small, then $\bar{E}/(\bar{s}+\bar{h})$ is similar to $\bar{E}/\bar{h}$. Thus, maximizing $\bar{E}/(\bar{s}+\bar{h})$ is much the same as maximizing $\bar{E}/h$, which is achieved, clearly, by including only the most profitable items in the diet. For instance, lions live more or less constantly in sight of their prey so that search time is negligible; handling time, on the other hand, and particularly pursuit time, can be long (and very energy consuming). Lions consequently specialize on prey that can be pursued most profitably: the immature, the lame and the old.

specialization should be greater in productive environments

3 Other things being equal, a predator should have a broader diet in an unproductive environment (where prey items are relatively rare and $\bar{s}$ is relatively large) than in a productive environment (where $\bar{s}$ is smaller). This prediction is broadly supported by the two examples shown in Figure 9.17: in experimental arenas, both bluegill sunfish (*Lepomis macrochirus*) and great tits (*Parus major*) had more specialized diets when prey density was higher. A related result has been reported from predators in their natural setting – brown and black bears (*Ursos arctos* and

U. americanus) feeding on salmon in Bristol Bay in Alaska. When salmon availability was high, bears consumed less biomass per captured fish, targeting energy-rich fish (those that had not spawned) or energy-rich body parts (eggs in females, brain in males). In essence their diet became more specialized when prey were abundant (Gende *et al.*, 2001).

4 Equation 9.1 depends on the profitability of the *i*th item (E_i/h_i), depends on the profitabilities of the items already in the diet ($\bar{E}/\bar{h}$) and depends on the search times for

the abundance of unprofitable prey types is irrelevant

items already in the diet ($\bar{s}$) and thus on their abundance. But it does not depend on the search time for the *i*th item, s_i. In other words, predators should ignore insufficiently profitable food types irrespective of their abundance. Re-examining the examples in Figure 9.17, we can see that these both refer to cases in which the optimal diet model does indeed predict that the least profitable items should be ignored completely. The foraging behavior was very similar to this prediction, but in both cases the animals consistently took slightly more than expected of the less profitable food types. In fact, this sort of discrepancy has been uncovered repeatedly, and there are a number of reasons why it may occur, which can be summarized crudely by noting that the animals are not omniscient. The optimal diet model, however, does not predict a perfect correspondence between observation and expectation. It predicts the sort of strategy that will be favored by natural selection, and says that the animals that come closest to this

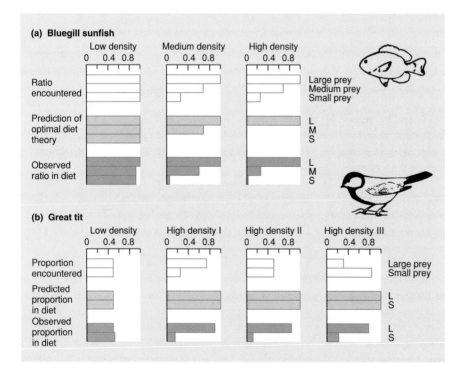

Figure 9.17 Two studies of optimal diet choice that show a clear but limited correspondence with the predictions of Charnov's (1976a) optimal diet model. Diets are more specialized at high prey densities; but more low profitability items are included than predicted by the theory. (a) Bluegill sunfish preying on different size classes of *Daphnia*: the histograms show ratios of encounter rates with each size class at three different densities, together with the predicted and observed ratios in the diet. (After Werner & Hall, 1974.) (b) Great tits preying on large and small pieces of mealworm. (After Krebs *et al.*, 1977.) The histograms in this case refer to the proportions of the two types of item taken. (After Krebs, 1978.)

strategy will be most favored. From this point of view, the correspondence between data and theory in Figure 9.17 seems much more satisfactory. Sih and Christensen (2001) reviewed 134 studies of optimal diet theory, focusing on the question of what factors might explain the ability of the theory to correctly predict diets. Contrary to their *a priori* prediction, forager groups (invertebrate versus ectothermic vertebrate versus endothermic vertebrate) did not differ in the likelihood of corroborating the theory. Their major conclusion was that while optimal diet theory generally works well for foragers that feed on immobile or relatively immobile prey (leaves, seeds, mealworms, zooplankton relative to fish), it often fails to predict diets of foragers that attack mobile prey (small mammals, fish, zooplankton relative to insect predators). This may be because variations among mobile prey in vulnerability (encounter rate and capture success) are often more important in determining predator diets than are variations in the active choices of predators (Sih & Christensen, 2001).

5 Equation 9.1 also provides a context for understanding the narrow specialization of predators that live in intimate association with their prey, especially where an individual predator is linked to an individual prey (e.g. many parasitoids and parasitic herbivores – and many parasites (see Chapter 12)). Since their whole lifestyle and life cycle are finely tuned to those of their prey (or host), handling time ($\bar{h}$) is low; but this precludes their being finely tuned to other prey species, for which, therefore, handling time is very high. Equation 9.1 will thus only apply within the specialist group, but not to any food item outside it.

On the other hand, polyphagy has definite advantages. Search costs ($\bar{s}$) are typically low – food is easy to find – and an individual is unlikely to starve because of fluctuations in the abundance of one type of food. In addition, polyphagous consumers can, of course, construct a balanced diet, and maintain this balance by varying preferences to suit altered circumstances, and can avoid consuming large quantities of a toxin produced by one of its food types. These are considerations ignored by Equation 9.1.

coevolution: predator–prey arms races?

Overall, then, evolution may broaden or restrict diets. Where prey exert evolutionary pressures demanding specialized morphological or physiological responses from the consumer, restriction is often taken to extremes. But where consumers feed on items that are individually inaccessible or unpredictable or lacking in certain nutrients, the diet often remains broad. An appealing and much-discussed idea is that particular pairs of predator and prey species have not only evolved but have coevolved. In other words, there has been an evolutionary 'arms race', whereby each improvement in predatory ability has been followed by an improvement in the prey's ability to avoid or resist the predator, which has been followed by a further improvement in

predatory ability, and so on. This may itself be accompanied, on a long-term, evolutionary timescale, by speciation, so that, for example, related species of butterfly are associated with related species of plants – all the species of the Heliconiini feed on members of the Passifloraceae (Ehrlich & Raven, 1964; Futuyma & May, 1992). To the extent that coevolution occurs, it may certainly be an additional force in favor of diet restriction. At present, however, hard evidence for predator–prey or plant–herbivore coevolution is proving difficult to come by (Futuyma & Slatkin, 1983; Futuyma & May, 1992).

There may seem, at first sight, to be a contradiction between the predictions of the optimal diet model and switching. In the latter, a consumer switches from one prey type to another as their relative densities change. But the optimal diet model suggests that the more profitable prey type should always be taken, irrespective of its density or the density of any alternative. Switching is presumed to occur, however, in circumstances to which the optimal diet model does not strictly apply. Specifically, switching often occurs when the different prey types occupy different microhabitats, whereas the optimal diet model predicts behavior within a microhabitat. Moreover, most other cases of switching involve a change in the profitabilities of items of prey as their density changes, whereas in the optimal diet model these are constants. Indeed, in cases of switching, the more abundant prey type is the more profitable, and in such a case the optimal diet model predicts specialization on whichever prey type is more profitable (that is, whichever is more abundant; in other words, switching).

9.5.4 Foraging in a broader context

It is worth stressing that foraging strategies will not always be strategies for simply maximizing feeding efficiency. On the contrary, natural selection will favor foragers that maximize their net benefits, and strategies will therefore

backswimmers forage suboptimally but avoid being preyed on . . .

often be modified by other, conflicting demands on the individuals concerned. In particular, the need to avoid predators will frequently affect an animal's foraging behavior.

This has been shown in work on foraging by nymphs of an aquatic insect predator, the backswimmer *Notonecta hoffmanni* (Sih, 1982). These animals pass through five nymphal instars (with I being the smallest and youngest, and V the oldest), and in the laboratory the first three instars are liable to be preyed upon by adults of the same species, such that the relative risk of predation from adults was:

I > II > III > IV = V ≅ no risk.

These risks appear to modify the behavior of the nymphs, in that they tend (both in the laboratory and in the field) to avoid the

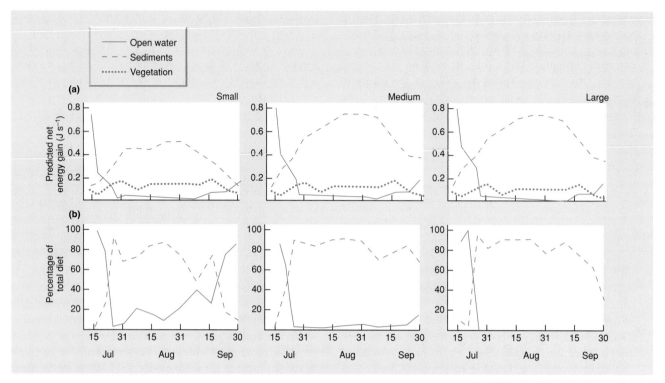

Figure 9.18 Seasonal patterns in (a) the predicted habitat profitabilities (net rate of energy gain) and (b) the actual percentage of the diet originating from each habitat, for three size classes of bluegill sunfish (*Lepomis macrochirus*). Piscivores were absent. (The 'vegetation' habitat is omitted from (b) for the sake of clarity – only 8–13% of the diet originated from this habitat for all size classes of fish.) There is good correspondence between the patterns in (a) and (b). (After Werner *et al.*, 1983b.)

central areas of water bodies, where the concentration of adults is greatest. In fact, the relative degree of avoidance was the same as the relative risk of predation from adults:

I > II > III > IV = V ≅ no avoidance.

Yet these central areas also contain the greatest concentration of prey items for the nymphs, and so, by avoiding predators, nymphs of instars I and II showed a reduction in feeding rate in the presence of adults (although those of instar III did not). The young nymphs displayed a less than maximal feeding rate as a result of their avoidance of predation, but an increased survivorship.

. . . as do certain fish

The modifying influence of predators on foraging behavior has also been studied by Werner *et al.* (1983b) working on bluegill sunfish. They estimated the net energy returns from foraging in three contrasting laboratory habitats – in open water, amongst water weeds and on bare sediment – and they examined how prey densities varied in comparable natural habitats in a lake through the seasons. They were then able to predict the time at which the sunfish should switch between different lake habitats so as to maximize their overall net energy returns. In the

absence of predators, three sizes of sunfish behaved as predicted (Figure 9.18). But in a further field experiment, this time in the presence of predatory largemouth bass, the small sunfish restricted their foraging to the water weed habitat (Figure 9.19) (Werner *et al.*, 1983a). Here, they were relatively safe from predation, although they could only achieve a markedly submaximal rate of energy intake. By contrast, the larger sunfish are more or less safe from predation by bass, and they continued to forage according to the optimal foraging predictions. In a similar vein, the nymphs of several species of algivorous mayflies largely restrict their feeding to the hours of darkness in streams that contain brown trout, reducing their overall feeding rates but also reducing the risk of predation (Townsend, 2003). In the case of mammals that feed at night, including mice, porcupines and hares, time spent feeding may be reduced in bright moonlight when predation risk is highest (Kie, 1999).

A foraging strategy is an integral part of an animal's overall pattern of behavior. The strategy is strongly influenced by the selective pressures favoring the maximization of feeding efficiency, but it may also be influenced by other, possibly conflicting demands. It is also worth pointing out one other thing. The places where animals occur, where they are maximally

predation and the realized niche

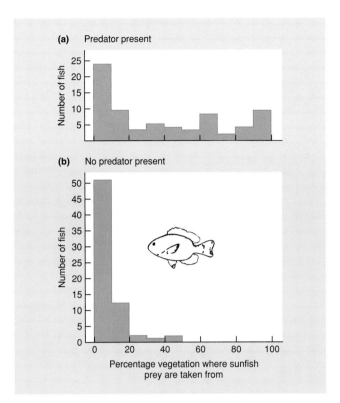

(a) Predator present

(b) No predator present

Percentage vegetation where sunfish
prey are taken from

Figure 9.19 (a) In contrast to Figure 9.18 and to (b), when largemouth bass (which prey on small bluegill sunfish) are present many sunfish take prey from areas where the percentage vegetation is high and where they are relatively protected from predation. (After Werner *et al.*, 1983a.)

abundant and where they choose to feed are all key components of their 'realized niches'. We saw in Chapter 8 that realized niches can be highly constrained by competitors. Here, we see that they can also be highly constrained by predators. This is also seen in the effects of predation by the barn owl (*Tyto alba*) on the foraging behavior of three heteromyid rodents, the Arizona pocket mouse (*Perognathus amplus*), Bailey's pocket mouse (*P. baileyi*) and Merriam's kangaroo rat (*Dipodomys merriami*) (Brown *et al.*, 1988). In the presence of owls, all three species moved to microhabitats where they were less at risk from owl predation and where they reduced their foraging activity. However they did so to varying extents, such that the way in which the microhabitat was partitioned between them was quite different in the presence and absence of owls.

9.6 Foraging in a patchy environment

food is patchily
distributed

For all consumers, food is distributed patchily. The patches may be natural and discrete physical objects: a bush

laden with berries is a patch for a fruit-eating bird; a leaf covered with aphids is a patch for a predatory ladybird. Alternatively, a 'patch' may only exist as an arbitrarily defined area in an apparently uniform environment; for a wading bird feeding on a sandy beach, different 10 m² areas may be thought of as patches that contain different densities of worms. In all cases though, a patch must be defined with a particular consumer in mind. One leaf is an appropriate patch for a ladybird, but for a larger and more active insectivorous bird, 1 m² of canopy or even a whole tree may represent a more appropriate patch.

Ecologists have been particularly interested in patch preferences of consumers where patches vary in the density of food or prey items they contain. There are many examples where predators show an 'aggregative response', spending more time in patches containing high densities (because these are the most profitable patches) (Figure 9.20a–d), although such direct density dependence is not always the case (Figure 9.20e). We deal with aggregative responses in more detail in Chapter 10 where their importance in population dynamics will be our focus, and particularly their potential to lend stability to predator–prey dynamics. For now, we concentrate on the behavior that leads to predator aggregation (Section 9.6.1), the optimal foraging approach to patch use (Section 9.6.2) and the distribution patterns that are likely to result when the opposing tendencies of predators to aggregate and to interfere with each other's foraging are both taken into account (Section 9.6.3).

9.6.1 Behavior that leads to aggregated distributions

There are various types of behavior underlying the aggregative responses of consumers, but they fall into two broad categories: those involved with the location of profitable patches, and the responses of consumers once within a patch. The first category includes all examples in which consumers perceive, at a distance, the existence of heterogeneity in the distribution of their prey.

locating a patch

Within the second category – responses of consumers within patches – there are two main aspects of behavior. The first is a change in the consumer's pattern of searching after encountering items of food. In particular, there is often a slowing down of movement and an increased rate of turning immediately following the intake of food, both of which lead to the consumer remaining in the vicinity of its last food item ('area-restricted search'). Alternatively, or in addition, consumers may simply abandon unprofitable patches more rapidly than they abandon profitable ones. Both types of behavior were evident when the carnivorous, net-spinning larva of the caddis-fly *Plectrocnemia conspersa* feeds on chironomid (midge) larvae in a laboratory stream. Caddis in their nets were provided with one prey item at the beginning of the experiment and then fed daily rations of

area-restricted search

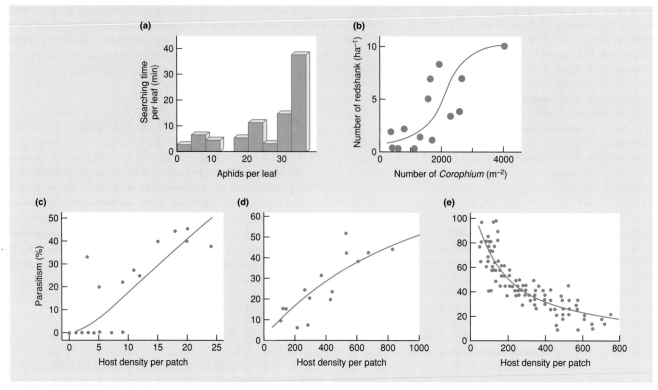

Figure 9.20 Aggregative responses: (a) coccinellid larvae (*Coccinella septempunctata*) spend more time on leaves with high densities of their aphid prey (*Brevicoryne brassicae*) (after Hassell & May, 1974); (b) redshank (*Tringa totanus*) aggregate in patches with higher densities of their amphipod prey (*Corophium volutator*) (after Goss-Custard, 1970); (c) direct density dependence when the parasitoid *Delia radicum* attacks *Trybliographa rapae*; and (d) direct density dependence when the parasitoid *Aspidiotiphagus citrinus* attacks *Fiorinia externa*. (e) But direct density dependence is not always the case: inverse density dependence when the parasitoid *Ooencyrtus kuwanai* attacks *Lymantria dispar*. ((c–e) after Pacala & Hassall, 1991.)

zero, one or three prey. The tendency to abandon the net was lowest at the higher feeding rates (Townsend & Hildrew, 1980). *Plectrocnemia*'s behavior in relation to prey patches also has an element of area-restricted search: the likelihood that it will spin a net in the first place depends on whether it happens to encounter a food item (which it can consume even without a net) (Figure 9.21a). Overall, therefore, a net is more likely to be constructed, and less likely to be abandoned, in a rich patch. These two behaviors account for a directly density-dependent aggregative response in the natural stream environment observed for much of the year (Figure 9.21b).

thresholds and giving-up times

The difference in the rates of abandonment of patches of high and low profitability can be achieved in a number of ways, but two are especially easy to envisage. A consumer might leave a patch when its feeding rate drops below a threshold level, or a consumer might have a giving-up time – it might abandon a patch whenever a particular time interval passes without the successful capture of food. Whichever mechanism is used, or indeed if the consumer simply uses area-restricted search, the consequences will be the same:

individuals will spend longer in more profitable patches, and these patches will therefore generally contain more consumers.

9.6.2 Optimal foraging approach to patch use

The advantages to a consumer of spending more time in higher profitability patches are easy to see. However, the detailed allocation of time to different patches is a subtle problem, since it depends on the precise differentials in profitability, the average profitability of the environment as a whole, the distance between the patches, and so on. The problem has been a particular focus of attention for optimal foraging theory. In particular, a great deal of interest has been directed at the very common situation in which foragers themselves deplete the resources of a patch, causing its profitability to decline with time. Amongst the many examples of this are insectivorous insects removing prey from a leaf, and bees consuming nectar from a flower.

Charnov (1976b) and Parker and Stuart (1976) produced similar models to predict the behavior of an optimal forager in such situations. They found that the optimal stay-time in a patch

Figure 9.21 (a) On arrival in a patch, fifth-instar *Plectrocnemia conspersa* larvae that encounter and eat a chironomid prey item at the beginning of the experiment ('fed') quickly cease wandering and commence net-building. Predators that fail to encounter a prey item ('unfed') exhibit much more widespread movement during the first 30 min of the experiment, and are significantly more likely to move out of the patch. (b) Directly density-dependent aggregative response of fifth-instar larvae in a natural environment expressed as mean number of predators against combined biomass of chironomid and stonefly prey per 0.0625 m² sample of streambed (n = 40). (After Hildrew & Townsend, 1980; Townsend & Hildrew, 1980.)

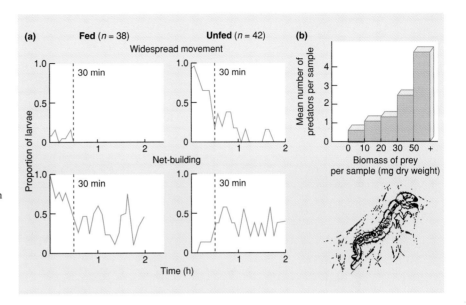

should be defined in terms of the rate of energy extraction experienced by the forager at the moment it leaves a patch (the 'marginal value' of the patch). Charnov called the results the 'marginal value theorem'. The models were formulated mathematically, but their salient features are shown in graphic form in Figure 9.22.

The primary assumption of the model is that an optimal forager will maximize its overall intake of a resource (usually energy) during a bout of foraging, taken as a whole. Energy will, in fact, be extracted in bursts because the food is distributed patchily; the forager will sometimes move between patches, during which time its intake of energy will be zero. But once in a patch, the forager will extract energy in a manner described by the curves in Figure 9.22a. Its initial rate of extraction will be high, but as time progresses and the resources are depleted, the rate of extraction will steadily decline. Of course, the rate will itself depend on the initial contents of the patch and on the forager's efficiency and motivation (Figure 9.22a).

| when should a forager leave a patch that it is depleting? |

The problem under consideration is this: at what point should a forager leave a patch? If it left all patches immediately after reaching them, then it would spend most of its time traveling between patches, and its overall rate of intake would be low. If it stayed in all patches for considerable lengths of time, then it would spend little time traveling, but it would spend extended periods in depleted patches, and its overall rate of intake would again be low. Some intermediate stay-time is therefore optimal. In addition, though, the optimal stay-time must clearly be greater for profitable patches than for unprofitable ones, and it must depend on the profitability of the environment as a whole.

Consider, in particular, the forager in Figure 9.22b. It is foraging in an environment where food is distributed patchily and

where some patches are more valuable than others. The average traveling time between patches is t_t. This is therefore the length of time the forager can expect to spend on average after leaving one patch before it finds another. The forager in Figure 9.22b has arrived at an average patch for its particular environment, and it therefore follows an average extraction curve. In order to forage optimally it must maximize its rate of energy intake not merely for its period in the patch, but for the whole period since its departure from the last patch (i.e. for the period $t_t + s$, where s is the stay-time in the patch).

If it leaves the patch rapidly then this period will be short ($t_t + s_{short}$ in Figure 9.22b). But by the same token, little energy will be extracted (E_{short}). The rate of extraction (for the whole period $t_t + s$) will be given by the slope of the line OS (i.e. $E_{short}/(t_t + s_{short})$). On the other hand, if the forager remains for a long period (s_{long}) then far more energy will be extracted (E_{long}); but, the overall rate of extraction (the slope of OL) will be little changed. To maximize the rate of extraction over the period $t_t + s$, it is necessary to maximize the slope of the line from O to the extraction curve. This is achieved simply by making the line a tangent to the curve (OP in Figure 9.22b). No line from O to the curve can be steeper, and the stay-time associated with it is therefore optimal (s_{opt}).

The optimal solution for the forager in Figure 9.22b, therefore, is to leave that patch when its extraction rate is equal to (tangential to) the slope

| how to maximize overall energy intake |

of OP, i.e. it should leave at point P. In fact, Charnov, and Parker and Stuart, found that the optimal solution for the forager is to leave all patches, irrespective of their profitability, at the same extraction rate (i.e. the same 'marginal value'). This extraction rate is given by the slope of the tangent to the average extraction curve (e.g. in Figure 9.22b), and it is therefore the maximum average overall rate for that environment as a whole.

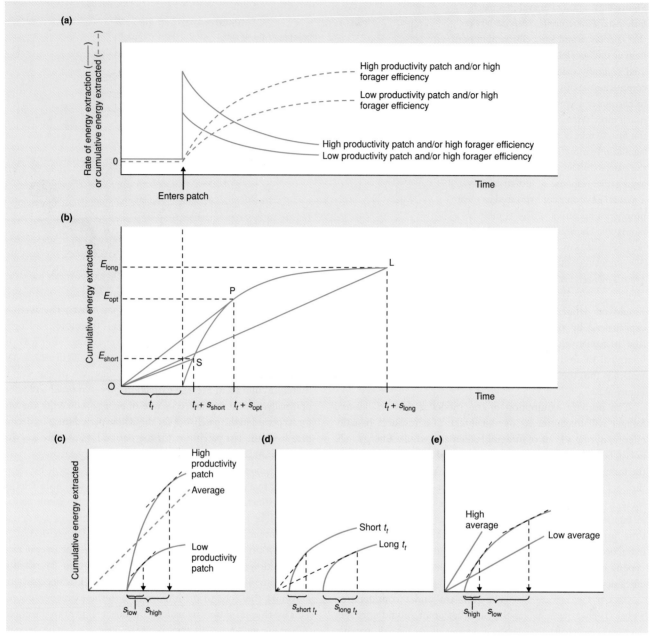

Figure 9.22   The marginal value theorem. (a) When a forager enters a patch, its rate of energy extraction is initially high (especially in a highly productive patch or where the forager has a high foraging efficiency), but this rate declines with time as the patch becomes depleted. The cumulative energy intake approaches an asymptote. (b) The options for a forager. The solid colored curve is cumulative energy extracted from an average patch, and t_t is the average traveling time between patches. The rate of energy extraction (which should be maximized) is energy extracted divided by total time, i.e. the slope of a straight line from the origin to the curve. Short stays in the patch (slope $= E_{short}/(t_t + s_{short})$) and long stays (slope $= E_{long}/(t_t + s_{long})$) both have lower rates of energy extraction (shallower slopes) than a stay (s_{opt}) which leads to a line just tangential to the curve. s_{opt} is therefore the optimum stay-time, giving the maximum overall rate of energy extraction. *All* patches should be abandoned at the *same* rate of energy extraction (the slope of the line OP). (c) Low productivity patches should be abandoned after shorter stays than high productivity patches. (d) Patches should be abandoned more quickly when traveling time is short than when it is long. (e) Patches should be abandoned more quickly when the average overall productivity is high than when it is low.

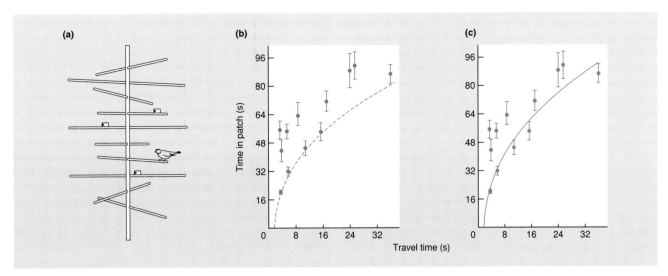

Figure 9.23 (a) An experimental 'tree' for great tits, with three patches. (b) Predicted optimal time in a patch plotted against traveling time (- - - - -), together with the observed mean points (± SE) for six birds, each in two environments. (c) The same data points, and the predicted time taking into account the energetic costs of traveling between patches. (After Cowie, 1977; from Krebs, 1978.)

predictions of the marginal value theorem . . .

The model therefore confirms that the optimal stay-time should be greater in more productive patches than in less productive patches (Figure 9.22c). Moreover, for the least productive patches (where the extraction rate is never as high as OP) the stay-time should be zero. The model also predicts that all patches should be depleted such that the final extraction rate from each is the same (i.e. the 'marginal value' of each is the same); and it predicts that stay-times should be longer in environments where the traveling time between patches is longer (Figure 9.22d) and that stay-times should be longer where the environment as a whole is less profitable (Figure 9.22e).

. . . supported by some experiments

Encouragingly, there is evidence from a number of cases that lends support to the marginal value theorem. In one of the first tests of the theory, Cowie (1977) considered the prediction set out in Figure 9.22d: that a forager should spend longer in each patch when the traveling time is longer. He used captive great tits in a large indoor aviary, and got the birds to forage for small pieces of mealworm hidden in sawdust-filled plastic cups – the cups were 'patches'. All patches on all occasions contained the same number of prey, but traveling time was manipulated by covering the cups with cardboard lids that varied in their tightness and therefore varied in the time needed to prize them off. Birds foraged alone, and Cowie used six in all, subjecting each to two habitats. One of these habitats always had longer traveling times (tighter lids) than the other. For each bird in each habitat Cowie measured the average traveling time and the curve of cumulative food intake within a patch. He then used the marginal value theorem to predict the optimal stay-time in habitats with different traveling

times, and compared these predictions with the stay-times he actually observed. As Figure 9.23 shows, the correspondence was quite close. It was closer still when he took account of the fact that there was a net loss of energy when the birds were traveling between patches.

Predictions of the marginal value theorem have also been examined through the behavior of the egg parasitoid, *Anaphes victus*, attacking the beetle *Listronotus oregonensis* in a laboratory setting (Boivin *et al.*, 2004). Patches differed in quality by virtue of the varying proportions of hosts already parasitized at the start of the experiment, and in line with the theorem's predictions, parasitoids stayed longer in the more profitable patches (Figure 9.24a). However, contrary to a further prediction, the marginal rate of fitness gain (the rate of progeny production in the final 10 min before leaving a patch) was greatest in the initially most profitable patches (Figure 9.24b).

As was the case with optimal diet theory, the risk of being preyed upon can be expected to modify the predicted outcomes of optimal patch use. With this in mind, Morris and Davidson (2000) compared the giving-up food extraction rates of white-footed mice (*Peromyscus leucopus*) in a forest habitat (where predation risk is low) and a forest-edge habitat (where predation risk is high). They provided 'patches' (containers) with 4 g of millet grain in 11 foraging sites in the two habitat types, and in both habitat types some sites were in relatively open situations and others were beneath shrubs. They then monitored the grain remaining at the time that the patches were abandoned on two separate days. Their results (Figure 9.25) supported the predictions that mice should abandon patches at

optimal patch use predictions are modified when there is a risk of being preyed upon

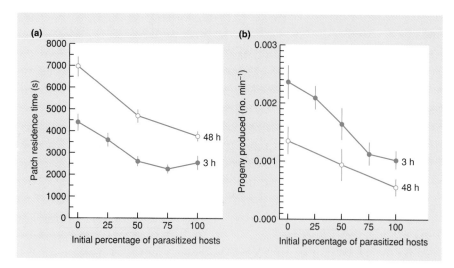

Figure 9.24 (a) When the parasitoid *Anaphes victus* attacked the beetle *Listronotus oregonensis* in patches of 16 hosts, a varying percentage of which had already been parasitized, parasitoids remained longer in the more profitable patches: those with the smaller percentage of parasitized hosts. (b) However, the marginal gain rate in fitness – the number of progeny produced per minute in the final 10 min before leaving a patch – was greatest in the initially most profitable patches. Bars represent standard errors. (After Boivin *et al.*, 2004.)

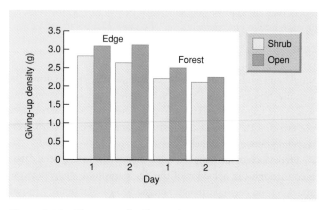

Figure 9.25 The mass of millet grain remaining (giving-up density, g) was higher in patches in the open (riskier) than in paired patches located under shrubs (safer), and was higher in forest-edge habitat (higher predation) than in forest (lower predation). (After Morris & Davidson, 2000.)

higher harvest rates in vulnerable edge habitats than in safe forest habitats, particularly in open situations (where predation risk is highest in each habitat).

predicted and observed behaviors do not correspond perfectly

A much fuller review of tests of the marginal value theorem is provided, for example, by Krebs and Kacelnik (1991). The picture this conveys is one of encouraging but not perfect correspondence – much like the balance of the results presented here. The main reason for the imperfection is that the animals, unlike the modelers, are not omniscient. As was clear in the case of the white-footed mice, they may need to spend time doing things other than foraging (e.g. avoiding predators). Foragers may also need to spend time learning about and sampling their environment, and are none the less likely to proceed in their foraging with imperfect information about the distribution of their hosts. For the parasitoids in Figure 9.24, for example, Boivin *et al.* (2004) suggest that they seem to base their assessment of overall habitat quality on the quality of the first patch they encounter; that is, they 'learn' but their learned assessment may still be wrong. Such a strategy would be adaptive, though, if there was considerable variation in quality between generations (so that each generation had to learn anew), but little variation in quality between patches within a generation (so that the first patch encountered was a fair indication of quality overall).

Nevertheless, in spite of their limited information, animals seem often to come remarkably close to the predicted strategy. Ollason (1980) developed a mechanistic model to account for this in the great tits studied by Cowie. Ollason's is a memory model. It assumes that an animal has a 'remembrance of past food', which Ollason likens to a bath of water without a plug. Fresh remembrance flows in every time the animal feeds. But remembrance is also draining away continuously. The rate of input depends on the animal's feeding efficiency and the productivity of the current feeding area. The rate of outflow depends on the animal's ability to memorize and the amount of remembrance. Remembrance drains away quickly, for example, when the amount is large (high water level) or the memorizing ability is poor (tall, narrow bath). Ollason's model simply proposes that an animal should stay in a patch until remembrance ceases to rise; an animal should leave a patch when its rate of input from feeding is slower than its rate of declining remembrance.

An animal foraging consistently with Ollason's model behaves in a way very similar to that predicted by the *mechanistic models of optimal foraging* marginal value theorem. This is shown for the case of Cowie's great tits in Figure 9.26. As Ollason himself remarks, this shows that to forage in a patchy environment in a way that approximates closely to optimality, an animal need not be omniscient, it does not need to sample and it does not need to perform numerical analyses to find the maxima of functions of many variables: all it

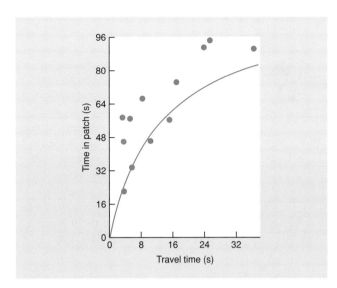

Figure 9.26 Cowie's (1977) great tit data (see Figure 9.23) compared to the predictions of Ollason's (1980) mechanistic memory model.

needs to do is to remember, and to leave each patch if it is not feeding as fast as it remembers doing. As Krebs and Davies (1993) point out, this is no more surprising than the observation that the same birds can fly without any formal qualification in aerodynamics.

Mechanistic models have also been developed and tested for a range of patterns of parasitoid attack (like that in Figure 9.24) (see Vos *et al.*, 1998; Boivin *et al.*, 2004). These highlight the important distinction between 'rule of thumb' behavior, where animals follow innate and unvarying rules, and learned behavior, where rules are subject to modification in the light of the forager's immediate experience. The weight of evidence suggests that learning plays at least some role in most foragers' decisions. There is an important distinction, too, between 'incremental' and 'decremental' behavior. With incremental behavior each successful attack in a patch increases the forager's chance of staying there. This is likely to be adaptive when there is considerable variation in quality between patches, because it encourages longer stay-times in better quality patches. With 'decremental' behavior each successful attack in a patch *de*creases the forager's chance of staying there. This is likely to be adaptive when all patches are of approximately the same quality, because it encourages foragers to leave depleted patches.

Thus, Ollason's model for great tits incorporated rule of thumb, incremental behavior. Boivin *et al.*, on the other hand, found their parasitoids to be exhibiting learned, decremental behavior: a parasitoid attacking a healthy host, for example, was 1.43 times more likely to leave a patch thereafter, and one rejecting a host that had already been attacked was 1.11 times more likely to leave. Vos *et al.* (1998), by contrast, found incremental behavior when the parasitoid *Cotesia glomerata* attacked its butterfly larva host, *Pieris brassicae*: each successful encounter increased its tendency to remain in a patch. For both the great tit and parasitoids, therefore, optimal foraging and mechanistic models are seen to be compatible and complementary in explaining how a predator has achieved its observed foraging pattern, and why that pattern has been favored by natural selection.

Finally, the principles of optimal foraging are also being applied to investigations of the foraging strategies of plants for nutrients (reviewed by Hutchings & de Kroon, 1994). When does it pay to produce long stolons moving rapidly from patch to patch? When does it pay to concentrate root growth within a limited volume, foraging from a patch until it is close to depletion? Certainly, it is good to see such intellectual cross-fertilization across the taxonomic divide.

optimal foraging in plants

9.6.3 Ideal free and related distributions: aggregation and interference

We can see, then, that consumers tend to aggregate in profitable patches where their expected rate of food consumption is highest. Yet we might also expect that consumers will compete and interfere with one another (discussed further in Chapter 10), thereby reducing their per capita consumption rate. It follows from this that patches that are initially most profitable become immediately less profitable because they attract most consumers. We might therefore expect the consumers to redistribute themselves, and it is perhaps not surprising that the observed patterns of predator distributions across prey patches vary substantially from case to case. But can we make some sense of this variation in pattern?

the ideal free distribution . . .

In an early attempt to do so, it was proposed that if a consumer forages optimally, the process of redistribution will continue until the profitabilities of all patches are equal (Fretwell & Lucas, 1970; Parker, 1970). This will happen because as long as there are dissimilar profitabilities, consumers should leave less profitable patches and be attracted to more profitable ones. Fretwell and Lucas called the consequent distribution the ideal free distribution: the consumers are 'ideal' in their judgement of profitability, and 'free' to move from patch to patch. Consumers were also assumed to be equal. Hence, with an ideal free distribution, because all patches come to have the same profitability, all consumers have the same consumption rate. There are some simple cases where consumers appear to conform to an ideal free distribution insofar as they distribute themselves in proportion to the profitabilities of different patches (e.g. Figure 9.27a), but even in such cases one of the underlying assumptions is likely to have been violated (e.g. Figure 9.27b – all consumers are not equal).

. . . is a balance between attractive and repellant forces

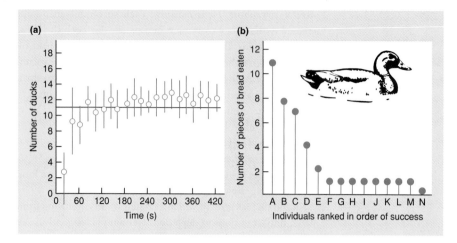

(a)

(b)

Figure 9.27   (a) When 33 ducks were fed pieces of bread at two stations around a pond (with a profitability ratio of 2 : 1), the number of ducks at the poorer station, shown here, rapidly approached one-third of the total, in apparent conformity with the predictions of ideal free theory. (b) However, contrary to the assumptions and other predictions of simple theory, the ducks were not all equal. (After Harper, 1982; from Milinski & Parker, 1991.)

incorporating a range of interference coefficients

The early ideas have been much modified taking account, for example, of unequal competitors (see Milinski & Parker, 1991; Tregenza, 1995, for reviews). In particular, ideal free theory was put in a more ecological context by Sutherland (1983) when he explicitly incorporated predator handling times and mutual interference amongst the predators. He found that predators should be distributed such that the proportion of predators in site i, p_i, is related to the proportion of prey (or hosts) in site i, h_i, by the equation:

$$p_i = k (h_i^{1/m}) \qquad (9.2)$$

where m is the coefficient of interference, and k is a 'normalizing constant' such that the proportions, p_i, add up to 1. It is now possible to see how the patch to patch distribution of predators might be determined jointly by interference and the selection by the predators of intrinsically profitable patches.

If there is no interference amongst the predators, then $m = 0$. All should exploit only the patch with the highest prey density (Figure 9.28), leaving lower density patches devoid of predators.

If there is a small or moderate amount of interference (i.e. $m > 0$, but $m < 1$ – a biologically realistic range), then high-density prey patches should still attract a disproportionate number of predators (Figure 9.28). In other words, there should be an aggregative response by the predators, which is not only directly density dependent, but actually accelerates with increasing prey density in a patch. Hence, the prey's risk of predation might itself be expected to be directly density dependent: the greatest risk of predation in the highest prey density patches (like the examples in Figure 9.20c, d).

With a little more interference ($m \approx 1$) the proportion of the predator population in a patch should still increase with the proportion of prey, but now it should do so more or less linearly rather than accelerating, such that the ratio of predators : prey is roughly the same in all patches (Figure 9.28, and, for example,

Figure 9.20a). Here, therefore, the risk of predation might be expected to be the same in all patches, and hence independent of prey density.

Finally, with a great deal of interference ($m > 1$) the highest density prey patches should have the lowest ratio of predators : prey (Figure 9.28). The risk of predation might therefore be expected to be greatest in the lowest prey density patches, and hence inversely density dependent (like the data in Figure 9.20e).

It is clear, therefore, that the range of patterns amongst the data in Figure 9.20 reflects a shifting balance between the forces of attraction and of repulsion. Predators are attracted to highly profitable patches; but they are repelled by the presence of other predators that have been attracted in the same way.

This description, however, of the relationship between the distribution of predators and the distribution of predation risk has been peppered with '*might be expected to*'s. The

pseudo-interference

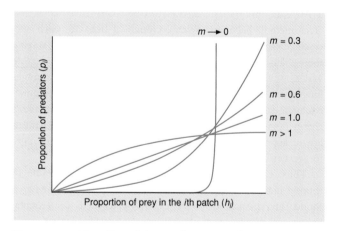

Figure 9.28   The effect of the interference coefficient, m, on the expected distribution of predators amongst patches of prey varying in the proportion of the total prey population they contain (and hence, in their 'intrinsic' profitability). (After Sutherland, 1983.)

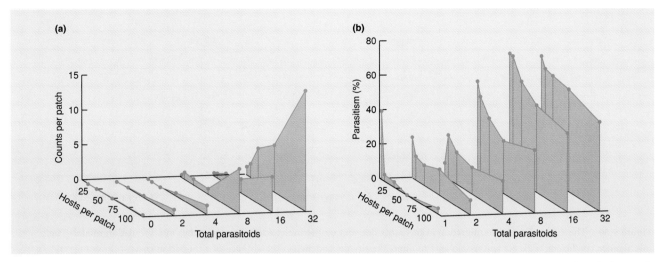

Fgiure 9.29 (a) The aggregative response of the egg parasitoid, *Trichogramma pretiosum*, which aggregates on patches with high densities of its host *Plodia interpunctella*. (b) The resultant distribution of ill effects: hosts on high-density patches are least likely to be parasitized. (After Hassell, 1982.)

reason is that the relationship also depends on a range of factors not so far considered. For example, Figure 9.29 shows a case where the parasitoid *Trichogramma pretiosum* aggregates in high-density patches of its moth host, but the risk of parasitism to the moth is greatest in low-density host patches. The explanation probably lies in time wasted by parasitoids in high-density host patches, dealing with already parasitized hosts that may still attract parasitoids because they are not physically removed from a patch (unlike preyed-upon prey) (Morrison & Strong, 1981; Hassell, 1982). Thus, earlier parasitoids in a patch may interfere indirectly with later arrivals, in that the previous presence of a parasitoid in a patch may reduce the effective rate at which later arrivals attack unparasitized hosts. This effect has been termed 'pseudo-interference' (Free *et al.*, 1977); its potentially important effects on population dynamics are discussed in Chapter 10.

| learning and migration | Expected patterns are modified further still if we incorporate learning by the predators, or the costs of migration between patches (Bernstein *et al.*, |

1988, 1991). With a realistic value of m $(= 0.3)$, the aggregative response of predators is directly density dependent (as expected) as long as the predators' learning response is strong relative to the rate at which they can deplete patches. But if their learning response is weak, predators may be unable to track the changes in prey density that result from patch depletion. Their distribution will then drift to one that is independent of the density of prey.

Similarly, when the cost of migration is low, the predators' aggregative response remains directly density dependent (with $m = 0.3$) (Figure 9.30a). When the cost of migration is increased, however, it still pays predators in the poorest patches to move, but for others the costs of migration can outweigh the potential gains of moving. For these, the distribution across prey patches

is random. This results in inverse density dependence in mortality rate between intermediate and good patches, and in a 'domed' relationship overall (Figure 9.30b). When the cost of migration is very high, it does not pay predators to move whatever patch they are in – mortality is inversely density dependent across all patches (Figure 9.30c).

Clearly, there is no shortage of potential causes for the wide range of types of distributions of predators, and of mortality rates, across prey patches (see Figures 9.20 and 9.29). Their consequences, in terms of population dynamics, are one of the topics dealt with in the chapter that follows. This highlights the crucial importance of forging links between behavioral and population ecology.

Summary

Predation is the consumption of one organism by another, in which the prey is alive when the predator first attacks it. There are two main ways in which predators can be classified. The first is 'taxonomic' – carnivores consume animals, herbivores consume plants, etc.– and the second is 'functional', in which true predators, grazers, parasitoids and parasites are distinguished.

The effects of herbivory on a plant depend on which herbivores are involved, which plant parts are affected, and the timing of attack relative to the plant's development. Leaf-biting, sap-sucking, mining, flower and fruit damage and root pruning can be expected to differ in the effect they have on the plant. Because the plant usually remains alive in the short term, the effects of herbivory are also crucially dependent on the response of the plant. The evolutionary selection pressure exerted by herbivores has led to a variety of plant physical and chemical defenses that

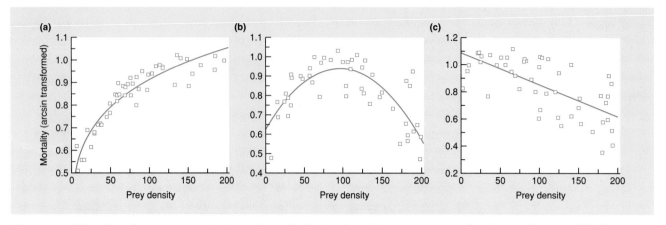

Figure 9.30 The effect of a cost to migration in predators distributing themselves across prey patches in a simulation model. The interference coefficient, m, is 0.3 and would lead to direct density dependence in the absence of a migration cost. (a) Low migration cost: direct density dependence is maintained. (b) Intermediate cost: a 'domed' relationship. (c) High cost: inverse density dependence. (After Bernstein *et al.*, 1991.)

resist attack. These may be present and effective continuously (constitutive defense) or increased production may be induced by attack (inducible defense). It is not straightforward to determine whether the supposed 'defenses' actually have measurable, negative effects on the herbivore and positive consequences for the plant, especially after the costs of mounting the response have been taken into account. We discuss the difficulties of revealing such effects and review the relationships between herbivory and plant survival and fecundity.

More generally, the immediate effect of predation on a population of prey is not always predictably harmful, first because the individuals that are killed are not always a random sample (and may be those with the lowest potential to contribute to the population's future) and second because of compensatory changes in the growth, survival or reproduction of the surviving prey (especially via reduced competition for a limiting resource). From the predator's point of view, an increase in the amount of food consumed can be expected to lead to increased rates of growth, development and birth, and decreased rates of mortality. However, there are a number of factors that complicate this simple relationship between consumption rate and consumer benefit.

Consumers can be classified on a continuum from monophagy (feeding on a single prey type) to polyphagy (many prey types). The preferences of many consumers are fixed – they are maintained irrespective of the relative availabilities of alternative food types. But many others switch their preference, such that food items are eaten disproportionately often when they are common. A mixed diet may be favored first because each food type contains a different undesirable toxic chemical. More generally, a generalist strategy would be favored if a consumer has more to gain than lose in accepting low-quality items, once encountered, rather than ignoring them and continuing to search. We discuss this in the context of optimal diet theory, the aim of which is to predict the foraging strategy to be expected under specified conditions.

Food is generally distributed patchily and ecologists have been particularly interested in patch preferences of consumers where patches vary in the density of food or prey items they contain. We describe the behaviors that lead to aggregated distributions and the nature of the distribution patterns that result. The advantages to a consumer of spending more time in higher profitability patches are easy to see. However, the detailed allocation of time to different patches is a subtle problem, depending on the precise differentials in profitability, the average profitability of the environment as a whole, the distance between patches, and so on. This is the domain of the theory of optimal patch use. The predictions of both optimal foraging and optimal patch use theory have to be modified when there is a simultaneous risk of a consumer being preyed upon.

Chapter 10
The Population Dynamics of Predation

10.1 Introduction: patterns of abundance and the need for their explanation

We turn now to the effects of predation on the population dynamics of the predator and its prey, where even a limited survey of the data reveals a varied array of patterns. There are certainly cases where predation has a profoundly detrimental effect on a prey population. For example, the 'vedalia' ladybird beetle (*Rodolia cardinalis*) is famous for having virtually eradicated the cottony cushion-scale insect (*Icerya purchasi*), a pest that threatened the California citrus industry in the late 1880s (see Section 15.2.5). On the other hand, there are many cases where predators and herbivores have no apparent effect on their prey's dynamics or abundance. For example, the weevil *Apion ulicis* has been introduced into many parts of the world in an attempt to control the abundance of gorse bushes (*Ulex europaeus*), and it has often become well established. The situation in Chile, however, is fairly typical, where, despite eating on average around half, and sometimes up to 94%, of the seeds produced, it has had no appreciable impact on gorse invasiveness (Norambuena & Piper, 2000).

There are also examples that appear to show predator and prey populations linked together by coupled oscillations in abundance (Figure 10.1), but there are many more examples in which predator and prey populations fluctuate in abundance apparently independently of one another.

It is clearly a major task for ecologists to develop an understanding of the patterns of predator–prey abundance, and to account for the differences from one example to the next. It is equally clear, though, that none of these predator and prey populations exist as isolated pairs, but rather as parts of multispecies systems, and that all these species are affected by environmental conditions. These broader issues of what determines a species' abundance are taken up again in Chapter 14. However, as with any complex process in science, we cannot understand the full complexity without a reasonable understanding of the components

– in this case, populations of predators and prey. Hence, this chapter deals with the consequences of predator–prey interactions for the dynamics of the populations concerned.

The approach will be firstly to use simple models to deduce the effects produced by different components of the interactions, teasing out the separate effects before seeking to understand those effects in combination. Then, field and experimental data will be examined to see whether the deductions appear to be supported or refuted. In fact, simple models are most useful when their predictions are *not* supported by real data – as long as the reason for the discrepancy can subsequently be discovered. Confirmation of a model's predictions provides consolidation; refutation with subsequent explanation is progress.

10.2 The basic dynamics of predator–prey and plant–herbivore systems: a tendency towards cycles

There have been two main series of models developed as attempts to understand predator–prey dynamics. Both will be examined here. The first (Section 10.2.1) is based on differential equations (and hence, applies most readily to populations in which breeding is continuous), but relies heavily on simple graphical models (Rosenzweig & MacArthur, 1963). The second (Section 10.2.3) uses difference equations to model host–parasitoid interactions with discrete generations. Despite this taxonomic limitation, these models have the advantage of having been subject to rigorous mathematical exploration. (We have also noted previously that there are a very large number of important parasitoid species.) Although the two series of models are explained separately, they have, of course, a common aim (to advance our understanding of predator–prey dynamics), and they can increasingly be seen as ends of a discrete-to-continuous spectrum of mathematical approaches.

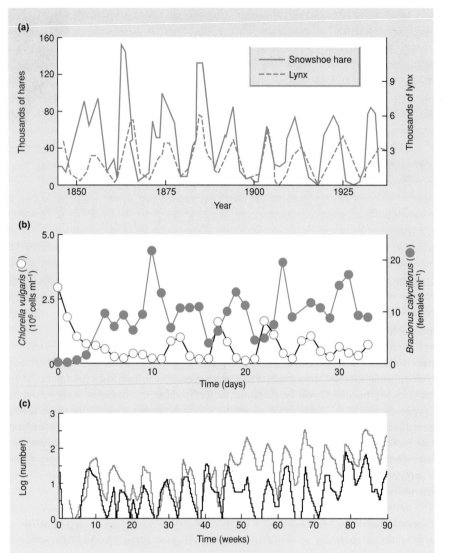

Figure 10.1 Coupled oscillations in the abundance of predators and prey. (a) The snowshoe hare (*Lepus americanus*) and the Canadian lynx (*Lynx canadensis*) as determined by the number of pelts lodged with the Hudson Bay Company. (After MacLulick, 1937.) (b) Parthenogenetic female rotifers, *Bracionus calyciflorus* (predators, ●), and unicellular green algae, *Chlorella vulgaris* (prey, ○) in laboratory cultures. (After Yoshida *et al.*, 2003). (c) The parasitoid *Venturia canescens* (——) and its moth host *Plodia interpunctella* (——) in laboratory cultures. (After Bjørnstad *et al.*, 2001.)

10.2.1 The Lotka–Volterra model

The simplest differential equation model is known (like the model of interspecific competition) by the name of its originators: Lotka–Volterra (Volterra, 1926; Lotka, 1932). This will serve as a useful point of departure. The model has two components: P, the numbers present in a predator (or consumer) population, and N, the numbers or biomass present in a prey or plant population.

We assume initially that in the absence of consumers the prey population increases exponentially (see Section 5.9):

$$dN/dt = rN. \qquad (10.1)$$

But prey individuals are removed by predators at a rate that depends on the frequency of predator–prey encounters. Encounters will increase with the numbers of predators (P) and the numbers of prey (N). However, the exact number encountered and successfully consumed will depend on the searching and attacking efficiency of the predator: a, sometimes also called the 'attack rate'. The consumption rate of prey will thus be aPN, and overall:

$$dN/dt = rN - aPN. \qquad (10.2)$$

the Lotka–Volterra prey equation

In the absence of prey, predator numbers in the model are assumed to decline exponentially through starvation:

$$dP/dt = -qP, \qquad (10.3)$$

where q is the predator mortality rate. This is counteracted by predator birth, the rate of which is assumed to depend on only two things: the rate at which food is consumed, aPN, and the predator's efficiency, f, at turning this food into predator offspring. Predator birth rate is therefore $faPN$, and overall:

the Lotka–Volterra
predator equation

$$dP/dt = faPN - qP. \qquad (10.4)$$

Equations 10.2 and 10.4 constitute the Lotka–Volterra model.

The properties of this model can be investigated by finding zero isoclines. Zero isoclines were described for models of two-species competition in Section 8.4.1. Here, there are separate zero isoclines for the predators and prey, both of which are drawn on a graph of prey density (x-axis) against predator density (y-axis). Each is a line joining those combinations of predator and prey density that lead either to an unchanging prey population ($dN/dt = 0$; prey zero isocline) or an unchanging predator population ($dP/dt = 0$; predator zero isocline). Having drawn, say, a prey zero isocline, we know that combinations to one side of it lead to prey decrease, and combinations to the other to prey increase. Thus, as we shall see, if we plot the prey and predator zero isoclines on the same figure, we can begin to determine the pattern of the dynamics of the joint predator–prey populations.

In the case of the prey (Equation 10.2), when:

$$dN/dt = 0, \; rN = aPN \qquad (10.5)$$

or:

$$P = r/a. \qquad (10.6)$$

properties revealed
by zero isoclines

Thus, since r and a are constants, the prey zero isocline is a line for which P itself is a constant (Figure 10.2a). Below it, predator abundance is low and the prey increase; above it, predator abundance is high and the prey decrease.

Likewise, for the predators (Equation 10.4), when:

$$dP/dt = 0, \; faPN = qP \qquad (10.7)$$

or:

$$N = q/fa. \qquad (10.8)$$

The predator zero isocline is therefore a line along which N is constant (Figure 10.2b). To the left, prey abundance is low and the predators decrease; to the right, prey abundance is high and the predators increase.

Putting the two isoclines together (Figure 10.2c) shows the behavior of joint populations. Predators increase in abundance when there are large numbers of prey, but this leads to an increased predation pressure on the prey, and thus to a decrease in prey abundance. This then leads to a food shortage for predators and a decrease in predator abundance, which leads to a relaxation of predation pressure and an increase in prey abundance, which leads to an increase in predator abundance, and so on (Figure 10.2d). Thus, predator and prey populations undergo 'coupled oscillations' in abundance, which continue indefinitely.

The Lotka–Volterra model, then, is useful in pointing to this underlying tendency for predator–prey interactions to generate fluctuations in the prey population tracked by fluctuations in the predator population. The detailed behavior of the model, however, should not be taken seriously, because the cycles it exhibits are 'structurally unstable', showing 'neutral stability'. That is, the populations would follow precisely the same cycles indefinitely, but only until some external influence shifted them to new values, after which they would follow new cycles indefinitely (Figure 10.2e). In practice, of course, environments are continually changing, and populations would continually be 'shifted to new values'. A population following the Lotka–Volterra model would, therefore, not exhibit regular cycles, but, because of repeated disturbance, fluctuate erratically. No sooner would it start one cycle than it would be diverted to a new one.

an underlying
tendency towards
coupled oscillations –
which are structurally
unstable in this case

For a population to exhibit regular and recognizable cycles, the cycles must themselves be stable: when an external influence changes the population level, there must be a tendency to return to the original cycle. In fact, as we shall see, predator–prey models (once we move beyond the very limiting assumptions of the Lotka–Volterra model) are capable of generating a whole range of abundance patterns: stable-point equilibria, multigeneration cycles, one-generation cycles, chaos, etc. – a range repeated in surveys of real populations. The challenge is to discover what light the models can throw on the behavior of real populations.

10.2.2 Delayed density dependence

The basic mechanism generating the coupled oscillations in these predator–prey interactions is a series of time-delayed 'numerical responses', i.e. changes in one species' abundance in response to the abundance of the other species. The first is a time delay between 'many prey' and 'many predators' (arising because the response of predator abundance to high prey abundance cannot occur instantaneously). There may then be another time delay between 'many predators' and 'few prey', and then between 'few prey' and 'few predators', and so on. In practice, therefore, even where coupled oscillations exist, their exact shape is likely to reflect the varying delays, and strengths, of the different numerical responses. Certainly, the shapes of apparent coupled oscillations in real populations are varied, and not all are symmetric like those generated by the Lotka–Volterra model (see Figure 10.1).

numerical responses

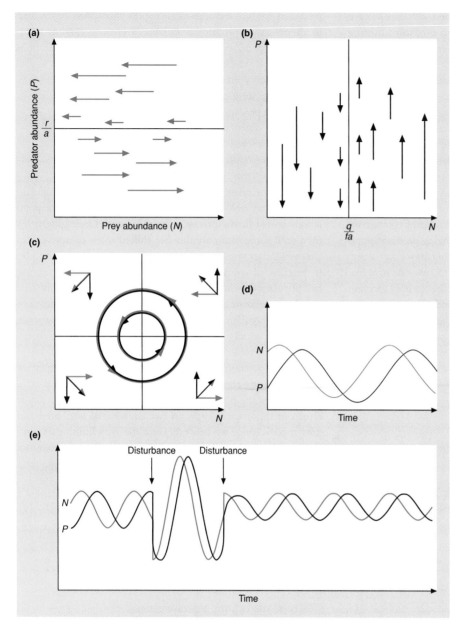

Figure 10.2 The Lotka–Volterra predator–prey model. (a) The prey zero isocline, with prey (N) increasing in abundance (arrows left to right) at lower predator densities (low P) and decreasing at higher predator densities. (b) The predator zero isocline, with predators increasing in abundance (arrows pointing upwards) at higher prey densities and decreasing at lower prey densities. (c) When the zero isoclines are combined, the arrows can also be combined, and these joint arrows progress in anticlockwise circles. In other words, the joint population moves with time from low predator/low prey (bottom left in (c)), to low predator/high prey (bottom right), to high predator/ high prey, to high predator/low prey and back to low predator/low prey. Note, however, that the lowest prey abundance ('9 o'clock') comes one-quarter of a cycle before the lowest predator abundance ('6 o'clock' – anticlockwise movement). These coupled cycles of predator–prey abundance, continuing indefinitely, are shown as numbers against time in (d). However, as shown in (e), these cycles exhibit neutral stability: they continue indefinitely if undisturbed, but each disturbance to a new abundance initiates a new, different series of neutrally stable cycles, around the same means but with a different amplitude.

the regulatory tendencies of delayed density dependence are relatively difficult to demonstrate

These responses are density dependent (see Section 5.2): they act to reduce the size of relatively large populations and allow relatively small populations to increase. Varley (1947) introduced the term 'delayed density dependence' to describe them. The strength of a delayed density-dependent effect is related not to the current abundance (that would be *direct* density dependence) but to abundance at some time in the past (i.e. the delay-length ago). Compared to direct density dependence, delayed density dependence is relatively difficult to demonstrate. To see this, we can examine the coupled oscillations produced by a particular parasitoid–host model, shown in Figure 10.3a (Hassell, 1985). The details of the model need not concern us, but note that the oscillations are damped: they get gradually smaller over time until a stable equilibrium is reached. The prey population, subject to delayed density dependence, is regulated in size by the predator. In Section 5.6, we demonstrated density dependence by plotting k values against the log of density; but when we plot the k values of predator-induced mortality

Figure 10.3 Delayed density dependence. (a) A parasitoid–host model followed over 50 generations: despite oscillations, the parasitoid has a regulatory effect on the host population. (b) For the same model, the *k* value of generation mortality plotted against the log of host density: no clear density-dependent relationship is apparent. (c) The points from (b) linked serially from generation to generation: they spiral in an anticlockwise direction – a characteristic of delayed density dependence. (After Hassell, 1985). (d) The *k* value of generation mortality plotted against the log of host density two generations previously: a clear delayed density-dependent relationship is again apparent.

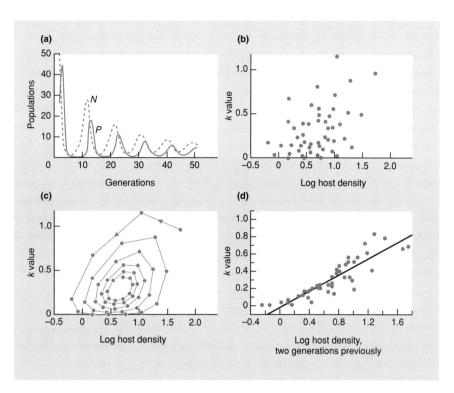

against the log of prey density in that generation (Figure 10.3b), no clear relationship is apparent. On the other hand, when the same points are linked together, each generation to the next (Figure 10.3c), they can be seen to describe an anticlockwise spiral. This spiraling is characteristic of delayed density dependence. Here, because the oscillations are damped, the points spiral inwards to the equilibrium point. Moreover, when we plot the *k* values of predator-induced mortality against the log of prey density *two generations previously* (Figure 10.3d), the delayed density dependence is clearly revealed by the positive relationship characteristic of density dependence in general. Indeed, the fact that a two-generation delay gives a better fitting relationship than delays that are either shorter or longer, tells us that two generations is our best estimate of the delay in this case.

The regulatory effects of delayed density dependence are relatively easy to reveal for the model population of Figure 10.3, because it is not subject to the fluctuations of a natural environment, it is not subject to the density-dependent attacks of any other predator, it is not subject to the inaccuracies of sampling error, and so on. Data of this quality, however, are rarely if ever available for natural or even experimental populations. We return to the question of uncovering and integrating delayed density-dependent effects into an overall account of what determines abundance in Chapter 14. For now, though, this discussion

highlights the relationship between 'regulation' and 'stability' in predator–prey interactions. Natural predator and prey populations tend to exhibit less violent and less regular fluctuations than those we have seen generated by the simplest models. Most of the rest of this chapter describes the search for explanations for these patterns and for the variations in dynamical pattern from case to case. A population that remains roughly constant in size provides evidence for the effects of both regulatory and stabilizing forces. The delayed density dependence of a predator–prey interaction 'regulates' in the sense of acting strongly on large populations and only weakly on small populations. But, as we have already seen, it can hardly be said, typically, to stabilize either population. What follows in this chapter is, therefore, in large part, a search for stabilizing forces that might complement the (delayed) regulatory forces that occur inherently in predator–prey interactions.

10.2.3 The Nicholson–Bailey model

Turning now to parasitoids, the basic model (Nicholson & Bailey, 1935) is again not so much realistic as a reasonable basis from which to start. Let H_t be the number of hosts, and P_t the number of parasitoids in generation t; r is the intrinsic rate of natural increase of the host. If H_a is the number of hosts attacked by parasitoids (in generation t), then, assuming no intraspecific competition amongst the hosts (exponential growth – see Section 4.7.1), and that each host can support only one parasitoid (commonly the case):

$$H_{t+1} = e^r(H_t - H_a), \qquad (10.9)$$

$$P_{t+1} = H_a. \qquad (10.10)$$

In other words, hosts that are not attacked reproduce, and those that are attacked yield not hosts but parasitoids.

To derive a simple formulation for H_a, let E_t be the number of host–parasitoid encounters in generation t. Then, if A is the parasitoid's searching efficiency:

$$E_t = AH_tP_t \qquad (10.11)$$

and:

$$E_t/H_t = AP_t. \qquad (10.12)$$

Note the similarity to the formulation in Equation 10.2. Remember, though, that we are dealing with parasitoids, and hence a single host can be encountered several times, although only one encounter leads to successful parasitization (i.e. only one parasitoid develops). Predators, by contrast, would remove their prey and prevent re-encounters. Thus, Equation 10.2 dealt with instantaneous rates, rather than numbers.

a model based on random encounters . . . If encounters are assumed to occur more or less at random, then the proportions of hosts that are encountered zero, one, two or more times are given by the successive terms in the appropriate 'Poisson distribution' (see any basic statistics textbook). The proportion not encountered at all, p_0, would be given by e^{-E_t/H_t}, and thus the proportion that *is* encountered (one or more times) is $1 - e^{-E_t/H_t}$. The number encountered (or attacked) is then:

$$H_a = H_t(1 - e^{-E_t/H_t}). \qquad (10.13)$$

Using this and Equation 10.12 to substitute into Equations 10.9 and 10.10 gives us:

$$H_{t+1} = H_t e^{(r - AP_t)} \qquad (10.14)$$

$$P_{t+1} = H_t(1 - e^{(-AP_t)}). \qquad (10.15)$$

. . . giving rise to (unstable) coupled oscillations This is the basic Nicholson–Bailey model of a host–parasitoid interaction. Its behavior is reminiscent of the Lotka–Volterra model but it is even less stable. An equilibrium combination of the two populations is a possibility, but even the slightest disturbance from this equilibrium leads to divergent coupled oscillations.

10.2.4 One-generation cycles

The coupled oscillations generated by the basic Lotka–Volterra and Nicholson–Bailey models are multigeneration cycles, i.e. there are several generations between successive peaks (or troughs), and such oscillations have lain at the heart of most attempts to understand cyclic predator–prey dynamics. However, other models of host–parasitoid (and host–pathogen) systems are able to generate coupled oscillations just *one* host generation in length (Knell, 1998; see, for example, Figure 10.1c). On the other hand, such 'generation cycles' can also occur in a population for reasons other than a predator–prey interaction – specifically as a result of competition between age classes within a population (Knell, 1998).

Predator–prey generation cycles occur essentially when the generation length of the consumer is roughly half that of its host – as it often is. Any small, chance peak in host abundance tends to generate a further peak in host abundance one host generation later. But any associated peak in consumer abundance occurs half a host generation length later, creating a trough in host abundance between the twin peaks. And this host trough creates a further host trough one generation later, but a consumer trough coinciding with the next host peak. Thus, the consumers have alternate 'feasts' and 'famines' that accentuate the originally small peaks and troughs in host abundance, and hence promote one-generation cycles (Figure 10.4).

10.2.5 Predator–prey cycles in nature: or are they?

The inherent tendency for predator–prey interactions to generate coupled oscillations in abundance might suggest an expectation of such oscillations in real populations. However, there are many important aspects of predator and prey ecology that have not been considered in the models derived so far; and as subsequent sections will show, these can greatly modify our expectations. Certainly, even if a population exhibits regular oscillations, this does not necessarily provide support for the Lotka–Volterra, Nicholson–Bailey or any other simple model. We saw cycles generated by intraspecific competition in Section 5.8, and we shall see several other routes to cycles in subsequent chapters (see also Kendall *et al.*, 1999). At this point, though, it is worth simply making the point that even when predators or prey exhibit regular cycles in abundance, it is never easy to demonstrate that these are *predator–prey* cycles.

hare and lynx: not the simple predator and prey they appear to be

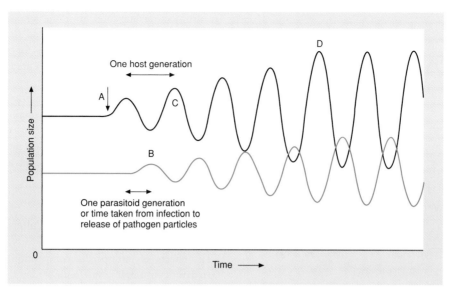

Figure 10.4   Schematic illustration of how a parasitoid or pathogen may generate coupled cycles in abundance in the host and itself that are approximately one host generation in length. For this, the parasitoid or pathogen must have a generation length approximately half that of the host. Any chance increase in host abundance (A) will first give rise to an increase in parasitoid abundance one parasitoid generation later (B), and also to an increase in host abundance one host generation later (C). But the parasitoid peak at B will also give rise to a coincident host trough, which will give rise to a parasitoid trough at C, reinforcing the host peak at that point. This mutual reinforcement will continue until by, say, D, persistent host generation-length cycles have become established. (After Knell, 1998; from Godfray & Hassell, 1989.)

The regular oscillations in the abundance of the snowshoe hare and the Canadian lynx shown in Figure 10.1a have often been held to epitomize predator–prey cycles. Recently, however, evidence has increasingly indicated that even this apparent exemplar is not as straightforward as it has seemed. Experimental manipulations carried out in the field are one powerful means of suggesting what forces are normally acting: if those forces are removed or exaggerated, is the cycle eliminated or enhanced? A whole series of coordinated field experiments has indicated that the cyclic hare is not simply a prey of the lynx (and other predators in the community), nor simply a predator of its plant food resources: the cycle can be understood only by taking account of its interactions both as a prey *and* as a predator (Krebs *et al.*, 2001). Furthermore, modern statistical analysis of the time series of abundances has tended to confirm this: the hare series carries a relatively complex 'signature', suggesting the influence of both its predators and its food, whereas the lynx series has a simpler signature, suggesting only the influence of its (hare) prey (Stenseth *et al.*, 1997; see also Section 14.5.2). What has so often been described as a predator-prey cycle seems rather to comprise one predator linked to a species that is both predator and prey.

moths and two
natural enemies

Apparently coupled one-generation cycles linking a moth host (*Plodia interpunctella*) and its parasitoid *Venturia canescens* were shown in Figure 10.1c. In this case, the dangers of jumping too readily to the conclusion

that these are predator–prey cycles are highlighted by the fact that the host also exhibits generation-length cycles when maintained alone, without any natural enemies, and also when maintained with another enemy, a granulovirus (Figure 10.5). It has been possible, however, to confirm that the cycles in Figure 10.1c are indeed coupled oscillations, using methods similar to those applied to the hare–lynx time series (Bjørnstad *et al.*, 2001). The host-alone cycles have within them the signature simply of intraspecific competition, and the virus seems to modulate this pattern but does not alter its basic structure (i.e. the patterns in Figure 10.5 are *not* predator–prey cycles). However, the host and parasitoid cycles in Figure 10.1c both carry the same, more complex signature that indicates a tightly coupled prey–predator interaction (see also Section 12.7.1).

We return to the question of cycles – indeed, some of the same cycles discussed above – in Section 14.6, as part of a more general exploration of how the whole range of biotic and abiotic factors come together to determine the level and pattern of a population's abundance.

10.3 Effects of crowding

The most obvious omission, perhaps, from the predator–prey interactions we have modeled so far has been any acknowledgement that prey abundance may be limited by other prey, and predator

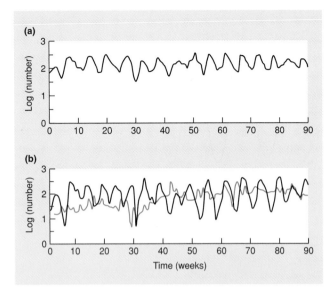

Figure 10.5 Host generation-length cycles in the moth *Plodia interpunctella* (a) alone (black line) and (b) with a granulovirus (colored line). These dynamics may be compared with those in Figure 10.1c. In spite of a superficial similarity in pattern, analysis indicates that those in (a) are generated by intraspecific competition; those in (b) are simply modulated versions of those in (a) and are therefore not predator–prey cycles. However, those in Figure 10.1c are predator–prey cycles. (After Bjornstad *et al.*, 2001.)

abundance by other predators. Prey are bound to be increasingly affected by intraspecific competition as their abundance increases; and predators, too, are likely to be limited at high densities by the availability of resting places, say, or safe refuges of their own, quite apart from their interaction with their most obvious resource, their prey.

mutual interference

More generally, predators have been assumed in the models discussed thus far to consume prey at a rate that depends only on prey abundance (in Equation 10.2, for example, the consumption rate per predator is simply aN). In reality, consumption rate will also often depend on the abundance of the predators themselves. Most obviously, food shortage – the abundance of prey *per predator* – will commonly result in a reduction in the consumption rate per individual as predator density increases. However, even when food is not limited, the consumption rate can be reduced by a number of processes known collectively as mutual interference (Hassell, 1978). For example, many consumers interact behaviorally with other members of their population, leaving less time for feeding and therefore depressing the overall feeding rate. For instance, humming-birds actively and aggressively defend rich sources of nectar. Alternatively, an increase in consumer density may lead to an increased rate of emigration, or of

consumers stealing food from one another (as do many gulls), or the prey themselves may respond to the presence of consumers and become less available for capture. All of these mechanisms give rise to a decline in predator consumption rate with predator density. Figure 10.6a, for example, shows significant reductions in consumption rate with abundance even at low densities of the crab *Carcinus aestuarii* foraging for the mussel *Musculista senhousia*; while Figure 10.6b shows that the kill rate of wolves, *Canis lupus*, preying on moose, *Alces alces*, in Isle Royale National Park, Michigan, USA, was lowest when there were most wolves.

10.3.1   Crowding in the Lotka–Volterra model

The effects of intraspecific competition, and of a decline in predator consumption rate with predator density, can be investigated by modifying the Lotka–Volterra isoclines. The details of incorporating intraspecific competition into the prey zero isocline are described by Begon *et al.* (1990), but the end result (Figure 10.7a) can be understood without reference to these details. At low prey densities there is no intraspecific competition, and the prey isocline is horizontal as in the Lotka–Volterra model. But as density increases, it is increasingly the case that prey densities below the isocline (prey increase) must be placed above the isocline (prey decrease) because of the effects of intraspecific competition. Hence, the isocline is increasingly lowered until it reaches the prey axis at the carrying capacity, K_N; that is, the prey can only just maintain themselves even in the absence of predators.

As we have seen, the predator isocline in the Lotka–Volterra model is vertical. This itself reflects the assumption that the ability of a predator population to increase in abundance is determined by the absolute abundance of prey, irrespective of the number of predators. If, however, mutual interference amongst the predators increases, then individual consumption rates will decline with predator abundance, and additional prey will be required to maintain a predator population of any given size. The predator zero isocline will depart increasingly from the vertical (Figure 10.7b). Moreover, at high densities, competition for other resources will put an upper limit on the predator population (a horizontal isocline) irrespective of prey numbers (Figure 10.7b).

crowding and the Lotka–Volterra isoclines

An alternative modification is to abandon altogether the assumption that consumption rate depends only on the absolute availability of prey, and assume ratio-dependent predation instead (Arditi & Ginzburg, 1989), although this alternative has itself been criticized (see Abrams, 1997; Vucetich *et al.*, 2002). In this case, the consumption rate depends on the ratio of prey to predators, and a particular ratio needs to be exceeded for the predators to increase in abundance: a

ratio-dependent predation

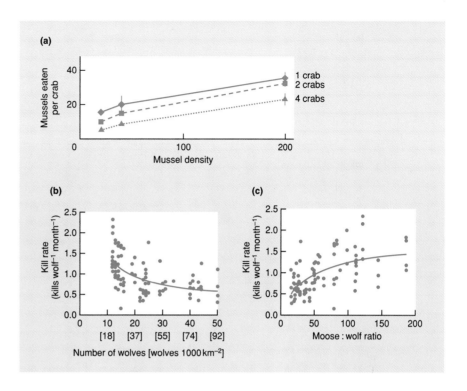

Figure 10.6 (a) Mutual interference amongst crabs, *Carcinus aestuarii*, feeding on mussels, *Musculista senhousia*. ♦ 1 crab; ■, 2 crabs; ▲, 4 crabs. The more crabs there were, the lower their per capita consumption rate. (After Mistri, 2003.) (b) Mutual interference amongst wolves, *Canis lupus*, preying on moose, *Alces alces*. (c) The same data but with wolf kill rate plotted against the moose : wolf ratio. The fitted curve assumes a dependence of kill rate on this ratio, but also that the wolves may become 'saturated' at high moose densities (see Section 10.4.2). This curve fits better than any for which kill rate depends on either predator density (e.g. (b)) or prey density. ((b, c) after Vucetich *et al.*, 2002.)

diagonal zero isocline passing through the origin (Figure 10.7c). Evidence of ratio-dependent predation is illustrated, for example, for the wolf–moose study in Figure 10.6c.

The likely effects of crowding in either population can now be deduced by combining the predator and prey isoclines (Figure 10.7d). Oscillations are still apparent for the most part, but these are no longer neutrally stable. Instead, they are damped so that they converge to a stable equilibrium. Predator–prey interactions in which either or both populations are substantially self-limited are likely, therefore, to exhibit patterns of abundance that are relatively stable, i.e. in which fluctuations in abundance are relatively slight.

crowding stabilizes dynamics

More particularly, when the predator is relatively inefficient, i.e. when many prey are needed to maintain a population of predators (curve (ii) in Figure 10.7d), the oscillations are damped quickly but the equilibrium prey abundance (N^*) is not much less than the equilibrium in the absence of predators (K_N). By contrast, when the predators are more efficient (curve (i)), N^* is lower and the equilibrium density of predators, P^*, is higher – but the interaction is less stable (the oscillations are more persistent). Moreover, if the predators are very strongly self-limited, then abundance may not oscillate at all (curve (iii)); but P^* will tend to be low, whilst N^* will tend to be not much less than K_N. Hence, for interactions where there is crowding, there appears to be a contrast between those in which predator density is low, prey abundance is little affected and the patterns of abundance are stable, and those in which predator

density is higher and prey abundance is more drastically reduced, but the patterns of abundance are less stable. (Figure 10.7d does not use ratio-dependent predation, but a predator isocline with a steeper slope in a ratio-dependent model (more efficient predation) can be equated, for present purposes, with an isocline rising from closer to the origin in the figure – that is, curve (i) rather than curve (ii).)

Essentially similar conclusions emerge from modifications of the Nicholson–Bailey model that incorporate either simple (logistic) crowding effects amongst the hosts or mutual interference amongst the predators (Hassell, 1978).

To quote examples of data proving the stabilizing influence of self-limitation on predator–prey dynamics would be difficult, simply because it would be all but impossible to compare the dynamics of matched populations with and without such self-limitation. On the other hand, populations of predators and prey with relatively stable dynamics are commonplace, as are the stabilizing forces of self-limitation we have discussed here. To take a more specific example, there are two groups of primarily herbivorous rodents that are widespread in the Arctic: the microtine rodents (lemmings and voles) and the ground squirrels. The microtines are renowned for their dramatic, cyclic fluctuations in abundance (see Chapter 14), but the ground squirrels have populations that remain remarkably constant from year to year, especially in open meadow and tundra habitats. There, significantly, they appear to be strongly self-limited by food availability, suitable burrowing habitat and their own spacing behavior (Karels & Boonstra, 2000).

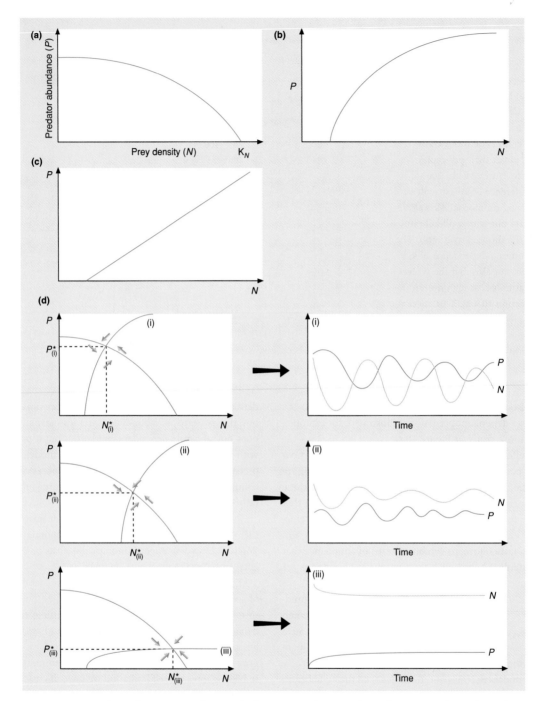

Figure 10.7   (a) A prey zero isocline subject to crowding. At the lowest prey densities this is the same as the Lotka–Volterra isocline, but when the density reaches the carrying capacity (K_N) the population can only just maintain itself even in the complete absence of predators. (b) A predator zero isocline subject to crowding (see text). (c) A predator zero isocline when there is prey : predator ratio dependent predation. (d) The prey zero isocline combined with the predator zero isoclines with increasing levels of crowding: (i), (ii) and (iii). P^* is the equilibrium abundance of predators, and N^* the equilibrium abundance of prey. Combination (i) is the least stable (most persistent oscillations) and has the most predators and least prey: the predators are relatively efficient. Less efficient predators, as in (ii), give rise to a lowered predator abundance, an increased prey abundance and less persistent oscillations. Strong predator self-limitation (iii) can eliminate oscillations altogether, but P^* is low and N^* is close to K_N.

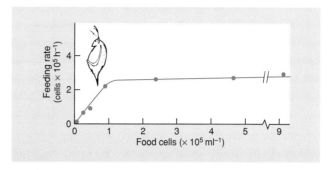

Figure 10.8   The type 1 functional response of *Daphnia magna* to different concentrations of the yeast *Saccharomyces cerevisiae*. (After Rigler, 1961.)

On a cautionary note, however, Umbanhowar *et al.* (2003), for example, failed to find evidence of mutual interference in a field study of the parasitoid *Tachinomyia similis* attacking its moth host *Orgyia vetusta*. The strength of mutual interference may often have been exaggerated by forcing predators to forage in artificial arenas at densities much higher than those they experience naturally. This is a useful reminder of the general point that an ecological force that is powerful in models or in the laboratory may none the less often be trivial, in practice, in natural populations. There can be little doubt, though, that self-limitation in its various forms frequently plays a key role in shaping predator–prey dynamics.

how important is mutual interference in practice?

10.4   Functional responses

Having examined the relationship between a predator's consumption rate and the predator's own abundance, above, we turn now to the effect on this consumption rate of the prey's abundance, the so-called functional response (Solomon, 1949). Below we describe the three main types of functional response (Holling, 1959), before considering how they might modify predator–prey dynamics.

10.4.1   The type 1 functional response

The most basic, 'type 1' functional response is that assumed by the Lotka–Volterra equations: consumption rate rises linearly with prey density (indicated by the constant, *a*, in Equation 10.2). An example is illustrated in Figure 10.8. The rate at which *Daphnia magna* consumed yeast cells rose linearly when the density of cells

varied, because the yeast cells were extracted by the *Daphnia* from a constant volume of water washed over their filtering apparatus, and the amount extracted therefore rose in line with food concentration. Above 10^5 cells ml^{-1}, however, the *Daphnia* could filter more cells but were unable to swallow all the food they filtered. They therefore ingested food at a maximum (plateau) rate irrespective of its concentration.

10.4.2   The type 2 functional response

The most frequently observed functional response is the 'type 2' response, in which consumption rate rises with prey density, but gradually decelerates until a plateau is reached at which consumption rate remains constant irrespective of prey density. (Realistically, even a type 1 response must have a plateau, as in the example above. The distinction is between the deceleration of a type 2 response and the linearity of the type 1 response.) Type 2 responses are shown for a carnivore, a herbivore and a parasitoid in Figure 10.9.

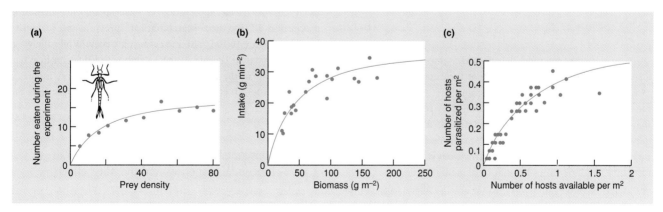

Figure 10.9   Type 2 functional responses. (a) Tenth-instar damselfly nymphs (*Ishnura elegans*) eating *Daphnia* of approximately constant size. (After Thompson, 1975.) (b) Wood bison (*Bison bison*) feeding on the sedge *Carex atherodes* presented at a range of sedge biomass densities. (After Bergman *et al.*, 2000.) (c) The parasitoid *Microplitis croceipes* attacking the tobacco budworm *Heliothis virescens*. (After Tillman, 1996.)

the type 2 response
and handling time

The type 2 response can be explained by noting that a predator has to
devote a certain handling time to each
prey item it consumes (i.e. pursuing, subduing and consuming
the prey item, and then preparing itself for further search). As
prey density increases, finding prey becomes increasingly easy.
Handling a prey item, however, still takes the same length of time,
and handling overall therefore takes up an increasing proportion
of the predator's time – until at high prey densities the predator
is effectively spending all of its time handling prey. The consumption rate therefore approaches and then reaches a maximum
(the plateau), determined by the maximum number of handling
times that can be fitted into the total time available.

We can derive a relationship between P_e (the number of
prey items eaten by a predator during a period of searching
time, T_s) and N, the density of those prey items (Holling, 1959).
P_e increases with the time available for searching, it increases with
prey density, and it increases with the searching efficiency or attack
rate of the predator, a. Thus:

Holling's type 2
response equation

$$P_e = aT_sN. \qquad (10.16)$$

However, the time available for searching will be less than the
total time, T, because of time spent handling prey. Hence, if T_h
is the handling time of each prey item, then T_hP_e is the total time
spent handling prey, and:

$$T_s = T - T_hP_e. \qquad (10.17)$$

Substituting this into Equation 10.16 we have:

$$P_e = a(T - T_hP_e)N \qquad (10.18)$$

or, rearranging:

$$P_e = aNT/1 + aT_hN. \qquad (10.19)$$

Note that the equation describes the amount eaten during
a specified period of time, T, and that the density of prey, N, is
assumed to remain constant throughout that period. In experiments, this can sometimes be guaranteed by replacing any prey
that are eaten, but more sophisticated models are required if prey
density is depleted by the predator. Such models are described
by Hassell (1978), who also discusses methods of estimating attack
rates and handling times from a set of data. (Trexler *et al.*, 1988,
discuss the general problem of fitting functional response curves
to sets of data.)

other routes to a
type 2 response

It would be wrong, however, to
imagine that the existence of a handling time is the only or the complete
explanation for all type 2 functional

responses. For instance, if the prey are of variable profitability,
then at high densities the diet may tend towards a decelerating number of highly profitable items (Krebs *et al.*, 1983); or a
predator may become confused and less efficient at high prey
densities.

10.4.3   The type 3 functional response

Type 3 functional responses are illustrated in Figure 10.10a–c.
At high prey densities they are similar to a type 2 response, and
the explanations for the two are the same. At low prey densities,
however, the type 3 response has an accelerating phase where an
increase in density leads to a more than linear increase in consumption rate. Overall, therefore, a type 3 response is 'S-shaped'
or 'sigmoidal'.

One important way in which a type
3 response can be generated is through
switching by the predator (see Section 9.5.2). The similarities
between Figures 9.15 and 10.10 are readily apparent. The difference is that discussions of switching focus on the density of a prey
type relative to the densities of alternatives, whereas functional
responses are based on only the absolute density of a single prey
type. In practice, though, absolute and relative densities are likely
to be closely correlated, and switching is therefore likely to lead
frequently to a type 3 functional response.

switching

More generally, a type 3 functional
response will arise whenever an
increase in food density leads to an
increase in the consumer's searching
efficiency, a, or a decrease in its handling
time, T_h, since between them these two determine consumption
rate (Equation 10.19). Thus, the small mammals in Figure 10.10a
appear to develop a search image for sawfly cocoons as they
become more abundant (increasing efficiency). The bluebottle
fly, *Calliphora vomitoria* (Figure 10.10b), spends an increasing
proportion of its time searching for 'prey' as prey density
increases (Figure 10.10d), also increasing efficiency. Whilst the wasp
Aphelinus thomsoni (Figure 10.10c) exhibits a reduction in mean
handling time as the density of its sycamore aphid prey increases
(Figure 10.10e). In each case, a type 3 functional response is the
result.

variations in
searching efficiency
or handling time

10.4.4   Consequences for population dynamics of
functional responses and the Allee effect

Different types of functional response
have different effects on population
dynamics. A type 3 response means a
low predation rate at low prey densities.
In terms of isoclines, this means that

type 3 responses
stabilize but may
be unimportant
in practice

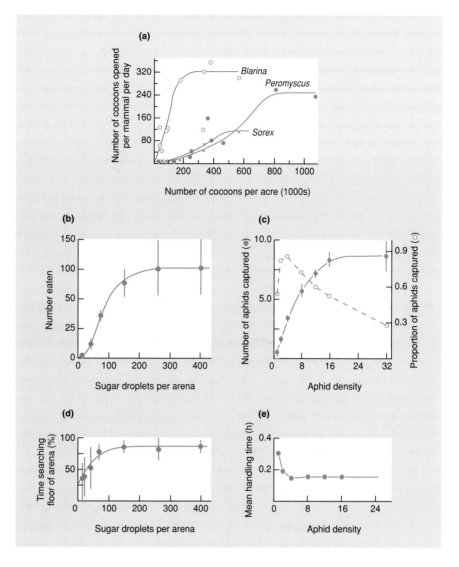

Figure 10.10 Type 3 (sigmoidal) functional responses. (a) The shrews *Sorex* and *Blarina* and the deer mouse *Peromyscus* responding to changing field densities of cocoons of the European pine sawfly, *Neodiprion sertifer*, in Ontario, Canada. (After Holling, 1959.) (b) The bluebottle fly, *Calliphora vomitoria*, feeding on sugar droplets. (After Murdie & Hassell, 1973.) (c) The wasp, *Aphelinus thomsoni*, attacking sycamore aphids, *Drepanosiphum platanoidis*: note the density-dependent increase in prey mortality rate at low prey densities (– – –) giving rise to the accelerating phase of the response curve (——). (After Collins *et al.*, 1981.) (d) The basis of the response in (b): searching efficiency of *C. vomitoria* increases with 'prey' (sugar droplet) density. (After Murdie & Hassell, 1973.) (e) The basis of the response in (c): handling time in *A. thomsoni* decreases with aphid density. (After Collins *et al.*, 1981.)

prey at low densities can increase in abundance virtually irrespective of predator density, and that the prey zero isocline will therefore rise vertically at low prey densities (Figure 10.11a). This could lend considerable stability to an interaction (Figure 10.11a, curve (i)), but for this the predator would have to be highly efficient at low prey densities (readily capable of maintaining itself), which contradicts the whole idea of a type 3 response (ignoring prey at low densities). Hence, curve (ii) in Figure 10.11a is likely to apply, and the stabilizing influence of the type 3 response may in practice be of little importance.

On the other hand, if a predator has a type 3 response to one particular type of prey because it switches its attacks amongst various prey types, then the population dynamics of the predator would be independent of the abundance of any particular prey type, and the vertical position of its zero isocline would therefore

be the same at all prey densities. As Figure 10.11b shows, this can lead potentially to the predators regulating the prey at a low and stable level of abundance.

An apparent example of this is provided by studies of vole cycles in Europe (Hanski *et al.*, 1991; see also Section 14.6.4). In subarctic Finnish Lapland, there are regular 4- or 5-year cycles, with a ratio of maximum : minimum vole densities generally exceeding 100. In southern Sweden small rodents show no regular multiannual cycles. But between the two, moving north to south in Fennoscandia, there is a gradient of decreasing regularity, amplitude and length of the cycle. Hanski *et al.* argue that this gradient is itself correlated with a gradient of increasing densities of generalist predators that switch between alternative

switching, stabilization and the voles of Fennoscandia

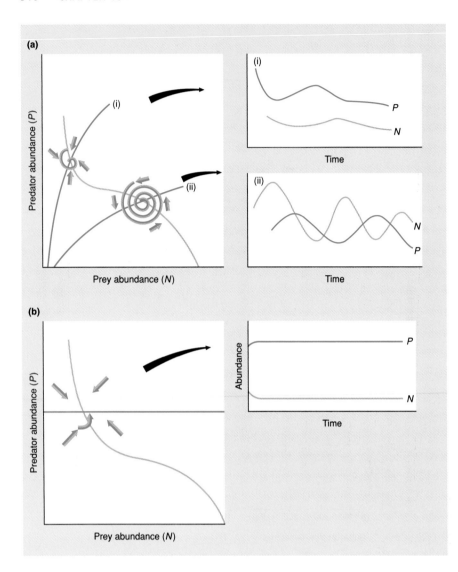

Figure 10.11 (a) The prey zero isocline is that which is appropriate when consumption rate is particularly low at low prey densities because of a type 3 functional response, an aggregative response (and partial refuge), an actual refuge or because of a reserve of plant material that is not palatable. With a relatively inefficient predator, predator zero isocline (ii) is appropriate and the outcome is not dissimilar from Figure 10.7. However, a relatively efficient predator will still be able to maintain itself at low prey densities. Predator zero isocline (i) will therefore be appropriate, leading to a stable pattern of abundance in which prey density is well below the carrying capacity and predator density is relatively high. (b) When a type 3 functional response arises because the predator exhibits switching behavior, the predator's abundance may be independent of the density of any particular prey type (main figure), and the predator zero isocline may therefore be horizontal (unchanging with prey density). This can lead to a stable pattern of abundance (inset) with prey density well below the carrying capacity.

prey as relative densities change (especially red foxes, badgers, domestic cats, buzzards, tawny owls and crows) and of specialist bird predators (especially other owl species and kestrels) that, being wide ranging in their activity, switch between alternative areas. In both cases, predator dynamics would be effectively independent of vole abundance, adding stability to the system in the manner of Figure 10.11b. In fact, Hanski *et al.* were able to go further in constructing a simple model of prey (voles) interacting with specialist predators (mustelids: stoats and weasels) and generalist (switching) predators. Their general contention was supported; as the number of generalist predators increased, oscillations in vole and mustelid abundance (which may or may not be the basis for the vole cycle) decreased in length and amplitude. Large enough densities of switching generalists stabilized the cycle entirely.

Turning to type 2 responses, if the predator has a response that reaches its plateau at relatively low prey densities (well below K_N), then the prey zero isocline has a hump, because there is a range of intermediate prey densities where the predators become less efficient with increasing prey density but the effects of competition amongst the prey are not intense. A hump will also arise here if the prey are subject to an 'Allee effect', where they have a disproportionately low rate of recruitment when their own density is low, perhaps because mates are difficult to find or because a 'critical number' must be exceeded before a resource can be properly exploited, i.e. there is inverse density dependence at low population densities (Courchamp *et al.*, 1999). If the predator

type 2 responses and the Allee effect destabilize – but not necessarily in practice

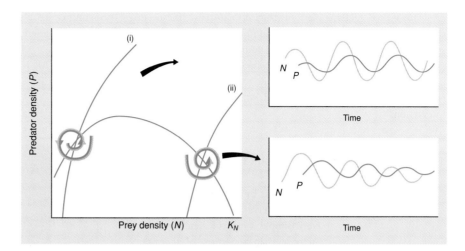

Figure 10.12 The possible effects of a prey isocline with a 'hump', either as a result of a type 2 functional response or an Allee effect. (i) If the predator is highly efficient, with its isocline crossing to the left of the hump, then the hump can be destabilizing, leading to the persistent oscillations of a limit cycle (inset). (ii) But if the predator is less efficient, crossing to the right of the hump, then the hump has little effect on the dynamics: the oscillations converge (inset).

isocline crosses to the right of the hump, then the population dynamics of the interaction will be little affected; but if the isocline crosses to the left of the hump, then the outcome will be persistent rather than convergent oscillations, i.e. the interaction will be *de*stabilized (Figure 10.12).

However, for a type 2 response to have this effect, predators would have to suffer serious reductions in their consumption rate at prey densities far below those at which the prey themselves suffer seriously from competition. This is unlikely. The potentially destabilizing effects of type 2 responses may also therefore be of little practical importance.

A destabilizing Allee effect has not apparently been established for any 'natural' predator–prey interaction. On the other hand, when we ourselves are the predator (for example, with exploited fisheries populations), we frequently have the ability (i.e. the technology) to maintain effective predation at low prey densities. If the prey population also exhibits an Allee effect, then the combination of this and persistent predation may all too readily drive a population towards extinction (Stephens & Sutherland, 1999; and see Section 15.3.5). That is, our isocline may cross that of the prey well to the left of their hump.

10.5 Heterogeneity, aggregation and spatial variation

Until now in this chapter, environmental heterogeneities, and the variable responses of predators and prey to such heterogeneities – all of which we saw in the previous chapter to be commonplace – have been ignored. We can ignore them no longer.

10.5.1 Aggregative responses to prey density

Because of the potential consequences for population dynamics, ecologists have been particularly interested in patch preferences where patches vary in the density of the food or prey items they contain (see Section 9.6). At one time it appeared, and was widely believed, that: (i) predators generally spent most time in patches containing high densities of prey (because these were the most profitable patches); (ii) most predators were therefore to be found in such patches; and (iii) prey in those patches were therefore most vulnerable to predation, whereas those in low-density patches were relatively protected and most likely to survive. Examples certainly exist to support the first two of these propositions (see Figure 9.20a–d), demonstrating an 'aggregative response' by the predators that is directly density dependent (predators spending most time in patches with high densities of prey such that prey and predator densities are positively correlated). However, this is not always the case. Furthermore, contrary to the third proposition, reviews of host–parasitoid interactions (e.g. Pacala & Hassell, 1991) have shown that prey (hosts) in high-density patches are not necessarily the most vulnerable to attack (direct density dependence): percentage parasitism may also be inversely density dependent or density independent between patches (see Figure 9.20e). Indeed, the reviews suggest that only around 50% of the studies examined show evidence of density dependence, and in only around 50% of these is the density dependence direct, as opposed to inverse. None the less, despite this variation in pattern, it remains true that the risk of predation often varies greatly between patches, and hence between individual prey.

do predators aggregate in high-density prey patches?

Many herbivores also display a marked tendency to aggregate, and many plants show marked variation in their risk of being attacked. The cabbage aphid (*Brevicoryne brassicae*) forms aggregates at two separate levels (Way & Cammell, 1970). Nymphs quickly form large groups when isolated on the surface of a single leaf, and populations on a single plant tend to be restricted to particular leaves. When aphids attack only one leaf

plants may be protected by the aggregative responses of herbivores

of a four-leaved cabbage plant (as they do naturally), the other three leaves survive; but if the same number of aphids are evenly spread over the four leaves, then all four leaves are destroyed (Way & Cammell, 1970). The aggregative behavior of the herbivores affords protection to the plant overall. But how might such heterogeneities influence the dynamics of predator–prey interactions?

10.5.2   Heterogeneity in the graphical model

refuges, partial
refuges and vertical
isoclines

We can start by incorporating into the Lotka–Volterra isoclines some relatively simple types of heterogeneity. Suppose that a portion of the prey population exists in a refuge: for example, shore snails packed into cracks in the cliff-face, away from marauding birds, or plants that maintain a reserve of material underground that cannot be grazed. In such cases, the prey zero isocline rises vertically at low prey densities (again, see Figure 10.11), since prey at low densities, hidden in their refuge, can increase in abundance irrespective of predator density.

Even if predators tend simply to ignore prey in low-density patches, as we have seen in some aggregative responses (see Section 9.6), this comes close to those prey being in a refuge, in the sense that the predators do not (rather than cannot) attack them. The prey may therefore be said to have a 'partial refuge', and this time the prey isocline can be expected to rise almost vertically at low prey abundances.

We saw above, when discussing type 3 functional responses, that such isoclines have a tendency to stabilize interactions. Early analyses of both the Lotka–Volterra and the Nicholson–Bailey systems (and early editions of this textbook) agreed with this conclusion: that spatial heterogeneities, and the responses of predators and prey to them, stabilize predator–prey dynamics, often at low prey densities (Beddington *et al.*, 1978). However, as we shall see next, subsequent developments have shown that the effects of heterogeneity are more complex than was previously supposed: the effects of heterogeneity vary depending on the type of predator, the type of heterogeneity, and so on.

10.5.3   Heterogeneity in the Nicholson–Bailey model

negative binomial
encounters . . .

Most progress has been made in untangling these effects in host–parasitoid systems. A good starting point is the model constructed by May (1978), in which he ignored precise details and argued simply that the distribution of host–parasitoid encounters was not random but aggregated. In particular, he assumed that this distribution could be described by a particular statistical model, the negative binomial. In this case

(in contrast to Section 10.2.3), the proportion of hosts not encountered at all is given by:

$$p_0 = \left[1 + \frac{AP_t}{k} \right]^{-k} \qquad (10.20)$$

where k is a measure of the degree of aggregation; maximal aggregation at $k = 0$, but a random distribution (recovery of the Nicholson–Bailey model) at $k = \infty$. If this is incorporated into the Nicholson–Bailey model (Equations 10.14 and 10.15), then we have:

$$H_{t+1} = H_t e^r \left[1 + \frac{AP_t}{k} \right]^{-k} \qquad (10.21)$$

$$P_{t+1} = H_t \left\{ 1 - \left[1 + \frac{AP_t}{k} \right]^{-k} \right\}. \qquad (10.22)$$

. . . that stabilize
dynamics

The behavior of a version of this model, which also includes a density-dependent host rate of increase, is illustrated in Figure 10.13, from which it is clear that the system is given a marked boost in stability by the incorporation of significant levels of aggregation ($k \leq 1$). Of particular importance is the existence of stable systems with low values of H^*/K; i.e. aggregation appears capable of generating stable host abundances well below the host's normal carrying capacity. This coincides with the conclusion drawn from Figure 10.11.

10.5.4   Aggregation of risk and spatial density dependence

pseudo-interference

How does this stability arise out of aggregation? The answer lies in what has been called 'pseudo-interference' (Free *et al.*, 1977). With mutual interference, as predator density increases, predators spend an increasing amount of time interacting with one another, and their attack rate therefore declines. With pseudo-interference, attack rate also declines with parasitoid density, but as a result of an increasing fraction of encounters being wasted on hosts that have already been attacked. The crucial point is that 'aggregation of risk' amongst hosts tends to increase the amount of pseudo-interference. At low parasitoid densities, a parasitoid is unlikely to have its attack rate reduced as a result of aggregation. But at higher parasitoid densities, parasitoids in aggregations (where most of them are) will increasingly be faced with host patches in which most or all of the hosts have already been parasitized. As parasitoid density increases, therefore, their effective attack rate (and hence their subsequent birth rate) declines rapidly – a *directly* density-dependent effect. This dampens both the natural oscillations in parasitoid density, and their impact on host mortality.

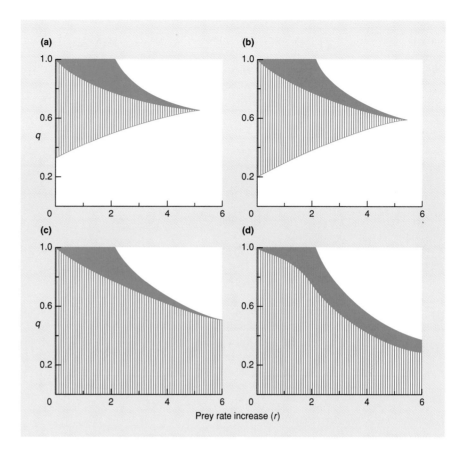

Figure 10.13 May's (1978) model of host–parasitoid aggregation, with host self-limitation incorporated, illustrates that aggregation can enhance stability and give rise to stability at low values of $q = H^*/K$. In the solid orange area there is an exponential approach to equilibrium; in the hatched area there is an oscillatory approach to equilibrium; outside these there is instability (the oscillations either diverge or are sustained). The four figures are for four values of k, the exponent of the negative binomial distribution in the model: (a) $k = \infty$: no aggregation, least stability; (b) $k = 2$; (c) $k = 1$; (d) $k = 0.1$: most aggregation. (After Hassell, 1978.)

aggregation of risk strengthens direct (temporal) density dependences

To summarize, aggregation of risk stabilizes host–parasitoid interactions by strengthening direct (not delayed) density dependencies that already exist (Taylor, 1993). The stabilizing powers of this spatial phenomenon, aggregation of risk, therefore arise not from any spatial density dependencies, but from its translation into direct, temporal density dependence.

But how does aggregation of risk relate to the aggregative responses of parasitoids? And do aggregative responses and aggregation of risk necessarily lead to enhanced stability? We can address these questions by examining Figure 10.14, bearing in mind from Section 9.6 that aggregated predators do not necessarily spend most time foraging in patches of high host density (spatial density dependence); foraging time can also be negatively correlated with host density (inverse density dependence) or independent of host density. Start with Figure 10.14a. The distribution of parasitoids over host patches follows a perfect, straight line density-dependent relationship. But, since the host : parasitoid ratio is therefore the same in each host patch, the risk is likely to be the same in each host patch, too. Thus, positive spatial density dependence does not necessarily lead to aggregation of risk and does not necessarily enhance stability. On the other hand, with a directly density-dependent relationship that

accelerates (Figure 10.14b), there does appear to be aggregation of risk, and this might well enhance stability (Hassell & May, 1973); but it turns out that whether or not it does so depends on the parasitoids' functional response (Ives, 1992a). With a type 1 response, assumed by most analyses, stability is enhanced. But with a more realistic type 2 response, initial increases in density-dependent aggregation from zero aggregation actually decrease aggregation of risk and are destabilizing. Only high levels of density-dependent aggregation are stabilizing.

Moreover, it is clear from Figures 10.14c and d that there can be considerable aggregation of risk with either inverse spatial density dependence or no spatial density dependence of any sort – and these would not be counteracted by a type 2 functional response. In partial answer to our two questions above, therefore, aggregative responses that are spatially density dependent are actually *least* likely to lead to aggregation of risk, and therefore least likely to enhance stability.

aggregative responses and aggregation of risk

In practice, of course, with real sets of data (like those in Figure 9.20), aggregation of risk will often arise from a combination of spatially density-dependent (direct or inverse) and density-independent responses (Chesson & Murdoch, 1986; Pacala & Hassell, 1991). Pacala, Hassell and coworkers have

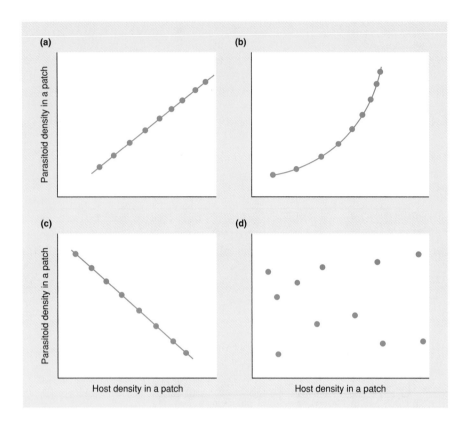

Figure 10.14 The aggregative responses of parasitoids and the aggregation of risk. (a) Parasitoids aggregate in high host-density patches, but the parasitoid : host ratio is the same in all patches (a perfect straight-line relationship), and hence the risk to hosts is apparently the same in all patches. (b) Parasitoid aggregation to high host-density patches now accelerates with increasing host density, and hosts in high-density patches are thus apparently at greater risk of parasitization: there is aggregation of risk. (c) With perfect inverse density dependence (i.e. parasitoid aggregation in *low* host-density patches) the hosts in the low-density patches are apparently at a much greater risk of parasitization: again there is aggregation of risk. (d) Even with no aggregative response (density independence) the hosts in some patches are apparently at a greater risk of parasitization (are subject to a higher parasitoid : host ratio) than others: here too there is aggregation of risk.

called the former the 'host density dependent' (HDD) component, and the latter the 'host density independent' (HDI) component, and have described methods by which, in real data sets like Figure 9.20, the aggregation of risk can be split between them. Interestingly, in an analysis of 65 data sets, representing 26 different host–parasitoid combinations (Pacala & Hassell, 1991), 18 appeared to have sufficient aggregation of risk to stabilize their interactions, but for 14 of these 18 cases, it was HDI variation that contributed most to the total, further weakening any imagined link between spatial density dependence and stability.

10.5.5 Heterogeneity in some continuous-time models

We have been pursuing parasitoids and hosts; and in doing so we have been retaining certain structural features in our analysis that we should now reconsider. In particular, our parasitoids have been assumed, in effect, to arrange themselves over host patches at the beginning of a generation (or whatever the time interval is between t and $t + 1$), and then to have to suffer the consequences of that arrangement until the beginning of the next generation. But suppose we move into continuous time – as appropriate for many parasitoids as it is for many other predators. Now, aggregation should be assumed to occur on a continuous basis, too. Predators in a depleted, or even a depleting, patch should leave

and redistribute themselves (see Section 9.6.2). The whole basis of pseudo-interference and hence stability, namely wasted predator attacks in high predator density patches, tends to disappear.

Murdoch and Stewart-Oaten (1989) went to, perhaps, the opposite extreme to the one we have been considering, by constructing a continuous-time model in which prey moved instantly into patches to replace prey that had been consumed, and predators moved instantly into patches to maintain a consistent pattern of predator–prey covariation over space. The effects on their otherwise neutrally stable Lotka–Volterra model contrast strongly with those we have seen previously. First, predator aggregation that is independent of local prey density now has *no* effect on either stability or prey density. However, predator aggregation that is directly dependent on local prey density has an effect that depends on the strength of this dependence – although it always lowers prey density (because predator efficiency is increased). If such density dependence is relatively weak (as Murdoch and Stewart-Oaten argue it usually is in practice), then stability is *decreased*. Only if it is stronger than seems typical in nature is stability increased.

Other, less 'extreme' continuous-time formulations (Ives, 1992b), or those that combine discrete generations with redistribution within generations (Rohani *et al.*, 1994), produce results

continuous redistribution of predators and prey

that are themselves intermediate between the 'Nicholson–Bailey extreme' and the 'Murdoch–Stewart-Oaten extreme'. It seems certain, however, that a preoccupation with models lacking within-generation movement has, in the past, led to a serious overestimation of the significance of aggregation to patches of high host density in stabilizing host–parasitoid interactions.

10.5.6   The metapopulation perspective

The continuous- and discrete-time approaches clearly differ, but they share a common perspective in seeing predator–prey interactions occurring within single populations, albeit populations with inbuilt variability. An alternative is a 'metapopulation' perspective (see Section 6.9), in which environmental patches support subpopulations that have their own internal dynamics, but are linked to other subpopulations by movement between patches.

A number of studies have investigated predator–prey metapopulation models, usually with unstable dynamics within patches. Mathematical difficulties have often limited analysis to two-patch models, where, if the patches are the same, and dispersal is uniform, stability is unaffected: patchiness and dispersal have no effect in their own right (Murdoch *et al.*, 1992; Holt & Hassell, 1993).

patch differences stabilize through asynchrony

Differences between the patches, however, tend, in themselves, to stabilize the interaction (Ives, 1992b; Murdoch *et al.*, 1992; Holt & Hassell, 1993). The reason is that any difference in parameter values between patches leads to asynchrony in the fluctuations in the patches. Inevitably, therefore, a population at the peak of its cycle tends to lose more by dispersal than it gains, a population at a trough tends to gain more than it loses, and so

on. Dispersal and asynchrony together, therefore, give rise to stabilizing temporal density dependence in net migration rates.

The situation becomes much more complex with the inclusion of aggregative behavior, since dispersal rates themselves become a much more complex function of both prey and predator densities. Aggregation appears to have two opposing effects (Murdoch *et al.*, 1992). It tends to increase the asynchrony between fluctuations in predator abundance (enhancing stability) but to reduce the asynchrony between prey fluctuations (decreasing stability). The balance between these forces appears to be sensitive to the strength of the aggregation, but perhaps even more sensitive to the assumptions built into the models (Godfray & Pacala, 1992; Ives, 1992b; Murdoch *et al.*, 1992). Aggregation may either stabilize or destabilize. In contrast to previous analyses, it has no clear effect on prey density since its stabilizing powers are not linked to predator efficiency.

an explicitly, and visually, spatial model

The treatment of a spatially heterogeneous predator–prey interaction as a problem in metapopulation dynamics was taken a stage further by Comins *et al.* (1992). They constructed computer models of an environment consisting of a patchwork of squares, which could actually be visualized as such (Figure 10.15). In each generation, two processes occurred in sequence. First, a fraction, μ_P, of predators, and a fraction, μ_N, of prey, dispersed from each square to the eight neighboring squares. At the same time, predators and prey from the eight neighboring squares were dispersing into the first square. Thus, for example, the dynamics for the density of prey, $N_{i,t+1}$, in square i in generation $t + 1$, was given by:

$$N_{i,t+1} = N_{i,t}(1 - \mu_N) + \mu_N \bar{N}_{i,t} \tag{10.23}$$

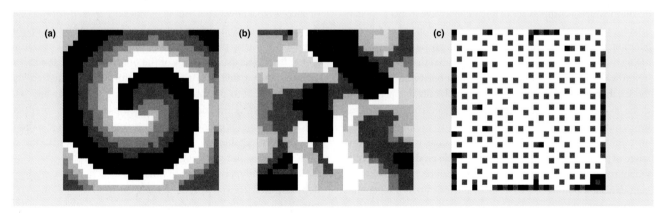

Figure 10.15   Instantaneous maps of population density for simulations of the dispersal model of Comins *et al.* (1992) with Nicholson–Bailey local dynamics. Different levels of shading represent different densities of hosts and parasitoids. Black squares represent empty patches; dark shades becoming paler represent patches with increasing host densities; light shades to white represent patches with hosts and increasing parasitoid densities. (a) Spirals: $\mu_N = 1$, $\mu_P = 0.89$; (b) spatial chaos: $\mu_i = 0.2$, $\mu_P = 0.89$; (c) a 'crystalline lattice': $\mu_N = 0.05$; $\mu_P = 1$. (After Comins *et al.*, 1992.)

or:

$$N_{i,t+1} = N_{i,t} + \mu_N(\bar{N}_{i,t} - N_{i,t}), \qquad (10.24)$$

where $\bar{N}_{i,t}$ is the mean density in the eight squares neighboring square i in generation t. The second phase then consisted of one generation of standard predator–prey dynamics, either following the Nicholson–Bailey equations or a discrete-time version of the Lotka–Volterra equations (May, 1973). Simulations were started with random prey and predator populations in a single patch, with all the other patches empty.

We know that within individual squares, if they existed in isolation, the dynamics would be unstable. But within the patchwork of squares as a whole, stable or at least highly persistent patterns can readily be generated (Figure 10.15). The general message is similar to the results that we have already seen: that stability can be generated by dispersal in metapopulations in which different patches are fluctuating asynchronously. Note especially, in this case, that a patch experiences a net gain in migrants when its density is lower than the mean of the eight patches with which it connects (Equation 10.24) but experiences a net loss when its density is higher – a kind of density dependence. Note, too, that the asynchrony arises in the present case because the population has spread from a single initial patch (all patches are, in principle, the same) and that it is maintained by dispersal being limited to the neighboring patches (rather than being a powerful equalizing force over all patches).

emergent spatial patterns

Moreover, the explicitly spatial aspects of this model have, quite literally, added another dimension to the results. Depending on the dispersal fractions and the host reproductive rate, a number of quite different spatial structures can be generated (although they tend to blur into one another) (Figure 10.15a–c). 'Spatial chaos' can occur, in which a complex set of interacting wave fronts are established, each one persisting only briefly. With somewhat different parameter values, and especially when both predator and prey are highly mobile, the patterns are more structured than chaotic, with 'spiral waves' rotating around almost immobile focal points. The model, therefore, makes the point very graphically that persistence at the level of a whole population does not necessarily imply either uniformity across the population or stability in individual parts of it. Static 'crystalline lattices' can even occur within a narrow range of parameter values, with highly mobile predators and rather sedentary prey, emphasizing that pattern can be generated internally within a population even in an intrinsically homogeneous environment.

Is there one general message that can be taken from this body of theory? Certainly, we cannot say 'aggregation does *this* or *that* to predator–prey interactions'. Rather, aggregation can have a variety of effects, and knowing which of these is likely will require detailed knowledge of predator and prey biology for the inter-

action concerned. In particular, the effects of aggregation have been seen to depend on the predator's functional response, the extent of host self-regulation, and so on – other features that we have examined in isolation. It is necessary, as stressed at the beginning of this chapter, in seeking to understand complex processes, to isolate conceptually the different components. But it is also necessary, ultimately, to recombine those components.

10.5.7 Aggregation, heterogeneity and spatial variation in practice

What, then, can be said about the role of spatial variation in practice? The stabilizing effects of heterogeneity were demonstrated famously, long ago, by Huffaker (1958; Huffaker *et al.*, 1963), who studied a system in which a predatory mite fed on a herbivorous mite, which fed on oranges interspersed amongst rubber balls in a tray. In the absence of its predator, the prey maintained a fluctuating but persistent population (Figure 10.16a); but if the predator was added during the early stages of prey population growth, it rapidly increased its own population size, consumed all of its prey and then became extinct itself (Figure 10.16b). The interaction was altered, however, when Huffaker made his microcosm more 'patchy' (creating, effectively, a metapopulation, though the term had not been coined at the time). He spread the oranges further apart, and partially isolated each one by placing a complex arrangement of vaseline barriers in the tray, which the mites could not cross. But he facilitated the dispersal of the prey by inserting a number of upright sticks from which they could launch themselves on silken strands carried by air currents. Dispersal between patches was therefore much easier for the prey than it was for the predators. In a patch occupied by both predators and prey, the predators consumed all the prey and then either became extinct themselves or dispersed (with a low rate of success) to a new patch. But in patches occupied by prey alone, there was rapid, unhampered growth accompanied by successful dispersal to new patches. In a patch occupied by predators alone, there was usually death of the predators before their food arrived. Each patch was therefore ultimately doomed to the extinction of both predators and prey. But overall, at any one time, there was a mosaic of unoccupied patches, prey–predator patches heading for extinction, and thriving prey patches; and this mosaic was capable of maintaining persistent populations of both predators and prey (Figure 10.16c).

Subsequently, others, too, have demonstrated the power of a metapopulation structure in promoting the persistence of coupled predator and prey populations when their dynamics

metapopulation effects in mites, beetles and ciliates

in individual subpopulations are unstable. Figure 10.17a, for example, shows this for a parasitoid attacking its beetle host. Figure 10.17b shows similar results for prey and predatory ciliates

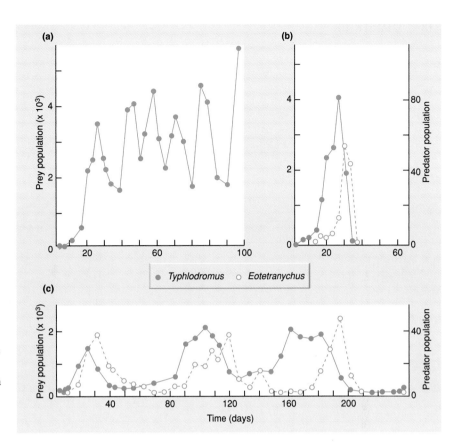

Figure 10.16 Hide and seek: predator–prey interactions between the mite *Eotetranychus sexmaculatus* and its predator, the mite *Typhlodromus occidentalis*. (a) Population fluctuations of *Eotetranychus* without its predator. (b) A single oscillation of the predator and prey in a simple system. (c) Sustained oscillations in a more complex system. (After Huffaker, 1958.)

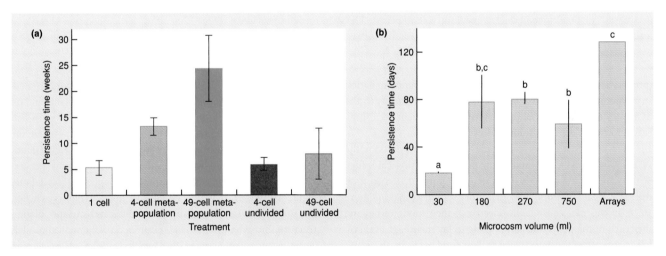

Figure 10.17 A metapopulation structure can increase the persistence of predator–prey interactions. (a) The parasitoid, *Anisopteromalus calandrae*, attacking its bruchid beetle host, *Callosobruchus chinensis*, living on beans either in small single 'cells' (short persistence time, left), or in combinations of cells (four or 49), which either had free access between them so that they effectively constituted a single population (persistence time not significantly increased, right), or had limited (infrequent) movement between cells so that they constituted a metapopulation of separate subpopulations (increased persistence time, center). Bars show standard errors. (After Bonsall *et al.*, 2002.) (b) The predatory ciliate, *Didinium nasutum*, feeding on the bacterivorous ciliate, *Colpidium striatum*, in bottles of various volumes, where persistence time varied little, except in the smallest populations (30 ml) where times were shorter, and also in 'arrays' of nine or 25 linked 30 ml bottles (metapopulations), where persistence was greatly prolonged: all populations persisted until the end of the experiment (130 days). Bars show standard errors; different letters above bars indicate treatments that were significantly different from one another ($P < 0.05$). (After Holyoak & Lawler, 1996.)

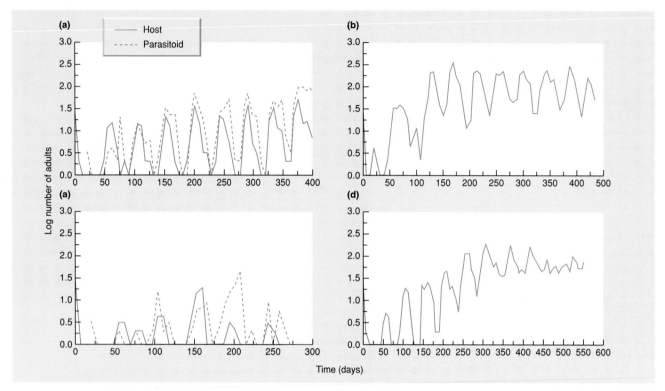

Figure 10.18   Long-term population dynamics in laboratory population cages of a host (*Plodia interpunctella*), with and without its parasitoid (*Venturia canescens*). (a) Host and parasitoid in deep medium, exhibiting coupled cycles in abundance, approximately one host generation in length. (b) The host alone in deep medium, exhibiting similar cycles. (c) Host and parasitoid in shallow medium, unable to persist. (d) The host alone in shallow medium, able to persist. The deep medium provides a refuge from attack for a proportion of the host population that is not present in the shallow medium (see Section 10.5.2). All data sets are selected from several replicates showing the same pattern. (After Begon *et al.*, 1995.)

(protists), where, in support of the role of a metapopulation structure, it was also possible to demonstrate asynchrony in the dynamics of individual subpopulations and frequent local prey extinctions and recolonizations (Holyoak & Lawler, 1996).

a refuge for a moth   A study providing support for the stabilizing powers of a physical refuge is illustrated in Figure 10.18, based on the same *Plodia–Venturia* host–parasitoid system as that in Figure 10.1c. In this case, hosts living deeper in their food are beyond the reach of the parasitoids attempting to lay their eggs in them. In the absence of this refuge, in a shallow food medium, this host–parasitoid interaction is unable to persist (Figure 10.18c), although the host alone does so readily (Figure 10.18d). With a refuge present, however, in a deeper food medium, the host and parasitoid can apparently persist together indefinitely (Figure 10.18a).

more mites: a   In fact, though, the distinctions metapopulation   between different types of spatial or a refuge?   heterogeneity may not be as clear cut in real systems as they are in mathematical models. Ellner *et al.* (2001), for example, examined a system of predatory mites, *Phytoseiulus*

persimilis, feeding on herbivorous mites, *Tetranychus urticae*, feeding on bean plants, *Phaseolus lunatus*. On individual plants and on a single 'continent' of 90 plants (Figure 10.19a), the system had no long-term persistence (Figure 10.19c). However, when the Styrofoam sheet supporting the plants was split into eight islands of 10 plants, connected by bridges that limited the mites' powers of dispersal (Figure 10.19b), persistence was apparently indefinite (Figure 10.19d, e). It would be easy to jump to the conclusion that stability was increased by the eight-island metapopulation structure. But when Ellner *et al.* examined mathematical models of the system that allowed the various aspects of the altered layout to be investigated one by one, they could detect no significant effect of such a structure. Instead, they suggested that the enhanced stability arose from a different aspect: a reduction in the predators' ability to detect and respond to prey outbreaks on individual plants – a prey 'refuge' effect that could arise in the absence of any explicit spatial structure.

One major problem in making pronouncements about the stabilizing role of aggregation of risk is that although, as we have seen, there have been wide-   real data confirm the complexity of natural systems

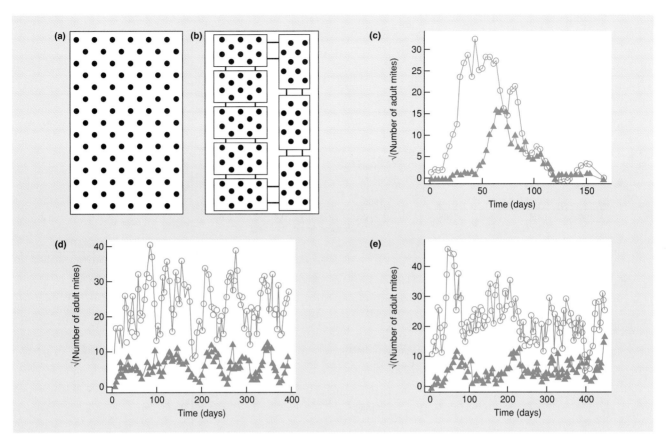

Figure 10.19 The population dynamics of the predatory mite, *Phytoseiulus persimilis*, and its herbivorous mite prey, *Tetranychus urticae*. They interacted either (a) on a single continent of 90 bean plants, the dynamics of which are shown in (c) (▲, predators; ○, prey), or (b) in a metapopulation of eight islands of 10 plants. For the latter, the dynamics of two replicates are shown in (d) and (e), where persistence (stability) is clearly enhanced. (After Ellner *et al.*, 2001.)

ranging surveys of the data on spatial distributions of attacks, these data generally come from studies of very short duration – often of only one generation. We do not know if the observed spatial patterns are typical for that interaction; nor do we know if the population dynamics show the degree of stability that the spatial patterns might seem to predict. One investigation that has examined population dynamics and spatial distributions over several generations is that of Redfern *et al.* (1992), who made a 7-year (seven-generation) study of two tephritid fly species that attack thistles and the guilds of parasitoids that attack the flies. For one host, *Terellia serratulae* (Figure 10.20a) there was evidence of year-to-year density dependence in the overall rate of parasitism (Figure 10.20b), but no strong evidence of significant levels of aggregation within generations, either overall (Figure 10.20c) or for parasitoid species individually. For the other species, *Urophora stylata* (Figure 10.20d), there was no apparent temporal density dependence but good evidence for the aggregation of risk (Figure 10.20e, f), and, to repeat a pattern we have seen before, most heterogeneity was contributed by the HDI component. It cannot be said, however, that the patterns of this study fit neatly,

overall, to the theory we have outlined. First, both hosts were attacked by several parasitoid species – not one, as assumed by most models. Second, the levels of aggregation (and to a lesser extent the HDI or HDD contributions) varied considerably and apparently randomly from year to year (Figure 10.20c, f): no one year was typical, and no single 'snap-shot' could have captured either interaction. Finally, while the relatively stable dynamics of *Terellia* may have reflected the more demonstrable direct density dependence in parasitism, this appeared to be quite unconnected to any differences in the aggregation of risk.

The effects of spatial heterogeneities on the stability of predator–prey dynamics are not only of purely scientific interest. They have also been the subject of lively debate (Hawkins & Cornell, 1999) in considering the prop-

spatial heterogeneity and the most effective biological control agents

erties and nature of biological control agents: natural enemies of a pest that are imported into an area, or otherwise aided and abetted, in order to control the pest (see Section 15.2.5). What is required of a good biological control agent is the ability to

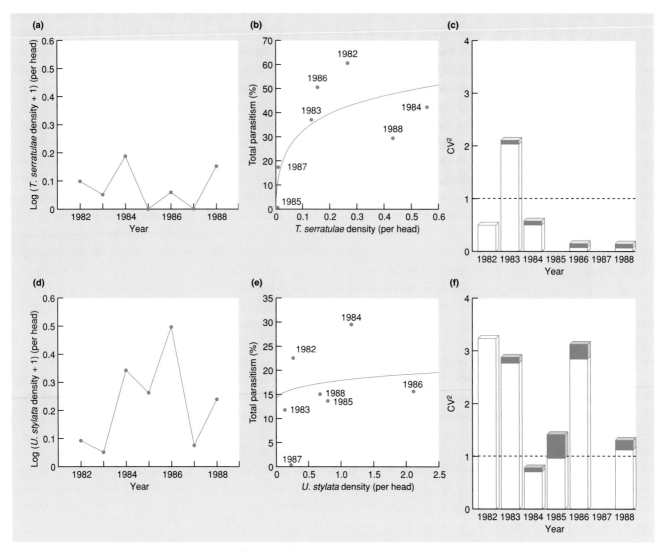

Figure 10.20 Attacks by parasitoids on tephrytid flies (*Terellia serratulae* and *Urophora stylata*) that attack thistle flower-heads. The dynamics of the populations are shown for *T. serratulae* in (a) and for *U. stylata* in (d). Temporal density dependence of parasitoid attacks on *T. serratulae* (b) is significant ($r^2 = 0.75$; $P < 0.05$), but for *U. stylata* (e) it is not ($r^2 = 0.44$; $P > 0.05$); both fitted lines take the form $y = a + b \log_{10}x$. However, whereas for *T. serratulae* (c) there is little aggregation of risk of parasitoid attack within years (measured as $CV^2 > 1$ for aggregation), with *U. stylata* (f) there is far more, most of which is HDI (no shading) rather than HDD (dark shading). (After Redfern *et al.*, 1992.)

reduce the prey (pest) to a stable abundance well below its normal, harmful level, and we have seen that some theoretical analyses suggest that this is precisely what aggregative responses help to generate. Establishing such a link in practice, however, has not proved easy. Murdoch *et al.* (1995), for example, noted that the California red scale, *Aonidiella aurantii*, an insect pest of citrus plants worldwide, appeared to be kept at low and remarkably stable densities in southern California by a parasitoid introduced to control it, *Aphytis mellitus*. The existence of a partial refuge from parasitization for the red scale seemed a plausible hypothesis for how this was achieved: on bark in the interior of the trees, rates of parasitism were very low and scale densities high, seemingly as a result of the activities of ants there that interfered with the searching parasitoids. Murdoch *et al.*, therefore, tested this hypothesis by a field experiment in which ants were removed from a number of trees. Parasitization rates in the refuge did increase, and scale abundance declined there (Figure 10.21), and there was some evidence that parasitization rates, and scale abundance, in the population as a whole were then more variable. But these effects were only slight and apparently short term, and there was certainly no evidence that scale abundance overall was increased by any diminution of the refuge effect.

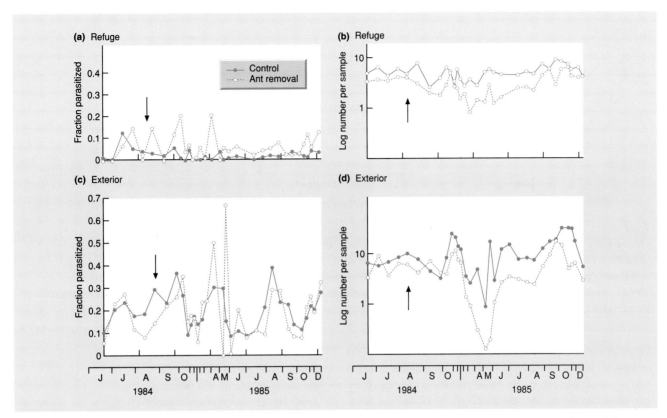

Figure 10.21 Results of a field experiment to test the hypothesis that the parasitoid *Aphytis mellitus* maintains the abundance of the California red scale, *Aonidiella aurantii*, at stable low levels because of a partial refuge from parasitization in the interior portions of citrus trees, where ants interfere with the parasitoids. When ants were removed from blocks of trees (time of removal indicated by the arrow), the fraction parasitized in the refuge tended to be higher (a), and scale abundance there was lower (b), but outside the refuge ('exterior') the fraction parasitized was only marginally more variable (c), and scale abundance was only more variable over one relatively brief period and tended to be lower than on control trees (d). (After Murdoch *et al.*, 1995.)

Moreover, Murdoch *et al.* (1985) had earlier argued that, in general, pest populations persist after successful biological control not as a result of aggregative responses, but because of the stochastic creation of host patches by colonization and their subsequent extinction when discovered by the agent: essentially, a metapopulation effect. Waage and Greathead (1988), however, suggested that a broader perspective could incorporate both aggregative responses and metapopulation effects. They proposed that scale insects and other homopterans, and mites (like Huffaker's), which may reproduce to have many generations within a patch, are often stabilized by asynchronies in the dynamics of different patches; whereas lepidopterans and hymenopterans, which typically occupy a patch for only part of a single generation, may often be stabilized by an aggregative response. In fact, though, with biological control, like predator–prey dynamics generally, building convincing links between patterns in population stability of natural populations and particular stabilizing mechanisms – or combinations of mechanisms – remains a challenge for the future.

10.6 Multiple equilibria: an explanation for outbreaks?

When predator and prey populations interact, there can sometimes be sudden changes in the abundance of one or both partners: outbreaks or crashes. Of course, this may reflect an equally sudden change in the environment, but ecologists working in a variety of fields have come to realize that there is not necessarily just one equilibrium combination of a predator and prey (about which there may or may not be oscillations). There can, instead, be 'multiple equilibria' or 'alternative stable states'.

Figure 10.22 is a model with multiple equilibria. The prey zero isocline has both a vertical section at low densities and a hump. This could reflect a type *a model with multiple equilibria* 3 functional response of a predator that also has a long handling time, or perhaps the combination of an aggregative response and an Allee effect in the prey. As a consequence, the predator zero

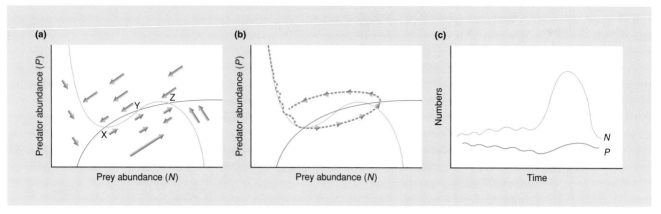

Figure 10.22 A predator–prey zero isocline model with multiple equilibria. (a) The prey zero isocline has a vertical section at low densities and a hump; the predator zero isocline can therefore cross it three times. Intersections X and Z are stable equilibria, but intersection Y is an unstable 'breakpoint' from which the joint abundances move towards either intersection X or intersection Z. (b) A feasible path that the joint abundances might take when subject to the forces shown in (a). (c) The same joint abundances plotted as numbers against time, showing that an interaction with characteristics that do not change can lead to apparent 'outbreaks' in abundance.

isocline crosses the prey zero isocline three times. The strengths and directions of the arrows in Figure 10.22a indicate that two of these points (X and Z) are fairly stable equilibria (although there are oscillations around each). The third point (Y), however, is unstable: populations near here will move towards either point X or point Z. Moreover, there are joint populations close to point X where the arrows lead to the zone around point Z, and joint populations close to point Z where the arrows lead back to the zone around point X. Even small environmental perturbations could put a population near point X on a path towards point Z, and vice versa.

The behavior of a hypothetical population, consistent with the arrows in Figure 10.22a, is plotted in Figure 10.22b on a joint abundance diagram, and in Figure 10.22c as a graph of numbers against time. The prey population, in particular, displays an 'eruption' in abundance, as it moves from a low-density equilibrium to a high-density equilibrium and back again. This eruption is in no sense a reflection of an equally marked change in the environment. It is, on the contrary, a pattern of abundance generated by the interaction itself (plus a small amount of environmental 'noise'), and in particular it reflects the existence of multiple equilibria. Similar explanations may be invoked to explain apparently complicated patterns of abundance in nature.

There are certainly examples of natural populations exhibiting outbreaks of abundance from levels that are otherwise low and apparently stable (Figure 10.23a), and there are other examples in which populations appear to alternate between two stable densities (Figure 10.23b). But it does not follow that each of these examples is necessarily an interaction with multiple equilibria.

In some cases, a plausible argument for multiple equilibria can be put forward. This is true, for instance, of Clark's (1964) work in Australia on the eucalyptus psyllid (*Cardiaspina albitextura*), a homopteran bug (Figure 10.23a). These insects appear to have a low-density equilibrium maintained by their natural predators (especially birds), and a much less stable high-density equilibrium reflecting intraspecific competition (the destruction of host tree foliage leading to reductions in fecundity and survivorship). Outbreaks from one to the other can occur when there is just a short-term failure of the predators to react to an increase in the density of adult psyllids. Similarly, the observation of two alternative equilibria in Figure 10.23b for the viburnum whitefly, *Aleurotrachelus jelinekii*, is reinforced by a model for that population which predicts the same pattern (Southwood *et al.*, 1989).

Alternative stable states have also been proposed for a number of plant–herbivore interactions, often where increased grazing pressure seems to have led to the 'collapse' of the vegetation from a high biomass to a much lower one, which is then stable in the sense that there is no return to the high biomass state even when grazing pressure is severely reduced (van de Koppel *et al.*, 1997). The grasslands of the Sahel region of Africa, grazed by livestock, and the arctic plants along the coast of Hudson Bay in Canada, grazed by geese, are both examples. The conventional explanation (Noy-Meir, 1975) has essentially been that depicted in Figure 10.22: when driven to a low biomass, plants may have very little material above ground and hence very limited powers of immediate regrowth. This is a classic 'Allee effect' – the prey

sudden changes in abundance: multiple equilibria – or sudden changes in the environment

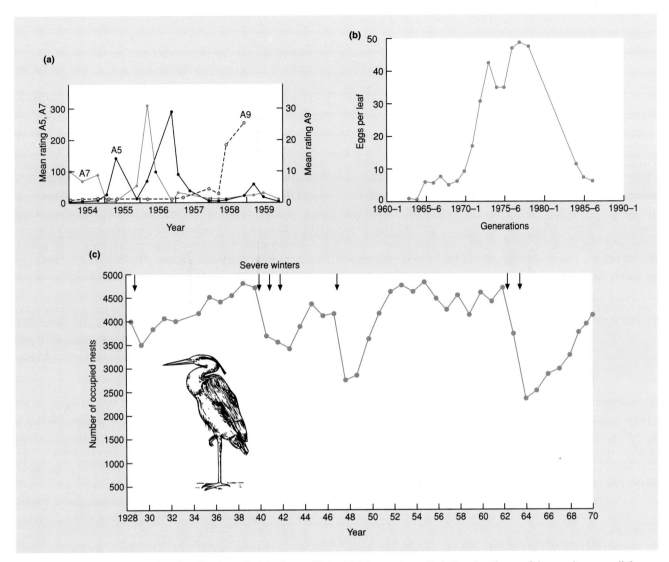

Figure 10.23 Possible examples of outbreaks and multiple equilibria. (a) Mean ratings of relative abundance of the eucalyptus psyllid, *Cardiaspina albitextura*, in three study areas in Australia (A5, A7 and A9). (After Clark, 1962.) (b) The mean number of eggs per leaf of the viburnum whitefly, *Aleurotrachelus jelinekii*, on a viburnum bush in Silwood Park, Berkshire, UK. No samples were taken between 1978 and 1979, and 1984 and 1985. (After Southwood *et al.*, 1989.) (c) Changes in the abundance of herons (*Ardea cinerea*) in England and Wales (measured by the number of nests occupied) are readily attributable to changes in environmental conditions (particularly severe winters). (After Stafford, 1971.)

suffering from too *low* an abundance – leading to a marked 'hump' in their isocline. It may also be, however, that the plants' problems at low biomass are compounded by soil deterioration – erosion, for example – introducing further positive feedback into the system: high grazing leading to low plant biomass, leading to poorer growing conditions, leading to lower plant biomass, leading to even poorer growing conditions, and so on (van de Koppel *et al.*, 1997).

On the other hand, there are many cases in which sudden changes in abundance are fairly accurate reflections of sudden changes in the environment or a food source. For instance, the number of herons nesting in England and Wales normally fluctuates around 4000–4500 pairs, but the population declines markedly after particularly severe winters (Figure 10.23c). This fish-eating bird is unable to find sufficient food when inland waters become frozen for long periods, but there is no suggestion that the lower population levels (2000–3000 pairs) are an alternative equilibrium. The population crashes are simply the result of density-independent mortality from which the herons rapidly recover.

10.7 Beyond predator–prey

The simplest mathematical models of predator–prey interactions produce coupled oscillations that are highly unstable. However, by adding various elements of realism to these models it is possible to reveal the features of real predator–prey relationships that are likely to contribute to their stability. A further insight provided by models is that predator–prey systems may exist in more than one stable state. We have seen that a variety of patterns in the abundance of predators and prey, both in nature and in the laboratory, are consistent with the conclusions derived from models. Unfortunately, we are rarely in a position to apply specific explanations to particular sets of data, because the critical experiments and observations to test the models have rarely been made. Natural populations are affected not just by their predators or their prey, but also by many other environmental factors that serve to 'muddy the waters' when direct comparisons are made with simple models.

Moreover, the attention of both modelers and data gatherers (not that the two need be different) is increasingly being directed away from single- or two-species systems, towards those in which three species interact. For example, a pathogen attacking a predator that attacks a prey, or a parasitoid and a pathogen both attacking a prey/host. Interestingly, in several of these systems, unexpected dynamical properties emerge that are not just the expected blend of the component two-species interactions (Begon *et al.*, 1996; Holt, 1997). We return to the problems of 'abundance' in a broader context in Chapter 14.

Summary

Predator and prey populations display a variety of dynamic patterns. It is a major task for ecologists to account for the differences from one example to the next.

A number of mathematical models illustrate an underlying tendency for predator and prey populations to undergo coupled oscillations (cycles) in abundance. We explain the Lotka–Volterra model, which is the simplest differential equation predator–prey model, and using zero isoclines we show that the coupled oscillations are structurally unstable in this case. The model also illustrates the role of delayed density-dependent numerical responses in generating the cycles. We explain, too, the Nicholson–Bailey host–parasitoid model, which also displays unstable oscillations.

In both these models, cycles are several prey (host) generations in length, but other models of host–parasitoid (and host–pathogen) systems are able to generate coupled oscillations just one host generation in length.

We ask whether there is good evidence for predator–prey cycles in nature, focusing especially on a hare–lynx system and a moth attacked by two natural enemies. Even when predators or prey exhibit regular cycles in abundance, it is never easy to demonstrate that these are predator–prey cycles.

We begin an examination of the effects on dynamics of factors missing from the simplest models by looking at crowding. For predators, the most important expression of this is mutual interference. We look at the effects of crowding in the Lotka–Volterra model, including ratio-dependent predation: crowding stabilizes the dynamics, although this effect is strongest when the predators are least efficient. Essentially similar conclusions emerge from modifications of the Nicholson–Bailey model. There is, though, little direct evidence for these effects in nature.

The functional response describes the effect of prey abundance on predator consumption rate. The three types of functional response are explained, including the role of handling time in generating type 2 responses, and of variations in handling time and searching efficiency in generating type 3 responses. We explain the consequences for predator–prey dynamics of the different types of functional responses and of the 'Allee effect' (lowered recruitment at low abundance). Type 2 responses tend to destabilize, and type 3 responses to stabilize, but these are not necessarily important in practice.

Predators often (but not always) exhibit an aggregative response. We examine the effects of refuges and partial refuges in the Lotka–Volterra model, suggesting that spatial heterogeneities, and the responses to them, stabilize predator–prey dynamics, often at low prey densities. However, further work, especially with host–parasitoid systems and the Nicholson–Bailey model, shows that the effects of heterogeneity are complex. Stability arises through 'aggregation of risk', strengthening direct density dependencies that already exist. But aggregative responses that are spatially density dependent are least likely to lead to aggregation of risk and least likely to enhance stability. Models with within-generation movement further undermine the significance of aggregative responses in stabilizing host–parasitoid interactions. A metapopulation perspective emphasizes that patch differences may stabilize through asynchrony, and also that predator–prey interactions may generate spatial as well as temporal patterns.

In practice, the stabilizing effects of metapopulation structure and of refuges have been demonstrated, and the general importance of responses to spatial heterogeneity in the choice of biocontrol agents has been the subject of lively debate.

Finally, predator–prey systems with more than one equilibrium combination of predators and prey are examined as a possible basis for prey (or predator) outbreaks.

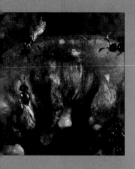

Chapter 11
Decomposers and Detritivores

11.1 Introduction

saprotrophs:
detritivores and
decomposers . . .

When plants and animals die, their bodies become resources for other organisms. Of course, in a sense, most consumers live on dead material – the carnivore catches and kills its prey, and the living leaf taken by a herbivore is dead by the time digestion starts. The critical distinction between the organisms in this chapter, and herbivores, carnivores and parasites, is that the latter all directly affect the rate at which their resources are produced. Whether it is lions eating gazelles, gazelles eating grass or grass parasitized by a rust fungus, the act of taking the resource harms the resource's ability to regenerate new resource (more gazelles or grass leaves). In contrast with these groups, saprotrophs (organisms that make use of dead organic matter) do not control the rate at which their resources are made available or regenerate; they are dependent on the rate at which some other force (senescence, illness, fighting, the shedding of leaves by trees) releases the resource on which they live. Exceptions exist among necrotrophic parasites (see Chapter 12) that kill and then continue to extract resources from the dead host. Thus, the fungus *Botrytis cinerea* attacks living bean leaves but continues this attack after the host's death. Similarly, maggots of the sheep blowfly *Lucilia cuprina* may parasitize and kill their host, whereupon they continue to feed on the corpse. In these cases the saprotroph can be said to have a measure of control over the supply of its food resource.

. . . do not generally
control their supply
of resources – 'donor
control'

We distinguish two groups of saprotrophs: decomposers (bacteria and fungi) and detritivores (animal consumers of dead matter). Pimm (1982) described the relationship that generally exists between decomposers or detritivores and their food as *donor controlled*: the donor (prey; i.e. dead organic matter) controls the density of the recipient (predator; i.e. decomposer or detritivore) but not the reverse. This is fundamentally different from truly interactive predator–prey interactions (see Chapter 10). However, while there is generally no direct feedback between decomposers/detritivores and the dead matter consumed (and thus donor-controlled dynamics apply), nevertheless it is possible to see an indirect 'mutualistic' effect through the release of nutrients from decomposing litter, which may ultimately affect the rate at which trees produce more litter. In fact, it is in nutrient recycling that decomposers and detritivores play their most fundamental role (see Chapter 19). More generally, of course, the food webs associated with decomposition are just like food webs based on living plants: they have a number of trophic levels, including predators of decomposers (microbivores) and of detritivores, and consumers of these predators, and exhibit a range of trophic interactions (not just donor controlled).

decomposition
defined

Immobilization occurs when an inorganic nutrient element is incorporated into an organic form – primarily during the growth of green plants. Conversely, decomposition involves the release of energy and the *mineralization* of chemical nutrients – the conversion of elements from an organic to inorganic form. Decomposition is defined as the gradual disintegration of dead organic matter and is brought about by both physical and biological agencies. It culminates with complex, energy-rich molecules being broken down by their consumers (decomposers and detritivores) into carbon dioxide, water and inorganic nutrients. Some of the chemical elements will have been locked up for a time as part of the body structure of the decomposer organisms, and the energy present in the organic matter will have been used to do work and is eventually lost as heat. Ultimately, the incorporation of solar energy in photosynthesis, and the immobilization of inorganic nutrients into biomass, is balanced by the loss of heat energy and organic nutrients when the organic matter is mineralized. Thus a given nutrient molecule may be successively immobilized and mineralized in a repeated round of nutrient cycling. We discuss the overall role played by decomposers and detritivores in the fluxes of energy and nutrients at the ecosystem level in Chapters 17 and 18. In the present chapter, we introduce the organisms involved and look in detail at the ways in which they deal with their resources.

decomposition . . .
. . . of dead
bodies, . . .

It is not only the bodies of dead animals and plants that serve as resources for decomposers and detritivores. Dead organic matter is continually produced during the life of both animals and plants and can be a major resource. Unitary organisms shed dead parts as they develop and grow – the larval skins of arthropods, the skins of snakes, the skin, hair, feathers and horn of other vertebrates. Specialist feeders are often associated with these cast-off resources. Among the fungi there are specialist decomposers of feathers and of horn, and there are arthropods that specialize on sloughed off skin. Human skin is a resource for the household mites that are omnipresent inhabitants of house dust and cause problems for many allergy sufferers.

. . . of shed parts of
organisms . . .

The continual shedding of dead parts is even more characteristic of modular organisms. Some polyps on a colonial hydroid or coral die and decompose, while other parts of the same genet continue to regenerate new polyps. Most plants shed old leaves and grow new ones; the seasonal litter fall onto a forest floor is the most important of all the sources of resource for decomposers and detritivores, but the producers do not die in the process. Higher plants also continually slough off cells from the root caps, and root cortical cells die as a root grows through the soil. This supply of organic material from roots produces the very resource-rich *rhizosphere*. Plant tissues are generally leaky, and soluble sugars and nitrogenous compounds also become available on the surface of leaves, supporting the growth of bacteria and fungi in the *phyllosphere*.

. . . and of feces

Finally, animal feces, whether produced by detritivores, microbivores, herbivores, carnivores or parasites, are a further category of resource for decomposers and detritivores. They are composed of dead organic material that is chemically related to what their producers have been eating.

The remainder of this chapter is in two parts. In Section 11.2 we describe the 'actors' in the saprotrophic 'play', and consider the relative roles of the bacteria and fungi on the one hand, and the detritivores on the other. Then, in Section 11.3, we consider, in turn, the problems and processes involved in the consumption by detritivores of plant detritus, feces and carrion.

11.2 The organisms

11.2.1 Decomposers: bacteria and fungi

If scavengers do not take a dead resource immediately it dies (such as hyenas consuming a dead zebra), the process of decomposition usually starts with colonization by bacteria and fungi. Other changes may occur at the same time: enzymes in the dead tissue may start to autolyze it and break down the carbohydrates and proteins into simpler, soluble forms. The dead material may also become leached by rainfall or, in an aquatic environment, may lose minerals and soluble organic compounds as they are washed out in solution.

Bacteria and fungal spores are omnipresent in the air and the water, and are usually present on (and often in) dead material before it is dead.

bacteria and fungi
are early colonists of
newly dead material

They usually have first access to a resource. These early colonists tend to use soluble materials, mainly amino acids and sugars that are freely diffusible. They lack the array of enzymes necessary for digesting structural materials such as cellulose, lignin, chitin and keratin. Many species of *Penicillium*, *Mucor* and *Rhizopus*, the so-called 'sugar fungi' in soil, grow fast in the early phases of decomposition. Together with bacteria having similar opportunistic physiologies, they tend to undergo population explosions on newly dead substrates. As the freely available resources are consumed, these populations collapse, leaving very high densities of resting stages from which new population explosions may develop when another freshly dead resource becomes available. They may be thought of as the opportunist 'r-selected species' among the decomposers (see Section 4.12). Another example is provided by the early colonizers of nectar in flowers, predominantly yeasts (simple sugar fungi); these may spread to the ripe fruit where they act on sugar in the juice to produce alcohol (as happens in the industrial production of wine and beer).

In nature, as in industrial processes such as the making of wine or sauerkraut, the activity of the early colonizers is dominated by the metabolism of

domestic and
industrial
decomposition

sugars and is strongly influenced by aeration. When oxygen is in free supply, sugars are metabolized to carbon dioxide by growing microbes. Under anaerobic conditions, fermentations produce a less complete breakdown of sugars to by-products such as alcohol and organic acids that change the nature of the environment for subsequent colonizers. In particular, the lowering of the pH by the production of acids has the effect of favoring fungal as opposed to bacterial activity.

Anoxic habitats are characteristic of waterlogged soils and, more particularly, of sediments of oceans and lakes. Aquatic sediments receive a continuous supply of dead organic matter from

aerobic and
anaerobic
decomposition
in nature

the water column above but aerobic decomposition (mainly by bacteria) quickly exhausts the available oxygen because this can only be supplied from the surface of the sediment by diffusion. Thus, at some depth, from zero to a few centimeters below the surface, depending mainly on the load of organic material, sediments are completely anoxic. Below this level are found a variety of bacterial types that employ different forms of anaerobic respiration

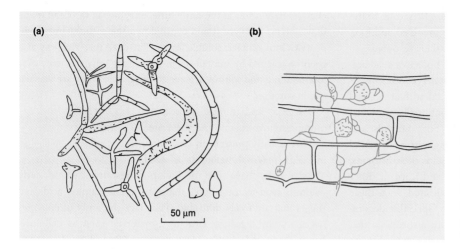

Figure 11.1 (a) Spores (conidia) of aquatic hyphomycete fungi from river foam. (b) Rhizomycelium of the aquatic fungus *Cladochytrium replicatum* within the epidermis of an aquatic plant. The circular bodies are zoosporangia. (After Webster, 1970.)

– that is, they use terminal inorganic electron acceptors other than oxygen in their respiratory process. The bacterial types occur in a predictable pattern with denitrifying bacteria at the top, sulfate-reducing bacteria next and methanogenic bacteria in the deepest zone. Sulfate is comparatively abundant in sea water and so the zone of sulfate-reducing bacteria is particularly wide (Fenchel, 1987b). In contrast, the concentration of sulfate in lakes is low, and methanogenesis plays a correspondingly larger role (Holmer & Storkholm, 2001).

A strong element of chance determines which species are the first to colonize newly dead material, but in some environments there are specialists with properties that enhance their chances of arriving early. Litter that falls into streams or ponds is often colonized by aquatic fungi (e.g. Hyphomycetes), which bear spores with sticky tips (Figure 11.1a) and are often of a curious form that seems to maximize their chance of being carried to and sticking to leaf litter. They may spread by growing from cell to cell within the tissues (Figure 11.1b).

decomposition of
more resistant tissues
proceeds more slowly

After the colonization of terrestrial litter by the 'sugar' fungi and bacteria, and perhaps also after leaching by rain or in the water, the residual resources are not diffusible and are more resistant to attack. In broad terms, the major components of dead terrestrial organic matter are, in a sequence of increasing resistance to decomposition: sugars < (less resistant than) starch < hemicelluloses, pectins and proteins < cellulose < lignins < suberins < cutins. Hence, after an initial rapid breakdown of sugar, decomposition proceeds more slowly, and involves microbial specialists that can use celluloses and lignins and break down the more complex proteins, suberin (cork) and cuticles. These are structural compounds, and their breakdown and metabolism depend on very intimate contact with the decomposers (most cellulases are surface enzymes requiring actual physical contact between the decomposer organism and its resource). The processes of

decomposition may now depend on the rate at which fungal hyphae can penetrate from cell to cell through lignified cell walls. In the decomposition of wood by fungi (mainly homobasidiomycetes), two major categories of specialist decomposers can be recognized: the brown rots that can decompose cellulose but leave a predominantly lignin-based brown residue, and the white rots that decompose mainly the lignin and leave a white cellulosic residue (Worrall *et al.*, 1997). The tough silicon-rich frustules of dead diatoms in the phytoplankton communities of lakes and oceans are somewhat analogous to the wood of terrestrial communities. The regeneration of this silicon is critical for new diatom growth, and decomposition of the frustules is brought about by specialized bacteria (Bidle & Azam, 2001).

The organisms capable of dealing with progressively more refractory compounds in terrestrial litter represent a natural succession starting with simple sugar fungi (mainly Phy-

succession of
decomposing
microorganisms

comycetes and Fungi Imperfecti), usually followed by septate fungi (Basidiomycetes and Actinomycetes) and Ascomycetes, which are slower growing, spore less freely, make intimate contact with their substrate and have more specialized metabolism. The diversity of the microflora that decomposes a fallen leaf tends to decrease as fewer but more highly specialized species are concerned with the last and most resistant remains.

The changing nature of a resource during its decomposition is illustrated in Figure 11.2a for beech leaf litter on the floor of a cool temperate deciduous forest in Japan. Polyphenols and soluble carbohydrates quickly disappeared, but the resistant structural holocellulose and lignin decomposed much more slowly. The fungi responsible for leaf decomposition follow a succession that is associated with the changing nature of the resource. The frequency of occurrence of early species, such as *Arthrinium* sp. (Figure 11.2b), was correlated with declines in holocellulose and soluble carbohydrate concentrations; Osono and Takeda (2001) suggest that they

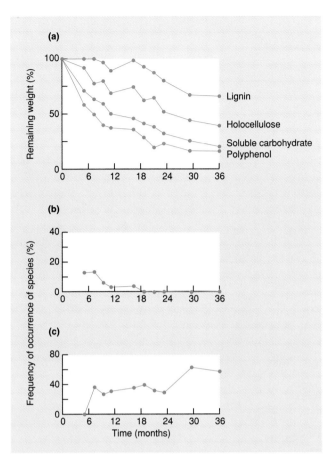

Figure 11.2 (a) Changes in the composition of beech (*Fagus crenata*) leaf litter (in mesh bags) during decomposition on a woodland floor in Japan over a 3-year period. Amounts are expressed as percentages of the starting quantities. (b, c) Changes in the frequency of occurrence of fungal species representative of: (b) early species (*Arthrinium* sp.) and (c) late species (*Mortierella ramanniana*). (After Osono & Takeda, 2001.)

depend on these components for their growth. Many late species, such as *Mortierella ramanniana*, seem to rely on sugars released by other fungi capable of decomposing lignin.

most microbial decomposers are relatively specialized

Individual species of microbial decomposer are not biochemically very versatile; most of them can cope with only a limited number of substrates. It is the diversity of species involved that allows the structurally and chemically complex tissues of a plant or animal corpse to be decomposed. Between them, a varied microbiota of bacteria and fungi can accomplish the complete degradation of dead material of both plants and animals. However, in practice they seldom act alone, and the process would be much slower and, moreover, incomplete, if they did so. The major factor that delays the decomposition of

organic residues is the resistance to decomposition of plant cell walls – an invading decomposer meets far fewer barriers in an animal body. The process of plant decomposition is enormously speeded up by any activity that grinds up and fragments the tissues, such as the chewing action of detritivores. This breaks open cells and exposes the contents and the surfaces of cell walls to attack.

11.2.2 Detritivores and specialist microbivores

specialist consumers of microbial organisms: microbivores

The microbivores are a group of animals that operate alongside the detritivores, and which can be difficult to distinguish from them. The name microbivore is reserved for the minute animals that specialize at feeding on microflora, and are able to ingest bacteria or fungi but exclude detritus from their guts. Exploitation of the two major groups of microflora requires quite different feeding techniques, principally because of differences in growth form. Bacteria (and yeasts) show a colonial growth form arising by the division of unicells, usually on the surface of small particles. Specialist consumers of bacteria are inevitably very small; they include free-living protozoans such as amoebae, in both soil and aquatic environments, and the terrestrial nematode *Pelodera*, which does not consume whole sediment particles but grazes among them consuming the bacteria on their surfaces. The majority of fungi, in contrast to most bacteria, are filamentous, producing extensively branching hyphae, which in many species are capable of penetrating organic matter. Some specialist consumers of fungi possess piercing, sucking stylets (e.g. the nematode *Ditylenchus*) that they insert into individual fungal hyphae. However, most fungivorous animals graze on the hyphae and consume them whole. In some cases, close mutualistic relationships exist between fungivorous beetles, ants and termites and characteristic species of fungi. These mutualisms are discussed in Chapter 13.

Note that microbivores consume a living resource and may not be subject to donor-controlled dynamics (Laakso *et al.*, 2000). In a study of decomposition of lake weed and phytoplankton in laboratory microcosms, Jurgens and Sala (2000) followed the fate of bacteria (decomposers) in the presence and absence of bacteria-grazing protists, namely *Spumella* sp. and *Bodo saltans* (microbivores). In the presence of the microbivores, there was a reduction of 50–90% in bacterial biomass and the bacterial community became dominated by large, grazer-resistant forms including filamentous bacteria.

The larger the animal, the less able it is to distinguish between microflora as food and the plant or animal detritus on which these are growing. In fact, the majority of the detritivorous animals involved in the decomposition of dead organic matter are generalist consumers, of both the detritus itself and the associated microfloral populations.

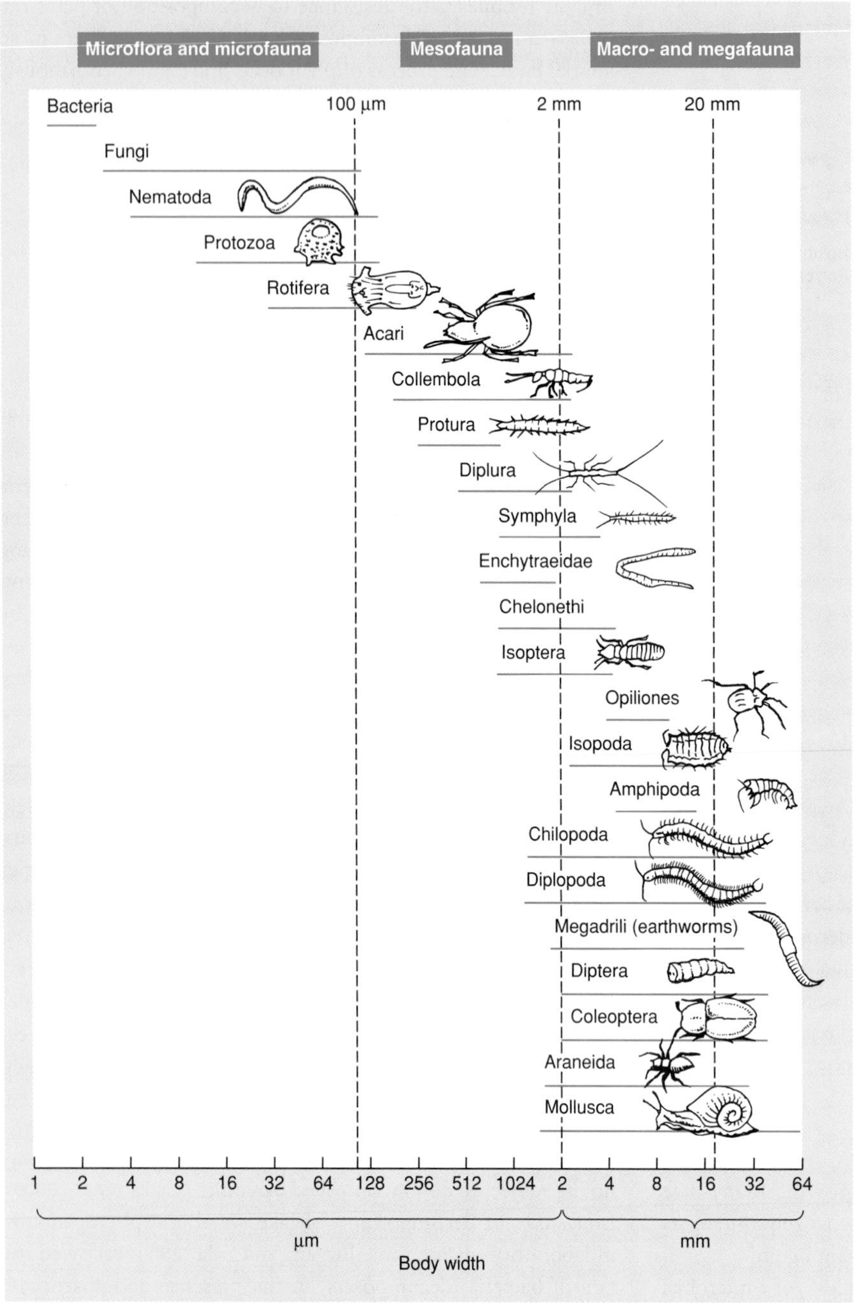

Figure 11.3   Size classification by body width of organisms in terrestrial decomposer food webs. The following groups are wholly carnivorous: Opiliones (harvest spiders), Chilopoda (centipedes) and Araneida (spiders). (After Swift *et al.*, 1979.)

classification of
decomposers . . .

. . . by size in
terrestrial
environments . . .

The protists and invertebrates that take part in the decomposition of dead plant and animal materials are a taxonomically diverse group. In terrestrial environments they are usually classified according to their size. This is not an arbitrary basis for classification, because size is an important feature for organisms that reach their resources by burrowing or crawling among cracks and crevices of litter or soil. The *microfauna* (including the specialist microbivores) includes protozoans, nematode worms and rotifers (Figure 11.3). The principal groups of the *mesofauna* (animals with a body width between 100 µm and 2 mm) are litter mites (Acari), springtails (Collembola) and pot worms (Enchytraeidae). The *macrofauna* (2–20 mm body width) and, lastly, the *megafauna* (> 20 mm) include woodlice (Isopoda), millipedes (Diplopoda), earthworms (Megadrili), snails and slugs (Mollusca) and the larvae of certain flies (Diptera) and beetles (Coleoptera). These animals are mainly responsible for the

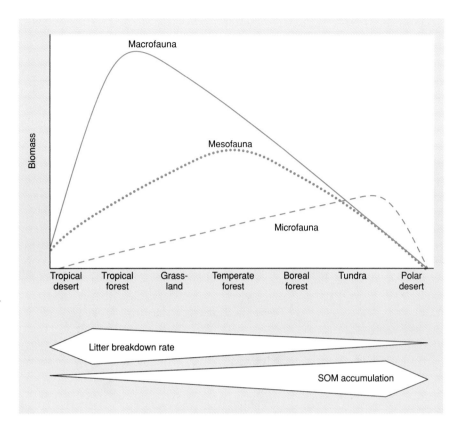

Figure 11.4 Patterns of latitudinal variation in the contribution of the macro-, meso- and microfauna to decomposition in terrestrial ecosystems. Soil organic matter (SOM) accumulation (inversely related to litter breakdown rate) is promoted by low temperatures and waterlogging, where microbial activity is impaired. (Swift *et al.*, 1979.)

initial shredding of plant remains. By their action, they may bring about a large-scale redistribution of detritus and thus contribute directly to the development of soil structure. It is important to note that the microfauna, with their short generation times, operate at the same scale as bacteria and can track bacterial population dynamics, whilst the mesofauna and the fungi they mainly depend on are both longer lived. The largest and longest lived detritivores, in contrast, cannot be finely selective in their diet, but choose patches of high decomposer activity (J. M. Anderson, personal communication).

Long ago, Charles Darwin (1888) estimated that earthworms in some pastures close to his house formed a new layer of soil 18 cm deep in 30 years, bringing about 50 tons ha^{-1} to the soil surface each year as worm casts. Figures of this order of magnitude have since been confirmed on a number of occasions. Moreover, not all species of earthworm put their casts above ground, so the total amount of soil and organic matter that they move may be much greater than this. Where earthworms are abundant, they bury litter, mix it with the soil (and so expose it to other decomposers and detritivores), create burrows (so increasing soil aeration and drainage) and deposit feces rich in organic matter. It is not surprising that agricultural ecologists become worried about practices that reduce worm populations.

Detritivores occur in all types of terrestrial habitat and are often found at remarkable species richness and in very great numbers.

Thus, for example, a square meter of temperate woodland soil may contain 1000 species of animals, in populations exceeding 10 million for nematode worms and protozoans, 100,000 for springtails (Collembola) and soil mites (Acari), and 50,000 or so for other invertebrates (Anderson, 1978). The relative importance of microfauna, mesofauna and macrofauna in terrestrial communities varies along a latitudinal gradient (Figure 11.4). The microfauna is relatively more important in the organic soils in boreal forest, tundra and polar desert. Here the plentiful organic matter stabilizes the moisture regime in the soil and provides suitable microhabitats for the protozoans, nematodes and rotifers that live in interstitial water films. The hot, dry, mineral soils of the tropics have few of these animals. The deep organic soils of temperate forests are intermediate in character; they maintain the highest mesofaunal populations of litter mites, springtails and pot worms. The majority of the other soil animal groups decline in numbers towards the drier tropics, where they are replaced by termites. Lower mesofaunal diversity in these tropical regions may be related to a lack of litter due to decomposition and consumption by termites, reflecting both low resource abundance and few available microhabitats (J. M. Anderson, personal communication).

On a more local scale, too, the nature and activity of the decomposer community depends on the conditions in which the organisms live. Temperature has a fundamental role in determining

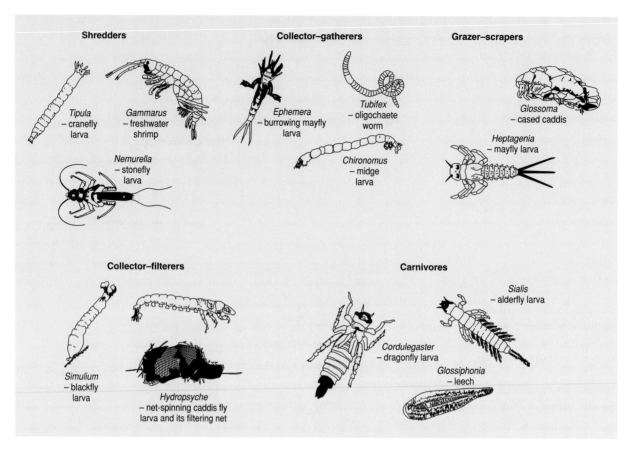

Figure 11.5 Examples of the various categories of invertebrate consumer in freshwater environments.

the rate of decomposition and, moreover, the thickness of water films on decomposing material places absolute limits on mobile microfauna and microflora (protozoa, nematode worms, rotifers and those fungi that have motile stages in their life cycles). In dry soils, such organisms are virtually absent. A continuum can be recognized from dry conditions through waterlogged soils to true aquatic environments. In the former, the amount of water and thickness of water films are of paramount importance, but as we move along the continuum, conditions change to resemble more and more closely those of the bed of an open-water community, where oxygen shortage, rather than water availability, may dominate the lives of the organisms.

... and by feeding mode in aquatic environments

In freshwater ecology the study of detritivores has been concerned less with the size of the organisms than with the ways in which they obtain their food. Cummins (1974) devised a scheme that recognizes four main categories of invertebrate consumer in streams. *Shredders* are detritivores that feed on coarse particulate organic matter (particles > 2 mm in size), and during feeding these serve to fragment the material. Very often

in streams, the shredders, such as cased caddis-fly larvae of *Stenophylax* spp., freshwater shrimps (*Gammarus* spp.) and isopods (e.g. *Asellus* spp.), feed on tree leaves that fall into the stream. *Collectors* feed on fine particulate organic matter (< 2 mm). Two subcategories of collectors are defined. *Collector–gatherers* obtain dead organic particles from the debris and sediments on the bed of the stream, whereas *collector–filterers* sift small particles from the flowing column of water. Some examples are shown in Figure 11.5. *Grazer–scrapers* have mouthparts appropriate for scraping off and consuming the organic layer attached to rocks and stones; this organic layer is comprised of attached algae, bacteria, fungi and dead organic matter adsorbed to the substrate surface. The final invertebrate category is *carnivores*. Figure 11.6 shows the relationships amongst these invertebrate feeding groups and three categories of dead organic matter. This scheme, developed for stream communities, has obvious parallels in terrestrial ecosystems (Anderson, 1987) as well as in other aquatic ecosystems. Earthworms are important shredders in soils, while a variety of crustaceans perform the same role on the sea bed. On the other hand, filtering is common among marine but not terrestrial organisms.

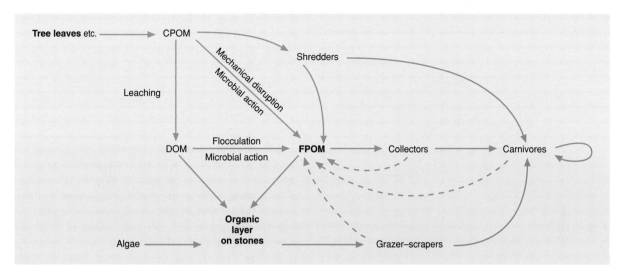

Figure 11.6 A general model of energy flow in a stream. A fraction of coarse particulate organic matter (CPOM) is quickly lost to the dissolved organic matter (DOM) compartment by leaching. The remainder is converted by three processes to fine particulate organic matter (FPOM): (i) mechanical disruption by battering; (ii) processing by microorganisms causing gradual break up; and (iii) fragmentation by the shredders. Note also that all animal groups contribute to FPOM by producing feces (dashed lines). DOM is also converted into FPOM by a physical process of flocculation or via uptake by microorganisms. The organic layer attached to stones on the stream bed derives from algae, DOM and FPOM adsorbed onto an organic matrix.

The feces and bodies of aquatic invertebrates are generally processed along with dead organic matter from other sources by shredders and collectors. Even the large feces of aquatic vertebrates do not appear to possess a characteristic fauna, probably because such feces are likely to fragment and disperse quickly as a result of water movement. Carrion also lacks a specialized fauna – many aquatic invertebrates are omnivorous, feeding for much of the time on plant detritus and feces with their associated microorganisms, but ever ready to tackle a piece of dead invertebrate or fish when this is available. This contrasts with the situation in the terrestrial environment, where both feces and carrion have specialized detritivore faunas (see Sections 11.3.3 and 11.3.5).

detritivore-dominated communities

Some animal communities are composed almost exclusively of detritivores and their predators. This is true not only of the forest floor, but also of shaded streams, the depths of oceans and lakes, and the permanent residents of caves: in short, wherever there is insufficient light for appreciable photosynthesis but nevertheless an input of organic matter from nearby plant communities. The forest floor and shaded streams receive most of their organic matter as dead leaves from trees. The beds of oceans and lakes are subject to a continuous settlement of detritus from above. Caves receive dissolved and particulate organic matter percolating down through soil and rock, together with windblown material and the debris of migrating animals.

11.2.3 The relative roles of decomposers and detritivores

The roles of the decomposers and detritivores in decomposing dead organic matter can be compared in a variety of ways. A comparison of numbers will reveal a predominance of bacteria. This is almost inevitable because we are counting individual cells. A comparison of biomass gives a quite different picture. Figure 11.7 shows the relative amounts of biomass represented in different groups involved in the decomposition of litter on a forest floor (expressed as the relative amounts of nitrogen present). For most of the year, decomposers (microorganisms) accounted for five to 10 times as much of the biomass as the detritivores. The biomass of detritivores varied less through the year because they are less sensitive to climatic change, and they were actually predominant during a period in the winter.

assessing the relative importance of decomposers and detritivores . . .

Unfortunately, the biomass present in different groups of decomposers is itself a poor measure of their relative importance in the process of decomposition. Populations of organisms with short lives and high activity may contribute more to the activities in the community than larger, long-lived, sluggish species (e.g. slugs!) that make a greater contribution to biomass.

Lillebo *et al.* (1999) attempted to distinguish the relative roles, in the

. . . in the decomposition of a salt marsh plant, . . .

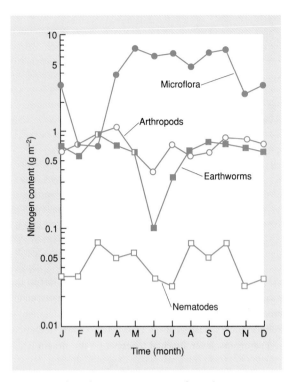

Figure 11.7 The relative importance in forest litter decomposition of microflora in comparison with arthropods, earthworms and nematodes, expressed in terms of their relative content of nitrogen – a measure of their biomass. Microbial activity is much greater than that of detritivores but the latter is more constant through the year. (After Ausmus *et al.*, 1976.)

decomposition of the salt marsh plant *Spartina maritima*, of bacteria, microfauna (e.g. flagellates) and macrofauna (e.g. the snail *Hydrobia ulvae*) by creating artificial communities in laboratory microcosms. At the end of the 99-day study, 32% of the biomass

of *Spartina* leaves remained in the bacteria treatment, whereas only 8% remained when the microfauna and macrofauna were also present (Figure 11.8a). Separate analyses of the mineralization of the carbon, nitrogen and phosphorus content of the leaves also revealed that bacteria were responsible for the majority of the mineralization, but that microfauna and particularly macrofauna enhanced the mineralization rates in the case of carbon and nitrogen (Figure 11.8b).

The decomposition of dead material is not simply due to the sum of the activities of microbes and detritivores: it is largely the result of interaction between the two. The shredding action of detritivores, such as the snail *Hydrobia ulvae* in the experiment of Lillebo *et al.* (1999), usually produces smaller particles with a larger surface area (per unit volume of litter) and thus increases the area of substrate available for microorganism growth. In addition, the activity of fungi may be stimulated by the disruption, through grazing, of competing hyphal networks. Moreover, the activity of both fungi and bacteria may be enhanced by the addition of mineral nutrients in urine and feces (Lussenhop, 1992).

The ways in which the decomposers and detritivores interact might be studied by following a leaf fragment through the process of decomposition, focusing attention on a part of the wall of a single cell. Initially, when the leaf falls to the ground, the piece of cell wall is protected from microbial attack because it lies within the plant tissue. The leaf is now chewed and the fragment enters the gut of, say, an isopod. Here it meets a new microbial flora in the gut and is acted on by the digestive enzymes of the isopod. The fragment emerges, changed by its passage through the gut. It is now part of the isopod's feces and is much more easily attacked by microorganisms, because it has been fragmented and partially digested. While microorganisms are colonizing, it may again be

... in a terrestrial environment, ...

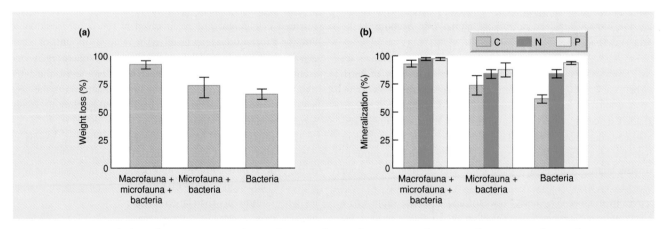

Figure 11.8 (a) Weight loss of *Spartina maritima* leaves during 99 days in the presence of: (i) macrofauna + microfauna + bacteria, (ii) microfauna + bacteria, or (iii) bacteria alone (mean ± SD). (b) Percentage of initial carbon, nitrogen and phosphorus content that was mineralized during 99 days in the three treatments. (After Lillebo *et al.*, 1999.)

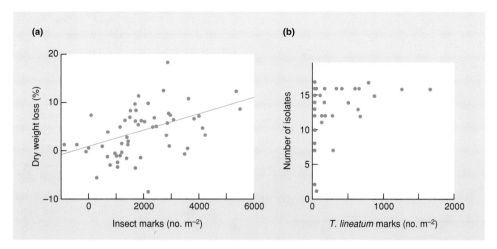

Figure 11.9 Relationships between (a) the decay of standard pieces of dead spruce wood over a 2.5-year period in Finland and the number of insect marks, and (b) the fungal infection rate (number of fungal isolates per standard piece of wood) and number of marks made by the beetle *Tripodendron lineatum*. Dry weight loss and number of insect marks in (a) were obtained by subtracting the values for each wood sample held in a permanently closed net cage from the corresponding value for its counterpart in a control cage that permitted insect entry. In some cases, the dry weight loss of the counterpart wood sample was lower, so the percentage weight loss was negative. This is possible because the number of insect visits does not explain all the variation in dry weight loss. (After Muller *et al.*, 2002.)

eaten, perhaps by a coprophagous springtail, and pass through the new environment of the springtail's gut. Incompletely digested fragments may again appear, this time in springtail feces, yet more easily accessible to microorganisms. The fragment may pass through several other guts in its progress from being a piece of dead tissue to its inevitable fate of becoming carbon dioxide and minerals.

... in a freshwater environment, ... Fragmentation by detritivores plays a key role in terrestrial situations because of the tough cell walls characteristic of vascular plant detritus. The same is true in many freshwater environments where terrestrial litter makes up most of the available detritus. In contrast, detritus at the lowest trophic level in marine environments consists of phytoplankton cells and seaweeds; the former present a high surface area without the need for physical disruption and the latter, lacking the structural polymers of vascular plant cell walls, are prone to fragmentation by physical factors. Rapid decomposition of marine detritus is probably less dependent on fragmentation by invertebrates; shredders are rare in the marine environment compared to its terrestrial and freshwater counterparts (Plante *et al.*, 1990).

... in dead wood ... Dead wood provides particular challenges to colonization by microorganisms because of its patchy distribution and tough exterior. Insects can enhance fungal colonization of dead wood by carrying fungi to their 'target' or by enhancing access of air-disseminated fungal propagules by making holes in the outer bark into the phloem and xylem. Muller *et al.* (2002)

distributed standard pieces of spruce wood (*Picea abies*) on a forest floor in Finland. After 2.5 years, the numbers of insect 'marks' (boring and gnawing) were recorded and were found to be correlated with dry weight loss of the wood (Figure 11.9a). This relationship comes about because of biomass consumption by the insects but also, to an unknown extent, by fungal action that has been enhanced by insect activity. Thus, fungal infection rate was always high when there were more than 400 marks per piece of wood made by the common ambrosia beetle *Tripodendron lineatum* (Figure 11.9b). This species burrows deeply into the sapwood and produces galleries about 1 mm in diameter. Some of the fungal species involved are likely to have been transmitted by the beetle (e.g. *Ceratocystis piceae*) but the invasion of other, air-disseminated types is likely to have been promoted by the galleries left by the beetle.

The enhancement of microbial respiration by the action of detritivores has also been reported in the decomposition of small mammal carcasses. ... and in small mammal carcasses Two sets of insect-free rodent carcasses weighing 25 g were exposed under experimental conditions in an English grassland in the fall. In one set the carcasses were left intact. In the other, the bodies were artificially riddled with tunnels by repeated piercing of the material with a dissecting needle to simulate the action of blowfly larvae in the carcass. The results of this experiment paralleled those of the wood decomposition study above; here, the tunnels enhanced microbial activity (Figure 11.10) by disseminating the microflora as well as increasing the aeration of the carcass.

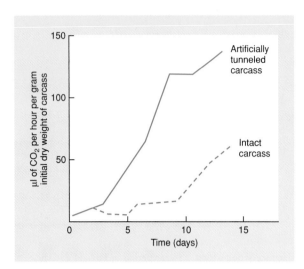

Figure 11.10   The evolution of carbon dioxide (CO_2), a measure of microbial activity, from carcasses of small mammals placed in 'respiration' cylinders and screened from insect attack. One set of carcasses was left intact, while the second set was pierced repeatedly with a dissecting needle to simulate the action of tunneling by blowfly larvae. (After Putman, 1978a.)

11.2.4   Ecological stoichiometry and the chemical composition of decomposers, detritivores and their resources

'ecological
stoichiometry' and
relations between
resources and
consumers

Ecological stoichiometry, defined by Elser and Urabe (1999) as the analysis of constraints and consequences in ecological interactions of the mass balance of multiple chemical elements (particularly the ratios of carbon to nitrogen and of carbon to phosphorus), is an approach that can shed light on the relations between resources and consumers. Many studies have focused on plant–herbivore relations (Hessen, 1997) but the approach is also important when considering decomposers, detritivores and their resources.

There is a great contrast between the chemical composition of dead plant tissue and that of the tissues of the heterotrophic organisms that consume and decompose it. While the major components of plant tissues, particularly cell walls, are structural polysaccharides, these are only of minor significance in the bodies of microorganisms and detritivores. However, being harder to digest than storage carbohydrates and protein, the structural chemicals still form a significant component of detritivore feces. Detritivore feces and plant tissue have much in common chemically, but the protein and lipid contents of detritivores and decomposers are significantly higher than those of plants and feces.

The rate at which dead organic matter decomposes is strongly dependent on its biochemical composition. This is because microbial tissue has very high nitrogen and phosphorus contents, indicative of high requirements for these nutrients. Roughly speaking, the stoichiometric ratios of carbon : nitrogen (C : N) and carbon : phosphorus (C : P) in decomposers are 10 : 1 and 100 : 1, respectively (e.g. Goldman *et al.*, 1987). In other words, a microbial population of 111 g can only develop if there is 10 g of nitrogen and 1 g of phosphorus available. Terrestrial plant material has much higher ratios, ranging from 19 to 315 : 1 for C : N and from 700 to 7000 : 1 for C : P (Enriquez *et al.*, 1993). Consequently, this material can support only a limited biomass of decomposer organisms and the whole pace of the decomposition process will itself be limited by nutrient availability. Marine and freshwater plants and algae tend to have ratios more similar to the decomposers (Duarte, 1992), and their rates of decomposition are correspondingly faster (Figure 11.11a). Figure 11.11b and c illustrate the strong relationships between initial nitrogen and phosphorus concentration in plant tissue and its decomposition rate for a wide range of plant detritus from terrestrial, freshwater and marine species.

decomposition rate
depends on . . .
. . . biochemical
composition . . .

The rate at which dead organic matter decomposes is also influenced by inorganic nutrients, especially nitrogen (as ammonium or nitrate), that are available from the environment. Thus, greater microbial biomass can be supported, and decomposition proceeds faster, if nitrogen is absorbed from outside. For example, grass litter decomposes faster in streams running through tussock grassland in New Zealand that has been improved for pasture (where the water is, in consequence, richer in nitrate) than in 'unimproved' settings (Young *et al.*, 1994).

. . . and mineral
nutrients in the
environment

One consequence of the capacity of decomposers to use inorganic nutrients is that after plant material is added to soil, the level of soil nitrogen tends to fall rapidly as it is incorporated into microbial biomass. The effect is particularly evident in agriculture, where the ploughing in of stubble can result in nitrogen deficiency of the subsequent crop. In other words, the decomposers compete with the plants for inorganic nitrogen. This raises a significant and somewhat paradoxical issue. We have noted that plants and decomposers are linked by an indirect *mutualism* mediated by nutrient recycling – plants provide energy and nutrients in organic form that are used by decomposers, and decomposers mineralize the organic material back to an inorganic form that can again be used by plants. However, stoichiometric constraints on carbon and nutrients also lead to *competition* between the plants and decomposers (usually for nitrogen in terrestrial communities, often

complex relationships
between decomposers
and living plants, . . .

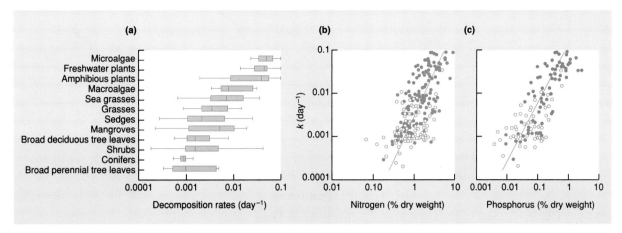

Figure 11.11 (a) Box plots showing the recorded decomposition rates of detritus from different sources. The decomposition rate is expressed as *k* (in log units per day), derived from the equation $W_t = W_0\,e^{-kt}$, which describes the loss in plant dry weight (*W*) with time (*t*) since the initiation of measurements. Boxes encompass the 25 and 75% quartiles of all data from the literature for each plant type. The central line represents the median and bars extend to the 95% confidence limits. The relationships between decomposition rate and the initial concentrations in the tissues (% dry weight) of (b) nitrogen and (c) phosphorus are also shown. Solid lines represent fitted regression lines and open and closed circles represent detritus decomposing on land and submersed, respectively. (After Enriquez *et al.*, 1993.)

for phosphorus in freshwater communities, and either nitrogen or phosphorus in marine communities).

... competition and mutualism

Daufresne and Loreau (2001) developed a model that incorporates both mutualistic and competitive relationships and posed the question 'what conditions must be met for plants and decomposers to coexist and for the ecosystem as a whole to persist?' Their model showed that the plant–decomposer system is generally persistent (both plant and decomposer compartments reach a stable positive steady state) only if decomposer growth is limited by the availability of carbon in the detritus – and this condition can only be achieved if the competitive ability of the decomposers for a limiting nutrient (e.g. nitrogen) was great enough, compared to that of plants, to maintain themselves in a state of carbon limitation. When decomposers were not competitive enough, they became nutrient-limited and the system eventually collapsed. Daufresne and Loreau (2001) note that the few experimental studies so far performed show bacteria can, in fact, outcompete plants for inorganic nutrients.

In contrast to terrestrial plants, the bodies of animals have nutrient ratios that are of the same order as those of microbial biomass; thus their decomposition is not limited by the availability of nutrients, and animal bodies tend to decompose much faster than plant material.

When dead organisms or their parts decompose in or on soil, they begin to acquire the C : N ratio of the decomposers. On the whole, if material with a nitrogen content of less than 1.2–1.3% is added to soil, any available ammonium ions are absorbed. If the material has a nitrogen content greater than 1.8%, ammonium

ions tend to be released. One consequence is that the C : N ratios of soils tend to be rather constant around values of 10; the decomposer system is in general remarkably homeostatic. However, in extreme situations, where the soil is very acid or waterlogged, the ratio may rise to 17 (an indication that decomposition is slow).

It should not be thought that the only activity of the microbial decomposers of dead material is to respire away the carbon and mineralize the remainder. A major consequence of microbial growth is the accumulation of microbial by-products, particularly fungal cellulose and microbial polysaccharides, which may themselves be slow to decompose and contribute to maintaining soil structure.

11.3 Detritivore–resource interactions

11.3.1 Consumption of plant detritus

Two of the major organic components of dead leaves and wood are cellulose and lignin. These pose considerable digestive problems for animal consumers, most of which are not capable of manufacturing the enzymatic machinery to deal with them. Cellulose catabolism (cellulolysis) requires *cellulase* enzymes. Without these, detritivores are unable to digest the cellulose component of detritus, and so cannot derive from it either energy to do work or the simpler chemical modules to use in their own tissue synthesis. Cellulases of animal origin have been definitely identified in remarkably few species, including a cockroach and some higher termites in the subfamily Nasutitermitinae (Martin, 1991) and the shipworm *Teledo navalis*, a marine bivalve mollusc

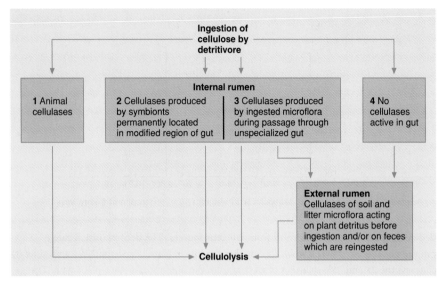

Figure 11.12 The range of mechanisms that detritivores adopt for digesting cellulose (cellulolysis). (After Swift *et al.*, 1979.)

that bores into the hulls of ships. In these organisms, cellulolysis poses no special problems.

most detritivores rely on microbial cellulases – they do not have their own

The majority of detritivores, lacking their own cellulases, rely on the production of cellulases by associated decomposers or, in some cases, protozoa. The interactions range from *obligate mutualism* between a detritivore and a specific and permanent gut microflora or microfauna, through *facultative mutualism*, where the animals make use of cellulases produced by a microflora that is ingested with detritus as it passes through an unspecialized gut, to animals that ingest the metabolic products of external cellulase-producing microflora associated with decomposing plant remains or feces (Figure 11.12).

woodlice rely on ingested microbial organisms

A wide range of detritivores appear to have to rely on the exogenous microbial organisms to digest cellulose. The invertebrates then consume the partially digested plant detritus along with its associated bacteria and fungi, no doubt obtaining a significant proportion of the necessary energy and nutrients by digesting the microflora itself. These animals, such as the springtail *Tomocerus*, can be said to be making use of an 'external rumen' in the provision of assimilable materials from indigestible plant remains. This process reaches a pinnacle of specialization in ambrosia beetles and in certain species of ants and termites that 'farm' fungus in specially excavated gardens (see Chapter 13).

cockroaches and termites rely on bacteria and protozoa

Clear examples of obligate mutualism are found amongst certain species of cockroach and termite that rely on symbiotic bacteria or protozoa for the digestion of structural plant polysaccharides. Nalepa *et al.* (2001) describe

the evolution of digestive mutualisms among the Dictyoptera (cockroaches and termites) from cockroach-like ancestors in the Upper Carboniferous that fed on rotting vegetation and relied on an 'external rumen'. The next stages involved progressive internalization of the microbiota associated with plant detritus, from indiscriminate coprophagy (feeding on feces of a variety of detritivorous species) through increasing levels of gregarious and social behavior that ensured neonates received appropriate innocula of gut biota. When proctodeal trophallaxis (the direct transfer of hindgut fluids from the rectal pouch of the parent to the mouth of the newborn young) evolved in certain cockroaches and lower termites, some microbes were captured and became ecologically dependent on the host. This specialized state ensured the direct transfer of the internal rumen, particularly those components that would degenerate if exposed to the external environment. In lower termites, such as *Eutermes*, symbiotic protozoa may make up more than 60% of the insect's body weight. The protozoa are located in the hindgut, which is dilated to form a rectal pouch. They ingest fine particles of wood, and are responsible for extensive cellulolytic activity, though bacteria are also implicated. Termites feeding on wood generally show effective digestion of cellulose but not of lignin, except for *Reticulitermes*, which has been reported to digest 80% or more of the lignin present in its food.

why no animal cellulases?

Given the versatility apparent in the evolutionary process, it may seem surprising that so few animals that consume plants can produce their own cellulase enzymes. Janzen (1981) has argued that cellulose is the master construction material of plants 'for the same reason that we construct houses of concrete in areas of high termite activity'. He views the use of cellulose, therefore, as a defense against attack, since higher organisms can rarely digest it unaided. From a different perspective,

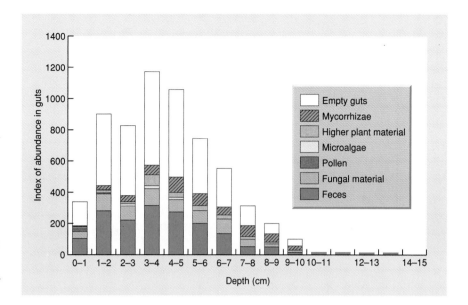

Figure 11.13 The distribution of gut content categories of springtails ($n = 6255$) (Collembola; all species combined) in relation to depth in the litter/soil of beech forests in Belgium. (After Ponge, 2000.)

it has been suggested that cellulolytic capacity is uncommon simply because it is a trait that is rarely advantageous for animals to possess (Martin, 1991). For one thing, diverse bacterial communities are commonly found in hindguts and this may have facilitated the evolution of symbiont-mediated cellulolysis. For another, the diets of plant-eaters generally suffer from a limited supply of critical nutrients, such as nitrogen and phosphorus, rather than of energy, which cellulolysis would release. This imposes the need for processing large volumes of material to extract the required quantities of nutrients, rather than extracting energy efficiently from small volumes of material.

detritus and microbial organisms are typically consumed together Because microbes, plant detritus and animal feces are often very intimately associated, there are inevitably many generalist consumers that ingest all these resources. In other words, many animals simply cannot manage to take a mouthful of one without the others. Figure 11.13 shows the various components of the gut contents of 45 springtail species (all species combined) collected at different depths in the litter and soil of beech forests in Belgium. Species that occurred in the top 2 cm lived in a habitat derived from beech leaves at various stages of microbial decomposition where microalgae, feces of slugs and woodlice, and pollen grains were also common. Their diets contained all the local components but little of the very abundant beech litter. At intermediate depths (2–4 cm) the springtails ate mainly spores and hyphae of fungi together with invertebrate feces (particularly the freshly deposited feces of enchytraeid pot worms). At the lowest depths, their diets consisted mainly of mycorrhizal material (the springtails browsed the fungal part of the fungal/plant root assemblage) and higher plant detritus (mainly derived from plant roots). There were clear interspecific differences

in both depth distributions and the relative importance of the different dietary components, and some species were more specialized feeders than others (e.g. *Isotomiella minor* ate only feces whereas *Willemia aspinata* ate only fungal hyphae). But most consumed more than one of the potential diet components and many were remarkably generalist (e.g. *Protaphorura eichhorni* and *Mesaphorura yosii*) (Ponge, 2000).

11.3.2 Consumption of fallen fruit

fruit-flies and rotten fruit Of course, not all plant detritus is so difficult for detritivores to digest. Fallen fruit, for example, is readily exploited by many kinds of opportunist feeders, including insects, birds and mammals. However, like all detritus, decaying fruits have associated with them a microflora, in this case mainly dominated by yeasts. Fruit-flies (*Drosophila* spp.) specialize at feeding on these yeasts and their by-products; and in fruit-laden domestic compost heaps in Australia, five species of fruit-fly show differing preferences for particular categories of rotting fruit and vegetables (Oakeshott *et al.*, 1982). *Drosophila hydei* and *D. immigrans* prefer melons, *D. busckii* specializes on rotting vegetables, while *D. simulans* is catholic in its tastes for a variety of fruits. The common *D. melanogaster*, however, shows a clear preference for rotting grapes and pears. Note that rotting fruits can be highly alcoholic. Yeasts are commonly the early colonists and the fruit sugars are fermented to alcohol, which is normally toxic, eventually even to the yeasts themselves. *D. melanogaster* tolerates such high levels of alcohol because it produces large quantities of alcohol dehydrogenase (ADH), an enzyme that breaks down ethanol to harmless metabolites. Decaying vegetables produce

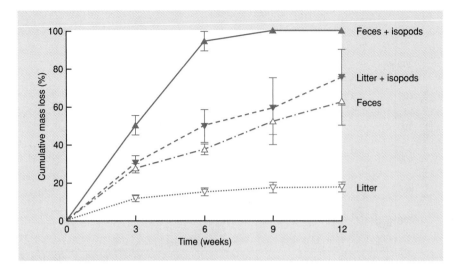

Figure 11.14 The cumulative mass loss of beech leaf litter and feces of grazing caterpillars (*Operophthera fagata*) in the presence and absence of feeding by isopods. Standard errors are shown. (After Zimmer & Topp, 2002.)

little alcohol, and *D. busckii*, which is associated with them, produces very little ADH. Intermediate levels of ADH were produced by the species preferring moderately alcoholic melons. The boozy *D. melanogaster* is also associated with winery wastes!

11.3.3 Feeding on invertebrate feces

isopods do best when they can eat their own feces

A large proportion of dead organic matter in soils and aquatic sediments may consist of invertebrate feces, which generalist detritivores often include in their diets. Some of the feces derive from grazing insects. In the laboratory, the feces of caterpillars of *Operophthera fagata* that had grazed leaves of beech (*Fagus sylvatica*) under the influence of leaching and microbial degradation decomposed faster than leaf litter itself; however, the decomposition rate was much enhanced when detritivorous isopods (*Porcellio scabar* and *Oniscus asellus*) fed on the feces (Figure 11.14). Thus, rates of decomposition and nutrient release into the soil from grazer feces can be increased through the feeding activity of coprophagous detritivores.

'coprophagy' may be more valuable when detrital quality is low

Feces of detritivores are common in many environments. It some cases, reingestion of feces may be critically important, by providing essential micronutrients or highly assimilable resources. In most cases, however, there are probably not marked nutritive benefits of feeding on feces compared with the detritus from which the feces were derived. Thus, the isopod *Porcellio scabar* gained no more from feeding on its feces, even when these were experimentally inoculated with microbes, than from feeding directly on the leaf litter of alder (*Alnus glutinosa*) (Kautz *et al.*, 2002). On the other hand, in the case of the less nutritionally preferred leaves of oak (*Quercus robur*), inoculated feces provided a small but significant increase in growth rate compared to the parent oak leaf material. Coprophagy may be more valuable when detrital quality is particularly low.

A remarkable story of coprophagy was unraveled in some small bog lakes in northeast England (MacLachlan *et al.*, 1979). These murky water bodies have restricted light penetration because of dissolved humic substances derived from the surrounding sphagnum peat, and they are characteristically poor in plant nutrients. Primary production is insignificant. The main organic input consists of poor-quality peat particles resulting from wave erosion of the banks. By the time the peat has settled from suspension it has been colonized, mainly by bacteria, and its caloric and protein contents have increased by 23 and 200%, respectively. These small particles are consumed by *Chironomus lugubris* larvae, the detritivorous young of a nonbiting chironomid midge. The feces the larvae produce become quite richly colonized by fungi, microbial activity is enhanced, and they would seem to constitute a high-quality food resource. But they are not reingested by *Chironomus* larvae, mainly because they are too large and too tough for its mouthparts to deal with. However, another common inhabitant of the lake, the small crustacean *Chydorus sphaericus*, finds chironomid feces very attractive. It seems always to be associated with them and probably depends on them for food. *Chydorus* clasps the chironomid fecal pellet just inside the valve of its carapace and rotates it while grazing the surface, causing gradual disintegration. In the laboratory, the presence of chydorids has been shown to speed up dramatically the breakdown of large *Chironomus* pellets to smaller particles. The final and most intriguing twist to the story is that the fragmented chironomid feces (mixed probably with chydorid feces) are now small enough to be used again by *Chironomus*. It is probable that *Chironomus lugubris* larvae grow faster when in the presence of *Chydorus sphaericus* because of

a midge and a cladoceran eat each other's feces

the availability of suitable fecal material to eat. The interaction benefits both participants.

11.3.4 Feeding on vertebrate feces

carnivore dung is attacked mainly by bacteria and fungi

The dung of carnivorous vertebrates is relatively poor-quality stuff. Carnivores assimilate their food with high efficiency (usually 80% or more is digested) and their feces retain only the least digestible components. In addition, carnivores are necessarily much less common than herbivores, and their dung is probably not sufficiently abundant to support a specialist detritivore fauna. What little research has been done suggests that decay is effected almost entirely by bacteria and fungi (Putman, 1983).

'autocoprophagy' among mammalian herbivores

In contrast, herbivore feces still contain an abundance of organic matter. Autocoprophagy (reingesting one's own feces) is quite a widespread habit among small to medium-sized mammalian herbivores, being reported from rabbits and hares, rodents, marsupials and a primate (Hirakawa, 2001). Many species produce soft and hard feces, and it is the soft feces that are usually reingested (directly from the anus), being rich in vitamins and microbial protein. If prevented from reingestion, many animals exhibit symptoms of malnutrition and grow more slowly.

herbivore dung supports its own characteristic detritivores

Herbivore dung is also sufficiently thickly spread in the environment to support its own characteristic fauna, consisting of many occasional visitors but with several specific dung-feeders. Dung removal varies both seasonally and spatially. In tropical and in warm temperate regions most activity occurs during summer rainfall, whereas in Mediterranean-type climates dung removal is highest during spring after the winter rainfall and again in mid-summer when temperatures are high (Davis, 1996). Dung removal also occurs at greater rates in unshaded situations and is faster on sand than on harder, more compacted clay soils (Davis, 1996). A wide range of animals are involved, including earthworms, termites and, in particular, beetles.

A good example of the predominant role of beetles is provided by elephant dung. Two main patterns of decay can be recognized, related to the wet and dry seasons. During the rains, within a few minutes of dung deposition the area is alive with beetles. The adult dung beetles feed on the dung but they also bury large quantities along with their eggs to provide food for the developing larvae. For example, the large African dung beetle, *Heliocopris dilloni*, carves a lump out of fresh dung and rolls this away for burying several meters from the original dung pile. Each beetle buries sufficient dung for several eggs. Once underground, a small quantity of dung is shaped into a cup, and lined with soil; a single egg is laid and then more dung is added to produce a sphere that is almost entirely covered with a thin layer of soil. A small area at the top of the ball, close to the location of the egg, is left clear of soil, possibly to facilitate gas exchange. After hatching, the larva feeds by a rotating action in the dung ball, excavating a hollow, and, incidentally, feeding on its own feces as well as the elephant's (Figure 11.15). When all the food supplied by its parents is used up, the larva covers the inside of its cell with a paste of its own feces, and pupates.

a diversity of dung beetles

The full range of tropical dung beetles in the family Scarabeidae vary in size from a few millimeters in length up to the 6 cm long *Heliocopris*. Not all remove dung and bury it at a distance from the dung pile. Some excavate their nests at various depths immediately below the pile, while others build nest chambers within the dung pile itself. Beetles in other families do not construct chambers but simply lay their eggs in the dung, and their larvae feed and grow within the dung mass until fully developed, when they move away to pupate in the soil. The beetles associated with elephant dung in the wet season may remove 100% of the dung pile. Any left may be processed by other detritivores such as flies and termites, as well as by decomposers.

Dung that is deposited in the dry season is colonized by relatively few beetles (adults emerge only in the rains). Some microbial activity is evident but this soon declines as the feces dry out. Rewetting during the rains stimulates more microbial activity but beetles do not exploit old dung. In fact a dung pile deposited in the dry season may persist for longer than 2 years, compared with 24 h or less for one deposited during the rains.

Australian cow dung poses a problem

Bovine dung has provided an extraordinary and economically very important problem in Australia. During the past two centuries the cow population increased from just seven individuals (brought over by the first English colonists in 1788) to 30 million or so. These produce some 300 million dung pats per day, covering as much as 6 million acres per year with dung. Deposition of bovine dung poses no particular problem elsewhere in the world, where bovines have existed for millions of years and have an associated fauna that exploits the fecal resources. However, the largest herbivorous animals in Australia, until European colonization, were marsupials such as kangaroos. The native detritivores that deal with the dry, fibrous dung pellets that these leave cannot cope with cow dung, and the loss of pasture under dung has imposed a huge economic burden on Australian agriculture. The decision was therefore made in 1963 to establish in Australia beetles of African origin, able to dispose of bovine dung in the most important places and under the most prevalent conditions where cattle are raised (Waterhouse, 1974); more than 20 species have been introduced (Doube *et al.*, 1991).

Figure 11.15 (a) An African dung beetle rolling a ball of dung. (Courtesy of Heather Angel.) (b) The larva of the dung beetle *Heliocopris* excavates a hollow as it feeds within the dung ball. (After Kingston & Coe, 1977.)

Adding to the problem, Australia is plagued by native bushflies (*Musca vetustissima*) and buffalo flies (*Haematobia irritans exigua*) that deposit eggs on dung pats. The larvae fail to survive in dung that has been buried by beetles, and the presence of beetles has been shown to be effective at reducing fly abundance (Tyndale-Biscoe & Vogt, 1996). Success depends on dung being buried within about 6 days of production, the time it takes for the fly egg (laid on fresh dung) to hatch and develop to the pupal stage. Edwards and Aschenborn (1987) surveyed the nesting behavior in southern Africa of 12 species of dung beetles in the genus *Onitis*. They concluded that *O. uncinatus* was a prime candidate for introduction to Australia for fly-control purposes, since substantial amounts of dung were buried on the first night after pad colonization. The least suitable species, *O. viridualus*, spent several days constructing a tunnel and did not commence burying until 6–9 days had elapsed.

11.3.5 Consumption of carrion

When considering the decomposition of dead bodies, it is helpful to distinguish three categories of organisms that attack carcasses. As before, both decomposers and invertebrate detritivores have a role to play. For example, the tenebrionid beetles *Argoporis apicalis* and *Cryptadius tarsalis* are particularly abundant on islands in the Gulf of California where large colonies of seabirds nest; here they feed on bird carcasses, as well as fish debris associated with the bird colonies (Sanchez-Pinero & Polis, 2000). In the case of carrion feeding, however, scavenging vertebrates are often also of considerable importance. Many carcasses of a size to make a single meal for one of a few of these scavenging detritivores will be removed completely within a very short time of death, leaving nothing for bacteria,

many carnivores are opportunistic carrion-feeders . . .

fungi or invertebrates. This role is played, for example, by arctic foxes and skuas in polar regions, by crows, gluttons and badgers in temperate areas, and by a wide variety of birds and mammals, including kites, jackals and hyenas, in the tropics.

... and vice versa

The chemical composition of the diet of carrion-feeders is quite distinct from that of other detritivores, and this is reflected in their complement of enzymes. Carbohydrase activity is weak or absent, but protease and lipase activity is vigorous. Carrion-feeding detritivores possess basically the same enzymatic machinery as carnivores, reflecting the chemical identity of their food. In fact, many species of carnivore (such as lions, *Panthera leo*) are also opportunistic carrion-feeders (DeVault & Rhodes, 2002) whilst classic carrion-feeders such as hyenas (*Crocuta crocuta*) sometimes operate as carnivores.

the arctic fox: a facultative carrion-feeder

Arctic foxes (*Alopex lagopus*) illustrate how the diet of facultative carrion-feeders can vary with food availability. Lemmings (*Dicrostonyx* and *Lemmus* spp.) are the live prey of foxes over much of their range and for much of the time (Elmhagen *et al.*, 2000). However, lemming populations go through dramatic population cycles (see Chapter 14), forcing the foxes to switch to alternative foods such as migratory birds and their eggs (Samelius & Alisauskas, 2000). In winter, marine foods become available when foxes can move onto the sea ice and scavenge carcasses of seals killed by polar bears. Roth (2002) investigated the extent to which foxes switched to carrion feeding in winter by comparing the ratios of carbon isotopes (^{13}C : ^{12}C) of suspected food (marine organisms have characteristically higher ratios than terrestrial organisms) and of fox hair (since carbon isotope signatures of predator tissue reflect the ratios of the prey consumed). Figure 11.16 shows that in three of the 4 years of the study the isotope signature of fox hair samples was much increased in winter, as expected if seal carrion was a major component of the diet. In the winter of 1994, however, a marked shift was not evident and it is of interest that lemming density was high at this time. It seems that foxes switched to seal carrion when the formation of sea ice made this possible, but only when alternative prey were not available.

seasonal variation in invertebrate and microbial activity

The relative roles played by decomposers, invertebrates and vertebrates are influenced by factors that affect the speed with which carcasses are discovered by scavengers in relation to the rate at which they disappear through microbial and invertebrate activity. This is illustrated for small rodent carcasses whose disappearance/decomposition was monitored in the Oxfordshire countryside in both the summer–fall and winter–spring periods (Figure 11.17). There are two points to note. First, the rate at which carcasses were removed was faster during the summer and fall, reflecting a greater scavenger activity at this time (presumably

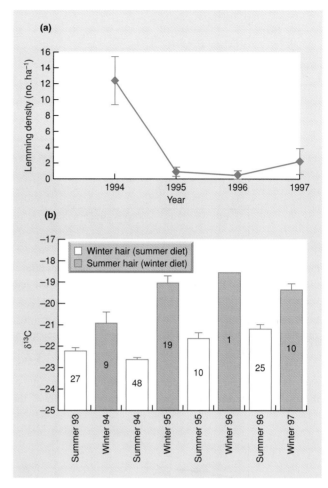

Figure 11.16 (a) Annual changes in lemming density in the summer, near Cape Churchill in Manitoba, Canada, and (b) carbon isotope ratios (mean ± SE) of fox hair in the winter (reflecting summer diet) and in the summer (reflecting winter diet). Numbers on the bars indicate sample sizes. (After Roth, 2002.)

because of higher scavenger population densities and/or higher feeding rates – these were not monitored in the study). Secondly, a greater percentage of the rodent bodies were removed in the winter–spring period, albeit over a longer timescale. At a time when microbial decay proceeds most slowly, all the carcasses persisted for long enough to be found by scavengers. During the summer and fall, decomposition was much more rapid and any carcass that was undiscovered for 7 or 8 days would have been largely decomposed and removed by bacteria, fungi and invertebrate detritivores.

specialist consumers of bone, hair and feathers

Certain components of animal corpses are particularly resistant to attack and are the slowest to disappear. However, some consumer species possess the enzymes to deal with them. For example, the blowfly larvae of *Lucilia* species produce a collagenase that can digest the collagen

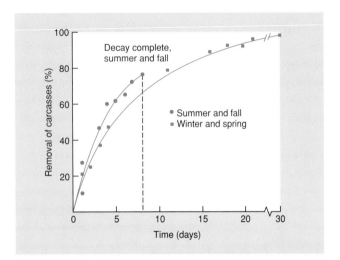

Figure 11.17 The rate of removal of small mammal corpses in the Oxfordshire (UK) countryside in two periods: summer–fall and winter–spring. (After Putman, 1983.)

and elastin present in tendons and soft bones. The chief constituent of hair and feathers, keratin, forms the basis of the diet of species characteristic of the later stages of carrion decomposition, in particular tineid moths and dermestid beetles. The midgut of these insects secretes strong reducing agents that break the resistant covalent links binding together peptide chains in the keratin. Hydrolytic enzymes then deal with the residues. Fungi in the family Onygenaceae are specialist consumers of horn and feathers. It is the corpses of larger animals that generally provide the widest variety of resources and thus attract the greatest diversity of carrion consumers (Doube, 1987). In contrast, the carrion community associated with dead snails and slugs consists of a relatively small number of sarcophagid and calliphorid flies (Kneidel, 1984).

One group of carrion-feeding invertebrates deserves special attention – the burying beetles (*Nicrophorus* spp.) (Scott, 1998). These species live exclusively on carrion on which they play out their extraordinary life history. Adult *Nicrophorus*, using their sensitive chemoreceptors, arrive at the carcass of a small mammal or bird within an hour or two of death. The beetle may tear flesh from the corpse and eat it or, if decomposition is sufficiently advanced, consume blowfly larvae instead. However, should a burying beetle arrive at a completely fresh corpse it sets about burying it where it lies, or may drag the body (many times its own weight) for several meters before starting to dig. It works beneath the corpse, painstakingly excavating and dragging the small mammal down little by little until it is completely underground (Figure 11.18). The various species of *Nicrophorus* vary in body size (and thus the size of corpse utilized), reproductive period (and

remarkable burying beetles

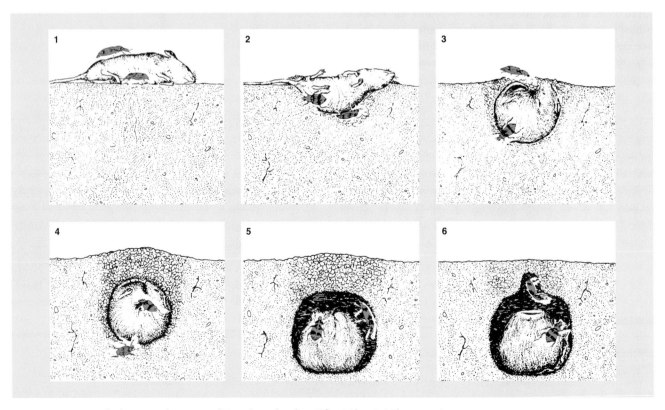

Figure 11.18 Burial of a mouse by a pair of *Nicrophorus* beetles. (After Milne & Milne, 1976.)

thus the season of activity), diel activity (some are diurnal, some crepuscular and some nocturnal) and the habitat they use (coniferous forest, hardwood forest, field, marsh or generalist) (Scott, 1998). Some species, such as *N. vespilloides*, only just cover the corpse, while others, including *N. germanicus*, may bury it to a depth of 20 cm. During the excavation, other burying beetles are likely to arrive. Competing individuals of the same or other species are fiercely repulsed, sometimes leading to the death of one combatant. A prospective mate, on the other hand, is accepted and the male and female work on together.

The buried corpse is much less susceptible to attack by other invertebrates than it was while on the surface. Additional protection is provided, under some circumstances, by virtue of a mutualistic relationship between the beetles and a species of mite, *Poecilochirus necrophori*, which invariably infests adult burying beetles, hitching a ride to a suitable carrion source. When the carcass is first buried the beetle systematically removes its hair and this clears it of virtually all the eggs of blowflies. However, if the carcass is buried only shallowly, flies will often lay more eggs and maggots will compete with the beetle larvae. It is now that the presence of mites has a beneficial effect. By piercing and consuming the fly eggs, the mites keep the carcass free of the beetle's competitors and dramatically improve beetle brood success (Wilson, 1986). Both adults, or sometimes just the female, remain in the chamber and provide parental care. A conical depression is prepared in the top of the meat-ball, into which droplets of partially digested meat are regurgitated. Older larvae are able to feed themselves but only when their offspring are ready to pupate do the adults force their way out through the soil and fly away.

carrion feeders on the sea bed

We have already noted that in freshwater environments carrion lack a specialized fauna. However, specialist carrion-feeders are found on the sea bed in very deep parts of the oceans. As the detritus sinks through very deep water all but the largest particles of organic matter are completely decomposed before they reach the bottom. In contrast, the occasional body of a fish, mammal or large invertebrate does settle on the sea bed. A remarkable diversity of scavengers exist there, though at low density, and these possess several characteristics that match a way of life in which meals are well spread out in space and time. For example, Dahl (1979) described several genera of deep sea gammarid crustaceans which, unlike their relatives at shallower depths and in fresh water, possess dense bundles of exposed chemosensory hairs that sense food, and sharp mandibles that can take large bites from carrion. These animals also have the capacity to gorge themselves far beyond what is normal in amphipods. Thus *Paralicella* possesses a soft body wall that can be stretched when feeding on a large meal so that the animal swells to two or three times its normal size, and *Hirondella* has a midgut that expands to fill almost the entire abdominal cavity and in which it can store meat.

11.4 Conclusion

Decomposer communities are, in their composition and activities, as diverse as or more diverse than any of the communities more commonly studied by ecologists. Generalizing about them is unusually difficult because the range of conditions experienced in their lives is so varied. As in all natural communities, the inhabitants not only have specialized requirements for resources and conditions, but their activities change the resources and conditions available for others. Most of this happens hidden from the view of the observer, in the crevices and recesses of soil and litter and in the depths of water bodies.

Despite these difficulties, some broad generalizations may be made.

1 Decomposers and detritivores tend to have low levels of activity when temperatures are low, aeration is poor, soil water is scarce and conditions are acid.

2 The structure and porosity of the environment (soil or litter) is of crucial importance, not only because it affects the factors listed in point 1 but because many of the organisms responsible for decomposition must swim, creep, grow or force their way through the medium in which their resources are dispersed.

3 The activities of the decomposers and detritivores are intimately interlocked, and may in some cases be synergistic. For this reason, it is very difficult to unravel their relative importance in the decomposition process.

4 Many of the decomposers and detritivores are specialists and the decay of dead organic matter results from the combined activities of organisms with widely different structures, forms and feeding habits.

5 Organic matter may cycle repeatedly through a succession of microhabitats within and outside the guts and feces of different organisms, as they are degraded from highly organized structures to their eventual fate as carbon dioxide and mineral nutrients.

6 The activity of decomposers unlocks the mineral resources such as phosphorus and nitrogen that are fixed in dead organic matter. The speed of decomposition will determine the rate at which such resources are released to growing plants (or become free to diffuse and thus to be lost from the ecosystem). This topic is taken up and discussed in Chapter 18.

7 Many dead resources are patchily distributed in space and time. An element of chance operates in the process of their colonization; the first to arrive have a rich resource to exploit, but the successful species may vary from dung pat to dung pat, and from corpse to corpse. The dynamics of competition between exploiters of such patchy resources require their own particular mathematical models (see Chapter 8). Because detritus is often an 'island' in a sea of quite different habitat, its study is conceptually similar to that discussed in Chapter 21 under the heading of island biogeography (see Section 21.5).

8 Finally, it may be instructive at this point to switch the emphasis away from the success with which decomposers and detritivores deal with their resources. It is, after all, the failure of organisms to decompose wood rapidly that makes the existence of forests possible! Deposits of peat, coal and oil are further testaments to the failures of decomposition.

Summary

We distinguish two groups of organisms that make use of dead organic matter (saprotrophs): decomposers (bacteria and fungi) and detritivores (animal consumers of dead matter). These do not control the rate at which their resources are made available or regenerate; they are dependent on the rate at which some other force (senescence, illness, fighting, the shedding of leaves by trees) releases the resource on which they live. They are donor controlled. Nevertheless, it is possible to see an indirect 'mutualistic' effect through the release of nutrients from decomposing litter, which may ultimately affect the rate at which trees produce more litter.

Immobilization occurs when an inorganic nutrient element is incorporated into an organic form – primarily during the growth of green plants. Conversely, decomposition involves the release of energy and the *mineralization* of chemical nutrients – the conversion of elements from an organic to inorganic form. Decomposition is defined as the gradual disintegration of dead organic matter and is brought about by both physical and biological agencies. It culminates, often after a reasonably predictable succession of colonizing decomposers, with complex energy-rich molecules being broken down into carbon dioxide, water and inorganic nutrients.

Most microbial decomposers are quite specialized, as are the tiny consumers of bacteria and fungi (microbivores), but detritivores are more often generalists. The larger the detritivore, the less able it is to distinguish between microbes as food and the detritus on which these are growing. We discuss the relative roles in decomposition of decomposers and detritivores in terrestrial, freshwater and marine environments.

The rate at which dead organic matter decomposes is strongly dependent on its biochemical composition and on the availability of mineral nutrients in the environment. Two of the major organic components of dead leaves and wood are cellulose and lignin. These pose considerable digestive problems for animal consumers, most of which are not capable of manufacturing the enzymatic machinery to deal with them. Most detritivores depend on microbial organisms to digest cellulose, in a variety of increasingly intimate associations. Dead fruit is a lot easier for detritivores to deal with.

Feces and carrion are abundant dead organic resources in all environments and, once again, microbial organisms and detritivores both play important roles. Many detritivores feed on feces, and the dung of herbivores (but not carnivores) supports its own characteristic fauna. Similarly, many carnivores are opportunistic feeders on carrion but there is also a specialized carrion-feeding fauna.

Decomposer communities are, in their composition and activities, as diverse as or more diverse than any of the communities more commonly studied by ecologists.

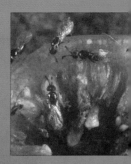

Chapter 12
Parasitism and Disease

12.1 Introduction: parasites, pathogens, infection and disease

Previously, in Chapter 9, we defined a parasite as an organism that obtains its nutrients from one or a very few host individuals, normally causing harm but not causing death immediately. We must follow this now with some more definitions, since there are a number of related terms that are often misused, and it is important not to do so.

When parasites colonize a host, that host is said to harbor an *infection*. Only if that infection gives rise to symptoms that are clearly harmful to the host should the host be said to have a *disease*. With many parasites, there is a presumption that the host can be harmed, but no specific symptoms have as yet been identified, and hence there is no disease. 'Pathogen' is a term that may be applied to any parasite that gives rise to a disease (i.e. is 'pathogenic'). Thus, measles and tuberculosis are infectious diseases (combinations of symptoms resulting from infections). Measles is the result of a measles virus infection; tuberculosis is the result of a bacterial (*Mycobacterium tuberculosis*) infection. The measles virus and *M. tuberculosis* are pathogens. But measles is not a pathogen, and there is no such thing as a tuberculosis infection.

Parasites are an important group of organisms in the most direct sense. Millions of people are killed each year by various types of infection, and many millions more are debilitated or deformed (250 million cases of elephantiasis at present, over 200 million cases of bilharzia, and the list goes on). When the effects of parasites on domesticated animals and crops are added to this, the cost in terms of human misery and economic loss becomes immense. Of course, humans make things easy for the parasites by living in dense and aggregated populations and forcing their domesticated animals and crops to do the same. One of the key questions we will address in this chapter is: 'to what extent are animals and plant populations *in general* affected by parasitism and disease?'

Parasites are also important numerically. An organism in a natural environment that does *not* harbor several species of parasite is a rarity. Moreover, many parasites and pathogens are host-specific or at least have a limited range of hosts. Thus, the conclusion seems unavoidable that more than 50% of the species on the earth, and many more than 50% of individuals, are parasites.

12.2 The diversity of parasites

The language and jargon used by plant pathologists and animal parasitologists are often very different, and there are important differences in the ways in which animals and plants serve as habitats for parasites, and in the way they respond to infection. But for the ecologist, the differences are less striking than the resemblances, and we therefore deal with the two together. One distinction that *is* useful, though, is that between microparasites and macroparasites (Figure 12.1) (May & Anderson, 1979).

Microparasites are small and often intracellular, and they multiply directly within their host where they are often extremely numerous. Hence, it is generally difficult, and usually inappropriate, to estimate precisely the number of microparasites in a host. The number of infected hosts, rather than the number of parasites, is the parameter usually studied. For example, a study of a measles epidemic will involve counting the number of cases of the disease, rather than the number of particles of the measles virus.

Macroparasites have a quite different biology: they grow but do not multiply in their host, and then produce specialized infective stages (microparasites do not do this) that are released to infect new hosts. The macroparasites of animals mostly live on the body or in the body cavities (e.g. the gut), rather than within the host cells. In plants, they are generally intercellular. It is usually possible to count or at least estimate the numbers of macroparasites in or on a host (e.g. worms in an intestine or lesions on a leaf), and the numbers of parasites as well as the numbers of infected hosts can be studied by the epidemiologist.

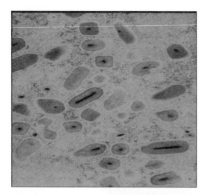

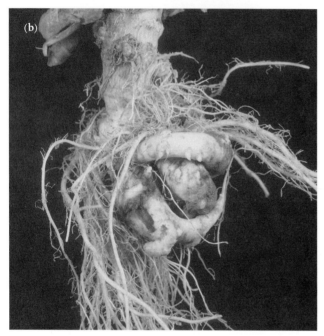

Figure 12.1 Plant and animal micro- and macroparasites. (a) An animal microparasite: particles of *the Plodia interpunctella* granulovirus (each within its protein coat) within a cell of their insect host. (b) A plant microparasite: 'club-root disease' of crucifers caused by multiplication of *Plasmodiophora brassicae*. (c) An animal macroparasite: a tapeworm. (d) A plant macroparasite: powdery mildew lesions. Reproduced by permission of: (a) Dr Caroline Griffiths; (b) Holt Studios/Nigel Cattlin; (c) Andrew Syred/Science Photo Library; and (d) Geoff Kidd/Science Photo Library.

direct and indirect life cycles: vectors

Cutting across the distinction between micro- and macroparasites, parasites can also be subdivided into those that are transmitted directly from host to host and those that require a vector or intermediate host for transmission and therefore have an indirect life cycle. The term 'vector' signifies an animal carrying a parasite from host to host, and some vectors play no other role than as a carrier; but many vectors are also intermediate hosts within which the parasite grows and/or multiplies. Indeed, parasites with indirect life cycles may elude the simple micro/macro distinction. For example, schistosome parasites spend part of their life cycle in a snail and part in a vertebrate (in some cases a human). In the snail, the parasite multiplies and so behaves as a microparasite, but in an infected human the parasite grows and produces eggs but does not itself multiply, and so behaves as a macroparasite.

12.2.1 Microparasites

Probably the most obvious microparasites are the bacteria and viruses that infect animals (such as the measles virus and the typhoid bacterium) and plants (e.g. the yellow net viruses of beet and tomato and the bacterial crown gall disease). The other major group of microparasites affecting animals is the protozoa (e.g. the trypanosomes that cause sleeping sickness and the *Plasmodium* species that cause malaria). In plant hosts some of the simpler fungi behave as microparasites.

The transmission of a microparasite from one host to another can be in some cases almost instantaneous, as in venereal disease and the short-lived infective agents carried in the water droplets of coughs and sneezes (influenza, measles, etc.). In other species the parasite may spend an extended dormant period 'waiting' for its new host. This is the case with the ingestion of food or water contaminated with the protozoan *Entamoeba histolytica*, which causes amoebic dysentery, and with the plant parasite *Plasmodiophora brassicae*, which causes 'club-root disease' of crucifers.

Alternatively, a microparasite may depend on a vector for its spread. The two most economically important groups of vector-transmitted protozoan parasites of animals are the trypanosomes, transmitted by various vectors including tsetse flies (*Glossina* spp.) and causing sleeping sickness in humans and nagana in domesticated (and wild) mammals, and the various species of *Plasmodium*, transmitted by anopheline mosquitoes and causing malaria. In both these cases, the flies act also as intermediate hosts, i.e. the parasite multiplies within them.

Many plant viruses are transmitted by aphids. In some 'nonpersistent' species (e.g. cauliflower mosaic virus), the virus is only viable in the vector for 1 h or so and is often borne only *on* the aphid's mouthparts. In other 'circulative' species (e.g. lettuce necrotic yellow virus), the virus passes from the aphid's gut to its circulatory system and thence to its salivary glands. Here, there is a latent period before the vector becomes infective, but it then remains infective for an extended period. Finally, there are 'propagative' viruses (e.g. the potato leaf roll virus) that multiply within the aphid. Nematode worms are also widespread vectors of plant viruses.

12.2.2 Macroparasites

The parasitic helminth worms are major macroparasites of animals. The intestinal nematodes of humans, for example, all of which are transmitted directly, are perhaps the most important human intestinal parasites, both in terms of the number of people infected and their potential for causing ill health. There are also many types of medically important animal macroparasites with indirect life cycles. For example, the tapeworms are intestinal parasites as adults, absorbing host nutrients directly across their body wall and proliferating eggs that are voided in the host's feces. The larval stages then proceed through one or two intermediate hosts before the definitive host (in these cases, the human) is reinfected. The schistosomes, as we have seen, infect snails and vertebrates alternately. Human schistosomiasis (bilharzia) affects the gut wall where eggs become lodged, and also affects the blood vessels of the liver and lungs when eggs become trapped there too. Filarial nematodes are another group of long-lived parasites of humans that all require a period of larval development in a blood-sucking insect. One, *Wucheria bancrofti*, does its damage (Bancroftian filariasis) by the accumulation of adults in the lymphatic system (classically, but only rarely, leading to elephantiasis). Larvae (microfilariae) are released into the blood and are ingested by mosquitoes, which also transmit more developed, infective larvae back into the host. Another filarial nematode, *Onchocerca volvulus*, which causes 'river blindness', is transmitted by adult blackflies (the larvae of which live in rivers, hence the name of the disease). Here, though, it is the microfilariae that do the major damage when they are released into the skin tissue and reach the eyes.

In addition, there are lice, fleas, ticks and mites and some fungi that attack animals. Lice spend all stages of their life cycle on their host (either a mammal or a bird), and transmission is usually by direct physical contact between host individuals, often between mother and offspring. Fleas, by contrast, lay their eggs and spend their larval lives in the 'home' (usually the nest) of their host (again, a mammal or a bird). The emerging adult then actively locates a new host individual, often jumping and walking considerable distances in order to do so.

Plant macroparasites include the higher fungi that give rise to the mildews, rusts and smuts, as well as the gall-forming and mining insects, and some flowering plants that are themselves parasitic on other plants.

Direct transmission is common amongst the fungal macroparasites of plants. For example, in the development of mildew on a crop of wheat, infection involves contact between a spore (usually wind dispersed) and a leaf surface, followed by penetration of the fungus into or between the host cells, where it begins to grow, eventually becoming apparent as a lesion of altered host tissue. This phase of invasion and colonization precedes an infective stage when the lesion matures and starts to produce spores.

Indirect transmission of plant macroparasites via an intermediate host is common amongst the rust fungi. For example, in black stem rust, infection is transmitted from an annual grass host (especially the cultivated cereals such as wheat) to the barberry shrub (*Berberis vulgaris*) and from the barberry back to wheat. Infections on the cereal are polycyclic, i.e. within a season spores may infect and form lesions that release spores that infect further cereal plants. It is this phase of intense

Figure 12.2   A cuckoo in the nest. Reproduced by permission of FLPA/Martin B. Withers.

multiplication by the parasite that is responsible for epidemic outbreaks of disease. On the other hand, the barberry is a long-lived shrub and the rust is persistent within it. Infected barberry plants may therefore serve as persistent foci for the spread of the rust into cereal crops.

holo- and hemiparasitic plants

Plants in a number of families have become specialized as parasites on other flowering plants. These are of two quite distinct types. *Holoparasites*, such as dodder (*Cuscuta* spp.), lack chlorophyll and are wholly dependent on the host plant for their supply of water, nutrients and fixed carbon. *Hemiparasites*, on the other hand, such as the mistletoes (*Phoraradendron* spp.), are photosynthetic but have a poorly developed root system of their own, or none at all. They form connections with the roots or stems of other species and draw most or all of their water and mineral nutrients from the host.

12.2.3   Brood and social parasitism

At first sight the presence of a section about cuckoos might seem out of place here. Mostly a host and its parasite come from very distant systematic groups (mammals and bacteria, fish and tapeworms, plants and viruses). In contrast, *brood parasitism* usually occurs between quite closely related species and even between members of the same species. Yet the phenomenon falls clearly within the definition of parasitism (a brood parasite 'obtains its nutrients from one or a few host individuals, *normally* causing harm but not causing death immediately'). Brood parasitism is well developed in social insects (sometimes then called social parasitism), where the parasites use workers of another, usually very closely related species to rear their progeny (Choudhary *et al.*, 1994). The phenomenon is best known, however, amongst birds.

the ecological importance of brood parasitic birds

Bird brood parasites lay their eggs in the nests of other birds (Figure 12.2), which then incubate and rear them. They usually depress the nesting success of the host. Amongst ducks, *intraspecific* brood parasitism appears to be most common. Most brood parasitism, however, is *interspecific*. About 1% of all bird species are brood parasites – including about 50% of the species of cuckoos, two genera of finches, five cowbirds and a duck (Payne, 1977). They usually lay only a single egg in the host's nest and may adjust the host's clutch size by removing one of its eggs. The developing parasite may evict the host's eggs or nestlings and harm any survivors by monopolizing parental care. There is therefore the potential for brood parasites to have profound effects on the population dynamics of the host species. However, the frequency of parasitized nests is usually very low (less than 3%), and some time ago Lack (1963) concluded that 'the cuckoo is an almost negligible cause of egg and nestling losses amongst English breeding birds'. None the less, some impression of the potential importance of brood parasites is apparent from the fact that magpies (*Pica pica*) in populations that coexist with great spotted cuckoos (*Clamator glandarius*) in Europe invest their reproductive effort into laying significantly larger clutches of eggs than those that live free of brood parasitism (Soler *et al.*, 2001) – but those eggs are smaller in compensation. The presumption that this is an evolutionary response to the losses they suffer due to the cuckoos is supported by the fact that magpies that lay larger parasitized clutches do indeed have a higher probability of successfully raising at least some of their own offspring.

host-specific
polymorphisms: gentes

Highly host-specific, polymorphic relationships have evolved among brood parasites. For instance, the cuckoo *Cuculus canorum* parasitizes many different host species, but there are different strains ('gentes') within the cuckoo species. Individual females of one strain favor just one host species and lay eggs that match quite closely the color and markings of the eggs of the preferred host. Thus, amongst cuckoo females there is marked differentiation between strains in their mitochondrial DNA, which is passed only from female to female, but not at 'microsatellite' loci within the nuclear DNA, which contains material from the male parents, who do not restrict matings to within their own strain (Gibbs *et al.*, 2000). It has long been suggested (Punnett, 1933) that this is possible because the genes controlling egg patterning are situated on the W chromosome, carried only by females. (In birds, unlike mammals, the females are the heterogametic sex.) This has now been established – though in great tits, *Parus major*, rather than in a species of brood parasite (Gosler *et al.*, 2000). Females produce eggs that resemble those of their mothers and maternal grandmothers (from whom they inherit their W chromosome) but not those of their paternal grandmothers. Of course, if female cuckoos lay eggs that look like those of the species with which they were reared, it is also necessary for them to lay their eggs, inevitably or at least preferentially, in the nests of that species. This is most likely to be the result of early 'imprinting' (i.e. a learned preference) within the nest (Teuschl *et al.*, 1998).

12.3 Hosts as habitats

The essential difference between the ecology of parasites and that of free-living organisms is that the habitats of parasites are themselves alive. A living habitat is capable of growth (in numbers and/or size); it is potentially reactive, i.e. it can respond *actively* to the presence of a parasite by changing its nature, developing immune reactions to the parasite, digesting it, isolating or imprisoning it; it is able to evolve; and in the case of many animal parasites, it is mobile and has patterns of movement that dramatically affect dispersal (transmission) from one habitable host to another.

12.3.1 Biotrophic and necrotrophic parasites

The most obvious response of a host to a parasite is for the whole host to die. Indeed, we can draw a distinction between parasites that kill and then continue life on the dead host (*necrotrophic parasites*) and those for which the host must be alive (*biotrophic parasites*). Necrotrophic parasites blur the tidy distinctions between parasites, predators and saprotrophs (see Section 11.1). Insofar as host death is often inevitable and sometimes quite rapid,

necrotrophic parasites are really predators, and once the host is dead they are saprotrophs. But for as long as the host is alive, necroparasites share many features with other types of parasite.

For a biotrophic parasite, the death of its host spells the end of its active life. Most parasites are biotrophic. *Lucilia cuprina*, the blowfly of sheep, however, is a necroparasite on an animal host. The fly lays eggs on the living host and the larvae (maggots) eat into its flesh and may kill it. The maggots continue to exploit the carcass after death but they are now detritivores rather than either parasites or predators. Necroparasites on plants include many that attack the vulnerable seedling stage and cause symptoms known as 'damping-off' of seedlings. *Botrytis fabi* is a typical fungal necroparasite of plants. It develops in the leaves of the bean *Vicia faba*, and the cells are killed, usually in advance of penetration. Spots and blotches of dead tissues form on the leaves and the pods. The fungus continues to develop as a decomposer, and spores are formed and then dispersed from the dead tissue, but not while the host tissue is still alive.

Most necroparasites can therefore be regarded as pioneer saprotrophs. They are one jump ahead of competitors because they can kill the host (or

necroparasites: pioneer saprotrophs

its parts) and so gain first access to the resources of its dead body. The response of the host to necroparasites is never very subtle. Amongst plant hosts, the most common response is to shed the infected leaves, or to form specialized barriers that isolate the infection. Potatoes, for example, form corky scabs on the tuber surface that isolate infections by *Actinomyces scabies*.

12.3.2 Host specificity: host ranges and zoonoses

We saw in the chapters on the interactions between predators and their prey that there is often a high degree of specialization of a particular predator species on a particular species of prey (monophagy). The specialization of parasites on one or a restricted range of host species is even more striking. For any species of parasite (be it tapeworm, virus, protozoan or fungus) the potential hosts are a tiny subset of the available flora and fauna. The overwhelming majority of other organisms are quite unable to serve as hosts: often, we do not know why.

There are, though, some patterns to this specificity. It seems, for example, that the more intimate a parasite's association with a particular host individual, the more likely it is to be restricted to a particular species of host. Thus, for example, most species of bird lice, which spend their entire lives on one host, exploit only one host species, whereas louse flies, which move actively from one host individual to another, can use several species of host (Table 12.1).

The delineation of a parasite's host range, however, is not always as straightforward as one might imagine.

natural and accidental hosts

Table 12.1 Specialization in ectoparasites that feed on birds and mammals. (After Price, 1980.)

Scientific name	Common name and lifestyle	Number of species	Percentage of species restricted to:		
			1 host	2 or 3 hosts	More than 3 hosts
Philopteridae	Bird lice (spend whole life on one host)	122	87	11	2
Streblidae	Blood-sucking flies (parasitize bats)	135	56	35	9
Oestridae	Botflies (females fly between hosts)	53	49	26	25
Hystrichopsyllidae	Fleas (jump between hosts)	172	37	29	34
Hippoboscidae	Louse flies (are highly mobile)	46	17	24	59

Species outside the host range are relatively easily characterized: the parasite cannot establish an infection within them. But for those inside the host range, the response may range from a serious pathology and certain death to an infection with no overt symptoms. What is more, it is often the 'natural' host of a parasite, i.e. the one with which it has coevolved, in which infection is asymptomatic. It is often 'accidental' hosts in which infection gives rise to a frequently fatal pathology. ('Accidental' is an appropriate word here, since these are often dead-end hosts, that die too quickly to pass on the infection, within which the pathogen cannot therefore evolve – and *to* which it cannot therefore be adapted.)

plague: a zoonotic infection with humans as accidental hosts

These issues take on not just parasitological but also medical importance in the case of *zoonotic infections*: infections that circulate naturally, and have coevolved, in one or more species of wildlife but also have a pathological effect on humans. A good example is bubonic and pneumonic plague: the human diseases caused by the bacterium *Yersinia pestis*. *Y. pestis* circulates naturally within populations of a number of species of wild rodent: for example, in the great gerbil, *Rhombomys opimus*, in the deserts of Central Asia, and probably in populations of kangaroo rats, *Dipodomys* spp., in similar habitats in southwestern USA. (Remarkably, little is known about the ecology of *Y. pestis* in the USA, despite its widespread nature and potential threat (see Biggins & Kosoy, 2001).) In these species, there are few if any symptoms in most cases of infection. There are, however, other species where *Y. pestis* infection is devastating. Some of these are closely related to the natural hosts. In the USA, populations of prairie dogs, *Cynomys* spp., also rodents, are regularly annihilated by epidemics of plague, and the disease is an important conservation issue. But there are also other species, only very distantly related to the natural hosts, where untreated plague is usually, and rapidly, fatal. Amongst these are humans. Why such a pattern of differential virulence so often occurs – low virulence in the coevolved host, high virulence in some unrelated hosts, but unable even to cause an infection in others – is an important unanswered question in host–pathogen biology. The issue of host–pathogen coevolution is taken up again in Section 12.8.

12.3.4 Habitat specificity within hosts

Most parasites are also specialized to live only in particular parts of their host. Malarial parasites live in the red blood cells of vertebrates. *Theileria* parasites of cattle, sheep and goats live in the lymphocytes of the mammal, and in the epithelial cells, and later in the salivary gland cells, of the tick that is the disease vector, and so on.

By transplanting parasites experimentally from one part of the host's body to another, it can be shown that many home in on target habitats. When nematode worms (*Nippostrongylus brasiliensis*) were transplanted from the jejunum into the anterior and posterior parts of the small intestine of rats, they migrated back to their original habitat (Alphey, 1970). In other cases, habitat search may involve growth rather than bodily movement. For instance, loose smut of wheat, the fungus *Ustilago tritici*, infects the exposed stigmas of wheat flowers and then grows as an extending filamentous system into the young embryo. Growth continues in the seedling, and the fungus mycelium keeps pace with the growth of the shoot. Ultimately, the fungus grows rapidly into the developing flowers and converts them into masses of spores.

parasites may search for habitats within their hosts

12.3.5 Hosts as reactive environments: resistance, recovery and immunity

Any reaction by an organism to the presence of another depends on it recognizing a difference between what is 'self' and what is 'not self'. In invertebrates, populations of phagocytic cells are responsible for much of a host's response to invaders, even to inanimate particles. In insects, hemocytes (cells in the hemolymph) isolate infective material by a variety of routes, especially encapsulation – responses that are accompanied by the production of a number of soluble compounds in the humoral system that recognize and respond to nonself material, some of which also operate at the midgut barrier in the absence of hemocytes (Siva-Jothy *et al.*, 2001).

invertebrates

Figure 12.3 The immune response. The mechanisms mediating resistance to infection can be divided into 'natural' or 'nonspecific' (left) and 'adaptive' (right), each composed of both cellular elements (lower half) and humoral elements (i.e. free in the serum or body fluids; upper half). The adaptive response begins when the immune system is stimulated by an antigen that is taken up and processed by a macrophage (MAC). The antigen is a part of the parasite, such as a surface molecule. The processed antigen is presented to T and B lymphocytes. T lymphocytes respond by stimulating various clones of cells, some of which are cytotoxic (NK, natural killer cells), as others stimulate B lymphocytes to produce antibodies. The parasite that bears the antigen can now be attacked in a variety of ways. PMN, polymorphonuclear neutrophil. (After Playfair, 1996.)

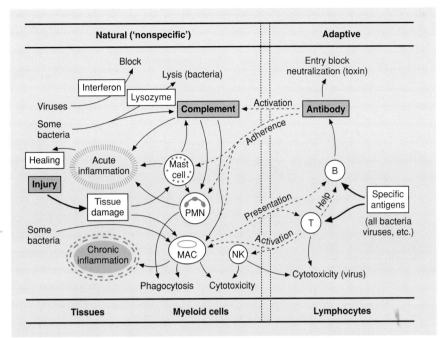

vertebrates: the immune response

In vertebrates there is also a phagocytic response to material that is not self, but their armory is considerably extended by a much more elaborate process: the immune response (Figure 12.3). For the ecology of parasites, an immune response has two vital features: (i) it may enable a host to recover from an infection; and (ii) it can give a once-infected host a 'memory' that changes its reaction if the parasite strikes again, i.e. the host has become immune to *re*infection. In mammals, the transmission of immunoglobulins to the offspring can sometimes even extend protection to the next generation.

For most viral and bacterial infections of vertebrates, the colonization of the host is a brief and transient episode in the host's life. The parasites multiply within the host and elicit a strong immunological response. By contrast, the immune responses elicited by many of the macroparasites and protozoan microparasites tend to be weaker. The infections themselves, therefore, tend to be persistent, and hosts may be subject to repeated reinfection.

contrasting responses to micro- and macroparasites

Indeed, responses to microparasites and helminths seem often to be dominated by different pathways within the immune system (MacDonald *et al.*, 2002), and these pathways can downregulate each other: helminth infection may therefore increase the likelihood of microparasitic infection and vice versa (Behnke *et al.*, 2001). Thus, for example, successful treatment of worm infections in patients that were also infected with HIV led to a significant drop in their HIV viral load (Wolday *et al.*, 2002).

plants

The modular structure of plants, the presence of cell walls and the absence of a true circulating system (such as blood or lymph) all make any form of immunological response an inefficient protection. There is no migratory population of phagocytes in plants that can be mobilized to deal with invaders. There is, however, growing evidence that higher plants possess complex systems of defense against parasites. These defenses may be constitutive – physical or biological barriers against invading organisms that are present whether the parasite is present or not – or inducible, arising in response to pathogenic attack (Ryan & Jagendorf, 1995; Ryan *et al.*, 1995). After a plant has survived a pathogenic attack, 'systematic acquired resistance' to subsequent attacks may be elicited from the host. For example, tobacco plants infected on one leaf with tobacco mosaic virus can produce local lesions that restrict the virus infection locally, but the plants then also become resistant to new infections not only by the same virus but to other parasites as well. In some cases the process involves the production of 'elicitins', which have been purified and shown to induce vigorous defense responses by the host (Yu, 1995).

the costliness of host defense

Central to our understanding of all host defensive responses to parasites is the belief that these responses are costly – that energy and material invested in the response must be diverted away from other important bodily functions – and that there must therefore be a trade-off between the response and other aspects of the life history: the more that is invested in one, the less can be invested

Table 12.2 Estimated energetic costs (percentage increase in resting metabolic rate relative to controls) made by various vertebrate hosts when mounting an immune response to a range of 'challenges' that induce such a response. (After Lochmiller & Derenberg, 2000.)

Species	Immune challenge	Cost (%)
Human	Sepsis	30
	Sepsis and injury	57
	Typhoid vaccination	16
Laboratory rat	Interleukin-1 infusion	18
	Inflammation	28
Laboratory mouse	Keyhole limpet hemocyanin injection	30
Sheep	Endotoxin	10–49

in the others. Evidence for this in vertebrates is reviewed by Lochmiller and Derenberg (2000), who illustrate, for example, the energetic price (in terms of an increase in resting metabolic rate) paid by a number of vertebrates when mounting an immune response (Table 12.2).

12.3.6 The consequences of host reaction: *S-I-R*

The variations in mechanisms used by different types of organism to fight infection are clearly interesting and important to parasitologists, medics and veterinarians. They are also important to ecologists working on particular systems, where an understanding of the overall biology is essential. But from the perspective of an ecological overview, the *consequences* for the hosts of these responses are more important, both at the whole organism and the population levels. First, these responses determine where individuals are on the spectrum from 'wholly susceptible' to 'wholly resistant' to infection – and if they become infected, where they are on the spectrum from being killed by infection to being asymptomatic. Second, in the case of vertebrates, the responses determine whether an individual still expresses a naive susceptibility or has acquired an immunity to infection.

These individual differences then determine, for a population, the structure of that population in terms of the numbers of individuals in the different classes. Many mathematical models of host–pathogen dynamics, for example, are referred to as *S-I-R* models, because they follow the changing numbers of susceptible, infectious and recovered (and immune) individuals in the population. The variations at the population level are then crucial in molding the features at the heart of ecology: the distributions and abundances of the organism concerned. We return to these questions of epidemic behavior in Section 12.4.2 and thereafter in this chapter.

12.3.7 Parasite-induced changes in growth and behavior

Some parasites induce a new *programed* change in the development of the host. The agromyzid flies and cecidomyid and cynipid wasps that form galls on higher plants are remarkable examples. The insects lay eggs in host tissue, which responds by renewed growth. The galls that are produced are the result of a morphogenetic response that is quite different from any structure that the plant normally produces. Just the presence, for a time, of the parasite egg may be sufficient to start the host tissue into a morphogenetic sequence that can continue even if the developing larva is removed. Amongst the gall-formers that attack oaks (*Quercus* spp.), each elicits a unique morphogenetic response from the host (Figure 12.4).

Fungal and nematode parasites of plants can also induce morphogenetic responses, such as enormous cell enlargement and the formation of nodules and other 'deformations'. After infection by the bacterium *Agrobacterium tumefaciens*, gall tissue can be recovered from the host plant that lacks the parasite but has now been set in its new morphogenetic pattern of behavior; it continues to produce gall tissue. In this case, the parasite has induced a genetic transformation of the host cells. Some parasitic fungi also 'take control' of their host plant and castrate or sterilize it. The fungus *Epichloe typhina*, which parasitizes grasses, prevents them from flowering and setting seed – the grass remains a vegetatively vigorous eunuch, leaving descendant parasites but no descendants of its own.

Most of the responses of modular organisms to parasites (and indeed other environmental stimuli) involve changes in growth and form, but in unitary organisms the response of hosts to infection more often involves a change in behavior: this often increases the chance of transmission of the parasite. In worm-infected hosts, irritation of the anus stimulates scratching, and parasite eggs are then carried from the fingers or claws to the mouth. Sometimes, the behavior of infected hosts seems to maximize the chance of the parasite reaching a secondary host or vector. Praying mantises have been observed walking to the edge of a river and apparently throwing themselves in, whereupon, within a minute of entering the water, a gordian worm (*Gordius*) emerges from the anus. This worm is a parasite of terrestrial insects but depends on an aquatic host for part of its life cycle. It seems that an infected host develops a hydrophilia that ensures that the parasite reaches a watery habitat. Suicidal mantises that are rescued will return to the riverbank and throw themselves in again.

galls

(sometimes dramatic) changes in host behavior

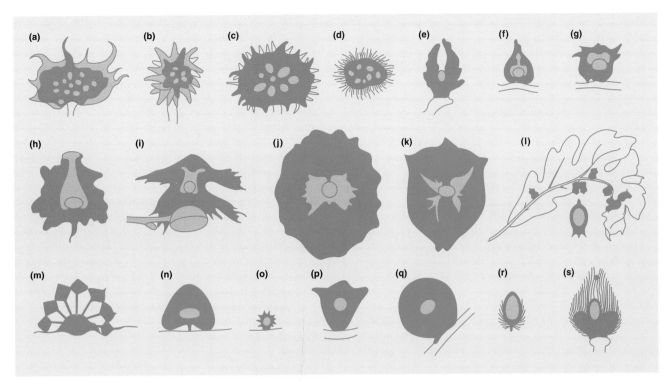

Figure 12.4 Galls formed by wasps of the genus *Andricus* on oaks (*Quercus petraea, Q. robur, Q. pubescens* or *Q. cerris*). Each figure shows a section through a gall induced by a different species of *Andricus*. The dark colored areas are the gall tissue and the central lighter areas are the cavities containing the insect larva. (From Stone & Cook, 1998.)

12.3.8   Competition within hosts

constant final yield?

Since hosts are the habitat patches for their parasites, it is not surprising that intra- and interspecific competition, observed in other species in other habitats, can also be observed in parasites within their hosts. There are many examples of the fitness of individual parasites decreasing within a host with increasing overall parasite abundance (Figure 12.5a), and of the overall output of parasites from a host reaching a saturation level (Figure 12.5b) reminiscent of the 'constant final yield' found in many plant monocultures subject to intraspecific competition (see Section 5.5.1).

competition or the immune response?

However, in vertebrates at least, we need to be cautious in interpreting such results simply as a consequence of intraspecific competition for limited resources, since the intensity of the immune reaction elicited from a host itself typically depends on the abundance of parasites. A rare attempt to disentangle these two effects utilized the availability of mutant rats lacking an effective immune response (Paterson & Viney, 2002). These and normal, control rats were subjected to experimental infection with a nematode, *Strongyloides*

ratti, at a range of doses. Any reduction in parasite fitness with dose in the normal rats could be due to intraspecific competition and/or an immune response that itself increases with dose; but clearly, in the mutant rats only the first of these is possible. In fact, there was no observable response in the mutant rats (Figure 12.6), indicating that at these doses, which were themselves similar to those observed naturally, there was no evidence of intraspecific competition, and that the pattern observed in the normal rats is entirely the result of a density-dependent immune response. Of course, this does not mean that there is never intraspecific competition amongst parasites within hosts, but it does emphasize the particular subtleties that arise when an organism's habitat is its reactive host.

We know from Chapter 8 that niche differentiation, and especially species having more effect on their own populations than on those of potential competitors, lies at the heart of our understanding of competitor coexistence. We noted earlier that parasites typically specialize on particular sites or tissues within their hosts, suggesting ample opportunity for niche differentiation. And in vertebrates at least, the specificity of the immune response also means that each parasite tends to have its greatest adverse effect on its own population. On the other hand, many parasites do have host tissues and resources in common; and it

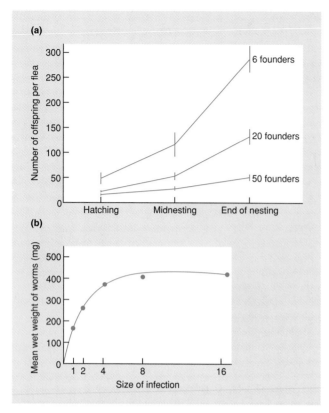

Figure 12.5 Density-dependent responses of parasites within their hosts. (a) The relationship between the number of fleas *Ceratophyllus gallinae* ('founders') added to the nests of blue tits and the number of offspring per flea (mean ± SE). The greater the density, the lower the reproductive rate of the fleas. This differential increased from an initial assessment at blue tit egg hatching, through to the end of the nestling period. (After Tripet & Richner, 1999.) (b) The mean weight of worms per infected mouse reaches a 'constant final yield' after deliberate infection at a range of levels with the tapeworm *Hymenolepis microstoma*. (After Moss, 1971.)

is easy to see that the presence of one parasite species may make a host less vulnerable to attack by a second species (for example, as a result of inducible responses in plants), or more vulnerable (simply because of the host's weakened state). All in all, it is no surprise that the ecology of parasite competition within hosts is a subject with no shortage of unanswered questions.

|interspecific competition amongst parasites| None the less, some evidence for interspecific competition amongst parasites comes from a study of two species of nematode, *Howardula aoronymphium* and *Parasitylenchus nearcticus*, that infect the fruit-fly *Drosophila recens* (Perlman & Jaenike, 2001). Of these, *P. nearcticus* is a specialist, being found only in *D. recens*, whereas *H. aoronymphium* is more of a generalist, capable of infect-

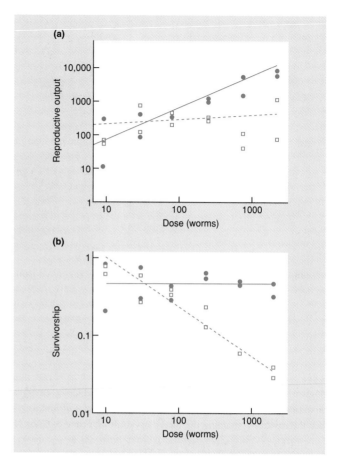

Figure 12.6 Host immune responses are necessary for density dependence in infections of the rat with the nematode *Strongyloides ratti*. (a) Overall reproductive output increases in line with the initial dose in mutant rats without an immune response (●; slope not significantly different from 1), but with an immune response (□) it is roughly independent of initial dose, i.e. it is regulated (slope = 0.15, significantly less than 1, P < 0.001). (b) Survivorship is independent of the initial dose in mutant rats without an immune response (●; slope not significantly different from 0), but with an immune response (□) it declines (slope = −0.62, significantly less than 0, P < 0.001). (After Paterson & Viney, 2002.)

ing a range of *Drosophila* species. In addition, *P. nearcticus* has the more profound effect on its host, typically sterilizing females, whereas *H. aoronymphium* seems to reduce host fecundity by only around 25% (though this itself represents a drastic reduction in host fitness). It is also apparent that whereas *H. aoronymphium* is profoundly affected by *P. nearcticus* when the two coexist within the same host in experimental infections (Figure 12.7a), this effect is not reciprocated (Figure 12.7b). Overall, therefore, competition is strongly asymmetric between the two parasites (as interspecific competition frequently is; see Section 8.3.3): the specialist *P. nearcticus* is both a more powerful exploiter of its host

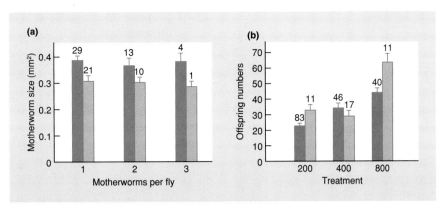

Figure 12.7 (a) Mean size ± SE (mm², longitudinal section area) of *Howardula aoronymphium* motherworms in 1-week-old hosts, *Drosophila recens*, in single and mixed infections. Size is a good index of fecundity in *H. aoronymphium*. The hosts contained either one, two or three *H. aoronymphium* motherworms, having been reared on a diet contaminated with either *H. aoronymphium* (dark bars) or mixed infections (*H. aoronymphium* and *Parasitylenchus nearcticus*; light bars). Size (fecundity) was consistently lower in mixed infections. (b) Number of *P. nearcticus* offspring (i.e. fecundity) ± SE, in single (dark bars) and mixed (light bars) infections. Numbers above the bars indicate sample sizes of flies; treatment numbers refer to the numbers of nematodes added to the diet. Fecundity was not reduced in mixed infections. (After Perlman & Jaenike, 2001.)

(reducing it to lower densities through its effect on fecundity) and stronger in interference competition. Coexistence between the species occurs, presumably, because the fly host provides the whole of both the fundamental and the realized niche of *P. nearcticus*, whereas it is only part of the realized niche of *H. aoronymphium*.

12.4 Dispersal (transmission) and dispersion of parasites amongst hosts

12.4.1 Transmission

hosts as islands

Janzen (1968) pointed out that we could usefully think of hosts as islands that are colonized by parasites. By using the same vocabulary, this brought host–parasite relationships into the same arena as MacArthur and Wilson's (1967) study of island biogeography (see Section 21.5). A human colonized by the malarial parasite is in a sense an inhabited island or patch. The chances of a mosquito vector carrying the parasite from one host to another correspond to the varying distances between different islands. Populations of parasites are thus maintained by the continual colonization of new host patches as old infected patches (hosts) die or become immune to new infection. The whole parasite population is then a 'metapopulation' (see Section 6.9), with each host supporting a subpopulation of the whole.

direct and indirect transmission; short- and long-lived agents

Different species of parasite are, of course, transmitted in different ways

between hosts. The most fundamental distinction, perhaps, is that between parasites that are transmitted directly from host to host and those that require a vector or intermediate host for transmission. Amongst the former, we should also distinguish between those where infection is by physical contact between hosts or by a very short-lived infective agent (borne, for example, in coughs and sneezes), and those where hosts are infected by long-lived infective agents (e.g. dormant and persistent spores).

We are largely familiar, through our own experience, with the nature of these distinctions amongst animal pathogens; but essentially the same patterns apply in plants. For example, many soil-borne fungal diseases are spread from one host plant to another by root contacts, or by the growth of the fungus through the soil from a base established on one plant, which gives it the resources from which to attack another. The honey fungus *Armillaria mellea* spreads through the soil as a bootlace-like 'rhizomorph' and can infect another host (usually a woody tree or shrub) where it meets their roots. In naturally diverse communities, such spread is relatively slow, but when plants occur as 'continents' of continuous interplant contacts, there are greatly increased opportunities for infection to spread. For diseases that are spread by wind, the foci of infection may become established at great distances from the origin; but the rate at which an epidemic develops locally is strongly dependent on the distance between individuals. It is characteristic of wind-dispersed propagules (spores, but also pollen and seeds) that the distribution achieved by dispersal is usually strongly 'leptokurtic': a few propagules go a very long way but the majority are deposited close to the origin.

12.4.2   Transmission dynamics

Transmission dynamics are in a very real sense the driving force behind the overall population dynamics of pathogens, but they are often the aspect about which we have least data (compared, say, to the fecundity of parasites or the death rate of infected hosts). We can, none the less, build a picture of the principles behind transmission dynamics (Begon *et al.*, 2002).

The rate of production of new infections in a population, as a result of transmission, depends on the per capita transmission rate (the rate of transmission per susceptible host 'target') and the number of susceptible hosts there are (which we can call S). In turn, per capita transmission rate is usually proportional, first, to the contact rate, k, between susceptible hosts and whatever it is that carries the infection. It also depends on the probability, p, that a contact that might transmit infection actually does so. Clearly, this probability depends on the infectiousness of the parasite, the susceptibility of the host, and so on. Putting these three components together we can say:

$$\text{the rate of production of new infections} = k \cdot p \cdot S. \quad (12.1)$$

the contact rate

The details of the contact rate, k, are different for different types of transmission.

- For parasites transmitted directly from host to host, we deal with the rate of contact between infected hosts and susceptible (uninfected) hosts.
- For hosts infected by long-lived infective agents that are isolated from hosts, it is the rate of contact between these and susceptible hosts.
- With vector-transmitted parasites we deal with the contact rate between host and vector (the 'host-biting rate'), and this goes to determine *two* key transmission rates: from infected hosts to susceptible vectors and from infected vectors to susceptible hosts.

But what is it that determines the per capita contact rate between susceptibles and infecteds? For long-lived infective agents, it is usually assumed that the contact rate is determined essentially by the density of these agents. For direct and vector-borne transmission, however, the contact rate needs to be broken down further into two components. The first is the contact rate between a susceptible individual and *all* other hosts (direct transmission) or all vectors; we can call this c. The second is then the proportion of those hosts or vectors that are infectious; we call this I/N, where I is the number of infecteds and N the total number of hosts (or vectors). Our expanded equation is now:

$$\text{the rate of production of new infections} = c \cdot p \cdot S \cdot (I/N). \quad (12.2)$$

We need to try to understand c and I/N in turn.

12.4.3   Contact rates: density- and frequency-dependent transmission

For most infections, it has often been assumed that the contact rate c increases in proportion to the density of the population, N/A, where A is the area occupied by the population, i.e. the denser the population, the more hosts come into contact with one another (or vectors contact hosts). Assuming for simplicity that A remains constant, the Ns in the equation then cancel, all the other constants can be combined into a single constant β, the 'transmission coefficient', and the equation becomes:

density-dependent transmission

$$\text{the rate of production of new infections} = \beta \cdot S \cdot I. \quad (12.3)$$

This, unsurprisingly, is known as *density-dependent transmission*.

On the other hand, it has long been asserted that for sexually transmitted diseases, the contact rate is constant: the frequency of sexual contacts is independent of population density. This time the equation becomes:

frequency-dependent transmission

$$\text{the rate of production of new infections} = \beta' \cdot S \cdot (I/N), \quad (12.4)$$

where the transmission coefficient again combines all the other constants but this time acquires a 'prime', β', because the combination of constants is slightly different. This is known as *frequency-dependent transmission*.

Increasingly, however, it has become apparent that the assumed simple correspondence between sexual transmission and frequency dependence on the one hand, and all other types of infection and density dependence on the other, is incorrect. For example, when density and frequency dependence were compared as descriptors of the transmission dynamics of cowpox virus, which is not sexually transmitted, in natural populations of bank voles (*Clethrionomys glareolus*), frequency dependence appeared, if anything, to be superior (Begon *et al.*, 1998). Frequency dependence appears to be a better descriptor than density dependence, too, for a number of (nonsexually transmitted) infections of insects (Fenton *et al.*, 2002). One likely explanation in such cases is that sexual contact is not the only aspect of behavior for which the contact rate varies little with population density: many social contacts, territory defense for instance, may come into the same category.

Secondly, $\beta \cdot S \cdot I$ and $\beta' \cdot S \cdot I/N$ are themselves increasingly recognized (e.g. McCallum *et al.*, 2001) as, at best, benchmarks against which real examples of transmission might be measured, rather than exact descriptors of the dynamics; or perhaps as ends of a spectrum along which real transmission terms could be assembled. For example, fitting the term $\beta S^x I^y$ to the transmission dynamics of granulovirus infection in larvae of the moth *Plodia interpunctella* revealed that the best fit was not to 'pure' density-dependent transmission, βSI, but to $\beta' S^{1.12} I^{0.14}$ (Figure 12.8).

ends of a spectrum

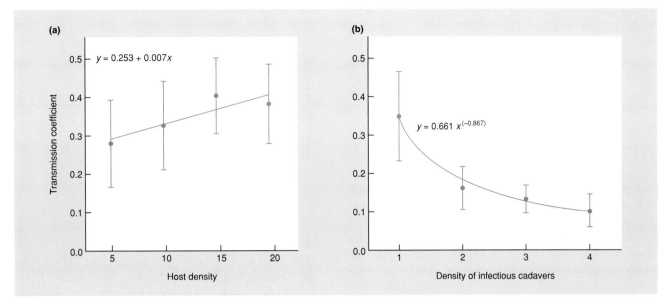

Figure 12.8 Estimating the transmission coefficient at various densities of (a) susceptible hosts and (b) infectious cadavers during the transmission of a granulovirus amongst moths, *Plodia interpunctella*, showed that the coefficient appeared to increase with the former and decrease with the latter. This is contrary to the expectations from density-dependent transmission (an apparently constant coefficient in both cases). (After Knell *et al.*, 1998.)

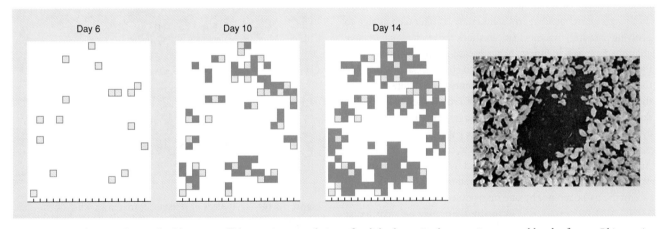

Figure 12.9 The spatial spread of damping-off disease in a population of radish plants, *Raphanus sativus*, caused by the fungus *Rhizoctonia solani*. Following initiation of the disease at isolated plants (light squares), the epidemic spreads rapidly to neighboring plants (dark squares), resulting in patches of damped-off plants (picture on the right). (Courtesy of W. Otten and C.A. Gilligan, Cambridge University.)

In other words, transmission was greater than expected (exponent greater than 1) at higher densities of susceptible hosts, probably because hosts at higher densities were short of food, moved more, and consumed more infectious material. But it was lower than expected (exponent less than 1) at higher densities of infectious host cadavers, probably because of strongly differential susceptibility amongst the hosts, such that the most susceptible become infected even at low cadaver densities, but the least susceptible remain uninfected even as cadaver density increases.

Turning from the contact rate, c, to the I/N term, there has usually been a simplifying assumption that this can

local hot spots

be based on numbers from the whole of a population. In reality, however, transmission typically occurs locally, between nearby individuals. In other words, use of such a term assumes either that all individuals in a population are intermingling freely with one another, or, slightly more realistically, that individuals are distributed approximately evenly across the population, so that all susceptibles are subject to roughly the same probability of a contact being with an infectious individual, I/N. The reality, however, is that there are likely to be hot spots of infection in a population, where I/N is high, and corresponding cool zones. Transmission, therefore, often gives rise to spatial waves of infection passing through a population (e.g. Figure 12.9), rather than simply the

overall rise in infection implied by a global transmission term like
β*SI*. This illustrates a very general point in modeling: that is, the
price paid in diminished realism when a complex process is
boiled down into a simple term (such as β*SI*). None the less,
as we shall see (and have seen previously in other contexts)
without such simple terms to help us, progress in understanding
complex processes would be impossible.

12.4.4   Host diversity and the spatial spread of disease

The further that hosts are isolated from one another, the more
remote are the chances that a parasite will spread between them.
It is perhaps no surprise, then, that the major disease epidemics
known amongst plants have occurred in crops that are not
islands in a sea of other vegetation, but 'continents' – large areas
of land occupied by one single species (and often by one single
variety of that species). Conversely, the spatial spread of an infection can be slowed down or even stopped by mixtures of susceptible
and resistant species or varieties (Figure 12.10). A rather similar
effect is described in Section 22.3.1.1, for Lyme disease in the United
States, where a variety of host species that are incompetent in
transmitting the spirochete pathogen 'dilute' transmission between
members of the most competent species.

   In agricultural practice, resistant cultivars offer a challenge to
evolving parasites: mutants that can attack the resistant strain have
an immediate gain in fitness. New, disease-resistant crop varieties
therefore tend to be widely adopted into commercial practice; but
they then often succumb, rather suddenly, to a different race of
the pathogen. A new resistant strain of crop is then used, and in
due course a new race of pathogen emerges. This 'boom and bust'
cycle is repeated endlessly and keeps the pathogen in a continually
evolving condition, and plant breeders in continual employment.
An escape from the cycle can be gained by the deliberate mixing
of varieties so that the crop is dominated neither by one virulent
race of the pathogen nor by one susceptible form of the crop itself.

the Janzen–Connell
effect

   In nature there may be a particular
risk of disease spreading from perennial
plants to seedlings of the same species
growing close to them. If this were
commonly the case, it could contribute to the species richness of
communities by preventing the development of monocultures.
This has been called the Janzen–Connell effect. In an especially
complete test for the effect, Packer and Clay (2000) showed for
black cherry, *Prunus serotina*, trees in a woodland in Indiana, first,
that seedlings were indeed less likely to survive close to their
parents (Figure 12.11a). Second, they showed that it was
something in the soil close to the parents that reduced survival
(Figure 12.11b), though this was only apparent at high seedling
density, and the effect could be removed by sterilizing the soil.
This suggests a pathogen, which high densities of seedlings, close
to the parent, amplify and transmit to other seedlings. In fact, dying

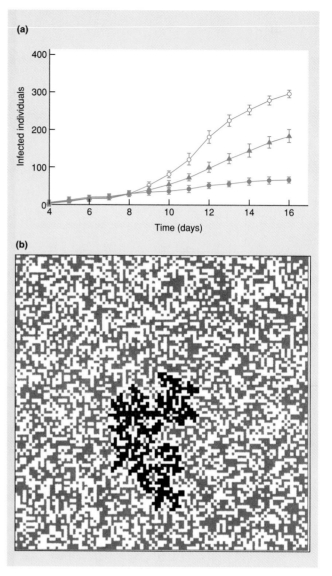

Figure 12.10   The effect of resistant forms in slowing down
the spread of damping-off epidemics caused by the fungus
Rhizoctonia solani. (a) Progress of epidemics in populations
following the introduction of *R. solani* into a susceptible
population (radish, *Raphanus sativus*: ○), a partially resistant
population (mustard, *Sinapsis alba*: ●) or a 50 : 50 mixture of the
two (▲). (b) A simulation showing that when 40% of the plants in
a population are of a resistant variety, the spread of a damping-off
epidemic following its introduction can be prevented. White
squares are resistant plants, black squares are infected, and gray
squares susceptible. Infection can only be transmitted to an
adjacent plant (sharing a 'side'). Here, the epidemic can spread
no further. (Courtesy of W. Otten, J. Ludlam and C.A. Gilligan,
Cambridge University.)

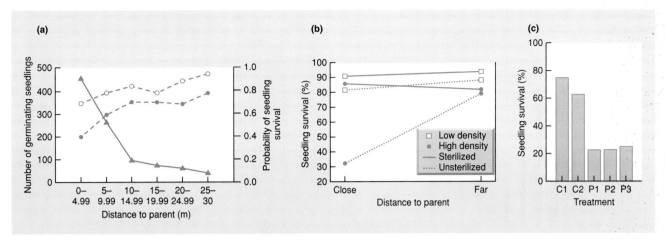

Figure 12.11 (a) The relationship between distance to parent, initial seedling germination (▲) and probability of seedling survival over time (dashed lines: ○, after 4 months; ●, after 16 months); $n = 974$ seedlings from beneath six trees. (b) The effect of distance from parent, seedling density and soil sterilization on seedling survival when seedlings were grown in pots containing soil collected close to or far from their parents. In high-density treatments, survival was significantly greater after the soil collected close to the tree was sterilized. ($P < 0.0001$). (c) Seedling survival in control and pathogen inoculation treatments ($n = 40$ per treatment). Control 1, potting mix only; control 2, 5 ml of sterile nutrient-rich fungal growth medium plus potting mix; P1, P2 and P3, three 5 ml replicates of pathogen inoculum plus potting mix. Survival was significantly lower in pathogen treatments compared with controls after 19 days ($X^2 = 13.8$, d.f. = 4, $P < 0.05$). (After Packer & Clay, 2000.)

seedlings were observed with the symptoms of 'damping off', and the damping-off fungus, *Pythium* sp., was isolated from dying seedlings and itself caused a significant reduction in seedling survival (Figure 12.11c).

12.4.5 The distribution of parasites within host populations: aggregation

Transmission naturally gives rise to an ever-changing dispersion of parasites within a population of hosts. But if we freeze the frame (or more correctly, carry out a cross-sectional survey of a population at one point in time), then we generate a distribution of parasites within the host population. Such distributions are rarely random. For any particular species of parasite it is usual for many hosts to harbor few or no parasites, and a few hosts to harbor many, i.e. the distributions are usually aggregated or clumped (Figure 12.12).

prevalence, intensity and mean intensity

In such populations, the mean density of parasites (mean number per host) may have little meaning. In a human population in which only one person is infected with anthrax, the mean density of *Bacillus anthracis* is a particularly useless piece of information. A more useful statistic, especially for microparasites, is the *prevalence* of infection: the proportion or percentage of a host population that is infected. On the other hand, infection *may* often vary in severity between individuals and is often clearly related to the number

of parasites that they harbor. The number of parasites in or on a particular host is referred to as the *intensity* of infection. The *mean intensity* of infection is then the mean number of parasites per host in a population (including those hosts that are not infected).

Aggregations of parasites within hosts may arise because individual hosts vary in their susceptibility to infection (whether due to genetic, behavioral or environmental factors), or because individuals vary in their exposure to parasites (Wilson *et al.*, 2002). The latter is especially likely to arise because of the local nature of transmission, and especially when hosts are relatively immobile. Infection then tends to be concentrated, at least initially, close to an original source of infection, and to be absent in individuals in areas that infection has yet to reach, or where it was previously but the hosts have recovered. It is clear, for example, even without explicit data on the distribution of parasites amongst hosts, that the parasites in Figure 12.9, at any one point in time, were aggregated at high intensities around the wave front – but absent ahead of and after it.

12.5 Effects of parasites on the survivorship, growth and fecundity of hosts

According to strict definition, parasites cause harm to their host. But it is not always easy to demonstrate this harm, which may be detectable only at some peculiarly sensitive stage of the host's life history or under particular circumstances (Toft & Karter, 1990). Indeed, there are examples of 'parasites' that feed on a host but

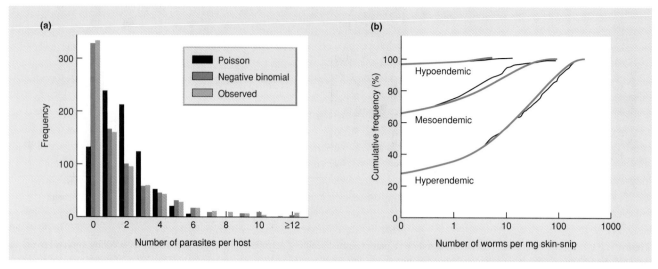

Figure 12.12 Examples of aggregated distributions of parasite numbers per host. (a) Crayfish, *Orconectes rusticus*, infected with the flatworm *Paragonimus kellicotti*. The distribution is significantly different from Poisson (random) ($X^2 = 723$, $P < 0.001$) but conforms well with a 'negative binomial', which is good at describing aggregated distributions: $X^2 = 12$, $P \approx 0.4$. (After Stromberg *et al.*, 1978; Shaw & Dobson, 1995.) (b) Distribution of *Onchocerca vulvulus* worms, which cause onchocerciasis or 'river blindness', in human Yanomami communities in southern Venezuela. Again the distributions, plotted as cumulative frequencies (black lines), conform well to a negative binomial distribution (colored lines), whether the typical intensity of infection is low (hypoendemic), moderate (mesoendemic) or high (hyperendemic). (After Vivas-Martinez *et al.*, 2000.)

appear to do it no harm. For example, in natural populations of Australia's sleepy lizard, *Tiliqua rugosa*, longevity was either not correlated or was positively associated with their load of ectoparasitic ticks (*Aponomma hydrosauri* and *Amblyomma limbatum*). There was no evidence that the ticks reduced host fitness (Bull & Burzacott, 1993).

There are of course, none the less, examples in which a detrimental effect of a parasite on host fitness has been demonstrated. Table 12.3, for example, shows one particular compilation of studies in which experimental manipulation of the loads of animal parasites revealed effects on either host fecundity or survival. (And while an effect on fecundity may seem less drastic than one on mortality, this seems less to be the case if one thinks of it as the death of potentially large numbers of offspring.)

On the other hand, the effects of parasites are often more subtle than a simple reduction in survival or fecundity. For example, the pied flycatcher

effects are often subtle . . .

Table 12.3 The impact of various parasites on the fecundity and survival of wild animals, as demonstrated through the experimental manipulation of parasite loads. (After Tompkins & Begon, 1999, where the original references may be found.)

Host	Parasite	Impact
Anderson's gerbil (*Gerbillus andersoni*)	*Synoternus cleopatrae* (flea)	Reduced survival
Barn swallow (*Hirundo rustica*)	*Ornithonyssus bursa* (mite)	Reduced fecundity
Cliff swallow (*Hirundo pyrrhonota*)	*Oeciacus vicarius* (bug)	Reduced fecundity
European starling (*Sturnus vulgaris*)	*Dermanyssus gallinae* (mite)	Reduced fecundity
	Ornithonyssus sylvarium (mite)	Reduced fecundity
Great tit (*Parus major*)	*Ceratophyllus gallinae* (flea)	Reduced fecundity
House martin (*Delichon urbica*)	*Oeciacus hirundinis* (bug)	Reduced fecundity
Pearly-eyed thrasher (*Margarops fuscatus*)	*Philinus deceptivus* (fly)	Reduced fecundity
Purple martin (*Progne subis*)	*Dermanyssus prognephilus* (mite)	Reduced fecundity
Red grouse (*Lagopus lagopus*)	*Trichostrongylus tenuis* (nematode)	Reduced fecundity
Snowshoe hare (*Lepus americanus*)	*Obeliscoides cuniculi* (nematode)	Reduced survival
Soay sheep (*Ovis aries*)	*Teladorsagia circumcincta* (nematode)	Reduced survival

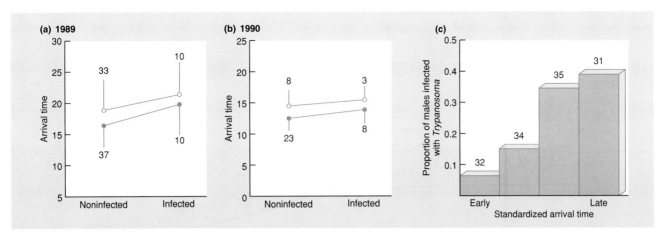

Figure 12.13 The mean date of arrival (1 = May 1) in Finland of male pied flycatchers (*Fidecula hypoleuca*) infected and uninfected with *Trypanosoma*: (a) 1989 and (b) 1990. ●, adult males; ○, yearling males. Sample sizes are indicated near the standard deviation bars. (c) The proportion of males infected with *Trypanosoma* amongst groups of migrants arriving in Finland at different times. (After Rätti *et al.*, 1993.)

(*Ficedula hypoleuca*) migrates from tropical West Africa to Finland to breed, and males that arrive early are particularly successful in finding mates. Males infected with the blood parasite *Trypanosoma* have shorter tails, tend to have shorter wings and arrive in Finland late and so presumably mate less often (Figure 12.13). Another example is provided by lice that feed on the feathers of birds and are commonly regarded as 'benign' parasites, with little or no effects on the fitness of their hosts. However, a long-term comparison of the effects of lice on feral rock doves (*Columba livia*) showed that the lice reduced the thermal protection given by the feathers and, in consequence, heavily infected birds incurred the costs of requiring higher metabolic rates to maintain their body temperatures (Booth *et al.*, 1993) and in the time that the birds spent in preening to keep the lice population under control.

In a similar vein, infection may make hosts more susceptible to predation. For example, postmortem examination of red grouse (*Lagopus lagopus scoticus*) showed that birds killed by predators carried significantly greater burdens of the parasitic nematode *Trichostrongylus tenuis* than the presumably far more random sample of birds that were shot (Hudson *et al.*, 1992a). Alternatively, the effect of parasitism may be to weaken an aggressive competitor and so allow weaker associated species to persist. For example, of two *Anolis* lizards that live on the Caribbean island of St Maarten, *A. gingivinus* is the stronger competitor and appears to exclude *A. wattsi* from most of the island. But the malarial parasite *Plasmodium azurophilum* very commonly affects *A. gingivinus* but rarely affects *A. wattsi*. Wherever the parasite infects *A. gingivinus*, *A. wattsi* is present; wherever the parasite is absent, only *A. gingivinus* occurs (Schall, 1992). Similarly, the holoparasitic plant, dodder (*Cuscuta salina*), which has a strong preference for *Salicornia* in a southern Californian salt marsh, is highly instrumental in determining the outcome of competition between *Salicornia* and other plant species within several zones of the marsh (Figure 12.14).

These latter examples make an important point. Parasites often affect their hosts not in isolation, but through an interaction with some other factor: infection may make a host more vulnerable to competition or predation; or competition or shortage of food may make a host more vulnerable to infection or to the effects of infection. This does not mean, however, that the parasites play only a supporting role. Both partners in the interaction may be crucial in determining not only the overall strength of the effect but also which particular hosts are affected.

Organisms that are resistant to parasites avoid the costs of parasitism, but, as with resistance to other natural enemies, resistance itself may carry a cost. This was tested with two cultivars of lettuce (*Lactuca sativa*), resistant or susceptible by virtue of two tightly linked genes to leaf root aphid (*Pemphigus bursarius*) and downy mildew (*Bremia lactucae*). The parasites were controlled by weekly applications of insecticides and fungicides. Resistant forms of lettuce bore fewer axillary buds than susceptibles (Figure 12.15), and this cost of resistance was most marked when the plants were making poor growth because of nutrient deficiency. In nature, hosts must always be caught between the costs of susceptibility and the costs of resistance.

Establishing that parasites have a detrimental effect on host characteristics of demographic importance is a critical first step in establishing that parasites influence the population and community dynamics of their hosts. But it is only a first step. A parasite may increase mortality, directly or indirectly, or decrease fecundity, without this affecting levels or patterns of abundance. The effect may simply be too trivial to have a measurable effect at the population level, or other factors and processes may act in

... affecting an interaction

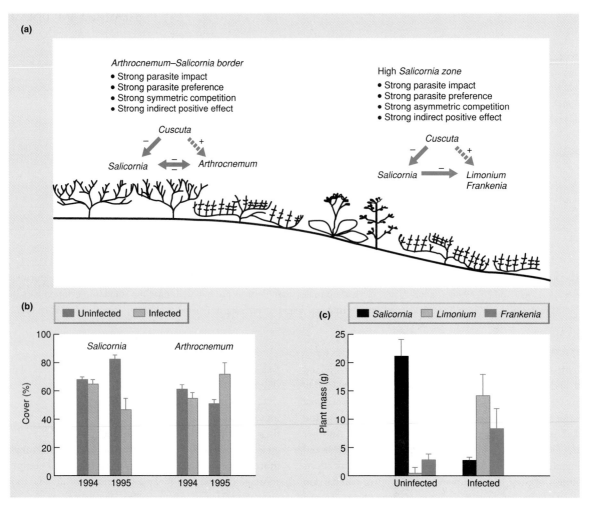

Figure 12.14 The effect of dodder, *Cascuta salina*, on competition between *Salicornia* and other species in a southern Californian salt marsh. (a) A schematic representation of the main plants in the community in the upper and middle zones of the marsh and the interactions between them (solid lines: direct effects; dashed lines: indirect effects). *Salicornia* (the relatively low-growing plant in the figure) is most attacked by, and most affected by, dodder (which is not itself shown in the figure). When uninfected, *Salicornia* competes strongly and symmetrically with *Arthrocnemum* at the *Arthrocnemum–Salicornia* border, and is a dominant competitor over *Limonium* and *Frankenia* in the middle (high *Salicornia*) zone. However, dodder significantly shifts the competitive balances. (b) Over time, *Salicornia* decreased and *Arthrocnemum* increased in plots infected with dodder. (c) Large patches of dodder suppress *Salicornia* and favor *Limonium* and *Frankenia*. (After Pennings & Callaway, 2002.)

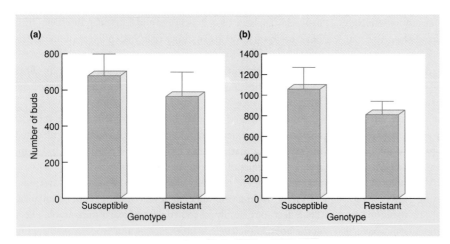

Figure 12.15 The number of buds produced by resistant and susceptible genotypes of two cultivars of lettuce, (a) and (b). Error bars are ±2 SE. (After Bergelson, 1994.)

a compensatory fashion – for example, loss to parasites may lead to a weakening of density-dependent mortality at a later stage in the life cycle. The effects of rare, devastating epidemics, whether in humans, other animals or plants, are easy to see; but for more typical, endemic parasites and pathogens, moving from the host-individual to the host-population level offers an immense challenge.

12.6   The population dynamics of infection

In principle, the sorts of conclusions that were drawn in Chapter 10 regarding the population dynamics of predator–prey and herbivore–plant interactions can be extended to parasites and hosts. Parasites harm individual hosts, which they use as a resource. The way in which this affects their populations varies with the densities of both parasites and hosts and with the details of the interaction. In particular, infected and uninfected hosts can exhibit compensatory reactions that may greatly reduce the effects on the host population as a whole. Theoretically, a range of outcomes can be predicted: varying degrees of reduction in host-population density, varying levels of parasite prevalence and various fluctuations in abundance.

effects on health or morbidity

With parasites, however, there are particular problems. One difficulty is that parasites often cause a reduction in the 'health' or 'morbidity' of their host rather than its immediate death, and it is therefore usually difficult to disentangle the effects of the parasites from those of other factors with which they interact (see Section 12.5). Another problem is that even when parasites cause a death, this may not be obvious without a detailed postmortem examination (especially in the case of microparasites). Also, the biologists who describe themselves as parasitologists have in the past tended to study the biology of their chosen parasite without much consideration of the effects on whole host populations; while ecologists have tended to ignore parasites. Plant pathologists and medical and veterinary parasitologists, for their part, generally study parasites with known severe effects that live typically in dense and aggregated populations of hosts, paying little attention to the more typical effects of parasites in populations of 'wildlife' hosts. Elucidation of the role of parasites in host-population dynamics is one of the major challenges facing ecology.

Here, we begin by looking at the dynamics of infection within host populations without considering any possible effects on the total abundance of hosts. This 'epidemiological' approach (Anderson, 1991) has especially dominated the study of human disease, where total abundance is usually considered to be determined by a whole spectrum of factors and is thus effectively independent of the prevalence of any one infection. Infection only affects the partitioning of this population into susceptible (uninfected), infected and other classes. We then take a more 'ecological' approach by considering the effects of parasites on host abundance

in a manner much more akin to conventional predator–prey dynamics.

12.6.1   The basic reproductive rate and the transmission threshold

In all studies of the dynamics of parasite populations or the spread of infection, there are a number of particularly key concepts. The first is the *basic reproductive rate*, R_0. For microparasites, because infected hosts are the unit of study, this is defined as the average number of new infections that would arise from a single infectious host introduced into a population of susceptible hosts. For macroparasites, it is the average number of established, reproductively mature offspring produced by a mature parasite throughout its life in a population of uninfected hosts.

R_0, the basic reproductive rate

The *transmission threshold*, which must be crossed if an infection is to spread, is then given by the condition $R_0 = 1$. An infection will eventually die out for $R_0 < 1$ (each present infection or parasite leads to less than one infection or parasite in the future), but an infection will spread for $R_0 > 1$.

the transmission threshold

Insights into the dynamics of infection can be gained by considering the various determinants of the basic reproductive rate. We do this in some detail for directly transmitted microparasites, and then deal more briefly with related issues for indirectly transmitted microparasites, and directly and indirectly transmitted macroparasites.

12.6.2   Directly transmitted microparasites: R_0 and the critical population size

For microparasites with direct, density-dependent transmission (see Section 12.4.3), R_0 can be said to increase with: (i) the average period of time over which an infected host remains infectious, L; (ii) the number of susceptible individuals in the host population, S, because greater numbers offer more opportunities for transmission of the parasite; and (iii) the transmission coefficient, β (see Section 12.4.3). Thus, overall:

$$R_0 = S\beta L. \qquad (12.5)$$

Note immediately that by this definition, the greater the number of susceptible hosts, the higher the basic reproductive rate of the infection (Anderson, 1982).

The transmission threshold can now be expressed in terms of a *critical population size*, S_T, where, because $R_0 = 1$ at that threshold:

the critical population size . . .

$$S_T = 1/(\beta L). \tag{12.6}$$

In populations with numbers of susceptibles less than this, the infection will die out ($R_0 < 1$). With numbers greater than this the infection will spread ($R_0 > 1$). (S_T is often referred to as the critical *community* size because it has mostly been applied to human 'communities', but this is potentially confusing in a wider ecological context.) These simple considerations allow us to make sense of some very basic patterns in the dynamics of infection (Anderson, 1982; Anderson & May, 1991).

... for different types of parasite

Consider first the kinds of population in which we might expect to find different sorts of infection. If microparasites are highly infectious (large βs), or give rise to long periods of infectiousness (large Ls), then they will have relatively high R_0 values even in small populations and will therefore be able to persist there (S_T is small). Conversely, if parasites are of low infectivity or have short periods of infectiousness, they will have relatively small R_0 values and will only be able to persist in large populations. Many protozoan infections of vertebrates, and also some viruses such as herpes, are persistent within individual hosts (large L), often because the immune response to them is either ineffective or short lived. A number of plant diseases, too, like club-root, have very long periods of infectiousness. In each case, the critical population size is therefore small, explaining why they can and do survive endemically even in small host populations.

On the other hand, the immune responses to many other human viral and bacterial infections are powerful enough to ensure that they are only very transient in individual hosts (small L), and they often induce lasting immunity. Thus, for example, a disease like measles has a critical population size of around 300,000 individuals, and is unlikely to have been of great importance until quite recently in human biology. However, it generated major epidemics in the growing cities of the industrialized world in the 18th and 19th centuries, and in the growing concentrations of population in the developing world in the 20th century. Around 900,000 deaths occur each year from measles infection in the developing world (Walsh, 1983).

12.6.3 Directly transmitted microparasites: the epidemic curve

The value of R_0 itself is also related to the nature of the *epidemic curve* of an infection. This is the time series of new cases following the introduction of the parasite into a population of hosts. Assuming there are sufficient susceptible hosts present for the parasite to invade (i.e. the critical population size, S_T, is exceeded), the initial growth of the epidemic will be rapid as the parasite sweeps through the population of susceptibles. But as these susceptibles either die or recover to immunity, their number, S, will decline, and so too therefore will R_0 (Equation 12.5). Hence, the rate of appearance of new cases will slow down and then decline. And if S falls below S_T and stays there, the infection will disappear – the epidemic will have ended. Two examples of epidemic curves, for Legionnaires' disease in Spain and for foot-and-mouth disease in the UK, are shown in Figure 12.16.

Not surprisingly, the higher the initial value of R_0, the more rapid will be the rise in the epidemic curve. But this will also lead

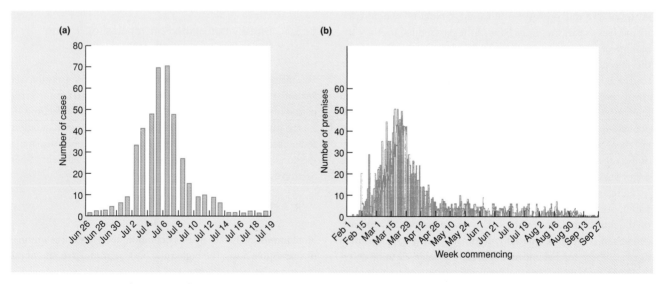

Figure 12.16 (a) An epidemic curve for an outbreak of Legionnaires' disease in Murcia, a municipality in southeastern Spain, in 2001. (After García-Fulgueiras *et al.*, 2003.) (b) An epidemic curve for an outbreak of foot-and-mouth disease (mostly affecting cattle and sheep) in the United Kingdom in 2001. Infected premises (farms) are shown, since infection was transmitted from farm to farm, and once infected, all the stock on that farm were destroyed. (After Gibbens & Wilesmith, 2002.)

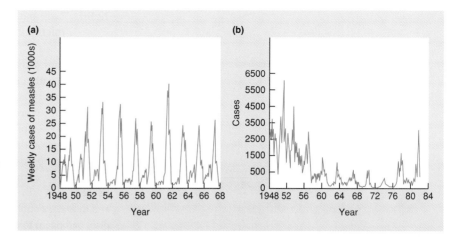

Figure 12.17 (a) Reported cases of measles in England and Wales from 1948 to 1968, prior to the introduction of mass vaccination. (b) Reported cases of pertussis (whooping cough) in England and Wales from 1948 to 1982. Mass vaccination was introduced in 1956. (After Anderson & May, 1991.)

to the more rapid removal of susceptibles from the population and hence to an earlier end to the epidemic: higher values of R_0 tend to give rise to shorter, sharper epidemic curves. Also, whether the infection disappears altogether (i.e. the epidemic simply ends) depends very largely on the rate at which new susceptibles either move into or are born into the population, since this determines how long the population remains below S_T. If this rate is too low, then the epidemic will indeed simply end. But a sufficiently rapid input of new susceptibles should prolong the epidemic, or even allow the infection to establish endemically in the population after the initial epidemic has passed.

12.6.4 Directly transmitted microparasites: cycles of infection

dynamic patterns of different types of parasite

This leads us naturally to consider the longer term patterns in the dynamics of different types of endemic infection. As described above, the immunity induced by many bacterial and viral infections reduces S, which reduces R_0, which therefore tends to lead to a decline in the incidence of the infection itself. However, in due course, and before the infection disappears altogether from the population, there is likely to be an influx of new susceptibles into the population, a subsequent increase in S and R_0, and so on. There is thus a marked tendency with such infections to generate a sequence from 'many susceptibles (R_0 high)', to 'high incidence', to 'few susceptibles (R_0 low)', to 'low incidence', to 'many susceptibles', etc. – just like any other predator–prey cycle. This undoubtedly underlies the observed cyclic incidence of many human diseases, with the differing lengths of cycle reflecting the differing characteristics of the diseases: measles with peaks every 1 or 2 years (Figure 12.17a), pertussis (whooping cough) every 3–4 years (Figure 12.17b), diphtheria every 4–6 years, and so on (Anderson & May, 1991).

By contrast, infections that do not induce an effective immune response tend to be longer lasting within individual hosts, but also tend not to give rise to the same sort of fluctuations in S and R_0. Thus, for example, protozoan infections tend to be much less variable (less cyclic) in their prevalence.

12.6.5 Directly transmitted microparasites: immunization programs

Recognizing the importance of critical population sizes also throws light on immunization programs, in which susceptible hosts are rendered nonsusceptible without ever becoming diseased (showing clinical symptoms), usually through exposure to a killed or attenuated pathogen. The direct effects here are obvious: the immunized individual is protected. But, by reducing the number of susceptibles, such programs also have the indirect effect of reducing R_0. Indeed, seen in these terms, the fundamental aim of an immunization program is clear – to hold the number of susceptibles below S_T so that R_0 remains less than 1. To do so is said to provide 'herd immunity'.

In fact, a simple manipulation of Equation 12.5 gives rise to a formula for the critical proportion of the population, p_c, that needs to be immunized in order to provide herd immunity (reducing R_0 to a maximum of 1, at most). If we define S_0 as the typical number of susceptibles prior to any immunization and note that S_T is the number still susceptible (not immunized) once the program to achieve $R_0 = 1$ has become fully established, then the proportion immunized is:

$$p_c = 1 - (S_T/S_0). \tag{12.7}$$

The formula for S_T is given in Equation 12.6, whilst that for S_0, from Equation 12.5, is simply $R_0/\beta L$, where R_0 is the basic reproductive rate of the infection prior to immunization. Hence:

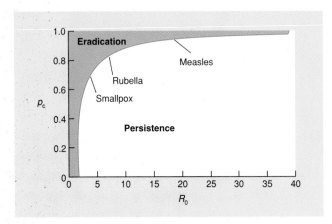

Figure 12.18 The dependence of the critical level of vaccination coverage required to halt transmission, p_c, on the basic reproductive rate, R_0, with values for some common human diseases indicated. (After Anderson & May, 1991.)

$$p_c = 1 - (1/R_0). \tag{12.8}$$

This reiterates the point that in order to eradicate a disease, it is not necessary to immunize the whole population – just a proportion sufficient to bring R_0 below 1. It also shows that this proportion will be higher the greater the 'natural' basic reproductive rate of the disease (without immunization). This general dependence of p_c on R_0 is illustrated in Figure 12.18, with the estimated values for a number of human diseases indicated on it. Note that smallpox, the only disease where in practice immunization seems to have led to eradication, has unusually low values of R_0 and p_c.

12.6.6 Directly transmitted microparasites: frequency-dependent transmission

Suppose, however, that transmission is frequency dependent (see Section 12.4.3), as it is likely to be, for example, with sexually transmitted diseases, where transmission occurs after an infected individual 'seeks out' (or is sought out by) a susceptible individual. Then there is no longer the same dependence on the number of susceptibles, and the basic reproductive rate is simply given by:

$$R_0 = \beta'L. \tag{12.9}$$

Here, there is apparently no threshold population size and such infections can therefore persist even in extremely small populations (where, to a first approximation, the chances of sexual contact for an infected host are the same as in large populations).

12.6.7 Crop pathogens: macroparasites viewed as microparasites

Most of plant pathology has been concerned with the dynamics of diseases within crops, and hence with the spread of a disease *within* a generation. Moreover, although most commonly studied plant pathogens are macroparasites in the sense we have defined them, they are typically treated like microparasites in that disease is monitored on the basis of some measure of disease severity – often, the proportion of the population infected (i.e. prevalence). We refer to y_t as the proportion affected by lesions at time t, and hence $(1 - y_t)$ is the proportion of the population without lesions and thus susceptible to infection. It is also usually necessary with plant pathogens to take explicit account of the latent period, length p, between the time when a lesion is initiated and the time when it becomes infectious (spore-forming) itself, in which state it remains for a further period l. Hence, the proportion of the population affected by *infectious* lesions at time t is $(y_{t-p} - y_{t-p-l})$. The rate of increase in the proportion of a plant population affected by lesions (Vanderplank, 1963; Zadoks & Schein, 1979; Gilligan, 1990) may thus be given by:

$$dy_t/dt = D(1 - y_t)(y_{t-p} - y_{t-p-l}), \tag{12.10}$$

which is essentially a βSI formulation, with D the plant pathologists' version of a transmission coefficient. This gives rise to S-shaped curves for the progress of a disease within a crop that broadly match the data derived from many crop–pathogen systems (Figure 12.19).

In the progress of such infections, plant pathologists recognize three phases.

1 The 'exponential' phase, when, although the disease is rarely detectable, rapid acceleration of parasite prevalence occurs. This is therefore the phase in which chemical control would be most effective, but in practice it is usually applied in phase 2. The exponential phase is usually considered arbitrarily to end at $y = 0.05$; about the level of infection at which a nonspecialist might detect that an epidemic was developing (the perception threshold).

2 The second phase, which extends to $y = 0.5$. (This is sometimes confusingly called the 'logistic' phase, although the whole curve is logistic.)

3 The terminal phase, which continues until y approaches 1.0. In this phase chemical treatment is virtually useless – yet it is at this stage that the greatest damage is done to the yield of a crop.

On the other hand, some crop diseases are not simply transmitted by the passive spread of infective particles from one host to another. For example, the anther smut fungus, *Ustilago violacea*, is spread between host plants of white campion, *Silene alba*, by

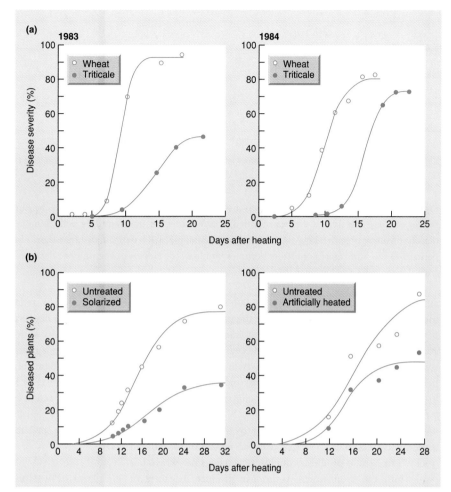

Figure 12.19 'S-shaped' curves of the progress of diseases through crops from an initial inoculum to an asymptotic proportion of the total population infected. (a) *Puccinia recondita* attacking wheat (cultivar Morocco) and triticale (a crop derived from the hybridization of wheat and rye) in 1983 and 1984. (b) *Fusarium oxysporum* attacking tomatoes in experiments comparing untreated and sterilized soil and untreated and artificially heated soil. (After Gilligan, 1990, in which the original data sources and methods of curve-fitting may be found.)

pollinating insects that adjust their flight distances to compensate for changes in plant density, such that the rate of transmission is effectively independent of host density (Figure 12.20a). However, this rate decreases significantly with the proportion of the population that is susceptible: transmission is frequency dependent (Figure 12.20b), favoring, as we have seen, persistence of the disease even in low-density populations. Of course, this is really just another case of frequency-dependent transmission in a sexually transmitted disease – except that sexual contact here is indirect rather than intimate.

12.6.8 Other classes of parasite

vector-borne infections

For microparasites that are spread from one host to another by a vector more generally (where the vector does not compensate for changes in host density as in the above example), the life cycle characteristics of both the host and vector enter into the calculation of R_0. In particular, the transmission threshold

($R_0 = 1$) is dependent on a ratio of vector : host numbers. For a disease to establish itself and spread, that ratio must exceed a critical level – hence, disease control measures are usually aimed directly at reducing the numbers of vectors, and are aimed only indirectly at the parasite. Many virus diseases of crops, and vector-transmitted diseases of humans and their livestock (malaria, onchocerciasis, etc.), are controlled by insecticides rather than chemicals directed at the parasite; and the control of all such diseases is of course crucially dependent on a thorough understanding of the vector's ecology.

The effective reproductive rate of a directly transmitted macroparasite (no intermediate host) is directly related to

directly transmitted macroparasites

the length of its reproductive period within the host (i.e. again, to L) and to its rate of reproduction (rate of production of infective stages). Both of these are subject to density-dependent constraints that can arise either because of competition between the parasites, or commonly because of the host's immune response (see Section 12.3.8). Their intensity varies with the distribution of the parasite population between its hosts and, as we have seen,

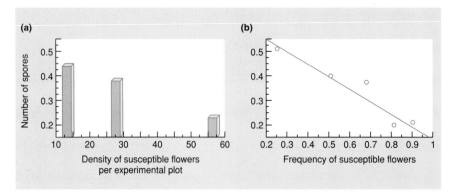

Figure 12.20 Frequency-dependent transmission of a sexually transmitted disease. The number of spores of *Ustilago violacea* deposited per flower of *Silene alba* ($\log_{10}(x + 1)$ transformed) where spores are transferred by pollinating insects. (a) The number is independent of the density of susceptible (healthy) flowers in experimental plots ($P > 0.05$) (and shows signs of decreasing rather than increasing with density, perhaps as the number of pollinators becomes limiting). (b) However, the number decreases with the frequency of susceptibles ($P = 0.015$). (After Antonovics & Alexander, 1992.)

aggregation of the parasites is the most common condition. This means that a very large proportion of the parasites exist at high densities where the constraints are most intense, and this tightly controlled density dependence undoubtedly goes a long way towards explaining the observed stability in prevalence of many helminth infections (such as hookworms and roundworms) even in the face of perturbations induced by climatic change or human intervention (Anderson, 1982).

Most directly transmitted helminths have an enormous reproductive capability. For instance, the female of the human hookworm *Necator* produces roughly 15,000 eggs per worm per day, whilst the roundworm *Ascaris* can produce in excess of 200,000 eggs per worm per day. The critical threshold densities for these parasites are therefore very low, and they occur and persist endemically in low-density human populations, such as hunter–gatherer communities.

indirectly transmitted macroparasites

Density dependence within hosts also plays a crucial role in the epidemiology of indirectly transmitted macroparasites, such as schistosomes. In this case, however, the regulatory constraints can occur in either or both of the hosts: adult worm survival and egg production are influenced in a density-dependent manner in the human host; but also, production of infective stages by the snail (intermediate host) is virtually independent of the number of (different) infective stages that penetrate the snail. Thus, levels of schistosome prevalence tend to be stable and resistant to perturbations from outside influences.

The threshold for the spread of infection depends directly on the abundance of both humans and snails (i.e. a product as opposed to the ratio that was appropriate for vector-transmitted microparasites). This is because transmission in both directions is by means of free-living infective stages. Thus, since it is inappropriate to reduce human abundance, schistosomiasis is often controlled by reducing snail numbers with molluscicides in an attempt to depress R_0 below unity (the transmission threshold). The difficulty with this approach, however, is that the snails have an enormous reproductive capacity, and they rapidly recolonize aquatic habitats once molluscicide treatment ceases. The limitations imposed by low snail numbers, moreover, are offset to an important extent by the long lifespan of the parasite in humans (L is large): the disease can remain endemic despite wide fluctuations in snail abundance.

12.6.9 Parasites in metapopulations: measles

With host–parasite dynamics, as with other areas of ecology, there is increasing recognition that populations cannot be seen as either homogeneous or isolated. Rather, hosts are usually distributed amongst a series of subpopulations, linked by dispersal between them, and which together comprise a 'metapopulation' (see Section 6.9). Thus, since the argument has already been made (see Section 12.4.1) that each host supports a subpopulation and a host population supports a metapopulation of parasites, host–parasite systems are typically metapopulations of metapopulations.

Such a perspective immediately changes our view of what is required of a host population if it is to support a persistent population of parasites. This is apparent from an analysis of the dynamics of measles in 60 towns and cities in England and Wales from 1944 to 1994: 60 subpopulations comprising an overall metapopulation (Figure 12.21) (Grenfell & Harwood, 1997). Taken as a whole, the metapopulation displayed regular cycles in the number of measles cases and measles was ever-present (Figure 12.21a), at least before widespread vaccination (c. 1968). But amongst the individual subpopulations, only the very largest were not liable to frequent 'stochastic fade-out' (disappearance of the disease when a few remaining infectious individuals fail to pass

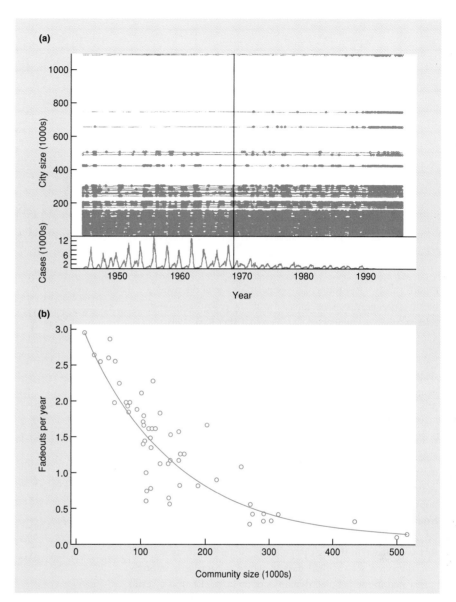

Figure 12.21 (a) The weekly measles notifications for 60 towns and cities in England and Wales, combined, are shown below for the period 1944–94. The vertical line indicates the start of mass vaccination around 1968. The data for the individual towns (town size on the vertical axis) are displayed above as a dot for each week without a measles notification. (b) Persistence of measles in these towns and cities in the prevaccination era (1944–67) as a function of population size. Persistence is measured inversely as the number of 'fade-outs' per year, where a fade-out here is defined as a period of three or more weeks without notification, to allow for the underreporting of cases. (After Grenfell & Harwood, 1997.)

it on), especially during the cycle troughs: the idea of a critical population size of around 300,000–500,000 is therefore well supported (Figure 12.21b). Thus, patterns of dynamics may be apparent, and persistence may be predictable, in a metapopulation taken as a whole. But in the individual subpopulations, especially if they are small, the patterns of dynamics and persistence are likely to be far less clear. The measles data set is unusual in that we have information both for the metapopulation and individual subpopulations. In many other cases, it is almost certain that the principle is similar but we have data only for the metapopulation (and do not appreciate the number of fade-outs in smaller parts of it), or we have data only for a subpopulation (and do not appreciate its links to other subpopulations within the larger metapopulation).

12.7 Parasites and the population dynamics of hosts

A key and largely unanswered question in population ecology is what role, if any, do parasites and pathogens play in the dynamics of their hosts? There are data (see Section 12.5) showing that parasites may affect host characteristics of demographic importance (birth and death rates), though even these data are relatively uncommon; and there are mathematical models showing that parasites have the *potential* to have a major impact on the dynamics of their hosts. But the point was also made earlier that it is a big step further to establish that dynamics are actually affected. There are certainly cases where a parasite or pathogen seems, by implication, to reduce the population size of its host. The

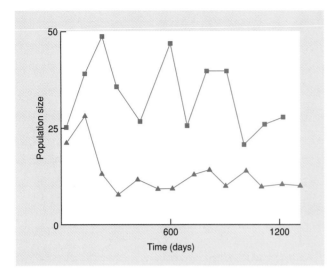

Figure 12.22 Depression of the population size of the flour beetle, *Tribolium castaneum*, infected with the protozoan parasite *Adelina triboli*: (■) uninfected, (▲) infected. (After Park, 1948.)

widespread and intensive use of sprays, injections and medicines in agricultural and veterinary practice all bear witness to the disease-induced loss of yield that would result in their absence. Data sets from controlled, laboratory environments showing reductions in host abundance by parasites have also been available for many years (Figure 12.22). However, good evidence from natural populations is extremely rare. Even when a parasite is present in one population but absent in another, the parasite-free population is certain to live in an environment that is different from that of the infected population; and it is likely also to be infected with some other parasite that is absent from or of low prevalence in the first population. Nevertheless, as we shall see, there are sets of field data in which a parasite is strongly implicated in the detailed dynamics of its host, either as a result of field-scale manipulations, or through using data on the effects of parasites on individual hosts in order to 'parameterize' mathematical models that can then be compared with field data.

12.7.1 Coupled (interactive) or modified host dynamics?

First, an important question, even when an effect of a parasite on host dynamics has been demonstrated, is whether the host and parasite interact, such that their dynamics are coupled in the manner usually envisaged for 'predator–prey' cycles, or whether the parasite simply modifies the underlying dynamics of the host, without there being any detectable feedback between host and parasite dynamics, and hence without any actual *interaction* between the two. This question has been addressed for the data shown in Figure 12.23 for the stored product moth *Plodia*

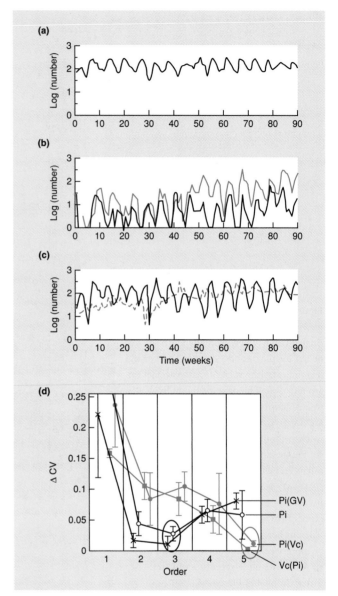

Figure 12.23 Dynamics of the host moth, *Plodia interpunctella* (———) alone (a), in the presence of the parasitoid *Venturia canescens* (———) (b), and in the presence of the *Plodia interpunctella* granulovirus (- - - - -) (c). The series show representative replicates of each treatment (out of three) for the first 90 weeks of the experiment. (d) Estimating the dimensionality or 'order' of the density dependence of the dynamics for each treatment (all replicates), which is predicted to increase with the number of interacting elements in the system. The lower the value of ΔCV, the better the 'fit'; error bars represent 1 SE. The best-fitting orders (circled) are three for the host alone (Pi) and the host in the presence of the virus (Pi(GV)), but five for the host in the presence of the parasitoid (Pi(Vc)), and five too for the parasitoid to which it is coupled (Vc(Pi)). (After Bjørnstad *et al.*, 2001.)

interpunctella and its granulovirus (PiGV) (touched on briefly in Section 10.2.5). The dynamics of the host in the presence and absence of the virus are different but only subtly so (Figure 12.23a, c), and detailed statistical analysis is required to try to understand the difference. Put simply, if host dynamics in infected populations are driven by an interaction between *Plodia* and PiGV, then the 'dimensionality' of those dynamics (essentially, the complexity of the statistical model required to describe them) should be greater than those of uninfected populations. In fact, although host fecundity was reduced and host development was slowed by the virus, and host abundance was more variable, the dimensionality of the dynamics was unaltered (Figure 12.23d): the virus modulated the vital rates of the host but did not interact with the host nor alter the underlying nature of its dynamics (Bjørnstad *et al.*, 2001). By contrast, when *Plodia* interacted with another natural enemy, the parasitoid *Venturia canescens*, the underlying pattern of 'generation cycles' (see Section 10.2.4) remained intact, but this time the dimensionality of the host dynamics was significantly increased (from dimension 3 to 5): the host and parasite interacted.

12.7.2 Red grouse and nematodes

Next we look at the red grouse – of interest both because it is a 'game' bird, and hence the focus of an industry in which British landowners charge for the right to shoot at it, and also because it is another species that often, although not always, exhibits regular cycles of abundance (Figure 12.24a). The underlying cause of these cycles has been disputed (Hudson *et al.*, 1998; Lambin *et al.*, 1999; Mougeot *et al.*, 2003), but one mechanism receiving strong support has been the influence of the parasitic nematode, *Trichostrongylus tenuis*, occupying the birds' gut ceca and reducing survival and breeding production (Figure 12.24b, c).

A model for this type of host–macroparasite interaction is described in Figure 12.25. Its analysis suggests that regular cycles both of host abundance and of mean number of parasites per host will be generated if:

$$\delta > \alpha k. \tag{12.11}$$

Here, δ is the parasite-induced reduction in host fecundity (relatively delayed density dependence: destabilizing), α is the parasite-induced host death rate (relatively direct density dependence: stabilizing), and k is the 'aggregation parameter' for the (assumed) negative binomial distribution of parasites amongst hosts. Cycles arise when the destabilizing effects of reduced fecundity outweigh the stabilizing effects of both increased mortality and the aggregation of parasites (providing a 'partial refuge' for the hosts) (see Chapter 10). Data from a cyclic study population in the north of England indicate that this condition is indeed satisfied. On the other hand, grouse populations that fail to show regular cycles or show them only very sporadically are often those in which the nematode cannot properly establish (S_T exceeds typical host abundance) (Dobson & Hudson, 1992; Hudson *et al.*, 1992b).

Such results from models are supportive of a role for the parasites in grouse cycles, but they fall short of the type of 'proof' that can come from a controlled experiment. A simple modification of the model in Figure 12.25, however, predicted that if a sufficient proportion (20%) of the population were treated for their nematodes with an anthelminthic, then the cycles would die out. This set the scene for a field-scale experimental manipulation designed to test the parasite's role (Hudson *et al.*, 1998). In two populations, the grouse were treated with anthelminthics in the expected years of two successive population crashes; in two others, the grouse were treated only in the expected year of one crash; while two further populations were monitored as unmanipulated controls. Grouse abundance was measured as 'bag records': the number of grouse shot. It is clear that the anthelminthic had an effect in the experiment (Figure 12.24d), and it is therefore equally clear that the parasites normally have an effect: that is, the parasites affected host dynamics. The precise nature of that effect, however, remains a matter of some controversy. Hudson and his colleagues themselves believed that the experiment demonstrated that the parasites were 'necessary and sufficient' for host cycles. Others felt that rather less had been fully demonstrated, suggesting for example that the cycles may have been reduced in amplitude rather than eliminated, especially as the very low numbers normally 'observed' in a trough (1 on their logarithmic scale equates to zero) are a result of there being no shooting when abundance is low (Lambin *et al.*, 1999; Tompkins & Begon, 1999). On the other hand, such controversy should not be seen as detracting from the importance of field-scale experiments in investigating the roles of parasites in the dynamics of host populations – nor, indeed, the roles of other factors. For example, a subsequent field manipulation supported the alternative hypothesis that red grouse cycles are the result of density-dependent changes in aggressiveness and the spacing behavior of males (Mougeot *et al.*, 2003). This system is examined again in a general discussion of cycles in Section 14.6.2.

12.7.3 Svarlbard reindeer and nematodes

Next, we stay with nematodes but switch to a mammal, the Svarlbard reindeer, *Rangifer tarandus plathyrynchus*, on the island of Svarlbard (Spitzbergen), north of Norway (Albon *et al.*, 2002). The system is attractive for its simplicity (the effects may be visible, uncluttered by complicating factors): (i) there are no mammalian herbivores competing with the reindeer for food; (ii) there are no mammalian predators; and (iii) the parasite community of the reindeer is itself very simple, dominated by two gastrointestinal nematodes, neither with an alternative host and only one of which, *Ostertagia gruehneri*, has a demonstrable pathogenic effect.

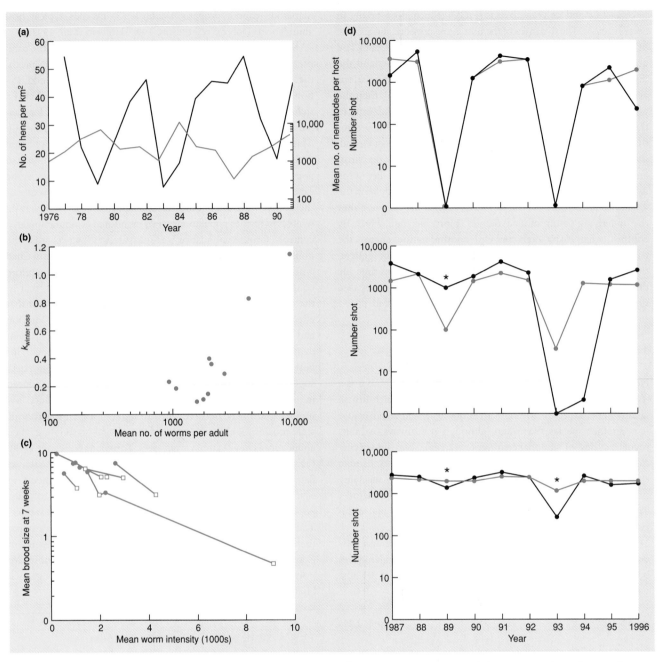

Figure 12.24 (a) Regular cycles in the abundance (breeding hens per km²) of red grouse (———) and the mean number of nematodes, *Trichostrongylus tenuis*, per host (———) at Gunnerside, UK. (b) *Trichostrongylus tenuis* reduces survival in the red grouse: over 10 years (1980–89) winter loss (measured as a *k* value) increased significantly ($P < 0.05$) with the mean number of worms per adult. (c) *T. tenuis* reduces fecundity in the red grouse: in each of 8 years, females treated with a drug to kill nematodes (●; representing mean values) had fewer worms and larger brood sizes (at 7 weeks) than untreated females (□). ((a–c) after Dobson & Hudson, 1992; Hudson *et al.*, 1992.) (d) Population changes of red grouse, as represented through bag records in two control sites (above), two populations with a single treatment each against nematodes (middle), and two populations with two treatments each (below). Asterisks represent the years of treatment when worm burdens in adult grouse were reduced by an anthelmintic. (After Hudson *et al.*, 1998.)

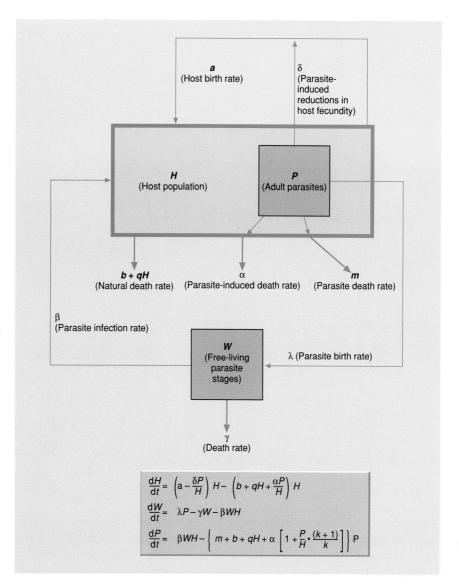

Figure 12.25 Flow diagram (above) depicting the dynamics of a macroparasitic infection such as the nematode *Trichostrongylus tenuis* in red grouse, where the parasite has free-living infective stages; and (below) the model equations describing those dynamics. Taking the equations in order, they describe: (i) hosts (H) increasing as a result of (density-independent) births (which, however, are reduced at a rate dependent on the average number of parasites per host, P/H), but decreasing as a result of deaths – both natural (density dependent) and induced by the parasite (again dependent on P/H); (ii) free-living parasite stages (W) increasing as a result of being produced by parasites in infected hosts, but decreasing both as a result of death and by being consumed by hosts; and (iii) parasites within hosts (P) increasing as a result of being consumed by hosts, but decreasing as a result of their own death within hosts, of the natural death of the hosts themselves and of disease-induced death of hosts. This final term is dependent on the distribution of parasites amongst hosts – here assumed to follow a negative binomial distribution, parameter k, accounting for the term in square brackets. (After Anderson & May, 1978; Dobson & Hudson, 1992.)

$$\frac{dH}{dt} = \left(a - \frac{\delta P}{H}\right)H - \left(b + qH + \frac{\alpha P}{H}\right)H$$

$$\frac{dW}{dt} = \lambda P - \gamma W - \beta WH$$

$$\frac{dP}{dt} = \beta WH - \left\{m + b + qH + \alpha\left[1 + \frac{P}{H}\cdot\frac{(k+1)}{k}\right]\right\}P$$

Over a period of 6 years, reindeer were treated with an anthelminthic each spring (April), and the effect of this on pregnancy rates 1 year later, as well as on subsequent calf production, was noted. Infection appeared to have no effect on survival, but untreated (i.e. infected) females had significantly lower pregnancy rates, after year-to-year variation had been accounted for ($X_1^2 = 4.92$, $P = 0.03$; Figure 12.26a), an effect that was maintained in the data on calf production. The extent of this effect increased significantly with increases in the abundance of the nematode in the previous fall ($F_{1,4} = 52.9$, $P = 0.002$; Figure 12.26b). Moreover, the abundance of the nematodes themselves was significantly and positively related to the density of reindeer 2 years earlier (Figure 12.26c). Hence, increases in host abundance appear to lead (after a delay) to increases in parasite abundance; increases in parasite abundance appear to lead (after a further delay) to

reductions in host fecundity; and reductions in host fecundity clearly have the potential to lead to reductions in host abundance.

In order to ask whether this circle was completed in practice, such that the parasite *did* regulate reindeer abundance, these various relationships, along with others, were fed as parameter values into a model of the reindeer–nematode interaction. Results are shown in Figure 12.26d. Three outcomes are possible: either the reindeer population is driven to extinction, or it shows unbounded exponential growth, or it is regulated to the numbers per square kilometer shown in the figure. Encouragingly, within the observed ranges of calf and old reindeer survival, the model predicts reindeer densities very much in line with those observed (around 1–3 km^{-2}). In the absence of an effect of the nematode on calf production, the model predicts unbounded growth. Thus, together, field experiments and observations, and a mathematical

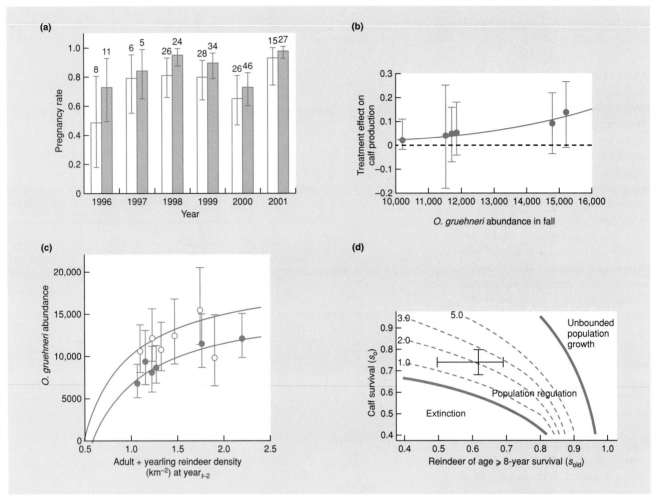

Figure 12.26 (a) The estimated pregnancy rate in April–May in controls (open bars) and reindeer treated with anthelminthics 12 months earlier (shaded bars). Numbers over the bars give the sample size of animals with pregnancy status determined. (b) The difference in the calf production of reindeer treated with anthelminthics in the previous April–May and controls, in relation to the estimated *Ostertagia gruehneri* abundance in October. (c) The estimated *Ostertagia gruehneri* abundance in October in relation (curvilinear regression) to adult and yearling reindeer summer density 2 years earlier at two sites: Colesdalen (●) and Sassendalen (○). Error bars in (a–c) give 95% confidence limits of the estimates. (d) Summary of the output from a model of the Svalbard reindeer population dynamics, using the range of possible values of annual calf survival and the annual survival of reindeer more than or equal to 8 years old. The bold lines give the boundaries between the parameter space where the host population becomes extinct, or is regulated, or shows unbounded growth. The dotted lines give the combination of parameter values in the regulated zone that give an average adult + yearling population density of 1, 2, 3 and 5 reindeer per km². The crossed bars indicate ranges of estimated values. (After Albon *et al.*, 2002.)

model, provide powerful support for a role of the nematodes in the dynamics of the Svarlbard reindeer.

12.7.4   Red foxes and rabies

We turn last to rabies: a directly transmitted viral disease of vertebrates, including humans, that attacks the central nervous system and is much feared both for the unpleasantness of its symptoms and the high probability of death once it has taken hold. In Europe, recent interest has focused on the interaction between rabies and the red fox (*Vulpes vulpes*). An epidemic in foxes spread westwards and southwards from the Polish–Russian border from the 1940s, and whilst the direct threat to humans is almost certainly slight, there is an economically significant transmission of rabies from foxes to cattle and sheep. The authorities in Great Britain have been especially worried about rabies since the disease has yet to cross the English Channel from mainland Europe, but there has been a strong desire to eliminate rabies from the European mainland too (Pastoret & Brochier, 1999). In this case,

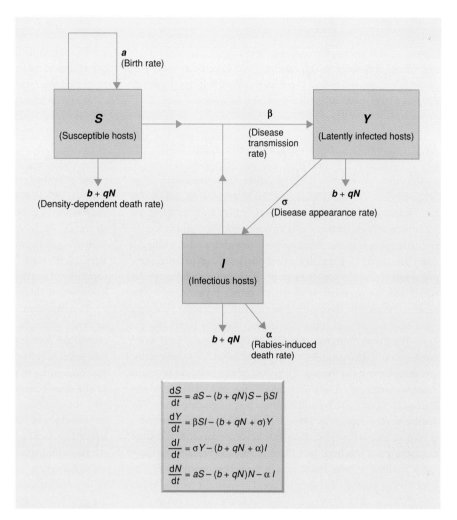

Figure 12.27 Flow diagram (above) depicting the dynamics of a rabies infection of a vertebrate host (such as the fox) and (below) the model equations describing those dynamics. Taking the equations in order, they describe: (i) susceptible hosts (S) increasing as a result of (density-independent) birth from the susceptible class only, but decreasing both as a result of natural (density-dependent) death and also by becoming infected through contact with infectious hosts; (ii) latently infected (noninfectious) hosts (Y) increasing as a result of susceptibles becoming infected, and decreasing both as a result of natural (density-dependent) death and also (as the rabies appears) by becoming converted into infectious hosts; and (iii) infectious hosts (I) increasing as a result of disease development in latently infected hosts, but decreasing as a result of natural and disease-induced mortality. Finally, the equation for the total host population (N = S + Y + I) is derived by summing the equations for S, Y and I. (After Anderson *et al.*, 1981.)

we look at the use of a model, first, to capture the observed host–pathogen dynamics in the field (and thus lend credibility to that model) and then to ask whether those dynamics can usefully be manipulated. That is: do we know enough about fox–rabies population dynamics to suggest how further spread of the disease might be prevented and how it might even be eliminated where it already exists?

A simple model of fox–rabies dynamics is described in Figure 12.27. This does indeed seem to capture the essence of the interaction successfully, since, with values for the various biological parameters taken from field data, the model predicts regular cycles of fox abundance and rabies prevalence, around 4 years in length – just like those found in a number of areas where rabies is established (Anderson *et al.*, 1981).

There are two methods that have a realistic chance of controlling rabies in foxes. The first is to kill numbers of them on a continuing basis, so as to hold their abundance below the rabies transmission threshold. The model suggests that this is around 1 km^{-2}, which is itself a helpful piece of information, given credence by the ability of the model to recreate observed

dynamics. As discussed much more fully in Chapter 15 (in the context of harvesting), the problem with repeated culls of this type is that by reducing density they relieve the pressure of intraspecific competition, leading to increases in birth rates and declines in natural death rates. Thus, culling becomes rapidly more problematic the greater the gap between the normal density and the target density (in this case, 1 km^{-2}). Culling may, therefore, be feasible with natural densities of around only 2 km^{-2}. However, since densities in, for instance, Great Britain often average 5 km^{-2} and may reach 50 km^{-2} in some urban areas, culls of a sufficient intensity will usually be unattainable. Culling will typically be of little practical use.

The second potential control method is vaccination – in this case, the placement of oral vaccine in baits to which the foxes are attracted. Such methods can reach around 80% of a fox population. Is that enough? The formula for answering this has already been given as Equation 12.7; the application of which suggests that vaccination should be successful at natural fox densities of up to 5 km^{-2}. Vaccination should therefore be successful, for example, throughout much of Great Britain, but appears to offer little hope of control in many urban areas. In fact, more than 20 years after the development of the model in Figure 12.27, rabies has still not spread to Great Britain, and the use of ever-improving oral vaccines appears to have halted the spread of rabies in Europe and indeed eliminated it from Belgium, Luxembourg and large parts of France (Pastoret & Brochier, 1999).

12.8   Coevolution of parasites and their hosts

myxomatosis

It may seem straightforward that parasites in a population select for the evolution of more resistant hosts, which in turn select for more infective parasites: a classic coevolutionary arms race. In fact, the process is not necessarily so straightforward, although there are certainly examples where the host and parasite drive one another's evolution. A most dramatic example involves the rabbit and the myxoma virus, which causes myxomatosis. The virus originated in the South American jungle rabbit *Sylvilagus brasiliensis*, where it causes a mild disease that only rarely kills the host. The South American virus, however, is usually fatal when it infects the European rabbit *Oryctolagus cuniculus*. In one of the greatest examples of the biological control of a pest, the myxoma virus was introduced into Australia in the 1950s to control the European rabbit, which had become a pest of grazing lands. The disease spread rapidly in 1950–51, and rabbit populations were greatly reduced – by more than 90% in some places. At the same time, the virus was introduced to England and France, and there too it resulted in huge reductions in the rabbit populations. The evolutionary changes that then occurred in Australia were followed

in detail by Fenner and his associates (Fenner & Ratcliffe, 1965; Fenner, 1983) who had the brilliant research foresight to establish baseline genetic strains of both the rabbits and the virus. They used these to measure subsequent changes in the virulence of the virus and the resistance of the host as they evolved in the field.

When the disease was first introduced to Australia it killed more than 99% of infected rabbits. This 'case mortality' fell to 90% within 1 year and then declined further (Fenner & Ratcliffe, 1965). The virulence of isolates of the virus sampled from the field was graded according to the survival time and the case mortality of control rabbits. The original, highly virulent virus (1950–51) was grade I, which killed 99% of infected laboratory rabbits. Already by 1952 most of the virus isolates from the field were the less virulent grades III and IV. At the same time, the rabbit population in the field was increasing in resistance. When injected with a standard grade III strain of the virus, field samples of rabbits in 1950–51 had a case mortality of nearly 90%, which had declined to less than 30% only 8 years later (Marshall & Douglas, 1961) (Figure 12.28).

This evolution of resistance in the European rabbit is easy to understand: resistant rabbits are obviously favored by natural selection in the presence of the myxoma virus. The case of the virus, however, is subtler. The contrast between the virulence of the myxoma virus in the European rabbit and its lack of virulence in the American host with which it had coevolved, combined with the attenuation of its virulence in Australia and Europe after its introduction, fit a commonly held view that parasites evolve toward becoming benign to their hosts in order to prevent the parasite eliminating its host and thus eliminating its habitat. This view, however, is quite wrong. The parasites favored by natural selection are those with the greatest fitness (broadly, the greatest

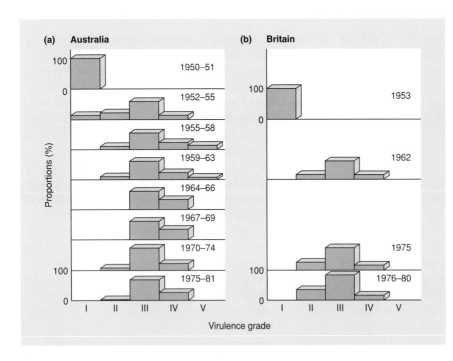

Figure 12.28   (a) The percentages in which various grades of myxoma virus have been found in wild populations of rabbits in Australia at different times from 1951 to 1981. Grade I is the most virulent. (After Fenner, 1983.) (b) Similar data for wild populations of rabbits in Great Britain from 1953 to 1980. (After May & Anderson, 1983; from Fenner, 1983.)

reproductive rate). Sometimes this is achieved through a decline in virulence, but sometimes it is not. In the myxoma virus, an initial decline in virulence was indeed favored – but further declines were not.

The myxoma virus is blood-borne and is transmitted from host to host by blood-feeding insect vectors. In Australia in the first 20 years after its introduction, the main vectors were mosquitoes (especially *Anopheles annulipes*), which feed only on live hosts. The problem for grade I and II viruses is that they kill the host so quickly that there is only a very short time in which the mosquito can transmit them. Effective transmission may be possible at very high host densities, but as soon as densities decline, it is not. Hence, there was selection against grades I and II and in favor of less virulent grades, giving rise to longer periods of host infectiousness. At the other end of the virulence scale, however, the mosquitoes are unlikely to transmit grade V of the virus because it produces very few infective particles in the host skin that could contaminate the vectors' mouthparts. The situation was complicated in the late 1960s when an alternative vector of the disease, the rabbit flea *Spilopsyllus cuniculi* (the main vector in England), was introduced to Australia. There is some evidence that more virulent strains of the virus may be favored when the flea is the main vector (see discussion in Dwyer *et al.*, 1990).

Overall, then, there has been selection in the rabbit–myxomatosis system not for decreased virulence as such, but for *increased transmissibility* (and hence increased fitness) – which happens in this system to be maximized at intermediate grades of virulence. Many parasites of insects rely on killing their host for effective transmission. In these, very high virulence is favored. In yet other cases, natural selection acting on parasites has clearly favored very low virulence: for example, the human herpes simplex virus may do very little tangible harm to its host but effectively gives it lifelong infectiousness. These differences reflect differences in the underlying host–parasite ecologies, but what the examples have in common is that there has been evolution toward increased parasite fitness.

bacteria and bacteriophages

In other cases, coevolution is more definitely antagonistic: increased resistance in the host and increased infectivity in the parasite. A classic example is the interaction between agricultural plants and their pathogens (Burdon, 1987), although in this case the resistant hosts are often introduced by human intervention. There may even be gene-for-gene matching, with a particular virulence allele in the pathogen selecting for a resistant allele in the host, which in turn selects for alleles other than the original allele in the pathogen, and so on. This, moreover, may give rise to polymorphism in the parasite and host, either as a result of different alleles being favored in different subpopulations, or because several alleles are simultaneously in a state of flux within their population, each being favored when they (and their matching allele in the other partner) is rare. In fact, such detailed processes have proved difficult to observe, but this has

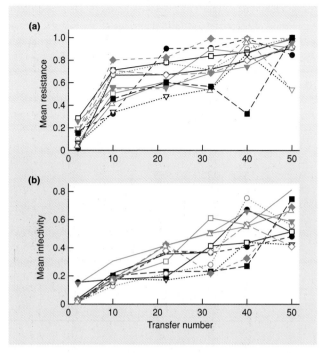

Figure 12.29   (a) Over evolutionary time (1 'transfer' ≈ 8 bacterial generations) bacterial resistance to phage increased in each of 12 bacterial replicates. 'Mean' resistance was the mean calculated over the 12 phage isolates from the respective time points. (b) Similarly, phage infectivity increased, where 'mean' infectivity was calculated over the 12 bacterial replicates. (After Buckling & Rainey, 2002.)

been done with a system comprising the bacterium *Pseudomonas fluorescens* and its viral parasite, the bacteriophage (or phage) SBW25φ2 (Buckling & Rainey, 2002).

Changes in both the host and parasite were monitored over evolutionary time, as 12 replicate coexisting populations of bacterium and phage were transferred from culture bottle to culture bottle. It is apparent that the bacteria became generally more resistant to the phage at the same time as the phage became generally more infective to the bacteria (Figure 12.29): each was being driven by the directional selection of an arms race. But this was only apparent because any given bacterial strain (from one of the 12 replicates) was tested against all 12 phage strains, and the phage strains were tested similarly. When, at the end of the experiment (Table 12.4), the resistance of each bacterial strain was tested against each phage strain in turn, it was clear that the bacteria were almost always *most* resistant (and often wholly resistant) to the phage strain with which they coevolved. There was therefore extensive evolutionary divergence amongst the strains – or subpopulations – and extensive polymorphism within the metapopulation as a whole.

Thus we close this chapter, appropriately, with another reminder that despite being relatively neglected by ecologists in the past, parasites are increasingly being recognized as major players in both the ecological and the evolutionary dynamics of their hosts.

Table 12.4   For each of 12 bacterial replicates (B1–B12) and their 12 respective phage replicates (φ1–φ12), entries in the table are the proportion of bacteria resistant to the phage at the end of a period of coevolution (50 transfers ≈ 400 bacterial generations). Coevolving pairs are shown along the diagonal in bold. Note that bacterial strains are usually most resistant to the phage strain with which they coevolved. (After Buckling & Rainey, 2002.)

Phage replicates	Bacterial replicates											
	B1	B2	B3	B4	B5	B6	B7	B8	B9	B10	B11	B12
φ1	**0.8**	0.9	1	1	1	1	1	1	0.85	0.85	0.75	0.65
φ2	0.1	**1**	0.3	1	0.85	0.25	1	1	0.85	0.9	0.8	0.65
φ3	0.75	0.75	**1**	1	1	0.9	1	1	0.85	0.9	0.9	0.65
φ4	0.15	0.9	0.8	**1**	0.85	0.6	0.6	1	0.85	1	0.85	0.35
φ5	0.25	0.9	1	1	**1**	0.9	1	0.8	0.85	1	0.8	0.65
φ6	0.2	1	0.85	0.8	0.75	**0.8**	0.85	0.9	0.85	0.75	0.45	0.25
φ7	0.2	0.75	0.6	1	0.4	0.45	**1**	0.9	0.85	1	0.75	0.35
φ8	0	0.95	0.55	0.95	0.35	0.25	0.8	**1**	0.85	1	0.7	0.25
φ9	0	0.7	0.55	0.45	0.7	0.35	1	1	**0.85**	1	0.5	0.1
φ10	0	0.7	0.9	0.7	0.55	0.9	1	1	0.7	**1**	0.5	0.4
φ11	0	0.5	0.9	0.75	0.7	1	1	0.95	0.75	1	**1**	0.35
φ12	0	0.15	0	0.1	0.65	0.35	1	1	0.7	0.8	0.85	**0.4**

Summary

We begin by defining parasite, infection, pathogen and disease. The diversity of animal and plant parasites is then outlined, based on the distinctions between micro- and macroparasites and between those with direct and those with indirect (vectored) life cycles. The particular case of social and brood parasites (e.g. cuckoos) is also described.

We explain the difference between biotrophic and necrotrophic parasites (pioneer saprotrophs), and we use a discussion of zoonoses (wildlife infections transmissible to man) to illustrate the nature of host specificity amongst parasites.

Hosts are reactive environments: they may resist, or recover, or (in vertebrates) acquire immunity. We describe the contrasting responses in vertebrates to micro- and macroparasites and contrast these in turn with the responses of plants to infection. The costliness of host defense against attack is emphasized. Parasites may also induce profound changes in host growth and behavior.

We explain why it may be difficult to distinguish the effects of intraspecific competition amongst parasites from parasite density-dependent host immune responses, and that patterns associated with interspecific competition are as observable amongst parasites as they are in other organisms.

The distinctions between different types of parasite transmission are outlined and a formal description of transmission dynamics is developed, using the form of the contact rate to distinguish between density- and frequency-dependent transmission, though it is emphasized that these may merely be ends of a spectrum. There may also be great spatial variation in the speed with which infection spreads, either as a result of infectious foci or because of spatial mixtures of susceptible and resistant species or varieties.

The distribution of parasites within host populations is usually aggregated. This makes it especially important to understand the distinctions between prevalence, intensity and mean intensity.

We discuss the effects of parasites on the survivorship, growth and fecundity of hosts. The effects are often subtle, affecting, for example, interactions of hosts with other species.

We then examine the dynamics of infection within host populations. Key concepts here are the basic reproductive rate, R_0, the transmission threshold ($R_0 = 1$) and the critical population size. These form a framework for directly transmitted microparasites that sheds light on the kinds of population in which we might expect to find different sorts of infection, on the nature of the epidemic curve of an infection, on the dynamic patterns of different types of parasite, and on the planning of immunization programs based on the principle of 'herd immunity'.

The dynamics are also outlined of pathogens attacking crops, of vector-borne infections and macroparasites, and of parasites infecting metapopulations of hosts.

We examine the role that parasites and pathogens play in the dynamics of their hosts. We address first the question of whether host and parasite dynamics are coupled, or whether the parasite simply modifies the underlying dynamics of the host, without there being any detectable feedback. A series of case studies then emphasizes that data supporting a role for parasites in the dynamics of their hosts are sparse and often liable to alternative interpretations.

Finally, we consider the coevolution of parasites and their hosts, stressing the absence of any 'cosy accommodation', but rather that the selective pressures in both cases – parasite and host – favor maximizing individual fitness.

Chapter 13

Symbiosis and Mutualism

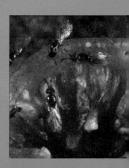

13.1 Introduction: symbionts, mutualists, commensals and engineers

No species lives in isolation, but often the association with another species is especially close: for many organisms, the habitat they occupy is an individual of another species. Parasites live within the body cavities or even the cells of their hosts; nitrogen-fixing bacteria live in nodules on the roots of leguminous plants; and so on. Symbiosis ('living together') is the term that has been coined for such close physical associations between species, in which a 'symbiont' occupies a habitat provided by a 'host'.

In fact, parasites are usually excluded from the category of symbionts, which is reserved instead for interactions where there is at least the suggestion of 'mutualism'. A mutualistic relationship is simply one in which organisms of different species interact to their mutual benefit. It usually involves the direct exchange of goods or services (e.g. food, defense or transport) and typically results in the acquisition of novel capabilities by at least one partner (Herre *et al.*, 1999). Mutualism, therefore, need not involve close physical association: mutualists need not be symbionts. For example, many plants gain dispersal of their seeds by offering a reward to birds or mammals in the form of edible fleshy fruits, and many plants assure effective pollination by offering a resource of nectar in their flowers to visiting insects. These are mutualistic interactions but they are not symbioses.

| mutualism: reciprocal exploitation not a cosy partnership | It would be wrong, however, to see mutualistic interactions simply as conflict-free relationships from which nothing but good things flow for both partners. Rather, current evolutionary |

thinking views mutualisms as cases of reciprocal exploitation where, none the less, each partner is a *net* beneficiary (Herre & West, 1997).

Nor are interactions in which one species provides the habitat for another necessarily either mutualistic (both parties benefit: '+ +') or parasitic (one gains, one suffers: '+ −'). In the first place, it may simply not be possible to establish, with solid data, that each of the participants either benefits or suffers. In addition, though, there are many 'interactions' between two species in which the first provides a habitat for the second, but there is no real suspicion that the first either benefits or suffers in any measurable way as a consequence. Trees, for example, provide habitats for the many species of birds, bats and climbing and scrambling animals that are absent from treeless environments. Lichens and mosses develop on tree trunks, and climbing plants such as ivy, vines and figs, though they root in the ground, use tree trunks as support to extend their foliage up into a forest canopy. Trees are therefore good examples of what have been called ecological or ecosystem 'engineers' (Jones *et al.*, 1994). By their very presence, they create, modify or maintain habitats for others. In aquatic communities, the solid surfaces of larger organisms are even more important contributors to biodiversity. Seaweeds and kelps normally grow only where they can be anchored on rocks, but their fronds are colonized in turn by filamentous algae, by tube-forming worms (*Spirorbis*) and by modular animals such as hydroids and bryozoans that depend on seaweeds for anchorage and access to resources in the moving waters of the sea.

More generally, many of these are likely to be examples of commensal 'interactions' (one partner gains, the other is neither harmed nor benefits: '+ 0'). Certainly, those cases where the harm to the host of a 'parasite' or the benefit to a 'mutualist' cannot be established should be classified as commensal or 'host–guest', bearing in mind that, like guests under other circumstances, they may be unwelcome when the hosts are ill or distressed! Commensals have received far less study than parasites and mutualists, though many of them have ways of life that are quite as specialized and fascinating.

Mutualisms themselves have often been neglected in the past compared to other types of interaction, yet mutualists compose most of the world's biomass. Almost all the plants that dominate

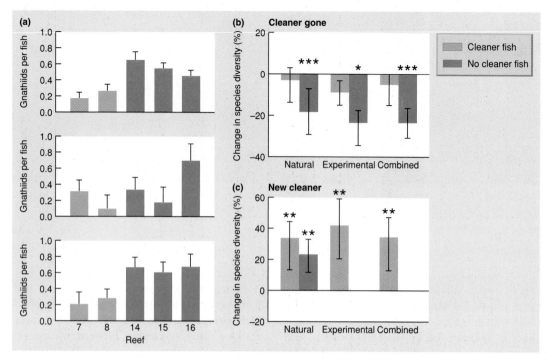

Figure 13.1   (a) Cleaner fish really do clean their clients. The mean number of gnathiid parasites per client (*Hemigymnus melapterus*) at five reefs, from three of which (14, 15 and 16) the cleaners (*Labroides dimidiatus*) were experimentally removed. In a 'long-term' experiment, clients without cleaners had more parasites after 12 days (upper panel: $F = 17.6$, $P = 0.02$). In a 'short-term' experiment, clients without cleaners did not have significantly more parasites at dawn after 12 h (middle panel: $F = 1.8$, $P = 0.21$), presumably because cleaners do not feed at night, but the difference was significant after a further 12 h of daylight (lower panel: $F = 11.6$, $P = 0.04$). Bars represent standard errors. (After Grutter, 1999.) (b) Cleaners increase reef fish diversity. The percentage change in the number of fish species present following natural or experimental loss of a cleaner fish, *L. dimidiatus*, from a reef patch (or the two treatments combined), in the short term (2–4 weeks, light bars) and the long term (4–20 months, dark bars). (c) The percentage change in the number of fish species present following natural or experimental immigration of a cleaner fish, *L. dimidiatus*, into a reef patch (or the two treatments combined), in the short term (2–4 weeks, light bars) and the long term (4–20 months, dark bars). The columns and error bars represent medians and interquartiles. *, $P < 0.05$; **, $P < 0.01$; ***, $P < 0.001$. (After Bshary, 2003.)

grasslands, heaths and forests have roots that have an intimate mutualistic association with fungi. Most corals depend on the unicellular algae within their cells, many flowering plants need their insect pollinators, and many animals carry communities of microorganisms within their guts that they require for effective digestion.

The rest of this chapter is organised as a progression. We start with mutualisms in which no intimate symbiosis is involved. Rather, the association is largely behavioral: that is, each partner behaves in a manner that confers a net benefit on the other. By Section 13.5, when we discuss mutualisms between animals and the microbiota living in their guts, we will have moved on to closer associations (one partner living within the other), and in Sections 13.6–13.10 we examine still more intimate symbioses in which one partner enters between or within another's cells. In Section 13.11 we interrupt the progression to look briefly at mathematical

models of mutualisms. Then, finally, in Section 13.12 – for completeness, though the subject is not strictly 'ecological' – we examine the idea that various organelles have entered into such intimate symbioses within the cells of their many hosts that it has ceased to be sensible to regard them as distinct organisms.

13.2   Mutualistic protectors

13.2.1   Cleaner and client fish

'Cleaner' fish, of which at least 45 species have been recognized, feed on ectoparasites, bacteria and necrotic tissue from the body surface of 'client' fish. Indeed, the cleaners often hold territories with 'cleaning stations' that their clients visit – and visit more often when they carry many parasites. The cleaners gain a food source and the clients are protected from infection. In fact, it has not always proved easy to establish that the clients benefit, but in experiments off Lizard Island on Australia's Great Barrier Reef, Grutter (1999) was able to do this for the cleaner fish, *Labroides dimidiatus*, which eats parasitic gnathiid isopods from its client fish, *Hemigymnus melapterus*. Clients had significantly (3.8 times) more parasites 12 days after cleaners were excluded from caged enclosures (Figure 13.1a, top panel); but even in the short term (up to 1 day), although removing cleaners, which only feed during daylight, had no effect when a check was made at dawn (middle panel), this led to there being significantly (4.5 times) more parasites following a further day's feeding (lower panel).

effects at the community level, too Moreover, further experiments using the same cleaner fish, but at a Red Sea reef in Egypt, emphasized the community-wide importance of these cleaner–client interactions (Bshary, 2003). When cleaners either left a reef patch naturally (so the patch had no cleaner) or were experimentally removed, the local diversity (number of species) of reef fish dropped dramatically, though this was only significant after 4–20 months, not after 2–4 weeks (Figure 13.1b). However, when cleaners either moved into a cleanerless patch naturally or were experimentally added, diversity increased significantly even within a few weeks (Figure 13.1c). Intriguingly, these effects applied not only to client species but to nonclients too.

In fact, several behavioral mutualisms are found amongst the inhabitants of tropical coral reefs (where the corals themselves are mutualists – see Section 13.7.1). The clown fish (*Amphiprion*), for example, lives close to a sea anemone (e.g. *Physobrachia*, *Radianthus*) and retreats amongst the anemone's tentacles when-ever danger threatens. Whilst within the anemone, the fish gains a covering of mucus that protects it from the anemone's stinging nematocysts (the normal function of the anemone slime is to prevent discharge of nematocysts when neighboring tentacles touch). The fish derives protection from this relationship, but the anemone also benefits because clown fish attack other fish that come near, including species that normally feed on the sea anemones.

13.2.2 Ant–plant mutualisms

The idea that there are mutualistic relationships between plants and ants was put forward by Belt (1874) after observing the behavior of aggressive ants on species of *Acacia* with swollen thorns in Central America. This relationship was later described more fully by Janzen (1967) for the Bull's horn acacia (*Acacia cornigera*) and its associated ant, *Pseudomyrmex ferruginea*. The plant bears hollow thorns that are used by the ants as nesting sites; its leaves have protein-rich 'Beltian bodies' at their tips (Figure 13.2) which the ants collect and use for food; and it has sugar-secreting nectaries on its vegetative parts that also attract the ants. The ants, for their part, protect these small trees from competitors by actively snipping off shoots of other species and also protect the plant from herbivores – even large (vertebrate) herbivores may be deterred.

In fact, ant–plant mutualisms appear to have evolved many times (even repeatedly in the same family of plants); and nectaries are present on *do the plants benefit?*

(a)

(b)

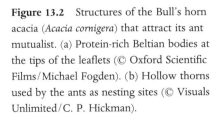

Figure 13.2 Structures of the Bull's horn acacia (*Acacia cornigera*) that attract its ant mutualist. (a) Protein-rich Beltian bodies at the tips of the leaflets (© Oxford Scientific Films/Michael Fogden). (b) Hollow thorns used by the ants as nesting sites (© Visuals Unlimited/C. P. Hickman).

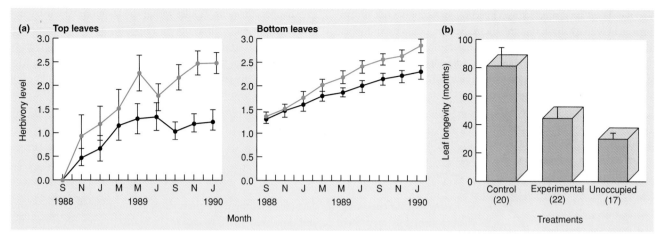

Figure 13.3 (a) The intensity of leaf herbivory on plants of *Tachigali myrmecophila* naturally occupied by the ant *Pseudomyrmex concolor* (●, n = 22) and on plants from which the ants had been experimentally removed (●, n = 23). Bottom leaves are those present at the start of the experiment and top leaves are those emerging subsequently. (b) The longevity of leaves on plants of *T. myrmecophila* occupied by *P. concolor* (control) and from which the ants were experimentally removed or from which the ants were naturally absent. Error bars ± standard error. (After Fonseca, 1994.)

the vegetative parts of plants of at least 39 families and in many communities throughout the world. Nectaries on or in flowers are easily interpreted as attractants for pollinators. But the role of extrafloral nectaries on vegetative parts is less easy to establish. They clearly attract ants, sometimes in vast numbers, but carefully designed and controlled experiments are necessary to show that the plants themselves benefit, such as the study of the Amazonian canopy tree *Tachigali myrmecophila*, which harbors the stinging ant *Pseudomyrmex concolor* in specialized hollowed-out structures (Figure 13.3). The ants were removed from selected plants;

these then bore 4.3 times as many phytophagous insects as control plants and suffered much greater herbivory. Leaves on plants that carried a population of ants lived more than twice as long as those on unoccupied plants and nearly 1.8 times as long as those on plants from which ants had been deliberately removed.

Mutualistic relationships, in this case between individual ant and plant species, should not, however, be viewed in isolation – a theme that will recur in this chapter. Palmer *et al.* (2000), for example, studied competition

competition amongst mutualistic ants

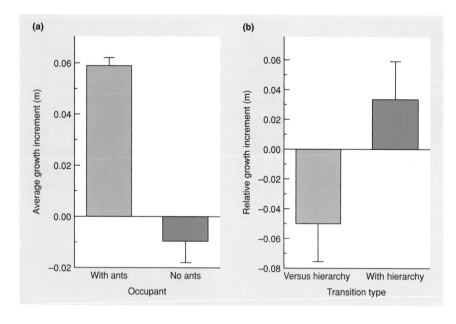

Figure 13.4 (a) Average growth increment was significantly greater (P < 0.0001) for *Acacia drepanolobium* trees continually occupied by ants (n = 651) than for uninhabited trees (n = 126). 'Continually occupied' trees were occupied by ant colonies at both an initial survey and one 6 months later. Uninhabited trees were vacant at the time of both surveys. (b) Relative growth increments were significantly greater (P < 0.05) for trees undergoing transitions in ant occupancy in the direction of the ants' competitive hierarchy (n = 85) than for those against the hierarchy (n = 48). Growth increment was determined relative to trees occupied by the same ant species when these ants were not displaced. Error bars show standard errors. (After Palmer *et al.*, 2000).

amongst four species of ant that have mutualistic relationships with *Acacia drepanolobium* trees in Laikipia, Kenya, nesting within the swollen thorns and feeding from the nectaries at the leaf bases. Experimentally staged conflicts and natural take-overs of plants both indicated a dominance hierarchy among the ant species. *Crematogaster sjostedti* was the most dominant, followed by *C. mimosae*, *C. nigriceps* and *Tetraponera penzigi*. Irrespective of which ant species had colonized a particular acacia tree, occupied trees tended to grow faster than unoccupied trees (Figure 13.4a). This confirmed the mutualistic nature of the interactions overall. But more subtly, changes in ant occupancy in the direction of the dominance hierarchy (take-over by a more dominant species) occurred on plants that grew faster than average, whereas changes in the opposite direction to the hierarchy occurred on plants that grew more slowly than average (Figure 13.4b).

These data therefore suggest that take-overs are rather different on fast and slow growing trees, though the details remain speculative. It may be, for example, that trees that grow fastest also produce ant 'rewards' at the greatest rate and are actively chosen by the dominant ant species; whereas slow growing trees are more readily abandoned by dominant species, with their much greater demands for resources. Alternatively, competitively superior ant species may be able to detect and preferentially colonize faster growing trees. What is clear is that these mutualistic interactions are not cosy relationships between pairs of species that we can separate from a more tangled web of interactions. The costs and benefits accruing to the different partners vary in space and time, driving complex dynamics amongst the competing ant species that in turn determine the ultimate balance sheet for the acacias.

Ant–plant interactions are reviewed by Heil and McKey (2003).

13.3 Culture of crops or livestock

13.3.1 Human agriculture

At least in terms of geographic extent, some of the most dramatic mutualisms are those of human agriculture. The numbers of individual plants of wheat, barley, oats, corn and rice, and the areas these crops occupy, vastly exceed what would have been present if they had not been brought into cultivation. The increase in human population since the time of hunter–gatherers is some measure of the reciprocal advantage to *Homo sapiens*. Even without doing the experiment, we can easily imagine the effect the extinction of humans would have on the world population of rice plants or the effect of the extinction of rice plants on the population of humans. Similar comments apply to the domestication of cattle, sheep and other mammals.

Similar 'farming' mutualisms have developed in termite and especially ant societies, where the farmers may protect individuals they exploit from competitors and predators and may even move or tend them.

13.3.2 Farming of insects by ants

Ants farm many species of aphids (homopterans) in return for sugar-rich secretions of honeydew. The 'flocks' *(farmed aphids: do they pay a price?)* of aphids benefit through suffering lower mortality rates caused by predators, showing increased feeding and excretion rates, and forming larger colonies. But it would be wrong, as ever, to imagine that this is a cosy relationship with nothing but benefits on both sides: the aphids are being manipulated – is there a price that they pay to be entered on the other side of the balance sheet (Stadler & Dixon, 1998)? This question has been addressed for colonies of the aphid *Tuberculatus quercicola* attended by the red wood ant *Formica yessensis* on the island of Hokkaido, northern Japan (Yao *et al.*, 2000). As expected, in the presence of predators, aphid colonies survived significantly longer when attended by ants than when ants were excluded by smearing ant repellent at the base of the oak trees on which the aphids lived (Figure 13.5a). However, there *were* also costs for the aphids: in an environment from which predators were excluded, and the effects of ant attendance on aphids could thus be viewed in isolation, ant-attended aphids grew less well and were less fecund than those where ants as well as predators were excluded (Figure 13.5b).

Another classic farming mutualism is that between ants and many species of lycaenid butterfly. In a number of cases, young lycaenid caterpillars feed *(ants and blue butterflies)* on their preferred food plants usually until their third or fourth instar, when they expose themselves to foraging ant workers that pick them up and carry them back to their nests – the ants 'adopt' them. There, the ants 'milk' a sugary secretion from a specialized gland of the caterpillars, and in return protect them from predators and parasitoids throughout the remainder of their larval and pupal lives. On the other hand, in other lycaenid–ant interactions the evolutionary balance is rather different. The caterpillars produce chemical signals mimicking chemicals produced by the ants, inducing the ants to carry them back to their nests and allowing them to remain there. Within the nests, the caterpillars may either act as social parasites ('cuckoos', see Section 12.2.3), being fed by the ants (e.g. the large-blue butterfly *Maculinea rebeli*, which feeds on the crossleaved gentian, *Gentiana cruciata*, and whose caterpillars mimic the larvae of the ant *Myrmica schenkii*), or they may simply prey upon the ants (e.g. another large-blue, *M. arion*, which feeds on wild thyme, *Thymus serpyllum*) (Elmes *et al.*, 2002).

13.3.3 Farming of fungi by beetles and ants

Much plant tissue, including wood, is unavailable as a direct source of food to most animals because they lack the enzymes that can digest cellulose and lignins (see Sections 3.7.2 and 11.3.1). However, many fungi possess these enzymes, and an

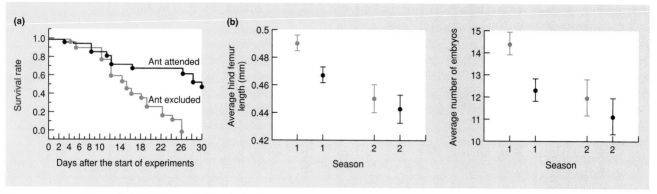

Figure 13.5 (a) Ant-excluded colonies of the aphid *Tuberculatus quercicola* were more likely to become extinct than those attended by ants ($X^2 = 15.9$, $P < 0.0001$). (b) But in the absence of predators, ant-excluded colonies perform better than those attended by ants. Shown are the averages for aphid body size (hind femur length; $F = 6.75$, $P = 0.013$) and numbers of embryos ($F = 7.25$, $P = 0.010$), $\pm$ SE, for two seasons (July 23 to August 11, 1998 and August 12 to August 31, 1998) in a predator-free environment. ●, ant-excluded treatment; ●, ant-attended treatment. (After Yao *et al.*, 2000.)

animal that can eat such fungi gains indirect access to an energy-rich food. Some very specialized mutualisms have developed between animal and fungal decomposers. Beetles in the group *Scolytidae* tunnel deep into the wood of dead and dying trees, and fungi that are specific for particular species of beetle grow in these burrows and are continually grazed by the beetle larvae. These 'ambrosia' beetles may carry inocula of the fungus in their digestive tract, and some species bear specialized brushes of hairs on their heads that carry the spores. The fungi serve as food for the beetle and in turn depend on it for dispersal to new tunnels.

Fungus-farming ants are found only in the New World, and the 210 described species appear to have evolved from a common ancestor: that is, the trait has appeared just once in evolution. The more 'primitive' species typically use dead vegetative debris as well as insect feces and corpses to manure their gardens; the genera *Trachymyrmex* and *Sericomyrmex* typically use dead vegetable matter; whereas species of the two most derived (evolutionarily 'advanced') genera, *Acromyrmex* and *Atta*, are 'leaf-cutters' using mostly fresh leaves and flowers (Currie, 2001). Leaf-cutting ants are the most remarkable of the fungus-farming ants. They excavate 2–3-liter cavities in the soil, and in these a basidiomycete fungus is cultured on leaves that are cut from neighboring vegetation (Figure 13.6). The ant colony may depend absolutely on the fungus for the nutrition of their larvae. Workers lick the fungus colonies and remove specialized swollen hyphae, which are aggregated into bite-sized 'staphylae'. These are fed to the larvae and this 'pruning' of the fungus may stimulate further fungal growth. The fungus gains from the association: it is both fed and dispersed by leaf-cutting ants and has never been found outside their nests. The reproductive female ant carries her last meal as a culture when she leaves one colony to found another.

Most phytophagous insects have very narrow diets – indeed, the vast majority of insect herbivores are strict monophages (see Section 9.5). The leaf-cutting ants are remarkable amongst insect herbivores in their polyphagy. Ants from a nest of *Atta cephalotes* harvest from 50 to 77% of the plant species in their neighborhood; and leaf-cutting ants generally may harvest 17% of total leaf production in tropical rainforest and be the ecologically dominant herbivores in the community. It is their polyphagy that gives them this remarkable status. In contrast to the *A. cephalotes* adults though, the larvae appear to be extreme dietary specialists, being restricted to nutritive bodies (gongylidia) produced by the fungus *Attamyces bromatificus*, which the adults cultivate and which decompose the leaf fragments (Cherrett *et al.*, 1989).

> leaf-cutting ants: remarkably polyphagous

Moreover, just as human farmers may be plagued by weeds, so fungus-farming ants have to contend with other species of fungi that may devastate their crop. Fungal pathogens of the genus *Escovopsis* are specialized (never found other than in fungus gardens) and virulent: in one experiment, nine of 16 colonies of the leaf-cutter *Atta colombica* that were treated with heavy doses of *Escovopsis* spores lost their garden within 3 weeks of treatment (Currie, 2001). But the ants have another mutualistic association to help them: a filamentous actinomycete bacterium associated with the surface of the ants is dispersed to new gardens by virgin queens on their nuptial flight, and the ants may even produce chemicals that promote the actinomycete's growth. For its part, the actinomycete produces an antibiotic with specialized and potent inhibitory effects against *Escovopsis*. It even appears to protect the ants themselves from pathogens and to promote the growth of the farmed fungi (Currie, 2001). *Escovopsis* therefore has ranged

> ants, farmed fungi and actinomycetes: a three-way mutualism

which often cleans itself and removes them if it can, but usually after carrying them some distance. In these cases the benefit is to the plant (which has invested resources in attachment mechanisms) and there is no reward to the animal.

Quite different are the true mutualisms between higher plants and the birds and other animals that feed on the fleshy fruits and disperse the seeds. Of course, for the relationship to be mutualistic it is essential that the animal digests only the fleshy fruit and not the seeds, which must remain viable when regurgitated or defecated. Thick, strong defenses that protect plant embryos are usually part of the price paid by the plant for dispersal by fruit-eaters. The plant kingdom has exploited a splendid array of morphological variations in the evolution of fleshy fruits (Figure 13.7).

Mutualisms involving animals that eat fleshy fruits and disperse seeds are seldom very specific to the species of animal concerned. Partly, this is because these mutualisms usually involve long-lived birds or mammals, and even in the tropics there are few plant species that fruit throughout the year and form a reliable food supply for any one specialist. But also, as will be apparent when pollination mutualisms are considered next, a more exclusive mutualistic link would require the plant's reward to be protected and denied to other animal species: this is much easier for nectar than for fruit. In any case, specialization by the animal is important in pollination, because interspecies transfers of pollen are disadvantageous, whereas with fruit and seed it is necessary only that they are dispersed away from the parent plant.

13.4.2 Pollination mutualisms

Most animal-pollinated flowers offer nectar, pollen or both as a reward to their visitors. Floral nectar seems to have no value to the plant other than as an attractant to animals and it has a cost to the plant, because the nectar carbohydrates might have been used in growth or some other activity.

Presumably, the evolution of specialized flowers and the involvement of animal pollinators have been favored because an animal may be able to recognize and discriminate between different flowers and so move pollen between different flowers of the same species but not to flowers of other species. Passive transfer of pollen, for example by wind or water, does not discriminate in this way and is therefore much more wasteful. Indeed, where the vectors and flowers are highly specialized, as is the case in many orchids, virtually no pollen is wasted even on the flowers of other species.

There are, though, costs that arise from adopting animals as mutualists in flower pollination. For example, animals carrying pollen may be responsible for the transmission of sexual diseases as well (Shykoff & Bucheli, 1995). The fungal pathogen *Microbotryum violaceum*, for example, is transmitted by pollinating visitors to the

(a)

(b)

Figure 13.6 (a) Partially excavated nest of the leaf-cutting ant *Atta vollenweideri* in the Chaco of Paraguay. The above-ground spoil heap excavated by the ants extended at least 1 m below the bottom of the excavation. (b) Queen of *A. cephalotes* (with an attendant worker on her abdomen) on a young fungus garden in the laboratory, showing the cell-like structure of the garden with its small leaf fragments and binding fungal hyphae. (Courtesy of J. M. Cherrett.)

against it not just two two-species mutualisms but a three-species mutualism amongst ants, farmed fungi and actinomycetes.

13.4   Dispersal of seeds and pollen

13.4.1   Seed dispersal mutualisms

Very many plant species use animals to disperse their seeds and pollen. About 10% of all flowering plants possess seeds or fruits that bear hooks, barbs or glues that become attached to the hairs, bristles or feathers of any animal that comes into contact with them. They are frequently an irritation to the animal,

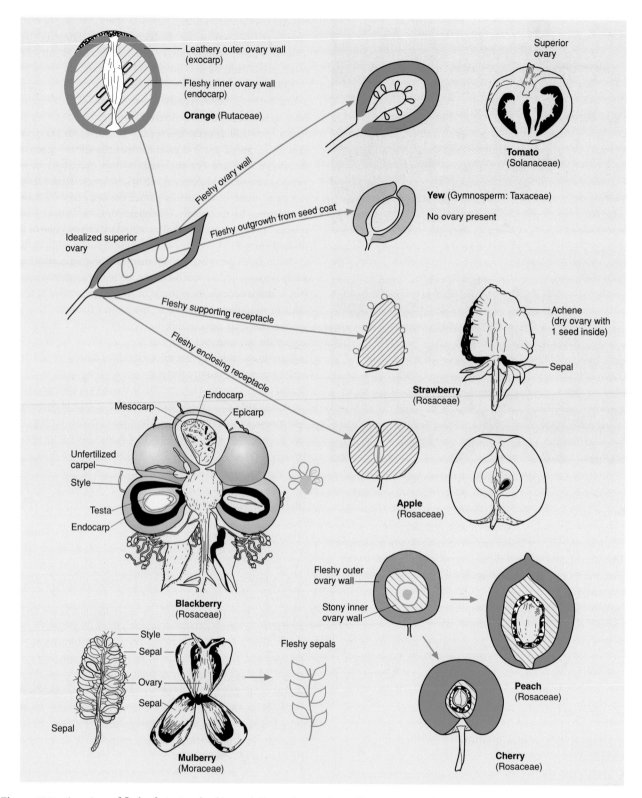

Figure 13.7 A variety of fleshy fruits involved in seed dispersal mutualisms illustrating morphological specializations that have been involved in the evolution of attractive fleshy structures.

(a)

(b)

Figure 13.8   Pollinators: (a) honeybee (*Apis mellifera*) on raspberry flowers, and (b) Cape sugarbird (*Promerops cafer*) feeding on *Protea eximia*. (Courtesy of Heather Angel.)

flowers of white campion (*Silene alba*) and in infected plants the anthers are filled with fungal spores.

insect pollinators: from generalists to ultraspecialists

Many different kinds of animals have entered into pollination liaisons with flowering plants, including humming-birds, bats and even small rodents and marsupials (Figure 13.8). However, the pollinators *par excellence* are, without doubt, the insects. Pollen is a nutritionally rich food resource, and in the simplest insect-pollinated flowers, pollen is offered in abundance and freely exposed to all and sundry. The plants rely for pollination on the insects being less than wholly efficient in their pollen consumption, carrying their spilt food with them from plant to plant. In more complex flowers, nectar (a solution of sugars) is produced as an additional or alternative reward. In the simplest of these, the nectaries are unprotected, but with increasing specialization the nectaries are enclosed in structures that restrict access to the nectar to just a few species of visitor. This range can be seen within the family Ranunculaceae. In the simple flower of *Ranunculus ficaria* the nectaries are exposed to all visitors, but in the more specialized flower of *R. bulbosus* there is a flap over the nectary, and in *Aquilegia* the nectaries have developed into long tubes and only visitors with long probosces (tongues) can reach the nectar. In the related *Aconitum* the whole flower is structured so that the nectaries are accessible only to insects of the right shape and size that are forced to brush against the anthers and pick up pollen. Unprotected nectaries have the advantage of a ready supply of pollinators, but because these pollinators are unspecialized they transfer much of the pollen to

the flowers of other species (though in practice, many generalists are actually 'sequential specialists', foraging preferentially on one plant species for hours or days). Protected nectaries have the advantage of efficient transfer of pollen by specialists to other flowers of the same species, but are reliant on there being sufficient numbers of these specialists.

Charles Darwin (1859) recognized that a long nectary, as in *Aquilegia*, forced a pollinating insect into close contact with the pollen at the nectary's mouth. Natural selection may then favor even longer nectaries, and as an evolutionary reaction, the tongues of the pollinator would be selected for increasing length – a reciprocal and escalating process of specialization. Nilsson (1988) deliberately shortened the nectary tubes of the long-tubed orchid *Platanthera* and showed that the flowers then produced many fewer seeds – presumably because the pollinator was not forced into a position that maximized the efficiency of pollination.

seasonality

Flowering is a seasonal event in most plants, and this places strict limits on the degree to which a pollinator can become an obligate specialist. A pollinator can only become completely dependent on specific flowers as a source of food if its life cycle matches the flowering season of the plant. This is feasible for many short-lived insects like butterflies and moths, but longer lived pollinators such as bats and rodents, or bees with their long-lived colonies, are more likely to be generalists, turning from one relatively unspecialized flower to another through the seasons or to quite different foods when nectar is unavailable.

Figure 13.9 Fig wasps on a developing fig. Reproduced by permission of Gregory Dimijian/Science Photo Library.

13.4.3 Brood site pollination: figs and yuccas

figs and fig wasps . . .   Not every insect-pollinated plant provides its pollinator with only a take-away meal. In a number of cases, the plants also provide a home and sufficient food for the development of the insect larvae (Proctor *et al.*, 1996). The best studied of these are the complex, largely species-specific interactions between figs (*Ficus*) and fig wasps (Figure 13.9) (Wiebes, 1979; Bronstein, 1988). Figs bear many tiny flowers on a swollen receptacle with a narrow opening to the outside; the receptacle then becomes the fleshy fruit. The best-known species is the edible fig, *Ficus carica*. Some cultivated forms are entirely female and require no pollination for fruit to develop, but in wild *F. carica* three types of receptacle are produced at different times of the year. (Other species are less complicated, but the life cycle is similar.) In winter, the flowers are mostly neuter (sterile female) with a few male flowers near the opening. Tiny females of the wasp *Blastophaga psenes* invade the receptacle, lay eggs in the neuter flowers and then die. Each wasp larva then completes its development in the ovary of one flower, but the males hatch first and chew open the seeds occupied by the females and then mate with them. In early summer the females emerge, receiving pollen at the entrance from the male flowers, which have only just opened.

The fertilized females carry the pollen to a second type of receptacle, containing neuter and female flowers, where they lay their eggs. Neuter flowers, which cannot set seed, have a short style: the wasps can reach to lay their eggs in the ovaries where they develop. Female flowers, though, have long styles so the wasps cannot reach the ovaries and their eggs fail to develop, but in laying these eggs they fertilize the flowers, which set seed. Hence, these receptacles generate a combination of viable seeds (that benefit the fig) and adult fig wasps (that obviously benefit the wasps, but also benefit the figs since they are the figs' pollinators).

Following another round of wasp development, fertilized females emerge in the fall, and a variety of other animals eat the fruit and disperse the seeds. The fall-emerging wasps lay their eggs in a third kind of receptacle containing only neuter flowers, from which wasps emerge in winter to start the cycle again.

This, then, apart from being a fascinating piece of natural history, is a good example of a mutualism in which the interests of the two participants   . . . show mutualism despite conflict   none the less appear not to coincide. Specifically, the optimal proportion of flowers that develop into fig seeds and fig wasps is different for the two parties, and we might reasonably expect to see a negative correlation between the two: seeds produced *at the expense* of wasps, and vice versa (Herre & West, 1997). In fact, detecting this negative correlation, and hence establishing the conflict of interest, has proved elusive for reasons that frequently apply in studies of evolutionary ecology. The two variables tend, rather, to be *positively* correlated, since both tend to increase with two 'confounding' variables: the overall size of fruit and the overall proportion of flowers in a fruit that are visited by wasps. Herre and West (1997), however, in analyzing data from nine species of New World figs, were able to over-come this in a way that is generally applicable in such situations. They controlled statistically for variation in the confounding variables (asking, in effect, what the relationship between seed and wasp numbers would be in a fruit of constant size in which a constant proportion of flowers was visited) and then were able to uncover a negative correlation. The fig and fig wasp mutualists *do* appear to be involved in an on-going evolutionary battle.

A similar, and similarly much studied, set of mutualisms occurs between the 35–50 species of *Yucca* plant that live in North and Central   yuccas and yucca moths   America and the 17 species of yucca moth, 13 of which are newly described since 1999 (Pellmyr & Leebens-Mack, 2000). A female moth uses specialized 'tentacles' to collect together pollen from several anthers in one flower, which she then takes to the flower of another inflorescence (promoting outbreeding) where she both lays eggs in the ovaries and carefully deposits the pollen, again using her tentacles. The development of the moth larvae requires successful pollination, since unpollinated flowers quickly die, but the larvae also consume seeds in their immediate vicinity, though many other seeds develop successfully. On completing their development, the larvae drop to the soil to pupate, emerging one or more years later during the yucca's flowering season. The reproductive success of an individual adult female moth is not, therefore, linked to that of an individual yucca plant in the same way as are those of female fig wasps and figs.

A detailed review of both seed dispersal and pollination mutualisms is given by Thompson (1995), who provides a thorough account of the processes that may lead to the evolution of such mutualisms.

13.5 Mutualisms involving gut inhabitants

Most of the mutualisms discussed so far have depended on patterns of behavior, where neither species lives entirely 'within' its partner. In many other mutualisms, one of the partners is a unicellular eukaryote or bacterium that is integrated more or less permanently into the body cavity or even the cells of its multicellular partner. The microbiota occupying parts of various animals' alimentary canals are the best known extracellular symbionts.

13.5.1 Vertebrate guts

The crucial role of microbes in the digestion of cellulose by vertebrate herbivores has long been appreciated, but it now appears that the gastrointestinal tracts of all vertebrates are populated by a mutualistic microbiota (reviewed in Stevens & Hume, 1998). Protozoa and fungi are usually present but the major contributors to these 'fermentation' processes are bacteria. Their diversity is greatest in regions of the gut where the pH is relatively neutral and food retention times are relatively long. In small mammals (e.g. rodents, rabbits and hares) the cecum is the main fermentation chamber, whereas in larger nonruminant mammals such as horses the colon is the main site, as it is in elephants, which, like rabbits, practice coprophagy (consume their own feces) (Figure 13.10). In ruminants, like cattle and sheep, and in kangaroos and other marsupials, fermentation occurs in specialized stomachs.

The basis of the mutualism is straightforward. The microbes receive a steady flow of substrates for growth in the form of food that has been eaten, chewed and partly homogenized. They live within a chamber in which pH and, in endotherms, temperature are regulated and anaerobic conditions are maintained. The vertebrate hosts, especially the herbivores, receive nutrition from food that they would otherwise find, literally, indigestible. The bacteria produce short-chain fatty acids (SCFAs) by fermentation of the host's dietary cellulose and starches and of the endogenous

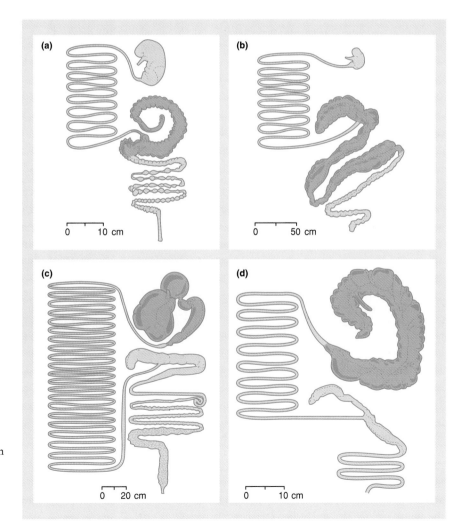

Figure 13.10 The digestive tracts of herbivorous mammals are commonly modified to provide fermentation chambers inhabited by a rich fauna and flora or microbes. (a) A rabbit, with a fermentation chamber in the expanded cecum. (b) A zebra, with fermentation chambers in both the cecum and colon. (c) A sheep, with foregut fermentation in an enlarged portion of the stomach, rumen and reticulum. (d) A kangaroo, with an elongate fermentation chamber in the proximal portion of the stomach. (After Stevens & Hume, 1998.)

Species	Function	Products
Bacteroides succinogenes	C, A	F, A, S
Ruminococcus albus	C, X	F, A, E, H, C
R. flavefaciens	C, X	F, A, S, H
Butyrivibrio fibrisolvens	C, X, PR	F, A, L, B, E, H, C
Clostridium lochheadii	C, PR	F, A, B, E, H, C
Streptococcus bovis	A, SS, PR	L, A, F
B. amylophilus	A, P, PR	F, A, S
B. ruminicola	A, X, P, PR	F, A, P, S
Succinimonas amylolytica	A, D	A, S
Selenomonas ruminantium	A, SS, GU, LU, PR	A, L, P, H, C
Lachnospira multiparus	P, PR, A	F, A, E, L, H, C
Succinivibrio dextrinosolvens	P, D	F, A, L, S
Methanobrevibacter ruminantium	M, HU	M
Methanosarcina barkeri	M, HU	M, C
Spirochete species	P, SS	F, A, L, S, E
Megasphaera elsdenii	SS, LU	A, P, B, V, CP, H, C
Lactobacillus sp.	SS	L
Anaerovibrio lipolytica	L, GU	A, P, S
Eubacterium ruminantium	SS	F, A, B, C

Table 13.1 A number of the bacterial species of the rumen, illustrating their wide range of functions and the wide range of products that they generate. (After Allison, 1984; Stevens & Hume, 1998.)

Functions: A, amylolytic; C, cellulolytic; D, dextrinolytic; GU, glycerol utilizing; HU, hydrogen utilizer; L, lipolytic; LU, lactate utilizing; M, methanogenic; P, pectinolytic; PR, proteolytic; SS, major soluble sugar fermenter; X, xylanolytic.
Products: A, acetate; B, butyrate; C, carbon dioxide; CP, caproate; E, ethanol; F, formate; H, hydrogen; L, lactate; M, methane P, propionate; S, succinate; V, valerate;.

carbohydrates contained in host mucus and sloughed epithelial cells. SCFAs are often a major source of energy for the host; for example, they provide more than 60% of the maintenance energy requirements for cattle and 29–79% of those for sheep (Stevens & Hume, 1998). The microbes also convert nitrogenous compounds (amino acids that escape absorption in the midgut, urea that would otherwise be excreted by the host, mucus and sloughed cells) into ammonia and microbial protein, conserving nitrogen and water; and they synthesize B vitamins. The microbial protein is useful to the host if it can be digested – in the intestine by foregut fermenters and following coprophagy in hindgut fermenters – but ammonia is usually not useful and may even be toxic to the host.

13.5.2 Ruminant guts

The stomach of ruminants comprises a three-part forestomach (rumen, reticulum and omasum) followed by an enzyme-secreting abomasum that is similar to the whole stomach of most other vertebrates. The rumen and reticulum are the main sites of fermentation, and the omasum serves largely to transfer material to the abomasum. Only particles with a volume of about 5 μl or less can pass from the reticulum into the omasum; the animal regurgitates and rechews the larger particles (the pro-

cess of rumination). Dense populations of bacteria (10^{10}–10^{11} ml^{-1}) and protozoa (10^5–10^6 ml^{-1} but occupying a similar volume to the bacteria) are present in the rumen. The bacterial communities of the rumen are composed almost wholly of obligate anaerobes – many are killed instantly by exposure to oxygen – but they perform a wide variety of functions (subsist on a wide variety of substrates) and generate a wide range of products (Table 13.1). Cellulose and other fibers are the important constituents of the ruminant's diet, and the ruminant itself lacks the enzymes to digest these. The cellulolytic activities of the rumen microflora are therefore of crucial importance. But not all the bacteria are cellulolytic: many subsist on substrates (lactate, hydrogen) generated by other bacteria in the rumen.

The protozoa in the gut are also a complex mixture of specialists. Most are holotrich ciliates and entodiniomorphs. A few can digest cellulose. The cellulolytic ciliates have intrinsic cellulases, although some other protozoa may use bacterial symbionts. Some consume bacteria: in their absence the number of bacteria rise. Some of the entodiniomorphs prey on other protozoa. Thus, the diverse processes of competition, predation and mutualism, and the food chains that characterize terrestrial and aquatic communities in nature, are all present within the rumen microcosm.

a complex community of mutualists

13.5.3 Refection

Eating feces is a taboo amongst humans, presumably through some combination of biological and cultural evolution in response to the health hazards posed by pathogenic microbes, including many that are relatively harmless in the hindgut but are pathogenic in more anterior regions. For many vertebrates, however, symbiotic microbes, living in the hindgut beyond the regions where effective nutrient absorption is possible, are a resource that is too good to waste. Thus coprophagy (eating feces) or refection (eating one's own feces) is a regular practice in many small herbivorous mammals. This is developed to a fine art in species such as rabbits that have a 'colonic separation mechanism' that allows them to produce separate dry, non-nutritious fecal pellets and soft, more nutritious pellets that they consume selectively. These contain high levels of SCFAs, microbial protein and B vitamins, and can provide 30% of a rabbit's nitrogen requirements and more B vitamins than it requires (Björnhag, 1994; Stevens & Hume, 1998).

13.5.4 Termite guts

Termites are social insects of the order Isoptera, many of which depend on mutualists for the digestion of wood. Primitive termites feed directly on wood, and most of the cellulose, hemicelluloses and possibly lignins are digested by mutualists in the gut (Figure 13.11), where the paunch (part of the segmented hindgut) forms a microbial fermentation chamber. However, the advanced termites (75% of all the species) rely much more heavily on their own cellulase (Hogan *et al.*, 1988), while a third group (the Macrotermitinae) cultivate wood-digesting fungi that the termites eat along with the wood itself, which the fungal cellulases assist in digesting.

Termites refecate, so that food material passes at least twice through the gut, and microbes that have reproduced during the first passage may be digested the second time round. The major group of microorganisms in the paunch of primitive termites are anaerobic flagellate protozoans. Bacteria are also present, but cannot digest cellulose. The protozoa engulf particles of wood and ferment the cellulose within their cells, releasing carbon dioxide and hydrogen. The principal products, subsequently absorbed by the host, are SCFAs (as in vertebrates) but in termites they are primarily acetic acid.

The bacterial population of the termite gut is less conspicuous than that of the rumen, but appears to play a part in two distinct mutualisms.

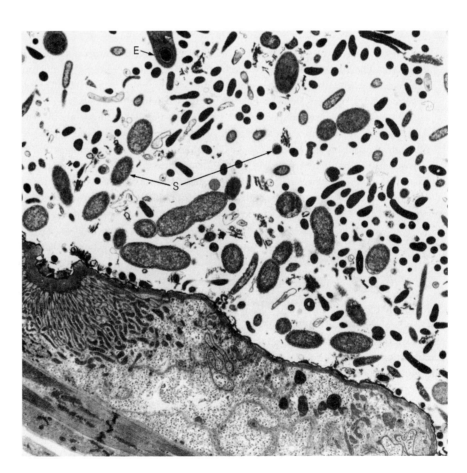

Figure 13.11 Electron micrograph of a thin section of the paunch of the termite *Reticulitermes flavipes*. Much of the flora is composed of aggregates of bacteria. Amongst them can be seen endospore-forming bacteria (E), spirochetes (S) and protozoa. (After Breznak, 1975.)

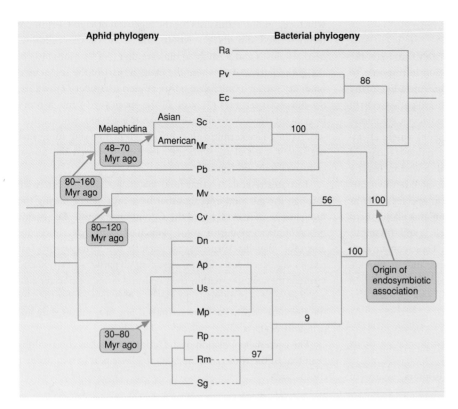

Figure 13.12 The phylogeny of selected aphids and their corresponding primary endosymbionts. Other bacteria are shown for comparison. The aphid phylogeny (after Heie, 1987) is shown on the left and the bacterial phylogeny on the right. Broken lines connect the associated aphids and bacteria. Three species of bacteria that are not endosymbionts are also shown in the phylogeny: Ec, *Escherichia coli*; Pv, *Proteus vulgaris*; Ra, *Ruminobacter amylophilus* (a rumen symbiont). The distances along the branches are drawn to be roughly proportionate to time. (After Moran *et al.*, 1993.) Aphid species: Ap, *Acyrthosiphon pisum*; Cv, *Chaitophorus viminalis*; Dn, *Diuraphis noxia*; Mp, *Myzus persicae*; Mr, *Melaphis rhois*; Mv, *Mindarus victoriae*; Pb, *Pemphigus betae*; Rm, *Rhopalosiphum maidis*; Rp, *Rhodalosiphon padi*; Sc, *Schlectendalia chinensis*; Sg, *Schizaphis graminum*; Us, *Uroleucon sonchi*.

1 Spirochetes tend to be concentrated at the surface of the flagellates. The spirochetes possibly receive nutrients from the flagellates, and the flagellates gain mobility from the movements of the spirochetes: a pair of mutualists living mutualistically within a third species.

2 Some bacteria in the termite gut are capable of fixing gaseous nitrogen – apparently the only clearly established example of nitrogen-fixing symbionts in insects (Douglas, 1992). Nitrogen fixation stops when antibacterial antibiotics are eaten (Breznak, 1975), and the rate of nitrogen fixation falls off sharply if the nitrogen content of the diet is increased.

13.6 Mutualism within animal cells: insect mycetocyte symbioses

In mycetocyte symbioses between microorganisms and insects, the maternally inherited microorganisms are found within the cytoplasm of specialized cells, mycetocytes, and the interaction is unquestionably mutualistic. It is required by the insects for the nutritional benefits the microorganisms bring, as key providers of essential amino acids, lipids and vitamins, and is required by the microorganisms for their very existence (Douglas, 1998). The symbioses are found in a wide variety of types of insect, and are universally or near-universally present in cockroaches, homopterans, bed bugs, sucking lice, tsetse flies, lyctid beetles and camponotid ants. They have evolved independently in different groups of microorganisms and their insect partners, but in effectively all cases the insects live their lives on nutritionally poor or unbalanced diets: phloem sap, vertebrate blood, wood and so on. Mostly the symbionts are various sorts of bacteria, although in some insects yeasts are involved.

Amongst these symbioses, most is known by far about the interactions between aphids and bacteria in the genus *Buchnera* (Douglas, 1998). The mycetocytes are found in the hemocoel of the aphids and the bacteria occupy around 60% of the mycetocyte cytoplasm. The bacteria cannot be brought into culture in the laboratory and have never been found other than in aphid mycetocytes, but the extent and nature of the benefit they bring to the aphids can be studied by removing the *Buchnera* by treating the aphids with antibiotics. Such 'aposymbiotic' aphids grow very slowly and develop into adults that produce few or no offspring. The most fundamental function performed by the bacteria is to produce essential amino acids that are absent in phloem sap from nonessential amino acids like glutamate, and antibiotic treatment confirms that the aphids cannot do this alone. In addition, though, the *Buchnera* seem to provide other benefits, since symbiotic aphids still outperform aposymbiotic aphids when the latter are provided with all the essential amino acids, but establishing further nutritional functions has proved elusive.

... provide an
ecological and
evolutionary link

The aphid–*Buchnera* interaction also provides an excellent example of how an intimate association between mutualists may link them at both the ecological and the evolutionary level. The *Buchnera* are transmitted transovarially, that is, they are passed by a mother to her offspring in her eggs. Hence, an aphid lineage supports a corresponding single *Buchnera* lineage, and this is no doubt the reason for the strictly congruent phylogenies of aphid and *Buchnera* species: each aphid species has its own *Buchnera* species (see, for example, Figure 13.12). Moreover, these molecular studies, which allow the *Buchnera* phylogeny to be reconstructed, also suggest that the aphids acquired *Buchnera* just once in their evolutionary history, apparently between 160 and 280 million years ago, after the divergence from the main aphid lineage of the only two aphid families not to have a mycetocyte symbiosis, the phylloxerids and the adelgids (Moran *et al.*, 1993). Providing a final twist, the only other aphids without *Buchnera* (in the family Hormaphididae) appear to have lost them secondarily in their evolutionary history, but they do instead host symbiotic yeasts (Douglas, 1998). It seems more likely that the yeasts competitively displaced the bacteria than that the bacteria were first lost and the yeasts subsequently acquired.

Lastly, Douglas (1998) also points out that whereas all Homoptera that feed on nutritionally deficient phloem sap have mycetocyte symbioses, including the aphids described above, those that have switched secondarily in their evolutionary history to feeding on intact plant cells have lost the symbiosis. This, then, is an illustration from a comparative, evolutionary perspective that even in clearly mutualistic symbioses like these, the benefit is a *net* benefit. Once the insects' requirements are reduced, as in a switch of diet, the balance of the costs and benefits of the symbionts is also changed. In this case, the costs clearly outweigh the benefits on a changed diet: those insects that *lost* their symbionts have been favored by natural selection.

13.7   Photosynthetic symbionts within aquatic invertebrates

Hydra and *Chlorella*

Algae are found within the tissues of a variety of animals, particularly in the phylum Cnidaria. In freshwater symbioses the algal symbiont is usually *Chlorella*. For example, in *Hydra viridis*, cells of *Chlorella* are present in large numbers (1.5×10^5 per hydroid) within the digestive cells of the endoderm. In the light, a *Hydra* receives photosynthates from the algae and 50–100% of its oxygen needs. It can also use organic food. Yet when a *Hydra* is maintained in darkness and fed daily with organic food, a reduced symbiotic population of algae is maintained for at least 6 months that can return to normal within 2 days of exposure to light (Muscatine & Pool, 1979). Thus,

armed with its symbionts, and depending on local conditions and resources, *Hydra* can behave both as an autotroph and a heterotroph. There must then be regulatory processes harmonizing the growth of the endosymbiont and its host (Douglas & Smith, 1984), as there must presumably be in all such symbioses. If this were not the case, the symbionts would either overgrow and kill the host or fail to keep pace and become diluted as the host grew.

marine plankton

There are many records of close associations between algae and protozoa in the marine plankton. For example, in the ciliate *Mesodinium rubrum*, 'chloroplasts' are present that appear to be symbiotic algae. The mutualistic consortium of protists and algae can fix carbon dioxide and take up mineral nutrients, and often forms dense populations known as 'red tides' (e.g. Crawford *et al.*, 1997). Extraordinarily high production rates have been recorded from such populations (in excess of $2 \, \mathrm{g \, m^{-3} \, h^{-1}}$ of carbon) – apparently the highest levels of primary productivity ever recorded for populations of aquatic microorganisms.

13.7.1   Reef-building corals and coral bleaching

We have already noted that mutualists dominate environments around the world in terms of their biomass. Coral reefs provide an important example: reef-building corals (another dramatic example of autogenic ecosystem engineering – see Section 13.1) are in fact mutualistic associations between heterotrophic Cnidaria and phototrophic dinoflagellate algae from the genus *Symbiodinium*. Coral reefs provide an illustration, too, of the potential vulnerability of even the most dominant of 'engineered' habitat features. There have been repeated reports of 'coral bleaching' since it was first described in 1984: the whitening of corals as a result of the loss of the endosymbionts and/or their photosynthetic pigments (Brown, 1997). Bleaching occurs mainly in response to unusually elevated temperatures (as seen at the Phuket study site, Thailand; Figure 13.13a), but also in response to high intensities of solar radiation and even disease. Thus, episodes of bleaching seem likely to become increasingly frequent as global temperatures rise (Figure 13.13a; see Section 2.8.2), which is a particular cause for concern, since some bleaching episodes have been followed by mass mortality of corals. This was apparent at Phuket, for example, associated with the bleaching episodes of 1991 and 1995 (Figure 13.13b). (On the other hand, a more catastrophic loss had occurred in 1987 as a result, not of bleaching, but of dredging activity, and the decline in cover in the early 1990s appeared to result from an *interaction* between bleaching and a variety of local human disturbances.)

bleaching and global warming

We clearly cannot be complacent about the effects of global warming on coral reefs – and there *are* likely always

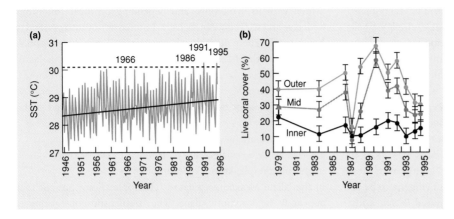

Figure 13.13 (a) Monthly mean sea surface temperatures (SSTs) for sea areas off Phuket, Thailand, from 1945 to 1995. The regression line for all points is shown ($P < 0.001$). The dashed line drawn at 30.11°C represents a tentative bleaching threshold. The years exceeding this are shown: bleaching was observed in 1991 and 1995 but not monitored prior to that. (b) Mean percentage coral cover ($\pm$ SE) on inner (——), mid (——) and outer (——) reef flats at Phuket, Thailand, over the period 1979–95. (After Brown, 1997.)

to be human disturbances with which bleaching effects can interact – but it is also apparent that reef corals are able to acclimate to the changed conditions that may induce bleaching and to recover from bleaching episodes. Their adaptability is illustrated by another study at Phuket. During the 1995 episode, it had been observed that bleaching in the coral *Goniastrea aspera* occurred predominantly on east- rather than west-facing surfaces. The latter normally suffer greater exposure to solar radiation, which also has a tendency to cause bleaching. This therefore suggests that tolerance to bleaching had been built up in the west-facing corals. Such a difference in tolerance was confirmed experimentally (Figure 13.14): there was little or no bleaching on the 'adapted' west-facing surfaces at high temperatures.

another mutualism extending beyond two species

Meanwhile, another study of coral bleaching adds to the growing realization that seemingly simple two-species mutualisms may be more complex and subtler than might be imagined. The ecologically dominant Caribbean corals *Montastraea annularis* and *M. faveolata* both host three quite separate 'species' or 'phylotypes' of *Symbiodinium* (denoted *A*, *B* and *C* and distinguishable only by genetic methods). Phylotypes *A* and *B* are common in shallower, high-irradiance habitats, whereas *C* predominates in deeper, lower irradiance sites – illustrated both by comparisons of colonies from different depths and of samples from different depths within a colony (Figure 13.14b). In the fall of 1995, following a prolonged period above the mean maximum summer temperature, bleaching occurred in *M. annularis* and *M. faveolata* in the reefs off Panama and elsewhere. Bleaching, however, was rare at the shallowest and the deepest sites, but was most apparent in shallower colonies at shaded sites and in deeper colonies at more exposed sites. A comparison of adjacent samples before and after bleaching provides an explanation (Figure 13.14c). The bleaching resulted from the selective loss of *Symbiodinium C*. It appears to have occurred at locations supporting *C* and one or both of the other two species, near the irradiance limit of *C* under non-

bleaching conditions. At shaded deep-water sites, dominated by *C*, the high temperatures in 1995 were not sufficient to push *C* into bleaching conditions. The shallowest sites were occupied by the species *A* and *B*, which were not susceptible to bleaching at these temperatures. Bleaching occurred, however, where *C* was initially present but was pushed beyond its limit by the increased temperature. At these sites, the loss of *C* was typically close to 100%, *B* decreased by around 14%, but *A* more than doubled in three of five instances.

It seems, therefore, first, that the coral–*Symbiodinium* mutualism involves a range of endosymbionts that allows the corals to thrive in a wider range of habitats than would otherwise be possible. Second, looking at the mutualism from the algal side, the endosymbionts must constantly be engaged in a competitive battle, the balance of which alters over space and time (see Section 8.5). Finally, bleaching (and subsequent recovery), and possibly also 'adaptation' of the type described above, may be seen as manifestations of this competitive battle: not breakdowns and reconstructions in a simple two-species association, but shifts in a complex symbiotic community.

13.8 Mutualisms involving higher plants and fungi

A wide variety of symbiotic associations are formed between higher plants and fungi. A very remarkable group of Ascomycete fungi, the Clavicipitaceae, grow in the tissues of many species of grass and a few species of sedge. The family includes species that are easily recognized as parasites (e.g. *Claviceps*, the ergot fungus, and *Epichloe*, the choke disease of grasses), others that are clearly mutualistic, and a large number where the costs and benefits are uncertain. The fungal mycelia characteristically grow as sparsely branched filaments running through intercellular spaces along the axis of leaves and stems, but they are not found in roots. Many of the symbiotic fungi produce powerful toxic alkaloids that

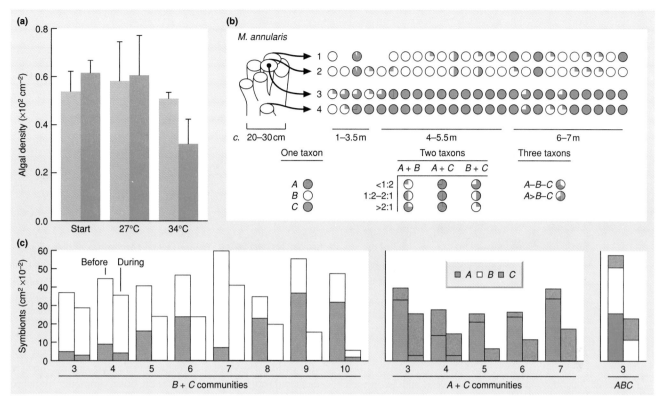

Figure 13.14 Coral acclimation and recovery in coral bleaching. (a) Algal density in western (light bars) and eastern (dark bars) cores of the coral *Goniastrea aspera* before and after exposure to elevated (34°C) and ambient (27°C) temperatures for 68 h. Mean values are shown; error bars show 1 SD (*n* = 5). (After Brown *et al.*, 2000.) (b) Symbiont communities in another coral, *Montastraea annularis*, collected in January 1995 off the coast of Panama. Each symbol represents a sample that contained the algal taxa *Symbiodinium A*, *B* or *C*, or mixtures of taxa summarized according to the code shown below. Columns in the data represent individual coral colonies (depth increases from left to right) and rows represent locations of higher (rows 1 and 2) and lower (rows 3 and 4) irradiance, as defined in the diagram to the left. (After Rowan *et al.*, 1997.) (c) Corresponding symbiont communities from close to the bleaching region of *Symbiodinium C* before (January 1995) and during (October 1995) an episode of coral bleaching. Densities of *A* (gray), *B* (white) and *C* (orange) before and during bleaching (left and right bars of each pair, respectively) in samples reported in *B* + *C* communities (3–10), *A* + *C* communities (3–7) and an *ABC* community. (After Rowan *et al.*, 1997.)

confer some protection from grazing animals (the evidence is reviewed in Clay, 1990) and, perhaps even more important, deter seed predators (Knoch *et al.*, 1993).

not roots but mycorrhizas

A quite different mutualism of fungi with higher plants occurs in roots. Most higher plants do not have roots, they have mycorrhizas – intimate mutualisms between fungi and root tissue. Plants of only a few families like the Cruciferae are the exception. Broadly, the fungal networks in mycorrhizas capture nutrients from the soil, which they transport to the plants in exchange for carbon. Many plant species can live without their mycorrhizal fungi in soils where neither nutrients nor water are ever limiting, but in the harsh world of natural plant communities, the symbioses, if not strictly obligate, are none the less 'ecologically obligate'. That is,

they are necessary if the individuals are to survive in nature (Buscot *et al.*, 2000). The fossil record suggests that the earliest land plants, too, were heavily infected. These species lacked root hairs, even roots in some cases, and the early colonization of the land may have depended on the presence of the fungi to make the necessary intimate contact between plants and substrates.

Generally, three major types of mycorrhiza are recognized. Arbuscular mycorrhizas are found in up to two-thirds of all plant species, including most nonwoody species and tropical trees. Ectomycorrhizal fungi form symbioses with many trees and shrubs, dominating boreal and temperate forests and also some tropical rainforests. Finally, ericoid mycorrhizas are found in the dominant species of heathlands including the northern hemisphere heaths and heathers (Ericaceae) and the Australian heaths (Epacridaceae).

Figure 13.15 Mycorrhiza of pine (*Pinus sylvestris*). The swollen, much branched structure is the modified rootlet enveloped in a thick sheath of fungal tissue. (Courtesy of J. Whiting; photograph by S. Barber.)

13.8.1 Ectomycorrhizas

An estimated 5000–6000 species of Basidiomycete and Ascomycete fungi form ectomycorrhizas (ECMs) on the roots of trees (Buscot *et al.*, 2000). Infected roots are usually concentrated in the litter layer of the soil. Fungi form a sheath or mantle of varying thickness around the roots. From there, hyphae radiate into the litter layer, extracting nutrients and water and also producing large fruiting bodies that release enormous numbers of wind-borne spores. The fungal mycelium also extends inwards from the sheath, penetrating between the cells of the root cortex to give intimate cell-to-cell contact with the host and establishing an interface with a large surface area for the exchange of photo-assimilates, soil water and nutrients between the host plant and its fungal partner. The fungus usually induces morphogenetic changes in the host roots, which cease to grow apically and remain stubby (Figure 13.15). Host roots that penetrate into the deeper, less organically rich layers of the soil continue to elongate.

The ECM fungi (see Buscot *et al.*, 2000 for a review) are effective in extracting the sparse and patchy supplies of phosphorus and especially nitrogen from the forest litter layer, and their high species diversity presumably reflects a corresponding diversity of niches in this environment (though this diversity of niches is very far from having been demonstrated). Carbon flows from the plant to the fungus, very largely in the form of the simple hexose sugars: glucose and fructose. Fungal consumption of these may represent up to 30% of the plants' net rate of photosynthate production. The plants, though, are often nitrogen-limited, since in the forest litter there are low rates of nitrogen mineralization (conversion from organic to inorganic forms), and inorganic

nitrogen is itself mostly available as ammonia. It is therefore crucial for forest trees that ECM fungi can access organic nitrogen directly through enzymic degradation, utilize ammonium as a preferred source of inorganic nitrogen, and circumvent ammonium depletion zones through extensive hyphal growth. None the less, the idea that this relationship between the fungi and their host plants is mutually exploitative rather than 'cosy' is emphasized by its responsiveness to changing circumstances. ECM growth is directly related to the rate of flow of hexose sugars from the plant. But when the direct availability of nitrate to the plants is high, either naturally or through artificial supplementation, plant metabolism is directed away from hexose production (and export) and towards amino acid synthesis. As a result the ECM degrades; the plants seem to support just as much ECM as they appear to need.

13.8.2 Arbuscular mycorrhizas

Arbuscular mycorrhizas (AMs) do not form a sheath but penetrate *within* the roots of the host, though they do not alter the host's root morphology. Roots become infected from mycelium present in the soil or from germ tubes that develop from asexual spores, which are very large and produced in small numbers – a striking contrast with the ECM fungi. Initially, the fungus grows between host cells but then enters them and forms a finely branched intracellular 'arbuscule'. The fungi responsible comprise a distinct phylum, the Glomeromycota (Schüßler *et al.*, 2001). Although originally divided into only about 150 species, suggesting a lack of host specificity (since there are vastly more species of hosts), modern genetic methods have uncovered a far greater diversity among the AM fungi, and there is increasing evidence of niche differentiation amongst them. For instance, when 89 root samples were taken from three grass species that co-occurred in the same plots in a field experiment, and their AM fungi were characterized using such a method – terminal restriction fragment length polymorphism – there was clear separation amongst the AM strains found on the different hosts (Figure 13.16).

a range of benefits?

There has been a tendency to emphasize facilitation of the uptake of phosphorus as the main benefit to plants from AM symbioses (phosphorus is a highly immobile element in the soil, which is therefore frequently limiting to plant growth), but the truth appears to be more complex than this. Benefits have been demonstrated, too, in nitrogen uptake, pathogen and herbivore protection, and resistance to toxic metals (Newsham *et al.*, 1995). Certainly, there are cases where the inflow of phosphorus is strongly related to the degree of colonization of roots by AM fungi. This has been shown for the bluebell, *Hyacinthoides non-scripta*, as colonization progresses during its phase of subterranean growth from August to February through to its above-ground photosynthetic phase thereafter (Figure 13.17a). Indeed, bluebells cultured without AM fungi are unable to take up phosphorus through their poorly branched system of roots (Merryweather & Fitter, 1995).

On the other hand, a factorial set of experiments examined the growth of the annual grass *Vulpia ciliata* ssp. *ambigua* at sites

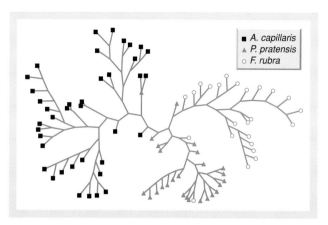

Figure 13.16 The similarity among 89 arbuscular mycorrhiza (AM) fungal communities taken from the roots of three coexisting grass species, *Agrostis capillaris*, *Poa pratensis* and *Festuca rubra*, assessed by terminal restriction fragment length polymorphism. Each terminal on the 'tree' is a different sample, with the grass species from which it originated shown. More similar samples are closer together on the tree. The similarity within, and the differentiation between, the AM fungal communities associated with different hosts are plainly apparent. (After Vandenkoornhuyse *et al.*, 2003.)

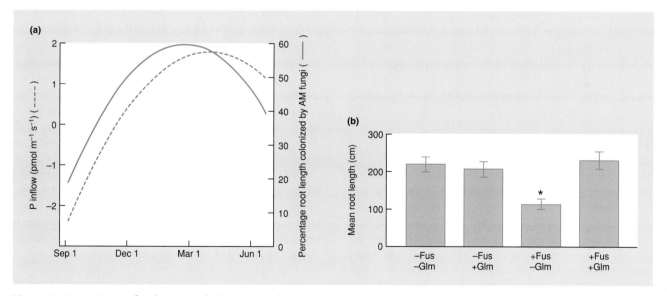

Figure 13.17 (a) Curves fitted to rates of phosphorus inflow (-----, left axis) and root colonization by arbuscular mycorrhiza (AM) fungi (——, right axis) in the bluebell, *Hyacinthoides non-scripta*, over a single growing season. (After Merryweather & Fitter, 1995; Newsham *et al.*, 1995.) (b) The effects of a factorial combination of *Fusarium oxysporum* (Fus) and an AM fungus, *Glomus* sp. (Glm), on the growth (root length) of *Vulpia* plants. Values are means of 16 replicates per treatment; bars show standard errors; the asterisk signifies a significant difference at $P < 0.05$ in a Fisher's pairwise comparison. (After Newsham *et al.*, 1994, 1995.)

in eastern England where there were large differences in the intensity of natural mycorrhizal infection (West *et al.*, 1993). In one treatment phosphate was applied, and in another the fungicide benomyl was used to control the fungal infection. Fecundity of the grass was scarcely affected by any of the treatments. An explanation was provided by a further set of experiments (Figure 13.17b) in which seedlings of *Vulpia* were grown with an AM fungus (*Glomus* sp.), with the pathogenic fungus *Fusarium oxysporum*, with both, and with neither. Growth was not enhanced by *Glomus* alone, but growth was harmed by *Fusarium* in the absence of *Glomus*. When both were present, growth returned to normal levels. Clearly, the mycorrhiza did not benefit the phosphorus-economy of the *Vulpia*, but it did protect it from the harmful effects of the pathogen. (In the previous experiment, benomyl presumably had no effect on performance because it controlled both mycorrhizal *and* pathogenic fungi.)

| it depends on the species | The key difference appears to be that *Vulpia*, unlike the bluebell, has a highly branched system of roots, and Newsham *et al.* (1995) go so far as to |

propose a continuum of AM function in relation to root architecture, with *Vulpia* and *Hyacinthoides* sitting towards the two extremes. Plants with finely branched roots have little need for supplementary phosphorus capture, but development of that same root architecture provides multiple points of entry for plant pathogens. In such cases AM symbioses are therefore likely to have evolved with an emphasis on plant protection. By contrast, root systems with few lateral and actively growing meristems are relatively invulnerable to pathogen attack, but these root systems are poor foragers for phosphorus. Here, AM symbioses are likely to have evolved with an emphasis on phosphorus capture. Of course, even this more sophisticated view of AM function is unlikely to be the whole story: other aspects of AM ecology, such as protection from herbivores and toxic metals, may well vary in ways unrelated to root architecture.

13.8.3 Ericoid mycorrhizas

Heathlands exist in environments characterized by soils with low levels of available plant nutrients, often as a result of regular fires in which, for example, up to 80% of the nitrogen that has accumulated between fires may be lost. It is unsurprising, therefore, that heathlands are dominated by many plants that have evolved an association with ericoid mycorrhizal fungi (Read, 1996). This enables them to facilitate the extraction of nitrogen and phosphorus from the superficial layers of detrital material generated by the plants. Indeed, the conservation of natural heathlands is threatened by nitrogen supplementation and fire control, which allow colonization and domination by grasses that would otherwise be unable to exist in these impoverished environments.

The ericoid mycorrhizal root itself is anatomically simple compared to other mycorrhizas, characterized by a reduction of its vascular and cortical tissues, by the absence of root hairs, and by the presence of swollen epidermal cells occupied by mycorrhizal fungi. As a result, the individual roots are delicate structures, often referred to as 'hair-roots'; collectively the hair-roots form a dense fibrous root system, the bulk of which is concentrated towards the surface of the soil profile (Pate, 1994). The fungi are effective, unlike the plants alone, in absorbing nitrate, ammonium and phosphate ions that have been mobilized by other decomposers in the soil (see Chapter 11), but crucially they are also 'saprotrophic'. They are therefore able to compete directly with the other decomposers in liberating nitrogen and phosphorus from the organic residues in which most of these elements are locked up in heathland ecosystems (Read, 1996). A mutualism can thus be seen, again, to be woven into a larger web of interactions: the symbiont enhances its contribution to the host by making a preemptive competitive strike for scarce inorganic resources, and its own competitive ability is presumably enhanced in turn by the physiological support provided by its host.

13.9 Fungi with algae: the lichens

| | mycobionts and phytobionts |

Of the 70,000 or so species of fungus that are known, approximately 20% are 'lichenized' (Palmqvist, 2000). Lichens are nutritionally specialized fungi (the so-called 'mycobiont' component) that have escaped from their normal way of life into a mutualistic association with a 'photobiont'. In around 90% of lichen species the photobiont is an alga, which provides carbon compounds to the mycobiont through photosynthesis. In some cases, the photobiont is a cyanobacterium, which may also provide fixed nitrogen to the association. In a relatively few, 'tripartite' lichen species (around 500) both an alga and a cyanobacterium are involved. Lichenized fungi belong to diverse taxonomic groups and the mutualistic algae to 27 different genera. Presumably, the lichen habit has evolved many times.

The photobionts are located extracellularly between the fungal hyphae, in a thin layer near the upper surface. Together, the two components form an integrated 'thallus' but the photobiont makes up only about 3–10% by weight. The advantage to the photobiont in the association, if any, has not been established clearly. All lichenized algal species, for example, can also occur free-living outside their association with their mycobiont. It may be that they are 'captured' by the fungus and exploited without any recompense. However, some of the species (e.g. of algal genus *Trebouxia*) are rare in their free-living form but very common in lichens, suggesting that there is something special about life in their mycobiont that they need. Moreover, since minerals, including nitrogen, are largely 'captured' from what is deposited

Figure 13.18   A variety of lichen species on a tree trunk. Reproduced by permission of Vaughan Fleming/Science Photo Library.

but synthesis is stimulated when the carbon supply is limiting (Palmqvist, 2000).

Lichenization, then, gives the mycobiont and the photobiont, between them, the functional role of higher plants, but in so doing it extends the ecological range of both partners onto substrata (rock surfaces, tree trunks) and into regions (arid, arctic and alpine) that are largely barred to higher plants. Indeed, it has been claimed that lichens dominate 8% of terrestrial communities, both in terms of abundance and species diversity. However, all lichens grow slowly: the colonizers of rock surfaces rarely extend faster than 1–5 mm year^{-1}. They are, though, very efficient accumulators of the mineral cations that fall or drip onto them, and this makes them particularly sensitive to environmental contamination by heavy metals and fluoride. Hence, they are amongst the most sensitive indicators of environmental pollution. The 'quality' of an environment in humid regions can be judged rather accurately from the presence or absence of lichen growth on tombstones and tree trunks.

One remarkable feature in the life of the lichenized fungi is that the growth form of the fungus is usually profoundly changed when the alga is present. When the fungi are cultured in isolation from the algae, they grow slowly in compact colonies, much like related free-living fungi; but in the presence of the algal symbionts they take on a variety of morphologies (Figure 13.18) that are characteristic of specific algal–fungal partnerships. In fact, the algae stimulate morphological responses in the fungi that are so precise that the lichens are classified as distinct species, and a cyanobacterium and an alga, for example, may elicit quite different morphologies from the same fungus.

remarkable morphological responses on the fungi

directly onto the lichen, often in rainwater and from the flow and drip down the branches of trees, and since the surface and biomass are largely fungal, the mycobiont must contribute the vast bulk of these minerals.

parallels with higher plants

Hence, the mutualistic pairs (and threesomes) in lichens provide two striking parallels with higher plants. There is a structural parallel: in plants, the photosynthetic chloroplasts (see also Section 13.12) are similarly concentrated close to light-facing surfaces. There is also a functional parallel. The economy of a plant relies on carbon produced largely in the leaves and nitrogen absorbed principally through the roots, with a relative shortage of carbon giving rise to shoot growth at the expense of roots, and a shortage of nitrogen leading to root growth at the expense of shoots. Likewise, in lichens, the synthesis of carbon-fixing photobiont cells is inhibited by a relative shortage of nitrogen in the mycobiont,

13.10   Fixation of atmospheric nitrogen in mutualistic plants

The inability of most plants and animals to fix atmospheric nitrogen is one of the great puzzles in the process of evolution, since nitrogen is in limiting supply in many habitats. However, the ability to fix nitrogen is widely though irregularly distributed amongst both the eubacteria ('true' bacteria) and the archaea (archaebacteria), and many of these have been caught up in tight mutualisms with systematically quite different groups of eukaryotes. Presumably such symbioses have evolved a number of times independently. They are of enormous ecological importance because of nitrogen's frequent importance (Sprent & Sprent, 1990).

The nitrogen-fixing bacteria that have been found in symbioses (not necessarily mutualistic) are members of the following taxa.

the range of nitrogen-fixing bacteria

1 Rhizobia, which fix nitrogen in the root nodules of most leguminous plants and just one nonlegume, *Parasponia* (a member of the family Ulmaceae, the elms). At least three genera are recognized: *Rhizobium*, *Bradyrhizobium* and *Azorhizobium*, which are so distinct that they should perhaps be in different families (Sprent & Sprent, 1990), and between them they may comprise 10^4 or more species.

2 Actinomycetes of the genus *Frankia*, which fix nitrogen in the nodules (actinorhiza) of a number of nonleguminous and mainly woody plants, such as alder (*Alnus*) and sweet gale (*Myrica*).

3 Azotobacteriaceae, which can fix nitrogen aerobically and are commonly found on leaf and root surfaces.

4 Bacillaceae, such as *Clostridium* spp., which occur in ruminant feces, and *Desulfotomaculum* spp., which fix nitrogen in mammalian guts.

5 Enterobacteriaceae, such as *Enterobacter* and *Citrobacter*, which occur regularly in intestinal floras (e.g. of termites) and occasionally on leaf surfaces and on root nodules.

6 Spirillaceae, such as *Spirillum lipiferum*, which is an obligate aerobe found on grass roots.

7 Cyanobacteria of the family Nostocaceae, which are found in association with a remarkable range (though rather few species) of flowering and nonflowering plants (see Section 13.10.3), and which we recently met as photobionts in lichens.

Of these, the association of the rhizobia with legumes is the most thoroughly studied, because of the huge agricultural importance of legume crops.

13.10.1 Mutualisms of rhizobia and leguminous plants

several steps to a liaison

The establishment of a liaison between rhizobia and legume plants proceeds by a series of reciprocating steps. The bacteria occur in a free-living state in the soil and are stimulated to multiply by root exudates and cells that have been sloughed from roots as they develop. These exudates are also responsible for switching on a complex set of genes in the rhizobia (*nod* genes) that control the process that induces nodulation in the roots of the host. In a typical case, a bacterial colony develops on the root hair, which then begins to curl and is penetrated by the bacteria. The host responds by laying down a wall that encloses the bacteria and forms an 'infection thread', within which the rhizobia proliferate extracellularly. This grows within the host root cortex, and the host cells divide in advance of it, beginning to form a nodule. Rhizobia in the infection thread cannot fix nitrogen, but some are released into the host meristem cells. There, surrounded by a host-derived peribacteroid membrane, they differentiate into 'bacteroids' that can fix nitrogen. In some species, those with 'indeterminate' growth like the rhizobia of the pea (*Pisum sativum*), the bacteroids themselves are unable to reproduce further. Only undifferentiated rhizobia are released back into the soil to associate with another root when the original root senesces. By contrast, in species with 'determinate' growth like those of the soybean (*Glycine max*), bacteroids survive root senescence and can then invade other roots (Kiers *et al.*, 2003).

A special vascular system develops in the host, supplying the products of photosynthesis to the nodule tissue and carrying away fixed nitrogen compounds (very often the amino acid asparagine) to other parts of the plant (Figure 13.19). The nitrogen-fixing nitrogenase enzyme accounts for up to 40% of the protein in the nodules and depends for its activity on a very low oxygen tension. A boundary layer of tightly packed cells within the nodule serves as a barrier to oxygen diffusion. A hemoglobin (leghemoglobin) is formed within the nodules, giving the active nodules a pink color. It has a high affinity for oxygen and allows the symbiotic bacteria to respire aerobically in the virtually anaerobic environment of the nodule. Indeed, wherever nitrogen-fixing symbioses occur, at least one of the partners has special structural (and usually also biochemical) properties that protect the anaerobic nitrogenase enzyme from oxygen, yet allow normal aerobic respiration to occur around it.

13.10.2 Costs and benefits of rhizobial mutualisms

The costs and benefits of this mutualism need to be considered carefully. From the plant's point of view, we need to compare the energetic costs of alternative processes by which supplies of fixed nitrogen might be obtained. The route for most plants is direct from the soil as nitrate or ammonium ions. The metabolically cheapest route is the use of ammonium ions, but in most soils ammonium ions are rapidly converted to nitrates by microbial activity (nitrification). The energetic cost of reducing nitrate from the soil to ammonia is about 12 mol of adenosine triphosphate (ATP) per mol of ammonia formed. The mutualistic process (including the maintenance costs of the bacteroids) is energetically slightly more expensive to the plant: about 13.5 mol of ATP. However, to the costs of nitrogen fixation itself we must also add the costs of forming and maintaining the nodules, which may be about 12% of the plant's total photosynthetic output. It is this that makes nitrogen fixation energetically inefficient. Energy, though, may be much more readily available for green plants than nitrogen. A rare and valuable commodity (fixed nitrogen) bought with a cheap currency (energy) may be no bad bargain. On the other hand, when a nodulated legume is provided with nitrates (i.e. when nitrate is not a rare commodity) nitrogen fixation declines rapidly.

The benefits to the rhizobia are more problematic from an evolutionary point of view, especially for those with indeterminate growth, where the rhizobia that have become bacteroids can

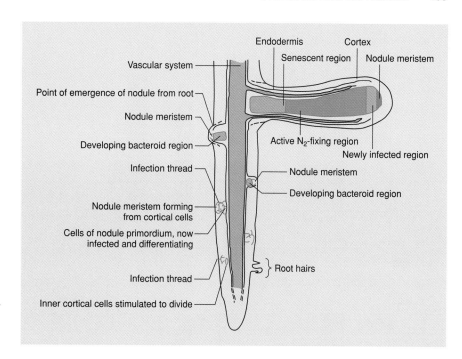

Figure 13.19 The development of the root nodule structure during the course of development of infection of a legume root by *Rhizobium*. (After Sprent, 1979.)

fix nitrogen but cannot reproduce. Hence, they cannot themselves benefit from the symbiosis, since 'benefit' must express itself, ultimately, as an increased reproductive rate (fitness). The rhizobia in the infection thread are capable of reproduction (and are therefore able to benefit), but they cannot fix nitrogen and are therefore not themselves involved in a mutualistic interaction. However, since the rhizobia are clonal, the bacteroids and the cells in the infection thread are all part of the same, single genetic entity. The bacteroids, therefore, by supporting the plant and generating a flow of photosynthates, can benefit the cells of the infection thread, and hence benefit the clone as a whole, in much the same way as the cells in a bird's wing can bring benefit, ultimately, to the cells that produce its eggs – and hence to the bird as a whole.

why no cheating?

One puzzle, though, since the rhizobia associated with a particular plant are typically a mixture of clones, is why individual clones do not 'cheat': that is, derive benefits from the plant, which itself derives benefit from the rhizobia in general, without themselves entering fully into the costly enterprise of fixing nitrogen. Indeed, we can see that this question of *cheating* applies to many mutualisms, once we recognize that they are, in essence, cases of mutual exploitation. There would be evolutionary advantage in exploiting without being exploited. Perhaps the most obvious answer is for the plant (in this case) to monitor the performance of the rhizobia and apply 'sanctions' if they cheat. This, clearly, will provide evolutionary stability to the mutualism by preventing cheats from escaping the interaction, and evidence for such sanctioning has indeed been found for a legume–rhizobium mutualism (Kiers *et al.*, 2003). A normally mutualistic rhizobium strain was prevented from cooperating (fixing nitrogen) by

growing its soybean host in an atmosphere in which air (80% nitrogen, 20% oxygen) was replaced with approximately 80% argon, 20% oxygen and only around 0.03% nitrogen, reducing the rate of nitrogen fixation to around 1% of normal levels. Thus, the rhizobium strain was forced to cheat. In experiments at the whole plant, the part-root and the individual nodule level, the reproductive success of the noncooperating rhizobia was decreased by around 50% (Figure 13.20). Noninvasive monitoring of the plants indicated that they were applying sanctions by withholding oxygen from the rhizobia. Cheating did not pay.

13.10.3 Nitrogen-fixing mutualisms in nonleguminous plants

The distribution of nitrogen-fixing symbionts in nonleguminous higher plants is patchy. A genus of actinomycete, *Frankia*, forms symbioses (actinorhiza) with members of at least eight families of flowering plants, almost all of which are shrubs or trees. The nodules are usually hard and woody. The best known hosts are the alder (*Alnus*), sea buckthorn (*Hippophaë*), sweet gale (*Myrica*), she-oak (*Casuarina*) and the arctic/alpine shrubs *Arctostaphylos* and *Dryas*. *Ceonothus*, which forms extensive stands in Californian chaparral, also develops *Frankia* nodules. Unlike rhizobia, the species of *Frankia* are filamentous and produce specialized vesicles and sporangia that release spores. Whilst the rhizobia rely on their host plant to protect their nitrogenase from oxygen, *Frankia* provides its own protection in the walls of the vesicles, which are massively thickened with as many as 50 monolayers of lipids.

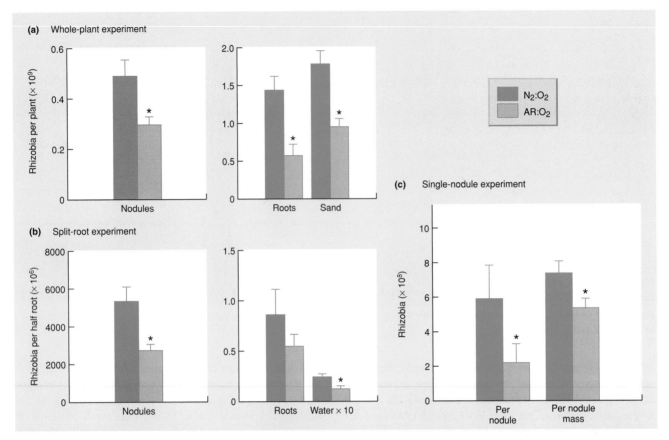

Figure 13.20 The number of rhizobia grew to much larger numbers when allowed to fix nitrogen in normal air ($N_2 : O_2$) than when prevented from doing so by manipulation of the atmosphere (Ar : O_2). (a) When the different treatments were applied at the whole plant level, there were greater numbers within the nodules (left; $P < 0.005$) and on the root surface (right; both $P < 0.01$) and in the surrounding sand ($P < 0.01$). $n = 11$ pairs; bars are standard errors. (b) When the different treatments were applied to different parts of the same root system, there were greater numbers within the nodules (left; $P < 0.001$) and for those in the surrounding water (right; $P < 0.01$), but not significantly so for those on the root surface. $n = 12$ plants; bars are standard errors. (c) When the different treatments were applied to individual nodules from the same root system, there were greater numbers on a per nodule basis ($P < 0.05$) and a per nodule mass basis ($P < 0.01$). $n = 6$ experiments; bars are standard errors. (After Keirs *et al.*, 2003.)

Cyanobacteria form symbioses with three genera of liverwort (*Anthoceros*, *Blasia* and *Clavicularia*), with one fern (the free-floating aquatic *Azolla*), with many cycads (e.g. *Encephalartos*) and with all 40 species of the flowering plant genus *Gunnera*, but with no other flowering plants. In the liverworts, the cyanobacteria *Nostoc* live in mucilaginous cavities and the plant reacts to their presence by developing fine filaments that maximize contact with it. *Nostoc* is found at the base of the leaves of *Gunnera*, in the lateral roots of many cycads, and in pouches in the leaves of *Azolla*.

13.10.4 Interspecific competition

The mutualisms of rhizobia and legumes (and other nitrogen-fixing mutualisms) must not be seen as isolated interactions between bacteria and their own host plants. In nature, legumes normally form mixed stands in association with nonlegumes. These are potential competitors with the legumes for fixed nitrogen (nitrates or ammonium ions in the soil). The nodulated legume sidesteps this competition by its access to a unique source of nitrogen. It is in this ecological context that nitrogen-fixing mutualisms gain their main advantage. Where nitrogen is plentiful, however, the energetic costs of nitrogen fixation often put the plants at a competitive *dis*advantage.

Figure 13.21, for example, shows the results of a classic experiment in which soybeans (*Glycine soja*, a legume) were grown in mixtures with *Paspalum*, a grass. The mixtures either received mineral nitrogen, or were inoculated with *Rhizobium*, or received both. The experiment

a classic 'replacement series'

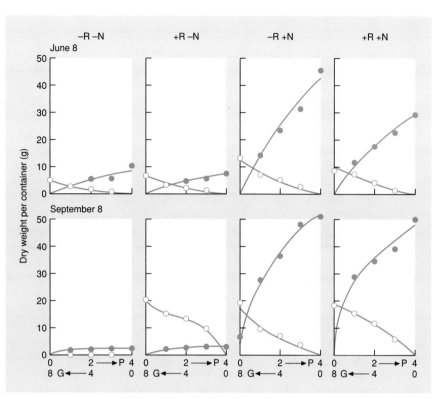

Figure 13.21 The growth of soybeans (*Glycine soja*, G, ○) and a grass (*Paspalum*, P, ●) grown alone and in mixtures with and without nitrogen fertilizer and with and without inoculation with nitrogen-fixing *Rhizobium*. The plants were grown in pots containing 0–4 plants of the grass and 0–8 plants of *Glycine*. The horizontal scale on each figure shows the mass of plants of the two species in each container. −R −N, no *Rhizobium*, no fertilizer; +R −N, inoculated with *Rhizobium* but no fertilizer; −R +N, no *Rhizobium* but nitrate fertilizer was applied; +R +N, inoculated with *Rhizobium* and nitrate fertilizer was supplied. (After de Wit *et al.*, 1966.)

was designed as a 'replacement series' (see Section 8.7.2), which allows us to compare the growth of pure populations of the grass and legume with their performances in the presence of each other. In the pure stands of soybean, yield was increased very substantially *either* by inoculation with *Rhizobium or* by application of fertilizer nitrogen, or by receiving both. The legumes can use either source of nitrogen as a substitute for the other. The grass, however, responded only to the fertilizer. Hence, when the species competed in the presence of *Rhizobium* alone, the legume contributed far more to the overall yield than did the grass: over a succession of generations, the legume would have outcompeted the grass. When they competed in soils supplemented with fertilizer nitrogen, however, whether or not *Rhizobium* was also present, it was the grass that made the major contribution: long term, it would have outcompeted the legume.

Quite clearly, then, it is in environments deficient in nitrogen that nodulated legumes have a great advantage over other species. But their activity raises the level of fixed nitrogen in the environment. After death, legumes augment the level of soil nitrogen on a very local scale with a 6–12-month delay as they decompose. Thus, their advantage is lost – they have improved the environment of their competitors, and the growth of associated grasses will be favored in these local patches. Hence, organisms that can fix atmospheric nitrogen can be thought of as locally suicidal. This is one reason why it is very difficult to grow

repeated crops of pure legumes in agricultural practice without aggressive grass weeds invading the nitrogen-enriched environment. It may also explain why leguminous herbs or trees usually fail to form dominant stands in nature.

Grazing animals, on the other hand, continually remove grass foliage, and the nitrogen status of a grass patch may again decline to a level at which the legume may once more be at a competitive advantage. In a stoloniferous legume, such as white clover, the plant is continually 'wandering' through the sward, leaving behind it local grass-dominated patches, whilst invading and enriching with nitrogen new patches where the nitrogen status has become low. The symbiotic legume in such a community not only drives its nitrogen economy but also some of the cycles that occur within its patchwork (Cain *et al.*, 1995).

13.10.5 Nitrogen-fixing plants and succession

An ecological succession (treated in much more detail in Chapter 17) is the directional replacement of species by other species at a site. A shortage of fixed nitrogen commonly hinders the earliest stages of the colonization of land by vegetation: the initial stages of a succession on open land. Some fixed nitrogen will be contributed in rain after thunderstorms, and some may be blown in from other more established areas, but nitrogen-fixing organisms such as bacteria, cyanobacteria and lichens are important pioneer colonizers. Higher plants with nitrogen-fixing symbionts, however, are rarely pioneers. The reason appears to be that open land is usually colonized first by plants with light, dispersible seeds. A legume seedling, however, depends on fixed nitrogen in its seed reserves and the soil before it can grow to a stage where it can nodulate and fix nitrogen for itself. It is likely, therefore, that only large-seeded legumes carry enough fixed nitrogen to carry them through the establishment phase, and species with such large seeds will not have the dispersibility needed to be pioneers (Grubb, 1986; see also Sprent & Sprent, 1990).

Finally, note that since symbiotic nitrogen fixation is energetically demanding, it is not surprising that most of the higher plant species that support nitrogen-fixing mutualists are intolerant of the shade that is characteristic of the late stages of successions. Higher plants with nitrogen-fixing mutualists are seldom in at the beginning of a succession and they seldom persist to the end.

13.11 Models of mutualisms

Several of the previous chapters on interactions have included a section on mathematical models. This is perhaps a good time to remind ourselves why this was – because the models, by separating essence from detail, were able to provide insights that would not be apparent from a catalog of actual examples. For modeling to be a success, then, it is imperative that the 'essence' is correctly identified. What is the essence of a mutualism? One might imagine it to be that each partner has a positive influence on the fitness of the other partner. At first sight, therefore, we might imagine that an appropriate model for a mutualistic interaction would simply replace the negative contributions in models of two-species competition (see Chapter 8) with positive contributions. However, such a model leads to absurd solutions in which both populations explode to unlimited size (May, 1981) because it places no limits on the carrying capacity of either species, which would therefore increase indefinitely. In practice, intraspecific competition for limiting resources must eventually determine a maximum carrying capacity for any mutualist population, even if the population of the partner mutualist is present in excess (Dean, 1983). Thus, a plant whose growth is limited by a shortage of fixed nitrogen may be released into faster growth by mutualism with a nitrogen-fixing partner, but its faster growth must soon become constrained by a shortage of some other limiting resource (e.g. water, phosphate, radiant energy).

This returns us to points made at the start of the chapter: that the essence of mutualism is subtler than 'mutual benefit'. Rather, instead of thinking of each partner as a benefit to the other, without qualification, it is better to think of each partner exploiting the other, with benefits to be gained but also costs to pay.

And to recognize, too, that the balance of benefits and costs can alter – with changing conditions, changing resource levels, the abundance of either partner, and the presence or abundance of other species. Even the simplest of models, therefore, would have to have terms not signifying a 'positive contribution', but terms that could be positive or negative according to the state of some other part of the model community: not simple at all compared to the models described, and found useful, in previous chapters.

In a sense, then, turning to models does prove helpful at this point. Models of predator–prey and competitor pairs in isolation capture an essence of predator–prey and competitive interactions. The fact that models of isolated pairs of 'mutualists' cannot do so re-emphasizes that mutualism is, in terms of population dynamics, *essentially* an interaction that should only be viewed within the broader context of a larger community. We have seen this earlier in the chapter, for example in ants and aphids in the presence and absence of aphid predators, in *Symbiodinium* taxa coexisting in coral, and in the legume–*Rhizobium* mutualism, which brings its great advantage to the legume when it is competing with some other plant (e.g. a grass) for limited nitrogen from the soil.

models of two-species mutualisms stress the importance of knowing the broader context

This point has been captured in a mathematical model, in which a two-species bee–plant pollination mutualism (Figure 13.22a) was examined embedded in a community that also contained another species of plant and a species of bird that preyed on the bees (Figure 13.22b) (Ringel *et al.*, 1996). The bees could either take nectar and pollen from the plants but fail to pollinate them (predator–prey) or could pollinate them successfully (mutualistic). The model of the simple mutualistic pair (Figure 13.22a) was, as described above, intrinsically unstable. The pair could only persist if the strength of intraspecific competition exceeded that of mutualism: the more mutualistic the interaction became, the more unstable it was. Taken at face value, such results seem to suggest that mutualisms will be rare (though we have seen they are not), since the conditions for their existence are restricted.

a bird, a bee and two plants

A quite different picture emerges, however, once the pair is embedded in the larger assemblage (Figure 13.22b). Using a variety of measures, one of which is illustrated and explained in Figure 13.22c, it was apparent that mutualisms tended to *increase* the chances of the assemblage's persistence. Clearly, there is no necessary paradox between the widespread occurrence of mutualistic interactions in nature and their effects in model assemblages of species. Equally clearly, though, while model assemblages are inevitably simple (e.g. just five species), the effects of mutualistic interactions in nature can easily be misjudged if they are *too* simple (i.e. the mutualistic pair alone).

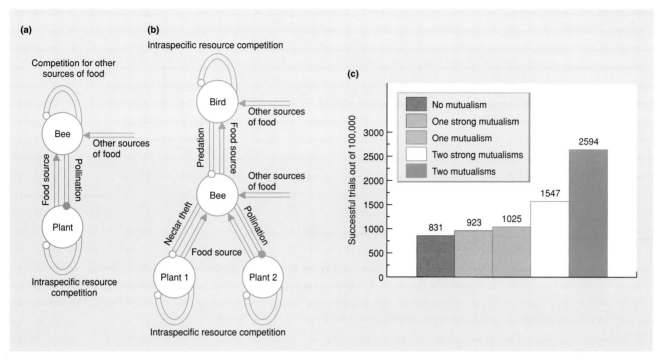

Figure 13.22 (a) A model two-species bee–plant mutualism. Both species are also subject to intraspecific competition. Filled arrowheads indicate a positive interaction, either a resource → consumer (pointed) or a pollination (round) interaction; open arrowheads indicate negative interactions, either consumer → resource or intraspecific competition. (b) The bee and plant embedded in a community with another plant and a bird predator of the bees. The plants suffer intraspecific competition but do not compete with one another. The birds suffer intraspecific competition but the bees do not. The bees take pollen and nectar from both plant species and either fail to pollinate them (predator–prey) or pollinate them successfully (mutualistic). In the figure, the interaction with plant 1 is predator–prey and that with plant 2 is mutualistic, but cases were examined in which neither, one or both were mutualistic. (c) Comparison of persistence in the possible assemblages in (b). Persistence of an assemblage is the maintenance of all species at positive population densities. The bars indicate the number that persisted when the dynamics of each assemblage were simulated 10,000 times, with the strengths of each interaction given by values generated randomly within defined bounds. In a 'strong mutualism', the strength of interaction could be up to twice that in a 'mutualism'. Mutualisms greatly increased the chances of persistence; two-tailed t-test of persistence versus no mutualism: one mutualism ($t = 4.52$, $P < 0.001$), one strong mutualism ($t = 2.21$, $P < 0.05$), two mutualisms ($t = 30.46$, $P < 0.001$), two strong mutualisms ($t = 14.78$, $P < 0.001$). (After Ringel et al., 1996.)

13.12 Evolution of subcellular structures from symbioses

We have seen in this chapter that there is remarkable variety in the types of association that may be regarded as symbiotic – many of them shown clearly to be mutualistic. They extend from patterns of behavior linking two very different organisms that spend parts of their lives apart, through the microbial communities of the vertebrate gut (strictly external to the body tissues), to the intercellular ectomycorrhizas and lichens, and the intracellular dinoflagellate algae of corals and mycetocyte bacteria of insects. We end this chapter by examining how an ecological interaction – mutualism – may lie at the heart of biological patterns operating on the longest evolutionary timescales.

It is now generally accepted that the origin of the various sorts of eukaryotes from more primitive ancestors has progressed at least in part through the *the serial endosymbiosis theory* inextricable merging of partners in a symbiosis. This view was championed especially by Margulis (1975, 1996) in the 'serial endosymbiosis theory' (Figure 13.23a). The aim is to understand the relationships between the three 'domains' of living organism: the archaebacteria or Archaea (many of them now 'extremophiles', living at high temperatures, low pHs and so on), the 'true' bacteria (Eubacteria) and the eukaryotes (Katz, 1998). One suggested first step (estimated to have occurred around 2 billion years ago) was the merger of archaeal and bacterial (spirochete) cells in an anaerobic symbiosis. The former brought its nucleocytoplasm and the latter brought its swimming motility, thus explaining the

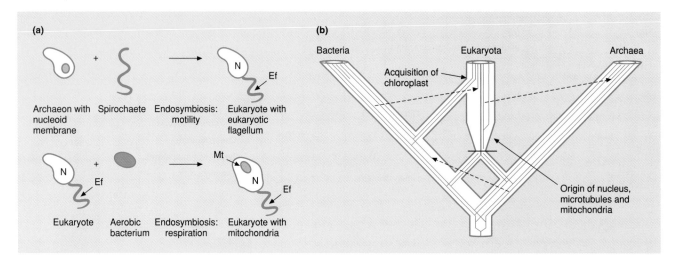

Figure 13.23   (a) The first two steps in the serial endosymbiosis theory for the origin of the eukaryotic cell. Ef, eukaryote flagellum; Mt, mitochondrion; N, nucleus. (b) A model for the origin of eukaryotes indicating a symbiosis between archaeal and bacterial lineages, and the possible simultaneous origin of nuclei, microtubules and mitochondria in eukaryotes. Bold lines represent lineage boundaries; pale lines are gene genealogies; broken arrows are possible lateral transfers of individual genes. (After Katz, 1998.)

chimeric nature – the mix of archaeal and bac-terial features – of the proteins and the genetic material of even the most primitive eukaryotes. Subsequently, some of these chimeras incorporated aerobic bacteria that were the forerunners of mitochondria, to become aerobic eukaryotes from which all other eukaryotes have evolved. Some of these later acquired phototrophic cyanobacteria that were the forerunners of chloroplasts, providing the stock from which the algae and higher plants evolved.

In fact, the serial endosymbiosis theory is merely one of several seeking to link the three domains and recreate the origins of the eukaryotes (Katz, 1998). A suggestion, for instance, that the most primitive eukaryotes have lost mitochondria, rather than never having had them, calls into question the whole sequential nature of eukaryote origins. It may also be that the 'lateral transfer' of individual genes (from one evolutionary lineage to another) has been more pervasive over evolutionary time than was previously imagined, so that the branching tree of life is in fact much more of a tangled web (Figure 13.23b). No doubt, as further evidence accumulates, these competing theories will themselves evolve further, both through progression and the lateral transfer of ideas. What they share, however, is the idea that mutualistic symbioses, beyond their ecological importance, lie at the heart of some of the most fundamental steps in evolution.

Summary

We start by distinguishing mutualism, symbiosis and commensalism and emphasizing that mutualism is best seen as reciprocal exploitation not a cosy partnership.

Mutualisms are examined in a progression: from those where the association is behavioral, through intimate symbioses in which one partner enters between or within another's cells, to those where organelles have entered into such intimate symbioses within the cells of their hosts that they cannot be regarded as distinct organisms.

'Cleaner' fish feed on ectoparasites, bacteria and necrotic tissue from the body surface of 'client' fish. The cleaners gain a food source and the clients are protected from infection. Many ant species protect plants from predators and competitors, while themselves feeding from specialized parts of the plants, though careful experiments are necessary to show that the plants themselves benefit.

Many species, including humans, culture crops or livestock from which they feed. Ants farm many species of aphids in return for sugar-rich secretions, though experiments demonstrate that there can be both costs and benefits for the aphids. Many ants and beetles farm fungi that give them access to otherwise indigestible plant material, and in some cases a three-way mutualism is established with actinomycetes that protect the fungi from pathogens.

Very many plant species use animals to disperse their seeds and pollen. We emphasize the importance of insect pollinators and the coevolutionary pressures generating a range from generalists to ultraspecialists. We also discuss brood site pollinations, of figs and yuccas, by fig wasps and yucca moths that rear their larvae in the fruits of the pollinated plant.

Many animals support a mutualistic microbiota within their guts, especially important in the digestion of cellulose. We outline the range of active sites, and the complex community of mutualists, within the guts of a variety of vertebrates and of termites, focusing especially on the ruminants and noting the importance in many cases of refection. We also describe insect mycetocyte symbioses, especially those between aphids and *Buchnera* species, through which microorganisms, mostly bacteria, living in specialized cells bring nutritional benefits to their insect hosts.

A number of aquatic invertebrates enter into mutualistic associations with photosynthetic algae, perhaps the most important of which are the reef-building corals. We focus especially on 'coral bleaching' – the whitening of corals as a result of the loss of the endosymbionts – and its possible relationship with global warming, and we emphasize the multi- (not two) species nature of these and many other mutualisms.

A wide variety of symbiotic associations are formed between higher plants and fungi. We concentrate on the mycorrhizas – intimate mutualisms between fungi and root tissues – possessed by most plants. We describe ectomycorrhizas, arbuscular mycorrhizas and ericoid mycorrhizas, noting the range of benefits that they can confer.

The biology is outlined of lichens, discussing the intimate associations between mycobiont fungi and phytobionts, mostly algae. Parallels with higher plants are particularly emphasized.

Mutualisms between plants and nitrogen-fixing bacteria are of enormous importance. We outline the range of these bacteria but focus mainly on the mutualisms of rhizobia and leguminous plants, describing the steps involved in establishing the liaison, the costs and benefits to both parties, and the role of the mutualism in determining the outcome of competition between legumes and other plants. This leads to a discussion of the part played by nitrogen-fixing plants in ecological successions.

We examine briefly some mathematical models of mutualisms, which re-emphasize the importance of looking beyond two focal species to the broader context.

Finally, we discuss the possibility that the origin of the various sorts of eukaryotes from more primitive ancestors has progressed at least in part through the inextricable merging of partners in mutualistic symbioses.

Chapter 14
Abundance

14.1 Introduction

Why are some species rare and others common? Why does a species occur at low population densities in some places and at high densities in others? What factors cause fluctuations in a species' abundance? These are crucial questions. To provide complete answers for even a single species in a single location, we might need, ideally, a knowledge of physicochemical conditions, the level of resources available, the organism's life cycle and the influence of competitors, predators, parasites, etc., as well as an understanding of how all these things influence abundance through their effects on the rates of birth, death and movement. In previous chapters, we have examined each of these topics separately. We now bring them together to see how we might discover which factors actually matter in particular examples.

counting is not enough

The raw material for the study of abundance is usually some estimate of population size. In its crudest form, this consists of a simple count. But this can hide vital information. As an example, picture three human populations containing identical numbers of individuals. One of these is an old people's residential area, the second is a population of young children, and the third is a population of mixed age and sex. No amount of attempted correlation with factors outside the population would reveal that the first was doomed to extinction (unless maintained by immigration), the second would grow fast but only after a delay, and the third would continue to grow steadily. More detailed studies, therefore, involve recognizing individuals of different age, sex, size and dominance and even distinguishing genetic variants.

estimates are usually deficient

Ecologists usually have to deal with estimates of abundance that are deficient. First, data may be misleading unless sampling is adequate over both space and time, and adequacy of either usually requires great

commitment of time and money. The lifetime of investigators, the hurry to produce publishable work and the short tenure of most research programs all deter individuals from even starting to conduct studies over extended periods of time. Moreover, as knowledge about populations grows, so the number of attributes of interest grows and changes; every study risks being out of date almost as soon as it begins. In particular, it is usually a technically formidable task to follow individuals in a population throughout their lives. Often, a crucial stage in the life cycle is hidden from view – baby rabbits within their warrens or seeds in the soil. It is possible to mark birds with numbered leg rings, roving carnivores with radiotransmitters or seeds with radioactive isotopes, but the species and the numbers that can be studied in this way are severely limited.

A large part of population theory depends on the relatively few exceptions where logistical difficulties have been overcome (Taylor, 1987). In fact, most

studied species may not be typical

of the really long-term or geographically extensive studies of abundance have been made of organisms of economic importance such as fur-bearing animals, game birds and pests, or the furry and feathered favorites of amateur naturalists. Insofar as generalizations emerge, we should treat them with great caution.

14.1.1 Correlation, causation and experimentation

Abundance data may be used to establish correlations with external factors (e.g. the weather) or correlations between features within the abundance data themselves (e.g. correlating numbers present in the spring with those present in the fall). Correlations may be used to predict the future. For example, high intensities of the disease 'late blight' in the canopy of potato crops usually occur 15–22 days after a period in which the minimum temperature is not less than 10°C and the relative humidity is more than 75% for two consecutive days. Such a correlation may alert the grower to the need for protective spraying.

Correlations may also be used to suggest, although not to prove, causal relationships. For example, a correlation may be demonstrated between the size of a population and its growth rate. The correlation may hint that it is the size of the population itself that causes the growth rate to change, but, ultimately, 'cause' requires a mechanism. It may be that when the population is high many individuals starve to death, or fail to reproduce, or become aggressive and drive out the weaker members.

density is an abstraction

In particular, as we have remarked previously, many of the studies that we discuss in this and other chapters have been concerned to detect 'density-dependent' processes, as if density itself is the cause of changes in birth rates and death rates in a population. But this will rarely (if ever) be the case: organisms do not detect and respond to the density of their populations. They usually respond to a shortage of resources caused by neighbors or to aggression. We may not be able to identify *which* individuals have been responsible for the harm done to others, but we need continually to remember that 'density' is often an abstraction that conceals what the world is like as experienced in the lives of real organisms.

Observing directly what is happening to the individuals may suggest more strongly still what causes a change in overall abundance. Incorporating observations on individuals into mathematical models of populations, and finding that the model population behaves like the real population, may also provide strong support for a particular hypothesis. But often, the acid test comes when it is possible to carry out a field experiment or manipulation. If we suspect that predators or competitors determine the size of a population, we can ask what happens if we remove them. If we suspect that a resource limits the size of a population, we can add more of it. Besides indicating the adequacy of our hypotheses, the results of such experiments may show that we ourselves have the power to determine a population's size: to reduce the density of a pest or weed, or to increase the density of an endangered species. Ecology becomes a predictive science when it can forecast the future: it becomes a management science when it can determine the future.

14.2 Fluctuation or stability?

Gilbert White's swifts

Perhaps the direct observations of abundance that span the greatest period of time are those of the swifts (*Micropus apus*) in the village of Selborne in southern England (Lawton & May, 1984). In one of the earliest published works on ecology, Gilbert White, who lived in the village, wrote of the swifts in 1778:

I am now confirmed in the opinion that we have every year the same number of pairs invariably; at least, the result of my inquiry has been exactly the same for a long time past. The number that I constantly find are eight pairs, about half of which reside in the church, and the rest in some of the lowest and meanest thatched cottages. Now, as these eight pairs – allowance being made for accidents – breed yearly eight pairs more, what becomes annually of this increase?

Lawton and May visited the village in 1983, and found major changes in the 200 years since White described it. It is unlikely that swifts had nested in the church tower for 50 years, and the thatched cottages had disappeared or had been covered with wire. Yet, the number of breeding pairs of swifts regularly to be found in the village was found to be 12. In view of the many changes that have taken place in the intervening centuries, this number is remarkably close to the eight pairs so consistently found by White.

Another example of a population showing relatively little change in adult numbers from year to year is seen in an 8-year study in Poland of the small, annual sand-dune plant *Androsace septentrionalis* (Figure 14.1a). Each year there was great flux within the population: between 150 and 1000 new seedlings per square meter appeared, but subsequent mortality reduced the population by between 30 and 70%. However, the population appears to be kept within bounds. At least 50 plants always survived to fruit and produce seeds for the next season.

The long-term study of nesting herons in the British Isles reported previously in Figure 10.23c reveals a picture of a bird population that has remained remarkably constant over long periods, but here, because repeated estimates were made, it is apparent that there were seasons of severe weather when the population declined precipitously before it subsequently recovered. By contrast, the mice in Figure 14.1b have extended periods of relatively low abundance interrupted by sporadic and dramatic irruptions.

14.2.1 Determination and regulation of abundance

Looking at these studies, and many others like them, some investigators have emphasized the apparent constancy of population sizes, while others have emphasized the fluctuations. Those who have emphasized constancy have argued that we need to look for stabilizing forces within populations to explain why they do not increase without bounds or decline to extinction. Those who have emphasized the fluctuations have looked to external factors, for example the weather, to explain the changes. Disagreements between the two camps dominated much of ecology in the middle third of the 20th century. By considering some of these

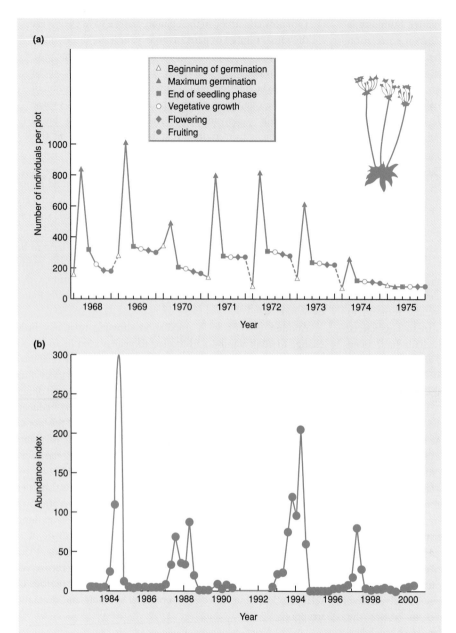

Figure 14.1 (a) The population dynamics of *Androsace septentrionalis* during an 8-year study. (After Symonides, 1979; a more detailed analysis of these data is given by Silvertown, 1982.) (b) Irregular irruptions in the abundance of house mice (*Mus domesticus*) in an agricultural habitat in Victoria, Australia, where the mice, when they irrupt, are serious pests. The 'abundance index' is the number caught per 100 trap-nights. In fall 1984 the index exceeded 300. (After Singleton *et al.*, 2001.)

arguments, it will be easier to appreciate the details of the modern consensus (see also Turchin, 2003).

distinguishing the determination and regulation of abundance

First, however, it is important to understand clearly the difference between questions about the ways in which abundance is *determined* and questions about the way in which abundance is *regulated*. Regulation is the tendency of a population to decrease in size when it is above a particular level, but to increase in size when below that level. In other words, regulation of a population can, by definition, occur only as a result of one or more density-dependent processes that act on rates of birth and/or death and/or movement. Various potentially density-dependent processes have been discussed in earlier chapters on competition, movement, predation and parasitism. We must look at regulation, therefore, to understand how it is that a population tends to remain within defined upper and lower limits.

On the other hand, the precise abundance of individuals will be determined by the combined effects of all the processes that affect a population, whether they are dependent or independent of density. Figure 14.2 shows this diagrammatically and very simply.

Figure 14.2 (a) Population regulation with: (i) density-independent birth and density-dependent death; (ii) density-dependent birth and density-independent death; and (iii) density-dependent birth and death. Population size increases when the birth rate exceeds the death rate and decreases when the death rate exceeds the birth rate. N^* is therefore a stable equilibrium population size. The actual value of the equilibrium population size is seen to depend on both the magnitude of the density-independent rate and the magnitude and slope of any density-dependent process. (b) Population regulation with density-dependent birth, b, and density-independent death, d. Death rates are determined by physical conditions which differ in three sites (death rates d_1, d_2 and d_3). Equilibrium population size varies as a result (N_1^*, N_2^*, N_3^*).

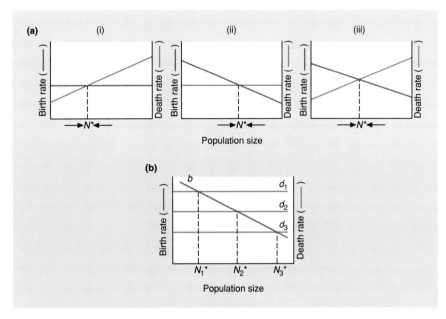

Here, the birth rate is density dependent, whilst the death rate is density independent but depends on physical conditions that differ in three locations. There are three equilibrium populations (N_1, N_2, N_3), which correspond to the three death rates, which in turn correspond to the physical conditions in the three environments. Variations in density-independent mortality like this were primarily responsible, for example, for differences in the abundance of the annual grass *Vulpia fasciculata* on different parts of a sand-dune environment in North Wales, UK. Reproduction was density dependent and regulatory, but varied little from site to site. However, physical conditions had strong density-independent effects on mortality (Watkinson & Harper, 1978). We must look at the determination of abundance, therefore, to understand how it is that a particular population exhibits a particular abundance at a particular time, and not some other abundance.

14.2.2 Theories of abundance

A. J. Nicholson

The 'stability' viewpoint usually traces its roots back to A. J. Nicholson, a theoretical and laboratory animal ecologist working in Australia (e.g. Nicholson, 1954), believing that density-dependent, biotic interactions play the main role in determining population size, holding populations in a state of balance in their environments. Nicholson recognized, of course, that 'factors which are uninfluenced by density may produce profound effects upon density' (see Figure 14.2), but he considered that density dependence 'is merely relaxed from time to time and subsequently resumed, and it remains the influence which adjusts population densities in relation to environmental favourability'.

Andrewartha and Birch

The other point of view can be traced back to two other Australian ecologists, Andrewartha and Birch (1954), whose research was concerned mainly with the control of insect pests in the wild. It is likely, therefore, that their views were conditioned by the need to predict abundance and, especially, the timing and intensity of pest outbreaks. They believed that the most important factor limiting the numbers of organisms in natural populations was the shortage of time when the rate of increase in the population was positive. In other words, populations could be viewed as passing through a repeated sequence of setbacks and recovery – a view that can certainly be applied to many insect pests that are sensitive to unfavorable environmental conditions but are able to bounce back rapidly. They also rejected any subdivision of the environment into Nicholson's density-dependent and density-independent 'factors', preferring instead to see populations as sitting at the center of an ecological web, where the essence was that various factors and processes interacted in their effects on the population.

no need for disagreement between the competing schools of thought

With the benefit of hindsight, it seems clear that the first camp was preoccupied with what regulates population size and the second with what determines population size – and both are perfectly valid interests. Disagreement seems to have arisen because of some feeling within the first camp that whatever regulates *also* determines; and some feeling in the second camp that the determination of abundance is, for practical purposes, all that really matters. It is indisputable, however, that no population can be absolutely free of regulation – long-term unrestrained population growth is unknown, and unrestrained declines to extinction are rare. Furthermore, any suggestion that density-dependent processes are rare or generally of only minor importance would be wrong. A very large number of studies have been made of various kinds of animals, especially of insects. Density dependence has by no means always been detected but is commonly seen when studies are continued for many generations. For instance, density dependence was detected in 80% or more of studies of insects that lasted more than 10 years (Hassell *et al.*, 1989; Woiwod & Hanski, 1992).

On the other hand, in the kind of study that Andrewartha and Birch focused on, weather was typically the major determinant of abundance and other factors were of relatively minor importance. For instance, in one famous, classic study of a pest, the apple thrips (*Thrips imaginis*), weather accounted for 78% of the variation in thrips numbers (Davidson & Andrewartha, 1948). To predict thrips abundance, information on the weather is of paramount importance. Hence, it is clearly not necessarily the case that whatever regulates the size of a population also determines its size for most of the time. And it would also be wrong to give regulation or density dependence some kind of preeminence. It may be occurring only infrequently or intermittently. And even when regulation is occurring, it may be drawing abundance toward a level that is itself changing in response to changing levels of resources. It is likely that no natural population is ever truly at equilibrium. Rather, it seems reasonable to expect to find some populations in nature that are almost always recovering from the last disaster (Figure 14.3a), others that are usually limited by an abundant resource (Figure 14.3b) or by a scarce resource (Figure 14.3c), and others that are usually in decline after sudden episodes of colonization (Figure 14.3d).

There is a very strong bias towards insects in the data sets available for the analysis of the regulation and determination of population size, and amongst these there is a preponderance of studies of pest species. The limited information from other groups suggests that terrestrial vertebrates may have significantly less variable populations than those of arthropods, and that populations of birds are more constant than those of mammals. Large terrestrial mammals seem to be regulated most often by their food supply, whereas in small mammals the single biggest cause of regulation seems to be the density-dependent exclusion of juveniles from breeding (Sinclair, 1989). For birds, food shortage and competition for territories and/or nest sites seem to be most important. Such generalizations, however, may be as much a reflection of biases in the species selected for study and of the neglect of their predators and parasites, as they are of any underlying pattern.

14.2.3 Approaches to the investigation of abundance

demographic, mechanistic and density approaches

Sibly and Hone (2002) distinguished three broad approaches that have been used to address questions about the determination and regulation of

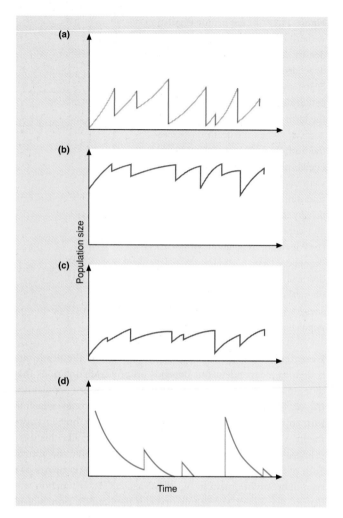

Figure 14.3 Idealized diagrams of population dynamics: (a) dynamics dominated by phases of population growth after disaster; (b) dynamics dominated by limitations on environmental carrying capacity – carrying capacity high; (c) same as (b) but carrying capacity low; and (d) dynamics within a habitable site dominated by population decay after more or less sudden episodes of colonization recruitment.

abundance. They did so having placed population growth rate at the center of the stage, since this summarizes the combined effects on abundance of birth, death and movement. The *demographic* approach (Section 14.3) seeks to partition variations in the overall population growth rate amongst the phases of survival, birth and movement occurring at different stages in the life cycle. The aim is to identify the most important phases. However, as we shall see, this begs the question 'Most important for what?' The *mechanistic* approach (Section 14.4) seeks to relate variations in growth rate directly to variations in specified factors – food, temperature, and so on – that might influence it. The approach itself can range from establishing correlations to carrying out field experiments. Finally, the *density* approach (Section 14.5) seeks to relate variations in growth rate to variations in density. This is a convenient framework for us to use in examining some of the wide variety of studies that have been carried out. However, as Sibly and Hone's (2002) survey makes clear, many studies are hybrids of two, or even all three, of the approaches. Lack of space will prevent us from looking at all of the different variants.

14.3 The demographic approach

14.3.1 Key factor analysis

key factors? or key phases?

For many years, the demographic approach was represented by a technique called *key factor analysis*. As we shall see, there are shortcomings in the technique and useful modifications have been proposed, but as a means of explaining important general principles, and for historical completeness, we start with key factor analysis. In fact, the technique is poorly named, since it begins, at least, by identifying key *phases* (rather than factors) in the life of the organism concerned.

the Colorado potato beetle

For a key factor analysis, data are required in the form of a series of life tables (see Section 4.5) from a number of different cohorts of the population concerned. Thus, since its initial development (Morris, 1959; Varley & Gradwell, 1968) it has been most commonly used for species with discrete generations, or where cohorts can otherwise be readily distinguished. In particular, it is an approach based on the use of *k* values (see Sections 4.5.1 and 5.6). An example, for a Canadian population of the Colorado potato beetle (*Leptinotarsa decemlineata*), is shown in Table 14.1 (Harcourt, 1971). In this species, 'spring adults' emerge from hibernation around the middle of June, when potato plants are breaking through the ground. Within 3

or 4 days oviposition (egg laying) begins, continuing for about 1 month and reaching its peak in early July. The eggs are laid in clusters on the lower leaf surface, and the larvae crawl to the top of the plant where they feed throughout their development, passing through four instars. When mature, they drop to the ground and pupate in the soil. The 'summer adults' emerge in early August, feed, and then re-enter the soil at the beginning of September to hibernate and become the next season's 'spring adults'.

The sampling program provided estimates of the population at seven stages: eggs, early larvae, late larvae, pupae, summer adults, hibernating adults and spring adults. One further category was included, 'females × 2', to take account of any unequal sex ratios amongst the summer adults. Table 14.1 lists these estimates for a single season. It also gives what were believed to be the main causes of death in each stage of the life cycle. In so doing, what is essentially a demographic technique (dealing with phases) takes on the mantle of a mechanistic approach (by associating each phase with a proposed 'factor').

mean *k* values: typical strengths of factors

The mean *k* values, determined for a single population over 10 seasons, are presented in the third column of Table 14.2. These indicate the relative strengths of the various factors that contribute to the total rate of mortality within a generation. Thus, the emigration of summer adults has by far the greatest proportional effect ($k_6 = 1.543$), whilst the starvation of older larvae, the frost-induced mortality of hibernating adults, the 'nondeposition' of eggs, the effects of rainfall on young larvae and the cannibalization of eggs all play substantial roles as well.

What this column of Table 14.2 does not tell us, however, is the relative importance of these factors as determinants of the year-to-year *fluctuations* in mortality. For instance, we can easily imagine a factor that repeatedly takes a significant toll from a population, but which, by remaining constant in its effects, plays

Table 14.1 Typical set of life table data collected by Harcourt (1971) for the Colorado potato beetle (in this case for Merivale, Canada, 1961–62).

Age interval	Numbers per 96 potato hills	Numbers 'dying'	'Mortality factor'	$Log_{10}N$	k value	
Eggs	11,799	2,531	Not deposited	4.072	0.105	(k_{1a})
	9,268	445	Infertile	3.967	0.021	(k_{1b})
	8,823	408	Rainfall	3.946	0.021	(k_{1c})
	8,415	1,147	Cannibalism	3.925	0.064	(k_{1d})
	7,268	376	Predators	3.861	0.024	(k_{1e})
Early larvae	6,892	0	Rainfall	3.838	0	(k_2)
Late larvae	6,892	3,722	Starvation	3.838	0.337	(k_3)
Pupal cells	3,170	16	*D. doryphorae*	3.501	0.002	(k_4)
Summer adults	3,154	126	Sex (52% ♀)	3.499	−0.017	(k_5)
♀ × 2	3,280	3,264	Emigration	3.516	2.312	(k_6)
Hibernating adults	16	2	Frost	1.204	0.058	(k_7)
Spring adults	14			1.146		
					2.926	(k_{total})

Mortality factor	k	Mean k value	Regression coefficient on k_{total}	b	r^2
Eggs not deposited	k_{1a}	0.095	−0.020	−0.05	0.27
Eggs infertile	k_{1b}	0.026	−0.005	−0.01	0.86
Rainfall on eggs	k_{1c}	0.006	0.000	0.00	0.00
Eggs cannibalized	k_{1d}	0.090	−0.002	−0.01	0.02
Eggs predation	k_{1c}	0.036	−0.011	−0.03	0.41
Larvae 1 (rainfall)	k_2	0.091	0.010	0.03	0.05
Larvae 2 (starvation)	k_3	0.185	0.136	0.37	0.66
Pupae (*D. doryphorae*)	k_4	0.033	−0.029	−0.11	0.83
Unequal sex ratio	k_5	−0.012	0.004	0.01	0.04
Emigration	k_6	1.543	0.906	2.65	0.89
Frost	k_7	0.170	0.010	0.002	0.02
	k_{total}	2.263			

Table 14.2 Summary of the life table analysis for Canadian Colorado beetle populations. *b* is the slope of the regression of each *k* factor on the logarithm of the numbers preceding its action; r^2 is the coefficient of determination. See text for further explanation. (After Harcourt, 1971.)

little part in determining the particular rate of mortality (and thus, the particular population size) in any 1 year. This can be assessed, however, from the next column of Table 14.2, which gives the regression coefficient of each individual *k* value on the total generation value, k_{total}.

regressions of *k* on
k_{total}: key factors

A mortality factor that is important in determining population changes will have a regression coefficient close to unity, because its *k* value will tend to fluctuate in line with k_{total} in terms of both size and direction (Podoler & Rogers, 1975). A mortality factor with a *k* value that varies quite randomly with respect to k_{total}, however, will have a regression coefficient close to zero. Moreover, the sum of all the regression coefficients within a generation will always be unity. The values of the regression coefficients will, therefore, indicate the relative strength of the association between different factors and the fluctuations in mortality. The largest regression coefficient will be associated with the *key factor* causing population change.

In the present example, it is clear that the emigration of summer adults, with a regression coefficient of 0.906, is the key factor. Other factors (with the possible exception of larval starvation) have a negligible effect on the changes in generation mortality, even though some have reasonably high mean *k* values. A similar conclusion can be drawn by simply examining graphs of the fluctuations in *k* values with time (Figure 14.4a).

Thus, whilst mean *k* values indicate the average strengths of various factors as causes of mortality in each generation, key factor analysis indicates their relative contribution to the yearly *changes* in generation mortality, and thus measures their importance as determinants of population size.

a role for factors in
regulation?

What, though, of population regulation? To address this, we examine the density dependence of each factor by plotting *k* values against $\log_{10}$ of the

numbers present before the factor acted (see Section 5.6). Thus, the last two columns in Table 14.2 contain the slopes (*b*) and coefficients of determination (r^2) of the various regressions of *k* values on their appropriate '$\log_{10}$ initial densities'. Three factors seem worthy of close examination. The emigration of summer adults (the key factor) appears to act in an overcompensating density-dependent fashion, since the slope of the regression (2.65) is considerably in excess of unity (see also Figure 14.4b). Thus, the key factor, although density dependent, does not so much regulate the population as lead to violent fluctuations in abundance (because of overcompensation). Indeed, the Colorado potato beetle–potato system would go extinct if potatoes were not continually replanted (Harcourt, 1971).

Also, the rate of larval starvation appears to exhibit undercompensating density dependence (although statistically this is not significant). An examination of Figure 14.4b, however, shows that the relationship would be far better represented not by a linear regression but by a curve. If such a curve is fitted to the data, then the coefficient of determination rises from 0.66 to 0.97, and the slope (*b* value) achieved at high densities would be 30.95 (although it is, of course, much less than this in the range of densities observed). Hence, it is quite possible that larval starvation plays an important part in regulating the population, prior to the destabilizing effects of pupal parasitism and adult emigration.

Key factor analysis has been applied to a great many insect populations, but to far fewer vertebrate or plant populations. Examples of these, though, are

wood frogs and an
annual plant

shown in Table 14.3 and Figure 14.5. In populations of the wood frog (*Rana sylvatica*) in three regions of the United States (Table 14.3), the larval period was the key phase determining abundance in each region (second data column), largely as a result of year-to-year variations in rainfall during the larval period. In low rainfall years, the ponds could dry out, reducing larval survival to catastrophic levels, sometimes as a result of a bacterial infection. Such

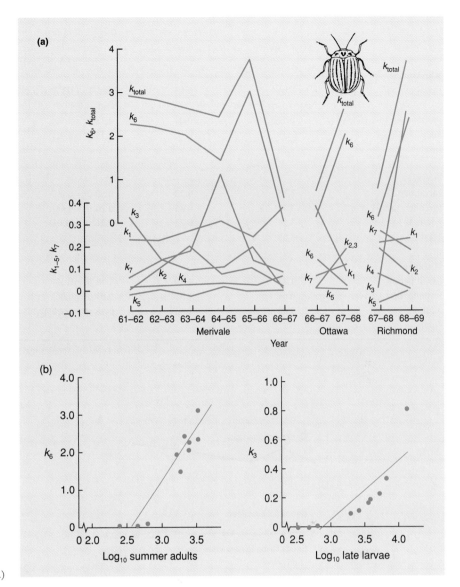

Figure 14.4 (a) The changes with time of the various *k* values of Colorado beetle populations at three sites in Canada. (After Harcourt, 1971.) (b) Density-dependent emigration by Colorado beetle 'summer' adults (slope = 2.65) (left) and density-dependent starvation of larvae (slope = 0.37) (right). (After Harcourt, 1971.)

mortality, however, was inconsistently related to the size of the larval population (one pond in Maryland, and only approaching significance in Virginia – third data column) and hence played an inconsistent part in regulating the sizes of the populations. Rather, in two of the regions it was during the adult phase that mortality was clearly density dependent and hence regulatory (apparently as a result of competition for food). Indeed, in two of the regions mortality was also most intense in the adult phase (first data column).

The key phase determining abundance in a Polish population of the sand-dune annual plant *Androsace septentrionalis* (Figure 14.5; see also Figure 14.1a) was found to be the seeds in the soil. Once

again, however, mortality did not operate in a density-dependent manner, whereas mortality of seedlings, which were not the key phase, was found to be density dependent. Seedlings that emerge first in the season stand a much greater chance of surviving.

Overall, therefore, key factor analysis (its rather misleading name apart) is useful in identifying important phases in the life cycles of study organisms. It is useful too in distinguishing the variety of ways in which phases may be important: in contributing significantly to the overall sum of mortality; in contributing significantly to variations in mortality, and hence in *determining* abundance; and in contributing significantly to the *regulation* of abundance by virtue of the density dependence of the mortality.

Age interval	Mean k value	Coefficient of regression on k_{total}	Coefficient of regression on log (population size)
Maryland			
Larval period	1.94	**0.85**	**Pond 1: 1.03 ($P = 0.04$)**
			Pond 2: 0.39 ($P = 0.50$)
Juvenile: up to 1 year	0.49	0.05	0.12 ($P = 0.50$)
Adult: 1–3 years	**2.35**	0.10	0.11 ($P = 0.46$)
Total	4.78		
Virginia			
Larval period	**2.35**	**0.73**	0.58 ($P = 0.09$)
Juvenile: up to 1 year	1.10	0.05	−0.20 ($P = 0.46$)
Adult: 1–3 years	1.14	0.22	**0.26 ($P = 0.05$)**
Total	4.59		
Michigan			
Larval period	1.12	**1.40**	1.18 ($P = 0.33$)
Juvenile: up to 1 year	0.64	1.02	0.01 ($P = 0.96$)
Adult: 1–3 years	**3.45**	−1.42	**0.18 ($P = 0.005$)**
Total	5.21		

Table 14.3 Key factor (or key phase) analysis for wood frog populations from three areas in the United States: Maryland (two ponds, 1977–82), Virginia (seven ponds, 1976–82) and Michigan (one pond, 1980–93). In each area, the phase with the highest mean k value, the key phase and any phase showing density dependence are highlighted in bold. (After Berven, 1995.)

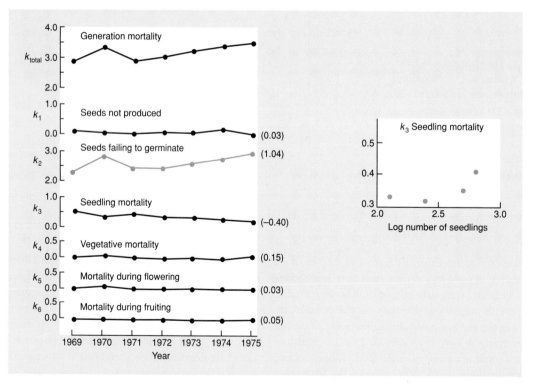

Figure 14.5 Key factor analysis of the sand-dune annual plant *Androsace septentrionalis*. A graph of total generation mortality (k_{total}) and of various k factors is presented. The values of the regression coefficients of each individual k value on k_{total} are given in brackets. The largest regression coefficient signifies the key phase and is shown as a colored line. Alongside is shown the one k value that varies in a density-dependent manner. (After Symonides, 1979; analysis in Silvertown, 1982.)

14.3.2 Sensitivities, elasticities and λ-contribution analysis

overcoming problems in key factor analysis

Although key factor analysis has been useful and widely used, it has been subject to persistent and valid criticisms, some technical (i.e. statistical) and some conceptual (Sibly & Smith, 1998). Important among these criticisms are: (i) the rather awkward way in which *k* values deal with fecundity: a value is calculated for 'missing' births, relative to the maximum possible number of births; and (ii) 'importance' may be inappropriately ascribed to different phases, because equal weight is given to all phases of the life history, even though they may differ in their power to influence abundance. This is a particular problem for populations in which the generations overlap, since mortalities (and fecundities) later in the life cycle are bound to have less effect on the overall rate of population growth than those occurring in earlier phases. In fact, key factor analysis was designed for species with discrete generations, but it has been applied to species with overlapping generations, and in any case, restricting it to the former is a limitation on its utility.

Sibly and Smith's (1998) alternative to key factor analysis, λ-contribution analysis, overcomes these problems. λ is the population growth rate (e^r) that we referred to as *R*, for example, in Chapter 4, but here we retain Sibly and Smith's notation. Their method, in turn, makes use of a weighting of life cycle phases taken from sensitivity and elasticity analysis (De Kroon *et al.*, 1986; Benton & Grant, 1999; Caswell, 2001; see also 'integral projection models', for example Childs *et al.* (2003)), which is itself an important aspect of the demographic approach to the study of abundance. Hence, we deal first, briefly, with sensitivity and elasticity analysis before examining λ-contribution analysis.

the population projection matrix revisited

The details of calculating sensitivities and elasticities are beyond our scope, but the principles can best be understood by returning to the population projection matrix, introduced in Section 4.7.3. Remember that the birth and survival processes in a population can be summarized in matrix form as follows:

$$\begin{bmatrix} p_0 & m_1 & m_2 & m_3 \\ g_0 & p_1 & 0 & 0 \\ 0 & g_1 & p_2 & 0 \\ 0 & 0 & g_2 & p_3 \end{bmatrix}$$

where, for each time step, m_x is the fecundity of stage *x* (into the first stage), g_x is the rate of survival and growth from stage *x* into the next stage, and p_x is the rate of persisting within stage *x*. Remember, too, that λ can be computed directly from this matrix. Clearly, the overall value of λ reflects the values of the various elements in the matrix, but their contribution to λ is not equal. The *sensitivity*, then, of each element (i.e. each biological process) is the amount by which λ would change for a given absolute change in the value of the matrix element, with the value of all the other elements held constant. Thus, sensitivities are highest for those processes that have the greatest power to influence λ.

sensitivity and elasticity

However, whereas survival elements (*g*s and *p*s here) are constrained to lie between 0 and 1, fecundities are not, and λ therefore tends to be more sensitive to absolute changes in survival than to absolute changes of the same magnitude in fecundity. Moreover, λ can be sensitive to an element in the matrix even if that element takes the value 0 (because sensitivities measure what *would* happen if there *was* an absolute change in its value). These shortcomings are overcome, though, by using the *elasticity* of each element to determine its contribution to λ, since this measures the proportional change in λ resulting from a proportional change in that element. Conveniently, too, with this matrix formulation the elasticities sum to 1.

elasticity analysis and the management of abundance

Elasticity analysis therefore offers an especially direct route to plans for the management of abundance. If we wish to increase the abundance of a threatened species (ensure λ is as high as possible) or decrease the abundance of a pest (ensure λ is as low as possible), which phases in the life cycle should be the focus of our efforts? Answer: those with the highest elasticities. For example, an elasticity analysis of the threatened Kemp's ridley sea turtle (*Lepidochelys kempi*) off the southern United States showed that the survival of older, especially subadult individuals was more critical to the maintenance of abundance than either fecundity or hatchling survival (Figure 14.6a). Therefore, 'headstarting' programs, in which eggs were reared elsewhere (Mexico) and imported, and which had dominated conservation practice through the 1980s, seem doomed to be a low-payback management option (Heppell *et al.*, 1996). Worryingly, headstarting programs have been widespread, and yet this conclusion seems likely to apply to turtles generally.

Elasticity analysis was applied, too, to populations of the nodding thistle (*Carduus nutans*), a noxious weed in New Zealand. The survival and reproduction of young plants were far more important to the overall population growth rate than those of older individuals (Figure 14.6b), but, discouragingly, although the biocontrol program in New Zealand had correctly targeted these phases through the introduction of the seed-eating weevil, *Rhinocyllus conicus*, the maximum observed levels of seed predation (*c.* 49%) were lower than those projected to be necessary to bring λ below 1 (69%) (Shea & Kelly, 1998). As predicted, the control program has had only limited success.

elasticity may say little about variations in abundance . . .

Thus, elasticity analyses are valuable in identifying phases and processes that are important in determining abundance, but they do so by focusing on typical or average values, and in that

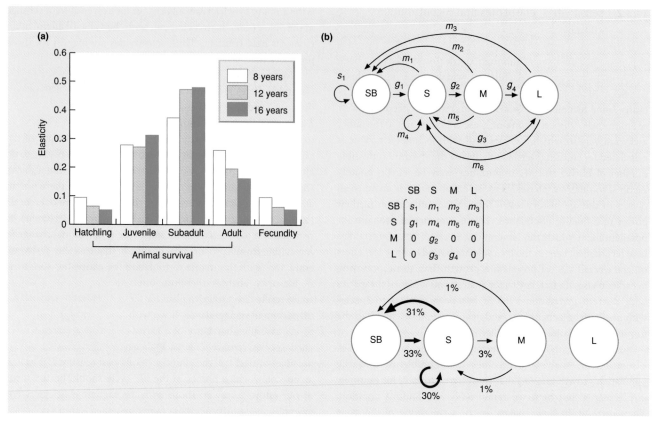

Figure 14.6 (a) Results of elasticity analyses for Kemp's ridley turtles (*Lepidochelys kempi*), showing the proportional changes in λ resulting from proportional changes in stage-specific annual survival and fecundity, on the assumption of three different ages of maturity. (After Heppell *et al.*, 1996.) (b) Top: diagrammatic representation of the life cycle structure of *Carduus nutans* in New Zealand, where SB is the seed bank and S, M and L are small, medium and large plants, and *s* is seed dormancy, *g* is growth and survival to subsequent stages, and *m* is the reproductive contribution either to the seed bank or to immediately germinating small plants. Middle: the population projection matrix summarizing this structure. Bottom: the results of an elasticity analysis for one population, in which the percentage changes in λ resulting from percentage changes in *s*, *g* and *r* are shown on the life cycle diagram. The most important transitions are shown in bold, and elasticities less than 1% are omitted altogether. (After Shea & Kelly, 1998.)

sense they seek to account for the typical size of a population. However, a process with a high elasticity may still play little part, in practice, in accounting for variations in abundance from year to year or site to site, if that process (mortality or fecundity) shows little temporal or spatial variation. There is even evidence from large herbivorous mammals that processes with high elasticity tend to vary little over time (e.g. adult female fecundity), whereas those with low elasticity (e.g. juvenile survival) vary far more (Gaillard *et al.*, 2000). The *actual* influence of a process on variations in abundance will depend on both elasticity *and* variation in the process. Gaillard *et al.* further suggest that the relative absence of variation in the 'important' processes may be a case of 'environmental canalisation': evolution, in the phases most important to fitness, of an ability to maintain relative constancy in the face of environmental perturbations.

In contrast to elasticity analyses, key factor analysis seeks specifically to understand temporal and spatial variations in abundance. The same is true of Sibly and Smith's λ-contribution analysis, to which we now return. We can note first that it deals with the contributions of the different phases not to an overall *k* value (as in key factor analysis) but to λ, a much more obvious determinant of abundance. It makes use of *k* values to quantify mortality, but can use fecundities directly rather than converting them into 'deaths of unborn offspring'. And crucially, the contributions of all mortalities and fecundities are weighted by their sensitivities. Hence, quite properly, where generations overlap, the chances of later phases being identified with a key factor are correspondingly lower in λ-contribution than in key factor

. . . but λ-contribution analysis does

analysis. As a result, λ-contribution analysis can be used with far more confidence when generations overlap. Subsequent investigation of density dependences proceeds in exactly the same way in λ-contribution analysis as in key factor analysis.

Table 14.4 contrasts the results of the two analyses applied to life table data collected on the Scottish island of Rhum between 1971 and 1983 for the red deer, *Cervus elaphus* (Clutton-Brock *et al.*, 1985). Over the 19-year lifespan of the deer, survival and birth rates were estimated in the following 'blocks': year 0, years 1 and 2, years 3 and 4, years 5–7, years 8–12 and years 13–19. This accounts for the limited number of different values in the k_x and m_x columns of the table, but the sensitivities of λ to these values are of course different for different ages (early influences on λ are more powerful), with the exception that λ is equally sensitive to mortality in each phase prior to first reproduction (since it is all 'death before reproduction'). The consequences of these differential sensitivities are apparent in the final two columns

of the table, which summarize the results of the two analyses by presenting the regression coefficients of each of the phases against k_{total} and λ_{total}, respectively. Key factor analysis identifies reproduction in the final years of life as the key factor and even identifies reproduction in the preceding years as the next most important phase. In stark contrast, in λ-contribution analysis, the low sensitivities of λ to birth in these late phases relegate them to relative insignificance – especially the last phase. Instead, survival in the earliest phase of life, where sensitivity is greatest, becomes the key factor, followed by fecundity in the 'middle years' where fecundity itself is highest. Thus, λ-contribution analysis combines the virtues of key factor and elasticity analyses: distinguishing the regulation and determination of abundance, identifying key phases or factors, while taking account of the differential sensitivities of growth rate (and hence abundance) to the different phases.

Table 14.4 Columns 1–4 contain life table data for the females of a population of red deer, *Cervus elaphus*, on the island of Rhum, Scotland, using data collected between 1971 and 1983 (Clutton-Brock *et al.*, 1985): x is age, l_x is the proportion surviving at the start of an age class, k_x, killing power, has been calculated using natural logarithms, and m_x, fecundity, refers to the birth of female calves. These data represent averages calculated over the period, the raw data having been collected both by following individually recognizable animals from birth and aging animals at death. The next two columns contain the sensitivities of λ, the population growth rate, to k_x and m_x in each age class. In the final two columns, the contributions of the various age classes have been grouped as shown. These columns show the contrasting results of a key factor analysis and a λ contribution analysis as the regression coefficients of k_x and m_x on k_{total} and λ_{total}, respectively, where λ_{total} is the deviation each year from the long-term average value of λ. (After Sibly & Smith, 1998, where details of the calculations may also be found.)

Age (years) at start of class, x	l_x	k_x	m_x	Sensitivity of λ to k_x	Sensitivity of λ to m_x	Regression coefficients of k_x, left, and m_x, right, on k_{total}	Regression coefficients of k_x, left, and m_x, right, on λ_{total}
0	1.00	0.45	0.00	−0.14	0.16	0.01, −	0.32, −
1	0.64	0.08	0.00	−0.14	0.09	0.01, −	0.14, −
2	0.59	0.08	0.00	−0.14	0.08		
3	0.54	0.03	0.22	−0.13	0.07	0.00, 0.05	0.03, 0.04
4	0.53	0.03	0.22	−0.11	0.06		
5	0.51	0.04	0.35	−0.10	0.05	−0.00, 0.03	0.08, 0.16
6	0.49	0.04	0.35	−0.08	0.05		
7	0.47	0.04	0.35	−0.07	0.04		
8	0.45	0.06	0.37	−0.05	0.04	0.01, 0.15	0.09, 0.12
9	0.42	0.06	0.37	−0.04	0.03		
10	0.40	0.06	0.37	−0.03	0.03		
11	0.38	0.06	0.37	−0.02	0.02		
12	0.35	0.06	0.37	−0.02	0.02		
13	0.33	0.30	0.30	−0.01	0.02	−0.05, 0.80	0.01, −0.00
14	0.25	0.30	0.30	−0.006	0.01		
15	0.18	0.30	0.30	−0.004	0.008		
16	0.14	0.30	0.30	−0.002	0.005		
17	0.10	0.30	0.30	−0.001	0.004		
18	0.07	0.30	0.30	−0.001	0.002		
19	0.06	0.30	0.30	−0.000	0.002		

14.4 The mechanistic approach

The previous section dealt with analyses directed at phases in the life cycle, but these often ascribe the effects occurring in particular phases to factors or processes – food, predation, etc. – known to operate during those phases. An alternative has been to study the role of particular factors in the determination of abundance directly, by relating the level or presence of the factor (the amount of food, the presence of predators) either to abundance itself or to population growth rate, which is obviously the proximate determinant of abundance. This mechanistic approach has the advantage of focusing clearly on the particular factor, but in so doing it is easy to lose sight of the relative importance of that factor compared to others.

14.4.1 Correlating abundance with its determinants

Figure 14.7, for example, shows four examples in which population growth rate increases with the availability of food. It also suggests that in general, such relationships are likely to level off at the highest food levels where some other factor or factors place an upper limit on abundance.

14.4.2 Experimental perturbation of populations

As we noted in the introduction to this chapter, correlations can be suggestive, but a much more powerful test of the importance of a particular factor is to manipulate that factor and monitor the population's response. Predators, competitors or food can be added or removed, and if they are important in determining abundance, this should be apparent in subsequent comparisons of control and manipulated populations. Examples are discussed below, when we examine what may drive the regular cycles of abundance exhibited by some species, but we should note straight away that field-scale experiments require major investments in time and effort (and money), and a clear distinction between controls and experimental treatments is inevitably much more difficult to achieve than in the laboratory or greenhouse.

One context in which predators have been added to a population is when biological control agents (natural enemies of a pest – see Section 15.2.5) have been released in attempts to control pests. However, because the motivation has been practical rather than intellectual, perfect experimental design has not usually been a priority. There have, for example, been many occasions when aquatic plants have undergone massive population explosions after their introduction to new habitats, creating significant economic problems by blocking navigation channels and irrigation pumps and upsetting local fisheries. The population explosions occur as the plants grow clonally, break up into fragments and become dispersed. The aquatic fern, *Salvinia molesta*, for instance, which originated in southeastern Brazil, has appeared since 1930 in various tropical and subtropical regions. It was first recorded in Australia in 1952 and spread very rapidly – under optimal conditions *Salvinia* has a doubling time of 2.5 days.

biological control: an experimental perturbation

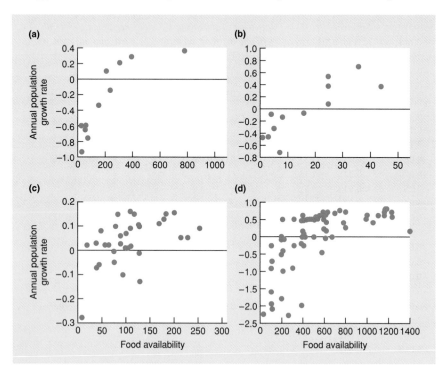

Figure 14.7 Increases in annual population grown rate ($r = \ln \lambda$) with the availability of food (pasture biomass (in kg ha^{-1}), except in (b) where it is vole abundance and in (c) where it is availability per capita. (a) Red kangaroo (from Bayliss, 1987). (b) Barn owl (after Taylor, 1994). (c) Wildebeest (from Krebs *et al.*, 1999). (d) Feral pig (from Choquenot, 1998). (After Sibly & Hone, 2002.)

Figure 14.8   Lake Moon Darra (North Queensland, Australia). (a) Covered by dense populations of the water fern (*Salvinia molesta*). (b) After the introduction of weevil (*Cyrtobagous* spp.). (Courtesy of P.M. Room.)

Significant pests and parasites appear to have been absent. In 1978, Lake Moon Darra (northern Queensland) carried an infestation of 50,000 tonnes fresh weight of *Salvinia* covering an area of 400 ha (Figure 14.8).

Amongst the possible control agents collected from *Salvinia*'s native range in Brazil, the black long-snouted weevil (*Cyrtobagous* sp.) was known to feed only on *Salvinia*. On June 3, 1980, 1500 adults were released in cages at an inlet to the lake and a further release was made on January 20, 1981. The weevil was free of any parasites or predators that might reduce its density and,

by April 1981, *Salvinia* throughout the lake had become dark brown. Samples of the dying weed contained around 70 adult weevils per square meter, suggesting a total population of 1000 million beetles on the lake. By August 1981, there was estimated to be less than 1 tonne of *Salvinia* left on the lake (Room *et al.*, 1981). This has been the most rapid success of any attempted biological control of one organism by the introduction of another, and it establishes the importance of the weevil in the persistently low abundance of *Salvinia* both after the weevil's introduction to Australia and in its native environment. It was a controlled experiment to the extent that other lakes continued to bear large populations of *Salvinia*.

treating red grouse for nematodes Both the power and the problems of field-scale experiments are further illustrated by an example we have already discussed in Section 12.7.2, in which a 'predator' (in this case a parasite) was not added but removed. When Hudson *et al.* (1998) treated cyclic populations of the red grouse *Lagopus lagopus* against the nematode *Trichostrongylus tenuis*, the extent of the grouse 'crash' was very substantially reduced, proving the importance of the nematodes, normally, in reducing grouse abundance, and justifying the effort that had gone into the manipulation. But as we have seen, in spite of this effort, controversy remained about whether the nematodes had been proved to be the cause of the cycles (in which case, the residual smaller crashes were dying echoes) or whether, instead, the experiment had only proved a role for the nematodes in determining a cycle's amplitude, leaving their role in cyclicity itself uncertain. Experiments are better than correlations, but when they involve ecological systems in the field, eliminating ambiguity can never be guaranteed.

14.5   The density approach

Correlations with density have not been altogether absent from the approaches we have considered so far, and indeed, density dependence played a central role in our discussions of the determinants of abundance (birth, death and movement) in earlier chapters. Some studies, however, have focused much more fixedly on density dependences in their own right. In particular, many such studies have been designed to seek evidence for both direct and *delayed* density dependence (see Section 10.2.2). It is a problem, for example, that conventional life table analyses may fail to detect delayed density dependence simply because they are not designed to do so (Turchin, 1990). An analysis of population time series for 14 species of forest insects detected direct density dependence clearly in only five, but it revealed delayed density dependence in seven of the remaining nine (Turchin, 1990). It may be that a similar proportion of populations classified, from their life tables, as lacking density dependence are actually subject to the delayed density dependence of a natural enemy.

14.5.1   Time series analysis: dissecting density dependence

abundance determination expressed as a time-lag equation A number of related approaches have sought to dissect the density-dependent 'structure' of species' population dynamics by a statistical analysis of time series of abundance. Abundance at

a given point in time may be seen as reflecting abundances at various times in the past. It reflects abundance in the immediate past in the obvious sense that the past abundance gave rise directly to the present abundance. It may also reflect abundance in the more distant past if, for example, that past abundance gave rise to an increased abundance of a predator, which in due course affected the present abundance (i.e. a delayed density dependence). In particular, and without going into technical details, the log of the abundance of a population at time t, X_t, can be expressed, at least approximately, as:

$$X_t = m + (1 + \beta_1)X_{t-1} + \beta_2 X_{t-2} + \ldots + \beta_d X_{t-d} + u_t, \qquad (14.1)$$

an equation that captures, in a particular functional form, the idea of present abundance being determined by past abundances (Royama, 1992; Bjørnstad *et al.*, 1995; see also Turchin & Berryman, 2000). Thus, m reflects the mean abundance around which there are fluctuations over time, β_1 reflects the strength of direct density dependence, and other βs reflect the strengths of delayed density dependences with various time lags up to a maximum d. Finally, u_t represents fluctuations from time-point to time-point imposed from outside the population, independent of density. It is easiest to understand this approach when the X_ts represent deviations from the long-term average abundance such that m disappears (the long-term average deviation from the mean is obviously zero). Then, in the absence of any density dependence (all βs zero) the abundance at time t will reflect simply the abundance at time $t - 1$ plus any 'outside' fluctuations u_t; while any regulatory tendencies will be reflected in β values of less than zero.

Fennoscandian microtines Applying this approach to a time series of abundance (i.e. a sequence of X_t values) the usual first step is to determine the statistical model (X_t as the dependent variable) with the optimal number of time lags: the one that strikes the best balance between accounting for the variations in X_t and not including too many lags. Essentially, additional lags are included as long as they account for a significant additional element of the variation. The β values in the optimal model may then shed light on the manner in which abundance in the population is regulated and determined. An example is illustrated in Figure 14.9, which summarizes analyses of 19 time series of microtine rodents (lemmings and voles) from various latitudes in Fennoscandia (Finland, Sweden and Norway) sampled once per year (Bjørnstad *et al.*, 1995). In almost all cases, the optimum number of lags was two, and so the analysis proceeded on the basis of these two lags: (i) direct density dependence; and (ii) density dependence with a delay of 1 year.

Figure 14.9a sets out the predicted dynamics, in general, of populations governed by these two density dependences (Royama, 1992). Remember that delayed density dependence is reflected in a value of β_2 less than 0, while direct density dependence is reflected

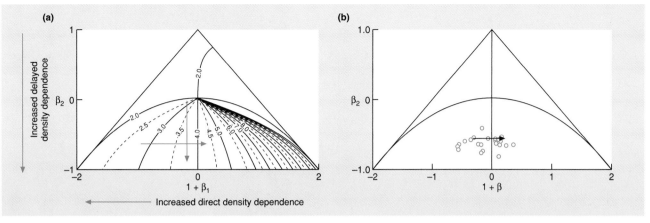

Figure 14.9 (a) The type of population dynamics generated by an autoregressive model (see Equation 14.1) incorporating direct density dependence, β_1, and delayed density dependence, β_2. Parameter values outside the triangle lead to population extinction. Within the triangle, the dynamics are either stable or cyclic and are always cyclic within the semicircle, with a period (length of cycle) as shown by the contour lines. Hence, as indicated by the arrows, the cycle period may increase as β_2 decreases (more intense delayed density dependence) and especially as β_1 increases (less intense direct density dependence). (b) The locations of the pairs of β_1 and β_2 values, estimated from 19 microtine rodent time series from Fennoscandia. The arrow indicates the trend of increasing latitude in the geographic origin of the time series, suggesting that a trend in cycle period with latitude, from around 3 to around 5 years, is the result of a decreased intensity of direct density dependence. (After Bjørnstad *et al.*, 1995.)

in a value of $(1 + \beta_1)$ less than 1. Thus, populations not subject to delayed density dependence tend not to exhibit cycles (Figure 14.9a), but β_2 values less than 0 generate cycles, the period (length) of which tends to increase both as delayed density dependence becomes more intense (down the vertical axis) and especially as direct density dependence becomes less intense (left to right on the horizontal axis).

support for the
specialist predation
hypothesis

The results of Bjørnstad *et al.*'s analysis are set out in Figure 14.9b. The estimated values of β_2 for the 19 time series showed no trend as latitude increased, but the β_1 values increased significantly. The points combining these pairs of βs, then, are shown in the figure, and the trend with increasing latitude is denoted by the arrow. It was known prior to the analysis, from the data themselves, that the rodents exhibited cycles in Fennoscandia and that the cycle length increased with latitude. The data in Figure 14.9b point to precisely the same patterns. But in addition, they suggest that the reasons lie in the structure of the density dependences: on the one hand, a strong delayed density dependence throughout the region, such as would result from the actions of specialist predators; and on the other hand, a significant decline with latitude in the intensity of direct density dependence, such as may result from an immediate shortage of food or the actions of generalist predators (see Figure 10.11b). As we shall see in Section 14.6.4 (see also Section 10.4.4), this in turn is supportive of the 'specialist predation' hypothesis for microtine cycles. The important point here, though, is the

illustration this example provides of the utility of such analyses, focusing on the abundances themselves, but suggesting underlying mechanisms.

14.5.2   Time series analysis: counting and characterizing lags

In other, related cases, the emphasis shifts to deriving the optimal statistical model because the number of lags in that model may provide clues as to how abundance is being determined. It may do so because Takens' theorem (see Section 5.8.5) indicates that a system that can be represented with three lags, for example, comprises three functional interacting elements, whereas two lags imply just two elements, and so on.

One example of this approach (another is described in Section 12.7.1)

hares and lynx . . .

is the study of Stenseth *et al.* (1997) of the hare–lynx system in Canada to which we have already referred briefly in Section 10.2.5. We noted there that the optimal model for the hare time series suggested three lags, whereas that for the lynx suggested two. The density dependences for these lags are illustrated in Figure 14.10a. For the hares, direct density dependence was weakly negative (remember that the slope shown is $1 + \beta_1$) and density dependence with a delay of 1 year was negligible, but there was significant density dependence with a delay of 2 years. For the lynx, direct density dependence was effectively absent, but there was strong density dependence with a delay of 1 year.

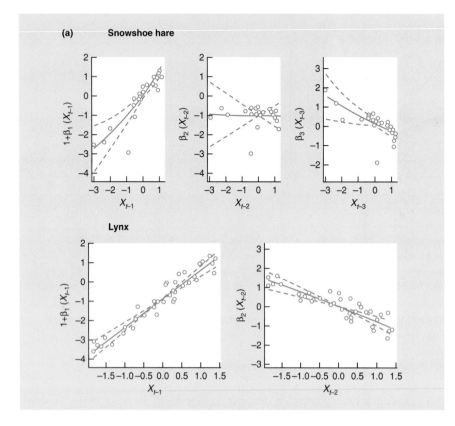

Figure 14.10 (a) Functions for the autoregressive equations (see Equation 14.1) for the snowshoe hare, above (three 'dimensions': direct density dependence and delays of 1 and 2 years), and the lynx, below (two dimensions: direct density dependence and a delay of 1 year). In each case, therefore, the slope indicates the estimated parameters, $1 + \beta_1$, β_2 and β_3, respectively, reflecting the intensity of density dependence. The 95% confidence intervals are also shown.

. . . display three and two dimensions, respectively

This, combined with a detailed knowledge of the whole community of which the hare and lynx are part (Figure 14.10b, c), provided justification for Stenseth *et al.* (1997) to go on to construct a three-equation model for the hares and a two-equation model for the lynx. Specifically, the model for the lynx comprised just the lynx and the hares, since the hares are by far the lynx's most important prey (Figure 14.10b). Whereas the model for the hares comprised the hares themselves, 'vegetation' (since hares feed relatively indiscriminately on a wide range of vegetation), and 'predators' (since a wide range of predators feed on the hares and even prey on one another in the absence of hares, adding a strong element of self-regulation within the predator guild as a whole) (Figure 14.10c).

Lastly, then, and again without going into technical details, Stenseth *et al.* were able to recaste the two- and three-equation models of the lynx and hare into the general, time-lag form of Equation 14.1. In so doing, they were also able to recaste the β values in the time-lag equations as appropriate combinations of the interaction strengths between and within the hares, the lynx, and so on. Encouragingly, they found that these combinations were entirely consistent with the slopes (i.e. the β values) in Figure 14.10a. Thus, the elements that appeared to determine hare and lynx abundance were first counted (three and two, respectively) and

then characterized. What we have here, therefore, is a powerful hybrid of a statistical (time series) analysis of densities and a mechanistic approach (incorporating into mathematical models knowledge of the specific interactions impinging on the species concerned).

Finally, note that related methods of time series analysis have been used in the search for chaos in ecological systems, as described in Section 5.8.5. The motivations in the two cases, of course, are somewhat different. The search for chaos, none the less, is, in a sense, an attempt to identify as 'regulated' populations that appear, at first glance, to be anything but.

14.5.3 Combining density dependence and independence – weather and ecological interactions

Seeking to dissect out the relative contributions of direct and delayed density dependences, however, could itself be seen as prejudging the determinants of abundance by focusing too much on density-dependent, as opposed to density-independent, processes. Other studies, then, have examined time series precisely with a view to understanding how density-dependent and -independent factors combine in generating particular patterns

multimammate rats in Tanzania

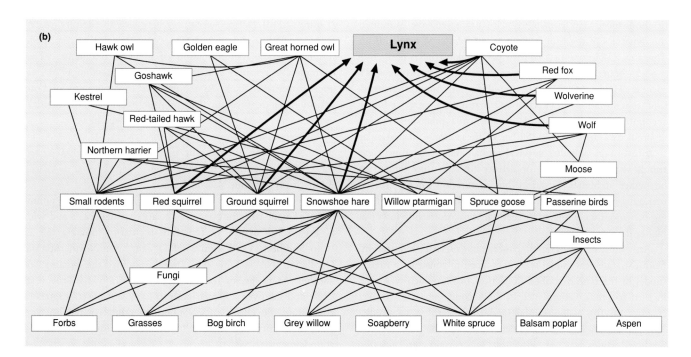

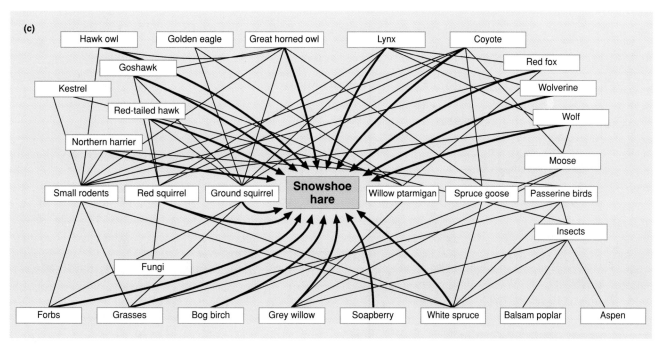

Figure 14.10   (*continued*) (b) The main species and groups of species in the boreal forest community of North America, with trophic interactions (who eats who) indicated by lines joining the species, and those affecting the lynx shown in bold. (c) The same community, but with the interactions of the snowshoe hare shown in bold. (After Stenseth *et al.*, 1997.)

of abundance. Leirs *et al.* (1997), for example, have examined the dynamics of the multimammate rat, *Mastomys natalensis*, in Tanzania. Using one part of their data to construct a predictive model (Figure 14.11a) and a second part to test that model's success (Figure 14.11b), they found first, in model construction, that variations in survival and maturation were far better accounted for by using both densities and preceding rainfall as predictors than by using either of these alone. In particular (Figure 14.11c), subadult survival probabilities showed no clear trends with either rainfall or density (though they tended to be higher at higher densities), but maturation rates increased markedly with rainfall (and were lowest at high densities following wet months), while adult survival was consistently lower at higher densities.

Estimates of demographic parameters (survival, maturation) from the statistical model were then used to construct a matrix model of the type described in Section 14.3.2, which was used in turn to predict abundance in the second, separate data set (Figure 14.11b), using rainfall and density there to predict 1 month ahead. The correspondence between the observed and predicted values was not perfect but was certainly encouraging

(Figure 14.11d). Hence, we can see here how the density, mechanistic (rainfall) and demographic approaches combine to provide insights into the determination of the rats' abundance. This example also reminds us that a proper understanding of abundance patterns is likely to require the incorporation of both density-dependent, biotic, deterministic effects and the density-independent, often stochastic effects of the weather.

Of course, not all the effects of the weather are wholly stochastic in the sense of being entirely unpredictable. Apart from obvious seasonal varia-

mice and the ENSO in Chile

tions, we saw in Section 2.4.1, for example, that there are a number of climatic patterns operating at large spatial scales and with at least some degree of temporal regularity, notably the El Niño–Southern Oscillation (ENSO) and the North Atlantic Oscillation (NAO). Lima *et al.* (1999) examined the dynamics of another rodent species, the leaf-eared mouse, *Phyllotis darwini*, in Chile and followed a similar path to Leirs and his colleagues in combining the effects of ENSO-driven rainfall variability and delayed density dependence in accounting for the observed abundance patterns.

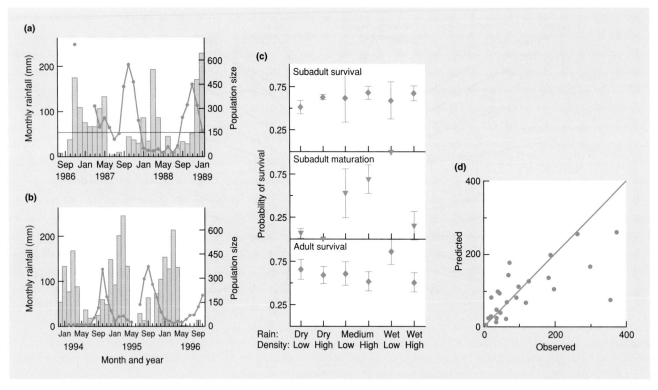

Figure 14.11 (a) Time series data for the multimammate rat (dots) and rainfall (bars) in Tanzania, used to derive a statistical model to predict rat abundance. (The horizontal line indicates the cut-off between 'high' and 'low' densities.) (b) Subsequent time series data used to test that model. (c) Estimates (± SE) from the model for the effects of density and rainfall on population size. (d) The relationship between predicted and observed population sizes in the test data ($r^2 = 0.49$, $P < 0.001$); the line of equality is also shown. (After Leirs *et al.*, 1997.)

14.6 Population cycles and their analysis

Regular cycles in animal abundance were first observed in the long-term records of fur-trading companies and gamekeepers. Cycles have also been reported from many studies of voles and lemmings and in certain forest Lepidoptera (Myers, 1988). Population ecologists have been fascinated by cycles at least since Elton drew attention to them in 1924. In part, this fascination is attributable to the striking nature of a phenomenon that is crying out for an explanation. But there are also sound scientific reasons for the preoccupation. First, cyclic populations, almost by definition, exist at different times at a wide range of densities. They therefore offer good opportunities (high statistical power) for detecting such density-dependent effects as might exist, and integrating these with density-independent effects in an overall analysis of abundance. Furthermore, regular cycles constitute a pattern with a relatively high ratio of 'signal' to 'noise' (compared, say, to totally erratic fluctuations, which may appear to be mostly noise). Since any analysis of abundance is likely to seek ecological explanations for the signal and attribute noise to stochastic perturbations, it is obviously helpful to know clearly which *is* signal and which is noise.

extrinsic and intrinsic factors

Explanations for cycles are usually classified as emphasizing either extrinsic or intrinsic factors. The former, acting from outside the population, may be food, predators or parasites, or some periodic fluctuation in the environment itself. Intrinsic factors are changes in the phenotypes of the organisms themselves (which might in turn reflect changes in genotype): changes in aggressiveness, in the propensity to disperse, in reproductive output, and so on. Below we examine studies on population cycles in three systems, all of which we have touched on previously: the red grouse (Section 14.6.2), the snowshoe hare (and lynx) (Section 14.6.3) and microtine rodents (Section 14.6.4). In each case, it will be important to bear in mind the problems of disentangling cause from effect; that is, of distinguishing factors that change density from those that merely vary *with* density. Equally, it will be important to try to distinguish the factors that affect density (albeit in a cyclic population) from those that actually impose a pattern of cycles (see also Berryman, 2002; Turchin, 2003).

14.6.1 Detecting cycles

The defining feature of a population cycle or oscillation is regularity: a peak (or trough) every *x* years. (Of course, *x* varies from case to case, and a certain degree of variation around *x* is inevitable; even in a '3-year cycle', the occasional interval of 2 or 4 years is to be expected.) The statistical methods applied to a time series, to determine whether the claim of 'cyclicity' can justifiably be made, usually involve the use of an autocorrelation function (Royama, 1992; Turchin & Hanski, 2001). This sets out the correlations between pairs of abundances one time interval apart, two time intervals apart, and so on (Figure 14.12a). The correlation between abundances just one time interval apart can often be high simply because one abundance has led directly to the next. Thereafter, a high positive correlation between pairs, for example, 4 years apart would indicate a regular cycle with a period of 4 years; while a further high *negative* correlation between pairs 2 years apart would indicate a degree of symmetry in the cycle: peaks and troughs typically 4 years apart; with peaks typically 2 years from troughs.

autocorrelation function analysis

It must be remembered, however, that it is not just the pattern of an autocorrelation function that is important but also its statistical significance. Even a single clear rise and fall in a relatively short time series may hint at a cycle (Figure 14.12b), but this pattern would need to be repeated in a much longer series before the autocorrelations were significant, and only then could a cycle be said to have been identified (and require explanation). It is no surprise that major investments in time and effort are required to study cycles in natural populations. Even where those investments have been made, the resulting 'ecological' time series are shorter than those commonly generated in, say, physics – and shorter than those probably envisaged by the statisticians who devised methods for analyzing them. Ecologists need always to be cautious in their interpretations.

14.6.2 Red grouse

The explanation for cycles in the dynamics of the red grouse (*Lagopus lagopus scoticus*) in the United Kingdom has been a matter of disagreement for decades. Some have emphasized an extrinsic factor, the parasitic nematode *Trichostrongylus tenuis* (Dobson & Hudson, 1992; Hudson *et al.*, 1998). Others have emphasized an intrinsic process through which increased density leads to more interactions between non-kin male birds and hence to more aggressive interactions. This leads in turn to wider territorial spacing and, with a delay because this is maintained into the next year, to reduced recruitment (Watson & Moss, 1980; Moss & Watson, 2001). Both viewpoints, therefore, rely on a delayed density dependence to generate the cyclic dynamics (see Section 10.2.2), though these are arrived at by very different means.

parasites?

We have already seen in Sections 12.7.4 and 14.4.2 that even field-scale experiments have been unable to determine the role of the nematodes with certainty. There seems little doubt that they reduce density, and the results of the experiment are consistent with them generating the cycles, too. But the results are also consistent with the nematodes determining the amplitude of the cycles but not generating them in the first place.

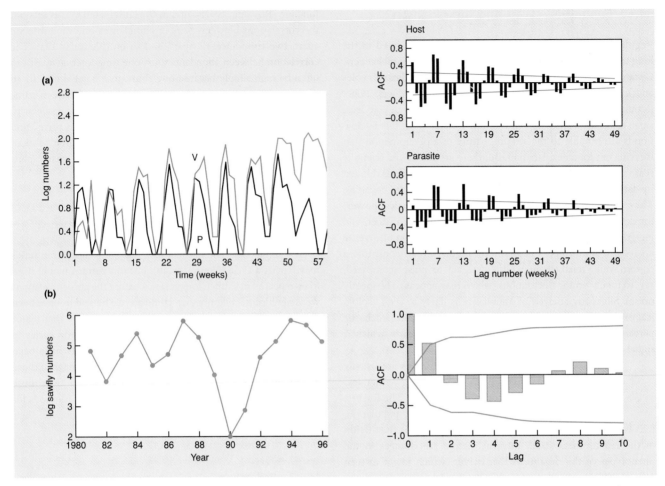

Figure 14.12 (a) Coupled oscillations in the abundance of the moth *Plodia interpunctella* and its parasitoid *Venturia canescens* (P and V, respectively) and, on the right, an autocorrelation function (ACF) analysis of those data (host above, parasitoid below). Sloping lines show the levels the bars must exceed for statistical significance ($P < 0.05$). The cycle periods (l) are 6–7 weeks, with significant correlations at l, $2l$, etc. and significant negative correlations at $0.5l$, $1.5l$, etc. (After Begon *et al.*, 1996.) (b) Time series for the abundance of the sawfly *Euura lasiolepis* (left) and an ACF analysis of those data (right). There is a hint of an 8-year cycle (positive correlation with a lag of 8 years; negative correlation with a lag of 4 years), but this does not come close to statistical significance (exceeding the lines). (After Turchin & Berryman, 2000; following Hunter & Price, 1998).

kinship, territories
and aggression?

In another field experiment, aspects of the alternative, 'kinship' or 'territorial behavior' hypothesis were tested (Mougeot *et al.*, 2003). In experimental areas, established males were given testosterone implants at the beginning of the fall, when territorial contests take place. This increased their aggressiveness (and hence the size of their territories) at densities that would not normally generate such aggression. By the end of the fall, it was clear that, relative to the control areas, the increased aggression of the older males had reduced the recruitment of the younger males: testosterone treatment had significantly reduced male densities and had particularly reduced the ratio of young (newly recruited) to old males, though there was no consistent effect on female densities (Figure 14.13a).

Moreover, in the following year, even though the direct effects of the testosterone had worn off, the young males had not returned (Figure 14.13a). Also, because young *relatives* had been driven out, levels of kinship were likely to be lower in experimental than in control areas. Hence, the kinship hypothesis predicts that recruitment and density in the experimental areas should have remained lower through the following year: that is, lower kinship leads to more aggression, which leads to larger territories, which leads to lower recruitment, which leads to lower density. These predictions were borne out (Figure 14.13b).

Thus, these results establish, at least, the potential for intrinsic processes to have (delayed) density-dependent effects on recruitment, and thus to generate cycles in the grouse. In a companion paper, Matthiopoulos *et al.* (2003) demonstrate how changes in aggressiveness can cause population cycles. As Mougeot *et al.* themselves note, though, the possibility remains that both the parasites and territorial behavior contribute to the observed cycles. Indeed, the two processes may interact: parasites, for example, reduce territorial behavior (Fox & Hudson, 2001). Certainly, it is far from guaranteed that either of the alternative explanations will ultimately be declared the 'winner'.

14.6.3   Snowshoe hares

The '10-year' hare and lynx cycles have also been examined in previous sections. We have seen, for example, from Stenseth *et al.*'s (1997) analysis of time series (see Section 14.5.2) that despite becoming a 'textbook' example of coupled predator–prey oscillations, the hare cycle appears in fact to be generated by interactions with both its food and its predators, both considered as guilds rather than as single species. The lynx cycle, on the other hand, does indeed appear to be generated by its interaction with the snowshoe hare.

This supports other results obtained by much more direct, experimental means, reviewed by Krebs *et al.* (2001). The demographic patterns underlying the hare cycle are relatively clear cut: both fecundity and survival begin to decline well before peak densities are reached, arriving at their minima around 2 years after density has started to decline (Figure 14.14).

First, we can ask: 'What role does the hares' interaction with their food play in these patterns?' A whole series of field experiments in which artificial food was added, or natural food was supplemented, or food quality was manipulated either by fertilizers or by cutting down trees to make high-quality twigs available, all pointed in the same direction. Food supplementation may improve individual condition and in some cases lead to higher densities, but food by itself seems to have no discernible influence on the cyclic pattern (Krebs *et al.*, 2001).

field-scale manipulations of food and/or predators

On the other hand, experiments in which either predators were excluded, or they were excluded and food was also supplemented, had much more dramatic effects. In the study by Krebs *et al.* (1995) carried out at Kluane Lake in the Yukon, Canada (Figure 14.15a), the combination of the two treatments all but eliminated the pattern of decline in survival over the cycle from 1988 to 1996, and predation played by far the major role in this.

Furthermore, food supplementation reduced slightly the initial decline in fecundity prior to peak densities (Figure 14.15b), but the combination of food supplementation and predator exclusion brought fecundity up to almost maximum levels at the phase of

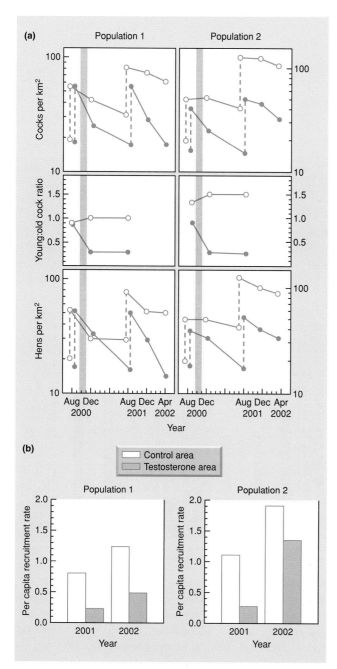

Figure 14.13   (a) Changes in grouse numbers (males (cocks); the young : old cock ratio; and females (hens)) in control (○) and testosterone-implant experimental areas (●) in two populations. The gray bar represents the period of time over which the males were given implants. (b) Per capita recruitment in the two populations was higher in the control areas than in the experimental areas, both in 2001, immediately after treatment, and 1 year later. (After Mougeot *et al.*, 2003.)

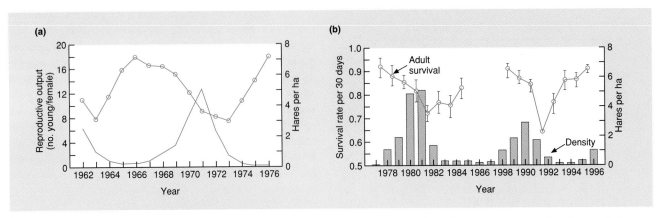

Figure 14.14 (a) Variation in reproductive output per year (dots) as density (continuous line) changes over a snowshoe hare cycle in central Alberta, Canada. (b) Variation in survival over two snowshoe hare cycles at Kluane Lake, Yukon, Canada. Too few hares were caught to estimate survival between 1985 and 1987. (After Krebs *et al.*, 2001; (a) following Cary & Keith, 1979.)

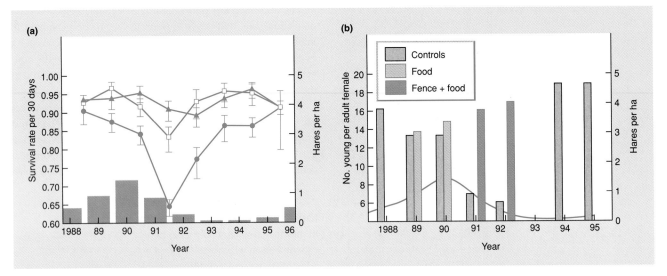

Figure 14.15 (a) Survival of radio-collared hares (with 90% confidence intervals) over a hare cycle from 1988 to 1996 at Kluane Lake, Yukon, Canada. The bars are densities; lines show the survival in controls (●) with mammalian predators excluded (□) and with mammalian predators excluded and food supplemented (▲). (b) Reproductive output over a hare cycle from 1988 to1995 at Kluane Lake (the line). It was possible to compare control values with those from treatments of food supplementation in 1989 and 1990, and with those where food was supplemented and mammalian predators excluded in 1991 and 1992. (After Krebs *et al.*, 2001; (a) following Krebs *et al.*, 1995.)

lowest fecundity following the density peak. Unfortunately, it was not possible to measure fecundity in a treatment where only food was supplemented – an example of the disappointments that almost inevitably accompany large field experiments – so the effects of food and predators cannot be disentangled. Of these, any effects of food shortage on fecundity would be easy to understand. It is also possible, though, that an increased frequency of interaction with predators could reduce fecundity through its physiological effects on hares (reduced energy or increased levels of stress-associated hormones).

Thus, these hard-won results from field experiments and the analyses of time series essentially agree in suggesting that the snowshoe hare cycle results from interactions with both its food and its predators, with the latter playing the dominant role. It is also noteworthy that, at least over some periods, there has been a high correlation between the hare cycle and the 10-year cycle of sunspot activity, which is known to affect broad weather patterns (Sinclair & Gosline, 1997). This type of extrinsic, abiotic factor was initially a strong candidate for playing a major role in driving population

sunspot cycles?

cycles generally (Elton, 1924). Subsequently, however, they have received little support. In the first place, many population cycles are of the wrong period and are also variable in period (see, for example, the microtine rodents in the next section). Second, the population cycles are often more pronounced than the extrinsic cycles that are proposed to be 'causing' them. Also, even when a correlation has been established, as in the present case, this simply begs the question of what links the two cycles: presumably it is climate acting on some combination of the factors we have already been considering – predators, food and intrinsic features of the population itself – although no mechanistic basis for such a link has been established.

Overall, then, the snowshoe hare work illustrates how a range of methodologies may come together in the search for an explanation of a cyclic pattern. It also provides a very sobering reminder of the logistical and practical difficulties – collecting long time series, undertaking large field experiments – that need to be accepted and overcome in order to build such explanations.

14.6.4 Microtine rodents: lemmings and voles

many microtines cycle – and many don't

There is no doubt that more effort has been expended overall in studying population cycles in microtine rodents (voles and lemmings) than in any other group of species. Cycle periods are typically 3 or 4 years, or much more rarely 2 or 5 years or even longer. These cyclic dynamics have been convincingly identified in a range of communities, including the following: voles (*Microtus* spp. and *Clethrionomys* spp.) in Fennoscandia (Finland, Norway and Sweden); lemmings (*Lemmus lemmus*) elsewhere in montane habitats in Fennoscandia; lemmings (*Lemmus* spp. and *Dicrostonyx* spp.) in the tundra of North America, Greenland and Siberia; voles (*Clethrionomys rufocanus*) in Hokkaido, northern Japan; common voles (*Microtus arvalis*) in central Europe; and field voles (*Microtus agrestis*) in northern England. On the other hand, there are many other microtine populations that show no evidence of multiannual cycles, including voles in southern Fennoscandia, southern England, elsewhere in Europe, and many locations in North America (Turchin & Hanski, 2001). It is also worth emphasizing that a quite different pattern, of irregular and spectacular irruptions in abundance and mass movement, is shown by just a few lemming populations, notably in Finnish Lapland. It is these whose suicidal behavior has been so grossly exaggerated (to say the least) in the name of film-makers' poetic license, unfairly condemning all lemmings to popular misconception (Henttonen & Kaikusalo, 1993).

trends in cyclicity

Over many decades, the same range of extrinsic and intrinsic factors have been proposed to explain microtine cycles as have been directed at population cycles generally. Given the variety of species and habitats, it is perhaps especially

unlikely in this case that there is a single all-encompassing explanation for all of the cycles. None the less, there are a number of features of the cycles that any explanation, or suite of explanations, must account for. First is the simple observation that some populations cycle while others do not. Also, there are cases (notably in Fennoscandia) where several coexisting species, often with apparently quite different ecologies, all cycle synchronously. And there are sometimes clear trends in cycle period, notably with increasing latitude (south to north) in Fennoscandia (see Section 14.5.1), where an explanation has been most intensively sought, but also for example in Hokkaido, Japan, where cyclicity increases broadly from southwest to northeast (Stenseth *et al.*, 1996), and in central Europe, where cyclicity increases from north to south (Tkadlec & Stenseth, 2001).

cycles result from a second-order process

A useful perspective from which to proceed is to acknowledge, as we have seen, that the rodent cycles are the result of a 'second-order' process (Bjørnstad *et al.*, 1995; Turchin & Hanski, 2001) (see Section 14.5.1); that is, they reflect the combined effects of a directly density-dependent process and a delayed density-dependent process. This immediately alerts us to the fact that, in principle at least, the direct and delayed processes need not be the same in every cyclic population: what is important is that two such processes act in conjunction.

We start with the 'intrinsic' theories. It is not surprising that voles and lemmings, which can achieve extremely high potential rates of population growth, should experience periods of overcrowding. Neither would it be surprising if overcrowding then produced changes in physiology and behavior. Mutual aggression (even fighting) might become more common and have consequences in the physiology, especially the hormonal balance, of the individuals. Individuals may grow larger, or mature later, under different circumstances. There might be increased pressure on some individuals to defend territories and on others to escape. Kin and non-kin might behave differently to one another when they are crowded. Powerful local forces of natural selection might be generated that favor particular genotypes (e.g. aggressors or escapists). These are responses that we easily recognize in crowded human societies, and ecologists have looked for the same phenomena when they try to explain the population behavior of rodents. All these effects have been found or claimed by rodent ecologists (e.g. Lidicker, 1975; Krebs, 1978; Gaines *et al.*, 1979; Christian, 1980). But it remains an open question whether any of them plays a critical role in explaining the behavior of rodent populations in nature.

dispersal, relatedness and aggression?

In the first place, we saw in Sections 6.6 and 6.7 the complexities of the relationships in rodents between density, dispersal, relatedness and ultimate survival and reproductive success. What is more, by no means all of this work has been carried out on species that exhibit cycles. Hence, there is little support for any universal rules, but there do seem to be tendencies for most dispersal to be natal (soon after birth), for males to disperse more than females, for effective dispersal (arriving, rather than simply traveling hopefully) to be more likely at lower densities, and for fitness to be greater the greater the relatedness of neighbors. This has led some in the field to argue that the 'jury is still out' (Krebs, 2003), but others have simply doubted any role for these processes in the regulation of rodent populations, especially in view of the frequent *inverse* density dependence (Wolff, 2003). Certainly, while variations between individuals may be associated with different phases of the cycle, this is very far from saying that they are *driving* the cycles. If individuals disperse more at particular cycle phases, say, or are larger, then this is likely to be a *response* either to a present or to a past level of food or space availability, or to predation pressure or infection intensity. That is, intrinsic variations are more likely to explain the detailed nature of responses, whereas extrinsic factors are more likely to explain the causes of the responses.

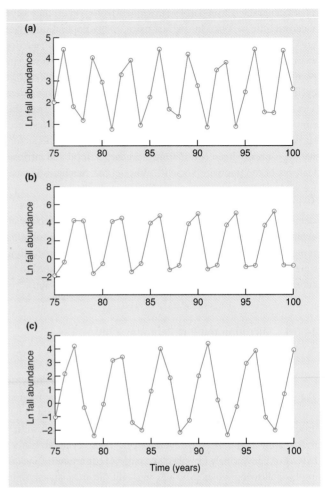

Figure 14.16 Behavior of Inchausti and Ginzburg's (1998) maternal effect model with differing values of the maximum yearly reproductive rate, *R*, and the maternal effect, *M*, through which the quality of daughters in one season is affected by the quality of mothers in the previous season (fall or spring). The simulations are given 75 'years' to settle into a regular pattern. (a) $R = 7.3$; $M = 15$. (b) $R = 4.4$; $M = 10$. (c) $R = 3.5$; $M = 5$. (After Inchausti & Ginzburg, 1998.)

None the less, in one case at least, an intrinsic cause has been proposed for the delayed density dependence. Inchausti and Ginzburg (1998) constructed a model with a 'maternal effect', in which mothers transmit their body condition phenotypically to their daughters, either from spring to fall or from fall to spring, and this in turn determines their per capita rate of growth. Thus, in this case the intrinsic quality of an individual is indeed a response to a past density, and hence to past resource availability, accounting for the delayed density dependence. Furthermore, when Inchausti and Ginzburg, focusing on Fennoscandia, fed what they believed to be reasonable values of population growth rate and the maternal effect into their model, both decreasing with latitude, they were able to recreate cycles with periods varying from 3 to 5 years (Figure 14.16). Turchin and Hanski (2001) criticized the parameter estimates (especially those of the growth rates) and claimed that the maternal effect model actually predicted 2-year cycles, at odds with those observed. Ergon *et al.* (2001) found with field voles, *Microtus agrestis*, from cyclic populations, that in transferring them between contrasting sites they rapidly took on characteristics appropriate to their new rather than their old populations – and certainly not those of their mothers. None the less, Inchausti and Ginzburg's results, set alongside the specialist predation hypothesis (see Section 14.5.1 and below), emphasize how the same pattern (here, the latitudinal gradient) might be achieved by quite different means. They also show that intrinsic theories remain 'in play' in the continuing search for an explanation for microtine cycles.

maternal effects?

the specialist predation hypothesis

Turning now to extrinsic factors, there are two main candidates: predators and food. (Parasites and pathogens excited Elton's interest immediately after his original, 1924 paper, but they were largely ignored subsequently until recent technical advances made their study a serious possibility. It remains to be seen what role, if any, they

play.) We have already made a start in examining predators in Sections 10.4.4 and 14.5.1. Their importance in microtine cycles, expressed as the '*specialist predation hypothesis*', has received considerable support since around 1990 from a series of mathematical models and field experiments, especially from workers focused on the cycles in Fennoscandia. The hypothesis, put simply, is that specialist predators are responsible for the delayed density dependence, whereas generalist predators, whose importance varies with latitude, are a major source of direct density dependence.

Early field experiments in which predators were removed (in Fennoscandia and elsewhere), although they

experimental support?

typically led to 2–3-fold increases in vole density, were subject to various criticisms of their experimental design: they were short term, or small scale, or they affected too many, or too few, of the predator species, and they often involved the erection of protective fences that are likely to have affected movements of the prey (voles) as well (Hanski *et al.*, 2001). Conclusive experiments may be a necessity, but this does not make them any easier! More recent experiments, too, give rise to some of the same misgivings. Klemola *et al.* (2000) excluded all the predators from four fenced (and net-roofed) exclosures in western Finland, 1 ha in size, for 2 years. The vole populations in the exclosures increased more than 20-fold in abundance compared to the control grids, until food shortages caused them to crash (Figure 14.17a). But the effects of specialists and generalists were inevitably combined in such a design; and while results such as these indicate an important role for predators in vole survival and abundance, they cannot prove

a role in causing (as opposed, say, to amplifying) the vole cycles. Korpimaki *et al.* (2002), worked in the same area but used four much larger unfenced areas (2.5–3 km²) for 3 years, reducing predator abundance over the summer but not the winter: mustelids (stoats and weasels) by trapping, and avian predators by removing natural and artificial nesting sites. Predator reduction increased vole density fourfold in the first (low) year; it accelerated an increase in density twofold in the second year; and it increased fall density twofold in the third (peak) year (Figure 14.17b). But again, specialists and generalists were not distinguished, and the temporal pattern of abundance was essentially unaltered.

The specialist predation model itself, which has been successively refined in a series of studies (the refinements are traced by Hanski *et al.*, 2001) has the following key features: (i) logistic population growth in the microtine prey, to reflect the directly density-dependent effects of food shortage on the microtines, preventing their populations from growing too large before specialist predators 'catch up'; (ii) specialist predators (weasels) with a population growth rate that declines as the ratio of specialist predators to prey increases; (iii) seasonal differences in the breeding of voles and weasels in the summer and winter; and (iv) generalist predators – generalist, switching mammals or wide ranging (nomadic) avian specialists that act in a directly density-dependent manner by responding immediately to changes in microtine density. Note, therefore, that the model includes both of the most-studied extrinsic factors: predators *and* food. Food provides the baseline direct density dependence; specialist predators provide the delayed density dependence. The generalist predators then provide a further source of direct density dependence that can be varied to mimic their known decline in abundance with latitude.

When the model is parameterized with field data from Fennoscandia, it can recreate an impressive number of the features of the observed dynamics. Cycles are of broadly the correct amplitude and period, and both the period and indeed the amplitude of the cycles increase with latitude as the density of generalist predators decreases, as observed in nature (Figure 14.18). A related model for the collared lemming, *Dicrostonyx groenlandicus*, preyed upon by one specialist predator (the stoat, *Mustela erminea*) and three generalists (Gilg *et al.*, 2003) was also able to recreate observed cycles in Greenland when parameterized with field data.

On the other hand, not all studies have conformed to the predictions of the specialist predation model. Lambin *et al.* (2000) described regular cycles of field voles in Kielder forest, northern England (55°N), with a period of 3–4 years and an approximately 10-fold difference between peak and trough densities (a difference of 1 on a log scale, such as on Figure 14.18). Yet, parameterizing the specialist predation model with the estimated intensity of generalist predation at this site would have predicted no cycles whatsoever – as would the latitude. What is more, a rigorous

supportive prediction?

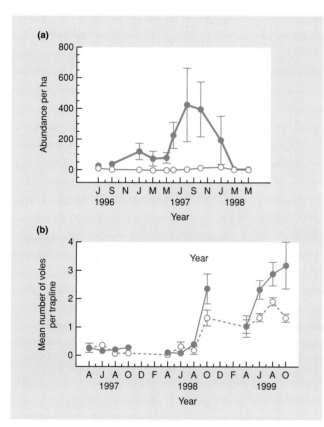

Figure 14.17 (a) Mean abundances of voles (±SE) from four small predator-exclosure grids (●) and four control grids (○) in western Finland. (After Klemola *et al.*, 2000.) (b) The density of voles (mean number of individuals caught per trapline, ±SE, in April, June, August and October) from four large predator-reduction sites (●) and four control sites (○) in western Finland. Predator reduction occurred only over the summer and vole densities tended to revert to control levels over the winter. (After Korpimaki *et al.*, 2002.)

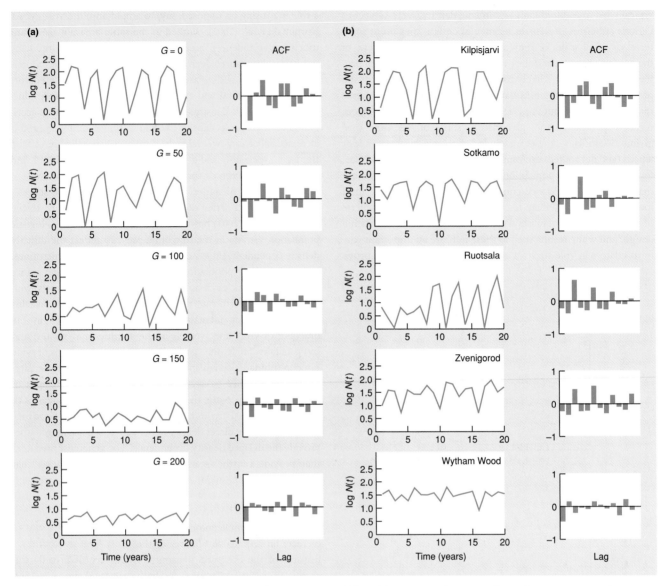

Figure 14.18 (a) Sample data generated by the specialist predation model, and the associated autocorrelation functions (ACFs), for various values of generalist predator abundance, G. As G increases, cycle period increases and cycle amplitude decreases, and at sufficiently high values the dynamics are sufficiently stabilized for the cycles to disappear altogether. (b) Comparable time series from five field sites: Kilpisjarvi (69°N; period = 5), Sotkamo (64°N; period = 4), Ruotsala (63°N; period = 3), Zvenigorod (57°N; period = 3) and Wytham Wood (51°N; no significant periodicity). (After Turchin & Hanski, 1997.)

program of reducing weasel numbers (i.e. specialist predators) at unfenced grids within the site (by about 60% in comparison with control sites) increased adult vole survival by about 25% but had no appreciable impact on the cyclic dynamics (Graham & Lambin, 2002).

Lambin and his collaborators conclude from these studies that generalist predators may not, after all, be responsible for the Fennoscandian gradient in cycle length; and that vole cycles need not be the result of the impact of specialist (i.e. weasel) predation (since they seem not to be at Kielder). Remember, too, that the results of time series analyses (see Section 14.5.1) and predator removal studies in Fennoscandia are consistent with the specialist predation hypothesis but do not prove it. In contrast, the response to these results of adherents of the specialist predation hypothesis (e.g. Korpimaki et al., 2003) has been to emphasize that the Kielder cycles are different from those in northern Fennoscandia (lower amplitude at Kielder, higher trough densities, less spatial synchrony and only one vole species involved). That is, they contend that the results at Kielder may tell us little or nothing about cycles in Fennoscandia. Though we may strive to avoid it, even rigorous studies are often open to alternative interpretations.

a role for food?

Turning finally to the role of food, both field observations and experiments suggest that it would be unwise to assume that the same forces act on voles and lemmings (Turchin & Batzli, 2001). In the first place, voles typically eat a range of vascular plants, including graminoids (grasses and sedges), whereas lemmings feed on a mixture of mosses and graminoids. Voles seem rarely to consume more than a few percent of available plant material (though, of course, the quality of the available food might be more important than its quantity – see, for example, Batzli, 1983); and food supplementation has typically failed to increase vole abundance (though experiments may have been foiled by the 'pantry effect' through which predators are attracted to high vole densities, counteracting the effects of supplementation). Lemmings, on the other hand, at peak densities, typically remove more than 50% and sometimes as much as 90–100% of the available vegetation.

Furthermore, through an analysis of models, Turchin and Batzli (2001) show clearly that the role that vegetation might play in cyclic dynamics depends critically on the nature of the vegetation itself, especially the vegetation's dynamics following significant consumption by herbivores. If the dynamics are logistic (i.e. S-shaped) then this may provide the delayed density dependence necessary to generate 'second-order' cycles in microtine abundance. But if the dynamics are 'regrowth' (i.e. a rapid initial response, decelerating until a saturation abundance is reached) then any density dependence will be direct rather than delayed. In this case, the microtine–food interaction may play an integral part in cyclic dynamics (as they do, for example, in the specialist predation hypothesis), but it cannot be the second-order driving force. Crucially, plants consumed by voles seem likely to exhibit rapid regrowth dynamics because of the large proportion of unconsumed plant parts, much of it underground. By contrast, mosses are, by their nature, wholly available to their consumers, and when lemmings devastate their vegetation they often grub underground for graminoid rhizomes and destroy these too. Lemming vegetation, therefore, is likely to exhibit logistic dynamics: rapid only after a slow start.

On this basis, Turchin and Batzli parameterized a model for microtines and a food supply with logistic growth, using what data were available for the brown lemming (*Lemmus sibiricus*) and its vegetation at Barrow, Alaska (Batzli, 1993). The results were encouraging although not perfect representations of the observed patterns: cycle amplitudes were too low (400- rather than 600-fold) and too long (6 rather than 4 years). On the other hand, uncertainty, and in some cases plain ignorance, surrounded several of the parameter estimates. The model can be 'tweaked' to generate the observed dynamics. Further painstaking work in the field,

especially to obtain winter parameter estimates from under the snow, will be required to determine whether such tweaking is justified by the truth about lemming biology.

The cycles of microtines have been studied for longer and with greater intensity than those of any other species, and have generated more theories to explain them, and more disagreements amongst disputing advocates. At the time of writing, a near-consensus appears to have been arrived at that a conjunction of direct and delayed density dependence is required to account for observed patterns; and most support is attracted to the contention that specialist predators provide the delayed density dependence, while food shortage and generalist predators provide the direct density dependence. All scientific 'conclusions' are provisional, however, and fashions change in science as in everything else. It remains to be seen how robust and universal currently fashionable explanations prove to be.

in conclusion

More generally, we started this chapter with a series of questions. Why are some species rare and others common? Why does a species occur at low population densities in some places and at high densities in others? What factors cause fluctuations in a species' abundance? Having reached the end of the chapter, it should be clear that none of these questions has a simple answer. We have seen for particular examples why a species is rare, or why another varies in abundance from place to place. But we must not expect the answer to be the same for every species – especially when we start a new study of a species that demands our attention, perhaps by its excessive abundance (a pest) or declining abundance (a target for conservation). It is crucial none the less that we have a clear idea of what the *possible* answers might be and how we might go about obtaining those answers. The aim of this chapter has been to examine those possibilities and how to distinguish them. In the next chapter, we turn explicitly to some of the pressing examples of populations whose abundance we need to understand in order to exert some measure of control – be they pests or natural resources that we wish to exploit.

Summary

We bring together topics from previous chapters, seeking to account for variations in abundance.

Ecologists may emphasize stability or fluctuations. To resolve these contrasting perspectives, it is necessary to distinguish clearly between factors that determine and those that regulate abundance. In doing so, we review historical conflicts between the viewpoints of Nicholson and Andrewartha and Birch. We then outline the demographic, mechanistic and density approaches to the investigation of abundance.

Starting with the demographic approach, we explain key factor analysis, its uses, but also its shortcomings. We therefore also explain λ-contribution analysis, which overcomes some of the problems with key factor analysis, and in developing this explanation we describe and apply elasticity analysis.

The mechanistic approach relates the level or presence of a factor (amount of food, presence of predators) either to abundance itself or to the population growth rate. This may be simply correlational, but may alternatively involve the experimental perturbation of populations. We note that the introduction of a biological control agent is one particular example of this.

Correlations with density are not absent from other approaches, but the density approach focuses on density dependences in their own right. We explain how time series analyses seek to dissect density dependences, especially the relative strengths of direct and delayed density dependence when abundance at a given point in time is expressed as reflecting abundances at various times in the past ('time lags'). We show, too, how related analyses may be valuable in counting and then characterizing the lags in an optimal description of a time series, and also in evaluating the respective contributions of density-dependent and -independent processes (especially weather) in determining abundance.

Regular, multigeneration cycles have in many ways, and for many years, been the benchmark against which ecologists have tested their ability to understand the determination of abundance. We explain how cycles may be identified within time series and then examine three case studies in detail.

Red grouse cycles illustrate the difficulties of distinguishing between alternative explanations – parasites and kinship/territorial behavior – both of which have support.

Work on cycles in snowshoe hares illustrates the coming together of detailed time series analyses and results obtained by much more direct, experimental means. It also provides a very sobering reminder of the logistical and practical difficulties that need to be accepted and overcome in order to build explanations.

More effort has been expended in studying population cycles in microtine rodents (voles and lemmings) than in any other group of species. We describe geographic trends in cyclicity and the need for an explanation to account for these, and we note that any such explanation must acknowledge that the cycles are the result of a 'second-order' process: a combination of a direct and a delayed density-dependent process. We then examine, in turn, three sets of explanations differing in their source of delayed density dependence: (i) 'intrinsic' theories, including maternal effects; (ii) the 'specialist predation hypothesis', supported by both mathematical models and field experiments, though both of these have also been subject either to criticism or contradictory evidence; and (iii) theories focused on food, which also have their problems.

We conclude by acknowledging that none of the questions posed at the beginning of the chapter has simple answers.

Chapter 15

Ecological Applications at the Level of Population Interactions: Pest Control and Harvest Management

15.1 Introduction

Humans are very much a part of all ecosystems. Our activities sometimes motivate us to drive towards extinction the species we identify as pests, to kill individuals of species we harvest for food or fiber while ensuring the persistence of their populations, and to prevent the extinction of species we believe to be endangered. The desired outcomes are very different for pest controllers, harvest managers and conservation ecologists, but all need management strategies based on the theory of population dynamics. Because much of the tool kit developed to manage endangered species is based on the dynamics of individual populations, we dealt with species conservation in Chapter 7 at the end of the first section of the book, which considered the ecology of individual organisms and single species populations. Pest controllers and harvest managers, on the other hand, mostly have to deal explicitly with multispecies interactions, and their work must be informed by the theory concerning population interactions covered in the book's second section (Chapters 8–14). Pest control and harvest management are the topics of the present chapter.

'sustainability' – an aim of both pest controllers and harvest managers

The importance of pest control and harvest management has grown exponentially as the human population has increased (see Section 7.1) and each touches on a different aspect of 'sustainability'. To call an activity 'sustainable' means that it can be continued or repeated for the foreseeable future. Concern has arisen, therefore, precisely because so much human activity is clearly unsustainable. We cannot continue to use the same pesticides if increasing numbers of pests become resistant to them. We cannot (if we wish to have fish to eat in future) continue to remove fish from the sea faster than the remaining fish can replace their lost companions.

Sustainability has thus become one of the core concepts – perhaps the core concept – in an ever-broadening concern for the fate of the earth and the ecological communities that occupy it. In defining sustainability we used the words 'foreseeable future'. We did so because, when an activity is described as sustainable, it is on the basis of what is known at the time. But many factors remain unknown or unpredictable. Things may take a turn for the worse (as when adverse oceanographic conditions damage a fishery already threatened by overexploitation) or some unforeseen additional problem may be discovered (resistance may appear to some previously potent pesticide). On the other hand, technological advances may allow an activity to be sustained that previously seemed unsustainable (new types of pesticide may be discovered that are more finely targeted on the pest itself rather than innocent bystander species). However, there is a real danger that we observe the many technological and scientific advances that have been made in the past and act on the faith that there will always be a technological 'fix' to solve our present problems, too. Unsustainable practices cannot be accepted simply from faith that future advances will make them sustainable after all.

The recognition of the importance of sustainability as a unifying idea in applied ecology has grown gradually, but there is something to be said for the claim that sustainability really came of age in 1991. This was when the Ecological Society of America published 'The sustainable biosphere initiative: an ecological research agenda', a 'call-to-arms for all ecologists' with a list of 16 co-authors (Lubchenco *et al.*, 1991). And in the same year, the World Conservation Union (IUCN), the United Nations Environment Programme and the World Wide Fund for Nature jointly published *Caring for the Earth. A Strategy for Sustainable Living* (IUCN/UNEP/WWF, 1991). The detailed contents of these documents are less important than their existence. They indicate a growing preoccupation with sustainability, shared by scientists, pressure groups and governments, and recognition that much of what we do is not sustainable. More recently, the emphasis has shifted from a purely ecological perspective to one that incorporates the social and economic conditions influencing sustainability

(Milner-Gulland & Mace, 1998) – this is sometimes referred to as the 'triple bottomline' of sustainability.

In this chapter we deal in turn with the application of population theory to the management of pests (Section 15.2) and harvests (Section 15.3). We have seen previously how the details of spatial structuring of populations can affect their dynamics (see Chapters 6 and 14). With this in mind, Section 15.4 presents examples of the application of a metapopulation perspective to pest control and harvest management.

We discussed in Chapter 7 how predicted global climate change is expected to affect species' distribution patterns. Such conclusions were based on the mapping of species' fundamental niches onto new global patterns of temperature and rainfall. We will not dwell on this phenomenon in the current chapter, but it should be noted that global change will also impact on population parameters, such as birth and death rates and the timing of breeding (e.g. Walther *et al.*, 2002; Corn, 2003), with implications for the population dynamics of pest and harvested (and endangered) species.

15.2   Management of pests

what is a pest?

A pest species is one that humans consider undesirable. This definition covers a multitude of sins: mosquitoes are pests because they carry diseases or because their bites itch; *Allium* spp. are pests because when harvested with wheat these weeds make bread taste of onions; rats and mice are pests because they feast on stored food; mustellids are pests in New Zealand because they are unwanted invaders that prey upon native birds and insects; garden weeds are pests for esthetic reasons. People want rid of them all.

15.2.1   Economic injury level and economic thresholds

economic injury level defines actual and potential pests

Economics and sustainability are intimately tied together. Market forces ensure that uneconomic practices are not sustainable. One might imagine that the aim of pest control is always total eradication of the pest, but this is not the general rule. Rather, the aim is to reduce the pest population to a level at which it does not pay to achieve yet more control (the economic injury level or EIL). Our discussion here is informed particularly by the theory covered in Chapter 14, which dealt with the combination of factors that determines a species' average abundance and fluctuations about that average. The EIL for a hypothetical pest is illustrated in Figure 15.1a: it is greater than zero (eradication is not profitable) but it is also below the typical, average abundance of the species. If the species was naturally self-limited to a density below the EIL, then it would never make economic sense to apply 'control' measures, and the

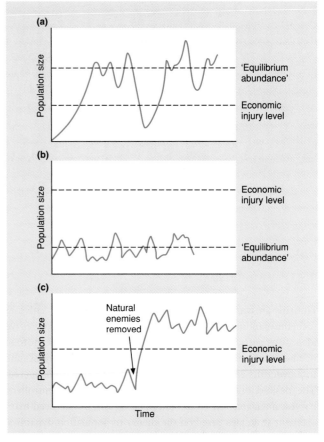

Figure 15.1   (a) The population fluctuations of a hypothetical pest. Abundance fluctuates around an 'equilibrium abundance' set by the pest's interactions with its food, predators, etc. It makes economic sense to control the pest when its abundance exceeds the economic injury level (EIL). Being a pest, its abundance exceeds the EIL most of the time (assuming it is not being controlled). (b) By contrast, a species that cannot be a pest fluctuates always below its EIL. (c) 'Potential' pests fluctuate normally below their EIL but rise above it in the absence of one or more of their natural enemies.

species could not, by definition, be considered a 'pest' (Figure 15.1b). There are other species, though, that have a carrying capacity in excess of their EIL, but have a typical abundance that is kept below the EIL by natural enemies (Figure 15.1c). These are potential pests. They can become actual pests if their enemies are removed.

When a pest population has reached a density at which it is causing economic injury, however, it is generally too late to start controlling it. More important, then, is the economic threshold (ET): the density of the pest at which action should be taken to prevent it reaching the EIL. ETs are predictions based either on cost–benefit analyses (Ramirez & Saunders, 1999) and detailed studies

the economic threshold – getting ahead of the pests

of past outbreaks, or sometimes on correlations with climatic records. They may take into account the numbers not only of the pest itself but also of its natural enemies. As an example, in order to control the spotted alfalfa aphid (*Therioaphis trifolii*) on hay alfalfa in California, control measures have to be taken at the times and under the following circumstances (Flint & van den Bosch, 1981):

1 In the spring when the aphid population reaches 40 aphids per stem.
2 In the summer and fall when the population reaches 20 aphids per stem, but the first three cuttings of hay are not treated if the ratio of ladybirds (beetle predators of the aphids) to aphids is one adult per 5–10 aphids or three larvae per 40 aphids on standing hay or one larva per 50 aphids on stubble.
3 During the winter when there are 50–70 aphids per stem.

15.2.2 Chemical pesticides, target pest resurgence and secondary pests

Chemical pesticides are a key part of the armory of pest managers but they have to be used with care because population theory (see, in particular, Chapter 14) predicts some undesirable responses to the application of a pesticide. Below we discuss the range of chemical pesticides and herbicides before proceeding to consider some undesirable consequences of their use.

15.2.2.1 Insecticides

insecticides and how they work

The use of *inorganics* goes back to the dawn of pest control and, along with the botanicals (below), they were the chemical weapons of the expanding army of insect pest managers of the 19th and early 20th century. They are usually metallic compounds or salts of copper, sulfur, arsenic or lead – and are primarily stomach poisons (i.e. they are ineffective as contact poisons) and they are therefore effective only against insects with chewing mouthparts. This, coupled with their legacy of persistent, broadly toxic metallic residues, has led now to their virtual abandonment (Horn, 1988).

Naturally occurring insecticidal plant products, or *botanicals*, such as nicotine from tobacco and pyrethrum from chrysanthemums, having run a course similar to the inorganics, have now also been largely superseded, particularly because of their instability on exposure to light and air. However, a range of *synthetic pyrethroids*, with much greater stability, such as permethrin and deltamethrin, have replaced other types of organic insecticide (described below) because of their relative selectivity against pests as opposed to beneficial species (Pickett, 1988).

Chlorinated hydrocarbons are contact poisons that affect nerve-impulse transmission. They are insoluble in water but show a high affinity for fats, thus tending to become concentrated in animal fatty tissue. The most notorious is DDT: a Nobel Prize was awarded for its rediscovery in 1948, but it was suspended from all but emergency uses in the USA in 1973 (although it is still being used in poorer countries). Others in use are toxaphene, aldrin, dieldrin, lindane, methoxychlor and chlordane.

Organophosphates are also nerve poisons. They are much more toxic (to both insects and mammals) than the chlorinated hydrocarbons, but are generally less persistent in the environment. Examples are malathion, parathion and diazinon.

Carbamates have a mode of action similar to the organophosphates, but some have a much lower mammalian toxicity. However, most are extremely toxic to bees (necessary for pollination) and parasitic wasps (the likely natural enemies of insect pests). The best-known carbamate is carbaryl.

Insect growth regulators are chemicals of various sorts that mimic natural insect hormones and enzymes, and hence interfere with normal insect growth and development. As such, they are generally harmless to vertebrates and plants, although they may be as effective against a pest's natural insect enemies as against the pest itself. The two main types that have been used effectively to date are: (i) chitin-synthesis inhibitors such as diflubenzuron, which prevent the formation of a proper exoskeleton when the insect molts; and (ii) juvenile hormone analogs such as methoprene, which prevent pest insects from molting into their adult stage, and hence reduce the population size in the next generation.

Semiochemicals are not toxins but chemicals that elicit a change in the behavior of the pest (literally 'chemical signs'). They are all based on naturally occurring substances, although in a number of cases it has been possible to synthesize either the semiochemicals themselves or analogs of them. Pheromones act on members of the same species; allelochemicals on members of another species. Sex-attractant pheromones are used commercially to control pest moth populations by interfering with mating (Reece, 1985), whilst the aphid alarm pheromone is used to enhance the effectiveness of a fungal pathogen against pest aphids in glasshouses in Great Britain by increasing the mobility of the aphids, and hence their rate of contact with fungal spores (Hockland *et al.*, 1986). These semiochemicals, along with the insect growth regulators, are sometimes referred to as 'third-generation' insecticides (following the inorganics and the organic toxins). Their development is relatively recent (Forrester, 1993).

15.2.2.2 Herbicides

Here, too, *inorganics* were once important although they have mostly been replaced, largely owing to the combined problems of persistence and nonspecificity. However, for these very reasons, borates for example, absorbed by plant roots and translocated to above-ground parts, are still sometimes used to provide semipermanent sterility to areas where no vegetation

the tool-kit of herbicides

of any sort is wanted. Others include a range of arsenicals, ammonium sulfamate and sodium chlorate (Ware, 1983).

More widely used are the *organic arsenicals*, for instance disodium methylarsonate. These are usually applied as spot treatments (since they are nonselective) after which they are translocated to underground tubers and rhizomes where they disrupt growth.

By contrast, the highly successful *phenoxy* or *hormone* weed killers, translocated throughout the plant, tend to be very much more selective. For instance, 2,4-D is highly selective against broad-leaved weeds, whilst 2,4,5-trichlorophenoxyethanoic acid (2,4,5-T) is used mainly to control woody perennials. They appear to act by inhibiting the production of enzymes needed for coordinated plant growth, leading ultimately to plant death.

The *substituted amides* have diverse biological properties. For example, diphenamid is largely effective against seedlings rather than established plants, and is therefore applied to the soil around established plants as a 'pre-emergence' herbicide, preventing the subsequent appearance of weeds. Propanil, on the other hand, has been used extensively on rice fields as a selective post-emergence agent.

The *nitroanilines* (e.g. trifluralin) are another group of soil-incorporated pre-emergence herbicides in very widespread use. They act, selectively, by inhibiting the growth of both roots and shoots.

The *substituted ureas* (e.g. monuron) are mostly rather nonselective pre-emergence herbicides, although some have post-emergence uses. Their mode of action is to block electron transport.

The *carbamates* were described amongst the insecticides, but some are herbicides, killing plants by stopping cell division and plant tissue growth. They are primarily selective, pre-emergence weed killers. One example, asulam, is used mostly for grass control amongst crops, and is also effective in reforestation and Christmas tree plantings.

The *thiocarbamates* (e.g. *S*-ethyl dipropylthiocarbamate) are another group of soil-incorporated pre-emergence herbicides, selectively inhibiting the growth of roots and shoots that emerge from weed seeds.

Amongst the *heterocyclic nitrogen* herbicides, probably the most important are the *triazines* (e.g. metribuzin). These are effective blockers of electron transport, mostly used for their post-emergence activity.

The *phenol derivatives*, particularly the nitrophenols such as 2-methyl-4,6-dinitrophenol, are contact chemicals with broad-spectrum toxicity extending beyond plants to fungi, insects and mammals. They act by uncoupling oxidative phosphorylation.

The *bipyridyliums* contain two important herbicides, diquat and paraquat. These are powerful, very fast acting contact chemicals of widespread toxicity that act by the destruction of cell membranes.

Finally worth mentioning is glyphosate (a *glyphosphate* herbicide): a nonselective, nonresidual, translocated, foliar-applied chemical, popular for its activity at any stage of plant growth and at any time of the year.

15.2.2.3   *Target pest resurgence*

A pesticide gets a bad name if, as is usually the case, it kills more species than just the one at which it is aimed. However, in the context of the sustainability of agriculture, the bad name is especially justified if it kills the pests' natural enemies and so contributes to undoing what it was employed to do. Thus, the numbers of a pest sometimes increase rapidly some time after the application of a pesticide. This is known as 'target pest resurgence' and occurs when the treatment kills both large numbers of the pest *and* large numbers of its natural enemies (an example is presented below in Figure 15.2). Pest individuals that survive the pesticide or that migrate into the area later find themselves with a plentiful food resource but few, if any, natural enemies. The pest population may then explode. Populations of natural enemies will probably eventually re-establish but the timing depends both on the relative toxicity of the pesticide to target and nontarget species and the persistence of the pesticide in the environment, something that varies dramatically from one pesticide to another (Table 15.1).

the pest bounces back because its enemies are killed

	Toxicity				
	Rat	Fish	Bird	Honeybee	Persistence
Permethrin (pyrethroid)	2	4	2	5	2
DDT (organochlorine)	3	4	2	2	5
Lindane (organochlorine)	3	3	2	4	4
Ethyl parathion (organophosphate)	5	2	5	5	2
Malathion (organophosphate)	2	2	1	4	1
Carbaryl (carbamate)	2	1	1	4	1
Diflubenzuron (chitin-synthesis inhibitor)	1	1	1	1	4
Methoprene (juvenile hormone analogue)	1	1	1	2	2
Bacillus thuringiensis	1	1	1	1	1

Table 15.1   The toxicity to nontarget organisms, and the persistence, of selected insecticides. Possible ratings range from a minimum of 1 (which may, therefore, include zero toxicity) to a maximum of 5. Most damage is done by insecticides that combine persistence with acute toxicity to nontarget organisms. This clearly applies, to an extent, to each of the first six (broad-spectrum) insecticides. (After Metcalf, 1982; Horn, 1988.)

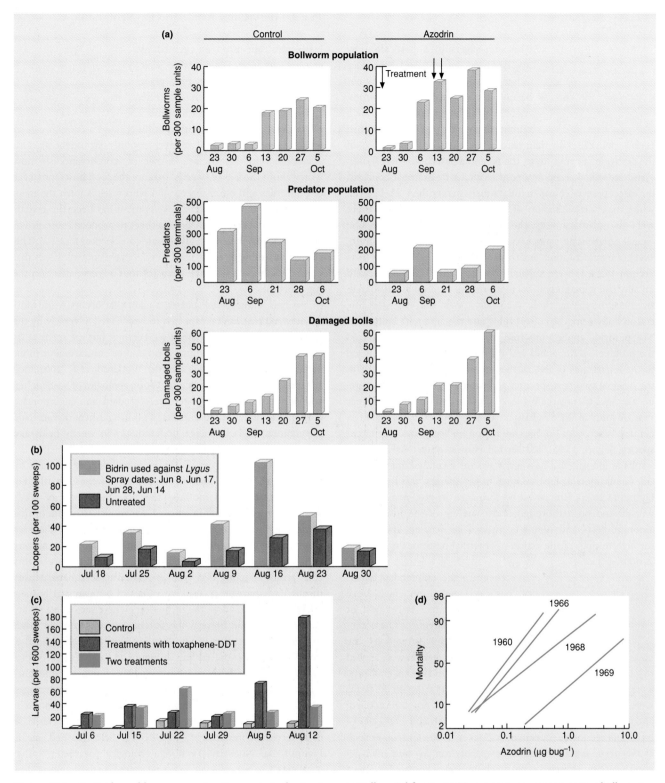

Figure 15.2 Pesticide problems amongst cotton pests in the San Joaquin Valley, California. (a) Target pest resurgence: cotton bollworms (*Heliothis zea*) resurged because the abundance of their natural predators was reduced – the number of damaged bolls was higher. (b) An increase in cabbage loopers (*Trichoplusia ni*) and (c) in beet army worms (*Spodoptera exigua*) were seen when plots were sprayed against the target lygus bugs (*Lygus hesperus*) – both are examples of secondary pest outbreaks. (d) Increasing resistance of lygus bugs to Azodrin®. (After van den Bosch *et al.*, 1971.)

15.2.2.4   *Secondary pests*

nonpests become
pests when their
enemies and
competitors are
killed

The after-effects of a pesticide may involve even more subtle reactions. When a pesticide is applied, it may not be only the target pest that resurges. Alongside the target are likely to be a number of potential pest species that had been kept in check by their natural enemies (see Figure 15.1c). If the pesticide destroys these, the potential pests become real ones – and are called secondary pests. A dramatic example concerns the insect pests of cotton in the southern part of the USA. In 1950, when mass dissemination of organic insecticides began, there were two primary pests: the Alabama leafworm and the boll weevil (*Anthonomus grandis*), an invader from Mexico (Smith, 1998). Organochlorine and organophosphate insecticides (see Section 15.2.2.1) were applied fewer than five times a year and initially had apparently miraculous results – cotton yields soared. By 1955, however, three secondary pests had emerged: the cotton bollworm, the cotton aphid and the false pink bollworm. The pesticide applications rose to 8–10 per year. This reduced the problem of the aphid and the false pink bollworm, but led to the emergence of five further secondary pests. By the 1960s, the original two pest species had become eight and there were, on average, an unsustainable 28 applications of insecticide per year. A study in the San Joaquin Valley, California, revealed target pest resurgence (in this case cotton bollworm was the target species; Figure 15.2a) and secondary pest outbreaks in action (cabbage loopers and beet army worms increased after insecticide application against another target species, the lygus bug; Figure 15.2b, c). Improved performance in pest management will depend on a thorough understanding of the interactions amongst pests and nonpests as well as detailed knowledge, through testing, of the action of potential pesticides against the various species.

mortality of
nontarget species
in general

Sometimes the unintended effects of pesticide application have been much less subtle than target pest or secondary pest resurgence. The potential for disaster is illustrated by the occasion when massive doses of the insecticide dieldrin were applied to large areas of Illinois farmland from 1954 to 1958 to 'eradicate' a grassland pest, the Japanese beetle. Cattle and sheep on the farms were poisoned, 90% of cats and a number of dogs were killed, and among the wildlife 12 species of mammals and 19 species of birds suffered losses (Luckman & Decker, 1960). Outcomes such as this argue for a precautionary approach in any pest management exercise. Coupled with much improved knowledge about the toxicity and persistence of pesticides, and the development of more specific and less persistent pesticides, such disasters should never occur again.

15.2.3   Herbicides, weeds and farmland birds

unintended effects
of the genetic
modification of crops
with herbicide
resistance

Herbicides are used in very large amounts and on a worldwide scale. They are active against pest plants and when used at commercial rates appear to have few significant effects on animals. Herbicide pollution of the environment did not, until relatively recently, arouse the passions associated with insecticides. However, conservationists now worry about the loss of 'weeds' that are the food hosts for larvae of butterflies and other insects and whose seeds form the main diet of many birds. A recent development has been the genetic modification of crops such as sugar beet to produce resistance to the nonselective herbicide glyphosate (see Section 15.2.2.2). This allows the herbicide to be used to effectively control weeds that normally compete with the crop without adverse affect on the sugar beet itself.

Fat hen (*Chenopodium album*), a plant that occurs worldwide, is one weed that can be expected to be affected adversely by the farming of genetically modified (GM) crops; but the seeds of fat hen are an important winter food source for farmland birds, including the skylark (*Alauda arvensis*). Watkinson *et al.* (2000) took advantage of the fact that the population ecologies of both fat hen and skylarks have been intensively studied and incorporated both into a model of the impacts of GM sugar beet on farmland populations. Skylarks forage preferentially in weedy fields and aggregate locally in response to weed seed abundance. Hence, the impact of GM sugar beet on the birds will depend critically on the extent to which high-density patches of weeds are affected. Watkinson *et al.* incorporated the possible effects of weed seed density on farming practice. Their model assumed: (i) that before the introduction of GM technology, most farms have a relatively low density of weed seeds, with a few farms having very high densities (solid line in Figure 15.3a); and (ii) the probability of a farmer adopting GM crops is related to seed bank density through a parameter ρ. Positive values of ρ mean that farmers are more likely to adopt the technology where seed densities are currently high and there is the potential to reduce yield losses to weeds. This leads to an increase in the relative abundance of low-density fields (dotted line in Figure 15.3a). Negative values of ρ indicate that farmers are more likely to adopt the technology where seed densities are currently low (intensively managed farms), perhaps because a history of effective weed control is correlated with a willingness to adopt new technology. This leads to a decreased frequency of low-density fields (dashed line in Figure 15.3a). Note that ρ is not an ecological parameter. Rather it reflects a socioeconomic response to the introduction of new technology. The way that farmers will respond is not self-evident and needs to be included as a variable in the model. It turns out that the relationship between current weed levels and uptake of the new technology (ρ) is as important to bird population

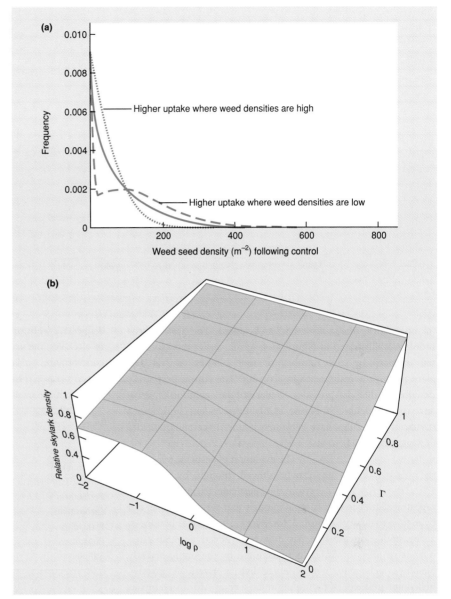

Figure 15.3 (a) Frequency distributions of mean seed densities across farms before the introduction of GM sugar beet (solid line), and in two situations where the technology has been adopted: where the technology is preferentially adopted on farms where weed density is currently high (dotted line) and where it is currently low (dashed line). (b) The relative density of skylarks in fields in winter (vertical axis; unity indicates field use before the introduction of GM crops) in relation to ρ (horizontal axis; positive values mean farmers are more likely to adopt GM technology where seed densities are currently high, negative values where seed densities are currently low) and to the approximate reduction in weed seed bank density due to the introduction of GM crops (Γ, third axis; realistic values are those less than 0.1). Note that the parameter space that real systems are expected to occupy is the 'slice' of the diagram nearest to you, where small positive or negative values of ρ give quite different skylark densities. (After Watkinson et al., 2000.)

density as the direct impact of the technology on weed abundance (Figure 15.3b), emphasizing the need for resource managers to think in terms of the triple bottomline of sustainability, with its ecological, social and economic dimensions.

15.2.4   Evolution of resistance to pesticides

evolved resistance: a widespread problem

Chemical pesticides lose their role in sustainable agriculture if the pests evolve resistance. The evolution of pesticide resistance is simply natural selection in action. It is almost certain to occur when vast numbers of individuals in a genetically variable population are killed in a systematic way by the pesticide. One or a few individuals may be unusually resistant (perhaps because they possess an enzyme that can detoxify the pesticide). If the pesticide is applied repeatedly, each successive generation of the pest will contain a larger proportion of resistant individuals. Pests typically have a high intrinsic rate of reproduction, and so a few individuals in one generation may give rise to hundreds or thousands in the next, and resistance spreads very rapidly in a population.

This problem was often ignored in the past, even though the first case of DDT resistance was reported as early as 1946

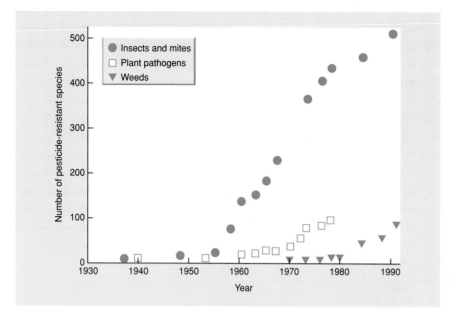

Figure 15.4   The increase in the number of arthropod (insects and mites), plant pathogens and weed species reported to be resistant to at least one pesticide. (After Gould, 1991.)

(in house-flies, *Musca domestica*, in Sweden). The scale of the problem is illustrated in Figure 15.4, which shows the exponential increases in the number of invertebrates, weeds and plant pathogens resistant to insecticides. The cotton pest study described earlier also provides evidence of the evolution of resistance to a pesticide (see Figure 15.2d). Even rodents and rabbits (*Oryctolagus cuniculus*) have evolved resistance to certain pesticides (Twigg *et al.*, 2002).

managing resistance

The evolution of pesticide resistance can be slowed, though, by changing from one pesticide to another, in a repeated sequence that is rapid enough that resistance does not have time to emerge (Roush & McKenzie, 1987). River blindness, a devastating disease that has now been effectively eradicated over large areas of Africa, is transmitted by the biting blackfly *Simulium damnosum*, whose larvae live in rivers. A massive helicopter pesticide spraying effort in several African countries (50,000 km of river were being treated weekly by 1999; Yameogo *et al.*, 2001) began with Temephos, but resistance appeared within 5 years (Table 15.2). Temephos was then replaced by another organophosphate, Chlorphoxim, but resistance rapidly evolved to this too. The strategy of using a range of pesticides on a rotational basis has prevented further evolution of resistance and by 1994 there were few populations that were still resistant to Temephos (Davies, 1994).

If chemical pesticides brought nothing but problems, however – if their use was intrinsically and acutely unsustainable – then they would already have fallen out of widespread use. This has not happened. Instead, their rate of production has increased rapidly. The ratio of cost to benefit for the individual producer has generally remained in favor of pesticide use. Moreover, in many poorer countries, the prospect of imminent mass starvation, or

Table 15.2   History of pesticide use against the aquatic larvae of blackflies, the vectors of river blindness in Africa. After early concentration on Temephos and Chlorphoxim, to which the insects became resistant, pesticides were used on a rotational basis to prevent the evolution of resistance. (After Davies, 1994.)

Name of pesticide	Class of chemical	History of use
Temephos	Organophosphate	1975 to present
Chlorphoxim	Organophosphate	1980–90
Bacillus thuringiensis H14	Biological insecticide	1980 to present
Permethrin	Pyrethroid	1985 to present
Carbosulfan	Carbamate	1985 to present
Pyraclofos	Organic phosphate	1991 to present
Phoxim	Organophosphate	1991 to present
Etofenprox	Pyrethroid	1994 to present

of an epidemic disease, are so frightening that the social and health costs of using pesticides have to be ignored. In general the use of pesticides is justified by objective measures such as 'lives saved', 'economic efficiency of food production' and 'total food produced'. In these very fundamental senses, their use may be described as sustainable. In practice, sustainability depends on continually developing new pesticides that keep at least one step ahead of the pests: pesticides that are less persistent, biodegradable and more accurately targeted at the pests.

15.2.5   Biological control

Outbreaks of pests occur repeatedly and so does the need to apply pesticides. But biologists can sometimes replace chemicals by

biological control:
the use of natural
enemies in a variety
of ways

another tool that does the same job and often costs a great deal less – biological control (the manipulation of the natural enemies of pests). Biological control involves the application of theory about interactions between species and their natural enemies (see Chapters 10, 12 and 14) to limit the population density of specific pest species. There are a variety of categories of biological control.

The first is the *introduction* of a natural enemy from another geographic area – very often the area in which the pest originated prior to achieving pest status – in order that the control agent should persist and thus maintain the pest, long term, below its economic threshold. This is a case of a desired invasion of an exotic species and is often called *classical biological control* or *importation*.

By contrast, *conservation biological control* involves manipulations that augment the density or persistence of populations of generalist natural enemies that are native to the pest's new area (Barbosa, 1998).

Inoculation is similar to introduction, but requires the periodic release of a control agent where it is unable to persist throughout the year, with the aim of providing control for only one or perhaps a few generations. A variation on the theme of inoculation is 'augmentation', which involves the release of an indigenous natural enemy in order to supplement an existing population, and is also therefore carried out repeatedly, typically to intercept a period of rapid pest population growth.

Finally, *inundation* is the release of large numbers of a natural enemy, with the aim of killing those pests present at the time, but with no expectation of providing long-term control as a result of the control agent's population increasing or maintaining itself. By analogy with the use of chemicals, agents used in this way are referred to as biological pesticides.

Insects have been the main agents of biological control against both insect pests (where parasitoids have been particularly useful) and weeds. Table 15.3 summarizes the extent to which they have been used and the proportion of cases where the establishment of an agent has greatly reduced or eliminated the need for other control measures (Waage & Greathead, 1988).

cottony cushion scale
insect: a classic case
of importation . . .

Probably the best example of 'classical' biological control is itself a classic. Its success marked the start of biological control in a modern sense. The cottony cushion scale insect, *Icerya purchasi*, was first discovered as a pest of Californian citrus orchards in 1868. By 1886 it had brought the citrus industry close to the point of destruction. Ecologists initiated a worldwide correspondence to try and discover the natural home and natural enemies of the scale, eventually leading to the importation to California of about 12,000 *Cryptochaetum* (a dipteran parasitoid) from Australia and 500 predatory ladybird beetles (*Rodolia cardinalis*) from Australia and New Zealand. Initially, the parasitoids seemed simply to have disappeared, but the predatory beetles underwent such a population explosion that all infestations of the scale insects in California were controlled by the end of 1890. Although the beetles have usually taken most or all of the credit, the long-term outcome has been that the beetles are instrumental in keeping the scale in check inland, but *Cryptochaetum* is the main agent of control on the coast (Flint & van den Bosch, 1981).

. . . illustrating
several general points

This example illustrates a number of important general points. Species may become pests simply because, by colonization of a new area, they escape the control of their natural enemies (the enemy release hypothesis) (Keane & Crawley, 2002). Biological control by importation is thus, in an important sense, restoration of the status quo for the specific predator–prey interaction (although the overall ecological context is certain to differ from what would have been the case where the pest and control agent originated). Biological control requires the classical skills of the taxonomist to find the pest in its native habitat, and particularly to identify and isolate its natural enemies. This may often be a difficult task – especially if the natural enemy has the desired effect of keeping the target species at a low carrying capacity, since both the target and the agent will then be rare in their natural habitat. Nevertheless, the rate of return on investment can be highly favorable. In the case of the cottony cushion scale, biological control has subsequently been transferred to 50 other countries and savings have been immense. In addition, this example illustrates the importance of establishing several, hopefully complementary, enemies to control a pest. Finally, classical biological control, like natural control, can be destabilized by chemicals. The first use of DDT in Californian citrus orchards in 1946–47 against the citricola scale *Coccus pseudomagnoliarum* led to an outbreak of the (by then) rarely seen cottony cushion scale when the DDT almost eliminated the ladybirds. The use of DDT was terminated.

conservation
biological control
of wheat aphids

Many pests have a diversity of natural enemies that already occur in their vicinity. For example, the aphid pests of wheat (e.g. *Sitobion avenae* or *Rhopalosiphum* spp.) are attacked by

Table 15.3 The record of insects as biological control agents against insect pests and weeds. (After Waage & Greathead, 1988.)

	Insect pests	Weeds
Control agent species	563	126
Pest species	292	70
Countries	168	55
Cases where agent has become established	1063	367
Substantial successes	421	113
Successes as a percentage of establishments	40	31

coccinellid and other beetles, heteropteran bugs, lacewings (Chrysopidae), syrphid fly larvae and spiders – all part of a large group of specialist aphid predators and generalists that include them in their diet (Brewer & Elliott, 2004). Many of these natural enemies overwinter in the grassy boundaries at the edge of wheat fields, from where they disperse and reduce aphid populations around the field edges. The planting of grassy strips within the fields can enhance these natural populations and the scale of their impact on aphid pests. This is an example of 'conservation biological control' in action (Barbosa, 1998).

inoculation against glasshouse pests

'Inoculation' as a means of biological control is widely practised in the control of arthropod pests in glasshouses, a situation in which crops are removed, along with the pests and their natural enemies, at the end of the growing season (van Lenteren & Woets, 1988). Two particularly important species of natural enemy used in this way are *Phytoseiulus persimilis*, a mite that preys on the spider mite *Tetranychus urticae*, a pest of cucumbers and other vegetables, and *Encarsia formosa*, a chalcid parasitoid wasp of the whitefly *Trialeurodes vaporariorum*, a pest in particular of tomatoes and cucumbers. By 1985 in Western Europe, around 500 million individuals of each species were being produced each year.

microbial control of insects via inundation

'Inundation' often involves the use of insect pathogens to control insect pests (Payne, 1988). By far the most widespread and important agent is the bacterium *Bacillus thuringiensis*, which can easily be produced on artificial media. After being ingested by insect larvae, gut juices release powerful toxins and death occurs 30 min to 3 days later. Significantly, there is a range of varieties (or 'pathotypes') of *B. thuringiensis*, including one specific against lepidoptera (many agricultural pests), another against diptera, especially mosquitos and blackflies (the vectors of malaria and onchocerciasis) and a third against beetles (many agricultural and stored product pests). *B. thuringiensis* is used inundatively as a microbial insecticide. Its advantages are its powerful toxicity against target insects and its lack of toxicity against organisms outside this narrow group (including ourselves and most of the pest's natural enemies). Plants, including cotton (*Gossypium hirsutum*), have been genetically modified to express the *B. thuringiensis* toxin (insecticidal crystal protein Cry1Ac). The survivorship of pink bollworm larvae (*Pectinaphora gossypiella*) on genetically modified cotton was 46–100% lower than on nonmodified cotton (Lui *et al.*, 2001). Concern has arisen about the widespread insertion of Bt into commercial genetically modified crops, because of the increased likelihood of the development of resistance to one of the most effective 'natural' insecticides available.

biological control is not always environmentally friendly

Biological control may appear to be a particularly environmentally friendly approach to pest control, but examples are coming to light where even carefully chosen and apparently successful introductions of biological control agents have impacted on nontarget species. For example, a seed-feeding weevil (*Rhinocyllus conicus*), introduced to North America to control exotic *Carduus* thistles, attacks more than 30% of native thistles (of which there are more than 90 species), reducing thistle densities (by 90% in the case of the Platte thistle *Cirsuim canescens*) with consequent adverse impacts on the populations of a native picture-winged fly (*Paracantha culta*) that feeds on thistle seeds (Louda *et al.*, 2003a). Louda *et al.* (2003b) reviewed 10 biological control projects that included the unusual but worthwhile step of monitoring nontarget effects and concluded that relatives of the target species were most likely to be attacked whilst rare native species were particularly susceptible. Their recommendations for management included the avoidance of generalist control agents, an expansion of host-specificity testing and the need to incorporate more ecological information when evaluating potential biological control agents.

15.2.6 Integrated pest management

IPM: an ecologically rather than chemically based philosophy

A variety of management implications of our understanding of pest population dynamics have been presented in previous sections. However, it is important to take a broader perspective and consider how all the different tools at the pest controller's disposal can be deployed most effectively, both to maximize the economic benefit of reducing pest density and to minimize the adverse environmental and health consequences. This is what integrated pest management (IPM) is intended to achieve. It combines physical control (for example, simply keeping invaders from arriving, keeping pests away from crops, or picking them off by hand when they arrive), cultural control (for example, rotating the crops planted in a field so pests cannot build up their numbers over several years), biological and chemical control, and the use of resistant varieties of crop. IPM came of age as part of the reaction against the unthinking use of chemical pesticides in the 1940s and 1950s.

IPM is ecologically based and relies heavily on natural mortality factors, such as weather and enemies, and seeks to disrupt the latter as little as possible. It aims to control pests below the EIL, and it depends on monitoring the abundance of pests and their natural enemies and using various control methods as complementary parts of an overall program. Broad-spectrum pesticides in particular, although not excluded, are used only very sparingly, and if chemicals are used at all it is in ways that minimize the costs and quantities used. The essence of the IPM approach is to make the control measures fit the pest problem, and no two problems are the same – even in adjacent fields. Thus, IPM often involves the development of computer-based expert systems that can be used by farmers to diagnose pest problems and suggest appropriate responses (Mahaman *et al.*, 2003).

IPM for the potato tuber moth

The caterpillar of the potato tuber moth (*Phthorimaea operculella*) commonly damages crops in New Zealand. An invader from a warm temperate subtropical country, it is most devastating when conditions are warm and dry (i.e. when the environment coincides closely with its optimal niche requirements – see Chapter 3). There can be as many as 6–8 generations per year and different generations mine leaves, stems and tubers. The caterpillars are protected both from natural enemies (parasitoids) and insecticides when in the tuber, so control must be applied to the leaf-mining generations. The IPM strategy for potato tuber moth (Herman, 2000) involves: (i) monitoring (female pheromone traps, set weekly from mid summer, are used to attract males, which are counted); (ii) cultural methods (the soil is cultivated to prevent soil cracking, soil ridges are molded up more than once and soil moisture is maintained); and (iii) the use of insecticides, but only when absolutely necessary (most commonly the organophosphate, methamidophos). Farmers follow the decision tree shown in Figure 15.5.

integration of IPM in sustainable farming systems

Implicit in the philosophy of IPM is the idea that pest control cannot be isolated from other aspects of food production and it is especially bound up with the means by which soil fertility is maintained and improved. These broader sustainable agricultural systems, including IFS (integrated farming systems) in the USA and LIFE (lower input farming and environment) in Europe (International Organisation for Biological Control, 1989; National Research Council, 1990), have advantages in terms of reduced environmental hazards. Even so, it is unreasonable to suppose that they will be adopted widely unless they are also sound in economic terms. In this context, Figure 15.6 shows the yields of apples from organic, conventional and integrated production systems in Washington State from 1994 to 1999 (Reganold *et al.*, 2001).

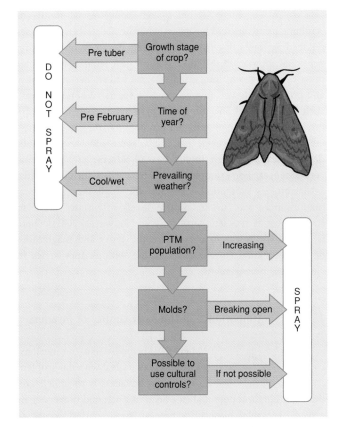

Figure 15.5 Decision flow chart for the integrated pest management of potato tuber moths (PTM) in New Zealand. Boxed phrases are questions (e.g. 'what is the growth stage of the crop?'), the words in the arrows are the farmer's answers to the questions (e.g. 'before the tuber has formed') and the recommended action is shown in the vertical box ('don't spray the crop'). Note that February is late summer in New Zealand. (After Herman, 2000.) Photograph © International Potato Center (CIP).

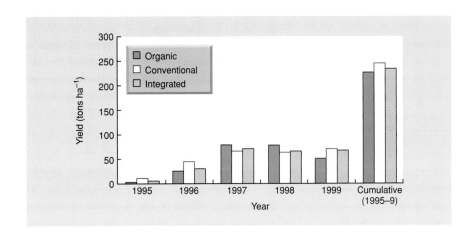

Figure 15.6 The fruit yields of three apple production systems. (From Reganold *et al.*, 2001.)

Organic management excludes such conventional inputs as synthetic pesticides and fertilizers whilst integrated farming uses reduced amounts of chemicals by integrating organic and conventional approaches. All three systems gave similar apple yields but the organic and integrated systems had higher soil quality and potentially lower environmental impacts. When compared with conventional and integrated systems, the organic system produced sweeter apples, higher profitability and greater energy efficiency. Note, however, that despite some widely held beliefs, organic farming is not totally free of adverse environmental consequences. For example, some approved pesticides are just as harmful as synthetic ones whilst the application of animal manure may lead to undesirable levels of nitrate runoff to streams just as synthetic fertilizers can (Trewavas, 2001). There is a need for research to compare the types and magnitudes of environmental consequences of the various approaches to agricultural management.

15.2.7 The importance of the early control of invaders

when a new pest invades . . .

Many pests begin life as exotic invaders. The best way to deal with the problem of potential invaders is to understand their immigration potential (see Section 7.4.2) and prevent their arrival by careful biosecurity processes at a nation's point of entry, or elsewhere on trade routes (Wittenberg & Cock, 2001). However, there are so many potential invaders that it is unrealistic to expect that they all will be prevented from arriving. Moreover, many arrivals will not establish, and many of those that do establish will do so without dramatic ecological consequences. Managers need to focus on the really problematic cases. Thus, the next step in an invader management strategy is to prioritize those that might arrive (or that have recently arrived) according to their likelihood of persisting, establishing large populations, spreading through the new area and causing significant problems. This is not an easy matter, but particular life history traits provide useful pointers (dealt with in Section 7.3.2). We will see in Chapter 22 that assessment of the potential to do harm at higher ecological levels (community/ecosystem) can also be helpful in prioritizing invaders for special attention (see Section 22.3.1).

. . . early control is best

The arrival of an exotic species with a high likelihood of becoming a significant invasive species should be a matter for urgent action, because this is the stage at which eradication is both feasible and easy to justify economically. Such campaigns sometimes rely on fundamental knowledge of population ecology. An example is the eradication of the South African sabellid polychaete worm, *Terebrasabella heterouncinata*, a parasite of abalone and other gastropods that became established near the outflow of an abalone aquaculture facility in California (Culver & Kuris, 2000).

Its population biology was understood sufficiently to know it was specific to gastropods, that two species of *Tegula* were its principal hosts in the area, and that large snails were most susceptible to the parasite. Volunteers removed 1.6 million large hosts, thereby reducing the density of susceptible hosts below that needed for parasite transmission (see Chapter 12), which became extinct.

However, in the words of Simberloff (2003), rapid responses to recent invaders will often 'resemble a blunderbuss attack rather than a surgical strike'. He notes, for example, that a string of successful eradications of small populations of weeds such as pampas grass (*Cortaderia selloana*) and ragwort (*Senecio jacobaea*) on New Zealand's offshore islands (Timmins & Braithwaite, 2002) were effective because of early action using brute-force methods. Similarly, the white-spotted tussock moth (*Orygyia thyellina*), discovered in a suburban region of Auckland, New Zealand, was eradicated (at a cost of US$5 million) using *Bacillus thuringiensis* spray (Clearwater, 2001). The only population biological information to hand was that females attracted males by pheromone, knowledge that was used to trap males and determine areas that needed respraying. Eradication of a recently established species known to be invasive elsewhere usually cannot and should not wait for new population studies to be performed.

Once invaders have established and spread through a new area and are determined to be pests, they are just another species at which the pest manager's armory must be directed.

15.3 Harvest management

harvesting aims to avoid over- and underexploitation

Harvesting of populations by people is clearly in the realm of predator–prey interactions and harvest management relies on the theory of predator–prey dynamics (see Chapters 10 and 14). When a natural population is exploited by culling or harvesting – whether this involves the removal of whales or fish from the sea, the capture of 'bushmeat' in the African savanna or the removal of timber from a forest – it is much easier to say what we want to avoid than precisely what we might wish to achieve. On the one hand, we want to avoid overexploitation, where too many individuals are removed and the population is driven into biological jeopardy, or economic insignificance or perhaps even to extinction. But harvest managers also want to avoid underexploitation, where far fewer individuals are removed than the population can bear, and a crop of food, for example, is produced which is smaller than necessary, threatening both the health of potential consumers and the livelihood of all those employed in the harvesting operation. However, as we shall see, the best position to occupy between these two extremes is not easy to determine, since it needs to combine considerations that are not only biological (the well-being of the exploited population) and economic (the profits being made

Figure 15.7 Fixed quota harvesting. The figure shows a single recruitment curve and three fixed quota harvesting curves: high quota (h_h), medium quota (h_m) and low quota (h_l). Arrows in the figure refer to changes to be expected in abundance under the influence of the harvesting rate to which the arrows are closest. ●, equilibria. At h_h the only 'equilibrium' is when the population is driven to extinction. At h_l there is a stable equilibrium at a relatively high density, and also an unstable breakpoint at a relatively low density. The MSY is obtained at h_m because it just touches the peak of the recruitment curve (at a density N_m): populations greater than N_m are reduced to N_m, but populations smaller than N_m are driven to extinction.

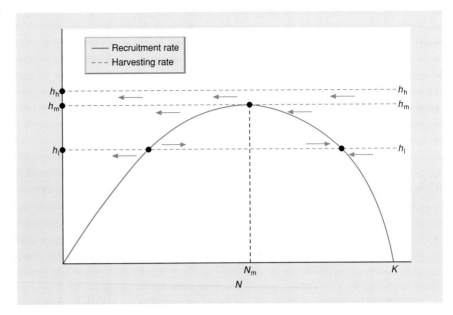

from the operation), but also social (local levels of employment and the maintenance of traditional lifestyles and human communities) (Hilborn & Walters, 1992; Milner-Gulland & Mace, 1998). We begin, though, with the biology.

15.3.1 Maximum sustainable yield

MSY: the peak of the net recruitment curve

The first point to grasp about harvesting theory is that high yields are obtained from populations held below, often well below, the carrying capacity. This fundamental pattern is captured by the model population in Figure 15.7. There, the natural net recruitment (or net productivity) of the population is described by an n-shaped curve (see Section 5.4.2). Recruitment rate is low when there are few individuals and low when there is intense intraspecific competition. It is zero at the carrying capacity (K). The density giving the highest net recruitment rate depends on the exact form of intraspecific competition. This density is $K/2$ in the logistic equation (see Section 5.9) but, for example, is only slightly less than K in many large mammals (see Figure 5.10d). Always, though, the rate of net recruitment is highest at an 'intermediate' density, less than K.

Figure 15.7 also illustrates three possible harvesting 'strategies', although in each case there is a fixed harvesting *rate*, i.e. a fixed number of individuals removed during a given period of time, or 'fixed quota'. When the harvesting and recruitment lines cross, the harvesting and recruitment rates are equal and opposite; the number removed per unit time by the harvester equals the number recruited per unit time by the population. Of particular interest is the harvesting rate h_m, the line that crosses (or, in fact, just touches) the recruitment rate curve at its peak. This is

the highest harvesting rate that the population can match with its own recruitment. It is known as the maximum sustainable yield (MSY), and as the name implies, it is the largest harvest that can be removed from the population on a regular and repeated (indeed indefinite) basis. It is equal to the maximum rate of recruitment, and it is obtained from the population by depressing it to the density at which the recruitment rate curve peaks.

The MSY concept is central to much of the theory and practice of harvesting. This makes the recognition of the following shortcomings in the concept all the more essential.

MSY has severe shortcomings . . .

1 By treating the population as a number of similar individuals, or as an undifferentiated biomass, it ignores all aspects of population structure such as size or age classes and their differential rates of growth, survival and reproduction. The alternatives that incorporate structure are considered below.
2 By being based on a single recruitment curve it treats the environment as unvarying.
3 In practice, it may be impossible to obtain a reliable estimate of the MSY.
4 Achieving an MSY is by no means the only, nor necessarily the best, criterion by which success in the management of a harvesting operation should be judged (see, for example, Section 15.3.9).

Despite all these difficulties, the MSY concept dominated resource management for many years in fisheries, forestry and wildlife exploitation. Prior to 1980, for example, there were 39 agencies for the management of marine fisheries, every

. . . but has been frequently used

one of which was required by its establishing convention to manage on the basis of an MSY objective (Clark, 1981). In many other areas, the MSY concept is still the guiding principle. Moreover, by assuming that MSYs are both desirable and attainable, a number of the basic principles of harvesting can be explained. Therefore, we begin here by exploring what can be learnt from analyses based on the MSY, but then look more deeply at management strategies for exploited populations by examining the various shortcomings of MSY in more detail.

15.3.2 Simple MSY models of harvesting: fixed quotas

fixed-quota harvesting is extremely risky . . .

The MSY density (N_m) is an equilibrium (gains = losses), but when harvesting is based on the removal of a fixed quota, as it is in Figure 15.7, N_m is a very fragile equilibrium. If the density exceeds the MSY density, then h_m exceeds the recruitment rate and the population declines towards N_m. This, in itself, is satisfactory. But if, by chance, the density is even slightly less than N_m, then h_m will once again exceed the recruitment rate. Density will then decline even further, and if a fixed quota at the MSY level is maintained, the population will decline until it is extinct. Furthermore, if the MSY is even slightly overestimated, the harvesting rate will always exceed the recruitment rate (h_h in Figure 15.7). Extinction will then follow, whatever the initial density. In short, a fixed quota at the MSY level might be desirable and reasonable in a wholly predictable world about which we had perfect knowledge. But in the real world of fluctuating environments and imperfect data sets, these fixed quotas are open invitations to disaster.

. . . whose dangers are illustrated by the Peruvian anchovy fishery

Nevertheless, a fixed-quota strategy has frequently been used. On a specified day in the year, the fishery (or hunting season) is opened and the cumulative catch logged. Then, when the quota (estimated MSY) has been taken, the fishery is closed for the rest of the year. An example of the use of fixed quotas is provided by the Peruvian anchovy (*Engraulis ringens*) fishery (Figure 15.8). From 1960 to 1972 this was the world's largest single fishery, and it constituted a major sector of the Peruvian economy. Fisheries experts advised that the MSY was around 10 million tonnes annually, and catches were limited accordingly. But the fishing capacity of the fleet expanded, and in 1972 the catch crashed. Overfishing seems at least to have been a major cause of the collapse, although its effects were compounded with the influences of profound climatic fluctuations. A moratorium on fishing would have been an ecologically sensible step, but this was not politically feasible: 20,000 people were dependent on the anchovy industry for employment. The stock took more than 20 years to recover (Figure 15.8).

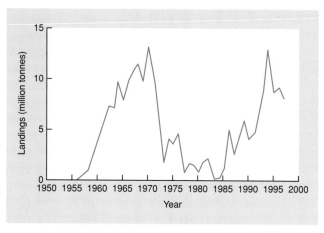

Figure 15.8 Landings of the Peruvian anchovy since 1950. (After Jennings *et al.*, 2001; data from FAO, 1995, 1998.)

15.3.3 A safer alternative: fixed harvesting effort

The risk associated with fixed quotas can be reduced if instead there is regulation of the harvesting effort. The yield from a harvest (H) can be thought of, simply, as being dependent on three things:

$$H = qEN. \tag{15.1}$$

Yield, H, increases with the size of the harvested population, N; it increases with the level of harvesting effort, E (e.g. the number of 'trawler-days' in a fishery or the number of 'gun-days'

regulating harvesting effort is less risky – but leads to a more variable catch

with a hunted population); and it increases with harvesting efficiency, q. On the assumption that this efficiency remains constant, Figure 15.9a depicts an exploited population subjected to three potential harvesting strategies differing in harvesting effort. Figure 15.9b then illustrates the overall relationship to be expected, in a simple case like this, between effort and average yield: there is an apparently 'optimum' effort giving rise to the MSY, E_m, with efforts both greater and less than this giving rise to smaller yields.

Adopting E_m is a much safer strategy than fixing an MSY quota. Now, in contrast to Figure 15.7, if density drops below N_m (Figure 15.9a), recruitment exceeds the harvesting rate and the population recovers. In fact, there needs to be a considerable overestimate of E_m before the population is driven to extinction (E_0 in Figure 15.9a). However, because there is a fixed effort, the yield varies with population size. In particular, the yield will be less than the MSY whenever the population size, as a result of natural fluctuations, drops below N_m. The appropriate reaction would be to reduce effort slightly or at least hold it steady whilst the population recovers. But an understandable (albeit misguided) reaction

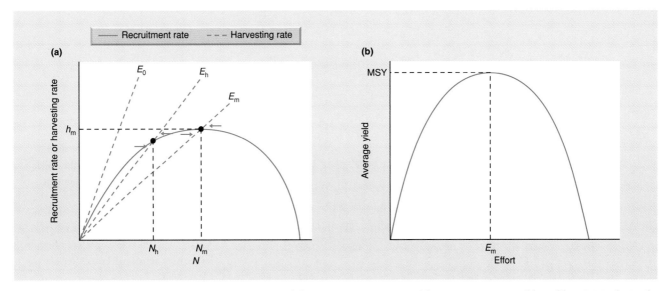

Figure 15.9 Fixed effort harvesting. (a) Curves, arrows and dots are as in Figure 15.7. The maximum sustainable yield (MSY) is obtained with an effort of E_m, leading to a stable equilibrium at a density of N_m with a yield of h_m. At a somewhat higher effort (E_h), the equilibrium density and the yield are both lower than with E_m but the equilibrium is still stable. Only at a much higher effort (E_0) is the population driven to extinction. (b) The overall relationship between the level of the fixed effort and average yield.

might be to compensate by increasing the effort. This, however, might depress population size further (E_h in Figure 15.9a); and it is therefore easy to imagine the population being driven to extinction as very gradual increases in effort chase an ever-diminishing yield.

There are many examples of harvests being managed by legislative regulation of effort, and this occurs in spite of the fact that effort usually defies precise measurement and control. For instance, issuing a number of gun licenses leaves the accuracy of the hunters uncontrolled; and regulating the size and composition of a fishing fleet leaves the weather to chance. Nevertheless, the harvesting of mule deer, pronghorn antelope and elks in Colorado was controlled by issuing a limited but varying number of hunting permits (Pojar, 1981). In the management of the important Pacific halibut stock, effort was limited by seasonal closures and sanctuary zones – although a heavy investment in fishery protection vessels was needed to make this work (Pitcher & Hart, 1982).

15.3.4 Other MSY approaches: harvesting a fixed proportion or allowing constant escapement

other MSY approaches: . . .

. . . harvesting a constant proportion . . .

Two further management strategies are based on the simple idea of availability of a surplus yield. First, a constant proportion of the population can be harvested (this is equivalent to fixing a hunting mortality rate and should have the same effect as harvesting at constant effort) (Milner-Gulland & Mace, 1998). Thus, in the Northwest Territories of Canada, 3–5% of the caribou and muskox populations can be killed each year (Gunn, 1998), a strategy that involves the expense of preharvest censuses so that numbers to be harvested can be calculated.

. . . or leaving a constant 'escapement' of breeding individuals

Another strategy leaves a fixed number of breeding individuals at the end of each hunting season (constant escapement), an approach that involves the even greater expense of continuous monitoring through the hunting season. Constant escapement is a particularly safe option because it rules out the accidental removal of all the breeding individuals before breeding has occurred. Constant escapement is particularly useful for annual species because they lack the buffer provided by immature individuals in longer lived species (Milner-Gulland & Mace, 1998). The Falkland Islands government uses a constant escapement strategy for the annual *Loligo* squid. Stock sizes are assessed weekly from mid-season onwards and the fishery is closed when the ratio of stocks in the presence and absence of fishing falls to 0.3–0.4. After 10 years of this management regime the squid fishery shows good signs of sustainability (Figure 15.10).

constant escapement seems to work best for alpine marmot hunting

Stephens *et al.* (2002) used simulation models to compare the outcomes for a population of alpine marmots (*Marmota marmota*) of fixed-quota, fixed-effort and threshold harvesting. In the latter case,

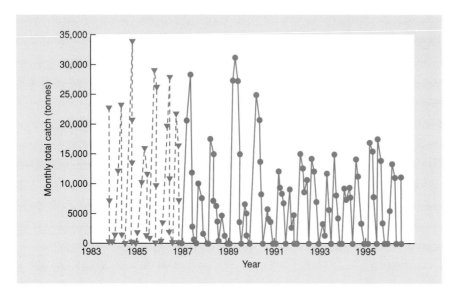

Figure 15.10 Monthly *Loligo* squid catches by licensed vessels in the Falkland Islands where a constant escapement management strategy is used. Note that there are two fishing seasons each year (February–May and August–October). The dotted lines (1984–86) represent estimated rather than actual catches. (After des Clers, 1998.)

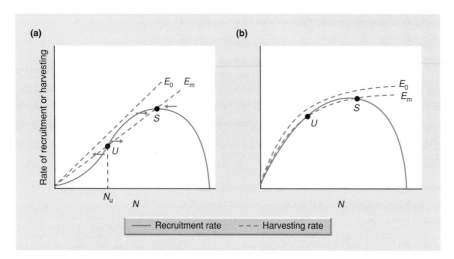

Figure 15.11 Multiple equilibria in harvesting. (a) When recruitment rate is particularly low at low densities, the harvesting effort giving the MSY (E_m) has not only a stable equilibrium (*S*) but also an unstable breakpoint (*U*) at a density below which the population declines to extinction. The population can also be driven to extinction by harvesting efforts (E_0) not much greater than E_m. (b) When harvesting efficiency declines at high densities, comments similar to those in (a) are appropriate.

harvesting only occurred during years in which the population exceeded a given threshold and exploitation continued until that threshold was reached (essentially a constant escapement approach). These social mammals are hunted in parts of Europe but the modeling was performed using extensive data available from a nonhunted population. They found that threshold harvesting provided the highest mean yields coupled with an acceptably low extinction risk. However, the introduction of error, associated with less frequent censuses (3-yearly rather than yearly), led to higher variance in yields and a much increased extinction probability (Stephens *et al.*, 2002). This emphasizes the importance of frequent censuses for constant escapement strategies to succeed.

15.3.5 Instability of harvested populations: multiple equilibria

Even with regulation of effort, harvesting near the MSY level may be courting disaster. The recruitment rate may be particularly low in the smallest populations (a pattern known as *depensation*; Figure 15.11a); for instance, the recruitment of young salmon is low at low densities because of intense predation from larger fish, and the recruitment of young whales may be low at low densities simply because of the reduced chances of males and females meeting to mate. However, depensation is apparently quite rare; Myers *et al.* (1995) detected it in only three of 128 fish stock

the problem of 'depensation'

data sets with 15 or more years available for analysis. Alternatively, harvesting efficiency may increase in small populations (Figure 15.11b). For instance, many clupeids (sardines, anchovies, herring) are especially prone to capture at low densities, because they form a small number of large schools that follow stereotyped migratory paths that the trawlers can intercept. With either depensation or higher harvesting efficiency at low density, small overestimates of E_m are liable to lead to overexploitation or even eventual extinction.

harvesting operations with multiple equilibria are susceptible to dramatic irreversible crashes

Even more important, however, is the fact that these interactions may have crucial 'multiple equilibria' (see Section 10.6). Note the two points where the harvesting line crosses the recruitment curve in Figure 15.11a. The point S is a stable equilibrium but the point U is an unstable 'breakpoint'. If the population is driven slightly below the MSY density, or even to a level slightly above N_u, a breakpoint, it returns to the MSY density (Figure 15.11a). But a marginally increased depression in density, to a level slightly below N_u, perhaps resulting from only a very small increase in effort, would make the harvesting rate greater than the recruitment rate. The population would be *en route* to extinction. Moreover, once the population is on this slippery slope, much more than a marginal reduction in effort is required to reverse the process. This is the crucial, practical point about multiple equilibria: a very slight change in behavior can lead to a wholly disproportionate change in outcome as the point of attraction in the system shifts from one stable state to another. Drastic changes in stock abundance can result from only small changes in harvesting strategy or small changes in the environment.

15.3.6 Instability of harvested populations: environmental fluctuations

It is tempting to attribute all fisheries' collapses simply to overfishing and human greed. Doing so, however, would be an unhelpful oversimplification. There is no doubt that fishing pressure often exerts a great strain on the ability of natural populations to sustain levels of recruitment that counteract overall rates of loss. But the immediate cause of a collapse – in 1 year rather than any other – is often the occurrence of unusually unfavorable environmental conditions. Moreover, when this is the case, the population is more likely to recover (once conditions have returned to a more favorable state) than it would be if the crash was the result of overfishing alone.

the anchoveta and the El Niño

The Peruvian anchovy (see Figure 15.8), prior to its major collapse from 1972 to 1973, had already suffered a dip in the upward rise in catches in the mid-1960s as a result of an 'El Niño event': the incursion of warm tropical water from the north severely reducing ocean upwelling, and hence productivity, within the cold Peruvian current coming from the south (see Section 2.4.1). By 1973, however, because fishing intensity had so greatly increased, the effects of a subsequent El Niño event were much more severe. Moreover, whilst the fishery showed some signs of recovery from 1973 to 1982, in spite of largely unabated fishing pressure, a further collapse occurred in 1983 associated with yet another El Niño event. Clearly, it is unlikely that the consequences of these natural perturbations to the usual patterns of current flow would have been so severe if the anchovy had not been exploited or had been only lightly fished. It is equally clear, though, that the history of the Peruvian anchovy fishery cannot be understood properly in terms simply of fishing, as opposed to natural events.

The three Norwegian and Icelandic herring fisheries also collapsed in the early 1970s and had certainly been subjected to increasing fishing intensities prior to that. Once again, however, an oceanic anomaly is implicated (Beverton, 1993). In the mid-1960s, a mass of cold, low-salinity water from the Arctic Basin formed north of Iceland. It drifted south until it became entrained in the Gulf Stream several years later, and then moved north again – although well to the east of its southward track. It eventually disappeared off Norway in 1982 (Figure 15.12a). Data for the number of 'recruits per spawner', essentially the birth rate, are illustrated in Figure 15.12b for the Norwegian springspawning and the Icelandic spring- and summer-spawning herring between 1947 and 1990, in terms of the difference each year between that year's value and the overall average. Also illustrated are the corresponding yearly temperature differentials in the Norwegian Sea, reflecting the southward and northward passage of the anomalous cold water body. There was a good correspondence between the cold water and poor recruitment in both the Icelandic and Norwegian stocks in the late 1960s and in the Norwegian stocks in 1979–81, the Icelandic stocks being then extinct (spring spawners) or too far west. It seems likely that the anomalous cold water led to unusually low recruitment, which was strongly instrumental in the crashes experienced by each of these fisheries.

herring and cold water

This cannot, however, account for all the details in Figure 15.12b – especially the succession of poor recruitment years in the Norwegian stocks in the 1980s. For this, a more complex explanation is required, probably involving other species of fish and perhaps alternative stable states (Beverton, 1993). None the less, it remains clear that whilst the dangers of overfishing should not be denied, these must be seen within the context of marked and often unpredictable natural variations. Given the likely effects of environmental conditions on the vital rates of harvested populations, a reliance on models with constant vital rates is even more risky. Engen *et al.* (1997) argue that the best harvesting strategies for such highly variable populations involve constant escapement (see Section 15.3.4).

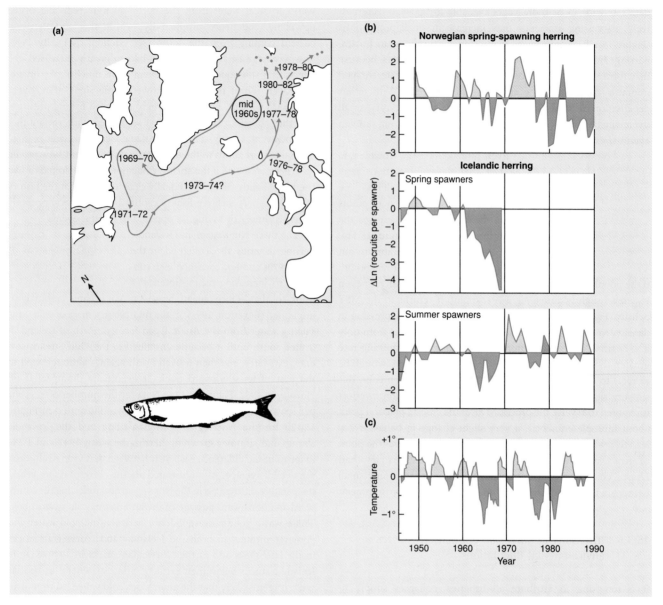

Figure 15.12 (a) The track of a large mass of cold, low-salinity water in the 1960s and 1970s, showing its presence in the Norwegian Sea both in the mid-1960s and the period 1977–82. (b) Annual differentials between overall averages and ln (recruits per spawner) for three herring stocks in the Norwegian Sea, and (c) the temperature in the Norwegian Sea. The Icelandic spring-spawning stock never recovered from its collapse in the early 1970s, preceded by low recruitment in the 1960s. (After Beverton, 1993.)

15.3.7 Recognizing structure in harvested populations: dynamic pool models

'dynamic pool' models recognize population structure

The simple models of harvesting that have been described so far are known as 'surplus yield' models. They are useful as a means of establishing some basic principles (like MSY), and they are good for investigating the possible consequences of different types of harvesting strategy.

But they ignore population structure, and this is a bad fault for two reasons. The first is that 'recruitment' is, in practice, a complex process incorporating adult survival, adult fecundity, juvenile survival, juvenile growth, and so on, each of which may respond in its own way to changes in density and harvesting strategy. The second reason is that most harvesting practices are primarily interested in only a portion of the harvested population (e.g. mature trees, or fish that are large enough to be saleable). The approach that attempts to take these complications into

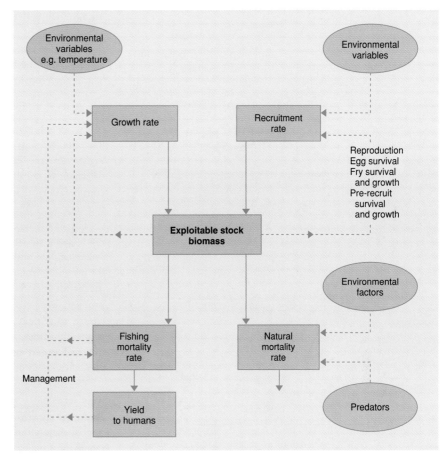

Figure 15.13 The dynamic pool approach to fishery harvesting and management, illustrated as a flow diagram. There are four main 'submodels': the growth rate of individuals and the recruitment rate into the population (which add to the exploitable biomass), and the natural mortality rate and the fishing mortality rate (which deplete the exploitable biomass). Solid lines and arrows refer to changes in biomass under the influence of these submodels. Dashed lines and arrows refer to influences either of one submodel on another, or of the level of biomass on a submodel or of environmental factors on a submodel. Each of the submodels can itself be broken down into more complex and realistic systems. Yield to humans is estimated under various regimes characterized by particular values inserted into the submodels. These values may be derived theoretically (in which case they are 'assumptions') or from field data. (After Pitcher & Hart, 1982.)

account involves the construction of what are called 'dynamic pool' models.

The general structure of a dynamic pool model is illustrated in Figure 15.13. The submodels (recruitment rate, growth rate, natural mortality rate and fishing rate of the exploited stock) combine to determine the exploitable biomass of the stock and the way this translates into a yield to the fishing community. In contrast to the surplus yield models, this biomass yield depends not only on the number of individuals caught but also on their size (past growth); whilst the quantity of exploitable (i.e. catchable) biomass depends not just on 'net recruitment' but on an explicit combination of natural mortality, harvesting mortality, individual growth and recruitment into catchable age classes.

There are many variants on the general theme (e.g. the submodels can be dealt with separately in each of the age classes and submodels can incorporate as much or as little information as is available or desirable). In all cases, though, the basic approach is the same. Available information (both theoretical and empirical) is incorporated into a form that reflects the dynamics of the structured population. This then allows the yield and the response of the population to different harvesting strategies to be estimated.

This in turn should allow a recommendation to the stock-manager to be formulated. The crucial point is that in the case of the dynamic pool approach, a harvesting strategy can include not only a harvesting intensity, but also a decision as to how effort should be partitioned amongst the various age classes.

A classic example of a dynamic pool model in action concerned the Arcto-Norwegian cod fishery, the most northerly of the Atlantic stocks (Garrod & Jones, 1974). The age class structure of the late 1960s was used to predict the medium-term effects on yield of different fishing intensities and different mesh sizes in the trawl. Some of the results are shown in Figure 15.14. The temporary peak after 5 or so years is a result of the very large 1969 year-class working through the population. Overall, however, it is clear that the best longer term prospects were predicted for a low fishing intensity and a large mesh size. Both of these give the fish more opportunity to grow (and reproduce) before they are caught, which is important because yield is measured in biomass, not simply in numbers. Higher fishing intensities and mesh sizes of 130 mm were predicted to lead to overexploitation of the stock.

dynamic pool models can lead to valuable recommendations . . .

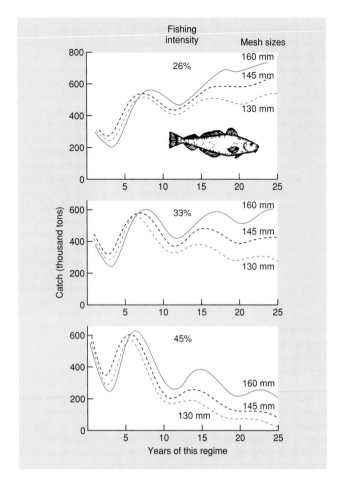

Figure 15.14 Garrod and Jones' (1974) predictions for the Arctic cod stock under three fishing intensities and with three different mesh sizes. (After Pitcher & Hart, 1982.)

15.3.8 Objectives for managing harvestable resources

If we treat the Garrod and Jones example as typical, then we might conclude that the biologist proposes – but the manager disposes. This is therefore an appropriate point at which to reconsider not only the objectives of harvesting programs, but also the criteria by which successful management should be judged and the role of ecologists in management overall. As Hilborn and Walters (1992) have pointed out, there are three alternative attitudes that ecologists can take, each of which has been popular but only one of which is sensible. Indeed, these are increasingly important considerations that apply not just to fisheries management but to every entry of ecologists into the public arena.

The first is to claim that ecological interactions are too complex, and our understanding and our data too poor, for pronouncements of any kind to be made (for fear of being wrong). The problem with this is that if ecologists *three attitudes for ecologists towards managers in the real world . . .* choose to remain silent because of some heightened sensitivity to the difficulties, there will always be some other, probably less qualified 'expert' ready to step in with straightforward, not to say glib, answers to probably inappropriate questions.

The second possibility is for ecologists to concentrate exclusively on ecology and arrive at a recommendation designed to satisfy purely ecological criteria. Any modification by managers or politicians of this recommendation is then ascribed to ignorance, inhumanity, political corruption or some other sin or human foible. The problem with this attitude is that it is simply unrealistic in any human activity to ignore social and economic factors.

The third alternative, then, is for ecologists to make ecological assessments that are as accurate and realistic *. . . but only one of them is sensible* as possible, but to assume that these will be incorporated with a broader range of factors when management decisions are made. Moreover, these assessments should themselves take account of the fact that the ecological interactions they address include humans as one of the interacting species, and humans are subject to social and economic forces. Finally, since ecological, economic and social criteria must be set alongside one another, choosing a single, 'best' option is likely to be seen by some involved in the decision as an opinion based on the proponent's particular set of values. It follows that a single recommendation is, in practice, far less useful in this discourse than laying out a series of possible plans of action with their associated consequences.

In the present context, therefore, we develop this third alternative by first looking beyond MSY to criteria that incorporate risk, economics, social consequences, and so on (Hilborn & Waters, 1992). We then briefly examine the means by which crucial parameters and variables are estimated in natural populations, since these, by determining the quality of available information, determine the degree of confidence with which recommendations can be made.

. . . but these may still be ignored Sadly, Garrod and Jones' recommendations were ignored by those with the power to determine fishing strategies. Mesh sizes were not increased until 1979, and then only from 120 to 125 mm. Fishing intensity never dropped below 45% and catches of 900,000 tonnes were taken in the late 1970s. Not surprisingly perhaps, surveys late in 1980 showed that these and other North Atlantic cod stocks were very seriously depleted as a result of overfishing. North Sea cod reach sexual maturity around the age of 4 years, but the species has been so heavily exploited that some 1 year olds are now harvested and 2 year olds are almost fully exploited, leaving only 4% of 1 year olds to survive to age 4 (Cook *et al.*, 1997).

Rattans (climbing spiny palms whose stems are used for weaving and furniture making in Southeast Asia) are threatened with overexploitation in a similar way, with harvesters cutting stems too young and reducing their ability to resprout (MacKinnon, 1998).

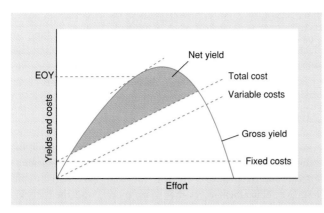

Figure 15.15 The economically optimum yield (EOY), that which maximizes 'profit', is obtained to the left of the peak of the yield-against-effort curve, where the difference between gross yield and total cost (fixed costs plus variable costs) is greatest. At this point, the gross yield and total cost lines have the same slope. (After Hilborn & Walters, 1992.)

15.3.9 Economic and social factors

the economically optimum yield – typically less than the MSY

Perhaps the most obvious shortcoming of a purely ecological approach is its failure to recognize that the exploitation of a natural resource is usually a business enterprise, in which the value of the harvest must be set against the costs of obtaining that harvest. Even if we distance ourselves from any preoccupation with 'profit', it makes no sense to struggle to obtain the last few tonnes of an MSY if the money spent in doing so could be much more effectively invested in some other means of food production. The basic idea is illustrated in Figure 15.15. We seek to maximize not total yield but net value – the difference between the gross value of the harvest and the sum of the fixed costs (interest payments on ships or factories, insurance, etc.) and the variable costs, which increase with harvesting effort (fuel, crew's expenses, etc.). This immediately suggests that the economically optimum yield (EOY) is less than the MSY, and is obtained through a smaller effort or quota. However, the difference between the EOY and the MSY is least in enterprises where most costs are fixed (the 'total cost' line is virtually flat). This is especially the case in high investment, highly technological operations such as deep-sea fisheries, which are therefore most prone to overfishing even with management aimed at economic optima.

discounting: liquidating stocks, or leaving them to grow?

A second important economic consideration concerns 'discounting'. This refers to the fact that in economic terms, each bird in the hand now (or each fish in the hold) is worth more than an equivalent bird or fish some time in the future. The reason is basically that the value of the current catch can be placed in the bank to accrue interest, so that its total value increases. In fact, a commonly used discount rate for natural resources is 10% per annum (90 fish now are as valuable as 100 fish in 1 year's time) despite the fact that the difference between the interest rates in the banks and the rate of inflation is usually only 2–5%. The economists' justification for this is a desire to incorporate 'risk'. A fish caught now has already been caught; one still in the water might or might not be caught – a bird in the hand really is worth two in the bush.

On the other hand, the caught fish is dead, whereas the fish still in the water can grow and breed (although it may also die). In a very real sense, therefore, each uncaught fish will be worth more than 'one fish' in the future. In particular, if the stock left in the water grows faster than the discount rate, as is commonly the case, then a fish put on deposit in the bank is not so sound an investment as a fish left on deposit in the sea. Nevertheless, even in cases like this, discounting provides an economic argument for taking larger harvests from a stock than would otherwise be desirable.

Moreover, in cases where the stock is less productive than the discount rate – for example, many whales and a number of long-lived fish – it seems to make sense, in purely economic terms, not only to overfish the stock, but actually to catch every fish ('liquidate the stock'). The reasons for not doing so are partly ethical – it would clearly be ecologically short sighted and a disdainful way of treating the hungry mouths to be fed in the future. But there are also practical reasons: jobs must be found for those previously employed in the fishery (or their families otherwise provided for), alternative sources of food must be found, and so on. This emphasizes, first, that a 'new economics' must be forged in which value is assigned not only to things that can be bought and sold – like fish and boats – but also to more abstract entities, like the continued existence of whales or other 'flagship species' (Hughey *et al.*, 2002). It also stresses the danger of an economic perspective that is too narrowly focused. The profitability of a fishery cannot sensibly be isolated from the implications that the management of the fishery has in a wider sphere.

social repercussions

'Social' factors enter in two rather separate ways into plans for the management of natural resources. First, practical politics might dictate, for instance, that a large fleet of small, individually inefficient boats is maintained in an area where there are no alternative means of employment. In addition, though, and of much more widespread importance, it is necessary for management plans to take full account of the way fishermen and harvesters will behave and respond to changing circumstances, rather than assuming that they will simply conform to the requirements for achieving either ecological or economic optima. Harvesting involves a predator–prey interaction: it makes no sense to base plans on the dynamics of the prey alone whilst simply ignoring those of the predator (us!).

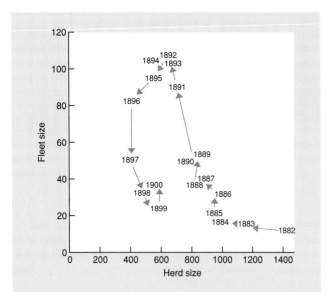

Figure 15.16 The fleet size of the North Pacific fur seal fishery (predators) responded to the size of the seal herd (prey) between 1882 and 1900 by exhibiting an anticlockwise predator–prey spiral. (After Hilborn & Walters, 1992; from data of Wilen, 1976, unpublished observations.)

harvester as predator: human behavior

The idea of the harvester as predator is reinforced in Figure 15.16, which shows a classic anticlockwise predator–prey spiral (see Chapter 10) for the North Pacific fur seal fishery in the last years of the 19th century. The figure illustrates a numerical response on the part of the predator – extra vessels enter the fleet when the stock is abundant, but leave when it is poor. But the figure also illustrates the inevitable time lag in this response. Thus, whatever a modeler or manager might propose, there is unlikely ever to be some perfect match, at an equilibrium, between stock size and effort. Moreover, whilst the sealers in the figure left

the fishery as quickly as they had entered it, this is by no means a general rule. The sealers were able to switch to fishing for halibut, but such switches are often not easy to achieve, especially where there has been heavy investment in equipment or long-standing traditions are involved. As Hilborn and Walters (1992) put it, 'Principle: the hardest thing to do in fisheries management is reduce fishing pressure'.

Switching is one aspect of a harvester's predatory behavior – its functional response (see Chapter 10). Harvesters will also generally 'learn' as there is an inevitable trend towards technological improvement. Even without this, harvesters usually improve their efficiency as they learn more about their stock – notwithstanding the assumptions of simple fixed-effort models.

15.3.10 Estimates from data: putting management into practice

The role of the ecologist in the management of a natural resource is in *stock assessment*: making quantitative predictions about the response of the biological population to alternative management choices and addressing questions like whether a given fishing intensity will lead to a decline in the size of the stock, whether nets of a given mesh size will allow the recruitment rate of a stock to recover, and so on. In the past, it has often been assumed that this can be done simply by careful monitoring. For example, as effort and yield increase in an expanding fishery, both are monitored, and the relationship between the two is plotted until it seems that the top of a curve like that in Figure 15.7 has been reached or just exceeded, identifying the MSY. This approach, however, is deeply flawed, as can be seen from Figure 15.17. In 1975, the International Commission for the Conservation of Atlantic Tunas (ICCAT) used the available data (1964–73) to plot the yield–effort relationship for the yellowfin tuna (*Thunnus albacares*) in the eastern Atlantic. They felt that they

monitoring effort and yield: the difficulties of 'finding the top of the curve'

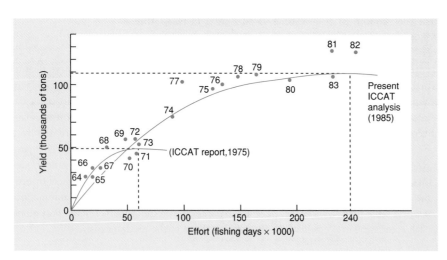

Figure 15.17 Estimated yield–effort relationships for the eastern Atlantic yellowfin tuna (*Thunnus albacares*) on the basis of the data for 1964–73 (ICCAT, 1975) and 1964–83 (ICCAT, 1985). (After Hunter *et al.*, 1986; Hilborn & Walters, 1992.)

had reached the top of the curve: a sustainable yield of around 50,000 tons (5.1×10^7 kg) and an optimum effort of about 60,000 fishing days. However, ICCAT were unable to prevent effort (and yield) rising further, and it soon became clear that the top of the curve had not been reached. A reanalysis using data up to 1983 suggested a sustainable yield of around 110,000 tons (1.1×10^8 kg) and an effort of 240,000 fishing days.

This illustrates what Hilborn and Walters (1992) describe as another principle: 'You cannot determine the potential yield from a fish stock without overexploiting it'. At least part of the reason for this is the tendency, already noted, for the variability in yield to increase as an MSY is approached. Furthermore, if we also recall the previously described difficulty in reducing fishing pressure, it is clear that in practice, managers are likely to have to wrestle with the combined challenge of estimation difficulties, ecological relationships (here, between yield and predictability) and socio-economic factors (here, concerning the regulation and reduction in effort). We have moved a long way from the simple fixed-effort models of Section 15.3.3.

The practical difficulties of parameter estimation are further illustrated in Figure 15.18, which displays the time series for total catch, fishing effort and catch per unit effort (CPUE) between 1969 and 1982 for yellowfin tuna for the whole Atlantic Ocean. As effort increased, CPUE declined – presumably, a reflection of a diminishing stock of fish. On the other hand, the catch continued to rise over this period, suggesting that perhaps the stock was not yet being overfished (i.e. the MSY had not yet been reached). These, then, are the data, and they come in probably the most commonly available form – a so-called 'one-way trip' time series. But can they suggest an MSY and can they suggest the effort required to achieve that MSY? Certainly, methods exist for performing the necessary calculations, but these methods require assumptions to be made about the underlying dynamics of the population.

estimates from catch and effort data: applying the Schaefer model

about the underlying dynamics of the population.

The most frequently used assumption describes the dynamics of the stock biomass, B, by:

$$\frac{dB}{dt} = rB\left(1 - \frac{B}{K}\right) - H \tag{15.2}$$

(Schaefer, 1954), which is simply the logistic equation of Chapter 5 (intrinsic rate of increase, r, carrying capacity, K) with a harvesting rate incorporated. The latter may itself be given, following Equation 15.1 (see Section 15.3.3), by $H = qEB$, where q is harvesting efficiency and E the harvesting effort. By definition:

$$CPUE = H/E = qB. \tag{15.3}$$

Hence:

$$B = CPUE/q \tag{15.4}$$

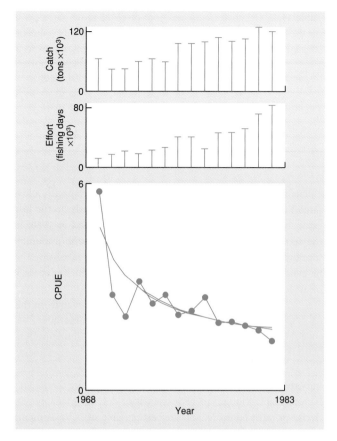

Figure 15.18 Changes in total catch, fishing effort and catch per unit effort (CPUE) between 1969 and 1982 for the yellowfin tuna (*Thussus albacares*) in the Atlantic Ocean. Also shown are three separate curves fitted to the CPUE time series by methods outlined in the text, the parameters of which are given in Table 15.4. (After Hilborn & Walters, 1992.)

and Equation 15.2 can be rewritten in terms of CPUE with either H or E as variables, and with r, q and K as parameters. For this model, the MSY is given by $rK/4$ and the effort required to achieve this by $r/2q$.

There are a number of methods of obtaining estimates of these parameters from field data, perhaps the best of which is the fitting of curves to time series (Hilborn & Walters, 1992).

time series analysis is best – but answers are still equivocal

However, when the time series is a one-way trip, as we have noted it often is, there is no unique 'best' set of parameter values. Table 15.4, for instance, shows the parameters for three separate curves fitted to the data in Figure 15.18, providing equally good fits (the same sum of squares), but with widely differing parameter values. There are, in effect, a large number of equally good alternative explanations for the data in Figure 15.18, in some of which, for example, the population has a low carrying capacity but a high intrinsic rate of increase and is being harvested

Table 15.4 Parameter estimates from three fits to the catch per unit effort (CPUE) time series for yellowfin tuna shown in Figure 15.18. r is the intrinsic rate of increase, K is the carrying capacity (equilibrium abundance in the absence of harvesting) and q is the harvesting efficiency. Effort is measured in fishing days; K and maximum sustainable yield (MSY) in tons. (After Hilborn & Walters, 1992.)

Fit number	r	K ($\times$ 1000)	q ($\times$ 10^{-7})	MSY ($\times$ 1000)	Effort at MSY ($\times$ 1000)	Sum of squares
1	0.18	2103	9.8	98	92	3.8
2	0.15	4000	4.5	148	167	3.8
3	0.13	8000	2.1	261	310	3.8

efficiently, whereas in others it has a high carrying capacity, a low rate of increase and is being harvested less efficiently. In the first case, the MSY had probably already been reached in 1980; in the second, catches could probably be doubled with impunity. Moreover, in each of these cases, the population is assumed to be behaving in conformity with Equation 15.2, which may itself be wide of the mark.

these uncertainties make ecologists all the more valuable

It is clear, therefore, even from this limited range of rather arbitrarily chosen examples, that there are immense limitations placed on stock assessments and management plans by inadequacies in both the available data and the means of analyzing them. This is not meant, though, to be a council of despair. Management decisions must be made, and the best possible stock assessments must form the basis – although not the sole basis – for these decisions. It is regrettable that we do not know more, but the problem would be compounded if we pretended that we did. Moreover, the ecological, economic and human behavioral analyses are important – as all analyses are – for identifying what we do *not* know, since, armed with this knowledge, we can set about obtaining whatever information is most useful. This has been formalized, in fact, in an 'adaptive management' approach, where, in an 'actively adaptive' strategy, a policy is sought which offers some balance between, on the one hand, probing for information (directed experimentation), and, on the other, exercising caution about losses in short-term yield and long-term overfishing (Hilborn & Walters, 1992). Indeed, there is a strong argument that says that the inadequacies in data and theory make the need for ecologists all the more profound: who else can appreciate the uncertainties and provide appropriately enlightened interpretations?

'dataless management' where no estimates are available?

However, to be realistic, managing most marine fisheries to achieve optimum yields will be very difficult to achieve. There are generally too few researchers to do the work and, in many parts of the world, no researchers at all. In these situations, a precautionary approach to fisheries management might involve locking away a proportion of a coastal or coral community in marine protected areas (Hall, 1998). The term *dataless management* has been applied to situations where local villagers follow simple prescriptions to make sustainability more likely. For example locals on the Pacific island of Vanuatu were provided with some simple principles of management for their trochus (*Tectus niloticus*) shellfishery (stocks should be harvested every 3 years and left unfished in between) with an apparently successful outcome (Johannes, 1998).

15.4 The metapopulation perspective in management

A repeated theme in previous chapters has been the spatial patchiness upon which population interactions are often played out. Managers need to understand the implications of such heterogeneous landscape structure when making their decisions. Various approaches are available to improve our understanding of populations in complex landscapes and we consider two in the following sections. First, landscapes with different degrees of habitat loss and fragmentation can be artificially created at a scale appropriate to populations of interest and their behavior can then be assessed in carefully controlled experiments (see Section 15.4.1 – in the context of biological control of pests). Second, simple deterministic models can throw light on the factors that need to be taken into account when managing populations in a habitat patchwork (see Section 15.4.2 – in the context of creating protected areas for fisheries management). We also saw earlier (see Section 7.5.6 – in the context of a reserve patchwork for an endangered species) how stochastic simulation models can be used to compare management scenarios where subpopulations exist in a metapopulation.

15.4.1 Biological control in a fragmented landscape

We know that spatial heterogeneity can stabilize predator–prey interactions (e.g. Chapter 10). However, the dynamics of pests and their biological control agents may become destabilized, resulting in pest outbreaks, if habitat change occurs at a scale that

natural enemy success may depend on predation efficiency in a patchy habitat

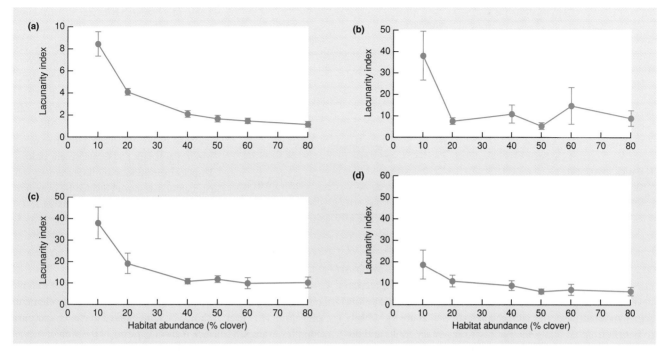

Figure 15.19   The distribution pattern (lacunarity index – a measure of aggregation) of (a) clover (i.e. habitat) and populations of (b) pest aphids, (c) an introduced ladybird beetle control agent (*Harmonia axyridis*) and (d) a native ladybird beetle (*Coleomegilla maculata*). In these experiments, clover plants were clumped together as opposed to being dispersed through the landscape. Error bars are ±1 SE. (After With *et al.*, 2002.)

interferes with the search behavior of a control agent (Kareiva, 1990).

With *et al.* (2002) created replicate landscapes (plots) of red clover (*Trifolium pratense*), each 16 × 16 m, that differed in terms of clover abundance (10, 20, 40, 50, 60 and 80% *T. pratense*). Their aim was to explore whether thresholds in landscape structure precipitate similar thresholds in the distribution of a pest aphid, *Acyrthosiphon pisum*, and to discover how landscape structure affects the search behavior of two ladybird beetle predators of aphids, one an introduced biocontrol agent, *Harmonia axyridis*, the other a native species, *Coleomegilla maculata*. Colonization by the aphids and beetles was by natural immigration to the outdoor plots.

Lacunarity is an index of aggregation derived from fractal geometry that quantifies the variability in the distribution of gap sizes (distances among clover patches in the landscape). The distribution of clover in the experimental landscapes showed a threshold at 20% habitat, indicating that gap sizes became greater and more variable below this level (Figure 15.19a). This threshold was mirrored by the aphids (Figure 15.19b) and was strongly tracked by the exotic control agent (*H. axyridis*) but not the native predator (*C. maculata*) (Figure 15.19c, d).

Although the native ladybird foraged more actively among stems within the clover cells, overall it was less mobile and moved less between clover cells in the landscape than the introduced

ladybird, which showed a greater tendency to fly (Table 15.5). With its greater mobility, the introduced species was more effective at tracking aphids when they occurred at low patch occupancy, a prerequisite for successful biological control (Murdoch & Briggs, 1996).

Findings such as these have implications both for the selection of effective biological control agents and for the design of agricultural systems, which may need to be managed to preserve habitat connectivity and thus enhance the efficiency of natural enemies and/or biological control agents (Barbosa, 1998).

15.4.2   Designing reserve networks for fisheries management

Over the last decade or so, coastal marine reserves or no-take zones have been promoted as a means of managing fisheries (e.g. Holland & Brazee, 1996). This is another example where an understanding of landscape structure, and metapopulation dynamics, will be necessary to devise management strategies. Probably the most fundamental questions of reserve design are the fraction of coastline that should be set aside and the appropriate size (and number) of reserves needed in relation to the

fishery management using no-take zones: metapopulation considerations

Scale and behavior measure	Introduced Harmonia axyridis	Native Coleomegilla maculata
Within clover cells		
Stems visited per minute	0.80 ± 0.05	1.20 ± 0.07
Between clover cells		
Cells visited per minute	0.22 ± 0.07	0.10 ± 0.04
Primary mode of movement	Fly	Crawl
Plot-wide movement		
Mean step length (m)	1.90 ± 0.21	1.10 ± 0.04
Displacement ratio	0.49 ± 0.05	0.19 ± 0.03

Table 15.5 Search behavior of introduced and native labybird beetles at different scales in experimental clover landscapes. Values are means ±1 SE. Each 16 × 16 m plot contains 256 cells (each 1 m²); clover cells are those cells in which clover was present. For individual ladybirds that made at least five cell transitions, plot-wide movements were quantified in terms of mean step length and displacement ratio. Displacement ratio is net displacement (straight-line distance) divided by overall path length. (After With *et al.*, 2002.)

dispersal potential of the target species. Hastings and Botsford (2003) developed a simple deterministic model to answer these questions for a hypothetical species with characteristics that are most likely to benefit from no-take zones: one with sedentary adults and dispersing larvae. Their approach is based on the idea that altering the spacing and width of reserves changes the fraction of larvae that are retained within or exported from reserves (Figure 15.20). It is, of course, larval export that provides the basis for a sustainable yield from nonreserve areas.

The MSY problem can be stated as 'fix the level of larval retention within reserves, F, to preserve the species, and adjust the fraction of coastline in reserves, c, to maximize the number of larvae that settle outside the reserves (available as yield)'. Note that because F remains constant (something the modelers have chosen to assume), changing c means changing the width of reserves. Suppose that a value of F of 0.35 is deemed necessary to maintain the species. The solid line in Figure 15.20b shows how c and reserve width need to change to maintain an F of 0.35. The mathematical details of the model need not concern us but it turns out that although the largest yield is obtained when the reserves are as small as possible (the arrow in Figure 15.20b), so that larval export to fished areas is maximized, the yield is only slightly reduced as the reserve configuration moves away from this optimum. Thus, Hastings and Botsford (2003) argue that practical considerations, such as making reserves large enough to be enforced, can be allowed to play a major part in reserve design, as long as reserves are not so large (beyond the 'shoulder' of the curve in Figure 15.20b) as to significantly depress yield.

Although the model is a gross simplification, particularly in terms of the lack of any uncertainty or temporal or spatial heterogeneity, it usefully highlights some general considerations of importance and provides a starting point for more sophisticated and species-specific models to address the question of whether reserve networks will be useful for fisheries management.

In each of the sections of this chapter we have sought to build on relatively simple concepts by gradually adding more elements of realism. However, it should be remembered that even our most complex examples still lack realism in terms of the web of species interactions within which the target species are embedded. In fact, many management solutions have to be focused at a higher level of ecological organization – multispecies communities and whole ecosystems. We deal with the ecology of communities and ecosystems in Chapters 16–21 before considering ecological applications at this ecological level in Chapter 22.

Summary

Sustainability is a core concept in an ever-broadening concern for the fate of the earth and the ecological communities that occupy it. In this chapter we deal with two key aspects of ecological management – the control of pests and the management of harvests of wild populations. Each depends on an understanding of population interactions (discussed in Chapters 8–14) and each has sustainability as a primary aim.

One might imagine that the goal of pest control is total eradication but this is generally restricted to cases where a new pest has invaded a region and a rapid effort is made to completely eliminate it. Usually, the aim is to reduce the pest population to a level at which it does not pay to achieve yet more control (the economic injury level or EIL). In this way, we can see that economics and sustainability are intimately tied together. When a pest population has reached a density at which it is causing economic injury, however, it is generally too late to start controlling it. More important, then, is the economic threshold (ET): the density of the pest at which action should be taken to prevent it reaching the EIL.

We describe the tool kit of chemical pesticides and herbicides. These are a key part of the armory of pest managers but they

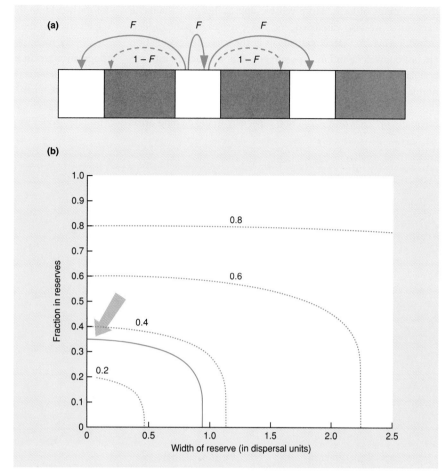

Figure 15.20   (a) Schematic representation of a network of marine reserves (white) and fished areas (gray). The fraction of coastline in the reserves is c, the fraction of larvae produced in the reserves is F, and the fraction of larvae produced in reserves that are exported is $1 - F$. (b) The combination of values of the fraction of coastline in reserves, c, and mean width of reserve (in units of mean dispersal distance) that yield a value of 0.35 for the fraction of larvae that are retained within reserves, F, along with similar combinations for other values of F. The arrow indicates the configuration that produces the maximum fishing yield outside the reserves. (After Hastings & Botsford, 2003.)

have to be used with care because of the possibility of 'target pest resurgence' (when treatment affects the pest's natural enemies more than the pest itself) and 'secondary pest outbreaks' (when natural enemies of 'potential' pests are strongly affected, allowing potential pests to become actual pests). Pests are also adept at evolving resistance to pesticides.

An alternative to chemical pesticides is biological manipulation of the natural enemies of the pests. Biological control may involve: (i) 'introduction', with the expectation of long-term persistence, of a natural enemy from another geographic area (often the one from where the pest originated); (ii) the manipulation of natural predators already present ('conservation biological control'); (iii) the periodic release of an agent that is unable to persist through the year but provides control for one or a few pest generations ('inoculation'); or (iv) the release of large numbers of enemies, which will not persist, to kill only those pests present at the time ('inundation', sometimes called, by analogy, biological pesticides). Biological control is by no means always environmentally friendly. Examples are coming to light where even carefully chosen and apparently successful introductions of

biological control agents have impacted on nontarget species, both by affecting nontarget species related to the pest and by affecting other species that interact in food webs with the nontarget species.

Integrated pest management (IPM) is a practical philosophy of pest management that is ecologically based but uses all methods of control, including chemicals, when appropriate. It relies heavily on natural mortality factors such as weather and natural enemies.

Whenever, a natural population is exploited by harvesting there is a risk of overexploitation. But harvesters also want to avoid underexploitation, where potential consumers are deprived and those who harvest are underemployed. Thus, as with many areas of applied ecology, there are important economic and sociopolitical perspectives to consider.

The concept of the maximum sustainable yield (MSY) has been a guiding principle in harvest management. We describe the different approaches to obtain an MSY – taking a fixed quota, regulating harvest effort, harvesting a constant proportion or allowing constant escapement – and we point out the shortcomings of each. More reliable approaches to sustainable harvesting are also discussed, including dynamic pool models (which recognize that all individuals in the harvested population are not equivalent and incorporate population structure into the population models) and approaches that explicitly incorporate economic factors (dealing with economically optimum yield, OEY, rather than simply MSY). We also note that no data are available for many of the world's fisheries, especially in developing areas of the world; in these cases, simple 'dataless' management principles may be the best that ecologists can propose.

Finally, many populations, including those of pests and harvested populations, exist in a heterogeneous environment, sometimes as metapopulations. Managers need to be aware of this possibility, for instance when determining which biological control agent to use in an agricultural landscape or when designing a network of 'no-take' zones as part of a fisheries management strategy.

Part 3
Communities and Ecosystems

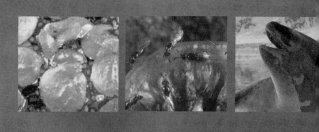

Introduction

In nature, areas of land and volumes of water contain assemblages of different species, in different proportions and doing different things. These communities of organisms have properties that are the sum of the properties of the individual denizens plus their interactions. The interactions are what make the community more than the sum of its parts. Just as it is a reasonable aim for a physiologist to study the behavior of different sorts of cells and tissues and then attempt to use a knowledge of their interactions to explain the behavior of a whole organism, so ecologists may use their knowledge of interactions between organisms in an attempt to explain the behavior and structure of a whole community. Community ecology, then, is the study of patterns in the structure and behavior of multispecies assemblages. Ecosystem ecology, on the other hand, is concerned with the structure and behavior of the same systems but with a focus on the flux of energy and matter.

We consider first the nature of the community. Community ecologists are interested in how groupings of species are distributed, and the ways these groupings can be influenced by both abiotic and biotic environmental factors. In Chapter 16 we start by explaining how the structure of communities can be measured and described, before focusing on patterns in community structure in space, in time and finally in a more complex, but more realistic spatiotemporal setting.

Communities, like all biological entities, require matter for their construction and energy for their activities. We examine the ways in which arrays of feeders and their food bind the inhabitants of a community into a web of interacting elements, through which energy (Chapter 17) and matter (Chapter 18) are moved. This ecosystem approach involves primary producers, decomposers and detritivores, a pool of dead organic matter, herbivores, carnivores and parasites *plus* the physicochemical environment that provides living conditions and acts both as a source and a sink for energy and matter. In Chapter 17, we deal with large-scale patterns in primary productivity before turning to the factors that limit productivity, and its fate, in terrestrial and aquatic settings. In Chapter 18, we consider the ways in which the biota accumulates, transforms and moves matter between the various components of the ecosystem.

In Chapter 19 we return to some key population interactions dealt with earlier in the book, and consider the ways that competition, predation and parasitism can shape communities. Then in Chapter 20 we recognize that the influence of a particular species often ramifies beyond a particular competitor, prey or host population, through the whole food web. The study of food webs lies at the interface of community and ecosystem ecology and we focus both on the population dynamics of interacting species in the community and on the consequences for ecosystem processes such as productivity and nutrient flux.

In Chapter 21 we attempt an overall synthesis of the factors, both abiotic and biotic, that determine species richness. Why the number of species varies from place to place, and from time to time, are interesting questions in their own right as well as being questions of practical importance. We will see that a full understanding of patterns in species richness has to draw on an understanding of all the ecological topics dealt with in earlier chapters of the book.

Finally, in the last of our trilogy of chapters dealing with the application of ecological theory, we consider in Chapter 22 the application of theory related to succession, food web ecology, ecosystem functioning and biodiversity. We conclude by recognizing that the application of ecological theory never proceeds in isolation – the sustainable use of natural resources requires that we also incorporate economic and sociopolitical perspectives.

To pursue an analogy we introduced earlier, the study of ecology at the community/ecosystem level is a little like making a study of watches and clocks. A collection can be made and the contents of each timepiece classified. We can recognize characteristics that they have in common in the way they are constructed and patterns in the way they behave. But to understand how they work, they must be taken to pieces, studied and put back together again. We will have understood the nature of natural communities when we *know* how to recreate those that we have, often inadvertently, taken to pieces.

Chapter 16
The Nature of the Community: Patterns in Space and Time

16.1 Introduction

Physiological and behavioral ecologists are concerned primarily with individual *organisms*. Coexisting individuals of a single species possess characteristics – such as density, sex ratio, age-class structure, rates of natality and immigration, mortality and emigration – that are unique to *populations*. We explain the behavior of a population in terms of the behavior of the individuals that comprise it. In their turn, activities at the population level have consequences for the next level up – that of the *community*. The community is an assemblage of species populations that occur together in space and time. Community ecology seeks to understand the manner in which groupings of species are distributed in nature, and the ways these groupings can be influenced by their abiotic environment (Part 1 of this textbook) and by interactions among species populations (Part 2). One challenge for community ecologists is to discern and explain patterns arising from this multitude of influences.

the search for rules of community assembly

In very general terms, the species that assemble to make up a community are determined by: (i) dispersal constraints; (ii) environmental constraints; and (iii) internal dynamics (Figure 16.1) (Belyea & Lancaster, 1999). Ecologists search for rules of community assembly, and we discuss these in this chapter and a number of others (particularly Chapters 19–21).

communities have collective properties . . .
. . . and emergent properties not possessed by the individual populations that comprise them

A community is composed of individuals and populations, and we can identify and study straightforward *collective* properties, such as species diversity and community biomass. However, we have already seen that organisms of the same and different species interact with each other in processes of mutualism, parasitism, predation and competition. The nature of the community is obviously more than just the sum of its constituent species. There are *emergent* properties that appear when the community is the focus of attention, as there are in other cases where we are concerned with the behavior of complex mixtures. A cake has emergent properties of texture and flavor that are not apparent simply from a survey of the ingredients. In the case of ecological communities, the limits to similarity of competing species (see Chapter 19) and the stability of the food web in the face of disturbance (see Chapter 20) are examples of emergent properties.

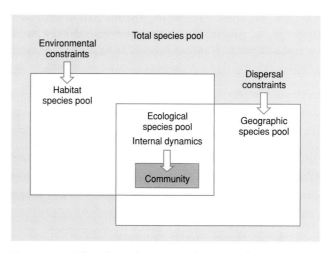

Figure 16.1 The relationships among five types of species pools: the total pool of species in a region, the geographic pool (species able to arrive at a site), the habitat pool (species able to persist under the abiotic conditions of the site), the ecological pool (the overlapping set of species that can both arrive and persist) and the community (the pool that remains in the face of biotic interactions). (Adapted from Belyea & Lancaster, 1999; Booth & Swanton, 2002.)

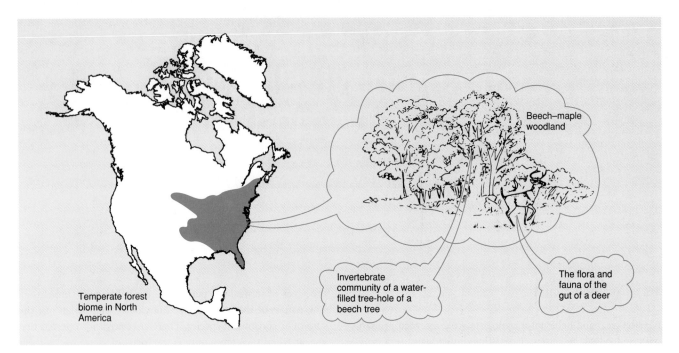

Figure 16.2 We can identify a hierarchy of habitats, nesting one into the other: a temperate forest biome in North America; a beech–maple woodland in New Jersey; a water-filled tree hole; or a mammalian gut. The ecologist may choose to study the community that exists on any of these scales.

Science at the community level poses daunting problems because the database may be enormous and complex. A first step is usually to search for patterns in the community's collective and emergent properties. Patterns are repeated consistencies, such as the repeated grouping of similar growth forms in different places, or repeated trends in species richness along different environmental gradients. Recognition of patterns leads, in turn, to the forming of hypotheses about the causes of these patterns. The hypotheses may then be tested by making further observations or by doing experiments.

A community can be defined at any scale within a hierarchy of habitats. At one extreme, broad patterns in the distribution of community types can be recognized on a global scale. The temperate forest biome is one example; its range in North America is shown in Figure 16.2. At this scale, ecologists usually recognize climate as the overwhelming factor that determines the limits of vegetation types. At a finer scale, the temperate forest biome in parts of New Jersey is represented by communities of two species of tree in particular, beech and maple, together with a very large number of other, less conspicuous species of plants, animals and microorganisms. Study of the community may be focused at this scale. On an even finer habitat scale, the characteristic invertebrate community that inhabits water-filled holes in beech trees may be studied, or the flora and fauna in the gut of a deer in the forest. Amongst these various scales of community study, no one is more legitimate than another. The scale appropriate for investigation depends on the sorts of questions that are being asked.

Community ecologists sometimes consider all of the organisms existing together in one area, although it is rarely possible to do this without a large team of taxonomists. Others restrict their attention within the community to a single taxonomic group (e.g. birds, insects or trees), or a group with a particular activity (e.g. herbivores or detritivores).

communities can be recognized at a variety of levels – all equally legitimate

The rest of this chapter is in six sections. We start by explaining how the structure of communities can be measured and described (Section 16.2). Then we focus on patterns in community structure: in space (Section 16.3), in time (Sections 16.4–16.6) and finally in a combined spatiotemporal setting (Section 16.7).

16.2 Description of community composition

One way to characterize a community is simply to count or list the species that are present. This sounds a straightforward procedure that enables us to describe and compare communities by their species 'richness' (i.e. the number of species present). In practice, though, it is often surprisingly difficult, partly because

species richness: the number of species present in a community

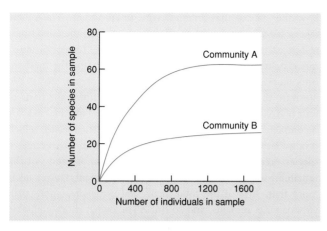

Figure 16.3 The relationship between species richness and the number of individual organisms from two contrasting hypothetical communities. Community A has a total species richness considerably in excess of community B.

of taxonomic problems, but also because only a subsample of the organisms in an area can usually be counted. The number of species recorded then depends on the number of samples that have been taken, or on the volume of the habitat that has been explored. The most common species are likely to be represented in the first few samples, and as more samples are taken, rarer species will be added to the list. At what point does one cease to take further samples? Ideally, the investigator should continue to sample until the number of species reaches a plateau (Figure 16.3). At the very least, the species richnesses of different communities should be compared on the basis of the same sample sizes (in terms of area of habitat explored, time devoted to sampling or, best of all, number of individuals or modules included in the samples). The analysis of species richness in contrasting situations figures prominently in Chapter 21.

16.2.1 Diversity indices

diversity incorporates richness, commonness and rarity

An important aspect of community structure is completely ignored, though, when the composition of the community is described simply in terms of the number of species present. It misses the information that some species are rare and others common. Consider a community of 10 species with equal numbers in each, and a second community, again consisting of 10 species, but with more than 50% of the individuals belonging to the most common species and less than 5% in each of the other nine. Each community has the same species richness, but the first, with a more 'equitable' distribution of abundances, is clearly more *diverse* than the second. Richness and equitablity combine to determine community diversity.

Knowing the numbers of individuals present in each species may not provide a full answer either. If the community is closely defined (e.g. the warbler community of a woodland), counts of the number of individuals in each species may suffice for many purposes. However, if we are interested in all the animals in the woodland, then their enormous disparity in size means that simple counts would be very misleading. There are also problems if we try to count plants (and other modular organisms). Do we count the number of shoots, leaves, stems, ramets or genets? One way round this problem is to describe the community in terms of the biomass per species per unit area.

Simpson's diversity index

The simplest measure of the character of a community that takes into account both the abundance (or biomass) patterns and the species richness, is Simpson's diversity index. This is calculated by determining, for each species, the proportion of individuals or biomass that it contributes to the total in the sample, i.e. the proportion is P_i for the ith species:

$$\text{Simpson's index, } D = \frac{1}{\displaystyle\sum_{i=1}^{S} P_i^2}, \tag{16.1}$$

where S is the total number of species in the community (i.e. the richness). As required, for a given richness, D increases with equitability, and for a given equitability, D increases with richness.

'equitability' or 'evenness'

Equitability can itself be quantified (between 0 and 1) by expressing Simpson's index, D, as a proportion of the maximum possible value D would assume if individuals were completely evenly distributed amongst the species. In fact, $D_{max} = S$. Thus:

$$\text{equitability, } E = \frac{D}{D_{max}} = \frac{1}{\displaystyle\sum_{i=1}^{S} P_i^2} \times \frac{1}{S}. \tag{16.2}$$

Shannon's diversity index

Another index that is frequently used and has essentially similar properties is the Shannon diversity index, H. This again depends on an array of P_i values. Thus:

$$\text{diversity, } H = -\sum_{i=1}^{S} P_i \ln P_i \tag{16.3}$$

and:

$$\text{equitability, } J = \frac{H}{H_{max}} = \frac{-\displaystyle\sum_{i=1}^{S} P_i \ln P_i}{\ln S}. \tag{16.4}$$

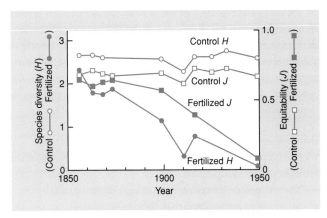

Figure 16.4 Species diversity (*H*) and equitability (*J*) of a control plot and a fertilized plot in the Rothamsteard 'Parkgrass' experiment. (After Tilman, 1982.)

An example of an analysis of diversity is provided by a uniquely long-term study that has been running since 1856 in an area of grassland at Rothamsted in England. Experimental plots have received a fertilizer treatment once every year, whilst control plots have not. Figure 16.4 shows how species diversity (*H*) and equitability (*J*) of the grass species changed between 1856 and 1949. Whilst the unfertilized area has remained essentially unchanged, the fertilized area has shown a progressive decline in diversity and equitability. One possible explanation may be that high nutrient availability leads to high rates of population growth and a greater chance of the most productive species coming to dominate and, perhaps, competitively exclude others.

16.2.2   Rank–abundance diagrams

Of course, attempts to describe a complex community structure by one single attribute, such as richness, diversity or equitability, can be criticized because so much valuable information is lost. A more complete picture of the distribution of species abundances in a community makes use of the full array of P_i values by plotting P_i against rank. Thus, the P_i for the most abundant species is plotted first, then the next most common, and so on until the array is completed by the rarest species of all. A rank–abundance diagram can be drawn for the number of individuals, or for the area of ground covered by different sessile species, or for the biomass contributed to a community by the various species.

rank–abundance models may be based on statistical or biological arguments

A range of the many equations that have been fitted to rank–abundance diagrams is shown in Figure 16.5. Two of these are statistical in origin (the log series and log-normal) with no foundation in any assumptions about

how the species may interact with one another. The others take some account of the relationships between the conditions, resources and species-abundance patterns (niche-orientated models) and are more likely to help us understand the mechanisms underlying community organization (Tokeshi, 1993). We illustrate the diversity of approaches by describing the basis of four of Tokeshi's niche-orientated models (see Tokeshi, 1993, for a complete treatment). The *dominance–preemption model*, which produces the least equitable species distribution, has successive species preempting a dominant portion (50% or more) of the remaining niche space; the first, most dominant species takes more than 50% of the total niche space, the next more than 50% of what remains, and so on. A somewhat more equitable distribution is represented by the *random fraction model*, in which successive species invade and take over an arbitrary portion of the niche space of any species previously present. In this case, irrespective of their dominance status, all species are subjected to niche division with equal probability. The *MacArthur fraction model*, on the other hand, assumes that species with larger niches are more likely to be invaded by new species; this results in a more equitable distribution than the random fraction model. Finally, the *dominance–decay model* is the inverse of the dominance–preemption model, in that the largest niche in an existing assemblage is always subject to a subsequent (random) division. Thus, in this model the next invading species is supposed to colonize the niche space of the species currently most abundant, yielding the most equitable species abundances of all the models.

Rank–abundance diagrams, like indices of richness, diversity and equitability, should be viewed as abstractions of the highly complex structure of communities that may be useful when

community indices are abstractions that may be useful when making comparisons

making comparisons. In principle, the idea is that finding the best fitting model should give us clues as to underlying processes, and perhaps as to how these vary from sample to sample. Progress so far, however, has been limited, both because of problems of interpretation and the practical difficulty of testing for the best fit between model and data (Tokeshi, 1993). However, some studies have successfully focused attention on a change in dominance/evenness relationships in relation to environmental change. Figure 16.5c shows how, assuming a geometric series can be appropriately applied, dominance steadily increased, whilst species richness decreased, during the Rothamsted long-term grassland experiment described above. Figure 16.5d shows how invertebrate species richness and equitability were both greater on an architecturally complex stream plant *Ranunculus yezoensis*, which provides more potential niches, than on a structurally simple plant *Sparganium emersum*. The rank–abundance diagrams of both are closer to the random fraction model than the MacArthur fraction model. Finally, Figure 16.5e shows how attached bacterial assemblages (biofilms), during colonization of

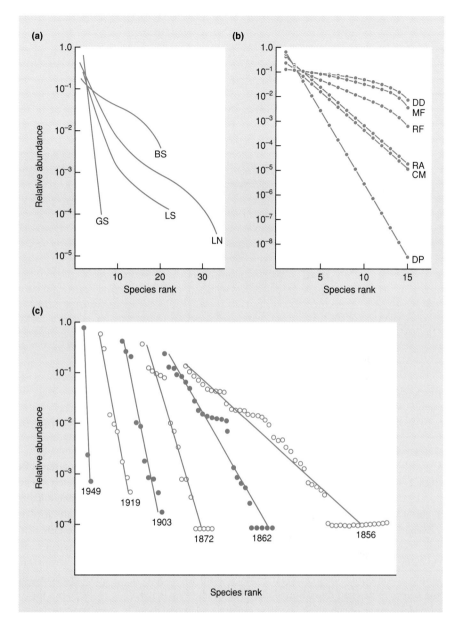

Figure 16.5 (a, b) Rank–abundance patterns of various models. Two are statistically orientated (LS and LN), whilst the rest can be described as niche orientated. (a) BS, broken stick; GS, geometric series; LN, log-normal; LS, log series. (b) CM, composite; DD, dominance decay; DP, dominance preempt; MF, MacArthur fraction; RA, random assortment; RF, random fraction. (c) Change in the relative abundance pattern (geometric series fitted) of plant species in an experimental grassland subjected to continuous fertilizer from 1856 to 1949. ((a–c) after Tokeshi, 1993.)

glass slides in a lake, change from a log-normal to a geometric pattern as the biofilm ages.

the energetics approach: an alternative to taxonomic description

Taxonomic composition and species diversity are just two of many possible ways of describing a community. Another alternative (not necessarily better but quite different) is to describe communities and ecosystems in terms of their standing crop and the rate of production of biomass by plants, and its use and conversion by heterotrophic microorganisms and animals. Studies that are orientated in this way may begin by describing the food web, and

then define the biomasses at each trophic level and the flow of energy and matter from the physical environment through the living organisms and back to the physical environment. Such an approach can allow patterns to be detected amongst communities and ecosystems that may have no taxonomic features in common. This approach will be discussed in Chapters 17 and 18.

Much recent research effort has been devoted to understanding the link between species richness and ecosystem functioning (productivity, decomposition and nutrient dynamics). Understanding the role of species richness in ecosystem processes has particular significance for how humans respond to biodiversity loss. We discuss this important topic in Section 21.7.

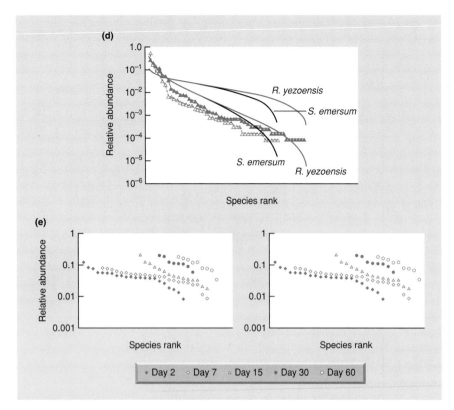

Figure 16.5 (*cont'd*) (d) Comparison of rank–abundance patterns for invertebrate species living on a structurally complex stream plant *Ranunculus yezoensis* (▲) and a simple plant *Sparganium emersum* (△); fitted lines represent the MacArthur fraction model (——, the upper one for *R. yezoensis* and the lower one for *S. emersum*) and the random fraction model (——, the upper one for *R. yezoensis* and the lower one for *S. emersum*). (After Taniguchi *et al.*, 2003.) (e) Rank–abundance patterns (based on a biomass index) for bacterial assemblages in lake biofilms of different ages (symbols from left to right represent days 2, 7, 15, 30, 60). (After Jackson *et al.*, 2001.)

16.3   Community patterns in space

16.3.1   Gradient analysis

Figure 16.6 shows a variety of ways of describing the distribution of vegetation used in a classic study in the Great Smoky Mountains (Tennessee), USA, where tree species give the vegetation its main character. Figure 16.6a shows the characteristic associations of the dominant trees on the mountainside, drawn as if the communities had sharp boundaries. The mountainside itself provides a range of conditions for plant growth, and two of these, altitude and moisture, may be particularly important in determining the distribution of the various tree species. Figure 16.6b shows the dominant associations graphed in terms of these two environmental dimensions. Finally, Figure 16.6c shows the abundance of each individual tree species (expressed as a percentage of all tree stems present) plotted against the single gradient of moisture.

species distributions along gradients end not with a bang but with a whimper

Figure 16.6a is a subjective analysis that acknowledges that the vegetation of particular areas differs in a characteristic way from that of other areas. It could be taken to imply that the various communities are sharply delimited. Figure 16.6b gives the same impression. Note that both Figure 16.6a and b are based on descriptions of the *vegetation*.

However, Figure 16.6c sharpens the focus by concentrating on the pattern of distribution of the individual *species*. It is then immediately obvious that there is considerable overlap in their abundance – there are no sharp boundaries. The various tree species are now revealed as being strung out along the gradient with the tails of their distributions overlapping. The results of this 'gradient analysis' show that the limits of the distributions of each species 'end not with a bang but with a whimper'. Many other gradient studies have produced similar results.

Perhaps the major criticism of gradient analysis as a way of detecting pattern in communities is that the choice of the gradient is almost always subjective. The investigator searches for some feature of the environment that appears to matter to the organisms and then organizes the data about the species concerned along a gradient of that factor. It is not necessarily the most appropriate factor to have chosen. The fact that the species from a community can be arranged in a sequence along a gradient of some environmental factor does not prove that this factor is the most important one. It may only imply that the factor chosen is more or less loosely correlated with whatever really matters in the lives of the species involved. Gradient analysis is only a small step on the way to the objective description of communities.

choice of gradient is almost always subjective

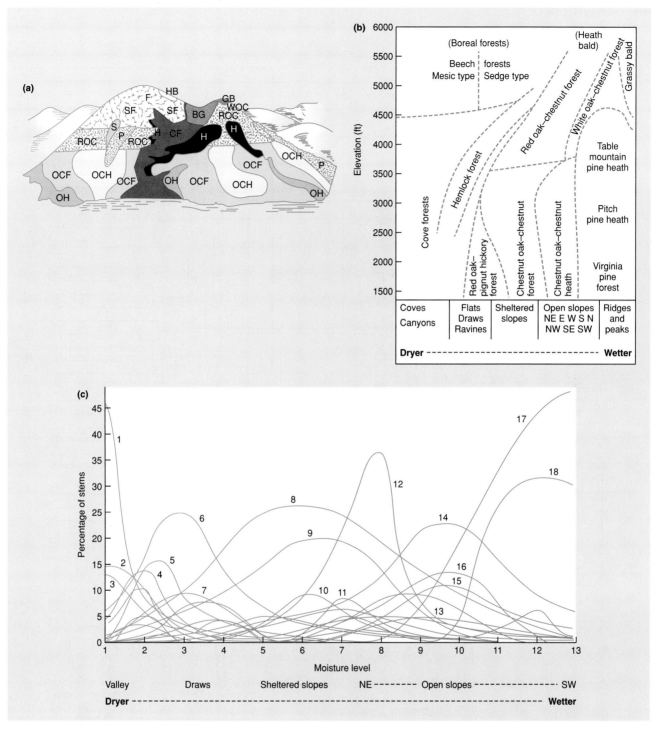

Figure 16.6 Three contrasting descriptions of distributions of the characteristic dominant tree species of the Great Smoky Mountains, Tennessee. (a) Topographic distribution of vegetation types on an idealized west-facing mountain and valley. (b) Idealized graphic arrangement of vegetation types according to elevation and aspect. (c) Distributions of individual tree populations (percentage of stems present) along the moisture gradient. Vegetation types: BG, beech gap; CF, cove forest; F, Fraser fir forest; GB, grassy bald; H, hemlock forest; HB, heath bald; OCF, chestnut oak–chestnut forest; OCH, chestnut oak–chestnut heath; OH, oak–hickory; P, pine forest and heath; ROC, red oak–chestnut forest; S, spruce forest; SF, spruce–fir forest; WOC, white oak–chestnut forest. Major species: 1, *Halesia monticola*; 2, *Aesculus octandra*; 3, *Tilia heterophylla*; 4, *Betula alleghaniensis*; 5, *Liriodendron tulipifera*; 6, *Tsuga canadensis*; 7, *B. lenta*; 8, *Acer rubrum*; 9, *Cornus florida*; 10, *Carya alba*; 11, *Hamamelis virginiana*; 12, *Quercus montana*; 13, *Q. alba*; 14, *Oxydendrum arboreum*; 15, *Pinus strobus*; 16, *Q. coccinea*; 17, *P. virginiana*; 18, *P. rigida*. (After Whittaker, 1956.)

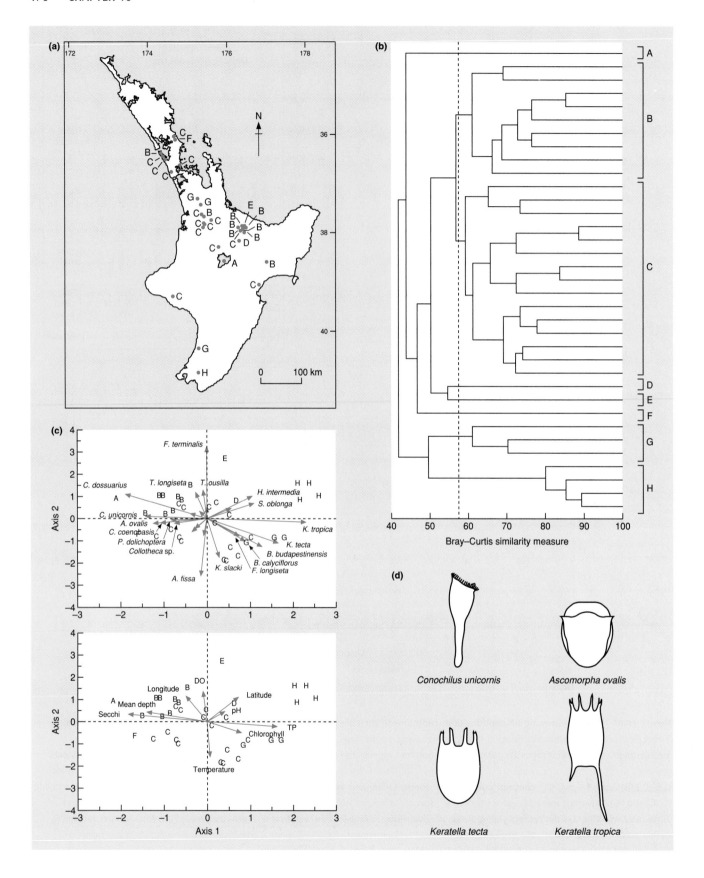

16.3.2 Classification and ordination of communities

Formal statistical techniques have been defined to take the subjectivity out of community description. These techniques allow the data from community studies to sort themselves, without the investigator putting in any preconceived ideas about which species tend to be associated with each other or which environmental variables correlate most strongly with the species distributions. One such technique is classification.

classification involves grouping similar communities together in clusters

Classification begins with the assumption that communities consist of relatively discrete entities. It produces groups of related communities by a process conceptually similar to taxonomic classification. In taxonomy, similar individuals are grouped together in species, similar species in genera, and so on. In community classification, communities with similar species compositions are grouped together in subsets, and similar subsets may be further combined if desired (see Ter Braak & Prentice, 1988, for details of the procedure).

The rotifer communities of a number of lakes in the North Island of New Zealand (Figure 16.7a) were subjected to a classification technique called cluster analysis (Duggan *et al.*, 2002). Eight clusters or classes were identified (Figure 16.7b), each based solely on the arrays of species present and their abundances. The spatial distribution of each class of rotifer community in the New Zealand lakes is shown in Figure 16.7a. Note that there is little consistent spatial relationship; communities in each class are dotted about the island. This illustrates one of the strengths of classification. Classification methods show the structure within a series of communities without the necessity of picking out some supposedly relevant environmental variable in advance, a procedure that is necessary for gradient analysis.

in ordination, communities are displayed on a graph so that those most similar in composition are closest together

Ordination is a mathematical treatment that allows communities to be organized on a graph so that those that are most similar in both species composition and relative abundance will appear closest together, whilst communities that differ greatly in the relative importance of a similar set of species, or that possess quite different species, appear far apart. Figure 16.7c shows the application of an ordination technique called canonical correspondence analysis (CCA) to the rotifer communities (Ter Braak & Smilauer 1998). CCA also allows the community patterns to be examined in terms of environmental variables. Obviously, the success of the method now depends on having sampled an appropriate variety of environmental variables. This is a major snag in the procedure – we may not have measured the qualities in the environment that are most relevant. The relationships between rotifer community composition and a variety of physicochemical factors are shown in Figure 16.7c. The link between classification and ordination can be gauged by noting that communities falling into classes A–H, derived from classification, are also fairly distinctly separated on the CCA ordination graph.

Community classes A and B tend to be associated with high water transparency ('Secchi depth'), whereas those in classes G and H are associated with high total phosphorus and chlorophyll

subsequently, it is necessary to ask what varies along the axes of the graph

concentrations; the other lake classes take up intermediate positions. Lakes that have been subject to a greater level of runoff of agricultural fertilizers or input of sewage are described as eutrophic. These tend to have high phosphorus concentrations, leading to higher chlorophyll levels and lower transparency (a greater abundance of phytoplankton cells). Evidently, the rotifer communities are strongly influenced by the level of eutrophication to which the lakes are subject. Species of rotifer that are characteristic of particularly eutrophic conditions, such as *Keratella tecta* and *K. tropica* (Figure 16.7d), were strongly represented in classes G and H, while those associated with more pristine conditions, such as *Conochilus unicornis* and *Ascomorpha ovalis*, were common in classes A and B.

The level of eutrophication, however, is not the only significant factor in explaining rotifer community composition. Class C communities, for example, while characteristic of intermediate phosphorus concentrations, can be differentiated along axis 2 according to dissolved oxygen concentration and lake temperature (themselves negatively related because oxygen solubility declines with increasing temperature).

What do these results tell us? First, and most specifically, the correlations with environmental factors, revealed by the analysis, give us some specific hypotheses to test about the relationship between community composition and underlying environmental factors. (Remember that correlation does not necessarily imply

ordination can generate hypotheses for subsequent testing

Figure 16.7 (*opposite*) (a) Thirty-one lakes in the North Island of New Zealand where rotifer communities (78 species in total) were sampled and described. (b) Results of cluster analysis (classification) on species composition data from the 31 lakes (based on the Bray–Curtis similarity measure); lake communities that are most similar cluster together and eight clusters are identified (A–H). (c) Results of canonical correspondence analysis (ordination). The positions in ordination space are shown for lake sites (shown as letters A–H corresponding to their classification), individual rotifer species (orange arrows in top panel) and environmental factors (orange arrows in lower panel). (d) Silhouettes of four of the rotifer species. (After Duggan *et al.*, 2002.)

causation. For example, dissolved oxygen and community composition may vary together because of a common response to another environmental factor. A direct causal link can only be proved by controlled experimentation.)

A second, more general point is relevant to the discussion of the nature of the community. The results emphasize that under a particular set of environmental conditions, a predictable association of species is likely to occur. It shows that community ecologists have more than just a totally arbitrary and ill-defined set of species to study.

16.3.3   Problems of boundaries in community ecology

are communities discrete entities with sharp boundaries?

There may be communities that are separated by clear, sharp boundaries, where groups of species lie adjacent to, but do not intergrade into, each other. If they exist, they are exceptional. The meeting of terrestrial and aquatic environments might appear to be a sharp boundary but its ecological unreality is emphasized by the otters or frogs that regularly cross it and the many aquatic insects that spend their larval lives in the water but their adult lives as winged stages on land or in the air. On land, quite sharp boundaries occur between the vegetation types on acidic and basic rocks where outcrops meet, or where serpentine (a term applied to a mineral rich in magnesium silicate) and nonserpentine rocks are juxtaposed. However, even in such situations, minerals are leached across the boundaries, which become increasingly blurred. The safest statement we can make about community boundaries is probably that they do not exist, but that some communities are much more sharply defined than others. The ecologist is usually better employed looking at the ways in which communities grade into each other, than in searching for sharp cartographic boundaries.

the community: not so much a superorganism . . .

In the first quarter of the 20th century there was considerable debate about the nature of the community. Clements (1916) conceived of the community as a sort of *superorganism* whose member species were tightly bound together both now and in their common evolutionary history. Thus, individuals, populations and communities bore a relationship to each other resembling that between cells, tissues and organisms.

In contrast, the *individualistic* concept devised by Gleason (1926) and others saw the relationship of coexisting species as simply the results of similarities in their requirements and tolerances (and partly the result of chance). Taking this view, community boundaries need not be sharp, and associations of species would be much less predictable than one would expect from the superorganism concept.

The current view is close to the individualistic concept. Results of direct gradient analysis, ordination and classification all indicate

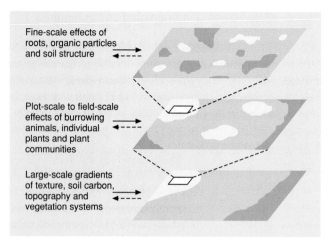

Figure 16.8   Determinants of spatial heterogeneity of communities of soil organisms including bacteria, fungi, nematodes, mites and collembolans. (After Ettema & Wardle, 2002.)

that a given location, by virtue mainly of its physical characteristics, possesses a reasonably predictable association of species. However, a given species that occurs in one predictable association is also quite likely to occur with another group of species under different conditions elsewhere.

A further point needs to be born in mind when considering the question of environmental patchiness and boundaries. Spatial heterogeneity in the distribution of communities can be viewed within a series of nested scales. Figure 16.8, for example, shows patterns in spatial heterogeneity in communities of soil organisms operating at scales from hectares to square millimeters (Ettema & Wardle, 2002). At the largest scale, these reflect patterns in environmental factors related to topography and the distribution of different plant communities. But at the other extreme, fine-scale patterns may be present as a result of the location of individual plant roots or local soil structure. The boundaries of patterns at these various scale are also likely to be blurred.

Whether or not communities have more or less clear boundaries is an important question, but it is not the fundamental consideration. Community ecology is the study of the *community level of organization* rather than of a spatially and temporally definable unit. It is concerned with the structure and activities of the multispecies assemblage, usually at one point in space and time. It is not necessary to have discrete boundaries between communities to study community ecology.

. . . more a level of organization

16.4   Community patterns in time

Just as the relative importance of species varies in space, so their patterns of abundance may change with time. In either case, a species will occur only where and when: (i) it is capable of reaching a location; (ii) appropriate conditions and resources exist there; and (iii) competitors, predators and parasites do not preclude it. A temporal sequence in the appearance and disappearance of species therefore seems to require that conditions, resources and/or the influence of enemies themselves vary with time.

For many organisms, and particularly short-lived ones, their relative importance in the community changes with time of year as the individuals act out their life cycles against a background of seasonal change. Sometimes community composition shifts because of externally driven physical change, such as the build up of silt in a coastal salt marsh leading to its replacement by forest. In other cases, temporal patterns are simply a reflection of changes in key resources, as in the sequence of heterotrophic organisms associated with fecal deposits or dead bodies as they decompose (see Figure 11.2). The explanation for such temporal patterns is relatively straightforward and will not concern us here. Nor will we dwell on the variations in abundance of species in a community from year to year as individual populations respond to a multitude of factors that influence their reproduction and survival (dealt with in Chapters 5, 6 and 8–14).

Our focus will be on patterns of community change that follow a disturbance, defined as a relatively discrete event that removes organisms (Townsend & Hildrew, 1994) or otherwise disrupts the community by influencing the availability of space or food resources, or by changing the physical environment (Pickett & White, 1985). Such disturbances are common in all kinds of community. In forests, they may be caused by high winds, lightning, earthquakes, elephants, lumberjacks or simply by the death of a tree through disease or old age. Agents of disturbance in grassland include frost, burrowing animals and the teeth, feet, dung or dead bodies of grazers. On rocky shores or coral reefs, disturbances may result from severe wave action during hurricanes, tidal waves, battering by logs or moored boats or the fins of careless scuba divers.

16.4.1 Founder-controlled and dominance-controlled communities

founder control: many species are equivalent in their ability to colonize

In response to disturbances, we can postulate two fundamentally different kinds of community response according to the type of competitive relationships exhibited by the component species – founder controlled and dominance controlled (Yodzis, 1986). *Founder-controlled* communities will occur if a large number of species are approximately equivalent in their ability to colonize an opening left by a disturbance, are equally well fitted to the abiotic environment and can hold the location until they die. In this case, the result of the disturbance is essentially a lottery. The winner is the species that happens to reach and establish itself in the disturbed

location first. The dynamics of founder-controlled communities are discussed in Section 16.7.4.

Dominance-controlled communities are those where some species are competitively superior to others so that an initial colonizer of an opening left by a disturbance cannot necessarily maintain its presence there. In these cases,

dominance control: some potential colonizers are competitively dominant

disturbances lead to reasonably predictable sequences of species because different species have different strategies for exploiting resources – early species are good colonizers and fast growers, whereas later species can tolerate lower resource levels and grow to maturity in the presence of early species, eventually outcompeting them. These situations are more commonly known by the term *ecological succession*, defined as the *nonseasonal, directional and continuous pattern of colonization and extinction on a site by species populations*.

16.4.2 Primary and secondary successions

Our focus is on successional patterns that occur on newly exposed landforms. If the exposed landform has not previously been influenced by a community, the sequence of species is referred to as a

primary succession: an exposed landform uninfluenced by a previous community

primary succession. Lava flows and pumice plains caused by volcanic eruptions (see Section 16.4.3), craters caused by the impact of meteors (Cockell & Lee, 2002), substrate exposed by the retreat of a glacier (Crocker & Major, 1955) and freshly formed sand dunes (see Section 16.4.4) are examples. In cases where the vegetation of an area has been partially or completely removed, but where well-developed soil and seeds and spores remain, the subsequent sequence of species is termed a secondary succession. The loss of trees locally as a result of disease, high winds, fire or felling may lead to secondary successions, as can cultivation followed by the abandonment of farmland (so-called old field successions – see Section 16.4.5).

Successions on newly exposed landforms typically take several hundreds of years to run their course. However, a precisely analagous process occurs amongst the animals and algae on

secondary succession: vestiges of a previous community are still present

recently denuded rock walls in the marine subtidal zone, and this succession takes only a decade or so (Hill *et al.*, 2002). The research life of an ecologist is sufficient to encompass a subtidal succession but not that following glacial retreat. Fortunately, however, information can sometimes be gained over the longer timescale. Often, successional stages in time are represented by community gradients in space. The use of historic maps, carbon dating or other techniques may enable the age of a community since exposure of the landform to be estimated. A series of communities currently in existence, but corresponding to different lengths of time since the onset of succession, can be inferred to reflect succession. However, whether or not different communities that are spread out in space really do represent various stages of succession must be judged with caution. We must remember, for example, that in northern temperate areas the vegetation we see may still be undergoing recolonization and responding to climatic change following the last ice age (see Chapter 1).

16.4.3   Primary succession on volcanic lava

facilitation: early successional species on volcanic lava pave the way for later ones

A primary succession on basaltic volcanic flows on Miyake-jima Island, Japan, was inferred from a known chronosequence (16, 37, 125 and >800 years old) (Figure 16.9a). In the 16-year-old flow, soil was very sparse and lacking in nitrogen; vegetation was absent except for a few small alder trees (*Alnus sieboldiana*). In the older plots, 113 taxa were recorded, including ferns, herbaceous perennials, lianas and trees. Of most significance in this primary succession were: (i) the successful colonization of the bare lava by the nitrogen-fixing alder; (ii) the facilitation (through improved nitrogen availability) of mid-successional *Prunus speciosa* and the late successional evergreen tree *Machilus thunbergii*; (iii) the formation of a mixed forest and the shading out of *Alnus* and *Prunus*; and (iv) finally, the replacement of *Machilus* by the longer lived *Castanopsis sieboldii* (Figure 16.9b).

16.4.4   Primary succession on coastal sand dunes

An extensive chronosequence of dune-capped beach ridges has been undertaken on the coast of Lake Michigan in the USA. Thirteen ridges of known

importance of seed availability rather than facilitation in sand dune succession

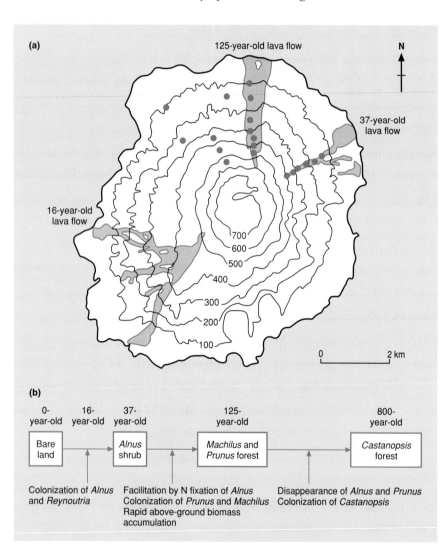

Figure 16.9   (a) Vegetation was described on 16-, 37- and 125-year-old lava flows on Miyake-jima Island, Japan. Analysis of the 16-year-old flow was nonquantitative (no sample sites shown). Sample sites on the other flows are shown as solid circles. Sites outside the three flows are at least 800 years old. (b) The main features of the primary succession in relation to lava age. (After Kamijo *et al.*, 2002.)

age (30–440 years old) show a clear pattern of primary succession to forest (Lichter, 2000). The dune grass *Ammophila breviligulata* dominates the youngest, still mobile dune ridge, but shrubby *Prunus pumila* and *Salix* spp. are also present. Within 100 years, these are replaced by evergreen shrubs such as *Juniperus communis* and by prairie bunch grass *Schizachrium scoparium*. Conifers such as *Pinus* spp., *Larix laricina*, *Picea strobus* and *Thuja occidentalis* begin colonizing the dune ridges after 150 years, and a mixed forest of *Pinus strobus* and *P. resinosa* develops between 225 and 400 years. Deciduous trees such as the oak *Quercus rubra* and the maple *Acer rubrum* do not become important components of the forest until 440 years.

It used to be thought that early successional dune species facilitated the later species by adding organic matter to the soil and increasing the availability of soil moisture and nitrogen (as in the volcanic primary succession). However, experimental seed addition and seedling transplant experiments have shown that later species are capable of germinating in young dunes (Figure 16.10a). While the more developed soil of older dunes may improve the performance of late successional species, their successful colonization of young dunes is mainly constrained by limited seed dispersal, together with seed predation by rodents (Figure 16.10b). *Ammophila* generally colonizes young, active dunes through horizontal vegetative growth. *Schizachrium*, one of the

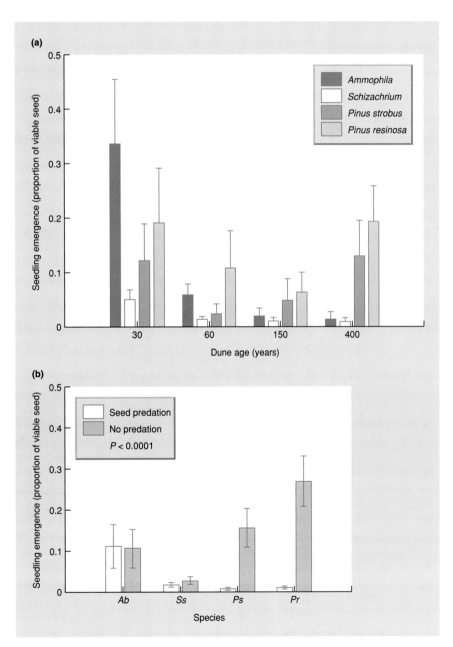

Figure 16.10 (a) Seedling emergence (means + SE) from added seeds of species typical of different successional stages on dunes of four ages. (b) Seedling emergence of the four species (Ab, *Ammophila breviligulata*, Ss, *Schizachrium scoparium*, Ps, *Pinus strobus*, Pr, *Pinus resinosa*) in the presence and absence of rodent predators of seeds (After Lichter, 2000.)

dominants of open dunes before forest development, has rates of germination and seedling establishment that are no better than *Pinus*, but its seeds are not preyed upon. Also, *Schizachrium* has the advantage of quickly reaching maturity and can continue to provide seeds at a high rate. These early species are eventually competitively excluded as trees establish and grow. Lichter (2000) considers that dune succession is better described in terms of the transient dynamics of colonization and competitive displacement, rather than the result of facilitation by early species (improving soil conditions) followed by competitive displacement.

16.4.5   Secondary successions in abandoned fields

abandoned old fields: succession to forest in North America . . .

Successions on old fields have been studied particularly along the eastern part of the USA where many farms were abandoned by farmers who moved west after the frontier was opened up in the 19th century (Tilman, 1987, 1988). Most of the precolonial mixed conifer–hardwood forest had been destroyed, but regeneration was swift. In many places, a series of sites that were abandoned for different, recorded periods of time are available for study. The typical sequence of dominant vegetation is: annual weeds, herbaceous perennials, shrubs, early successional trees and late successional trees.

. . . but to grassland in China

Old-field succession has also been studied in the productive Loess Plateau in China, which for millennia has been affected by human activities so that few areas of natural vegetation remain. The Chinese government has launched some conservation projects focused on the recovery of damaged ecosystems. A big question mark is whether the climax vegetation of the Plateau will prove to be grassland steppe or forest. Wang (2002) studied the vegetation at four plots abandoned by farmers for known periods of time (3, 26, 46 and 149 years). He was able to age some of his plots in an unusual manner. Graveyards in China are sacred and human activities are prohibited in their vicinity – gravestone records indicated how long ago the older areas had been taken out of agricultural production. Of a total of 40 plant species identified, several were considered dominant at the four successional stages (in terms of relative abundance and relative ground cover). In the first stage (recently abandoned farmland) *Artemesia scoparia* and *Seraria viridis* were most characteristic, at 26 years *Lespedeza davurica* and *S. viridis* dominated, at 46 years *Stipa bungeana*, *Bothriochloa ischaemun*, *A. gmelinii* and *L. davurica* were most important, while at 149 years *B. ischaemun* and *A. gmelinii* were dominant (Figure 16.11). The early successional species were annuals and biennials with high seed production. By 26 years, the perennial herb *L. davurica*, with its ability to spread laterally by vegetative means and a well-developed root system, had replaced *A. scoparia*.

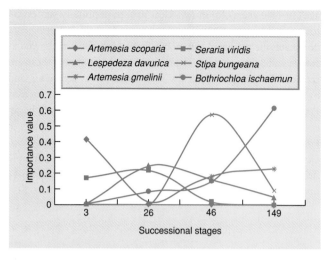

Figure 16.11   Variation in the relative importance of six species during an old-field succession on the Loess Plateau in China. (After Wang, 2002.)

The 46-year-old plot was characterized by the highest species richness and diverse life history strategies, dominated by perennial lifestyles. The dominance of *B. ischaemun* at 149 years was related to its perennial nature, ability to spread clonally and high competitive ability. As in Tilman's (1987, 1988) North American studies, soil nitrogen content increased during the succession and may have facilitated some species in the succession. Wang concludes that the grass *B. ischaemun* is the characteristic climax species in this Loess Plateau habitat, and thus the vegetation seems likely to succeed to steppe grassland rather than forest.

16.5   Species replacement probabilities during successions

forest succession can be represented as a tree-by-tree replacement model . . .

A model of succession developed by Horn (1981) sheds some light on the successional process. Horn recognized that in a hypothetical forest community it would be possible to predict changes in tree species composition given two things. First, one would need to know for each tree species the probability that, within a particular time interval, an individual would be replaced by another of the same species or of a different species. Second, an initial species composition would have to be assumed.

Horn considered that the proportional representation of various species of saplings established beneath an adult tree reflected the probability of an individual tree's replacement by each of those species. Using this information, he estimated the probability, after 50 years, that a site now occupied by a given species will be taken over by another species or will still be occupied by

Table 16.1 A 50-year tree-by-tree transition matrix from Horn (1981). The table shows the probability of replacement of one individual by another of the same or different species 50 years hence.

| | Occupant 50 years hence | | | |
Present occupant	Grey birch	Blackgum	Red maple	Beech
Grey birch	0.05	0.36	0.50	0.09
Blackgum	0.01	0.57	0.25	0.17
Red maple	0.0	0.14	0.55	0.31
Beech	0.0	0.01	0.03	0.96

Table 16.2 The predicted percentage composition of a forest consisting initially of 100% grey birch. (After Horn, 1981.)

| | Age of forest (years) | | | | | | |
Species	0	50	100	150	200	∞	Data from old forest
Grey birch	100	5	1	0	0	0	0
Blackgum	0	36	29	23	18	5	3
Red maple	0	50	39	30	24	9	4
Beech	0	9	31	47	58	86	93

the same species (Table 16.1). Thus, for example, there is a 5% chance that a location now occupied by grey birch will still support grey birch in 50 years' time, whereas there is a 36% chance that blackgum will take over, a 50% chance for red maple and 9% for beech.

Beginning with an observed distribution of the canopy species in a stand in New Jersey in the USA known to be 25 years old, Horn modeled the changes in species composition over several centuries. The process is illustrated in simplified form in Table 16.2 (which deals with only four species out of those present). The progress of this hypothetical succession allows several predictions to be made. Red maple should dominate quickly, whilst grey birch disappears. Beech should slowly increase to predominate later, with blackgum and red maple persisting at low abundance. All these predictions are borne out by what happens in the real succession (final column).

... that predicts a stable species composition and the time taken to reach it

The most interesting feature of Horn's so-called Markov chain model is that, given enough time, it converges on a stationary, stable composition that is independent of the initial composition of the forest. The outcome is inevitable (it depends only on the matrix of replacement probabilities) and will be achieved whether the starting point is 100% grey birch or 100% beech, 50% blackgum and 50% red maple, or any other combination (as long as adjacent areas provide a source of seeds of species not initially present). Korotkov et al. (2001) have used a similar Markov modeling approach to predict the time it should take to reach the climax state from any other stage in old-field successions culminating in mixed conifer–broadleaf forest in central Russia. From field abandonment to climax is predicted to take 480–540 years, whereas a mid-successional stage of birch forest with spruce undergrowth should take 320–370 years to reach the climax.

Since Markov models seem to be capable of generating quite accurate predictions, they may prove to be a useful tool in formulating plans for forest management. However, the models are simplistic and the assumption that transition probabilities remain constant in space and over time and are not affected by historic factors, such as initial biotic conditions and the order of arrival of species, are likely to be wrong in many cases (Facelli & Pickett, 1990). Hill et al. (2002) addressed the question of spatiotemporal variation in species replacement probabilities in a subtidal community succession including sponges, sea anemones, polychaetes and encrusting algae. In this case, the predicted successions and endpoints were similar whether replacement probabilities were averaged or were subject to realistic spatial or temporal variation. And the outcomes of all three models were very similar to the observed community structure (Figure 16.12).

16.6 Biological mechanisms underlying successions

Despite the advantages of simple Markov models, a theory of succession should ideally not only predict but also explain. To do this, we need to consider the *biological* basis for the replacement values in the model, and here we have to turn to alternative approaches.

an ideal theory of succession should predict *and* explain

16.6.1 Competition–colonization trade-off and successional niche mechanisms

Rees et al. (2001) drew together a diversity of experimental, comparative and theoretical approaches to produce some generalizations about vegetation dynamics. Early successional plants have a series of correlated traits, including high fecundity, effective dispersal, rapid growth when resources are abundant, and poor growth and survival when resources are scarce. Late successional species usually have the opposite traits, including an ability to grow, survive and compete when resources are scarce. In the absence of disturbance,

a trade-off between colonization and competitive ability?

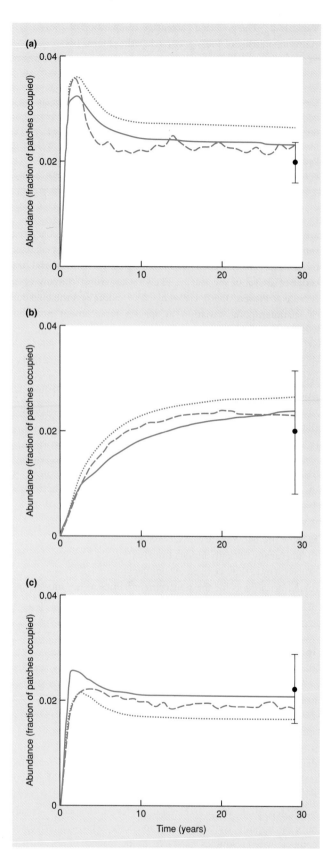

late successional species eventually outcompete early species, because they reduce resources beneath the levels required by the early successional species. Early species persist for two reasons: (i) because their dispersal ability and high fecundity permits them to colonize and establish in recently disturbed sites before late successional species can arrive; or (ii) because rapid growth under resource-rich conditions allows them to temporarily outcompete late successional species even if they arrive at the same time. Rees and his colleagues refer to the first mechanism as a *competition–colonization trade-off* and the second as the *successional niche* (early conditions suit early species because of their niche requirements). The competition–colonization trade-off is strengthened by a further physiological inevitability. Huge differences in per capita seed production among plant species are inversely correlated to equally large variations in seed size; plants producing tiny seeds tend to produce many more of them than plants producing large seeds (see Section 4.8.5). Thus, Rees *et al.* (2001) point out that small-seeded species are good colonists (many propagules) but poor competitors (small seed food reserves), and vice versa for large-seeded species.

16.6.2 Facilitation

Cases of competition–colonization trade-offs and/or successional niche relations are prominent in virtually every succession that has been described, including all those in the previous section. In addition, we have seen cases where early species may change the abiotic environment in ways (e.g. increased soil nitrogen) that make it easier for later species to establish and thrive. Thus, *facilitation* has to be added to the list of phenomena underlying some successions. We cannot say how common this state of affairs is. However, the converse is by no means uncommon; thus, many plant species alter the environment in a way that makes it more, rather than less, suitable for themselves (Wilson & Agnew, 1992). Thus, for example, woody vegetation can trap water from fog or ameliorate frosts, improving the conditions for growth of the species concerned, whilst grassy swards can intercept surface flowing water and grow better in the moister soil that is created.

the importance of facilitation – but not always

Figure 16.12 (*left*) Simulated recovery dynamics (Markov chain models) of three of the species that make up a subtidal community starting from 100% bare rock for spatially varying, time varying or homogeneous replacement probabilities: (a) the bryozoan *Crisia eburnea*, (b) the sea anenome *Metridium senile* and (c) encrusting coralline algae. The points at the end of each plot (±95% confidence intervals) are the observed abundances at a site in the Gulf of Maine, USA. (After Hill *et al.*, 2002.)

16.6.3   Interactions with enemies

an important role for seed predation?

Rees *et al.* (2001) point out that it follows from the competition–colonization trade-off that recruitment of competitively dominant plants should be determined largely by the rate of arrival of their seeds. This means that herbivores that reduce seed production are more likely to reduce the density of dominant competitors than of subordinates. Recall that this is just what happened in the sand-dune study described in Section 16.4.4. In a similar vein, Carson and Root (1999) showed that by removing insect predators of seeds, the meadow goldenrod (*Solidago altissima*), which normally appears about 5 years into an old-field succession, became dominant after only 3 years. This happened because release from seed predation allowed it to outcompete earlier colonists more quickly.

Thus, apart from competition–colonization trade-off, successional niche and facilitation, we have to add a fourth mechanism – interactions with enemies – if we are to fully understand plant successions. Experimental approaches, such as that employed to understand the role of seed predators, have also shown that the nature of soil food webs (Gange & Brown, 2002), the presence and disturbance of litter (Ganade & Brown, 2002), and the presence of mammals that consume vegetation (Cadenasso *et al.*, 2002) sometimes play roles in determining successional sequences.

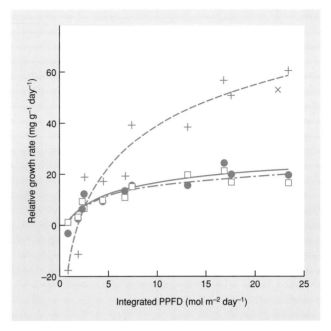

Figure 16.13   Relative growth rate (during the July–August 1994 growing season) of trembling aspen (+), northern red oak (●) and sugar maple (□) in relation to photosynthetic photon flux density (PPFD). (After Kaelke *et al.*, 2001.)

16.6.4   Resource-ratio hypothesis

Tilman's resource-ratio hypothesis emphasizes changing competitive abilities

A further example of a successional niche being responsible for species replacement is worth highlighting. Trembling aspen (*Populus tremuloides*) is a tree that appears earlier in successions in North America than northern red oak (*Quercus rubra*) or sugar maple (*Acer saccharum*). Kaelke *et al.* (2001) compared the growth of seedlings of all three species when planted along a gradient of light availability ranging from forest understory (2.6% of full light) to small clearings (69% of full light). The aspen outgrew the others when relative light availability exceeded 5%. However, there was a rank reversal in relative growth rate in deep shade; here the oak and maple, typical of later stages of succession, grew more strongly and survived better than aspen (Figure 16.13). In his *resource-ratio* hypothesis of succession, Tilman (1988) places strong emphasis on the role of changing relative competitive abilities of plant species as conditions slowly change with time. He hypothesized that species dominance at any point in a terrestrial succession is strongly influenced by the relative availability of two resources: not just by light (as demonstrated by Kaelke *et al.*, 2001) but also by a limiting soil nutrient (often nitrogen). Early in succession, the habitat experienced by seedlings has low nutrient but high light availability. As

a result of litter input and the activities of decomposer organisms, nutrient availability increases with time – this can be expected to be particularly marked in primary successions that begin with a very poor soil (or no soil at all). But total plant biomass also increases with time and, in consequence, light penetration to the soil surface decreases. Tilman's ideas are illustrated in Figure 16.14 for five hypothetical species. Species A has the lowest requirement for the nutrient and the highest requirement for light at the soil surface. It has a short, prostrate growth form. Species E, which is the superior competitor in high-nutrient, low-light habitats, has the lowest requirement for light and the highest for the nutrient. It is a tall, erect species. Species B, C and D are intermediate in their requirements and each reaches its peak abundance at a different point along the soil nutrient–light gradient. There is scope for further experimental testing of Tilman's hypothesis.

16.6.5   Vital attributes

Noble and Slatyer (1981) were also interested in defining the qualities that determine the place of a species in a succession. They called these properties *vital attributes*. The two most important relate to: (i) the method of recovery after

beyond just competitive ability: Noble and Slatyer's 'vital attributes'

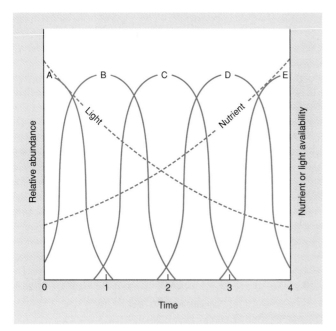

Figure 16.14 Tilman's (1988) resource-ratio hypothesis of succession. Five hypothetical plant species are assumed to be differentiated in their requirements for a limiting soil nutrient and light. During the succession, the habitat starts with a nutrient-poor soil but high light availability, changing gradually into a habitat with a rich soil but low availability of light at the soil surface. Relative competitive abilities change as conditions vary, and first one species and then another comes to dominate.

disturbance (four classes are defined: vegetative spread, V; seedling pulse from a seed bank, S; seedling pulse from abundant dispersal from the surrounding area, D; no special mechanism with just moderate dispersal from only a small seed bank, N); and (ii) the ability of individuals to reproduce in the face of competition (defined in terms of tolerance T at one extreme and intolerance I at the other). Thus, for example, a species may be classed as SI if disturbance releases a seedling pulse from a seed bank, and if the plants are intolerant of competition (being unable to germinate or grow in competition with older or more advanced individuals of either their own or another species). Seedlings of such a species could establish themselves only immediately after a disturbance, when competitors are rare. Of course, a seedling pulse fits well with such a pioneer existence. An example is the annual *Ambrosia artemisiifolia* which often figures early in old-field successions. In contrast, the American beech (*Fagus grandifolia*) could be classed as VT (being able to regenerate vegetatively from root stumps, and tolerant of competition since it is able to establish itself and reproduce in competition with older or more advanced individuals of either its own or another species) or NT (if no stumps remain, it would invade slowly via seed dispersal). In either case, it would eventually displace other species and form part of the

'climax' vegetation. Noble and Slatyer argue that it should be possible to classify all the species in an area according to these two vital attributes (to which relative longevity might be added as a third). Given this information, quite precise predictions about successional sequences should be possible.

Lightning-induced fires produce regular and natural disturbances in many ecosystems in arid parts of the world and two fire-response syndromes, analogous to two of Noble and Slatyer's disturbance recovery classes, can be identified. Resprouters have massive, deeply penetrating root systems, and survive fires as individuals, whereas reseeders are killed by the fire but re-establish through heat-stimulated germination and growth of seedlings (Bell, 2001). The proportion of species that can be classified as resprouters is higher in forest and shrubland vegetation of southwest Western Australia (Mediterranean-type climate) than in more arid areas of the continent. Bell suggests that this is because the Western Australian communities have been subject to more frequent fires than other areas, conforming to the hypothesis that short intervals between fires (averaging 20 years or less in many areas of Western Australia) promote the success of resprouters. Longer intervals between fires, on the other hand, allow fuel loads to build up so that fires are more intense, killing resprouters and favoring the reseeding strategy.

The consideration of vital attributes from an evolutionary point of view suggests that certain attributes are likely to occur together more often than by chance. We can envisage two alternatives that might increase the fitness of an organism in a succession (Harper, 1977), either: (i) the species reacts to the competitive selection pressures and evolves characteristics that enable it to persist longer in the succession, i.e. it responds to *K* selection; or (ii) it may develop more efficient mechanisms of escape from the succession, and discover and colonize suitable early stages of succession elsewhere, i.e. it responds to *r* selection (see Section 4.12). Thus, from an evolutionary point of view, good colonizers can be expected to be poor competitors and vice versa. This is evident in Table 16.3, which lists some physiological characteristics that tend to go together in early and late successional plants.

r and *K* species and succession

16.6.6 The role of animals in successions

The structure of communities and the successions within them have most often been treated as essentially botanical matters. There are obvious reasons for this. Plants commonly provide most of the biomass and the physical structure of communities; moreover, plants do not hide or run away and this makes it rather easy to assemble species lists, determine abundances and detect change. The massive contribution that plants make to

necromass and the late successional role of trees

Table 16.3 Physiological characteristics of early and late successional plants. (After Bazzaz, 1979.)

Attribute	Early successional plants	Late successional plants
Seed dispersal in time	Well dispersed	Poorly dispersed
Seed germination:		
enhanced by		
light	Yes	No
fluctuating temperatures	Yes	No
high NO_3^-	Yes	No
inhibited by		
far-red light	Yes	No
high CO_2 concentration	Yes	No?
Light saturation intensity	High	Low
Light compensation point	High	Low
Efficiency at low light	Low	High
Photosynthetic rates	High	Low
Respiration rates	High	Low
Transpiration rates	High	Low
Stomatal and mesophyll resistances	Low	High
Resistance to water transport	Low	High
Recovery from resource limitation	Fast	Slow
Resource acquisition rates	Fast	Slow?

determining the character of a community is not just a measure of their role as the primary producers, it is also a result of their slowness to decompose. The plant population not only contributes biomass to the community, but is also a major contributor of *necromass*. Thus, unless microbial and detritivore activity is fast, dead plant material accumulates as leaf litter or as peat. Moreover, the dominance of trees in so many communities comes about because they accumulate dead material; the greater part of a tree's trunk and branches is dead. The tendency in many habitats for shrubs and trees to succeed herbaceous vegetation comes largely from their ability to hold leaf canopies (and root systems) on an extending skeleton of predominantly dead support tissue (the heart wood).

animals are often affected by, but may also affect, successions

Animal bodies decompose much more quickly, but there are situations where animal remains, like those of plants, can determine the structure and succession of a community. This happens when the animal skeleton resists decomposition, as is the case in the accumulation of calcified skeletons during the growth of corals. A coral reef, like a forest or a peat bog, gains its structure, and drives its successions, by accumulating its dead past. Reef-forming corals, like forest trees, gain their dominance in their respective communities by holding their assimilating parts progressively higher on predominantly dead support. In both cases, the organisms have an almost overwhelming effect on the abiotic environment, and they 'control' the lives of other organisms within it. The coral reef community (dominated by an animal, albeit one with a plant symbiont) is as structured, diverse and dynamic as a tropical rainforest.

The fact that plants dominate most of the structure and succession of communities does not mean that animals always follow the communities that plants dictate. This will often be the case, of course, because the plants provide the starting point for all food webs and determine much of the character of the physical environment in which animals live. But it is also sometimes the animals that determine the nature of the plant community. We have already seen how seed-eating insects and rodents can slow successions in old fields and sand dunes by causing a higher seed mortalilty of later successional species. A particularly dramatic example of a role for animals, and on a much larger scale, comes from the savanna at Ndara in Kenya. The vegetation in savannas is often held in check by grazers. The experimental exclusion of elephants from a plot of savanna led to a more than threefold increase in the density of trees over a 10-year period (work by Oweyegha-Afundaduula, reported in Deshmukh, 1986).

More often though, animals are passive followers of successions amongst the plants. This is certainly the case for passerine bird species in an old-field succession (Figure 16.15). Arbuscular mycorrhizal fungi (see Section 13.8.2), which show a clear sequence of species replacement in the soils associated with an old-field succession (Johnson *et al.*, 1991), may also be passive followers of the plants. But this does not mean that the birds, which eat seeds, or the fungi, which affect plant growth and survival, do not influence the succession in its course. They probably do.

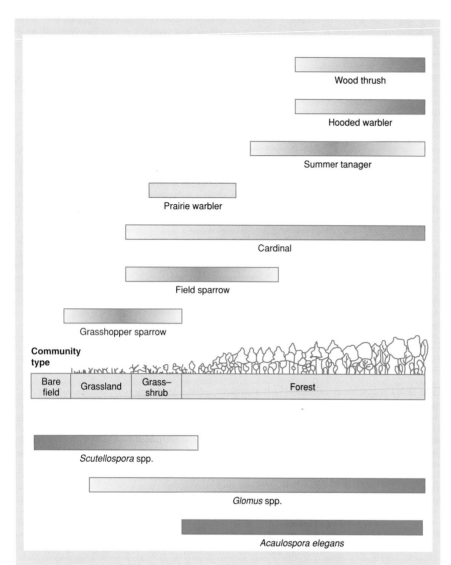

Figure 16.15   Top: bird species distributions along a plant succession gradient in the Piedmont region of Georgia, USA. Differential shading indicates relative abundance of the birds. (After Johnston & Odum, 1956; from Gathreaux, 1978.) Bottom: distributions of vesicular–arbuscular mycorrhizae in the soils associated with an old-field succession in Minnesota. Differential shading indicates relative abundance of spores of species in the genera *Scutellospora*, *Glomus* and *Acaulospora*. (After Johnson *et al.*, 1991).

16.6.7   Concept of the climax

Do successions come to an end? It is clear that a stable equilibrium will occur if individuals that die are replaced on a one-to-one basis by young of the same species. At a slightly more complex level, Markov models (see Section 16.5) tell us that a stationary species composition should, in theory, occur whenever the replacement probabilities (of one species by itself or by any one of several others) remain constant through time.

The concept of the climax has a long history. One of the earliest students of succession, Frederic Clements (1916), is associated with the idea that a single climax will dominate in any given climatic region, being the endpoint of all successions, whether they happened to start from a sand dune, an abandoned old field or even a pond filling in and progressing towards a terrestrial climax.

This *monoclimax* view was challenged by many ecologists, amongst whom Tansley (1939) was prominent. The *polyclimax* school of thought recognized that a local climax may be governed by one factor or a combination of factors: climate, soil conditions, topography, fire and so on. Thus, a single climatic area could easily contain a number of specific climax types. Later still, Whittaker (1953) proposed his climax pattern hypothesis. This conceives a continuity of climax types, varying gradually along environmental gradients and not necessarily separable into discrete climaxes. (This is an extension of Whittaker's approach to gradient analysis of vegetation, discussed in Section 16.3.1.)

In fact, it is very difficult to identify a stable climax community in the field.

climaxes may be approached rapidly – or, so slowly that they are rarely ever reached

Usually, we can do no more than point out that the rate of change of succession slows down to the point where any change is imperceptible to us. In this context, the subtidal rockface succession illustrated in Figure 16.12 is unusual in that convergence to a climax took only a few years. Old-field successions might take 100–500 years to reach a 'climax', but in that time the probabilities of further fires or hurricanes are so high that a process of succession may rarely go to completion. If we bear in mind that forest communities in northern temperate regions, and probably also in the tropics, are still recovering from the last glaciation (see Chapter 1), it is questionable whether the idealized climax vegetation is often reached in nature.

16.7 Communities in a spatiotemporal context: the patch dynamics perspective

the idea of a successional mosaic

A forest, or a rangeland, that appears to have reached a stable community structure when studied on a scale of hectares, will always be a mosaic of miniature successions. Every time a tree falls or a grass tussock dies, an opening is created in which a new succession starts. One of the most seminal papers in the history of ecology was entitled 'Pattern and process in the plant community' (Watt, 1947). Part of the pattern of a community is caused by the dynamic processes of deaths, replacements and microsuccessions that the broad view may conceal. Thus, although we can point to patterns in community composition in space (see Section 16.3) and in time (see Section 16.4), it is often more meaningful to consider space and time together.

disturbance . . . gaps . . . dispersal . . . recruitment

We have already seen that disturbances that open up gaps are common in all kinds of community. The formation of gaps is obviously of considerable significance to sessile or sedentary species that have a requirement for open space, but gaps have also proved to be important for mobile species such as invertebrates on the beds of streams (Matthaei & Townsend, 2000). The patch dynamics concept of communities views the habitat as patchy, with patches being disturbed and recolonized by individuals of various species. Implicit in the patch dynamics view is a critical role for disturbance as a reset mechanism (Pickett & White, 1985). A single patch without migration is, by definition, a closed system, and any extinction caused by disturbance would be final. However, extinction within a patch in an open system is not necessarily the end of the story because of the possibility of reinvasion from other patches.

Fundamental to the patch dynamics perspective is recognition of the importance of migration between habitat patches. This may involve adult individuals, but very often the process of most significance is the dispersal of immature propagules (seeds, spores, larvae) and their recruitment to populations within habitat patches. The order of arrival and relative recruitment levels of individual species may determine or modify the nature and outcome of population interactions in the community (Booth & Brosnan, 1995).

In Section 16.4.1 we identified two fundamentally different kinds of situations within communities: those in which some species are strongly competitively superior are *dominance controlled* (equivalent to succession) and those in which all species have similar competitive abilities are *founder controlled*. Within the patch dynamics framework, the dynamics of these two situations are different and we deal with them in turn.

16.7.1 Dominance-controlled communities

dominance control and succession

In patch dynamics models where some species are competitively superior to others, the effect of the disturbance is to knock the community back to an earlier stage of succession (Figure 16.16). The open space is colonized by one or more of a group of opportunistic, early successional species (p_1, p_2, etc., in Figure 16.16). As time passes, more species invade, often those with poorer powers of dispersal. These eventually reach maturity, dominating mid-succession (m_1, m_2, etc.) and many or all of the pioneer species are driven to extinction. Later still, the community regains the climax stage when the most efficient competitors (c_1, c_2, etc.) oust their neighbors. In this sequence, diversity starts at a low level, increases at the mid-successional stage and usually declines again at the climax. The gap essentially undergoes a minisuccession.

disturbance scale and phasing

Some disturbances are synchronized, or phased, over extensive areas. A forest fire may destroy a large tract of a climax community. The whole area then proceeds through a more or less synchronous succession, with diversity increasing through the early colonization phase and falling again through competitive exclusion as the climax is approached. Other disturbances are much smaller and produce a patchwork of habitats. If these disturbances are unphased, the resulting community comprises a mosaic of patches at different stages of succession. A climax mosaic, produced by unphased disturbances, is much richer in species than an extensive area undisturbed for a very long period and occupied by just one or a few dominant climax species. Towne (2000) monitored the plant species that established in prairie grassland where large ungulates had died (mainly bison, *Bos bison*). Scavengers remove most of the body tissue but copious amounts of body fluids and decomposition products seep into the soil. The flush of nutrients combined with death of the previous vegetation produces a competitor-free, disturbed area where resources are unusually abundant. The patches are also exceptional because the soil has not been disturbed (as it would be after a ploughed field is abandoned or a badger makes a burrow); thus,

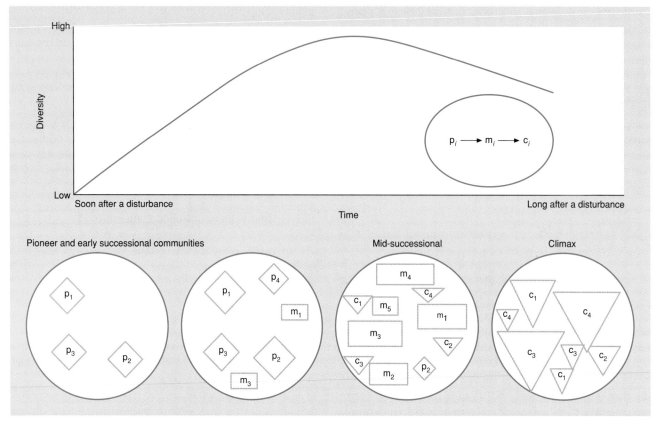

Figure 16.16 Hypothetical minisuccession in a gap. The occupancy of gaps is reasonably predictable. Diversity begins at a low level as a few pioneer (p_i) species arrive; reaches a maximum in mid-succession when a mixture of pioneer, mid-successional (m_i) and climax (c_i) species occur together; and drops again as competitive exclusion by the climax species takes place.

the colonizing plants do not derive from the local seed bank. The unusual nature of the disturbed patches means that many of the pioneer species are rare in the prairie as a whole, and carcass sites contribute to species diversity and community heterogeneity for many years.

16.7.2 Frequency of gap formation

Connell's 'intermediate disturbance hypothesis'

The influence that disturbances have on a community depends strongly on the frequency with which gaps are opened up. In this context, the intermediate disturbance hypothesis (Connell, 1978; see also the earlier account by Horn, 1975) proposes that the highest diversity is maintained at intermediate levels of disturbance. Soon after a severe disturbance, propagules of a few pioneer species arrive in the open space. If further disturbances occur frequently, gaps will not progress beyond the pioneer stage in Figure 16.16, and the diversity of the community as a whole will be low. As the interval between disturbances increases, the diversity will

also increase because time is available for the invasion of more species. This is the situation at an intermediate frequency of disturbance. At very low frequencies of disturbance, most of the community for most of the time will reach and remain at the climax, with competitive exclusion having reduced diversity. This is shown diagrammatically in Figure 16.17, which plots the pattern of species richness to be expected as a result of unphased high, intermediate and low frequencies of gap formation, in separate patches and for the community as a whole.

The influence of the frequency of gap formation was studied in southern California by Sousa (1979a, 1979b), in an intertidal algal community associated with boulders of various sizes. Wave action disturbs small boulders more often than large ones. Using a sequence of photographs, Sousa estimated the probability that a given boulder would be moved during the course of 1 month. A class of mainly small boulders (which required a force of less than 49 Newtons to move them) had a monthly probability of movement of 42%. An intermediate class (which required a force of 50–294 N) had a much smaller monthly probability of movement,

boulders on a rocky shore that vary in disturbability . . .

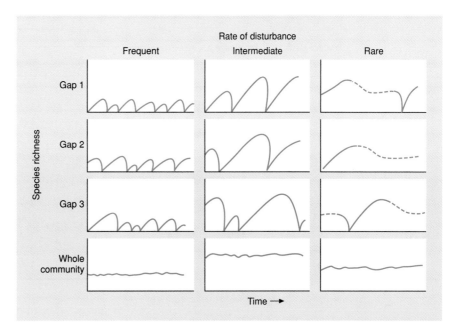

Figure 16.17 Diagrammatic representation of the time course of species richness in three gaps, and in the community as a whole, at three frequencies of disturbance. The disturbance is unphased. Dashed lines indicate the phase of competitive exclusion as the climax is approached.

9%. Finally, the class of mainly large boulders (which required a force >294 N) moved with a probability of only 0.1% per month. The 'disturbability' of the boulders had to be assessed in terms of the force required to move them, rather than simply in terms of top surface area, because some rocks which appeared to be small were actually stable portions of larger, buried boulders, and a few large boulders with irregular shapes moved when a relatively small force was applied. The three classes of boulder (<49, 50–294 and >294 N) can be viewed as patches exposed to a decreasing frequency of disturbance when waves caused by winter storms overturn them.

Species richness increased during early stages of succession through a process of colonization by the pioneer green alga *Ulva* spp. and various other algae, but declined again at the climax because of competitive exclusion by the perennial red alga *Gigartina canaliculata*. It is important to note that the same succession occurred on small boulders that had been artificially made stable. Thus, variations in the communities associated with the surfaces of boulders of different size were not simply an effect of size, but rather of differences in the frequency with which they were disturbed.

. . . provide support for the hypothesis

Communities on unmanipulated boulders in each of the three size/disturbability classes were assessed on four occasions. Table 16.4 shows that the percentage of bare space decreased from small to large boulders, indicating the effects of the greater frequency of disturbance of small boulders. Mean species richness was lowest on the regularly disturbed small boulders. These were dominated most commonly by *Ulva* spp. (and barnacles, *Chthamalus fissus*). The highest levels of species richness were consistently recorded on the intermediate boulder class. Most held mixtures of three to five abundant species from all successional stages. The largest boulders had a lower mean species richness than the intermediate class, although a monoculture was achieved on only a few boulders. *G. canaliculata* covered most of the rock surfaces.

These results offer strong support for the intermediate disturbance hypothesis as far as frequency of appearance of gaps is concerned. However, we must be careful not to lose sight of the fact that this is a highly stochastic process. By chance, some small boulders were not overturned during the period of study. These few were dominated by the climax species *G. canaliculata*. Conversely, two large boulders in the May census had been overturned, and these became dominated by the pioneer *Ulva*. On average, however, species richness and species composition followed the predicted pattern.

This study deals with a single community conveniently composed of identifiable patches (boulders) that become gaps (when overturned by waves) at short, intermediate or long intervals. Recolonization occurs mainly from propagules derived from other patches in the community. Because of the pattern of disturbance, this mixed boulder community is more diverse than would be one with only large boulders.

Disturbances in small streams often take the form of bed movements during periods of high discharge. Because

further support from a study of streams

of differences in flow regimes and in the substrates of stream beds, some stream communities are disturbed more frequently and to a larger extent than others. This variation was assessed in 54 stream sites in the Taieri River in New Zealand (Townsend *et al.*, 1997) by recording the frequency at which at least 40% (chosen arbitrarily) of the bed moved and the average percentage that

Census date	Boulder class (N)	Percentage bare space	Species richness		
			Mean	Standard error	Range
November 1975	< 49	78.0	1.7	0.18	1–4
	50–294	26.5	3.7	0.28	2–7
	> 294	11.4	2.5	0.25	1–6
May 1976	< 49	66.5	1.9	0.19	1–5
	50–294	35.9	4.3	0.34	2–6
	> 294	4.7	3.5	0.26	1–6
October 1976	< 49	67.7	1.9	0.14	1–4
	50–294	32.2	3.4	0.40	2–7
	> 294	14.5	2.3	0.18	1–6
May 1977	< 49	49.9	1.4	0.16	1–4
	50–294	34.2	3.6	0.20	2–5
	> 294	6.1	3.2	0.21	1–5

Table 16.4 Seasonal patterns in bare space and species richness on boulders in each of three classes, categorized according to the force (in Newtons) required to move them. (After Sousa, 1979b.)

moved (assessed on five occasions during 1 year, using painted particles of sizes characteristic of the stream bed in question). The pattern of richness of insect species conformed to the intermediate disturbance hypothesis (Figure 16.18). It is likely that low richness at high frequencies and intensities of disturbance reflects the inability of many species to persist in such situations. Whether low richness at low frequencies and intensities of disturbance is due to competitive exclusion, as proposed in the intermediate disturbance hypothesis, remains to be tested.

16.7.3 Formation and filling of gaps

Gaps of different sizes may influence community structure in different ways because of contrasting mechanisms of *influence of gap size . . .* recolonization. The centers of very large gaps are most likely to be colonized by species producing propagules that travel relatively great distances. Such mobility is less important in small gaps, since most recolonizing propagules will be produced by adjacent established individuals. The smallest gaps of all may be filled simply by lateral movements of individuals around the periphery.

Intertidal beds of mussels provide excellent opportunities to study the processes of formation and filling-in of gaps. In the absence of disturbance, mussel beds may persist as extensive monocultures. More often, they are an ever-changing mosaic of many species that inhabit gaps formed by the action of waves. Gaps can appear virtually anywhere, and may exist for years as islands in a sea of mussels. The size of these gaps at the time of formation ranges from the dimensions of a single mussel to hundreds of square meters. In general, a mussel or group of mussels becomes infirm or damaged through disease, predation, old age or, most often, the effects of storm waves or battering by logs. Gaps begin to fill as soon as they are formed.

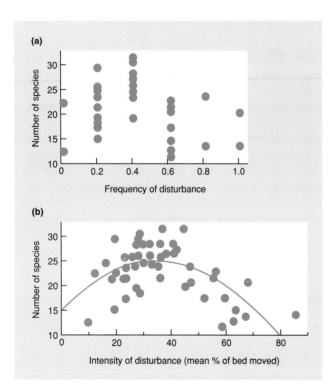

Figure 16.18 Relationship between invertebrate species richness and: (a) frequency of disturbance – assessed as the number of occasions in 1 year when more than 40% of the bed moved (analysis of variance significant at $P < 0.0001$), and (b) intensity of disturbance – average percentage of the bed that moved (polynomial regression fitted, relationship significant at $P < 0.001$) assessed at 54 stream sites in the Taieri River, New Zealand. The patterns are essentially the same; intensity and frequency of disturbance are strongly correlated. (After Townsend *et al.*, 1997.)

Table 16.5 Measures of area, perimeter and perimeter : area ratio for the experimental gaps created in two experiments on semiexposed shores in southeast Brazil. (From Tanaka & Magalhaes, 2002.)

	Area (cm²)	Perimeter (cm)	Perimeter : area ratio
Patch size effects			
Square	25	20	0.8
Square	100	40	0.2
Square	400	80	0.2
Patch shape effects			
Square	100.0	40.0	0.4
Circle	78.5	31.4	0.4
Rectangle	112.5	45.0	0.4
Sector	190.1	78.6	0.4

. . . and gap shape

In their experimental study of mussel beds of *Brachidontes solisianus* and *B. darwinius* in Brazil, Tanaka and Magalhaes (2002) aimed to determine the differential effects of patch size and perimeter : area ratio on the dynamics of succession. In an experiment on one moderately exposed shoreline, they created square gaps with different areas (because of identical shapes, the bigger squares had smaller perimeter : area ratios) (Table 16.5). On a nearby and physically very similar shore, they created patches of four different shapes and chose areas for each that produced identical perimeter : area ratios (Figure 16.19a). Note that a circle has the most perimeter per unit area of any shape. The gap sizes were within the range observed for natural gaps, which did not differ on the two shores (Figure 16.19b).

colonization of gaps . . .
. . . in mussel beds, . . .

Higher densities of the herbivorous limpet *Collisella subrugosa* occurred in the small gaps in the first 6 months after gap formation (Figure 16.19c). Small gaps, compared to medium and large gaps, were also most quickly colonized by lateral migration of the two mussel species, but with *B. darwinius* predominating. The larger gaps had higher densities of the barnacle *Chthamalus bisinuatus* and sheltered more limpets at their edges, while central areas had more *Brachiodontes* recruited from larvae after 6 months (Figure 16.19d). The gaps with identical perimeter : area ratios showed very similar patterns of colonization despite their different sizes, emphasizing that colonization dynamics are mainly determined by distance from adjacent sources of colonists.

The limpet is probably associated with patch edges because here they are less vulnerable to visually hunting predators. The negative relationship between distributions of the limpet and the barnacle may be due to the former dislodging the latter from the substrate. Tanaka and Magalhaes conclude that the mussel

B. darwinius is a more effective colonist of disturbed patches than *B. solisianus*, and suggest that *B. darwinius* would gradually come to dominate the whole of the shoreline if it were not for occasional massive recruitment events of *B. solisianus*.

. . . in grassland . . .

The pattern of colonization of gaps in mussel beds is repeated in almost every detail in the colonization of gaps in grassland caused by burrowing animals or patches killed by urine. Initially, leaves lean into the gap from plants outside it. Then colonization begins by clonal spread from the edges, and a very small gap may close up quickly. In larger gaps, new colonists may enter as dispersed seed, or germinate from the seed bank in the soil. Over 2–3 years the vegetation begins to acquire the character that it had before the gap was formed.

. . . and in mangrove forest

The gaps produced in forests vary greatly in size. Lightning-induced gaps in mangrove forest in the Dominican Republic, for example, range from 200 to 1600 m² or more (Figure 16.20). Lightning almost always kills groups of trees in a 20–30 m circle, and the trees remain as standing dead for several years. In a forest dominated by red mangrove *Rhizophora mangle* and white mangrove *Laguncularia racemosa*, and with some black mangrove *Avicennia germinans*, Sherman *et al.* (2000) compared the performance of the three species in lightning gaps and under forest canopy. Seedling density did not differ in gaps and intact forest, but sapling density and the growth rates of all three species were much higher in the gaps (Table 16.6). However, gap regeneration was dominated by *R. mangle* because its mortality rate was much lower in gaps than was the case for the other species. Sherman *et al.* (2000) note that the peat mat on the forest floor usually collapses after lightning damage, resulting in increased levels of standing water. They suggest that the success of *R. mangle* in gaps is due to their higher tolerance of flooding conditions.

Organisms other than plants can also be overrepresented in gaps. In a study of tropical rainforest in Costa Rica, Levey (1988) found that nectarivorous and frugivorous birds were much more abundant in treefall gaps, reflecting the fact that understory plants in gaps tend to produce more fruit over a longer period than conspecifics fruiting under a closed canopy.

16.7.4 Founder-controlled communities

founder-controlled communities: a competitive lottery not a predictable succession

In the dominance-controlled communities discussed in Section 16.7.1 there was the familiar *r* and *K* selection dichotomy in which colonizing ability and competitive status are inversely related. In founder-controlled communities, on the other hand, all species are both good colonists and essentially equal competitors; thus, within a patch opened by disturbance, a competitive lottery rather than a predictable succession is to

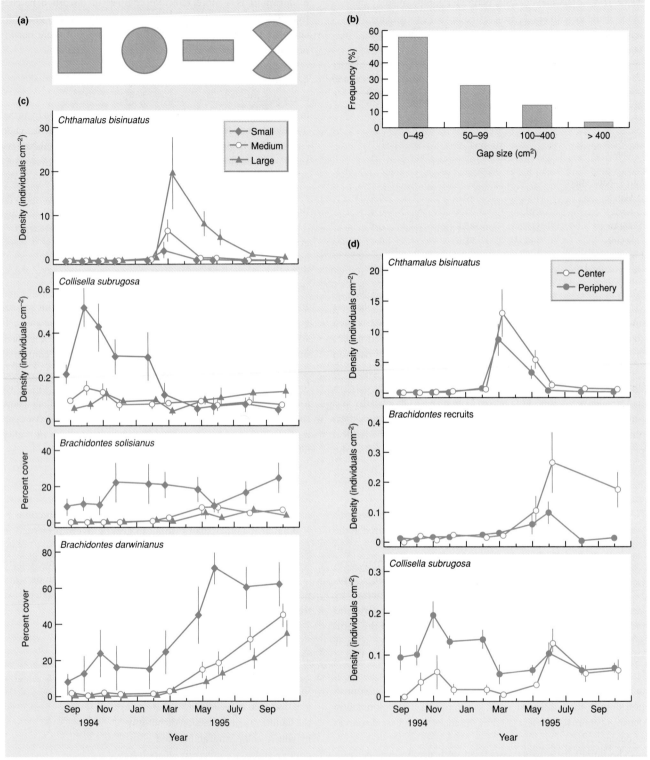

Figure 16.19 (a) The four shapes used in patch-shape experiments: square, circle, rectangle and 'sector' (see Table 16.5). (b) Size distribution of natural gaps in the mussel beds. (c) Mean abundances (±SE) of four colonizing species in experimentally cleared small, medium and large square gaps. (d) Recruitment of three species at the periphery (within 5 cm of the gap edge) and in the center of 400 cm² square gaps. (After Tanaka & Magalhaes, 2002.)

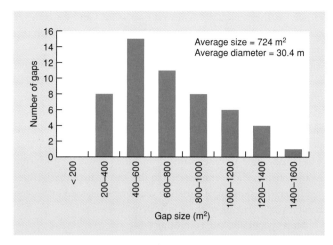

Figure 16.20 Frequency distribution of gaps created by lightning in a tropical mangrove forest in the Dominican Republic. (After Sherman *et al.*, 2000.)

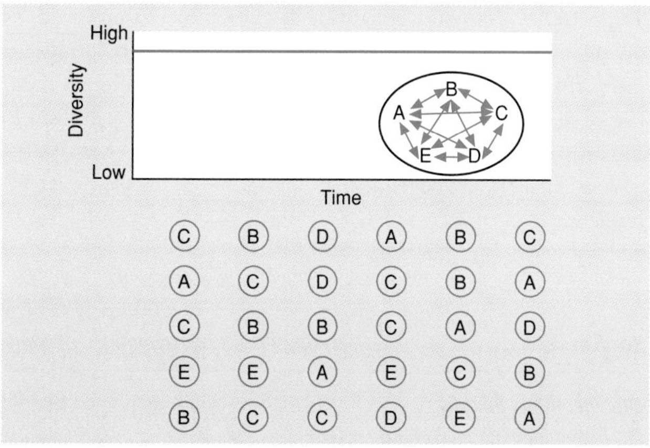

Figure 16.21 Hypothetical competitive lottery: occupancy of gaps which periodically become available. Each of species A–E is equally likely to fill a gap, regardless of the identity of its previous occupant. Species richness remains high and relatively constant.

be expected. If a large number of species are approximately equivalent in their ability to invade gaps, are equally tolerant of the abiotic conditions and can hold the gaps against all comers during their lifetime, then the probability of competitive exclusion may be much reduced in an environment where gaps are appearing continually and randomly. A further condition for coexistence is that the number of young that invade and occupy the gaps should not be consistently greater for parent populations that produce more offspring, otherwise the most productive species would come to monopolize space even in a continuously disturbed environment.

fish coexisting on coral reefs

If these idealized conditions are met, it is possible to envisage how the occupancy of a series of gaps will change through time (Figure 16.21). On each occasion that an organism dies (or is killed) the gap is reopened for invasion. All conceivable replacements are possible and species richness will be maintained at a high level. Some tropical reef communities of fish may conform to this model (Sale, 1977, 1979). They are extremely diverse. For example, the number of species of fish on the Great Barrier Reef ranges from 900 in the south to 1500 in the north, and more than 50 resident species

may be recorded on a single patch of reef 3 m in diameter. Only a proportion of this diversity is likely to be attributable to resource partitioning of food and space – indeed, the diets of many of the coexisting species are very similar. In this community, vacant living space seems to be a crucial limiting factor, and it is generated unpredictably in space and time when a resident dies or is killed. The lifestyles of the species match this state of affairs. They breed often, sometimes year-round, and produce numerous clutches of dispersive eggs or larvae. It can be argued that the species compete in a lottery for living space in which larvae are the tickets, and the first arrival at the vacant space wins the site, matures quickly and holds the space for its lifetime.

Three species of herbivorous pomacentrid fish co-occur on the upper slope of Heron Reef, part of the Great Barrier Reef off eastern Australia. Within rubble patches, the available space is occupied by a series of contiguous and usually nonoverlapping territories, each up to 2 m² in area, held by individuals of *Eupomacentrus apicalis*, *Plectroglyphidodon lacrymatus* and *Pomacentrus wardi*. Individuals hold territories throughout their juvenile and adult life and defend them against a broad range of chiefly herbivorous species, including conspecifics. There seems to be no particular tendency for space initially held by one species to be

Table 16.6 Initial size, and growth and mortality rates over a 1-year period of saplings of three mangrove species in lightning-induced gaps and under intact forest canopy. (After Sherman *et al.*, 2000.)

	Initial sapling diameter (cm ± SE)		Growth rate–diameter increment (cm ± SE)		Mortality (%)	
	Gaps	*Canopy*	*Gaps*	*Canopy*	*Gaps*	*Canopy*
Rhizophora mangle	1.9 ± 0.06	2.3 ± 0.06	0.58 ± 0.03	0.09 ± 0.01	9	16
Laguncularia racemosa	1.7 ± 0.11	1.8 ± 0.84	0.46 ± 0.04	0.11 ± 0.06	32	40
Avicennia germinans	1.3 ± 0.25	1.7 ± 0.45	0.51 ± 0.04	–	56	88

Table 16.7 Numbers of individuals of each species observed occupying sites, or parts of sites, that had been vacated during the immediately prior interperiod between censuses through the loss of residents of each species. The sites vacated through loss of 120 residents have been reoccupied by 131 fish; the species of the new occupant is not dependent on the species of the previous resident.

	Reoccupied by:		
Resident lost	E. apicalis	P. lacrymatus	P. wardi
Eupomacentrus apicalis	9	3	19
Plectroglyphidodon lacrymatus	12	5	9
Pomacentrus wardi	27	18	29

taken up, following mortality, by the same species. Nor is any successional sequence of ownership evident (Table 16.7). *P. wardi* both recruited and lost individuals at a higher rate than the other two species, but all three species appear to have recruited at a sufficient level to balance their rates of loss and maintain a resident population of breeding individuals.

plants in grassland or forest

Thus, the maintenance of high reef diversity depends, at least in part, on the unpredictability of the supply of living space; and as long as all species win some of the time and in some places, they will continue to put larvae into the plankton, and hence, into the lottery for new sites. An analogous situation has been postulated for the highly diverse chalk grasslands of Great Britain (Grubb, 1977) and even for trees in temperate and tropical forest gaps (Busing & Brokaw, 2002). Any small gap that appears is rapidly exploited, by a seed in grassland and very often by a sapling in a forest gap. In these cases, the tickets in the lottery are saplings or seeds (either in the act of dispersal or as components of a persistent seed bank in the soil). Which seeds or saplings develop to established plants, and therefore which species comes to occupy the gap, may depend on a strong random element since many species overlap in their requirements for successful growth. The successful plant rapidly establishes itself and retains the patch for its lifetime, in a similar way to the reef fish described above.

16.8 Conclusions: the need for a landscape perspective

founder and dominance control as a continuum of possibilities

The lottery hypothesis and the notion of the founder-controlled community were important steps in the development of our understanding of the range of community dynamics that can occur.

However, these should be viewed not as hard and fast rules to which some communities are subject, but rather as extremes on a continuum from dominance to founder control. Real communities may be closer to one or other end of this continuum, but in reality component species or component patches may be dominance controlled or founder controlled within the same community. Syms and Jones (2000), for example, acknowledge that more than half of within-reef variation in fish species composition in their study of patch reefs in the Great Barrier Reef was attributable to unexplained, and thus possibly stochastic, factors such as those emphasized in the lottery hypothesis. But a significant proportion of variation could be explained by specific habitat requirements of the constituent species.

importance of a 'landscape ecology' perspective

More generally, no community is truly the homogeneous, temporally invariant system described by simple Lotka–Volterra mathematics and exemplified by laboratory microcosms, although some are less variable than others. In most real communities, population dynamics will be spatially distributed and temporal variation will be present. In a closed system, composed of a single patch, species extinctions can occur for two very different reasons: (i) as a result of biotic instability caused by competitive exclusion, overexploitation and other strongly destabilizing species interactions; or (ii) as a result of environmental instability caused by unpredictable disturbances and changes in conditions. By integrating unstable patches of either of these types into the open system of a larger landscape (consisting of many patches out of phase with each other), persistent species-rich communities can result (DeAngelis & Waterhouse, 1987). This is the principal message to emerge from the patch dynamics perspective, and its larger scale counterpart, 'landscape ecology' (Wiens *et al.*, 1993), stressing the importance of the spatial scale at which we view communities and the open nature of most of them. Note the strong link between the patch dynamics view of community organization and metapopulation theory, which deals with the effects on the dynamics of populations of dividing them into fragments (see Section 6.9). In a model that combines extinction–colonization dynamics (the metapopulation approach) with the dynamics of patch succession, Amarasekare and Possingham (2001) show that persistence of a species in the landscape depends: (i) on the net rate at which suitable patches arise relative to the species' colonization ability; as well as (ii) the longevity of the dormant stages (e.g. seed bank) relative to disturbance frequency.

multiple classes of disturbance . . .

Future development of ideas about patch dynamics is likely to concern the consequences of multiple classes of disturbance. Steinauer and Collins (2001) have made a start by showing that disturbances caused by urine deposition and grazing by bison (*Bos bison*) interact with each other. The abundance of four common grass species, and of all of them combined, increased on urine patches in ungrazed prairie grassland. However, the abundance of the grass *Andropogon gerardii*, and all grasses combined, decreased on urine patches in grazed prairie. The changed dynamics reflect the fact that bison preferentially graze on urine patches. In addition, grazed areas initiated on urine patches tend to expand well beyond the area of urine deposition, increasing the size and severity of disturbance by grazing.

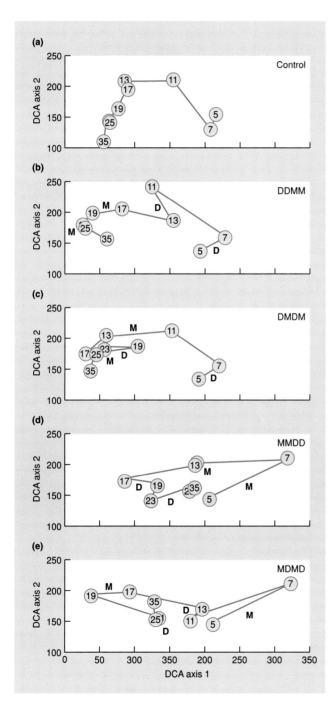

(a) Control

(b) DDMM

(c) DMDM

(d) MMDD

(e) MDMD

Finally, just as we can readily see how community dynamics may vary according to the order in which colonizing species happen to appear after a disturbance, it is equally the case that the order in which different kinds of disturbance occur may affect the outcome. Fukami (2001) addressed this issue by introducing two classes of disturbance (drought or the addition of predatory mosquito larvae) in various sequences to laboratory microsoms of protists and small metazoans (analogous to natural communities in water-filled bamboo stumps). Different disturbance sequences drove the microcosms into different successional trajectories, sometimes leading to divergence in final community composition (both in terms of species richness and relative abundance of the component species). This is illustrated graphically by ordination diagrams (see Section 16.3.2) that show the sequence of positions of communities in the same ordination space through experiments where disturbances were imposed in different sequences (Figure 16.22). It may often be the case that a knowledge of the disturbance history will be needed to predict the responses of communities to disturbances imposed in the future (such as global climate change).

... may interact to determine community patterns

Summary

The community is an assemblage of species populations that occur together in space and time. Community ecology seeks to understand the manner in which groupings of species are distributed in nature, and how they are influenced by their abiotic environment and by species interactions.

We begin by explaining how the structure of communities can be measured and described, in terms of species composition, species richness, diversity, equitability (evenness) and rank–abundance diagrams.

Figure 16.22 (*left*) Temporal changes in species composition and relative abundance of microcosms composed of a specific mix of protists and metazoans. The change is expressed in ordination plots based on a procedure called detrended correspondence analysis (DCA). (Recall that ordination is a mathematical treatment that allows communities to be organized on a graph so that those that are most similar in species composition and relative abundance appear closest together, whilst communities that differ greatly in the relative importance of a similar set of species, or that possess quite different species, appear far apart.) Data points are the mean ordination scores on different days in the experiment (from day 5 to day 35). The letter D indicates periods of drought disturbance, and the letter M, mosquito disturbance. (a–e) The results of the control and disturbances imposed in different sequences. (After Fukami, 2001.)

The assessment of community patterns in space has progressed from subjective 'gradient analysis' to objective mathematical approaches ('classification' and 'ordination') that permit relationships between community composition and abiotic factors to be systematically explored. We note that most communities are not delimited by sharp boundaries, where one group of species is abruptly replaced by another. Moreover a given species that occurs in one predictable association is also quite likely to occur with another group of species under different conditions elsewhere.

Just as the relative importance of species varies in space, so their patterns of abundance may change with time. A particular species can occur where it is capable of reaching a location, appropriate conditions and resources exist, and competitors, predators and parasites do not preclude it. A temporal sequence in the appearance and disappearance of species therefore requires that conditions, resources and/or the influence of enemies themselves vary with time. We emphasize and explain patterns of community change that follow a disturbance. Sometimes these patterns are predictable (succession; dominance control), in other cases highly stochastic (founder control).

Although we can discern and often explain patterns in community composition in space and in time, it is often more meaningful to consider space and time together. The patch dynamics concept of communities views the landscape as patchy, with patches being disturbed and recolonized by individuals of various species. Implicit in this view are critical roles for disturbance as a reset mechanism, and of migration between habitat patches. The community dynamics of patchy landscapes are strongly influenced by the frequency of gap formation and the sizes and shapes of these gaps in relation to the colonization and competitive properties of the species concerned.

Chapter 17
The Flux of Energy through Ecosystems

17.1 Introduction

All biological entities require matter for their construction and energy for their activities. This is true not only for individual organisms, but also for the populations and communities that they form in nature. The intrinsic importance of fluxes of energy (this chapter) and of matter (see Chapter 18) means that community processes are particularly strongly linked with the abiotic environment. The term *ecosystem* is used to denote the biological community *together with* the abiotic environment in which it is set. Thus, ecosystems normally include primary producers, decomposers and detritivores, a pool of dead organic matter, herbivores, carnivores and parasites *plus* the physicochemical environment that provides the living conditions and acts both as a source and a sink for energy and matter. Thus, as is the case with all chapters in Part 3 of this book, our treatment calls upon knowledge of individual organisms in relation to conditions and resources (Part 1) together with the diverse interactions that populations have with one another (Part 2).

Lindemann laid the foundations of ecological energetics A classic paper by Lindemann (1942) laid the foundations of a science of ecological energetics. He attempted to quantify the concept of food chains and food webs by considering the efficiency of transfer between trophic levels – from incident radiation received by a community through its capture by green plants in photosynthesis to its subsequent use by herbivores, carnivores and decomposers. Lindemann's paper was a major catalyst for the International Biological Programme (IBP), which, with a view to human welfare, aimed to understand the biological basis of productivity of areas of land, fresh waters and the seas (Worthington, 1975). The IBP provided the first occasion on which biologists throughout the world were challenged to work together towards a common end. More recently, a further pressing issue has again galvanized the community of ecologists into action. Deforestation, the burning of fossil fuels and other pervasive human influences

are causing dramatic changes to global climate and atmospheric composition, and can be expected in turn to influence patterns of productivity on a global scale. Much of the current work on productivity has a prime objective of providing the basis for predicting the effects of changes in climate, atmospheric composition and land use on terrestrial and aquatic ecosystems (aspects that will be dealt with in Chapter 22).

progressive improvements in technology to assess productivity The decades since Lindemann's classic work have seen a progressive improvement in technology to assess productivity. Early calculations in terrestrial ecosystems involved sequential measurements of biomass of plants (usually just the aboveground parts) and estimates of energy transfer efficiency between trophic levels. In aquatic ecosystems, production estimates relied on changes in the concentrations of oxygen or carbon dioxide measured in experimental enclosures. Increasing sophistication in the measurement, *in situ*, of chlorophyll concentrations and of the gases involved in photosynthesis, coupled with the development of satellite remote-sensing techniques, now permit the extrapolation of local results to the global scale (Field *et al.*, 1998). Thus, satellite sensors can measure vegetation cover on land and chlorophyll concentrations in the sea, from which rates of light absorption are calculated and, based on our understanding of photosynthesis, these are converted to estimates of productivity (Geider *et al.*, 2001).

some definitions: standing crop and biomass, . . . Before proceeding further it is necessary to define some new terms. The bodies of the living organisms within a unit area constitute a *standing crop* of biomass. By *biomass* we mean the mass of organisms per unit area of ground (or per unit area or unit volume of water) and this is usually expressed in units of energy (e.g. $J\ m^{-2}$) or dry organic matter (e.g. $t\ ha^{-1}$) or carbon (e.g. $g\ C\ m^{-2}$). The great bulk of the biomass in communities is almost always formed by plants, which are the primary producers of biomass because of their almost unique ability to fix carbon in photosynthesis. (We have to say 'almost unique' because bacterial photosynthesis and chemosynthesis may also contribute to forming new biomass.) Biomass includes the whole bodies of the organisms even though parts of them may be dead. This needs to be borne in mind, particularly when considering woodland and forest communities in which the bulk of the biomass is dead heartwood and bark. The living fraction of biomass represents active capital capable of generating interest in the form of new growth, whereas the dead fraction is incapable of new growth. In practice we include in biomass all those parts, living or dead, which are attached to the living organism. They cease to be biomass when they fall off and become litter, humus or peat.

. . . primary and secondary productivity, autotrophic respiration, . . .

The *primary productivity* of a community is the rate at which biomass is produced per unit area by plants, the primary producers. It can be expressed either in units of energy (e.g. $J\,m^{-2}\,day^{-1}$) or dry organic matter (e.g. $kg\,ha^{-1}\,year^{-1}$) or carbon (e.g. $g\,C\,m^{-2}\,year^{-1}$). The total fixation of energy by photosynthesis is referred to as *gross primary productivity* (GPP). A proportion of this is respired away by the plants (autotrophs) and is lost from the community as respiratory heat (RA – *autotrophic respiration*). The difference between GPP and RA is known as *net primary productivity* (NPP) and represents the actual rate of production of new biomass that is available for consumption by heterotrophic organisms (bacteria, fungi and animals). The rate of production of biomass by heterotrophs is called *secondary productivity*.

. . . net ecosystem productivity, and heterotrophic and ecosystem respiration

Another way to view energy flux in ecosystems involves the concept of *net ecosystem productivity* (NEP, using the same units as GPP or NPP). This acknowledges that the carbon fixed in GPP can leave the system as inorganic carbon (usually carbon dioxide) via either autotrophic respiration (RA) or, after consumption by heterotrophs, via *heterotrophic respiration* (RH)—the latter consisting of respiration by bacteria, fungi and animals. Total *ecosystem respiration* (RE) is the sum of RA and RH. NEP then is equal to GPP – RE. When GPP exceeds RE, the ecosystem is fixing carbon faster than it is being released and thus acts as a carbon sink. When RE exceeds GPP, carbon is being released faster than it is fixed and the ecosystem is a net carbon source. That the rate of ecosystem respiration can exceed GPP may seem paradoxical. However, it is important to note that an ecosystem can receive organic matter from sources other than its own photosynthesis – via the import of dead organic matter that has been produced elsewhere. Organic matter produced by photosynthesis within an ecosystem's boundaries is known as *autochthonous*, whereas that imported from elsewhere is called *allochthonous*.

In what follows we deal first with large-scale patterns in primary productivity (Section 17.2) before considering the factors that limit productivity in terrestrial (Section 17.3) and aquatic (Section 17.4) settings. We then turn to the fate of primary productivity and consider the flux of energy through food webs (Section 17.5), placing particular emphasis on the relative importance of grazer and decomposer systems (we return to food webs and their detailed population interactions in Chapter 20). We finally turn to seasonal and longer term variations in energy flux through ecosystems.

17.2 Patterns in primary productivity

The net primary production of the planet is estimated to be about 105 petagrams of carbon per year (1 $Pg = 10^{15}\,g$) (Geider *et al.*, 2001). Of this, 56.4 $Pg\,C\,year^{-1}$ is produced in terrestrial ecosystems and 48.3 $Pg\,C\,year^{-1}$ in aquatic ecosystems (Table 17.1). Thus, although oceans

primary productivity depends on, but is not solely determined by, solar radiation

Marine	NPP	Terrestrial	NPP
Tropical and subtropical oceans	13.0	Tropical rainforests	17.8
Temperate oceans	16.3	Broadleaf deciduous forests	1.5
Polar oceans	6.4	Mixed broad/needleleaf forests	3.1
Coastal	10.7	Needleleaf evergreen forests	3.1
Salt marsh/estuaries/seaweed	1.2	Needleleaf deciduous forests	1.4
Coral reefs	0.7	Savannas	16.8
		Perennial grasslands	2.4
		Broadleaf shrubs with bare soil	1.0
		Tundra	0.8
		Desert	0.5
		Cultivation	8.0
Total	48.3	Total	56.4

Table 17.1 Net primary production (NPP) per year for major biomes and for the planet in total (in units of petagrams of C). (From Geider *et al.*, 2001.)

cover about two-thirds of the world's surface, they account for less than half of its production. On the land, tropical rainforests and savannas account between them for about 60% of terrestrial NPP, reflecting the large areas covered by these biomes and their high levels of productivity. All biological activity is ultimately dependent on received solar radiation but solar radiation alone does not determine primary productivity. In very broad terms, the fit between solar radiation and productivity is far from perfect because incident radiation can be captured efficiently only when water and nutrients are available and when temperatures are in the range suitable for plant growth. Many areas of land receive abundant radiation but lack adequate water, and most areas of the oceans are deficient in mineral nutrients.

17.2.1 Latitudinal trends in productivity

the productivity of forests, grasslands and lakes follows a latitudinal pattern

In the forest biomes of the world a general latitudinal trend of increasing productivity can be seen from boreal, through temperate, to tropical conditions (Table 17.2). However, there is also considerable variation, much of it due to differences in water availability, local topography and associated variations in microclimate. The same latitudinal trend (and local variations) exists in the above-ground productivity of grassland communities (Figure 17.1). Note the considerable differences in the relative importance of above-ground and below-ground productivity in the different grassland biomes. It is technically difficult to estimate below-ground productivity and early reports of NPP often ignored or underestimated the true values. As far as aquatic communities are concerned, a latitudinal trend is clear in lakes (Brylinski & Mann, 1973) but not in the oceans, where productivity may more often be limited by a shortage of nutrients – very high productivity occurs in marine communities where there are upwellings of nutrient-rich waters, even at high latitudes and low temperatures.

Table 17.2 Gross primary productivity (GPP) of forests at various latitudes in Europe and North and South America, estimated as the sum of net ecosystem productivity and ecosystem respiration (calculated from CO_2 fluxes measured in the forest canopies – only one estimate for tropical forest was included by the reviewers). (From data in Falge *et al.*, 2002.)

Forest type	Range of GPP estimates (g C m⁻² year⁻¹)	Mean of estimates (g C m⁻² year⁻¹)
Tropical rainforest	3249	3249
Temperate deciduous	1122–1507	1327
Temperate coniferous	992–1924	1499
Cold temperate deciduous	903–1165	1034
Boreal coniferous	723–1691	1019

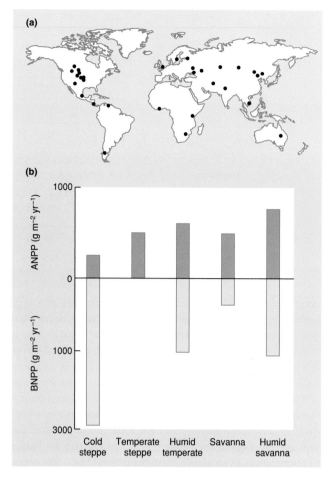

Figure 17.1 (a) The location of 31 grassland study sites included in this analysis. (b) Above-ground net primary productivity (ANPP) and below-ground net primary productivity (BNPP) for five categories of grassland biomes (BNPP not available for temperate steppe). The values in each case are averages for 4–8 grassland studies. The technique involved summing increments in the biomass of live plants, standing dead matter and litter between successive samples in the study period (average 6 years). (From Scurlock *et al.*, 2002.)

The overall trends with latitude suggest that radiation (a resource) and temperature (a condition) may often limit the productivity of communities. But other factors frequently constrain productivity within even narrower limits.

17.2.2 Seasonal and annual trends in primary productivity

The large ranges in productivity in Table 17.2 and the wide confidence intervals in Figure 17.1 emphasize the

productivity shows considerable temporal variation

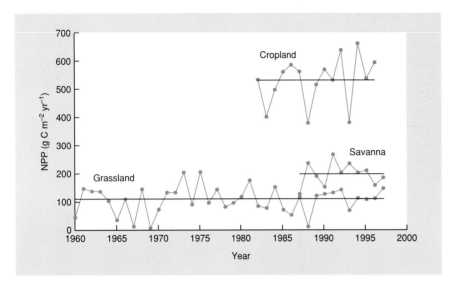

Figure 17.2 Interannual variation in net primary productivity (NPP) in a grassland in Queensland, Australia (above-ground NPP), a cropland in Iowa, USA (total above- and below-ground NPP) and a tropical savanna in Senegal (above-ground NPP). Black horizontal lines show the mean NPP for the whole study period. (After Zheng *et al.*, 2003.)

considerable variation that exists within a given class of ecosystems. It is important to note also that productivity varies from year to year in a single location (Knapp & Smith, 2001). This is illustrated for a temperate cropland, a tropical grassland and a tropical savanna in Figure 17.2. Such annual fluctuations no doubt reflect year-to-year variation in cloudless days, temperature and rainfall. At a smaller temporal scale, productivity reflects seasonal variations in conditions, particularly in relation to the consequences of temperature for the length of the growing season. For example, the period when daily GPP is high persists for longer in temperate than in boreal situations (Figure 17.3). Moreover, the growing season is more extended but the amplitude of seasonal change is smaller in evergreen coniferous forests than in their deciduous counterparts (where the growing season is curtailed by the shedding of leaves in the fall).

17.2.3 Autochthonous and allochthonous production

autochthonous and allochthonous production . . .

All biotic communities depend on a supply of energy for their activities. In most terrestrial systems this is contributed *in situ* by the photosynthesis of green plants – this is autochthonous production. Exceptions exist, however, particularly where colonial animals deposit feces derived from food consumed at a distance from the colony (e.g. bat colonies in caves, seabirds on coastland) – guano is an example of allochthonous organic matter (dead organic material formed outside the ecosystem).

. . . vary in systematic ways in lakes, rivers and estuaries

In aquatic communities, the autochthonous input is provided by the photosynthesis of large plants and attached algae in shallow waters (littoral zone) and by microscopic phytoplankton in the open water. However, a substantial proportion of the organic matter in aquatic communities comes from allochthonous material that arrives in rivers, via groundwater or is blown in by the wind. The relative importance of the two autochthonous sources (littoral and planktonic) and the allochthonous source of organic material in an aquatic system depends on the dimensions of the body of water and the types of terrestrial community that deposit organic material into it.

A small stream running through a wooded catchment derives most of its energy input from litter shed by surrounding vegetation (Figure 17.4). Shading from the trees prevents any significant growth of planktonic or attached algae or aquatic higher plants. As the stream widens further downstream, shading by trees is restricted to the margins and autochthonous primary production increases. Still further downstream, in deeper and more turbid waters, rooted higher plants contribute much less, and the role of the microscopic phytoplankton becomes more important. Where large river channels are characterized by a flood plain, with associated oxbow lakes, swamps and marshes, allochthonous dissolved and particulate organic may be carried to the river channel from its flood plain during episodes of flooding (Junk *et al.*, 1989; Townsend 1996).

The sequence from small, shallow lakes to large, deep ones shares some of the characteristics of the river continuum just discussed (Figure 17.5). A small lake is likely to derive quite a large proportion of its energy from the land because its periphery is large in relation to its area. Small lakes are also usually shallow, so internal littoral production is more important than that by phytoplankton. In contrast, a large, deep lake will derive only limited organic matter from outside (small periphery relative to lake surface area) and littoral production, limited to the shallow margins, may also be low. The organic inputs to the community may then be due almost entirely to photosynthesis by the phytoplankton.

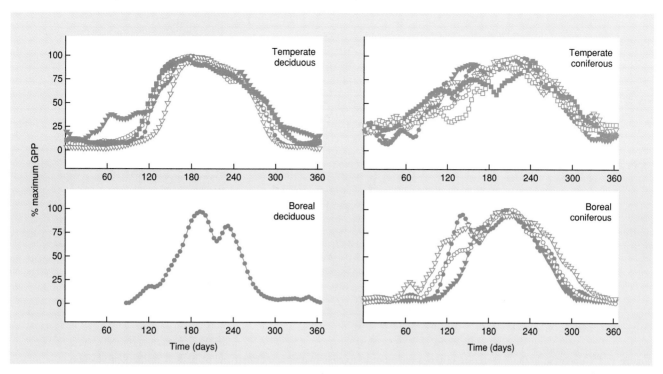

Figure 17.3 Seasonal development of maximum daily gross primary productivity (GPP) for deciduous and coniferous forests in temperate (Europe and North America) and boreal locations (Canada, Scandinavia and Iceland). The different symbols in each panel relate to different forests. Daily GPP is expressed as the percentage of the maximum achieved in each forest during 365 days of the year. (After Falge *et al.*, 2002.)

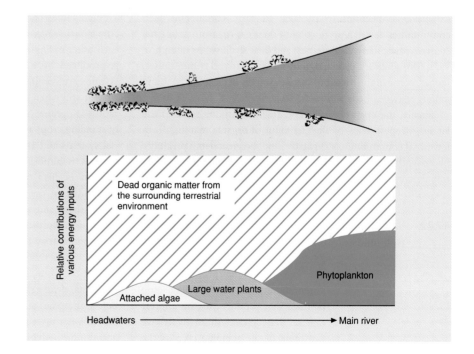

Figure 17.4 Longitudinal variation in the nature of the energy base in stream communities.

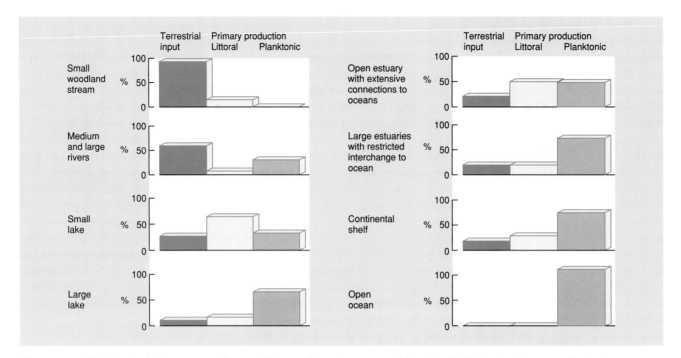

Figure 17.5 Variation in the importance of terrestrial input of organic matter and littoral and planktonic primary production in contrasting aquatic communities.

Estuaries are often highly productive systems, receiving allochthonous material and a rich supply of nutrients from the rivers that feed them. The most important autochthonous contribution to their energy base varies. In large estuarine basins, with restricted interchange with the open ocean and with small marsh peripheries relative to basin area, phytoplankton tend to dominate. By contrast, seaweeds dominate in some open basins with extensive connections to the sea. In turn, continental shelf communities derive a proportion of their energy from terrestrial sources (particularly via estuaries) and their shallowness often provides for significant production by littoral seaweed communities. Indeed, some of the most productive systems of all are to be found among seaweed beds and reefs.

Finally, the open ocean can be described in one sense as the largest, deepest 'lake' of all. The input of organic material from terrestrial communities is negligible, and the great depth precludes photosynthesis in the darkness of the sea bed. The phytoplankton are then all-important as primary producers.

17.2.4　Variations in the relationship of productivity to biomass

NPP : B ratios are very low in forests and very high in aquatic communities

We can relate the productivity of a community to the standing crop biomass that produces it (the interest rate on the capital). Alternatively, we can think of the standing crop as the biomass that is sustained by the productivity (the capital resource that is sustained by earnings). Overall, there is a dramatic difference in the total biomass that exists on land (800 Pg) compared to the oceans (2 Pg) and fresh water (< 0.1 Pg) (Geider *et al.*, 2001). On an areal basis, biomass on land ranges from 0.2 to 200 kg m^{-2}, in the oceans from less than 0.001 to 6 kg m^{-2} and in freshwater biomass is generally less than 0.1 kg m^{-2} (Geider *et al.*, 2001). The average values of net primary productivity (NPP) and standing crop biomass (B) for a range of community types are plotted against each other in Figure 17.6. It is evident that a given value of NPP is produced by a smaller biomass when nonforest terrestrial systems are compared with forests, and the biomass involved is smaller still when aquatic systems are considered. Thus NPP : B ratios (kilograms of dry matter produced per year per kilogram of standing crop) average 0.042 for forests, 0.29 for other terrestrial systems and 17 for aquatic communities. The major reason for this is almost certainly that a large proportion of forest biomass is dead (and has been so for a long time) and also that much of the living support tissue is not photosynthetic. In grassland and scrub, a greater proportion of the biomass is alive and involved in photosynthesis, though half or more of the biomass may be roots. In aquatic communities, particularly where productivity is due mainly to phytoplankton, there is no support tissue, there is no need for roots to absorb water and nutrients, dead cells do not accumulate (they are usually eaten before they die) and the photosynthetic output per kilogram of biomass is thus very high indeed. Another factor that helps to account for high NPP : B ratios in phytoplankton communities is

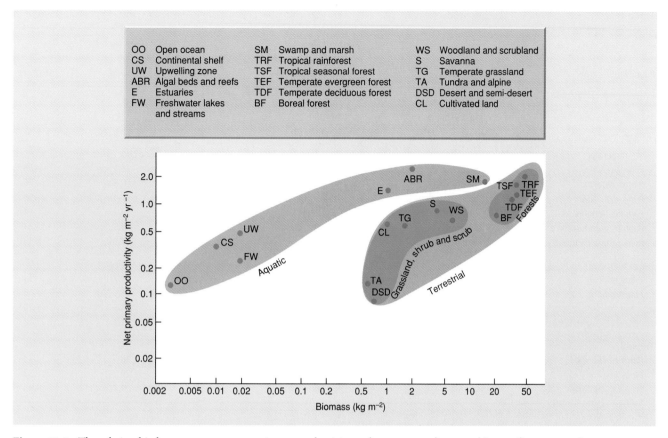

OO	Open ocean	SM	Swamp and marsh	WS	Woodland and scrubland
CS	Continental shelf	TRF	Tropical rainforest	S	Savanna
UW	Upwelling zone	TSF	Tropical seasonal forest	TG	Temperate grassland
ABR	Algal beds and reefs	TEF	Temperate evergreen forest	TA	Tundra and alpine
E	Estuaries	TDF	Temperate deciduous forest	DSD	Desert and semi-desert
FW	Freshwater lakes and streams	BF	Boreal forest	CL	Cultivated land

Figure 17.6 The relationship between average net primary productivity and average standing crop biomass for a range of ecosystems. (Based on data in Whittaker, 1975.)

the rapid turnover of biomass (turnover times of biomass in oceans and fresh waters average 0.02–0.06 years, compared to 1–20 years on land; Geider *et al.*, 2001). The annual NPP shown in the figure is actually produced by a number of overlapping phytoplankton generations, while the standing crop biomass is only the average present at an instant.

NPP : B ratios tend to decrease during successions

Ratios of NPP to biomass tend to decrease during successions. This is because the early successional pioneers are rapidly growing herbaceous species with relatively little support tissue (see Section 16.6). Thus, early in the succession the NPP : B ratio is high. However, the species that come to dominate later are generally slow growing, but eventually achieve a large size and come to monopolize the supply of space and light. Their structure involves considerable investment in nonphotosynthesizing and dead support tissues, and as a consequence their NPP : B ratio is low.

When attention is focused on trees, a common pattern is for above-ground NPP to reach a peak early in succession and then gradually decline by as much as 76%, with a mean reduction of 34% (Table 17.3). The reductions are no doubt partly due to a shift from photosynthesizing to respiring tissues. In addition,

nutrient limitation may become more significant later in the succession or the longer branches and taller stems of older trees may increase resistance to the transpiration stream and thus limit photosynthesis (Gower *et al.*, 1996). Trees characteristic of different stages in succession show different patterns of NPP with stand age. In a subalpine coniferous forest, for example, the early successional whitebark pine (*Pinus albicaulis*) reached a peak above-ground NPP at about 250 years and then declined, whereas the late successional, shade-tolerant subalpine fir (*Abies lasiocarpa*) continued towards a maximum beyond 400 years (Figure 17.7). The late successional species allocated almost twice as much biomass to leaves as its early successional counterpart, and maintained a high photosynthesis : respiration ratio to a greater age (Callaway *et al.*, 2000).

17.3 Factors limiting primary productivity in terrestrial communities

Sunlight, carbon dioxide (CO_2), water and soil nutrients are the resources required for primary production on land, while temperature, a condition, has a strong influence on the rate

Table 17.3   Above-ground net primary productivity (ANPP) for forest age sequences in contrasting biomes. (After Gower *et al.*, 1996.)

Biome/species	Location	Range of stand ages, in years (no. of stands shown in brackets)	ANPP (t dry mass ha⁻¹ year⁻¹)		
			Peak	*Oldest*	*% change*
Boreal					
Larix gmelinii	Yakutsk, Siberia	50–380 (3)	4.9	2.4	−51
Picea abies	Russia	22–136 (10)	6.2	2.6	−58
Cold temperate					
Abies baisamea	New York, USA	0–60 (6)	3.2	1.1	−66
Pinus contorta	Colorado, USA	40–245 (3)	2.1	0.5	−76
Pinus densiflora	Mt Mino, Japan	16–390 (7)	16.1	7.4	−54
Populus tremuloides	Wisconsin, USA	8–83 (5)	11.1	10.7	−4
Populus grandidentata	Michigan, USA	10–70	4.6	3.5	−24
Pseudotsuga menziesii	Washington, USA	22–73 (4)	9.9	5.1	−45
Warm temperate					
Pinus elliottii	Florida, USA	2–34 (6)	13.2	8.7	−34
Pinus radiata	Puruki, NZ (Tahi)	2–6 (5)	28.5	28.5	0
	(Rue)	2–7 (6)	29.2	23.5	−20
	(Toru)	2–8 (7)	31.1	31.1	0
Tropical					
Pinus caribaea	Afaka, Nigeria	5–15 (4)	19.2	18.5	−4
Pinus kesiya	Meghalaya, India	1–22 (9)	30.1	20.1	−33
Tropical rainforest	Amazonia	1–200 (8)	13.2	7.2	−45

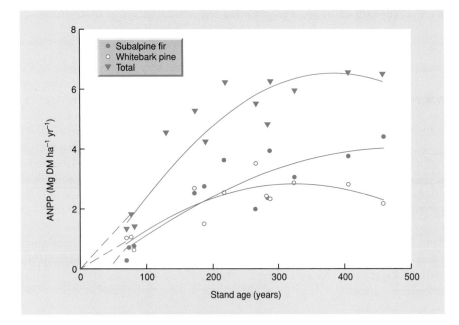

Figure 17.7   Annual above-ground net primary productivity (ANPP) (Mg dry matter ha⁻¹ year⁻¹) in stands of different ages in a subalpine coniferous forest in Montana, USA: early successional whitebark pine, late successional subalpine fir, and total ANPP. (After Callaway *et al.*, 2000.)

of photosynthesis. CO_2 is normally present at a level of around 0.03% of atmospheric gases. Turbulent mixing and diffusion prevent the CO_2 concentration from varying much from place to place, except in the immediate neighborhood of a leaf, and

CO_2 probably plays little role in determining differences between the productivities of different communities (although global increases in CO_2 concentration are expected to have profound effects (e.g. DeLucia *et al.*, 1999). On the other hand, the quality and quantity of light, the availability of water and nutrients, and temperature all vary dramatically from place to place. They are all candidates for the role of limiting factor. Which of them actually sets the limit to primary productivity?

17.3.1   Inefficient use of solar energy

terrestrial communities use radiation inefficiently

Depending on location, something between 0 and 5 joules of solar energy strikes each square meter of the earth's surface every minute. If all this were converted by photosynthesis to plant biomass (that is, if photosynthetic efficiency were 100%) there would be a prodigious generation of plant material, one or two orders of magnitude greater than recorded values. However, much of this solar energy is unavailable for use by plants. In particular, only about 44% of incident shortwave radiation occurs at wavelengths suitable for photosynthesis. Even when this is taken into account, though, productivity still falls well below the maximum possible. Photosynthetic efficiency has two components – the efficiency with which light is intercepted by leaves and the efficiency with which intercepted light is converted by photosynthesis to new biomass (Stenberg *et al.*, 2001). Figure 17.8 shows the range in overall net photosynthetic efficiencies (percentage of incoming photosynthetically active radiation (PAR) incorporated into above-ground NPP) in seven coniferous forests, seven deciduous forests and eight desert communities studied as part of the International Biological Programme (see Section 17.1). The conifer communities had the highest efficiencies, but these were only between 1 and 3%. For a similar level of incoming radiation, deciduous forests achieved 0.5–1%, and, despite their greater energy income, deserts were able to convert only 0.01–0.2% of PAR to biomass.

productivity may still be limited by a shortage of PAR

However, the fact that radiation is not used efficiently does not in itself imply that it does not limit community productivity. We would need to know whether at increased intensities of radiation the productivity increased or remained unchanged. Some of the evidence given in Chapter 3 shows that the intensity of light during part of the day is below the optimum for canopy photosynthesis. Moreover, at peak light intensities, most canopies still have their lower leaves in relative gloom, and would almost certainly photosynthesize faster if the light intensity were higher. For C_4 plants a saturating intensity of radiation never seems to be reached, and the implication is that productivity may in fact be limited by a shortage of PAR even under the brightest natural radiation.

There is no doubt, however, that what radiation is available would be used more efficiently if other resources were in abundant supply. The much higher values of community productivity recorded from agricultural systems bear witness to this.

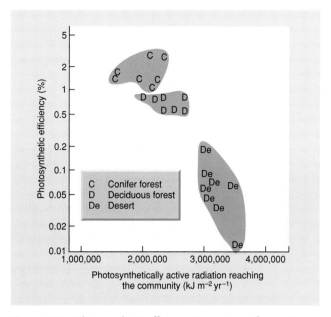

Figure 17.8   Photosynthetic efficiency (percentage of incoming photosynthetically active radiation converted to above-ground net primary productivity) for three sets of terrestrial communities in the USA. (After Webb *et al.*, 1983.)

17.3.2   Water and temperature as critical factors

shortage of water may be a critical factor

The relationship between the NPP of a wide range of ecosystems on the Tibetan Plateau and both precipitation and temperature is illustrated in Figure 17.9. Water is an essential resource both as a constituent of cells and for photosynthesis. Large quantities of water are lost in transpiration – particularly because the stomata need to be open for much of the time for CO_2 to enter. It is not surprising that the rainfall of a region is quite closely correlated with its productivity. In arid regions, there is an approximately linear increase in NPP with increase in precipitation, but in the more humid forest climates there is a plateau beyond which productivity does not continue to rise. Note that a large amount of precipitation is not necessarily equivalent to a large amount of water available for plants; all water in excess of field capacity will drain away if it can. A positive relationship between productivity and mean annual temperature can also be seen in Figure 17.9. However, the pattern can be expected to be complex because, for example, higher temperatures are associated with rapid water loss through evapotranspiration; water shortage may then become limiting more quickly.

interaction of temperature and precipitation

To unravel the relationships between productivity, rainfall and temperature, it is more instructive to concentrate on a single ecosystem

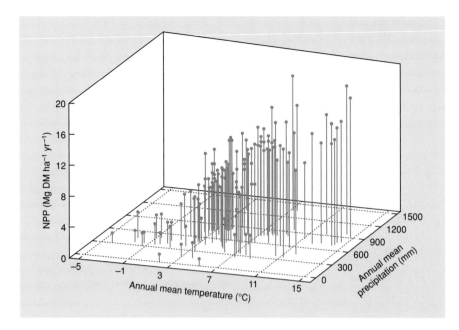

Figure 17.9   Relationship between total net primary productivity (Mg dry matter ha^{-1} year^{-1}) and annual precipitation and temperature for ecosystems on the Tibetan Plateau. The ecosystems include forests, woodlands, shrublands, grasslands and desert. (After Luo *et al.*, 2002.)

type. Above-ground NPP was estimated for a number of grassland sites along two west-to-east precipitation gradients in the Argentinian pampas. One of these gradients was in mountainous country and the other in the lowlands. Figure 17.10 shows the relationship between an index of above-ground NPP (ANPP) and precipitation and temperature for the two sets of sites. There are strong positive relationships between ANPP and precipitation but

the slopes of the relationships differed between the two environmental gradients (Figure 17.10a).

The relationships between ANPP and temperature are similar for two further environmental gradients (both north-to-south elevation transects) in Figure 17.10b – both show a hump-shaped pattern. This probably results from the overlap of two effects of increasing temperature: a positive effect on the length of the

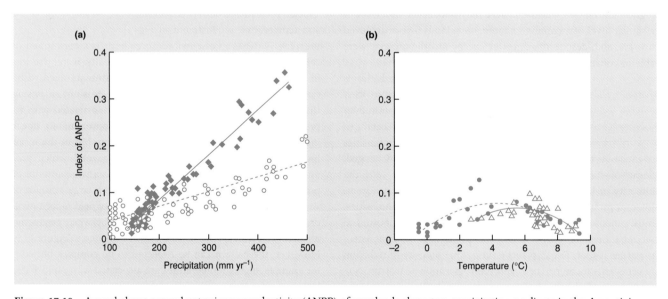

Figure 17.10   Annual above-ground net primary productivity (ANPP) of grasslands along two precipitation gradients in the Argentinian pampas. NPP is shown as an index based on satellite radiometric measurements with a known relationship to absorbed photosynthetically active radiation in plant canopies. (a) NPP in relation to annual precipitation. (b) NPP in relation to annual mean temperature. Open circles and diamonds represent sites along precipitation gradients in the lowland and mountainous regions respectively. Closed circles and triangles represent sites along two elevation transects. (After Jobbagy *et al.*, 2002.)

growing season and a negative effect through increased evapo-transpiration at higher temperatures. Because temperature is the main constraint on productivity at the cool end of the gradients, an increase in NPP is observed as we move from the coolest to warmer sites. However, there is a temperature value above which the growing season does not lengthen and the dominating effect of increasing temperature is now to increase evapo-transpiration, thus reducing water availability and curtailing NPP (Epstein *et al.*, 1997).

productivity and the structure of the canopy   Water shortage has direct effects on the rate of plant growth but also leads to the development of less dense vegetation. Vegetation that is sparse intercepts less light (much of which falls on bare ground). This wastage of solar radiation is the main cause of the low productivity in many arid areas, rather than the reduced photosynthetic rate of drought-affected plants. This point is made by comparing the productivity per unit weight of leaf biomass instead of per unit area of ground for the studies shown in Figure 17.8. Coniferous forest produced 1.64 g g^{-1} year^{-1}, deciduous forest 2.22 g g^{-1} year^{-1} and desert 2.33 g g^{-1} year^{-1}.

17.3.3 Drainage and soil texture can modify water availability and thus productivity

There was a notable difference in the slopes of the graphs of NPP against precipitation for the mountainous and lowland sites in Figure 17.10. The slope was much lower in the mountainous case and it seems likely that the steeper terrain in this region resulted in a higher rate of water runoff from the land and, thus, a lower efficiency in the use of precipitation (Jobbagy *et al.*, 2002).

soil texture can influence productivity   A related phenomenon has been observed when forest production on sandy, well-drained soils is compared with soils consisting of finer particle sizes. Data are available for the accumulation through time of forest biomass at a number of sites where all the trees had been removed by a natural disturbance or human clearance. For forests around the world, Johnson *et al.* (2000) have reported the relationship between above-ground biomass accumulation (a rough index of ANPP) and accumulated growing season degree-days (stand age in years × growing season temperature × growing season as a proportion of the year). In effect, 'growing season degree-days' combine the time for which the stand has been accumulating biomass with the average temperature at the site in question. Figure 17.11 shows that productivity of broadleaf forests is generally much lower, for a given value for growing season degree-days, when the forest is on sandy soil. Such soils have less favorable soil-moisture-holding capacities and this accounts in some measure for their poorer productivity. In addition, however, nutrient retention may be lower in coarse soils, further

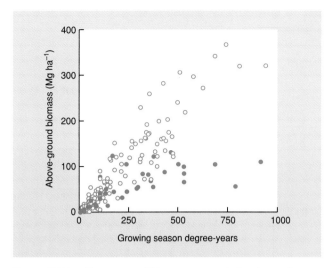

Figure 17.11   Above-ground biomass accumulation (a rough index of NPP) expressed as megagrams (= 10^6 g) per hectare in relation to accumulated growing season degree-days in broadleaf forest stands growing on sandy or nonsandy soils. ○, nonsandy soils; ●, sandy soils. (After Johnson *et al.*, 2000.)

reducing productivity compared to soils with finer texture. This was confirmed by Reich *et al.* (1997) who, in their compilation of data for 50 North American forests, found that soil nitrogen availability (estimated as annual net nitrogen mineralization rate) was indeed lower in sandier soils and, moreover, that ANPP was lower per unit of available nitrogen in sandy situations.

17.3.4 Length of the growing season

The productivity of a community can be sustained only for that period of the year when the plants have photosynthetically active foliage. Deciduous trees have a self-imposed limit on the period when they bear foliage. In general, the leaves of deciduous species photosynthesize fast and die young, whereas evergreen species have leaves that photosynthesize slowly but for longer (Eamus, 1999). Evergreen trees hold a canopy throughout the year, but during some seasons they may barely photosynthesize at all or may even respire faster than they photosynthesize. Evergreen conifers tend to dominate in nutrient-poor and cold conditions, perhaps because in other situations their seedlings are outcompeted by their faster growing deciduous counterparts (Becker, 2000).

The latitudinal patterns in forest productivity seen earlier (see Table 17.2) are largely the result of differences in the number of days when there is active photosynthesis. In this context, Black *et al.* (2000) measured net ecosystem pro-   length of the growing season: a pervasive influence on productivity

ductivity (NEP) in a boreal deciduous forest in Canada for 4 years. First leaf emergence occurred considerably earlier in 1998 when

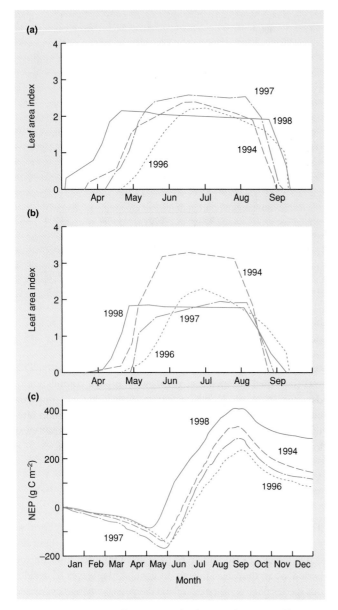

Figure 17.12 Seasonal patterns in leaf area index (area of leaves divided by ground area beneath the foliage) of (a) overstory aspen (*Populus tremuloides*) and (b) understory hazelnut (*Corylus cornuta*) in a boreal deciduous forest during four study years with contrasting spring temperatures. (c) Cumulative net ecosystem productivity (NEP). (After Black *et al.*, 2000.)

the April/May temperature was warmest (9.89°C) and a month later in 1996 when the April/May temperature was coldest (4.24°C) (Figure 17.12a, b). Equivalent spring temperatures in 1994 and 1997 were 6.67 and 5.93°C. The difference in the length of the growing season in the four study years can be gauged from the pattern of cumulative NEP (Figure 17.12c). During winter and early spring, NEP was negative because ecosystem respiration exceeded gross

ecosystem productivity. NEP became positive earlier in warmer years (particularly 1998) so that overall total carbon sequestered by the ecosystem in the four years was 144, 80, 116 and 290 g C m^{-2} year^{-1} for 1994, 1996, 1997 and 1998, respectively.

In our earlier discussion of the study of Argentinian pampas communities (see Figure 17.10) we noted that higher NPP was not only directly affected by precipitation and temperature but was partly determined by length of the growing season. Figure 17.13 shows that the start of the growing season was positively related to mean annual temperature (paralleling the boreal forest study above), whereas the end of the growing season was determined partly by temperature but also by precipitation (it ended earlier where temperatures were high and precipitation was low). Again we see a complex interaction between water availability and temperature.

17.3.5 Productivity may be low because mineral resources are deficient

No matter how brightly the sun shines and how often the rain falls, and no matter how equable the temperature is, productivity must be low if there is no soil in a terrestrial community, or if the soil is deficient in essential mineral nutrients. The geological conditions that determine slope and aspect also determine whether a soil forms, and they have a large, though not wholly dominant, influence on the mineral content of the soil. For this reason, a mosaic of different levels of community productivity develops within a particular climatic regime. Of all the mineral nutrients, the one that has the most pervasive influence on community productivity is fixed nitrogen (and this is invariably partly or mainly biological, not geological, in origin, as a result of nitrogen fixation by microorganisms). There is probably no agricultural system that does not respond to applied nitrogen by increased primary productivity, and this may well be true of natural vegetation as well. Nitrogen fertilizers added to forest soils almost always stimulate forest growth.

the crucial importance of nutrient availability

The deficiency of other elements can also hold the productivity of a community far below that of which it is theoretically capable. A classic example is deficiency of phosphate and zinc in South Australia, where the growth of commercial forest (Monterey pine, *Pinus radiata*) is made possible only when these nutrients are supplied artificially. In addition, many tropical systems are primarily limited by phosphorus.

17.3.6 Résumé of factors limiting terrestrial productivity

The ultimate limit on the productivity of a community is determined by the amount of incident radiation that it receives – without this, no photosynthesis can occur.

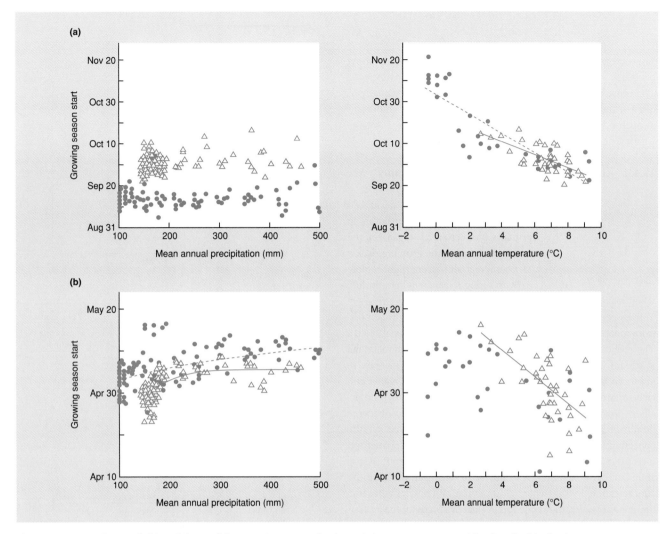

Figure 17.13 (a) Start and (b) end dates of the growing season for Argentinian pampas communities described in Section 17.3.2. Circles represent sites along the precipitation gradient in the mountainous region and triangles represent sites along the lowland gradient. (After Jobbagy *et al.*, 2002.)

Incident radiation is used inefficiently by all communities. The causes of this inefficiency can be traced to: (i) shortage of water restricting the rate of photosynthesis; (ii) shortage of essential mineral nutrients, which slows down the rate of production of photosynthetic tissue and its effectiveness in photosynthesis; (iii) temperatures that are lethal or too low for growth; (iv) an insufficient depth of soil; (v) incomplete canopy cover, so that much of the incident radiation lands on the ground instead of on foliage (this may be because of seasonality in leaf production and leaf shedding *or* because of defoliation by grazing animals, pests and diseases); and (vi) the low efficiency with which leaves photosynthesize – under ideal conditions, efficiencies of more than 10% (of PAR) are hard to achieve even

in the most productive agricultural systems. However, most of the variation in primary productivity of world vegetation is due to factors (i) to (v), and relatively little is accounted for by intrinsic differences between the photosynthetic efficiencies of the leaves of the different species.

In the course of a year, the productivity of a community may (and probably usually will) be limited by a succession of the factors (i) to (v). In a grassland community, for instance, the primary productivity may be far below the theoretical maximum because the winters are too cold and light intensity is low, the summers are too dry, the rate of nitrogen mobilization is too slow, and for periods grazing animals may reduce the standing crop to a level at which much incident light falls on bare ground.

17.4   Factors limiting primary productivity in aquatic communities

The factors that most frequently limit the primary productivity of aquatic environments are the availability of light and nutrients. The most commonly limiting nutrients are nitrogen (usually as nitrate) and phosphorus (phosphate), but iron can be important in open ocean environments.

17.4.1   Limitation by light and nutrients in streams

in small forest streams, light and nutrients interact to determine productivity

Streams flowing through deciduous forests undergo marked transitions in primary production by algae on the stream bed during the growing season as conditions shift from light-replete early in spring to severely light-limited when leaves develop on the overhanging trees. In a stream in Tennessee, leaf emergence reduced PAR reaching the stream bed from more than 1000 to less than 30 μmol m^{-2} s^{-1} (Hill *et al.*, 2001). The reduction in PAR was paralleled by an equally dramatic fall in stream GPP (Figure 17.14). This is despite a large increase in photosynthetic efficiency from less than 0.3 to 2%; the higher efficiencies arose both because existing taxa acclimated physiologically to low irradiances and because more efficient taxa became dominant later in the season. Intriguingly, as PAR levels fell, the concentration of both nitrate (Figure 17.14a) and phosphate rose. It seems that nutrients limited primary production when PAR was abundant early in spring, with uptake by the algae reducing the concentration in the water at this time. When light became limiting, however, the reduction in algal productivity meant that less of the available nutrients were removed from the supply in the flowing water.

17.4.2   Nutrients in lakes

productivity in lakes . . .

. . . shows a pervasive role for nutrients . . .

Like streams, lakes receive nutrients by the weathering of rocks and soils in their catchment areas, in the rainfall and as a result of human activity (fertilizers and sewage input). They vary considerably in nutrient availability. A study of 12 Canadian lakes shows a clear relationship between gross primary productivity (GPP) and phosphorus concentration and demonstrates the importance of nutrients in limiting lake productivity (Figure 17.15). Note that GPP easily exceeded ecosystem respiration in most lakes, emphasizing the overriding importance of autochthonous production in these lakes. The outlier in the top right corner of Figure 17.15b was atypical of the study sites because it received sewage effluent; here the allochthonous input of organic matter led to a higher consumption than production of organic carbon in the lake.

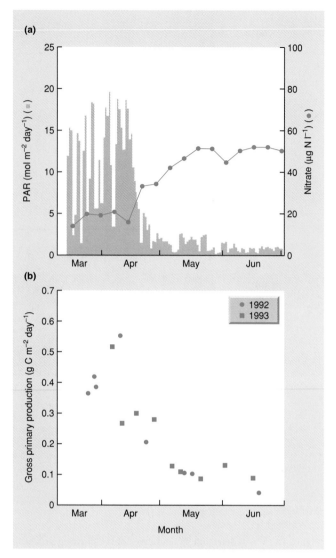

Figure 17.14   (a) Photosynthetically active radiation (PAR) reaching the bed of a Tennessee stream (bars) and stream water nitrate concentration (circles) during the spring of 1992 (the patterns were very similar in 1993). (b) Gross primary productivity in the stream during the spring in 1992 and 1993 (calculated on the basis of whole stream diurnal changes in oxygen concentration). (After Hill *et al.*, 2001.)

It is worth noting that the balance of radiant energy relative to the availability of key nutrients can affect C : N : P ratios (stoichiometry) in the tissues of primary producers. Thus, Sterner *et al.* (1997b) found in some phosphorus-deficient Canadian lakes that the availability of PAR relative to total phosphorus (PAR : TP) affected the balance of carbon fixation and phosphorus uptake in algal communities and, thereby, caused variations in C : P ratios in

. . . whose availability may interact with radiant energy to affect algal 'stoichiometry' (C : N : P ratios)

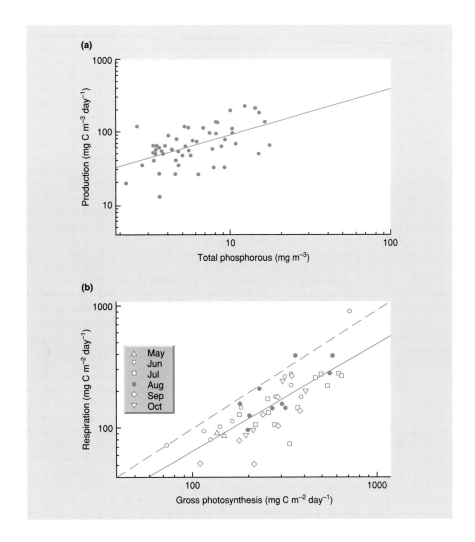

Figure 17.15 (a) Relationship between the gross primary productivity of phytoplankton (microscopic plants) in the open water of some Canadian lakes and phosphorus concentration. (b) The relationship between ecosystem respiration and gross photosynthesis measured on various dates in the study lakes. The dashed line shows where respiration equals GPP. The solid line shows the regression line for the relationship. Metabolic measurements were made in bottles in the laboratory at lake temperatures on depth-integrated water samples taken from the field. (After Carignan *et al.*, 2000.)

living algal cells and algal detritus. The zooplankton that consume live algae and the decomposers and detritivores that depend on algal detritus each have specific nutrient requirements, and these are very different from the nutrient ratios in algae. Thus, the changes in algal stoichiometry noted by Sterner *et al.* have consequences for heterotrophic metabolism and productivity. We consider elsewhere how such imbalances between the stoichiometry of plant tissue and of its consumers affect food web interactions, decomposition and nutrient cycling (see Sections 11.2.4, 17.5.4 and 18.2.5).

17.4.3 Nutrients and the importance of upwellings in oceans

rich supplies of nutrients in marine environments . . .

. . . from estuaries . . .

In the oceans, locally high levels of primary productivity are associated with high nutrient inputs from two sources. First, nutrients may flow continuously into coastal shelf regions from estuar-

ies. An example is provided in Figure 17.16. Productivity in the inner shelf region is particularly high both because of high nutrient concentrations and because the relatively clear water provides a reasonable depth within which net photosynthesis is positive (the *euphotic zone*). Closer to land, the water is richer in nutrients but is highly turbid and its productivity is less. The least productive zones are on the outer shelf (and open ocean) where it might be expected that primary productivity would be high because the water is clear and the euphotic zone is deep. Here, however, productivity is low because of the extremely low concentrations of nutrients.

Ocean upwellings are a second source of high nutrient concentrations. These occur on continental shelves where the wind is consistently parallel to, or at a slight angle to, the coast. As a result, water moves offshore and is replaced by cooler, nutrient-rich water originating from the bottom, where nutrients have been accumulating by sedimentation. Strong upwellings can also occur adjacent to submarine ridges, as well as in areas of very strong currents. Where it reaches the surface,

. . . and upwellings

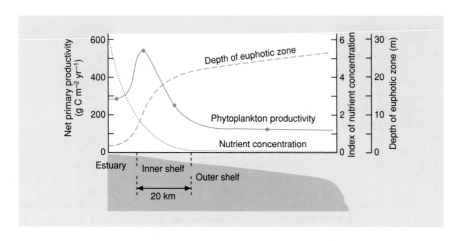

Figure 17.16 Variation in phytoplankton net primary productivity, nutrient concentration and euphotic depth on a transect from the coast of Georgia, USA, to the edge of the continental shelf. (After Haines, 1979.)

the nutrient-rich water sets off a bloom of phytoplankton production. A chain of heterotrophic organisms takes advantage of the abundant food, and the great fisheries of the world are located in these regions of high productivity.

iron as a limiting factor in oceans

Recently, iron has been identified as a limiting nutrient that potentially affects about one-third of the open ocean (Geider *et al.*, 2001). Iron, which is very insoluble in seawater, is ultimately derived from wind-blown particulate material, and large areas of ocean receive insufficient amounts. When iron is added experimentally to ocean areas, massive blooms of phytoplankton can result (Coale *et al.*, 1996); such blooms are also likely to occur when large storms supply land-derived iron to the oceans.

While nutrients are the most influential factors for local ocean productivity, temperature and PAR also play a role at a larger scale (Figure 17.17).

temperature and PAR also affect productivity

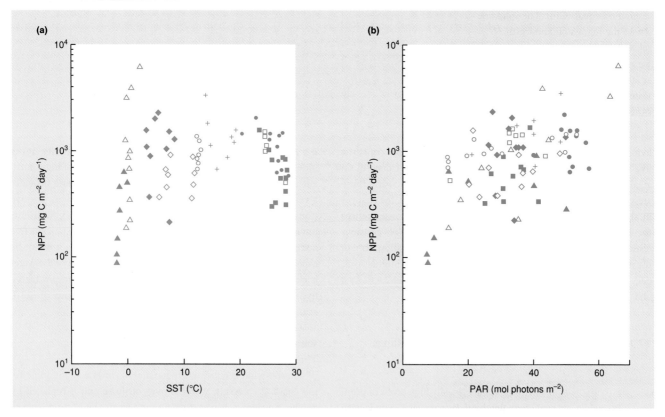

Figure 17.17 Relationships between daily depth-integrated estimates of net primary production (NPP) and: (a) sea surface temperature (SST), and (b) above-water daily photosynthetically available radiation (PAR). The different symbols relate to different data sets from various oceans. (After Campbell *et al.*, 2002.)

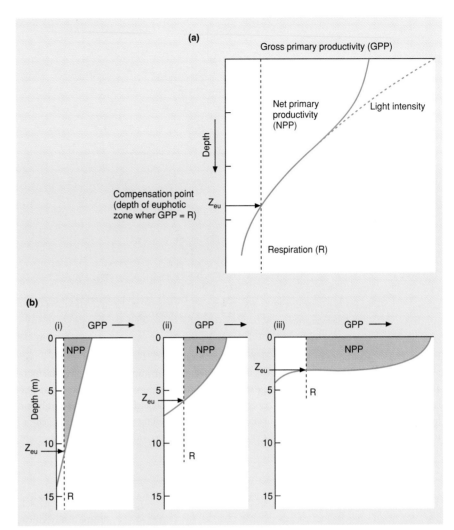

Figure 17.18 (a) The general relationship with depth, in a water body, of gross primary productivity (GPP), respiratory heat loss (R) and net primary productivity (NPP). The compensation point (or depth of the euphotic zone, eu) occurs at the depth (Z_{eu}) where GPP just balances R and NPP is zero. (b) Total NPP increases with nutrient concentration in the water (lake iii > ii > i). Increasing fertility itself is responsible for greater biomasses of phytoplankton and a consequent decrease in the depth of the euphotic zone.

This has significance for our ability to estimate ocean primary productivity because sea surface temperature and PAR (together with surface chlorophyll concentration, another factor correlated with NPP) can be measured using satellite telemetry.

17.4.4 Productivity varies with depth in aquatic communities

phytoplankton productivity varies with depth

Although the concentration of a limiting nutrient usually determines the productivity of aquatic communities on an areal basis, in any given water body there is also considerable variation with depth as a result of attenuation of light intensity. Figure 17.18a shows how GPP declines with depth. The depth at which GPP is just balanced by phytoplankton respiration, R, is known as the compensation point. Above this, NPP is positive. Light is absorbed by water molecules as well as by dissolved and particulate matter, and it declines exponentially with depth. Near the surface, light is superabundant, but at greater depths its supply is limited and light intensity ultimately determines the extent of the euphotic zone. Very close to the surface, particularly on sunny days, there may even be photoinhibition of photosynthesis. This seems to be due largely to radiation being absorbed by the photosynthetic pigments at such a rate that it cannot be used via the normal photosynthetic channels, and it overflows into destructive photo-oxidation reactions.

The more nutrient-rich a water body is, the shallower its euphotic zone is likely to be (Figure 17.18b). This is not really a paradox. Water bodies with higher nutrient concentrations usually possess a greater biomasses of phytoplankton that absorb light and reduce its availability at greater depth. (This is exactly analogous to the shading influence of the tree canopy in a forest, which may remove up to 98% of the radiant energy before it can reach the ground layer vegetation or, as we saw above, a stream bed.) Even quite shallow lakes, if sufficiently fertile, may be devoid of water weeds on the bottom because of shading by phytoplankton. The relationships shown in Figure 17.18a and b are derived from lakes but the pattern is qualitatively similar in ocean environments (Figure 17.19).

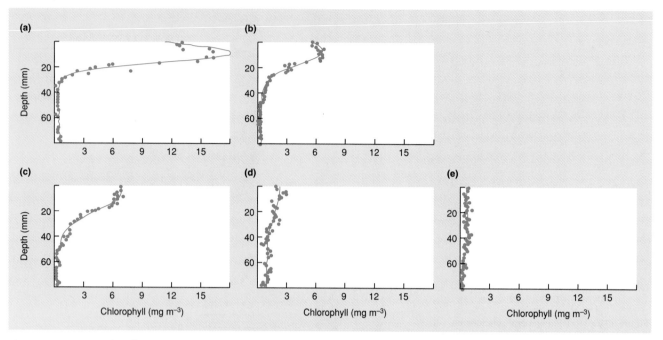

Figure 17.19 Examples of vertical chlorophyll profiles recorded in the ocean off the coast of Namibia. Example (a) is typical of locations associated with ocean upwelling: as cold upwelled water warms up, a surface phytoplankton bloom develops, reducing light penetration and thus productivity in deeper water. Example (b) illustrates how peak abundance can shift to deeper water as a surface bloom in an upwelling area depletes the nutrient concentrations there. The surface phytoplankton bloom in example (c) is less dramatic than in (a) (perhaps reflecting lower nutrient concentrations in the upwelling water); as a result, chlorophyll concentration remains relatively high to a greater depth. Examples (d) and (e) are for locations where nutrient concentrations are much lower. (After Silulwane *et al.*, 2001.)

17.5   The fate of energy in ecosystems

Secondary productivity is defined as the rate of production of new biomass by heterotrophic organisms. Unlike plants, heterotrophic bacteria, fungi and animals cannot manufacture from simple molecules the complex, energy-rich compounds they need. They derive their matter and energy either directly by consuming plant material or indirectly from plants by eating other heterotrophs. Plants, the primary producers, comprise the first trophic level in a community; primary consumers occur at the second trophic level; secondary consumers (carnivores) at the third, and so on.

17.5.1   Relationships between primary and secondary productivity

there is a general positive relationship between primary and secondary productivity

Since secondary productivity depends on primary productivity, we should expect a positive relationship between the two variables in communities. Turning again to the stream study described

in Section 17.4.1, recall that primary productivity declined dramatically during the summer when a canopy of tree leaves above the stream shaded out most of the incident radiation. A principal grazer of the algal biomass is the snail *Elimia clavaeformis*. Figure 17.20a shows how the growth rate of individual snails in the stream was lowest in the summer; there was a statistically significant positive relationship between snail growth and monthly stream bed PAR (Hill *et al.*, 2001). Figure 17.20b–d illustrates the general relationship between primary and secondary productivity in aquatic and terrestrial examples. Secondary productivity by zooplankton, which principally consume phytoplankton cells, is positively related to phytoplankton productivity in a range of lakes in different parts of the world (Figure 17.20b). The productivity of heterotrophic bacteria in lakes and oceans also parallels that of phytoplankton (Figure 17.20c); they metabolize dissolved organic matter released from intact phytoplankton cells or produced as a result of 'messy feeding' by grazing animals. Figure 17.20d shows how the productivity of *Geospiza fortis* (one of Darwin's finches), measured in terms of average brood size on an island in the Galápagos archipelago, is related to annual rainfall, itself an index of primary productivity.

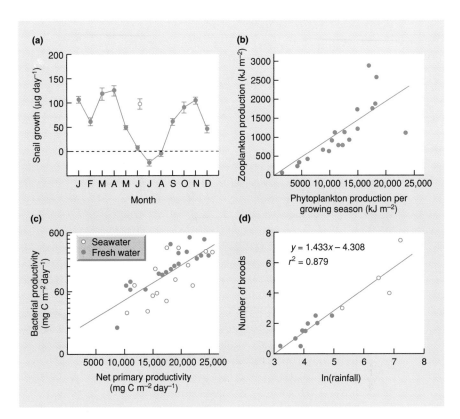

Figure 17.20 (a) Seasonal pattern of snail growth (mean increase in weight of individually marked snails during a month on the stream bed ± SE). The open circle represents growth at a nearby unshaded stream site in June. (After Hill *et al.*, 2001.) (b) Relationship between primary and secondary productivity for zooplankton in lakes. (After Brylinsky & Mann, 1973.) (c) Relationship between bacterial and phytoplankton productivity in fresh water and seawater. (After Cole *et al.* 1988.) (d) Mean clutch size of *Geospiza fortis* in relation to annual rainfall (positively related to primary productivity); the open circles are for particularly wet years when El Niño weather events occurred. (After Grant *et al.*, 2000.).

most primary productivity does not pass through the grazer system

A general rule in both aquatic and terrestrial ecosystems is that secondary productivity by herbivores is approximately an order of magnitude less than the primary productivity upon which it is based. This is a consistent feature of all grazer systems: that part of the trophic structure of a community that depends, at its base, on the consumption of *living* plant biomass (in the ecosystem context we use 'grazer' in a different sense to its definition in Chapter 9). It results in a pyramidal structure in which the productivity of plants provides a broad base upon which a smaller productivity of primary consumers depends, with a still smaller productivity of secondary consumers above that. Trophic levels may also have a pyramidal structure when expressed in terms of density or biomass. (Elton (1927) was the first to recognize this fundamental feature of community architecture and his ideas were later elaborated by Lindemann (1942).) But there are many exceptions. Food chains based on trees will certainly have larger numbers (but *not* biomass) of herbivores per unit area than of plants, while chains dependent on phytoplankton production may give inverted pyramids of biomass, with a highly productive but small biomass of short-lived algal cells maintaining a larger biomass of longer lived zooplankton.

The productivity of herbivores is invariably less than that of the plants on which they feed. Where has the missing energy gone? First, not all of the plant biomass produced is consumed alive by herbivores. Much dies without being grazed and supports the decomposer community (bacteria, fungi and detritivorous animals). Second, not all plant biomass eaten by herbivores (nor herbivore biomass eaten by carnivores) is assimilated and available for incorporation into consumer biomass. Some is lost in feces, and this also passes to the decomposers. Third, not all energy that has been assimilated is actually converted to biomass. A proportion is lost as respiratory heat. This occurs both because no energy conversion process is ever 100% efficient (some is lost as unusable random heat, consistent with the second law of thermodynamics) and also because animals do work that requires energy, again released as heat. These three energy pathways occur at all trophic levels and are illustrated in Figure 17.21.

17.5.2 Possible pathways of energy flow through a food web

Figure 17.22 provides a complete description of the trophic structure of a community. It consists of the grazer system pyramid of productivity, but with two additional elements of realism.

alternative pathways that energy can trace through the community

Most importantly, it adds a *decomposer system* – this is invariably coupled to the grazer system in communities. Secondly, it recognizes that there are subcomponents of each trophic level in each

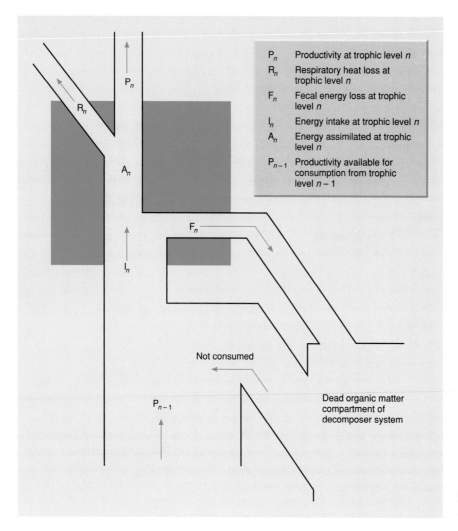

Figure 17.21 The pattern of energy flow through a trophic compartment.

The legend in Figure 17.21 reads:

P_n Productivity at trophic level n

R_n Respiratory heat loss at trophic level n

F_n Fecal energy loss at trophic level n

I_n Energy intake at trophic level n

A_n Energy assimilated at trophic level n

P_{n-1} Productivity available for consumption from trophic level $n-1$

Not consumed

Dead organic matter compartment of decomposer system

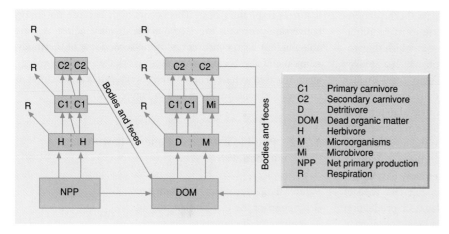

The legend in Figure 17.22 reads:

C1   Primary carnivore
C2   Secondary carnivore
D   Detritivore
DOM Dead organic matter
H   Herbivore
M   Microorganisms
Mi   Microbivore
NPP Net primary production
R   Respiration

Figure 17.22 A generalized model of trophic structure and energy flow through a food web. (After Heal & MacLean, 1975.)

subsystem that operate in different ways. Thus a distinction is made between microbes and detritivores that occupy the same trophic level and utilize dead organic matter, and between consumers of microbes (microbivores) and of detritivores. Displayed in Figure 17.22 are the possible routes that a joule of energy, fixed in net primary production, can take as it is dissipated on its path through the community. A joule of energy may be consumed and assimilated by a herbivore that uses part of it to do work and loses it as respiratory heat. Or it might be consumed by a herbivore and later assimilated by a carnivore that dies and enters the dead organic matter compartment. Here, what remains of the joule may be assimilated by a fungal hypha and consumed by a soil mite, which uses it to do work, dissipating a further part of the joule as heat. At each consumption step, what remains of the joule may fail to be assimilated and may pass in the feces to be dead organic matter, or it may be assimilated and respired, or assimilated and incorporated into the growth of body tissue (or the production of offspring – as in the case of broods of the bird in Figure 17.20d). The body may die and what remains of the joule enter the dead organic matter compartment, or it may be captured alive by a consumer in the next trophic level where it meets a further set of possible branching pathways. Ultimately, each joule will have found its way out of the community, dissipated as respiratory heat at one or more of the transitions in its path along the food chain. Whereas a molecule or ion may cycle endlessly through the food chains of a community, energy passes through just once.

The possible pathways in the grazer and decomposer systems are the same, with one critical exception – feces and dead bodies are lost to the grazer system (and enter the decomposer system), but feces and dead bodies from the decomposer system are simply sent back to the dead organic matter compartment at its base. This has a fundamental significance. The energy available as dead organic matter may finally be completely metabolized – and all the energy lost as respiratory heat – even if this requires several circuits through the decomposer system. The exceptions to this are situations where: (i) matter is exported out of the local environment to be metabolized elsewhere, for example detritus washed out of a stream; and (ii) local abiotic conditions are very unfavorable to decomposition processes, leaving pockets of incompletely metabolized high-energy matter, otherwise known as oil, coal and peat.

17.5.3 The importance of transfer efficiencies in determining energy pathways

the relative importance of energy pathways depends on three transfer efficiencies: . . .

The proportions of net primary production that flow along each of the possible energy pathways depend on *transfer efficiencies* in the way energy is used and passed from one step to the next. A knowledge of the values of just three categories of transfer efficiency is all that is required to predict the pattern of energy flow. These are *consumption efficiency* (CE) *assimilation efficiency* (AE) and *production efficiency* (PE).

. . . consumption efficiency, . . .

consumption efficiency,
$$CE = I_n/P_{n-1} \times 100.$$

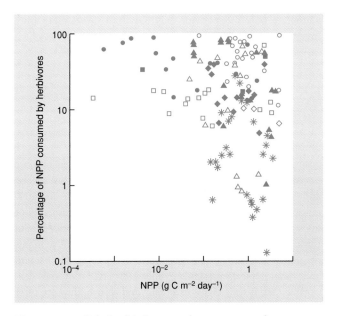

Figure 17.23 Relationship between the percentage of net primary production (NPP) consumed by herbivores and net primary productivity. ○, phytoplankton; ●, benthic microalgae; □, macroalgal beds; ◆, freshwater macrophyte meadows; ■, seagrass meadows; ▲, marshes; △, grasslands; ◇, mangroves; *, forests. (Data from a number of sources, compiled by Cebrian, 1999.)

Repeated in words, CE is the percentage of total productivity available at one trophic level (P_{n-1}) that is actually consumed ('ingested') by a trophic compartment 'one level up' (I_n). For primary consumers in the grazer system, CE is the percentage of joules produced per unit time as NPP that finds its way into the guts of herbivores. In the case of secondary consumers, it is the percentage of herbivore productivity eaten by carnivores. The remainder dies without being eaten and enters the decomposer chain.

Various reported values for the consumption efficiencies of herbivores are shown in Figure 17.23. Most of the estimates are remarkably low, usually reflecting the unattractiveness of much plant material because of its high proportion of structural support tissue, but sometimes also as a consequence of generally low herbivore densities (because of the action of their natural enemies). The consumers of microscopic plants (microalgae growing on beds or free-living phytoplankton) can achieve greater densities, have less structural tissue to deal with and account for a greater percentage of primary production. Median values for consumption efficiency are less than 5% in forests, around 25% in grasslands and more than 50% in phytoplankton-dominated communities. We know much less about the consumption efficiencies of carnivores feeding on their prey, and any estimates are speculative. Vertebrate predators may consume 50–100% of production from vertebrate prey but perhaps only 5% from invertebrate prey. Invertebrate predators consume perhaps 25% of available invertebrate prey production.

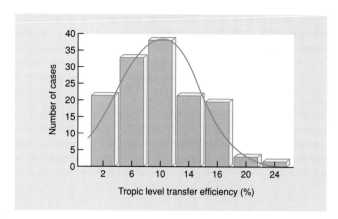

Figure 17.24 Frequency distribution of trophic-level transfer efficiencies in 48 trophic studies of aquatic communities. There is considerable variation among studies and among trophic levels. The mean is 10.13 % (SE = 0.49). (After Pauly & Christensen, 1995.)

. . . assimilation efficiency . . .

assimilation efficiency,
$$AE = A_n/I_n \times 100.$$

Assimilation efficiency is the percentage of food energy taken into the guts of consumers in a trophic compartment (I_n) that is assimilated across the gut wall (A_n) and becomes available for incorporation into growth or to do work. The remainder is lost as feces and enters the base of the decomposer system. An 'assimilation efficiency' is much less easily ascribed to microorganisms. Food does not enter an invagination of the outside world passing through the microorganism's body (like the gut of a higher organism) and feces are not produced. In the sense that bacteria and fungi typically assimilate effectively 100% of the dead organic matter they digest externally and absorb, they are often said to have an 'assimilation efficiency' of 100%.

Assimilation efficiencies are typically low for herbivores, detritivores and microbivores (20–50%) and high for carnivores (around 80%). In general, animals are poorly equipped to deal with dead organic matter (mainly plant material) and living vegetation, no doubt partly because of the very widespread occurrence of physical and chemical plant defenses, but mainly as a result of the high proportion of complex structural chemicals such as cellulose and lignin in their make-up. As Chapter 11 describes, however, many animals contain a symbiotic gut microflora that produces cellulase and aids in the assimilation of plant organic matter. In one sense, these animals have harnessed their own personal decomposer system. The way that plants allocate production to roots, wood, leaves, seeds and fruits influences their usefulness to herbivores. Seeds and fruits may be assimilated with efficiencies as high as 60–70%, and leaves with about 50% efficiency, while the assimilation efficiency for wood may be as low as 15%. The animal food of carnivores (and detritivores such as vultures that consume animal carcasses) poses less of a problem for digestion and assimilation.

. . . and production efficiency . . .

production efficiency,
$$PE = P_n/A_n \times 100.$$

Production efficiency is the percentage of assimilated energy (A_n) that is incorporated into new biomass (P_n). The remainder is entirely lost to the community as respiratory heat. (Energy-rich secretory and excretory products, which have taken part in metabolic processes, may be viewed as production, P_n, and become available, like dead bodies, to the decomposers.)

Production efficiency varies mainly according to the taxonomic class of the organisms concerned. Invertebrates in general have high efficiencies (30–40%), losing relatively little energy in respiratory heat and converting more assimilate to production.

Amongst the vertebrates, ectotherms (whose body temperature varies according to environmental temperature) have intermediate values for PE (around 10%), whilst endotherms, with their high energy expenditure associated with maintaining a constant temperature, convert only 1–2% of assimilated energy into production. The small-bodied endotherms have the lowest efficiencies, with the tiny insectivores (e.g. wrens and shrews) having the lowest production efficiencies of all. On the other hand, microorganisms, including protozoa, tend to have very high production efficiencies. They have short lives, small size and rapid population turnover. Unfortunately, available methods are not sensitive enough to detect population changes on scales of time and space relevant to microorganisms, especially in the soil. In general, efficiency of production increases with size in endotherms and decreases very markedly in ectotherms.

trophic level transfer efficiency,
$$TLTE = P_n/P_{n-1} \times 100.$$

. . . which combine to give trophic level transfer efficiency

The overall trophic transfer efficiency from one trophic level to the next is simply CE × AE × PE. In the period after Lindemann's (1942) pioneering work, it was generally assumed that trophic transfer efficiencies were around 10%; indeed some ecologists referred to a 10% 'law'. However, there is certainly no law of nature that results in precisely one-tenth of the energy that enters a trophic level transferring to the next. For example, a compilation of trophic studies from a wide range of freshwater and marine environments revealed that trophic level transfer efficiencies varied between about 2 and 24%, although the mean was 10.13% (Figure 17.24).

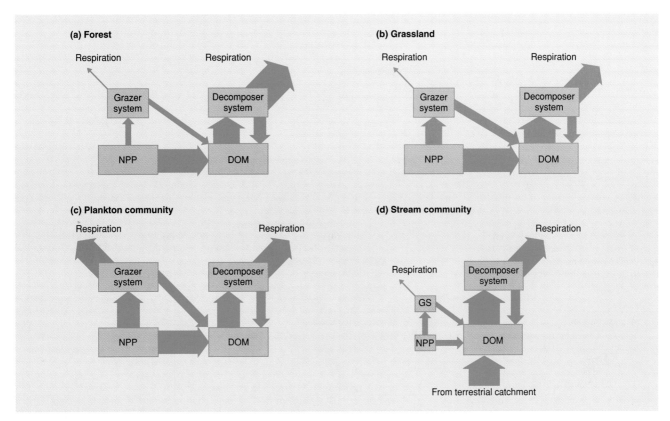

Figure 17.25   General patterns of energy flow for: (a) a forest, (b) a grassland, (c) a marine plankton community, and (d) the community of a stream or small pond. The relative sizes of the boxes and arrows are proportional to the relative magnitudes of compartments and flows. DOM, dead organic matter; NPP, net primary production.

17.5.4 Energy flow through contrasting communities

relative roles of grazer and decomposer systems in contrasting communities

Given accurate values for net primary productivity (NPP) in an ecosystem, and CE, AE and PE for the various trophic groupings shown in the model in Figure 17.22, it should be possible to predict and understand the relative importance of the different possible energy pathways. Perhaps not surprisingly, no study has incorporated all ecosystem compartments and all transfer efficiencies of the component species. However, some generalizations are possible when the gross features of contrasting systems are compared (Figure 17.25). Thus, the decomposer system is probably responsible for the majority of secondary production, and therefore respiratory heat loss, in every community in the world. The grazer system has its greatest role in plankton communities, where a large proportion of NPP is consumed alive and assimilated at quite a high efficiency. Even here, though, it is now clear that very high densities of heterotrophic bacteria in the plankton community subsist on dissolved organic molecules excreted by phytoplankton cells, perhaps consuming more than 50% of primary productivity as 'dead' organic matter

in this way (Fenchel, 1987a). The grazer system holds little sway in terrestrial communities because of low herbivore consumption and assimilation efficiencies, and it is almost nonexistent in many small streams and ponds simply because primary productivity is so low. The latter depend for their energy base on dead organic matter produced in the terrestrial environment that falls or is washed or blown into the water. The deep-ocean benthic community has a trophic structure very similar to that of streams and ponds (all can be described as heterotrophic communities). In this case, the community lives in water too deep for photosynthesis to be appreciable or even to take place at all, but it derives its energy base from dead phytoplankton, bacteria, animals and feces that sink from the autotrophic community in the euphotic zone above. From a different perspective, the ocean bed is equivalent to a forest floor beneath an impenetrable forest canopy.

We can move from the relatively gross generalizations above to consider in Figure 17.26 a greater range of terrestrial and aquatic ecosystems (data compiled from over 200 published reports by Cebrian, 1999). Figure 17.26a first shows the range of values for NPP

grazer consumption efficiencies are highest where plants have low C : N and C : P ratios

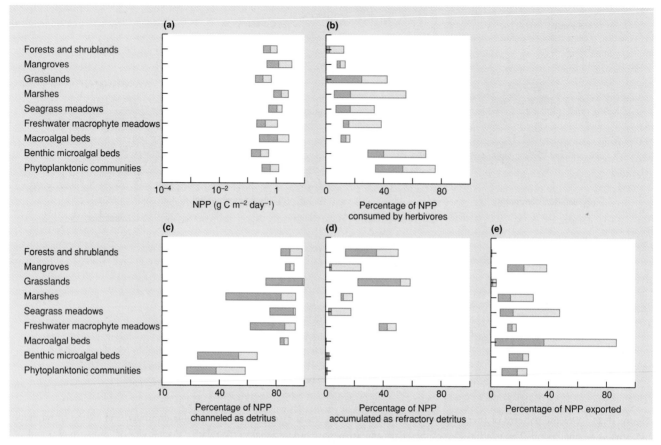

Figure 17.26 Box plots showing for a range of ecosystem types: (a) net primary productivity (NPP), (b) percentage of NPP consumed by detritivores, (c) percentage of NPP channeled as detritus, (d) percentage of NPP accumulated as refractory detritus, and (e) percentage of NPP exported. Boxes encompass 25 and 75% quartiles and the central lines represents the median of a number of studies. (After Cebrian, 1999.)

in a variety of terrestrial and aquatic ecosystems. Figure 17.26b re-emphasizes how consumption efficiency by grazers is particularly low in ecosystems where plant biomass contains considerable support tissue and relatively low amounts of nitrogen and phosphorus (i.e. forests, shrublands and mangroves). Plant biomass not consumed by herbivores becomes detritus and contributes by far the largest proportion to the dead organic matter (DOM) box in Figure 17.25. Not surprisingly, the percentage of NPP destined to be detritus is highest in forests and lowest in phytoplankton and benthic microalgal communities (Figure 17.26c). Plant biomass from terrestrial communities is not only unpalatable to herbivores, it is also relatively more difficult for decomposers and detritivores to deal with. Thus, Figure 17.26d shows that a greater proportion of primary production accumulates as refractory detritus (persisting for more than a year) in forests, shrublands, grasslands and freshwater macrophyte meadows. Finally, Figure 17.26e shows the percentage of NPP that is exported out of the systems. The values are generally modest (medians of 20% or less) indicating that, in most cases, the majority of biomass

produced in an ecosystem is consumed or decomposed there. The most obvious exceptions are mangroves and, in particular, macroalgal beds (which often inhabit rocky shores), where relatively large proportions of plant biomass are displaced and moved away by storm and tidal action.

In general then, communities composed of plants whose stoichiometry represents a higher nutritional status (higher nitrogen and phosphorus concentrations, i.e. lower C : N and C : P) lose a higher percentage to herbivores, produce a smaller proportion of detritus, experience faster decomposition rates and, in consequence, accumulate less refractory detritus and have smaller stores of dead organic carbon (Cebrian, 1999).

The presentation of information in Figure 17.26 emphasizes spatial patterns in the way energy moves through the world's ecosystems. However, we should not lose sight of the temporal patterns that exist in the balance between production and consumption

temporal patterns in the balance between production and consumption of organic matter

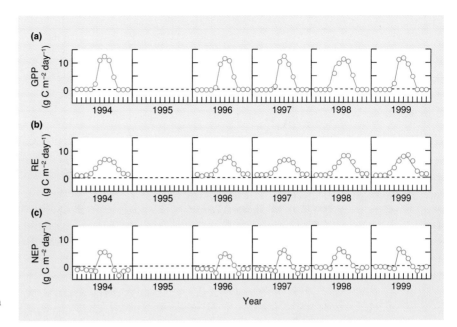

Figure 17.27 Monthly mean values for: (a) gross primary productivity (GPP), (b) ecosystem respiration (RE), and (c) net ecosystem productivity (NEP) in a Canadian aspen forest. (After Arain *et al.*, 2002.)

of organic matter. Figure 17.27 shows how GPP, RE (the sum of autotrophic and heterotrophic respiration) and net ecosystem productivity (NEP) varied seasonally during 5 years of study of a boreal aspen (*Populus tremuloides*) forest in Canada. Total annual GPP (the area under the GPP curves in Figure 17.27a) was highest in 1998 when the temperature was high (probably the result of an El Niño event – see below) and lowest in 1996 when the temperature was particularly low. Annual variations in GPP (e.g. 1419 g C m^{-2} in 1998, 1187 g C m^{-2} in 1996) were large compared to variations in RE (1132 g C m^{-2} and 1106 g C m^{-2}, respectively) because the occurrence of warm springs caused photosynthesis to increase faster than respiration. This led overall to higher values of NEP in warmer years (290 g C m^{-2} in 1998, 80 g C m^{-2} in 1996). Note how NEP is negative (RE exceeds GPP and carbon stores are being used by the community) except in the summer months when GPP consistently exceeds RE. At this site, the cumulative annual values for NEP were always positive, indicating that more carbon is fixed than is respired each year and the forest is a carbon sink. However, this is not true for all ecosystems every year (Falge *et al.*, 2002).

consequences of the
ENSO for ecosystem
energetics

The aspen forest discussed above is by no means the only ecosystem where annual variations in energy flux may be due to climatic cycles such as the El Niño–Southern Oscillation (ENSO; see also Section 2.4.1). ENSO events occur sporadically but typically occur every 3–6 years. During such events, the temperature may be significantly higher in some locations and lower in others and, just as significantly, rainfall can be 4–10 times higher in some areas (Holmgren *et al.*, 2001). The El Niño has been correlated with dramatic changes in aquatic ecosystems (even leading to the collapse of fisheries; Jordan, 1991). More recently, it has become obvious that the El Niño can cause major changes on land too. Figure 17.28 shows the annual variation in caterpillar numbers on the Galápagos Islands in a standard census conducted in various years since 1977, plotted on the same graph as annual rainfall. The remarkably strong correlation comes about because of the dependence of caterpillar numbers on primary productivity, which itself is considerably higher in wet years. We saw in Figure 17.20d how the total number of broods of the finch *Geospiza fortis* was much greater in the four ENSO years (open circles in that figure). This reflects the much greater production in very wet years of the seeds, fruits and caterpillars that they feed on. Not only do the finches increase the number of broods, but also the size of their clutches and the probability of successful rearing to the fledging stage.

Our growing knowledge of the impact of ENSO events on energy flux through ecosystems suggests that the predicted changes in extreme weather events expected as a result of human-induced global climate change will profoundly alter ecosystem processes in many parts of the world, a topic to which we will return in Chapter 22.

But next we turn to the flux of matter through ecosystems, recognizing that the rate at which resources are supplied and used by autotrophs and heterotrophs depends fundamentally on the supply of nutrients (Chapter 18). We shall see later how ecosystem productivity helps determine the consequences of competitive and predator–prey interactions for community composition (Chapter 19), food web ecology (Chapter 20) and species richness (Chapter 21).

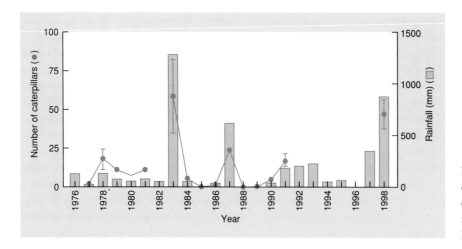

Figure 17.28 Annual variation in mean caterpillar numbers (± SE; ●) in a standard census against a histogram of annual rainfall on the Galápagos island of Daphne Major. (After Grant *et al.*, 2000.)

Summary

The term ecosystem is used to denote the biological community (primary producers, decomposers, detritivores, herbivores, etc.) together with the abiotic environment in which it is set. Lindemann laid the foundations of a science of ecological energetics by considering the efficiency of transfer between trophic levels – from incident radiation received by a community through its capture by green plants in photosynthesis to its subsequent use by heterotrophs. This is the topic of the present chapter.

The bodies of the living organisms within a unit area constitute a standing crop of biomass. Primary productivity is the rate at which biomass is produced per unit area by plants. The total fixation of energy by photosynthesis is gross primary productivity (GPP), a proportion of which is respired by the plants as autotrophic respiration (RA). The difference between GPP and RA is net primary productivity (NPP) and represents the actual rate of production of new biomass that is available for consumption by heterotrophic organisms. The rate of production of biomass by heterotrophs is secondary productivity, and their respiration is heterotrophic respiration (HE). Net ecosystem productivity (NEP) is GPP minus total respiration (RA + RH).

We discuss the broad patterns in primary productivity across the face of the globe and in relation to seasonal and annual variations in conditions, and note that primary productivity : biomass ratios are higher in aquatic than terrestrial communities.

The factors that limit terrestrial primary productivity are solar energy (and particularly its inefficient use by plants), water and temperature (and their complex interactions), soil texture and drainage, and mineral nutrient availability. The length of the growing season is particularly influential. In aquatic environments, primary productivity depends in particular on the availability of solar radiation (with strong patterns related to water depth) and nutrients (especially important are human inputs to lakes, estuarine inputs to oceans and ocean upwelling zones).

Unlike plants, heterotrophic bacteria, fungi and animals cannot manufacture from simple molecules the complex, energy-rich compounds they need. They derive their matter and energy either directly by consuming plant material or indirectly from plants by eating other heterotrophs. There is a general positive relationship between primary and secondary productivity in ecosystems, but most primary production passes, when dead, through the detritus system rather than as living material through the grazing system. The pathways traced by energy through communities are determined by three energy transfer efficiencies (consumption, assimilation and production efficiencies). Grazer consumption efficiencies are highest where plants have little structural support tissue and low C : N and C : P ratios. We discuss temporal patterns in the balance between primary productivity and its consumption by heterotrophs, and show that broad climatic patterns (such as El Niño) can profoundly influence ecosystem energetics.

Chapter 18
The Flux of Matter through Ecosystems

18.1 Introduction

Chemical elements and compounds are vital for the processes of life. Living organisms expend energy to extract chemicals from their environment, they hold on to them and use them for a period, then lose them again. Thus, the activities of organisms profoundly influence the patterns of flux of chemical matter in the biosphere. Physiological ecologists focus their attention on how individual organisms obtain and use the chemicals they need (see Chapter 3). However, in this chapter, as in the last, we change the emphasis and consider the ways in which the biota on an area of land, or within a volume of water, accumulates, transforms and moves matter between the various components of the ecosystem. The area that we choose may be that of the whole globe, a continent, a river catchment or simply a square meter.

18.1.1 Relationships between energy flux and nutrient cycling

The great bulk of living matter in any community is water. The rest is made up mainly of carbon compounds (95% or more) and this is the form in which energy is accumulated and stored. The energy is ultimately dissipated when the carbon compounds are oxidized to carbon dioxide (CO_2) by the metabolism of living tissue or of its decomposers. Although we consider the fluxes of energy and carbon in different chapters, the two are intimately bound together in all biological systems.

Carbon enters the trophic structure of a community when a simple molecule, CO_2, is taken up in photosynthesis. If it becomes incorporated in net primary productivity, it is available for consumption as part of a molecule of sugar, fat, protein or, very often, cellulose. It follows exactly the same route as energy, being successively consumed, defecated, assimilated and perhaps incorporated into secondary productivity somewhere within one of the trophic compartments. When the high-energy molecule in which the carbon is resident is finally used to provide energy for work, the energy is dissipated as heat (as we have discussed in Chapter 17) and the carbon is released again to the atmosphere as CO_2. Here, the tight link between energy and carbon ends.

Once energy is transformed into heat, it can no longer be used by living organisms to do work or to fuel the synthesis of biomass. (Its only possible role is momentary, in helping to maintain a high body temperature.) The heat is eventually lost to the atmosphere and can never be recycled. In contrast, the carbon in CO_2 can be used again in photosynthesis. Carbon, and all other nutrient elements (e.g. nitrogen, phosphorus, etc.) are available to plants as simple inorganic molecules or ions in the atmosphere (CO_2), or as dissolved ions in water (nitrate, phosphate, potassium, etc.). Each can be incorporated into complex organic carbon compounds in biomass. Ultimately, however, when the carbon compounds are metabolized to CO_2, the mineral nutrients are released again in simple inorganic form. Another plant may then absorb them, and so an individual atom of a nutrient element may pass repeatedly through one food chain after another. The relationship between energy flow and nutrient cycling is illustrated in Figure 18.1.

By its very nature, then, each joule of energy can be *used* only once, whereas chemical nutrients, the building blocks of biomass, simply change the form of molecule of which they are part (e.g. nitrate-N to protein-N to nitrate-N). They can be used again, and repeatedly recycled. Unlike the energy of solar radiation, nutrients are not in unalterable supply, and the process of locking some up into living biomass reduces the supply remaining to the rest of the community. If plants, and their consumers, were not eventually decomposed, the supply of nutrients would become exhausted and life on the planet would cease. The activity of heterotrophic organisms is crucial in bringing about nutrient

energy cannot be cycled and reused; matter can . . .

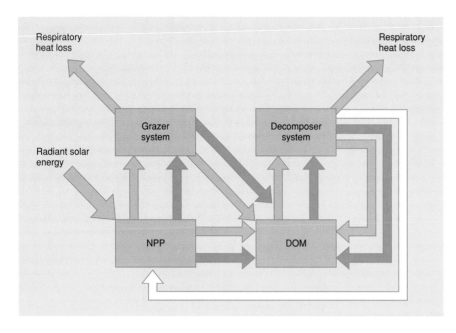

Figure 18.1 Diagram to show the relationship between energy flow (pale arrows) and nutrient cycling. Nutrients locked in organic matter (dark arrows) are distinguished from the free inorganic state (white arrow). DOM, dead organic matter; NPP, net primary production.

cycling and maintaining productivity. Figure 18.1 shows the release of nutrients in their simple inorganic form as occurring only from the decomposer system. In fact, some is also released from the grazer system. However, the decomposer system plays a role of overwhelming importance in nutrient cycling.

... but nutrient cycling is never perfect

The picture described in Figure 18.1 is an oversimplification in one important respect. Not all nutrients released during decomposition are necessarily taken up again by plants. Nutrient recycling is never perfect and some nutrients are exported from land by runoff into streams (ultimately to the ocean) and others, such as nitrogen and sulfur, that have gaseous phases, can be lost to the atmosphere. Moreover, a community receives additional supplies of nutrients that do not depend directly on inputs from recently decomposed matter – minerals dissolved in rain, for example, or derived from weathered rock.

18.1.2 Biogeochemistry and biogeochemical cycles

the 'bio' in biogeochemistry

We can conceive of pools of chemical elements existing in compartments. Some compartments occur in the *atmosphere* (carbon in CO_2, nitrogen as gaseous nitrogen, etc.), some in the rocks of the *lithosphere* (calcium as a constituent of calcium carbonate, potassium in feldspar) and others in the *hydrosphere* – the water in soil, streams, lakes or oceans (nitrogen in dissolved nitrate, phosphorus in phosphate, carbon in carbonic acid, etc.). In all these cases the elements exist in an inorganic form. In contrast, living organisms (the biota) and dead and decaying bodies

can be viewed as compartments containing elements in an organic form (carbon in cellulose or fat, nitrogen in protein, phosphorus in adenosine triphosphate, etc.). Studies of the chemical processes occurring within these compartments and, more particularly, of the fluxes of elements between them, comprise the science of biogeochemistry.

Many geochemical fluxes would occur in the absence of life, if only because all geological formations above sea level are eroding and degrading. Volcanoes release sulfur into the atmosphere whether there are organisms present or not. On the other hand, organisms alter the rate of flux and the differential flux of the elements by extracting and recycling some chemicals from the underlying geochemical flow (Waring & Schlesinger, 1985). The term biogeochemistry is apt.

The flux of matter can be investigated at a variety of spatial and temporal scales. Ecologists interested in the gains, uses and losses of nutrients by the community of a small pond or a hectare of grassland can focus on local pools of chemicals. They need not concern themselves with the contribution to the nutrient budget made by volcanoes or the possible fate of nutrients leached from land to eventually be deposited on the ocean floor. At a larger scale, we find that the chemistry of streamwater is profoundly influenced by the biota of the area of land it drains (its catchment area; see Section 18.2.4) and, in turn, influences the chemistry and biota of the lake, estuary or sea into which it flows. We deal with the details of nutrient fluxes through terrestrial and aquatic ecosystems in Sections 18.2 and 18.3. Other investigators are interested in the global scale. With their broad brush they paint a picture of the contents and fluxes of the largest conceivable compartments – the entire

biogeochemistry can be studied at different scales

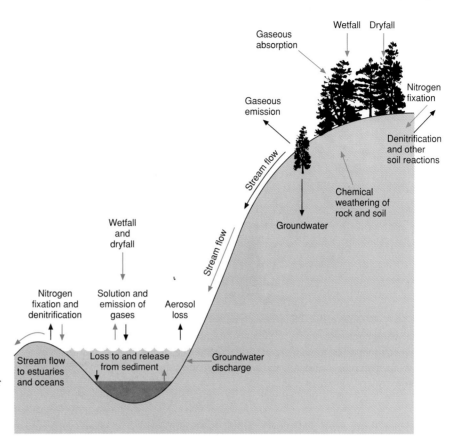

Figure 18.2 Components of the nutrient budgets of a terrestrial and an aquatic system. Note how the two communities are linked by stream flow, which is a major output from the terrestrial system and a major input to the aquatic one. Inputs are shown in color and outputs in black.

atmosphere, the oceans as a whole, and so on. Global biogeochemical cycles will be discussed in Section 18.4.

18.1.3   Nutrient budgets

Nutrients are gained and lost by ecosystems in a variety of ways (Figure 18.2). We can construct a nutrient budget by identifying and measuring all the processes on the credit and debit sides of the equation. For some nutrients, in some ecosystems, the budget may be more or less in balance.

inputs sometimes balance outputs . . . but not always

In other cases, the inputs exceed the outputs and nutrients accumulate in the compartments of living biomass and dead organic matter. This is especially obvious during community succession (see Section 17.4).

Finally, outputs may exceed inputs if the biota is disturbed by an event such as fire, massive defoliation (such as that caused by a plague of locusts) or large-scale deforestation or crop harvesting by people. Another important source of loss in terrestrial systems occurs where mineral export (e.g. of base cations due to acid rain) exceeds replenishment from weathering.

The components of nutrient budgets are discussed below.

18.2   Nutrient budgets in terrestrial communities

18.2.1   Inputs to terrestrial communities

Weathering of parent bedrock and soil is generally the dominant source of nutrients such as calcium, iron, magnesium, phosphorus and potassium, which may then be taken up via the roots of

nutrient inputs . . . . . . from the weathering of rock and soil, . . .

plants. Mechanical weathering is caused by processes such as freezing of water and the growth of roots in crevices. However, much more important to the release of plant nutrients are chemical weathering processes. Of particular significance is carbonation, in which carbonic acid (H_2CO_3) reacts with minerals to release ions, such as calcium and potassium. Simple dissolution of minerals in water also makes nutrients available from rock and soil, and so do hydrolytic reactions involving organic acids released by the ectomycorrhizal fungi (see Section 13.8.1) associated with plant roots (Figure 18.3).

Atmospheric CO_2 is the source of the carbon content of terrestrial communities. Similarly, gaseous nitrogen from

. . . from the atmosphere, . . .

the atmosphere provides most of the nitrogen content of communities. Several types of bacteria and blue-green algae possess

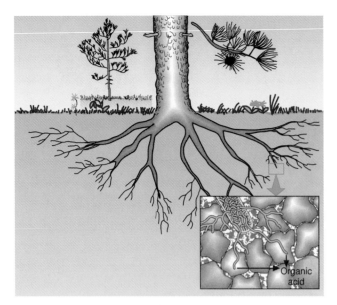

Figure 18.3 Ectomycorrhizal fungi associated with tree roots can mobilize phosphorus, potassium, calcium and magnesium from solid mineral substrates through organic acid secretion, and these nutrients then become available to the host plant via the fungal mycelium. (After Landeweert *et al.*, 2001.)

the enzyme nitrogenase and convert atmospheric nitrogen to soluble ammonium (NH_4^+) ions, which can then be taken up through the roots and used by plants. All terrestrial ecosystems receive some available nitrogen through the activity of free-living bacteria, but communities containing plants such as legumes and alder trees (*Alnus* spp.), with their root nodules containing symbiotic nitrogen-fixing bacteria (see Section 13.10), may receive a very substantial proportion of their nitrogen in this way. More than 80 kg ha^{-1} year^{-1} of nitrogen was supplied to a stand of alder by biological nitrogen fixation, for example, compared with 1–2 kg ha^{-1} year^{-1} from rainfall (Bormann & Gordon, 1984); and nitrogen fixation by legumes can be even more dramatic: values in the range 100–300 kg ha^{-1} year^{-1} are not unusual.

... as wetfall and dryfall, ...   Other nutrients from the atmosphere become available to communities as *wetfall* (in rain, snow and fog) or *dryfall* (settling of particles during periods without rain, and gaseous uptake). Rain is not pure water but contains chemicals derived from a number of sources: (i) trace gases, such as oxides of sulfur and nitrogen; (ii) aerosols produced when tiny water droplets from the oceans evaporate in the atmosphere and leave behind particles rich in sodium, magnesium, chloride and sulfate; and (iii) dust particles from fires, volcanoes and windstorms, often rich in calcium, potassium and sulfate. The constituents of rainfall that serve as nuclei for raindrop formation make up the *rainout* component, whereas other constituents, both particulate and gaseous, are cleansed from the atmosphere as the rain falls – these are the *washout* component (Waring & Schlesinger,

1985). The nutrient concentrations in rain are highest early in a rainstorm, but fall subsequently as the atmosphere is progressively cleansed. Snow scavenges chemicals from the atmosphere less effectively than rain, but tiny fog droplets have particularly high ionic concentrations. Nutrients dissolved in precipitation mostly become available to plants when the water reaches the soil and can be taken up by the plant roots. However, some are absorbed by leaves directly.

Dryfall can be a particularly important process in communities with a long dry season. In four Spanish oak forests (*Quercus pyrenaica*) situated along a rainfall gradient, for example, dryfall sometimes accounted for more than half of the atmospheric input to the tree canopy of magnesium, manganese, iron, phosphorus, potassium, zinc and copper (Figure 18.4). For most elements, the importance of dryfall was more marked in forests in drier environments. However, dryfall was not insignificant for forests in wetter locations. Figure 18.4 also plots for each nutrient the annual forest demand (annual increase in above-ground biomass multiplied by the mineral concentration in the biomass). Annual deposition of many elements in wetfall and dryfall was much greater than needed to satisfy demand (e.g. Cl, S, Na, Zn). But for other elements, and especially for forests in dryer environments, annual atmospheric inputs more or less matched demand (e.g. P, K, Mn, Mg) or were inadequate (N, Ca). Of course element deficits would be greater if root productivity had been taken into account, and other sources of nutrient input must be particularly significant for a number of these elements.

While we may conceive of wetfall and dryfall inputs arriving vertically, part of the pattern of nutrient income to a forest depends on its ability to intercept horizontally driven air-borne nutrients. This was demonstrated for mixed deciduous forests in New York State when the aptly named Weathers *et al.* (2001) showed that inputs of sulfur, nitrogen and calcium at the forest edge were 17–56% greater than in its interior. The widespread tendency for forests to become fragmented as a result of human activities is likely to have had unexpected consequences for their nutrient budgets because more fragmented forests have a greater proportion of edge habitat.

Streamwater plays a major role in the output of nutrients from terrestrial ecosystems (see Section 18.3). However, *... from hydrological inputs ...* in a few cases, stream flow can provide a significant input to terrestrial communities when, after flooding, material is deposited in floodplains.

Last, and by no means least, human activities contribute significant inputs of nutrients to many communities. For *... and from human activities* example, the amounts of CO_2 and oxides of nitrogen and sulfur in the atmosphere have been increased by the burning of fossil fuels, and the concentrations of nitrate and phosphate in streamwater have been raised by agricultural practices and sewage disposal. These changes have far-reaching consequences, which will be discussed later.

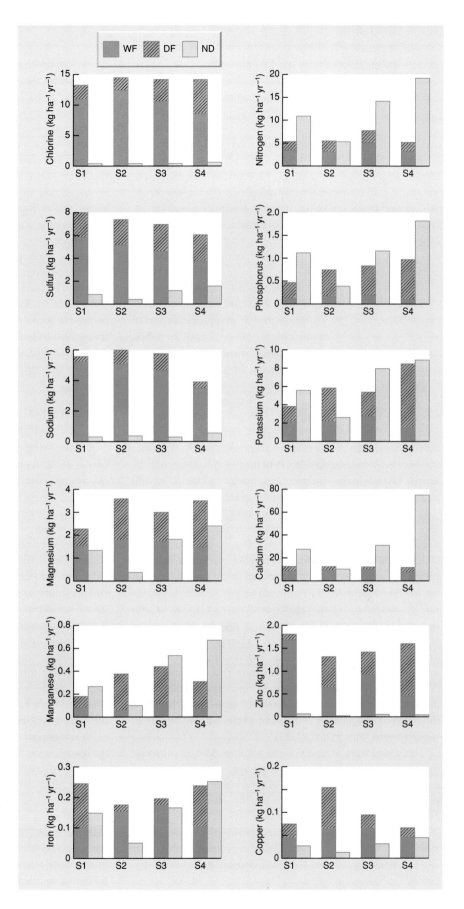

Figure 18.4 Annual atmospheric deposition as wetfall (WF) and dryfall (DF) compared to annual nutrient demand (ND; to account for above-ground tree growth) for four oak forests along a rainfall gradient (S1 wettest, S4 driest) in Spain. (After Marcos & Lancho, 2002.)

18.2.2   Outputs from terrestrial communities

nutrients can
be lost ...

A particular nutrient atom may be taken up by a plant that is then eaten by a herbivore which then dies and is decomposed, releasing the atom back to the soil from where it is taken up through the roots of another plant. In this manner, nutrients may circulate within the community for many years. Alternatively, the atom may pass through the system in a matter of minutes, perhaps without interacting with the biota at all. Whatever the case, the atom will eventually be lost through one of the variety of processes that remove nutrients from the system (see Figure 18.2). These processes constitute the debit side of the nutrient budget equation.

... to the
atmosphere ...

Release to the atmosphere is one pathway of nutrient loss. In many communities there is an approximate annual balance in the carbon budget; the carbon fixed by photosynthesizing plants is balanced by the carbon released to the atmosphere as CO_2 from the respiration of plants, microorganisms and animals. Other gases are released through the activities of anaerobic bacteria. Methane is a well-known product of the soils of bogs, swamps and floodplain forests, produced by bacteria in the waterlogged, anoxic zone of wetland soils. However, its net flux to the atmosphere depends on the rate at which it is produced in relation to its rate of consumption by aerobic bacteria in the shallower, unsaturated soil horizons, with as much as 90% consumed before it reaches the atmosphere (Bubier & Moore, 1994). Methane may be of some importance in drier locations too. It is produced by fermentation in the anaerobic stomachs of grazing animals, and even in upland forests, periods of heavy rainfall may produce anaerobic conditions that can persist for some time within microsites in the organic layer of the soil (Sexstone et al., 1985). In such locations, bacteria such as Pseudomonas reduce nitrate to gaseous nitrogen or N_2O in the process of denitrification. Plants themselves may be direct sources of gaseous and particulate release. For example, forest canopies produce volatile hydrocarbons (e.g. terpenes) and tropical forest trees emit aerosols containing phosphorus, potassium and sulfur (Waring & Schlesinger, 1985). Finally, ammonia gas is released during the decomposition of vertebrate excreta and has been shown to be a significant component in the nutrient budget of many systems (Sutton et al., 1993).

Other pathways of nutrient loss are important in particular instances. For example, fire can turn a very large proportion of a community's carbon into CO_2 in a very short time. The loss of nitrogen as volatile gas can be equally dramatic: during an intense wild fire in a conifer forest in northwest USA, 855 kg ha^{-1} (equal to 39% of the pool of organic nitrogen) was lost in this way (Grier, 1975). Substantial losses of nutrients also occur when foresters or farmers harvest and remove their trees and crops.

For many elements, the most important pathway of loss is in stream flow. The water that drains from the soil of a terrestrial community, via the groundwater, into a stream carries a load of nutrients that is partly dissolved and partly particulate. With the exception of iron and phosphorus, which are not mobile in soils, the loss of plant nutrients is predominantly in solution. Particulate matter in stream flow occurs both as dead organic matter (mainly tree leaves) and as inorganic particles. After rainfall or snowmelt the water draining into streams is generally more dilute than during dry periods, when the concentrated waters of soil solution make a greater contribution. However, the effect of high volume more than compensates for lower concentrations in wet periods. Thus, total loss of nutrients is usually greatest in years when rainfall and stream discharge are high. In regions where the bedrock is permeable, losses occur not only in stream flow but also in water that drains deep into the groundwater. This may discharge into a stream or lake after a considerable delay and at some distance from the terrestrial community.

... and to
groundwater
and streams

18.2.3   Carbon inputs and outputs may vary with forest age

Law et al. (2001) compared patterns of carbon storage and flux in a young (clear cut 22 years previously) and an old forest (not previously logged, trees from 50 to 250 years old) of ponderosa pine (Pinus ponderosa) in Oregon, USA. Their results are summarized in Figure 18.5.

Total ecosystem carbon content (vegetation, detritus and soil) of the old forest was about twice that of its young counterpart. There were notable differences in percentage carbon stored

an old forest is a
net sink for carbon
(input greater than
output) ...

in living biomass (61% in old, 15% in young) and in dead wood on the forest floor (6% in old, 26% in young). These differences reflect the influence of soil organic matter and woody debris in the young forest derived from the prelogged period of its history. As far as living biomass is concerned, the old forest contained more than 10 times as much as the young forest, with the biggest difference in the wood component of tree biomass.

Below-ground primary productivity differed little between the two forests but because of a much lower above-ground net primary productivity (ANPP) in the young forest, total net primary productivity (NPP) was 25% higher in the old forest. Shrubs accounted for 27% of ANPP in the young forest, but only 10% in the old forest. Heterotrophic respiration (decomposers, detritivores and other animals) was somewhat lower in the old forest than NPP, indicating that this forest is a net sink for carbon. In the young forest, however, heterotrophic respiration exceeded NPP making this site a net source of CO_2 to the atmosphere. In

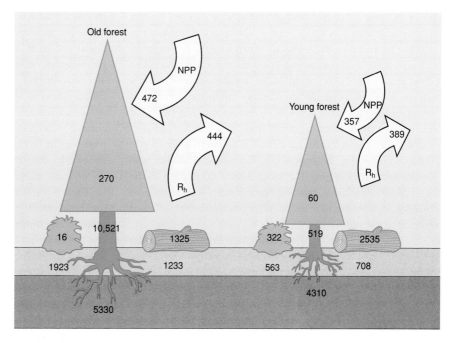

Figure 18.5 Annual carbon budgets for an old and a young ponderosa pine forest. Carbon storage figures are in g C m^{-2} while net primary productivity (NPP) and heterotrophic respiration (R_h) are in g C m^{-2} year^{-1} (arrows). The numbers above ground represent carbon storage in tree foliage, in the remainder of forest biomass, in understory plants, and in dead wood on the forest floor. The numbers just below the ground surface are for tree roots and litter. The lowest numbers are for soil carbon. (After Law *et al.*, 2001.)

... whereas a young forest is a net carbon source (output greater than input)

both forests, respiration from the soil community accounted for 77% of total heterotrophic respiration.

These results provide a good illustration of the pathways, stores and fluxes of carbon in forest communities. They also serve to emphasize that nutrient inputs and outputs are by no means always in balance in ecosystems.

18.2.4 Importance of nutrient cycling in relation to inputs and outputs

the movement of water links terrestrial and aquatic communities

Because many nutrient losses from terrestrial communities are channeled through streams, a comparison of the chemistry of streamwater with that of incoming precipitation can reveal a lot about the differential uptake and cycling of chemical elements by the terrestrial biota. Just how important is nutrient cycling in relation to the through-put of nutrients? Is the amount of nutrients cycled per year small or large in comparison with external supplies and losses? The most thorough study of this question has been carried out by Likens and his associates in the Hubbard Brook Experimental Forest, an area of temperate deciduous forest drained by small streams in the White Mountains of New Hampshire, USA. The catchment area – the extent of terrestrial environment drained by a particular stream – was taken as the unit of study because of the role that streams play in nutrient export. Six small catchments were defined and their outflows were monitored. A

Table 18.1 Annual nutrient budgets for forested catchments at Hubbard Brook (kg ha^{-1} year^{-1}). Inputs are for dissolved materials in precipitation or as dryfall. Outputs are losses in streamwater as dissolved material plus particulate organic matter. (After Likens *et al.*, 1971.)

	NH_4^+	NO_3^-	K^+	Ca^{2+}	Mg^{2+}	Na^+
Input	2.7	16.3	1.1	2.6	0.7	1.5
Output	0.4	8.7	1.7	11.8	2.9	6.9
Net change*	+2.3	+7.6	−0.6	−9.2	−2.2	−5.4

* Net change is positive when the catchment gains matter and negative when it loses it.

network of precipitation gauges recorded the incoming amounts of rain, sleet and snow. Chemical analyses of precipitation and streamwater made it possible to calculate the amounts of various nutrients entering and leaving the system, and these are shown in Table 18.1. A similar pattern is found each year. In most cases, the output of chemical nutrients in stream flow is greater than their input from rain, sleet and snow. The source of the excess chemicals is parent rock and soil, which are weathered and leached at a rate of about 70 g m^{-2} year^{-1}.

In almost every case, the inputs and outputs of nutrients are small in comparison with the amounts held in biomass and recycled within the system. Nitrogen, for example, was added to the system not only in precipitation

Hubbard Brook – forest inputs and outputs are small compared to internal cycling

(6.5 kg ha^{-1} year^{-1}) but also through atmospheric nitrogen fixation by microorganisms (14 kg ha^{-1} year^{-1}). (Note that denitrification by other microorganisms, releasing nitrogen to the atmosphere, will also have been occurring but was not measured.) The export in streams of only 4 kg ha^{-1} year^{-1} emphasizes how securely nitrogen is held and cycled within the forest biomass. Stream output represents only 0.1% of the total nitrogen standing crop held in living and dead forest organic matter. Nitrogen was unusual in that its net loss in stream runoff was less than its input in precipitation, reflecting the complexity of inputs and outputs and the efficiency of its cycling. However, despite the net loss to the forest of other nutrients, their export was still low in relation to the amounts bound in biomass. In other words, relatively efficient recycling is the norm.

deforestation uncouples cycling and leads to a loss of nutrients

In a large-scale experiment, all the trees were felled in one of the Hubbard Brook catchments and herbicides were applied to prevent regrowth. The overall export of dissolved inorganic nutrients from the disturbed catchment then rose to 13 times the normal rate (Figure 18.6). Two phenomena were responsible. First, the enormous reduction in transpiring surfaces (leaves) led to 40% more precipitation passing through the groundwater to be discharged to the streams, and this increased outflow caused greater rates of leaching of chemicals and weathering of rock and soil. Second, and more significantly, deforestation effectively broke the within-system nutrient cycling by uncoupling the decomposition process from the plant uptake process. In the absence of nutrient uptake in the spring, when the deciduous trees would have started production, the inorganic nutrients released by decomposer activity were available to be leached in the drainage water.

The main effect of deforestation was on nitrate-N, emphasizing the normally efficient cycling to which inorganic nitrogen is subject. The output of nitrate in streams increased 60-fold after the disturbance. Other biologically important ions were also leached faster as a result of the uncoupling of nutrient cycling mechanisms (potassium: 14-fold increase; calcium: sevenfold increase; magnesium: fivefold increase). However, the loss of sodium, an element of lower biological significance, showed a much less dramatic change following deforestation (2.5-fold increase). Presumably it is cycled less efficiently in the forest and so uncoupling had less effect.

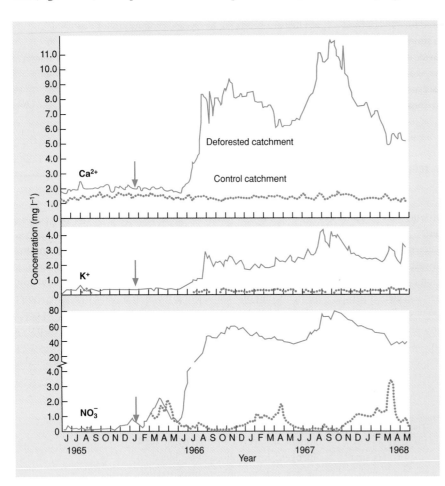

Figure 18.6 Concentrations of ions in streamwater from the experimentally deforested catchment and a control catchment at Hubbard Brook. The timing of deforestation is indicated by arrows. Note that the 'nitrate' axis has a break in it. (After Likens & Borman, 1975.)

18.2.5 Some key points about nutrient budgets in terrestrial ecosystems

diversity of patterns of nutrient input and output

The examples discussed above have illustrated that ecosystems do not generally have balanced inputs and outputs of nutrients. However, in many cases (as in the Hubbard Brook Forest) nutrients such as nitrogen are cycled quite tightly, and inputs and outputs are small compared to stored pools. For carbon too, fluxes may be small compared to storage, but note that tight cycling is not the rule in this case; the carbon molecules in respired CO_2 will rarely be the same ones taken up by photosynthesis (because of the huge pool of CO_2 involved).

We have also seen that nutrient budgets of a single category of ecosystem can differ dramatically, either because of internal properties (the age of trees in the pine forests in Section 18.2.3) or external factors (the dryness of the climate in the oak forests in Figure 18.4). Similarly, in a semiarid grassland in Colorado, nitrogen availability to grass plants adjacent to actively growing roots was greater in months when there was more rainfall (Figure 18.7).

decomposition and nutrient flux . . .

. . . influenced by stoichiometry . . .

Many other factors influence nutrient flux rates and stores. For example, the stoichiometry of elements in foliage (and thus in detritus when the leaves die) can influence decomposition rates and nutrient flux (see Section 11.2.4). There is a theoretical critical detritus C : N ratio of 30 : 1 above which bacteria and fungi are nitrogen-limited, when they then take up exogenous ammonium and nitrate ions from the soil, competing with plants for these resources (Daufresne & Loreau,

2001). When the C : N ratio is below 30 : 1, the microbes are carbon-limited and decomposition increases soil inorganic nitrogen, which may in turn increase plant nitrogen uptake (Kaye & Hart, 1997). In general, plants are most often nitrogen-limited and microbes carbon-limited, and whilst microbes are more significant in the control of nitrogen cycling, it is the plants that regulate carbon inputs which control microbial activity (Knops *et al.*, 2002).

A quite different chemical property of foliage may have an equally dramatic effect. Polyphenols are a very widely distributed class of secondary metabolites in plants that often provide protection against attack; their evolution is usually interpreted in terms of defense against herbivores. However, the polyphenols in detritus can also influence the flux of soil nutrients (Hattenschwiler & Vitousek, 2000). Different classes of polyphenols have been found to affect fungal spore germination and hyphal growth. They have also been shown to inhibit nitrifying bacteria and to suppress or, in some cases, stimulate symbiotic nitrogen-fixing bacteria. Finally, polyphenols may restrict the activity and abundance of soil detritivores. Overall, polyphenols may tend to reduce decomposition rates (as they decrease herbivory rates) with important consequences for nutrient fluxes, but more work is needed on this topic (Hattenschwiler & Vitousek, 2000).

. . . and plant defense chemicals

18.3 Nutrient budgets in aquatic communities

When attention is switched from terrestrial to aquatic communities, there are several important distinctions to be made. In particular, aquatic systems receive the bulk of their supply of nutrients from stream inflow (see Figure 18.2). In stream and river communities, and also in lakes with a stream outflow, export in outgoing stream water is a major factor. By contrast, in lakes without an outflow (or where this is small relative to the volume of the lake), and also in oceans, nutrient accumulation in permanent sediments is often the major export pathway.

18.3.1 Streams

nutrient 'spiraling' in streams

We noted, in the case of Hubbard Brook, that nutrient cycling within the forest was great in comparison to nutrient exchange through import and export. By contrast, only a small fraction of available nutrients take part in biological interactions in stream and river communities (Winterbourn & Townsend, 1991). The majority flows on, as particles or dissolved in the water, to be discharged into a lake or the sea. Nevertheless, some nutrients do cycle from an inorganic form in streamwater to an organic form in biota to an inorganic form in streamwater, and so on. But because of the inexorable transport downstream,

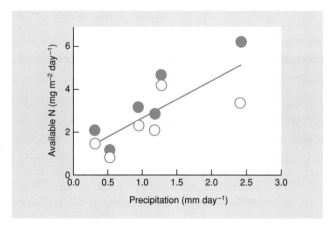

Figure 18.7 Nitrogen available to actively growing roots of the bunchgrass *Bouteloua gracilis* in shortgrass steppe ecosystems in relation to precipitation in the study period. The values for the six sampling periods are the averages of eight replicate plots. ●, downslope plots; ○, upslope plots (up to 11 m further up the same hillslope). (After Hook & Burke, 2000.)

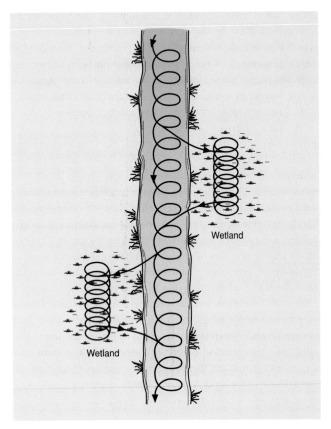

Figure 18.8 Nutrient spiraling in a river channel and adjacent wetland areas. (After Ward, 1988.)

the displacement of nutrients is better represented as a spiral (Elwood *et al.*, 1983), where fast phases of inorganic nutrient displacement alternate with periods when the nutrient is locked in biomass at successive locations downstream (Figure 18.8). Bacteria, fungi and microscopic algae, growing on the substratum of the stream bed, are mainly responsible for the uptake of inorganic nutrients from streamwater in the biotic phase of spiraling. Nutrients, in organic form, pass on through the food web via invertebrates that graze and scrape microbes from the substratum (grazer–scrapers – see Figure 11.5). Ultimately, decomposition of the biota releases inorganic nutrient molecules and the spiral continues. The concept of nutrient spiraling is equally applicable to 'wetlands', such as backwaters, marshes and alluvial forests, which occur in the floodplains of rivers. However, in these cases spiraling can be expected to be much tighter because of reduced water velocity (Prior & Johnes, 2002).

A dramatic example of spiraling occurs when the larvae of blackflies (collector–filterers; see Figure 11.5) use their modified mouthparts to filter out and consume fine particulate organic matter which otherwise would be carried downstream. Because of very high densities (sometimes as many as 600,000 blackfly larvae per square meter of river bed) a massive quantity of fine particulate matter may be converted by the larvae into fecal pellets (estimated at 429 t dry mass of fecal pellets per day in a Swedish river; Malmqvist *et al.*, 2001). Fecal pellets are much larger than the particulate food of the larvae and so are much more likely to settle out on the river bed, especially in slower flowing sections of river (Figure 18.9). Here they provide organic matter as food for many other detritivorous species.

18.3.2 Lakes

In lakes, it is usually the phytoplankton and their consumers, the zooplankton, which play the key roles in nutrient cycling. However, most lakes are interconnected with each other by rivers, and standing stocks of nutrients are determined only partly by processes within the lakes. Their position with respect to other water bodies in the landscape can also have a marked effect on nutrient status. This is well illustrated for a series of lakes connected by a river that ultimately flows into Toolik Lake in arctic Alaska (Figure 18.10a).

nutrient flux in lakes: important roles for plankton and lake position

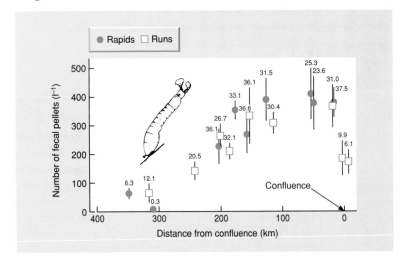

Figure 18.9 Downstream trends in the Vindel River in Sweden (shown as distance from the confluence with the larger Ume River) in the concentration of fecal pellets (number of fecal pellets per liter ± SE) of blackfly larvae (family Simuliidae). The generally lower concentrations in the 'runs' reflect the higher probability of pellets settling to the river bed in these sections compared to the 'rapids' sections. The numbers above the error bars are percentages of the mass of total organic matter in the flowing water (seston) made up of fecal pellets. (After Malmqvist *et al.*, 2001.)

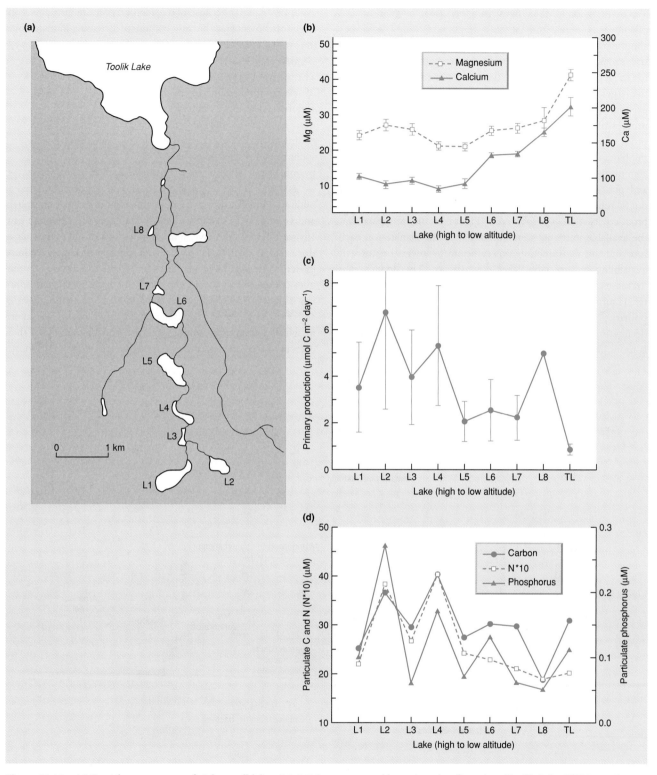

Figure 18.10 (a) Spatial arrangement of eight small lakes (L1–L8) interconnected by a river that flows into Toolik Lake (TL) in arctic Alaska. (b) Mean values, averaged over all sampling occasions during 1991–97 (±SE), for magnesium (Mg) and calcium (Ca) concentrations in the study lakes. (c) Pattern in primary productivity down the lake chain. (d) Mean values for carbon (C), nitrogen (N) and phosphorus in particulate form. (After Kling *et al.*, 2000.)

The main reason for the downstream increase in magnesium and calcium was increased weathering (Figure 18.10b). This comes about because a greater proportion of the water entering downstream lakes has been in intimate contact with the parent rock for longer; put another way, the higher concentrations reflect the larger catchment areas that feed the downstream lakes. The pattern for calcium and magnesium may also partly reflect progressive evaporative concentration with longer residence times of water in the system as well as material processing by the biota in streams and lakes as the water moves downstream. The nutrients that generally limit production in lakes, nitrogen and phosphorus, were in very low concentrations and could not be reliably measured. However, the downstream decrease in productivity that was observed (Figure 18.10c) suggests that the available nutrients were consumed by the plankton in each lake and this consumption was sufficient to lower the nutrient availability in successive lakes downstream. The downstream decrease of nitrogen, phosphorus and carbon in particulate matter (Figure 18.10d) simply reflects the lower downstream rates of primary productivity. Note that it is unusual to have a downstream decline in productivity. In less pristine conditions, productivity is more likely to increase in a downstream direction (e.g. Kratz

et al., 1997), partly because of the addition of more nutrients from larger catchment areas but also because of increasing human inputs in lowland areas through fertilizer application and sewage.

Many lakes in arid regions, lacking a stream outflow, lose water only by evaporation. The waters of these endorheic lakes (internal flow) are thus more concentrated than their freshwater counterparts, being particularly rich in sodium (with values up to 30,000 mg l^{-1} or more) but also in other nutrients such as phosphorus (up to 7000 µg l^{-1} or more). Saline lakes should not be considered as oddities; globally, they are just as abundant in terms of numbers and volume as freshwater lakes (Williams, 1988). They are usually very fertile and have dense populations of blue-green algae (for example, *Spirulina platensis*), and some, such as Lake Nakuru in Kenya, support huge aggregations of plankton-filtering flamingoes (*Phoeniconaias minor*). No doubt, the high level of phosphorus is due in part to the concentrating effect of evaporation. In addition, there may be a tight nutrient cycle in lakes such as Nakuru in which continuous flamingo feeding and the supply of their excreta to the sediment creates circumstances where phosphorus

> saline lakes lose water only by evaporation, and have high nutrient concentrations

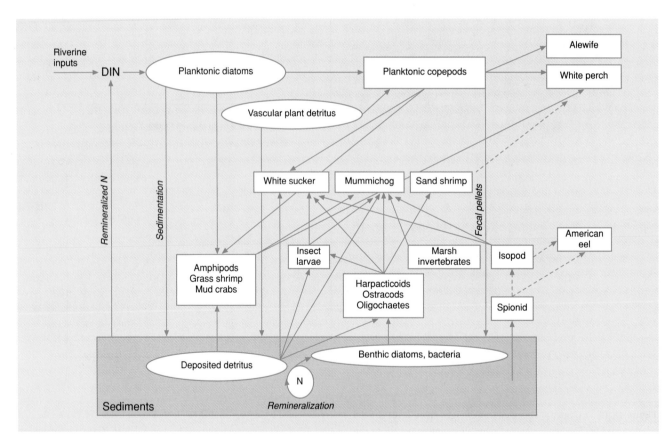

Figure 18.11 Conceptual model of nitrogen (N) flux through the food web of the upper Parker River estuary, Massachusetts, USA. Dashed arrows indicate suspected pathways. DIN, dissolved inorganic nitrogen. (After Hughes *et al.*, 2000.)

is continuously regenerated from the sediment to be taken up again by phytoplankton (Moss, 1989).

18.3.3   Estuaries

nutrient flux in estuaries: roles for planktonic and benthic organisms . . .

In estuaries, both planktonic organisms (as in lakes) and benthic organisms (as in rivers) are significant in nutrient flux. Hughes *et al.* (2000) introduced tracer levels of a rare isotope of nitrogen (as nitrate-containing ^{15}N) into the water of an estuary in Massachusetts, USA, to study how nitrogen derived from the catchment area is used and transformed in the estuarine food web. They focused their study on the upper, low salinity part of the estuary where water derived from the river catchment first meets the saline influence of tidal seawater. The planktonic centric diatom *Actinocyclus normanii* turned out to be the primary vector of nitrogen to some benthic organisms (large crustaceans) and particularly pelagic organisms (planktonic copepods and juvenile fishes). Certain components of the sedimentary biota received a small proportion of their nitrogen via the centric diatom (10–30%; e.g. pennate diatoms, harpacticoid copepods, oligochaete worms, bottom-feeding fishes such as mummichog, *Fundulus heteroclitus*, and sand shrimps). But many others obtained almost all their nitrogen from a pathway based on plant detritus. The patterns of nitrogen flow through this estuarine food web are shown in Figure 18.11. The relative importance of nutrient fluxes through the grazer and decomposer systems can be expected to vary from estuary to estuary.

. . . and human activities

The chemistry of estuarine (and coastal marine) water is strongly influenced by features of the catchment area through which the rivers have been flowing, and human activities play a major role in determining the nature of the water supplied. In a revealing comparison, van Breeman (2002) describes the forms of nitrogen in water at the mouths of rivers in North and South America. In the North American case, where the river flows through a largely forested region but has been subject to considerable human impact (fertilizer input, logging, acid precipitation, etc.), nitrogen was almost exclusively exported to estuaries and the sea in inorganic form (only 2% organic). In contrast, a pristine South American river, subject to very little human impact, exported 70% of its nitrogen in organic form. In Australian rivers too, pristine forested catchments export little nitrogen or phosphorus, and the predominant form of nitrogen is organic. As human population density increases (greater agricultural runoff and sewage) and forests are cleared (less tight retention of nutrients), however, the export to river mouths of both nitrogen and phosphorus increases and the predominant form of nitrogen changes to inorganic (Figure 18.12).

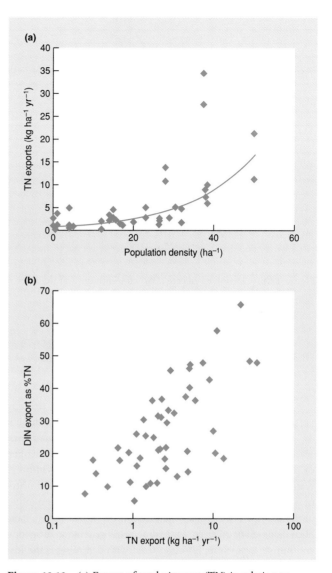

Figure 18.12   (a) Export of total nitrogen (TN) in relation to population density in 24 catchment areas near Sydney, Australia. (b) Rivers with low TN export rates (more pristine) contain nitrogen predominantly in organic form and the percentage of TN that is inorganic increases with TN. DIN, dissolved inorganic nitrogen. (After Harris, 2001.)

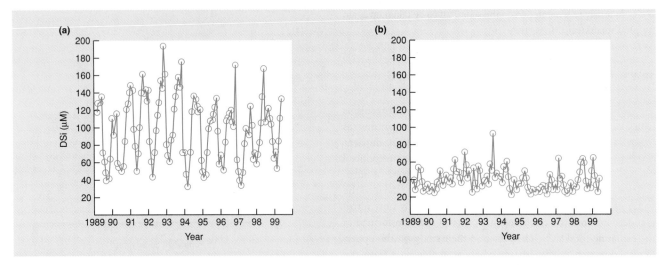

Figure 18.13   Dissolved silicate (DSi) concentrations at the river mouths of (a) the nondammed River Kalixalven and (b) the dammed River Lulealven. (Humborg *et al.*, 2002).

18.3.4   Continental shelf regions of the oceans

coastal regions of oceans are influenced by their terrestrial catchment areas . . .

The nutrient budgets of coastal regions of oceans, like estuaries, are strongly influenced by the nature of catchment areas that supply the water, via rivers, to the sea. Concentrations of nitrogen or phosphorus may limit productivity in these areas as in other water bodies, but a further human-induced effect on the chemistry of riverwater has special significance for planktonic communities in the oceans. Today, more than 25% of the world's rivers have been dammed or diverted (for hydroelectric generation, irrigation and human water supply). Associated with damming is the loss of upper soils and vegetation through inundation, loss of soil through shoreline erosion, and underground channeling of water through tunnels. These effects reduce the contact of water with vegetated soil and, therefore, reduce weathering. Figure 18.13 illustrates the patterns of export of dissolved silicate, an essential component of the cells of planktonic diatoms in the sea, for a dammed river and a freely flowing river in Sweden. The export of silicate was dramatically lower in the dammed case. The possible ecological effects of silicate reduction to nutrient fluxes and productivity in the sea may become particularly significant in East Asia, where major rivers are being dammed at accelerating rates (Milliman, 1997).

. . . and local upwelling

Another important mechanism of nutrient enrichment in coastal regions is local upwelling, bringing high nutrient concentrations from deep to shallow water where they fuel primary productivity, often producing phytoplankton blooms. Three categories of upwelling have been described and studied off the east coast of Australia: (i) wind-driven upwellings in response to seasonal north and northeasterly breezes; (ii) upwelling driven by the encroachment of the East Australian Current (EAC) onto the continental shelf; and (iii) upwelling caused by the separation of the EAC from the coast. Figure 18.14 provides examples of the distribution of nitrate concentrations associated with each mechanism. Wind-driven upwellings (generally considered to be the dominant mechanism globally) are not persistent or massive in scale. The highest nitrate concentrations are generally associated with encroachment upwellings, whilst separation upwellings are the most widespread along the coast of New South Wales.

18.3.5   Open oceans

We can view the open ocean as the largest of all endorheic 'lakes' – a huge basin of water supplied by the world's rivers and losing water only by evaporation. Its great size, in comparison to the input from rain and rivers, leads to a remarkably constant chemical composition.

the open ocean: an important role for plankton . . .

We considered biologically mediated transformations of carbon in terrestrial ecosystems in Section 18.2.3. Figure 18.15 illustrates the same thing but for the open ocean. The main transformers of dissolved inorganic carbon (essentially CO_2) are the small phytoplankton, which recycle CO_2 in the euphotic zone, and the larger plankton, which generate the majority of the carbon flux in particulate and dissolved organic form to the deep ocean floor. Figure 18.16 shows that, in general, only a small proportion of carbon fixed near the

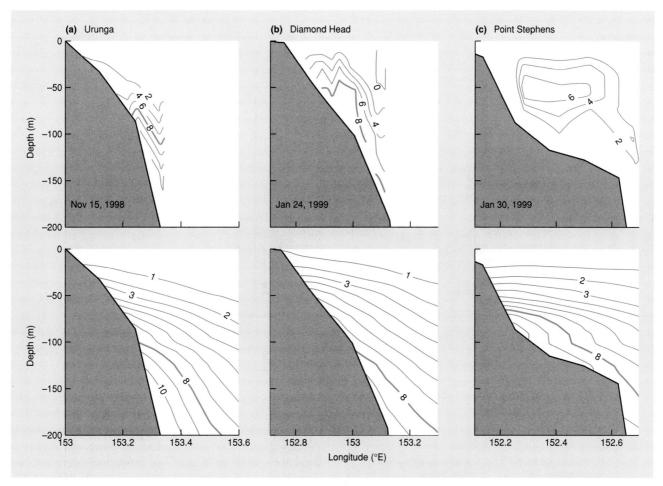

Figure 18.14 Contours of nitrate concentration during upwelling events along the New South Wales coast at: (a) Urunga (wind-driven), (b) Diamond Head (encroachment-driven), and (c) Point Stephens (separation-driven). The bottom graph in each case shows the mean nitrate concentrations that can be taken as characteristic of these sites in the absence of an upwelling event. Maximum concentration is 10 μmol l^{-1}. The contour interval is 1 or 2 μmol l^{-1} and the thick orange line represents 8 μmol l^{-1}. (After Roughan & Middleton, 2002.)

surface finds its way to the ocean bed. What reaches the ocean floor is consumed by the deep-sea biota, some is remineralized into dissolved organic form by decomposers, and a small proportion becomes buried in the sediment.

. . . which may follow a seasonal pattern

Just as we saw in terrestrial ecosystems, marked seasonal and interannual differences in nutrient flux and availability can be detected in the deep ocean. Thus, Figure 18.17a shows how chlorophyll *a* concentrations varied during the spring bloom at a site in the North Atlantic, reflecting a succession of dominant phytoplankton species. Large diatoms bloomed first, consuming almost all the available silicate (Figure 18.17b). Subsequently, a bloom of small flagellates used up the remaining nitrate. Over a longer timescale, a remarkable shift in the relative abundance of organic nitrogen and phosporus has been witnessed in the North Pacific.

The ocean has traditionally been viewed as nitrogen-limited, but, when nitrogen limitation is extreme, nitrogen-fixing taxa such as *Trichodesmium* spp. grow over large areas and bring into play the inexhaustible pool of dissolved N$_2$ in the ocean. This has led to a decade-long shift in the N : P ratio in suspended particulate organic matter (Figure 18.17c). Under these circumstances, phosphorus, iron or some other nutrient will eventually limit productivity.

About 30% of the world's oceans have long been known to have low productivity despite high concentrations of nitrate. The hypothesis that this paradox was due to the iron limitation of phytoplankton productivity has been tested in locations as different as the eastern equatorial Pacific and the open polar Southern Ocean (Boyd, 2002). Large infusions of dissolved iron

iron as a factor limiting ocean primary productivity?

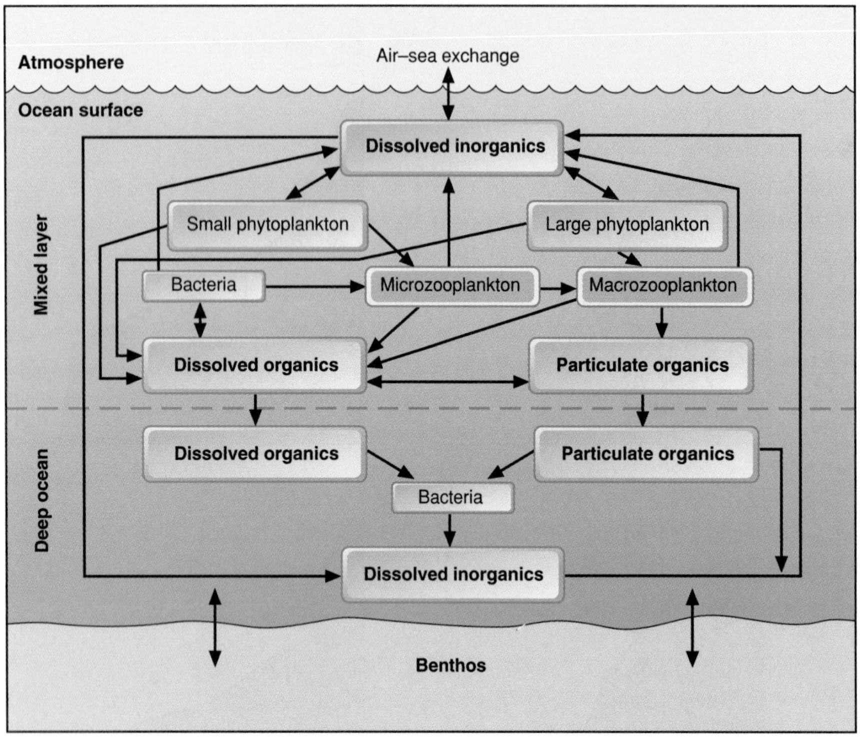

Figure 18.15 Biologically mediated transformations of carbon in the open ocean. (After Fasham *et al.*, 2001.)

at sites in each ocean led in both cases to dramatic increases in primary productivity and decreases in nitrate and silicate, as these were taken up during algal production (the results are expressed as nitrate removal in Figure 18.18). Bacterial productivity tripled within a few days in both cases, and rates of herbivory by micrograzers (flagellates and ciliates) also increased, but less so in the polar situation (where dominance by a grazer-resistant, highly silicified diatom probably suppressed grazing). The metazoan community, dominated by copepods, showed relatively little change in either situation.

It is an intriguing thought that in places such as the eastern equatorial Pacific or polar Southern Ocean, blooms in productivity might sometimes be caused by long-distance wind transport of land-derived, iron-rich particles. This would mirror, but on a very different scale, the high productivity associated with inputs of land-derived, nutrient-rich water from rivers.

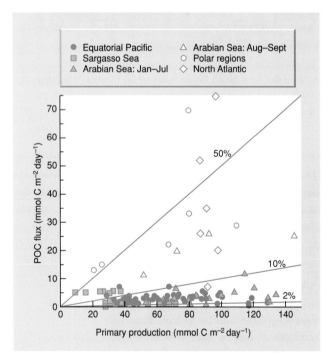

Figure 18.16 Relationship between the export of particulate organic carbon (POC) to the ocean depths, recorded at 100 m, and ocean primary productivity in the world's oceans. (After Buesseler, 1998.)

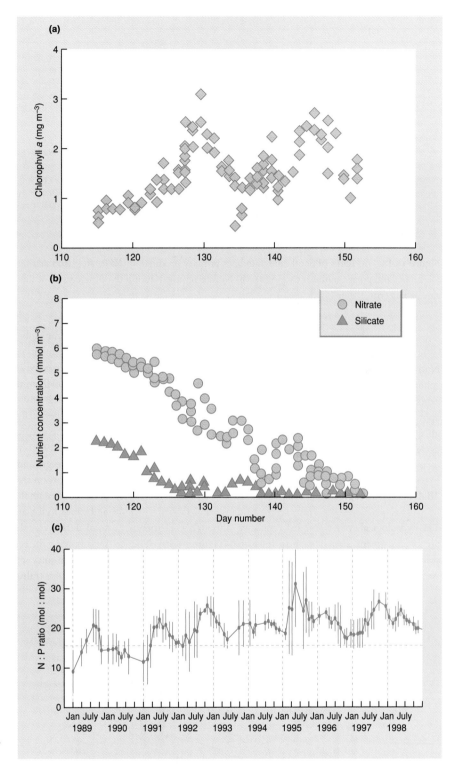

Figure 18.17 Patterns in (a) chlorophyll *a* and (b) silicate and nitrate concentrations during a spring bloom in the North Atlantic. Day number is days since January 1. (After Fasham *et al.*, 2001.) (c) Shift in the ratio of N : P in suspended particulate matter measured in the North Pacific Gyre. (After Karl, 1999.)

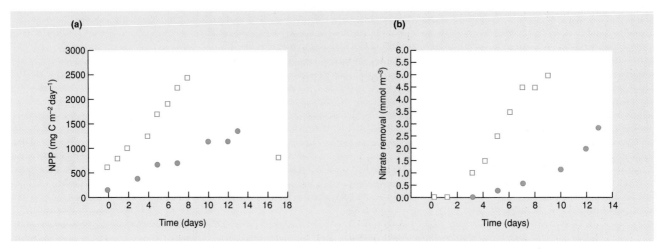

Figure 18.18   (a) Rates of depth-integrated net primary production (NPP) after iron addition at sites in the eastern equatorial Pacific Ocean (□) and polar Southern Ocean (●). (b) Nitrate removal during the time course of the two experiments. Note that silicate followed similar patterns. (After Boyd, 2002.)

18.4   Global biogeochemical cycles

Nutrients are moved over vast distances by winds in the atmosphere and by the moving waters of streams and ocean currents. There are no boundaries, either natural or political. It is appropriate, therefore, to conclude this chapter by moving to an even larger spatial scale to examine global biogeochemical cycles.

18.4.1   Hydrological cycle

The hydrological cycle is simple to conceive (although its elements are by no means always easy to measure) (Figure 18.19). The principal source of water is the oceans; radiant energy makes water evaporate into the atmosphere, winds distribute it over the

surface of the globe, and precipitation brings it down to earth (with a net movement of atmospheric water from oceans to continents), where it may be stored temporarily in soils, lakes and icefields. Loss occurs from the land through evaporation and transpiration or as liquid flow through stream channels and groundwater aquifers, eventually to return to the sea. The major pools of water occur in the oceans (97.3% of the total for the biosphere; Berner & Berner, 1987), the ice of polar ice caps and glaciers (2.06%), deep in the groundwater (0.67%) and in rivers and lakes (0.01%). The proportion that is in transit at any time is very small – water draining through the soil, flowing along rivers and present as clouds and vapor in the atmosphere constitutes only about 0.08% of the total. However, this small percentage plays a crucial role, both by supplying the requirements for survival of living organisms and for community productivity, and because so many chemical nutrients are transported with the water as it moves.

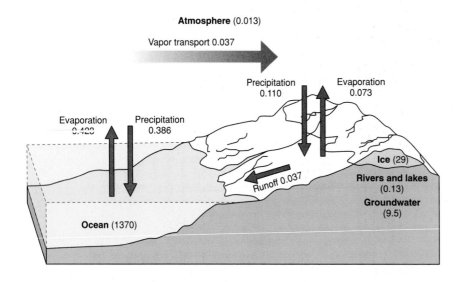

Figure 18.19   The hydrological cycle showing fluxes and sizes of reservoirs ($\times 10^6$ km^3). Values in parentheses represent the size of the various reservoirs. (After Berner & Berner, 1987.)

The hydrological cycle would proceed whether or not a biota was present. However, terrestrial vegetation can modify to a significant extent the fluxes that occur. Plants live between two counterflowing movements of water (McCune & Boyce, 1992). One moves within the plant, proceeding from the soil into the roots, up through the stem and out from the leaves as transpiration. The other is deposited on the canopy as precipitation from where it may evaporate or drip from the leaves or flow down the stem to the soil. In the absence of vegetation, some of the incoming water would evaporate from the ground surface but the rest would enter the stream flow (via surface runoff and groundwater discharge). Vegetation can intercept water at two points on this journey, preventing some from reaching the stream and causing it to move back into the atmosphere by: (i) catching some in foliage from where it may evaporate; and (ii) preventing some from draining from the soil water by taking it up in the transpiration stream.

We have seen on a small scale how cutting down the forest in a catchment in Hubbard Brook can increase the throughput to streams of water together with its load of dissolved and particulate matter. It is small wonder that large-scale deforestation around the globe, usually to create new agricultural land, can lead to the loss of topsoil, nutrient impoverishment and increased severity of flooding.

Another major perturbation to the hydrological cycle will be global climate change resulting from human activities (see Section 18.4.6). The predicted temperature increase, with its concomitant changes to wind and weather patterns, can be expected to affect the hydrological cycle by causing some melting of polar caps and glaciers, by changing patterns of precipitation and by influencing the details of evaporation, transpiration and stream flow.

18.4.2   A general model of global nutrient flux

The world's major abiotic reservoirs for nutrients are illustrated in Figure 18.20. The biotas of both terrestrial and aquatic habitats obtain some of their nutrient elements predominantly via the weathering of rock. This is the case, for example, for phosphorus. Carbon and nitrogen, on the other hand, derive mainly from the atmosphere – the first from CO_2 and the second from gaseous nitrogen, fixed by microorganisms in the soil and water. Sulfur derives from both atmospheric and lithospheric sources. In the following sections we consider phosphorus, nitrogen, sulfur and carbon in turn, and ask how human activities upset the global biogeochemical cycles of these biologically important elements.

18.4.3   Phosphorus cycle

The principal stocks of phosphorus occur in the water of the soil, rivers, lakes and oceans and in rocks and ocean sediments. The phosphorus

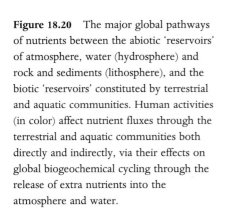

Figure 18.20   The major global pathways of nutrients between the abiotic 'reservoirs' of atmosphere, water (hydrosphere) and rock and sediments (lithosphere), and the biotic 'reservoirs' constituted by terrestrial and aquatic communities. Human activities (in color) affect nutrient fluxes through the terrestrial and aquatic communities both directly and indirectly, via their effects on global biogeochemical cycling through the release of extra nutrients into the atmosphere and water.

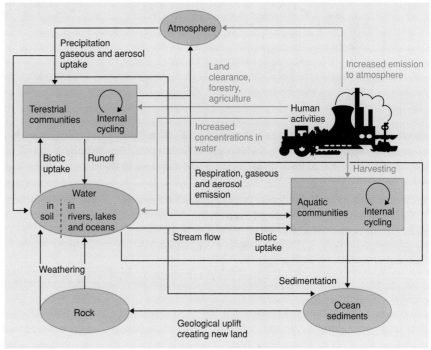

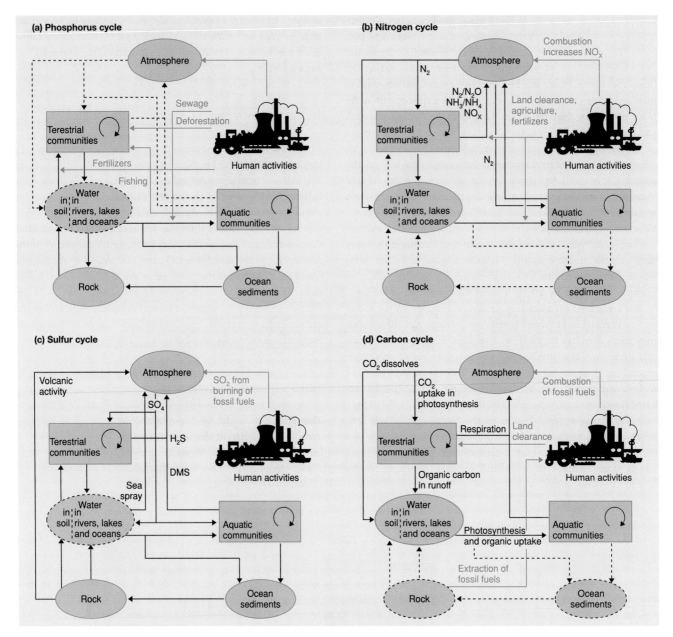

Figure 18.21 The main pathways of nutrient flux (black) and the perturbations caused by human activities (color) for four important nutrient elements: (a) phosphorus, (b) nitrogen, (c) sulfur (DMS, dimethylsufide), and (d) carbon. Insignificant compartments and fluxes are represented by dashed lines. (Based on the model illustrated in Figure 18.10, where further details can be found.)

cycle may be described as an 'open' cycle because of the general tendency for mineral phosphorus to be carried from the land inexorably to the oceans, mainly in rivers, but also to smaller extents in groundwater, or via volcanic activity and atmospheric fallout, or through abrasion of coastal land. The cycle may alternatively be termed a 'sedimentary cycle' because ultimately phosphorus becomes incorporated in ocean sediments (Figure 18.21a). We can unravel an intriguing story that starts in a terrestrial catchment area. A typical phosphorus atom, released from the rock by

chemical weathering, may enter and cycle within the terrestrial community for years, decades or centuries before it is carried via groundwater into a stream, where it takes part in the nutrient spiraling described in Section 18.3.1. Within a short time of entering the stream (weeks, months or years), the atom is carried to the ocean. It then makes, on average, about 100 round trips between the surface and deep waters, each lasting perhaps 1000 years. During each trip, it is taken up by organisms that live at the ocean surface, before eventually settling into the deep again.

On average, on its 100th descent (after 10 million years in the ocean) it fails to be released as soluble phosphorus, but instead enters the bottom sediment in particulate form. Perhaps 100 million years later, the ocean floor is lifted up by geological activity to become dry land. Thus, our phosphorus atom will eventually find its way back via a river to the sea, and to its existence of cycle (biotic uptake and decomposition) within cycle (ocean mixing) within cycle (continental uplift and erosion).

human activities contribute the majority of phosphorus in inland waters . . .

Human activities affect the phosphorus cycle in a number of ways. Marine fishing transfers about 50 Tg (1 teragram = 10^{12} g) of phosphorus from the ocean to the land each year. Since the total oceanic pool of phosphorus is around 120 Pg (1 petagram = 10^{15} g), this reverse flow has negligible consequences for the ocean compartment. However, phosphorus from the fish catch will eventually move back through the rivers to the sea and, thus, fishing contributes indirectly to increased concentrations in inland waters. More than 13 Tg of phosphorus are dispersed annually over agricultural land as fertilizer (some derived from the marine fish catch) and a further 2 or 3 Tg as an additive to domestic detergents. Much of the former reaches the aquatic system as agricultural runoff, whereas the latter arrives in domestic sewage. In addition, deforestation and many forms of land cultivation increase erosion in catchment areas and contribute to artificially high amounts of phosphorus in runoff water. All told, human activities have almost doubled the inflow of phosphorus to the oceans above that which occurs naturally (Savenko, 2001).

. . . and cause eutrophication

An increase to phosphorus input to the oceans on this scale is likely to have increased productivity to some extent, but as the more concentrated water passes through rivers, estuaries, coastal waters and particularly lakes, its influence can be particularly profound. This is because phosphorus is often the nutrient whose supply limits aquatic plant growth. In many lakes worldwide, the input of large quantities of phosphorus from agricultural runoff and sewage and also of nitrogen (mainly as runoff from agricultural land) produces ideal conditions for high phytoplankton productivity. In such cases of cultural eutrophication (enrichment), the lake water becomes turbid because of dense populations of phytoplankton (often the blue-green species), and large aquatic plants are outcompeted and disappear along with their associated invertebrate populations. Moreover, decomposition of the large biomass of phytoplankton cells may lead to low oxygen concentrations, which kill fish and invertebrates. The outcome is a productive community, but one with low biodiversity and low esthetic appeal. The remedy is to reduce nutrient input; for example, by altering agricultural practices and by diverting sewage, or by chemically 'stripping' phosphorus from treated sewage before it is discharged. Where phosphate loading has been reduced in deep lakes,

such as Lake Washington in North America, a reversal of the trends described above may occur within a few years (Edmonson, 1970). In shallow lakes, however, phosphorus stored in the sediment may continue to be released and the physical removal of some of the sediment may be called for (Moss et al., 1988).

The effects of agricultural runoff and sewage discharge are localized, in the sense that only those waters that drain the catchment area concerned are affected. But the problem is pervasive and worldwide.

18.4.4   Nitrogen cycle

the nitrogen cycle has an atmospheric phase of overwhelming importance

The atmospheric phase is predominant in the global nitrogen cycle, in which nitrogen fixation and denitrification by microbial organisms are by far the most important (Figure 18.21b). Atmospheric nitrogen is also fixed by lightning discharges during storms and reaches the ground as nitric acid dissolved in rainwater, but only about 3–4% of fixed nitrogen derives from this pathway. Organic forms of nitrogen are also widespread in the atmosphere, some of which results from the reaction of hydrocarbons and oxides of nitrogen in polluted air masses. In addition, amines and urea are naturally injected as aerosols or gases from terrestrial and aquatic ecosystems; and a third source consists of bacteria and pollen (Neff et al., 2002). While the atmospheric phase produces by far the most important input of nitrogen, there is also evidence that nitrogen from certain geological sources may fuel local productivity in terrestrial and freshwater communities (Holloway et al., 1998; Thompson et al., 2001). The magnitude of the nitrogen flux in stream flow from terrestrial to aquatic communities may be relatively small, but it is by no means insignificant for the aquatic systems involved. This is because nitrogen is one of the two elements (along with phosphorus) that most often limits plant growth. Finally, there is a small annual loss of nitrogen to ocean sediments.

In a model for the terrestrial part of the biosphere, nitrogen fixation accounts for the input of 211 Tg N year^{-1}. This is the predominant annual source of nitrogen and can be compared with the total amount stored in terrestrial vegetation and soil of 296 Pg year^{-1} (280 Pg year^{-1} of which is in the soil, and 90% of this in organic form) (Lin et al., 2000).

humans impact on the nitrogen cycle in diverse ways

Human activities have a variety of far-reaching effects on the nitrogen cycle. Deforestation, and land clearance in general, leads to substantial increases in nitrate flux in the stream flow and N_2O losses to the atmosphere (see Section 18.2.2). In addition, technological processes yield fixed nitrogen as a by-product of internal combustion and in the production of fertilizers. The agricultural practice of planting legume crops, with their root nodules containing nitrogen-fixing bacteria, contributes further to nitrogen fixation. In fact, the amount of fixed nitrogen produced by these human activities is of the same order of magnitude as that produced by natural nitrogen fixation. The production of nitrogenous fertilizers (more than 50 Tg year^{-1}) is of particular significance because an appreciable proportion of fertilizer added to land finds its way into streams and lakes. The artificially raised concentrations of nitrogen contribute to the process of cultural eutrophication of lakes.

Human activities impinge on the atmospheric phase of the nitrogen cycle too. For example, fertilization of agricultural soils leads to increased runoff as well as an increase in denitrification, and the handling and spreading of manure in areas of intensive animal husbandry releases substantial amounts of ammonia to the atmosphere. Atmospheric ammonia (NH_3) is increasingly recognized as a major pollutant when it is deposited downwind of livestock farming areas (Sutton *et al.*, 1993). Since many plant communities are adapted to low nutrient conditions, an increased input of nitrogen can be expected to cause changes to community composition. Lowland heathland is particularly sensitive to nitrogen enrichment (this is a terrestrial counterpart to lake eutrophication) and, for example, more than 35% of former Dutch heathland has now been replaced by grassland (Bobbink *et al.*, 1992). Further sensitive communities include calcareous grasslands and upland herb and bryophyte floras, where declines in species richness have been recorded (Sutton *et al.*, 1993). The vegetation of some other terrestrial communities may be less sensitive, because it may reach a stage where nitrogen is not limited. Increased nitrogen deposition to forests, for example, can be expected to result initially in increased forest growth, but at some point the system becomes 'nitrogen-saturated' (Aber, 1992). Further increases in nitrogen deposition can be expected to 'break through' into drainage, with raised concentrations of nitrogen in stream runoff contributing to eutrophication of downstream lakes.

nitrogen and acid rain

There is clear evidence of increased NH_3 emissions during the past few decades and current estimates indicate that these account for 60–80% of anthropogenic nitrogen input to European ecosystems, at least in localized areas around livestock operations (Sutton *et al.*, 1993). The other 20–40% derives from oxides of nitrogen (NO_x), resulting from combustion of oil and coal in power stations, and from industrial processes and traffic emissions. Atmospheric NO_x is converted, within days, to nitric acid, which contributes, together with NH_3, to the acidity of precipitation within and downwind of industrial regions. Sulfuric acid is the other culprit, and we outline the consequences of acid rain in the next section, after dealing with the global sulfur cycle.

18.4.5 Sulfur cycle

In the global phosphorus cycle we have seen that the lithospheric phase is predominant (Figure 18.21a), whereas the nitrogen cycle has an atmospheric phase of overwhelming importance (Figure 18.21b). Sulfur, by contrast, has atmospheric and lithospheric phases of similar magnitude (Figure 18.21c).

the sulfur cycle has atmospheric and lithospheric phases of similar magnitude

Three natural biogeochemical processes release sulfur to the atmosphere: (i) the formation of the volatile compound dimethylsulfide (DMS) (by enzymatic breakdown of an abundant compound in phytoplankton – dimethylsulfonioproprionate); (ii) anaerobic respiration by sulfate-reducing bacteria; and (iii) volcanic activity. Total biological release of sulfur to the atmosphere is estimated to be 22 Tg S year^{-1}, and of this more than 90% is in the form of DMS. Most of the remainder is produced by sulfur bacteria that release reduced sulfur compounds, particularly H_2S, from waterlogged bog and marsh communities and from marine communities associated with tidal flats. Volcanic production provides a further 7 Tg S year^{-1} to the atmosphere (Simo, 2001). A reverse flow from the atmosphere involves oxidation of sulfur compounds to sulfate, which returns to earth as both wetfall and dryfall.

The weathering of rocks provides about half the sulfur draining off the land into rivers and lakes, the remainder deriving from atmospheric sources. On its way to the ocean, a proportion of the available sulfur (mainly dissolved sulfate) is taken up by plants, passed along food chains and, via decomposition processes, becomes available again to plants. However, in comparison to phosphorus and nitrogen, a much smaller fraction of the flux of sulfur is involved in internal recycling in terrestrial and aquatic communities. Finally, there is a continuous loss of sulfur to ocean sediments, mainly through abiotic processes such as the conversion of H_2S, by reaction with iron, to ferrous sulfide (which gives marine sediments their black color).

sulfur and acid rain

The combustion of fossil fuels is the major human perturbation to the global sulfur cycle (coal contains 1–5% sulfur and oil contains 2–3%). The SO_2 released to the atmosphere is oxidized and converted to sulfuric acid in aerosol droplets, mostly less than 1 μm in size. Natural and human releases of sulfur to the atmosphere are of similar magnitude and together account for 70 Tg S year^{-1} (Simo, 2001). Whereas natural inputs are spread fairly evenly over the globe, most human inputs are concentrated in and around industrial zones in northern Europe and eastern North America, where they can contribute up to 90% of the total (Fry & Cooke, 1984). Concentrations decline progressively downwind from sites of production, but they can still be high at distances of several hundred kilometers. Thus, one nation can export its SO_2 to other countries; concerted international political action is required to alleviate the problems that arise.

Water in equilibrium with CO_2 in the atmosphere forms dilute carbonic acid with a pH of about 5.6. However, the pH of acid precipitation (rain or snow) can average well below 5.0, and values as low as 2.4 have been recorded in Britain, 2.8 in

Scandinavia and 2.1 in the USA. The emission of SO_2 often contributes most to the acid rain problem, though together NO_x and NH_3 account for 30–50% of the problem (Mooney et al., 1987; Sutton et al., 1993).

We saw earlier how a low pH can drastically affect the biotas of streams and lakes (see Chapter 2). Acid rain (see Section 2.8) has been responsible for the extinction of fish in thousands of lakes, particularly in Scandinavia. In addition, a low pH can have far-reaching consequences for forests and other terrestrial communities. It can affect plants directly, by breaking down lipids in foliage and damaging membranes, or indirectly, by increasing leaching of some nutrients from the soil and by rendering other nutrients unavailable for uptake by plants. It is important to note that some perturbations to biogeochemical cycles arise through indirect, 'knock-on' effects on other biogeochemical components. For example, alterations in the sulfur flux in themselves are not always damaging to terrestrial and aquatic communities, but the effect of sulfate's ability to mobilize metals such as aluminum, to which many organisms are sensitive, may indirectly lead to changes in community composition. (In another context, sulfate in lakes can reduce the ability of iron to bind phosphorus, releasing the phosphorus and increasing phytoplankton productivity (Caraco, 1993).)

Provided that governments show the political will to reduce emissions of SO_2 and NO_x (for example, by making use of techniques already available to remove sulfur from coal and oil), the acid rain problem should be controllable. Indeed reductions in sulfur emissions have occurred in various parts of the world.

18.4.6   Carbon cycle

opposing forces of photosynthesis and respiration drive the global carbon cycle

Photosynthesis and respiration are the two opposing processes that drive the global carbon cycle. It is predominantly a gaseous cycle, with CO_2 as the main vehicle of flux between the atmosphere, hydrosphere and biota. Historically, the lithosphere played only a minor role; fossil fuels lay as dormant reservoirs of carbon until human intervention in recent centuries (Figure 18.21d).

Terrestrial plants use atmospheric CO_2 as their carbon source for photosynthesis, whereas aquatic plants use dissolved carbonates (i.e. carbon from the hydrosphere). The two subcycles are linked by exchanges of CO_2 between the atmosphere and oceans as follows:

$$\text{atmospheric } CO_2 \rightleftharpoons \text{dissolved } CO_2$$
$$CO_2 + H_2O \rightleftharpoons H_2CO_3 \text{ (carbonic acid)}.$$

In addition, carbon finds its way into inland waters and oceans as bicarbonate resulting from weathering (carbonation) of calcium-rich rocks such as limestone and chalk:

$$CO_2 + H_2O + CaCO_3 \rightleftharpoons CaH_2(CO_3)_2.$$

Respiration by plants, animals and microorganisms releases the carbon locked in photosynthetic products back to the atmospheric and hydrospheric carbon compartments.

The concentration of CO_2 in the atmosphere has increased from about 280 parts per million (ppm) in 1750 to more than 370 ppm today and it is still rising. The pattern of increase recorded at the Mauna Loa Observatory in Hawaii since 1958 is shown in Figure 18.22. (Note the cyclical decreases in CO_2 associated with higher rates of photosynthesis during summer in the northern hemisphere – reflecting the fact that most of the world's landmass is north of the equator.)

CO$_2$ in the atmosphere has increased significantly because of . . .

We discussed this increase in atmospheric CO_2, and the associated exaggeration in the greenhouse effect, in Sections 2.9.1 and 2.9.2, but armed with a more comprehensive appreciation of carbon budgets, we can now revisit this subject. The principal causes of the increase has been the combustion of fossil fuels and, to a much smaller extent, the kilning of limestone to produce cement (the latter produces less than 2% of that produced by fossil fuel burning). Together, during the period 1980–95, these accounted for a net increase in the atmosphere averaging 5.7 ($\pm$ 0.5) Pg C year^{-1} (Houghton, 2000).

. . . the combustion of fossil fuels . . .

Land-use change has caused a further 1.9 ($\pm$ 0.2) Pg of carbon to enter the atmosphere each year. The exploitation of tropical forest causes a significant release of CO_2, but the precise effect depends on whether forest is cleared for permanent agriculture, shifting agriculture or timber production. The burning that follows most forest clearance quickly converts some of the vegetation to CO_2, while decay of the remaining vegetation releases CO_2 over a more extended period. If forests have been cleared to provide for permanent agriculture, the carbon content of the soil is reduced by decomposition of the organic matter, by erosion and sometimes by mechanical removal of the topsoil. Clearance for shifting agriculture has similar effects, but the regeneration of ground flora and secondary forest during the fallow period sequesters a proportion of the carbon originally lost. Shifting agriculture and timber extraction involve 'temporary' clearance in which the net release of CO_2 per unit area is significantly less than is the case for 'permanent' clearance for agriculture or pasture. Changes to land use in non-tropical terrestrial communities seem to have had a negligible effect on the net release of CO_2 to the atmosphere.

. . . and exploitation of tropical forest

some of the extra CO$_2$ dissolves in the oceans . . . or is taken up by terrestrial plants

The total amount of carbon released each year to the atmosphere

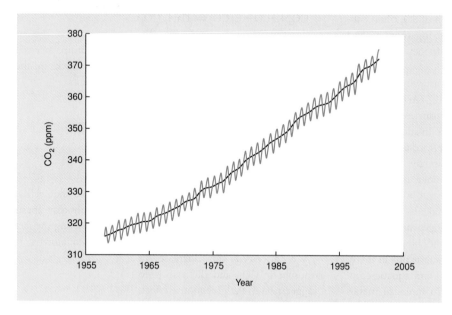

Figure 18.22 Concentration of atmospheric carbon dioxide (CO_2) at the Mauna Loa Observatory, Hawaii, showing the seasonal cycle (resulting from changes in photosynthetic rate) and the long-term increase that is due largely to the burning of fossil fuels. (Courtesy of the Climate Monitoring and Diagnostics Laboratory of the National Oceanic and Atmospheric Administration.)

by human activities (7.6 Pg C year^{-1}; see Section 2.9.1) can be compared with the 100–120 Pg C year^{-1} released naturally by respiration of the world's biota (Houghton, 2000). Where does the extra CO_2 go? The observed increase in atmospheric CO_2 accounts for 3.2 ($\pm$ 1.0) Pg C year^{-1} (i.e. 42% of the human inputs). Much of the rest, 2.1 ($\pm$ 0.6) Pg C year^{-1}, dissolves in the oceans. This leaves 2.3 Pg C year^{-1}, which is generally put down to a residual terrestrial sink, the magnitude, location and causes of which are uncertain, but are believed to involve increased terrestrial productivity in northern mid-latitude regions (i.e. part of the increase in CO_2 may serve to 'fertilize' terrestrial communities and be assimilated into extra biomass) and the recovery of forests from earlier disturbances (Houghton, 2000).

accurate prediction of future changes in carbon emissions is a pressing matter

There is considerable year-to-year variation in the estimates of CO_2 sources and sinks, and of the increase in the atmosphere (Figure 18.23). Indeed, this variation is what allowed standard errors to be placed on average values in the previous paragraphs. The declines in atmospheric increase in CO_2 between 1981 and 1982 followed dramatic rises in oil prices, while the declines in 1992 and 1993 followed the economic collapse of the Soviet Union. In 1997–98 (not shown in Figure 18.23), a remarkable wildfire in a small part of the globe doubled the growth rate of CO_2 in the atmosphere. Massive forest fires in Indonesia produced a carbon emission of about 1 Pg in just a few weeks. The burned areas included vast deposits of peat, which lost 25–85 cm of their depth during the fire, and most of the released carbon came from this source rather than the burning of wood. The fires in Indonesia were particularly serious due to a combination of circumstances – drought caused by the 1997–98 El Niño event, the thickness of peat present, and particular logging practices that allowed the vegetation and soil to dry out (Schimel & Baker, 2002). The accurate prediction of future changes in carbon emissions is a pressing matter, but it will be a difficult task because so many variables – climatic, political and sociological – impinge on the carbon balance. We return to the many dimensions of the ecological challenges facing mankind at the very end of the book (see Section 22.5.3).

Summary

Living organisms expend energy to extract chemicals from their environment, hold on to them and use them for a period, and then lose them again. In this chapter, we consider the ways in which the biota on an area of land, or within a volume of water, accumulates, transforms and moves matter between the various living and abiotic components of the ecosystem. Some abiotic compartments occur in the *atmosphere* (carbon in carbon dioxide, nitrogen as gaseous nitrogen), some in the rocks of the *lithosphere* (calcium, potassium) and others in the *hydrosphere* – the water of soils, streams, lakes or oceans (nitrogen in dissolved nitrate, phosphorus in phosphate).

Nutrient elements are available to plants as simple inorganic molecules or ions and can be incorporated into complex organic carbon compounds in biomass. Ultimately, however, when the carbon compounds are metabolized to carbon dioxide, the mineral nutrients are released again in simple inorganic form. Another plant may then absorb them, and so an individual atom of a nutrient element may pass repeatedly through one food chain after another. By its very nature, each joule of energy in a high-energy

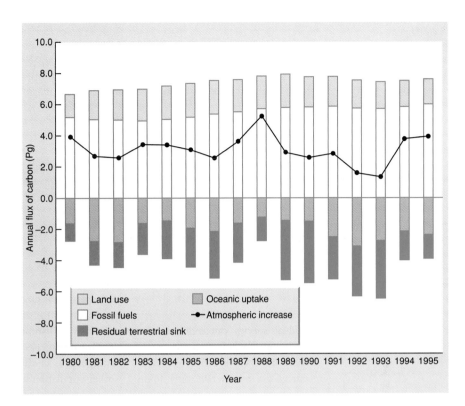

Figure 18.23 Annual variations in the atmospheric increase in carbon dioxide (circles and black line) and in carbon released (histograms above the midline) or accumulated (histograms below the midline) in the global carbon cycle from 1980 to 1995. (After Houghton, 2000.)

compound can be used only once, whereas chemical nutrients can be used again, and repeatedly recycled (although nutrient cycling is never perfect).

We discuss the ways that nutrients are gained and lost in ecosystems and note that inputs and outputs of a given nutrient may be in balance. However, this is by no means always so, in which case the ecosystem is a net source or sink for the nutrient in question. We discuss the components of nutrient budgets, and the factors affecting inputs and outputs, in forests, streams, lakes, estuaries and oceans.

Because nutrients are moved over vast distances by winds in the atmosphere and by the moving waters of streams and ocean currents, we conclude the chapter by examining global biogeochemical cycles. The principal source of water in the hydrological cycle is the oceans; radiant energy makes water evaporate into the atmosphere, winds distribute it over the surface of the globe, and precipitation brings it down to earth. Phosphorus derives mainly from the weathering of rocks (lithosphere); its cycle may be described as sedimentary because of the general tendency for mineral phosphorus to be carried from the land inexorably to the oceans where ultimately it becomes incorporated in sediments. The sulfur cycle has an atmospheric phase and a lithospheric phase of similar magnitude. In contrast, the atmospheric phase is predominant in both the carbon and nitrogen cycles. Photosynthesis and respiration are the two opposing processes that drive the global carbon cycle while nitrogen fixation and denitrification by microbial organisms are of particular importance in the nitrogen cycle. Human activities contribute significant inputs of nutrients to ecosystems and disrupt local and global biogeochemical cycles.

Chapter 19
The Influence of Population Interactions on Community Structure

19.1 Introduction

interspecific competition may determine which, and how many, species can coexist

Individual species can influence the composition of whole communities in a variety of ways. Every species provides resources for others that prey upon or parasitize them, but some species (trees, for example) provide a wide range of resources that are used by a large number of consumer species (discussed in Chapter 3). Thus, oak trees can be very influential in determining the composition and diversity of the communities of which they are part by providing acorns, leaves, stemwood and roots for their specialist herbivores as well as a similar range of dead organic materials that are exploited by detritivores and decomposers (see Chapter 11). Species may also help determine community composition and diversity by influencing conditions (see Chapter 2). Thus, large plants create microhabitats that encompass the niche requirements of many smaller plants and animals, whilst large animals provide a range of conditions on and in their bodies that are exploited by a variety of parasites (see Chapter 12). During succession we have also seen that some early colonizers facilitate the entry of later species by changing conditions in a way that favors the latter (see Chapter 16). We will not dwell further on these processes.

The current chapter pays particular attention to the way that competition, predation and parasitism can shape communities. The ideas we present reflect a debate that has been central to ecology for the last four decades. As we explain below, there are sound theoretical reasons for expecting interspecific competition to be important in shaping communities by determining which, and how many, species can coexist. Indeed, the prevalent view amongst ecologists in the 1970s was that competition was of overriding importance (MacArthur, 1972; Cody, 1975). Subsequently, the conventional wisdom has moved away from this monolithic view to one giving more prominence to nonequilibrial and stochastic

factors, such as physical disturbance and inconstancy in conditions (see Chapter 16), and to an important role for predation and parasitism (e.g. Diamond & Case, 1986; Gee & Giller, 1987; Hudson & Greenman, 1998). We first consider the role of interspecific competition in theory and practice before proceeding to the other population interactions that in some communities and for some organisms make competition much less influential.

19.2 Influence of competition on community structure

The view that interspecific competition plays a central and powerful role in the shaping of communities was first fostered by the competitive exclusion principle (see Chapter 8) which says that if two or more species compete for the same limiting resource, then all but one of them will be driven to extinction. More sophisticated variants of the principle, namely the concepts of limiting similarity, optimum similarity and niche packing (see Chapter 8), have suggested a limit to the similarity of competing species, and thus, a limit to the number of species that can be fitted into a particular community before niche space is fully saturated. Within this theoretical framework, interspecific competition is obviously important, because it excludes particular species from some communities, and determines precisely which species coexist in others. The crucial question, however, is: 'how important are such theoretical effects in the real world?'

19.2.1 Prevalence of current competition in communities

There is no argument about whether competition *sometimes* affects community structure; nobody doubts that it

competition is not always of overriding importance

does. Equally, nobody claims that competition is of overriding importance in each and every case. In a community where the species are competing with one another on a day-to-day or minute-by-minute basis, and where the environment is homogeneous, it is indisputable that competition will have a powerful effect on community structure. Suppose instead, though, that other factors prevent competition from progressing to competitive exclusion, by depressing densities or periodically reversing competitive superiority. In this context, Hutchinson (1961) noted how phytoplankton communities generally exhibit high diversity despite limited opportunities for resource partitioning (his 'paradox of the plankton') and suggested that short-term fluctuations in conditions (e.g. temperature) or resources (light or nutrients) might prevent competitive exclusion and permit high diversity. Floder et al. (2002) tested this hypothesis by comparing species diversity of inocula of natural phytoplankton communities in microcosms maintained at high (100 μmol photons m^{-2} s^{-1}) or low (20 μmol photons m^{-2} s^{-1}) light levels with diversities achieved when light levels were periodically switched from high to low and back again every 1, 3, 6 or 12 days in a 49-day experiment. As predicted by Hutchinson, diversities were higher under fluctuating conditions, where interspecific competition was less likely to proceed to competitive exclusion (Figure 19.1).

literature reviews suggest competition is widespread . . .

Perhaps the most direct way of determining the importance of competition in practice is from the results of experimental field manipulations, in which one species is removed from or added to the community, and the responses of the other species are monitored. Two important surveys of field experiments

on interspecific competition were published in 1983. Schoener (1983) examined the results of all the experiments he could find – 164 studies in all – and noted that approximately equal numbers of studies had dealt with terrestrial plants, terrestrial animals and marine organisms, but that studies of freshwater organisms amounted to only about half the number in the other groups. Amongst the terrestrial studies, however, most were concerned with temperate regions and mainland populations and relatively few dealt with phytophagous (plant-eating) insects. Any conclusions were therefore bound to be subject to the limitations imposed by what ecologists had chosen to look at. Nevertheless, Schoener found that approximately 90% of the studies had demonstrated the existence of interspecific competition, and that the figures were 89, 91 and 94% for terrestrial, freshwater and marine organisms, respectively. Moreover, if he looked at single species or small groups of species (of which there were 390) rather than at whole studies which may have dealt with several groups of species, 76% showed effects of competition at least sometimes, and 57% showed effects in all the conditions under which they were examined. Once again, terrestrial, freshwater and marine organisms gave very similar figures. Connell's (1983) review was less extensive than Schoener's: 72 studies in six major journals, dealing with a total of 215 species and 527 different experiments. Interspecific competition was demonstrated in most of the studies, in more than half of the species, and in approximately 40% of the experiments. In contrast to Schoener, Connell found that interspecific competition was more prevalent in marine than in terrestrial organisms, and also that it was more prevalent in large than in small organisms.

Taken together, Schoener's and Connell's reviews certainly seem to indicate that active, current interspecific competition is widespread. Its percentage occurrence amongst species is admittedly lower than its percentage occurrence amongst whole studies. However, this is to be expected, since, for example, if four species are arranged along a single niche dimension and all adjacent species competed with each other, this would still be only three out of six (or 50%) of all possible pairwise interactions.

Connell also found, however, that in studies of just one pair of species, interspecific competition was almost always apparent, whereas with more species the prevalence dropped markedly (from more than 90% to less than 50%). This can be explained to some extent by the argument outlined above, but it may also indicate biases in the particular pairs of species studied, and in the studies that are actually reported (or accepted by journal editors). It is highly likely that many pairs of species are chosen for study because they are 'interesting' (because competition between them is suspected) and if none is found this is simply not reported. Judging the prevalence of competition from such studies is rather like judging the prevalence of debauched clergymen from the 'gutter press'. Bias in the choice of studies is a real problem, only partially alleviated

. . . but are the data biased?

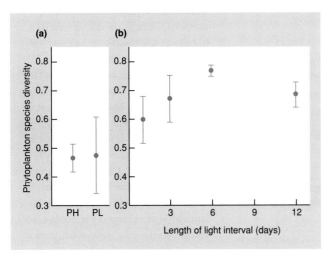

Figure 19.1 Average phytoplankton species diversity (Shannon diversity index, ± SE) at the end of 49-day experiments (a) under constant light conditions and (b) under fluctuating light intensity. PH, permanent high light intensity; PL, permanent low light intensity. (After Floder et al., 2002.)

in investigations of larger groups of species when a number of 'negatives' can be conscientiously reported alongside one or a few 'positives'. Thus the results of surveys, such as those by Schoener and Connell, exaggerate, to an unknown extent, the frequency of competition.

the strength of competition is likely to vary from community to community

As previously noted, phytophagous insects were poorly represented in Schoener's data, but reviews of phytophagous insects have tended to suggest either that competition is relatively rare in the group overall (Strong *et al.*, 1984) or at least in certain types of phytophagous insects, for example 'leaf-biters' (Denno *et al.*, 1995). There are also examples of 'vacant niches' for phytophagous insects: feeding sites or feeding modes on a widespread plant that are utilized by insects in one part of the world, but not in another part of the world where the native insect fauna is different (Figure 19.2). This failure to saturate the niche space also argues against a powerful role for interspecific competition. On a more general level, it has been suggested that herbivores *as a whole* are seldom food-limited, and are therefore not likely to compete for common resources (Hairston *et al.*, 1960; Slobodkin *et al.*, 1967; see Section 20.2.5). The bases for this suggestion are the observations that green plants are normally abundant and largely intact, they are rarely devastated, and most herbivores are scarce most of the time. Schoener

found the proportion of herbivores exhibiting interspecific competition to be significantly lower than the proportions of plants, carnivores or detritivores.

Taken overall, therefore, current interspecific competition has been reported in studies on a wide range of organisms and in some groups its incidence may be particularly obvious, for example amongst sessile organisms in crowded situations. However, in other groups of organisms, interspecific competition may have little or no influence. It appears to be relatively rare among herbivores generally, and particularly rare amongst some types of phytophagous insect.

19.2.2   Structuring power of competition

Even when competition is potentially intense, the species concerned may nevertheless coexist. This has been highlighted in theoretical studies of model communities in which species compete for patchy and ephemeral resources, and in which species themselves have aggregated distributions, with each species distributed independently of the others (e.g. Atkinson & Shorrocks, 1981; Shorrocks & Rosewell, 1987; see Chapter 8). The species exhibited 'current competition' (like that in Schoener's and Connell's surveys), in

Atkinson and Shorrock's simulations

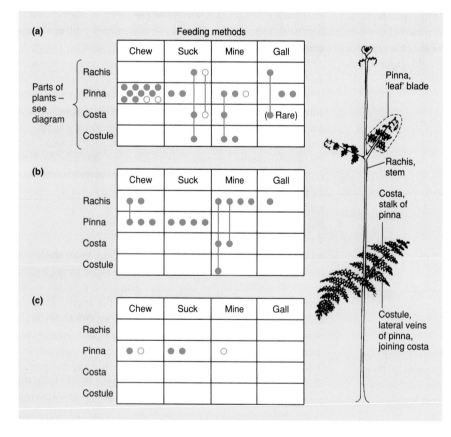

Figure 19.2 Feeding sites and feeding methods of herbivorous insects attacking bracken (*Pteridium aquilinum*) on three continents. (a) Skipwith Common in northern England; the data are derived from both a woodland and a more open site. (b) Hombrom Bluff, a savanna woodland in Papua New Guinea. (c) Sierra Blanca in the Sacramento Mountains of New Mexico and Arizona, USA; as at Skipwith, the data here are derived from both an open and a wooded site. Each bracken insect exploits the frond in a characteristic way. Chewers live externally and bite large pieces out of the plant; suckers puncture individual cells or the vascular system; miners live inside tissues; and gall-formers do likewise but induce galls. Feeding sites are indicated on the diagram of the bracken frond. Feeding sites of species exploiting more than one part of the frond are joined by lines. •, open and woodland sites; ○, open sites only. (After Lawton, 1984.)

that the removal of one species led to an increase in the abundance of others. But, despite the fact that competition coefficients were high enough to lead to competitive exclusion in a uniform environment, the patchy nature of the environment and the aggregative behaviors of individuals of the species made coexistence possible without any niche differentiation. Thus, even if interspecific competition is actually affecting the abundance of populations, it need not determine the species composition of the community. In a field study of 60 insect taxa (Diptera and Hymenoptera) that exploit the patchy resources provided by 66 mushroom taxa, Wertheim *et al.* (2000) found insect coexistence could be explained by intraspecific aggregation in the manner described above, while resource partitioning was judged not to contribute detectably to biodiversity.

the ghost of competition past

On the other hand, even when interspecific competition is absent or difficult to detect, this does not necessarily mean that it is unimportant as a structuring force. Species may not compete at present because selection in the past favored an avoidance of competition, and thus a differentiation of niches (Connell's 'ghost of competition past' – see Chapter 8). Alternatively, unsuccessful competitors may already have been driven to extinction; the present, observed species may then simply be those that are able to exist because they compete very little or not at all with other species. Furthermore, species may compete only rarely (perhaps during population outbreaks), or only in localized patches of especially high density, but the results of such competition may be crucial to their continued existence at a particular location. In all of these cases, interspecific competition must be seen as a powerful influence on community structure, affecting which species coexist and the precise nature of those species. Yet, this influence will not be reflected in the level of current competition. It is clear that the intensity of current competition may sometimes be linked only weakly to the structuring power of competition within the community.

expectations from competition theory

This weak link has led a number of community ecologists to carry out studies on competition that do not rely on the existence of current competition. The approach has been first to predict what a community *should* look like if interspecific competition was shaping it or had shaped it in the past, and then to examine real communities to see whether they conform to these predictions.

The predictions themselves emerge readily from conventional competition theory (see Chapter 8).

1 Potential competitors that coexist in a community should, at the very least, exhibit niche differentiation (see Section 19.2.3).
2 This niche differentiation will often manifest itself as morphological differentiation (see Section 19.2.4).
3 Within any one community, potential competitors with little or no niche differentiation would be unlikely to coexist. Their

distributions in space should therefore be negatively associated: each should tend to occur only where the other is absent (see Section 19.2.5).

In the following sections, we will discuss studies that deal with the documentation of patterns consistent with a role for competition in structuring communities.

19.2.3 Evidence from community patterns: niche differentiation

The various types of niche differentiation in animals and plants were outlined in Chapter 8. On the one hand, resources may be utilized differentially. This may express itself directly within a single habitat, or as a difference in microhabitat, geographic distribution or temporal appearance if the resources are themselves separated spatially or temporally. Alternatively, species and their competitive abilities may differ in their responses to environmental conditions. This too can express itself as either microhabitat, geographic or temporal differentiation, depending on the manner in which the conditions themselves vary.

19.2.3.1 Niche complementarity

In one study of niche differentiation and coexistence, a number of species of anenome fish were examined near Madang in Papua New Guinea (Elliott & Mariscal, 2001). This region has the highest reported species richness of both anenome fishes (nine) and their host anenomes (10).

evidence from community patterns . . .
. . . in anenome fishes in Papua New Guinea, . . .

Each individual anenome is typically occupied by individuals of just one species of anenome fish because the residents aggressively exclude intruders (although aggressive interactions are less frequently observed between anenome fish of very different sizes). Anenomes seem to be a limiting resource for the fishes because almost all the anenomes were occupied, and when some were transplanted to new sites they were quickly colonized and the abundance of adult fish increased. Surveys at three replicate reef sites in four zones (nearshore, mid-lagoon, outer barrier reef and offshore: Figure 19.3a) showed that each anenome fish was primarily associated with a particular species of anenome and each showed a characteristic preference for a particular zone (Figure 19.3b). Different anenome fish that lived with the same anenome were typically associated with different zones. Thus, *Amphiprion percula* occupied the anenome *Heteractis magnifica* in nearshore zones, while *A. perideraion* occupied *H. magnifica* in offshore zones. Elliott and Mariscal concluded that coexistence of the nine anenome fishes on the limited anenome resource was possible because of the differentiation of their niches, together with the ability of small anenome fish species (*A. sandaracinos* and *A.*

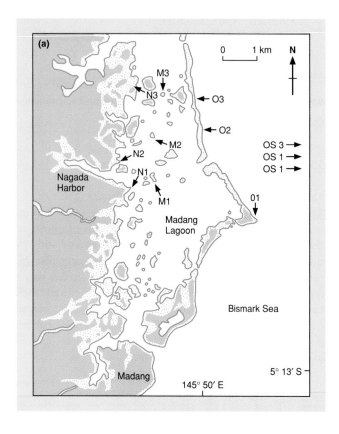

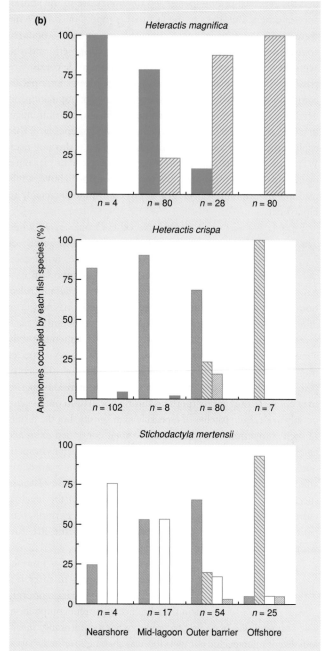

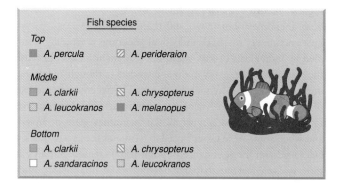

Figure 19.3 (a) Map showing the location of three replicate study sites in each of four zones within and outside Madang Lagoon (N, nearshore; M, mid-lagoon; O, outer barrier reef; OS, offshore reef). The white areas indicate water, heavy stippling represents coral reef and light stippling represents land. (b) The percentage of three common species of anenome (*Heteractis magnifica*, *H. crispa* and *Stichodactyla mertensii*) occupied by different anemone fish species (*Amphiprion* spp., in key on left) in each of four zones. The number of anenomes censused in each zone is shown by *n*. (After Elliott & Mariscal, 2001.)

leucokranos) to cohabit the same anenome with larger species. The pattern is consistent with what would be expected of communities molded by competition (specifically predictions 1 and 3 above).

Two further points, illustrated by the anenome fish, are worth highlighting. First, they can be considered to be a guild, in that they are a group of species that exploit the same class of environmental resources in a similar way (Root, 1967). If interspecific competition is to occur at all, or if it has occurred in the past, then it will be most likely to occur, or to have occurred, within guilds. But this does *not* mean that guild members necessarily compete or have necessarily competed: the onus is on ecologists to demonstrate that this is the case.

The second point about the anenome fish is that they demonstrate *niche complementarity*. That is, within the guild as a whole, niche differentiation involves several niche dimensions, and species that occupy a similar position along one dimension (anenome species used) tend to differ along another dimension (zone occupied). Complementary differentiation along several dimensions has also been reported for guilds as diverse as lizards (Schoener, 1974), bumblebees (Pyke, 1982), bats (McKenzie & Rolfe, 1986), rainforest carnivores (Ray & Sunquist, 2001) and tropical trees (Davies *et al.*, 1998), as described next.

19.2.3.2 Niche differentiation in space

. . . in trees in Borneo, . . .

Trees vary in their capacity to use resources such as light, water and nutrients. A study in Borneo of 11 tree species in the genus *Macaranga* showed marked differentiation in light requirements, from extremely high light-demanding species such as *M. gigantea* to shade-tolerant species such as *M. kingii* (Figure 19.4a). The average light levels intercepted by the crowns of these trees tended to increase as they grew larger, but the ranking of the species did not change. The shade-tolerant species were smaller (Figure 19.4b) and persisted in the understory, rarely establishing in disturbed microsites (e.g. *M. kingii*), in contrast to some of the larger, high-light species that are pioneers of large forest gaps (e.g. *M. gigantea*). Others were associated with intermediate light levels and can be considered small gap specialists (e.g. *M. trachyphylla*). The *Macaranga* species were also differentiated along a second niche gradient, with some species being more common on clay-rich soils and others on sand-rich soils (Figure 19.4b). This differentiation may be based on nutrient availability (generally higher in clay soils) and/or soil moisture availability (possibly lower in the clay soils because of thinner root mats and humus layers). As with the anenome fish, there is evidence of niche complementarity among the *Macaranga* species. Thus, species with similar light requirements differed in terms of preferred soil textures, especially in the case of the shade-tolerant species.

The apparent niche partitioning by *Macaranga* species was partly related to horizontal heterogeneity in resources (light levels in relation to gap size, distribution of soil types) and partly to vertical heterogeneity (height achieved, depth of root mat).

Ectomycorrhizal fungi have also been shown to exploit resources differentially in the vertical plane in the floor of a pine (*Pinus resinosa*) forest. Until recently, it was not possible to study the *in situ* distribution of ectomycorrhizal hyphae, but now DNA analyses allow the identification of putative species (even in the absence of species names) and permit their distributions to be compared. The forest soil had a well-developed litter layer above a fermentation layer (the F layer) and a thin humified layer (the H layer), with mineral soil beneath (the B horizon). Of the 26 species separated by the DNA analysis, some were very largely restricted to the litter layer (group A in Figure 19.5), others to the F layer (group D), the H layer (group E) or the B horizon (group F). The remaining species were more general in their distributions (groups B and C).

19.2.3.3 Niche differentiation in time

Intense competition may, in theory, be avoided by partitioning resources in horizontal or vertical space, as in the examples above, *or* in time (Kronfeld-Schor & Dayan, 2003), for example, by a staggering of life cycles through the year. It is notable that two species of mantids, which feature as predators in many parts of the world, commonly coexist both in Asia and North America. *Tenodera sinensis* and *Mantis religiosa* have life cycles that are 2–3 weeks out of phase. To test the hypothesis that this asynchrony serves to reduce interspecific competition, the timing of their egg hatch was experimentally synchronized in replicated field enclosures (Hurd & Eisenberg, 1990). *T. sinensis*, which normally hatches earlier, was unaffected by *M. religiosa*. In contrast, the survival and body size of *M. religiosa* declined in the presence of *T. sinensis*. Because these mantids are both competitors for shared resources and predators of each other, the outcome of this experiment probably reflects a complex interaction between the two processes.

. . . in preying mantids in North America . . .

In plants too, resources may be partitioned in time. Thus, tundra plants growing in nitrogen-limited conditions in Alaska were differentiated in their timing of nitrogen uptake, as well as the soil depth from which it was extracted and the chemical form of nitrogen used. To trace how tundra species differed in their uptake of different nitrogen sources, McKane *et al.* (2002) injected three chemical forms labeled with the rare isotope ^{15}N (inorganic ammonium, nitrate and organic glycine) at two soil depths (3 and 8 cm) on two occasions (June 24 and August 7) in a $3 \times 2 \times 2$ factorial design. Concentration of the ^{15}N tracer was measured in each of five common tundra plants in 3–6 replicates of each treatment 7 days after application. The five plants proved to be well differentiated in

. . . and in tundra plants in Alaska

their use of nitrogen sources (Figure 19.6). Cottongrass (*Eriophorum vaginatum*) and the cranberry bush (*Vaccinium vitis-idaea*) both relied on a combination of glycine and ammonium, but cranberry obtained more of these forms early in the growing season and at a shallower depth than cottongrass. The evergreen shrub *Ledum palustre* and the dwarf birch (*Betula nana*) used mainly ammonium but *L. palustre* obtained more of this form early in the season while the birch exploited it later. Finally, the grass *Carex bigelowii* was the only species to use mainly nitrate. Here, niche complementarity can be seen along three niche dimensions and differences in timing of use may help explain the coexistence of these species on a limited resource.

19.2.3.4   Niche differentiation – apparent or real? Null models

the aim of demonstrating that patterns are not generated merely by chance

Many cases of apparent resource partitioning have been reported. It is likely, however, that studies failing to detect such differentiation have tended to go unpublished. It is always possible, of course, that these 'unsuccessful' studies are flawed and incomplete, and that they have failed to deal with the relevant niche dimensions; but a number have been sufficiently beyond reproach to raise the possibility that in certain groups resource partitioning is not an important feature. Strong (1982) studied a group of hispine beetles (Chrysomelidae) that commonly coexist as adults in the rolled leaves of *Heliconia* plants. These long-lived tropical beetles are closely related, eat the same food and occupy the same habitat. They would appear to be good candidates for demonstrating resource partitioning. Yet, Strong could find no evidence of segregation, except in the case of just one of the 13 species studied, which was segregated weakly from a number of others. The beetles lack any aggressive behavior, either within or between species; their host specificity does not change as a function of co-occupancy of leaves with other species that might be competitors; and the levels of food and habitat are commonly not limiting for these beetles, which suffer heavily from parasitism and predation. In these species, resource partitioning associated with interspecific competition does not appear to structure the community. As we have seen, this may well be true of many phytophagous insect communities. Plant

Figure 19.4 (*right*) (a) Percentage of individuals in each of five crown illumination classes for 11 *Macaranga* species (sample sizes in parentheses). (b) Three-dimensional distribution of the 11 species with respect to maximum height, the proportion of stems in high light levels (class 5 in (a)) and proportion of stems in sand-rich soils. Each species of *Macaranga* is denoted by a single letter: G, *gigantean*; W, *winkleri*; H, *hosei*; Y, *hypoleuca*; T, *triloba*; B, *beccariana*; A, *trachyphylla*; K, *kingii*; U, *hullettii*; V, *havilandii*; L, *lamellata*. (After Davies *et al.*, 1998.)

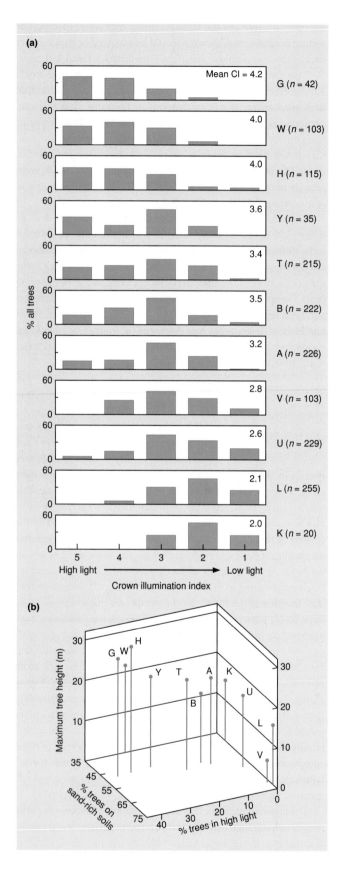

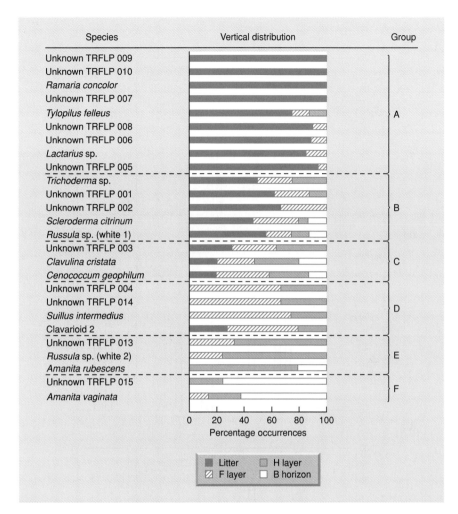

Figure 19.5 The vertical distribution of 26 ectomycorrhizal fungal species in the floor of a pine forest determined by DNA analysis. Most have not been named formally but are shown as a code (TRFLP, terminal restriction fragment length polymorphism). The vertical distribution histograms show the percentage of occurrences of each species in litter, the F layer, the H layer and the B horizon. (After Dickie *et al.*, 2002.)

studies involving taxa as diverse as phytoplankton (see Figure 19.1) and trees (Brokaw & Busing, 2000) have similarly failed to provide evidence consistent with a strong role for niche partitioning in promoting coexistence and species diversity. Whilst patterns consistent with a niche differentiation hypothesis are reasonably widespread, they are by no means universal.

null hypotheses are intended to ensure statistical rigor Moreover, a number of workers, notably Simberloff and Strong, have criticized what they see as a tendency to interpret 'mere differences' as confirming the importance of interspecific competition. Such reports beg the question of whether the differences are large enough or regular enough to be different from what might be found at random among a set of species. This problem led to an approach known as *null model analysis* (Gotelli, 2001). Null models are models of actual communities that retain certain of the characteristics of their real counterparts, but reassemble the components at random, specifically excluding the consequences of biological interactions. In fact, such analyses are attempts to

follow a much more general approach to scientific investigation, namely the construction and testing of *null hypotheses*. The idea (familiar to most readers in a statistical context) is that the data are rearranged into a form (the null model) representing what the data would look like in the absence of the phenomenon under investigation (in this case species interactions, particularly interspecific competition). Then, if the actual data show a significant statistical difference from the null hypothesis, the null hypothesis is rejected and the action of the phenomenon under investigation is strongly inferred. Rejecting (or falsifying) the absence of an effect is reckoned to be better than confirming its presence, because there are well-established statistical methods for testing whether things are significantly different (allowing falsification) but none for testing whether things are 'significantly similar'.

Lawlor (1980) looked at 10 North American lizard communities, consisting of 4–9 species, for which he had estimates of the amounts of each of a null model of food resource use in lizard communities . . .

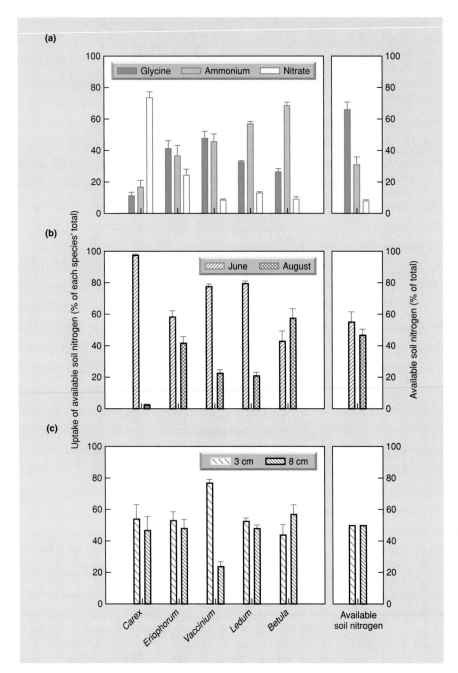

Figure 19.6 Mean uptake of available soil nitrogen (± SE) in terms of (a) chemical form, (b) timing of uptake and (c) depth of uptake by the five most common species in tussock tundra in Alaska. Data are expressed as the percentage of each species' total uptake (left panels) or as the percentage of the total pool of nitrogen available in the soil (right panels). (After McKane *et al.*, 2002.)

20 food categories consumed by each species in each community (data from Pianka, 1973). A number of null models of these communities were created (see below), which were then compared with their real counterparts in terms of their patterns of overlap in resource use. If competition is or has been a significant force in determining community structure, the niches should be spaced out, and overlap in resource use in the real communities should be less than predicted by the null models.

Lawlor's analysis was based on the 'electivities' of the consumer species, where the electivity of species i for resource k was the proportion of the diet of species i which consisted of resource k. Electivities therefore ranged from 0 to 1. These electivities were in turn used to calculate, for each pair of species in a community, an index of resource-use overlap, which itself varied between 0 (no overlap) and 1 (complete overlap). Finally, each community was characterized by a single value: the mean resource overlap for all pairs of species present.

... based on four
'reorganization
algorithms'

The null models were of four types, generated by four 'reorganization algorithms' (RA1–RA4, Figure 19.7). Each retained a different aspect of the structure of the original community whilst randomizing the remaining aspects of resource use.

RA1 retained the minimum amount of original community structure. Only the original number of species and the original number of resource categories were retained. Observed electivities (including zeros) were replaced in every case by random values between 0 and 1. This meant that there were far fewer zeros than in the original community. The niche breadth of each species was therefore increased.

RA2 replaced all electivities, *except zeros*, with random values. Thus, the qualitative degree of specialization of each consumer was retained (i.e. the number of resources consumed to any extent by each species was correct).

RA3 retained not only the original qualitative degree of specialization but also the original consumer niche breadths. No randomly generated electivities were used. Instead, the original

sets of values were rearranged. In other words, for each consumer, all electivities, both zeros and non-zeros, were randomly reassigned to the different resource types.

RA4 reassigned only the non-zero electivities. Of all the algorithms, this one retained most of the original community structure.

Each of the four algorithms was applied to each of the 10 communities. In every one of these 40 cases, 100 'null model' communities were generated and the corresponding 100 mean values of resource overlap were calculated. If competition were important in the real community, these mean overlaps should exceed the real community value. The real community was therefore considered to have a *significantly* lower mean overlap than the null model ($P < 0.05$) if five or fewer of the 100 simulations gave mean overlaps less than the real value.

The results are shown in Figure 19.7. Increasing the niche breadths of all consumers (RA1) resulted in the highest mean overlaps (significantly higher the lizards appear to pass the test . . .

than the real communities). Rearranging the observed non-zero electivities (RA2 and RA4) also always resulted in mean overlaps

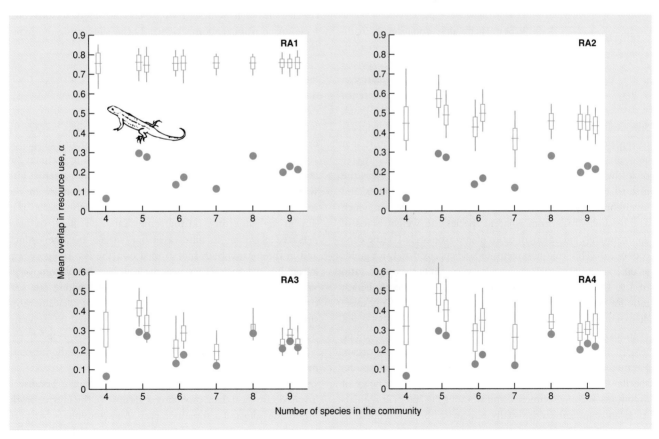

Figure 19.7 The mean indices of resource-use overlap for each of Pianka's (1973) 10 North American lizard communities are shown as solid circles. These can be compared, in each case, with the mean (horizontal line), standard deviation (vertical rectangle) and range (vertical line) of mean overlap values for the corresponding set of 100 randomly constructed communities. The analysis was performed using four different reorganization algorithms (RAs), as described in the text. (After Lawlor, 1980.)

that were significantly higher than those actually observed. With RA3, on the other hand, where all electivities were reassigned, the differences were not always significant. But in all communities, the algorithm mean was higher than the observed mean. In the case of these lizard communities, therefore, the observed low overlaps in resource use suggest that the niches are segregated, and that interspecific competition plays an important role in community structure.

... whereas grassland ants do not

A study similar to that in Figure 19.7 concerned spatial and temporal niche partitioning in grassland ant communities in Oklahoma (Albrecht & Gotelli, 2001). In this case, there was little evidence of niche partitioning on a seasonal basis. However, on a smaller spatial scale, at individual bating stations there was significantly less spatial niche overlap than expected by chance. This pattern of results – sometimes a role for competition is confirmed, sometimes not – has been the general conclusion from the null model approach.

19.2.4 Evidence from morphological patterns

Hutchinson's 'rule' about size ratios of coexisting species applied to brachiopods

Where niche differentiation is manifested as morphological differentiation, the spacing out of niches can be expected to have its counterpart in regularity in the degree of morphological difference between species belonging to a guild. Specifically, a common feature claimed for animal guilds that appear to segregate strongly along a single-resource dimension is that adjacent species tend to exhibit regular differences in body size or in the size of feeding structures. Hutchinson (1959) was the first to catalog many examples, drawn from vertebrates and invertebrates, of sequences of potential competitors in which average individuals from adjacent species had weight ratios of approximately 2.0 or length ratios of approximately 1.3 (the cube root of 2.0). This 'rule' also seems to hold approximately for guilds as different as coexisting cuckoo-doves (mean body weight ratio of 1.9; Diamond, 1975), bumblebees (mean proboscis length ratio for worker bees of 1.32; Pyke, 1982), weasels (mean canine diameter ratio of between 1.23 and 1.50; Dayan et al., 1989) and even fossil brachiopods (between 1.48 and 1.57 for body outline length, an index of the size of the brachiopod's feeding organ; Hermoyian et al., 2002). Models of competition do not predict specific values for size ratios that might apply across a range of organisms and environments, and whether the apparent regularity is an empirical quirk usually remains to be determined. In the case of the brachiopod community (Figure 19.8), however, Hermoyian et al. (2002) built 100,000 null models that each drew four species at random from the complete strophomenide brachiopod fossil fauna (74 taxa) and calculated size ratios between adjacent species. On the basis of their results, they

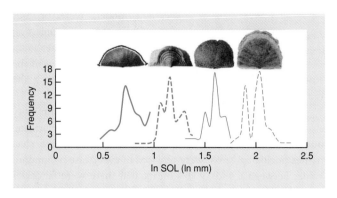

Figure 19.8 Distributions of strophomenide body outline length (SOL) of samples of four coexisting species of strophomenide brachiopods collected from a late Ordovician (*c.* 448–438 million years before present) marine sediment in Indiana, USA. The species shown, from left to right, are *Eochonetes clarksvillensis*, *Leptaena richmondensis*, *Strophomena planumbona* and *Rafinesquina alternata*. (After Hermoyian *et al.*, 2002.)

rejected the null hypothesis ($P < 0.03$) that the observed ratios could have arisen from randomly selected taxa, supporting the hypothesis of limiting similarity.

are extinctions more likely for very similar competitors?

If interspecific competition does in fact shape a community, it will often do so through a process of selective extinction. Species that are too similar will simply fail to persist together. The detailed records of ornithologists from the six main Hawaiian islands during the period 1860–1980 allowed Moulton and Pimm (1986) to estimate, at least to the nearest decade, when each species of passerine bird was introduced and if and when it became extinct. In the records, overall, there were 18 pairs of congeneric species present at the same time on the same island. Of these, six pairs persisted together; in nine cases one species became extinct; and in three cases both species died out (the last category was ignored in the analysis because the outcome is not compatible with pairwise competitive exclusion). In cases where one species became extinct the species pair was morphologically more similar than in cases where both species persisted: the average percentage difference in bill length was 9 and 22%, respectively. This statistically significant result is consistent with the competition hypothesis.

Moulton and Pimm's approach was informative because it invoked historical data, providing a glimpse of the elusive workings of the 'ghost of competition past'. An evolutionary perspective has been even more explicitly incorporated by the use of 'cladistic analysis', which allows us to reconstruct phylogenies (evolutionary trees) based on similarities and differences between species in their DNA molecules and/or in morphological (or other biologically meaningful) characteristics.

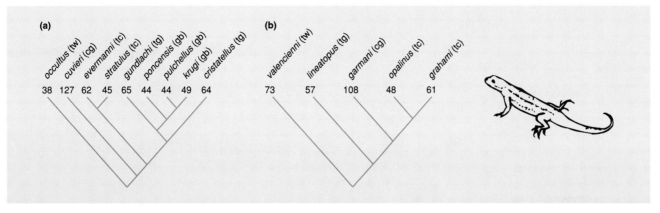

Figure 19.9 Phylogenies of lizards in the genus *Anolis* (a) on Puerto Rico and (b) on Jamaica. For each species, size (snout–vent length, mm) and ecomorph are shown: cg, crown-giant; gb, grass–bush; tc, trunk–crown; tg, trunk–ground; tw, twig. (After Losos, 1992.)

evidence from the divergent evolution of lizards on islands

The results of such an analysis of the *Anolis* lizards of Puerto Rico (Figure 19.9a) are consistent with the hypothesis of divergent evolution in body size (Losos, 1992). The two-species stage in evolution (the first, lowermost node in Figure 19.9a) was composed of species with markedly different snout–vent lengths (SVL – a standard index of size for lizards) of approximately 38 and 64 mm (*A. occultus* and the ancestor of all the remaining types, respectively) whilst sizes at the three-species stage (the next node) were 38, 64 and 127 mm. In Jamaica, on the other hand (Figure 19.9b), no such pattern is observed; the two- and three-species stages were composed of species of similar size (61 and 73 mm, then 57, 61 and 73 mm SVL). However, the phylogenies of the two islands show remarkable consistency when viewed from the point of view of patterns in 'ecomorphs' – each distinct in morphology, ecology and behavior. On both islands, the two-morph stage was composed of a short-legged twig ecomorph, which crawls slowly on narrow supports on the periphery of trees, and a generalist ancestral species. At the three-morph stage, too, both islands possessed the same assembly – a twig ecomorph, one specialized at foraging in the tree crown and a trunk–ground type, the latter being robust and long-legged and using its jumping and running abilities to forage on the ground. At the four-morph stage the patterns were again identical, each having added a trunk–crown type. Only at the five-morph stage was there a difference – the grass–bush morph was the last to evolve on Puerto Rico, but its counterpart has never appeared on Jamaica (Figure 19.10). Note that on each island a morph usually consists of a single species of *Anolis*, but Puerto Rico has several trunk–ground and grass–bush species. This phylogenetic analysis is consistent with the hypothesis that the faunal assembly on both Puerto Rico and Jamaica has occurred via sequential microhabitat partitioning, with morphological differences perhaps being related to differences in microhabitat utilization. Extending this work to further islands, Losos *et al.* (1998) confirmed that adaptive radiation in similar environments can produce strikingly similar evolutionary outcomes.

19.2.5 Evidence from negatively associated distributions

A number of studies have used patterns in distribution as evidence for the importance of interspecific competition. Foremost amongst these is Diamond's (1975) survey of the land birds living on the islands of the Bismarck Archipelago off the coast of New Guinea. The most striking evidence comes from distributions that Diamond refers to as 'checkerboard'. In these, two or more ecologically similar species (i.e. members of the same guild) have mutually exclusive but interdigitating distributions such that any one island supports only one of the species (or none at all). Figure 19.11 shows this for two small, ecologically similar cuckoo-dove species: *Macropygia mackinlayi* and *M. nigrirostris*.

evidence from 'checkerboard' distributions in . . . . . . island birds . . .

A null model approach to the analysis of distributional differences involves comparing the pattern of species co-occurrences at a suite of locations with that which would be expected by chance. An excess of negative associations would then be consistent with a role for competition in determining community structure.

Thorough censuses of both native and exotic (introduced) plants occurring on 23 small islands in Lake Manapouri in the South Island of New Zealand (Wilson, 1988b), were the basis for computing a standard index of association for every pair of species:

. . . and native and exotic plants on islands in a lake

$$d_{ik} = (O_{ik} - E_{ik})/\mathrm{SD}_{ik} \qquad (19.1)$$

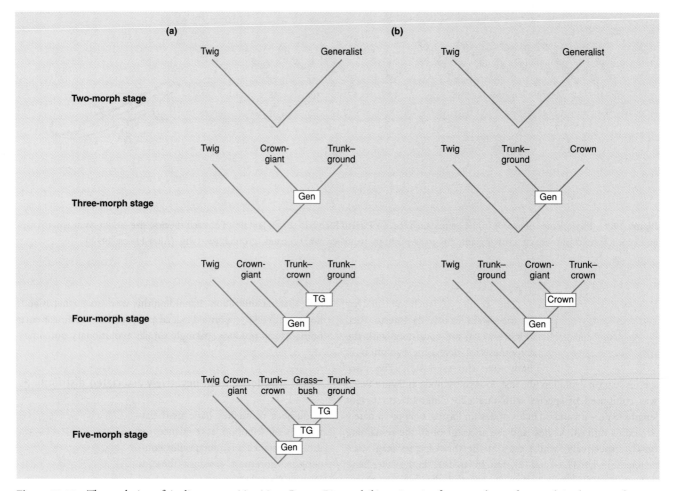

Figure 19.10 The evolution of *Anolis* communities (a) on Puerto Rico and (b) on Jamaica for two-, three-, four- and, in the case of Puerto Rico, five-ecomorph communities. Labels at nodes in the trees are the estimated ecological characteristics of the ancestors. (After Losos, 1992.)

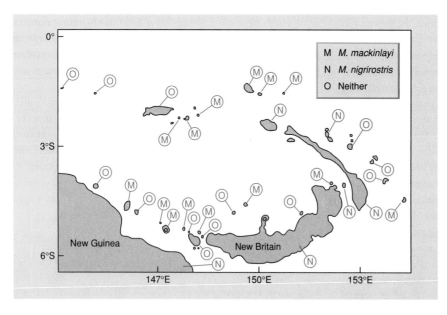

Figure 19.11 Checkerboard distribution of two small *Macropygia* cuckoo-dove species in the Bismarck region. Islands whose pigeon faunas are known are designated as M (*M. mackinlayi* resident), N (*M. nigrirostris* resident) or O (neither species resident). Note that most islands have one of these species, no island has both and some islands have neither. (After Diamond, 1975.)

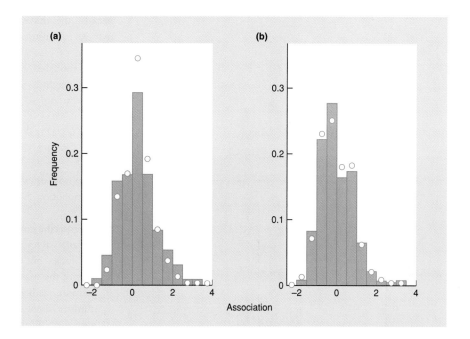

Figure 19.12 A comparison between the observed values of association between pairs of (a) native plant species and (b) exotic plant species on islands in Lake Manapouri (histograms), and the distributions expected on the basis of a neutral model (○). (After Wilson, 1988b.)

where d_{ik} is the difference between the observed (O_{ik}) and the expected (E_{ik}) number of islands shared by species i and k, expressed in terms of the standard deviation of the expected number (SD_{ik}).

The resulting sets of association values for the real communities of native and exotic species are presented as histograms in Figure 19.12. These can be compared with null model communities in which island species richnesses and species frequencies of occurrence were fixed at those observed, but species occurrences on islands were randomized (Wilson, 1987). One thousand randomizations were performed, yielding a mean frequency in each d_{ik} category (the circles in Figure 19.12). The analysis of native plants showed an excess of negative associations (highly statistically significant for the bottom four categories) and of positive associations (highly significant for the top five categories), with a corresponding deficit of associations near zero. In contrast, the analysis of exotic plants showed no significant departure from the null model.

In the case of native species, the excess of negative associations is consistent with the action of competitive exclusion, and this is particularly likely for the woody species. However, we cannot rule out an explanation based on a tendency of particular pairs of species to occur in different habitats, which themselves are not represented on every island (Wilson, 1988b). The most likely explanation for the excess of positive associations amongst native plants is a tendency for certain species to occur in the same habitats. The agreement of exotic species with the null model may reflect their generally weedy status and effective colonization abilities, or it may indicate that the exotics have not yet reached an equilibrium distribution (Wilson, 1988b).

The number of checkerboard pairs in a community can be readily calculated by counting the number of unique pairs of species that never co-occur. A less strict version of Diamond's assembly rule that 'some pairs of species never coexist' can be assessed using the C score of Stone and Roberts (1990). This index also measures the degree to which species co-occur but does not require perfect segregation between species. The C score is calculated for each pair of species as $(R_i - S)(R_j - S)$ where R_i and R_j are the number of sites where species i and j occur, and S is the number of sites in which both species co-occur. This score is then averaged over all possible pairs of species in the matrix. For a community structured by competitive interactions, the number of checkerboard pairs should be greater and the C score should be larger than expected by chance.

Gotelli and McCabe (2002) checked the generality of negatively associated distributions (in support of a structuring role for competition) in a meta-analysis of various taxonomic groups in 96 data sets that reported the distribution of species assemblages across sets of replicated sites. For every real data set, 1000 randomized versions were prepared and an index of association d_{ik} was computed (as in Wilson, 1988b) but Gotelli and McCabe called this index the standardized effect size (SES). The results of this analysis for all 96 data sets together support the predictions that the C score and the number of checkerboard pairs should be larger than expected by chance (Figure 19.13a, b). The null hypothesis in each case is that the mean SES should be zero (real communities not different from simulated communities) and that 95% of the values should lie between −2.0 and +2.0. The null hypothesis can be rejected in both cases. Figure 19.13c shows that

comparison of taxonomic groups

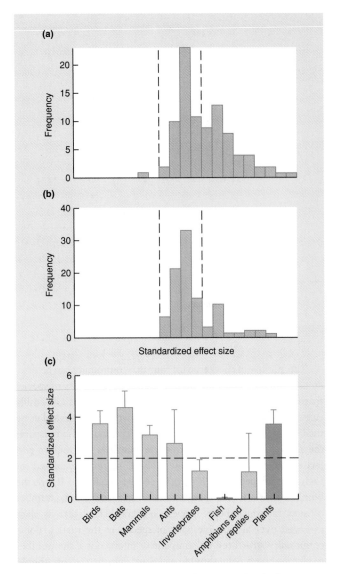

Figure 19.13 Frequency histograms for standardized effect sizes measured for 96 presence–absence matrices taken from the literature in the case of (a) the C score and (b) the number of species pairs forming perfect checkerboard distributions. (c) Standardized effect sizes for the C score for different taxonomic groups. The dashed line indicates an effects size of 2.0, which is the approximate 5% significance level. (After Gotelli & McCabe, 2002.)

plants and homeothermic vertebrates tend to have higher SESs for the C score, indicating stronger tendencies towards negative species associations than the poikilotherms have (invertebrates, fish and reptiles), with the exception of ants.

Gotelli and McCabe (2002) do not go so far as to claim they have performed a definitive test of the role of competition. They note that some species may exhibit 'habitat checkerboards'

because they have affinities to nonoverlapping habitats. Others may reveal 'historical checkerboards', co-occurring infrequently because of restricted dispersal since allopatric speciation (i.e. having speciated in different places). However, these results add further weight to a widespread role for competition in structuring communities.

19.2.6 Appraisal of the role of competition

We can now draw a number of conclusions about the evidence for competition discussed in this section.

1 Interspecific competition is a possible and indeed a plausible explanation for many aspects of the organization of many communities – but it is not often a proven explanation.
2 One of the main reasons for this is that active, current competition has been studied and demonstrated in only a small number of communities. Its actual prevalence overall can be judged only imperfectly from the results and considerations discussed above.
3 As an alternative to current competition, the ghost of competition past can always be invoked to account for present-day patterns. But it can be invoked so easily because it is impossible to observe directly and therefore is difficult to disprove.
4 The communities chosen for study may not be typical. The ecologists observing them have usually been specifically interested in competition, and they may have selected appropriate, 'interesting' systems. Studies that fail to show niche differentiation may often have been considered 'unsuccessful' and are likely to have gone largely unreported.
5 The community patterns uncovered, even where they appear to support the competition hypothesis, often have alternative explanations. For example, species that have negatively associated distributions may have recently speciated allopatrically, and their distributions may still be expanding into one another's ranges.
6 The recurring alternative explanation to competition as the cause of community patterns is that these have arisen simply by chance. Niche differentiation may occur because the various species have evolved independently into specialists, and their specialized niches happen to be different. Even niches arranged along a resource dimension at random are bound to differ to some extent. Similarly, species may differ in their distribution because each has been able, independently, to colonize and establish itself in only a small proportion of the habitats that are suitable for it. Ten blue and 10 red balls thrown at random into 100 boxes are almost certain to end up with different distributions. Hence, competition cannot be inferred from mere 'differences' alone. But, what sorts of differences *do* allow the action of competition to be inferred? This is the domain of the null model approach.

7   The aim of the null model approach, whether applied to niche differentiation, morphological patterns or negatively associated distribution patterns, is undoubtedly worthy. We need to guard against the temptation to see competition in a community simply because we are looking for it. On the other hand, the approach is bound to be of limited use unless it is applied to groups (usually guilds) within which competition may be expected. In its favor, the null model approach concentrates the minds of investigators, and it can stop them from jumping to conclusions too readily. Ultimately, though, it can never take the place of a detailed understanding of the field ecology of the species in question, or of manipulative experiments designed to reveal competition by increasing or reducing species abundances (Law & Watkinson, 1989). It can only be part of the community ecologist's armory.

8   Interspecific competition is certain to vary in importance from community to community: it has no single, general role. For example, it appears frequently to be important in vertebrate communities, particularly those of stable, species-rich environments, and in communities dominated by sessile organisms such as plants and corals; whilst, for example, in some phytophagous insect communities it is less often important. A challenge for the future is to understand why some guilds show evidence for a role for competition, such as regularity in size ratios, whilst others do not (Hopf et al., 1993).

9   Finally, we should not lose sight of the fact that community organization in field studies is almost certain to be influenced by more than one kind of population interaction; for example the anenome fish (see Section 19.2.3.1) and ectomycorrhizal fungal cases (see Section 19.2.3.2) both involved mutualism as well as competition, and the mantids in Section 19.2.3.3 were intraguild predators as well as competitors. The interaction between predation and competition can be particularly influential, as we shall see in Section 19.4.

19.3   Equilibrium and nonequilibrium views of community organization

It is possible to conceive of a world with just one species of plant (or herbivore) with supreme performance over an enormous range of tolerance. In this scenario the most competitive species (the one that is most efficient at converting limited resources into descendants) would be expected to drive all less competitive species to extinction. The species richness we witness in real communities is a clear demonstration of the failure of evolution to produce such supreme species. An extension of this competitive argument holds that diversity can be explained through a partitioning of resources amongst competing species whose requirements do not overlap completely, as discussed in detail in Section 19.2. However, this argument rests on two assumptions that are not necessarily always valid.

The first assumption is that the organisms are actually competing, which in turn implies that resources are limiting. But there are many situations where physical disturbances, such as storms on a rocky shore or frequent fires, may hold down the densities of populations so that resources are not limiting and individuals do not compete for them. The role of physical disturbances, and the associated patch dynamics view of communities, were discussed in Chapter 16. In an exactly analogous manner, the action of predators or parasites is often a disturbance in the 'normal' course of a competitive interaction; the resulting mortality may open up a gap for colonization in a way that is sometimes indistinguishable from that of battering by waves on a rocky shore or a hurricane in a forest.

The second assumption is that when competition is operating and resources are in limited supply, one species will inevitably exclude another. But in the real world, when no year is exactly like another, and no square centimeter of ground exactly the same as the next, the process of competitive exclusion may not proceed to its monotonous end. Any force that continually changes direction at least delays, and may prevent, an equilibrium or a stable conclusion being reached. Any force that simply interrupts the process of competitive exclusion may prevent extinction and enhance species richness.

A basic distinction can thus be made between *equilibrium* and *nonequilibrium* theories. An equilibrium theory, like the one concerned with niche differentiation, helps us to focus attention on the properties of a system at an equilibrium point – time and variation are not the central concern. A nonequilibrium theory, on the other hand, is concerned with the transient behavior of a system away from an equilibrium point, and specifically focuses our attention on time and variation. Of course, it would be naive to think that any real community has a precisely definable equilibrium point, and it is wrong to ascribe this view to researchers who are associated with equilibrium theories. The truth is that investigators who focus attention on equilibrium points have in mind that these are merely states towards which systems tend to be attracted, but about which there may be greater or lesser fluctuation. In one sense, therefore, the contrast between equilibrium and nonequilibrium theories is a matter of degree. However, this difference of focus is instructive in unraveling the important role of temporal heterogeneity in communities.

Thus, predators and parasites, like physical disturbances, can interrupt the process of competitive exclusion, influence profoundly the outcome of competitive processes, and impose their own order on community organization. Predation and parasitism can also affect community structure through the process of 'apparent competition' (see Section 8.6), where one or more prey or host species suffers from the actions of predators or parasites that are sustained by the presence of other species of prey or hosts. We turn to predation and parasitism in the next two sections.

19.4   The influence of predation on community structure

19.4.1   Effects of grazers

Lawn-mowers are relatively unselective predators capable of maintaining a close-cropped sward of vegetation. Darwin (1859) was the first to notice that the mowing of a lawn could maintain a higher richness of species than occurred in its absence. He wrote that:

> If turf which has long been mown, and the case would be the same with turf closely browsed by quadrupeds, be let to grow, the most vigorous plants gradually kill the less vigorous, though fully grown plants; thus out of 20 species growing on a little plot of mown turf (3 feet by 4 feet) nine species perished from the other species being allowed to grow up freely.

grazing can increase plant species richness (exploiter-mediated coexistence) . . .

Grazing animals are usually more choosy than lawn-mowers, and this is clearly demonstrated by the occurrence in the neighborhood of rabbit (*Oryctolagus cuniculus*) burrows of plants which for chemical or physical reasons are unacceptable as food to the rabbits (including the poisonous deadly nightshade *Atropa belladonna* and the stinging nettle *Urtica dioica*). Nevertheless, many grazers seem to have a similar general effect to lawn-mowers. Thus, in one experiment, grazing by oxen (*Bos taurus*) and zebu cows (*Bos taurus indicus*) in natural pasture in the Ethiopian highlands was manipulated to provide a no-grazing control and four grazing intensity treatments (several replicates of each) in two sites. Figure 19.14 shows how the mean number of plant species varied in the sites in October, the period when plant productivity was at its highest (Mwendera *et al.*, 1997). Significantly more species occurred at intermediate levels of grazing than where there was no grazing or heavier grazing (*P* < 0.05). In the ungrazed plots, several highly competitive plant species, including the grass *Bothriochloa insculpta*, accounted for 75–90% of ground cover. At intermediate levels of grazing, however, the cattle apparently kept the aggressive, competitively dominant grasses in check and allowed a greater number of plant species to persist. But at very high intensities of grazing, species numbers were reduced as the cattle were forced to turn from heavily grazed, preferred plant species to less preferred species, driving some to extinction. Where grazing pressure was particularly intense, grazing-tolerant species such as *Cynodon dactylon* became dominant.

. . . but not always

The composition of plant communities in different grazing regimes clearly depends on a variety of species traits. First, competitively superior species can be expected to dominate in the absence of grazing. A particularly striking example

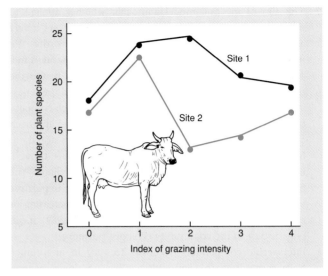

Figure 19.14   Mean species richness of pasture vegetation in plots subjected to different levels of cattle grazing in two sites in the Ethiopian highlands in October. 0, no grazing; 1, light grazing; 2, moderate grazing; 3, heavy grazing; 4, very heavy grazing (estimated according to cattle stocking rates). (After Mwendera *et al.*, 1997.)

has been provided by Paine (2002), who reported that the exclusion of macroherbivores (urchins, chitons and limpets) from a North American rocky intertidal zone caused the multi-species kelp community to collapse to a virtual monoculture of *Alaria marginata*; this was 10 times more productive than its grazed counterpart (86.0 versus 8.6 kg wet mass m^{-2} year^{-1}). Second, we have seen that plant species with physical or chemical characteristics that deter grazers are likely to be more strongly represented in grazed locations. Bullock *et al.* (2001) have also noted that while certain dominant grasses decreased in importance in response to sheep grazing, most dicotyledonous species increased in abundance, at least at certain times of year. Moreover, summer grazing produced an increased representation of plant species best able to colonize gaps.

When predation promotes the coexistence of species amongst which there would otherwise be competitive exclusion (because the densities of some or all of the species are reduced

exploiter-mediated coexistence is more likely in nutrient-rich situations

to levels at which competition is relatively unimportant), this is known generally as 'exploiter-mediated coexistence'. Many examples of this phenomenon have been reported, such as that in Figure 19.14, but grazer-mediated coexistence is far from universal. Proulx and Mazumder (1998) performed a meta-analysis of 44 reports of the effects of grazing on plant species richness from lake, stream, marine, grassland and forest ecosystems. Their conclusion was that the outcome was strongly related to whether the studies had been performed in nutrient-rich or nutrient-poor situations. All

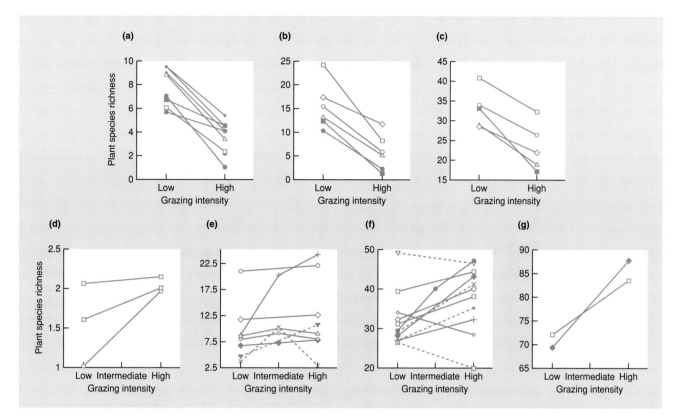

Figure 19.15 (a–c) Species richness under contrasting grazing pressure (low or high) in nonenriched or nutrient-poor ecosystems. The different lines show the results of different aquatic or terrestrial studies and are presented in three panels simply for clarity. (d–g) Species richness under contrasting grazing pressure (low, intermediate or high) in various enriched or nutrient-rich ecosystems. (After Proulx & Mazumder, 1998.)

19 studies from nonenriched or nutrient-poor ecosystems exhibited significantly lower species richness under high grazing than under low grazing (Figure 19.15a–c). In contrast, 14 of 25 comparisons from enriched or nutrient-rich ecosystems showed significantly higher species richness under high grazing (indicating grazer-mediated coexistence) (Figure 19.15d–g). Nine of the remaining 11 nutrient-rich studies showed no difference with grazing regime whilst two showed a decline in species richness. The lack of grazer-mediated coexistence in unproductive situations may reflect the poor growth potential of the less competitive species that, in nutrient-rich circumstances, would be released from competitive domination as a result of grazing.

community responses to grazing depend on productivity . . .

Osem *et al.* (2002) focused on the interactive effects of grazing and productivity in a study of annual herbaceous plant communities in Mediterranean semiarid rangeland in Israel. They recorded the response of the community to protection from sheep grazing in four neighboring topographic situations – south-facing slopes, north-facing slopes, hilltops and wadi (dry stream) shoulders (Figure 19.16). Annual above-ground primary productivity was measured each year for 4 years at the peak

season in the four fenced subplots per site and was found to be typical of semiarid ecosystems (10–200 g dry matter m^{-2}) except on wadi shoulders (up to 700 g dry matter m^{-2}). The measured values were taken to represent 'potential' productivity in the adjacent grazed subplots. Grazing only increased plant species richness in the most productive site (wadi) (Figure 19.16d). In the other, less productive sites, species richness was unaffected or declined with grazing. These results are consistent with those reported by Proulx and Mazumder (1998) and support the long-standing proposal of Huston (1979) that grazing should change diversity in opposite ways in resource-poor and resource-rich ecosystems.

Figure 19.17a and b plots species richness in relation to potential productivity individually for all subplots and all years (because precipitation and productivity varied both spatially and temporally) for both grazed and ungrazed locations. Under grazing, species richness was positively related to productivity over the whole range measured. In the absence of grazers, however, a positive relationship only occurred in low-productivity sites. Osem *et al.* (2002) hypothesize (Figure 19.17c) that at low productivity, plant growth and diversity are limited by the soil resources of water and nutrients, while at higher productivity (with its associated larger

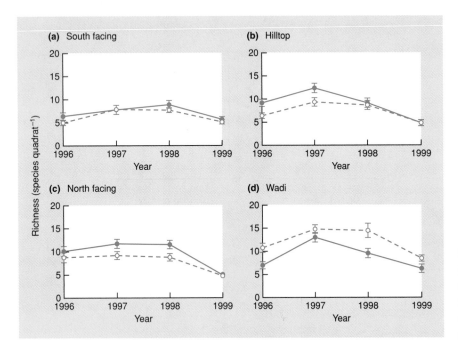

Figure 19.16 Species richness (per 20 × 20 cm quadrats) in four topographic sites in Israel in April: (a) south-facing slopes, (b) hilltops, (c) north-facing slopes, and (d) wadis. ●, ungrazed subplots; ○, grazed subplots. (After Osem *et al.*, 2002.)

biomass) competition is predominantly for the canopy resource of light. Thus, in the low productivity range, richness was either unaffected or reduced by grazing, probably because of plant removal and trampling. In the high productivity wadi sites, however, species richness continued to increase with grazing, most likely because of a reduction in light competition through removal of the palatable larger species.

... and plant species traits

Taken overall, then, the way that plant species richness responds to grazing depends partly on grazing intensity, but also on the evolutionary history of the plant community and thus the particular plant species traits that are represented, as well as the primary productivity of the ecosystem in question. An increase in species richness in response to grazing can be expected if grazers feed preferentially on competitively dominant species, a prediction that has received support in situations as diverse as cattle grazing in Ethiopia (see above) and periwinkles (*Littorina littorea*) feeding on algae in rocky tide pools (Lubchenco, 1978). Conversely, a reduction in species richness can be expected if the preferred food plants are competitively inferior, as was the case for periwinkles feeding on algae on emergent substrata in Lubchenco's study.

19.4.2 The effect of carnivores

predator-mediated coexistence on a rocky shore . . .

The rocky intertidal zone also provided the location for pioneering work by Paine (1966) on the influence of a top carnivore on community structure.

The starfish *Pisaster ochraceus* preys on sessile filter-feeding barnacles and mussels, and also on browsing limpets and chitons and a small carnivorous whelk. These species, together with a sponge and four macroscopic algae, form predictable associations on rocky shores of the Pacific coast of North America. Paine removed all the starfish from a typical piece of shoreline about 8 m long and 2 m deep, and continued to exclude them for several years. At irregular intervals, the density of invertebrates and the cover of benthic algae were assessed in the experimental area and in an adjacent control site. The latter remained unchanged during the study. Removal of *P. ochraceus*, however, had dramatic consequences. Within a few months, the barnacle *Balanus glandula* settled successfully. Later, barnacles were crowded out by mussels (*Mytilus californianus*), and eventually the site became dominated by the latter. All but one of the species of algae disappeared, apparently through lack of space, and the browsers tended to move away, partly because space was limited and partly due to lack of suitable food. Overall, the removal of starfish led to a reduction in the number of species from 15 to eight. The main influence of the starfish *Pisaster* appears to be to make space available for competitively subordinate species. It cuts a swathe free of barnacles and, most importantly, free of the dominant mussels that would otherwise outcompete other invertebrates and algae for space. Once again, there is exploiter-mediated coexistence. Note that this argument applies specifically to the primary space occupiers, such as mussels, barnacles and macroalgae. In contrast, the number of less conspicuous species associated with living and dead mussel shells would be expected to increase in the bed that develops after *Pisaster* removal (more than 300 species of animals and plants occur in mussel beds; Suchanek, 1992).

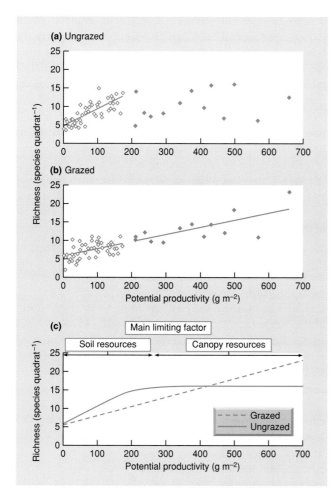

Figure 19.17 Relationship between annual above-ground productivity (measured in ungrazed subplots) and species richness in (a) ungrazed and (b) grazed subplots. Open symbols represent low productivity subplots (< 200 g dry matter m⁻²; hilltop, south- and north-facing slopes in all years plus wadi in the dry 1999 season). Closed symbols represent high productivity subplots (> 200 g dry matter m⁻²; wadi sites in years other than 1999). (c) Conceptual model of the relationship between productivity and species richness in grazed and ungrazed semiarid Mediterranean rangeland. (After Osem *et al.*, 2002.)

Experiments similar to those of Paine have been performed in the more challenging environment of hydrothermal vents at a depth of 2500 m in the eastern tropical Pacific Ocean (Micheli *et al.*, 2002). Colonization of replicate recruitment substrates (10 cm basalt cubes) was monitored for 5 months at increasing distances from the vent in three sites in the presence and absence (exclusion cages) of predators (fish and crabs). In terms of reduced prey abundance (particularly two gastropods endemic to vents – the limpet *Lepetodrilus elevatus* and the snail *Cyathermia naticoides*), the effects of predation were strongest near the vent where productivity and the overall abundance of invertebrates were greatest. Species richness, which generally declined with distance from the vent, was usually lower in the presence of predators (but only statistically significantly so at the Worm Hole site – Figure 19.18). The reason for a lack of predator-mediated coexistence is unknown.

... but not in hydrothermal vent communities

Turning now to terrestrial ecosystems, in a study of nine Scandinavian islands, pigmy owls (*Glaucidium passerinum*) occurred on only four of the islands, and the pattern of occurrence of three species of passerine birds in the genus *Parus* showed a striking relationship with this distribution (Table 19.1). The five islands without the predatory owl were home to only one species, the coal tit (*Parus ater*). However, in the presence of the owl, the coal tit was always joined by two larger tit species, the willow tit (*P. montanus*) and the crested tit (*P. cristatus*). Kullberg and Ekman (2000) argue that the smaller coal tit is superior in exploitation competition for food. The two larger species, however, have an advantage via interference competition for foraging sites close to the trunk of trees where they are safer from predators; in other words the larger species are less affected than the coal tit by predation from the owl. It seems that the owl may be responsible for predator-mediated coexistence, by reducing the competitive dominance enjoyed by coal tits in its absence.

predator-mediated coexistence among passerine birds ...

However, an increase in species richness with predation is by no means universal in terrestrial ecosystems. Spiller and Schoener (1998) reviewed a

... but not among communities of insects or spiders

Table 19.1 Area, distance to mainland and occurrence of breeding pairs of pygmy owls and three species of tit. (After Kullberg & Ekman, 2000.)

Island	Area (km²)	Distance to mainland (km)	Pygmy owl	Coal tit	Willow tit	Crested tit
Åland	970	50	+	+	+	+
Ösel	3000	15	+	+	+	+
Dagö	989	10	+	+	+	+
Karlö	200	7	+	+	+	+
Gotland	3140	85		+		
Öland	1345	4		+		
Bornholm	587	35		+		
Hanö	2.2	4		+		
Visingsö	30	6		+		

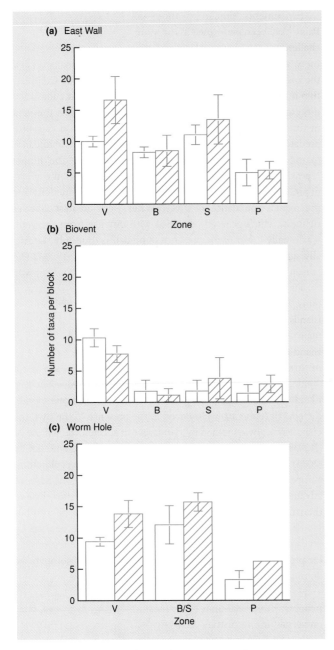

Figure 19.18 Patterns of invertebrate species richness (worms in the order Vestimentifera (class Pogonophora), worms in the class Polychaeta, gastropods, bivalves and crustaceans) per implanted recruitment substrate after 5 months at three sites with two experimental treatments: (a) East Wall, (b) Biovent, and (c) Worm Hole. The results are presented for four zones whose boundaries are based on water temperature and dominant benthic invertebrates (at increasing distances from the hydrothermal vent: Vestimentiferan (V), bivalve (B), suspension-feeder (S) and periphery (P)). The two middle zones were combined at the Worm Hole site. Experimental treatments: □, uncaged; ▨, caged to exclude mobile predators – fish and crabs. (After Micheli *et al.*, 2002.)

number of studies involving birds preying upon grasshoppers, rodents upon carabid beetles and lizards upon web spiders and concluded that these predators usually reduced prey richness or had no effect. In their own study in the Bahamas, they censused spider populations at 2-month intervals for 4.5 years in enclosures (three replicates) containing or lacking lizards. Species richness was dramatically increased by the exclusion of lizards (mainly *Anolis sagrei*) at high and medium levels in the vegetation (Figure 19.19a). The lizards preyed preferentially upon rare species of spiders (Figure 19.19b), resulting in increased dominance of the already abundant *Metapeira datona*, a species whose relative invulnerability to predation is probably due to its small size and habit of living in a suspended retreat rather than in the middle of the web.

As was the case with grazers, the manner in which prey species richness responds to predation no doubt depends partly on predation intensity, partly on

predator dietary preferences may modify the outcome

ecosystem productivity, and partly on the particular characteristics of the prey species. Once again we have seen increases in prey species richness where carnivores feed preferentially on competitively dominant prey (starfish feeding on mussels, pygmy owls feeding on coal tits) and a decrease where the preferred prey are competitively inferior (lizards feeding on spiders).

Another reason for contrasting effects of consumers on lower trophic levels relates to their prey-selection behavior. They seldom simply take potential prey species from a community in turn, bringing each to extinction

frequency-dependent selection may sometimes enhance diversity

before turning to the next. Selection is moderated by the time or energy spent in search for the preferred prey (see Chapter 9), and many species take a mixed diet. However, others switch sharply from one type of prey to another, taking disproportionately more of the most common acceptable types of prey. In theory, such behavior could lead to the coexistence of a large number of relatively rare species (a frequency-dependent form of exploiter-mediated coexistence). In this context, there is evidence that predation on the seeds of tropical trees is often more intense where the seeds are more dense (beneath and near the adult that produced them) (Connell, 1979); the herbivorous butterfly *Battus philenor* forms search images for leaf shape when foraging for its two larval host plants, and concentrates on whichever happens to be the more common (Rausher, 1978); the freshwater zooplanktivorous fish *Rutilus rutilus* switches from large planktonic waterfleas, its preferred prey, to small sediment-dwelling waterfleas when the density of the former falls below about 40 per liter (Townsend *et al.*, 1986); and piscivorous coral reef fish (*Cephalopholis boenak* and *Pseudochromis fuscus*) concentrate on highly abundant cardinal fish (mainly *Apogon fragilis*) when these are present, leaving recruits of many other fish species relatively unmolested (Webster & Almany, 2002). However, such frequency-

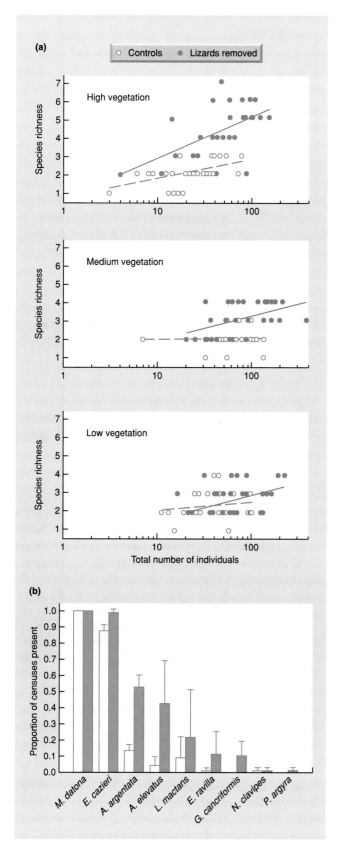

dependent selection is not a general rule and may not be common. For one thing, some species are so highly specialized that switching is not an option – giant pandas are specialists on bamboo shoots and specialization in diet is equally extreme amongst many phytophagous insects. Moreover, in other cases a predator may be sustained by one prey type whilst exterminating others. This has been claimed for the introduced snake *Boiga irregularis*, on the small island of Guam, midway between Japan and New Guinea. Coincident with its arrival in the early 1950s, and its subsequent spread through Guam, most of the 18 native bird species have declined dramatically and seven are now extinct. Savidge (1987) argues that by including abundant small lizards in its diet, *B. irregularis* has maintained high densities whilst exterminating the more vulnerable bird species.

19.5 Influence of parasitism on community structure

The incidence of a parasite, like that of other types of exploiter, may determine whether or not a host species occurs in an area. Thus, the extinction of nearly

parasites may drive vulnerable host species extinct

50% of the endemic bird fauna of the Hawaiian Islands has been attributed in part to the introduction of bird pathogens such as malaria and bird pox (van Riper *et al.*, 1986); and changes in the distribution of the North American moose (*Alces alces*) have been associated with the parasitic nematode *Pneumostrongylus tenuis* (Anderson, 1981). Probably the largest single change wrought in the structure of communities by a parasite has been the destruction of the chestnut (*Castanea dentata*) in North American forests, where it had been a dominant tree over large areas until the introduction of the fungal pathogen *Endothia parasitica*, probably from China.

Like grazers and carnivores, parasites can cause more subtle effects too. In many streams in Michigan, USA, larvae of the herbivorous caddis-fly *Glossosoma nigrior* play a key role in the community

a microparasite with subtle direct and indirect effects in a stream community

because their foraging maintains attached algae at very low levels, with negative consequences for most other stream herbivores

Figure 19.19 (*left*) (a) Spider species richness plotted against total number of individuals (all censuses) in the presence and absence of lizards at three heights in the vegetation. For a given number of individuals, enclosures without lizards (●) contained a greater number of spider species than enclosures with lizards (○) except low in the vegetation. (b) Mean proportion of censuses in which each web spider was recorded per enclosure in the presence (□) and absence of lizards (■). Error bars ± SD. (After Spiller & Schoener, 1998.)

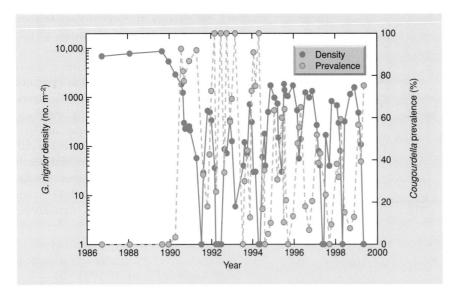

Figure 19.20 *Glossosoma nigrior* density in Seven Mile Creek, Michigan, and the percentage of the population infected (prevalence) by *Cougourdella*. (After Kohler, 1992.)

(Kohler, 1992). *G. nigrior* is subject to sporadic outbreaks of a highly specific microsporidian microparasite, *Cougourdella*, which result in dramatic whole stream reductions in *G. nigrior* density that may be maintained for years. In Seven Mile Creek, for example, the mean *G. nigrior* density was 6285 per m² in the 10 generations before a parasite outbreak in 1990 but it averaged 164 per m² for the next decade (Figure 19.20). The decline of *G. nigrior* leads to increased abundance of its food resource (Figure 19.21a). As a result, several herbivores (Figure 19.21b–d), including a species that was previously absent or extremely rare (Figure 19.21e), increased in abundance after the streams had experienced a parasite-caused decline in *G. nigrior*. Thus, by reducing the abundance of the competitively dominant herbivore, the parasite increased herbivore equitability (one aspect of species diversity) and may have been responsible for an increase in species richness. This example, therefore, has the hallmarks of parasite-mediated coexistence The parasite was also responsible for further effects – increased algal abundance seems to have resulted in more fine particulate dead organic matter (through sloughing off of algal cells) fueling an increase in the density of filter-feeders (Figure 19.21f), and the increase in abundance of vulnerable herbivore species (*G. nigrior* is relatively invulnerable to predators) led to increased densities of predaceous caddis-flies (*Rhyacophila manistee*) and stoneflies (*Paragnetina media*) (Figure 19.21g).

parasite-mediated coexistence in Caribbean lizards . . .

In terrestrial ecosystems, too, there are apparent examples of parasite-mediated coexistence. For example, the malarial parasite *Plasmodium azurophilum* infects two *Anolis* lizards on the Caribbean island of St Martin. One of the lizards, thought to be the competitive dominant, is widespread throughout the island while the other is only found in a limited area. Schall (1992) reported that the superior competitor was much more likely to be infected

by the parasite and, intriguingly, the two species only coexisted where the parasite was present. Once again, though, this is far from a universal pattern. For example, the invading grey squirrel (*Sciurus carolensis*) is displacing the resident red squirrel (*S. vulgaris*) throughout much of its range in Britain. At least part of the reason seems to be that the invader has brought with it a parapox virus that has little discernible effect on the grey squirrel but a dramatic adverse effect on the health of the native red squirrel (Tompkins *et al.*, 2003).

Brood parasites (see Section 12.2.3), such as brown-headed cowbirds (*Molothrus ater*), might also be expected

. . . but not in songbirds

to affect the composition or richness of the communities in which they operate. De Groot and Smith (2001) made use of a cowbird reduction program in a pine (*Pinus banksiana*) forest in Michigan (designed to protect one of the cowbird's hosts, the endangered Kirtland's warbler *Dendroica kirtlandii*) to investigate whether the songbird community as a whole was affected by a reduction in the density of the brood parasite. Their results provided no support for parasite-mediated coexistence, nor was there a change in community composition or an increase in the representation of songbird species known to be unsuitable as hosts for cowbirds.

Parasites may sometimes influence community composition not by altering the outcome of competitive interactions but through an impact on a key member of the community that acts as an ecosystem engineer (*sensu* Jones *et al.*, 1994, 1997). The juvenile stages of the

parasites that influence species that are themselves strong community interactors

trematode *Curuteria australis* encyst in the foot of cockles, *Austrovenus stutchburyi*, and impair the burrowing ability of the cockles. This results in heavily infected cockles remaining stranded at the surface of the sediments, where they are easy prey

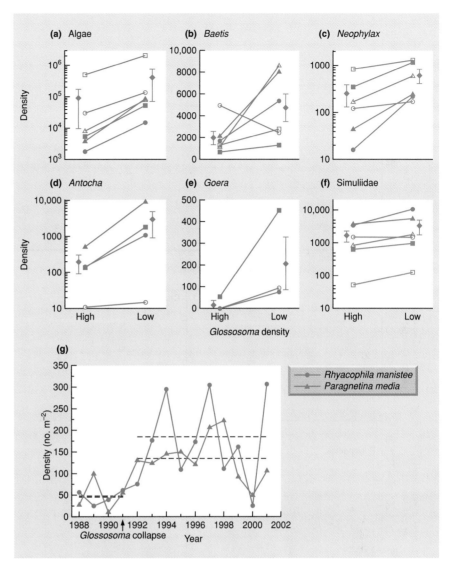

Figure 19.21 Mean densities of (a) attached algae (cells cm^{-2}), (b–e) herbivorous insects (number m^{-2}) and (f) filter-feeders (number m^{-2}), in relation to *Glossosoma nigrior* density (high, before a parasite outbreak; low, during a parasite outbreak) in six streams. Lines connect data points for each of the six streams; the points with error bars ($\pm$ 1 SE) are the overall means. (g) Predator densities before and after a parasite-induced reduction of *G. nigrior* in Silver Creek (dashed lines show mean densities before and after the collapse). (After Kohler & Wiley, 1997.)

for oystercatchers, the trematode's definitive host (Thomas & Poulin, 1998; Mouritsen, 2002). Cockles, the dominant bivalves in New Zealand soft-sediment intertidal zones, are normally buried 2–3 cm under the sediment surface. But in areas of intense parasitism, large numbers protrude from the sediment or lie on its surface, increasing surface heterogeneity and changing patterns of water flow and sedimentation. Mouritsen and Poulin (2005) manipulated the density of surface cockles by creating plots with 30 or 100 surface cockles added to compare with control plots with naturally few cockles at the surface. After 6 months, there were significantly more species of macrofaunal invertebrates (polychaetes, molluscs, crustaceans, etc.) in the treatments with added surface cockles and the densities of a variety of taxa were greater in these experimental plots (Figure 19.22).

19.6 Appraisal of the effects of predators and parasites

1 Selective predators are likely to act to enhance species richness in a community if their preferred prey are competitively dominant and in situations where community productivity is high. It seems likely that there is some general correlation between palatability to predators and high growth rates. If the production of chemical and physical defenses by prey requires a sacrifice of resources used in growth and reproduction, we might expect species that are competitive dominants in the absence of predators (and hence, which devote resources to competition rather than defense) to suffer excessively from their presence. Thus, selective predators may often enhance

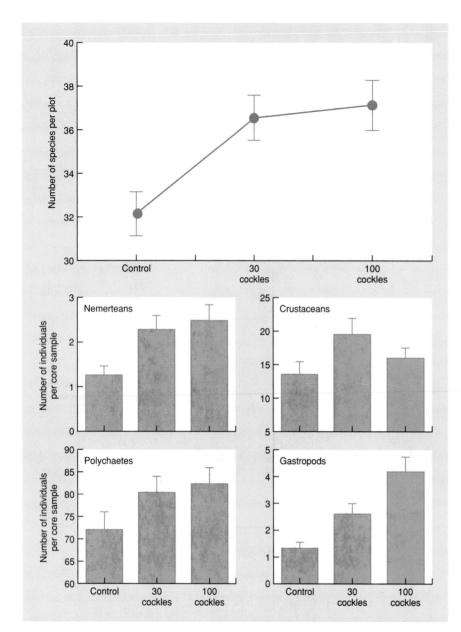

Figure 19.22 Manipulation of the density of cockles at the sediment surface, mimicking variation in infection levels by the trematode *Curtuteria australis*, and its effect on intertidal communities. Mean (± 1 SE) species richness of macrofaunal invertebrate species per plot and mean density of several invertebrate taxa in three experimental treatments (0, 30 and 100 cockles added to a 1 m² plot). All means are derived from five core samples taken per plot, with seven plots per treatment. (After Mouritsen & Poulin, 2005)

species richness. If the predators act in a frequency-dependent manner their action should be even stronger. Even very generalist predators may increase community diversity through exploiter-mediated coexistence, because if prey are attacked simply in proportion to their abundance, it will be those species that are assimilating resources and producing biomass and offspring most rapidly (the competitive dominants) that will be most abundant, and will therefore be most severely set back by predation. Note, however, that predators seem just as frequently to cause a reduction in species richness, or to have no effect.

2 An intermediate intensity of predation is most likely to be associated with high prey species richness, since too low an intensity may not prevent competitive exclusion of inferior prey species, whilst too high an intensity may itself drive preferred prey to extinction. (Note, however, that 'intermediate' is difficult to define *a priori*.)

3 The role of predators and parasites in shaping community structure may often be least significant in communities where physical conditions are more severe, variable or unpredictable (Connell, 1975). In sheltered coastal sites, predation appears to be a dominant force shaping community structure

(Paine, 1966), but in exposed rocky tidal communities where there is direct wave action, predators seem to be scarce and to have a negligible influence on community structure (Menge & Sutherland, 1976; Menge *et al.*, 1986). Deep-sea hydrothermal vents provide an exception to these generalizations, probably because the physically severe circumstances close to the vents also engender very high levels of productivity.

4 The effects of animals on a community often extend far beyond just those due to the cropping of their prey. Burrowing animals (such as earthworms, rabbits and porcupines) and mound-builders (ants and termites) – and parasitized cockles – all create disturbances and act as ecosystem engineers (by modifying the physical structure of the environment) (Wilby *et al.*, 2001). Their activities provide local heterogeneities, including sites for new colonists to become established and for microsuccessions to take place. Larger grazing animals introduce a mosaic of nutrient-rich patches, as a result of dunging and urinating, in which the local balance of other species is profoundly changed. Even the footprint of a cow in a wet pasture may so change the microenvironment that it is now colonized by species that would not be present were it not for the disturbance (Harper, 1977). The predator is just one of the many agents disturbing community equilibrium.

5 Carnivores that also feed at other trophic levels (omnivores) may have particularly far-reaching consequences for the community. For example, omnivorous freshwater crayfish can influence the composition of plants (which they consume), herbivores and carnivores (which they consume or with which they compete), and even detritivores because their extreme omnivory includes feeding on dead plant and animal material (Usio & Townsend, 2002, 2004). Moreover, they can also act as ecosystem engineers by dislodging animals and detritus as they move or burrow through the substrate (Statzner *et al.*, 2000).

19.7 Pluralism in community ecology

It would be wrong to replace one monolithic view of community organization (the overriding importance of competition and niche differentiation) with another (the overriding importance of forces such as predation and disturbance that make competition much less influential). Certainly, communities structured by competition are not a general rule, but neither necessarily are communities structured by any single agency. Most communities are probably organized by a mixture of forces – competition, predation, disturbance and recruitment – although their relative importance may vary systematically, with competition and predation figuring more prominently in communities where recruitment levels are high (Menge & Sutherland, 1987) and in less disturbed environments (Menge & Sutherland, 1976; Townsend, 1991).

communities are not necessarily structured by a single biotic process

In an elegant series of experiments, Wilbur (1987) investigated the interactions between competition, predation and disturbance as they influenced four species of frog and toad that occur in North American ponds. In the absence of predators, tadpoles of *Scaphiopus holbrooki* were competitively dominant whilst, at the opposite extreme, *Hyla chrysoscelis* had a very low competitive status (Figure 19.23a). The presence of predatory salamanders, *Notophthalmus viridescens*, did not alter the total number of tadpoles reaching metamorphosis, but relative abundances were shifted because *S. holbrooki*, the competitive dominant, was selectively eaten (Figure 19.23b). Finally, Wilbur subjected his tadpole communities, in the presence and absence of predators, to water loss, to simulate a natural drying regime (disturbance). The influence of competition was to slow growth and retard the

physical conditions can moderate the effects of predators and parasites

Figure 19.23 (a) Relative abundance of tadpoles of each of four species introduced at high density into ponds (initial), and the relative abundances of metamorphs at the end of the experiment (final). (b) Number of metamorphs of four species in the absence and presence of predatory salamanders, and in ponds which persist or which dry up 100 days into the experiment. (After Wilbur, 1987; Townsend, 1991a.)

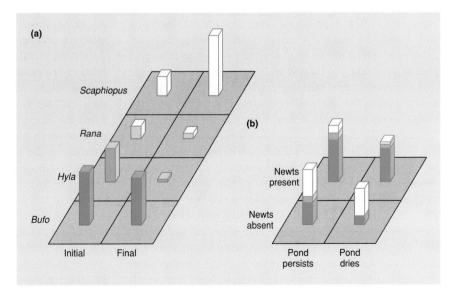

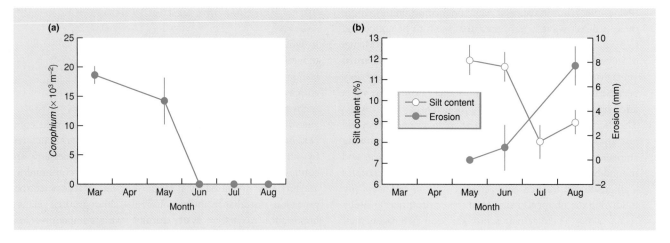

Figure 19.24 Extermination of a dense intertidal amphipod population by microphallid trematodes and the consequent changes in sediment characteristics and topography of the mudflat. (a) Mean density (± SE) of *Corophium volutator*. (b) Mean silt content (particles < 63 μm) and substrate erosion. (c) The topography of the *Corophium* bed before the parasite-induced *C. volutator* die-off. (d) The topography of the flat a few months after the disappearance of amphipods. (After Mouritsen & Poulin, 2002 and Mouritsen *et al.*, 1998. Reproduced by permission of K. Mouritsen)

timing of metamorphosis, thus increasing the risk of desiccation in drying ponds. *S. holbrooki* had the shortest larval period and made up a greater proportion of metamorphs in the drying experiment without predators. The presence of predators ameliorated the impact of competition, allowing surviving tadpoles of several species to grow rapidly enough to metamorphosize before the ponds dried up.

The consequences of parasitism may also be moderated by physical conditions. The mud snail *Hydrobia ulvae* and the amphipod *Corophium volutator* dominate the benthic macrofaunal community on intertidal mudflats in the Danish Wadden Sea. These two species serve as first and second intermediate hosts, respectively, to microphallid trematodes, with sandpipers (*Calidris* spp.) as definitive hosts. The trematode eggs are expelled in the bird's droppings and the detritus-feeding snails accidentally eat them. The parasite larvae hatch and reproduce inside the snail, releasing into the water on a daily basis vast numbers of swimming cercariae that seek out an amphipod. As a consequence of a temperature-dependent release of cercariae from the snails, the parasites cause intensity-dependent mortality in the amphipod hosts, which itself increases rapidly with increasing temperature (Mouritsen & Jensen,

1997). *C. volutator* normally increases rapidly during spring and summer, commonly achieving densities exceeding 80,000 individuals per m² in early fall (Mouritsen *et al.*, 1997). Because these amphipods make permanent U-shaped burrows that stabilize the substrate, and because of their patchy distribution, *Corophium*-dominated mudflats have a characteristic topography with a mosaic of elevated plateaux (high-density patches) and sediment depressions (low-density patches) (Mouritsen *et al.*, 1998). In this state, the *Corophium* bed is very stable even during strong onshore gales. But during the spring of 1990, ambient temperatures were unusually high and so was the prevalence of microphallid infections in the snail population, resulting in a massive release of cercariae from the snails and, within 5 weeks, the complete collapse of the amphipod population (Figure 19.24a) (Jensen & Mouritsen, 1992). As the sediment-stabilizing amphipods disappeared, the plateaux of the former *Corophium* bed (which covered about 80 ha) were subject to significant erosion (Figure 19.24b). The characteristic mudflat topography eventually vanished (Figure 19.26c, d) with dramatic consequences for many other mudflat macroinvertebrates, including species of nemertinea, polychaeta, gastropoda, bivalvia and crustacea.

remember that some biotic interactions are positive in their effects

We began this chapter by noting the diversity of ways in which a single species can affect communities and ecosystems. It would be wrong to finish it with the impression that competition, predation and parasitism are the principal population interactions that determine community organization. Facilitation is also of major significance, though once again its significance varies with physical conditions. Thus, the presence of a canopy of the seaweed *Ascophyllum nodosum* at its upper intertidal boundary in communities in the Gulf of Maine reduced maximum daily rock temperatures by 5–10°C and evaporative losses by an order of magnitude, with positive outcomes for recruitment, growth and survivorship of a range of benthic organisms (Bertness *et al.*, 1999). In fact, nearly half of the recorded population interactions in this zone were positive (facilitation) rather than negative (competitive or predatory). On the other hand, at *A. nodosum*'s lower boundary, rather than ameliorating physical conditions (these are not so severe deeper in the intertidal zone) the seaweed canopy provided excellent conditions for herbivores and carnivores and consumer pressure was severe.

Positive interactions among terrestrial plant species have also been demonstrated in many communities (Wilson & Agnew, 1992; Jones *et al.*, 1994). Plants sometimes benefit their neighbors by reducing the likelihood of consumption by herbivores. Thus, Callaway *et al.* (2000) examined the role played by two competitively dominant and highly unpalatable plants, the physically defended thistle *Cirsium obvalatum* and the chemically defended *Veratrum lobelianum*. Both have invaded grazed meadows in the central Caucasus in the Republic of Georgia. Forty four percent (15/34) of all species in the study were rare (< 1.0% cover) in open meadow, but occurred at significantly higher covers under *C. obvalatum* and *V. lobelianum* (i.e. within a 60 × 60 cm plot containing one of the unpalatable species). Eight species were only found under an unpalatable species, and the communities associated with them had 78–128% more species in flower or fruit than in the open meadow sites. It seems that tasty species may avoid being eaten, and grow and reproduce better, if they associate with an unpalatable neighbor.

Finally, we have seen how the effects of predators and parasites are not restricted to their prey/hosts or even just those species with which they or their prey compete. Sometimes the effects extend beyond a single, or adjacent, trophic level to spread throughout the food web. This was the case, for example, for starfish (see Section 19.4.2), parasitized caddis larvae (see Section 19.5) and omnivorous crayfish (see Section 19.6). We turn our attention to the complex workings of whole food webs in the next chapter.

Summary

Individual species can influence the composition of whole communities in a variety of ways. In this chapter we pay particular attention to the manner in which competition, predation and parasitism can shape communities.

The view that interspecific competition plays a central and powerful role in the shaping of communities was first fostered by the competitive exclusion principle, with its implication of a limit to the similarity of competing species, and thus a limit to the number of species that can be fitted into a particular community before niche space is fully saturated. There is no argument about whether competition sometimes affects community structure, nor does it play an overriding role in every case. Thus, other factors may prevent competition from progressing to competitive exclusion, by depressing densities or periodically reversing competitive superiority. Moreover, even when competition is intense, the species may coexist if they have aggregated distributions, with each species distributed independently of the others.

Evidence from community studies of niche differentiation for important resources in space and time is consistent with a role for competition in determining community composition. However, the documentation of mere differences among species is insufficient. The approach has been to build null models of actual communities that retain certain of the characteristics of their real counterparts (in terms of diets, feeding morphologies or distributions of coexisting species) but reassemble the components at random, excluding the consequences of competition. Comparisons of predicted and observed patterns have sometimes supported a role for competition, but by no means in all cases.

Grazing animals sometimes increase plant species richness (exploiter-mediated coexistence) by interrupting the process of competitive exclusion, thus imposing their own order on community composition. Plant coexistence is more likely to be fostered by grazers in nutrient-rich situations, and where the preferred food plants would otherwise be competitively superior to less preferred ones.

Carnivorous animals may, likewise, increase the species richness of prey. This has been recorded for rocky shore invertebrates and woodland bird communities, but not for deep-sea vent communities or in terrestrial insect and spider studies. The outcome for species richness in the face of predation again depends on a number of factors, including the pattern of dietary preference and the relative competitive status of the prey.

The incidence of a parasite, like that of other types of exploiter, may determine whether or not a host species occurs in an area; parasites can cause more subtle effects too, by influencing species that are themselves strong interactors or ecosystem engineers in terrestrial, freshwater and marine communities. Parasites are sometimes responsible for exploiter-mediated coexistence.

Communities are not necessarily structured by a single biotic process and the role of consumers in shaping community structure can be expected to be modified according to abiotic conditions. Biotic effects may often be least significant in communities where physical conditions are more severe, variable or unpredictable.

Chapter 20
Food Webs

20.1 Introduction

In the previous chapter we began to consider how population interactions can shape communities. Our focus was on interactions between species occupying the same trophic level (interspecific competition) or between members of adjacent trophic levels. It has already become clear, however, that the structure of communities cannot be understood solely in terms of direct interactions between species. When competitors exploit living resources, the interaction between them necessarily involves further species – those whose individuals are being consumed – while a recurrent effect of predation is to alter the competitive status of prey species, leading to the persistence of species that would otherwise be competitively excluded (consumer-mediated coexistence).

In fact, the influence of a species often ramifies even further than this. The effects of a carnivore on its herbivorous prey may also be felt by any plant population upon which the herbivore feeds, by other predators and parasites of the herbivore, by other consumers of the plant, by competitors of the herbivore and of the plant, and by the myriad of species linked even more remotely in the food web. This chapter is about food webs. In essence, we are shifting the focus to systems usually with at least three trophic levels and 'many' (at least more than two) species.

The study of food webs lies at the interface of community and ecosystem ecology. Thus, we will focus both on the population dynamics of interacting species in the community (species present, connections between them in the web, and interaction strengths) and on the consequences of these species interactions for ecosystem processes such as productivity and nutrient flux.

First, we consider the incidental effects – repercussions further away in the food web – when one species affects the abundance of another (Section 20.2). We examine indirect, 'unexpected' effects in general (Section 20.2.1) and then specifically the effects of 'trophic cascades' (Sections 20.2.3 and 20.2.4). This leads naturally to the question of when and where the control of food webs is 'top-down' (the abundance, biomass or diversity at lower trophic levels depends on the effects of consumers, as in a trophic cascade) or 'bottom-up' (a dependence of community structure on factors acting from lower trophic levels, such as nutrient concentration and prey availability) (Section 20.2.5). We then pay special attention to the properties and effects of 'keystone' species – those with particularly profound and far-reaching consequences elsewhere in the food web (Section 20.2.6).

Second, we consider interrelationships between food web structure and stability (Sections 20.3 and 20.4). Ecologists are interested in community stability for two reasons. The first is practical – and pressing. The stability of a community measures its sensitivity to disturbance, and natural and agricultural communities are being disturbed at an ever-increasing rate. It is essential to know how communities react to such disturbances and how they are likely to respond in the future. The second reason is less practical but more fundamental. The communities we actually see are, inevitably, those that have persisted. Persistent communities are likely to possess properties conferring stability. The most fundamental question in community ecology is: 'Why are communities the way they are?' Part of the answer is therefore likely to be: 'Because they possess certain stabilizing properties'.

20.2 Indirect effects in food webs

20.2.1 'Unexpected' effects

The removal of a species (experimentally, managerially or naturally) can be a powerful tool in unraveling the workings of a food web. If a predator species is removed, we expect an increase in the density of its prey. If a competitor species is removed, we expect an increase in the success of species with which it competes. Not surprisingly, there are plenty of examples of such expected results.

Sometimes, however, removing a species may lead to a *de*crease in competitor abundance, or the removal of a predator may lead to a decrease in prey abundance. Such unexpected effects arise when direct effects are less important than the effects that occur through indirect pathways. Thus, the removal of a species might increase the density of one competitor, which in turn causes another competitor to decline. Or the removal of a predator might increase the abundance of a prey species that is competitively superior to another, leading to a decrease in the density of the latter. In a survey of more than 100 experimental studies of predation, more than 90% demonstrated statistically significant results, and of these about one in three showed unexpected effects (Sih *et al.*, 1985).

These indirect effects are brought especially into focus when the initial removal is carried out for some managerial reason – either the biological control of a pest (Cory & Myers, 2000) or the eradication of an exotic, invader species (Zavaleta *et al.*, 2001)

– since the deliberate aim is to solve a problem, not create further, unexpected problems.

For example, there are many islands on which feral cats have been allowed

mesopredators

to escape domestication and now threaten native prey, especially birds, with extinction. The 'obvious' response is to eliminate the cats (and conserve their island prey), but as a simple model developed by Courchamp *et al.* (1999) explains, the programs may not have the desired effect, especially where, as is often the case, rats have also been allowed to colonize the island (Figure 20.1). The rats ('mesopredators') typically both compete with and prey upon the birds. Hence, removal of the cats ('superpredators'), which normally prey upon the rats as well as the birds, is likely to increase not decrease the threat to the birds once predation pressure on the mesopredators is removed. Thus, introduced cats on Stewart Island, New Zealand preyed upon an endangered flightless parrot, the kakapo, *Strigops habroptilus* (Karl & Best, 1982);

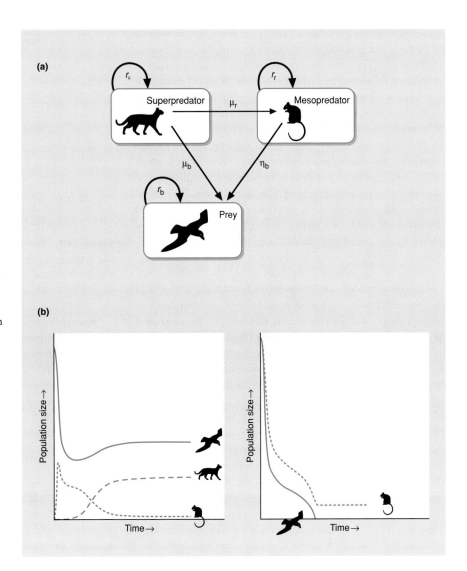

Figure 20.1 (a) Schematic representation of a model of an interaction in which a 'superpredator' (such as a cat) preys both on 'mesopredators' (such as rats, for which it shows a preference) at a per capita rate μ_r, and on prey (such as birds) at a per capita rate μ_b, while the mesopredator also attacks prey at a per capita rate η_b. Each species also recruits to its own population at net per capita rates r_c, r_r and r_b. (b) The output of the model with realistic parameter values: with all three species present, the superpredator keeps the mesopredator in check and all three species coexist (left); but in the absence of the superpredator, the mesopredator drives the prey to extinction (right). (After Courchamp *et al.*, 1999.)

but controlling cats alone would have been risky, since their preferred prey are three species of introduced rats, which, unchecked, could pose far more of a threat to the kakapo. In fact, Stewart Island's kakapo population was translocated to smaller offshore islands where exotic mammalian predators (like rats) were absent or had been eradicated.

Further indirect effects, though not really 'unexpected', have occurred following the release of the weevil, *Rhinocyllus conicus*, as a biological control agent of exotic thistles, *Carduus* spp., in the USA (Louda *et al.*, 1997). The beetle also attacks native thistles in the genus *Cirsium* and reduces the abundance of a native picture-winged fly, *Paracantha culta*, which feeds on thistle seeds – the weevil indirectly harms species that were never its intended target.

20.2.2 Trophic cascades

The indirect effect within a food web that has probably received most attention is the so-called trophic cascade (Paine, 1980; Polis *et al.*, 2000). It occurs when a predator reduces the abundance of its prey, and this cascades down to the trophic level below, such that the prey's own resources (typically plants) increase in abundance. Of course, it need not stop there. In a food chain with four links, a top predator may reduce the abundance of an intermediate predator, which may allow the abundance of a herbivore to increase, leading to a decrease in plant abundance.

The Great Salt Lake of Utah in the USA provides a natural experiment that illustrates a trophic cascade. There, what is essentially a two-level trophic system (zooplankton–phytoplankton) is augmented by a third trophic level (a predatory insect, *Trichocorixa verticalis*) in unusually wet years when salinity is lowered (Wurtsbaugh, 1992). Normally, the zooplankton, dominated by a brine shrimp (*Artemia franciscana*), are capable of keeping phytoplankton biomass at a low level, producing high water clarity. But when salinity declined from above 100 g l^{-1} to 50 g l^{-1} in 1985, *Trichochorixa* invaded and *Artemia* biomass was reduced from 720 to 2 mg m^{-3}, leading to a massive increase in the abundance of phytoplankton, a 20-fold increase in chlorophyll *a* concentration and a fourfold decrease in water clarity (Figure 20.2).

Another example of a trophic cascade, but also of the complexity of indirect effects, is provided by a 2-year experiment in which bird predation pressure was manipulated in an intertidal community on the northwest coast of the USA, in order to determine the effects of the birds on three limpet species (prey) and their algal food (Wootton, 1992). Glaucous-winged gulls (*Larus glaucescens*) and oystercatchers (*Haematopus bachmani*) were excluded by means of wire cages from large areas (each 10 m²) in which limpets were common. Overall, limpet biomass was much lower in the presence of birds, and the effects of bird predation cascaded down to the plant trophic level, because grazing pressure on the fleshy algae was reduced. In addition, the birds freed up space for algal colonization through the removal of barnacles (Figure 20.3).

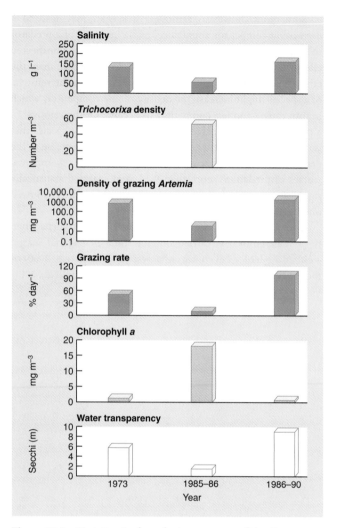

Figure 20.2 Variation in the pelagic ecosystem of the Great Salt Lake during three periods that differed in salinity. (After Wurtsbaugh, 1992.)

It also became evident, however, that while birds reduced the abundance of one of the limpet species, *Lottia digitalis*, as might have been expected, they increased the abundance of a second limpet species (*L. strigatella*) and had no effect on the third, *L. pelta*. The reasons are complex and go well beyond the direct effects of consumption of limpets. *L. digitalis*, a light-colored limpet, tends to occur on light-colored goose barnacles (*Pollicipes polymerus*), whilst dark *L. pelta* occurs primarily on dark Californian mussels (*Mytilus californianus*). Both limpets show strong habitat selection for these cryptic locations. Predation by gulls reduced the area covered by goose barnacles (to the detriment of *L. digitalis*), leading through competitive release to an increase in the area covered by mussels (benefiting *L. pelta*). The third species, *L. strigatella*, is competitively inferior to the others and increased in density because of competitive release.

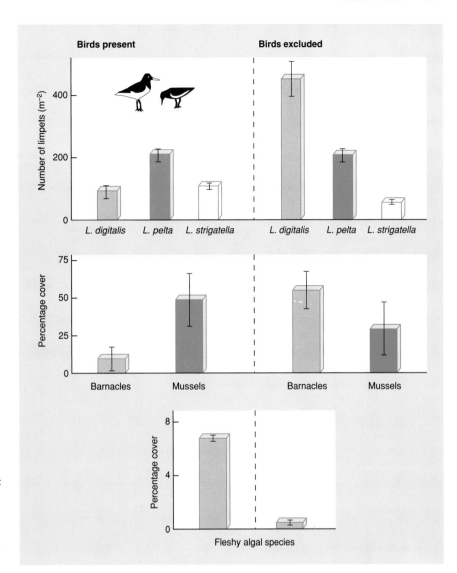

Figure 20.3 When birds are excluded from the intertidal community, barnacles increase in abundance at the expense of mussels, and three limpet species show marked changes in density, reflecting changes in the availability of cryptic habitat and competitive interactions as well as the easing of direct predation. Algal cover is much reduced in the absence of effects of birds on intertidal animals (means ± SE are shown). (After Wootton, 1992.)

20.2.3 Four trophic levels

In a four-level trophic system, if it is subject to trophic cascade, we might expect that the abundances of the top carnivores and the herbivores are positively correlated, as are those of the primary carnivores and the plants. This is precisely what was found in an experimental study of the food web in Eel River, northern California (Figure 20.4a) (Power, 1990). Large fish (roach, *Hesperoleucas symmetricus*, and steelhead trout, *Oncorhynchus mykiss*) reduced the abundance of fish fry and invertebrate predators, allowing their prey, tuft-weaving midge larvae (*Pseudochironomus richardsoni*) to attain high density and to exert intense grazing pressure on filamentous algae (*Cladophora*), whose biomass was thus kept low.

Support for the expected pattern also comes from the tropical lowland forests of Costa Rica and a study of *Tarsobaenus* beetles preying on *Pheidole* ants that prey on a variety of herbivores that attack ant-plants, *Piper cenocladum* (though the detailed trophic interactions are slightly more complex than this – Figure 20.5a). A descriptive study at a number of sites showed precisely the alternation of abundances expected in a four-level trophic cascade: relatively high abundances of plants and ants associated with low levels of herbivory and beetle abundance at three sites, but low abundances of plants and ants associated with high levels of herbivory and beetle abundance at a fourth (Figure 20.5b). Moreover, when beetle abundance was manipulated experimentally at one of the sites, ant and plant abundance were significantly higher, and levels of herbivory lower, in the absence of beetles than in their presence (Figure 20.5c).

On the other hand, in a four-level trophic stream community in New Zealand (brown trout (*Salmo trutta*),

four levels can act like three

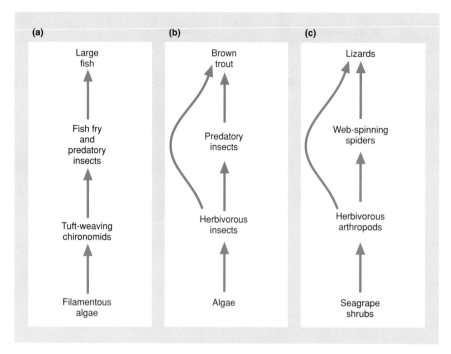

Figure 20.4 Three examples of food webs, each with four trophic levels. (a) The absence of omnivory (feeding at more than one trophic level) in this North American stream community means it functions as a four-level trophic system. On the other hand, web (b) from a New Zealand stream community and web (c) from a terrestrial Bahamanian community both function as three-level trophic webs. This is because of the strong direct effects of omnivorous top predators on herbivores and their less influential effects on intermediate predators. (After Power, 1990; Flecker & Townsend, 1994; Spiller & Schoener, 1994, respectively.)

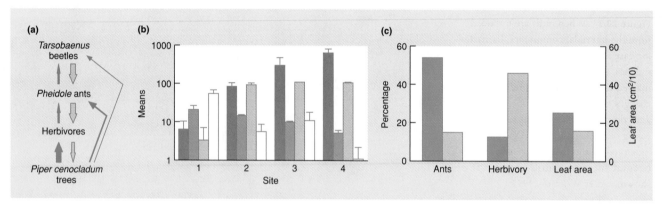

Figure 20.5   (a) Schematic representation of a four-level food chain in Costa Rica. Pale arrows denote mortality and dark arrows a contribution to the consumer's biomass; arrow breadth denotes their relative importance. Both (b) and (c) show evidence of a trophic cascade flowing down from the beetles: with positive correlations between the beetles and herbivores and between the ants and trees. (b) The relative abundance of ant-plants (■), abundance of ants (■) and of beetles (■), and strength of herbivory (□) at four sites. Means and standard errors are shown; the units of measurement are various and are given in the original references. (c) The results of an experiment at site 4 when replicate enclosures were established without beetles (■) and with beetles (■). Units are: ants, % of plant petioles occupied; herbivory, % of leaf area eaten; leaf area, cm² per 10 leaves. (After Letourneau & Dyer, 1998a, 1998b; Pace *et al.*, 1999.)

predatory invertebrates, grazing invertebrates and algae), the presence of the top predator did not lead to reduced algal biomass, because the fish influenced not only the predatory invertebrates but also directly affected the activity of the herbivorous species at the next trophic level down (Figure 20.4b) (Flecker & Townsend, 1994). They did this both by consuming grazers and by con-straining the foraging behavior of the survivors (McIntosh & Townsend, 1994). A similar situation has been

reported for a four-level trophic terrestrial community in the Bahamas, consisting of lizards, web spiders, herbivorous arthropods and seagrape shrubs (*Coccoloba uvifera*) (Figure 20.4c) (Spiller & Schoener, 1994). The results of experimental manipulations indicated a strong interaction between top predators (lizards) and herbivores, but a weak effect of lizards on spiders. Consequently, the net effect of top predators on plants was positive and there was less leaf damage in the presence of lizards. These four-level trophic communities have a trophic cascade, but it functions as if they had only three levels.

20.2.4 Cascades in all habitats? Community- or species-level cascades?

are trophic cascades all wet?

So much of the discussion of trophic cascades, including their original identification, has been based on aquatic (either marine or freshwater) examples that the question has seriously been asked 'are trophic cascades all wet?' (Strong, 1992). As pointed out by Polis *et al.* (2000), however, in order to answer this question we should recognize a distinction between community- and species-level cascades (Polis, 1999). In the former, the predators in a community, as a whole, control the abundance of the herbivores, such that the plants, as a whole, are released from control by the herbivores. But in a species-level cascade, increases in a particular predator give rise to decreases in particular herbivores and increases in particular plants, without this affecting the whole community. Thus, Schmitz *et al.* (2000), in apparent contradiction of the 'all cascades are wet' proposition, reviewed a total of 41 studies in terrestrial habitats demonstrating trophic cascades; but Polis *et al.* (2000) pointed out that all of these referred only to subsets of the communities of which they were part – that is, they were essentially species-level cascades. Moreover, the measures of plant performance in these studies were typically short term and small scale (for instance, 'leaf damage' as in the lizard–spider–herbivore–seagrape example above) rather than broader scale responses of significance to the whole community, such as plant biomass or productivity.

Polis *et al.* (2000) proposed, then, that community-level cascades are most likely to occur in systems with the following characteristics: (i) the habitats are relatively discrete and homogeneous; (ii) the prey population dynamics (including those of the primary producers) are uniformly fast relative to those of their consumers; (iii) the common prey tend to be uniformly edible; and (iv) the trophic levels tend to be discrete and species interactions strong, such that the system is dominated by discrete trophic chains.

If this proposition is correct, then community-level cascades are most likely in pelagic communities of lakes and in benthic communities of streams and rocky shores (all 'wet') and perhaps in agricultural communities. These tend to be discrete, relatively simple communities, based on fast-growing plants often dominated by a single taxon (phytoplankton, kelp or an agricultural crop). This is not to say (as the Schmitz *et al.* (2000) review confirms) that such forces are absent in more diffuse, species-rich systems, but rather that patterns of consumption are so differentiated that their overall effects are buffered. From the point of view of the whole community, such effects may be represented as trophic trickles rather than cascades.

Certainly, the accumulating evidence seems to support a pattern of overt community-level cascades in simple, especially wet, communities, and much more limited cascades embedded within a broader web in more diverse, especially terrestrial, communities. It remains to be seen, however, whether this reflects some underlying realities or simply differences in the practical difficulties of manipulating and studying cascades in different habitats. An attempt to decide whether there are real differences between aquatic and terrestrial food webs was forced to conclude that there is little evidence, either empirical or theoretical, to either support or refute the idea (Chase, 2000).

20.2.5 Top-down or bottom-up control of food webs? Why is the world green?

We have seen that trophic cascades are normally viewed 'from the top', starting at the highest trophic level. So, in a three-level trophic community, we think of the predators controlling the abundance of the grazers and say that the grazers are subject to 'top-down control'. Reciprocally, the predators are subject to bottom-up control (abundance determined by their resources): a standard predator–prey interaction. In turn, the plants are also subject to bottom-up control, having been released from top-down control by the effects of the predators on the grazers. Thus, in a trophic cascade, top-down and bottom-up control alternate as we move from one trophic level to the next.

But suppose instead that we start at the other end of the food chain, and assume that the plants are controlled bottom-up by competition for their resources. It is still possible for the herbivores to be limited by competition for plants – *their* resources – and for the predators to be limited by competition for herbivores. In this scenario, all trophic levels are subject to bottom-up control (also called 'donor control'), because the resource controls the abundance of the consumer but the consumer does not control the abundance of the resource. The question has therefore arisen: 'Are food webs – or are particular *types* of food web – dominated by either top-down or bottom-up control?' (Note again, though, that even when top-down control 'dominates', top-down and bottom-up control are expected to alternate from trophic level to trophic level.)

top-down, bottom-up and cascades

Clearly, this is linked to the issues we have just been dealing with. Top-down control should dominate in systems with powerful community-level trophic cascades. But in systems where trophic cascades, if they exist at all, are limited to the species level, the community as a whole could be dominated by top-down or bottom-up control. Also, there are some communities that tend, inevitably, to be dominated by bottom-up control, because consumers have little or no influence on the supply of their food resource. The most obvious group of organisms to which this applies is the detritivores (see Chapter 11), but consumers of nectar and seeds are also likely to come into this category (Odum & Biever, 1984) and few of the multitude of rare phytophagous insects are likely to have any impact upon the abundance of their host plants (Lawton, 1989).

why is the world green? . . .

The widespread importance of top-down control, foreshadowing the idea of the trophic cascade, was first advocated in a famous paper by Hairston *et al.* (1960), which asked 'Why is the world green?' They answered, in effect, that the world is green because top-down control predominates: green plant biomass accumulates because predators keep herbivores in check. The argument was later extended to systems with fewer or more than three trophic levels (Fretwell, 1977; Oksanen *et al.*, 1981).

. . . or is it prickly and bad tasting?

Murdoch (1966), in particular, challenged these ideas. His view, described by Pimm (1991) as 'the world is prickly and tastes bad', emphasized that even if the world is green (assuming it is), it does not necessarily follow that the herbivores are failing to capitalize on this because they are limited, top-down, by their predators. Many plants have evolved physical and chemical defenses that make life difficult for herbivores (see Chapter 3). The herbivores may therefore be competing fiercely for a limited amount of palatable and unprotected plant material; and their predators may, in turn, compete for scarce herbivores. A world controlled from the bottom-up may still be green.

Oksanen (1988), moreover, has argued that the world is not always green – particularly if the observer is standing in the middle of a desert or on the northern coast of Greenland. Oksanen's contention (see also Oksanen *et al.*, 1981) is that: (i) in extremely unproductive or 'white' ecosystems, grazing will be light because there is not enough food to support effective populations of herbivores: both the plants and the herbivores will be limited bottom-up; (ii) at the highest levels of plant productivity, in 'green' ecosystems, there will also be light grazing because of top-down limitation by predators (as argued by Hairston *et al.*, 1960); but (iii) between these extremes, ecosystems may be 'yellow', where plants are top-down limited by grazers because there are insufficient herbivores to support effective populations of predators. The suggestion, then, is that productivity shifts the balance between top-down and bottom-up control by altering the lengths of food chains. This still remains to be critically tested.

an influence of primary productivity?

There are also suggestions that the level of primary productivity may be influential in other ways in determining whether top-down or bottom-up control is predominant. Chase (2003) examined the effect of nutrient concentrations on a freshwater web comprising an insect predator, *Belostoma flumineum*, feeding on two species of herbivorous snails, *Physella girina* and *Helisoma trivolvis*, in turn feeding on macrophytes and algae within a larger food web including zooplankton

and phytoplankton. At the lowest nutrient concentrations, the snails were dominated by the smaller *P. gyrina*, vulnerable to predation, and the predator gave rise to a trophic cascade extending to the primary producers. But at the highest concentrations, the snails were dominated by the larger *H. trivolvis*, relatively invulnerable to predation, and no trophic cascade was apparent (Figure 20.6). This study, therefore, also lends support to Murdoch's proposition that the 'world tastes bad', in that invulnerable herbivores gave rise to a web with a relative dominance of bottom-up control. Overall, though, we see again that the elucidation of clear patterns in the predominance of top-down or bottom-up control remains a challenge for the future.

20.2.6 Strong interactors and keystone species

Some species are more intimately and tightly woven into the fabric of the food web than others. A species whose removal would produce a significant effect (extinction or a large change in density) in at least one other species may be thought of as a strong interactor. Some strong interactors would lead, through their removal, to significant changes spreading throughout the food web – we refer to these as *keystone species*.

A keystone is the wedge-shaped block at the highest point of an arch that locks the other pieces together. Its early use in food web architecture referred to a top predator (the starfish *Pisaster* on a rocky shore; see Paine (1966) and Section 19.4.2) that has an indirect beneficial effect on a suite of inferior competitors by depressing the abundance of a superior competitor. Removal of the keystone predator, just like the removal of the keystone in an arch, leads to a collapse of the structure. More precisely, it leads to extinction or large changes in abundance of several species, producing a community with a very different species composition and, to our eyes, an obviously different physical appearance.

what is a keystone species?

It is now usually accepted that keystone species can occur at other trophic levels (Hunter & Price, 1992). Use of the term has certainly broadened since it was first coined (Piraino *et al.*, 2002), leading some to question whether it has any value at all. Others have defined it more narrowly – in particular, as a species whose impact is 'disproportionately large relative to its abundance' (Power *et al.*, 1996). This has the advantage of excluding from keystone status what would otherwise be rather trivial examples, especially 'ecological dominants' at lower trophic levels, where one species may provide the resource on which a whole myriad of other species depend – for example, a coral, or the oak trees in an oak woodland. It is certainly more challenging and more useful to identify species with disproportionate effects.

Semantic quibbles aside, it remains important to acknowledge that while all species no doubt influence the structure of their communities to a degree, some are far more influential than

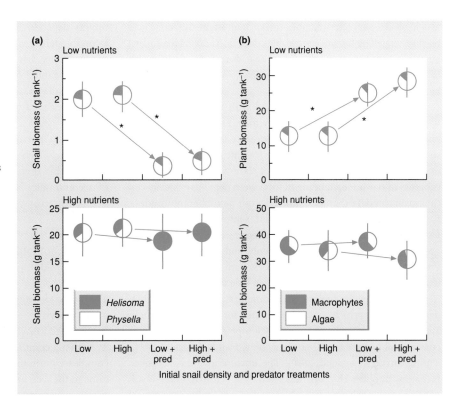

Figure 20.6 Top-down control, but only with low productivity. (a) Snail biomass and (b) plant biomass in experimental ponds with low or high nutrient treatments (vertical bars are standard errors). With low nutrients, the snails were dominated by *Physella* (vulnerable to predation) and the addition of predators led to a significant decline (indicated by *) in snail biomass and a consequent increase in plant biomass (dominated by algae). But with high nutrients, *Helisoma* snails (less vulnerable to predation) increased their relative abundance, and the addition of predators led neither to a decline in snail biomass nor to an increase in plant biomass (often dominated by macrophytes). (After Chase, 2003.)

others. Indeed, various indices have been proposed to measure this influence (Piraino *et al.*, 2002); for example, the 'community importance' of a species is the percentage of other species lost from the community after its removal (Mills *et al.*, 1993). Also, recognizing the concept of keystone species and attempting to identify them are both important from a practical point of view because keystone species are likely to have a crucial role in conservation: changes in their abundance will, by definition, have significant repercussions for a whole range of other species. Inevitably, though, the dividing line between keystone species and the rest is not clear cut.

keystone species can occur throughout the food web

In principle, keystone species can occur throughout the food web. Jones *et al.* (1997) point out that it need not even be their trophic role that makes them important, but rather that they act as 'ecological engineers' (see Section 13.1). Beavers, for example, in cutting down a tree and building a dam, create a habitat on which hundreds of species rely. Keystone mutualists (Mills *et al.*, 1993) may also exert influence out of proportion to their abundance: examples include a pollinating insect on which an ecologically dominant plant relies, or a nitrogen-fixing bacterium supporting a legume and hence the whole structure of a plant community and the animals reliant on it. Certainly, keystone species are limited neither to top predators nor consumers mediating coexistence amongst their prey. For example,

lesser snow geese (*Chen caerulescens caerulescens*) are herbivores that breed in large colonies in coastal brackish and freshwater marshes along the west coast of Hudson Bay in Canada. At their nesting sites in spring, before the onset of above-ground growth of vegetation, adult geese grub for the roots and rhizomes of graminoid plants in dry areas and eat the swollen bases of sedge shoots in wet areas. Their activity creates bare patches (1–5 m²) of peat and sediment. Since there are few pioneer plant species able to recolonize these patches, recovery is very slow. Furthermore, in ungrubbed brackish marshes, intense grazing by high densities of geese later in the summer is essential in establishing and maintaining grazing 'lawns' of *Carex* and *Puccinellia* (Kerbes *et al.*, 1990). It seems reasonable to consider the lesser snow goose as a keystone (herbivore) species.

20.3 Food web structure, productivity and stability

Any ecological community can be characterized by its *structure* (number of species, interaction strength within the food web, average length of food chains, etc.), by certain *quantities* (especially biomass and the rate of production of biomass, which we can summarize as 'productivity') and by its temporal *stability* (Worm & Duffy, 2003). In the remainder of this chapter, we examine some of the interrelationships between these three.

Much of the very considerable recent interest in this area has been generated by the understandable concern to know what might be the consequences of the inexorable decline in biodiversity (a key aspect of structure) for the stability and productivity of biological communities.

We will be particularly concerned with the effects of food web structure (food web complexity in this section; food chain length and a number of other measures in Section 20.4) on the stability of the structure itself and the stability of community productivity. It should be emphasized at the outset, however, that progress in our understanding of food webs depends critically on the quality of data that are gathered from natural communities. Recently, several authors have called this into doubt, particularly for earlier studies, pointing out that organisms have often been grouped into taxa extremely unevenly and sometimes at the grossest of levels. For example, even in the same web, different taxa may have been grouped at the level of kingdom (plants), family (Diptera) and species (polar bear). Some of the most thoroughly described food webs have been examined for the effects of such an uneven resolution by progressively lumping web elements into coarser and coarser taxa (Martinez, 1991; Hall & Raffaelli, 1993, Thompson & Townsend, 2000). The uncomfortable conclusion is that most food web properties seem to be sensitive to the level of taxonomic resolution that is achieved. These limitations should be borne in mind as we explore the evidence for food web patterns in the following sections.

First, however, it is necessary to define 'stability', or rather to identify the various different types of stability.

20.3.1 What do we mean by 'stability'?

resilience and resistance

Of the various aspects of stability, an initial distinction can be made between the resilience of a community (or any other system) and its resistance. *Resilience* describes the speed with which a community returns to its former state after it has been perturbed and displaced from that state. *Resistance* describes the ability of the community to avoid displacement in the first place. (Figure 20.7 provides a figurative illustration of these and other aspects of stability.)

local and global stability

The second distinction is between local stability and global stability. *Local stability* describes the tendency of a community to return to its original state (or something close to it) when subjected to a small perturbation. *Global stability* describes this tendency when the community is subjected to a large perturbation.

dynamic fragility and robustness

A third aspect is related to the local/global distinction but concentrates more on the environment of the community. The stability of any community depends on the environment in which it exists, as well as on the densities and characteristics of the component species. A community that is stable only within a narrow range of environmental conditions, or for only a very limited range of species' characteristics, is said to be *dynamically fragile*. Conversely, one that is stable within a wide range of conditions and characteristics is said to be *dynamically robust*.

Lastly, it remains for us to specify the aspect of the community on which we will focus. Ecologists have often taken a demographic approach. They have concentrated on the *structure* of a community. However, it is also possible to focus on the stability of ecosystem processes, especially *productivity*.

20.3.2 Community complexity and the 'conventional wisdom'

The connections between food web structure and food web stability have preoccupied ecologists for at least half a century. Initially, the 'conventional wisdom' was that increased complexity within a community leads to increased stability; that is, more complex communities are better able to remain structurally the same in the face of a disturbance such as the loss of one or more species. Increased complexity, then as now, was variously taken to mean more species, more interactions between species, greater average strength of interaction, or some combination of all of these things. Elton (1958) brought together a variety of empirical and theoretical observations in support of the view that more complex communities are more stable (simple mathematical models are inherently unstable, species-poor island communities are liable to invasion, etc.). Now, however, it is clear his assertions were mostly either untrue or else liable to some other plausible interpretation. (Indeed, Elton himself pointed out that more extensive analysis was necessary.) At about the same time, MacArthur (1955) proposed a more theoretical argument in favor of the conventional wisdom. He suggested that the more possible pathways there were by which energy passed through a community, the less likely it was that the densities of constituent species would change in response to an abnormally raised or lowered density of one of the other species.

20.3.3 Complexity and stability in model communities: populations

The conventional wisdom, however, has by no means always received support, and has been undermined in particular by the analysis of mathematical models. A watershed study was that by May (1972). He constructed model food webs comprising a number of species, and examined the way in which the population size of each species changed in the neighborhood of its equilibrium abundance (i.e. the *local* stability of *individual populations*).

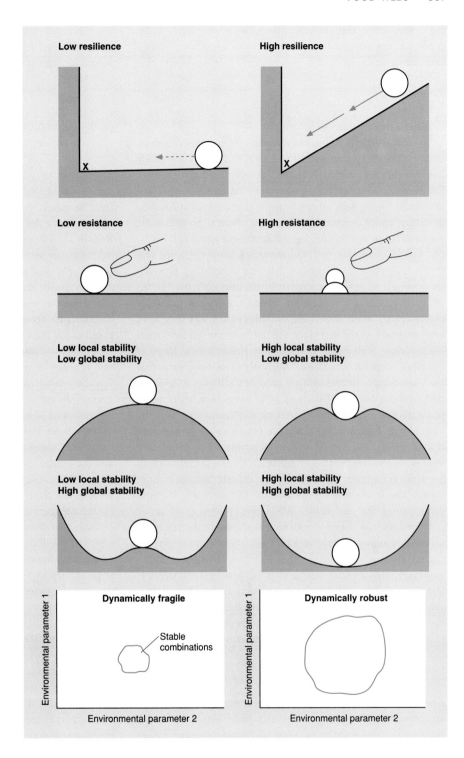

Figure 20.7 Various aspects of stability, used in this chapter to describe communities, illustrated here in a figurative way. In the resilience diagrams, X marks the spot from which the community has been displaced.

Each species was influenced by its interaction with all other species, and the term β_{ij} was used to measure the effect of species j's density on species i's rate of increase. The food webs were 'randomly assembled', with all self-regulatory terms (β_{ii}, β_{jj}, etc.) set at -1, but all other β values distributed at random, including a certain number of zeros. The webs could then be described by three parameters: S, the number of species; C, the 'connectance' of the web (the fraction of all possible pairs of species that interacted directly, i.e. with β_{ij} non-zero); and β, the average 'interaction strength' (i.e. the average of the non-zero β values, disregarding sign). May found that these food webs were only likely to be stable (i.e. the populations would return to equilibrium after a small disturbance) if:

$$\beta(SC)^{1/2} < 1. \tag{20.1}$$

Otherwise, they tended to be unstable.

In other words, increases in the number of species, in connectance and in interaction strength all tend to increase instability (because they increase the left-hand side of the inequality above). Yet each of these represents an increase in complexity. Thus, this model (along with others) suggests that complexity leads to *instability*, and it certainly indicates that there is no necessary, unavoidable connection linking stability to complexity.

many models defy the conventional wisdom

Other studies, however, have suggested that this connection between complexity and instability may be an artefact arising out of the particular characteristics of the model communities or the way they have been analyzed. In the first place, randomly assembled food webs often contain biologically unreasonable elements (e.g. loops of the type: A eats B eats C eats A). Analyses of food webs that are constrained to be reasonable (Lawlor, 1978; Pimm, 1979) show that whilst stability still declines with complexity, there is no sharp transition from stability to instability (compared with the inequality in Equation 20.1). Second, if systems are 'donor controlled' (i.e. $\beta_{ij} > 0$, $\beta_{ji} = 0$), stability is unaffected by or actually increases with complexity (DeAngelis, 1975). And the relationship between complexity and stability in models becomes more complicated if attention is focused on the resilience of those communities that *are* stable. While the proportion of stable communities may decrease with increased complexity, resilience *within* this subset (a crucial aspect of stability) may increase (Pimm, 1979).

Finally, though, the relationship between species richness and the variability of populations appears to be affected in a very general way by the relationship between the mean (m) and variance (s^2) of abundance of individual populations over time (Tilman, 1999). This relationship can be denoted as:

$$s^2 = cm^z, \tag{20.2}$$

where c is a constant and z is the so-called scaling coefficient. There are grounds for expecting values of z to lie between 1 and 2 (Murdoch & Stewart-Oaten, 1989) and most observed values seem to do so (Cottingham *et al.*, 2001). In this range, population variability increases with species richness (Figure 20.8) – a connection between complexity and population instability, as found in May's original model.

Overall, therefore, most models indicate that population stability tends to decrease as complexity increases. This is sufficient to undermine the conventional wisdom prior to 1970. However, the conflicting results amongst the models at least suggest that no single relationship will be appropriate in all communities. It would be wrong to replace one sweeping generalization with another.

20.3.4 Complexity and stability in model communities: whole communities

The effects of complexity, especially species richness, on the stability of aggregate properties of whole communities, such as their biomass or productivity, seem rather more straightforward, at least from a theoretical point of view (Cottingham *et al.*, 2001). Broadly, in richer communities, the dynamics of these aggregate properties are *more* stable. In the first place, as long as the fluctuations in different populations are not perfectly correlated, there is an inevitable 'statistical averaging' effect when populations are added together – when one goes up, another is going down – and this tends to increase in effectiveness as richness (the number of populations) increases.

This effect interacts in turn with the variance to mean relationship of Equation 20.2. As richness increases, average abundance tends to decrease,

aggregate properties are more stable in richer communities

and the value of z in Equation 20.2 determines how the variance in abundance changes with this. Specifically, the greater the value of z, the greater the proportionate decrease in variance, and the greater the increase in stability with increasing richness (Figure 20.8). Only in the rare and probably unrealistic case of z being less than 1 (variance *increases* proportionately as mean abundance declines) is the statistical averaging effect absent.

Note that the related topic of the relationship between richness and productivity – in so far as this is different from the relationship between richness and the *stability* of productivity – is picked up in the next chapter (see Section 21.7), which is devoted to species richness.

20.3.5 Complexity and stability in practice: populations

Even if complexity and population instability are connected in models, this does not mean that we should necessarily expect to see the same association

what should we expect to see in nature?

in real communities. For one thing, the range and predictability of environmental conditions will vary from place to place. In a stable and predictable environment, a community that is dynamically fragile may still persist. However, in a variable and unpredictable environment, only a community that is dynamically robust will be able to persist. Hence, we might expect to see: (i) complex and fragile communities in stable and predictable environments, and simple and robust communities in variable and unpredictable

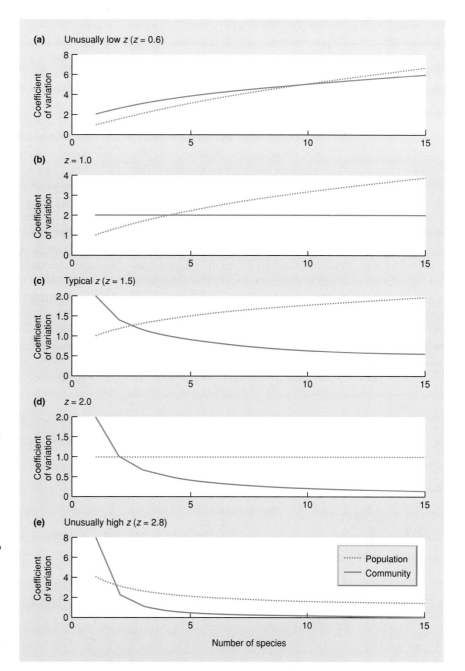

Figure 20.8 The effect of species richness (number of species) on the temporal variability (coefficient of variation, CV) of population size and aggregate community abundance, in model communities in which all species are equally abundant and have the same CV, for various values of the scaling coefficient, z, in the relationship between the mean and variance of abundance (Equation 20.2). (a) $z = 0.6$, an unusually low value. (b) $z = 1.0$, the lower end of typical values. (c) $z = 1.5$, a typical value. (d) $z = 2.0$, the upper end of typical values. (e) $z = 2.8$, an unusually high value. (After Cottingham *et al.*, 2001.)

environments; but (ii) approximately the same recorded stability (in terms of population fluctuations, etc.) in all communities, since this will depend on the inherent stability of the community combined with the variability of the environment. Moreover, we might expect manmade perturbations to have their most profound effects on the dynamically fragile, complex communities of stable environments, which are relatively unused to perturbations, but least effect on the simple, robust communities of variable environments, which have previously been subjected to natural perturbations.

It is also worth noting that there is likely to be an important parallel between the properties of a community and the properties of its constituent populations. In stable environments, populations will be subject to a relatively high degree of K selection (see Section 4.12); in variable environments they will be subject to a relatively high degree of r selection. The K-selected populations (high competitive ability, high inherent survivorship but low reproductive output) will be *resistant* to perturbations, but once perturbed will have little capacity to recover

connections to r and K

(low resilience). The *r*-selected populations, by contrast, will have little resistance but a higher resilience. The forces acting on the component populations will therefore reinforce the properties of their communities, namely fragility (low resilience) in stable environments and robustness in variable ones.

A number of studies have examined the relationship between S, C and β in real communities, following the prediction summarized in Equation 20.1. The argument they use runs as follows. The communities we observe must be stable – otherwise we would not be able to observe them. If communities are only stable for $\beta(SC)^{1/2} < 1$ (or at least when the left-hand side of the inequality is low), then increases in S will lead to decreased stability unless there are compensatory decreases in C and/or β. It is usually assumed, for want of evidence, that β is constant (though ecologists are rising to the challenge of quantifying interaction strengths – e.g. Benke *et al.*, 2001). Thus, communities with more species will only retain stability if there is an associated decline in average connectance, C. We should therefore observe a negative correlation between S and C. A group of 40 food webs was gleaned from the literature by Briand (1983), including terrestrial, freshwater and marine examples. For each community, a single value for

> what is the evidence from real communities?

connectance was calculated as the total number of identified interspecies links as a proportion of the total possible number. Connectance is plotted against S in Figure 20.9a. As predicted, connectance decreases with species number.

However, the data in Briand's compilation were not collected for the purpose of quantitative study of food web properties. Moreover, the level of taxonomic resolution varied substantially from web to web. More recent studies, in which food webs have been much more rigorously documented, indicate that C may decrease with S (as predicted) (Figure 20.9b), that C may be independent of S (Figure 20.9c) or may even increase with S (Figure 20.9d). Thus, no single relationship between complexity and stability receives consistent support from food web analyses.

> connectance decreases with species richness – except when it doesn't

Might other hypotheses do better in accounting for the recorded patterns in connectance? Morphological, physiological and behavioral features restrict the number of types of prey that a consumer can exploit. If each species is adapted to feed on a fixed number of other species, then SC turns out to be constant (Warren, 1994), and C should decrease with increasing S. But if each species feeds on anything whose characteristics fall within

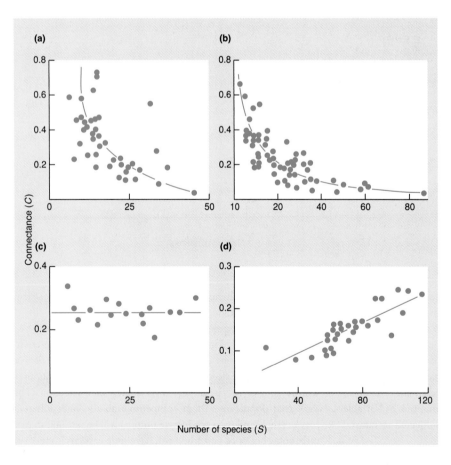

Figure 20.9 The relationships between connectance (C) and species richness (S). (a) For a compilation from the literature of 40 food webs from terrestrial, freshwater and marine environments. (After Briand, 1983.) (b) For a compilation of 95 insect-dominated webs from various habitats. (After Schoenly *et al.*, 1991.) (c) For seasonal versions of a food web for a large pond in northern England, varying in species richness from 12 to 32. (After Warren, 1989.) (d) For food webs from swamps and streams in Costa Rica and Venezuela. (After Winemiller, 1990.) ((a–d) after Hall & Raffaelli, 1993.)

Table 20.1 The influence of nutrient addition on species richness, equitability (*H*/ln *S*) and diversity (Shannon's index, *H*) in two fields; and grazing by African buffalo on species diversity in two areas of vegetation. (After McNaughton, 1977.)

	Control plots	Experimental plots	Statistical significance
Nutrient addition			
Species richness per 0.5 m² plot			
Species-poor plot	20.8	22.5	NS
Species-rich plot	31.0	30.8	NS
Equitability			
Species-poor plot	0.660	0.615	NS
Species-rich plot	0.793	0.740	$P < 0.05$
Diversity			
Species-poor plot	2.001	1.915	NS
Species-rich plot	2.722	2.532	$P < 0.05$
Grazing			
Species diversity			
Species-poor plot	1.069	1.357	NS
Species-rich plot	1.783	1.302	$P < 0.005$

NS, not significant.

the range to which it is adapted, then as richness increases, so too will the likely number within the acceptable range. In this more realistic case, connectance would be roughly constant. Moreover, if webs are made up of specialists, overall connectance will be low, whereas webs composed of generalists will have high connectance. The proportion of specialists may change with richness. Thus, the inconsistency of pattern may simply reflect a diversity of forces acting on different webs.

The prediction that populations in richer communities are less stable when disturbed can also be investigated experimentally. One classic study, for example, monitored the resistance in two grassland communities (McNaughton, 1978). In the first, plant nutrients were added to the soil of a community in New York State; in the second, the action of grazing animals was manipulated in the Serengeti. In both cases, the treatment was applied to species-rich and species-poor plant communities, and in both, disturbance reduced the diversity of the former but not the latter (Table 20.1). This was consistent with the prediction, but the effects, while significant, were relatively slight.

Similarly, Tilman (1996) pooled data for 39 common plant species from 207 grassland plots in Cedar Creek Natural History Area, Minnesota, over an 11-year period. He found that variation in the biomass of individual species increased significantly, but only very weakly, with the richness of the plots (Figure 20.10a).

Finally, there have been a number of studies directed at the question of whether the level of 'perceived stability' of natural populations (interannual variation in abundance) varies with the richness or complexity of the community. Leigh (1975) for herbivorous vertebrates, Bigger (1976) for crop pests and Wolda (1978) for insects, all failed to find evidence that it did so.

Overall, therefore, like the theoretical studies, empirical studies hint at decreased population stability (increased variability) in more complex communities, but the effect seems to be weak and inconsistent.

no consistent answers

20.3.6 Complexity and stability in practice: whole communities

Turning to the aggregate, whole community level, evidence is largely consistent in supporting the prediction that increased richness in a community increases stability (decreases variability), though a number of studies have failed to detect any consistent relationship (Cottingham *et al.*, 2001; Worm & Duffy, 2003).

First, returning to McNaughton's (1978) studies of US and Serengeti grasslands, the effects of perturbations were quite different when viewed in ecosystem (as opposed to population) terms. The addition of fertilizer significantly increased primary productivity in the species-poor field in New York State (+53%), but only slightly and insignificantly changed productivity in the species-rich field (+16%); and grazing in the Serengeti significantly reduced the standing crop biomass in the species-poor grassland (−69%), but only slightly reduced that of the species-rich field (−11%). Similarly, in Tilman's (1996) Minnesota grasslands, in contrast to the weak negative effect found at the population level, there was a strong positive effect of richness on the stability of community biomass (Figure 20.10b).

data support the models: aggregates are more stable in richer communities

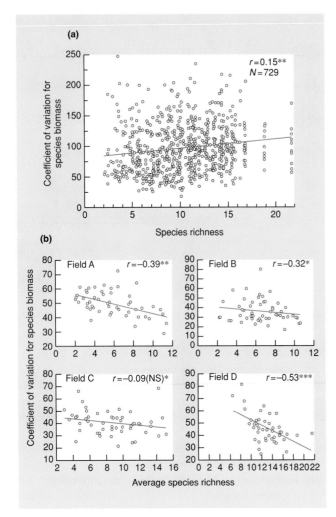

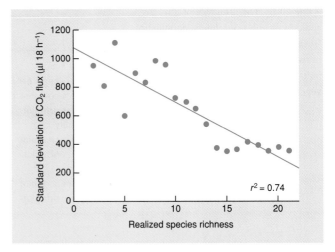

Figure 20.11   Variation (i.e. 'instability') in productivity (standard deviation of carbon dioxide flux) declined with species richness in microbial communities observed over a 6-week period. Richness is described as 'realized' because it refers to the number of species present at the time of the observation, irrespective of the number of species with which the community was initiated. (After McGrady-Steed *et al.*, 1997.)

Figure 20.10   (a) The coefficient of variation of population biomass for 39 plant species from plots in four fields in Minnesota over 11 years (1984–94) plotted against species richness in the plots. Variation increased with richness but the slope was very shallow. (b) The coefficient of variation for community biomass in each plot plotted against species richness for each of the four fields. Variation consistently decreased with richness. In both cases, regression lines and correlation coefficients are shown. *, $P < 0.05$; **, $P < 0.01$; ***, $P < 0.001$. (After Tilman, 1996.)

McGrady-Steed *et al.* (1997) manipulated richness in aquatic microbial communities (producers, herbivores, bacterivores and predators) and found that variation in another ecosystem measure, carbon dioxide flux (a measure of community respiration) also declined with richness (Figure 20.11). On the other hand, in an experimental study of small grassland communities perturbed by an induced drought, Wardle *et al.* (2000) found detailed community composition to be a far better predictor of stability than overall richness.

Studies of the response of a community to a perturbation (e.g. McNaughton, 1978) or of variations in the community in response to year-to-year variations in the environment (e.g. Tilman, 1996), are focused largely on the resistance of communities to change. A quite different perspective examines the resilience of communities to perturbations in ecosystem characteristics such as the energy or nutrient levels contained within them. O'Neill (1976), for example, considered the community as a three-compartment system consisting of active plant tissue (P), heterotrophic organisms (H) and inactive dead organic matter (D). The rate of change in the standing crop in these compartments depends on transfers of energy between them (Figure 20.12a). Inserting real data from six communities representing tundra, tropical forest, temperate deciduous forest, a salt marsh, a freshwater spring and a pond, O'Neill subjected the *models* of these communities to a standard perturbation: a 10% decrease in the initial standing crop of active plant tissue. He then monitored the rates of recovery towards equilibrium, and plotted these as a function of the energy input per unit standing crop of living tissue (Figure 20.12b).

The pond system, with a relatively low standing crop and a high rate of biomass turnover, was the most resilient. Most of its plant populations have short lives and rapid rates of population increase. The salt marsh and forests had intermediate values, whilst tundra had the lowest resilience. There is a clear relationship

importance of the nature – not just the richness – of the community

cycling rather than energy flow. Here too, then, stability seems more influenced by the nature of the species in the community than by simple measures such as overall richness.

20.3.7 The number of species or their identity? Keystones again

Indeed, it is clear that the whole concept of a keystone species (see Section 20.2.6) is itself a recognition of the fact that the effects of a disturbance on structure or function are likely to depend very much on the precise nature of the disturbance – that is, on *which* species are lost. Reinforcement of this idea is provided by a simulation study carried out by Dunne *et al.* (2002), in which they took 16 published food webs and subjected them to the sequential removal of species according to one of four criteria: (i) removing the most connected species first; (ii) randomly removing species; (iii) removing the most connected species first excluding basal species (those having predators but no prey); and (iv) removing the least connected species first. The stability of the webs was then judged by the number of secondary extinctions that resulted from the simulated removals, such extinctions occurring when species were left with no prey (and so basal species were subject to primary but not secondary extinction). In the first place, the robustness of community composition in the face of species loss increased with connectance of the communities – further support for an increase in community stability with complexity. Overall, however, it is also clear that secondary extinctions followed most rapidly when the most connected species were removed, and least rapidly when the least connected species were removed, with random removals lying between the two (Figure 20.13). There were, moreover, some interesting exceptions when, for example, the removal of a least connected species led to a rapid cascade of secondary extinctions because it was a basal species with a single predator, which was itself preyed upon by a wide variety of species. This, finally in this section, reminds us that the idiosyncrasies of individual webs are likely always to undermine the generality of any 'rules' even if such rules can be agreed on.

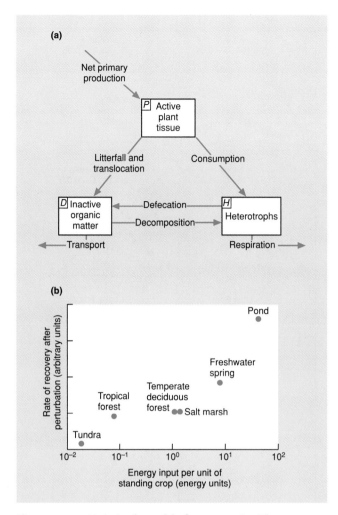

Figure 20.12   (a) A simple model of a community. The three boxes represent components of the system and arrows represent transfers of energy between the system components. (b) The rate of recovery (index of resilience) after perturbation (as a function of energy input per unit standing crop) for models of six contrasting communities. The pond community was most resilient to perturbation, tundra least so. (After O'Neill, 1976.)

20.4 Empirical patterns in food webs: the number of trophic levels

In the previous section, we examined very general aspects of food web structure – richness, complexity – and related them to the stability of food webs. In this section, we examine some more specific aspects of structure and ask, first, if there are detectable repeated patterns in nature, and second whether we can account for them. We deal first, at greatest length, with the number of trophic levels, and then turn to omnivory and the extent to which food webs are compartmentalized.

between resilience and energy input per unit standing crop. This seems to depend in part on the relative importance of heterotrophs in the system. The most resilient system, the pond, had a biomass of heterotrophs 5.4 times that of autotrophs (reflecting the short life and rapid turnover of phytoplankton, the dominant plants in this system), whilst the least resilient tundra had a heterotroph : autotroph ratio of only 0.004. Thus, the flux of energy through the system has an important influence on resilience. The higher this flux, the more quickly will the effects of a perturbation be 'flushed' from the system. An exactly analogous conclusion has been reached by DeAngelis (1980), but for nutrient

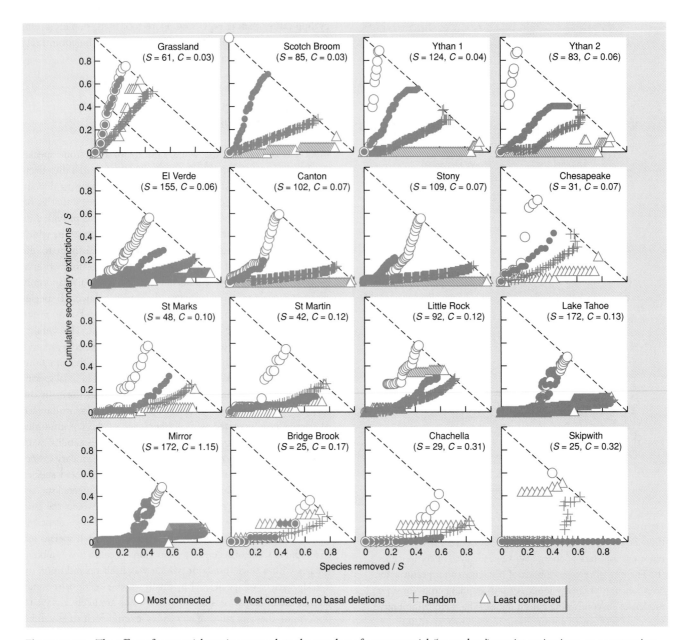

Figure 20.13 The effect of sequential species removal on the number of consequential ('secondary') species extinctions, as a proportion of the total number of species originally in the web, S, for each of 16 previously described food webs. The four different rules for species removal are described in the key. Robustness of the webs (the tendency *not* to suffer secondary extinctions) increased with the connectance of the webs, C (regression coefficients for the four rules: -0.62 (NS), 1.16 ($P < 0.001$), 1.01 ($P < 0.001$) and 0.47 ($P < 0.005$)). Overall, though, robustness was lowest when the most connected species were removed first and highest when the least connected were removed first. The origins of the webs are described in Dunne *et al.* (2002). (After Dunne *et al.*, 2002.)

food chain length

A fundamental feature of any food web is the number of trophic links in the pathways that run from basal species to top predators. Variations in the number of links have usually been investigated by examining *food chains*, defined as sequences of species running from a basal species to a species that feeds on it, to another species that feeds on the second, and so on up to a top predator (fed on by no other species). This does not imply a belief that communities are organized as linear chains (as opposed to more diffuse webs); rather, individual chains are identified purely as a means

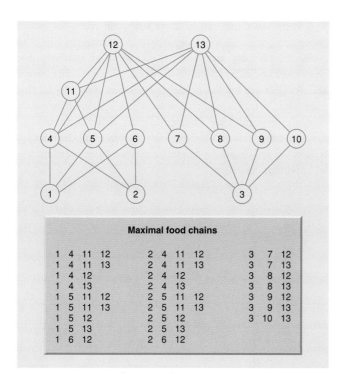

	Maximal food chains							
1	4	11	12	2	4	11	12	
1	4	11	13	2	4	11	13	
1	4	12		2	4	12		
1	4	13		2	4	13		
1	5	11	12	2	5	11	12	
1	5	11	13	2	5	11	13	
1	5	12		2	5	12		
1	5	13		2	5	13		
1	6	12		2	6	12		

(maximal food chains, third column:)

3	7	12
3	7	13
3	8	12
3	8	13
3	9	12
3	9	13
3	10	13

Figure 20.14 Community matrix for an exposed intertidal rocky shore in Washington State, USA. The pathways of all possible maximal food chains are listed. 1, detritus; 2, plankton; 3, benthic algae; 4, acorn barnacles; 5, *Mytilus edulis*; 6, *Pollicipes*; 7, chitons; 8, limpets; 9, *Tegula*; 10, *Littorina*; 11, *Thais*; 12, *Pisaster*; 13, *Leptasterias*. (After Briand, 1983.)

of trying to quantify the number of links. Food chain length has been defined in various ways (Post, 2002), and in particular has sometimes been used to describe the number of species in the chain, and sometimes (as here) the number of links. For instance, starting with basal species 1 in Figure 20.14, we can trace four possible trophic pathways via species 4 to a top predator: 1–4–11–12, 1–4–11–13, 1–4–12 and 1–4–13. This provides four food chain lengths: 3, 3, 2 and 2. Figure 20.14 lists a total of 21 further chains, starting from basal species 1, 2 and 3. The average of all the possible food chain lengths is 2.32. Adding one to this gives us the number of trophic levels that can be assigned to the food web. Almost all communities described have consisted of between two and five trophic levels, and most of these have had three or four. What sets the limit on food chain length? And how can we account for variations in length?

parasites are usually ignored

In addressing these questions, we will conform to a bias that has pervaded investigations of food chain length – a bias in favor of predators and against parasites. Thus, when a food chain is described as having four trophic levels, these would typically be a plant, a herbivore, a predator that eats the herbivore, and a top predator that eats the intermediate predator. Assume the top predator is an eagle. Even without collecting the data, it is all but certain that the eagle is attacked by parasites (perhaps fleas), which are themselves attacked by pathogens. But the convention is to describe the chain as having four trophic levels. Indeed, descriptions of food webs generally have paid little attention to parasites. There is little doubt that this neglect will have to be rectified (Thompson *et al.*, 2005).

20.4.1 Productivity? Productive space? Or just space?

It has long been argued that energetic considerations set a limit to the number of trophic levels that an environment can support. Of the radiant energy that reaches the earth, only a small fraction is fixed by photosynthesis and made available as either live food for herbivores or dead food for detritivores. Indeed, the amount of energy available for consumption is considerably less than that fixed by the plants, because of work done by the plants (in growth and maintenance) and because of losses due to inefficiencies in all energy-conversion processes (see Chapter 17). Thereafter, each feeding link amongst heterotrophs is characterized by the same phenomenon: at most 50%, sometimes as little as 1%, and typically around 10% of energy consumed at one trophic level is available as food to the next. The observed pattern of just three or four trophic levels could arise, therefore, simply because a viable population of predators at a further trophic level could not be supported by the available energy.

greater primary productivity supports more trophic levels? . . .

The most obvious testable predictions stemming from this hypothesis are, first, systems with greater primary productivity (e.g. at lower latitudes) should be able to support a larger number of trophic levels; and second, systems where energy is transferred more efficiently (e.g. based on insects rather than vertebrates) should also have more trophic levels. However, these predictions have received little support from natural systems. For instance, an analysis of 32 published food webs in habitats ranging from desert and woodland to Arctic lakes and tropical seas found no difference in the length of food chains when 22 webs from low-productivity habitats (less than 100 g of carbon m^{-2} $year^{-1}$) were compared with 10 webs from high-productivity habitats (greater than 1000 g m^{-2} $year^{-1}$). The median food chain length was 2.0 in both cases (Briand & Cohen, 1987). Moreover, a survey of 95 insect-dominated webs revealed first that food chains in tropical webs were no longer than those from (presumably) less productive temperate and desert situations, but also that these food chains composed of insects were no longer than those involving vertebrates (Schoenly *et al.*, 1991).

On the other hand, a number of studies on a much smaller scale (e.g. in a group of streams; Townsend *et al.*, 1998) or where resource availability has been manipulated experimentally, have shown food chain length to decrease with decreased productivity, especially when the decreases take productivity below around 10 g carbon m^{-2} $year^{-1}$ (Post, 2002). For example, in an experiment using water-filled containers as analogs of natural tree-holes, a 10-fold or 100-fold reduction from a 'natural' level of energy input (leaf litter) reduced maximal food chain length by one link, because in this simple community of mosquitoes, midges, beetles and mites, the principal predator – a chironomid midge *Anatopynia pennipes* – was usually absent from the less productive habitats (Jenkins *et al.*, 1992). This suggests that the simple productivity argument may indeed apply in the least productive environments (the most unproductive deserts, the deepest reaches of caves). However, establishing this is likely to prove difficult, since there are other reasons for expecting top predators to be absent from such environments (their size, their isolation, etc.; Post, 2002).

> . . . or should it
> be *total* available
> energy?

In fact, though, the simple productivity argument may have been misguided in the first place: what matters in an ecological community is not the energy available per unit area but the *total* available energy, that is, productivity per unit area multiplied by the space (or volume) occupied by the ecosystem – the 'productive space' hypothesis (Schoener, 1989). A very small and isolated habitat, for example, no matter how productive locally, is unlikely to provide enough energy for viable populations at higher trophic levels. A number of studies appear to support the productive space hypothesis, in that the number of trophic levels is positively correlated with the total available energy – an example is shown in Figure 20.15a. On the other hand, the rare attempts that have been made to determine the separate contributions of ecosystem size and local productivity have detected an effect from size but not from productivity (e.g. Figure 20.15b).

Results like these may indicate that total energy is indeed important but is far more dependent on ecosystem size than productivity per unit area. But they may mean, alternatively, that ecosystem size affects food chain length by some other means and available energy has no detectable effect (Post, 2002). One possibility is that ecosystem size affects species richness (it certainly does so – see Chapter 21) and richer webs tend to support longer chains. Unsurprisingly, richness and chain length tend to be associated. Untangling causation from correlation is an important challenge.

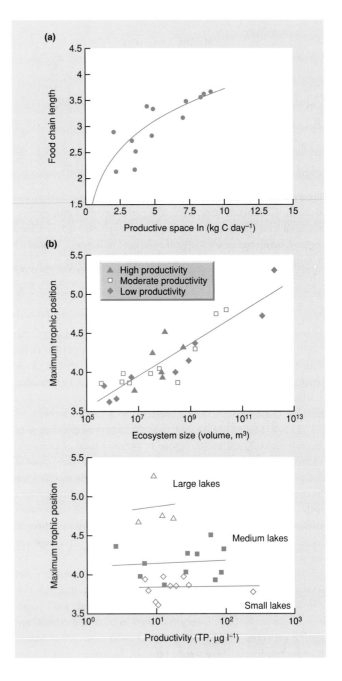

Figure 20.15 (*right*) (a) The food chain length (FCL) increases with productive space for the food webs of 14 lakes in Ontario and Quebec; productive space (PS) = productivity × lake area; FCL = $2.94PS^{0.21}$, $r^2 = 0.48$. (After Vander Zanden *et al.*, 1999.) (b) Relationships between maximum trophic position and ecosystem size (above) or productivity (below) for 25 lakes in northeastern North America. The maximum trophic position increased with ecosystem size apparently independently of whether productivity was low (2–11 μg l^{-1} total phosphorus (TP)), moderate (11–30 μg l^{-1} TP) or high (30–250 μg l^{-1} TP). However, when small (3×10^5 to 3×10^7 m^3), medium (3×10^7 to 3×10^9 m^3) and large lakes (3×10^9 to 3×10^{12} m^3) were examined separately, the maximum trophic position did not vary with productivity. The maximum trophic position is the trophic position (FCL + 1) of the species with the highest average trophic position in each of the lake food webs. (After Post *et al.*, 2000.)

If available energy is found ultimately to have no effect on food chain length, it should perhaps be borne in mind that species richness is usually significantly higher in productive regions (see Chapter 21), and that each consumer probably feeds on only a limited range of species at a lower trophic level. Hence, the amount of energy flowing up a single food chain in a productive region (a large amount of energy, but divided amongst many subsystems) may not be very different from that flowing up a single food chain in an unproductive region (having been divided amongst fewer subsystems).

20.4.2 Dynamic fragility of model food webs

Another popular idea has been that the length of food chains is limited by the lowered stability (especially resilience) of longer chains. In turn, we might then expect food chains to be shorter in environments subject to greater disturbance, where only the most stable food chains could persist. In particular, when Pimm and Lawton (1977) examined variously structured four-species Lotka–Volterra models (Figure 20.16a), webs with more trophic levels had return times after a perturbation that were substantially longer than those with fewer levels. Because less resilient systems are unlikely to persist in an inconstant environment, it was argued that only systems with few trophic levels will commonly be found in nature. However, these models had self-limitation (effectively, intraspecific competition) only at the lowest trophic level, and food chain length and the proportion of self-limited species was therefore confounded (Figure 20.16a). When a wider range of food webs was examined with self-limitation distributed more systematically (Figure 20.16b–e) (Sterner *et al.*, 1997a), there was a weak but significant *increase* in stability in longer food chains when the number of species and the number of self-limited species were held constant. Overall, there is no convincing case for dynamic fragility affecting the length of food chains significantly.

20.4.3 Constraints on predator design and behavior

There may also be evolutionary constraints on the anatomy or behavior of predators that limit the lengths of food chains. To feed on prey at a given trophic level, a predator has to be large enough, maneuverable enough and fierce enough to effect a capture. In general, predators are larger than their prey (not true, though, of grazing insects and parasites), and body size tends to increase (and density to decrease) at successive trophic levels (Cohen *et al.*, 2003). There may well be a limit above which design constraints rule out another link in the food chain. It may be impossible to design a predator that is both fast enough to catch an eagle and big and fierce enough to kill it.

Also, consider the arrival in a community of a new carnivore species. Would it do best to feed on the herbivores or the

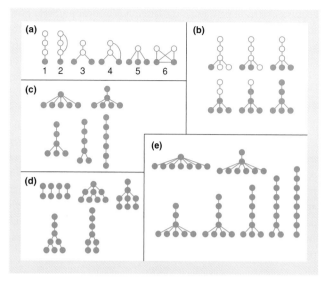

Figure 20.16 Sets of model food webs, the dynamics of which were examined to determine the effect of food chain length on stability having accounted for variations in the number of species and the number with self-limitation (●). (a) The original set examined by Pimm and Lawton (1977). (b) Six-species, four-level webs with varying degrees of self-limitation. (c) Six-species webs of self-limited species with varying numbers of trophic levels and species concentrated in the basal level. (d) Eight-species webs of self-limited species with varying numbers of trophic levels and species dispersed among the levels. (e) Eight-species webs of self-limited species with varying numbers of trophic levels and species concentrated in the basal level. (After Sterner *et al.*, 1997a.)

carnivores already there? The herbivores are more abundant and less well protected. The advantage to feeding low down in the food chain can readily be seen. Of course, if all species did this, competition would intensify, and feeding higher in the food chain could reduce competition. But it is difficult to imagine a top predator sticking religiously to a rule that it should prey only on the trophic level immediately below it, especially as the prey there are likely to be larger, fiercer and rarer than species at lower levels. Overall, theoretical explorations (Hastings & Conrad, 1979) suggest that an evolutionarily stable food chain length (one that would be optimal for predator fitness) would be around two (three trophic levels). Such arguments, however, have rather little to offer by way of explanation for the variations in food chain length.

Thus, there are complete answers to neither of our original questions (see p. 595). The constraints on predators are likely to set some general upper limit on the lengths of many food chains. Food chains are likely to be atypically short in especially unproductive environments. Food chain length seems to increase with increases in productive space, but it is unclear whether this is an association with the total energy available in an ecosystem or with ecosystem size alone – and if the latter, it is unclear precisely how size comes to determine food chain length. The two longest established hypotheses – energy per unit area and dynamic fragility – have, if anything, the least support.

are the data simply not good enough?

Finally, it is important to note that, as with connectance, estimates of food chain length are sensitive to the degree of taxonomic resolution. This may be why many of the more recently documented webs have longer than average chain lengths ranging from five to seven (Hall & Raffaelli, 1993). Moreover, if a well-resolved large web is progressively simplified by lumping taxa together (in a manner analogous to earlier studies), the estimate of food chain length declines (Martinez, 1993). There is clearly a need for rigorous studies of many more food webs before acceptable generalizations can be reached.

20.4.4   Omnivory

Technically, an omnivore is an animal that takes prey from more than one trophic level. Compilations of early descriptions of food webs indicated that omnivores are usually uncommon; this was taken to support expectations from simple model communities, where omnivory is destabilizing (Pimm, 1982). It was argued that in cases of omnivory, intermediate species both compete with and are preyed upon by top species and in consequence are unlikely to persist long. A more complex and realistic model incorporated 'life history omnivory', in which different life history stages of a species feed on different trophic levels, as when tadpoles are herbivores and adult frogs and toads are carnivores (Pimm & Rice, 1987). Life history omnivory also reduces stability, but much less than single life stage omnivory does. Intriguingly, omnivory is not destabilizing in donor-control models, and omnivores are common in decomposer food webs (Walter, 1987; Usio & Townsend, 2001; Woodward & Hildrew, 2002), to which donor-control dynamics can be applied.

In fact, an increasing number of studies indicate that omnivory is not uncommon at all, and that earlier indications of its rarity were an artefact of the webs being only poorly described (Polis & Strong, 1996; Winemiller, 1996). For example, Sprules and Bowerman (1988) found omnivory to be common in plankton food webs in North American glacial lakes, having identified all their zooplankton to species level and produced webs that were much more reliable as a result (Figure 20.17). Polis (1991) found similar results in his detailed study of a desert sand community. What is more, later modeling studies have undermined the whole suggestion that omnivory is inherently destabilizing. Dunne *et al.*'s (2002) simulation study detected no relationship between the level of omnivory and the stability of webs to species removal, while other models indicate that omnivory may in fact stabilize food webs (McCann & Hastings, 1997). It is sobering to note that

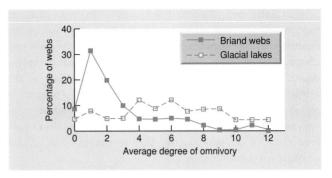

Figure 20.17   The prevalence of omnivory in glacial lakes in northeast North America (Sprules & Bowerman, 1988) is much greater than that observed in Briand's set of food webs (see Figure 20.9a). The degree of omnivory in a web is quantified as the number of closed omnivorous links divided by the number of top predators. A closed omnivorous link exists when a feeding path can be traced to a prey more than one trophic level away, and from that prey back to the predator through at least one other prey occupying an intermediate trophic level.

theoretical and empirical studies have managed to march in step twice in quick succession, but to quite different tunes. It reminds us that both sorts of study can only ever be as good as the assumptions on which they are inevitably based.

20.4.5   Compartmentalization

A food web is compartmentalized if it is organized into subunits within which interactions are strong, but between which interactions are weak. (The most perfectly compartmentalized community possesses only linear food chains.) Do food webs tend to be compartmentalized?

Not surprisingly, in studies where habitat divisions are major and unequivocal, there is a clear tendency for compartments to map onto habitats. For instance, Figure 20.18 shows the results of a classic study describing the major interactions within and between three interconnected habitats on Bear Island in the Arctic Ocean (Summerhayes & Elton, 1923). There is a significantly smaller number of interactions between habitats than would be expected by chance (Pimm & Lawton, 1980).

On the other hand, when habitat divisions are subtler, the evidence for compartments is typically poor, and there are even greater difficulties in providing a clear demonstration of compartments (or the lack of them) *within* habitats. Early analyses, certainly, suggested that food webs within habitats are only as compartmentalized as would be expected by chance alone (Pimm & Lawton, 1980; Pimm, 1982). More recently, though, promising methodological advances have been made that seem capable of identifying compartments within larger webs, especially when the

Figure 20.18 The major interactions within and between three interconnected habitats on Bear Island in the Arctic Ocean. 1, plankton; 2, marine animals; 3, seals; 4a, plants; 4b, dead plants; 5, worms; 6, geese; 7, Collembola; 8, Diptera; 9, mites; 10, Hymenoptera; 11, seabirds; 12, snow bunting; 13, purple sandpiper; 14, ptarmigan; 15, spiders; 16, ducks and divers; 17, Arctic fox; 18, skua and glaucous gull; 19, planktonic algae; 20a, benthic algae; 20b, decaying matter; 21, protozoa a; 22, protozoa b; 23, invertebrates a; 24, Diptera; 25, invertebrates b; 26, microcrustacean; 27, polar bear. (After Pimm & Lawton, 1980.)

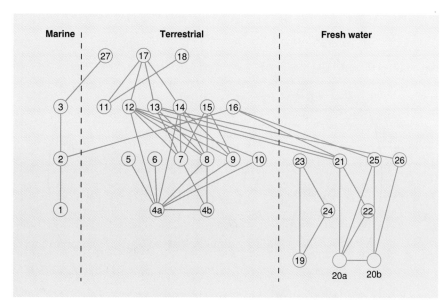

taxonomic resolution within the web is high and the strengths of interactions between the species can be weighted (Krause *et al.*, 2002). Interestingly, the methods lean heavily on ideas from sociology, where the aim is to identify social cliques within a broader society. An example is shown in Figure 20.19. Also, an alternative perspective has been to emphasize that what have been described as distinct food webs in different habitats may often be linked by 'spatial subsidies' – crucial flows of energy and materials (Polis *et al.*, 1997) – as, for example, when lake fish that normally prey upon other fish in the pelagic (open water) food web, switch to quite different prey in the benthic food web when their preferred prey are scarce (Schindler & Scheuerell, 2002). That is, what might seem to be separate webs are in fact compartments within a larger web.

Since no clear consensus has emerged that food webs are more compartmentalized than would be expected by chance alone, it would be inappropriate to argue that compartmentalization has been 'favored' because compartmentalized webs persist. None the less, since the earliest theoretical studies (e.g. May, 1972), a consensus *has* emerged that communities will have increased stability if they are compartmentalized, and it is easy to see why this might be so. In the first place, a disturbance to a compartmentalized web tends to be contained within the disturbed compartment, limiting the overall extent of the effects in the wider web. In addition, though, spatial subsidies between compartments will tend to buffer individual compartments against the worst excesses of disturbances within them. For instance, in the example above, piscivorous fish, when their preferred prey are rare, may switch to the benthos rather than driving populations of those preferred prey to extinction. The apparent contradiction between these two justifications of the stabilizing properties of compartmentalization can be resolved if we emphasize the first where a

seemingly unified web is in fact a series of semidetached compartments, and emphasize the second where seemingly separate webs are in fact coupled. Thus, it may be that an intermediate degree of compartmentalization is the most stable.

This chapter closes, then, with a tone that has pervaded much of it: suggestive but uncertain. Further progress, though, is essential. One standard answer of ecologists to the layman's question 'What does it matter if we lose *that* species?' is, quite rightly, 'But you must also consider the wider effects of that loss; losing that species may affect the whole food web of which it is part'. The need for further understanding of those wider effects is intense.

answers are uncertain – but it is important that we discover them

Summary

In this chapter, we shift the focus to systems that usually have at least three trophic levels and with 'many' species.

We describe 'unexpected' effects in food webs, where, for example, the removal of a predator may lead to a decrease in prey abundance.

The indirect effect within food webs that has received most attention is the trophic cascade. We discuss cascades in systems with three and four trophic levels, and address the question of whether cascades are equally common in all types of habitat, requiring a distinction to be made between community- and species-level cascades. We ask whether food webs, or particular types of food web, are dominated by either top-down (trophic cascade) or bottom-up control. We then define and discuss the importance of keystone species.

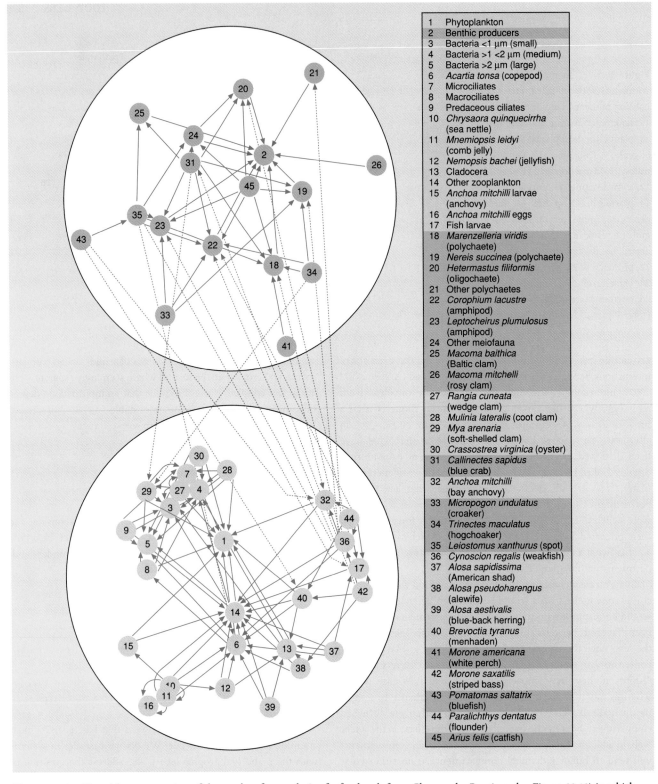

Figure 20.19 Pictorial representation of the results of an analysis of a food web from Chesapeake Bay (see also Figure 20.13) in which interactions between the 45 taxa were quantified and the taxa assigned to compartments (the number of which was not predetermined) in such a way as to maximize the differential between the connectance within compartments (in this case 0.0099) and that between compartments (in this case 0.000087, more than two orders of magnitude lower). Food webs may be considered compartmentalized if that differential is sufficiently large. Arrows represent interactions and point from predator to prey: solid color, within compartments; dashed lines, between compartments. (After Krause *et al.*, 2002.)

Any ecological community can be characterized by its structure, its productivity and its temporal stability. The variety of meanings of 'stability' is outlined, distinguishing resilience and resistance, local and global stability, and dynamic fragility and robustness.

For many years, the 'conventional wisdom' was that more complex communities were more stable. We describe the simple mathematical models that first undermined this view. We show how, in general, the effects of food web complexity on population stability in model systems has been equivocal, whereas for aggregate properties of whole model communities, such as their biomass or productivity, complexity (especially species richness) tends consistently to enhance stability.

In real communities, too, evidence is equivocal at the population level, including both studies that have examined the relationships between species richness and connectance and those that have manipulated richness experimentally. Again, turning to the aggregate, whole community level, evidence is largely consistent in supporting the prediction that increased richness increases stability (decreases variability). We stress, though, the importance of the nature, not just the richness, of a community in these regards, returning to the importance of keystone species.

Limitations and patterns in food chain length are discussed. We examine the evidence that food chain length is limited by productivity, by 'productive space' (productivity compounded by the extent of the community) or simply by 'space' – but that evidence is inconclusive. We examine, too, the arguments that food chain length is limited by dynamic fragility (ultimately unconvincing) or by constraints on predator design and behavior. There is a clear need for rigorous studies of many more food webs before acceptable generalizations can be reached.

We examine work linking the prevalence of omnivory and its effect on food web stability, noting that earlier work found omnivory to be rare and destabilizing, whereas later work found it common and with no consistent effect on stability.

Finally, we ask whether food webs tend to be more compartmentalized than would be expected by chance. As long as habitat divisions are subtle, the evidence for compartments is typically poor, and there are even greater difficulties in demonstrating compartments (or the lack of them) within habitats. There is, though, a clear consensus from theoretical studies that communities will have increased stability if they are compartmentalized.

Chapter 21
Patterns in Species Richness

21.1 Introduction

hot spots of species richness

Why the number of species varies from place to place, and from time to time, are questions that present themselves not only to ecologists but to anybody who observes and ponders the natural world. They are interesting questions in their own right – but they are also questions of practical importance. A remarkable 44% of the world's plant species and 35% of vertebrate species (other than fish) are endemic to just 25 separate 'hot spots' occupying a small proportion of the earth's surface (Myers *et al.*, 2000). Knowledge of the spatial distribution of species richness is a prerequisite for prioritizing conservation efforts both at a large scale (setting global priorities) and at a regional and local scale (setting national priorities). This aspect of conservation planning will be discussed in Section 22.4.

biodiversity and species richness

It is important to distinguish between *species richness* (the number of species present in a defined geographical unit – see Section 16.2) and *biodiversity*. The term biodiversity makes frequent appearances in both the popular media and the scientific literature – but it often does so without an unambiguous definition. At its simplest, biodiversity is synonymous with species richness. Biodiversity, though, can also be viewed at scales smaller and larger than the species. For example, we may include genetic diversity within species, recognizing the value of conserving genetically distinct subpopulations and subspecies. Above the species level, we may wish to ensure that species without close relatives are afforded special protection, so that the overall evolutionary variety of the world's biota is maintained as large as possible. At a larger scale still, we may include in biodiversity the variety of community types present in a region – swamps, deserts, early and late stages in a woodland succession, and so on. Thus, 'biodiversity' may itself, quite reasonably, have a diversity of meanings. Yet it is necessary to be specific if the term is to be of any practical use.

In this chapter we restrict our attention to species richness, partly because of its fundamental nature but mainly because so many more data are available for this than for any other aspect of biodiversity. We will address several questions. Why do some communities contain more species than others? Are there patterns or gradients of species richness? If so, what are the reasons for these patterns? There are plausible answers to the questions we ask, but these answers are by no means conclusive. Yet this is not so much a disappointment as a challenge to ecologists of the future. Much of the fascination of ecology lies in the fact that many of the problems are blatant, whereas the solutions are not. We will see that a full understanding of patterns in species richness must draw on our knowledge of all the ecological topics dealt with so far in this book.

the question of scale: macroecology

As with other areas of ecology, scale is a paramount feature in discussions of species richness; explanations for patterns usually have both smaller and larger scale components. Thus, the number of species living on a boulder in a river will reflect local influences such as the range of microhabitats provided (on the surface, in crevices and beneath the boulder) and the consequences of species interactions taking place (competition, predation, parasitism). However, larger scale influences of both a spatial and temporal nature will also be important. Thus, species richness may be large on our boulder because the regional pool of species is itself large (in the river as a whole or, at a still larger scale, in the geographic region) or because there has been a long interlude since the boulder was last turned over by a flood (or since the region was last glaciated). Comparatively more emphasis has been placed on local as opposed to regional questions in ecology, prompting Brown and Maurer (1989) to designate a subdiscipline of ecology as *macroecology* – to deal explicitly with understanding distribution and abundance at large spatial and temporal scales. Geographic patterns in species richness are a principal focus of macroecology (e.g. Gaston & Blackburn, 2000; Blackburn & Gaston, 2003).

21.1.1 Four types of factor affecting species richness

geographic factors There are a number of factors to which the species richness of a community can be related, and these are of several different types. First, there are factors that can be referred to broadly as 'geographic', notably latitude, altitude and, in aquatic environments, depth. These have often been correlated with species richness, as we shall discuss below, but presumably they cannot be causal agents in their own right. If species richness changes with latitude, then there must be some other factor changing with latitude, exerting a direct effect on the communities.

factors correlated with latitude A second group of factors does indeed show a tendency to be correlated with latitude (or altitude or depth), but they are not perfectly correlated. To the extent that they are correlated at all, they may play a part in explaining latitudinal and other gradients. But because they are not perfectly correlated, they serve also to blur the relationships along these gradients. Such factors include climatic variability, the input of energy, the productivity of the environment, and possibly the 'age' of the environment and the 'harshness' of the environment.

factors that are independent of latitude A further group of factors vary geographically but quite independently of latitude (or altitude, island location or depth). They therefore tend to blur or counteract relationships between species richness and other factors. This is true of the amount of physical disturbance a habitat experiences, the isolation of the habitat and the extent to which it is physically and chemically heterogeneous.

biotic factors Finally, there is a group of factors that are biological attributes of a community, but are also important influences on the structure of the community of which they are part. Notable amongst these are the amount of predation or parasitism in a community, the amount of competition, the spatial or architectural heterogeneity generated by the organisms themselves and the successional status of a community. These should be thought of as 'secondary' factors in that they are themselves the consequences of influences outside the community. Nevertheless, they can all play powerful roles in the final shaping of community structure.

A number of these factors have been discussed in previous chapters (disturbance and successional status in Chapter 16, competition, predation and parasitism in Chapter 19). In this chapter we continue by examining the relationships between species richness and factors that can be thought of as exerting an influence in their own right. We do this first by considering factors whose variation is primarily spatial (productivity, spatial heterogeneity, environmental harshness – Section 21.3) and, second, those whose variation is primarily temporal (climatic variation and environmental age – Section 21.4). We will then be in a position to consider patterns in species richness related to habitat area and remoteness (island patterns – Section 21.5), before moving to gradients in species richness related to latitude, altitude, depth, succession and position in the fossil record (Section 21.6). In Section 21.7, we take a different tack by asking whether variations in species richness themselves have consequences for the functioning of ecosystems (e.g. productivity, decomposition rate and nutrient cycling). We begin, though, by constructing a simple theoretical framework (following MacArthur (1972), probably the greatest macroecologist, although he did not use the term) to help us think about variations in species richness.

21.2 A simple model of species richness

To try to understand the determinants of species richness, it will be useful to begin with a simple model. Assume, for simplicity, that the resources available to a community can be depicted as a one-dimensional continuum, R units long (Figure 21.1). Each species uses only a portion of this resource continuum, and these portions define the *niche breadths* (n) of the various species: the average niche breadth within the community is $\bar{n}$. Some of these niches overlap, and the overlap between adjacent species can be measured by a value o. The average niche overlap within the community is then $\bar{o}$. With this simple background, it is possible to consider why some communities should contain more species than others.

First, for given values of $\bar{n}$ and $\bar{o}$, a community will contain more species the larger the value of R, i.e. the greater the range of resources (Figure 21.1a). This is true when the community is dominated by competition and the species 'partition' the resources (see Section 19.2). But, it will also presumably be true when competition is relatively unimportant. Wider resource spectra provide the means for existence of a wider range of species, whether or not those species interact with one another.

a model incorporating niche breadth, niche overlap and resource range

Second, for a given range of resources, more species will be accommodated if $\bar{n}$ is smaller, i.e. if the species are more specialized in their use of resources (Figure 21.1b).

Alternatively, if species overlap to a greater extent in their use of resources (greater $\bar{o}$), then more may coexist along the same resource continuum (Figure 21.1c).

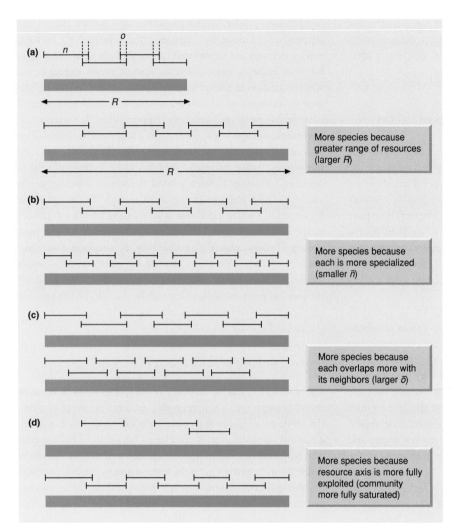

Figure 21.1 A simple model of species richness. Each species utilizes a portion *n* of the available resources (*R*), overlapping with adjacent species by an amount *o*. More species may occur in one community than in another (a) because a greater range of resources is present (larger *R*), (b) because each species is more specialized (smaller average *n*), (c) because each species overlaps more with its neighbors (larger average *o*), or (d) because the resource dimension is more fully exploited. (After MacArthur, 1972.)

Finally, a community will contain more species the more fully saturated it is; conversely, it will contain fewer species when more of the resource continuum is unexploited (Figure 21.1d).

21.2.1 The relationship between local and regional species richness

local vs regional richness – saturated or unsaturated communities?

One way to assess the degree to which communities are saturated with species is to plot the relationship between local species richness (assessed on a spatial scale where all the species could encounter each other in a community) and regional species richness (the number of species in the regional pool that could theoretically colonize the community). Local species richness is sometimes referred to as α richness (or α diversity) and regional species richness as γ richness. If communities are saturated with species

(i.e. the niche space is fully utilized), local richness will reach an asymptote in its relationship with regional richness (Figure 21.2a). This appears to be the case for the Brazilian ground-dwelling ant communities studied by Soares *et al.* (2001) (Figure 21.2b). Similar patterns have been described for aquatic and terrestrial plant groups, fish, mammals and parasites, but nonsaturating patterns have just as often been described for a variety of taxa, including fish (Figure 21.2c), insects, birds, mammals, reptiles, molluscs and corals (reviewed by Srivastava, 1999). Local regional richness plots provide a useful tool for addressing the question of community saturation, but they must be used with caution. For example, Loreau (2000) points out that the nature of the relationship depends on the way that total richness (γ) is partitioned between within-community (α) and between-community richness (β), and this is a matter of the scale at which different communities are distinguished from one another. In other words, researchers might erroneously include within a single community several habitats that should be considered as different communities, or, alternatively,

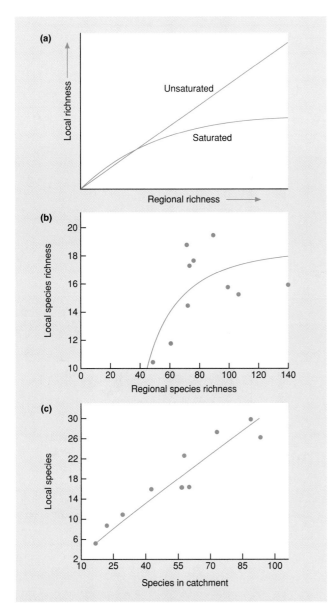

they may study local communities at an inappropriately small scale (e.g. 1 m² quadrats may have been too small to be 'local' communities in the ground-dwelling ant study of Soares *et al.*, 2001).

21.2.2 Species interactions and the simple model of species richness

We can also consider the relationship between the model in Figure 21.1 and two important kinds of species interactions described in previous chapters – interspecific competition and predation (see especially Chapter 19). If a community is dominated by interspecific competition, the resources are likely to be fully exploited. Species richness will then depend on the range of available resources, the extent to which species are specialists and the permitted extent of niche overlap (see Figure 21.1a–c).

Predation, on the other hand, is capable of exerting contrasting effects. First, we know that predators can exclude certain prey species; in the absence of these species the community may then be less than fully saturated, in the sense that some available resources may be unexploited (see Figure 21.1d). In this way, predation may reduce species richness. Second, though, predation may tend to keep species below their carrying capacities for much of the time, reducing the intensity and importance of direct interspecific competition for resources. This may then permit much more niche overlap and a greater richness of species than in a community dominated by competition (see Figure 21.1c). Finally, predation may generate richness patterns similar to those produced by competition when prey species compete for 'enemy-free space' (see Chapter 8). Such 'apparent competition' means that invasion and the stable coexistence of prey are favored by prey being sufficiently different from other prey species already present. In other words, there may be a limit to the similarity of prey that can coexist (equivalent to the presumed limits to similarity of coexisting competitors).

the role of competition

the role of predation

21.3 Spatially varying factors that influence species richness

21.3.1 Productivity and resource richness

For plants, the productivity of the environment can depend on whichever nutrient or condition is most limiting to growth (dealt with in detail in Chapter 17). Broadly speaking, the productivity of the environment for animals follows the same trends as for plants, both as a result of the changes in resource levels at the base of the food chain, and as a result of the changes in critical conditions such as temperature.

variations in productivity

Figure 21.2 (a) In a saturated community, local richness is expected to increase with regional richness at very low levels of regional richness, but to quickly reach an upper limit. In an unsaturated community, on the other hand, local richness is expected to be a constant proportion of regional richness. (After Srivastava, 1999.) (b) Asymptotic relationship between local richness of litter-dwelling ant communities in 1 m² quadrats in 10 forest remnants in Brazil in relation to the size of the regional species pool (assumed to be the total number of species in the forest remnant concerned). (After Soares *et al.*, 2001.) (c) Nonasymptotic relationship between local species richness (number recorded over equal-sized areas of a river bed) and regional species pools (the number of species present in the entire drainage basin from which the local sample was drawn). (After Rosenzweig & Ziv, 1999.)

If higher productivity is correlated with a wider *range* of available resources, then this is likely to lead to an increase in species richness (see Figure 21.1a). However, a more productive environment may have a higher rate of supply of resources but not a greater variety of resources. This might lead to more individuals per species rather than more species. Alternatively again, it is possible, even if the overall variety of resources is unaffected, that rare resources in an unproductive environment may become abundant enough in a productive environment for extra species to be added, because more specialized species can be accommodated (see Figure 21.1b).

increased
productivity might
lead to . . .
. . . increased
richness . . .

In general, then, we might expect species richness to increase with productivity – a contention that is supported by an analysis of the species richness of trees in North America in relation to a crude measure of available environmental energy, *potential* evapotranspiration (PET). This is the amount of water that would evaporate or be transpired from a saturated surface (Figure 21.3a). However, while energy (heat and light) is necessary for tree functioning, plants also depend critically on actual water availability; energy and water availability inevitably interact, since higher energy inputs lead to more evapotranspiration and a greater requirement for water (Whittaker *et al.*, 2003). Thus, in a study of southern African trees, species richness increased with water availability (annual rainfall), but first increased and then decreased with available energy (PET) (Figure 21.3b). We present and discuss further hump-shaped relationships later in this section.

When the North American work (Figure 21.3a) was extended to four vertebrate groups, species richness was found to be correlated to some extent with tree species richness itself. However, the best correlations were consistently with PET (Figure 21.4). Why should animal species richness be positively correlated with crude atmospheric energy? The answer is not known with any certainty, but it may be because for an ectotherm, such as a reptile, extra atmospheric warmth would enhance the intake and utilization of food resources. While for an endotherm, such as a bird, the extra warmth would mean less expenditure of resources in maintaining body temperature and more available for growth and reproduction. In both cases, then, this could lead to faster individual and population growth and thus to larger populations. Warmer environments might therefore allow species with narrower niches to persist and such environments may therefore support more species in total (see Figure 21.1b) (Turner *et al.*, 1996).

Sometimes there seems to be a direct relationship between animal species richness and plant productivity. This was the case, for example, for the relationship between bird species richness and mean annual net primary productivity in South Africa (van Rensburg *et al.*, 2002). In the cases of seed-eating rodents and seed-eating ants in the southwestern deserts of the United States,

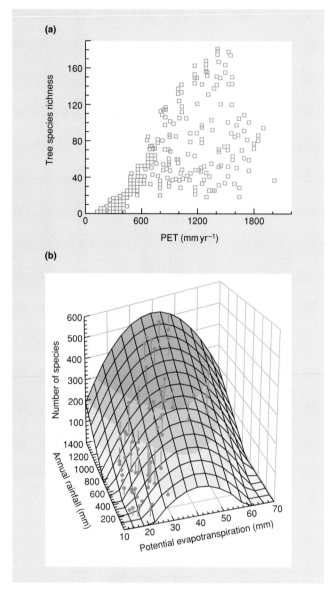

Figure 21.3 (a) Species richness of trees in North America, north of the Mexican border (in which the continent has been divided into 336 quadrats following lines of latitude and longitude) in relation to potential evapotranspiration (PET). (After Currie & Paquin, 1987; Currie, 1991.) (b) Species richness of southern African trees (in 25,000 km² cells) as a function of annual rainfall and PET. The surface describes the regression model between species richness, annual rainfall and PET, and the stalks show the residual variation associated with each data point. (After Whittaker *et al.*, 2003; data from O'Brien, 1993.)

Brown and Davidson (1977) recorded strong positive correlations between species richness and precipitation. In arid regions it is well established that mean annual precipitation is closely related to plant productivity and thus to the amount of seed resource

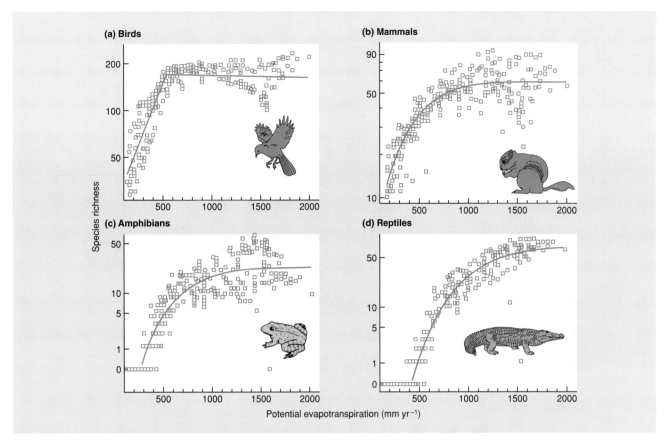

Figure 21.4   Species richness of (a) birds, (b) mammals, (c) amphibians, and (d) reptiles in North America in relation to potential evapotranspiration. (After Currie, 1991.)

available. It is particularly noteworthy that in species-rich sites, the communities contained more species of very large ants (which consume large seeds) and more species of very small ants (which take small seeds) (Davidson, 1977). It seems that either the range of sizes of seeds is greater in the more productive environments (see Figure 21.1a) or the abundance of seeds becomes sufficient to support extra consumer species with narrower niches (see Figure 21.1b).

... or decreased richness ...

On the other hand, an increase in diversity with productivity is by no means universal, as noted in the unique Parkgrass experiment which started in 1856 at Rothamstead in England (see Section 16.2.1). A 3.2 ha (8-acre) pasture was divided into 20 plots, two serving as controls and the others receiving a fertilizer treatment once a year. While the unfertilized areas remained essentially unchanged, the fertilized areas showed a progressive decline in species richness (and diversity).

Such declines have long been recognized. Rosenzweig (1971) referred to them as illustrating the 'paradox of enrichment'. One possible resolution of the paradox is that high productivity leads to high rates of population growth, bringing about the extinction of some of the species present because of a speedy conclusion to any potential competitive exclusion. At lower productivity, the environment is more likely to have changed before competitive exclusion is achieved. An association between high productivity and low species richness has been found in several other studies of plant communities (reviewed by Cornwell & Grubb, 2003).

It is perhaps not surprising, then, that several studies have demonstrated both an increase and a decrease in richness with increasing productivity – that is, that species richness may be highest at intermediate levels of productivity.

... or an increase then a decrease (hump-shaped relationships)

Species richness is low at the lowest productivities because of a shortage of resources, but also declines at the highest productivities where competitive exclusions speed rapidly to their conclusion. For instance, there are humped curves when the species richness of desert rodents is plotted against precipitation (and thus productivity) along a gradient in Israel (Abramsky & Rosenzweig, 1983), when the species richness of central European plants is plotted against soil nutrient supply (Cornwell & Grubb,

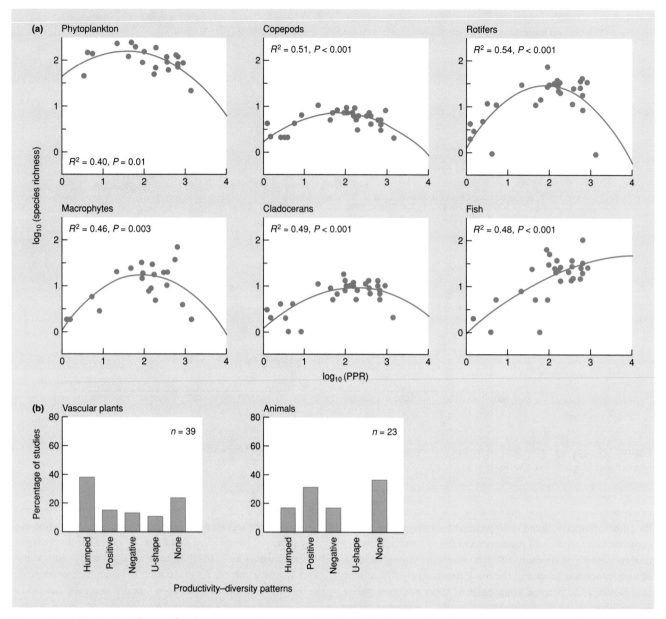

Figure 21.5   (a) Species richness of various taxonomic groups in lakes in North America plotted against gross primary productivity (PPR), with fitted quadratic regression lines (all significant at $P < 0.01$). (After Dodson *et al.*, 2000.) (b) Percentage of published studies on plants and animals showing various patterns of relationship between species richness and productivity. (After Mittelbach *et al.*, 2001.)

2003) and when the species richness of various taxonomic groups is plotted against gross primary productivity in the open water zones of lakes in North America (Figure 21.5a). An analysis of a wide range of such studies found that when communities of the same general type (e.g. tallgrass prairie) but differing in productivity were compared (Figure 21.5b), a positive relationship was the most common finding in animal studies (with fair numbers of humped and negative relationships), whereas with plants, humped relationships were the most common, with smaller numbers of

positives and negatives (and even some unexplained U-shaped curves). When Venterink *et al.* (2003) assessed the relationship between plant species richness and plant productivity in 150 European wetland sites that differed in the nutrient that was limiting productivity (nitrogen, phosphorus or potassium), they found hump-shaped patterns for nitrogen- and phosphorus-limited sites but species richness declined monotonically with productivity in potassium-limited sites. Clearly, increased productivity can and does lead to increased or decreased species richness – or both.

productivity may affect species richness in combination with other factors

Productivity often, perhaps always, exerts its influence on species richness in combination with other factors. Thus, we saw earlier how grazer-mediated coexistence was most likely to occur in nutrient-rich situations where plant productivity is high, whereas grazing in nutrient-poor, unproductive settings was associated with a reduction in plant richness (see Section 19.4). Moreover, disturbance (dealt with in Chapter 16) can also interact with nutrient supply (productivity) to determine species richness patterns. Wilson and Tilman (2002) monitored for 8 years the effects of four levels each of disturbance (different amounts of annual tilling) and nitrogen addition (in a complete factorial design) on species richness in agricultural fields that had been abandoned 30 years previously. Species richness showed a hump-shaped relationship with disturbance in the zero nitrogen and lowest nitrogen addition treatments because over time, at intermediate disturbance levels, annual plants colonized plots that would otherwise have become dominated by perennials. However, there was no relationship between species richness and disturbance in the high nitrogen treatments, where clearly competitively dominant species emerged even in disturbed plots (Figure 21.6). The higher nutrient levels were presumably sufficient to support rapid growth of competitive dominants, and to lead to competitive exclusion of subordinates between disturbance episodes.

21.3.2 Spatial heterogeneity

We have already seen how the patchy nature of an environment, coupled with aggregative behavior, can lead to coexistence of competing species (see Section 8.5.5). In addition, environments that are more spatially heterogeneous can be expected to accommodate extra species because they provide a greater variety of microhabitats, a greater range of microclimates, more types of places to hide from predators and so on. In effect, the extent of the resource spectrum is increased (see Figure 21.1a).

richness and heterogeneity in an abiotic environment

In some cases, it has been possible to relate species richness to the spatial heterogeneity of the abiotic environment. For instance, a study of plant species growing in 51 plots alongside the Hood River, Canada, revealed a positive relationship between species richness and an index of spatial heterogeneity (based, among other things, on the number of categories of substrate, slope, drainage regimes and soil pH present) (Figure 21.7a).

animal richness related to plant spatial heterogeneity

Most studies of spatial heterogeneity, though, have related the species richness of animals to the structural diversity of

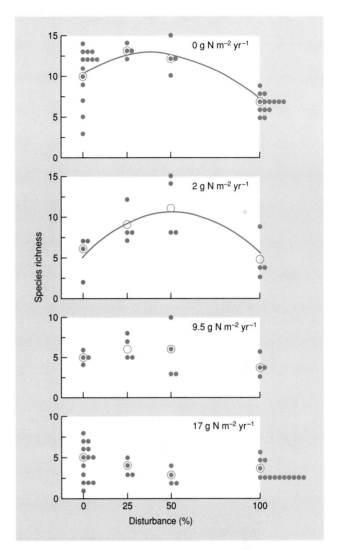

Figure 21.6 Species richness in old fields in Minnesota, USA, after 8 years across four levels of disturbance (quantified in terms of the percentage of bare ground produced by annual tilling) at four levels of nitrogen addition. Dots are values from replicate plots (1 m^2) and open circles are treatment means. Regression lines are shown only for significant relationships ($P < 0.05$). (After Wilson & Tilman, 2002.)

the plants in their environment (Figure 21.7b–d), occasionally as a result of experimental manipulation of the plants, as with the spiders in Figure 21.7b, but more commonly through comparisons of different natural communities (Figure 21.7c, d). However, whether spatial heterogeneity arises intrinsically from the abiotic environment or is provided by other biological components of the community, it is capable of promoting an increase in species richness.

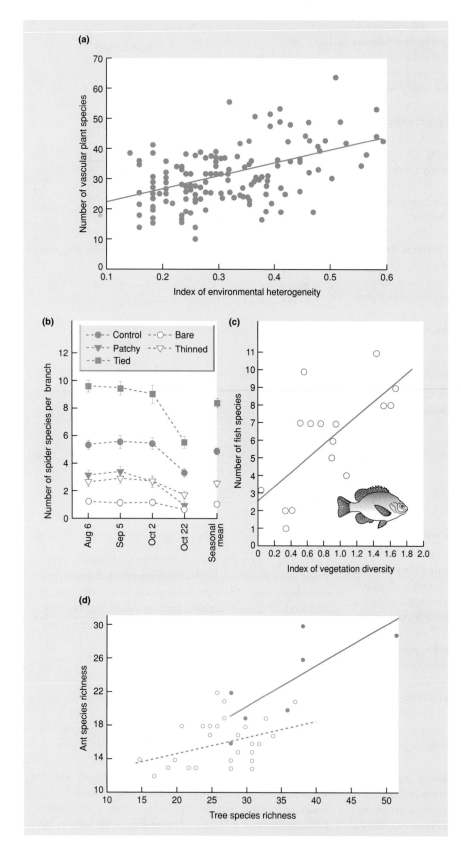

Figure 21.7 Relationship between the number of plants per 300 m² plot beside the Hood River, Northwest Territories, Canada, and an index (ranging from 0 to 1) of spatial heterogeneity in abiotic factors associated with topography and soil. (After Gould & Walker, 1997.) (b) In an experimental study, the number of spider species living on Douglas fir branches increases with their structural diversity. Those 'bare', 'patchy' or 'thinned' were less diverse than normal ('control') by virtue of having needles removed; those 'tied' were more diverse because their twigs were entwined together. (After Halaj *et al.*, 2000.) (c) Relationships between animal species richness and an index of structural diversity of vegetation for freshwater fish in 18 Wisconsin lakes. (After Tonn & Magnuson, 1982.) (d) Relationship between arboreal ant species richness in two regions of Brazilian savanna and the species richness of trees (a surrogate for spatial heterogeneity). ○, Distrito Federal; ●, Paraopeba region. (After Ribas *et al.*, 2003.)

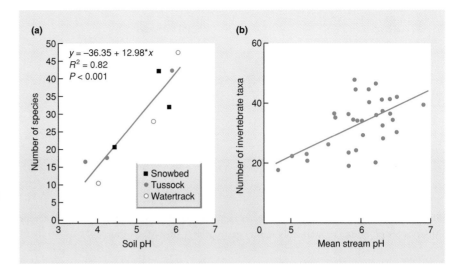

Figure 21.8 (a) The number of plant species per 72 m² sampling unit in the Alaskan Arctic tundra increases with pH. (After Gough *et al.*, 2000.) (b) The number of taxa of invertebrates in streams in Ashdown Forest, southern England, increases with the pH of the streamwater. (After Townsend *et al.*, 1983.)

21.3.3   Environmental harshness

what is harsh?

Environments dominated by an extreme abiotic factor – often called harsh environments – are more difficult to recognize than might be immediately apparent. An anthropocentric view might describe as extreme both very cold and very hot habitats, unusually alkaline lakes and grossly polluted rivers. However, species have evolved and live in all such environments, and what is very cold and extreme for us must seem benign and unremarkable to a penguin in the Antarctic.

We might try to get around the problem of defining environmental harshness by 'letting the organisms decide'. An environment may be classified as *extreme* if organisms, by their failure to live there, show it to be so. But if the claim is to be made – as it often is – that species richness is lower in extreme environments, then this definition is circular, and it is designed to prove the very claim we wish to test.

Perhaps the most reasonable definition of an extreme condition is one that requires, of any organism tolerating it, a morphological structure or biochemical mechanism that is not found in most related species and is costly, either in energetic terms or in terms of the compensatory changes in the organism's biological processes that are needed to accommodate it. For example, plants living in highly acidic soils (low pH) may be affected directly through injury by hydrogen ions or indirectly via deficiencies in the availability and uptake of important resources such as phosphorus, magnesium and calcium. In addition, aluminum, manganese and heavy metals may have their solubility increased to toxic levels, and mycorrhizal activity and nitrogen fixation may be impaired. Plants can only tolerate low pH if they have specific structures or mechanisms allowing them to avoid or counteract these effects.

Environments that experience a low pH can thus be considered harsh, and the mean number of plant species recorded per sampling unit in a study in the Alaskan Arctic tundra was indeed lowest in soils of low pH (Figure 21.8a). Similarly, the species richness of benthic stream invertebrates in the Ashdown Forest (southern UK) was markedly lower in the more acidic streams (Figure 21.8b). Further examples of extreme environments that are associated with low species richness include hot springs, caves and highly saline water bodies such as the Dead Sea. The problem with these examples, however, is that they are also characterized by other features associated with low species richness such as low productivity and low spatial heterogeneity. In addition, many occupy small areas (caves, hot springs) or areas that are rare compared to other types of habitat (only a small proportion of the streams in southern England are acidic). Hence extreme environments can often be seen as small and isolated islands. We will see in Section 21.5.1 that these features, too, are usually associated with low species richness. Although it appears reasonable that intrinsically extreme environments should as a consequence support few species, this has proved an extremely difficult proposition to establish.

are harsh environments the cause of low species richness?

21.4   Temporally varying factors that influence species richness

Temporal variation in conditions and resources may be predictable or unpredictable and operate on timescales from minutes through to centuries and millennia. All may influence species richness in profound ways.

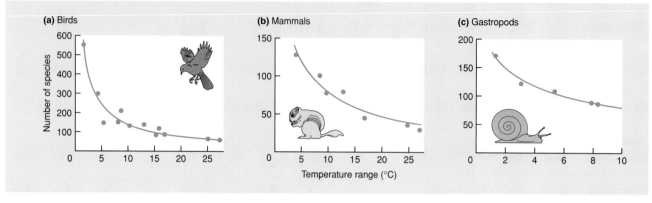

Figure 21.9 Relationships between species richness and the range of monthly mean temperatures at sites along the west coast of North America for (a) birds, (b) mammals and (c) gastropods. (After MacArthur, 1975.)

21.4.1 Climatic variation

temporal niche differentiation in seasonal environments

The effects of climatic variation on species richness depend on whether the variation is predictable or unpredictable (measured on timescales that matter to the organisms involved). In a predictable, seasonally changing environment, different species may be suited to conditions at different times of the year. More species might therefore be expected to coexist in a seasonal environment than in a completely constant one (see Figure 21.1a). Different annual plants in temperate regions, for instance, germinate, grow, flower and produce seeds at different times during a seasonal cycle; while phytoplankton and zooplankton pass through a seasonal succession in large, temperate lakes with a variety of species dominating in turn as changing conditions and resources become suitable for each.

specialization in nonseasonal environments

On the other hand, there are opportunities for specialization in nonseasonal environments that do not exist in seasonal environments. For example, it would be difficult for a long-lived obligate fruit-eater to exist in a seasonal environment when fruit is available for only a very limited portion of the year. But such specialization is found repeatedly in nonseasonal, tropical environments where fruit of one type or another is available continuously.

climatic instability may increase or decrease richness . . .

Unpredictable climatic variation (climatic instability) could have a number of effects on species richness: (i) stable environments may be able to support specialized species that would be unlikely to persist where conditions or resources fluctuated dramatically (see Figure 21.1b); (ii) stable environments are more likely to be saturated with species (see Figure 21.1d); and (iii) theory suggests that a higher degree of niche overlap will be found in more stable environments (see Figure 21.1c). All these processes could increase species richness. On the other hand, populations in a stable environment are more likely to reach their carrying capacities, the community is more likely to be dominated by competition, and species are therefore more likely to be excluded by competition (where $\bar{o}$ is smaller, see Figure 21.1c).

. . . but there is no good evidence either way

Some studies have seemed to support the notion that species richness increases as climatic variation decreases. For example, there is a significant negative relationship between species richness and the range of monthly mean temperatures for birds, mammals and gastropods that inhabit the west coast of North America (from Panama in the south to Alaska in the north) (Figure 21.9). However, this correlation does not prove causation, since there are many other things that change between Panama and Alaska. There is no established relationship between climatic instability and species richness.

21.4.2 Environmental age: evolutionary time

variable recovery from an ancient disturbance?

It has also often been suggested that communities that are 'disturbed' even on very extended timescales may none the less lack species because they have yet to reach an ecological or an evolutionary equilibrium. Thus communities may differ in species richness because some are closer to equilibrium and are therefore more saturated than others (see Figure 21.1d).

unchanging tropics and recovering temperate zones?

For example, many have argued that the tropics are richer in species than are more temperate regions at least in part because the tropics have existed over long and uninterrupted periods of evolutionary time, whereas the temperate regions are still recovering from the Pleistocene glaciations. It seems, however, that the long-term stability of the tropics has in the past been greatly exaggerated by ecologists. Whereas the climatic and biotic zones of the temperate region moved toward the equator during the glaciations, the tropical forest appears to have contracted to a limited number of small refuges surrounded by grasslands. A simplistic contrast between the unchanging tropics and the disturbed and recovering temperate regions is therefore untenable.

A comparison between the two polar regions may be more instructive. Both Arctic and Antarctic marine environments are cold, seasonal and strongly influenced by ice but their histories are quite different. The Arctic basin lost its fauna when covered by thick permanent ice at the height of the last glaciation and recolonization is underway; whereas a shallow water fauna has existed around the Antarctic since the mid-Palaeozoic (Clarke & Crame, 2003). Today the two polar faunas contrast markedly, the Arctic being depauperate and the Antarctic rich, most likely reflecting the importance of their histories.

21.5 Habitat area and remoteness: island biogeography

larger islands contain more species: contrasting explanations

It is well established that the number of species on islands decreases as island area decreases. Such a *species–area* rela-

tionship is shown in Figure 21.10a for terrestrial vascular plants on islands in the Stockholm Archipelago, Sweden.

'Islands', however, need not be islands of land in a sea of water. Lakes are islands in a 'sea' of land, mountain tops are high-altitude islands in a low-altitude ocean, gaps in a forest canopy where tree have fallen are islands in a sea of trees, and there can be islands of particular geological types, soil types or vegetation types surrounded by dissimilar types of rock, soil or vegetation. Species–area relationships can be equally apparent for these types of islands (Figure 21.10b–d).

The relationship between species richness and habitat area is one of the most consistent of all ecological patterns. However, the pattern raises an important question: 'Is the impoverishment of species on islands more than would be expected in comparably small areas of mainland?' In other words, does the characteristic isolation of islands contribute to their impoverishment of species? These are important questions for an understanding of community structure since there are many oceanic islands, many lakes, many mountaintops, many woodlands surrounded by fields, many isolated trees, and so on.

21.5.1 MacArthur and Wilson's 'equilibrium' theory

Probably the most obvious reason why larger areas should contain more species is that larger areas typically encompass more

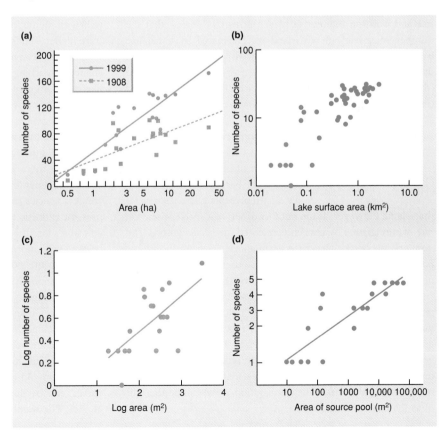

Figure 21.10 Species–area relationships. (a) Plants on islands east of Stockholm, Sweden: ●, survey completed in 1999 after grazing and hay-making had ceased; ■ survey completed in 1908 when intensive agriculture was practised. (After Lofgren & Jerling, 2002.) (b) Birds inhabiting lakes in Florida. (After Hoyer & Canfield, 1994.) (c) Bats inhabiting different-sized caves in Mexico. (After Brunet & Medellin, 2001.) (d) Fish living in Australian desert springs that have source pools of different sizes. (After Kodric-Brown & Brown, 1993.)

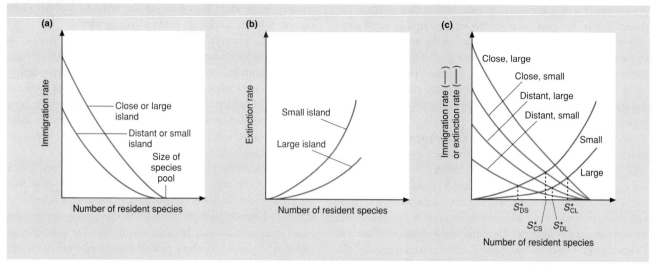

Figure 21.11 MacArthur and Wilson's (1976) equilibrium theory of island biogeography. (a) The rate of species immigration on to an island, plotted against the number of resident species on the island, for large and small islands and for close and distant islands. (b) The rate of species extinction on an island, plotted against the number of resident species on the island, for large and small islands. (c) The balance between immigration and extinction on small and large and on close and distant islands. In each case, S^* is the equilibrium species richness; C, close; D, distant; L, large; S, small.

different types of habitat. However, MacArthur and Wilson (1967) believed this explanation to be too simple. In their *equilibrium theory of island biogeography*, they argued: (i) that island size and isolation themselves played important roles – that the number of species on an island is determined by a balance between immigration and extinction; (ii) that this balance is dynamic, with species continually going extinct and being replaced (through immigration) by the same or by different species; and (iii) that immigration and extinction rates may vary with island size and isolation.

MacArthur and Wilson's immigration curves . . .

Taking immigration first, imagine an island that as yet contains no species at all. The rate of immigration of *species* will be high, because any colonizing individual represents a species new to that island. However, as the number of resident species rises, the rate of immigration of new, unrepresented species diminishes. The immigration rate reaches zero when all species from the source pool (i.e. from the mainland or from other nearby islands) are present on the island in question (Figure 21.11a).

The immigration graph is drawn as a curve, because immigration rate is likely to be particularly high when there are low numbers of residents and many of the species with the greatest powers of dispersal are yet to arrive. In fact, the curve should really be a blur rather than a single line, since the precise curve will depend on the exact sequence in which species arrive, and this will vary by chance. In this sense, the immigration curve can be thought of as the *most probable curve*.

The exact immigration curve will depend on the degree of remoteness of the island from its pool of potential colonizers (Figure 21.11a). The curve will always reach zero at the same point (when all members of the pool are resident), but it will generally have higher values on islands close to the source of immigration than on more remote islands, since colonizers have a greater chance of reaching an island the closer it is to the source. It is also likely that immigration rates will generally be higher on a large island than on a small island, since the larger island represents a larger target for the colonizers (Figure 21.11a).

The rate of species extinction on an island (Figure 21.11b) is bound to be zero when there are no species there, and it will generally be low when there are few species. However, as the number of resident species rises, the extinction rate is assumed by the theory to increase, probably at a more than proportionate rate. This is thought to occur because with more species, competitive exclusion becomes more likely, and the population size of each species is on average smaller, making it more vulnerable to chance extinction. Similar reasoning suggests that extinction rates should be higher on small than on large islands as population sizes will typically be smaller on small islands (Figure 21.11b). As with immigration, the extinction curves are best seen as 'most probable' curves.

. . . and extinction curves

In order to see the net effect of immigration and extinction, their two curves can be superimposed (Figure 21.11c). The number of species where

the balance between immigration and extinction

the curves cross (S^*) is a dynamic equilibrium and should be the characteristic species richness for the island in question. Below S^*, richness increases (immigration rate exceeds extinction rate); above S^*, richness decreases (extinction exceeds immigration). The theory, then, makes a number of predictions:

1 The number of species on an island should eventually become roughly constant through time.
2 This should be a result of a continual *turnover* of species, with some becoming extinct and others immigrating.
3 Large islands should support more species than small islands.
4 Species number should decline with the increasing remoteness of an island.

predictions of equilibrium theory are not all exclusive to this theory

Note, though, that several of these predictions could also be made without any reference to the equilibrium theory. An approximate constancy of species number would be expected if richness were determined simply by island type. Similarly, a higher richness on larger islands would be expected as a consequence of larger islands having more habitat types. One test of the equilibrium theory, therefore, would be whether richness increases with area at a rate greater than could be accounted for by increases in habitat diversity alone (see Section 21.5.2).

The effect of island remoteness can be considered quite separately from the equilibrium theory. Merely recognizing that many species are limited in their dispersal ability, and have not yet colonized all islands, leads to the prediction that more remote islands are less likely to be saturated with potential colonizers (see Section 21.5.3). However, the final prediction arising from the equilibrium theory – constancy as a result of turnover – is truly characteristic of the equilibrium theory (see Section 21.5.4).

21.5.2 Habitat diversity alone – or a separate effect of area?

an example where habitat diversity is paramount

The most fundamental question in island biogeography, then, is whether there is an 'island effect' as such, or whether islands simply support few species because they are small areas containing few habitats. Does richness increase with area at a rate *greater* than could be accounted for by increases in habitat diversity alone? Some studies have attempted to partition species–area variation on islands into that which can be entirely accounted for in terms of habitat diversity, and that which remains and must be accounted for by island area in its own right. For beetles on the Canary Islands, the relationship between species richness and habitat diversity (as measured by plant species richness) is much stronger

than that with island area, and this is particularly marked for the herbivorous beetles, presumably because of their particular food plant requirements (Figure 21.12a).

On the other hand, in a study of a variety of animal groups living on the Lesser Antilles island in the West Indies, the variation in species richness from island to island was partitioned, statistically, into that attributable to island area alone, that attributable to habitat diversity alone, that attributable to correlated variation between area and habitat diversity (and hence not attributable to either alone), and that attributable to neither. For reptiles and amphibians (Figure 21.12b), like the beetles of the Canary Islands, habitat diversity was far more important than island area. But for bats, the reverse was the case, and for birds and butterflies, both area itself and habitat diversity had important parts to play.

partitioning variation between habitat diversity and island area itself

experimental reductions in mangrove island area

An experiment was carried out to try to separate the effects of habitat diversity and area on some small mangrove islands in the Bay of Florida (Simberloff, 1976). These islands consist of pure stands of the mangrove species *Rhizophora mangle*, which support communities of insects, spiders, scorpions and isopods. After a preliminary faunal survey, some islands were reduced in size – by means of a power saw. Habitat diversity was not affected, but arthropod species richness on three islands none the less diminished over a period of 2 years (Figure 21.13). A control island, the size of which was unchanged, showed a slight *increase* in richness over the same period, presumably as a result of random events.

species–area graphs for islands and comparable mainland areas

Another way of trying to distinguish a separate effect of island area is to compare species–area graphs for islands with those for arbitrarily defined areas of mainland. The species–area relationships for mainland areas should be due almost entirely to habitat diversity (together with any 'sampling' effect involving increased probabilities of detecting rare species in larger areas). All species will be well able to 'disperse' between mainland areas, and the continual flow of individuals across the arbitrary boundaries will therefore mask local extinctions (i.e. what would be an extinction on an island is soon reversed by the exchange of individuals between local areas). An arbitrarily defined area of mainland should thus contain more species than an otherwise equivalent island, and this is usually interpreted as meaning that the slopes of the species–area graphs for islands should be steeper than those for mainland areas (since the effect of island isolation should be most marked on small islands, where extinctions are most likely). The difference between the two types of graph would then be attributable to the island effect in its own right. Table 21.1 shows that despite considerable variation, the island graphs do typically have steeper slopes.

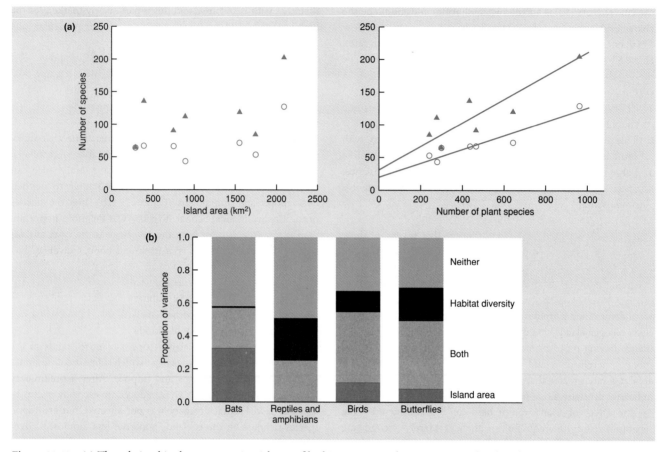

Figure 21.12   (a) The relationships between species richness of herbivorous (○) and carnivorous (▲) beetles of the Canary Islands and both island area and plant species richness. (After Becker, 1992.) (b) Proportion of variance, for four animal groups, in species richness among islands in the Lesser Antilles related uniquely to island area, uniquely to habitat diversity, to correlated variation between area and habitat diversity and unexplained by either. (After Ricklefs & Lovette, 1999.)

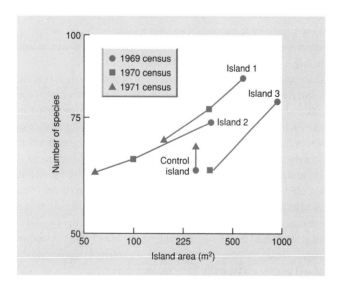

Note that a reduced number of species per unit area on islands should also lead to a lower value for the intercept on the *S*-axis of the species–area graph. Figure 21.14a illustrates both an increased slope and a reduced value for the intercept for the species–area graph for ant species on isolated Pacific islands, compared with the graph for progressively smaller areas of the very large island of New Guinea. Figure 21.14b gives a similar relationship for reptiles on islands off the coast of South Australia.

Figure 21.13   (*left*) The effect on the number of arthropod species of artificially reducing the size of mangrove islands. Islands 1 and 2 were reduced in size after both the 1969 and 1970 censuses. Island 3 was reduced only after the 1969 census. The control island was not reduced, and the change in its species richness was attributable to random fluctuations. (After Simberloff, 1976.)

Table 21.1 Values of the slope z, of species–area curves ($\log S = \log C + z \log A$, where S is species richness, A is area and C is a constant giving the number of species when A has a value of 1), for arbitrary areas of mainland, oceanic islands and habitat islands. (After Preston, 1962; May, 1975b; Gorman, 1979; Browne, 1981; Matter *et al.*, 2002; Barrett *et al.*, 2003; Storch *et al.*, 2003.)

Taxonomic group	Location	z
Arbitrary areas of mainland		
Birds	Central Europe	0.09
Flowering plants	England	0.10
Birds	Neoarctic	0.12
Savanna vegetation	Brazil	0.14
Land plants	Britain	0.16
Birds	Neotropics	0.16
Oceanic islands		
Birds	New Zealand islands	0.18
Lizards	Californian islands	0.20
Birds	West Indies	0.24
Birds	East Indies	0.28
Birds	East Central Pacific	0.30
Ants	Melanesia	0.30
Land plants	Galápagos	0.31
Beetles	West Indies	0.34
Mammals	Scandinavian islands	0.35
Habitat islands		
Zooplankton (lakes)	New York State	0.17
Snails (lakes)	New York State	0.23
Fish (lakes)	New York State	0.24
Birds (Paramo vegetation)	Andes	0.29
Mammals (mountains)	Great Basin, USA	0.43
Terrestrial invertebrates (caves)	West Virginia	0.72

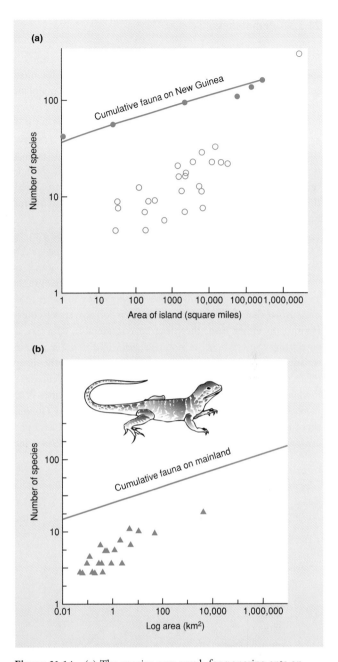

Figure 21.14 (a) The species–area graph for ponerine ants on various Moluccan and Melanesian islands compared with a graph for different-sized sample areas on the very large island of New Guinea. (After Wilson, 1961.) (b) The species–area graph for reptiles on islands off the coast of South Australia compared with the mainland species–area relationship. In this case, the islands were formed within the last 10,000 years as a result of rising sea level. (After Richman *et al.*, 1988.)

plant extinction and immigration rates in relation to island size

Overall, therefore, studies like this suggest a separate area effect (larger islands are larger targets for colonization; populations on larger islands have a lower risk of extinction) beyond a simple correlation between area and habitat diversity. Lofgren and Jerling (2002) were able to quantify plant extinction rates and immigration rates on islands of different sizes in the Stockholm Archipelago (see Figure 21.10a) by comparing species lists in their survey (1996–99) with those reported by J. W. Hamner from the period 1884–1908. In the intervening time, 93 new species appeared while 20 species disappeared from the islands. Many of the newcomers were trees, bushes and shade-tolerant shrubs, reflecting succession after the cessation of cattle grazing and hay-making in the 1960s. Despite the confounding effect of succession, and as predicted, extinction rate was negatively correlated and immigration rate positively correlated with island size.

21.5.3   Remoteness

It follows from the above argument that the island effect and the species impoverishment of an island should be greater for more remote islands. (Indeed, the comparison of islands with mainland areas is only an extreme example of a comparison of islands varying in remoteness, since local mainland areas can be thought of as having minimal remoteness.) Remoteness, however, can mean two things. First, it can simply refer to the degree of physical isolation. Alternatively, a single island can also itself vary in remoteness, depending on the type of organism being considered: the same island may be remote from the point of view of land mammals but not from the point of view of birds.

bird species richness on islands decreases with 'remoteness'

The effects of remoteness can be demonstrated either by plotting species richness against remoteness itself, or by comparing the species–area graphs of groups of islands (or for groups of organisms) that differ in their remoteness (or powers of colonization). In either case, there can be considerable difficulty in extricating the effects of remoteness from all the other characteristics by which two islands may differ. Nevertheless, the direct effect of remoteness can be seen in Figure 21.15 for nonmarine, lowland birds on tropical islands in the southwest Pacific. With increasing distance from the large source island of New Guinea, there is a decline in the number of species, expressed as a percentage of the number present on an island of similar area but

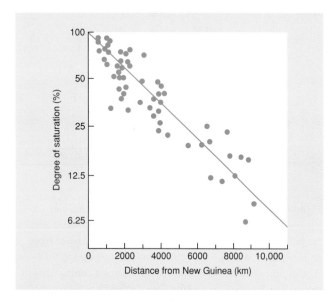

Figure 21.15 The number of resident, nonmarine, lowland bird species on islands more than 500 km from the larger source island of New Guinea expressed as a proportion of the number of species on an island of equivalent area but close to New Guinea, and plotted as a function of island distance from New Guinea. (After Diamond, 1972.)

close to New Guinea. Species richness decreases exponentially with distance, approximately halving every 2600 km. The species–area graph in Figure 21.16a also shows that remote islands of a given size possess fewer species than their counterparts close to a land mass. In addition, Figure 21.16b contrasts the species–area graphs of two classes of organisms in two regions: the relatively remote Azores (in the Atlantic, far to the west of Portugal) and the Channel Islands (close to the north coast of France). Whereas the Azores are indeed far more remote than the Channel Islands from the point of view of the birds, the two island groups are apparently equally remote for ferns, which are particularly good dispersers because of their light, wind-blown spores. Thus, on the basis of all these examples, the species impoverishment caused by the island effect does indeed appear to increase as the degree of isolation of the island increases. Note, also, that a multiple regression analysis of Lofgren and Jerling's 1999 Stockholm Archipelago database (see Figure 21.10a) demonstrated the overriding effect of island area on plant species richness (73% of variation explained), but distance to the nearest island also contributed significantly, explaining a further 17% of variation.

A more transient but none the less important reason for the species impoverishment of islands, especially remote islands, is the fact that many lack species that they could potentially support, simply because there has been insufficient time for the species to colonize. An example is the island of Surtsey, which emerged in 1963 as a result of a volcanic eruption (Fridriksson, 1975). The new island, 40 km southwest of Iceland, was reached by bacteria and fungi, some seabirds, a fly and the seeds of several beach plants within 6 months of the start of the eruption. Its first established vascular plant was recorded in 1965, and the first moss colony in 1967. By 1973, 13 species of vascular plant and more than 66 mosses had become established (Figure 21.17). Colonization is continuing still. The general importance of this example is that the communities of many islands can be understood *neither* in terms of simple habitat suitability *nor* as a characteristic equilibrium richness. Rather, they stress that many island communities have not reached equilibrium and are certainly not fully 'saturated' with species.

21.5.4   Which species? Turnover

MacArthur and Wilson's equilibrium theory predicts not only a characteristic species turnover . . . species richness for an island, but also a *turnover* of species in which new species continually colonize whilst others become extinct. This implies a significant degree of chance regarding precisely which species are present at any one time. However, studies of turnover itself are rare, because communities have to be followed over a period of time (usually difficult and costly). Good studies of turnover are rarer still, because it is necessary to count every species on every occasion so as to avoid 'pseudo-immigrations'

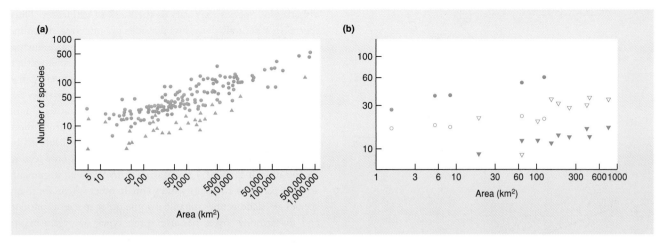

Figure 21.16 Remoteness increases the species impoverishment of islands. (a) A species–area plot for the land birds of individual islands in tropical and subtropical seas. ▲, islands more than 300 km from the next largest land mass or the very remote Hawaiian and Galápagos archipelagos; ●, islands less than 300 km from source. (b) Species–area plots in the Azores and the Channel Islands for land and freshwater breeding birds (▼, Azores; ●, Channel Islands) and for native ferns (▽, Azores; ○, Channel Islands). The Azores are more remote for birds but not for ferns. (After Williamson, 1981.)

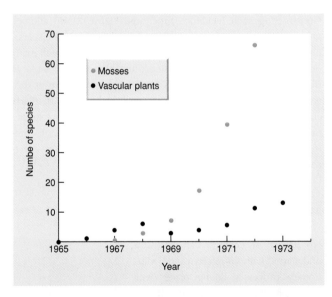

Figure 21.17 The number of species of mosses and vascular plants recorded on the new island of Surtsey from 1965 to 1973. (After Fridriksson, 1975.)

and 'pseudo-extinctions'. Indeed, any results are bound to be underestimates of actual turnover, because an observer cannot be everywhere all the time.

. . . is relatively high for temperate woodland birds . . . One revealing study involved censuses from 1949 to 1975 of the breeding birds in a small oak wood (Eastern Wood) in southern England. In all, 44 species bred in the wood over this period, and 16 of them bred every year. The number breeding in any one year varied between 27 and 36, with an average of 32 species. The immigration and extinction 'curves' are shown in Figure 21.18. Their most obvious feature is the scattering of points as compared with the assumed simplicity of the MacArthur–Wilson model. Nevertheless, whilst the positive correlation in the extinction graph is statistically insignificant, the negative correlation in the immigration graph is highly significant; and the two lines do seem to cross at roughly 32 species, with three new immigrants and three extinctions each year. There is clearly a considerable turnover of species, and consequently considerable year-to-year variation in the bird community of Eastern Wood despite its approximately constant species richness.

In contrast, a long-term study (surveys in 1954, 1976 and annually from 1984 to 1990) of the 15-strong bird community on tropical Guana Island, *. . . but not for birds on a tropical island* revealed no such turnover – no new species established and only one went extinct, as a result of habitat destruction (Mayer & Chipley, 1992). The position of Guana Island within an archipelago of numerous small islands may reduce the likelihood of local extinctions if there is continuous dispersal from island to island. On the other hand, it is conceivable that tropical birds really do have lower turnover rates – because they are more often sedentary, have lower adult mortality and are more often resident, as opposed to migratory (Mayer & Chipley, 1992).

Experimental evidence of turnover and indeterminacy is provided by the work of Simberloff and Wilson (1969), who exterminated the invertebrate fauna on a series of small mangrove islands in the Florida Keys and monitored recolonization. Within about 200 days, species richness had stabilized around the level

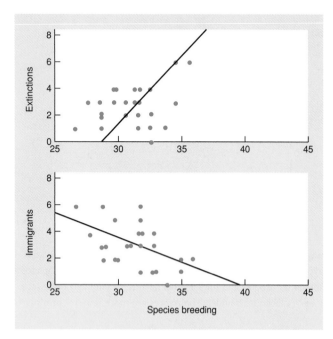

Figure 21.18 Immigration and extinction of breeding birds at Eastern Wood, UK. The line in the extinction diagram is at 45°. The line in the immigration diagram is the calculated regression line with a slope of −0.38. (After Beven, 1976; from Williamson, 1981.)

prior to defaunation, but with many differences in species composition. Since then, the rate of turnover of species on the islands has been estimated as 1.5 extinctions and colonizations per year (Simberloff, 1976).

Thus, the idea that there is a turnover of species leading to a characteristic equilibrium richness on islands, but an indeterminacy regarding particular species, appears to be correct – at least approximately.

21.5.5 Which species? Disharmony

some taxa are better suited to reach islands and persist there . . .

It has long been recognized – for example by Hooker in 1866 – that one of the main characteristics of island biotas is 'disharmony', that is, the relative proportions of different taxa are not the same on islands as they are on the mainland. We have already seen from the species–area relationships in Figure 21.16 that groups of organisms with good powers of dispersal (like ferns and, to a lesser extent, birds) are more likely to colonize remote islands than are groups with relatively poor powers of dispersal (most mammals).

. . . or vary in their risk of extinction

However, variation in dispersal ability is not the only factor leading to

disharmony. Species may vary in their risk of extinction. Thus, species that naturally have low densities per unit area are bound to have only small populations on islands, and a chance fluctuation in a small population is quite likely to eliminate it altogether. Vertebrate predators, which generally have relatively small populations, are notable for their absence on many islands. For example, the birds on the Atlantic island of Tristan da Cunha have no bird, mammal or reptile predators apart from those released by humans. Specialist predators are also liable to be absent from islands because their immigration can only lead to colonization if their prey have arrived first. Similar arguments apply to parasites, mutualists and so on. In other words, for many species an island is only suitable if some other species is present, and disharmony arises because some types of organism are more 'dependent' than others.

The development by Diamond (1975) of *incidence functions* and *assembly rules* for the birds of the Bismark Archipelago is probably the fullest attempt to understand island communities by combining ideas on dispersal and extinction differentials with those on sequences of arrival and habitat suitability. Constructing such incidence functions (Figure 21.19) allowed Diamond to contrast 'supertramp' species (high rates of dispersal but a poorly developed ability to persist in communities with many other species), with 'high *S*' species (only able to persist on large islands with many other species), and to contrast these in turn with intermediate categories. Such work illustrates particularly clearly that it takes far more than a count of the number of species present to characterize the community of an island. Island communities are not merely impoverished – the impoverishment affects particular types of organism disproportionately.

incidence functions and assembly rules

21.5.6 Which species? Evolution

No aspect of ecology can be fully understood without reference to evolutionary processes taking place over evolutionary timescales, and this is particularly true for an understanding of island communities. On isolated islands, the rate at which new species evolve may be comparable with or even faster than the rate at which they arrive as new colonists. Clearly, the communities of many islands will be incompletely understood by reference only to ecological processes.

evolution rate on islands may be faster than colonization rate

One widespread illustration of this is the very common occurrence, especially on 'oceanic' islands, of *endemic* species (i.e. species that are found nowhere else). Almost all the species of *Drosophila* on Hawaii, for example (see Section 1.4.1), are endemics (apart from cosmopolitan urban 'pests') as are most of the species

endemism – more likely on remote islands (and for poor dispersers) . . .

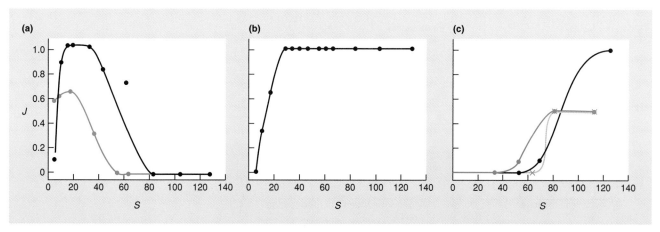

Figure 21.19 Incidence functions for various species in the Bismarcks in which *J*, the proportion of islands occupied by a given species, is plotted against *S*, a measure of island 'size' (actually the total number of bird species present). (a) Incidence functions for two 'supertramps': ●, flycatcher *Monarcha cinerascens*; ●, honeyeater *Myzomela pammelaena*. (b) Incidence function for the pigeon *Chalcophaps stephani*, a competent colonizer and, apparently, an effective competitor. (c) Incidence functions for three species that are restricted to larger islands: ●, hawk *Henicopernis longicauda*; ●, rail *Rallina tricolor*; ×, heron *Butorides striatus*. (After Diamond, 1975.)

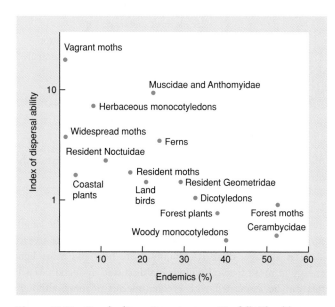

Figure 21.20 Poorly dispersing groups on Norfolk Island have a higher proportion of endemic species, and are more likely to contain species that have reached Norfolk Island from either New Caledonia or New Zealand than species from Australia, which is further away. The converse holds for good dispersers. (After Holloway, 1977.)

of land birds on the island of Tristan da Cunha. A more complete illustration of the balance between colonization and the evolution of endemics is provided by the animals and plants of Norfolk Island (Figure 21.20). This small island (about 70 km²) is

approximately 700 km from New Caledonia and New Zealand, but about 1200 km from Australia, and the ratio of Australian species to New Zealand and New Caledonian species within a group can therefore be used as a measure of that group's dispersal ability. As Figure 21.20 shows, the proportion of endemics on Norfolk Island is highest in groups with poor dispersal ability and lowest in groups with good dispersal ability.

In a similar vein, Lake Tanganyika, one of the ancient and deep Great Rift lakes of Africa, contains 214 species of cichlid fish, many of which show exquisite specializations in the manner and location of their feeding. Of these 214 species, 80% are endemic. With an estimated age of the lake of 9–12 million years, together with evidence that the various endemic groups diverged some 3.5–5 million years ago, it is likely that this uniquely diverse, endemic fish fauna evolved within the lake from a single ancestral lineage (Meyer, 1993). By contrast, Lake Rudolph, which has only been an isolated water body for 5000 years, since its connection to the Nile system was broken, contains only 37 species of cichlid of which only 16% are endemic (Fryer & Iles, 1972).

. . . and in more ancient ecosystems

21.6 Gradients of species richness

Sections 21.3–21.5 demonstrated how difficult explanations for variations in species richness are to formulate and test. It is easier to describe patterns, especially gradients, in species richness. These are discussed next. Explanations for these, too, however, are often very uncertain.

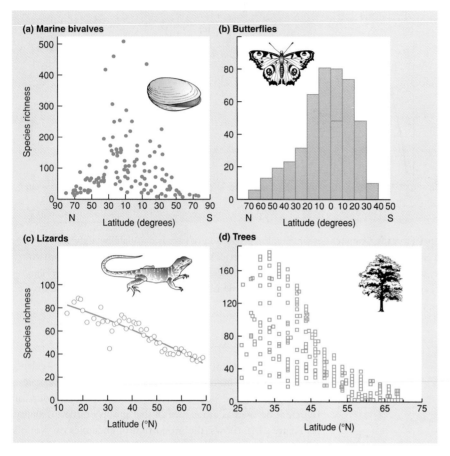

Figure 21.21 Latitudinal patterns in species richness in: (a) marine bivalves (after Flessa & Jablonski, 1995); (b) swallowtail butterflies (after Sutton & Collins, 1991); (c) quadruped mammals in North America (after Rosenzweig & Sandlin 1997); and (d) trees in North America (after Currie & Paquin, 1987.)

21.6.1 Latitudinal gradients

richness decreases with latitude

One of the most widely recognized patterns in species richness is the increase that occurs from the poles to the tropics. This can be seen in a wide variety of groups, including trees, marine invertebrates, mammals and lizards (Figure 21.21). The pattern can be seen, moreover, in terrestrial, marine and freshwater habitats.

a diversity of explanations: predation, . . .

A number of explanations related to our discussions in Section 21.3 and 21.4 have been put forward for the general latitudinal trend in species richness, but not one of these is without problems. In the first place, the richness of tropical communities has been attributed to a greater intensity of predation and to more specialized predators (Janzen, 1970; Connell, 1971; Clark & Clark, 1984). More intense predation could reduce the importance of competition, permitting greater niche overlap and promoting higher richness (see Figure 21.1c). However, even if predation is more intense in the tropics, which is far from certain, it cannot readily be forwarded as the root cause of tropical richness, since this begs the question of what gives rise to the richness of the predators themselves.

Second, increasing species richness may be related to an increase in productivity as one moves from the poles to the equator. The length of the growing season increases from the poles to the tropics and, on average, there is certainly more heat and more light energy in more tropical regions. As discussed in Section 21.3.1, this can be associated with greater species richness, although increased productivity in at least some cases has been associated with reduced richness.

. . . productivity, . . .

. . . nutrient supply, . . .

Moreover, light and heat are not the only determinants of plant productivity. Tropical soils often have lower concentrations of plant nutrients than temperate soils. The species-rich tropics might therefore be seen, in this sense, as reflecting their *low* productivity. In fact, tropical soils are poor in nutrients because most of the nutrients are locked up in the large tropical biomass. A productivity argument might therefore have to run as follows. The light, temperature and water regimes of the tropics lead to high biomass communities but not necessarily to diverse communities. This, though, leads to nutrient-poor soils and perhaps a wide range of light regimes from the forest floor to the canopy far above. These in turn lead to high plant species richness and thus to high animal species richness. There is certainly no *simple* 'productivity explanation' for the latitudinal trend in richness.

... climate ...

Some ecologists have invoked the climate of low latitudes as a reason for their high species richness. Specifically, equatorial regions are generally less seasonal than temperate regions, and this may allow species to be more specialized (i.e. have narrower niches, see Figure 21.1b). The greater evolutionary 'age' of the tropics has also been proposed as a reason for their greater species richness (Flenley, 1993), and another line of argument suggests that the repeated fragmentation and coalescence of tropical forest refugia promoted genetic differentiation and speciation, accounting for much of the high richness in tropical regions (Connor, 1986). These ideas, too, are plausible but very far from proven.

... or area?

A final idea, the area hypothesis of Terborgh (1973), is worth highlighting. The area of the tropical zone is much greater than that of the other latitudinal zones, and Rosenzweig (2003) has claimed that more area means more species. Note that in such enormous geographic areas the focus is not on a balance between immigration and extinction (as it was for islands in Section 21.5.1) but between speciation and extinction. Species inhabiting more extensive regions (i.e. tropical species) can, in consequence, have larger geographic ranges. Rosenzweig (2003) argues that species with larger ranges (and consequently larger population sizes) are both less likely to go extinct (see Section 7.5) and more likely to speciate (allopatrically, because of a greater likelihood that their range will be bisected by a barrier). If it is true that extinction rates are lower and speciation rates are higher in regions of greater spatial extent, such regions should also have higher equilibrium species richnesses. However, the evidence for the underlying assumptions is scant.

Overall, therefore, the latitudinal gradient lacks an unambiguous explanation. This is hardly surprising. The components of a possible explanation – trends with area, productivity, climatic stability and so on – are themselves understood only in an incomplete and rudimentary way, and the latitudinal gradient intertwines these components with one another, and with other, often opposing forces: isolation, harshness and so on.

21.6.2 Gradients with altitude and depth

decreasing, increasing or hump-shaped richness relationships with altitude

A decrease in species richness with altitude, analogous to that observed with latitude, has frequently been reported in terrestrial environments (e.g. Figure 21.22a, b). On the other hand, some have reported a monotonic increase with altitude (e.g. Figure 21.22c) while about half the studies of altitudinal species richness have described hump-shaped patterns (e.g. Figure 21.22d) (Rahbek, 1995).

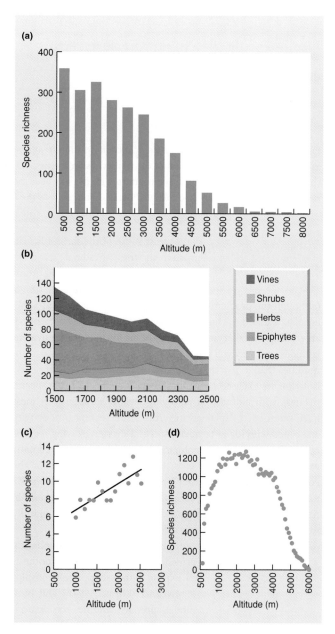

Figure 21.22 Relationships between species richness and altitude for: (a) breeding birds in the Nepalese Himalayas (after Hunter & Yonzon, 1992); (b) plants in the Sierra Manantlán, Mexico (after Vázquez & Givnish, 1998); (c) ants in Lee Canyon in the Spring Mountains of Nevada, USA (after Sanders *et al.*, 2003); and (d) flowering plants in the Nepalese Himalayas (after Grytnes & Vetaas, 2002).

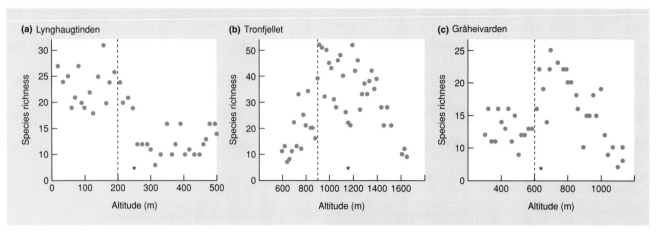

Figure 21.23 Scatter plots of species richness in relation to altitude for three transects in Norway. In each case the treeline is shown as a dashed line and the midpoint of the transect as an asterisk. (a) Lynghaugtinden shows a monotonic decline in richness with altitude. (b) Tronfjellet shows a hump-shaped pattern with its peak near the midpoint of the transect. (c) Gråheivarden shows an increase in richness just above the treeline followed by a decline towards the mountain top. (After Grytnes, 2003.)

again, a diversity of potential explanations

At least some of the factors instrumental in the latitudinal trend in richness are also likely to be important as explanations for altitudinal trends (although the problems in explaining the latitudinal trend apply equally to altitude). Thus, high-altitude communities almost invariably occupy smaller areas than those in lowlands at equivalent latitudes, and they will usually be more isolated from similar communities than in lowland sites. Therefore the effects of area and isolation are likely to contribute to observed decreases in species richness with altitude. In addition, declines in species richness have often been explained in terms of decreasing productivity associated with lower temperatures and shorter growing seasons at higher altitude, or physiological stress associated with climatic extremes near mountain tops. Indeed, the explanation for the converse, positive relationship between ant diversity and altitude in Figure 21.22c, is that precipitation increased with altitude in this case, resulting in higher productivity and less physiologically extreme conditions at higher altitude.

'hard boundaries' and hump-shaped relationships

The concept of 'hard boundaries' provides the basis for a hypothesis to explain hump-shaped relationships (Colwell & Hurtt, 1994). This null model approach assumes the random placement of species between an upper hard boundary (mountain top) and a lower hard boundary (valley bottom) and predicts a symmetric humped relationship in the middle of the gradient (which tapers most steeply as the boundaries are approached). Grytnes and Vetaas (2002) modeled the altitudinal pattern in Himalayan flowering plants and found that the actual distribution (Figure 21.22d) fitted best to a model combining hard boundaries with an underlying monotonic decline in richness with altitude.

In a revealing study of altitudinal transects in Norway, Grytnes (2003) reported a variety of patterns in vascular plant richness. The most northerly of the transects, at Lynghaugtinden, showed a monotonic decline, conforming best to the hypothesis relating declining area to increasing altitude (Figure 21.23a). Tronfjellet, on the other hand, had a pattern broadly consistent with the hard boundary hypothesis, peaking in richness in the middle of the altitudinal range and with steep declines near the boundaries (Figure 21.23b). Enriching the picture even further, Gråheivarden, the most southerly transect, revealed a pattern consistent with a third, 'mass effect' hypothesis. This concerns the establishment of species in sites where a self-maintaining population could not exist, via a spilling over of taxa from an adjacent biotic zone. The Gråheivarden transect supported the mass effect prediction of increased species richness near the treeline, where forest and alpine communities abut (Figure 21.23c).

patterns with depth in aquatic environments

In aquatic environments, the change in species richness with depth shows strong similarities to the terrestrial gradient with altitude. In larger lakes, the cold, dark, oxygen-poor abyssal depths contain fewer species than the shallow surface waters. Likewise, in marine habitats, plants are confined to the photic zone (where they can photosynthesize), which rarely extends below 30 m. In the open ocean, therefore, there is a rapid decrease in richness with depth, reversed only by the variety of often bizarre animals living on the ocean floor. Interestingly, however, in coastal regions the effect of depth on the species richness of benthic (bottom-dwelling) animals produces a peak of richness at about 1000 m, possibly reflecting higher environmental predictability there (Figure 21.24). At greater depths, beyond the continental slope, species richness declines again, probably because of the extreme paucity of food resources in abyssal regions.

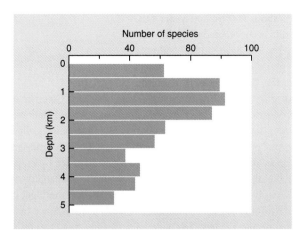

Figure 21.24 Depth gradient in species richness of the megabenthos (fish, decapods, holothurians and asteroids) in the ocean southwest of Ireland. (After Angel, 1994.)

21.6.3 Gradients during community succession

a hump-shaped richness relationship during succession . . .

We saw earlier (see, for example, Section 16.7.1) how, in community successions, if they run their full course, the number of species first increases (because of colonization) but eventually decreases (because of competition). This is most firmly established for plants, but the few studies that have been carried out on animals in successions indicate, at least, a parallel increase in species richness in the early stages of succession. Figure 21.25 illustrates this for birds following shifting cultivation in a tropical rainforest in northeast India, and for insects associated with old-field successions.

To a certain extent, the successional gradient is a necessary consequence of the gradual colonization of an area by species from surrounding communities that are at later successional stages; that is, later stages are more fully saturated with species (see Figure 21.1d). However, this is a small part of the story, since succession involves a process of replacement of species and not just the mere addition of new ones.

Indeed, as with the other gradients in species richness, there is something of a cascade effect with succession: one process that increases richness kick-starts a second, which feeds into a third, and so on. The earliest species will be those that are the best colonizers and the best competitors for open space. They immediately provide resources (and introduce heterogeneity) that were not previously present. For example, the earliest plants generate resource-depletion zones in the soil that inevitably increase the spatial heterogeneity of plant nutrients. The plants themselves provide a new variety of microhabitats, and for the animals that might feed on them they provide a much greater range of food resources (see Figure 21.1a). The increase in herbivory and predation may then feed back to promote further increases in species richness (predator-mediated coexistence: see Figure 21.1c), which provides further resources and more heterogeneity, and so on. In addition, temperature, humidity and wind speed are much less variable (over time) within a forest than

. . . caused by a cascade of effects?

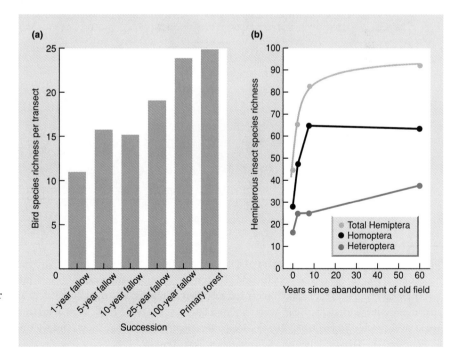

Figure 21.25 The increase in species richness during successions. (a) Birds following shifting cultivation in a tropical rainforest in northeast India. (After Shankar Raman *et al.*, 1998.) (b) Hemipterous insects following an old-field succession. (After Brown & Southwood, 1983.)

in an exposed early successional stage, and the enhanced constancy of the environment may provide a stability of conditions and resources that permits specialist species to build up populations and persist (see Figure 21.1b). As with the other gradients, the interaction of many factors makes it difficult to disentangle cause from effect. But with the successional gradient of richness, the tangled web of cause and effect appears to be of the essence.

21.6.4   Patterns in taxon richness in the fossil record

Finally, it is of interest to take the processes that are believed to be instrumental in generating present-day gradients in richness and apply them to trends occurring over much longer timespans. The imperfection of the fossil record has always been the greatest impediment to the paleontological study of evolution. Nevertheless, some general patterns have emerged, and our knowledge of six important groups of organisms is summarized in Figure 21.26.

Until about 600 million years ago, the world was populated virtually only by bacteria and algae, but then almost all the phyla of marine invertebrates entered the fossil record within the space of only a few million years (Figure 21.26a). Given that the introduction of a higher trophic level can increase richness at a lower level, it can be argued that the first single-celled herbivorous protist was probably instrumental in the Cambrian explosion in species richness. The opening up of space by cropping of the algal monoculture, coupled with the availability of recently evolved eukaryotic cells, may have caused the biggest burst of evolutionary diversification in earth's history. Since that time, taxonomic richness has increased steadily but erratically (Figure 21.26a), with five so-called mass extinctions and many smaller ones. Analysis of the pattern of 'recovery' peaks following extinction peaks indicates that the average recovery time is 10 million years (Kirchner & Weil, 2000).

Cambrian explosion: exploiter-mediated coexistence?

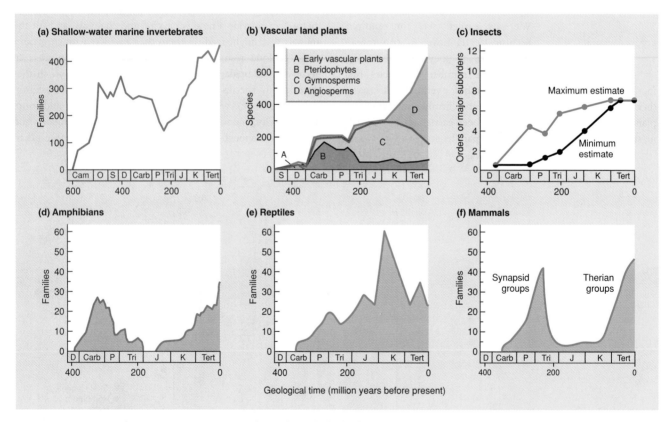

Figure 21.26   Curves showing patterns in taxon richness through the fossil record. (a) Families of shallow-water invertebrates. (After Valentine, 1970.) (b) Species of vascular land plants in four groups: early vascular plants, pteridophytes, gymnosperms and angiosperms. (After Niklas *et al.*, 1983.) (c) Major orders and suborders of insects. The minimum values are derived from definite fossil records; the maximum values include 'possible' records. (From Strong *et al.*, 1984.) (d–f) Vertebrate families of amphibians, reptiles and mammals, respectively. (After Webb, 1987.) Key to geological periods: Cam, Cambrian; O, Ordovician; S, Silurian; D, Devonian; Carb, Carboniferous; P, Permian; Tri, Triassic; J, Jurassic; K, Cretaceous; Tert, Tertiary.

Permian decline:
a species–area
relationship?

The dramatic decline in the number of families of shallow-water invertebrates as a result of the mass extinction at the end of the Permian (Figure 21.26a) could have been a result of the coalescence of the earth's continents to produce the single supercontinent of Pangaea. The joining of the continents produced a marked reduction in the area occupied by shallow seas (which occur around the periphery of continents) and thus a marked decline in the area of habitat available to shallow-water invertebrates. Moreover, at this time the world was subject to a prolonged period of global cooling in which huge quantities of water were locked up in enlarged polar caps and glaciers, causing a widespread reduction of warm shallow sea environments. Thus, a species–area relationship may be invoked to account for the reduction in richness of this fauna. In a related context, Rosenzweig (2003) reports a significant positive relationship when the number of fossil plant species of the northern hemisphere from various periods in the earth's history is plotted against the total area of the land mass during the period in question (11 periods with nonoverlapping lists of fossil species).

competitive
displacement among
the major plant
groups?

The analysis of fossil remains of vascular land plants (Figure 21.26b) reveals four distinct evolutionary phases: (i) a Silurian–mid-Devonian proliferation of early vascular plants; (ii) a subsequent Late Devonian–Carboniferous radiation of fern-like lineages; (iii) the appearance of seed plants in the late Devonian and the adaptive radiation to a gymnosperm-dominated flora; and (iv) the appearance and rise of the flowering plants (angiosperms) in the Cretaceous and Tertiary. It seems that after an initial invasion of the land, made possible by the appearance of roots, the diversification of each plant group coincided with a decline in species numbers of the previously dominant group. In two of the transitions (early plants to gymnosperms, and gymnosperms to angiosperms), this pattern may reflect the competitive displacement of older, less specialized taxa by newer and presumably more specialized taxa.

coevolution as a
driver of richness
increase?

The first undoubtedly phytophagous insects are known from the Carboniferous. Thereafter, modern orders appeared steadily (Figure 21.26c) with the Lepidoptera (butterflies and moths) arriving last on the scene, at the same time as the rise of the angiosperms. Reciprocal evolution and counterevolution between plants and herbivorous insects has almost certainly been, and still is, an important mechanism driving the increase in richness observed in both land plants and insects through their evolution.

extinctions of large
animals: prehistoric
overexploitation?

Toward the end of the last ice age, the continents were much richer in large animals than they are today. For example, Australia was home to many genera of giant marsupials; North America had its mammoths, giant ground sloths, and more than 70 other genera of large mammals; and New Zealand and Madagascar were home to giant flightless birds, the moas (Dinorthidae) and elephant bird (*Aepyornis*), respectively. Over the past 50,000 years or so, a major loss of this biotic diversity has occurred over much of the globe. The extinctions particularly affected large terrestrial animals (Figure 21.27a); they were more pronounced in some parts of the world than others; and they occurred at different times in different places (Figure 21.27b). The extinctions broadly mirror patterns of human migration. Thus, the arrival in Australia of ancestral aborigines occurred 40,000 or more years ago; stone spear points became abundant throughout the United States about 11,500 years ago; and humans have been in both Madagascar and New Zealand for about 1000 years. It seems likely, therefore, that the arrival of efficient human hunters led to the rapid overexploitation of vulnerable and profitable large prey. Africa, where humans originated, shows much less evidence of loss, perhaps because coevolution of large animals alongside early humans provided ample time for them to develop effective defenses (Owen-Smith, 1987).

The Pleistocene extinctions herald the modern age, in which the influence upon natural communities of human activities has been increasing dramatically (see Chapters 7, 15 and 22).

21.7 Species richness and ecosystem functioning

switching focus: how
does species richness
influence ecosystem
functioning?

In this penultimate chapter section, rather than seeking to discern and explain patterns in species richness we switch focus to address the consequences of variations in species richness for ecosystem functioning. Specifically, we deal with productivity, decomposition and the flux of nutrients and water (discussed more fully in Chapters 11, 17 and 18). Understanding the role of biodiversity in ecosystem processes is important both for fundamental and practical reasons, because it has implications for how humans respond to biodiversity loss. We have already discussed the effects of richness on the stability of ecosystem functioning (see Section 20.3.6). Here we present examples of studies from various ecosystem types that reveal relationships between species richness and the ecosystem processes themselves, before proceeding to consider several hypotheses to account for such relationships.

21.7.1 Positive relationships between species richness and ecosystem functioning

increased species
richness resulting
in . . .

. . . higher
productivity, . . .

As part of an international research effort, standard protocols were used at eight European field sites to investigate the effect of reduction in grassland species richness on primary productivity

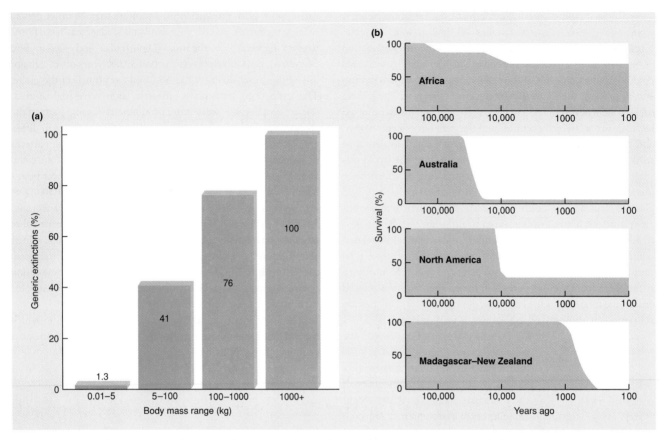

Figure 21.27   (a) The percentage of genera of large mammalian herbivores that have gone extinct in the last 130,000 years is strongly size dependent (data from North and South America, Europe and Australia combined). (After Owen-Smith, 1987.) (b) Percentage survival of large animals on three continents and two large islands (New Zealand and Madagascar). (After Martin, 1984.)

(measured as accumulation of above-ground biomass) by synthesizing grassland communities with different numbers of species of grasses, nitrogen-fixing legumes and other herbaceous species (herbs). While the detailed results differed among sites, there was an overall log-linear reduction of average productivity with loss of species (Figure 21.28a). For a given species richness, there was also a decline in productivity with a reduction in the number of functional groups (grasses, legumes, herbs) (Figure 21.28b).

. . . faster
decomposition . . .

Jonsson and Malmqvist (2000) studied the effect on decomposition of species richness of the larvae of three stonefly species that feed on tree leaves falling into streams. Every replicate had 12 stonefly larvae present – 12 of one species, six of each of two species, or four of each of three species (with 10 replicates of all possible combinations). The rate of loss of leaf mass during a 46-day mesocosm experiment was positively related to species richness (Figure 21.28c).

. . . and reduced
nutrient loss

The microbial decomposition of soil organic matter releases ammonium ions (nitrogen mineralization).

Seven years after commencing a replicated manipulation of grassland species richness in Minnesota, USA, Zak et al. (2003) found that mineralization rates in soil samples were positively related to plant species richness (Figure 21.28d). Nutrient flux was also found to be related to species richness of submersed macrophytes in mesocosms simulating wetland communities – the uptake of phosphorus by algae growing on the surface of the macrophytes was greater (and total phosphorus loss from the mesocosms was reduced) when more macrophytes were present (Figure 21.28e).

21.7.2 Contrasting explanations for richness–ecosystem process relationships

In an intense and sometimes acrimonious debate (Kaiser, 2000; Loreau et al., 2001), three principal hypotheses have been advanced to account for positive relationships between species richness and ecosystem functioning.

contrasting
hypotheses

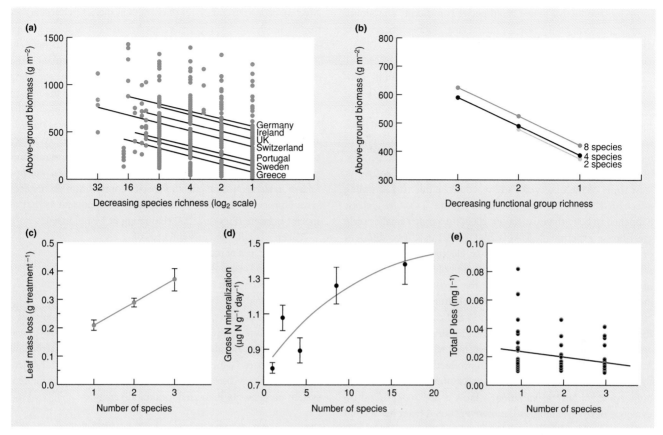

Figure 21.28 (a) Primary productivity (measured as above-ground biomass accumulation after 2 years) versus species richness in a large number of simulated grassland assemblages across Europe (regression lines are shown for each country). (After Hector *et al.*, 1999.) (b) Primary productivity versus functional group richness in the European grasslands combined. (After Hector *et al.*, 1999.) (c) Decomposition rate (loss of leaf mass) versus number of stream-dwelling shredding stonefly species present. (After Jonsson & Malmqvist, 2000.) (d) Gross nitrogen mineralization (per gram of soil) versus plant species richness in a 7-year grassland manipulation. (After Zak *et al.*, 2003.) (e) Rate of loss of phosphorus from mesocosms containing one, two or three submersed macrophyte species, simulating wetland communities. (After Engelhardt & Ritchie, 2002.)

complementarity . . .

On one hand, if species show niche differentiation (see Chapter 8) they may use resources in complementary ways, utilizing a greater proportion of available resources (see Figure 21.1d) and contributing to a higher level of ecosystem productivity (or decomposition or nutrient cycling). This is the *complementarity hypothesis*.

. . . and facilitation . . .

A second idea, the *facilitation hypothesis*, recognizes that some species may have positive effects on the ecosystem role played by other species. For example, some submersed wetland macrophytes facilitate colonization by algae more than others do (Figure 21.28d) (Engelhardt & Ritchie, 2002).

. . . predict 'overyielding' and point to the value of conserving biodiversity . . .

Both the complementarity and facilitation hypotheses predict 'over-

yielding', where productivity or decomposition rates in multi-species communities are faster than in communities with fewer species. As long as either of these hypotheses applies, a case can be made for the need for management to conserve biodiversity to maintain ecosystem functioning.

On the other hand, it may be that positive relationships between richness and functioning are artefacts of the species that happen to be assembled together in experiments. The so-called *sampling effect hypothesis* suggests that the more species are present in an assemblage, the more likely it is to contain, by chance, a highly competitive or productive species. Thus, species-rich communities may on average be more productive because they are more likely to contain an especially productive species. In this case, overyielding will not be seen (a monoculture of the productive species will

. . . but the sampling effect hypothesis does not

be no less productive than a multispecies community containing it) and there will be no case for conserving biodiversity *per se* to maintain ecosystem functioning.

some studies have not revealed overyielding . . .

The multinational study of Hector *et al.* (1999) produced results consistent with the complementarity hypothesis because grassland productivity was greater when more functional plant types were present, something that is likely to be a reflection of niche differentiation. However, the work was criticized because of a failure to test properly for overyielding and because at least some of the observed patterns could be due to a sampling effect of whether the nitrogen-fixing legume *Trifolium pratense* happened to be in the mix (Kaiser, 2000). In a much smaller scale greenhouse experiment, Mikola *et al.* (2002) used two different experimental designs. In the first, like Hector *et al.* (2002), they selected plant species at random from a pool to produce a range of richnesses (richness design). In the second, richness levels included deliberately replicated monocultures, bicultures, tricultures and six-species mixes (richness and composition design). In both cases, there was a positive relationship between richness and productivity (total shoot mass); but whereas in the richness design 34% of total variation in productivity was explained by species richness, in the richness and composition design, this only explained 16% of variation. Mikola *et al.* (2002) found no evidence of overyielding and, moreover, noted from the second experimental design that productivity was greatly affected by the presence of one species, *Trifolium hybridum*, another nitrogen-fixing legume. Both observations are consistent with the sampling effect hypothesis.

. . . while other studies have

However, in another field-scale experiment, designed so that the most productive monocultures could be compared with multispecies plots that included the most productive species, Tilman *et al.* (2001) gathered evidence of overyielding, finding that many high-richness plots had greater productivity that the single best-performing monoculture. Moreover, there was clearly 'overyielding' of decomposition when more stonefly species were present in Jonsson and Malmqvist's (2000) stream experiment (see Figure 21.28c).

It is clear that the results of a range of experimental manipulations of species richness differ in the extent to which the three hypotheses apply, and it should be noted that these are by no means mutually exclusive. As further studies accumulate, generalizations can be expected to emerge. For example, it may be that complementarity will be most prominent in situations where niche differentiation is most marked.

a tri-trophic level experiment showing overyielding

In the studies discussed above, the emphasis has been on manipulating the richness of a single trophic level (plants or detritivores). In contrast, Downing and Leibold (2002) investigated the effect on ecosystem processes of changes in species richness across trophic levels (one, three or five species in each of three groups – macrophytes, benthic grazers and invertebrate predators – in field mesocosms mimicking ponds). They aimed to disentangle the effects of species richness from species composition (the sampling effect) by nesting and replicating seven particular combinations of species within each richness level. Ecosystem productivity (mainly by periphyton, phytoplankton and microorganisms) was significantly greater at the highest richness level than at the two lower ones; ecosystem respiration showed a similar but nonsignificant pattern; while decomposition (weight loss of tree leaves) was not related to richness. The effect of species composition on the ecosystem processes was at least as statistically significant as that of species richness (Figure 21.29) (Downing & Leibold, 2002).

practical importance of biodiversity loss (or gain?)

Taken overall, the consequences of the on-going loss of biodiversity can be expected to be complex and difficult to predict unless compositional changes are also accounted for, and this will be particularly so in the context of whole food webs. Paradoxically, however, while biodiversity is generally declining globally, local biodiversity is commonly increasing because of the arrival of invaders (Sax & Gaines, 2003). Thus, a more meaningful objective will in many cases be to determine the consequences for ecosystem processes of increased local biodiversity.

21.8 Appraisal of patterns in species richness

richness patterns: generalizations and exceptions

There are many generalizations that can be made about the species richness of communities. We have seen how richness may peak at intermediate levels of available environmental energy or of disturbance frequency, and how richness declines with a reduction in island area or an increase in island remoteness. We find also that species richness decreases with increasing latitude, and declines or shows a hump-backed relationship with altitude or depth in the ocean. It increases with a rise in spatial heterogeneity but may decrease with an increase in temporal heterogeneity (increased climatic variation). It increases, at least initially, during the course of succession and with the passage of evolutionary time. However, for many of these generalizations important exceptions can be found, and for most of them the current explanations are not entirely adequate.

bidirectional relationships between richness and ecosystem characteristics

The results of descriptive surveys of species richness may, at first glance, seem at odds with the outcomes of experimental manipulations. Many experiments, for example, have shown that increasing the number of species leads to more productivity (see Section 21.7). On the other hand, we saw in Section 21.3.1 that

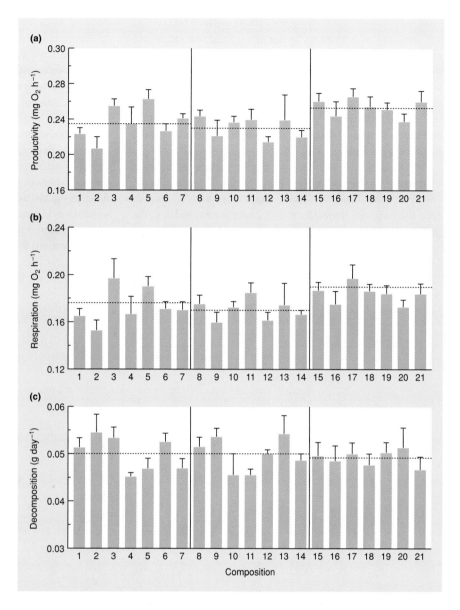

Figure 21.29 The response of (a) ecosystem productivity, (b) ecosystem respiration and (c) decomposition to species composition nested in species richness. Cases 1–7 are unique species combinations at low richness (one species each of macrophyte, benthic grazer and invertebrate predator), 8–14 at intermediate richness (three species of each group) and 15–21 at high richness (five species of each group). Average productivity was significantly higher in the high richness case (the dotted lines show the overall means for richness classes) but there were no significant richness effects on respiration or decomposition. The variability within richness classes reveals the strong effect of species composition on all the ecosystem processes. (After Downing & Leibold, 2002.)

more productive environments may contain more or fewer species than unproductive environments. It is important to understand that relationships such as these are bidirectional; changes in biodiversity can be both a cause and a consequence of changes in productivity (Worm & Duffy, 2003), further complicating the picture.

the pressing need to understand biodiversity and its significance

Unraveling richness patterns is one of the most difficult and challenging areas of modern ecology. No single mechanism is likely to adequately explain a particular pattern, and patterns at local scales are likely to be influenced by processes operating at both local and regional scales. Clear, unambiguous predictions and tests of ideas are often very

difficult to devise and will require great ingenuity of future generations of ecologists. Because of the increasing importance of recognizing and conserving the world's biodiversity, though, it is crucial that we come to understand thoroughly these patterns in species richness. We assess the adverse effects of human activities, and how they may be remedied, in Chapter 22.

Summary

Why do some communities contain more species than others? Are there patterns or gradients of species richness? If so, what are the reasons for these patterns? There are plausible answers to the questions we ask, but these answers are by no means conclusive.

In simple terms, the number of species that can be packed into a community is determined by the size of the realized niches and the extent to which they overlap, in relation to the range of available resources. Competition and predation can modify the outcome in predictable ways. In addition, a community will contain more species the more fully saturated it is; a phenomenon that can be addressed by plotting the relationship between local diversity and regional diversity (the number of species that could theoretically colonize).

We describe the influence on species richness of a range of spatially varying factors (productivity, spatial heterogeneity, environmental harshness) and temporally varying factors (climatic variation, environmental age, habitat area) and describe patterns of richness that increase, decrease or show hump-backed relationships with these factors. Interactions among factors (e.g. productivity with grazing or disturbance) are frequently involved in determining patterns. We pay particular attention to island biogeography theory and the interaction between immigration and extinction rates in determining species richness – in relation to island area and remoteness.

Next we turn to gradients in species richness, drawing on examples relating to latitude, altitude, depth, succession and evolutionary history. Explanations for these patterns invoke all the factors discussed earlier.

In the final section, rather than seeking to discern and explain patterns in species richness we switch focus to address the consequences of variations in species richness for ecosystem functioning, discussing productivity, decomposition and the flux of nutrients in turn. Understanding the role of biodiversity in ecosystem processes is important for practical reasons, because it has implications for how humans should respond to biodiversity loss.

Chapter 22
Ecological Applications at the Level of Communities and Ecosystems: Management Based on the Theory of Succession, Food Webs, Ecosystem Functioning and Biodiversity

22.1 Introduction

This is the last of the trilogy of chapters dealing with the application of ecological theory. In the first, Chapter 7, we considered how our understanding at the level of individual organisms and of single populations – related to niche theory, life history theory, dispersal behavior and intraspecific competition – can provide solutions to a multitude of practical problems. The second, Chapter 15, used the theory of the dynamics of interacting populations to guide the control of pests and the sustainable harvesting of wild populations. This final synthesis recognizes that individuals and populations exist in a web of species interactions embedded in a network of energy and nutrient flows. Thus, we deal with the application of theory related to succession (Chapter 16), food webs and ecosystem functioning (Chapters 17–20) and biodiversity (Chapter 21).

application of community and ecosystem theory

Community composition is hardly ever static and, as we saw in Chapter 16, some temporal patterns are quite predictable. Management objectives, on the other hand, often seem to require stasis – the annual production of an agricultural crop, the restoration of a particular combination of species or the long-term survival of an endangered species. Management will sometimes be ineffective in these situations if managers fail to take into account underlying successional processes (see Section 22.2).

We turn to the application of theory about food webs and ecosystem functioning in Section 22.3. Every species of concern to managers has its complement of competitors, mutualists, predators and parasites, and an appreciation of such complex interactions is often needed to guide management action (see Section 22.3.1). Farmers seek to maximize economic returns by manipulating ecosystems with irrigation and by applying fertilizers. But nutrient runoff from farm land, together with treated or untreated human sewage, can upset the functioning of aquatic ecosystems through the process of cultural eutrophication (nutrient enrichment), increasing productivity, changing abiotic conditions and altering species composition. Our understanding of lake ecosystem functioning has provided guidelines for 'biomanipulation' of lake food webs to reverse some of the adverse effects of human activities (see Section 22.3.2). Moreover, knowledge of terrestrial ecosystem functioning can help determine optimal farm practices, where crop productivity involves minimal input of nutrients (see Section 22.3.3). The setting of ecosystem restoration objectives (and the ability to monitor whether these are achieved) requires the development of tools to measure 'ecosystem health', a topic we deal with in Section 22.3.4.

So much of the planet's surface is used for, or adversely affected by, human habitation, industry, mining, food production and harvesting, that one of our most pressing needs is to plan and set aside networks of reserved land. The augmentation of existing reserves by further areas needs to be done in a systematic way to ensure that biodiversity objectives are achieved at minimal cost (because resources are always limited). Section 22.4 describes how our knowledge of patterns of species richness (see

Chapter 21) can be used to design networks of reserves, whether specifically for conservation (see Section 22.4.1) or for multiple uses, such as harvesting, tourism and conservation combined (see Section 22.4.2).

> ecological applications often involve economic and sociopolitical considerations

Finally, in Section 22.5 we deal with a reality that applied ecologists cannot ignore. The application of ecological theory never proceeds in isolation. First, there are inevitably economic considerations – how can farmers maximize production while minimizing costs and adverse ecological consequences; how can we set economic values for biodiversity and ecosystem functioning so that these can be evaluated alongside profits from forestry or mining; how can returns be maximized from the limited funds available for conservation? These issues are discussed in Section 22.5.1. Second, there are almost always sociopolitical considerations (see Section 22.5.2) – what methods can be used to reconcile the desires of all interested parties, from farmers and harvesters to tourism operators and conservationists; should the requirements for sustainable management be set in law or encouraged by education; how can the needs and perspectives of indigenous people be taken into account? These issues come together in the so-called triple bottom line of sustainability, with its ecological, economic and sociopolitical perspectives (see Section 22.5.3).

22.2 Succession and management

22.2.1 Managing succession in agroecosystems

Gardeners and farmers alike devote considerable effort to fighting succession by planting desired species and weeding out unwanted competitors.

> farmers often have to resist successional processes

In an attempt to maintain the characteristics of an early successional stage – growing a highly productive annual grass – arable farmers are forced to resist the natural succession to herbaceous perennials (and beyond, to shrubs and trees; see Section 16.4.5). Menalled et al. (2001) compared the impact of four agricultural management systems on the weed communities that developed in Michigan, USA, over a period of 6 years (consisting of two rotations from corn to soybean to wheat). Above-ground weed biomass and species richness were lowest in the conventional system (high external chemical input of synthetic fertilizer and herbicides, ploughed), intermediate in the no-till system (high external chemical input, no ploughing) and highest in the low-input (low external chemical input, ploughed) and organic systems (no external chemical input, ploughed) (Figure 22.1). A widely varying mixture of monocot (grass) and dicot species were represented in the conventional treatment and an equally unpredictable set of annual grasses dominated the no-till treatment.

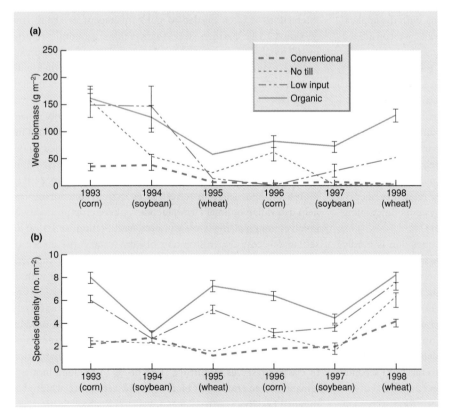

Figure 22.1 (a) Weed biomass and (b) weed species richness in four agricultural management treatments (see key; six replicate 1 ha plots in each treatment) over a period of 6 years consisting of two rotations of corn (*Zea mays*) to soybean (*Glycine max*) to wheat (*Triticum aestivum*). (After Menalled *et al.*, 2001.)

On the other hand, the weed communities of the low-input and organic treatments were more constant: an annual dicot (*Chenopodium album*) and two perennial weeds (*Trifolium pratense* and *Elytrigia repens*) were the dominant species under these conditions. Menalled *et al.* (2001) point out the potential advantages of a management system that fosters a more predictable weed community because control treatments can then be designed specifically against the species concerned.

benzoin 'gardening' in Sumatra – rapid reversion to forest

Other forms of agricultural 'gardening' pose fewer problems in the way they interrupt succession. Benzoin is an aromatic resin, used to make incense, flavoring and medicinal products, which for hundreds of years has been tapped from the bark of tropical trees in the genus *Styrax*. Benzoin still provides a significant income to many villagers in Sumatra who plant benzoin gardens (*S. paralleloneurum*) after clearing the understory in 0.5–3.0 ha areas of montane broadleaf forest. Two years later, farmers thin all the larger trees to allow light to reach the saplings (the thinnings are left in the garden) and annual tapping begins after 8 years. Yields typically decline after 30 years but resin may be harvested for up to 60 years before the garden is left to return to the forest. Garcia-Fernandez *et al.* (2003) identified three categories of garden: G1 was the most plantation-like, with intensive thinning and high densities of *S. paralleloneurum* trees, and G3 was the most forest-like. Total tree species richness was high in plots of primary (pristine) and 'secondary' forest (30–40 years after gardening had ceased) and also in the gardens, except for the most intensely managed situation where richness was significantly lower (but nevertheless with an average of 26 tree species) (Figure 22.2a). As predicted by succession theory (see Section 16.4), climax species typical of mature forest were most common in primary forest and there was a more even mix of pioneer and mid-successional tree species in secondary forest and in the least intensively managed gardens (G3) (Figure 22.2.b). However, gardens with an intermediate or high intensity of management were dominated by mid-successional trees (mainly because benzoin trees are in this class). It is not unusual for indigenous people to be aware of a wide range of uses for forest plants. Figure 22.2c shows the representation in the garden and forest plots of trees in each of four classes: no known use (12%), subsistence use (food, fiber or medicine; 42%), local market use (23%) and international market use (23%). The international category dominated in intensively managed gardens (i.e. benzoin and its products) whereas trees in the subsistence and local market categories were well represented in less intensively managed gardens and in primary and secondary forest. Although benzoin management requires competing vegetation to be trimmed, tree species richness remains quite high even in the most intensively managed gardens. This traditional form of forest gardening maintains a diverse community whose structure allows rapid recovery to a forest community when tapping ceases. It represents a good balance between development and conservation.

Fire is an important resource management tool for Australian aboriginal people such as the clan who own the Dukaladjarranj area of northeastern Arnhem Land (Figure 22.3a). Burning, to provide green forage for game animals, is planned by custodians (aboriginal people with special responsibilities for the land) and focuses initially on dry grasses on higher ground, moving progressively to moister sites as these dry out with the passage of the season. Each fire is typically of low intensity and small in extent, producing a patchy mosaic of burned and unburned areas and thus a diversity of habitats at different successional stages (see Section 16.7.1). Towards the end of the dry season, when it is very hot and dry, burning ceases except in controllable situations such as the reburning of previously burnt areas. In a collaboration between indigenous people and professional ecologists, Yibarbuk *et al.* (2001) lit experimental fires to assess their impact on the flora and fauna. They found that burned sites attracted large kangaroos and other favored game and that important plant foods, such as yams, remained abundant (results that would have hardly been a surprise to the indigenous collaborators) (Figure 22.3b). Fire-sensitive vegetation in decline elsewhere, such as *Callitris intratropica* woodlands and sandstone heath dominated by myrtaceaous and proteaceaous shrubs, remained well represented in the study area. In addition, the Dukaladjarranj area compares favorably with the Kakadu National Park, a conservation area with high vertebrate and plant diversity. Thus, Dukaladjarranj contains several rare species and a number of others that have declined in unmanaged areas and, moreover, the representation of exotic plant and animal invaders was remarkably low. The traditional regime, with its many small, low-intensity fires, contrasts dramatically with the more typical contemporary pattern of intensive, uncontrolled fires near the end of the dry season. These blaze across vast areas of western and central Arnhem Land (sometimes covering more than 1 million ha) that are unoccupied and unmanaged, and regularly find their way onto the western rim of the Arnhem Land plateau and into Kakadu and Nitmiluk National Parks (Figure 22.3a). It seems that continued aboriginal occupation of the study area and the maintenance of traditional fire management practices limits the accumulation of fuel (in fire-promoting grass species and in litter), reducing the likelihood of massive fires that can eliminate fire-sensitive vegetation types. A return to indigenous-style burning seems to hold promise for the restoration and conservation of threatened species and communities in these landscapes (Marsden-Smedley & Kirkpatrick, 2000) and provides important clues for the management of fire-prone areas in other parts of the world.

aboriginal burning regime provides resources and maintains biota

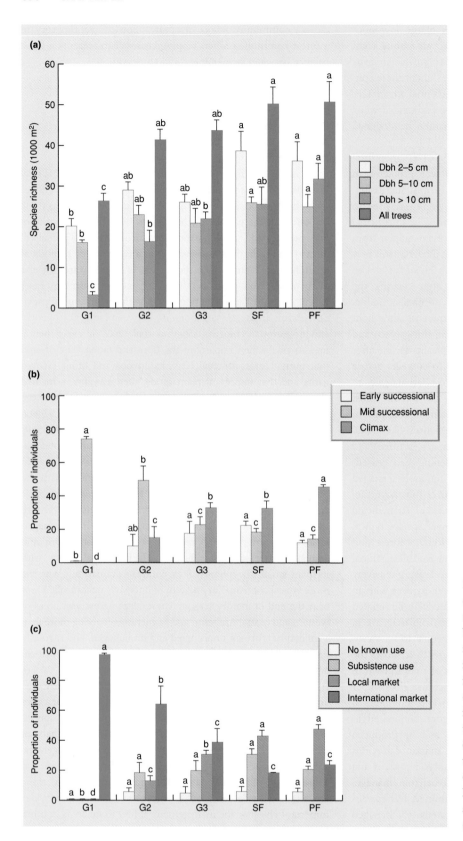

Figure 22.2 (a) Tree species richness in different tree size classes (Dbh is diameter at breast height) in three categories of benzoin garden (G1, most intensely managed; G2, intermediate; G3, least intensively managed) and in secondary forest (SF; 30–40 years after abandonment of benzoin gardens) and in primary forest (PF). (b) Percentage of individual trees in three successional categories. (c) Percentage of individual trees in various utility categories. Each data point is based on three replicate 1 ha plots. Different letters above each type of bar indicate statistically significant differences. (After Garcia-Fernandez et al., 2003.)

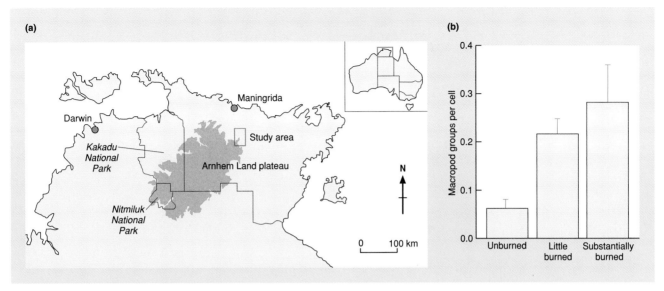

Figure 22.3 (a) Location of the fire management study area near the northeastern end of the Arnhem Plateau in the Northern Territory of Australia; the position of two National Parks is also shown. (b) Mean number (+2 SE) of kangaroo groups sighted during a helicopter survey of 0.25 km² plots with different recent burning histories. (After Yibarbuk *et al.*, 2001.)

22.2.2 Managing succession for restoration

restoration sometimes needs no intervention . . .

The goal of restoration ecology is often a relatively stable successional stage (Prach *et al.*, 2001) and ideally a climax. Once an undesirable land use ceases, managers need not intervene if they are prepared to wait for natural succession to run its course. Thus, abandoned rice fields in mountainous central Korea proceed from an annual grass stage (*Alopecurus aequalis*), through forbs (*Aneilema keisak*), rushes (*Juncus effusus*) and willows (*Salix koriyanagi*), to reach a species-rich and stable alder woodland community (*Alnus japonica*) within 10–50 years (Figure 22.4) (Lee *et al.*, 2002). Succession cannot always be counted on to promote habitat restoration, especially if natural sources of seeds are small and distant, but this was not the case here. In fact, the only active intervention worth considering is the dismantling of artificial rice paddy levees to accelerate, by a few years, the early stages of succession.

. . . but may be hastened by species introductions

Meadow grasslands subject to agricultural intensification, including the application of artificial fertilizers and herbicides and heavy grazing regimes, have dramatically fewer plant species than grasslands under historic 'traditional' management. The restoration of biodiversity in these situations involves a secondary succession that typically takes more than 10 years; it can be achieved by returning to a traditional regime without

mineral fertilizer in which hay is cut in mid-July and cattle are grazed in the fall (Smith *et al.*, 2003). However, in contrast to the mountain rice field case discussed above, meadow community recovery in lowland England by natural colonization from seed rain or the seed bank is a slow and unreliable process (Pywell *et al.*, 2002). Fortunately, recovery can be speeded up by sowing a species-rich mixture of seeds of desirable plants adapted to the prevailing conditions. Thus, in a 4-year study comparing species richness of grasses and forbs in plots that were unsown (natural regeneration from cereal stubble) or sown with a species-rich seed mixture (containing more than 25 species), the sown plots had twice as many established species in years 1 and 2 than naturally regenerating plots (means of 26.4 and 22.0 compared with 10.4 and 11.3, respectively). By year 4 there was little difference in species richness (22.0 versus 18.7) but the sown treatment had a species composition that included late successional grassland species and was much closer to that found in local nonintensively farmed grasslands (Pywell *et al.*, 2002).

restoration timetable for salt marsh animal communities

Restoration objectives often include recovery not just of plants but of the animal components of communities too. Tidal salt marshes are much rarer than they once were because of drainage and tidal interference through the installation of tide gates, culverts and dykes. The restoration of tidal action (by removing tide gates, etc.) and thus of links between the marshes, estuaries and the larger coastal system along the Long Island Sound shoreline of Connecticut, USA, led to the recovery of salt marsh vegetation, including *Spartina*

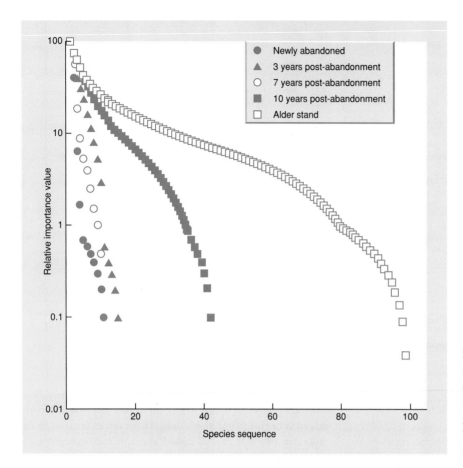

Figure 22.4 Rank–abundance diagram of plant species grouped by site age (time since abandonment of rice paddy field). Importance values are the relative ground cover of the plant species present. The alder stand was 50 years old. (After Lee *et al.*, 2002.)

alterniflora, *S. patens* and *Distichlis spicata*. Recovery was relatively fast (increasing at a rate of 5% of total area per year) where tidal flooding was frequent (i.e. at lower elevations and with higher soil watertables) but was otherwise slow (about 0.5% of total area per year). In the fast recovery sites, it took 10–20 years to achieve 50% coverage of specialist salt marsh plants. Characteristic salt marsh animals followed a similar timetable. Thus, in five sites in marshes at Barn Island that have been recovering for known periods (and for which nearby reference marshes are available for comparison), the high marsh snail *Melampus bidentatus* only achieved densities comparable to reference conditions after 20 years (Figure 22.5a). The bird community also took 10–20 years to reach a community composition similar to reference circumstances. Marsh generalists that forage and breed both in uplands and tidal wetlands (such as song sparrows *Melospiza melodia* and red-winged blackbirds *Agelaius phoeniceus*) dominated early in the restoration sequence, to be replaced later by marsh specialists such as marsh wrens *Cistothorus palustris*, snowy egrets *Egretta thula* and spotted sandpipers *Actitis macularia*) (Figure 22.5b). Typical fish communities in restoration salt marsh creeks recovered more quickly, within 5 years. It seems that the restoration of a natural tidal regime sets marshes on trajectories towards restoration of full

ecological functioning, although this generally takes one or more decades. The process can probably be speeded up if managers plant salt marsh species.

22.2.3 Managing succession for conservation

Some endangered animal species are associated with a particular stage of succession and their conservation then depends on an understanding of the successional sequence; intervention may be required to maintain their habitat at an appropriate successional stage. An intriguing example is provided by a giant New Zealand insect, the weta *Deinacrida mahoenuiensis* (Orthoptera; Anostostomatidae). This species, which was believed extinct after being formerly widespread in forest habitats, was discovered in the 1970s in an isolated patch of gorse (*Ulex europaeus*). Ironically, in New Zealand gorse is an introduced weed that farmers spend much time and effort attempting to control. Its dense, prickly sward provides a refuge for the giant weta against other introduced

understanding succession is crucial for the conservation of a rare insect

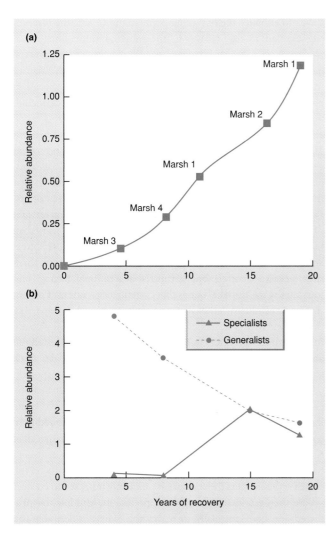

Figure 22.5   (a) Relative abundance of the snail *Melampus bidentatus* (expressed as mean density in the restoration area divided by density in a nearby reference marsh) in five sites in four marshes at Barn Island, Connecticut, that differ in the period since a natural tidal regime was restored. A relative abundance of 1.0 indicates a full recovery of this species. (b) Relative abundance (recovering/reference) of birds considered as salt marsh specialists (▲) and salt marsh generalists (●) on Barn Island marshes plotted against years of restoration at the time the counts were conducted. Again a relative abundance of 1.0 indicates full restoration of the specialist or generalist guild. (After Warren *et al.*, 2002.)

pests, particularly rats but also hedgehogs, stoats and possums, which readily captured wetas in their original forest home. New Zealand's Department of Conservation purchased this important patch of gorse from the landowner who insisted that cattle should be permitted to overwinter in the reserve. Conservationists were unhappy about this but the cattle sub-

sequently proved to be part of the weta's salvation. By opening up paths through the gorse, the cattle provided entry for feral goats that browse the gorse, producing a dense hedge-like sward and preventing the gorse habitat from succeeding to a stage inappropriate to the wetas. This story involves a single endangered endemic insect together with a whole suite of introduced pests (gorse, rats, goats, etc.) and introduced domestic animals (cattle). Before the arrival of people in New Zealand, the island's only land mammals were bats, and New Zealand's endemic fauna has proved to be extraordinarily vulnerable to the mammals that arrived with people. However, by maintaining gorse succession at an early stage, the grazing goats provide a habitat in which the weta can escape the attentions of the rats and other predators.

22.3   Food webs, ecosystem functioning and management

22.3.1   Management guided by food web theory

Studies that unravel the complex interactions in food webs (dealt with in Chapter 20) can provide key information for managers on issues as diverse as minimizing human disease risk, setting objectives for marine protected areas or predicting invaders with the most potential to disrupt ecosystem functioning.

understanding food webs for management . . .

22.3.1.1   *Lyme disease*

Lyme disease, which if untreated can damage the heart and nervous system and lead to a type of arthritis, each year affects tens of thousands of people around the world. It is caused by a spirochete bacterium (*Borrelia burgdorferi*) carried by ticks in the genus *Ixodes*. The ticks take 2 years to pass through four developmental stages, involving a succession of vertebrate hosts. Eggs are laid in the spring and uninfected larvae take a single blood meal from a host (usually a small mammal or bird) before dropping off and molting into the overwintering nymphal stage. Infected hosts transmit the spirochete to the larval ticks, which remain infective throughout their lives (i.e. after they have molted into nymphs and subsequently into adults). Next year the nymph seeks a host in the spring/early summer for another single blood meal; this is the most risky stage for human infection because the nymphs are small and difficult to detect and attach to hosts at a time of peak human recreation in forests and parks. Between 1 and 40% of nymphs carry the spirochete in Europe and the USA (Ostfeld & Keesing, 2000). The nymph drops off and molts into an adult that takes a final blood meal and reproduces on a third host, often a larger mammal such as a deer.

. . . of disease . . .

The most abundant small mammal host in the eastern USA, and by far the most competent transmitter of the spirochete, is the white-footed mouse (*Peromyscus leucopus*). Jones *et al.* (1998) added acorns, a preferred food of the mice, to the floor of an oak forest to simulate one of the occasional crop masting years that occur, and found mice numbers increased the following year and that the prevalence of spirochete infection in nymphal black-legged ticks (*Ixodes scapularis*) increased 2 years after acorn addition. It seems that despite the complexity of the food web of which the spirochetes are part, it may be possible to predict high-risk years for transmission to humans well in advance by monitoring the acorn crop. Of further interest to managers is evidence that outbreaks of pest moths, whose caterpillars can cause massive defoliation of forest, may be more likely to occur 1 year after very poor acorn crops, when mice, which also feed on moth pupae, are rare (Jones *et al.*, 1998).

A final point about disease transmission is worth emphasizing. The potential mammal, bird and reptile hosts of ticks show a great variation in the efficiency with which they are competent to transmit the spirochete to the tick. Ostfeld and Keesing (2000) hypothesized that a high species richness of potential hosts would result in lower disease prevalence in humans if the high transmission efficiency of the key species (such as white-footed mice) is diluted by the presence of a multitude of less competent species. (Note that what really matters is whether the total number of individuals of the more competent species is 'swamped' by a large number of individuals of the less competent ones; relative abundance is important as well as species richness.) Ostfeld and Keesing produced evidence in favor of their hypothesis in the form of a negative relationship between disease cases and small mammal host richness in 10 regions of the USA. Unfortunately, cases of Lyme disease were concentrated in more northerly states, where species richness was lower, suggesting that both disease and mammal richness follow a latitudinal pattern. Thus, whether the link between the two is causal or incidental remains to be determined. This is an important question, however, because a negative relationship between host diversity and disease transmission for vector-borne diseases (including Chagas' disease, plague and Congo hemorrhagic fever) would provide one more reason for managers to act to maintain biodiversity.

22.3.1.2   Management for an abalone fishery

... and of both harvested shellfish and a charismatic top predator

Sometimes biodiversity can be too high to achieve particular management objectives! Commercial and recreational fisheries for abalones (gastropods in the family Haliotidae) are prone to collapse through overfishing. Adult abalones do not move far and the protection of broodstock in reserved portions of their coastal marine habitat has potential for promoting the export of planktonic larvae to enhance the harvested populations outside the reserves (see Section 15.4.2). However, the most common function of marine-protected areas is the conservation of biodiversity, and the question arises whether protected areas can serve both fisheries management and biodiversity objectives. A keystone species in coastal habitats along the Pacific coast of North America, including those in California, is the sea otter (*Enhydra lutris*), hunted almost to extinction in the 18th and 19th centuries but increasingly widespread as a result of protected status. Sea otters eat abalones, and valuable fisheries for red abalone (*Haliotis rufescens*) developed while sea otters were rare; now there is concern that the fisheries will be unsustainable in the presence of sea otters. Fanshawe *et al.* (2003) compared the population characteristics of abalone in sites along the Californian coast that varied in harvest intensity and sea otter presence: two sites lacked sea otters and had been 'no-take' abalone zones for 20 years or more, three sites lacked sea otters but permitted recreational fishing, and four sites were 'no-take' zones that contained sea otters. The aim was to determine whether marine-protected areas can help make the abalone fishery sustainable when all links in the food web are fully restored. Sea otters and recreational harvest influenced red abalone populations in similar ways but the effects were very much stronger where sea otters were present. Red abalone populations in protected areas had substantially higher densities (15–20 abalone per 20 m^2) than in areas with sea otters (< 4 per 20 m^2), while harvested areas generally had intermediate densities. In addition, 63–83% of individual abalones in protected areas were larger than the legal harvesting limit of 178 mm, compared with 18–26% in harvested areas and less than 1% in sea otter areas. Finally, in the presence of sea otters the abalones were mainly restricted to crevices where they are least vulnerable to predation. Multiple-use protected areas are not likely to be feasible where a desirable top predator feeds intensively on prey targeted by a fishery. Fanshawe *et al.* (2003) recommend separate single-purpose categories of protected area, but this may not work in the long term either; the maintenance of the *status quo* when sea otters are expanding their range is likely eventually to require culling of the otters, something that may prove politically unacceptable.

22.3.1.3   Invasions by salmonid fish in streams and lakes

Just as sea otters alter the behavior of their abalone prey, so the introduced brown trout (*Salmo trutta*) in New Zealand changes the behavior of herbivorous invertebrates (including

food web and ecosystem consequences of invading fish

nymphs of the mayfly *Deleatidium* spp.) that graze algae on the beds of invaded streams – daytime activity is significantly reduced in the presence of trout (Townsend, 2003). Brown trout rely principally on vision to capture prey, whereas the native fish they have replaced (*Galaxias* spp.) rely on mechanical cues. The hours of

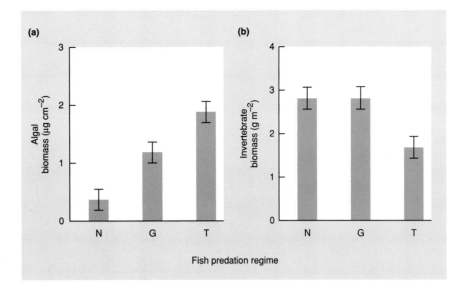

Figure 22.6 (a) Total algal biomass (chlorophyll *a*) and (b) invertebrate biomass (± SE) for an experiment performed in the summer in a small New Zealand stream. G, *Galaxias* present; N, no fish; T, trout present. (After Flecker & Townsend, 1994.)

darkness thus provide a refuge against trout predation analogous to the crevices occupied by the abalone. That an exotic predator such as trout has direct effects on *Galaxias* distribution or mayfly behavior is not surprising, but the influence also cascades to the plant trophic level. Three treatments were established in artificial flow-through channels placed in a real stream – no fish, *Galaxias* present or trout present, at naturally occurring densities. After 12 days, algal biomass was highest where trout were present (Figure 22.6a), partly because of a reduction in grazer biomass (Figure 22.6b) but also because of a reduction of grazing (only feeding at night) by the grazers that remain. This trophic cascade also changed the rate at which radiant energy was captured by the algae (annual net primary production was six times greater in a trout stream than in a neighboring *Galaxias* stream; Huryn, 1998) and, this in turn, resulted in more efficient cycling of nitrogen, the limiting nutrient in these streams (Simon *et al.*, 2004). Thus, important elements of ecosystem functioning, namely energy flux (see Chapter 17) and nutrient flux (see Chapter 18), were altered by the invading trout.

|managers should beware invaders that link ecosystem compartments in new ways|

Other salmonids, including rainbow trout (*Oncorhyncus mykiss*), have invaded many fishless lakes in North America where a similar increase in plant (phytoplankton) biomass has been recorded. A fish-induced reduction in benthic and planktonic grazers is partly responsible, but Schindler *et al.* (2001) argue that the main reason for increased primary production is that trout feed on benthic and littoral invertebrates and then, via their excretion, transfer phosphorus (the limiting nutrient) into the open water habitat of the phytoplankton. In their

review of the impacts of these and other freshwater invaders on community and ecosystem functioning, Simon and Townsend (2003) conclude that biosecurity managers should pay particular attention to invaders that have a novel method of resource acquisition or a broad niche that links previously unlinked ecosystem compartments.

22.3.1.4 Conflicting hypotheses about invasions

|where do invaders fit into food webs?|

A widely cited hypothesis in invasion biology related to population and food web interactions (see Chapters 19 and 20) and species richness (see Chapter 21) is that species-rich communities are more resistant to invasion than species-poor communities. This is because resources are more fully utilized in the former and competitors and predators are more likely to be present that can exclude potential invaders (Elton, 1958). On this basis, as invaders accumulate in an ecosystem, the rate of further invasions should be reduced (Figure 22.7a). But the opposite has also been postulated – the 'invasional meltdown' hypothesis (Figure 22.7b) (Simberloff & Von Holle, 1999). This argues that the rate of invasions will actually increase with time, partly because the disruption of native species promotes further invasions and partly because some invaders have facilitative rather than negative effects on later arrivals. Ricciardi's (2001) review of invasions of the Great Lakes of North America reveals a pattern that conforms closely to the meltdown hypothesis (Figure 22.7c). Among interactions between pairs of invaders, it is usually competition (−/−) and predation (+/−) that are given prominence. Ricciardi's review is unusual because it also accounted for mutualisms (+/+), commensalisms (+/0) and

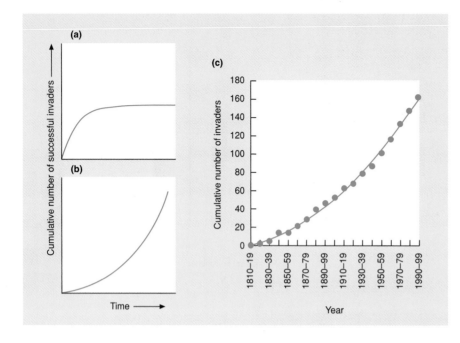

Figure 22.7 Predicted temporal trends in the cumulative number of successful invasions according to (a) the biotic resistance hypothesis and (b) the invasional meltdown hypothesis. (c) Cumulative number of invaders of the Great Lakes of North America – the pattern conforms to the invasional meltdown hypothesis. (After Ricciardi, 2001.)

amensalisms (−/0). There were 101 pairwise interactions in all, three cases of mutualism, 14 of commensalism, four of amensalism, 73 of predation (herbivory, carnivory and parasitism) and seven of competition. Thus, about 17% of reported cases involved one invader facilitating the success of another, whether directly or indirectly. An example of direct facilitation is the provision by invading dreissenid mussels of food in the form of fecal deposits and of increased habitat heterogeneity that favor further invaders such as the amphipod *Echinogammarus ischnus* (Stewart *et al.*, 1998). Indirect facilitation occurred in the 1950s and 1960s when the parasitic sea lamprey *Petromyzon marinus* suppressed native predatory salmonid fish to the benefit of invading fish such as *Alosa pseudoharengus* (Ricciardi, 2001). In addition, one-third of the cases of predation in Ricciardi's analysis could be said to involve 'facilitation' because a newcomer benefitted from a previously established invader. We do not know how widely the invasional meltdown hypothesis applies in different ecosystems, but the history of the Great Lakes suggests that it would generally be unwarranted for managers to take no further action just because several invaders were already established.

22.3.2 Managing eutrophication by manipulating lake food webs

which lakes can be managed to reverse nutrient enrichment?

The excess input of nutrients (particularly phosphorus; Schindler, 1977) from sources such as sewage and agricultural runoff has caused many 'healthy' oligotrophic lakes (low nutrients, low plant productivity with abundant macrophytes, and clear water) to switch to a eutrophic condition. Here, high nutrient inputs lead to high phytoplankton productivity (sometimes dominated by bloom-forming toxic species), making the water turbid and, in the worst situations, leading to anoxia and fish kills (see Section 18.4.3). In some cases the obvious management response of reducing phosphorus input (by sewage diversion, for example) may cause rapid and complete reversal. Lake Washington provides a success story in this *reversible* category (Edmondson, 1991), which includes lakes that are deep, cold and rapidly flushing and lakes that have only been briefly subject to cultural eutrophication (Carpenter *et al.*, 1999). At the other end of the scale are lakes that seem to be *irreversible* because the minimum attainable rate of phosphorus input, or phosphorus recycling from accumulated reserves in lake sediment, is too high to allow the switch back to oligotrophy. This applies particularly to lakes in phosphorus-rich regions (e.g. related to soil chemistry) and lakes that have received very high phosphorus inputs over an extended period. In an intermediate category, which Carpenter *et al.* (1999) refer to as *hysteretic* lakes, eutrophication can be reversed by combining the control of phosphorus inputs with interventions such as chemical treatment to immobilize phosphorus in the sediment or a biological intervention known as biomanipulation. Our discussion focuses on this final category because it depends on a knowledge of interactions in food webs (see Chapter 20) between piscivorous fish, planktivorous fish, herbivorous zooplankton and phytoplankton to guide the management of lakes towards a particular ecosystem endpoint (Mehner *et al.*, 2002).

The primary aim of biomanipulation is to improve water quality by lowering phytoplankton density and thus increasing water clarity. The approach involves increasing the grazing of zooplankton on phytoplankton via a reduction in the biomass of zooplanktivorous fish (by fishing them out or by increasing piscivorous fish biomass). Major successes have occurred in shallow lakes where nutrient levels are not too excessive (Meijer *et al.*, 1999). Lathrop *et al.* (2002) were more ambitious than most in attempting to biomanipulate the relatively large and deep eutrophic Lake Mendota in Wisconsin, USA. They combined the management objective of improving water quality with one of augmenting the recreational fishery for piscivorous walleye (*Stizostedion vitreum*) and northern pike (*Esox lucius*). In total, more than 2 million fingerlings of the two species were stocked beginning in 1987, and piscivore biomass rapidly responded and stabilized at 4–6 kg ha^{-1} (Figure 22.8a). The combined biomass of zooplanktivorous fish declined, as expected, from 300–600 kg ha^{-1} prior to biomanipulation to 20–40 kg ha^{-1} in subsequent years. The reduction in predation pressure on zooplankton (Figure 22.8b) led, in turn, to a switch from small zooplanktivorous grazers (*Daphnia galeata mendotae*) to the larger and more efficient grazer *D. pulicaria*. In many years when *D. pulicaria* were dominant, their high grazing pressure reduced phytoplankton density and increased water clarity (Figure 22.8c). The desired response would probably have been more emphatic had there not been an increase in phosphorus concentrations during the biomanipulation period, mainly as a result of increased agricultural and urban runoff. Lathrop *et al.* (2002) conclude that the favorable biomanipulation state of high grazing pressure should see further improvements as new management actions to reduce phosphorus inputs take effect.

Cultural eutrophication has equally dramatic effects in rivers, estuaries and marine ecosystems. Coastal eutrophication has become a major cause for concern. The United Nations Environment Program (UNEP) has reported that 150 sea areas worldwide are now regularly starved of oxygen as a result of the decomposition of algal blooms fueled particularly by nitrogen from agricultural runoff of fertilizers and sewage from large cities (UNEP, 2003).

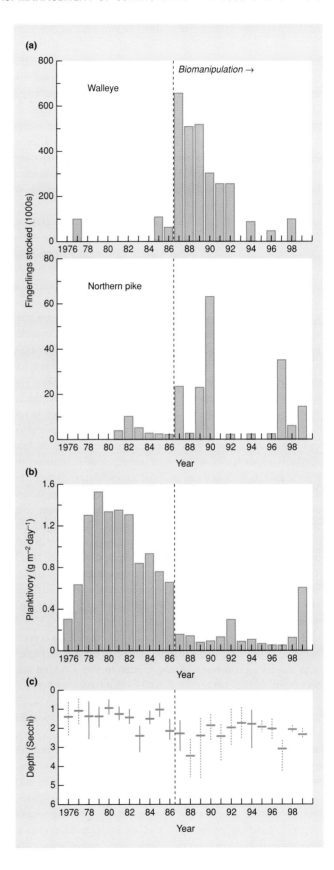

Figure 22.8 (*right*) (a) Fingerlings of two piscivorous fish stocked in Lake Mendota; the major biomanipulation effort started in 1987. (b) Estimates of zooplankton biomass consumed by zooplanktivorous fish per unit area per day. The principal zooplanktivore species were *Coregonus artedi*, *Perca flavescens* and *Morone chrysops*. (c) Mean and range during summer of the maximum depth at which a Secchi disc is visible (a measure of water clarity); dotted vertical lines are for periods when the large and efficient grazer *Daphnia pulicaria* was dominant. (After Lathrop *et al.*, 2002.)

22.3.3 Managing ecosystem processes in agriculture

Intensive land use is not only associated with phosphorus pollution but also with an increase in the amount of the nitrate that leaches into the groundwater and thence into rivers and lakes, affecting food webs and ecosystem functioning (see Section 18.4.4). The excess nitrate also finds its way into drinking water where it is a health hazard, potentially contributing to the formation of carcinogenic nitrosamines and in young children to a reduction in the oxygen-carrying capacity of the blood. The Environmental Protection Agency in the United States recommends a maximum concentration of nitrate of 10 mg l^{-1}.

problems with nutrient enrichment of land

Pigs, cattle and poultry are the three major nitrogen contributors in industrialized agriculture feedlots. The nitrogen-rich waste from factory-farmed poultry is easily dried and forms a transportable, inoffensive and valuable fertilizer for crops and gardens. In contrast, the excreta from cattle and pigs are 90% water and have an unpleasant smell. A commercial unit for fattening 10,000 pigs produces as much pollution as a town of 18,000 inhabitants. The law in many parts of the world increasingly restricts the discharge of agricultural slurry into watercourses. The simplest practice returns the material to the land as semisolid manure or as sprayed slurry. This dilutes its concentration in the environment to what might have occurred in a more primitive and sustainable type of agriculture and converts pollutant into fertilizer. However, if nitrate ions are not taken up again by plants, rainfall leaches them into the groundwater. In fact, the disassociation of livestock and crops in farms specializing in one or the other, rather than mixed farms, has made a major contribution to nitrate pollution of waterways. For example, the centralization of livestock production in the USA has tended to occur in regions that produce little crop feed (Mosier et al., 2002). Thus, for example, of the 11 Tg of nitrogen excreted in animal waste in the USA in 1990 only 34% was returned to cropped fields. Much of the remainder will eventually have found its way into waterways.

Most of the fixed nitrogen in natural communities is present in the vegetation and in the organic fraction of the soil. As organisms die they contribute organic matter to the soil, and this decomposes to release carbon dioxide so that the ratio of carbon to nitrogen falls; when the ratio approaches 10 : 1, nitrogen begins to be released from the soil organic matter as ammonium ions. In aerated regions of the soil, the ammonium ions become oxidized to nitrite and then to nitrate ions, which are leached by rainfall down the soil profile. Both the processes of organic matter decomposition and the formation of nitrates are usually fastest in the summer, when natural vegetation is growing most quickly. Nitrates may then be absorbed by the growing vegetation as fast as they are formed – they are not present in the soil long enough for significant quantities to be leached out of the

plants' rooting zone and lost to the community. Natural vegetation most often is a 'nitrogen-tight' ecosystem.

In contrast, there are several reasons why nitrates leach more easily from agricultural land and managed forests than from natural vegetation.

1 For part of the year agricultural land carries little or no living vegetation to absorb nitrates (and for many years forest biomass is below its maximum).
2 Crops and managed forests are usually monocultures that can capture nitrates only from their own rooting zones, whereas natural vegetation often has a diversity of rooting systems and depths.
3 When straw and forestry waste are burned, the organic nitrogen within them is returned to the soil as nitrates.
4 When agricultural land is used for grazing animals their metabolism speeds up the rate at which carbon is respired, reduces the C : N ratio, and increases nitrate formation and leaching.
5 Nitrogen in agriculture fertilizer is usually applied only once or twice a year rather than being steadily released as it is during the growth of natural vegetation; it is therefore more readily leached and finds its way into drainage waters.

Because nitrogen is not efficiently recycled on agricultural land or in managed forests, repeated cropping leads to losses of nitrogen from the

the problem is getting worse

ecosystem and thus to decreasing crop productivity. To maintain crop yields the available nitrogen has to be supplemented with fertilizer nitrogen, some of which is obtained by mining potassium nitrate in Chile and Peru, but the majority comes from the energy-expensive industrial process of nitrogen-fixation, in which nitrogen is catalytically combined with hydrogen under high pressure to form ammonia and, in turn, nitrate. Nitrogen fertilizers are applied in agriculture either as nitrates or as urea or ammonium compounds (which are oxidized to nitrates). However, it is wrong to regard artificial fertilization as the only practice that leads to nitrate pollution; nitrogen fixed by crops of legumes such as alfalfa, clover, peas and beans also finds its way into nitrates that leach into drainage water. Figure 22.9 shows how the amounts of synthetic fertilizer and nitrogen-fixing crops have increased in the last 50 years, and the dramatic increases are set to continue over the next half century (Tilman et al., 2001), particularly in developing countries.

A variety of approaches are available to tackle the problems of nitrate in drinking water and eutrophication, for example by maintaining ground cover of vegetation year-round, by practising mixed cropping rather than monoculture, by integrating animal and crop production and more generally returning organic matter to the soil, by maintaining low

management of the nutrient enrichment of land

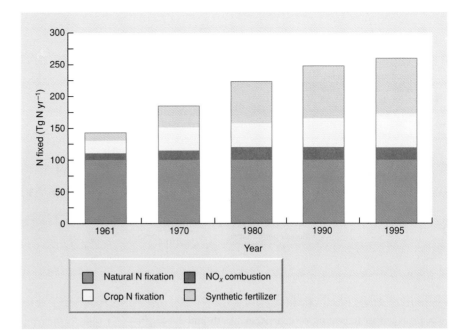

Figure 22.9 Estimates of global nitrogen fixation for representative years since 1961 in four categories. Natural nitrogen fixation remained constant but fixation by crops and in the production of synthetic fertilizer both increased dramatically. NO_x combustion refers to the oxidation of atmospheric nitrogen when fossil fuels are burnt; NO_x is deposited in downwind ecosystems. (After Galloway *et al.*, 1995.)

stocking levels, by matching nitrogen supply to crop demand and by using advanced 'controlled release' fertilizers (Mosier *et al.*, 2002). The role played by nitrogen-fixing symbionts (both fungal arbuscular mycorrhizae and bacterial rhizobia) is of particular interest. Root symbionts do not augment crop productivity consistently. Rather, different species, or the same species under different soil conditions, can range from acting parasitically (when they act as a sink for plant resources in the relationship) to mutualistic (when they significantly enhance plant performance). Kiers *et al.* (2002) argue that research is needed to determine how farm management practices, including fertilization, ploughing and crop rotation, influence the short-term responses and, over a slightly longer timeframe, the evolution of nitrogen-fixing symbionts. Such knowledge would help identify management regimes to enhance mutualistic rather than parasitic interactions.

22.3.4 Ecosystem health and its assessment

characterizing the state of degraded ecosystems – an analogy with human health

Many ecosystems around the world have been degraded by human activities. Using an analogy with human health, managers frequently describe ecosystems as 'unhealthy' if their community structure (species richness, species composition, food web architecture – see Chapters 16, 20 and 21) or ecosystem functioning (productivity, nutrient dynamics, decomposition – see Chapters 17 and 18) has been fundamentally upset by human pressures. Aspects of ecosystem health are sometimes reflected directly in human health (nitrogen content in groundwater and thus drinking water, toxic algae in lakes and oceans, species richness of animal hosts that transmit human diseases in oak forests) but also in natural processes (ecosystem services) that people value, such as flood control, the availability of wild food (including hunted animals and gathered fungi and plants) and recreational opportunities. Management strategies are often framed in the context of *pressure* (human actions), *state* (resulting community structure and ecosystem functioning) and management *response* (Figure 22.10) (Fairweather, 1999). Just as physicians use indicators in their assessment of human health (body temperature, blood pressure, etc.), ecosystem managers need ecosystem health indicators to help set priorities for action and to determine the extent to which their interventions have been successful.

The ponderosa pine forests (*Pinus ponderosa*) of the western USA can be used to illustrate the relationship between pressure, state and response *ecosystem health of a forest* (Rapport *et al.*, 1998). A variety of human influences are at play but Yazvenko and Rapport (1997) consider the most important pressure has been fire suppression (just as we saw in the Australian ecosystem described in Section 22.2.1, ponderosa pine forests evolved in a situation where periodic natural fires occurred). With fire suppression, the state of the forest has shifted towards decreased productivity and increased tree mortality, changed patterns of nutrient cycling, and an increased rate and magnitude of outbreaks of tree pests and diseases. These changed properties

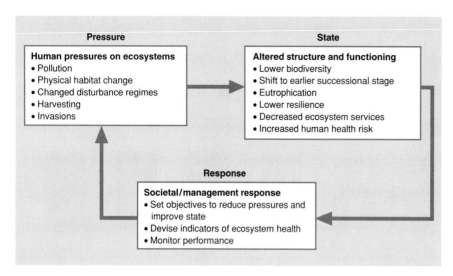

Figure 22.10 The linkage between *pressures* caused by human activities, *state* in terms of community composition and ecosystem processes, and management *response*. Adverse effects on ecosystems sometimes involve processes with clear value in human terms; such impacted ecosystem services include reduced recreational opportunities, poor water quality, diminished natural flood control, negative impacts on harvestable wildlife and on biodiversity generally.

can be taken as indicators of ecosystem health and successful restoration (response) will be evident when the indicators reverse the trends.

ecosystem health
of a river

River health has been measured in a number of ways, from assessment of abiotic evidence of pressures (e.g. nutrient concentrations and sediment loads), through community composition to ecosystem functioning (such as the rate of decomposition of leaves of overhanging vegetation that fall naturally into rivers; Gessner & Chauvet, 2002). Some health indexes include more than one of these indicators; in other cases managers rely on a single measure. In New Zealand, for example, river managers use the macroinvertebrate community index (MCI) (Stark, 1993). This is based on the presence or absence of certain types of river invertebrates that differ in their ability to tolerate pollution; healthy streams with abundant species that are intolerant of pollution have high values of MCI (120 or above) whereas unhealthy streams have values as low as 80 or less. Figure 22.11a shows the relationship, for sites on the Kakaunui River on the east coast of New Zealand's South Island, between MCI and the percentage of the catchment area that has been developed (for pasture or urban development; here land development is the pressure).

ecosystem health as a
social construct

We should not lose sight of the fact that the concept of ecosystem health is generally a social construct. A healthy ecosystem is one that the community believes to be healthy and different social groups hold different ideas about this (e.g. anglers consider that a river is healthy if it contains many big representatives of preferred fish species; parents if their children do not get sick swimming in the river; conservationists if native species are abundant). The Kakaunui River is within the territory of a Maori group who wished to develop a tool so their perceptions of river health could be taken into account by managers. Their Cultural Stream Health Measure (CSHM) includes components related to the extent to which the surrounding catchment area, the riparian zone, the banks and the stream bed appear impacted by human activities. The CSHM (Figure 22.11b) turned out to be strongly correlated with the MCI despite the fact that it included no invertebrate component.

22.4 Biodiversity and management

22.4.1 Selecting conservation areas

Producing individual species survival plans may be the best way to deal with species recognized to be in deep trouble and identified to be of special importance (e.g. keystone species, evolutionarily unique species, charismatic large animals that are easy to 'sell' to the public). However, there is no possibility that all endangered species could be dealt with one at a time. For instance, the US Fish and Wildlife Service calculated it would need to spend about $4.6 billion over 10 years to fully recover all gazetted species in the USA (US Department of the Interior, 1990), whereas the annual budget for 1993 was $60 million (Losos, 1993). In the face of such funding shortfalls, there has been a growing trend towards multispecies rather than single-species protection plans, but this carries a risk that the specific requirements of endangered species will receive insufficient attention. Thus, an analysis of USA cases showed that species in multispecies plans were significantly more likely to exhibit declining population trends (Boersma et al., 2001). For this reason, Clark and Harvey (2002) advocate the grouping together of species according to the threats they face. Despite

multispecies or
single-species
management plans?

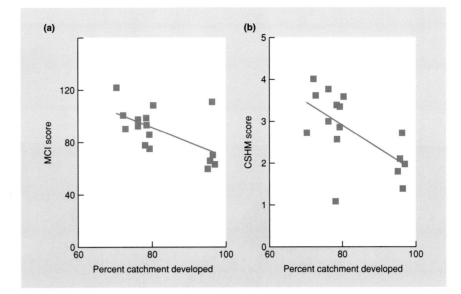

Figure 22.11 Relationships between percentage development of the catchment area of sites in the Kakaunui River (for pasture and urban use) and (a) the macroinvertebrate community index (MCI), commonly used by river managers in New Zealand, and (b) the Maori Cultural Stream Health Measure (CSHM). (After Townsend *et al.*, 2004.)

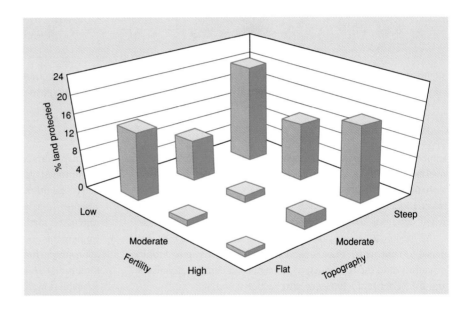

Figure 22.12 Protected areas in southwest Australia are most often situated in steeply sloping and poorly productive areas that are not in demand for agriculture or urban development. (After Pressey, 1995; Bibby, 1998.)

some shortcomings, however, we can generally expect to conserve the greatest biodiversity if we protect whole communities by setting aside protected areas.

protected areas: limits to growth

Protected areas of various kinds (national parks, nature reserves, multiple-use management areas, etc.) grew in number and area through the 20th century, with the greatest expansion occurring since 1970. However, the 4500 protected areas in existence in 1989 still only represented 3.2% of the world's land area. At best, and given the political will, perhaps 6% of land area may eventually be provided protection – the rest would be considered necessary to provide the natural resources needed by the human population (Primack,

1993). Understandably, but nevertheless disturbingly, reserves have often been established on land that no one else wants (Figure 22.12). Areas of high species richness and distributions of endangered plant and animal species often overlap with human population centers (Figure 22.13). Thus, although protection of wilderness is of value and relatively easy, conserving maximum diversity will require greater focus on areas of high human value.

Priorities for marine conservation, which have lagged behind terrestrial efforts, are now being urgently addressed. In taxonomic terms, most of the world's biota is found in the sea (32 of the 33 known animal phyla are marine, 15 exclusively so) and marine communities are

priorities for marine protected areas

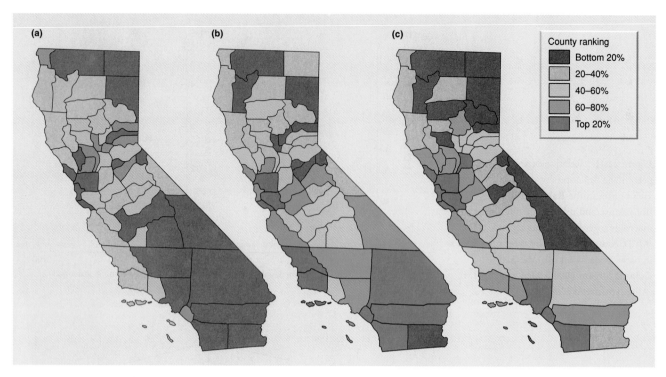

Figure 22.13   Counties of California ranked according to: (a) plant species richness (number per 2.59 km² sample area); (b) the proportion of plant species listed as threatened or endangered; and (c) human population density. (After Dobson *et al.*, 2001.)

subject to a number of potentially adverse influences, including overfishing, habitat disruption and pollution from land-based activities. There are some fundamental distinctions between marine and terrestrial ecosystems that need to be borne in mind when designing marine reserves. Most prominent among these is the greater 'openness' of marine areas, with long-distance dispersal of nutrients, organic and inorganic matter, planktonic organisms and the reproductive propagules of benthic organisms and fish (Carr *et al.*, 2003; see also Section 15.4.2).

systematic approach to conservation planning

The overall aim of conservation areas, whether terrestrial or marine, is to represent the biota of each region in a way that separates the biodiversity from the processes that threaten it. Margules and Pressey (2000) recommend the following steps for systematic conservation planning.

1   Compile data on biodiversity and on the distribution of rare and endangered species in the planning region.
2   Identify conservation goals and set explicit conservation targets for species and community types as well as quantitative targets for minimum reserve size and connectivity.
3   Review existing conservation areas to measure the extent to which quantitative goals have already been achieved and

identify imminent threats to underrepresented species and community types.
4   Select additional conservation areas to augment existing reserves in a way that best achieves the conservation goals (discussed further below).
5   Implement conservation actions having decided the most appropriate form of management for each area and having established an implementation timetable if resources are not available for all actions to be carried out at once.
6   Maintain the required values of conservation areas and monitor key indicators that will reflect management success, modifying management as required.

We know that the biotas of different locations vary in species richness (centers of diversity – see Section 21.1), the extent to which the biota is unique (centers of endemism) and the extent to which the biota is endangered (hot spots of extinction, for example because of imminent habitat destruction). One or more of these criteria could be used to prioritize potential areas for protection (Figure 22.14). Moreover, if we were to give less weight to the 'existence' value of species (every species equal) and more weight to the potential value of species that may provide future

centers of diversity, endemism, extinction and utility

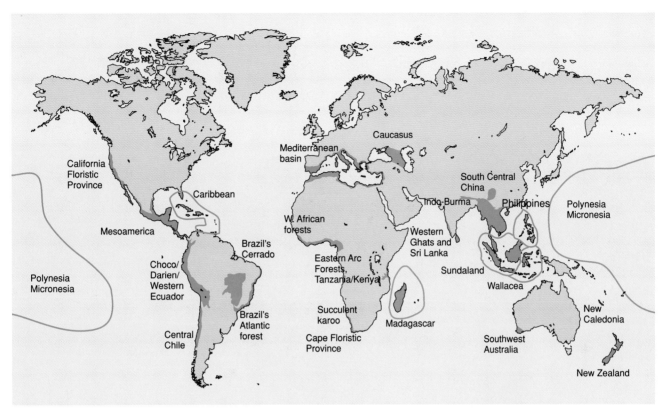

Figure 22.14 The global distribution of biodiversity hot spots where exceptional concentrations of endemic species are undergoing exceptional loss of habitat. As many as 44% of all species of the earth's vascular plants and 35% of its vertebrates are confined to 25 hot spots that make up only 1.4% of its land surface. (After Myers *et al.*, 2000.)

benefit (for food, domestication, medical products, etc.) we could prioritize locations that contained more species likely to be useful (centers of utility).

the key concepts of complementarity and irreplaceability

But, biodiversity encompasses more than just species richness. The selection of new areas should also try to ensure protection of representatives of as many types of community and ecosystem as possible. Two key principles here are *complementarity* and *irreplaceability* (Pressey *et al.*, 1993).

With limited resources, the ideal strategy is to assess the content of candidate areas and to proceed in a stepwise fashion, selecting at each step the site that is most complementary to the others in the features it contains. A number of algorithms are now available to carry out this procedure efficiently. For example, one algorithm lays more stress on the degree of uniqueness of the community or land system, while another lays more stress on the average rareness of the land systems present in the different locations (Figure 22.15a).

A related but subtly different approach identifies irreplaceability as a fundamental measure of the conservation value of a site.

Irreplaceability is an index of the potential contribution that a site will make to a defined conservation goal and the extent to which the options for conservation are lost if the site is lost (Figure 22.15b). The notions of complementarity and irreplaceability can equally well be applied to strategies designed to maximize species richness. However, complementarity algorithms for species richness should be implemented with care because they have a tendency to select areas that are at the margins of species ranges more often than would be expected by chance (Araujo & Williams, 2001), and rare species could do less well at the margins than in the centers of their ranges.

A perhaps rather surprising application of island biogeography theory (see Section 21.5) is in nature conservation. This is because many conserved areas and nature reserves are surrounded by

design of nature reserves: clues from island biogeography theory

an 'ocean' of habitat made unsuitable, and therefore hostile, by people. Can the study of islands in general provide us with 'design principles' that can be used in the planning of nature reserves? The answer is a cautious 'yes' (Soulé, 1986); some general points can be made.

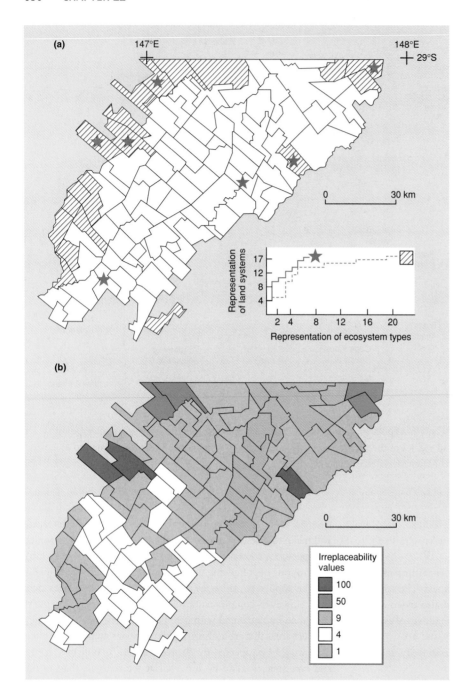

Figure 22.15 (a) A map of 95 pastoral holdings in New South Wales, Australia, showing two sets of holdings needed to represent all 17 ecosystem types at least once. Stars indicate a minimum set identified by a complementarity algorithm that selects sites with unique ecosystems, and then proceeds stepwise to select the site with the rarest unrepresented ecosystem type. Shading indicates the set required if all holdings are scored according to the average rareness of the ecosystem types they contain. (b) A landscape of conservation value for each holding derived by predicting irreplaceability levels. (After Pressey *et al.*, 1993.)

1 One problem that conservation managers sometimes face is whether to construct one large reserve or several small ones adding up to the same total area (sometimes referred to as the SLOSS (single large or several small) debate). If each of the small reserves supported the same species, then it would be preferable to construct the larger reserve in the expectation of conserving more species (this recommendation derives from the species–area relationships discussed in Section 21.5.1).

2 On the other hand, if the region as a whole is heterogeneous, then each of the small reserves may support a different group of species and the total conserved might exceed that in a large reserve. In fact, collections of small islands tend to contain more species than a comparable area composed of one or a few large islands. The pattern is similar for habitat islands and, most significantly, for national parks. Thus, several small parks contained more species than larger ones of the same area in studies of mammals and birds in East African parks, of mammals

Table 22.1 Activities permitted or prohibited for each of four planned levels of protection (from left to right in order of decreasing protection) for the Asinara Island National Marine Reserve of Italy. (After Villa *et al.*, 2002.)

Category	Activity	*No-take, no-entry*	*Entry, no-take*	*General reserve*	*Partial reserve*
Research	Nondestructive research	Aa	Aa	A	A
Sea access	Sailing	P	L	A	A
	Motor boating	P	P	L	L
	Swimming	P	P	A	A
Staying	Anchorage	P	P	L	L
	Mooring	P	L	Aa	A
Recreation	Diving	P	L	Aa	A
	Guided tours	P	L	Aa	A
	Recreational fishing	P	P	L	A
Exploitation	Artisanal	P	P	L	L
	Sport	P	P	P	L
	Scuba	P	P	P	P
	Commercial fishing	P	P	P	P

A, allowed without authorization; Aa, allowed upon authorization; L, subject to specific limitations; P, prohibited.

and lizards in Australian reserves, and of large mammals in national parks in the USA (Quinn & Harrison, 1988). It seems likely that habitat heterogeneity is a general feature of considerable importance in determining species richness.

3 A point of particular significance is that local extinctions are common events (see Section 7.5), and so recolonization of habitat fragments is critical for the survival of fragmented populations. Thus, we need to pay particular attention to the spatial relationships amongst fragments, including the provision of dispersal corridors. There are potential disadvantages – for example, corridors could increase the correlation among fragments of catastrophic effects such as the spread of fire or disease – but the arguments in favor are persuasive. Indeed, high recolonization rates (even if this means conservation managers themselves moving organisms around) may be indispensable to the success of conservation of endangered metapopulations (see Section 15.5.3). Note especially that human fragmentation of the landscape, producing subpopulations that are more and more isolated, is likely to have had the strongest effect on populations with naturally low rates of dispersal. Thus, the widespread declines of the world's amphibians may be due, at least in part, to their poor potential for dispersal (Blaustein *et al.*, 1994).

22.4.2 Multipurpose reserve design

managing for multiple objectives – beyond conservation

Many of the new generation of marine protected areas are designed as multiple-use reserves, accommodating many different users (environmentalists, cultural harvesters, commercial fishers, tourism operators, etc.) (Airame *et al.*, 2003). It is clear, too, that conservation and sustainable use on land (forestry, agriculture) can often proceed hand in hand as long as the planning has a scientific basis and the negotiated objectives are clear (Margules & Pressey, 2000).

A good example of multipurpose design is provided by Villa *et al.* (2002), who used a systematic approach to design one of the first marine reserve zoning plans in Italy. They involved all the different interest groups (fishing, recreation, conservation) in defining priorities, and used a GIS (geographic information system) to map marine areas for different uses and degrees of protection. Italian law recognizes reserves with three levels of protection: 'integral' reserves (only available for research), 'general' reserves and the less restrictive 'partial' reserves. Villa *et al.*'s starting point was to accept 'partial' and 'general' reserves but to split 'integral' reserves into two categories: no-entry, no-take zones (where only nondestructive research is permitted) and public entry, no-take zones, which allow visitors a full experience of the reserve, apart from exploitation. Permitted activities for the four categories are shown in Table 22.1.

The next step was to produce maps of 27 variables important to one or more interest groups. These included fish diversity, fish nursery areas, sites used by life history stages of key species (e.g. limpets, sea mammals, marine birds), archeological interest, suitability for various forms of fishing (e.g. traditional artisanal, commercial), suitability for various recreational activities (e.g. snorkeling, whale watching), tourist infrastructure and pollution status. Planning sessions with each interest group yielded weightings or relative importance values for the variables. Taking these into account, five higher level maps were produced (using an approach developed for economic analysis and urban planning known as multiple-criteria analysis): (i) the natural value of the marine environment (NVM – aggregating values related to biodiversity, rarity and crucial habitats such as nursery areas); (ii) the natural value of the coastal environment (NVC – aggregating values related to endemic coastal species including seabirds, and habitat suitable for the reintroduction of turtles and seals);

(iii) the recreational activity value (RAV – aggregating values for all recreational activities); (iv) the commercial resource value (CRV – aggregating traditional fishing sites plus other suitable areas); and (v) the ease of access value (EAV – aggregating marine access routes and harbors). Aggregated maps for NVM, NVC and RAV are shown in Figure 22.16a–c.

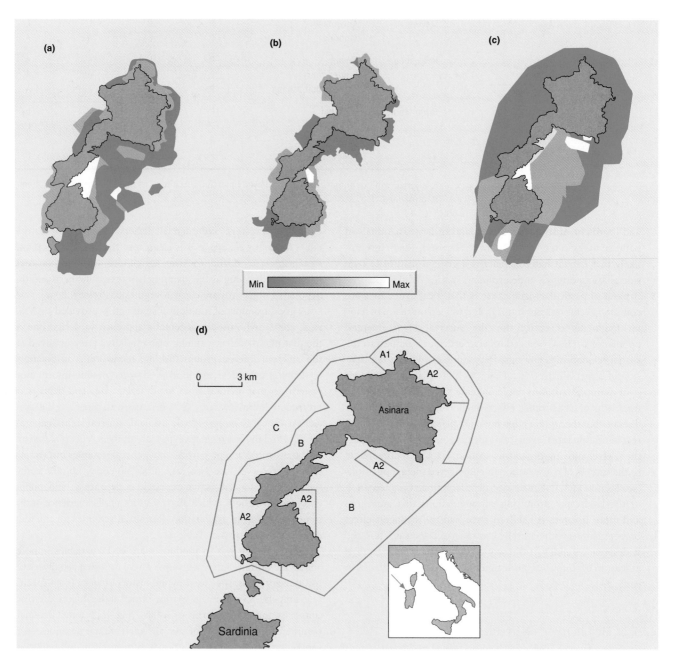

Figure 22.16   Maps of the natural value of (a) the marine environment (NVM), (b) the coastal environment (NVC) and (c) recreational activity value (RAV) for areas around Asinara Island (the island land area is shown in the center in gray). Lighter shades of color represent higher values. (d) Final zoning plan for the Asinara Island National Marine Reserve. A1, no-entry, no-take; A2, entry, no-take; B, general reserve; C, partial reserve. The inset map shows the location of the reserve in relation to the mainland of Italy. (After Villa *et al.*, 2002.)

The final stage was the production of a zoning plan. The researchers sought to avoid complex zoning that would make management and enforcement difficult and paid particular attention to the views of the various interest groups to reduce remaining conflicts to a minimum. The final plan (Figure 22.16d) had one no-entry, no-take zone (reflecting biological importance and relative remoteness), four entry, no-take zones to protect specific values such as endangered species (reflecting biological value but with easy access), two general reserve zones (to protect sensitive benthic assemblages, such as seagrass meadows that suffer little from the permitted activities; see Table 22.1) and one partial reserve zone as a buffer for adjacent reserve zones (in an area where traditional fishing practices are compatible with conservation). The zoning proposal also identified three channels providing maximum boat access with minimal environmental disturbance.

22.5 Triple bottom line of sustainability

The main emphasis up to this point has been on the use of ecological theory to help solve environmental problems and establish strategies that are likely to be sustainable in the long term. However, we have already come across a variety of examples where ecological aspects of sustainability cannot be divorced from economic (e.g. limited funds for conservation action) or social aspects (e.g. related to disease risk or the importance of involving diverse interest groups, including indigenous peoples, in resource management). Similar examples were also encountered in the two earlier chapters dealing with ecological applications (e.g. Sections 7.2.3, 7.5.5.2, 7.5.6, 15.2.1, 15.2.3 and 15.3.9). Here we deal more explicitly with the economic and sociopolitical threads of environmental sustainability.

22.5.1 Economic perspective

importance of the economic perspective

The importance of economics in resource management is obvious for activities such as harvest management (see Section 15.3), agricultural management (including pest control; see Sections 15.2 and 22.2.1) and the use of scarce funds when planning conservation management and protected areas (see Sections 7.5 and 22.4). When it comes to conservation of species, biodiversity or ecosystems, however, it is more difficult to assign economic value to the entities to be conserved. It is necessary to do this because of the economic arguments in favor of human activities that make conservation a necessity: agriculture, the felling of trees, the harvesting of wild animal populations, the exploitation of minerals, the burning of fossil fuels, irrigation, the discharge of wastes and so on. While there are not really arguments against conservation, the case for conservation will be most likely to be effective if framed in cost–benefit terms because governments determine their policies against a background of the money they have to spend and the priorities accepted by their electorates.

how can species be assigned economic value?

We first consider how individual species can be valued. There are three main components: (i) the direct value of the products that are harvested; (ii) the indirect value where aspects of biodiversity bring economic benefit without the need to consume the resource; and (iii) the ethical value.

Many species are recognized as having actual direct value as living resources; many more species are likely to have a potential value which as yet remains untapped (Miller, 1988). Wild meat and plants remain a vital resource in many parts of the world, whilst most of the world's food is derived from plants that were originally domesticated from wild plants in tropical and semiarid regions. In the future, wild strains of these species may be needed for their genetic diversity in attempts to breed for improved yield, pest resistance, drought resistance and so on, and quite different species of plants and animals may be found that are appropriate for domestication. In another context, we have seen in Section 15.2 the potential benefits that could come from natural enemies if they can be used as biological control agents for pest species. Most natural enemies of most pests remain unstudied and often unrecognized. Finally, about 40% of the prescription and nonprescription drugs used throughout the world have active ingredients extracted from plants and animals. Aspirin, probably the world's most widely used drug, was derived originally from the leaves of the tropical willow, *Salix alba*. The nine-banded armadillo (*Dasypus novemcinctus*) has been used to study leprosy and prepare a vaccine for the disease; the Florida manatee (*Trichechus manatus*), an endangered mammal, is being used to help understand hemophilia. These are by no means isolated cases and a large-scale worldwide search is underway to discover organisms with new medicinal applications. The vast majority of the world's animals and plants have yet to be screened – the potential value of any that go extinct can never be realized. By conserving species, we maintain their option value – the potential to provide benefit in the future.

Nonconsumptive, indirect economic value is sometimes relatively easy to calculate. For example, a multitude of wild insect species are responsible for pollinating crop plants. The value of these pollinators could be assigned either by calculating the extent to which the insects increase the value of the crop or by the expenditure necessary to hire hives of honeybees to do the pollinating (Primack, 1993). In a related context, the monetary value of recreation and ecotourism, often called amenity value, is becoming ever more considerable. On a smaller scale, a multitude of natural history films, books and educational programs are 'consumed' annually without harming the wildlife upon which they are based.

The final category is ethical value. Many people believe that there are ethical grounds for conservation, arguing that every species is of value in its own right, and would be of equal value even if people were not here to appreciate or exploit it. From this perspective even species with no conceivable economic value require protection.

Of these three main reasons for conserving biodiversity, the first two, direct and indirect economic value, have a truly objective basis. The third, ethics, on the other hand, is subjective and is faced with the problem that a subjective reason will inevitably carry less weight with those not committed to the conservationist cause.

valuing the functioning of ecosystems: 'ecosystem services'

It is clear that assigning a value to species is not always straightforward. However, even more ingenuity is required to assign value to benefits that accrue to people from natural ecosystems as a whole – ecosystem services such as the production of wild species for food, fiber and pharmaceuticals, maintenance of the chemical quality of natural waters, buffering of communities against floods and droughts, ecosystem resistance to pest invasion, protection and maintenance of soils, regulation of local and global climate, the breakdown of organic and inorganic wastes, recreational opportunities, etc. The value of all ecosystem services was estimated globally at US$33 trillion per year (Costanza *et al.*, 1997), updated for the year 2000 to US$38 trillion per year, an amount that is similar to the gross national product of all the world's economies (Balmford *et al.*, 2002).

Such gross estimates are fraught with difficulty and have been criticized, partly because of the assumption that limited local knowledge can be safely extrapolated to a global sum as though demand and value are the same in different parts of the world. Balmford *et al.* (2002) argue that the value of retaining habitat in a relatively undisturbed condition would be best determined by estimating the differences in benefit from relatively intact and exploited versions of a particular ecosystem. They go beyond the mere calculation of private benefit to the exploiters to incorporate the dollar values of diverse public benefits of ecosystem services. The results of three case studies are presented in Figure 22.17. In each case, the estimates of private benefit and ecosystem services are for 30–50-year periods.

including ecosystem services in the valuation of natural resources

The first case study deals with tropical forest in Cameroon and compares low-impact forestry, conversion to small-scale agriculture and conversion to oil palm and rubber plantations. The value of all ecosystem services combined was highest under sustainable forestry; here ecosystem services included the control of sedimentation, flood prevention, carbon sequestration by the vegetation (i.e. contributing to a reduction of carbon dioxide in the atmosphere and thus

counteracting global warming) and a range of species values (see Section 7.5). Overall the total economic value (combining private benefit with the value of ecosystem services, expressed as net present value – NPV) over 32 years for low-impact forestry was 18% greater than for small-scale farming, while plantations actually made a net loss when both private benefit and ecosystem services were taken into account.

Analysis of a mangrove ecosystem in Thailand showed that the private benefit from shrimp farming shrank almost to nothing when the economics took into account the loss of ecosystem services from timber and nontimber products, charcoal, offshore fisheries and storm protection associated with the natural ecosystem (Figure 22.17b). The total value of intact mangroves exceeded that of shrimp farming by 70%.

Finally, the draining of freshwater marshes often produces private benefit (sometimes, as in this Canadian example, in large part because of drainage subsidies provided by the government). However, ecosystem services from intact wetland include hunting, trapping and angling and when the dollar values of these are taken into account the overall economic value of intact wetland exceeded converted land by about 60% (Figure 22.17c).

These analyses prompted Balmford *et al.* (2002) to suggest that a large-scale expansion of the world's network of protected areas (costing as much as US$45 billion per year) would actually represent a 'strikingly good bargain' in comparison to the US$38 trillion per year that ecosystem services may be worth.

22.5.2 Social perspectives

In his analysis of the history of fisheries, Pitcher (2001) points out how successive technological advances have driven the inexorable decline in abundance, diversity and representation of high-value species in catches (Figure 22.18). He identified three stages that can be recognized during depletion episodes: the first stage is ecological, comprising depletion and local extinction; the second is economic, comprising a positive feedback loop between increased catching power and depletion, driven by the need to repay money; and the third is social, comprising a shifting baseline in what each generation considers acceptable (or primal) abundance and diversity. It is possible to devise sustainable regimes at any stage but this has not often happened. At the current stage, the question arises should managers simply devise a sustainable management policy or actually attempt to rebuild the fishery? Pitcher challenges communities to attempt a 'back to the future' strategy, in which models of past ecosystems (constructed on the basis of local and traditional environmental knowledge) are subjected to economic comparison with current and alternative ecosystems. He suggests that large no-take reserves and the reintroduction of high-value species will figure prominently in the restoration of such historic ecosystems.

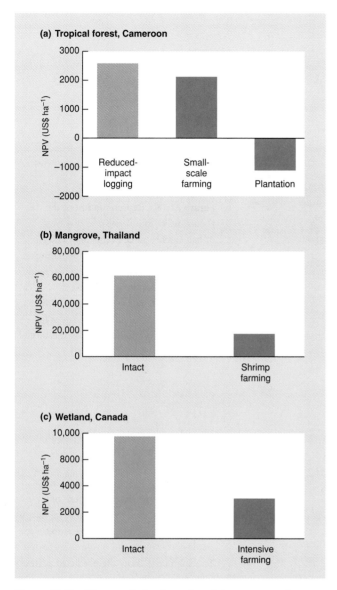

Figure 22.17 The marginal values of retaining or converting natural habitats expressed as net present value (NPV; in terms of US$ in the year 2000). (a) Tropical forest in Cameroon – estimates for three land uses over a 32-year time period, using a discount rate of 10%. Discounting allows for the fact that in economic terms each tree (or fish or bird) in the hand now is worth more than an equivalent bird some time in the future (see Section 15.3.8). The discount rate used was that adopted by the original researchers. (b) Mangrove in Thailand – estimates for intact mangrove forest and for conversion to shrimp farming over a 30-year period with a 6% discount rate. (c) Wetland in Canada – estimates for intact wetland and for conversion to intensive farming over a 50-year period with a discount rate of 4%. (After Balmford *et al.*, 2002; from original studies by G. Yaron, S. Sathirathai and W. van Vuuren & P. Roy, respectively.)

Managers can benefit from a coalescence of the economic approach of Balmford *et al.* (2002) and the social approach of Villa *et al.* (2002), where diverse local interest groups were involved in developing a management strategy. Aboriginal people can be expected to play a central role in sustainability developments in their territories not least because of their extensive knowledge of both the contemporary and historical situation. We have referred frequently in this chapter to lessons to be learnt from indigenous people and the importance of their involvement in resource management (benzoin gardening in Sumatra, fire management by Australian aborigines, Maori development of river health indicators). Maori have also been one of the groups, along with commercial and recreational fishers, tourism operators and environmentalists, comprising the Guardians of Fiordland's Fisheries and Marine Environment (GOFF). Over 3 years, they developed a zoning plan for New Zealand's Fiordland area on the west coast of New Zealand's South Island (Teirney, 2003). This was an entirely bottom-up effort by the local community (not directed top-down by governmental or nongovernmental agencies) and the diverse groups have worked face-to-face from the beginning. While challenging to manage (a skilled facilitator was involved), this approach provides a model for minimizing conflict, stimulating reciprocal learning and formulating objectives for sustainable ecosystem use that have proved difficult to achieve by top-down means. The New Zealand government has committed itself to implement the GOFF plan.

community action . . . and the role of aboriginal people

22.5.3 Putting it all together

In the past, the importance of ecosystem services was only appreciated after they had been lost. However, as ecological understanding has increased and now the economic significance is appreciated, sociopolitical change has become evident in a number of ways. In Costa Rica, for example, the government has been paying landowners since 1997 for ecosystem services such as carbon sequestration, protection of catchment areas, biodiversity and scenic beauty (payments of about US$50 ha^{-1}, which come mainly from taxes on fossil fuels) (Daily *et al.*, 2000). Private enterprise has also begun to respond. Thus, a company called Earth Sanctuaries Ltd became the world's first conservation company to go public when it was listed on the Australian Stock Market. It bought and restored land, earning income from tourism and wildlife sales. The company lobbied and won a change in Australian accounting law so that it could include its rare native animals as assets (Daily *et al.*, 2000). Such approaches, involving far-reaching political change, require price tags to be placed on natural ecosystems.

applying a triple bottom line approach . . .

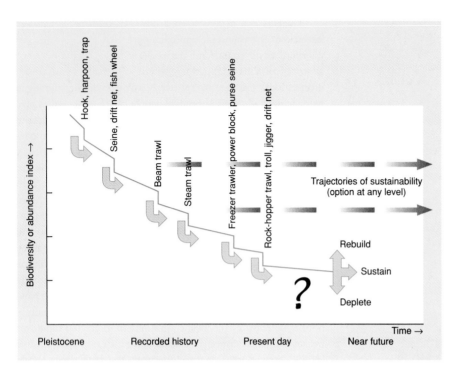

Figure 22.18 Representation of the reduction in abundance and diversity of fish catches since prehistory. The downward steps depict serial depletion as new fishing technologies are invented. Horizontal gray arrows represent sustainable management regimes, which in theory could be devised at any stage. Future options are indicated by the three-way arrow. (After Pitcher, 2001.)

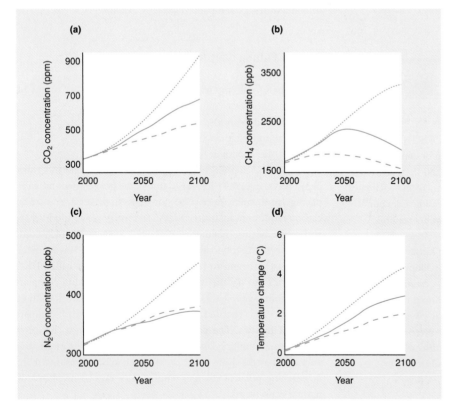

Figure 22.19 Predicted changes in concentration in the atmosphere of (a) carbon dioxide, (b) methane, (c) nitrous oxide and (d) predicted temperature changes to 2100 based on three scenarios. The solid lines show the predicted patterns for a future world of very rapid economic growth, a global population that peaks midcentury, the rapid increase of more efficient technologies, and a population that does not rely heavily on any one particular energy source. The dotted lines show patterns for a similar scenario but one where energy use is fossil-fuel intensive (as it has been until now). The dashed lines are for a more optimistic and sustainable scenario with a similar pattern of population growth but with a rapid change toward a service and information economy, with reductions in the use of materials and the introduction of clean and resource-efficient technologies. (After IPCC, 2001.)

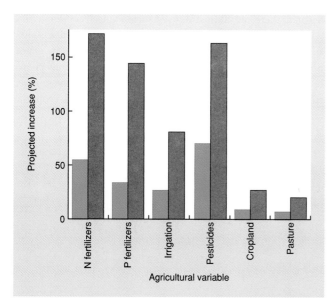

Figure 22.20 Projected increases in nitrogen (N) and phosphorus (P) fertilizers, irrigated land, pesticide use and total areas under crops and pasture by the years 2020 (orange bars) and 2050 (gray bars). (From Laurance, 2001; data from Tilman *et al.*, 2001.)

The range of problems facing the human race in the early years of a new millennium are unprecedented, and most of those problems are – in the broadest sense – ecological. Philosophers may have contemplated 'Man's place in the world' for generations, but the question has now taken on a new and much more practical meaning. The luxury of asking 'What does it all mean?' is being replaced by the urgent 'What are we to do?' The closing sections of this book have made the point that ecologists cannot address this question alone – and nobody would let us, even if we wanted to! But equally, the question cannot be addressed without the intimate involvement of those with deep, scientific, ecological understanding. Ecologists of the future face two challenges, equally urgent: to advance our science, and to involve our science thoroughly in local, national and global policies. We must believe that those challenges will be met: to doubt it would only paralyze us.

Summary

In this last of the trilogy of chapters (Chapters 7, 15 and 22), we deal with the application of theory related to succession, food webs, ecosystem functioning and biodiversity.

Managers need to be aware that community composition is hardly ever static. Management objectives that seem to require stasis – the annual production of an agricultural crop, the restoration of a particular combination of species, the long-term survival of an endangered species – are likely to fail unless succession is taken into account.

Every species of concern to managers has its complement of competitors, mutualists, predators and parasites, and an appreciation of such complex interactions is often needed to guide management action in diverse fields including human disease, conservation, harvesting and biosecurity.

Nutrient runoff from agricultural land, together with treated or untreated human sewage, can upset the functioning of aquatic ecosystems through the process of cultural eutrophication, increasing productivity, changing abiotic conditions and altering species composition. One potential solution is the 'biomanipulation' of lake food webs to reverse some of the adverse effects of nutrient enrichment. Moreover, knowledge of terrestrial ecosystem functioning can help determine optimal farm practices, where crop productivity involves minimal input of nutrients. The setting of ecosystem restoration objectives (and the ability to monitor whether these are achieved) requires the development of tools to measure the 'ecosystem health' of terrestrial and aquatic environments.

Much of the planet's surface is used for, or adversely affected by, human habitation, industry, mining, food production and harvesting. Thus, there is a pressing need to use our knowledge of the distribution of biodiversity to design networks of reserved land and water, whether specifically for conservation or for multiple uses, such as harvesting, tourism and conservation combined.

...to global climate change...

As with other pressing problems where the application of ecological knowledge is important, dealing with future climate change also requires a triple bottom line approach that brings together ecological, economic and social perspectives for a sustainable future. Estimates of future greenhouse gas emissions, the concentrations to be expected in the atmosphere, and the resulting changes to global temperature vary considerably. Figure 22.19 shows predicted patterns of increase, and in some cases eventual decreases, based on a variety of scenarios related to conceivable values for population increase, potential changes in the use of various energy sources, and likely technological advances.

...and to increasing agricultural development

A further example of predicted global change concerns the significant threats posed to ecosystems around the world by increasing agricultural development. Given the projected increase in human population, the associated impacts of increased erosion, unsustainability of water supply, salinization and desertification, excess plant nutrients finding their way into waterways, and the unwanted consequences of chemical pesticides will all increase over the next 50 years as more land is converted to grow crops and pasture (Figure 22.20). To control the environmental impacts of agricultural expansion, we will need scientific and technological advances as well as the implementation of effective government policies. Once again sustainability requires its three faces – ecology, economics and sociopolitics.

We finish by emphasizing a reality that applied ecologists cannot ignore. The application of ecological theory never proceeds in isolation. First, there are inevitably economic considerations: how can farmers maximize production while minimizing adverse ecological consequences; how can biodiversity and ecosystem functioning be evaluated alongside profits from forestry or mining; how can returns be maximized from limited conservation dollars? Second, there are almost always sociopolitical considerations: what methods can be used to reconcile interested parties; should sustainable management be set in law or encouraged by education; how can the needs and perspectives of indigenous people be taken into account? These issues come together in the so-called triple bottom line of sustainability, with its ecological, economic and sociopolitical perspectives.

References

Aber, J.D. (1992) Nitrogen cycling and nitrogen saturation in temperate forest ecosystems. *Trends in Ecology and Evolution*, 7, 220–224.

Aber, J.D. & Federer, C.A. (1992) A generalised, lumped-parameter model of photosynthesis, evapotranspiration and net primary production in temperate and boreal forest ecosystems. *Oecologia*, 92, 463–474.

Abrahamson, W.G. (1975) Reproductive strategies in dewberries. *Ecology*, 56, 721–726.

Abrams, P. (1976) Limiting similarity and the form of the competition coefficient. *Theoretical Population Biology*, 8, 356–375.

Abrams, P. (1983) The theory of limiting similarity. *Annual Review of Ecology and Systematics*, 14, 359–376.

Abrams, P.A. (1990) Ecological vs evolutionary consequences of competition. *Oikos*, 57, 147–151.

Abrams, P. (1997) Anomalous predictions of ratio-dependent models of predation. *Oikos*, 80, 163–171.

Abramsky, Z. & Rosenzweig, M.L. (1983) Tilman's predicted productivity–diversity relationship shown by desert rodents. *Nature*, 309, 150–151.

Abramsky, Z. & Sellah, C. (1982) Competition and the role of habitat selection in *Gerbillus allenbyi* and *Meriones tristrami*: a removal experiment. *Ecology*, 63, 1242–1247.

Adams, E.S. (2001) Approaches to the study of territory size and shape. *Annual Review of Ecology and Systematics*, 32, 277–303.

Agrawal, A.A. (1998) Induced responses to herbivory and increased plant performance. *Science*, 279, 1201–1202.

Airame, S., Dugan, J.E., Lafferty, K.D., Leslie, H., McArdle, D.A. & Warner, R.R. (2003) Applying ecological criteria to marine reserve design: a case study from the California Channel Islands. *Ecological Applications*, 13 (Suppl.), S170–S184.

Akçakaya, H.R. (1992) Population viability analysis and risk assessment. In: *Proceedings of Wildlife 2001: Populations* (D.R. McCullough, ed.), pp. 148–158. Elsevier, Amsterdam.

Albertson, F.W. (1937) Ecology of mixed prairie in west central Kansas. *Ecological Monographs*, 7, 481–547.

Albon, S.D., Stien, A., Irvine, R.J., Langvatn, R., Ropstad, E. & Halvorsen, O. (2002) The role of parasites in the dynamics of a reindeer population. *Proceedings of the Royal Society of London, Series B*, 269, 1625–1632.

Albrecht, M. & Gotelli, N.J. (2001) Spatial and temporal niche partitioning in grassland ants. *Oecologia*, 126, 134–141.

Alexander, R. McN. (1991) Optimization of gut structure and diet for higher vertebrate herbivores. In: *The Evolutionary Interaction of Animals and Plants* (J.L. Harper & J.H. Lawton, eds), pp. 73–79. The Royal Society, London; also in *Philosophical Transactions of the Royal Society of London, Series B*, 333, 249–255.

Allen, K.R. (1972) Further notes on the assessment of Antarctic fin whale stocks. *Report of the International Whaling Commission*, 22, 43–53.

Allison, M.J. (1984) Microbiology of the rumen and small and large intestines. In: *Dukes' Physiology of Domestic Animals*, 10th edn (M.J. Swenson, ed.), pp. 340–350. Cornell University Press, Ithaca, NY.

Alonso, A., Dallmeier, F., Granek, E. & Raven, P. (2001) *Connecting with the Tapestry of Life*. Smithsonian Institution/Monitoring and Assessment of Biodiversity Program and President's Committee of Advisors on Science and Technology, Washington, DC.

Alphey, T.W. (1970) Studies on the distribution and site location of *Nippostrongylus brasiliensis* within the small intestine of laboratory rats. *Parasitology*, 61, 449–460.

Amarasekare, P. & Possingham, H. (2001) Patch dynamics and metapopulation theory: the case of successional species. *Journal of Theoretical Biology*, 209, 333–344.

Anderson, J.M. (1978) Inter- and intrahabitat relationships between woodland Cryptostigmata species diversity and diversity of soil and litter micro-habitats. *Oecologia*, 32, 341–348.

Anderson, J.M. (1987) Forest soils as short, dry rivers: effects of invertebrates on transport processes. *Verhandlungen der Gesellschaft fur Okologie*, 17, 33–45.

Anderson, R.M. (1981) Population ecology of infectious disease agents. In: *Theoretical Ecology: principles and applications*, 2nd edn (R.M. May, ed.), pp. 318–355. Blackwell Scientific Publications, Oxford.

Anderson, R.M. (1982) Epidemiology. In: *Modern Parasitology* (F.E.G. Cox, ed.), pp. 205–251. Blackwell Scientific Publications, Oxford.

Anderson, R.M. (1991) Populations and infectious diseases: ecology or epidemiology? *Journal of Animal Ecology*, 60, 1–50.

Anderson, R.M. & May, R.M. (1978) Regulation and stability of host-parasite population interactions. I. Regulatory processes. *Journal of Animal Ecology*, 47, 219–249.

Anderson, R.M. & May, R.M. (1991) *Infectious Diseases of Humans: dynamics and control*. Oxford University Press, Oxford.

Anderson, R.M., Jackson, H.C., May, R.M. & Smith, A.D.M. (1981) Population dynamics of fox rabies in Europe. *Nature*, 289, 765–771.

Andrewartha, H.G. & Birch, L.C. (1954) *The Distribution and Abundance of Animals*. University of Chicago Press, Chicago.

Andrews, P., Lorde, J.M. & Nesbit Evans, E.M. (1979) Patterns of ecological diversity in fossil and mammalian faunas. *Biological Journal of the Linnean Society*, 11, 177–205.

Angel, M.V. (1994) Spatial distribution of marine organisms: patterns and processes. In: *Large Scale Ecology and Conservation Biology* (P.J. Edwards, R.M. May & N.R. Webb, eds), pp. 59–109. Blackwell, Oxford.

Angermeier, P.L. (1995) Ecological attributes of extinction-prone species: loss of freshwater fishes of Virginia. *Conservation Biology*, 9, 143–158.

Antonovics, J. & Alexander, H.M. (1992) Epidemiology of anther-smut infection of *Silene alba* (= *Silene latifolia*) caused by *Ustilago violacea*: patterns of spore deposition in experimental populations. *Proceedings of the Royal Society of London, Series B*, 250, 157–163.

Antonovics, J. & Bradshaw, A.D. (1970) Evolution in closely adjacent plant populations. VIII. Clinical patterns at a mine boundary. *Heredity*, 25, 349–362.

ANZECC (Australia and New Zealand Environment and Conservation Council) (1998) *National Koala Conservation Strategy*. Environment Australia, Canberra.

Arain, M.A., Black, T.A., Barr, A.G. *et al.* (2002) Effects of seasonal and interannual climate variability on net ecosystem productivity of boreal deciduous and conifer forests. *Canadian Journal of Forestry Research*, 32, 878–891.

Araujo, M.B. & Williams, P.H. (2001) The bias of complementarity hotspots toward marginal populations. *Conservation Biology*, 15, 1710–1720.

Arditi, R. & Ginzburg, L.R. (1989) Coupling in predator–prey dynamics: ratio dependence. *Journal of Theoretical Biology*, 139, 311–326.

Armbruster, P. & Lande, R. (1992) A population viability analysis for African elephant (*Loxodonta africana*): how big should reserves be? *Conservation Biology*, 7, 602–610.

Ashmole, N.P. (1971) Sea bird ecology and the marine environment. In: *Avian Biology*, Vol. 1 (D.S. Farner & J.R. King, eds), pp. 224–286. Academic Press, New York.

Aston, J.L. & Bradshaw, A.D. (1966) Evolution in closely adjacent plant populations. II. *Agrostis stolonifera* in maritime habitats. *Heredity*, 21, 649–664.

Atkinson, D., Ciotti, B.J. & Montagnes, D.J.S. (2003) Protists decrease in size linearly with temperature: *ca.* 2.5% °C^{-1}. *Proceedings of the Royal Society of London, Series B*, 270, 2605–2611.

Atkinson, W.D. & Shorrocks, B. (1981) Competition on a divided and ephemeral resource: a simulation model. *Journal of Animal Ecology*, 50, 461–471.

Audesirk, T. & Audesirk, G. (1996) *Biology: life on earth*. Prentice Hall, Upper Saddle River, NJ.

Ausmus, B.S., Edwards, N.T. & Witkamp, M. (1976) Microbial immobilisation of carbon, nitrogen, phosphorus and potassium: implications for forest ecosystem processes. In: *The Role of Terrestrial and Aquatic Organisms in Decomposition Processes* (J.M. Anderson & A. MacFadyen, eds), pp. 397–416. Blackwell Scientific Publications, Oxford.

Ayal, Y. (1994) Time-lags in insect response to plant productivity: significance for plant–insect interactions in deserts. *Ecological Entomology*, 19, 207–214.

Ayala, F.J. (1969) Evolution of fitness. IV. Genetic evolution of inter-specific competitive ability in *Drosophila*. *Genetics*, 61, 737–47.

Bach, C.E. (1994) Effects of herbivory and genotype on growth and survivorship of sand-dune willow (*Salix cordata*). *Ecological Entomology*, 19, 303–309.

Badger, M.R., Andrews, T.J., Whitney, S.M. *et al.* (1997) The diversity and coevolution of Rubisco, plastids, pyrenoids, and chloroplast-based CO_2-concentrating mechanisms in algae. *Canadian Journal of Botany*, 76, 1052–1071.

Baker, A.J.M. (2002) The use of tolerant plants and hyperaccumula-tors. In: *The Restoration and Management of Derelict Land: modern approaches* (M.H. Wong & A.D. Bradshaw, eds), pp. 138–148. World Scientific Publishing, Singapore.

Bakker, K. (1961) An analysis of factors which determine success in competition for food among larvae of *Drosophila melanogaster*. *Archieves Néerlandaises de Zoologie*, 14, 200–281.

Baldwin, I.T. (1998) Jasmonate-induced responses are costly but benefit plants under attack in native populations. *Proceedings of the National Academy of Science of the USA*, 95, 8113–8118.

Balmford, A., Bruner, A., Cooper, P. *et al.* (2002) Economic reasons for conserving wild nature. *Science*, 297, 950–953.

Balmford, A., Green, R.E. & Jenkins, M. (2003) Measuring the changing state of nature. *Trends in Ecology and Evolution*, 18, 326–330.

Baltensweiler, W., Benz, G., Bovey, P. & Delucchi, V. (1977) Dynamics of larch budmoth populations. *Annual Review of Ecology and Systematics*, 22, 79–100.

Barbosa, P. (ed.) (1998) *Conservation Biological Control*. Academic Press, San Diego.

Baross, J.A. & Deming, J.W. (1995) Growth at high temperatures: isolation and taxonomy, physiology, ecology. In: *Microbiology of Deep-Sea Hydrothermal Vent Habitats* (D.M. Karl, ed.), pp. 169–218. CRC Press, New York.

Barrett, K., Wait, D. & Anderson, W.B. (2003) Small island bio-geography in the Gulf of California: lizards, the subsidized island biogeography hypothesis, and the small island effect. *Journal of Biogeography*, 30, 1575–1581.

Bascompte, J. & Sole, R.V. (1996) Habitat fragmentation and extinction thresholds in spatially explicit models. *Journal of Animal Ecology*, 65, 465–473.

Batzli, G.O. (1983) Responses of arctic rodent populations to nutritional factors. *Oikos*, 40, 396–406.

Batzli, G.O. (1993) Food selection in lemmings. In: *The Biology of Lemmings* (N.C. Stenseth & R.A. Ims, eds), pp. 281–302. Academic Press, London.

Bayliss, P. (1987) Kangaroo dynamics. In: *Kangaroos, their Ecology and Management in the Sheep Rangelands of Australia* (G. Caughley, N. Shepherd & J. Short, eds), pp. 119–134. Cambridge University Press, Cambridge, UK.

Bazzaz, F.A. (1979) The physiological ecology of plant succession. *Annual Review of Ecology and Systematics*, 10, 351–371.

Bazzaz, F.A. (1990) The response of natural ecosystems to the rising global CO_2 levels. *Annual Review of Ecology and Systematics*, 21, 167–196.

Bazzaz, F.A. & Williams, W.E. (1991) Atmospheric CO_2 concentrations within a mixed forest: implications for seedling growth. *Ecology*, 72, 12–16.

Bazzaz, F.A., Miao, S.L. & Wayne, P.M. (1993) CO_2-induced growth enhancements of co-occurring tree species decline at different rates. *Oecologia*, 96, 478–482.

Beaumont, L.J. & Hughes, L. (2002) Potential changes in the distributions of latitudinally restricted Australian butterfly species in response to climate change. *Global Change Biology*, 8, 954–971.

Becker, P. (1992) Colonization of islands by carnivorous and herbivorous Heteroptera and Coleoptera: effects of island area, plant species richness, and 'extinction' rates. *Journal of Biogeography*, 19, 163–171.

Becker, P. (2000) Competition in the regeneration niche between conifers and angiosperms: Bond's slow seedling hypothesis. *Functional Ecology*, 14, 401–412.

Beddington, J.R., Free, C.A. & Lawton, J.H. (1978) Modelling biological control: on the characteristics of successful natural enemies. *Nature*, 273, 513–519.

Beebee, T.J.C. (1991) Purification of an agent causing growth inhibition in anuran larvae and its identification as a unicellular unpigmented alga. *Canadian Journal of Zoology*, 69, 2146–2153.

Begon, M. (1976) Temporal variations in the reproductive condition of *Drosophila obscura* Fallén and *D. subobscura* Collin. *Oecologia*, 23, 31–47.

Begon, M. & Bowers, R.G. (1995) Beyond host–pathogen dynamics. In: *Ecology of Infectious Diseases in Natural Populations* (B.T. Grenfell & A.P. Dobson, eds), pp. 478–509. Cambridge University Press, Cambridge, UK.

Begon, M. & Wall, R. (1987) Individual variation and competitor coexistence: a model. *Functional Ecology*, 1, 237–241.

Begon, M., Bennett, M., Bowers, R.G., French, N.P., Hazel, S.M. & Turner, J. (2002) A clarification of transmission terms in host–microparasite models: numbers, densities and areas. *Epidemiology and Infection*, 129, 147–153.

Begon, M., Feore, S.M., Bown, K., Chantrey, J., Jones, T. & Bennett, M. (1998) Population and transmission dynamics of cowpox virus in bank voles: testing fundamental assumptions. *Ecology Letters*, 1, 82–86.

Begon, M., Firbank, L. & Wall, R. (1986) Is there a self-thinning rule for animal populations? *Oikos*, 46, 122–124.

Begon, M., Harper, J.L. & Townsend, C.R. (1990) *Ecology: individuals, populations and communities*, 2nd edn. Blackwell Scientific Publications, Oxford.

Begon, M., Sait, S.M. & Thompson, D.J. (1995) Persistence of a predator–prey system: refuges and generation cycles? *Proceedings of the Royal Society of London, Series B*, 260, 131–137.

Begon, M., Sait, S.M. & Thompson, D.J. (1996) Predator–prey cycles with period shifts between two- and three-species systems. *Nature*, 381, 311–315.

Behnke, J.M., Bayer, A., Sinski, E. & Wakelin, D. (2001) Interactions involving intestinal nematodes of rodents: experimental and field studies. *Parasitology*, 122, S39–S49.

Bell, D.T. (2001) Ecological response syndromes in the flora of southwestern Western Austalia: fire resprouters versus reseeders. *The Botanical Review*, 67, 417–441.

Bell, G. & Koufopanou, V. (1986) The cost of reproduction. *Oxford Surveys in Evolutionary Biology*, 3, 83–131.

Bellows, T.S. Jr. (1981) The descriptive properties of some models for density dependence. *Journal of Animal Ecology*, 50, 139–156.

Belt, T. (1874) *The Naturalist in Nicaragua*. J.M. Dent, London.

Belyea, L.R. & Lancaster, J. (1999) Assembly rules within a contingent ecology. *Oikos*, 86, 402–416.

Benke, A.C., Wallace, J.B., Harrison, J.W. & Koebel, J.W. (2001) Food web quantification using secondary production analysis: predaceous invertebrates of the snag habitat in a subtropical river. *Freshwater Biology*, 46, 329–346.

Bennet, K.D. (1986) The rate of spread and population increase of forest trees during the postglacial. *Philosophical Transactions of the Royal Society of London, Series B*, 314, 523–531; and in *Quantitative Aspects of the Ecology of Biological Invasions* (H. Kornberg & M.H. Williamson, eds), pp. 21–27. The Royal Society, London.

Benson, J.F. (1973a) The biology of Lepidoptera infesting stored products, with special reference to population dynamics. *Biological Reviews*, 48, 1–26.

Benson, J.F. (1973b) Population dynamics of cabbage root fly in Canada and England. *Journal of Applied Ecology*, 10, 437–446.

Benton, T.G. & Grant, A. (1999) Elasticity analysis as an important tool in evolutionary and population ecology. *Trends in Ecology and Evolution*, 14, 467–471.

Bergelson, J.M. (1985) A mechanistic interpretation of prey selection by *Anax junius* larvae (Odonata: Aeschnidae). *Ecology*, 66, 1699–1705.

Bergelson, J. (1994) The effects of genotype and the environment on costs of resistance in lettuce. *American Naturalist*, 143, 349–359.

Berger, J. (1990) Persistence of different-sized populations: an empirical assessment of rapid extinctions in bighorn sheep. *Conservation Biology*, 4, 91–98.

Bergman, C.M., Fryxell, J.M. & Gates, C.C. (2000) The effect of tissue complexity and sward height on the functional response of wood bison. *Functional Ecology*, 14, 61–69.

Berner, E.K. & Berner, R.A. (1987) *The Global Water Cycle: geochemistry and environment*. Prentice-Hall, Englewood Cliffs, NJ.

Bernstein, C., Kacelnik, A. & Krebs, J.R. (1988) Individual decisions and the distribution of predators in a patchy environment. *Journal of Animal Ecology*, 57, 1007–1026.

Bernstein, C., Kacelnik, A. & Krebs, J.R. (1991) Individual decisions and the distribution of predators in a patchy environment. II. The influence of travel costs and structure of the environment. *Journal of Animal Ecology*, 60, 205–225.

Berry, J.A. & Björkman, O. (1980) Photosynthetic response and adaptation to temperature in higher plants. *Annual Review of Plant Physiology*, 31, 491–543.

Berryman, A.A. (ed.) (2002) *Population Cycles: the case for trophic interactions*. Oxford University Press, Oxford.

Bertness, M.D., Leonard, G.H., Levin, J.M., Schmidt, P.R. & Ingraham, A.O. (1999) Testing the relative contribution of positive and negative interactions in rocky intertidal communities. *Ecology*, 80, 2711–2726.

Berven, K.A. (1995) Population regulation in the wood frog, *Rana sylvatica*, from three diverse geographic localities. *Australian Journal of Ecology*, 20, 385–392.

Beven, G. (1976) Changes in breeding bird populations of an oak-wood on Bookham Common, Surrey, over twenty-seven years. *London Naturalist*, 55, 23–42.

Beverton, R.J.H. (1993) The Rio Convention and rational harvesting of natural fish resources: the Barents Sea experience in context. In: *Norway/UNEP Expert Conference on Biodiversity* (O.T. Sandlund & P.J. Schei, eds), pp. 44–63. DN/NINA, Trondheim.

Bibby, C.J. (1998) Selecting areas for conservation. In: *Conservation Science and Action* (W.J. Sutherland, ed.), pp. 176–201. Blackwell Science, Oxford.

Bidle, K.D. & Azam, F. (2001) Bacterial control of silicon regeneration from diatom detritus: significance of bacterial ectohydrolases and species identity. *Limnology and Oceanography*, 46, 1606–1623.

Bigger, M. (1976) Oscillations of tropical insect populations. *Nature*, 259, 207–209.

Biggins, D.E. & Kosoy, M.Y. (2001) Influences of introduced plague on North American mammals: implications from ecology of plague in Asia. *Journal of Mammalogy*, 82, 906–916.

Bignell, D.E. (1989) Relative assimilations of carbon-14-labeled microbial tissues and carbon-14-labeled plant fiber ingested with leaf litter by the millipede *Glomeris marginata* under experimental conditions. *Soil Biology and Biochemistry*, 21, 819–828.

Björnhag, G. (1994) Adaptations in the large intestine allowing small animals to eat fibrous food. In: *The Digestive System in Mammals. Food, Form and Function* (D.J. Chivers & P. Langer, eds), pp. 287–312. Cambridge University Press, Cambridge, UK.

Bjørnstad, O.N. & Grenfell, B.T. (2001) Noisy clockwork: time series analysis of population fluctuations in animals. *Science*, 293, 638–643.

Bjørnstad, O.N., Falck, W. & Stenseth, N.C. (1995) A geographic gradient in small rodent density fluctuations: a statistical modelling approach. *Proceedings of the Royal Society of London, Series B*, 262, 127–133.

Bjørnstad, O.N., Sait, S.M., Stenseth, N.C., Thompson, D.J. & Begon, M. (2001) The impact of specialized enemies on the dimensionality of host dynamics. *Nature*, 409, 1001–1006.

Black, J.N. (1963) The interrelationship of solar radiation and leaf area index in determining the rate of dry matter production of swards of subterranean clover (*Trifolium subterraneum*). *Australian Journal of Agricultural Research*, 14, 20–38.

Black, T.A., Chen, W.J., Barr, A.G. et al. (2000) Increased carbon sequestration by a boreal deciduous forest in years with a warm spring. *Geophysical Research Letters*, 27, 1271–1274.

Blackburn, T.M. & Gaston, K.J. (eds) (2003) *Macroecology: concepts and consequences*. Blackwell Publishing, Oxford.

Blackford, J.C., Allen, J.I. & Gilbert, F.J. (2004) Ecosystem dynamics at six contrasting sites: a generic modelling study. *Journal of Marine Systems*, 52, 191–215.

Blaustein, A.R., Wake, D.B. & Sousa, W.P. (1994) Amphibian declines: judging stability, persistence, and susceptibility of populations to local and global extinctions. *Conservation Biology*, 8, 60–71.

Blueweiss, L., Fox, H., Kudzma, V., Nakashima, D., Peters, R. & Sams, S. (1978) Relationships between body size and some life history parameters. *Oecologia*, 37, 257–272.

Bobbink, R., Boxman, D., Fremstad, E., Heil, G., Houdijk, A. & Roelofs, J. (1992) Critical loads for nitrogen eutrophication of terrestrial and wetland ecosystems based upon changes in vegetation and fauna. In: *Critical Loads for Nitrogen* (P. Grennfelt & E. Thornelof, eds), pp. 111–160. Nordic Council of Ministers, Copenhagen.

Boden, T.A., Kanciruk, P. & Fartell, M.P. (1990) *Trends '90. A Compendium of Data on Global Change*. Carbon Dioxide Analysis Center, Oak Ridge National Laboratory, Oak Ridge, TN.

Boersma, P.D., Kareiva, P., Fagan, W.F., Clark, J.A. & Hoekstra, J.M. (2001) How good are endangered species recovery plans? *BioScience*, 51, 643–650.

Boivin, G., Fauvergue, X. & Wajnberg, E. (2004) Optimal patch residence time in egg parasitoids: innate versus learned estimate of patch quality. *Oecologia*, 138, 640–647.

Bojorquez-Tapia, L.A., Brower, L.P., Castilleja, G. et al. (2003) Mapping expert knowledge: redesigning the monarch butterfly biosphere reserve. *Conservation Biology*, 17, 367–379.

Bolker, B.M., Pacala, S.W. & Neuhauser, C. (2003) Spatial dynamics in model plant communities: what do we really know. *American Naturalist*, 162, 135–148.

Bonnet, X., Lourdais, O., Shine, R. & Naulleau, G. (2002) Reproduction in a typical capital breeder: costs, currencies and complications in the aspic viper. *Ecology*, 83, 2124–2135.

Bonsall, M.B. & Hassell, M.P. (1997) Apparent competition structures ecological assemblages. *Nature*, 388, 371–372.

Bonsall, M.B., French, D.R. & Hassell, M.P. (2002) Metapopulation structure affects persistence of predator–prey interactions. *Journal of Animal Ecology*, 71, 1075–1084.

Booth, B.D. & Swanton, C.J. (2002) Assembly theory applied to weed communities. *Weed Science*, 50, 2–13.

Booth, D.J. & Brosnan, D.M. (1995) The role of recruitment dynamics in rocky shore and coral reef fish communities. *Advances in Ecological Research*, 26, 309–385.

Booth, D.T., Clayton, D.H. & Block, B.A. (1993) Experimental demonsration of the energetic cost of parasitism in free-ranging hosts. *Proceedings of the Royal Society of London, Series B*, 253, 125–129.

Boots, M. & Begon, M. (1993) Trade-offs with resistance to a granulosis virus in the Indian meal moth, examined by a laboratory evolution experiment. *Functional Ecology*, 7, 528–534.

Bormann, B.T. & Gordon, J.C. (1984) Stand density effects in young red alder plantations: productivity, photosynthate partitioning and nitrogen fixation. *Ecology*, 65, 394–402.

Bossenbroek, J.M., Kraft, C.E. & Nekola, J.C. (2001) Prediction of long-distance dispersal using gravity models: zebra mussel invasion in inland lakes. *Ecological Applications*, 11, 1778–1788.

Boyce, M.S. (1984) Restitution of *r*- and *K*-selection as a model of density-dependent natural selection. *Annual Review of Ecology and Systematics*, 15, 427–447.

Boyd, P.W. (2002) The role of iron in the biogeochemistry of the Southern Ocean and equatorial Pacific: a comparison of *in situ* iron enrichments. *Deep-Sea Research II*, 49, 1803–1821.

Bradshaw, A.D. (1987) The reclamation of derelict land and the ecology of ecosystems. In: *Restoration Ecology* (W.R. Jordan III, M.E. Gilpin & J.D. Aber, eds), pp. 53–74. Cambridge University Press, Cambridge, UK.

Bradshaw, A.D. (2002) Introduction – an ecological perspective. In: *The Restoration and Management of Derelict Land: modern approaches* (M.H. Wong & A.D. Bradshaw, eds), pp. 1–6. World Scientific Publishing, Singapore.

Branch, G.M. (1975) Intraspecific competition in *Patella cochlear* Born. *Journal of Animal Ecology*, 44, 263–281.

Brewer, M.J. & Elliott, N.C. (2004) Biological control of cereal aphids in North America and mediating effects of host plant and habitat manipulations. *Annual Review of Entomology*, 49, 219–242.

Breznak, J.A. (1975) Symbiotic relationships between termites and their intestinal biota. In: *Symbiosis* (D.H. Jennings & D.L. Lee, eds), pp. 559–580. Symposium 29, Society for Experimental Biology, Cambridge University Press, Cambridge, UK.

Briand, F. (1983) Environmental control of food web structure. *Ecology*, 64, 253–263.

Briand, F. & Cohen, J.E. (1987) Environmental correlates of food chain length. *Science*, 238, 956–960.

Brittain, J.E. & Eikeland, T.I. (1988) Invertebrate drift – a review. *Hydrobiologia*, 166, 77–93.

Brokaw, N. & Busing, R.T. (2000) Niche versus chance and tree diversity in forest gaps. *Trends in Ecology and Evolution*, 15, 183–188.

Bronstein, J.L. (1988) Mutualism, antagonism and the fig–pollinator interaction. *Ecology*, 69, 1298–1302.

Brook, B.W., O'Grady, J.J., Chapman, A.P., Burgman, M.A., Akcakaya, H.R. & Frankham, R. (2000) Predictive accuracy of population viability analysis in conservation biology. *Nature*, 404, 385–387.

Brookes, M. (1998) The species enigma. *New Scientist*, June 13, 1998.

Brooks, R.R. (ed.) (1998) *Plants that Hyperaccumulate Heavy Metals*. CAB International, Wallingford, UK.

Brower, J.E., Zar, J.H. & van Ender, C.N. (1998) *Field and Laboratory Methods for General Ecology*, 4th edn. McGraw-Hill, Boston.

Brower, L.P. & Corvinó, J.M. (1967) Plant poisons in a terrestrial food chain. *Proceedings of the National Academy of Science of the USA*, 57, 893–898.

Brown, B.E. (1997) Coral bleaching: causes and consequences. *Coral Reefs*, 16, S129–S138.

Brown, E., Dunne, R.P., Goodson, M.S. & Douglas, A.E. (2000) Bleaching patterns in reef corals. *Nature*, 404, 142–143.

Brown, H.T. & Escombe, F. (1990) Static diffusion of gases and liquids in relation to the assimilation of carbon and translocation in plants. *Philosophical Transactions of the Royal Society of London, Series B*, 193, 223–291.

Brown, J.H. & Davidson, D.W. (1977) Competition between seed-eating rodents and ants in desert ecosystems. *Science*, 196, 880–882.

Brown, J.H. & Kodric-Brown, A. (1977) Turnover rates in insular biogeography: effect of immigration and extinction. *Ecology*, 58, 445–449.

Brown, J.H. & Maurer, B.A. (1989) Macroecology: the division of food and space among species on continents. *Science*, 243, 1145–1150.

Brown, J.S., Kotler, B.P., Smith, R.J. & Wirtz, W.O. III (1988) The effects of owl predation on the foraging behaviour of heteromyid rodents. *Oecologia*, 76, 408–415.

Brown, K.M. (1982) Resource overlap and competition in pond snails: an experimental analysis. *Ecology*, 63, 412–422.

Brown, V.K. & Southwood, T.R.E. (1983) Trophic diversity, niche breadth, and generation times of exopterygote insects in a secondary succession. *Oecologia*, 56, 220–225.

Browne, R.A. (1981) Lakes as islands: biogeographic distribution, turnover rates, and species composition in the lakes of central New York. *Journal of Biogeography*, 8, 75–83.

Brunet, A.K. & Medellín, R.A. (2001) The species–area relationship in bat assemblages of tropical caves. *Journal of Mammalogy*, 82, 1114–1122.

Brylinski, M. & Mann, K.H. (1973) An analysis of factors governing productivity in lakes and reservoirs. *Limnology and Oceanography*, 18, 1–14.

Bshary, R. (2003) The cleaner wrasse, *Labroides dimidiatus*, is a key organism for reef fish diversity at Ras Mohammed National Park, Egypt. *Journal of Animal Ecology*, 72, 169–176.

Bubier, J.L. & Moore, T.R. (1994) An ecological perspective on methane emissions from northern wetlands. *Trends in Ecology and Evolution*, 9, 460–464.

Buchanan, G.A., Crowley, R.H., Street, J.E. & McGuire, J.A. (1980) Competition of sicklepod (*Cassia obtusifolia*) and redroot pigweed (*Amaranthus retroflexus*) with cotton (*Gossypium hirsutum*). *Weed Science*, 28, 258–262.

Buckling, A. & Rainey, P.B. (2002) Antagonistic coevolution between a bacterium and a bacteriophage. *Proceedings of the Royal Society of London, Series B*, 269, 931–936.

Buesseler, K.O. (1998) The decoupling of production and particulate export in the surface ocean. *Global Biogeochemical Cycles*, 12, 297–310.

Bull, C.M. & Burzacott, D. (1993) The impact of tick load on the fitness of their lizard hosts. *Oecologia*, 96, 415–419.

Bullock, J.M., Franklin, J., Stevenson, M.J. *et al.* (2001) A plant trait analysis of responses to grazing in a long-term experiment. *Journal of Applied Ecology*, 38, 253–267.

Bullock, J.M., Mortimer, A.M. & Begon, M. (1994a) Physiological integration among tillers of *Holcus lanatus*: age-dependence and responses to clipping and competition. *New Phytologist*, 128, 737–747.

Bullock, J.M., Mortimer, A.M. & Begon, M. (1994b) The effect of clipping on interclonal competition in the grass *Holcus lanatus* – a response surface analysis. *Journal of Ecology*, 82, 259–270.

Bullock, J.M., Moy, I.L., Pywell, R.F., Coulson, S.J., Nolan, A.M. & Caswell, H. (2002) Plant dispersal and colonization processes at local and landscape scales. In: *Dispersal Ecology* (J.M. Bullock, R.E. Kenward & R.S. Hails, eds), pp. 279–302. Blackwell Science, Oxford.

Burdon, J.J. (1987) *Diseases and Plant Population Biology.* Cambridge University Press, Cambridge, UK.

Buscot, F., Munch, J.C., Charcosset, J.Y., Gardes, M., Nehls, U. & Hampp, R. (2000) Recent advances in exploring physiology and biodiversity of ectomycorrhizas highlight the functioning of these symbioses in ecosystems. *FEMS Microbiology Reviews*, 24, 601–614.

Busing, R.T. & Brokaw, N. (2002) Tree species diversity in temperate and tropical forest gaps: the role of lottery recruitment. *Folia Geobotanica*, 37, 33–43.

Buss, L.W. (1979) Byrozoan overgrowth interactions – the interdependence of competition for food and space. *Nature*, 281, 475–477.

Cadenasso, M.L., Pickett, S.T.A. & Morin, P.J. (2002) Experimental test of the role of mammalian herbivores on old field succession: community structure and seedling survival. *Journal of the Torrey Botanical Society*, 129, 228–237.

Cain, M.L., Pacala, S.W., Silander, J.A. & Fortin, M.-J. (1995) Neighbourhood models of clonal growth in the white clover *Trifolium repens. American Naturalist*, 145, 888–917.

Caldwell, M.M. & Richards, J.H. (1986) Competing root systems: morphology and models of absorption. In: *On the Economy of Plant Form and Function* (T.J. Givnish, ed.), pp. 251–273. Cambridge University Press, Cambridge, UK.

Callaghan, T.V. (1976) Strategies of growth and population dynamics of plants: 3. Growth and population dynamics of *Carex bigelowii* in an alpine environment. *Oikos*, 27, 402–413.

Callaway, R.M., Kikvidze, Z. & Kikodze, D. (2000) Facilitation by unpalatable weeds may conserve plant diversity in overgrazed meadows in the Caucasus mountains. *Oikos*, 89, 275–282.

Cammell, M.E., Tatchell, G.M. & Woiwod, I.P. (1989) Spatial pattern of abundance of the black bean aphid, *Aphis fabae*, in Britain. *Journal of Applied Ecology*, 26, 463–472.

Campbell, J., Antoine, D., Armstrong, R. *et al.* (2002) Comparison of algorithms for estimating ocean primary productivity from surface chlorophyll, temperature, and irradiance. *Global Biogeochemical Cycles*, 16, 91–96.

Caraco, N.F. (1993) Disturbance of the phosphorus cycle: a case of indirect effects of human activity. *Trends in Ecology and Evolution*, 8, 51–54.

Caraco, T. & Kelly, C.K. (1991) On the adaptive value of physiological integration of clonal plants. *Ecology*, 72, 81–93.

Cardillo, M. & Bromham, L. (2001) Body size and risk of extinction in Australian mammals. *Conservation Biology*, 15, 1435–1440.

Carignan, R., Planas, D. & Vis, C. (2000) Planktonic production and respiration in oligotrophic Shield lakes. *Limnology and Oceanography*, 45, 189–199.

Carne, P.B. (1969) On the population dynamics of the eucalypt-defoliating chrysomelid *Paropsis atomaria* OI. *Australian Journal of Zoology*, 14, 647–672.

Carpenter, S.R., Ludwig, D. & Brock, W.A. (1999) Management of eutrophication for lakes subject to potentially irreversible change. *Ecological Applications*, 9, 751–771.

Carr, M.H., Neigel, J.E., Estes, J.A., Andelman, S., Warner, R.R. & Largier, J.L. (2003) Comparing marine and terrestrial ecosystems: implications for the design of coastal marine reserves. *Ecological Applications*, 13 (Suppl.), S90–S107.

Carson, H.L. & Kaneshiro, K.Y. (1976) *Drosophila* of Hawaii: systematics and ecological genetics. *Annual Review of Ecology and Systematics*, 7, 311–345.

Carson, W.P. & Root, R.B. (1999) Top-down effects of insect herbivores during early succession: influence on biomass and plant dominance. *Oecologia*, 121, 260–272.

Cary, J.R. & Keith, L.B. (1979) Reproductive change in the 10-year cycle of snowshoe hares. *Canadian Journal of Zoology*, 57, 375–390.

Caswell, H. (2001) *Matrix Population Models*, 2nd edn. Sinauer, Sunderland, MA.

Caughley, G. (1994) Directions in conservation biology. *Journal of Animal Ecology*, 63, 215–244.

Cebrian, J. (1999) Patterns in the fate of production in plant communities. *American Naturalist*, 154, 449–468.

Charlesworth, D. & Charlesworth, B. (1987) Inbreeding depression and its evolutionary consequences. *Annual Review of Ecology and Systematics*, 18, 237–268.

Charnov, E.L. (1976a) Optimal foraging: attack strategy of a mantid. *American Naturalist*, 110, 141–151.

Charnov, E.L. (1976b) Optimal foraging: the marginal value theorem. *Theoretical Population Biology*, 9, 129–136.

Charnov, E.L. & Krebs, J.R. (1974) On clutch size and fitness. *Ibis*, 116, 217–219.

Chase, J.M. (2000) Are there real differences among aquatic and terrestrial food webs? *Trends in Ecology and Evolution*, 15, 408–412.

Chase, J.M. (2003) Experimental evidence for alternative stable equilibria in a benthic pond food web. *Ecology Letters*, 6, 733–741.

Cherrett, J.M., Powell, R.J. & Stradling, D.J. (1989) The mutualism between leaf-cutting ants and their fungus. In: *Insect/Fungus Interactions* (N. Wilding, N.M. Collins, P.M. Hammond & J.F. Webber, eds), pp. 93–120. Royal Entomological Society Symposium No. 14. Academic Press, London.

Chesson, P. & Murdoch, W.W. (1986) Aggregation of risk: relationships among host–parasitoid models. *American Naturalist*, 127, 696–715.

Childs, D.Z., Rees, M., Rose, K.E., Grubb, P.J. & Elner, S.P. (2003) Evolution of complex flowering strategies: an age- and size-structured integral projection model. *Proceedings of the Royal Society of London, Series B*, 270, 1829–1838.

Choquenot, D. (1998) Testing the relative influence of intrinsic and extrinsic variation in food availability on feral pig populations in Australia's rangelands. *Journal of Animal Ecology*, 67, 887–907.

Choudhary, M., Strassman, J.E., Queller, D.C., Turilazzi, S. & Cervo, R. (1994) Social parasites in polistine wasps are monophyletic: implications for sympatric speciation. *Proceedings of the Royal Society of London, Series B*, 257, 31–35.

Christian, J.J. (1980) Endocrine factors in population regulation. In: *Biosocial Mechanisms of Population Regulation* (M.N. Cohen, R.S. Malpass & H.G. Klein, eds), pp. 55–115. Yale University Press, New Haven, CT.

Clapham, W.B. (1973) *Natural Ecosystems*. Collier–Macmillan, New York.

Clark, C.W. (1981) Bioeconomics. In: *Theoretical Ecology: principles and applications*, 2nd edn. (R.M. May, ed.), pp. 387–418. Blackwell Scientific Publications, Oxford.

Clark, C.W. & Mangel, M. (2000) *Dynamic State Variable Models in Ecology*. Oxford University Press, New York.

Clark, D.A. & Clark, D.B. (1984) Spacing dynamics of a tropical rain forest tree: evaluation of the Janzen–Connell model. *American Naturalist*, 124, 769–788.

Clark, J.A. & Harvey, E. (2002) Assessing multi-species recovery plans under the endangered species act. *Ecological Applications*, 12, 655–662.

Clark, L.R. (1962) The general biology of *Cardiaspina albitextura* (Psyllidae) and its abundance in relation to weather and parasitism. *Australian Journal of Zoology*, 10, 537–586.

Clark, L.R. (1964) The population dynamics of *Cardiaspina albitextura* (Psyllidae). *Australian Journal of Zoology*, 12, 362–380.

Clarke, A. (2004) Is there a universal temperature dependence of metabolism? *Functional Ecology*, 18, 252–256.

Clarke, A. & Crame, J.A. (2003) The importance of historical processes in global patterns of diversity. In: *Macroecology: concepts and consequences* (T.M. Blackburn & K.J. Gaston, eds), pp. 130–151. Blackwell Publishing, Oxford.

Clarke, B.C. & Partridge, L. (eds) (1988) Frequency dependent selection. *Philosophical Transactions of the Royal Society of London, Series B*, 319, 457–645.

Clay, K. (1990) Fungal endophytes of grasses. *Annual Review of Ecology and Systematics*, 21, 275–297.

Clearwater, J.R. (2001) Tackling tussock moths: strategies, timelines and outcomes of two programs for eradicating tussock moths from Auckland suburbs. In: *Eradication of Island Invasives: practical actions and results achieved* (M. Clout & D. Veitch, eds). Conference of the Invasive Species Specialist Group of the World Conservation Union (IUCN) Species Survival Commission. UICN, Auckland, New Zealand.

Clements, F.E. (1916) *Plant Succession: analysis of the development of vegetation*. Carnegie Institute of Washington Publication No. 242. Washington, DC.

Clinchy, M., Haydon, D.T. & Smith, A.T. (2002) Pattern does not equal process: what does patch occupancy really tell us about metapopulation dynamics? *American Naturalist*, 159, 351–362.

Clobert, J., Wolff, J.O., Nichols, J.D., Danchin, E. & Dhondt, A.A. (2001) Introduction. In: *Dispersal* (J. Clobert, E. Danchin, A.A. Dhondt & J.D. Nichols, eds), pp. xvii–xxi. Oxford University Press, Oxford.

Clutton-Brock, T.H. & Harvey, P.H. (1979) Comparison and adaptation. *Proceedings of the Royal Society of London, Series B*, 205, 547–565.

Clutton-Brock, T.H., Major, M., Albon, S.D. & Guinness, F.E. (1987) Early development and population dynamics in red deer. I. Density-dependent effects on juvenile survival. *Journal of Animal Ecology*, 56, 53–67.

Clutton-Brock, T.H., Major, M. & Guinness, F.E. (1985) Population regulation in male and female red deer. *Journal of Animal Ecology*, 54, 831–836.

Coale, K.H., Johnson, K.S., Fitzwater, S.E. *et al.* (1996) A massive phytoplankton bloom induced by an ecosystem-scale iron fertilization experiment in the Equatorial Pacific Ocean. *Nature*, 383, 495–501.

Cockell, C.S. & Lee, P. (2002) The biology of impact craters – a review. *Biological Reviews*, 77, 279–310.

Cody, M.L. (1975) Towards a theory of continental species diversities. In: *Ecology and Evolution of Communities* (M.L. Cody & J.M. Diamond, eds), pp. 214–257. Belknap, Cambridge, MA.

Cohen, J.E., Jonsson, T. & Carpenter, S.R. (2003) Ecological community description using food web, species abundance, and body-size. *Proceedings of the National Academy of Science of the USA*, 100, 1781–1786.

Cole, J.J., Findlay, S. & Pace, M.L. (1988) Bacterial production in fresh and salt water ecosystems: a cross-system overview. *Marine Ecology Progress Series*, 4, 1–10.

Collado-Vides, L. (2001) Clonal architecture in marine macroalgae: ecological and evolutionary perspectives. *Evolutionary Ecology*, 15, 531–545.

Collins, M.D., Ward, S.A. & Dixon, A.F.G. (1981) Handling time and the functional response of *Aphelinus thomsoni*, a predator and parasite of the aphid, *Drepanosiphum platanoidis*. *Journal of Animal Ecology*, 50, 479–487.

Colwell, R.K. & Hurtt, G.C. (1994) Non-biological gradients in species richness and a spurious Rapoport effect. *American Naturalist*, 144, 570–595.

Comins, H.N., Hassell, M.P. & May, R.M. (1992) The spatial dynamics of host–parasitoid systems. *Journal of Animal Ecology*, 61, 735–748.

Compton, S.G. (2001) sailing with the wind: dispersal by small flying insects. In: *Dispersal Ecology* (J.M. Bullock, R.E. Kenward & R.S. Hails, eds), pp. 113–133. Blackwell Science, Oxford.

Connell, J.H. (1961) The influence of interspecific competition and other factors on the distribution of the barnacle *Chthamalus stellatus*. *Ecology*, 42, 710–723.

Connell, J.H. (1970) A predator–prey system in the marine intertidal region. I. *Balanus glandula* and several predatory species of *Thais*. *Ecological Monographs*, 40, 49–78.

Connell, J.H. (1971) On the role of natural enemies in preventing competitive exclusion in some marine animals and in rain forest trees. In: *Dynamics of Populations* (P.J. den Boer & G.R. Gradwell, eds), pp. 298–310. *Proceedings of the Advanced Study Institute in Dynamics of Numbers in Populations, Oosterbeck*. Centre for Agricultural Publishing and Documentation, Wageningen.

Connell, J.H. (1975) Some mechanisms producing structure in natural communities: a model and evidence from field experiments. In: *Ecology and Evolution of Communities* (M.L. Cody & J.M. Diamond, eds), pp. 460–490. Belknap, Cambridge, MA.

Connell, J.H. (1978) Diversity in tropical rainforests and coral reefs. *Science*, 199, 1302–1310.

Connell, J.H. (1979) Tropical rain forests and coral reefs as open nonequilibrium systems. In: *Population Dynamics* (R.M. Anderson, B.D. Turner & L.R. Taylor, eds), pp. 141–163. Blackwell Scientific Publications, Oxford.

Connell, J.H. (1980) Diversity and the coevolution of competitors, or the ghost of competition past. *Oikos*, 35, 131–138.

Connell, J.H. (1983) On the prevalence and relative importance of interspecific competition: evidence from field experiments. *American Naturalist*, 122, 661–696.

Connell, J.H. (1990) Apparent versus 'real' competition in plants. In: *Perspectives on Plant Competition* (J.B. Grace & D. Tilman, eds), pp. 9–26. Academic Press, New York.

Connor, E.F. (1986) The role of Pleistocene forest refugia in the evolution and biogeography of tropical biotas. *Trends in Ecology and Evolution*, 1, 165–169.

Cook, L.M., Dennis, R.L.H. & Mani, G.S. (1999) Melanic morph frequency in the peppered moth in the Manchester area. *Proceedings of the Royal Society of London, Series B*, 266, 293–297.

Cook, R.M., Sinclair, A. & Stefansson, G. (1997) Potential collapse of North Sea cod stocks. *Nature*, 385, 521–522.

Coomes, D.A., Rees, M., Turnbull, L. & Ratcliffe, S. (2002) On the mechanisms of coexistence among annual-plant species, using neighbourhood techniques and simulation models. *Plant Ecology*, 163, 23–38.

Corn, P.S. (2003) Amphibian breeding and climate change: importance of snow in the mountains. *Conservation Biology*, 17, 622–625.

Cornell, H.V. & Hawkins, B.A. (2003) Herbivore responses to plant secondary compounds: a test of phytochemical coevolution theory. *American Naturalist*, 161, 507–522.

Cornwell, W.K. & Grubb, P.J. (2003) Regional and local patterns in plant species richness with respect to resource availability. *Oikos*, 100, 417–428.

Cortes, E. (2002) Incorporating uncertainty into demographic modeling: application to shark populations and their conservation. *Conservation Biology*, 16, 1048–1062.

Cory, J.S. & Myers, J.H. (2000) Direct and indirect ecological effects of biological control. *Trends in Ecology and Evolution*, 15, 137–139.

Costantino, R.F., Desharnais, R.A., Cushing, J.M. & Dennis, B. (1997) Chaotic dynamics in an insect population. *Science*, 275, 389–391.

Costanza, R., D'Arge, R., de Groot, R. *et al.* (1997) The value of the world's ecosystem services and natural capital. *Nature* 387, 253–260.

Cotrufo, M.F., Ineson, P., Scott, A. *et al.* (1998) Elevated CO_2 reduces the nitrogen concentration of plant tissues. *Global Change Biology*, 4, 43–54.

Cottingham, K.L., Brown, B.L. & Lennon, J.T. (2001) Biodiversity may regulate the temporal variability of ecological systems. *Ecology Letters*, 4, 72–85.

Courchamp, F., Clutton-Brock, T. & Grenfell, B. (1999) Inverse density dependence and the Allee effect. *Trends in Ecology and Evolution*, 14, 405–410.

Courchamp, F., Langlais, M. & Sugihara, G. (1999) Cats protecting birds: modeling the mesopredator release effect. *Journal of Animal Ecology*, 68, 282–292.

Cowie, R.J. (1977) Optimal foraging in great tits *Parus major*. *Nature*, 268, 137–139.

Cox, C.B., Healey, I.N. & Moore, P.D. (1976) *Biogeography*, 2nd edn. Blackwell Scientific Publications, Oxford.

Crawford, D.W., Purdie, D.A., Lockwood, A.P.M. & Weissman, P. (1997) Recurrent red-tides in the Southampton Water estuary caused by the phototrophic ciliate *Mesodinium rubrum*. *Estuarine, Coastal and Shelf Science*, 45, 799–812.

Crawley, M.J. (1983) *Herbivory: the dynamics of animal–plant interactions*. Blackwell Scientific Publications, Oxford.

Crawley, M.J. (1986) The structure of plant communities. In: *Plant Ecology* (M.J. Crawley, ed.), pp. 1–50. Blackwell Scientific Publications, Oxford.

Crawley, M.J. (1989) Insect herbivores and plant population dynamics. *Annual Review of Entomology*, 34, 531–564.

Crawley, M.J. & May, R.M. (1987) Population dynamics and plant community structure: competition between annuals and perennials. *Journal of Theoretical Biology*, 125, 475–489.

Crocker, R.L. & Major, J. (1955) Soil development in relation to vegetation and surface age at Glacier Bay, Alaska. *Journal of Ecology*, 43, 427–448.

Cronk, Q.C.B. & Fuller, J.L. (1995) *Plant Invaders*. Chapman & Hall, London.

Culver, C.S. & Kuris, A.M. (2000) The apparent eradication of a locally established introduced marine pest. *Biological Invasions*, 2, 245–253.

Cummins, K.W. (1974) Structure and function of stream ecosystems. *Bioscience*, 24, 631–641.

Currie, C.R. (2001) A community of ants, fungi, and bacteria: a multilateral approach to studying symbiosis. *Annual Review of Microbiology*, 55, 357–380.

Currie, D.J. (1991) Energy and large-scale patterns of animal and plant species richness. *American Naturalist*, 137, 27–49.

Currie, D.J. & Paquin, V. (1987) Large-scale biogeographical patterns of species richness in trees. *Nature*, 39, 326–327.

Daan, S., Dijkstra, C. & Tinbergen, J.M. (1990) Family planning in the kestrel (*Falco tinnunculus*): the ultimate control of covariation of laying date and clutch size. *Behavior*, 114, 83–116.

Dahl, E. (1979) Deep-sea carrion feeding amphipods: evolutionary patterns in niche adaptation. *Oikos*, 33, 167–175.

Daily, G.C., Soderqvist, T., Aniyar, S. *et al.* (2000) The value of nature and the nature of value. *Science*, 289, 395–396.

Darwin, C. (1859) *The Origin of Species by Means of Natural Selection*, 1st edn. John Murray, London.

Darwin, C. (1888) *The Formation of Vegetable Mould Through the Action of Worms*. John Murray, London.

Daufresne, T. & Loreau, M. (2001) Ecological stoichiometry, primary producer–decomposer interactions, and ecosystem persistence. *Ecology*, 82, 3069–3082.

Davidson, D.W. (1977) Species diversity and community organization in desert seed-eating ants. *Ecology*, 58, 711–724.

Davidson, D.W., Samson, D.A. & Inouye, R.S. (1985) Granivory in the Chihuahuan Desert: interactions within and between trophic levels. *Ecology*, 66, 486–502.

Davidson, J. & Andrewartha, H.G. (1948) The influence of rainfall, evaporation and atmospheric temperature on fluctuations in the size of a natural population of *Thrips imaginis* (Thysanoptera). *Journal of Animal Ecology*, 17, 200–222.

Davies, J.B. (1994) Sixty years of onchocerciasis vector control – a chronological summary with comments on eradication, reinvasion, and insecticide resistance. *Annual Review of Entomology*, 39, 23–45.

Davies, K.F., Margules, C.R. & Lawrence, J.F. (2000) Which traits of species predict population declines in experimental forest fragments. *Ecology*, 81, 1450–1461.

Davies, N.B. (1977) Prey selection and social behaviour in wagtails (Aves: Motacillidae). *Journal of Animal Ecology*, 46, 37–57.

Davies, N.B. & Houston, A.I. (1984) Territory economics. In: *Behavioural Ecology: an evolutionary approach*, 2nd edn (J.R. Krebs & N.B. Davies, eds), pp. 148–169. Blackwell Scientific Publications, Oxford.

Davies, S.J., Palmiotto, P.A., Ashton, P.S., Lee, H.S. & Lafrankie, J.V. (1998) Comparative ecology of 11 sympatric species of *Macaranga* in Borneo: tree distribution in relation to horizontal and vertical resource heterogeneity. *Journal of Ecology*, 86, 662–673.

Davis, A.L.V. (1996) Seasonal dung beetle activity and dung dispersal in selected South African habitats: implications for pasture improvement in Australia. *Agriculture, Ecosystems and Environment*, 58, 157–169.

Davis, M.B. (1976) Pleistocene biogeography of temperate deciduous forests. *Geoscience and Man*, 13, 13–26.

Davis, M.B. & Shaw, R.G. (2001) Range shifts and adaptive responses to quarternary climate change. *Science*, 292, 673–679.

Davis, M.B., Brubaker, L.B. & Webb, T. III (1973) Calibration of absolute pollen inflex. In: *Quaternary Plant Ecology* (H.J.B. Birks & R.G. West, eds), pp. 9–25. Blackwell Scientific Publications, Oxford.

Dayan, T., Simberloff, E., Tchernov, E. & Yom-Tov, Y. (1989) Inter- and intraspecific character displacement in mustelids. *Ecology*, 70, 1526–1539.

De Groot, K.L. & Smith, J.N.M. (2001) Community-wide impacts of a generalist brood parasite, the brown-headed cowbird (*Molothrus ater*). *Ecology*, 82, 868–881.

de Jong, G. (1994) The fitness of fitness concepts and the description of natural selection. *Quarterly Review of Biology*, 69, 3–29.

De Kroon, H., Plaisier, A., Van Groenendael, J. & Caswell, H. (1986) Elasticity: the relative contribution of demographic parameters to population growth rate. *Ecology*, 67, 1427–1431.

de Wet, N., Ye, W., Hales, S., Warrick, R., Woodward, A. & Weinstein, P. (2001) Use of a computer model to identify potential hotspots for dengue fever in New Zealand. *New Zealand Medical Journal*, 114, 420–422.

de Wit, C.T. (1960) On competition. *Verslagen van landbouwkundige Onderzoekingen*, 660, 1–82.

de Wit, C.T. (1965) Photosynthesis of leaf canopies. *Verslagen van Landbouwkundige Onderzoekingen*, 663, 1–57.

de Wit, C.T., Tow, P.G. & Ennik, G.C. (1966) Competition between legumes and grasses. *Verslagen van landbouwkundige Onderzoekingen*, 112, 1017–1045.

Dean, A.M. (1983) A simple model of mutualism. *American Naturalist*, 121, 409–417.

DeAngelis, D.L. (1975) Stability and connectance in food web models. *Ecology*, 56, 238–243.

DeAngelis, D.L. (1980) Energy flow, nutrient cycling and ecosystem resilience. *Ecology*, 61, 764–771.

DeAngelis, D.L. & Waterhouse, J.C. (1987) Equilibrium and non-equilibrium concepts in ecological models. *Ecological Monographs*, 57, 1–21.

Deevey, E.S. (1947) Life tables for natural populations of animals. *Quarterly Review of Biology*, 22, 283–314.

DeLucia, E.H., Hamilton, J.G., Naidu, S.L. *et al.* (1999) Net primary production of a forest ecosystem under experimental CO_2 enrichment. *Science*, 284, 1177–1179.

Delworth, T.L. Stouffer, R.J., Dixon, K.W. *et al.* (2002) Review of simulations of climate variability and change with the GFDL R30 coupled climate model. *Climate Dynamics*, 19, 555–574.

Denno, R.F., McClure, M.S. & Ott, J.R. (1995) Interspecific interactions in phytophagous insects: competition reexamined and resurrected. *Annual Review of Entomology*, 40, 297–331.

des Clers, S. (1998) Sustainability of the Falkland Islands *Loligo* squid fishery. In: *Conservation of Biological Resources* (E.J. Milner-Gulland & R. Mace, eds), pp. 225–241. Blackwell Science, Oxford.

Deshmukh, I. (1986) *Ecology and Tropical Biology*. Blackwell Scientific Publications, Oxford.

DeVault, T.L. & Rhodes, O.E. (2002) Identification of vertebrate scavengers of small mammal carcasses in a forested landscape. *Acta Theriologica*, 47, 185–192.

Dezfuli, B.S., Volponi, S., Beltrami, I. & Poulin, R. (2002) Intra- and interspecific density-dependent effects on growth in helminth parasites of the cormorant *Phalacrocorax carbo sinensis*. *Parasitology*, 124, 537–544.

Diamond, J.M. (1972) Biogeographic kinetics: estimation of relaxation times for avifaunas of South-West Pacific islands. *Proceedings of the National Academy of Science of the USA*, 69, 3199–3203.

Diamond, J.M. (1975) Assembly of species communities. In: *Ecology and Evolution of Communities* (M.L. Cody & J.M. Diamond, eds), pp. 342–444. Belknap, Cambridge, MA.

Diamond, J.M. (1983) Taxonomy by nucleotides. *Nature*, 305, 17–18.

Diamond, J.M. (1984) 'Normal' extinctions of isolated populations. In: *Extinctions* (M.H. Nitecki, ed.), pp. 191–245. University of Chicago Press, Chicago.

Diamond, J. & Case, T.J. (eds) (1986) *Community Ecology*. Harper & Row, New York.

Dickie, I.A., Xu, B. & Koide, R.T. (2002) Vertical niche differentiation of ectomycorrhizal hyphae in soil as shown by T-RFLP analysis. *New Phytologist*, 156, 527–535.

Dieckmann, U., Law, R. & Metz, J.A.J. (2000) *The Geometry of Ecological Interactions: simplifying spatial complexity*. Cambridge University Press, Cambridge, UK.

Dixon, A.F.G. (1998) *Aphid Ecology*. Chapman & Hall, London.

Dobson, A.P. & Hudson, P.J. (1992) Regulation and stability of a free-living host–parasite system: *Trichostrongylus tenuis* in red grouse. II. Population models. *Journal of Animal Ecology*, 61, 487–498.

Dobson, A.P., Rodriguez, J.P. & Roberts, W.M. (2001) Synoptic tinkering: integrating strategies for large-scale conservation. *Ecological Applications*, 11, 1019–1026.

Dodson, S.I., Arnott, S.E. & Cottingham, K.L. (2000) The relationship in lake communities between primary productivity and species richness. *Ecology*, 81, 2662–2679.

Doube, B.M. (1987) Spatial and temporal organization in communities associated with dung pads and carcasses. In: *Organization of Communities: past and present* (J.H.R. Gee & P.S. Giller, eds), pp. 255–280. Blackwell Scientific Publications, Oxford.

Doube, B.M., Macqueen, A., Ridsill-Smith, T.J. & Weir, T.A. (1991) Native and introduced dung beetles in Australia. In: *Dung Beetle Ecology* (I. Hanski & Y. Cambefort, eds), pp. 255–278. Princeton University Press, Princeton, NJ.

Douglas, A. & Smith, D.C. (1984) The green hydra symbiosis. VIII. Mechanisms in symbiont regulation. *Proceedings of the Royal Society of London, Series B*, 221, 291–319.

Douglas, A.E. (1992) Microbial brokers of insect–plant interactions. In: *Proceedings of the 8th Symposium on Insect–Plant Relationships* (S.B.J. Menken, J.H. Visser & P. Harrewijn, eds), pp. 329–336. Kluwer Academic Publishing, Dordrecht.

Douglas, E. (1998) Nutritional interactions in insect–microbial symbioses: aphids and their symbiotic bacteria *Buchnera*. *Annual Review of Entomology*, 43, 17–37.

Downing, A.L. & Leibold, M.A. (2002) Ecosystem consequences of species richness and composition in pond food webs. *Nature*, 416, 837–840.

Drès, M. & Mallet, J. (2001) Host races in plant-feeding insects and their importance in sympatric speciation. *Philosophical Transactions of the Royal Society of London, Series B*, 357, 471–492.

Duarte, C.M. (1992) Nutrient concentration of aquatic plants: patterns across species. *Limnology and Oceanography*, 37, 882–889.

Ducrey, M. & Labbé, P. (1985) Étude de la régénération naturelle contrölée en fôret tropicale humide de Guadeloupe. I. Revue bibliographique, milieu naturel et élaboration d'un protocole expérimental. *Annales Scientifiques Forestière*, 42, 297–322.

Ducrey, M. & Labbé, P. (1986) Étude de la régénération naturelle contrölée en fôret tropicale humide de Guadeloupe. II. Installation et croissance des semis après les coupes d'ensemencement. *Annales Scientifiques Forestière*, 43, 299–326.

Duggan, I.C., Green, J.D. & Shiel, R.J. (2002) Distribution of rotifer assemblages in North Island, New Zealand, lakes: relationshipsto environmental and historical factors. *Freshwater Biology*, 47, 195–206.

Dulvy, N.K. & Reynolds, J.D. (2003) Predicting extinction vulnerability in skates. *Conservation Biology*, 16, 440–450.

Dunne, J.A., Williams, R.J. & Martinez, N.J. (2002) Network structure and biodiversity loss in food webs: robustness increases with connectance. *Ecology Letters*, 5, 558–567.

Dwyer, G., Levin, S.A. & Buttel, L. (1990) A simulation model of the population dynamics and evolution of myxomatosis. *Ecological Monographs*, 60, 423–447.

Dytham, C. (1994) Habitat destruction and competitive coexistence: a cellular model. *Journal of Animal Ecology*, 63, 490–491.

Eamus, D. (1999) Ecophysiological traits of deciduous and evergreen woody species in the seasonally dry tropics. *Trends in Ecology and Evolution*, 14, 11–16.

Ebert, D., Zschokke-Rohringer, C.D. & Carius, H.J. (2000) Dose effects and density-dependent regulation in two microparasites of *Daphnia magna*. *Oecologia*, 122, 200–209.

Edmonson, W.T. (1970) Phosphorus, nitrogen and algae in Lake Washington after diversion of sewage. *Science*, 169, 690–691.

Edmonson, W.T. (1991) *The Uses of Ecology: Lake Washington and beyond*. University of Washington Press, Seattle, WA.

Edwards, P.B. & Aschenborn, H.H. (1987) Patterns of nesting and dung burial in *Onitis* dung beetles: implications for pasture productivity and fly control. *Journal of Applied Ecology*, 24, 837–851.

Ehleringer, J.R. & Monson, R.K. (1993) Evolutionary and ecological aspects of photosynthetic pathway variation. *Annual Review of Ecology and Systematics*, 24, 411–439.

Ehleringer, J.R., Sage, R.F., Flanagan, L.B. & Pearcy, R.W. (1991) Climate change and the evolution of C_4 photosynthesis. *Trends in Ecology and Evolution*, 6, 95–99.

Ehrlich, P. & Raven, P.H. (1964) Butterflies and plants: a study in coevolution. *Evolution*, 18, 586–608.

Eis, S., Garman, E.H. & Ebel, L.F. (1965) Relation between cone production and diameter increment of douglas fir (*Pseudotsuga menziesii* (Mirb). Franco), grand fir (*Abies grandis* Dougl.) and western white pine (*Pinus monticola* Dougl.). *Canadian Journal of Botany*, 43, 1553–1559.

Elliott, J.K. & Mariscal, R.N. (2001) Coexistence of nine anenome-fish species: differential host and habitat utilization, size and recruitment. *Marine Biology*, 138, 23–36.

Elliott, J.M. (1984) Numerical changes and population regulation in young migratory trout *Salmo trutta* in a Lake District stream 1966–83. *Journal of Animal Ecology*, 53, 327–350.

Elliott, J.M. (1993) The self-thinning rule applied to juvenile sea-trout, *Salmo trutta*. *Journal of Animal Ecology*, 62, 371–379.

Elliott, J.M. (1994) *Quantitative Ecology and the Brown Trout*. Oxford University Press, Oxford.

Ellison, A.M. & Gotelli, N.J. (2002) Nitrogen availability alters the expression of carnivory in the northern pitcher plant, *Sarracenia purpurea*. *Proceedings of the National Academy of Sciences of the USA*, 99, 4409–4412.

Ellner, S.P., McCauley, E., Kendall, B.E. *et al.* (2001) Habitat structure and population persistence in an experimental community. *Nature*, 412, 538–543.

Elmes, G.W., Akino, T., Thomas, J.A., Clarke, R.T. & Knapp, J.J. (2002) Interspecific differences in cuticular hydrocarbon profiles of *Myrmica* ants are sufficiently consistent to explain host specificity by *Maculinea* (large blue) butterflies. *Oecologia*, 130, 525–535.

Elmhagen, B., Tannerfeldt, M., Verucci, P. & Angerbjorn, A. (2000) The arctic fox (*Alopex lagopus*): an opportunistic specialist. *Journal of Zoology*, 251, 139–149.

Elner, R.W. & Hughes, R.N. (1978) Energy maximisation in the diet of the shore crab *Carcinus maenas* (L.). *Journal of Animal Ecology*, 47, 103–116.

Elser, J.J. & Urabe, J. (1999) The stoichiometry of consumer-driven nutrient recycling: theory, observations, and consequences. *Ecology*, 80, 735–751.

Elton, C.S. (1924) Periodic fluctuations in the number of animals: their causes and effects. *British Journal of Experimental Biology*, 2, 119–163.

Elton, C. (1927) *Animal Ecology*. Sidgwick & Jackson, London.

Elton, C. (1933) *The Ecology of Animals*. Methuen, London.

Elton, C.S. (1958) *The Ecology of Invasions by Animals and Plants*. Methuen, London.

Elwood, J.W., Newbold, J.D., O'Neill, R.V. & van Winkle, W. (1983) Resource spiralling: an operational paradigm for analyzing lotic ecosystems. In: *Dynamics of Lotic Ecosystems* (T.D. Fontaine & S.M. Bartell, eds), pp. 3–28. Ann Arbor Science Publishers, Ann Arbor, MI.

Emiliani, C. (1966) Isotopic palaeotemperatures. *Science*, 154, 851–857.

Engelhardt, K.A.M. & Ritchie, M.E. (2002) The effect of aquatic plant species richness on wetland ecosystem processes. *Ecology*, 83, 2911–2924.

Engen, S., Lande, R. & Saether, B.-E. (1997) Harvesting strategies for fluctuating populations based on uncertain population estimates. *Journal of Theoretical Ecology*, 186, 201–212.

Enquist, B.J., Brown, J.H. & West, G.B. (1998) Allometric scaling of plant energetics and population density. *Nature*, 395, 163–165.

Enriquez, S., Duarte, C.M. & Sand-Jensen, K. (1993) Patterns in decomposition rates among photosynthetic organisms: the importance of detritus C : N : P content. *Oecologia*, 94, 457–471.

Ens, B.J., Kersten, M., Brenninkmeijer, A. & Hulscher, J.B. (1992) Territory quality, parental effort and reproductive success of oystercatchers (*Haematopus ostralegus*). *Journal of Animal Ecology*, 61, 703–715.

Epstein, H.E., Lauenroth, W.K. & Burke, I.C. (1997) Effects of temperature and soil texture on ANPP in US Great Plains. *Ecology*, 78, 2628–2631.

Ergon, T., Lambin, X. & Stenseth, N.C. (2001) Life history traits of voles in a fluctuating population respond to the immediate environment. *Nature*, 411, 1041–1043.

Ericsson, G., Wallin, K., Ball, J.P. & Broberg, M. (2001) Age-related reproductive effort and senescence in free-ranging moose, *Alces alces*. *Ecology*, 82, 1613–1620.

Erwin, T.L. (1982) Tropical forests: their richness in Coleoptera and other arthropod species. *Coleopterists Bulletin*, 36, 74–75.

Ettema, C.H. & Wardle, D.A. (2002) Spatial soil ecology. *Trends in Ecology and Evolution*, 17, 177–183.

Facelli, J.M. & Pickett, S.T.A. (1990) Markovian chains and the role of history in succession. *Trends in Ecology and Evolution*, 5, 27–30.

Fahrig, L. & Merriam, G. (1994) Conservation of fragmented populations. *Conservation Biology*, 8, 50–59.

Fairweather, P.G. (1999) State of environment indicators of 'river health': exploring the metaphor. *Freshwater Biology*, 41, 211–220.

Fajer, E.D. (1989) The effects of enriched CO_2 atmospheres on plant–insect–herbivore interactions: growth responses of larvae of the specialist butterfly, *Junonia coenia* (Lepidoptera: Nymphalidae). *Oecologia*, 81, 514–520.

Falge, E., Baldocchi, D., Tenhunen, J. *et al.* (2002) Seasonality of ecosystem respiration and gross primary production as derived from FLUXNET measurements. *Agricultural and Forest Meteorology*, 113, 53–74.

Fanshawe, S., Vanblaricom, G.R. & Shelly, A.A. (2003) Restored top carnivores as detriments to the performance of marine protected areas intended for fishery sustainability: a case study. *Conservation Biology*, 17, 273–283.

FAO (1995) *World Fishery Production 1950–93*. Food and Agriculture Organization, Rome.

FAO (1999) *The State of World Fisheries and Aquaculture 1998*. Food and Agriculture Organization, Rome.

Fasham, M.J.R., Balino, B.M. & Bowles, M.C. (2001) A new vision of ocean biogeochemistry after a decade of the Joint Global Ocean Flux Study (JGOFS). *Ambio Special Report*, 10, 4–31.

Feeny, P. (1976) Plant apparency and chemical defence. *Recent Advances in Phytochemistry*, 10, 1–40.

Fenchel, T. (1987a) *Ecology – Potentials and Limitations*. Ecology Institute, Federal Republic of Germany.

Fenchel, T. (1987b) Patterns in microbial aquatic communities. In: *Organization of Communities: past and present* (J.H.R. Gee & P.S. Giller, eds), pp. 281–294. Blackwell Scientific Publications, Oxford.

Fenner, F. (1983) Biological control, as exemplified by smallpox eradication and myxomatosis. *Proceedings of the Royal Society of London, Series B*, 218, 259–285.

Fenner, F. & Ratcliffe, R.N. (1965) *Myxomatosis*. Cambridge University Press, London.

Fenton, A., Fairbairn, J.P., Norman, R. & Hudson, P.J. (2002) Parasite transmission: reconciling theory and reality. *Journal of Animal Ecology*, 71, 893–905.

Field, C.B., Behrenfeld, M.J., Randerson, J.T. & Falkowski, P.G. (1998) Primary production of the biosphere: integrating terrestrial and oceanic components. *Science*, 281, 237–240.

Fieldler, P.L. (1987) Life history and population dynamics of rare and common Mariposa lilies (*Calochortus* Pursh: Liliaceae). *Journal of Ecology*, 75, 977–995.

Firbank, L.G. & Watkinson, A.R. (1985) On the analysis of competition within two-species mixtures of plants. *Journal of Applied Ecology*, 22, 503–517.

Firbank, L.G. & Watkinson, A.R. (1990) On the effects of competition: from monocultures to mixtures. In: *Perspectives on Plant Competition* (J.B. Grace & D. Tilman, eds), pp. 165–192. Academic Press, New York.

Fischer, M. & Matthies, D. (1998) Effects of population size on performance in the rare plant *Gentianella germanica*. *Journal of Ecology*, 86, 195–204.

Fisher, R.A. (1930) *The Genetical Theory of Natural Selection*. Clarendon Press, Oxford.

FitzGibbon, C.D. (1990) Anti-predator strategies of immature Thomson's gazelles: hiding and the prone response. *Animal Behaviour*, 40, 846–855.

FitzGibbon, C.D. & Fanshawe, J. (1989) The condition and age of Thomson's gazelles killed by cheetahs and wild dogs. *Journal of Zoology*, 218, 99–107.

Flashpohler, D.J., Bub, B.R. & Kaplin, B.A. (2000) Application of conservation biology research to management. *Conservation Biology*, 14, 1898–1902.

Flecker, A.S. & Townsend, C.R. (1994) Community-wide consequences of trout introduction in New Zealand streams. *Ecological Applications*, 4, 798–807.

Flenley, J. (1993) The origins of diversity in tropical rain forests. *Trends in Ecology and Evolution*, 8, 119–120.

Flessa, K.W. & Jablonski, D. (1995) Biogeography of recent marine bivalve mollusks and its implications of paleobiogeography and the geography of extinction: a progress report. *Historical Biology*, 10, 25–47.

Flint, M.L. & van den Bosch, R. (1981) *Introduction to Integrated Pest Management*. Plenum Press, New York.

Floder, S., Urabe, J. & Kawabata, Z. (2002) The influence of fluctuating light intensities on species composition and diversity of natural phytoplankton communities. *Oecologia*, 133, 395–401.

Flower, R.J., Rippey, B., Rose, N.L., Appleby, P.G. & Battarbee, R.W. (1994) Palaeolimnological evidence for the acidification and contamination of lakes by atmospheric pollution in western Ireland. *Journal of Ecology*, 82, 581–596.

Fonseca, C.R. (1994) Herbivory and the long-lived leaves of an Amazonian ant-tree. *Journal of Ecology*, 82, 833–842.

Fonseca, D.M. & Hart, D.D. (1996) Density-dependent dispersal of black fly neonates is mediated by flow. *Oikos*, 75, 49–58.

Ford, E.B. (1940) Polymorphism and taxonomy. In: *The New Systematics* (J. Huxley, ed.), pp. 493–513. Clarendon Press, Oxford.

Ford, E.B. (1975) *Ecological Genetics*, 4th edn. Chapman & Hall, London.

Forrester, N.W. (1993) Well known and some not so well known insecticides: their biochemical targets and role in IPM and IRM programmes. In: *Pest Control and Sustainable Agriculture* (S. Corey, D. Dall & W. Milne, eds), pp. 28–34. CSIRO, East Melbourne.

Foster, B.L. & Tilman, D. (2003) Seed limitation and the regulation of community structure in oak savanna grassland. *Journal of Ecology*, 91, 999–1007.

Fowler, S.V. & Lawton, J.H. (1985) Rapidly induced defenses and talking trees: the devil's advocate position. *American Naturalist*, 126, 181–195.

Fox, A. & Hudson, P.J. (2001) Parasites reduce territorial behaviour in red grouse (*Lagopus lagopus scoticus*). *Ecology Letters*, 4, 139–143.

Fox, C.J. (2001) Recent trends in stock-recruitment of Blackwater herring (*Clupea harengus* L.) in relation to larval production. *ICES Journal of Marine Science*, 58, 750–762.

Foy, C.L. & Inderjit (2001) Understanding the role of allelopathy in weed interference and declining plant diversity. *Weed Technology*, 15, 873–878.

Franklin, I.A. (1980) Evolutionary change in small populations. In: *Conservation Biology, an Evolutionary–Ecological Perspective* (M.E. Soulé & B.A. Wilcox, eds), pp. 135–149. Sinauer Associates, Sunderland, MA.

Franklin, I.R. & Frankham, R. (1998) How large must populations be to retain evolutionary potential. *Animal Conservation*, 1, 69–73.

Franks, F., Mathias, S.F. & Hatley, R.H.M. (1990) Water, temperature and life. *Philosophical Transactions of the Royal Society of London, Series B*, 326, 517–533; also in *Life at Low Temperatures* (R.M. Laws & F. Franks, eds), pp. 97–117. The Royal Society, London.

Freckleton, R.P. & Watkinson, A.R. (2001) Nonmanipulative determination of plant community dynamics. *Trends in Ecology and Evolution*, 16, 301–307.

Free, C.A., Beddington, J.R. & Lawton, J.H. (1977) On the inadequacy of simple models of mutual interference for parasitism and predation. *Journal of Animal Ecology*, 46, 543–554.

Fretwell, S.D. (1977) The regulation of plant communities by food chains exploiting them. *Perspectives in Biology and Medicine*, 20, 169–185.

Fretwell, S.D. & Lucas, H.L. (1970) On territorial behaviour and other factors influencing habitat distribution in birds. *Acta Biotheoretica*, 19, 16–36.

Fridriksson, S. (1975) *Surtsey: evolution of life on a volcanic island*. Butterworths, London.

Fry, G.L.A. & Cooke, A.S. (1984) Acid deposition and its implications for nature conservation in Britain. *Focus on Nature Conservation*, No. 7. Nature Conservancy Council, Attingham Park, Shrewsbury, UK.

Fryer, G. & Iles, T.D. (1972) *The Cichlid Fishes of the Great Lakes of Africa*. Oliver & Boyd, Edinburgh.

Fukami, T. (2001) Sequence effects of disturbance on community structure. *Oikos*, 92, 215–224.

Fussmann, G.F. & Heber, G. (2002) Food web complexity and chaotic population dynamics. *Ecology Letters*, 5, 394–401.

Futuyma, D.J. (1983) Evolutionary interactions among herbivorous insects and plants. In: *Coevolution* (D.J. Futuyma & M. Slatkin, eds), pp. 207–231. Sinauer Associates, Sunderland, MA.

Futuyma, D.J. & May, R.M. (1992) The coevolution of plant–insect and host–parasite relationships. In: *Genes in Ecology* (R.J. Berry, T.J. Crawford & G.M. Hewitt, eds), pp. 139–166. Blackwell Scientific Publications, Oxford.

Futuyma, D.J. & Slatkin, M. (eds) (1983) *Coevolution*. Sinauer, Sunderland, MA.

Gaillard, J.-M., Festa-Bianchet, M., Yoccoz, N.G., Loison, A. & Toïgo, C. (2000) Temporal variation in fitness components and population dynamics of large herbivores. *Annual Review of Ecology and Systematics*, 31, 367–393.

Gaines, M.S., Vivas, A.M. & Baker, C.L. (1979) An experimental analysis of dispersal in fluctuating vole populations: demographic parameters. *Ecology*, 60, 814–828.

Galloway, J.N., Schlesinger, W.H., Levy, H., Michaels, A. & Schnoor, J.L. (1995) Nitrogen fixation: anthropogenic enhancement–environmental response. *Global Biogeochemical Cycles*, 9, 235–252.

Galloway, L.F. & Fenster, C.B. (2000) Population differentiation in an annual legume: local adaptation. *Evolution*, 54, 1173–1181.

Ganade, G. & Brown, V.K. (2002) Succession in old pastures of central Amazonia: role of soil fertility and plant litter. *Ecology*, 83, 743–754.

Gandon, S. & Michalakis, Y. (2001) Multiple causes of the evolution of dispersal. In: *Dispersal* (J. Clobert, E. Danchin, A.A. Dhondt & J.D. Nichols, eds), pp. 155–167. Oxford University Press, Oxford.

Gange, A.C. & Brown, V.K. (2002) Soil food web components affect plant community structure during early succession. *Ecological Research*, 17, 217–227.

Garcia-Fernandez, C., Casado, M.A. & Perez, M.R. (2003) Benzoin gardens in North Sumatra, Indonesia: effects of management on tree diversity. *Conservation Biology*, 17, 829–836.

García-Fulgueiras, A., Navarro, C., Fenoll, D. *et al.* (2003) Legionnaires' disease outbreak in Murcia, Spain. *Emerging Infectious Diseases*, 9, 915–921.

Garrod, D.J. & Jones, B.W. (1974) Stock and recruitment relationships in the N.E. Atlantic cod stock and the implications for management of the stock. *Journal Conseil International pour l'Exploration de la Mer*, 173, 128–144.

Gaston, K.J. & Blackburn, T.M. (2000) *Pattern and Process in Macroecology*. Blackwell Science, Oxford.

Gathreaux, S.A. (1978) The structure and organization of avian communities in forests. In: *Proceedings of the Workshop on Management of Southern Forests for Nongame Birds* (R.M. DeGraaf, ed.), pp. 17–37. Southern Forest Station, Asheville, NC.

Gause, G.F. (1934) *The Struggle for Existence*. Williams & Wilkins, Baltimore (reprinted 1964 by Hafner, New York).

Gause, G.F. (1935) Experimental demonstration of Volterra's periodic oscillation in the numbers of animals. *Journal of Experimental Biology*, 12, 44–48.

Gavloski, J.E. & Lamb, R.J. (2000a) Specific impacts of herbivores: comparing diverse insect species on young plants. *Environmental Entomology*, 29, 1–7.

Gavloski, J.E. & Lamb, R.J. (2000b) Compensation for herbivory in cruciferous plants: specific responses in three defoliating insects. *Environmental Entomology*, 29, 1258–1267.

Gee, J.H.R. & Giller, P.S. (eds) (1987) *Organization of Communities: past and present*. Blackwell Scientific Publications, Oxford.

Geider, R.J., Delucia, E.H., Falkowski, P.G. *et al.* (2001) Primary productivity of planet earth: biological determinants and physical constraints in terrestrial and aquatic habitats. *Global Change Biology*, 7, 849–882.

Geiger, R. (1955) *The Climate Near the Ground*. Harvard University Press, Cambridge, MA.

Gende, S.M., Quinn, T.P. & Willson, M.F. (2001) Consumption choice by bears feeding on salmon. *Oecologia*, 127, 372–382.

Gessner, M.O. & Chauvet, E. (2002) A case for using litter breakdown to assess functional stream integrity. *Ecological Applications*, 12, 498–510.

Gianoli, E. & Neimeyer, H.M. (1997) Lack of costs of herbivory-induced defenses in a wild wheat: integration of physiological and ecological approaches. *Oikos*, 80, 269–275.

Gibbens, J.C. & Wilesmith, J.W. (2002) Temporal and geographical distribution of cases of foot-and-mouth disease during the early weeks of the 2001 epidemic in Great Britain. *Veterinary Record*, 151, 407–412.

Gibbs, H.L., Sorenson, M.D., Marchetti, K., Brooke, M. de L., Davies, N.B. & Nakamura, H. (2000) Genetic evidence for female host-specific races of the common cuckoo. *Nature*, 407, 183–186.

Gilg, O., Hanski, I. & Sittler, B. (2003) Cyclic dynamics in a simple vertebrate predator–prey community. *Science*, 302, 866–868.

Gillespie, J.H. (1977) Natural selection for variances in offspring numbers: a new evolutionary principle. *American Naturalist*, 111, 1010–1014.

Gilligan, C.A. (1990) Comparison of disease progress curves. *New Phytologist*, 115, 223–242.

Gillooly, J.F., Brown, J.H., West, G.B., Savage, V.M. & Charnov, E.L. (2001) Effects of size and temperature on metabolic rate. *Science*, 293, 2248–2251.

Gillooly, J.F., Charnov, E.L., West, G.B., Savage, V.M. & Brown, J.H. (2002) Effects of size and temperature on developmental time. *Nature*, 417, 70–73.

Gilman, M.P. & Crawley, M.J. (1990) The cost of sexual reproduction in ragwort (*Senecio jacobaea* L.). *Functional Ecology*, 4, 585–589.

Glawe, G.A., Zavala, J.A., Kessler, A., Van Dam, N.M. & Baldwin, I.T. (2003) Ecological costs and benefits correlated with trypsin proteinase inhibitor production in *Nicotinia attenuata*. *Ecology*, 84, 79–90.

Gleason, H.A. (1926) The individualistic concept of the plant association. *Torrey Botanical Club Bulletin*, 53, 7–26.

Godfray, H.C.J. (1987) The evolution of clutch size in invertebrates. *Oxford Surveys in Evolutionary Biology*, 4, 117–154.

Godfray, H.C.J. (1994) *Parasitoids: behavioral and evolutionary ecology*. Princeton University Press, Princeton, NJ.

Godfray, H.C.J. & Crawley, M.J. (1998) Introduction. In: *Conservation Science and Action* (W.J. Sutherland, ed.), pp. 39–65. Blackwell Science, Oxford.

Godfray, H.C.J. & Hassell, M.P. (1989) Discrete and continuous insect populations in tropical environments. *Journal of Animal Ecology*, 58, 153–174.

Godfray, H.C.J. & Pacala, S.W. (1992) Aggregation and the population dynamics of parasitoids and predators. *American Naturalist*, 140, 30–40.

Goldman, J.C., Caron, D.A. & Dennett, M.R. (1987) Regulation of gross growth efficiency and ammonium regeneration in bacteria by substrate C : N ratio. *Limnology and Oceanography*, 32, 1239–1252.

Gomez, J.M. & Gonzalez-Megias, A. (2002) Asymmetrical interactions between ungulates and phytophagous insects: being different matters. *Ecology*, 83, 203–211.

Gorman, M.L. (1979) *Island Ecology*. Chapman & Hall, London.

Gosler, A.G., Barnett, P.R. & Reynolds, S.J. (2000) Inheritance and variation in eggshell patterning in the great tit *Parus major*. *Proceedings of the Royal Society of London, Series B*, 267, 2469–2473.

Goss-Custard, J.D. (1970) Feeding dispersion in some over-wintering wading birds. In: *Social Behaviour in Birds and Mammals* (J.H. Crook, ed.), pp. 3–34. Academic Press, New York.

Gotelli, N.J. (2001) Research frontiers in null model analysis. *Global Ecology and Biogeography*, 10, 337–343.

Gotelli, N.J. & McCabe, D.J. (2002) Species co-occurrence: a meta-analysis of J.M. Diamond's assembly rules model. *Ecology*, 83, 2091–2096.

Gough, L. Shaver, G.R., Carroll, J., Royer, D.L. & Laundre, J.A. (2000) Vascular plant species richness in Alaskan arctic tundra: the importance of soil pH. *Journal of Ecology*, 88, 54–66.

Gould, F. (1991) The evolutionary potential of crop pests. *American Scientist*, 79, 496–507.

Gould, S.J. (1966) Allometry and size in ontogeny and phylogeny. *Biological Reviews*, 41, 587–640.

Gould, W.A. & Walker, M.D. (1997) Landscape-scale patterns in plant species richness along an arctic river. *Canadian Journal of Botany*, 75, 1748–1765.

Gower, S.T., McMurtrie, R.E. & Murty, D. (1996) Aboveground net primary production declines with stand age: potential causes. *Trends in Ecology and Evolution*, 11, 378–382.

Grace, J.B. & Wetzel, R.G. (1998) Long-term dynamics of *Typha* populations. *Aquatic Botany*, 61, 137–146.

Graham, I. & Lambin, X. (2002) The impact of weasel predation on cyclic field-vole survival: the specialist predator hypothesis contradicted. *Journal of Animal Ecology*, 71, 946–956.

Grant, P.R., Grant, R., Keller, L.F. & Petren, K. (2000) Effects of El Nino events on Darwin's finch productivity. *Ecology*, 81, 2442–2457.

Gray, S.M. & Robinson, B.W. (2001) Experimental evidence that competition between stickleback species favours adaptive character divergence. *Ecology Letters*, 5, 264–272.

Greene, D.F. & Calogeropoulos, C. (2001) Measuring and modelling seed dispersal of terrestrial plants. In: *Dispersal Ecology* (J.M. Bullock, R.E. Kenward & R.S. Hails, eds), pp. 3–23. Blackwell Science, Oxford.

Greenwood, P.J. (1980) Mating systems, philopatry and dispersal in birds and mammals. *Animal Behaviour*, 28, 1140–1162.

Greenwood, P.J., Harvey, P.H. & Perrins, C.M. (1978) Inbreeding and dispersal in the great tit. *Nature*, 271, 52–54.

Grenfell, B. & Harwood, J. (1997) (Meta)population dynamics of infectious diseases. *Trends in Ecology and Evolution*, 12, 395–399.

Grier, C.C. (1975) Wildfire effects on nutrient distribution and leaching in a coniferous forest ecosystem. *Canadian Journal of Forest Research*, 5, 599–607.

Griffith, D.M. & Poulson, T.M. (1993) Mechanisms and consequences of intraspecific competition in a carabid cave beetle. *Ecology*, 74, 1373–1383.

Griffiths, R.A., Denton, J. & Wong, A.L.-C. (1993) The effect of food level on competition in tadpoles: interference mediated by protothecan algae? *Journal of Animal Ecology*, 62, 274–279.

Grime, J.P., Hodgson, J.G. & Hunt, R. (1988) *Comparative Plant Ecology: a functional approach to common British species*. Unwin-Hyman, London.

Grubb, P. (1977) The maintenance of species richness in plant communities: the importance of the regeneration niche. *Biological Reviews*, 52, 107–145.

Grubb, P.J. (1986) The ecology of establishment. In: *Ecology and Design in Landscape* (A.D. Bradshaw, D.A. Goode & E. Thorpe, eds), pp. 83–97. Symposia of the British Ecological Society, No. 24. Blackwell Scientific Publications, Oxford.

Grutter, A.S. (1999) Cleaner fish really do clean. *Nature*, 398, 672–673.

Grytnes, J.A. (2003) Species-richness patterns of vascular plants along seven altitudinal transects in Norway. *Ecography*, 26, 291–300.

Grytnes, J.A. & Vetaas, O.R. (2002) Species richness and altitude: a comparison between null models and interpolated plant species richness along the Himalayan altitudinal gradient, Nepal. *American Naturalist*, 159, 294–304.

Guiñez, R. & Castilla, J.C. (2001) An allometric tridimensional model of self-thinning for a gregarious tunicate. *Ecology*, 82, 2331–2341.

Gunn, A. (1998) Caribou and muskox harvesting in the Northwest Territories. In: *Conservation of Biological Resources* (E.J. Milner-Gulland & R. Mace, eds), pp. 314–330. Blackwell Science, Oxford.

Haefner, P.A. (1970) The effect of low dissolved oxygen concentrations on temperature–salinity tolerance of the sand shrimp, *Crangon septemspinosa*. *Physiological Zoology*, 43, 30–37.

Haines, E. (1979) Interaction between Georgia salt marshes and coastal waters: a changing paradigm. In: *Ecological Processes in Coastal and Marine Systems* (R.J. Livingston, ed.). Plenum Press, New York.

Hainsworth, F.R. (1981) *Animal Physiology*. Addison-Wesley, Reading, MA.

Hairston, N.G., Smith, F.E. & Slobodkin, L.B. (1960) Community structure, population control, and competition. *American Naturalist*, 44, 421–425.

Halaj, J., Ross, D.W. & Moldenke, A.R. (2000) Importance of habitat structure to the arthropod food-web in Douglas-fir canopies. *Oikos*, 90, 139–152.

Haldane, J.B.S. (1949) Disease and evolution. *La Ricerca Scienza*, 19 (Suppl.), 3–11.

Hall, S.J. (1998) Closed areas for fisheries management – the case consolidates. *Trends in Ecology and Evolution*, 13, 297–298.

Hall, S.J. & Raffaelli, D.G. (1993) Food webs: theory and reality. *Advances in Ecological Research*, 24, 187–239.

Hamilton, W.D. (1971) Geometry for the selfish herd. *Journal of Theoretical Biology*, 31, 295–311.

Hamilton, W.D. & May, R.M. (1977) Dispersal in stable habitats. *Nature*, 269, 578–581.

Hansen, J., Ruedy, R., Glasgoe, J. & Sato, M. (1999) GISS analysis of surface temperature change. *Journal of Geophysical Research*, 104, 30997–31022.

Hanski, I. (1991) Single-species metapopulation dynamics: concepts, models and observations. In: *Metapopulation Dynamics* (M.E. Gilpin & I. Hanski, eds), pp. 17–38. Academic Press, London.

Hanski, I. (1994) A practical model of metapopulation dynamics. *Journal of Animal Ecology*, 63, 151–162.

Hanski, I. (1996) Metapopulation ecology. In: *Population dynamics in ecological space and time* (O.E. Rhodes Jr., R.K. Chesser & M.H. Smith, eds), pp. 13–43. University of Chicago Press, Chicago.

Hanski, I. (1999) *Metapopulation Ecology*. Oxford University Press, Oxford.

Hanski, I. & Gyllenberg, M. (1993) Two general metapopulation models and the core-satellite hypothesis. *American Naturalist*, 142, 17–41.

Hanski, I. & Simberloff, D. (1997) The metapopulation approach, its history, conceptual domain, and application to conservation. In: *Metapopulation Biology* (I.A. Hanski & M.E. Gilpin, eds), pp. 5–26. Academic Press, San Diego, CA.

Hanski, I., Hansson, L. & Henttonen, H. (1991) Specialist predators, generalist predators, and the microtine rodent cycle. *Journal of Animal Ecology*, 60, 353–367.

Hanski, I., Henttonen, H., Korpimaki, E., Oksanen, L. & Turchin, P. (2001) Small-rodent dynamics and predation. *Ecology*, 82, 1505–1520.

Hanski, I., Pakkala, T., Kuussaari, M. & Lei, G. (1995) Metapopulation persistence of an endangered butterfly in a fragmented landscape. *Oikos*, 72, 21–28.

Harcourt, D.G. (1971) Population dynamics of *Leptinotarsa decemlineata* (Say) in eastern Ontario. III. Major population processes. *Canadian Entomologist*, 103, 1049–1061.

Harper, D.G.C. (1982) Competitive foraging in mallards: 'ideal free ducks'. *Animal Behaviour*, 30, 575–584.

Harper, J.L. (1955) The influence of the environment on seed and seedling mortality. VI. The effects of the interaction of soil moisture content and temperature on the mortality of maize grains. *Annals of Applied Biology*, 43, 696–708.

Harper, J.L. (1961) Approaches to the study of plant competition. In: *Mechanisms in Biological Competition* (F.L. Milthorpe, ed.), pp. 1–39. Symposium No. 15, Society for Experimental Biology. Cambridge University Press, Cambridge, UK.

Harper, J.L. (1977) *The Population Biology of Plants*. Academic Press, London.

Harper, J.L. & White, J. (1974) The demography of plants. *Annual Review of Ecology and Systematics*, 5, 419–463.

Harper, J.L., Jones, M. & Sackville Hamilton, N.R. (1991) The evolution of roots and the problems of analysing their behaviour. In: *Plant Root Growth: an ecological perspective* (D. Atkinson, ed.), pp. 3–22. Special Publication of the British Ecological Society, No. 10. Blackwell Scientific Publications, Oxford.

Harper, J.L., Rosen, R.B. & White, J. (eds) (1986) The growth and form of modular organisms. *Philosophical Transactions of the Royal Society of London, Series B*, 313, 1–250.

Harris, G.P. (2001) Biogeochemistry of nitrogen and phosphorus in Australian catchments, rivers and estuaries: effects of land use and flow regulation and comparisons with global patterns. *Marine and Freshwater Research*, 52, 139–149.

Harrison, S. & Taylor, A.D. (1997) Empirical evidence for metapopulations. In: *Metapopulation Biology* (I.A. Hanski & M.E. Gilpin, eds), pp. 27–42. Academic Press, San Diego, CA.

Hart, A.J., Bale, J.S., Tullett, A.G., Worland, M.R. & Walters, K.F.A. (2002) Effects of temperature on the establishment potential of the predatory mite *Amblyseius californicus* McGregor (Acari: Phytoseiidae) in the UK. *Journal of Insect Physiology*, 48, 593–599.

Harvey, P.H. (1996) Phylogenies for ecologists. *Journal of Animal Ecology*, 65, 255–263.

Harvey, P.H. & Pagel, M.D. (1991) *The Comparative Method in Evolutionary Biology*. Oxford University Press, Oxford.

Harvey, P.H. & Zammuto, R.M. (1985) Patterns of mortality and age at first reproduction in natural populations of mammals. *Nature*, 315, 319–320.

Hassell, M.P. (1978) *The Dynamics of Arthropod Predator–Prey Systems*. Princeton University Press, Princeton, NJ.

Hassell, M.P. (1982) Patterns of parasitism by insect parasitoids in patchy environments. *Ecological Entomology*, 7, 365–377.

Hassell, M.P. (1985) Insect natural enemies as regulating factors. *Journal of Animal Ecology*, 54, 323–334.

Hassell, M.P. & May, R.M. (1973) Stability in insect host–parasite models. *Journal of Animal Ecology*, 43, 567–594.

Hassell, M.P. & May, R.M. (1974) Aggregation of predators and insect parasites and its effect on stability. *Journal of Animal Ecology*, 43, 567–594.

Hassell, M.P., Latto, J. & May, R.M. (1989) Seeing the wood for the trees: detecting density dependence from existing life-table studies. *Journal of Animal Ecology*, 58, 883–892.

Hastings, A. & Botsford, L.W. (2003) Comparing designs of marine reserves for fisheries and for biodiversity. *Ecological Applications*, 13 (Suppl.), S65–S70.

Hastings, A., Hom, C.L., Ellner, S., Turchin, P. & Godfray, H.C.J. (1993) Chaos in ecology: is mother nature a strange attractor? *Annual Review of Ecology and Systematics*, 24, 1–33.

Hastings, H.M. & Conrad, M. (1979) Length and evolutionary stability of food chains. *Nature*, 282, 838–839.

Hattenschwiler, S. & Vitousek, P.M. (2000) The role of polyphenols in terrestrial ecosystem nutrient cycling. *Trends in Ecology and Evolution*, 15, 238–243.

Hawkins, B.A. & Cornell, H.V. (eds) (1999) *Theoretical Approaches to Biological Control*. Cambridge University Press, Cambridge, UK.

Heal, O.W. & MacLean, S.F. (1975) Comparative productivity in ecosystems – secondary productivity. In: *Unifying Concepts in Ecology* (W.H. van Dobben & R.H. Lowe–McConnell, eds), pp. 89–108. Junk, The Hague.

Heal, O.W., Menault, J.C. & Steffen, W.L. (1993) *Towards a Global Terrestrial Observing System (GTOS): detecting and monitoring change in terrestrial ecosystems*. MAB Digest 14 and IGBP Global Change Report 26, UNESCO, Paris and IGBP, Stockholm.

Hearnden, M., Skelly, C. & Weinstein, P. (1999) Improving the surveillance of mosquitoes with disease-vector potential in New Zealand. *New Zealand Public Health Report*, 6, 25–28.

Hector, A., Shmid, B., Beierkuhnlein, C. *et al.* (1999) Plant diversity and productivity experiments in European grasslands. *Science*, 286, 1123–1127.

Heed, W.B. (1968) Ecology of Hawaiian Drosophiladae. *University of Texas Publications*, 6861, 387–419.

Heie, O.E. (1987) Palaeontology and phylogeny. In: *Aphids: their biology, natural enemies and control. World Crop Pests*, Vol. 2A (A.K. Minks & P. Harrewijn, eds), pp. 367–391. Elsevier, Amsterdam.

Heil, M. & McKey, D. (2003) Protective ant–plant interactions as model systems in ecological and evolutionary research. *Annual Review of Ecology, Evolution and Systematics*, 34, 425–453.

Hendon, B.C. & Briske, D.D. (2002) Relative herbivory tolerance and competitive ability in two dominant : subordinate pairs of perennial grasses in a native grassland. *Plant Ecology*, 160, 43–51.

Hengeveld, R. (1990) *Dynamic Biogeography*. Cambridge University Press, Cambridge, UK.

Henttonen, H. & Kaikusalo, A. (1993) Lemming movements. In: *The Biology of Lemmings* (N.C. Stenseth & R.A. Ims, eds), pp. 157–186. Academic Press, London.

Heppell, S.S., Crowder, L.B. & Crouse, D.T. (1996) Models to evaluate headstarting as a management tool for long-lived turtles. *Ecological Applications*, 6, 556–565.

Herman, T.J.B. (2000) Developing IPM for potato tuber moth. *Commercial Grower*, 55, 26–28.

Hermoyian, C.S., Leighton, L.R. & Kaplan, P. (2002) Testing the role of competition in fossil communities using limiting similarity. *Geology*, 30, 15–18.

Herre, E.A. & West, S.A. (1997) Conflict of interest in a mutualism: documenting the elusive fig wasp–seed trade-off. *Proceedings of the Royal Society of London, Series B*, 264, 1501–1507.

Herre, E.A., Knowlton, N., Mueller, U.G. & Rehner, S.A. (1999) The evolution of mutualisms: exploring the paths between conflict and cooperation. *Trends in Ecology and Evolution*, 14, 49–53.

Herrera, C.M., Jordano, P., Guitian, J. & Traveset, A. (1998) Annual variability in seed production by woody plants and the masting concept: reassessment of principles and relationship to pollination and seed dispersal. *American Naturalist*, 152, 576–594.

Hessen, D.O. (1997) Stoichiometry in food webs. Lotka revisited. *Oikos*, 79, 195–200.

Hestbeck, J.B. (1982) Population regulation of cyclic mammals: the social fence hypothesis. *Oikos*, 39, 157–163.

Hilborn, R. & Walters, C.J. (1992) *Quantitative Fisheries Stock Assessment*. Chapman & Hall, New York.

Hildrew, A.G. & Townsend, C.R. (1980) Aggregation, interference and the foraging by larvae of *Plectrocnemia conspersa* (Trichoptera: Polycentropodidae). *Animal Behaviour*, 28, 553–560.

Hildrew, A.G., Townsend, C.R., Francis, J. & Finch, K. (1984) Cellulolytic decomposition in streams of contrasting pH and its relationship with invertebrate community structure. *Freshwater Biology*, 14, 323–328.

Hill, M.F., Witman, J.D. & Caswell, H. (2002) Spatio-temporal variation in Markov chain models of subtidal community succession. *Ecology Letters*, 5, 665–675.

Hill, W.R., Mulholland, P.J. & Marzolf, E.R. (2001) Stream ecosystem responses to forest leaf emergence in spring. *Ecology*, 82, 2306–2319.

Hirakawa, H. (2001) Coprophagy in leporids and other mammalian herbivores. *Mammal Review*, 31, 61–80.

Hockland, S.H., Dawson, G.W., Griffiths, D.C., Maples, B., Pickett, J.A. & Woodcock, C.M. (1986) The use of aphid alarm pheremone (*E*-β-farnesene) to increase effectiveness of the entomophilic fungus *Verticillium lecanii* in controlling aphids on chrysanthemums under glass. In: *Fundamental and Applied Aspects of Invertebrate Pathology* (R.A. Sampson, J.M. Vlak & D. Peters, eds), p. 252. Foundation of the Fourth International Colloquium of Invertebrate Pathology, Wageningen.

Hodgkinson, K.C. (1992) Water relations and growth of shrubs before and after fire in a semi-arid woodland. *Oecologia*, 90, 467–473.

Hogan, M.E., Veivers, P.C., Slaytor, M. & Czolij, R.T. (1988) The site of cellulose breakdown in higher termites (*Nasutitermes walkeri* and *Nasutitermes exitosus*). *Journal of Insect Physiology*, 34, 891–899.

Holland, D.S. & Brazee, R.J. (1996) Marine reserves for fishery management. *Marine Resource Economics*, 11, 157–171.

Holling, C.S. (1959) Some characteristics of simple types of predation and parasitism. *Canadian Entomologist*, 91, 385–398.

Holmer, M. & Storkholm, P. (2001) Sulphate reduction and sulphur cycling in lake sediments: a review. *Freshwater Biology*, 46, 431–451.

Holmgren, M., Scheffer, M., Ezcurra, E., Gutierrez, J.R. & Mohren, M.J. (2001) El Nino effects on the dynamics of terrestrial ecosystems. *Trends in Ecology and Evolution*, 16, 89–94.

Holloway, J.D. (1977) *The Lepidoptera of Norfolk Island, their Biogeography and Ecology*. Junk, The Hague.

Holloway, J.M., Dahlgren, R.A., Hansen, B. & Casey, W.H. (1998) Contribution of bedrock nitrogen to high nitrate concentrations in stream water. *Nature*, 395, 785–788.

Holt, R.D. (1977) Predation, apparent competition and the structure of prey communities. *Theoretical Population Biology*, 12, 197–229.

Holt, R.D. (1984) Spatial heterogeneity, indirect interactions, and the coexistence of prey species. *American Naturalist*, 124, 377–406.

Holt, R.D. (1997) Community modules. In: *Multitrophic Interactions in Terrestrial Ecosystems* (A.C. Gange & V.K. Brown, eds), pp. 333–349. Blackwell Science, Oxford.

Holt, R.D. & Hassell, M.P. (1993) Environmental heterogeneity and the stability of host–parasitoid interactions. *Journal of Animal Ecology*, 62, 89–100.

Holway, D.A. & Suarez, A.V. (1999) Animal behaviour: an essential component of invasion biology. *Trends in Ecology and Evolution*, 14, 328–330.

Holyoak, M. & Lawler, S.P. (1996) Persistence of an extinction-prone predator–prey interaction through metapopulation dynamics. *Ecology*, 77, 1867–1879.

Hook, P.B. & Burke, I.C. (2000) Biogeochemistry in a shortgrass landscape: control by topography, soil texture, and microclimate. *Ecology*, 81, 2686–2703.

Hopf, F.A., Valone, T.J. & Brown, J.H. (1993) Competition theory and the structure of ecological communities. *Evolutionary Ecology*, 7, 142–154.

Horn, D.S. (1988) *Ecological Approach to Pest Management*. Elsevier, London.

Horn, H.S. (1975) Markovian processes of forest succession. In: *Ecology and Evolution of Communities* (M.L. Cody & J.M. Diamond, eds), pp. 196–213. Belknap, Cambridge, MA.

Horn, H.S. (1981) Succession. In: *Theoretical Ecology: principles and applications* (R.M. May, ed.), pp. 253–271. Blackwell Scientific Publications, Oxford.

Houghton, R.A. (2000) Interannual variability in the global carbon cycle. *Journal of Geophysical Research*, 105, 20121–20130.

Hoyer, M.V. & Canfield, D.E. (1994) Bird abundance and species richness on Florida lakes: influence of trophic status, lake morphology and aquatic macrophytes. *Hydrobiologia*, 297, 107–119.

Hu, S., Firestone, M.K. & Chapin, F.S. III (1999) Soil microbial feedbacks to atmospheric CO_2 enrichment. *Trends in Ecology and Evolution*, 14, 433–437.

Hudson, P. & Greenman, J. (1998) Competition mediated by parasites: biological and theoretical progress. *Trends in Ecology and Evolution*, 13, 387–390.

Hudson, P.J., Dobson, A.P. & Newborn, D. (1992a) Do parasites make prey vulnerable to predation? Red grouse and parasites. *Journal of Animal Ecology*, 61, 681–692.

Hudson, P.J., Newborn, D. & Dobson, A.P. (1992b) Regulation and stability of a free-living host–parasite system: *Trichostrongylus tenuis* in red grouse. I. Monitoring and parasite reduction experiments. *Journal of Animal Ecology*, 61, 477–486.

Hudson, P.J., Dobson, A.P. & Newborn, D. (1998) Prevention of population cycles by parasite removal. *Science*, 282, 2256–2258.

Huffaker, C.B. (1958) Experimental studies on predation: dispersion factors and predator–prey oscillations. *Hilgardia*, 27, 343–383.

Huffaker, C.B., Shea, K.P. & Herman, S.G. (1963) Experimental studies on predation. *Hilgardia*, 34, 305–330.

Hughes, J.E., Deegan, L.A., Peterson, B.J., Holmes, R.M. & Fry, B. (2000) Nitrogen flow through the food web in the oligohaline zone of a New England estuary. *Ecology*, 81, 433–452.

Hughes, L. (2000) Biological consequences of global warming: is the signal already apparent. *Trends in Ecology and Evolution*, 15, 56–61.

Hughes, R.N. (1989) *A Functional Biology of Clonal Animals*. Chapman & Hall, London.

Hughes, R.N. & Croy, M.I. (1993) An experimental analysis of frequency-dependent predation (switching) in the 15-spined stickleback, *Spinachia spinachia*. *Journal of Animal Ecology*, 62, 341–352.

Hughes, R.N. & Griffiths, C.L. (1988) Self-thinning in barnacles and mussels: the geometry of packing. *American Naturalist*, 132, 484–491.

Hughes, T.P. & Connell, J.H. (1987) Population dynamics based on size or age? A reef-coral analysis. *American Naturalist*, 129, 818–829.

Hughes, T.P., Ayre, D. & Connell, J.H. (1992) The evolutionary ecology of corals. *Trends in Ecology and Evolution*, 7, 292–295.

Hughey, K.F.D., Cullen, R. & Moran, E. (2002) Integrating economics into priority setting and evaluation in conservation management. *Conservation Biology*, 17, 93–103.

Huisman, J. (1999) Population dynamics of light-limited phytoplankton: microcosm experiments. *Ecology*, 80, 202–210.

Humborg, C., Blomqvist, S., Avsan, E., Bergensund, Y. & Smedberg, E. (2002) Hydrological alterations with river damming in northern Sweden: implications for weathering and river biogeochemistry. *Global Biogeochemical Cycles*, 16, 1–13.

Hunter, J.R., Argue, A.W., Bayliff, W.H. *et al.* (1986) *The Dynamics of Tuna Movement: an evaluation of past and future research*. FAO Fisheries Technical Paper No. 277. Food and Agriculture Organization of the United Nations, Rome.

Hunter, M.D. & Price, P.W. (1992) Playing chutes and ladders: heterogeneity and the relative roles of bottom-up and top-down forces in natural communities. *Ecology*, 73, 724–732.

Hunter, M.D. & Price, P.W. (1998) Cycles in insect populations: delayed density dependence or exogenous driving variables? *Ecological Entomology*, 23, 216–222.

Hunter, M.L. & Yonzon, P. (1992) Altitudinal distributions of birds, mammals, people, forests, and parks in Nepal. *Conservation Biology*, 7, 420–423.

Hurd, L.E. & Eisenberg, R.M. (1990) Experimentally synchronized phenology and interspecific competition in mantids. *American Midland Naturalist*, 124, 390–394.

Huryn, A.D. (1998) Ecosystem level evidence for top-down and bottom-up control of production in a grassland stream system. *Oecologia*, 115, 173–183.

Husband, B.C. & Barrett, S.C.H. (1996) A metapopulation perspective in plant population biology. *Journal of Ecology*, 84, 461–469.

Huston, M. (1979) A general hypothesis of species diversity. *American Naturalist*, 113, 81–102.

Hutchings, M.J. (1983) Ecology's law in search of a theory. *New Scientist*, 98, 765–767.

Hutchings, M.J. & de Kroon, H. (1994) Foraging in plants: the role of morphological plasticity in resource acquisition. *Advances in Ecological Research*, 25, 159–238.

Hutchinson, G.E. (1957) Concluding remarks. *Cold Spring Harbour Symposium on Quantitative Biology*, 22, 415–427.

Hutchinson, G.E. (1959) Homage to Santa Rosalia, or why are there so many kinds of animals? *American Naturalist*, 93, 145–159.

Hutchinson, G.E. (1961) The paradox of the plankton. *American Naturalist*, 95, 137–145.

IGBP (International Geosphere–Biosphere Programme) (1990) *Global Change. Report No. 12. The Initial Core Projects. The International Geosphere–Biosphere Programme: a study of global change of the International Council of Scientific Unions (ICSU)*. IGBP, Stockholm, Sweden.

Ims, R. & Yoccoz, N. (1997) Studying transfer processes in metapopulations: emigration, migration and colonization. In: *Metapopulation Biology* (I. Hanski & M. Gilpin, eds), pp. 247–265. Academic Press, San Diego, CA.

Inchausti, P. & Ginzburg, L.R. (1998) Small mammal cycles in northern Europe: patterns and evidence for a maternal effect hypothesis. *Journal of Animal Ecology*, 67, 180–194.

Inghe, O. (1989) Genet and ramet survivorship under different mortality regimes: a cellular automata model. *Journal of Theoretical Biology*, 138, 257–270.

Inglesfield, C. & Begon, M. (1983) The ontogeny and cost of migration of *Drosophila subobscura* Collin. *Biological Journal of the Linnean Society*, 19, 9–15.

Interlandi, S.J. & Kilham, S.S. (2001) Limiting resources and the regulation of diversity in phytoplankton communities. *Ecology*, 82, 1270–1282.

International Organisation for Biological Control (1989) *Current Status of Integrated Farming Systems Research in Western Europe* (P. Vereijken & D.J. Royle, eds). IOBC WPRS Bulletin 12(5).

IPCC (2001) *Third Assessment Report of the Intergovernmental Panel on Climate Change*. Working Group 1, Intergovernmental Panel on Climate Change, Geneva.

IUCN/UNEP/WWF (1991) *Caring for the Earth. A Strategy for Sustainable Living*. Gland, Switzerland.

Ives, A.R. (1992a) Density-dependent and density-independent parasitoid aggregation in model host–parasitoid systems. *American Naturalist*, 140, 912–937.

Ives, A.R. (1992b) Continuous-time models of host–parasitoid interactions. *American Naturalist*, 140, 1–29.

Jackson, C.R., Churchill, P.F. & Roden, E.E. (2001) Successional changes in bacterial assemblage structure during epilithic biofilm development. *Ecology*, 82, 555–566.

Jackson, S.T. & Weng, C. (1999) Late Quaternary extinction of a tree species in eastern North America. *Proceedings of the National Academy of Sciences of the USA*, 96, 13847–13852.

Jamieson, I.G. & Ryan, C.J. (2001) Island takahe: closure of the debate over the merits of introducing Fiordland takahe to predator-free islands. In: *The Takahe: fifty years of conservation management and research* (W.G. Lee & I.G. Jamieson, eds), pp. 96–113. University of Otago Press, Dunedin, New Zealand.

Janis, C.M. (1993) Tertiary mammal evolution in the context of changing climates, vegetation and tectonic events. *Annual Review of Ecology and Systematics*, 24, 467–500.

Jannasch, H.W. & Mottl, M.J. (1985) Geomicrobiology of deep-sea hydrothermal vents. *Science*, 229, 717–725.

Janzen, D.H. (1967) Interaction of the bull's-horn acacia (*Acacia cornigera* L.) with an ant inhabitant (*Pseudomyrmex ferruginea* F. Smith) in eastern Mexico. *University of Kansas Science Bulletin*, 47, 315–558.

Janzen, D.H. (1968) Host plants in evolutionary and contemporary time. *American Naturalist*, 102, 592–595.

Janzen, D.H. (1970) Herbivores and the number of tree species in tropical forests. *American Naturalist*, 104, 501–528.

Janzen, D.H. (1980) Specificity of seed-eating beetles in a Costa Rican deciduous forest. *Journal of Ecology*, 68, 929–952.

Janzen, D.H. (1981) Evolutionary physiology of personal defence. In: *Physiological Ecology: an evolutionary approach to resource use* (C.R. Townsend & P. Calow, eds), pp. 145–164. Blackwell Scientific Publications, Oxford.

Janzen, D.H., Juster, H.B. & Bell, E.A. (1977) Toxicity of secondary compounds to the seed-eating larvae of the bruchid beetle *Callosobruchus maculatus*. *Phytochemistry*, 16, 223–227.

Jeffries, M.J. & Lawton, J.H. (1984) Enemy-free space and the structure of ecological communities. *Biological Journal of the Linnean Society*, 23, 269–286.

Jeffries, M.J. & Lawton, J.H. (1985) Predator–prey ratios in communities of freshwater invertebrates: the role of enemy free space. *Freshwater Biology*, 15, 105–112.

Jenkins, B., Kitching, R.L. & Pimm, S.L. (1992) Productivity, disturbance and food web structure at a local spatial scale in experimental container habitats. *Oikos*, 65, 249–255.

Jennings, S., Kaiser, M.J. & Reynolds, J.D. (2001) *Marine Fisheries Ecology*. Blackwell Science, Oxford.

Jensen, K.T. & Mouritsen, K.M. (1992) Mass mortality in two common soft-bottom invertebrates, *Hydrobia ulvae* and *Corophium volutator* – the possible role of trematodes. *Helgoländer Meeresuntersuchungen*, 46, 329–339.

Jobbagy, E.G., Sala, O.E. & Paruelo, J.M. (2002) Patterns and controls of primary production in the Patagonian steppe: a remote sensing approach. *Ecology*, 83, 307–319.

Johannes, R.E. (1998) Government-supported village-based management of marine resources in Vanuatu. *Ocean Coastal Management*, 40, 165–186.

Johnson, C.G. (1967) International dispersal of insects and insect-borne viruses. *Netherlands Journal of Plant Pathology*, 73 (Suppl. 1), 21–43.

Johnson, C.M., Zarin, D.J. & Johnson, A.H. (2000) Post-disturbance aboveground biomass accumulation in global secondary forests. *Ecology*, 81, 1395–1401.

Johnson, N.C., Zak, D.R., Tilman, D. & Pfleger, F.L. (1991) Dynamics of vesicular–arbuscular mycorrhizae during old field succession. *Oecologia*, 86, 349–358.

Johnston, D.W. & Odum, E.P. (1956) Breeding bird populations in relation to plant succession on the piedmont of Georgia. *Ecology*, 37, 50–62.

Jones, C.G., Lawton, J.H. & Shachak, M. (1994) Organisms as ecosystem engineers. *Oikos*, 69, 373–386.

Jones, C.G., Lawton, J.H. & Schachak, M. (1997) Positive and negative effects of organisms as physical ecosystem engineers. *Ecology*, 78, 1946–1957.

Jones, C.G., Ostfeld, R.S., Richard, M.P., Schauber, E.M. & Wolff, J.O. (1998) Chain reactions linking acorns to gypsy moth outbreaks and Lyme disease risk. *Science*, 279, 1023–1026.

Jones, M. & Harper, J.L. (1987) The influence of neighbours on the growth of trees. I. The demography of buds in *Betula pendula*. *Proceedings of the Royal Society of London, Series B*, 232, 1–18.

Jones, M., Mandelik, Y. & Dayan, T. (2001) Coexistence of temporally partitioned spiny mice: roles of habitat structure and foraging behaviour. *Ecology*, 82, 2164–2176.

Jones, W.T. (1988) Density-related changes in survival of philopatric and dispersing kangaroo rats. *Ecology*, 69, 1474–1478.

Jones, W.T., Waser, P.M., Elliott, L.F., Link, N.E. & Bush, B.B. (1988) Philopatry, dispersal, and habitat saturation in the banner-tailed kangaroo rat, *Dipodomys spectabilis*. *Ecology*, 69, 1466–1473.

Jonsson, M. & Malmqvist, B. (2000) Ecosystem process rate increases with animal species richness: evidence from leaf-eating, aquatic insects. *Oikos*, 89, 519–523.

Jordan, R.S. (1991) Impact of ENSO events on the southeastern Pacific region with special reference to the interaction of fishing and climatic variability. In: *ENSO Teleconnections Linking Worldwide Climate Anomalies: scientific basis and societal impacts* (M. Glantz, ed.), pp. 401–430, Cambridge University Press, Cambridge, UK.

Jowett, I.G. (1997) Instream flow methods: a comparison of approaches. *Regulated Rivers: Research and Management*, 13, 115–127.

Juniper, S.K., Tunnicliffe, V. & Southward, E.C. (1992) Hydrothermal vents in turbidite sediments on a Northeast Pacific spreading centre: organisms and substratum at an ocean drilling site. *Canadian Journal of Zoology*, 70, 1792–1809.

Junk, W.J., Bayley, P.B. & Sparks, R.E. (1989) The flood–pulse concept in river-floodplain systems. *Canadian Special Publications in Fisheries and Aquatic Sciences*, 106, 110–127.

Jurgens, K. & Sala, M.M. (2000) Predation-mediated shifts in size distribution of microbial biomass and activity during detritus decomposition. *Oikos*, 91, 29–40.

Jutila, H.M. (2003) Germination in Baltic coastal wetland meadows: similarities and differences between vegetation and seed bank. *Plant Ecology*, 166, 275–293.

Kaelke, C.M., Kruger, E.L. & Reich, P.B. (2001) Trade-offs in seedling survival, growth, and physiology among hardwood species of contrasting successional status along a light availability gradient. *Canadian Journal of Forestry Research*, 31, 1602–1616.

Kaiser, J. (2000) Rift over biodiversity divides ecologists. *Science*, 289, 1282–1283.

Kamijo, T., Kitayama, K., Sugawara, A., Urushimichi, S. & Sasai, K. (2002) Primary succession of the warm-temperate broad-leaved forest on a volcanic island, Miyake-jima, Japan. *Folia Geobotanica*, 37, 71–91.

Kaplan, R.H. & Salthe, S.N. (1979) The allometry of reproduction: an empirical view in salamanders. *American Naturalist*, 113, 671–689.

Karban, R. & Baldwin, I.T. (1997) *Induced Responses to Herbivory*. University of Chicago Press, Chicago.

Karban, R., Agrawal, A.A., Thaler, J.S. & Adler, L.S. (1999) Induced plant responses and information content about risk of herbivory. *Trends in Ecology and Evolution*, 14, 443–447.

Kareiva, P. (1990) Population dynamics in spatially complex environments: theory and data. *Philosophical Transactions of the Royal Society of London, Series B*, 330, 175–190.

Karels, T.J. & Boonstra, R. (2000) Concurrent density dependence and independence in populations of arctic ground squirrels. *Nature*, 408, 460–463.

Karl, B.J. & Best, H.A. (1982) Feral cats on Stewart Island: their foods, and their effects on kakapo. *New Zealand Journal of Zoology*, 9, 287–294.

Karl, D. (1999) A sea of change: biogeochemical variability in the North Pacific subtropical gyre. *Ecosystems*, 2, 181–214.

Katz, L.A. (1998) Changing perspectives on the origin of eukaryotes. *Trends in Ecology and Evolution*, 13, 493–498.

Kautz, G., Zimmer, M. & Topp, W. (2002) Does *Porcellio scabar* (Isopoda: Oniscidea) gain from coprophagy? *Soil Biology and Biochemistry*, 34, 1253–1259.

Kawano, K. (2002) Character displacement in giant rhinoceros beetles. *American Naturalist*, 159, 255–271.

Kaye, J.P. & Hart, S.C. (1997) Competition for nitrogen between plants and soil microorganisms. *Trends in Ecology and Evolution*, 12, 139–142.

Kays, S. & Harper, J.L. (1974) The regulation of plant and tiller density in a grass sward. *Journal of Ecology*, 62, 97–105.

Keane, R.M. & Crawley, M.J. (2002) Exotic plant invasions and the enemy release hypothesis. *Trends in Ecology and Evolution*, 17, 164–170.

Keddy, P.A. (1982) Experimental demography of the sand-dune annual, *Cakile edentula*, growing along an environmental gradient in Nova Scotia. *Journal of Ecology*, 69, 615–630.

Keddy, P.A. & Shipley, B. (1989) Competitive heirarchies in herbaceous plant communities. *Oikos*, 54, 234–241.

Keeling, M. (1999) Spatial models of interacting populations. In: *Advanced Ecological Theory* (J. McGlade, ed.), pp. 64–99. Blackwell Science, Oxford.

Kelly, C.A. & Dyer, R.J. (2002) Demographic consequences of inflorescence-feeding insects for *Liatris cylindrica*, an iteroparous perennial. *Oecologia*, 132, 350–360.

Kelly, D. (1994) The evolutionary ecology of mast seeding. *Trends in Ecology and Evolution*, 9, 465–470.

Kelly, D., Harrison, A.L., Lee, W.G., Payton, I.J., Wilson, P.R. & Schauber, E.M. (2000) Predator satiation and extreme mast seeding in 11 species of *Chionochloa* (Poaceae). *Oikos*, 90, 477–488.

Kendall, B.E., Briggs, C.J., Murdoch, W.W. *et al.* (1999) Why do populations cycle? A synthesis of statistical and mechanistic modeling approaches. *Ecology*, 80, 1789–1805.

Kerbes, R.H., Kotanen, P.M. & Jefferies, R.L. (1990) Destruction of wetland habitats by lesser snow geese: a keystone species on the west coast of Hudson Bay. *Journal of Applied Ecology*, 27, 242–258.

Kery, M., Matthies, D. & Fischer, M. (2001) The effect of plant population size on the interactions between the rare plant *Gentiana cruciata* and its specialized herbivore *Maculinea rebeli*. *Journal of Ecology*, 89, 418–427.

Kessler, A. & Baldwin, I.T. (2004) Herbivore-induced plant vaccination. Part I. The orchestration of plant defences in nature and their fitness consequences in the wild tobacco. *Plant Journal*, 38, 639–649.

Kettlewell, H.B.D. (1955) Selection experiments on industrial melanism in the Lepidoptera. *Heredity*, 9, 323–342.

Khalil, M.A.K. (1999) Non-CO_2 greenhouse gases in the atmosphere. *Annual Review of Energy and the Environment*, 24, 645–661.

Kicklighter, D.W., Bruno, M., Donges, S. *et al.* (1999) A first-order analysis of the potential role of CO_2 fertilization to affect the global carbon budget: a comparison of four terrestrial biosphere models. *Tellus*, 51B, 343–366.

Kie, J.G. (1999) Optimal foraging and risk of predation: effects on behaviour and social structure in ungulates. *Journal of Mammalogy*, 80, 1114–1129.

Kiers, E.T., Rousseau, R.A., West, S.A. & Denison, R.F. (2003) Host sanctions and the legume–rhizobium mutualism. *Nature*, 425, 78–81.

Kiers, E.T., West, S.A. & Denison, R.F. (2002) Mediating mutualisms: farm management practices and evolutionary changes in symbiont co-operation. *Journal of Applied Ecology*, 39, 745–754.

Kimura, M. & Weiss, G.H. (1964) The stepping stone model of population structure and the decrease of genetic correlation with distance. *Genetics*, 49, 561–576.

Kingsland, S.E. (1985) *Modeling Nature*. University of Chicago Press, Chicago.

Kingston, T.J. & Coe, M.J. (1977) The biology of a giant dung-beetle (*Heliocorpis dilloni*) (Coleoptera: Scarabaeidae). *Journal of Zoology*, 181, 243–263.

Kinnaird, M.F. & O'Brien, T.G. (1991) Viable populations for an endangered forest primate, the Tana River crested mangabey (*Cercocebus galeritus galeritus*). *Conservation Biology*, 5, 203–213.

Kira, T., Ogawa, H. & Shinozaki, K. (1953) Intraspecific competition among higher plants. I. Competition–density–yield inter-relationships in regularly dispersed populations. *Journal of the Polytechnic Institute, Osaka City University*, 4(4), 1–16.

Kirchner, J.W. & Weil, A. (2000) Delayed biological recovery from extinctions throughout the fossil record. *Nature*, 404, 177–180.

Kirk, J.T.O. (1994) *Light and Photosynthesis in Aquatic Ecosystems*. Cambridge University Press, Cambridge, UK.

Kitting, C.L. (1980) Herbivore–plant interactions of individual limpets maintaining a mixed diet of intertidal marine algae. *Ecological Monographs*, 50, 527–550.

Klemola, T., Koivula, M., Korpimaki, E. & Norrdahl, K. (2000) Experimental tests of predation and food hypotheses for population cycles of voles. *Proceedings of the Royal Society of London, Series B*, 267, 352–356.

Klemow, K.M. & Raynal, D.J. (1981) Population ecology of *Melilotu-salba* in a limestone quarry. *Journal of Ecology*, 69, 33–44.

Kling, G.W., Kipphut, G.W., Miller, M.M. & O'Brien, W.J. (2000) Integration of lakes and streams in a landscape perspective: the importance of material processing on spatial patterns and temporal coherence. *Freshwater Biology*, 43, 477–497.

Knapp, A.K. & Smith, M.D. (2001) Variation among biomes in temporal dynamics of aboveground primary production. *Science*, 291, 481–484.

Kneidel, K.A. (1984) Competition and disturbance in communities of carrion breeding Diptera. *Journal of Animal Ecology*, 53, 849–865.

Knell, R.J. (1998) Generation cycles. *Trends in Ecology and Evolution*, 13, 186–190.

Knell, R.J., Begon, M. & Thompson, D.J. (1998) Transmission of *Plodia interpunctella* granulosis virus does not conform to the mass action model. *Journal of Animal Ecology*, 67, 592–599.

Knoch, T.R., Faeth, S.H. & Arnott, D.L. (1993) Endophytic fungi alter foraging and dispersal by desert seed-harvesting ants. *Oecologia*, 95, 470–473.

Knops, J.M.H., Bradley, K.L. & Wedin, D.A. (2002) Mechanisms of plant species impacts on ecosystem nitrogen cycling. *Ecology Letters*, 5, 454–466.

Kodric-Brown, A. & Brown, J.M. (1993) Highly structured fish communities in Australian desert springs. *Ecology*, 74, 1847–1855.

Koenig, W.D. & Knops, J.M.H. (1998) Scale of mast-seeding and tree-ring growth. *Nature*, 396, 225–226.

Kohler, S.L. (1992) Competition and the structure of a benthic stream community. *Ecological Monographs*, 62, 165–188.

Kohler, S.L. & Wiley, M.J. (1997) Pathogen outbreaks reveal large-scale effects of competition in stream communities. *Ecology*, 78, 2164–2176.

Koller, D. & Roth, N. (1964) Studies on the ecological and physiological significance of amphicarpy in *Gymnarrhena micrantha* (Compositae). *American Journal of Botany*, 51, 26–35.

Korotkov, V.N., Logofet, D.O. & Loreau, M. (2001) Succession in mixed boreal forest in Russia: Markov models and non-Markov effects. *Ecological Modelling*, 142, 25–38.

Korpimaki, E., Klemola, T., Norrdahl, K. *et al.* (2003) Vole cycles and predation. *Trends in Ecology and Evolution*, 18, 494–495.

Korpimaki, E., Norrdahl, K., Klemola, T., Pettersen, T. & Stenseth, N.C. (2002) Dynamic effects of predators on cyclic voles: field experimentation and model extrapolation. *Proceedings of the Royal Society of London, Series B*, 269, 991–997.

Kosola, K.R., Dickmann, D.I. & Parry, D. (2002) Carbohydrates in individual poplar fine roots: effects of root age and defoliation. *Tree Physiology*, 22, 741–746.

Kozlowski, J. (1993) Measuring fitness in life-history studies. *Trends in Ecology and Evolution*, 7, 155–174.

Kraft, C.E. & Johnson, L.E. (2000) Regional differences in rates and patterns of North American inland lake invasions by zebra mussels (*Dreissena polymorpha*). *Canadian Journal of Fisheries and Aquatic Sciences*, 57, 993–1001.

Kratz, T.K., Webster, K.E., Bowser, C.J., Magnuson, J.J. & Benson, B.J. (1997) The influence of landscape position on lakes in northern Wisconsin. *Freshwater Biology*, 37, 209–217.

Krause, A.E., Frank, K.A., Mason, D.M., Ulanowicz, R.E. & Taylor, W.W. (2002) Compartments revealed in food-web structure. *Nature*, 426, 282–285.

Krebs, C.J. (1972) *Ecology*. Harper & Row, New York.

Krebs, C.J. (1999) *Ecological Methodology*, 2nd edn. Addison-Welsey Educational, Menlo Park, CA.

Krebs, C.J. (2003) How does rodent behaviour impact on population dynamics? In: *Rats, Mice and People: rodent biology and management* (G.R. Singleton, L.A. Hynds, C.J. Krebs & D.J. Spratt, eds), pp. 117–123. Australian Centre for International Agricultural Research, Canberra.

Krebs, C.J., Boonstra, R., Boutin, S. & Sinclair, A.R.E. (2001) What drives the 10-year cycle of snowshoe hares? *Bioscience*, 51, 25–35.

Krebs, C.J., Boutin, S., Boonstra, R. *et al.* (1995) Impact of food and predation on the snowshoe hare cycle. *Science*, 269, 1112–1115.

Krebs, C.J., Sinclair, A.R.E., Boonstra, R., Boutin, S., Martin, K. & Smith, J.N.M. (1999) Community dynamics of vertebrate herbivores: how can we untangle the web? In: *Herbivores: between plants and predators* (H. Olff, V.K. Brown & R.H. Drent, eds), pp. 447–473. Blackwell Science, Oxford.

Krebs, J.R. (1971) Territory and breeding density in the great tit, *Parus major* L. *Ecology*, 52, 2–22.

Krebs, J.R. (1978) Optimal foraging: decision rules for predators. In: *Behavioural Ecology: an evolutionary approach* (J.R. Krebs & N.B. Davies, eds), pp. 23–63. Blackwell Scientific Publications, Oxford.

Krebs, J.R. & Davies, N.B. (1993) *An Introduction to Behavioural Ecology*, 3rd edn. Blackwell Scientific Publications, Oxford.

Krebs, J.R. & Kacelnik, A. (1991) Decision-making. In: *Behavioural Ecology: an evolutionary approach*, 3rd edn (J.R. Krebs & N.B. Davies, eds), pp. 105–136. Blackwell Scientific Publications, Oxford.

Krebs, J.R., Erichsen, J.T., Webber, M.I. & Charnov, E.L. (1977) Optimal prey selection in the great tit (*Parus major*). *Animal Behaviour*, 25, 30–38.

Krebs, J.R., Stephens, D.W. & Sutherland, W.J. (1983) Perspectives in optimal foraging. In: *Perspectives in Ornithology* (A.H. Brush & G.A. Clarke, Jr., eds), pp. 165–216. Cambridge University Press, New York.

Kreitman, M., Shorrocks, B. & Dytham, C. (1992) Genes and ecology: two alternative perspectives using *Drosophila*. In: *Genes in Ecology* (R.J. Berry, T.J. Crawford & G.M. Hewitt, eds), pp. 281–312. Blackwell Scientific Publications, Oxford.

Kriticos, D.J., Sutherst, R.W., Brown, J.R., Adkins, S.W. & Maywald, G.F. (2003) Climate change and the potential distribution of an invasive alien plant: *Acacia nilotica* spp. *indica* in Australia. *Journal of Applied Ecology*, 40, 111–124.

Kronfeld-Schor, N. & Dayan, T. (2003) Partitioning of time as an ecological resource. *Annual Review of Ecology, Evolution and Systematics*, 34, 153–181.

Kubanek, J., Whalen, K.E., Engel, S. *et al.* (2002) Multiple defensive roles for triterpene glycosides from two Caribbean sponges. *Oecologia*, 131, 125–136.

Kullberg, C. & Ekman, J. (2000) Does predation maintain tit community diversity? *Oikos*, 89, 41–45.

Kunert, G. & Weisser, W.W. (2003) The interplay between density- and trait-mediated effects in predator–prey interactions: a case study in aphid wing polymorphism. *Oecologia*, 135, 304–312.

Kunin, W.E. & Gaston, K.J. (1993) The biology of rarity: patterns, causes and consequences. *Trends in Ecology and Evolution*, 8, 298–301.

Kuno, E. (1991) Some strange properties of the logistic equation defined with r and K: inherent defects or artifacts? *Researches on Population Ecology*, 33, 33–39.

Laakso, J., Setala, H. & Palojarvi, A. (2000) Influence of decomposer food web structure and nitrogen availability on plant growth. *Plant and Soil*, 225, 153–165.

Labbé, P. (1994) Régénération après passage du cyclone Hugo en forêt dense humide de Guadeloupe. *Acta Ecologica*, 15, 301–315.

Lack, D. (1947) The significance of clutch size. *Ibis*, 89, 302–352.

Lack, D. (1963) Cuckoo hosts in England. (With an appendix on the cuckoo hosts in Japan, by T. Royama.) *Bird Study*, 10, 185–203.

Lacy, R.C. (1993) VORTEX: a computer simulation for use in population viability analysis. *Wildlife Research*, 20, 45–65.

Lambin, X. & Krebs, C.J. (1993) Influence of female relatedness on the demography of Townsend's vole populations in spring. *Journal of Animal Ecology*, 62, 536–550.

Lambin, X. & Yoccoz, N.G. (1998) The impact of population kinstructure on nestling survival in Townsend's voles, *Microtus townsendii*. *Journal of Animal Ecology*, 67, 1–16.

Lambin, X., Aars, J. & Piertney, S.B. (2001) Dispersal, intraspecific competition, kin competition and kin facilitation: a review of the empirical evidence. In: *Dispersal* (J. Clobert, E. Danchin, A.A. Dhondt & J.D. Nichols, eds), pp. 110–122. Oxford University Press, Oxford.

Lambin, X., Krebs, C.J., Moss, R., Stenseth, N.C. & Yoccoz, N.G. (1999) Population cycles and parasitism. *Science* (technical comment), 286, 2425a.

Lambin, X., Petty, S.J. & MacKinnon, J.L. (2000) Cyclic dynamics in field vole populations and generalist predation. *Journal of Animal Ecology*, 69, 106–118.

Lande, R. (1993) Risks of population extinction from demographic and environmental stochasticity, and random catastrophes. *American Naturalist*, 142, 911–927.

Lande, R. & Barrowclough, G.F. (1987) Effective population size, genetic variation, and their use in population management. In: *Viable Populations for Conservation* (M.E. Soulé, ed.), pp. 87–123. Cambridge University Press, Cambridge, UK.

Landeweert, R., Hoffland, E., Finlay, R.D., Kuyper, T.W. & van Breemen, N. (2001) Linking plants to rocks: ectomycorrhizal fungi mobilize nutrients from minerals. *Trends in Ecology and Evolution*, 16, 248–254.

Larcher, W. (1980) *Physiological Plant Ecology*, 2nd edn. Springer-Verlag, Berlin.

Lathrop, R.C., Johnson, B.M., Johnson, T.B. *et al.* (2002) Stocking piscivores to improve fishing and water clarity: a synthesis of the Lake Mendota biomanipulation project. *Freshwater Biology*, 47, 2410–2424.

Laurance, W.F. (2001) Future shock: forecasting a grim fate for the Earth. *Trends in Ecology and Evolution*, 16, 531–533.

Law, B.E., Thornton, P.E., Irvine, J., Anthoni, P.M. & van Tuyl, S. (2001) Carbon storage and fluxes in ponderosa pine forests at different developmental stages. *Global Climate Change*, 7, 755–777.

Law, R. & Watkinson, A.R. (1987) Response-surface analysis of two-species competition: an experiment on *Phleum arenarium* and *Vulpia fasciculata*. *Journal of Ecology*, 75, 871–886.

Law, R. & Watkinson, A.R. (1989) Competition. In: *Ecological Concepts* (J.M. Cherrett, ed.), pp. 243–284. Blackwell Scientific Publications, Oxford.

Lawler, I.R., Foley, W.J. & Eschler, B.M. (2000) Foliar concentration of a single toxin creates habitat patchiness for a marsupial folivore. *Ecology*, 81, 1327–1338.

Lawler, S.P. Morin, P.J. (1993) Temporal overlap, competition, and priority effects in larval anurans. *Ecology*, 74, 174–182.

Lawlor, L.R. (1978) A comment on randomly constructed ecosystem models. *American Naturalist*, 112, 445–447.

Lawlor, L.R. (1980) Structure and stability in natural and randomly constructed competitive communities. *American Naturalist*, 116, 394–408.

Lawton, J.H. (1984) Non-competitive populations, non-convergent communities, and vacant niches: the herbivores of bracken. In: *Ecological Communities Conceptual Issues and the Evidence* (D.R. Strong, D. Simberloff, L.G. Abele & A.B. Thistle, eds), pp. 67–100. Princeton University Press, Princeton, NJ.

Lawton, J.H. (1989) Food webs. In: *Ecological Concepts* (J.M. Cherrett, ed.), pp. 43–78. Blackwell Scientific Publications, Oxford.

Lawton, J.H. & May, R.M. (1984) The birds of Selborne. *Nature*, 306, 732–733.

Lawton, J.H. & Woodroffe, G.L. (1991) Habitat and the distribution of water voles: why are there gaps in a species' range? *Journal of Animal Ecology*, 60, 79–91.

Le Cren, E.D. (1973) Some examples of the mechanisms that control the population dynamics of salmonid fish. In: *The Mathematical Theory of the Dynamics of Biological Populations* (M.S. Bartlett & R.W. Hiorns, eds), pp. 125–135. Academic Press, London.

Lee, C.-S., You, Y.-H. & Robinson, G.R. (2002) Secondary succession and natural habitat restoration in abandoned rice fields of central Korea. *Restoration Ecology*, 10, 306–314.

Lee, W.G. & Jamieson, I.G. (eds) (2001) *The Takahe: fifty years of conservation management and research.* University of Otago Press, Dunedin, New Zealand.

Leigh, E. (1975) Population fluctuations and community structure. In: *Unifying Concepts in Ecology* (W.H. van Dobben & R.H. Lowe–McConnell, eds), pp. 67–88. Junk, The Hague.

Leirs, H., Stenseth, N.C., Nichols, J.D., Hines, J.E., Verhagen, R. & Verheyen, W. (1997) Stochastic seasonality and non-linear density dependent factors regulate population size in an African rodent. *Nature*, 389, 176–180.

Lekve, K., Ottersen, G., Stenseth, N.C. & Gjøsæter, J. (2002) Length dynamics in juvenile coastal Skagerrak cod: effects of biotic and abiotic processes. *Ecology*, 83, 1676–1688.

Lennartsson, T., Nilsson, P. & Tuomi, J. (1998) Induction of over-compensation in the field gentian, *Gentianella campestris*. *Ecology*, 79, 1061–1072.

Lessells, C.M. (1991) The evolution of life histories. In: *Behavioural Ecology*, 3rd edn (J.R. Krebs & N.B. Davies, eds), pp. 32–68. Blackwell Scientific Publications, Oxford.

Letourneau, D.K. & Dyer, L.A. (1998a) Density patterns of *Piper* ant-plants and associated arthropods: top-predator trophic cascades in a terrestrial system? *Biotropica*, 30, 162–169.

Letourneau, D.K. & Dyer, L.A. (1998b) Experimental test in a lowland tropical forest shows top-down effects through four trophic levels. *Ecology*, 79, 1678–1687.

Leverich W.J. & Levin, D.A. (1979) Age-specific survivorship and reproduction in *Phlox drummondii*. *American Naturalist*, 113, 881–903.

Levey, D.J. (1988) Tropical wet forest treefall gaps and distributions of understorey birds and plants. *Ecology*, 69, 1076–1089.

Levins, R. (1968) *Evolution in Changing Environments*. Princeton University Press, Princeton, NJ.

Levins, R. (1969) Some demographic and genetic consequences of environmental heterogeneity for biological control. *Bulletin of the Entomological Society of America*, 15, 237–240.

Levins, R. (1970) Extinction. In: *Lectures on Mathematical Analysis of Biological Phenomena*, pp. 123–138. Annals of the New York Academy of Sciences, Vol. 231.

Lewontin, R.C. & Levins, R. (1989) On the characterization of density and resource availability. *American Naturalist*, 134, 513–524.

Lichter, J. (2000) Colonization constraints during primary succession on coastal Lake Michigan sand dunes. *Journal of Ecology*, 88, 825–839.

Lidicker, W.Z. Jr. (1975) The role of dispersal in the demography of small mammal populations. In: *Small Mammals: their productivity and population dynamics* (K. Petruscwicz, F.B. Golley & L. Ryszkowski, eds), pp. 103–128. Cambridge University Press, New York.

Likens, G.E. (1992) *The Ecosystem Approach: its use and abuse. Excellence in Ecology*, Book 3. Ecology Institute, Oldendorf-Luhe, Germany.

Likens, G.E. & Bormann, F.G. (1975) An experimental approach to New England landscapes. In: *Coupling of Land and Water Systems* (A.D. Hasler, ed.), pp. 7–30. Springer-Verlag, New York.

Likens, G.E. Bormann, F.H., Pierce, R.S. & Fisher, D.W. (1971) Nutrient–hydrologic cycle interaction in small forested watershed ecosystems. In: *Productivity of Forest Ecosystems* (P. Duvogneaud, ed.), pp. 553–563. UNESCO, Paris.

Lillebo, A.I., Flindt, M.R., Pardal, M.A. & Marques, J.C. (1999) The effect of macrofauna, meiofauna and microfauna on the degradation of *Spartina maritima* detritus from a salt marsh area. *Acta Oecologica*, 20, 249–258.

Lima, M., Keymer, J.E. & Jaksic, F.M. (1999) El Niño–Southern Oscillation-driven rainfall variability and delayed density dependence cause rodent outbreaks in western South America: linking demography and population dynamics. *American Naturalist*, 153, 476–491.

Lin, B., Sakoda, A., Shibasaki, R., Goto, N. & Suzuki, M. (2000) Modelling a global biogeochemical nitrogen cycle in terrestrial ecosystems. *Ecological Modelling*, 135, 89–110.

Lindemann, R.L. (1942) The trophic–dynamic aspect of ecology. *Ecology*, 23, 399–418.

Lochmiller, R.L. & Derenberg, C. (2000) Trade-offs in evolutionary immunology: just what is the cost of immunity? *Oikos*, 88, 87–98.

Lofgren, A. & Jerling, L. (2002) Species richness, extinction and immigration rates of vascular plants on islands in the Stockholm Archipelago, Sweden, during a century of ceasing management. *Folia Geobotanica*, 37, 297–308.

Loik, M.E. & Nobel, P.S. (1993) Freezing tolerance and water relations of *Opuntia fragilis* from Canada and the United States. *Ecology*, 74, 1722–1732.

Loladze, I. (2002) Rising atmospheric CO_2 and human nutrition: toward globally imbalanced plant stoichiometry? *Trends in Ecology and Evolution*, 17, 457–461.

Long, S.P., Humphries, S. & Falkowski, P.G. (1994) Photoinhibition of photosynthesis in nature. *Annual Review of Plant Physiology and Plant Molecular Biology*, 45, 633–662.

Lonsdale, W.M. (1990) The self-thinning rule: dead or alive? *Ecology*, 71, 1373–1388.

Lonsdale, W.M. & Watkinson, A.R. (1983) Light and self-thinning. *New Phytologist*, 90, 431–435.

Loreau, M. (2000) Are communities saturated? On the relationship between α, β and γ diversity. *Ecology Letters*, 3, 73–76.

Loreau, M., Naeem, S., Inchausti, P. *et al.* (2001) Biodiversity and ecosystem functioning: current knowledge and future challenges. *Science*, 294, 804–808.

Losos, E. (1993) The future of the US Endangered Species Act. *Trends in Ecology and Evolution*, 8, 332–336.

Losos, J.B. (1992) The evolution of convergent structure in Caribbean *Anolis* communities. *Systematic Biology*, 41, 403–420.

Losos, J.B., Jackman, T.R., Larson, A., de Queiroz, K. & Rodriguez-Schettino, L. (1998) Contingency and determinism in replicated adaptive radiations of island lizards. *Science*, 279, 2115–2118.

Lotka, A.J. (1932) The growth of mixed populations: two species competing for a common food supply. *Journal of the Washington Academy of Sciences*, 22, 461–469.

Loucks, C.J., Zhi, L., Dinerstein, E., Dajun, W., Dali, F. & Hao, W. (2003) The giant pandas of the Qinling Mountains, China: a case study in designing conservation landscapes for elevational migrants. *Conservation Biology*, 17, 558–565.

Louda, S.M. (1982) Distribution ecology: variation in plant recruitment over a gradient in relation to insect seed predation. *Ecological Monographs*, 52, 25–41.

Louda, S.M. & Rodman, J.E. (1996) Insect herbivory as a major factor in the shade distribution of a native crucifer (*Cardamine cordifolia* A. Gray, bittercress). *Journal of Ecology*, 84, 229–237.

Louda, S.M., Arnett, A.E., Rand, T.A. & Russell, F.L. (2003a) Invasiveness of some biological control insects and adequacy of their ecological risk assessment and regulation. *Conservation Biology*, 17, 73–82.

Louda, S.M., Kendall, D., Connor, J. *et al.* (1997) Ecological effects of an insect introduced for the biological control of weeds. *Science*, 277, 1088–1990.

Louda, S.M., Pemberton, R.W., Johnson, M.T. & Follett, P.A. (2003b) Non-target effects – the Achilles' heel of biological control? Retrospective analyses to reduce risk associated with biocontrol introductions. *Annual Review of Entomology*, 48, 365–396.

Lovett Doust, J., Schmidt, M. & Lovett Doust, L. (1994) Biological assessment of aquatic pollution: a review, with emphasis on plants as biomonitors. *Biological Reviews*, 69, 147–186.

Lovett Doust, L. & Lovett Doust, J. (1982) The battle strategies of plants. *New Scientist*, 95, 81–84.

Lowe, V.P.W. (1969) Population dynamics of the red deer (*Cervus elaphus* L.) on Rhum. *Journal of Animal Ecology*, 38, 425–457.

Lubchenco, J. (1978) Plant species diversity in a marine intertidal community: importance of herbivore food preference and algal competitive abilities. *American Naturalist*, 112, 23–39.

Lubchenco, J., Olson, A.M., Brubaker, L.B. *et al.* (1991) The sustainable biosphere initiative: an ecological research agenda. *Ecology*, 72, 371–412.

Luckmann, W.H. & Decker, G.C. (1960) A 5-year report on observations in the Japanese beetle control area of Sheldon, Illinois. *Journal of Economic Entomology*, 53, 821–827.

Lui, Y.B., Tabashnik, B.E., Dennehy, T.J. *et al.* (2001) Effects of Bt cotton and Cry1Ac toxin on survival and development of pink bollworm (Lepidoptera : Gelechiidae). *Journal of Economic Entomology*, 94, 1237–1242.

Lukens, R.J. & Mullany, R. (1972) The influence of shade and wet on southern corn blight. *Plant Disease Reporter*, 56, 203–206.

Luo, T., Li, W. & Zhu, H. (2002) Estimated biomass and productivity of natural vegetation on the Tibetan Plateau. *Ecological Applications*, 12, 980–997.

Lussenhop, J. (1992) Mechanisms of microarthropod–microbial interactions in soil. *Advances in Ecological Research*, 23, 1–33.

MacArthur, J.W. (1975) Environmental fluctuations and species diversity. In: *Ecology and Evolution of Communities* (M.L. Cody & J.M. Diamond, eds), pp. 74–80. Belknap, Cambridge, MA.

MacArthur, R.H. (1955) Fluctuations of animal populations and a measure of community stability. *Ecology*, 36, 533–536.

MacArthur, R.H. (1962) Some generalized theorems of natural selection. *Proceedings of the National Academy of Science of the USA*, 48, 1893–1897.

MacArthur, R.H. (1972) *Geographical Ecology*. Harper & Row, New York.

MacArthur, R.H. & Levins, R. (1964) Competition, habitat selection and character displacement in a patchy environment. *Proceedings of the National Academy of Sciences*, 51, 1207–1210.

MacArthur, R.H. & Levins, R. (1967) The limiting similarity, convergence and divergence of coexisting species. *American Naturalist*, 101, 377–385.

MacArthur, R.H. & Pianka, E.R. (1966) On optimal use of a patchy environment. *American Naturalist*, 100, 603–609.

MacArthur, R.H. & Wilson, E.O. (1967) *The Theory of Island Biogeography*. Princeton University Press, Princeton, NJ.

McCallum, H., Barlow, N. & Hone, J. (2001) How should pathogen transmission be modelled? *Trends in Ecology and Evolution*, 16, 295–300.

McCann, K. & Hastings, A. (1997) Re-evaluating the omnivory–stability relationship in food webs. *Proceedings of the Royal Society of London, Series B*, 264, 1249–1254.

McCune, D.C. & Boyce, R.L. (1992) Precipitation and the transfer of water, nutrients and pollutants in tree canopies. *Trends in Ecology and Evolution*, 7, 4–7.

MacDonald, A.S., Araujo, M.I. & Pearce, E.J. (2002) Immunology and parasitic helminth infections. *Infection and Immunity*, 70, 427–433.

Mace, G.M. (1994) An investigation into methods for categorizing the conservation status of species. In: *Large-Scale Ecology and Conservation Biology* (P.J. Edwards, R.M. May & N.R. Webb, eds), pp. 293–312. Blackwell Scientific Publications, Oxford.

Mace, G.M. & Lande, R. (1991) Assessing extinction threats: toward a reevaluation of IUCN threatened species categories. *Conservation Biology*, 5, 148–157.

McGrady-Steed, J., Harris, P.M. & Morin, P.J. (1997) Biodiversity regulates ecosystem predictability. *Nature*, 390, 162–165.

Mack, R.N., Simberloff, D., Lonsdale, W.M., Evans, H., Clout, M. & Bazzaz, F.A. (2000) Biotic invasions: causes, epidemiology, global consequences and control. *Ecological Applications*, 10, 689–710.

McIntosh, A.R. & Townsend, C.R. (1994) Interpopulation variation in mayfly anti-predator tactics: differential effects of contrasting fish predators. *Ecology*, 75, 2078–2090.

McKane, R.B., Johnson, L.C., Shaver, G.R. *et al.* (2002) Resource-based niches provide a basis for plant species diversity and dominance in arctic tundra. *Nature*, 415, 68–71.

McKay, J.K., Bishop, J.G., Lin, J.-Z., Richards, J.H., Sala, A. & Mitchell-Olds, T. (2001) Local adaptation across a climatic gradient despite small effective population size in the rare sapphire rockcress. *Proceedings of the Royal Society of London, Series B*, 268, 1715–1721.

McKenzie, N.L. & Rolfe, J.K. (1986) Structure of bat guilds in the Kimberley mangroves, Australia. *Journal of Animal Ecology*, 55, 401–420.

McKey, D. (1979) The distribution of secondary compounds within plants. In: *Herbivores: their interaction with secondary plant metabolites* (G.A. Rosenthal & D.H. Janzen, eds), pp. 56–134. Academic Press, New York.

MacKinnon, K. (1998) Sustainable use as a conservation tool in the forests of South-East Asia. In: *Conservation of Biological Resources*

(E.J. Milner-Gulland & R. Mace, eds), pp. 174–192. Blackwell Science, Oxford.

McKone, M.J., Kelly, D. & Lee, W.G. (1998) Effect of climate change on mast-seeding species: frequency of mass flowering and escape from specialist insect seed predators. *Global Change Biology*, 4, 591–596.

MacLachlan, A.J., Pearce, L.J. & Smith, J.A. (1979) Feeding interactions and cycling of peat in a bog lake. *Journal of Animal Ecology*, 48, 851–861.

MacLulick, D.A. (1937) Fluctuations in numbers of the varying hare (*Lepus americanus*). *University of Toronto Studies, Biology Series*, 43, 1–136.

McMahon, T. (1973) Size and shape in biology. *Science*, 179, 1201–1204.

McNaughton, S.J. (1975) r- and k-selection in *Typha*. *American Naturalist*, 109, 251–261.

McNaughton, S.J. (1977) Diversity and stability of ecological communities: a comment on the role of empiricism in ecology. *American Naturalist*, 111, 515–525.

McNaughton, S.J. (1978) Stability and diversity of ecological communities. *Nature*, 274, 251–253.

Maguire, L.A., Seal, U.S. & Brussard, P.F. (1987) Managing critically endangered species: the Sumatran rhino as a case study. In: *Viable Populations for Conservation* (M.E. Soulé, ed.), pp. 141–158. Cambridge University Press, Cambridge, UK.

Mahaman, B.D., Passam, H.C., Sideridis, A.B. & Yialouris, C.P. (2003) DIARES-IPM: a diagnostic advisory rule-based expert system for integrated pest management in solanaceous crop systems. *Agricultural Systems*, 76, 1119–1135.

Malmqvist, B., Wotton, R.S. & Zhang, Y. (2001) Suspension feeders transform massive amounts of seston in large northern rivers. *Oikos*, 92, 35–43.

Marchetti, M.P. & Moyle, P.B. (2001) Effects of flow regime on fish assemblages in a regulated California stream. *Ecological Applications*, 11, 530–539.

Marcos, G.M., & Lancho, J.F.G. (2002) Atmospheric deposition in oligotrophic *Quercus pyrenaica* forests: implications for forest nutrition. *Forest Ecology and Management*, 171, 17–29.

Margules, C.R. & Pressey, R.L. (2000) Systematic conservation planning. *Nature*, 405, 243–253.

Margulis, L. (1975) Symbiotic theory of the origin of eukaryotic organelles. In: *Symbiosis* (D.H. Jennings & D.L. Lee, eds), pp. 21–38. Symposium 29, Society for Experimental Biology. Cambridge University Press, Cambridge, UK.

Margulis, L. (1996) Archaeal–eubacterial mergers in the origin of Eukarya: phylogenetic classification of life. *Proceedings of the National Academy of Sciences of the USA*, 93, 1071–1076.

Maron, J.L., Combs, J.K. & Louda, S.M. (2002) Convergent demographic effects of insect attack on related thistles in coastal vs continental dunes. *Ecology*, 83, 3382–3392.

Marsden-Smedley, J.B. & Kirkpatrick, J.B. (2000) Fire management in Tasmania's Wilderness World Heritage Area: ecosystem restoration using indigenous-style fire regimes. *Ecological Management and Restoration*, 1, 195–203.

Marshall, I.D. & Douglas, G.W. (1961) Studies in the epidemiology of infectious myxomatosis of rabbits. VIII. Further observations on changes in the innate resistance of Australian wild rabbits exposed to myxomatosis. *Journal of Hygiene*, 59, 117–122.

Martin, M.M. (1991) The evolution of cellulose digestion in insects. In: *The Evolutionary Interaction of Animals and Plant* (W.G. Chaloner, J.L. Harper & J.H. Lawton, eds), pp. 105–112. The Royal Society, London; also in *Philosophical Transactions of the Royal Society of London, Series B*, 333, 281–288.

Martin, P.R. & Martin, T.E. (2001) Ecological and fitness consequences of species coexistence: a removal experiment with wood warblers. *Ecology*, 82, 189–206.

Martin, P.S. (1984) Prehistoric overkill: the global model. In: *Quaternary Extinctions: a prehistoric revolution* (P.S. Martin & R.G. Klein, eds), pp. 354–403. University of Arizona Press, Tuscon, AZ.

Martinez, N.D. (1991) Artefacts or attributes? Effects of resolution on the Little Rock Lake food web. *Ecological Monographs*, 61, 367–392.

Martinez, N.D. (1993) Effects of resolution on food web structure. *Oikos*, 66, 403–412.

Marzusch, K. (1952) Untersuchungen über di Temperaturabhängigkeit von Lebensprozessen bei Insekten unter besonderer Berücksichtigung winter-schlantender Kartoffelkäfer. *Zeitschrift für vergleicherde Physiologie*, 34, 75–92.

Matter, S.F., Hanski, I. & Gyllenberg, M. (2002) A test of the metapopulation model of the species–area relationship. *Journal of Biogeography*, 29, 977–983.

Matthaei, C.D. & Townsend, C.R. (2000) Long-term effects of local disturbance history on mobile stream invertebrates. *Oecologia*, 125, 119–126.

Matthaei, C.D., Peacock, K.A. & Townsend, C.R. (1999) Scour and fill patterns in a New Zealand stream and potential implications for invertebrate refugia. *Freshwater Biology*, 42, 41–58.

Matthiopoulos, J., Moss, R., Mougeot, F., Lambin, X. & Redpath, S.M. (2003) Territorial behaviour and population dynamics in red grouse *Lagopus lagopus scoticus*. II. Population models. *Journal of Animal Ecology*, 72, 1083–1096.

May, R.M. (1972) Will a large complex system be stable? *Nature*, 238, 413–414.

May, R.M. (1973) On relationships among various types of population models. *American Naturalist*, 107, 46–57.

May, R.M. (1975a) Biological populations obeying difference equations: stable points, stable cycles and chaos. *Journal of Theoretical Biology*, 49, 511–524.

May, R.M. (1975b) Patterns of species abundance and diversity. In: *Ecology and Evolution of Communities* (M.L. Cody & J.M. Diamond, eds), pp. 81–120. Belknap, Cambridge, MA.

May, R.M. (1976) Estimating r: a pedagogical note. *American Naturalist*, 110, 496–499.

May, R.M. (1978) Host–parasitoid systems in patchy environments: a phenomenological model. *Journal of Animal Ecology*, 47, 833–843.

May, R.M. (1981) Models for two interacting populations. In: *Theoretical Ecology: principles and applications*, 2nd edn (R.M. May, ed.), pp. 78–104. Blackwell Scientific Publications, Oxford.

May, R.M. (1990) How many species? *Philosophical Transactions of the Royal Society of London, Series B*, 330, 293–304.

May, R.M. & Anderson, R.M. (1979) Population biology of infectious diseases. *Nature*, 280, 455–461.

May, R.M. & Anderson, R.M. (1983) Epidemiology and genetics in the coevolution of parasites and hosts. *Proceedings of the Royal Society of London, Series B*, 219, 281–313.

Mayer, G.C. & Chipley, R.M. (1992) Turnover in the avifauna of Guana Island, British Virgin Islands. *Journal of Animal Ecology*, 61, 561–566.

Maynard Smith, J. (1972) *On Evolution*. Edinburgh University Press, Edinburgh.

Maynard Smith, J. & Slatkin, M. (1973) The stability of predator–prey systems. *Ecology*, 54, 384–391.

Mehner, T., Benndorf, J., Kasprzak, P. & Koschel, R. (2002) Biomanipulation of lake ecosystems: successful applications and expanding complexity in the underlying science. *Freshwater Biology*, 47, 2453–2465.

Meijer, M.-L., de Boois, I., Scheffer, M., Portielje, R. & Hosper, H. (1999) Biomanipulation in shallow lakes in the Netherlands: an evaluation of 18 case studies. *Hydrobiologia*, 408/9, 13–30.

Menalled, F.D., Gross, K.L. & Hammond, M. (2001) Weed aboveground and seedbank community responses to agricultural management systems. *Ecological Applications*, 11, 1586–1601.

Menge, B.A. & Sutherland, J.P. (1976) Species diversity gradients: synthesis of the roles of predation, competition, and temporal heterogeneity. *American Naturalist*, 110, 351–369.

Menge, B.A. & Sutherland, J.P. (1987) Community regulation: variation in disturbance, competition, and predation in relation to environmental stress and recruitment. *American Naturalist*, 130, 730–757.

Menge, B.A., Lubchenco, J., Gaines, S.D. & Ashkenas, L.R. (1986) A test of the Menge–Sutherland model of community organization in a tropical rocky intertidal food web. *Oecologia*, 71, 75–89.

Menges, E.S. (2000) Population viability analyses in plants: challenges and opportunities. *Trends in Ecology and Evolution*, 15, 51–56.

Menges, E.S. & Dolan, R.W. (1998) Demographic viability of populations of *Silene regia* in midwestern prairies: relationships with fire management, genetic variation, geographic location, population size and isolation. *Journal of Ecology*, 86, 63–78.

Merryweather, J.W. & Fitter, A.H. (1995) Phosphorus and carbon budgets: mycorrhizal contribution in *Hyacinthoides non-scripta* (L.) Chouard ex Rothm. under natural conditions. *New Phytologist*, 129, 619–627.

Metcalf, R.L. (1982) Insecticides in pest management. In: *Introduction to Insect Pest Management*, 2nd edn (R.L. Metcalf & W.L. Luckmann, eds), pp. 217–277. Wiley, New York.

Meyer, A. (1993) Phylogenetic relationships and evolutionary processes in East African cichlid fishes. *Trends in Ecology and Evolution*, 8, 279–284.

Micheli, F., Peterson, C.H., Mullineaux, L.S. *et al.* (2002) Predation structures communities at deep-sea hydrothermal vents. *Ecological Monographs*, 72, 365–382.

Mikola, J., Salonen, V. & Setala, H. (2002) Studying the effects of plant species richness on ecosystem functioning: does the choice of experimental design matter? *Oecologia*, 133, 594–598.

Milinski, M. & Parker, G.A. (1991) Competition for resources. In: *Behavioural Ecology: an evolutionary approach*, 3rd edn (J.R. Krebs & N.B. Davies, eds), pp. 137–168. Blackwell Scientific Publications, Oxford.

Miller, G.T. Jr. (1988) *Environmental Science*, 2nd edn. Wadsworth, Belmont.

Milliman, J.D. (1997) Blessed dams or damned dams? *Nature*, 386, 325–326.

Mills, J.A., Lavers, R.B. & Lee, W.G. (1984) The takahe – a relict of the Pleistocene grassland avifauna of New Zealand. *New Zealand Journal of Ecology*, 7, 57–70.

Mills, L.S., Soule, M.E. & Doak, D.F. (1993) The keystone-species concept in ecology and conservation. *Bioscience*, 43, 219–224.

Milne, L.J. & Milne, M. (1976) The social behaviour of burying beetles. *Scientific American*, August, 84–89.

Milner-Gulland, E.J. & Mace, R. (1998) *Conservation of Biological Resources*. Blackwell Science, Oxford.

Minorsky, P.V. (1985) An heuristic hypothesis of chilling injury in plants: a role for calcium as the primary physiological transducer of injury. *Plant Cell and Environment*, 8, 75–94.

Mistri, M. (2003) Foraging behaviour and mutual interference in the Mediterranean shore crab, *Carcinus aestuarii*, preying upon the immigrant mussel *Musculista senhousia*. *Estuarine, Coastal and Shelf Science*, 56, 155–159.

Mittelbach, G.G., Steiner, C.F., Scheiner, S.M. *et al.* (2001) What is the observed relationship between species richness and productivity? *Ecology*, 82, 2381–2396.

Moilanen, A., Smith, A.T. & Hanski, I. (1998) Long-term dynamics in a metapopulation of the American pika. *American Naturalist*, 152, 530–542.

Montagnes, D.J.S., Kimmance, S.A. & Atkinson, D. (2003) Using Q_{10}, can growth rates increase linearly with temperature? *Aquatic Microbial Ecology*, 32, 307–313.

Montague, J.R., Mangan, R.L. & Starmer, W.T. (1981) Reproductive allocation in the Hawaiian Drosophilidae: egg size and number. *American Naturalist*, 118, 865–871.

Mooney, H.A. & Gulmon, S.L. (1979) Environmental and evolutionary constraints on the photosynthetic pathways of higher plants. In: *Topics in Plant Population Biology* (O.T. Solbrig, S. Jain, G.B. Johnson & P.H. Raven, eds), pp. 316–337. Columbia University Press, New York.

Mooney, H.A., Vitousek, P.M. & Matson, P.A. (1987) Exchange of materials between terrestrial ecosystems and the atmosphere. *Science*, 238, 926–932.

Moran, N.A., Munson, M.A., Baumann, P. & Ishikawa, H. (1993) A molecular clock in endosymbiotic bacteria is calibrated using the insect hosts. *Proceedings of the Royal Society of London, Series B*, 253, 167–171.

Moreau, R.E. (1952) The place of Africa in the palaearctic migration system. *Journal of Animal Ecology*, 21, 250–271.

Morris, C.E. (2002) Self-thinning lines differ with fertility level. *Ecological Research*, 17, 17–28.

Morris, D.W. & Davidson, D.L. (2000) Optimally foraging mice match patch use with habitat differences in fitness. *Ecology*, 81, 2061–2066.

Morris, R.F. (1959) Single-factor analysis in population dynamics. *Ecology*, 40, 580–588.

Morris, R.J., Lewis, O.T. & Godfray, C.J. (2004) Experimental evidence for apparent competition in a tropical forest food web. *Nature*, 428, 310–313.

Morrison, G. & Strong, D.R. Jr. (1981) Spatial variations in egg density and the intensity of parasitism in a neotropical chrysomelid (*Cephaloleia consanguinea*). *Ecological Entomology*, 6, 55–61.

Morrow, P.A. & Olfelt, J.P. (2003) Phoenix clones: recovery after long-term defoliation-induced dormancy. *Ecology Letters*, 6, 119–125.

Mosier, A.R., Bleken, M.A., Chaiwanakupt, P. *et al.* (2002) Policy implications of human-accelerated nitrogen cycling. *Biogeochemistry*, 57/58, 477–516.

Moss, B. (1989) *Ecology of Fresh Waters: man and medium*, 2nd edn. Blackwell Scientific Publications, Oxford.

Moss, B., Balls, H., Booker, I., Manson, K. & Timms, M. (1988) Problems in the construction of a nutrient budget for the River Bure and its Broads (Norfolk) prior to its restoration from eutrophication. In: *Algae and the Aquatic Environment* (F.E. Round, ed.), pp. 326–352. Biopress, Bristol, UK.

Moss, G.D. (1971) The nature of the immune response of the mouse to the bile duct cestode, *Hymenolepis microstoma*. *Parasitology*, 62, 285–294.

Moss, R. & Watson, A. (2001) Population cycles in birds of the grouse family (Tetraonidae). *Advances in Ecological Research*, 32, 53–111.

Mothershead, K. & Marquis, R.J. (2000) Fitness impacts of herbivory through indirect effects on plant-pollinator interactions in *Oenothera macrocarpa*. *Ecology*, 81, 30–40.

Mougeot, F., Redpath, S.M., Leckie, F. & Hudson, P.J. (2003) The effect of aggressiveness on the population dynamics of a territorial bird. *Nature*, 421, 737–739.

Mougeot, F., Redpath, S.M., Moss, R., Matthiopoulos, J. & Hudson, P.J. (2003) Territorial behaviour and population dynamics in red grouse, *Lagopus lagopus scoticus*. I. Population experiments. *Journal of Animal Ecology*, 72, 1073–1082.

Moulton, M.P. & Pimm, S.L. (1986) The extent of competition in shaping an introduced avifauna. In: *Community Ecology* (J. Diamond & T.J. Case, eds), pp. 80–97. Harper & Row, New York.

Mouritsen, K.N. (2002) The parasite-induced surfacing behaviour in the cockle *Austrovenus stutchburyi*: a test of an alternative hypothesis and identification of potential mechanisms. *Parasitology*, 124, 521–528.

Mouritsen, K.N. & Jensen, K.T. (1997) Parasite transmission between soft-bottom invertebrates: temperature mediated infection rates and mortality in *Corophium volutator*. *Marine Ecology Progress Series*, 151, 123–134.

Mouritsen, K.N. & Poulin, R. (2002) Parasitism, community structure and biodiversity in intertidal ecosystems. *Parasitology*, 124, S101–S117.

Mouritsen, K.N. & Poulin, R. (2005) Parasite boosts biodiversity and changes animal community structure by trait-mediated indirect effects. *Oikos*, 108, 344–350.

Mouritsen, K.N., Jensen, T. & Jensen K.T. (1997) Parasites on an intertidal *Corophium*-bed: factors determining the phenology of microphallid trematodes in the intermediate host population of the mud-snail *Hydrobia ulvae* and the amphipod *Corophium volutator*. *Hydrobiologia*, 355, 61–70.

Mouritsen, K.N., Mouritsen, L.T. & Jensen, K.T. (1998) Change of topography and sediment characteristics on an intertidal mud-flat following mass-mortality of the amphipod *Corophium volutator*. *Journal of Marine Biology Association of the United Kingdom*, 78, 1167–1180.

Müller, H.J. (1970) Food distribution, searching success and predator-prey models. In: *The Mathematical Theory of the Dynamics of Biological Populations* (R.W. Hiorns, ed.), pp. 87–101. Academic Press, London.

Muller, M.M., Varama, M., Heinonen, J. & Hallaksela, A. (2002) Influence of insects on the diversity of fungi in decaying spruce wood in managed and natural forests. *Forest Ecology and Management*, 166, 165–181.

Murdie, G. & Hassell, M.P. (1973) Food distribution, searching success and predator-prey models. In: *The Mathematical Theory of the Dynamics of Biological Populations* (R.W. Hiorns, ed.), pp. 87–101. Academic Press, London.

Murdoch, W.W. (1966) Community structure, population control and competition – a critique. *American Naturalist*, 100, 219–226.

Murdoch, W.W. & Briggs, C.J. (1996) Theory for biological control: recent developments. *Ecology*, 77, 2001–2013.

Murdoch, W.W. & Stewart-Oaten, A. (1975) Predation and population stability. *Advances in Ecological Research*, 9, 1–131.

Murdoch, W.W. & Stewart-Oaten, A. (1989) Aggregation by parasitoids and predators: effects on equilibrium and stability. *American Naturalist*, 134, 288–310.

Murdoch, W.W., Avery, S. & Smith, M.E.B. (1975) Switching in predatory fish. *Ecology*, 56, 1094–1105.

Murdoch, W.W., Briggs, C.J., Nisbet, R.M., Gurney, W.S.C. & Stewart-Oaten, A. (1992) Aggregation and stability in meta-population models. *American Naturalist*, 140, 41–58.

Murdoch, W.W., Luck, R.F., Swarbrick, S.L., Walde, S., Yu, D.S. & Reeve, J.D. (1995) Regulation of an insect population under biological control. *Ecology*, 76, 206–217.

Murton, R.K., Isaacson, A.J. & Westwood, N.J. (1966) The relationships between wood pigeons and their clover food supply and the mechanism of population control. *Journal of Applied Ecology*, 3, 55–93.

Murton, R.K., Westwood, N.J. & Isaacson, A.J. (1974) A study of wood-pigeon shooting: the exploitation of a natural animal population. *Journal of Applied Ecology*, 11, 61–81.

Muscatine, L. & Pool, R.R. (1979) Regulation of numbers of intracellular algae. *Proceedings of the Royal Society of London, Series B*, 204, 115–139.

Mwendera, E.J., Saleem, M.A.M & Woldu, Z. (1997) Vegetation response to cattle grazing in the Ethiopian Highlands. *Agriculture, Ecosystems and Environment*, 64, 43–51.

Myers, J.H. (1988) Can a general hypothesis explain population cycles of forest Lepidoptera. *Advances in Ecological Research*, 18, 179–242.

Myers, J.H., Mittermeier, R.A., Mittermeier, C.G., da Fonseca, G.A.B. & Kent, J. (2000) Biodiversity hotspots for conservation priorities. *Nature*, 403, 853–858.

Myers, R.A. (2001) Stock and recruitment: generalizations about maximum reproductive rate, density dependence, and variability using meta-analytic approaches. *ICES Journal of Marine Science*, 58, 937–951.

Myers, R.A., Barrowman, N.J., Hutchings, S.A. & Rosenberg, A.A. (1995) Population dynamics of exploited fish stocks at low population levels. *Science*, 269, 1106–1108.

Nalepa, C.A., Bignell, D.E. & Bandi, C. (2001) Detritivory, coprophagy, and the evolution of digestive mutualisms in Dictyoptera. *Insectes Sociaux*, 48, 194–201.

National Research Council (1990) *Alternative Agriculture*. National Academy of Sciences, Academy Press, Washington, DC.

Nedergaard, J. & Cannon, B. (1990) Mammalian hibernation. *Philosophical Transactions of the Royal Society of London, Series B*, 326, 669–686; also in *Life at Low Temperatures* (R.M. Laws & F. Franks, eds), pp. 153–170. The Royal Society, London.

Neff, J.C., Holland, E.A., Dentener, F.J., McDowell, W.H. & Russell, K.M. (2002) The origin, composition and rates of organic nitrogen deposition: a missing piece of the nitrogen cycle? *Biogeochemistry*, 57/58, 99–136.

NERC (1990) *Our Changing Environment*. Natural Environment Research Council, London. (NERC acknowledges the significant contribution of Fred Pearce to the document.)

Neubert, M.G. & Caswell, H. (2000) Demography and dispersal: calculation and sensitivity analysis of invasion speed for structured populations. *Ecology*, 81, 1613–1628.

Neumann, R.L. (1967) Metabolism in the Eastern chipmunk (*Tamias striatus*) and the Southern flying squirrel (*Glaucomys volans*) during the winter and summer. In: *Mammalian Hibernation III* (K.C. Fisher, A.R. Dawe, C.P. Lyman, E. Schönbaum & F.E. South, eds), pp. 64–74. Oliver & Boyd, Edinburgh and London.

Newsham, K.K., Fitter, A.H. & Watkinson, A.R. (1994) Root pathogenic and arbuscular mycorrhizal mycorrhizal fungi determine fecundity of asymptomatic plants in the field. *Journal of Ecology*, 82, 805–814.

Newsham, K.K., Fitter, A.H. & Watkinson, A.R. (1995) Multifunctionality and biodiversity in arbuscular mycorrhizas. *Trends in Ecology and Evolution*, 10, 407–411.

Newton, I. & Rothery, P. (1997) Senescence and reproductive value in sparrowhawks. *Ecology*, 78, 1000–1008.

Nicholson, A.J. (1954) An outline of the dynamics of animal populations. *Australian Journal of Zoology*, 2, 9–65.

Nicholson, A.J. & Bailey, V.A. (1935) The balance of animal populations. *Proceedings of the Zoological Society of London*, 3, 551–598.

Niklas, K.J., Tiffney, B.H. & Knoll, A.H. (1983) Patterns in vascular land plant diversification. *Nature*, 303, 614–616.

Nilsson, L.A. (1988) The evolution of flowers with deep corolla tubes. *Nature*, 334, 147–149.

Noble, J.C. & Slatyer, R.O. (1981) Concepts and models of succession in vascular plant communities subject to recurrent fire. In: *Fire and the Australian Biota* (A.M. Gill, R.H. Graves & I.R. Noble, eds). Australian Academy of Science, Canberra.

Noble, J.C., Bell, A.D. & Harper, J.L. (1979) The population biology of plants with clonal growth. I. The morphology and structural demography of *Carex arenaria*. *Journal of Ecology*, 67, 983–1008.

Nolan, A.M., Atkinson, P.M. & Bullock, J.M. (1998) Modelling change in the lowland heathlands of Dorset, England. In: *Innovationsin GIS 5* (S. Carver, ed.), pp. 234–243. Taylor & Francis, London.

Normabuena, H. & Piper, G.L. (2000) Impact of *Apion ulicis* Forster on *Ulex europaeus* L. seed dispersal. *Biological Control*, 17, 267–271.

Norton, I.O. & Sclater, J.G. (1979) A model for the evolution of the Indian Ocean and the breakup of Gondwanaland. *Journal of Geophysical Research*, 84, 6803–6830.

Noy-Meir, I. (1975) Stability of grazing systems: an application of predator–prey graphs. *Journal of Ecology*, 63, 459–483.

Nunes, S., Zugger, P.A., Engh, A.L., Reinhart, K.O. & Holekamp, K.E. (1997) Why do female Belding's ground squirrels disperse away from food resources. *Behavioural Ecology and Sociobiology*, 40, 199–207.

Nye, P.H. & Tinker, P.B. (1977) *Solute Movement in the Soil–Root System*. Blackwell Scientific Publications, Oxford.

O'Brien, E.M. (1993) Climatic gradients in woody plant species richness: towards an explanation based on an analysis of southern Africa's woody flora. *Journal of Biogeography*, 20, 181–198.

O'Neill, R.V. (1976) Ecosystem persistence and heterotrophic regulation. *Ecology*, 57, 1244–1253.

Oakeshott, J.G., May, T.W., Gidson, J.B. & Willcocks, D.A. (1982) Resource partitioning in five domestic *Drosophila* species and its relationship to ethanol metabolism. *Australian Journal of Zoology*, 30, 547–556.

Obeid, M., Machin, D. & Harper, J.L. (1967) Influence of density on plant to plant variations in fiber flax, *Linum usitatissimum*. *Crop Science*, 7, 471–473.

Odum, E.P. & Biever, L.J. (1984) Resource quality, mutualism, and energy partitioning in food chains. *American Naturalist*, 124, 360–376.

Ødum, S. (1965) Germination of ancient seeds; floristical observations and experiments with archaeologically dated soil samples. *Dansk Botanisk Arkiv*, 24, 1–70.

Oedekoven, M.A. & Joern, A. (2000) Plant quality and spider predation affects grasshoppers (Acrididae): food-quality-dependent compensatory mortality. *Ecology*, 81, 66–77.

Ogden, J. (1968) *Studies on Reproductive Strategy with Particular Reference to Selected Composites*. PhD thesis, University of Wales, UK.

Oinonen, E. (1967) The correlation between the size of Finnish bracken (*Pteridium aquilinum* (L.) Kuhn) clones and certain periods of site history. *Acta Forestalia Fennica*, 83, 3–96.

Oksanen, L. (1988) Ecosystem organisation: mutualism and cybernetics of plain Darwinian struggle for existence. *American Naturalist*, 131, 424–444.

Oksanen, L., Fretwell, S., Arruda, J. & Niemela, P. (1981) Exploitation ecosystems in gradients of primary productivity. *American Naturalist*, 118, 240–261.

Oksanen, T.A., Koskela, E. & Mappes, T. (2002) Hormonal manipulation of offspring number: maternal effort and reproductive costs. *Evolution*, 56, 1530–1537.

Ollason, J.G. (1980) Learning to forage – optimally? *Theoretical Population Biology*, 18, 44–56.

Ormerod, S.J. (2002) The uptake of applied ecology. *Journal of Applied Ecology*, 39, 1–7.

Ormerod, S.J. (2003) Restoration in applied ecology: editor's introduction. *Journal of Applied Ecology*, 40, 44–50.

Orshan, G. (1963) Seasonal dimorphism of desert and Mediterranean chamaephytes and its significance as a factor in their water economy. In: *The Water Relations of Plants* (A.J. Rutter & F.W. Whitehead, eds), pp. 207–222. Blackwell Scientific Publications, Oxford.

Osawa, A. & Allen, R.B. (1993) Allometric theory explains self-thinning relationships of mountain beech and red pine. *Ecology*, 74, 1020–1032.

Osem, Y., Perevolotsky, A. & Kigel, J. (2002) Grazing effect on diversity of annual plant communities in a semi-arid rangeland: interactions with small-scale spatial and temporal variation in primary productivity. *Journal of Ecology*, 90, 936–946.

Osmundson, D.B., Ryel, R.J., Lamarra, V.L. & Pitlick, J. (2002) Flow–sediment–biota relations: implications for river regulation effects on native fish abundance. *Ecological Applications*, 12, 1719–1739.

Osono, T. & Takeda, H. (2001) Organic chemical and nutrient dynamics in decomposing beech leaf litter in relation to fungal ingrowth and succession during 3-year decomposition processes in a cool temperate deciduous forest in Japan. *Ecological Research*, 16, 649–670.

Ostfeld, R.S. & Keesing, F. (2000) Biodiversity and disease risk: the case of Lyme disease. *Conservation Biology*, 14, 722–728.

Ottersen, G., Planque, B., Belgrano, A., Post, E., Reid, P.C. & Stenseth, N.C. (2001) Ecological effects of the North Atlantic Oscillation. *Oecologia*, 128, 1–14.

Owen-Smith, N. (1987) Pleistocene extinctions: the pivotal role of megaherbivores. *Paleobiology*, 13, 351–362.

Pacala, S.W. (1997) Dynamics of plant communities. In: *Plant Ecology* (M.J. Crawley, ed.), pp. 532–555. Blackwell Science, Oxford.

Pacala, S.W. & Crawley, M.J. (1992) Herbivores and plant diversity. *American Naturalist*, 140, 243–260.

Pacala, S.W. & Hassell, M.P. (1991) The persistence of host–parasitoid associations in patchy environments. II. Evaluation of field data. *American Naturalist*, 138, 584–605.

Pace, M.L., Cole, J.J., Carpenter, S.R. & Kitchell, J.F. (1999) Trophic cascades revealed in diverse ecosystems. *Trends in Ecology and Evolution*, 14, 483–488.

Packer, A. & Clay, K. (2000) Soil pathogens and spatial patterns of seedling mortality in a temperate tree. *Nature*, 404, 278–281.

Paine, R.T. (1966) Food web complexity and species diversity. *American Naturalist*, 100, 65–75.

Paine, R.T. (1979) Disaster, catastrophe and local persistence of the sea palm *Postelsia palmaeformis*. *Science*, 205, 685–687.

Paine, R.T. (1980) Food webs: linkage, interaction strength, and community infrastructure. *Journal of Animal Ecology*, 49, 667–685.

Paine, R.T. (1994) *Marine Rocky Shores and Community Ecology: an experimentalist's perspective*. Ecology Institute, Oldendorf/Luhe, Germany.

Paine, R.T. (2002) Trophic control of production in a rocky intertidal community. *Science*, 296, 736–739.

Palmblad, I.G. (1968) Competition studies on experimental populations of weeds with emphasis on the regulation of population size. *Ecology*, 49, 26–34.

Palmer, T.M., Young, T.P., Stanton, M.L. & Wenk, E. (2000) Short-term dynamics of an acacia ant community in Laikipia, Kenya. *Oecologia*, 123, 425–435.

Palmqvist, K. (2000) Carbon economy in lichens. *New Phytologist*, 148, 11–36.

Park, T. (1948) Experimental studies of interspecific competition. I. Competition between populations of the flour beetle *Tribolium confusum* Duval and *Tribolium castaneum* Herbst. *Ecological Monographs*, 18, 267–307.

Park, T. (1954) Experimental studies of interspecific competiton. II. Temperature, humidity and competition in two species of *Tribolium*. *Physiological Zoology*, 27, 177–238.

Park, T. (1962) Beetles, competition and populations. *Science*, 138, 1369–1375.

Park, T., Mertz, D.B., Grodzinski, W. & Prus, T. (1965) Cannibalistic predation in populations of flour beetles. *Physiological Zoology*, 38, 289–321.

Parker, G.A. (1970) The reproductive behaviour and the nature of sexual selection in *Scatophaga stercoraria* L. (Diptera: Scatophagidae) II. The fertilization rate and the spatial and temporal relationships of each sex around the site of mating and oviposition. *Journal of Animal Ecology*, 39, 205–228.

Parker, G.A. (1984) Evolutionarily stable strategies. In: *Behavioral Ecology: an evolutionary approach*, 2nd edn (J.R. Krebs & N.B. Davies, eds), pp. 30–61. Blackwell Scientific Publications, Oxford.

Parker, G.A. & Stuart, R.A. (1976) Animal behaviour as a strategy optimizer: evolution of resource assessment strategies and optimal emigration thresholds. *American Naturalist*, 110, 1055–1076.

Parmesan, C. & Yohe, G. (2003) A globally coherent fingerprint of climate change impacts across natural systems. *Nature*, 421, 38–42.

Partridge, L. & Farquhar, M. (1981) Sexual activity reduces lifespan of male fruitflies. *Nature*, 294, 580–581.

Pastoret, P.P. & Brochier, B. (1999) Epidemiology and control of fox rabies in Europe. *Vaccine*, 17, 1750–1754.

Pate, J.S. (1994) The mycorrhizal association: just one of many nutrient acquiring specializations in natural ecosystems. *Plant and Soil*, 159, 1–10.

Paterson, S. & Viney, M.E. (2002) Host immune responses are necessary for density dependence in nematode infections. *Parasitology*, 125, 283–292.

Pauly, D. & Christensen, V. (1995) Primary production required to sustain global fisheries. *Nature*, 374, 255–257.

Pavia, H. & Toth, G.B. (2000) Inducible chemical resistance to herbivory in the brown seaweed *Ascophyllum nodosum*. *Ecology*, 81, 3212–3225.

Payne, C.C. (1988) Pathogens for the control of insects: where next? *Philosophical Transactions of the Royal Society of London, Series B*, 318, 225–248.

Payne, R.B. (1977) The ecology of brood parasitism in birds. *Annual Review of Ecology and Systematics*, 8, 1–28.

Pearcy, R.W., Björkman, O., Caldwell, M.M., Keeley, J.E., Monson, R.K. & Strain, B.R. (1987) Carbon gain by plants in natural environments. *Bioscience*, 37, 21–29.

Pearl, R. (1928) *The Rate of Living*. Knopf, New York.

Pellmyr, O. & Leebens-Mack, J. (1999) Reversal of mutualism as a mechanism for adaptive radiation in yucca moths. *American Naturalist*, 156, S62–S76.

Penn, A.M., Sherwin, W.B., Gordon, G., Lunney, D., Melzer, A. & Lacy, R.C. (2000) Demographic forecasting in koala conservation. *Conservation Biology*, 14, 629–638.

Pennings, S.C. & Callaway, R.M. (2002) Parasitic plants: parallels and contrasts with herbivores. *Oecologia*, 131, 479–489.

Perlman, S.J. & Jaenike, J. (2001) Competitive interactions and persistence of two nematode species that parasitize *Drosophila recens*. *Ecology Letters*, 4, 577–584.

Perrins, C.M. (1965) Population fluctuations and clutch size in the great tit, *Parus major* L. *Journal of Animal Ecology*, 34, 601–647.

Perry, J.N., Smith, R.H., Woiwod, I.P. & Morse, D.R. (eds) (2000) *Chaos in Real Data*. Kluwer, Dordrecht.

Peters, R.H. (1983) *The Ecological Implications of Body Size*. Cambridge University Press, Cambridge, UK.

Petit, J.R., Jouzel, J., Raynaud, D. *et al.* (1999) Climate and atmospheric history of the past 420,000 years from the Vostok ice core, Antarctica. *Nature*, 399, 429–436.

Petranka, J.W. (1989) Chemical interference competition in tadpoles: does it occur outside laboratory aquaria? *Copeia*, 1989, 921–930.

Petren, K. & Case, T.J. (1996) An experimental demonstration of exploitation competition in an ongoing invasion. *Ecology*, **77**, 118–132.

Petren, K., Grant, B.R. & Grant, P.R. (1999) A phylogeny of Darwin's finches based on microsatellite DNA variation. *Proceedings of the Royal Society of London, Series B*, 266, 321–329.

Pettifor, R.A., Perrins, C.M. & McCleery, R.H. (2001) The individual optimization of fitness: variation in reproductive output, including clutch size, mean nestling mass and offspring recruitment, in manipulated broods of great tits *Parus major*. *Journal of Animal Ecology*, 70, 62–79.

Pianka, E.R. (1970) On *r*- and *k*-selection. *American Naturalist*, 104, 592–597.

Pianka, E.R. (1973) The structure of lizard communities. *Annual Review of Ecology and Systematics*, 4, 53–74.

Pickett, J.A. (1988) Integrating use of beneficial organisms with chemical crop protection. *Philosophical Transactions of the Royal Society of London, Series B*, 318, 203–211.

Pickett, S.T.A. & White, P.S. (eds) (1985) *The Ecology of Natural Disturbance as Patch Dynamics*. Academic Press, New York.

Piersma, T. & Davidson, N.C. (1992) The migrations and annual cycles of five subspecies of knots in perspective. In: *The Migration of Knots, Wader Study Group Bulletin 64, Supplement, April 1992*, pp. 187–197. Joint Nature Conservation Committee, Publications Branch, Peterborough, UK.

Pimentel, D., Lach, L., Zuniga, R. & Morrison, D. (2000) Environmental and economic costs of nonindigenous species in the United States. *BioScience*, 50, 53–65.

Pimm, S.L. (1979) Complexitiy and stability: another look at MacArthur's original hypothesis. *Oikos*, 33, 351–357.

Pimm, S.L. (1982) *Food Webs*. Chapman & Hall, London.

Pimm, S.L. (1991) *The Balance of Nature: ecological issues in the conservation of species and communities*. University of Chicago Press, Chicago and London.

Pimm, S.L. & Lawton, J.H. (1977) The number of trophic levels in ecological communities. *Nature*, 275, 542–544.

Pimm, S.L. & Lawton, J.H. (1980) Are food webs divided into compartments? *Journal of Animal Ecology*, 49, 879–898.

Pimm, S.L. & Rice, J.C. (1987) The dynamics of multi-species, multi-life-stage models of aquatic food webs. *Theoretical Population Biology*, 32, 303–325.

Piraino, S., Fanelli, G. & Boero, F. (2002) Variability of species' roles in marine communities: change of paradigms for conservation priorities. *Marine Biology*, 140, 1067–1074.

Pisek, A., Larcher, W., Vegis, A. & Napp-Zin, K. (1973) The normal temperature range. In: *Temperature and Life* (H. Precht, J. Christopherson, H. Hense & W. Larcher, eds), pp. 102–194. Springer-Verlag, Berlin.

Pitcher, T.J. (2001) Fisheries managed to rebuild ecosystems? Reconstructing the past to salvage the future. *Ecological Applications*, 11, 601–617.

Pitcher, T.J. & Hart, P.J.B. (1982) *Fisheries Ecology*. Croom Helm, London.

Plante, C.J., Jumars, P.A. & Baross, J.A. (1990) Digestive associations between marine detritivores and bacteria. *Annual Review of Ecology and Systematics*, 21, 93–127.

Playfair, J.H.L. (1996) *Immunology at a Glance*, 6th edn. Blackwell Science, Oxford.

Podoler, H. & Rogers, D.J. (1975) A new method for the identification of key factors from life-table data. *Journal of Animal Ecology*, 44, 85–114.

Pojar, T.M. (1981) A management perspective of population modelling. In: *Dynamics of Large Mammal Populations* (D.W. Fowler & T.D. Smith, eds), pp. 241–261. Wiley-Interscience, New York.

Polis, G.A. (1991) Complex trophic interactions in deserts: an empirical critique of food-web theory. *American Naturalist*, 138, 123–155.

Polis, G.A. (1999) Why are parts of the world green? Multiple factors control productivity and the distribution of biomass. *Oikos*, 86, 3–15.

Polis, G.A. & Strong, D. (1996) Food web complexity and community dynamics. *American Naturalist*, 147, 813–846.

Polis, G.A., Anderson, W.B. & Holt, R.D. (1997) Towards an integration of landscape and food web ecology: the dynamics of spatially subsidized food webs. *Annual Review of Ecology and Systematics*, 28, 289–316.

Polis, G.A., Sears, A.L.W., Huxel, G.R., Strong, D.R. & Maron, J. (2000) When is a trophic cascade a trophic cascade? *Trends in Ecology and Evolution*, 15, 473–475.

Ponge, J.-F. (2000) Vertical distribution of Collembola (Hexapoda) and their food resources in organic horizons of beech forests. *Biology and Fertility of Soils* 32, 508–522.

Pope, S.E., Fahrig, L. & Merriam, H.G. (2000) Landscape complementation and metapopulation effects on leopard frog populations. *Ecology*, 81, 2498–2508.

Possingham, H.P., Andelman, S.J., Burgman, M.A., Medellin, R.A., Master, L.L. & Keith, D.A. (2002) Limits to the use of threatened species lists. *Trends in Ecology and Evolution*, 17, 503–507.

Post, D.M., Pace, M.L. & Hairston, N.G. Jr. (2000) Ecosystem size determines food-chain length in lakes. *Nature*, 405, 1047–1049.

Post, R.M. (2002) The long and the short of food-chain length. *Trends in Ecology and Evolution*, 17, 269–277.

Poulson, M.E. & DeLucia, E.H. (1993) Photosynthetic and structural acclimation to light direction in vertical leaves of *Silphium terebinthaceum*. *Oecologia*, 95, 393–400.

Power, M.E. (1990) Effects of fish in river food webs. *Science*, 250, 411–415.

Power, M.E., Tilman, D., Estes, J.A. *et al.* (1996) Challenges in the quest for keystones. *Bioscience*, 46, 609–620.

Prach, K., Bartha, S., Joyce, C.A., Pysek, P., van Diggelen, R. & Wiegleb, G. (2001) The role of spontaneous vegetation succession in ecosystem restoration: a perspective. *Applied Vegetation Science*, 4, 111–114.

Prance, G.T. (1987) Biogeography of neotropical plants. In: *Biogeography and Quaternary History of Tropical America* (T.C. Whitmore & G.T. Prance, eds), pp. 46–65. Oxford Monographs on Biogeography, No. 3. Clarendon Press, Oxford.

Praw, J.C. & Grant, J.W.A. (1999) Optimal territory size in the convict cichlid. *Behaviour*, 136, 1347–1363.

Pressey, R.L. (1995) Conservation reserves in New South Wales: crown jewels or leftovers? *Search*, 26, 47–51.

Pressey, R.L., Humphries, C.J., Margules, C.R., Vane-Wright, R.I. & Williams, P.H. (1993) Beyond opportunism: key principles for systematic reserve selection. *Trends in Ecology and Evolution*, 8, 124–128.

Preston, F.W. (1962) The canonical distribution of commoness and rarity. *Ecology*, 43, 185–215, 410–432.

Preszler, R.W. & Price, P.W. (1993) The influence of *Salix* leaf abscission on leaf-miner survival and life-history. *Ecological Entomology*, 18, 150–154.

Price, P.W. (1980) *Evolutionary Biology of Parasites*. Princeton University Press, Princeton, NJ.

Primack, R.B. (1993) *Essentials of Conservation Biology*. Sinauer Associates, Sunderland, MA.

Prior, H. & Johnes, P.J. (2002) Regulation of surface water quality in a cretaceous chalk catchment, UK: an assessment of the relative importance of instream and wetland processes. *Science of the Total Environment*, 282/283, 159–174.

Proctor, M., Yeo, P. & Lack, A. (1996) *The Natural History of Pollination*. Harper Collins, London.

Proulx, M. & Mazumder, A. (1998) Reversal of grazing impact on plant species richness in nutrient-poor vs nutrient-rich ecosystems. *Ecology*, 79, 2581–2592.

Pulliam, H.R. (1988) Sources, sinks and population regulation. *American Naturalist*, 132, 652–661.

Pulliam, H.R. & Caraco, T. (1984) Living in groups: is there an optimal group size? In: *Behavioural Ecology: an evolutionary approach*, 2nd edn (J.R. Krebs & N.B. Davies, eds), pp. 122–147. Blackwell Scientific Publications, Oxford.

Punnett, R.C. (1933) Inheritance of egg-colour in the parasitic cuckoos. *Nature*, 132, 892.

Putman, R.J. (1978) Patterns of carbon dioxide evolution from decaying carrion. Decomposition of small carrion in temperate systems. *Oikos*, 31, 47–57.

Putman, R.J. (1983) *Carrion and Dung: the decomposition of animal wastes*. Edward Arnold, London.

Pyke, G.H. (1982) Local geographic distributions of bumblebees near Crested Butte, Colorado: competition and community structure. *Ecology*, 63, 555–573.

Pywell, R.F., Bullock, J.M., Hopkins, A. *et al.* (2002) Restoration of species-rich grassland on arable land: assessing the limiting processes using a multi-site experiment. *Journal of Applied Ecology*, 39, 294–309.

Pywell, R.F., Bullock, J.M., Roy, D.B., Warman, L., Walker, K.J. & Rothery, P. (2003) Plant traits as predictors of performance in ecological restoration. *Journal of Applied Ecology*, 40, 65–77.

Quinn, J.F. & Harrison, S.P. (1988) Effects of habitat fragmentation and isolation on species richness – evidence from biogeographic patterns. *Oecologia*, 75, 132–140.

Raffaelli, D. & Hawkins, S. (1996) *Intertidal Ecology*. Kluwer, Dordrecht.

Rahbek, C. (1995) The elevational gradient of species richness: a uniform pattern? *Ecography*, 18, 200–205.

Rainey, P.B. & Trevisano, M. (1998) Adaptive radiation in a heterogeneous environment. *Nature*, 394, 69–72.

Ramirez, O.A. & Saunders, J.L. (1999) Estimating economic thresholds for pest control: an alternative procedure. *Journal of Economic Entomology*, 92, 391–401.

Randall, M.G.M. (1982) The dynamics of an insect population throughout its altitudinal distribution: *Coleophora alticolella* (Lepidoptera) in northern England. *Journal of Animal Ecology*, 51, 993–1016.

Ranta, E., Kaitala, V., Alaja, S. & Tesar, D. (2000) Nonlinear dynamics and the evolution of semelparous and iteroparous reproductive strategies. *American Naturalist*, 155, 294–300.

Rapport, D.J., Costanza, R. & McMichael, A.J. (1998) Assessing ecosystem health. *Trends in Ecology and Evolution*, 13, 397–402.

Rätti, O., Dufva, R. & Alatalo, R.V. (1993) Blood parasites and male fitness in the pied flycatcher. *Oecologia*, 96, 410–414.

Raunkiaer, C. (1934) *The Life Forms of Plants*. Oxford University Press, Oxford. (Translated from the original published in Danish, 1907.)

Raup, D.M. (1978) Cohort analysis of generic survivorship. *Paleobiology*, 4, 1–15.

Rausher, M.D. (1978) Search image for leaf shape in a butterfly. *Science*, 200, 1071–1073.

Rausher, M.D. (2001) Co-evolution and plant resistance to natural enemies. *Nature*, 411, 857–864.

Raushke, E., Haar, T.H. von der, Bardeer, W.R. & Paternak, M. (1973) The annual radiation of the earth–atmosphere system during 1969–70 from Nimbus measurements. *Journal of the Atmospheric Science*, 30, 341–346.

Ray, J.C. & Sunquist, M.E. (2001) Trophic relations in a community of African rainforest carnivores. *Oecologia*, 127, 395–408.

Read, A.F. & Harvey, P.H. (1989) Life history differences among the eutherian radiations. *Journal of Zoology*, 219, 329–353.

Read, D.J. (1996) The structure and function of the ericoid mycorrhizal root. *Annals of Botany*, 77, 365–374.

Redfern, M., Jones, T.H. & Hassell, M.P. (1992) Heterogeneity and density dependence in a field study of a tephritid–parasitoid interaction. *Ecological Entomology*, 17, 255–262.

Reece, C.H. (1985) The role of the chemical industry in improving the effectiveness of agriculture. *Philosophical Transactions of the Royal Society of London, Series B*, 310, 201–213.

Reed, D.H., O'Grady, J.J., Brook, B.W., Ballou, J.D. & Frankham, R. (2003) Estimates of minimum viable population sizes for vertebrates and factors influencing those estimates. *Biological Conservation*, 113, 23–34.

Rees, M., Condit, R., Crawley, M., Pacala, S. & Tilman, D. (2001) Long-term studies of vegetation dynamics. *Science*, 293, 650–655.

Reeve, J.D., Rhodes, D.J. & Turchin, P. (1998) Scramble competition in the southern pine beetle, *Dendroctonus frontalis*. *Ecological Entomology*, 23, 433–443.

Reganold, J.P., Glover, J.D., Andrews, P.K. & Hinman, H.R. (2001) Sustainability of three apple production systems. *Nature*, 410, 926–929.

Reich, P.B., Grigal, D.F., Aber, J.D. & Gower, S.T. (1997) Nitrogen mineralization and productivity in 50 hardwood and conifer stands on diverse soils. *Ecology*, 78, 335–347.

Reid, W.V. & Miller, K.R. (1989) *Keeping Options Alive: the scientific basis for conserving biodiversity.* World Resources Insititute, Washington, DC.

Rejmanek, M. & Richardson, D.M. (1996) What attributes make some plant species more invasive? *Ecology*, 77, 1655–1660.

Reunanen, P., Monkkonen, M. & Nikula, A. (2000) Managing boreal forest landscapes for flying squirrels. *Conservation Biology*, 14, 218–227.

Reznick, D.N. (1982) The impact of predation on life history evolution in Trinidadian guppies: genetic basis of observed life history patterns. *Evolution*, 36, 1236–1250.

Reznick, D.N. (1985) Cost of reproduction: an evaluation of the empirical evidence. *Oikos*, 44, 257–267.

Reznick, D.N., Bryga, H. & Endler, J.A. (1990) Experimentally induced life history evolution in a natural population. *Nature*, 346, 357–359.

Rhoades, D.F. & Cates, R.G. (1976) Towards a general theory of plant antiherbivore chemistry. *Advances in Phytochemistry*, 110, 168–213.

Ribas, C.R., Schoereder, J.H., Pic, M. & Soares, S.M. (2003) Tree heterogeneity, resource availability, and larger scale processes regulating arboreal ant species richness. *Austral Ecology*, 28, 305–314.

Ribble, D.O. (1992) Dispersal in a monogamous rodent, *Peromyscus californicus*. *Ecology*, 73, 859–866.

Ricciardi, A. (2001) Facilitative interactions among aquatic invaders: is an 'invasional meltdown' occurring in the Great Lakes? *Canadian Journal of Fisheries and Aquatic Science*, 58, 2513–2525.

Ricciardi, A. & MacIsaac, H.J. (2000) Recent mass invasion of the North American Great Lakes by Ponto-Caspian species. *Trends in Ecology and Evolution*, 15, 62–65.

Richards, O.W. & Waloff, N. (1954) Studies on the biology and population dynamics of British grasshoppers. *Anti-Locust Bulletin*, 17, 1–182.

Richman, A.D., Case, T.J. & Schwaner, T.D. (1988) Natural and unnatural extinction rates of reptiles on islands. *American Naturalist*, 131, 611–630.

Rickards, J., Kelleher, M.J. & Storey, K.B. (1987) Strategies of freeze avoidance in larvae of the goldenrod gall moth *Epiblema scudderiana*: winter profiles of a natural population. *Journal of Insect Physiology*, 33, 581–586.

Ricklefs, R.E. & Lovette, I.J. (1999) The role of island area *per se* and habitat diversity in the species–area relationships of four Lesser Antillean faunal groups. *Journal of Animal Ecology*, 68, 1142–1160.

Ridley, M. (1993) *Evolution.* Blackwell Science, Boston.

Rigler, F.H. (1961) The relation between concentration of food and feeding rate of *Daphnia magna* Straus. *Canadian Journal of Zoology*, 39, 857–868.

Riis, T. & Sand-Jensen, K. (1997) Growth reconstruction and photosynthesis of aquatic mosses: influence of light, temperature and carbon dioxide at depth. *Journal of Ecology*, 85, 359–372.

Ringel, M.S., Hu, H.H. & Anderson, G. (1996) The stability and persistence of mutualisms embedded in community interactions. *Theoretical Population Biology*, 50, 281–297.

Robertson, J.H. (1947) Responses of range grasses to different intensities of competition with sagebrush (*Artemisia tridentata* Nutt.). *Ecology*, 28, 1–16.

Robinson, S.P., Downton, W.J.S. & Millhouse, J.A. (1983) Photosynthesis and ion content of leaves and isolated chloroplasts of salt-stressed spinach. *Plant Physiology*, 73, 238–242.

Rohani, P., Godfray, H.C.J. & Hassell, M.P. (1994) Aggregation and the dynamics of host–parasitoid systems: a discrete-generation model with within-generation redistribution. *American Naturalist*, 144, 491–509.

Rombough, P. (2003) Modelling developmental time and temperature. *Nature*, 424, 268–269.

Room, P.M., Harley, K.L.S., Forno, I.W. & Sands, D.P.A. (1981) Successful biological control of the floating weed *Salvinia*. *Nature*, 294, 78–80.

Room, P.M., Maillette, L. & Hanan, J.S. (1994) Module and metamer dynamics and virtual plants. *Advances in Ecological Research*, 25, 105–157.

Root, R. (1967) The niche exploitation pattern of the blue-grey gnatcatcher. *Ecological Monographs*, 37, 317–350.

Rose, M.R., Service, P.M. & Hutchinson, E.W. (1987) Three approaches to trade-offs in life-history evolution. In: *Genetic Constraints on Evolution* (V. Loeschke, ed.). Springer, Berlin.

Rosenthal, G.A., Dahlman, D.L. & Janzen, D.H. (1976) A novel means for dealing with L-canavanine, a toxic metabolite. *Science*, 192, 256–258.

Rosenzweig, M.L. (1971) Paradox of enrichment: destabilization of exploitation ecosystems in ecological time. *Science*, 171, 385–387.

Rosenzweig, M.L. (2003) How to reject the area hypothesis of latitudinal gradients. In: *Macroecology: concepts and consequences* (T.M. Blackburn & K.J. Gaston, eds), pp. 87–106. Blackwell Publishing, Oxford.

Rosenzweig, M.L. & MacArthur, R.H. (1963) Graphical representation and stability conditions of predator–prey interactions. *American Naturalist*, 97, 209–223.

Rosenzweig, M.L. & Sandlin, E.A. (1997) Species diversity and latitudes: listening to area's signal. *Oikos*, 80, 172–176.

Rosenzweig, M.L. & Ziv, Y. (1999) The echo pattern in species diversity: pattern and process. *Ecography*, 22, 614–628.

Ross, K., Cooper, N., Bidwell, J.R. & Elder, J. (2002) Genetic diversity and metal tolerance of two marine species: a comparison between populations from contaminated and reference sites. *Marine Pollution Bulletin*, 44, 671–679.

Roth, J.D. (2002) Temporal variability in arctic fox diet as reflected in stable-carbon isotopes: the importance of sea ice. *Oecologia*, 133, 70–77.

Roughan, M. & Middleton, J.H. (2002) A comparison of observed upwelling mechanisms off the east coast of Australia. *Continental Shelf Research*, 22, 2551–2572.

Roush, R.T. & McKenzie, J.A. (1987) Ecological genetics of insecticide and acaricide resistance. *Annual Review of Entomology*, 32, 361–380.

Rowan, R., Knowlton, N., Baker, A. & Jara, J. (1997) Landscape ecology of algal symbionts creates variation in episodes of coral bleaching. *Nature*, 388, 265–269.

Rowe, C.L. (2002) Differences in maintenance energy expenditure by two estuarine shrimp (*Palaemonetes pugio* and *P. vulgaris*) that may permit partitioning of habitats by salinity. *Comparative Biochemistry and Physiology Part A*, 132, 341–351.

Royama, T. (1992) *Analytical Population Dynamics*. Chapman & Hall, London.

Ruiters, C. & McKenzie, B. (1994) Seasonal allocation and efficiency patterns of biomass and resources in the perennial geophyte *Sparaxis grandiflora* subspecies *fimbriata* (Iridaceae) in lowland coastal Fynbos, South Africa. *Annals of Botany*, 74, 633–646.

Rundle, H.D., Nagel, L., Boughman, J.W. & Schluter, D. (2000) Natural selection and parallel speciation in sympatric sticklebacks. *Science*, 287, 306–308.

Ryan, C.A. & Jagendorf, A. (1995) Self defense by plants. *Proceedings of the National Academy of Science of the USA*, 92, 4075.

Ryan, C.A., Lamb, C.J., Jagendorf, A.T. & Kolattukudy, P.E. (eds) (1995) Self-defense by plants: induction and signalling pathways. *Proceedings of the National Academy of Science of the USA*, 92, 4075–4205.

Sackville Hamilton, N.R., Matthew, C. & Lemaire, G. (1995) In defence of the −3/2 boundary line: a re-evaluation of self-thinning concepts and status. *Annals of Botany*, 76, 569–577.

Sackville Hamilton, N.R., Schmid, B. & Harper, J.L. (1987) Life history concepts and the population biology of clonal organisms. *Proceedings of the Royal Society of London, Series B*, 232, 35–57.

Sale, P.F. (1977) Maintenance of high diversity in coral reef fish communities. *American Naturalist*, 111, 337–359.

Sale, P.F. (1979) Recruitment, loss and coexistence in a guild of territorial coral reef fishes. *Oecologia*, 42, 159–177.

Salisbury, E.J. (1942) *The Reproductive Capacity of Plants*. Bell, London.

Salonen, K., Jones, R.I. & Arvola, L. (1984) Hypolimnetic retrieval by diel vertical migrations of lake phytoplankton. *Freshwater Biology*, 14, 431–438.

Saloniemi, I. (1993) An environmental explanation for the character displacement pattern in *Hydrobia* snails. *Oikos*, 67, 75–80.

Salt, D.E., Smith, R.D. & Raskin, I. (1998) Phytoremediation. *Annual Review of Plant Physiology and Plant Molecular Biology*, 49, 643–668.

Samelius, G. & Alisauskas, R.T. (2000) Foraging patterns of arctic foxes at a large arctic goose colony. *Arctic*, 53, 279–288.

Sanchez-Pinero, F. & Polis, G.A. (2000) Bottom-up dynamics of allochthonous input: direct and indirect effects of seabirds on islands. *Ecology*, 81, 3117–3132.

Sanders, N.J., Moss, J. & Wagner, D. (2003) Patterns of ant species richness along elevational gradients in an arid ecosystem. *Global Ecology and Biogeography*, 12, 93–102.

Saunders, M.A. (1999) Earth's future climate. *Philosophical Transactions of the Royal Society of London, Series A*, 357, 3459–3480.

Savenko, V.S. (2001) Global hydrological cycle and geochemical balance of phosphorus in the ocean. *Oceanology*, 41, 360–366.

Savidge, J.A. (1987) Extinction of an island forest avifauna by an introduced snake. *Ecology*, 68, 660–668.

Sax, D.F. & Gaines, S.D. (2003) Species diversity: from global decreases to local increases. *Trends in Ecology and Evolution*, 18, 561–566.

Schaefer, M.B. (1954) Some aspects of the dynamics of populations important to the management of marine fisheries. *Bulletin of the Inter-American Tropical Tuna Commission*, 1, 27–56.

Schaffer, W.M. (1974) Optimal reproductive effort in fluctuating environments. *American Naturalist*, 108, 783–790.

Schaffer, W.M. & Kot, M. (1986) Chaos in ecological systems: the coals that Newcastle forgot. *Trends in Ecology and Evolution*, 1, 58–63.

Schall, J.J. (1992) Parasite-mediated competition in *Anolis* lizards. *Oecologia*, 92, 58–64.

Schimel, D. & Baker, D. (2002) The wildfire factor. *Nature*, 420, 29–30.

Schindler, D.E. (1977) Evolution of phosphorus limitation in lakes. *Science*, 195, 260–262.

Schindler, D.E. & Scheuerell, M.D. (2002) Habitat coupling in lake ecosystems. *Oikos*, 98, 177–189.

Schindler, D.E., Knapp, K.A. & Leavitt, P.R. (2001) Alteration of nutrient cycles and algal production resulting from fish introductions into mountain lakes. *Ecosystems*, 4, 308–321.

Schluter, D. (2001) Ecology and the origin of species. *Trends in Ecology and Evolution*, 16, 372–380.

Schluter, D. & McPhail, J.D. (1992) Ecological character displacement and speciation in sticklebacks. *American Naturalist*, 140, 85–108.

Schluter, D. & McPhail, J.D. (1993) Character displacement and replicate adaptive radiation. *Trends in Ecology and Evolution*, 8, 197–200.

Schlyter, F. & Anderbrant, O. (1993) Competition and niche separation between two bark beetles: existence and mechanisms. *Oikos*, 68, 437–447.

Schmidt-Nielsen, K. (1984) *Scaling: why is animal size so important?* Cambridge University Press, Cambridge, UK.

Schmitt, R.J. (1987) Indirect interactions between prey: apparent competition, predator aggregation, and habitat segregation. *Ecology*, 68, 1887–1897.

Schmitz, O.J., Hamback, P.A. & Beckerman, A.P. (2000) Trophic cascades in terrestrial systems: a review of the effects of carnivore removals on plants. *American Naturalist*, 155, 141–153.

Schoener, T.W. (1974) Resource partitioning in ecological communities. *Science*, 185, 27–39.

Schoener, T.W. (1983) Field experiments on interspecific competition. *American Naturalist*, 122, 240–285.

Schoener, T.W. (1989) Food webs from the small to the large. *Ecology*, 70, 1559–1589.

Schoenly, K., Beaver, R.A. & Heumier, T.A. (1991) On the trophic relations of insects: a food-web approach. *American Naturalist*, 137, 597–638.

Schultz, J.C. (1988) Plant responses induced by herbivory. *Trends in Ecology and Evolution*, 3, 45–49.

Schüßler, A., Schwarzott, D. & Walker, C. (2001) A new fungal phylum, the *Glomeromycota*: phylogeny and evolution. *Mycological Research*, 105, 1413–1421.

Schwarz, C.J. & Seber, G.A.F. (1999) Estimating animal abundance: review III. *Statistical Science* 14, 427–456.

Scott, M.P. (1998) The ecology and behaviour of burying beetles. *Annual Review of Entomology*, 43, 595–618.

Scurlock, J.M.O., Johnson, K. & Olson, R.J. (2002) Estimating net primary productivity from grassland biomass dynamic measurements. *Global Change Biology*, 8, 736–753.

Semere, T. & Froud-Williams, R.J. (2001) The effect of pea cultivar and water stress on root and shoot competition between vegetative plants of maize and pea. *Journal of Applied Ecology*, 38, 137–145.

Severini, M., Baumgärtner, J. & Limonta, L. (2003) Parameter estimation for distributed delay based population models from laboratory data: egg hatching of *Oulema duftschmidi*. *Ecological Modelling*, 167, 233–246.

Sexstone, A.Y., Parkin, T.B. & Tiedje, J.M. (1985) Temporal response of soil denitrification rates to rainfall and irrigation. *Soil Science Society of America Journal*, 49, 99–103.

Shankar Raman, T., Rawat, G.S. & Johnsingh, A.J.T. (1998) Recovery of tropical rainforest avifauna in relation to vegetation succession following shifting cultivation in Mizoram, north-east India. *Journal of Applied Ecology*, 35, 214–231.

Shaw, D.J. & Dobson, A.P. (1995) Patterns of macroparasite abundance and aggregation in wildlife populations: a quantitative review. *Parasitology*, 111, S111–S133.

Shea, K. & Chesson, P. (2002) Community ecology theory as a framework for biological invasions. *Trends in Ecology and Evolution*, 17, 170–177.

Shea, K. & Kelly, D. (1998) Estimating biocontrol agent impact with matrix models: *Carduus nutans* in New Zealand. *Ecological Applications*, 8, 824–832.

Sherman, R.E., Fahey, T.J. & Battles, J.J. (2000) Small-scale disturbance and regeneration dynamics in a neotropical mangrove forest. *Journal of Ecology*, 88, 165–178.

Sherratt, T.N. (2002) The coevolution of warning signals. *Proceedings of the Royal Society of London, Series B*, 269, 741–746.

Shirley, B.W. (1996) Flavonoid biosynthesis: 'new' functions for an old pathway. *Trends in Plant Science*, 1, 377–382.

Shorrocks, B. & Rosewell, J. (1987) Spatial patchiness and community structure: coexistence and guild size of drosophilids on ephemeral resources. In: *Organization of Communities: past and present* (J.H.R. Gee & P.S. Giller, eds), pp. 29–52. Blackwell Scientific Publications, Oxford.

Shorrocks, B., Rosewell, J. & Edwards, K. (1990) Competition on a divided and ephemeral resource: testing the assumptions. II. Association. *Journal of Animal Ecology*, 59, 1003–1017.

Shykoff, J.A. & Bucheli, E. (1995) Pollinator visitation patterns, floral rewards and the probability of transmission of *Micro-botryum violaceum*, a venereal disease of plants. *Journal of Ecology*, 83, 189–198.

Sibly, R.M. & Calow, P. (1983) An integrated approach to life-cycle evolution using selective landscapes. *Journal of Theoretical Biology*, 102, 527–547.

Sibly, R.M. & Hone, J. (2002) Population growth rate and its determinants: an overview. *Philosophical Transactions of the Royal Society of London, Series B*, 357, 1153–1170.

Sibly, R.M. & Smith, R.H. (1998) Identifying key factors using λ-contribution analysis. *Journal of Animal Ecology*, 67, 17–24.

Sih, A. (1982) Foraging strategies and the avoidance of predation by an aquatic insect, *Notonecta hoffmanni*. *Ecology*, 63, 786–796.

Sih, A. & Christensen, B. (2001) Optimal diet theory: when does it work, and when and why does it fail? *Animal Behaviour*, 61, 379–390.

Sih, A., Crowley, P., McPeek, M., Petranka, J. & Strohmeier, K. (1985) Predation, competition and prey communities: a review of field experiments. *Annual Review of Ecology and Systematics*, 16, 269–311.

Silulwane, N.F., Richardson, A.J., Shillington, F.A. & Mitchell-Innes, B.A. (2001) Identification and classification of vertical chlorophyll patterns in the Benguela upwelling system and Angola-Benguela front using an artificial neural network. *South African Journal of Marine Science*, 23, 37–51.

Silvertown, J.W. (1980) The evolutionary ecology of mast seeding in trees. *Biological Journal of the Linnean Society*, 14, 235–250.

Silvertown, J.W. (1982) *Introduction to Plant Population Ecology*. Longman, London.

Silvertown, J.W., Franco, M., Pisanty, I. & Mendoza, A. (1993) Comparative plant demography – relative importance of life-cycle components to the finite rate of increase in woody and herbaceous perennials. *Journal of Ecology*, 81, 465–476.

Silvertown, J.W., Holtier, S., Johnson, J. & Dale, P. (1992) Cellular automaton models of interspecific competition for space – the effect of pattern on process. *Journal of Ecology*, 80, 527–533.

Simberloff, D.S. (1976) Experimental zoogeography of islands: effects of island size. *Ecology*, 57, 629–648.

Simberloff, D. (1998) Small and declining populations. In: *Conservation Science and Action* (W.J. Sutherland, ed.), pp. 116–134. Blackwell Science, Oxford.

Simberloff, D. (2003) How much information on population biology is needed to manage introduced species? *Conservation Biology*, 17, 83–92.

Simberloff, D. & Von Holle, B. (1999) Positive interactions of non-indigenous species: invasional meltdown? *Biological Invasions*, 1, 21–32.

Simberloff, D.S. & Wilson, E.O. (1969) Experimental zoogeography of islands: the colonization of empty islands. *Ecology*, 50, 278–296.

Simberloff, D., Dayan, T., Jones, C. & Ogura, G. (2000) Character displacement and release in the small Indian mongoose, *Herpestes javanicus*. *Ecology*, 81, 2086–2099.

Simo, R. (2001) Production of atmospheric sulfur by oceanic plankton: biogeochemical, ecological and evolutionary links. *Trends in Ecology and Evolution*, 16, 287–294.

Simon, K.S. & Townsend, C.R. (2003) The impacts of freshwater invaders at different levels of ecological organisation, with emphasis on ecosystem consequences. *Freshwater Biology*, 48, 982–994.

Simon, K.S., Townsend, C.R., Biggs, B.J.F., Bowden, W.B. & Frew, R.D. (2004) Habitat-specific nitrogen dynamics in New Zealand streams containing native or invasive fish. Ecosystems, 7, 1–16.

Simpson, G.G. (1952) How many species? *Evolution*, 6, 342.

Sinclair, A.R.E. (1975) The resource limitation of trophic levels in tropical grassland ecosystems. *Journal of Animal Ecology*, 44, 497–520.

Sinclair, A.R.E. (1989) The regulation of animal populations. In: *Ecological Concepts* (J. Cherrett, ed.), pp. 197–241. Blackwell Scientific Publications, Oxford.

Sinclair, A.R.E. & Gosline J.M. (1997) Solar activity and mammal cycles in the Northern Hemisphere. *American Naturalist*, 149, 776–784.

Sinclair, A.R.E. & Norton-Griffiths, M. (1982) Does competition or facilitation regulate migrant ungulate populations in the Serengeti? A test of hypothesis. *Oecologia*, 53, 354–369.

Sinervo, B. (1990) The evolution of maternal investment in lizards: an experimental and comparative analysis of egg size and its effects on offspring performance. *Evolution*, 44, 279–294.

Sinervo, B., Svensson, E. & Comendant, T. (2000) Density cycles and an offspring quantity and quality game driven by natural selection. *Nature*, 406, 985–988.

Singleton, G., Krebs, C.J., Davis, S., Chambers, L. & Brown, P. (2001) Reproductive changes in fluctuating house mouse populations in southeastern Australia. *Proceedings of the Royal Society of London, Series B*, 268, 1741–1748.

Siva-Jothy, M.T., Katsubaki, Y., Hooper, R.E. & Plaistow, S.J. (2001) Investment in immune function under chronic and acute immune challenge in an insect. *Physiological Entomology*, 26, 1–5.

Skogland, T. (1983) The effects of dependent resource limitation on size of wild reindeer. *Oecologia*, 60, 156–168.

Slobodkin, L.B., Smith, F.E. & Hairston, N.G. (1967) Regulation in terrestrial ecosystems, and the implied balance of nature. *American Naturalist*, 101, 109–124.

Smith, F.D.M., May, R.M., Pellew, R., Johnson, T.H. & Walter, K.R. (1993) How much do we know about the current extinction rate? *Trends in Ecology and Evolution*, 8, 375–378.

Smith, J.W. (1998) Boll weevil eradication: area-wide pest management. *Annals of the Entomological Society of America*, 91, 239–247.

Smith, R.S., Shiel, R.S., Bardgett, R.D. *et al.* (2003) Soil microbial community, fertility, vegetation and diversity as targets in the restoration management of a meadow grassland. *Journal of Applied Ecology*, 40, 51–64.

Smith, S.D., Dinnen-Zopfy, B. & Nobel, P.S. (1984) High temperature responses of North American cacti. *Ecology*, 6, 643–651.

Snaydon, R.W. (1996) Above-ground and below-ground interactions in intercropping. In: *Dynamics and Nitrogen in Cropping Systems of the Semi-arid Tropics* (O. Ito, C. Johansen, J.J. Adu-Gyamfi, K. Katayama, J.V.D.K. Kumar Rao & T.J. Rego, eds), pp. 73–92. JIRCAS, Kasugai, Japan.

Snell, T.W. (1998) Chemical ecology of rotifers. *Hydrobiologia*, 388, 267–276.

Snyman, A. (1949) The influence of population densities on the development and oviposition of *Plodia interpunctella* Hubn. (Lepidoptera). *Journal of the Entomological Society of South Africa*, 12, 137–171.

Soares, S.M., Schoereder, J.H. & DeSouza, O. (2001) Processes involved in species saturation of ground-dwelling ant communities (Hymenoptera, Formicidae). *Austral Ecology*, 26, 187–192.

Soderquist, T.R. (1994) The importance of hypothesis testing in reintroduction biology: examples from the reintroduction of the carnivorous marsupial *Phascogale tapoatafa*. In: *Reintroduction Biology of Australian and New Zealand Fauna* (M. Serena, ed.), pp. 156–164. Beaty & Sons, Chipping Norton, UK.

Sol, D. & Lefebvre, L. (2000) Behavioural flexibility predicts invasion success in birds introduced to New Zealand. *Oikos*, 90, 599–605.

Solbrig, O.T. & Simpson, B.B. (1974) Components of regulation of a population of dandelions in Michigan. *Journal of Ecology*, 62, 473–486.

Soler, J.J., Martinez, J.G., Soler, M. & Moller, A.P. (2001) Life history of magpie populations sympatric or allopatric with the brood parasitic great spotted cuckoo. *Ecology*, 82, 1621–1631.

Solomon, M.E. (1949) The natural control of animal populations. *Journal of Animal Ecology*, 18, 1–35.

Sommer, U. (1990) Phytoplankton nutrient competition – from laboratory to lake. In: *Perspectives on Plant Competition* (J.B. Grace & D. Tilman, eds), pp. 193–213. Academic Press, New York.

Sorrell, B.K., Hawes, I., Schwarz, A.-M. & Sutherland, D. (2001) Interspecfic differences in photosynthetic carbon uptake, photosynthate partitioning and extracellular organic carbon release by deep-water characean algae. *Freshwater Biology*, 46, 453–464.

Soulé, M.E. (1986) *Conservation Biology: the science of scarcity and diversity*. Sinauer Associates, Sunderland, MA.

Sousa, M.E. (1979a) Experimental investigation of disturbance and ecological succession in a rocky intertidal algal community. *Ecological Monographs*, 49, 227–254.

Sousa, M.E. (1979b) Disturbance in marine intertidal boulder fields: the nonequilibrium maintenance of species diversity. *Ecology*, 60, 1225–1239.

South, A.B., Rushton, S.P., Kenward, R.E. & Macdonald, D.W. (2002) Modelling vertebrate dispersal and demography in real landscapes: how does uncertainty regarding dispersal behaviour influence predictions of spatial population dynamics. In: *Dispersal Ecology* (J.M. Bullock, R.E. Kenward & R.S. Hails, eds), pp. 327–349. Blackwell Science, Oxford.

Southwood, T.R.E. & Henderson, P.H. (2000) *Ecological Methods*, 3rd edn. Blackwell Science, Oxford.

Southwood, T.R.E., Hassell, M.P., Reader, P.M. & Rogers, D.J. (1989) The population dynamics of the viburnum whitefly (*Aleurotrachelus jelinekii*). *Journal of Animal Ecology*, 58, 921–942.

Speed, M.P. (1999) Batesian, quasi-Batesian or Müllerian mimicry? Theory and data in mimicry research. *Evolutionary Ecology*, 13, 755–776.

Speed, M. & Ruxton, G.D. (2002) Animal behaviour: evolution of suicidal signals. *Nature*, 416, 375.

Spiller, D.A. & Schoener, T.W. (1994) Effects of a top and intermediate predators in a terrestrial food web. *Ecology*, 75, 182–196.

Spiller, D.A. & Schoener, T.W. (1998) Lizards reduce spider species richness by excluding rare species. *Ecology*, 79, 503–516.

Sprent, J.I. (1979) *The Biology of Nitrogen Fixing Organisms*. McGraw Hill, London.

Sprent, J.I. & Sprent, P. (1990) *Nitrogen Fixing Organisms: pure and applied aspects*. Chapman & Hall, London.

Sprules, W.G. & Bowerman, J.E. (1988) Omnivory and food chain length in zooplankton food webs. *Ecology*, 69, 418–426.

Srivastava, D.S. (1999) Using local-regional richness plots to test for species saturation: pitfalls and potentials. *Journal of Animal Ecology*, 68, 1–16.

Staaland, H., White, R.G., Luick, J.R. & Holleman, D.F. (1980) Dietary influences on sodium and potassium metabolism of reindeer. *Canadian Journal of Zoology*, 58, 1728–1734.

Stadler, B. & Dixon, A.F.G. (1998) Costs of ant attendance for aphids. *Journal of Animal Ecology*, 67, 454–459.

Stafford, J. (1971) Heron populations of England and Wales 1928–70. *Bird Study*, 18, 218–221.

Stark, J.D. (1993) Performance of the Macroinvertebrate Community Index: effects of sampling method, sample replication, water depth, current velocity, and substratum on index values. *New Zealand Journal of Marine and Freshwater Research*, 27, 463–478.

Statzner, B.E., Fievet, J.Y. Champagne, R., Morel, D. & Herouin, E. (2000) Crayfish as geomorphic agents and ecosystem engineers: biological behavior affects sand and gravel erosion in experimental streams. *Limnology and Oceanography*, 45, 1030–1040.

Stauffer, B. (2000) Long term climate records from polar ice. *Space Science Reviews*, 94, 321–336.

Stearns, S.C. (1977) The evolution of life history traits. *Annual Review of Ecology and Systematics*, 8, 145–171.

Stearns, S.C. (1983) The impact of size and phylogeny on patterns of covariation in the life history traits of mammals. *Oikos*, 41, 173–187.

Stearns, S.C. (1992) *The Evolution of Life Histories*. Oxford University Press, Oxford.

Stearns, S.C. (2000) Life history evolution: successes, limitations, and prospects. *Naturwissenschaften*, 87, 476–486.

Steinauer, E.M. & Collins, S.L. (2001) Feedback loops in ecological hierarchies following urine deposition in tallgrass prairie. *Ecology*, 82, 1319–1329.

Steingrimsson, S.O. & Grant, J.W.A. (1999) Allometry of territory-size and metabolic rate as predictors of self-thinning in young-of-the-year Atlantic salmon. *Journal of Animal Ecology*, 68, 17–26.

Stemberger, R.S. & Gilbert, J.J. (1984) Spine development in the rotifer *Keratella cochlearis*: induction by cyclopoid copepods and *Asplachna*. *Freshwater Biology*, 14, 639–648.

Stenberg, P., Palmroth, S., Bond, B.J., Sprugel, D.G. & Smolander, H. (2001) Shoot structure and photosynthetic efficiency along the light gradient in Scots pine canopy. *Tree Physiology*, 21, 805–814.

Stenseth, N.C., Bjørnstad, O.N. & Saihto, T. (1996) A gradient from stable to cyclic populations of *Clethrionomys rufocanus* in Hokkaido, Japan. *Proceedings of the Royal Society of London, Series B*, 263, 1117–1126.

Stenseth, N.C., Falck, W., Bjørnstad, O.N. & Krebs, C.J. (1997) Population regulation in snowshoe hare and Canadian lynx: asymmetric food web configurations between hare and lynx. *Proceedings of the National Academy of Sciences of the USA*, 94, 5147–5152.

Stenseth, N.C., Ottersen, G., Hurrell, J.W. *et al.* (2003) Studying climate effects on ecology through the use of climate indices: the North Atlantic Oscillation, El Nino Southern Oscillation and beyond. *Proceedings of the Royal Society of London, Series B*, 270, 2087–2096.

Stephens, D.W. & Krebs, J.R. (1986) *Foraging Theory*. Princeton University Press, Princeton, NJ.

Stephens, G.R. (1971) The relation of insect defoliation to mortality in Connecticut forests. *Connecticut Agricultural Experimental Station Bulletin*, 723, 1–16.

Stephens, P.A. & Sutherland, W.J. (1999) Consequences of the Allee effect for behaviour, ecology and conservation. *Trends in Ecology and Evolution*, 14, 401–405.

Stephens, P.A., Frey-Roos, F., Arnold, W. & Sutherland, W.J. (2002) Sustainable exploitation of social species: a test and comparison of models. *Journal of Applied Ecoology*, 39, 629–642.

Sterner, R.W., Bajpai, A. & Adams, T. (1997a) The enigma of food chain length: absence of theoretical evidence for dynamic constraints. *Ecology*, 78, 2258–2262.

Sterner, R.W., Elser, J.J., Fee, E.J., Guildford, S.J. & Chrzanowski, T.H. (1997b) The light : nutrient ratio in lakes: the balance of energy and materials affects ecosystem structure and processes. *American Naturalist*, 150, 663–684.

Stevens, C.E. & Hume, I.D. (1998) Contributions of microbes in vertebrate gastrointestinal tract to production and conservation of nutrients. *Physiological Reviews*, 78, 393–426.

Stewart, T.W., Miner, J.G. & Lowe, R.L. (1998) Macroinvertebrate communities on hard substrates in western Lake Erie: structuring effects of *Dreissena*. *Journal of Great Lakes Research*, 24, 868–879.

Stoll, P. & Prati, D. (2001) Intraspecific aggregation alters competitive interactions in experimental plant communities. *Ecology*, 82, 319–327.

Stoll, P., Weiner, J., Muller-Landau, H., Muller, E. & Hara, T. (2002) Size symmetry of competition alters biomass–density relationships. *Proceedings of the Royal Society of London, Series B*, 269, 2191–2195.

Stone, G.M. & Cook, J.M. (1998) The structure of cynipid oak galls: patterns in the evolution of an extended phenotype. *Proceedings of the Royal Society of London, Series B*, 265, 979–988.

Stone, L. & Roberts, A. (1990) The checkerboard score and species distributions. *Oecologia*, 85, 74–79.

Stolp, H. (1988) *Microbial Ecology*. Cambridge University Press, Cambridge, UK.

Storch, D., Sizling, A.L. & Gaston, K.J. (2003) Geometry of the species area relationship in central European birds: testing the mechanism. *Journal of Animal Ecology*, 72, 509–519.

Storey, K.B. (1990) Biochemical adaptation for cold hardiness in insects. *Philosophical Transactions of the Royal Society of London, Series B*, 326, 635–654; also in *Life at Low Temperatures* (R.M. Laws & F. Franks, eds), pp. 119–138. The Royal Society, London.

Stowe, L.G. & Teeri, J.A. (1978) The geographic distribution of C_4 species of the Dicotyledonae in relation to climate. *American Naturalist*, 112, 609–623.

Strauss, S.Y. & Agrawal, A.A. (1999) The ecology and evolution of plant tolerance to herbivory. *Trends in Ecology and Evolution*, 14, 179–185.

Strauss, S.Y., Irwin, R.E. & Lambrix, V.M. (2004) Optimal defence theory and flower petal colour predict variation in the secondary chemistry of wild radish. *Journal of Ecology*, 92, 132–141.

Strauss, S.Y., Rudgers, J.A., Lau, J.A. & Irwin, R.E. (2002) Direct and ecological costs of resistance to herbivory. *Trends in Ecology and Evolution*, 17, 278–285.

Strobeck, C. (1973) N species competition. *Ecology*, 54, 650–654.

Stromberg, P.C., Toussant, M.J. & Dubey, J.P. (1978) Population biology of *Paragonimus kellicotti* metacercariae in central Ohio. *Parasitology*, 77, 13–18.

Strong, D.R. Jr. (1982) Harmonious coexistence of hispine bettles on *Heliconia* in experimental and natural communities. *Ecology*, 63, 1039–1049.

Strong, D.R. (1992) Are trophic cascades all wet? Differentiation and donor-control in speciose ecosystems. *Ecology*, 73, 747–754.

Strong, D.R. Jr., Lawton, J.H. & Southwood, T.R.E. (1984) *Insects on Plants: community patterns and mechanisms*. Blackwell Scientific Publications, Oxford.

Suchanek, T.H. (1992) Extreme biodiversity in the marine environment: mussel bed communities of *Mytilus californianus*. *Northwest Environmental Journal*, 8, 150.

Summerhayes, V.S. & Elton, C.S. (1923) Contributions to the ecology of Spitsbergen and Bear Island. *Journal of Ecology*, 11, 214–286.

Sunderland, K.D., Hassall, M. & Sutton, S.L. (1976) The population dynamics of *Philoscia muscorum* (Crustacea, Oniscoidea) in a dune grassland ecosystem. *Journal of Animal Ecology*, 45, 487–506.

Susarla, S., Medina, V.F. & McCutcheon, S.C. (2002) Phytoremediation: an ecological solution to organic chemical contamination. *Ecological Engineering*, 18, 647–658.

Sutherland, W.J. (1983) Aggregation and the 'ideal free' distribution. *Journal of Animal Ecology*, 52, 821–828.

Sutherland, W.J. (1998) The importance of behavioural studies in conservation biology. *Animal Behaviour*, 56, 801–809.

Sutherland, W.J., Gill, J.A. & Norris, K. (2002) Density-dependent dispersal in animals: concepts, evidence, mechanisms and consequences. In: *Dispersal Ecology* (J.M. Bullock, R.E. Kenward & R.S. Hails, eds), pp. 134–151. Blackwell Science, Oxford.

Sutton, M.A., Pitcairn, C.E.R. & Fowler, D. (1993) The exchange of ammonia between the atmosphere and plant communities. *Advances in Ecological Research*, 24, 302–393.

Sutton, S.L. & Collins, N.M. (1991) Insects and tropical forest conservation. In: *The Conservation of Insects and their Habitats* (N.M. Collins & J.A. Thomas, eds), pp. 405–424. Academic Press, London.

Swift, M.J., Heal, O.W. & Anderson, J.M. (1979) *Decomposition in Terrestrial Ecosystems*. Blackwell Scientific Publications, Oxford.

Symonides, E. (1979) The structure and population dynamics of psammophytes on inland dunes. II. Loose-sod populations. *Ekologia Polska*, 27, 191–234.

Symonides, E. (1983) Population size regulation as a result of intra-population interactions. I. The effect of density on the survival and development of individuals of *Erophila verna* (L.). *Ecologia Polska*, 31, 839–881.

Syms, C. & Jones, G.P. (2000) Disturbance, habitat structure, and the dynamics of a coral-reef fish community. *Ecology*, 81, 2714–2729.

Szarek, S.R., Johnson, H.B. & Ting, I.P. (1973) Drought adaption in *Opuntia basilaris*. Significance of recycling carbon through Crassulacean acid metabolism. *Plant Physiology*, 52, 539–541.

Tamm, C.O. (1956) Further observations on the survival and flowering of some perennial herbs. *Oikos*, 7, 274–292.

Tanaka, M.O. & Magalhaes, C.A. (2002) Edge effects and succession dynamics in *Brachidontes* mussel beds. *Marine Ecology Progress Series*, 237, 151–158.

Taniguchi, H., Nakano, S. & Tokeshi, M. (2003) Habitat complexity, patch size and the abundance of ephiphytic invertebrates on plants. *Freshwater Biology*, 00, 00–00.

Taniguchi, Y. & Nakano, S. (2000) Condition-specific competition: implications for the altitudinal distribution of stream fishes. *Ecology*, 81, 2027–2039.

Tansley, A.G. (1917) On competition between *Galium sylvestre* Poll. (*G. asperum* Schreb.) on different types of soil. *Journal of Ecology*, 5, 173–179.

Tansley, A.G. (1939) *The British Islands and their Vegetation*. Cambridge University Press, Cambridge, UK.

Taylor, A.D. (1993) Heterogeneity in host–parasitoid interactions: the CV^2 rule. *Trends in Ecology and Evolution*, 8, 400–405.

Taylor, I. (1994) *Barn Owls. Predator–Prey Relationships and Conservation*. Cambridge University Press, Cambridge, UK.

Taylor, L.R. (1987) Objective and experiment in long-term research. In: *Long-Term Studies in Ecology* (G.E. Likens, ed.), pp. 20–70. Springer-Verlag, New York.

Taylor, R.B., Sotka, E. & Hay, M.E. (2002) Tissue-specific induction of herbivore resistance: seaweed response to amphipod grazing. *Oecologia*, 132, 68–76.

Teeri, J.A. & Stowe, L.G. (1976) Climatic patterns and the distribution of C$_4$ grasses in North America. *Oecologia*, 23, 112.

Teirney, L.D. (2003) *Fiordland Marine Conservation Strategy: Te Kaupapa Atawhai o Te Moana o Atawhenua.* Guardians of Fiordland's Fisheries and Marine Environment Inc., Te Anau, New Zealand.

Téllez-Valdés, O. & Dávila-Aranda, P. (2003) Protected areas and climate change: a case study of the cacti in the Tehuacán-Cuicatlán Biosphere Reserve, Mexico. *Conservation Biology*, 17, 846–853.

Ter Braak, C.J.F. & Prentice, I.C. (1988) A theory of gradient analysis. *Advances in Ecological Research*, 18, 272–317.

Ter Braak, C.J.F. & Smilauer, P. (1998) CANOCO for Windows version 4.02. Wageningen, the Netherlands.

Terborgh, J. (1973) On the notion of favorableness in plant ecology. *American Naturalist*, 107, 481–501.

Teuschl, Y., Taborsky, B. & Taborsky, M. (1998) How do cuckoos find their nests? The role of habitat imprinting. *Animal Behaviour*, 56, 1425–1433.

Thomas, C.D. (1990) What do real population dynamics tell us about minimum viable population sizes? *Conservation Biology*, 4, 324–327.

Thomas, C.D. & Harrison, S. (1992) Spatial dynamics of a patchily distributed butterfly species. *Journal of Applied Ecology*, 61, 437–446.

Thomas, C.D. & Jones, T.M. (1993) Partial recovery of a skipper butterfly (*Hesperia comma*) from population refuges: lessons for conservation in a fragmented landscape. *Journal of Animal Ecology*, 62, 472–481.

Thomas, C.D., Thomas, J.A. & Warren, M.S. (1992) Distributions of occupied and vacant butterfly habitats in fragmented landscapes. *Oecologia*, 92, 563–567.

Thomas, F. & Poulin, R. (1998) Manipulation of a mollusc by a trophically transmitted parasite: convergent evolution of phylogenetic inheritance? *Parasitology*, 116, 431–436.

Thomas, S.C. & Weiner, J. (1989) Growth, death and size distribution change in an *Impatiens pallida* population. *Journal of Ecology*, 77, 524–536.

Thompson, D.J. (1975) Towards a predator–prey model incorporating age-structure: the effects of predator and prey size on the predation of *Daphnia magna* by *Ischnura elegans*. *Journal of Animal Ecology*, 44, 907–916.

Thompson, D.J. (1989) Sexual size dimorphism in the damselfly *Coenagrion puella* (L.). *Advances in Odonatology*, 4, 123–131.

Thompson, J.N. (1988) Evolutionary ecology of the relationship between oviposition preference and performance of offspring in phytophagous insects. *Entomologia Experimentia et Applicata*, 47, 3–14.

Thompson, J.N. (1995) *The Coevolutionary Process.* University of Chicago Press, Chicago.

Thompson, R.M. & Townsend, C.R. (2000) Is resolution the solution?: the effect of taxonomic resolution on the calculated properties of three stream food webs. *Freshwater Biology*, 44, 413–422.

Thompson, R.M., Mouritsen, K.N. & Poulin, R. (2005) Importance of parasites and their life cycle characteristics in determining the structure of a large marine food web. *Journal of Animal Ecology*, 74, 77–85.

Thompson, R.M., Townsend, C.R., Craw, D., Frew, R. & Riley, R. (2001) (Further) links from rocks to plants. *Trends in Ecology and Evolution*, 16, 543.

Thórhallsdóttir, T.E. (1990) The dynamics of five grasses and white clover in a simulated mosaic sward. *Journal of Ecology*, 78, 909–923.

Tillman, P.G. (1996) Functional response of *Microplitis croceipes* and *Cardiochiles nigriceps* (Hymenoptera: Braconidae) to variation in density of tobacco budworm (Lepidoptera: Noctuidae). *Environmental Entomology*, 25, 524–528.

Tilman, D. (1977) Resource competition between planktonic algae: an experimental and theoretical approach. *Ecology*, 58, 338–348.

Tilman, D. (1982) *Resource Competition and Community Structure.* Princeton University Press, Princeton, NJ.

Tilman, D. (1986) Resources, competition and the dynamics of plant communities. In: *Plant Ecology* (M.J. Crawley, ed.), pp. 51–74. Blackwell Scientific Publications, Oxford.

Tilman, D. (1987) Secondary succession and the pattern of plant dominance along experimental nitrogen gradients. *Ecological Monographs*, 57, 189–214.

Tilman, D. (1988) *Plant Strategies and the Dynamics and Structure of Plant Communities.* Princeton University Press, Princeton, NJ.

Tilman, D. (1990) Mechanisms of plant competition for nutrients: the elements of a predictive theory of competition. In: *Perspectives on Plant Competition* (J.B. Grace & D. Tilman, eds), pp. 117–141. Academic Press, New York.

Tilman, D. (1996) Biodiversity: population versus ecosystem stability. *Ecology*, 77, 350–363.

Tilman, D. (1999) The ecological consequences of changes in biodiversity: a search for general principles. *Ecology*, 80, 1455–1474.

Tilman, D. & Wedin, D. (1991a) Plant traits and resource reduction for five grasses growing on a nitrogen gradient. *Ecology*, 72, 685–700.

Tilman, D. & Wedin, D. (1991b) Dynamics of nitrogen competition between successional grasses. *Ecology*, 72, 1038–1049.

Tilman, D., Fargione, J., Wolff, B. *et al.* (2001) Forecasting agriculturally driven global environmental change. *Science*, 292, 281–284.

Tilman, D., Mattson, M. & Langer, S. (1981) Competition and nutrient kinetics along a temperature gradient: an experimental test of a mechanistic approach to niche theory. *Limnology and Oceanography*, 26, 1020–1033.

Tilman, D., Reich, P.B., Knops, J., Wedin, D., Meilke, T. & Lehman, C. (2001) Diversity and productivity in a long-term grassland experiment. *Science*, 294, 843–845.

Timmermann, A., Oberhuber, J., Bacher, A., Esch, M., Latif, M. & Roeckner, E. (1999) Increased El Niño frequency in a climate model forced by future greenhouse warming. *Nature*, 398, 694–697.

Timmins, S.M. & Braithwaite, H. (2002) Early detection of new invasive weeds on islands. In: *Turning the Tide: eradication of invasive species* (D. Veitch & M. Clout, eds), pp. 311–318. Invasive Species Specialist Group of the World Conservation Union (IUCN), Auckland, New Zealand.

Tinbergen, L. (1960) The natural control of insects in pinewoods. 1: factors influencing the intensity of predation by songbirds. *Archives néerlandaises de Zoologie*, 13, 266–336.

Tjallingii, W.F. & Hogen Esch, Th. (1993) Fine structure of aphid stylet routes in plant tissues in correlation with EPG signals. *Physiological Entomology*, 18, 317–328.

Tkadlec, E. & Stenseth, N.C. (2001) A new geographical gradient in vole population dynamics. *Proceedings of the Royal Society of London, Series B*, 268, 1547–1552.

Toft, C.A. & Karter, A.J. (1990) Parasite–host coevolution. *Trends in Ecology and Evolution*, 5, 326–329.

Tokeshi, M. (1993) Species abundance patterns and community structure. *Advances in Ecological Research*, 24, 112–186.

Tompkins, D.M. & Begon, M. (1999) Parasites can regulate wildlife populations. *Parasitology Today*, 15, 311–313.

Tompkins, D.M., White, A.R., Boots, M. (2003) Ecological replacement of native red squirrels by invasive greys driven by disease. *Ecology Letters*, 6, 189–196.

Tonn, W.M. & Magnuson, J.J. (1982) Patterns in the species composition and richness of fish assemblages in northern Wisconsin lakes. *Ecology*, 63, 137–154.

Towne, E.G. (2000) Prairie vegetation and soil nutrient responses to ungulate carcasses. *Oecologia*, 122, 232–239.

Townsend, C.R. (1991) Community organisation in marine and freshwater environments. In: *Fundamentals of Aquatic Ecology* (R.S.K. Barnes & K.H. Mann, eds), pp. 125–144. Blackwell Scientific Publications, Oxford.

Townsend, C.R. (1996) Concepts in river ecology: pattern and process in the catchment hierarchy. *Archiv fur Hydrobiologie*, 113 (Suppl. 10), 3–21.

Townsend, C.R. (2003) Individual, population, community and ecosystem consequences of a fish invader in New Zealand streams. *Conservation Biology*, 17, 38–47.

Townsend, C.R. & Hildrew, A.G. (1980) Foraging in a patchy environment by a predatory net-spinning caddis larva: a test of optimal foraging theory. *Oecologia*, 47, 219–221.

Townsend, C.R. & Hildrew, A.G. (1994) Species traits in relation to a habitat templet for river systems. *Freshwater Biology*, 31, 265–275.

Townsend, C.R., Hildrew, A.G. & Francis, J.E. (1983) Community structure in some southern English streams: the influence of physiochemical factors. *Freshwater Biology*, 13, 521–544.

Townsend, C.R., Scarsbrook, M.R. & Dolédec, S. (1997) The intermediate disturbance hypothesis, refugia and bio-diversity in streams. *Limnology and Oceanography*, 42, 938–949.

Townsend, C.R., Thompson, R.M., Macintosh, A.R., Kilroy, C., Edwards, E. & Scarsbrook, M.R. (1998) Disturbance, resource supply and food-web architecture in streams. *Ecology Letters*, 1, 200–209.

Townsend, C.R., Tipa, G., Teirney, L.D. & Niyogi, D.K. (2004) Development of a tool to facilitate participation of Maori in the management of stream and river health. *Ecology and Health*, 1, 184–195.

Townsend, C.R., Winfield, I.J., Peirson, G. & Cryer, M. (1986) The response of young roach *Rutilus rutilus* to seasonal changes in abundance of microcrustacean prey: a field demonstration of switching. *Oikos*, 46, 372–378.

Tracy, C.R. (1976) A model of the dynamic exchanges of water and energy between a terrestrial amphibian and its environment. *Ecological Monographs*, 46, 293–326.

Tregenza, T. (1995) Building on the ideal free distribution. *Advances in Ecological Research*, 26, 253–307.

Trewavas, A. (2001) Urban myths of organic farming: organic agriculture began as an idealogy, but can it meet today's needs? *Nature*, 410, 409–410.

Trewick, S.A. & Worthy, T.H. (2001) Origins and prehistoric ecology of takahe based on morphometric, molecular, and fossil data. In: *The Takahe: fifty years of conservation management and research* (W.G. Lee & I.G. Jamieson, eds), pp. 31–48. University of Otago Press, Dunedin, New Zealand.

Trexler, J.C., McCulloch, C.E. & Travis, J. (1988) How can the functional response best be determined? *Oecologia*, 76, 206–214.

Tripet, F. & Richner, H. (1999) Density dependent processes in the population dynamics of a bird ectoparasite *Ceratophyllus gallinae*. *Ecology*, 80, 1267–1277.

Turchin, P.D. (1990) Rarity of density dependence or population regulation with lags? *Nature*, 344, 660-663.

Turchin, P. (2003) *Complex Population Dynamics*. Princeton University Press, Princeton, NJ.

Turchin, P. & Batzli, G.O. (2001) Availability of food and the population dynamics of arvicoline rodents. *Ecology*, 82, 1521–1534.

Turchin, P. & Berryman, A.A. (2000) Detecting cycles and delayed density dependence: a comment on Hunter and Price (1998). *Ecological Entomology*, 25, 119–121.

Turchin, P. & Hanski, I. (1997) An empirically based model for latitudinal gradient in vole population dynamics. *American Naturalist*, 149, 842–874.

Turchin, P. & Hanski, I. (2001) Contrasting alternative hypotheses about rodent cycles by translating them into parameterized models. *Ecology Letters*, 4, 267–276.

Turesson, G. (1922a) The species and variety as ecological units. *Hereditas*, 3, 100–113.

Turesson, G. (1922b) The genotypical response of the plant species to the habitat. *Hereditas*, 3, 211–350.

Turkington, R. & Harper, J.L. (1979) The growth, distribution and neighbour relationships of *Trifolium repens* in a permanent pasture. IV. Fine scale biotic differentiation. *Journal of Ecology*, 67, 245–254.

Turkington, R. & Mehrhoff, L.A. (1990) The role of competition in structuring pasture communities. In: *Perspectives on Plant Competition* (J.B. Grace & D. Tilman, eds), pp. 307–340. Academic Press, New York.

Turner, J.R.G., Lennon, J.J. & Greenwood, J.J.D. (1996) Does climate cause the global biodiversity gradient? In: *Aspects of the Genesis and Maintenance of Biological Diversity* (M. Hochberg, J. Claubert & R. Barbault, eds), pp. 199–220. Oxford University Press, London and New York.

Twigg, L.E., Martin, G.R. & Lowe, T.J. (2002) Evidence of pesticide resistance in medium-sized mammalian pests: a case study with 1080 poison and Australian rabbits. *Journal of Applied Ecology*, 39, 549–560.

Tyndale-Biscoe, M. & Vogt, W.G. (1996) Population status of the bush fly, *Musca vetustissima* (Diptera: Muscidae), and native dung beetles (Coleoptera: Scarabaeninae) in south-eastern Australia in relation to establishment of exotic dung beetles. *Bulletin of Entomological Research*, 86, 183–192.

Uchmanski, J. (1985) Differentiation and frequency distributions of body weights in plants and animals. *Philosophical Transactions of the Royal Society of London, Series B*, 310, 1–75.

Umbanhowar, J., Maron, J. & Harrison, S. (2003) Density dependent foraging behaviors in a parasitoid lead to density dependent parasitism of its host. *Oecologia*, 137, 123–130.

UNEP (2003) *Global Environmental Outlook Year Book 2003*. United Nations Environmental Program, GEO Section, PO Box 30552, Nairobi, Kenya.

United Nations (1999) *The World at Six Billion*. United Nations Population Division, Department of Economic and Social Affairs, United Nations Secretariat, New York.

US Department of the Interior (1990) *Audit Report: the Endangered Species Program*. US Fish and Wildlife Service Report 90–98.

Usio, N. & Townsend, C.R. (2001) The significance of the crayfish *Paranephrops zealandicus* as shredders in a New Zealand headwater stream. *Journal of Crustacean Biology*, 21, 354–359.

Usio, N. & Townsend, C.R. (2002) Functional significance of crayfish in stream food webs: roles of omnivory, substrate heterogeneity and sex. *Oikos*, 98, 512–522.

Usio, N. & Townsend, C.R. (2004) Roles of crayfish: consequences of predation and bioturbation for stream invertebrates. *Ecology*, 85, 807–822.

Valentine, J.W. (1970) How many marine invertebrate fossil species? A new approximation. *Journal of Paleontology*, 44, 410–415.

Valladares, V.F. & Pearcy, R.W. (1998) The functional ecology of shoot architecture in sun and shade plants of *Heteromeles arbutifolia* M. Roem., a Californian chaparral shrub. *Oecologia*, 114, 1–10.

Van Breeman, N. (2002) Natural organic tendency. *Nature*, 415, 381–382.

van de Koppel, J., Rietkerk, M. & Weissing, F.J. (1997) Catastrophic vegetation shifts and soil degradation in terrestrial grazing systems. *Trends in Ecology and Evolution*, 12, 352–356.

van den Bosch, R., Leigh, T.F., Falcon, L.A., Stern, V.M., Gonzales, D. & Hagen, K.S. (1971) The developing program of integrated control of cotton pests in California. In: *Biological Control* (C.B. Huffaker, ed.), pp. 377–394. Plenum Press, New York.

Van der Juegd, H.P. (1999) *Life history decisions in a changing environment: a long-term study of a temperate barnacle goose population*. PhD thesis, Uppsala University, Uppsala.

van Leneteren, J.C. & Woets, J. (1988) Biological and integrated pest control in greenhouses. *Annual Review of Entomology*, 33, 239–269.

van Rensburg, B.J., Chown, S.L. & Gaston, K.J. (2002) Species richness, environmental correlates, and spatial scale: a test using South African birds. *American Naturalist*, 159, 566–577.

van Riper, C., van Riper, S.G., Goff, M.L. & Laird, M. (1986) The epizootiology and ecological significance of malaria in Hawaiian land birds. *Ecological Monographs*, 56, 327–344.

Vandenkoornhuyse, P., Ridgway, K.P., Watson, I.J., Fitter, A.H. & Young, J.P.W. (2003) Co-existing grass species have distinctive arbuscular mycorrhizal communities. *Molecular Ecology*, 12, 3085–3095.

Vander Zanden, M.J., Shuter, B.J., Lester, L. & Rasmussen, J.B. (1999) Patterns of food chain length in lakes: a stable isotope study. *American Naturalist*, 154, 406–416.

Vandermeest, D.B., Van Lear, D.H. & Clinton, B.D. (2002) American chestnut as an allelopath in the southern Appalachians. *Forest Ecology and Management*, 165, 173–181.

Vanderplank, J.E. (1963) *Plant Diseases: epidemics and control*. Academic Press, New York.

Varley, G.C. (1947) The natural control of population balance in the knapweed gall-fly (*Urophora jaceana*). *Journal of Animal Ecology*, 16, 139–187.

Varley, G.C. & Gradwell, G.R. (1968) Population models for the winter moth. *Symposium of the Royal Entomological Society of London*, 9, 132–142.

Varley, G.C. & Gradwell, G.R. (1970) Recent advances in insect population dynamics. *Annual Review of Entomology*, 15, 1–24.

Vaughan, N., Lucas, E.-A, Harris, S. & White, P.C.L. (2003) Habitat associations of European hares *Lepus europaeus* in England and Wales: implications for farmland management. *Journal of Applied Ecology*, 40, 163–175.

Vázquez, G.J.A. & Givnish, T.J. (1998) Altitudinal gradients in tropical forest composition, structure, and diversity in the Sierra de Manantlán. *Journal of Ecology*, 86, 999–1020.

Venterink, H.O., Wassen, M.J., Verkroost, A.W.M. & de Ruiter, P.C. (2003) Species richness–productivity patterns differ between N-, P- and K-limited wetlands. *Ecology*, 84, 2191–2199.

Verhulst, P.F. (1838) Notice sur la loi que la population suit dans son accroissement. *Correspondances Mathématiques et Physiques*, 10, 113–121.

Villa, F., Tunesi, L. & Agardy, T. (2002) Zoning marine protected areas through spatial multiple-criteria analysis: the case of the Asinara Island National Marine Reserve of Italy. *Conservation Biology*, 16, 515–526.

Vivas-Martinez, S., Basanez, M.-G., Botto, C. *et al.* (2000) Amazonian onchocerciasis: parasitological profiles by host-age, sex, and endemicity in southern Venezuela. *Parasitology*, 121, 513–525.

Volk, M., Niklaus, P.A. & Korner, C. (2000) Soil moisture effects determine CO_2 responses of grassland species. *Oecologia*, 125, 380–388.

Volterra, V. (1926) Variations and fluctuations of the numbers of individuals in animal species living together. (Reprinted in 1931. In: R.N. Chapman, *Animal Ecology*. McGraw Hill, New York.)

Vos, M., Hemerik, L. & Vet, L.E.M. (1998) Patch exploitation by the parasitoids *Cotesia rubecula* and *Cotesia glomerata* in multi-patch environments with different host distributions. *Journal of Animal Ecology*, 67, 774–783.

Vucetich, J.A., Peterson, R.O. & Schaefer, C.L. (2002) The effect of prey and predator densities on wolf predation. *Ecology*, 83, 3003–3013.

Waage, J.K. & Greathead, D.J. (1988) Biological control: challenges and opportunities. *Philosophical Transactions of the Royal Society of London, Series B*, 318, 111–128.

Wallace, J.B. & O'Hop, J. (1985) Life on a fast pad: waterlily leaf beetle impact on water lilies. *Ecology*, 66, 1534–1544.

Wallis, G.P. (1994) Population genetics and conservation in New Zealand: a hierarchical synthesis and recommendations for the 1990s. *Journal of the Royal Society of New Zealand*, 24, 143–160.

Walsh, J.A. (1983) Selective primary health care: strategies for control of disease in the developing world. IV. Measles. *Reviews of Infectious Diseases*, 5, 330–340.

Walter, D.E. (1987) Trophic behaviour of 'mycophagous' micro-arthropods. *Ecology*, 68, 226–228.

Walther, G.-R., Post, E., Convey, P. *et al.* (2002) Ecological responses to recent climate change. *Nature*, 416, 389–395.

Wang, G.-H. (2002) Plant traits and soil chemical variables during a secondary vegetation succession in abandoned fields on the Loess Plateau. *Acta Botanica Sinica*, 44, 990–998.

Ward, J.V. (1988) Riverine–wetland interactions. In: *Freshwater Wetlands and Wildlife* (R.R. Sharitz & J.W. Gibbon, eds), pp. 385–400. Office of Science and Technology Information, US Department of Energy, Oak Ridge, TN.

Wardle, D.A., Bonner, K.I. & Barker, G.M. (2000) Stability of ecosystem properties in response to above-ground functional group richness and composition. *Oikos*, 89, 11–23.

Ware, G.W. (1983) *Pesticides Theory and Application*. W.H. Freeman, New York.

Waring, R.H. & Schlesinger, W.H. (1985) *Forest Ecosystems: concepts and management*. Academic Press, Orlando, FL.

Warren, P.H. (1989) Spatial and temporal variation in the structure of a freshwater food web. *Oikos*, 55, 299–311.

Warren, P.H. (1994) Making connections in food webs. *Trends in Ecology and Evolution*, 9, 136–141.

Warren, R.S., Fell, P.E., Rozsa, R. *et al.* (2002) Salt marsh restoration in Connecticut: 20 years of science and management. *Restoration Ecology*, 10, 497–513.

Waser, N.M. & Price, M.V. (1994) Crossing distance effects in *Delphinium nelsonii*: outbreeding and inbreeding depression in progeny fitness. *Evolution*, 48, 842–852.

Waterhouse, D.F. (1974) The biological control of dung. *Scientific American*, 230, 100–108.

Watkinson, A.R. (1984) Yield–density relationships: the influence of resource availability on growth and self-thinning in populations of *Vulpia fasciculata*. *Annals of Botany*, 53, 469–482.

Watkinson, A.R. & Harper, J.L. (1978) The demography of a sand dune annual: *Vulpia fasciculata*. I. The natural regulation of populations. *Journal of Ecology*, 66, 15–33.

Watkinson, A.R., Freckleton, R.P., Robinson, R.A. & Sutherland, W.J. (2000) Predictions of biodiversity response to genetically modified herbicide-tolerant crops. *Science*, 289, 1554–1557.

Watson, A. & Moss, R. (1980) Advances in our understanding of the population dynamics of red grouse from a recent fluctuation in numbers. *Areda*, 68, 103–111.

Watt, A.S. (1947) Pattern and process in the plant community. *Journal of Ecology*, 35, 1–22.

Way, M.J. & Cammell, M. (1970) Aggregation behaviour in relation to food utilization by aphids. In: *Animal Populations in Relation to their Food Resource* (A. Watson, ed.), pp. 229–247. Blackwell Scientific Publications, Oxford.

Weathers, K.C., Caldenasso, M.L. & Pickett, S.T.A. (2001) Forest edges as nutrient and pollutant concentrators: potential synergisms between fragmentation, forest canopies and the atmosphere. *Conservation Biology*, 15, 1506–1514.

Weaver, J.E. & Albertson, F.W. (1943) Resurvey of grasses, forbs and underground plant parts at the end of the great drought. *Ecological Monographs*, 13, 63–117.

Webb, S.D. (1987) Community patterns in extinct terrestrial invertebrates. In: *Organization of Communities: past and present* (J.H.R. Gee & P.S. Giller, eds), pp. 439–468. Blackwell Scientific Publications, Oxford.

Webb, W.L., Lauenroth, W.K., Szarek, S.R. & Kinerson, R.S. (1983) Primary production and abiotic controls in forests, grasslands and desert ecosystems in the United States. *Ecology*, 64, 134–151.

Webster, J. (1970) *Introduction to Fungi*. Cambridge University Press, Cambridge, UK.

Webster, M.S. & Almany, G.R. (2002) Positive indirect effects in a coral reef fish community. *Ecology Letters*, 5, 549–557.

Wegener, A. (1915) *Entstehung der Kontinenter und Ozeaner*. Samml. Viewig, Braunschweig. English translation (1924) *The Origins of Continents and Oceans*. Translated by J.G.A. Skerl. Metheun, London.

Weiner, J. (1986) How competition for light and nutrients affects size variability in *Ipomoea tricolor* populations. *Ecology*, 67, 1425–1427.

Weiner, J. (1990) Asymmetric competition in plant populations. *Trends in Ecology and Evolution*, 5, 360–364.

Weiner, J. & Thomas, S.C. (1986) Size variability and competition in plant monocultures. *Oikos*, 47, 211–222.

Weller, D.E. (1987) A reevaluation of the −3/2 power rule of plant self-thinning. *Ecological Monographs*, 57, 23–43.

Weller, D.E. (1990) Will the real self-thinning rule please stand up? A reply to Osawa and Sugita. *Ecology*, 71, 1204–1207.

Werner, E.E., Gilliam, J.F., Hall, D.J. & Mittlebach, G.G. (1983a) An experimental test of the effects of predation risk on habitat use in fish. *Ecology*, 64, 1540–1550.

Werner, E.E., Mittlebach, G.G., Hall, D.J. & Gilliam, J.F. (1983b) Experimental tests of optimal habitat use in fish: the role of relative habitat profitability. *Ecology*, 64, 1525–1539.

Werner, H.H. & Hall, D.J. (1974) Optimal foraging and the size selection of prey by the bluegill sunfish *Lepomis macrohirus*. *Ecology*, 55, 1042–1052.

Werner, P.A. & Platt, W.J. (1976) Ecological relationships of co-occurring golden rods (*Solidago*: Compositae). *American Naturalist*, 110, 959–971.

Wertheim, B., Sevenster, J.G., Eijs, I.E.M & van Alphen, J.J.M. (2000) Species diversity in a mycophagous insect community: the case of spatial aggregation vs. resource partitioning. *Journal of Animal Ecology*, 69, 335–351.

Wesson, G. & Wareing, P.F. (1969) The induction of light sensitivity in weed seeds by burial. *Journal of Experimental Biology*, 20, 413–425.

West, G.B., Brown, J.H. & Enquist, B.J. (1997) A general model for the origin of allometric scaling laws in biology. *Science*, 276, 122–126.

West, H.M., Fitter, A.H. & Watkinson, A.R. (1993) Response of *Vulpia ciliata* ssp. *ambigua* to removal of mycorrhizal infection and to phosphate application under natural conditions. *Journal of Ecology*, 81, 351–358.

Westphal, M.I., Pickett, M., Getz, W.M. & Possingham, H.P. (2003) The use of stochastic dynamic programming in optimal landscape reconstruction for metapopulations. *Ecological Applications*, 13, 543–555.

Wharton, D.A. (2002) *Life at the Limits: organisms in extreme environments*. Cambridge University Press, Cambridge, UK.

White, J. (1980) Demographic factors in populations of plants. In: *Demography and Evolution in Plant Populations* (O.T. Solbrig, ed.), pp. 21–48. Blackwell Scientific Publications, Oxford.

White, T.C.R. (1993) *The Inadequate Environment: nitrogen and the abundance of animals*. Springer, Berlin.

Whittaker, R.H. (1953) A consideration of climax theory: the climax as a population and pattern, *Ecological Monographs*, 23, 41–78

Whittaker, R.H. (1956) Vegetation of the Great Smoky Mountains. *Ecological Monographs*, 23, 41–78.

Whittaker, R.H. (1975) *Communities and Ecosystems*, 2nd edn. Macmillan, London.

Whittaker, R.J., Willis, K.J. & Field, R. (2003) Climatic–energetic explanations of diversity: a macroscopic perspective. In: *Macroecology: concepts and consequences* (T.M. Blackburn & K.J. Gaston, eds), pp. 107–129. Blackwell Publishing, Oxford.

Wiebes, J.T. (1979) Coevolution of figs and their insect pollinators. *Annual Review of Ecology and Systematics*, 10, 1–12.

Wiens, J.A., Stenseth, N.C., Van Horne, B. & Ims, R.A. (1993) Ecological mechanisms and landscape ecology. *Oikos*, 66, 369–380.

Wilbur, H.M. (1987) Regulation of structure in complex systems: experimental temporary pond communities. *Ecology*, 68, 1437–1452.

Wilby, A., Shachak, M. & Boeken, B. (2001) Integration of ecosystem engineering and trophic effects of herbivores. *Oikos*, 92, 436–444.

Williams, G.C. (1966) *Adaptation and Natural Selection*. Princeton University Press, Princeton, NJ.

Williams, K.S., Smith, K.G. & Stephen, F.M. (1993) Emergence of 13-year periodical cicadas (Cicacidae: *Magicicada*): phenology, mortality and predator satiation. *Ecology*, 74, 1143–1152.

Williams, W.D. (1988) Limnological imbalances: an antipodean viewpoint. *Freshwater Biology*, 20, 407–420.

Williamson, M.H. (1972) *The Analysis of Biological Populations*. Edward Arnold, London.

Williamson, M.H. (1981) *Island Populations*. Oxford University Press, Oxford.

Williamson, M. (1999) Invasions. *Ecography*, 22, 5–12.

Wilson, D.S. (1986) Adaptive indirect effects. In: *Community Ecology* (J. Diamond & T.J. Case, eds), pp. 437–444. Harper & Row, New York.

Wilson, E.O. (1961) The nature of the taxon cycle in the Melanesian ant fauna. *American Naturalist*, 95, 169–193.

Wilson, J.B. (1987) Methods for detecting non-randomness in species co-occurrences: a contribution. *Oecologia*, 73, 579–582.

Wilson, J.B. (1988a) Shoot competition and root competition. *Journal of Applied Ecology*, 25, 279–296.

Wilson, J.B. (1988b) Community structure in the flora of islands in Lake Manapouri, New Zealand. *Journal of Ecology*, 76, 1030–1042.

Wilson, J.B. (1999) Guilds, functional types and ecological groups. *Oikos*, 86, 507–522.

Wilson, J.B. & Agnew, A.D. (1992) Positive-feedback switches in plant communities. *Advances in Ecological Research*, 23, 263–333.

Wilson, K., Bjørnstad, O.N., Dobson, A.P. *et al.* (2002) Heterogeneities in macroparasite infections: patterns and processes. In: *The Ecology of Wildlife Diseases* (P.J. Hudson, A. Rizzoli, B.T. Grenfell, H. Heesterbeek & A.P. Dobson, eds), pp. 6–44. Oxford University Press, Oxford.

Wilson, S.D. & Tilman, D. (2002) Quadratic variation in old-field species richness along gradients of disturbance and nitrogen. *Ecology*, 83, 492–504.

Winemiller, K.O. (1990) Spatial and temporal variation in tropical fish trophic networks. *Ecological Monographs*, 60, 331–367.

Winemiller, K. (1996) Factors driving temporal and spatial variation in aquatic floodplain food webs. In: *Food Webs: integration of patterns and dynamics* (G.A. Polis & K.O. Winemiller, eds), pp. 298–312. Chapman & Hall, New York.

Winterbourn, M.J. & Townsend, C.R. (1991) Streams and rivers: one-way flow systems. In: *Fundamentals of Aquatic Ecology* (R.S.K. Barnes & K.H. Mann, eds), pp. 230–244. Blackwell Scientific Publications, Oxford.

With, K.A., Pavuk, D.M., Worchuck, J.L., Oates, R.K. & Fisher, J.L. (2002) Threshold effects of landscape structure on biological control in agroecosystems. *Ecological Applications*, 12, 52–65.

Wittenberg, R. & Cock, M.J.W. (2001) *Invasive Alien Species: a toolkit of best prevention and management practices*. CAB International, Oxford.

Woiwod, I.P. & Hanski, I. (1992) Patterns of density dependence in moths and aphids. *Journal of Animal Ecology*, 61, 619–629.

Wolda, H. (1978) Fluctuations in abundance of tropical insects. *American Naturalist*, 112, 1017–1045.

Wolday, D., Mayaan, S., Mariam, Z.G. *et al.* (2002) Treatment of intestinal worms is associated with decreased HIV plasma viral load. *Journal of Acquired Immune Deficiency Syndromes*, 31, 56–62.

Wolff, J.O. (1997) Population regulation in mammals: an evolutionary perspective. *Journal of Animal Ecology*, 66, 1–13.

Wolff, J.O. (2003) Density-dependence and the socioecology of space use in rodents. In: *Rats, Mice and People: rodent biology and management* (G.R. Singleton, L.A. Hynds, C.J. Krebs & D.J. Spratt, eds), pp. 124–130. Australian Centre for International Agricultural Research, Canberra.

Wolff, J.O., Schauber, E.M. & Edge, W.D. (1997) Effects of habitat loss and fragmentation on the behavior and demography of gray-tailed voles. *Conservation Biology*, 11, 945–956.

Woodward, F.I. (1987) *Climate and Plant Distribution*. Cambridge University Press, Cambridge, UK.

Woodward, F.I. (1990) The impact of low temperatures in controlling the geographical distribution of plants. *Philosophical Transactions of the Royal Society of London, Series B*, 326, 585–593; also in *Life at Low Temperatures* (R.M. Laws & F. Franks, eds), pp. 69–77. The Royal Society, London.

Woodward, F.I. (1994) Predictions and measurements of the maximum photosynthetic rate, Amax, at the global scale. In: *Ecophysiology: photosynthesis* (E.D. Schulze & M.M. Caldwell, eds), pp. 491–509. Springer-Verlag, Berlin.

Woodward, G. & Hildrew, A.G. (2002) Body-size determinants of niche overlap and intraguild predation within a complex food web. *Journal of Animal Ecology*, 71, 1063–1074.

Woodwell, G.M., Whittaker, R.H. & Houghton, R.A. (1975) Nutrient concentrations in plants in the Brookhaven oak pine forest. *Ecology*, 56, 318–322.

Wootton, J.T. (1992) Indirect effects, prey susceptibility, and habitat selection: impacts of birds on limpets and algae. *Ecology*, 73, 981–991.

Worland, M.R. & Convey, P. (2001) Rapid cold hardening in Antarctic microarthropods. *Functional Ecology*, 15, 515–524.

Worm, B. & Duffy, J.E. (2003) Biodiversity, productivity and stability in real food webs. *Trends in Ecology and Evolution*, 18, 628–632.

Worrall, J.W., Anagnost, S.E. & Zabel, R.A. (1997) Comparison of wood decay among diverse lignicolous fungi. *Mycologia*, 89, 199–219.

Worthen, W.B. & McGuire, T.R. (1988) A criticism of the aggregation model of coexistence: non-independent distribution of dipteran species on ephemeral resources. *American Naturalist*, 131, 453–458.

Worthington, E.B. (ed.) (1975) *Evolution of I.B.P.* Cambridge University Press, Cambridge, UK.

Wurtsbaugh, W.A. (1992) Food-web modification by an invertebrate predator in the Great Salt Lake (USA). *Oecologia*, 89, 168–175.

Wynne-Edwards, V.C. (1962) *Animal Dispersion in Relation to Social Behaviour*. Oliver & Boyd, Edinburgh.

Yako, L.A., Mather, M.E. & Juanes, F. (2002) Mechanisms for migration of anadromous herring: an ecological basis for effective conservation. *Ecological Applications*, 12, 521–534.

Yameogo, L., Crosa, G., Samman, J. *et al.* (2001) Long-term assessment of insecticide treatments in West Africa: aquatic entomofauna. *Chemosphere*, 44, 1759–1773.

Yao, I., Shibao, H. & Akimoto, S. (2000) Costs and benefits of ant attendance to the drepanosiphid aphid *Tuberculatus quercicola*. *Oikos*, 89, 3–10.

Yazvenko, S.B. & Rapport, D.J. (1997) The history of Ponderosa pine pathology: implications for management. *Journal of Forestry*, 95, 16–20.

Yibarbuk, D., Whitehead, P.J., Russell-Smith, J. *et al.* (2001) Fire ecology and aboriginal land management in central Arnhem Land, northern Australia: a tradition of ecosystem management. *Journal of Biogeography*, 28, 325–343.

Yoda, K., Kira, T., Ogawa, H. & Hozumi, K. (1963) Self thinning in overcrowded pure stands under cultivated and natural conditions. *Journal of Biology, Osaka City University*, 14, 107–129.

Yodzis, P. (1986) Competiton, mortality and community structure. In: *Community Ecology* (J. Diamond & T.J. Case, eds), pp. 480–491. Harper & Row, New York.

Yoshida, T., Jones, L.E., Ellner, S.P., Fussmann, G.F. & Hairston, N.G. Jr. (2003) Rapid evolution drives ecological dynamics in a predator–prey system. *Nature*, 424, 303–306.

Young, R.G., Huryn, A.D. & Townsend, C.R. (1994) Effects of agricultural development on processing of tussock leaf litter in high country New Zealand streams. *Freshwater Biology*, 32, 413–428.

Young, T.P. (1990) Evolution of semelparity in Mount Kenya lobelias. *Evolutionary Ecology*, 4, 157–172.

Young, T.P. & Augspurger, C.K. (1991) Ecology and evolution of long-lived semelparous plants. *Trends in Ecology and Evolution*, 6, 285–289.

Yu, L.M. (1995) Elicitins from *Phytophthora* and basic resistance in tobacco. *Proceedings of the National Academy of Science of the USA*, 92, 4088–4094.

Zadoks, J.S. & Schein, R.D. (1979) *Epidemiology and Disease Management*. Oxford University Press, Oxford.

Zahn, R. (1994) Fast flickers in the tropics. *Nature (London)*, 372, 621–622.

Zak, D.R., Holmes, W.E., White, D.C., Peacock, A.D. & Tilman, D. (2003) Plant diversity, soil microbial communities, and ecosystem function: are there any links? *Ecology*, 84, 2042–2050.

Zavala, J.A., Paankar, A.G., Gase, K. & Baldwin, I.T. (2004) Constitutive and inducible trypsin proteinase inhibitor production incurs large fitness costs in *Nicotinia attenuata*. *Proceedings of the National Academy of Sciences of the USA*, 101, 1607–1612.

Zavaleta, E.S., Hobbs, R.J. & Mooney, H.A. (2001) Viewing invasive species removal in a whole-ecosystem context. *Trends in Ecology and Evolution*, 16, 454–459.

Zeide, B. (1987) Analysis of the 3/2 power law of self-thinning. *Forest Science*, 33, 517–537.

Zheng, D., Prince, S. & Wright, R. (2003) Terrestrial net primary production estimates in 0.5 degree grid cells from field observations – a contribution to global biogeochemical modelling. *Global Change Biology*, 9, 46–64.

Zheng, D.W., Bengtsson, J. & Agren, G.I. (1997) Soil food webs and ecosystem processes: decomposition in donor-control and Lotka–Volterra systems. *American Naturalist*, 145, 125–148.

Ziemba, R.E. & Collins, J.P. (1999) Development of size structure in tiger salamanders: the role of intraspecific interference. *Oecologia*, 120, 524–529.

Zimmer, M. & Topp, W. (2002) The role of coprophagy in nutrient release from feces of phytophagous insects. *Soil Biology and Biochemistry*, 34, 1093–1099.

Organism index

Subject index